The Neurofeedback Book
Second Edition

Also by Michael and Lynda Thompson

Setting up for Clinical Success with Procomp+/Biograph

Also by William Sears and Lynda Thompson

The A.D.D. Book: New Understandings, New Approaches to Parenting Your Child

The Neurofeedback Book
Second Edition

Michael Thompson, M.D., & *Lynda Thompson, Ph.D.*

Consultants: James Thompson, Ph.D., David Hagedorn, Ph.D.,
Andrea Reid-Chung, M.A., C.Psych.Assoc., Tammy Binder, M.D.

Published by:
The Association for Applied Psychophysiology and Biofeedback, 2015

© copyright ADD Centres Ltd. / Biofeedback Institute of Toronto; 2015, Dr. Thompson;
www.addcentre.com

PUBLISHED BY:

The Association for Applied Psychophysiology and Biofeedback, 2015

COPYRIGHT © 2015

by

ADD Centres Ltd./Biofeedback Institute of Toronto

All rights reserved.

Second Edition

Drawing by Amanda Reeves and Maya Berenkey

Disclaimer: This text is intended only as a guide for professionals who are learning about the field of neurofeedback. It is published for teaching purposes only and none of this constitutes medical or psychological advice. None of the suggestions are to be taken as treatment protocols.

Professionals will learn about treatments from their colleagues in appropriate professional learning situations. This book is only a broad overview and guideline that will hopefully assist readers with an introduction to the field.

Note to Readers: The first edition of **The Neurofeedback Book** was published in 2003 as a working document and people were invited to give feedback to improve it and thus allow for a better referenced, more accurate and complete second edition. This second edition is still incomplete because the field of applied neuroscience is constantly changing. Thus the reader is encouraged, once again, to share any suggestions or corrections so that we can further improve this text for an eventual third edition. Thank you in advance for your feedback.

ISBN: 978-0-9842979-0-0

Table of Contents – Broad Overview

Preface i
Acknowledgments iii
A Word from the Authors v
Overview of the Book v
Introduction to the First Edition by Joel Lubar ix

PART ONE
What Neurofeedback Is and the Science Behind It 1

Section I	Definition, Description and Overview of Biofeedback, Neurofeedback, Learning Theory and Applications	2
Section II	Origin of the Electroencephalogram (EEG)	17
Section III	The EEG: Frequencies and Normal and Abnormal Waveforms	32
Section IV	Measuring the EEG: Instrumentation and Electronics	51
Section V	Neuroanatomical Structures, Connections and Neurochemistry	68

PART TWO
Functional Neuroanatomy Organized with Reference to Networks, Lobes of the Brain, 10-20 Sites, and Brodmann Areas 123

Section VI	Introduction/Orientation with Diagrams	124
Section VII	Functional Networks and Behavior	139
Section VIII	**Functions Related to Lobes of the Brain, International 10-20 Sites, and Brodmann Areas**	156
Appendices	....	243

PART THREE
Introduction to Assessment and Intervention 247

Section IX	Basics of Assessment (Electrode Placement, Identification of Artifacts, Effects of Medication and Toxic Substances on EEG, Interpretation of the EEG using QEEG)	248
Section X	Autonomic Nervous System (ANS) and Skeletal Muscle Tone (EMG) Assessment	359
Section XI	Fundamentals of Intervention (Electrode Placement, Bandwidths, Combining Operant and Classical Conditioning, Single-Channel Assessment and Session Graphing, Advanced Terminology (Phase Shift, Non-linear, Chaos, Neuromodulation…))	383
Section XII	Combined NFB + BFB Intervention Fundamentals	445
Section XIII	Intervention Summary (Basic Stages of Training and Tracking Progress)	467
Section XIV	Alpha-Theta Therapy (Combining NFB and BFB with Psychotherapy and Relaxation) .	475
Section XV	Metacognitive Strategies	483

PART FOUR
Assessment and Intervention: Advanced Procedures 521

Section XVI	Evoked Potentials (ERPs)	521
Section XVII	Using LORETA Mathematics in Assessment	543

PART FIVE
Other Treatment Modalities that can be Combined with Neurofeedback and Biofeedback 565

Section XVIII	Heart Rate Variability (HRV) Training	565
Section XIX	Transcranial Direct Current Stimulation (tDCS)	590
Section XX	Cranial Electrical Stimulation (CES)	591
Section XXI	Passive Infrared Hemoencephalography (pIR HEG) and Near Infrared Spectrophotometry Hemoencephalography (NIRS HEG)	611
Section XXII	Slow Cortical Potential (SCP) Neurofeedback	616

PART SIX
Assessment and Intervention with Specific Disorders and Syndromes
(Additions to the First Edition of *The Neurofeedback Book*) 619

Section XXIII	Concussion/Mild Traumatic Brain Injury (TBI)	620
Section XXIV	Affect Networks – Depression	641
Section XXV	Asperger's Syndrome and Autistic Spectrum Disorders (ASDs)	648
Section XXVI	Attention Deficits and Their Brain Correlates	660
Section XXVII	Seizure Disorders	676
Section XXVIII	Movement Disorders (Brief Additions for Parkinson's, Dystonia, Tourette), and Fibromyalgia	681
Section XXIX	Post-Traumatic Stress Disorder (PTSD)	684
Section XXX	Optimal Performance Training	691

PART SEVEN
Efficacy, Statistics, Research Design and Multiple Choice Questions 693

Section XXXI	Efficacy Criteria Used in the Field of Neurofeedback	693
Section XXXII	Brief Overview of Statistics and Research Design	697
Section XXXIII	Multiple Choice Questions	708

References	773
Index	805

Preface to *The Neurofeedback Book, Second Edition*

There have been impressive advances in the area of applied neuroscience and brain computer interface training since *The Neurofeedback Book* was published over a decade ago. Though a number of excellent books have been published in the interim—most of them edited texts with different contributors sharing their expertise in specific areas of the field—there is still no other text that brings together in one tome an overview of all the information relevant to the effective practice of neurofeedback. Thus there was demand for a second edition that would still include the fundamentals but also add updated information concerning advancements in theory, practice and research related to neurofeedback. This Second Edition follows in the spirit of the first by presenting the knowledge relevant to clinical practice in a format that is accessible to people just entering the field, though it also has enough depth to be helpful to seasoned providers of biofeedback services. The basics regarding history of the field, neuroanatomy, generation of the EEG, electronics and instrumentation for the measurement of psychophysiological variables have not changed, there has been additional research, enhancements in theoretical understanding, new (or improved) clinical interventions, and advances in the technology used in our field. This is too broad a field to cover completely, and we are not exhaustive in giving references and covering topics that are not central or necessary to carry out the fundamentals of good educational and/or clinical neurofeedback and biofeedback training.

When it comes to clinical approaches, there is still diversity in the field. We have chosen to present the training methods that are evidence-based. Therefore we some equipment and training methods that may be promising and may be effective for some individuals have been omitted because they do not yet have sufficient research support to meet the criterion of being evidence-based. In this we follow the lead of the Biofeedback Certification International Alliance. Indeed, since we teach BCIA-accredited courses in Neurofeedback, you will find nearly all the material in the BCIA Blueprint of Knowledge covered here, albeit with somewhat different organization and some additions. When material goes beyond the fundamentals but was deemed to be of interest to some practitioners, we put this material in *italics*.

The original textbook presented the basics required to do clinical practice, including both neurofeedback (NFB) and general biofeedback (BFB), augmented by the addition of coaching in metacognitive strategies. The metacognitive component remains a feature of our approach to doing NFB training sessions. It is achieved by doing part of each training session on task. Examples of tasks are described in the chapter on metacognition, which is largely unchanged from the first edition. The fundamentals regarding EEG generation and electronics for measuring this electrical activity have similarly remained the same, so those chapters are also largely unchanged.

Since the first edition was published, however, more has become known about neural networks and the functional connectivity of different areas of the brain. So there is a greatly expanded section on functional neuroanatomy, which is discussed in the context of the International 10-20 Placement System for electrodes and Brodmann area (BA) functions. With the wider use of Low Resolution Electromagnetic Tomography (LORETA) for source localization when doing 19-channel quantitative EEG (abbreviated to either QEEG or qEEG), these BAs have become more relevant to neurofeedback work.

There are also new additions to the section on assessment. The new material explains how LORETA analysis, event-related potentials (ERPs), and heart rate variability (HRV) have become an integral part of assessments done for complex problems, as in cases in which people have suffered head injuries.

In addition, several new treatment techniques are covered, such as Z-score NFB, which may be carried out with multiple surface electrode sites, and LORETA Z-score NFB, which is a program that uses 19-channel EEG with simultaneous LORETA analysis combined with comparisons to database norms. Also, methods such as transcranial direct current stimulation (tDCS) and passive infrared hemoelectroencephalography (pIR HEG) have been added because these methods now have some research support. In describing how these techniques can be integrated into a basic clinical practice, we emphasize how a careful, complete assessment of each patient leads to a multimodal approach and the prescription of specific treatment modalities. The descriptions of these techniques will be followed by clinical examples that expand the approaches given in the original text. Thus, *The Neurofeedback Book, Second Edition* has a considerable amount of new information to keep practitioners current.

The authors will attempt, as was done in the original edition, to explain each area in simple terms. Complex theoretical ideas, formulations, or mathematics are not addressed, but for those who wish to have more depth, there are references provided.

In short, the additions to *The Neurofeedback Book, Second Edition* include the following topics: Cortical Brodmann Areas and their Primary Functions; Neural Networks; Cortex-Basal Ganglia-Thalamus-Cortex Loops that Underlie Neural Networks; Autonomic Nervous System as it is Involved in the Control of Heart Rate Variability; Amygdala to Hypothalamus to Brain Stem Connections and their Relation to Afferent and Efferent Connections to the Heart.

Note: Some items are repeated a number of times in different parts of this book. In part this is done because repetition is good for learning. In part it is just to avoid the reader having to search, find, and flip back to a previous section in order to review something each time it is being discussed. The reader will notice that the terms *client* and *patient* are used interchangeably. *Client* is well understood in North America, but it is not a term that is used in clinical settings in many countries in Europe and Asia. *Client* is the broader term because it applies to people without a diagnosed clinical condition who are training for optimal performance. Perhaps the term *student* could also have been used because the techniques require that the trainer be a coach and a facilitator. It is a learning and training approach that involves educating the client. Indeed, the authors have always wanted to see NFB + BFB incorporated into school programs.

Introductory Statement

The field of combined EEG neurofeedback and peripheral biofeedback has its foundations in functional neuroanatomy and neurophysiology. Quantitative electroencephalography (QEEG), whether single-channel or multi-channel, is a means of quantifying EEG data using high speed computers in order to distinguish, both with statistics and images, patterns that are outside normal, neurotypical, database means. Low resolution electromagnetic tomography (LORETA) allows the practitioner to find the source of different frequencies deep within the cortex. This data, when combined with the practitioner's knowledge of the dominant functions of different Brodmann areas and their relationship to neural networks (Thompson et al., 2011, 2015), can assist the practitioner in deciding whether a specific frequency at that site may correlate with a patient's symptoms. Thus these data can help the clinician decide whether or not a pattern correlates with a cognitive, motor, sensory, or affect problem for a specific patient. Or, for that matter, whether it matches a strength or special competence that person has.

The QEEG can then be used to give almost instantaneous visual, auditory, and sometimes even tactile feedback to inform the client about whether or not they are meeting their objectives for moving their EEG patterns to a more optimal level. Note the deliberate omission in this discussion of the word *normalize*. What might be an optimal change for one client may not be the best for another. The simplest analogy is IQ. Moving a client with an IQ of 85 to the average of the population would be admirable. It would be a ridiculous goal for a client with an IQ of 130.

The goal of combined NFB + HRV training is to influence the central nervous system (CNS) in a manner that will optimize the performance of a particular client. It is important for the practitioner to have a reasonable understanding of how, neuro-anatomically, this might be expected to occur. For this reason, current additions to *The Neurofeedback Book* place considerable emphasis on functional neuroanatomy.

The reader will note that in our description of good clinical practice, NFB is never a stand-alone procedure. We recognize that researchers feel a need to do research on individual items such as NFB without other therapeutic or educational inputs, but we feel they should clearly state in their reporting that they are not doing research on NFB as it is done in standard clinical practice. In practice it is always combined with the work of a good therapist/teacher/coach/trainer who adds specific metacognitive strategies and tasks, plus, as appropriate, other biofeedback techniques such as heart rate variability training. There is also consideration of other factors such as diet, sleep, and exercise. The implications of positive or negative research findings should be evaluated in this context.

On the other hand, large case series that include both subjective (such as questionnaires and school reports) and objective measures (such as standardized testing like the WISC or WAIS) may better reflect the value of these combined approaches to the clients. So we encourage clinicians to collect data in a systematic way and to publish the results they achieve.

Acknowledgements

Our mentors and teachers in this field are extraordinary scientists and wonderful individuals – intelligent, innovative, rigorous and also friendly. One of them should have written this book, but they all are too busy doing cutting-edge work, so it got left to us to interpret a field they created.

Our training in biochemistry, physiology, medicine and psychology gave a background to neurology, neuroanatomy, the EEG, biofeedback and learning, but our introduction to the use of the EEG for biofeedback training (neurofeedback) came from Joel Lubar.

In 1992 Dr. Lubar was giving one of his classic introductory workshops in Fort Lauderdale on "A Non-medication Approach to Treating ADD," hosted by the Biofeedback Society of Florida. February in Florida instead of Toronto? That seemingly easy choice led to our discovery of the field. Then came more training with Tom Allen in Florida, Joel and Judy Lubar in Tennessee, Frank and Mary Diets in Arizona, and Susan and Siegfried Othmer in California. Within the year ADD Centres Ltd. was born out of our joint practices that dealt to a large extent with learning disabilities, Attention Deficit Disorder and children who had Asperger's syndrome and autism. Then Barry Sterman's address at the Northwestern Biofeedback Society meeting in Banff in November of 1993 made it clear there was plenty more to master. We took further training in full-cap assessment techniques taught by George Fitzsimmons, a psychologist in the Department of Education at the University of Alberta. He was also taught by Joel Lubar, and had started the first neurofeedback center in Canada at the University of Alberta, called the Neuronal Re-regulation Programme.

Since then there has been regular attendance at meetings of the Association for Applied Psychophysiology and Biofeedback, the Society for Neuronal Regulation (now the International Society for Neurofeedback and Research), Future Health's Winter Brain meetings, plus pleasant forays to the Biofeedback Foundation of Europe (now the Biofeedback Federation of Europe) and the World Congress of Psychophysiology.

The warmth of people in this field and the generous way that everyone shares their knowledge is perhaps the most impressive feature of the neurofeedback community. In addition to Joel Lubar and Barry Sterman, pioneers like Joe Kamiya, Tom Budzynski and Peter Rosenfeld were welcoming and ready to share knowledge from the first time we met them in Key West at an early Rob Kall Winter Brain meeting. George Fuller von Bozay modeled combining biofeedback with neurofeedback, which has become a mainstay of our approach. Jay Gunkelman added an element of fun and constructive drive that, along with the lectures by Bob Thatcher, made us dig out Michael's neuroanatomy and electroencephalography textbooks, plus his neurology notes and diagrams from a year spent dissecting the brain. Vendors and manufacturers are, likewise, great sources of information. Frank Diets was unfailingly patient in teaching about things electronic in response to Michael's calls. His wife, Mary, always assisted us in understanding the fine points of biofeedback. Hal Myers, Larry Klein and their staff at Thought Technology have assisted at every turn. Hal and Larry are indefatigable proponents of our field and add Canadian content, as does the ever innovative Paul Swingle. Lexicor staff contributed significantly to education and trained our son James in 19-channel intricacies, which he later applied when doing his doctoral work at Penn State concerning concussion in athletes. There are also behind-the-scenes heroes like Francine Butler, the former Executive Director of AAPB, and Judy Crawford, Director of Certification (and heart and soul) of BCIA. As we got to the point that we could share what we had learned, we presented papers about our clinical work and started being invited to give workshops. We have been privileged to teach in Canada, the United States, Australia, Israel, Switzerland, Germany, Norway, China, Holland, South Korea, Mexico, Italy, Belgium, United Kingdom (Wales), Austria, South Africa, Poland, Turkey, Sweden, Egypt, Taiwan, and even in international waters during a SABA meeting that cruised to Alaska. Our thanks go to all those professionals whose questions and discussion have continually pushed us to revise our concepts and techniques.

For this text, Barry Sterman reviewed the section on QEEG and contributed some brain map examples. Jay Gunkelman also shared EEG examples and his extensive experience in reading EEGs. Joel Lubar wrote a section on LORETA and kindly contributed the introduction to the first edition of the book. Bonny Beuret, at that time Director of the Lerninstitut Basel in Switzerland, deserves particular thanks for her reading of the original manuscript and helpful editing. The talented Amanda Reeves produced the drawings and worked on typing several early drafts of the manuscript. Our children must have a special thanks for putting up with their parents being buried in manuscripts for years. They have each made their own unique contributions. James wrote the chapter on Statistics and Research design and has worked with us at international meetings presenting on the training of athletes and executives. Dr. Kate Thompson (now Dr. McCubbin) helped with scientific references, and Dr. Aaron Thompson put us straight on what learning strategies are useful in high school and university settings.

This book represents our inchoate strivings to understand a fascinating and fast-moving field and to share that knowledge. The relatively new discipline of neurofeedback spans diverse domains of knowledge: neuroscience, anatomy, physiology, electroencephalography, psychology, learning theory, biofeedback, electronics and electrical engineering, measuring devices, physics, computers, statistics and research design, therapeutic interventions, stress management, sports psychology, metacognition, sleep research, pharmacology, nutrition, and the list goes on. If you feel a bit overwhelmed, that is appropriate. After more than two decades in the field, we still do, too.

Yet you should not feel discouraged. It is possible to learn enough of the necessary bits from each of the complementary disciplines to make a start at applying your knowledge to help people learn to self regulate. Then you keep learning more from clients and from colleagues. We gratefully acknowledge what clients, in particular, have taught us. Our hope and purpose in writing this book is that your learning curve will be a little easier and more efficient than ours. We have tried to give you "the basics" in one place. In return, we expect that you will be contributing your experience to the field in the decades to come.

Additional Thanks and Acknowledgements for the Second Edition

Michael Thompson would like to give a special thanks to Lynda, and their children and grandchildren, for their patience in putting up with his hyperfocus as he sat on the porch at the lake reading articles, tracking updates, and writing this text, year after year, from the time the original book was submitted for publication.

Michael and Lynda would like to acknowledge the assistance of all staff at the ADD Centre and Biofeedback Institute of Toronto for their help and support over the last 22 years. The "ADD Centre family" is extensive, and each member knows the unique contributions they have made. Particular thanks are given to Andrea Reid-Chung, our Clinical Coordinator, for taking over the organization of the Centre, including international mentoring and staff training, and for her constant attention to the quality of service people receive. Lena Santhirasegaram has, for more than a dozen years, been the heart and soul of the Toronto branch of our operations. She also deserves special mention. Thanks are given to Amanda Reeves (now an internationally recognized Canadian artist living in San Francisco) for the original drawings, and to Bojana Knezevic and Maya Berenkey for updates on the figures. Special, painstaking help with references was done by Tanushree Bhandari.

It goes without saying that we are constantly learning from and grateful to our patients. They have demonstrated how very basic neurofeedback training done even at a single site could have a profound effect on many different aspects of their day-to-day functioning. We thank our clients for all their help and for their permission to use their data without personal identification in publications.

Thanks is also extended to all those professionals who always responded immediately to queries and technical questions. James Thompson and David Hagedorn of Evoke Neuroscience, Frank Diets, Bob Thatcher, Barry Sterman, Jay Gunkleman, Joel Lubar, Merlyn Hurd, Judy Crawford, and many, many others in the field of neurofeedback and biofeedback always kindly responded to our questions. With respect to basics of heart rate variability training, special thanks are given to Richard Gevirtz, Paul Lehrer, Evgeny and Bronya Vaschillo, Donald Moss, Eric Peper, David Hagedorn, and to staff at Thought Technology with particular recognition of Frank DiGregario, Marc Saab and Didier Compatalade.

We also thank all those professionals and organizations who have invited us to present over a hundred workshops on five continents. Participants coming all the way to Canada to our annual workshops have extended our reach to various other countries beyond the ones where we have taught, including France, Portugal, Spain, India, Pakistan, Iran, Columbia, and Brazil. Our thanks go to all those professionals whose questions and discussion have continually pushed us to keep learning ourselves.

A Word from the Authors

Welcome to the field of neurofeedback. Once you enter, you will never be bored, and you will never stop learning. There is always more to master and it keeps one both curious and humble.

Sharing Our Bias With You

Biofeedback and neurofeedback are tools that assist individuals to learn self-regulation skills that improve their functioning. These tools may be used in two general ways. The first is an educational or training use. The candidates for this are students, athletes and persons who wish to optimize their performance in school, jobs, sports, or even interpersonal relationships. This educational intervention is done by the individual, most often with the help of a teacher, trainer, or coach who has specialized training in biofeedback and/or neurofeedback. The second use is a therapeutic one. The feedback tools assist the therapist to help a client or patient to manage symptoms or overcome a disorder, disease, condition (e.g., substance abuse) or an emotional difficulty. In this case, a health care professional is helping a patient and the biofeedback and neurofeedback are one part of the treatment plan.

We chose the term *training* for feedback because the analogy is that neurofeedback equipment is a lot like exercise equipment. We are facilitating exercise for the brain. An exercise machine may be used by an individual at home, in a gym or at school with the assistance of a trainer or coach. You certainly need to know something about the equipment and how to use it, plus the individual doing training should have a training regimen designed by someone knowledgeable at the outset. That is like the first educational use of neurofeedback. Analogous to the second use would be exercise equipment in a hospital or clinic when a physiotherapist is doing rehabilitation. The use depends on the presenting situation: a person is doing exercise either for self-improvement or because they are a patient requiring treatment. Similarly, neurofeedback is used either for training clients or for treating patients.

Just as with an exercise machine, one must not attribute undue power to the use of biofeedback or neurofeedback equipment. The goals of the user may often be achieved using other techniques: meditation, relaxation, yoga, cognitive training, and metacognitive strategies, to name just a few. EEG biofeedback, however, is usually more efficient. Training can be done quite rapidly using modern technology as compared to these other techniques. Neurofeedback can also be combined with other interventions.

For the most part our focus is on an educational approach both at our ADD Centre and in this book. Thus we talk of trainers who have a coaching role, rather than therapists, when describing who is helping the student or client learn self-regulation. Topics such as alpha-theta training, which is appropriately used by therapists, will receive less discussion in this book than will approaches that improve attention and concentration. Tension, distractibility, lack of ability to quickly shift between broad and narrow focus, poor organization and inability to sustain concentration all interfere in work, school and sports. Helping people to have more flexible brains and to be able to produce a state that is calm, alert and focused is what we specialize in and do every day at our ADD Centre. We *add* to a person's self-regulatory skills using neurofeedback. How we do this and why we do this (that is, the scientific underpinnings) is the subject of this book.

Overview of this Book

This book is written to assist you in understanding the basis of neurofeedback and the fundamentals of how to do EEG biofeedback. It does NOT detail how to use different NFB instruments. That is done in the manuals provided by manufacturers of the equipment and in workshops and during mentoring.

Note from the outset that we do not yet know the precise mechanisms by which EEG biofeedback works. Ours is an empirical field based on observed outcomes. There is a lot of science behind it. As the late Dr. John Basmajian stressed in his address at the AAPB convention in 2000, we do not have to be apologetic in the least about our field because it is solidly research based. (Basmajian was, himself, a biofeedback pioneer who demonstrated in the 1960s that you could do operant conditioning of single motor neuron units.) Understand that our knowledge is still growing and much of what is written today may be obsolete in a few years' time.

Part One begins with questions. What is biofeedback? Why use an EEG? What kind of learning takes place? How is the EEG produced? What can be observed with the EEG? How does the EEG instrument detect and display this information? Neuroanatomy related to neurofeedback is covered and includes the following: the synapse, nerve conduction, the structure of the cortex, the fundamentals of pyramidal cells, inhibitory cells, the influence of subcortical structures on the EEG, and some aspects of what is known about the functions of the basal ganglia and the lobes of the brain.

Part Two, in this second edition, consists of additions, including a section published by AAPB as a separate monograph called *Functional Neuroanatomy*. In this part the concept of neural networks is introduced. This is followed by functions of each of the Brodmann areas with reference to the networks that include that Brodmann area. The Brodmann areas are organized according to the lobes of the brain for ease of learning and understanding their functions and network connections.

Part Three attempts to answer the question of how and why one does biofeedback (BFB) combined with neurofeedback (NFB). It includes how to do an NFB assessment, artifact the data and carry out NFB training. It also includes a brief BFB stress assessment detailing what sensors are used, what they represent, and how to carry out a combined NFB + BFB training session. Included is mention of other techniques that may accelerate the learning process, with particular emphasis on a detailed look at metacognition.

Part Four in this Second Edition consists of further details for carrying out basic EEG assessments and the fundamentals that underlie additions to our assessment and treatment options. This section will include graphing results of training sessions and the removal of artifacts – some of which were not described in the first edition. It has sections on additions to assessment that include two older techniques with new ways of incorporating them to improve our assessments: namely, LORETA and Evoked Potentials. This is followed by a short description of terms that a practitioner may encounter in lectures and academic papers, such as phase shift and phase lock, chaos theory and nonlinear mathematics, and Independent Component Analysis (ICA).

Part Five covers additions to intervention that include LORETA Z-score NFB, Heart Rate Variability Training, Transcranial Direct Current Stimulation, Passive Infrared Hemoencephalography, and Slow Cortical Potential Training.

Part Six expands on discussions of important disorders. These include concussion/mild traumatic brain injury, depression with anxiety, Asperger's syndrome and autistic spectrum disorders, attention-deficit/hyperactivity disorder, and optimal performance training.

Part Seven contains information about research design and statistics. It was written by James Thompson when he was a graduate student at Penn State. He is closer to these topics than his parents. It was updated to include efficacy criteria for evaluating research agreed upon by a joint committee of the AAPB and the ISNR. Understanding basic concepts in these areas is vital, not only for doing research but also for evaluating intelligently the research done by others. We want everyone in the field of neurofeedback to continue in the scientific vein in which our field was first conceived. This Part concludes with practice multiple choice questions and answers based on the material in this text.

Current and Future Status of Neurofeedback

Our understanding of where neurofeedback fits in the larger realm of neuroscience and how it works is still developing. There is still a lack of awareness of the field and some professionals who are critical, though most are simply unaware. In the Thompson family, we compare it to the status of Sir Edward Jenner's work 200 years ago. Edward Jenner's research and application of vaccination against smallpox culminated in the publication of his findings in 1798. It was a fine example of tried and true empirical research but it was not initially accepted by the medical authorities of his time. The Royal Society, of which he was a member, wrote to him and apparently suggested he stop his writing about vaccination because it was damaging his excellent reputation that was based on his solid observations concerning the cuckoo bird. Despite criticism in England, Jenner was invited to Russia to vaccinate the Czar and his family. Vaccination became accepted in continental Europe and eventually in Great Britain; indeed, vaccination against smallpox became mandatory in 1853. Smallpox was eventually eradicated worldwide in the late 20th century.

This portrait of Sir Edward Jenner hangs in the home of the authors.

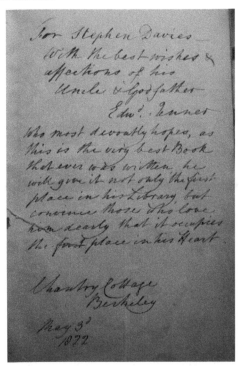

This inscription is on the first page of the Bible given by Jenner to his nephew in 1822, a year before he died. The excerpt reflects a personal side of this great man. The Bible and the portrait are in the personal collection of Michael Thompson, a descendant of Jenner's sister.

Public ridicule concerning vaccination included cartoons such as the one below.

1802 caricature entitled "The Cow Pock – Publications of ye Anti-Vaccine Society" (Public Domain; James Gillray – Library of Congress, Prints and Photographs Division). It seems that some patients feared that vaccination with cowpox to confer immunity against smallpox would make them sprout cow-like appendages.

Our field is still in the new-idea-being-resisted stage, but it will eventually become mainstream for the simple reason that it works and is based on reasonable neuroanatomical theory and objective EEG observations and research. We all look forward to neuroscience advancements leading to increased understanding of how neurofeedback works and the most efficient ways to use this valuable tool.

Self-regulation skills will become a major part of health maintenance in the 21st century. Two powerful reasons underlie this assertion. First, self-regulation aided by biofeedback and neurofeedback is effective in managing a wide range of conditions, many of which do not respond readily to traditional medical approaches. Secondly, the cost to society is much less since the focus is on promoting health with self-regulation rather than being passive in receiving help when things go wrong. Self-regulation is a learning method that engenders a long-term positive change. Pharmaceutical methods, for the most part, are only effective as long as the medication is continued and are costly in the long term. Self-regulation is health-oriented, not illness-oriented.

Why isn't neurofeedback better accepted in some quarters? Why is neurofeedback largely ignored? One reason is that it is outside the scope of most healthcare professionals' training. It is an enormous conceptual shift for many with biochemical or neurological training and experience to move from a psychopharmacological model to an educational model using biofeedback. If you have no experience with something, you may not even perceive it. (The indigenous people reportedly did not see Cortez' ships in the bay because they had no prior experience of ships that could transport people across vast oceans.) Accept that neurofeedback is not yet on the radar screen for most people, but be assured that awareness is growing. The internet and word-of-mouth are spreading the word faster than the scientific publications, though thankfully they are increasing, too.

Research is important for the respectability of our field. In the area of ADHD, for example, the American Pediatric Society in its November 2012 efficacy review gave biofeedback Level 1 efficacy – the highest level in their ranking system. This was a milestone that was based on a review of the research literature. It will make physicians more comfortable in mentioning neurofeedback as a treatment option for ADHD, especially for patients who suffer side effects with medications, for nonresponders to medications, and for those who are simply more comfortable with a nondrug approach.

We all have a responsibility to do things right so that the word that spreads is favorable. Aim for success with every client. Keep in mind the advice to promise less and deliver more.

Notes to the Reader

The terms *EEG biofeedback*, *Brain-Computer Interface Training*, and *neurofeedback* (NFB) are used interchangeably in this text. Other terms in common usage are *neurotherapy* and, simply, *brain wave biofeedback*. Brain-Computer Interface training (BCI) is a term that sounds more modern and sidesteps the erroneously held view of some grant reviewers that biofeedback is something from the 1970s that faded away. Whatever the term, the process involves monitoring brain electrical activity, displaying it as an electroencephalogram (EEG), and giving the client instant information, thus feedback, about what their brain is doing.

The term *biofeedback* (BFB) is used when the information given back to the client involves autonomic nervous system modalities or muscle electrical activity. Muscle contractions are measured and shown with an electromyogram (EMG) tracing. The other modalities commonly used are temperature, heart rate, respiration rate, and skin conductance, which is also called electrodermal response (EDR). Formerly the inverse of EDR, galvanic skin response (GSR) was the term used.

We suggest that you, yourself, use metacognitive strategies while reading this book. In other words, go beyond (*meta*) regular thinking and perceiving (*cognition*) and be aware of how you are thinking, how you best learn and remember new material. To start off right, apply active reading strategies. **Before opening the rest of this book,** first think about what you would like to know and how you would design such a text. Second, scan the major headings, then the subheadings, and think about how your design matches or differs from the one we decided to use. Third, review the intent of the book, as described here in this "Word from the Authors." Fourth, begin to read in whatever chapter seems most useful to you. Throughout your reading, please generate questions and try to find the answers, either here or in other texts. These steps encourage emotional involvement that helps you remember the material. Make sure you have a personal reason for bothering to read this text.

In keeping with the government's policy in our province, in the text we have, for the most part, adopted the use of *they, their, themselves* instead of *he/she, his/her, himself/herself* even when the sentence would normally require the singular. For example, you may find a phrase such as, "the client changed *their* theta wave amplitude," which is grammatically incorrect but has now become the preferred terminology in Ontario, Canada because it is gender neutral. We have primarily used American spelling, respecting the roots of neurofeedback being in the USA.

Feedback is Invited

This text is a short primer about a broad field. For this reason everything is simplified. Simplification may be overdone, and errors do occur. Also, we put the criterion of "doneness" ahead of perfection. So, when you find errors, please let us know by email to addcentre@gmail.com or by writing to us at The ADD Centre, 50 Village Centre Place, Mississauga, ON L4Z 1V9 Canada. We will attempt to put corrections that we make from your suggestions onto our website: www.addcentre.com. Your feedback will also contribute to any updated editions of this text.

An Introduction to *The Neurofeedback Book*

Joel F. Lubar Ph.D.,
BCIA-Senior Fellow, BCIA-EEG, Fellow ISNR
Professor, University of Tennessee
Co-Director,
Southeastern Biofeedback and Neurobehavioral Institute
2003

I am very pleased to have this opportunity to write the Foreword for Drs. Michael and Lynda Thompson's new book *The Neurofeedback Book*. Neurofeedback has been around a long time, since the 1960s when it was only called EEG biofeedback. We have waited more than 40 years for the advent of a true textbook of neurofeedback. *The Neurofeedback Book* fulfills this need beautifully and in great detail. This book is written for people at all levels of knowledge in the field, from beginners to the very advanced. The writing style is clear and straightforward.

The Thompsons' book begins with basic science dealing with the definitions of biofeedback and neurofeedback and a rather detailed discussion regarding EEG in terms of its generation, and its normal and abnormal manifestations. The genesis of the EEG in terms of its emergence from activity both at the cellular level and intracortical dynamics and thalamocortical pacemakers is very complex. This text does a fine job of sorting these out and elaborating them in Part One. There is a very detailed coverage of the terminology that is germane to the EEG literature, such as *coherence, phase, asymmetry, synchrony*, and basic terms such as *frequency, magnitude different waveforms,* and many others. The majority of the material that one would need in preparation for certification in the area of EEG biofeedback is contained in this volume. The basic neurophysiology and basic science behind EEG is well outlined both in terms of its historical importance and the present state of knowledge. The book contains numerous excellent illustrations, in black and white and in color, showing different types of EEG phenomena including topographic brain maps and comparisons of clinical cases with normative databases.

There is a wealth of clinical information contained in the volume dealing with most of the common disorders that are currently treated both medically and with behavioral medicine approaches including neurofeedback. There are illustrations of different types of feedback displays from a variety of different instruments. Considerable effort is expended in discussions of clinical procedures including assessment, psychophysiological profiling, combining conventional biofeedback with neurofeedback, and tracking patient progress. The Thompsons describe in detail their metacognitive approach for working with attentional and learning disorders.

One important part of the book is the detailed list of references and the whole section of multiple choice questions that are tied to the BCIA blueprint areas for EEG Biofeedback Certification. This again emphasizes the value of *The Neurofeedback Book* as a primer for new certificants and as a review for those seeking recertification or additional certifications in the field. In summary, *The Neurofeedback Book* should be read and reread by all in the field and should be made available for University courses in psychophysiology and behavioral medicine. It provides a wealth of material for patients in treatment using neurofeedback approaches and for educating professionals in health care who might refer patients for treatment or who might be considering embarking on the neurofeedback-neurotherapy endeavor.

PART ONE

What Neurofeedback Is and The Science Behind It

This section contains a brief overview of definitions, learning theory, origin of the EEG, instrumentation and neuroanatomy. In each of these areas it is expected that the reader will have had undergraduate courses. This section is meant only to give a short review of the area as it pertains to the use of neurofeedback. The reader is encouraged to review their own basic textbook on neuroanatomy. The section on instrumentation and electronics goes into detail that is not necessary for most neurofeedback practitioners, but is important in understanding the differential amplifier and the importance of impedance.

SECTION I
What Neurofeedback Is and the Science Behind It

What is biofeedback in general and EEG biofeedback in particular? *Definition, Description and Overview of Biofeedback, Learning Theory, and Neurofeedback Applications.*

What Is Biofeedback?

Biofeedback is the use of instrumentation to mirror psychophysiological processes of which the individual is not normally aware, and which may be brought under voluntary control (George Fuller, 1984).

Think of *bio* as referring to biology, the science of life, and all those dynamic processes that are going on all the time in our bodies. The brain with its >100 billion neurons runs this whole dynamic organization. The nerves carry the brain's messages to every corner of the human body. Through the action of neurotransmitters, neuromodulators and neurohormones, every cell in the body can be influenced by the brain. When you give the brain information, you influence the system. Biofeedback simply means *feeding* information *back* to the individual who generated the bio-signals in the first place.

One example of *feeding back bio*logical information is heart rate variability biofeedback. When the heart runs faster, something in the nervous system is related to this increase. In this case, it is a portion of the autonomic nervous system called the *sympathetic* nervous system. In our body there is always a balance of stimulating and slowing, excitation and inhibition. In the example of heart rate, slowing of the heart corresponds to its being released from sympathetic stimulation, which speeded it up. The parasympathetic system (more specifically, the vagus nerve, which has branches to many of our internal organs) takes over control of heart rate, inhibiting or slowing down the heart.

Biofeedback for this modality involves using an instrument that monitors heart rate variability and displays this virtually instantly (in "real time") to the client whose cardiac activity is being measured. The display is the feedback and it usually has both a visual and an auditory component.

Biofeedback implies more than just a passive measurement. It implies an active involvement of the client. Biofeedback is done so that the client will become actively involved in controlling their own physiology; hence the term *applied psychophysiology*.

EEG biofeedback (or neurofeedback) is based on two basic tenets: that brain electrical activity – the electroencephalogram or EEG – reflects mental states, and that the activity can be trained. Electrical activity being produced in the brain can be displayed on a computer screen virtually instantly (within approximately 50 to 100 milliseconds with modern equipment). The display is in the form of a line with a mixture of waves. Most people are familiar with the electrocardiogram (EKG) that represents the electrical activity of the heart. The EEG is similar but less regular. It looks rather like waves on the surface of a lake. Like waves on a lake, what we observe is always a complex sum: small little waves with low amplitude and little force or power, like those produced by a gentle breeze, run at a high frequency, whereas larger waves (high amplitude, more power), like the waves produced by a ferry boat, run at a slower frequency. The little ripples change amplitude and frequency with each gust of wind – and are thus desynchronous. The larger waves, however, roll by in a more regular, synchronous fashion. We have already noted that there is a different generator for each of these types of waves – those from the ferry boat and those from the wind. Indeed we could imagine a smaller motor boat going by and producing a regular, synchronous, wave of a little higher frequency but less power than those from the much larger ferry boat. The little ripples may ride on top of the larger waves and the surface activity is constantly changing. This analogy can be kept in mind when watching the EEG on the computer screen.

Similarly, in the case of the EEG, waves which come from different generators (cortex, thalamus) are different in frequency. The raw EEG contains all these different frequencies in a single wavy line, faster waves riding on top of slower waves.

EEG biofeedback involves recording this information using electrodes placed on the scalp and displaying it, that is *feeding it back,* on a computer screen. As the clients alter their own mental state, the amplitudes of various brain wave frequencies are changing. Clients see this change as it is reflected by various displays on the computer monitor, and attempt to alter their brain wave pattern to achieve a predefined goal. In this manner, the client learns to self-regulate. It is a learned normalization of EEG patterns (Sterman).

To summarize, advanced electronics and mathematical computations have made it possible to convert EEG patterns into images on a computer monitor. Learning to change the computer image reflects self-regulation of the EEG. Self-regulation of the EEG requires that the client self regulate underlying mental states that were responsible for that EEG pattern. When the EEG patterns reflect system changes in thalamus-basal ganglia-cortical interchanges, the person is actually learning self-regulation of this complex dynamic neural system.

It is well established in science that a positive reward for a behavior is followed by an increase in the probability of that behavior recurring (Edward Thorndike's Law of Effect). The production of particular brain wave patterns is the behavior we reward when doing EEG biofeedback. The reward is information about success using sounds and visual displays provided by a computer. Rewarding a behavior (or a sequence of neurophysiological occurrences) therefore 'shapes' the contributing components of that sequence in a way that results in an increased frequency of that behavior sequence recurring (Sterman, 2000). Shaping is done through a process which is termed *operant conditioning*.

The term *operant conditioning* originally reflected the fact that the behaviors being trained resulted in a series of learned responses which constituted an operation or observed action on the environment. Advances in techniques demonstrated that internal changes such as skin temperature or heart rate could also be influenced in this manner. External rewards could thus influence physiological changes in the body (Sterman, 2000). The operations were no longer on the external environment, and a new term for this type of work emerged. After much discussion in the 1960s it came to be called *biofeedback.*

When we reward changes in neuronal behavior based on measurement of the electroencephalogram (EEG), which reflects changes in neuronal activity, we use the terms *EEG biofeedback* or *neurofeedback*. The evidence that EEG biofeedback can produce significant and sustained physiological changes was well documented as far back as the early 1970s. (For references. see the review by M. Barry Sterman, entitled EEG Markers for Attention Deficit Disorder: Pharmacological and Neurofeedback Applications. *Child Study Journal*, Vol. 30, No. 1, 2000).

Biofeedback is **not a new invention**. Biofeedback is a universal, natural, biological process. A simple example is learning to ride a bicycle. If a child is seven or eight years old he can learn, usually in half an hour, how to ride, put his bike down for the winter, and then ride again without relearning in the spring. How is this possible? The answer is *natural neurofeedback.* Instead of a biofeedback instrument on the desk, we have one inside our body, in the inner ear, called the vestibular apparatus. It consists of fluid inside tiny semicircular canals which detect movement in all directions. This information is instantly sent to the brain along the auditory pathways (just like neurofeedback displays are sent by the optic and auditory pathways). The brain assimilates the data and, far faster than one can consciously think about what is happening, adjusts the muscles. The result is that the child is balanced on the bicycle. The means for learning was a kind of internal neurofeedback. Other ways of gaining control over brain states that have been practiced for centuries include yoga, meditation, and martial arts.

Treatment *versus* Training
Treatment, for the most part, implies passivity. Taking a medication or undergoing a surgical procedure are examples of passive involvement of the patient. Learning (training) is an active process that requires some motivation and repetition of exercises.

What Can Be Measured?
Concerning Biofeedback
In many forms of biofeedback we measure functioning of the autonomic (sympathetic and parasympathetic) nervous system. You may think of *autonomic* as meaning 'automatic.' Decades ago, Western scientists thought that this portion of our nervous system, which governed the actions of our

internal organs (the heart, blood vessels, lungs, gastrointestinal system, bladder, and so on), was not under conscious control. On the other hand, Eastern practitioners in places like India and China have practiced controlling aspects of these systems for thousands of years. As a Greek once said, "Nothing is new under the sun." We have simply added electronic measurements to help the learning go faster.

Western science made a leap forward when it, too, recognized that humans are capable of consciously learning how to self-regulate much of their own physiology. It became clear that we could regulate autonomic biological functions such as peripheral skin temperature, electrodermal responses (sweating), heart rate, and the synchrony of heart rate with respiration, which is called respiratory sinus arrhythmia (RSA). In addition, we now use the term *biofeedback* to refer to control of muscle tension (EMG). How to assess and self-regulate each of these physiological functions is dealt with in a later section.

Concerning Neurofeedback

In *neurofeedback* we measure the frequency and amplitude of different brain waves. These are recorded by means of small electrodes (sensors) placed on the surface of the scalp, using a highly conductive electrode paste. The electrode records evidence of electrical activity produced by the underlying neurons (nerve cells) in the brain. This recording is called an *electroencephalogram (EEG)*. *Electro,* because you are measuring electricity (the potential difference between two electrodes), *encephalo,* which refers to brain, and *gram,* which means writing. Older machines use pens that draw the brain waves on paper that is rolled past the pen. Current technology uses a computer display of the waves. The raw EEG shows the morphology (shape) of the waves, the amplitude (how high the waves are from peak to trough) and the frequency (how many waves there are in one second). Waves with different frequencies occur together, often with faster waves riding on slower waves. Different EEG patterns correspond to different mental states. For example, there are different patterns for sleep compared to being awake, for focused concentration and problem solving compared to drifting off and day dreaming, for impulsive, hyperactive states compared to calm, reflective states, and so on.

A quantitative electroencephalogram (QEEG) involves not just recording the EEG but doing measurements; that is, quantifying data concerning the amount of electrical activity occurring at particular frequencies (say, 4 Hz), or across defined frequency bands (say, 4-8 Hz). The electrical activity is usually expressed either as amplitude, measured in microvolts (μV) or millionths of a volt, or as power, measured in picowatts (pW). The raw EEG shows brain waves, amplitude and wave forms seen as a function of time.

The QEEG uses computer algorithms that transform this raw EEG into quantitative displays that assist the clinician to recognize deviations from normal. You can do a simple QEEG using three leads. The electrodes comprise a positive (+ve) lead, a negative (–ve) lead and a ground. The ground is not truly a ground wire in modern instruments; it performs a type of electrical or instrument housekeeping that results in a good quality recording.

The EEG instrument (the electroencephalograph) measures the potential difference between the +ve and the –ve leads. The positive lead is called the *active* lead and is usually placed over the area you wish to record and measure. The –ve lead, called a *reference* electrode, is usually placed over a relatively inactive area electrically, such as the mastoid bone or an earlobe. This is called a referential placement. It is also possible to measure the potential difference between two active electrode sites on the scalp. This kind of sequential recording gives much lower amplitudes as compared to a referential recording.

The potential difference between the two active sites is also dependent on the ***phase*** of the waveforms being measured and compared. Imagine that you were measuring and comparing the amplitude of two waves running at nine cycles (or waves) per second. Cycles-per-second (cps) is called Hertz (Hz), named after Heinrich R. Hertz, a German physicist who died in 1894. If the waves were in phase with both going up together and one of these waves registered +6 μV and the other +4 μV, the difference would be 2 μV. On the other hand, if the waves were going in different directions, one up and one down, if the first measured +6 μV and the second –4 μV, the difference between the two sites would be much higher, namely, 10 μV. So changes in amplitude can be achieved when using a sequential placement (both +ve and –ve electrodes are over cortical sites) either by changes in amplitude at either site or by a change in phase of similar waveforms at the two sites. It is harder to interpret what is happening but Lubar points out that it gives the brain more ways of learning the task. (The task being to change the amplitude; for example, reduce theta activity or increase sensorimotor activity.)

This same kind of comparison or measurement between pairs of electrodes can be done for many more leads placed at different locations on the scalp. Usually 19 leads at active sites are used, and it is called a *full cap* assessment. This term arises from the fact that this type of assessment may utilize a soft, thin, cloth cap (a popular one was developed by Marvin Sams), which looks very like a swimmer's cap. The cap has tiny electrodes sewn into it. The data from a full cap assessment can be quantified in various ways. The clinician can look at power, relative power or percent power (percent of a certain bandwidth compared to the total power of all bandwidths), coherence, comodulation and phase. (These terms will be discussed later.) It can also be compared to a normal database, with several from which to choose. This is discussed in the section on assessment in Part II. There are centers where work is being done to improve the spatial information from an EEG recording that uses even more electrodes, sometimes over 200.

Another, newer, experimental method of describing information concerning electrical activity is called LORETA (low resolution electro-magnetic tomography assessment). This is a mathematical processthat looks at surface EEG information and infers what activity is occurring in areas a little deeper in the cortex. It was first developed by Roberto Pasqual-Marquis in Zurich. At this time these data appear to be correlating very well with MRI (magnetic resonance imagery) findings. However, LORETA is very sensitive to many kinds of artifact.

We are now able to use LORETA-derived information to guide neurofeedback interventions. Section VII will describe LORETA Z-score neurofeedback.

Note that EEG, though it lacks the spatial resolution possible with imaging techniques such as MRI or PET, has the best temporal resolution. You can see what the brain is doing over time very accurately. EEG also has the advantage of being noninvasive, whereas other mapping techniques often involve injections of radioactive material. With the PET (positron emission tomography) scan, a radioactive form of oxygen is injected. Positrons are given off. These collide with electrons. The result is two gamma rays or photons that are detected by the scanner, which calculates their source. Metabolically active regions of the brain have an increased demand for oxygen and thus regions of increased (or decreased) blood flow can be identified. These other techniques are well accepted by the scientific community, and EEG data correspond well to the other measures. With Attention Deficit Disorder, for example, the slowing of the EEG in the central and frontal regions parallels decreased glucose metabolism shown on PET scans and decreased blood flow seen on SPECT scans in the frontal region.

Event-Related Potentials (ERPs)

An ERP is a measure of brain electrical activity that occurs as a response to a specific stimulus. The EEG, on the other hand, is a measurement of spontaneous and ongoing activity in the brain. ERPs are usually thought of as being time-locked to a specific stimulus. There are interesting aspects to this. For example, ERPs have been found at the exact time when a stimulus was expected but when there was no actual external stimulus present (Sutton, Teuting, Zubin & John, 1967). The definition of an ERP as proposed by Vaughn in 1969 stated that the ***ERP is a brain response that shows a stable time relationship to actual or anticipated stimuli.***

In North America there is not a lot of overlap between people using ERPs and those doing neurofeedback, but research in these two ways of looking at brain electrical activity are complementary. Recently, however, recordings using Evoke Neuroscience equipment include 19-channel EEG, ERPs, and heart rate variability (HRV) measurements in every assessment. The ERP research literature is much larger than the literature on neurofeedback and, because conditions are carefully controlled, it has more scientific respect. Most often ERPs are measured at Fz, Cz and Pz (see the 10-20 electrode system diagram for these locations) and amplitude and scalp distribution are among the variables measured. Amplitudes of one common measurement called the P300 are usually highest parietally and lowest in the frontal region. Research has shown that ERPs can distinguish between different clinical conditions so they are used in diagnosis. The most common application is the use of ERPs by audiologists to test hearing since the presence of the response indicates the brain has responded to the sound even if the person cannot give a verbal response.

Most ERPs are only made visible by averaging many, many samples (at least 20, usually more than a hundred and sometimes several thousand samples). Specific ERPs, in a given individual, come a set time interval after the stimulus and are always the same waveform. When sufficient samples are averaged, the ERP deflections are consistent and will remain, whereas other brain activity will be random and will therefore cancel out. Vaughn suggests four types of ERPs: sensory, motor, long-latency potentials and steady-potential shifts. The sensory ERPs are those evoked by sight, sound, smell and touch. Auditory ERPs occur with a negative peak at about 80-90 ms, and a positive peak at about 170 ms, after the stimulus. These together are called the N1-P2 complex. It occurs in the auditory cortex in the temporal lobe (Vaughn & Arezzo, 1988). Motor ERPs precede and accompany motor movement, and are proportional in amplitude to the strength and speed of muscle contraction. They are seen in the precentral area (motor cortex).

Long-latency potentials reflect subjective responses to expected or unexpected stimuli. They run between 250 and 750 ms after the stimulus. The most often mentioned ERP is a positive response called the P300. It comes approximately 300 ms after an oddball stimulus, although it can be later, depending on variables such as age and processing speed. Children with ADD tend to have a slower P300 than their non-ADD age peers. An *oddball* refers to a meaningful stimulus which is different than the other stimuli in a series. An example is a high tone in a series of low tones or hearing your name when a list of names is being read aloud. The P300 (sometimes shortened to just "P3") indicates that the brain has noticed something. The P300 was apparently discovered by Sutton, Barron and Zubin in 1965. The orienting response is also seen as an ERP. A switch in attention evokes what can be termed P3a. Engagement operations may evoke parietal P3b responses. Disengagement may evoke a frontal-central P3b response.

One important negative long-latency potential is the N400 (Kutas & Hillyard, 1980). It occurs as a response to unexpected endings in sentences or other semantic deviations. The lyrics of the song, *Oh Suzanna,* would presumably evoke a series of N400 responses: "It rained all night the day I left, the weather it was dry. The sun so hot, I froze to death. Suzanna don't you cry."

An example of a *steady-potential shift* is one that occurs after a person is told that they must wait for a signal (warning) and then respond to an event. It is a kind of anticipation response. It is seen as a negative shift that occurs between the warning signal and the event. This type of steady-potential shift is called a *contingent negative variation (CNV)* (Walter, Cooper, Aldridge, McCallum & Winter, 1964).

ERPs have been investigated as an aid to diagnosis. For example, using go/no-go paradigms, there are response differences between ADHD children and normals. In a 'go' condition the subject performs an action in response to a cue. A green light is an example of a cue that says you can cross the street. A 'go' stimulus produces alpha desynchronization. In a 'no-go' condition the subject withholds acting in response to a cue that indicates he is not to act. A red light at a street corner is an example of a cue that says you must withhold a response and not cross the street. In this case the subject must suppress a prepared action. There is motor inhibition. Following a 'no-go' stimulus there is an initial desynchronization followed by synchronization in the frontal and occipital areas.

These ERP responses are impaired in ADHD. The ERP amplitudes are higher in normal subjects. It has been demonstrated that 20 sessions of beta training, in subjects diagnosed with ADHD, can result in a dramatic increase in the ERP response (Grin-Yatsenko & Kropotov, 2001). At the time of writing, Yuri Kropotov's group in St. Petersburg, Russia, were doing work using ERPs that should help elucidate the areas of the cortex that are involved in discrimination tasks. Peter Rosenfeld, at Northwestern University near Chicago, USA, has been working on the use of ERPs in lie detection. He has found a different scalp distribution regarding the amplitude of the P300 when someone lies. Interestingly, when the amplitudes of P300 are graphed at Fz-Cz and Pz there is a straight line when subjects are telling the truth and a crooked line when lies are told (Rosenfeld, 1998).

ERPs can also be used to demonstrate the effects of injury. For example, Kropotov has shown that an auditory ERP will decrease if there is an injury in the left temporal-parietal area (auditory cortex) but will increase if the damage is in the frontal area. This increase indicates a loss of inhibition from the frontal to the temporal lobe.

Event-Related Desynchronization (ERD)

Event-related desynchronization (ERD) refers to the observation that increased cognitive or sensory workload results in a decrease in rhythmic slow wave activity and an increase in desynchronized beta activity. When the task is completed there is **post-reinforcement synchronization (PRS)** of the EEG. M. Barry Sterman describes these patterns in his work concerning EEG measurements in Top Gun pilots. He notes that this synchronization phase is self-rewarding. It seems almost as if the brain rewards itself with a little rest – a burst of synchronous alpha waves – after completing a task. Sterman also found that a pilot would shift from faster brain wave activity (beta) to alpha when he was on overload; for example, attempting an impossible landing in the simulator. Thus alpha can reflect different things under different conditions – in these instances, either a very brief rest or giving up. Nothing is ever simple regarding the human brain. It is best to learn early that you must live with a bit of ambiguity in the field of neurofeedback.

Slow Cortical Potentials (SCPs)

The major work in this area is carried out in Europe, principally by researchers such as Nils Birbaumer and his colleagues at the University of Tübingen, Germany, and by John Gruzelier of the Department of Psychology at Goldsmiths University, formerly at the Imperial College of Medicine in London. Very few practitioners in North America are working with SCPs. Some of the equipment used for assessment and for neurofeedback in North America can measure these slow cortical potentials, such as the Biograph-Infinti from Thought Technology. SCPs are very slow waves that indicate the shift between positive and negative. These shifts underlie the electrical activity we are usually measuring. ERPs are described in more detail in Section XXII of this book.

There is great interest in this meticulous work. Gruzelier examined slow cortical potentials in patients with schizophrenia. Birbaumer has been able to teach subjects with severe amyotrophic lateral sclerosis (ALS or Lou Gehrig's disease), who could not speak or move or otherwise communicate, to consciously make a positive or negative shift like an on-off switch, and thus communicate. He has used this to assist them to mark letters of the alphabet and thus slowly write sentences. He has also demonstrated that a shift to electro-positivity can decrease voltages of the normal AC (alternating current) activity of the brain. It may even halt an epileptic discharge.

Early Findings with the EEG

For further description of the history of the field of neurofeedback the reader should consult Jim Robbins' book *A Symphony in the Brain*.

History

The earliest measurements of electrical patterns in the brain were done with animals. These findings, using a string galvanometer to measure the activity, were reported in 1875 by a British scientist named Richard Caton. In the 1920s, Hans Berger, a German psychiatrist, made detailed recordings and observations using his own son as the subject. He observed a pattern of uniform electrical waves in humans that he labelled as the first order waves. This came to be known as *alpha rhythm*, with reference to the first letter of the Greek alphabet. He also observed periods when these waves were absent, the pattern of waves was smaller and desynchronous. This pattern was called *beta*. His studies indicated that when people had their eyes closed the alpha rhythm was prominent, but when they opened them it was greatly reduced, thus linking the alpha rhythm to the brain resting. Berger's observations, published in 1929, are still valid today. He gave not only the Greek letter designations to the field, he also coined the word *electroencephalogram* and the abbreviation *EEG*. His findings were replicated by two British scientists, Adrian and Matthews, in 1934, which brought the field of EEG into the English scientific literature.

In 1958, psychologist Joe Kamiya, using careful scientific methods, demonstrated that individuals could correctly identify when they were producing alpha waves, but subjects were not able to say precisely how they were making that discrimination. Kamiya had one subject say "A" or "B" to indicate if he was in the state. By the third day he scored 400 correct guesses in a row! Kamiya noted later that he was fortunate in finding a subject so attuned to his own mental state as it increased his motivation to do more studies. This early finding is important when considering neurofeedback, which requires individuals to change their mental state according to feedback based on the brain waves they are producing. Research investigating the EEG as it relates to consciousness and awareness has continued for almost half a century. Thomas Hardt, for example, has done EEG recordings with Zen Buddhist monks in Japan, and continues to research questions of consciousness at his center in San Francisco.

M. Barry Sterman, working with cats in the late 1960s at the University of California Los Angeles, demonstrated that they could be trained using a method called *operant conditioning* to increase a specific spindle-like brain wave pattern which ran at a frequency between 12 to 19 cycles per second. He gave the spindle-like activity between 12 and 15 Hz the name *sensorimotor rhythm (SMR)*.

We are grateful to Dr. Maurice Barry Sterman for allowing us to use the picture below, which was taken in his lab during his brilliant experiments that for the first time demonstrated that brain waves could be operantly conditioned.

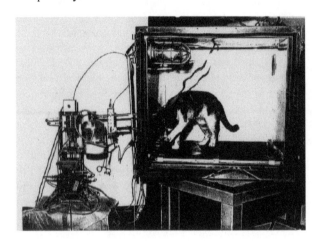

Closely following on this discovery was his observation that cats that had increased their SMR activity were resistant to seizures caused by exposure to hydrazine, a toxic chemical used in rocket fuel. This hydrazine had been causing seizures in air force personnel exposed to it during rocket fuelling operations. He tried the same operant conditioning to increase SMR in human patients with epilepsy and found that it decreased the frequency, duration and severity of seizures, and sometimes totally controlled their seizures. This effect has been successfully reproduced in other laboratories, as reviewed in Sterman's article published in *Clinical Electroencephalography,* January 2000. (That whole issue of *Clinical EEG* is devoted to neurofeedback and it is definitely recommended reading.)

Another physiological psychologist who was already working with EEG, Joel Lubar, came from the University of Tennessee to UCLA on a National Science Foundation grant to work for an academic year with Sterman. The observation had been made that many seizure disorder patients had hyperactivity and became calmer with SMR training. The question arose as to whether SMR training might be beneficial for children with hyperactivity. Margaret Shouse, a graduate student with Lubar at the time, decided to direct her doctoral thesis effort to this question. Using many outcome measures, she found that a significant number of the children she provided with SMR training objectively benefited. Building on earlier work concerning the EEG and on these findings, they published a paper on the neurofeedback treatment of an ADHD child (Shouse & Lubar, 1976, 1979).

Lubar has continued his work on operant conditioning for ADHD at the University of Tennessee. He discovered that measuring the theta/beta ratio was a key to differentiating between normal and ADHD clients. Joel Lubar and his wife Judith, a social worker, have now amassed over 35 years of experience using the EEG for assessment and treatment. The Lubars have taught hundreds of professionals and students, and are responsible for much of the research in this field, particularly with respect to using neurofeedback for people with ADD.

Some Applications that Use the EEG

Clinical Electroencephalography

The primary (medical) use of the EEG is **not for most readers of this text.** *Clinical encephalography* is used to detect and analyze transient events in the EEG that are abnormal. These events have a known clinical significance. An example is spike and wave activity that may be observed in a seizure disorder. This is largely the domain of neurologists and electroencephalographers.

The role of the neurofeedback provider is fundamentally different from that of a neurologist. Neurology is a medical specialty wherein the EEG is used for the detection of abnormalities such as seizure disorders, space-occupying lesions (tumors and aneurisms) and arteriovenous malformations. The neurofeedback practitioner's interest usually lies in an entirely different area, that is, normal EEG waves and variations on normal. What is the foreground for our work is just background for the neurologist. Our assumption is that the neurofeedback client has their own medical practitioner and will always have their medical problems handled through that professional or by referral to a medical specialist. Neurofeedback training can be a helpful adjunct to medical treatment. It does not replace it.

Assessment Using Quantitative Electroencephalography

The second use of the EEG is to distinguish patterns that indicate a person is an appropriate candidate for neurofeedback training. Patterns should correspond to the history given by the client. This work involves the use of quantitative electroencephalography (QEEG). This describes the spectral characteristics of the EEG. It may reveal differences from normal databases that would not easily be perceived by visual inspection of the raw EEG. The differences usually involve comparing amplitudes of different frequency bands to expected values and / or examining the communication between different areas of the brain. This information is used for planning a neurofeedback (NFB) training regimen.

Another use of the brain maps generated by QEEG methods is to look at patterns found in different diagnostic groups. The late E. Roy John and Leslie Prichip at New York University in Manhattan have done this in collaboration with psychiatrists at Bellevue Psychiatric Hospital, and have published extensively (John, 1988). They can predict medication response in those with depression, for example, thus increasing the chances of choosing the correct class of antidepressant (SSRI versus tricyclic antidepressants) at the outset of treatment.

NFB practitioners observe the EEG waveforms and distinguish electrical frequency patterns for the purpose of setting up a training program to meet their clients' objectives and assist them to gain a degree of self-regulation. Normal brain wave patterns have been found to correspond to various mental states. We have detailed this in the section on "States of Consciousness and Band Widths." In brief, it has been found that delta activity, 0.5-3 Hz, is found in sleep and also in conjunction with learning disabilities and brain injury. Theta waves, 4-7 Hz, are seen in drowsy states, which are also states in which some quite creative thoughts may occur. Low alpha, 8-10 Hz, is found in dissociative states, some kinds of meditation, and tuning out from external stimuli (daydreaming). High alpha, 11-12 Hz, can be found associated with creative reflection as well as relaxed calm states of optimal performance. Sensorimotor rhythm frequencies, 13-15 Hz. imply being motorically calm with reflection before action. Low beta waves, 16-20 Hz, are associated with singular focus, external orientation and problem solving, while higher beta frequencies may be found in association with anxiety (above 20 Hz) and rumination (around 30 Hz). Note that there is overlap in the frequency bands (theta may be defined as 3-7, 4-7, or 4-8 Hz, for example). There are also shifts with age, moving to the right along the frequency spectrum (frequency on a horizontal spectrum increases as you move from left to right; thus 7 Hz in a child may be 8 Hz activity in an adult). Indeed, in young children, for example, alpha wave forms may have a frequency of 7 Hz.

Correlations of Bandwidths to Mental States (Cz)

Frequency Bands	Correlations
0.5-3 Hz Delta	Movement or eye-blink artifact. Brain damage. Learning disabilities. The dominant frequency in infants.
3-5 Hz Low Theta	Tuned out. Sleepy.
6-7 Hz High Theta	Internal orientation, important in memory recall, can be very creative but may not recall ideas for very long after emerging from this mental state unless these ideas are consciously worked on and developed. Not focused on external learning stimuli such as reading or listening. The dominant frequency in young children.
7.5-8.5 Hz	Visualization.
8-10 (or 11) Hz Low Alpha	Internally oriented and may be observed in some types of meditation. It is possible, but rare, to have a dissociative experience when totally in this state. Adults (eyes closed) have alpha as the dominant frequency.
12 Hz (11-13 Hz) High Alpha	Can correlate with a very alert broad awareness state. This can be a readiness state seen especially in high-level athletes. Persons with high intelligence often demonstrate a higher peak alpha frequency.
13-15 Hz SMR	When this corresponds to sensory motor rhythm (only over the central cortex: C3, Cz, C4) it can correlate with decreased motor and sensory activity combined with a mental state that maintains alertness and focus. Appears to correlate with a calm state, decreased anxiety and impulsivity. It may also correlate with a decrease in involuntary motor activity.
16-20 Hz Beta	Correlates with active problem-solving cognitive activity. It requires more beta when you are learning a task than when you have mastered it.
19-23 Hz	This may correlate with emotional intensity, including anxiety.
24-36 Hz	Can correlate with multi-tasking and be seen in persons who have high intelligence. It can also correlate with ruminating, which may be negative in a person who is depressed.
~ 27 Hz (Elevated in the mid 20s)	May correlate with family history of addiction
38-42 Hz Sheer (Gamma)	Cognitive activity – related to attention and increasing it may help to improve learning disability. It is also referred to as a "binding" rhythm. It may also be seen at the moment of correcting balance.
44-58 Hz	Reflects the effect of muscle activity on the EEG.
60 Hz (50 in Europe and Australia)	Usually electrical interference.

In the above table the frequency brackets used for each name are approximate. To emphasize this we have used different bandwidths throughout the text in this book. The practitioner should always state explicitly the frequency he is training.

The heavy black lines separate four regions that we talk about in the training section of this text. For example, below 10 Hz the waves are usually referred to as slow waves. (Fast waves are those above 10 Hz.) Above 19 Hz is often referred to as high-beta. In training, both the slow waves and the high-beta are often discouraged.

Learning Self-Regulation

The third use of the EEG is for learning self-regulation of brain wave patterns through operant conditioning. Using the information contained in the EEG, the person is given feedback and the reward is information about successfully producing the desired patterns in the EEG. For example, we may ask a client with symptoms of Attention Deficit Disorder to practice holding a mental state wherein they decrease theta (and/or low alpha) while at the same time raising SMR (and/or low beta). The mental state which corresponds to this is calm, alert, focused concentration. This will be discussed in detail in Section XII and XXVI.

Learning Theory and Neurofeedback

There are two basic learning paradigms, Operant Conditioning (Instrumental Learning) and Classical (Pavlovian) Conditioning. Both are relevant to neurofeedback.

Operant Conditioning or Instrumental Learning

This type of learning is based on **The Law of Effect,** which can be simply stated: When you reward behavior you increase the likelihood of its recurrence.

This law was first stated by Edward Thorndike in 1911. He mainly studied cats in puzzle boxes where the cats had to figure out how to get out of the box to get food that was visible just outside the box. He concluded that responses to a particular situation that were followed by satisfaction were more likely to occur when the animal was again in that situation. (That is, rewarded responses were more likely to be repeated than were responses that were followed by discomfort. In this case, things that did not work resulted in the cat staying hungry). This is also known as Trial and Error Learning since Thorndike's cats tried a lot of things that did not work, like mewing and scratching, before they figured out how to get out of the box by pulling on a string or stepping on a treadle. On subsequent trials they did not bother with the other behavior but immediately did the thing that got them out of the box and to the food.

Skinner took Thorndike's Law of Effect and refined it by introducing the idea of operant classes. An *operant* is a response that operates on the environment. Skinner thus emphasizes the function of a behavior. Having a temper tantrum or smiling nicely are in the same class of operants if they both produce attention from the parent. Skinner's operants are voluntary behavior, and this distinguishes these responses from classically conditioned reflexive responses (see below). In what came to be known as Skinner Boxes, a pigeon would be trained to peck at a disc or a rat would learn to press a lever (operant behaviors) with food as a reinforcer. Further experiments established the importance of schedules of reinforcement; for example, a variable reinforcement schedule is more resistant to extinction than a continuous reinforcement schedule. (Hence the problem with gambling where an occasional big payoff is highly reinforcing and results in behavior that is hard to eliminate.)

Skinner and other behaviorists also introduced concepts of secondary reinforcement, shaping and chaining. They investigated how to apply the principles when training animals and also in human learning. When shaping behavior, you reward successive approximations to the behavior; for example, you might reward your dog for lying down as a first step in training him to roll over. In general, operant conditioning can be used for the learning of responses that are under voluntary control. Motivation is a factor and the reward must be meaningful or desired.

Operant conditioning occurs frequently in everyday life. When a young child is asked to do 10 math questions that the child considers very boring, receiving an external reward each time they finish a question (with a double reward if it is correct) may help. If rewards are abruptly stopped, the math behavior may quickly extinguish. If the child is just occasionally rewarded (partial reinforcement) then the math homework completion will be more resistant to extinction. The child finds that they finish the homework more quickly and get out to play. Soon the child may finish the work quickly and correctly just with the knowledge that they will then have earned free time. The play time is a secondary reinforcer. This may turn into what parents call a good habit as they grow older. The essential factor in operant conditioning is that when you reward behavior you increase the likelihood of its recurrence.

Classical Conditioning or Autonomic Conditioning

Classical conditioning is a term that refers to another type of learning. It was originally described by Pavlov in Russia as a conditioned, or learned, reflex. Pavlov had studied the reflex involved when a dog salivates at the presentation of food. He paired ringing a bell with the delivery of meat powder, and the dog "learned" to salivate at the sound of the bell. The food was an unconditioned stimulus that produced an unconditioned response (salivation). When the conditioned stimulus, the bell, was paired with the presentation of meat powder it came to elicit an almost identical conditioned response of salivation. Pavlov also did *second order conditioning* by illuminating a light just before the bell, without any meat. The light, too, came to elicit salivation.

True classical conditioning can only be done when there is a reflex response to begin with, so it is really restricted to autonomic nervous system responses, and does not apply to new behavior. Motivation is largely irrelevant.

Emotional conditioning, which is a subset of classical conditioning, occurs when any gut reaction, from anxiety to relaxation, is paired with a neutral object. An example would be a person who liked to fly developing a fear of flying after a particularly rough flight that terrified them. In a similar vein, a young child (or the family dog for that matter) may get excited and run to the front door when he hears it open because that sound has been paired with father coming home to play. In this case father's arrival home is an unconditioned stimulus that elicits increased excitement/autonomic arousal in the child and the family dog.

John Watson conducted an (in)famous experiment that demonstrated acquisition and generalization of a fear response. He conditioned fear of a white rat in little Albert, an 11-month old child who loved to touch and explore things, by making a loud noise as Albert reached for the rat. The fear generalized to other white furry objects (rabbits, cotton, Santa mask, Watson's white hair). In **classical** conditioning the conditioned stimulus *automatically elicits* a conditioned response after it has been paired a sufficient number of times with an unconditioned stimulus that elicits an autonomic response. This is why motivation is irrelevant in this kind of learning. Watson also coined the term "behaviorism," which he used in a lecture in 1912, though it was Skinner and operant conditioning, rather than Watson with his work using classical conditioning, who became known as the Great Behaviorist.

What Kind of Learning Occurs with EEG Biofeedback?

Operant Conditioning

In Sterman's seminal work with cats back in the 1960s, the production of brain wave patterns, which came to be known as sensorimotor rhythm, was rewarded with milk and chicken broth. In our work with EEG biofeedback operant conditioning occurs when the client is rewarded for finding a mental state which results in his meeting the thresholds which have been set for designated slow and fast waves. This is rewarded with visual and auditory feedback, usually using a game-like display. There may be secondary rewards of praise or tokens, which may be exchanged for little rewards. It seems that the human brain will learn with information about success as the reward. Soon the person can get into the desired mental state quite rapidly. It is similar to training the person to hit a tennis serve. At first it is awkward and difficult. If they practice exactly the same swing many times, it becomes automatic. With motor training, coaches estimate that between 1500 and 5000 correct repetitions of a movement are needed for it to be automatic. With neurofeedback the general rule of thumb (at least for managing ADD symptoms) is a minimum of 40 training sessions.

In *operant conditioning* of brain waves, the student operates on the display by changing their mental state until rewards are received. The student practices this many times. After enough practice, moving to that state becomes almost automatic. At that juncture, our job in neurofeedback is to facilitate the transfer of this skill to other situations, such as the classroom or when doing homework. For this a second step of pairing, the desired mental state with doing an academic task can be helpful. This second step is hypothesized to involve *classical conditioning*.

The basic principle is that when you reward production of a particular brain wave pattern with auditory and visual feedback, then that information acts as a reward and you increase the likelihood of recurrence of that brain-wave activity. The human brain will work for information.

Classical Conditioning

Classical conditioning occurs when the desired mental state of focused concentration is paired with carrying out an academic task during the neurofeedback training session. This is done by having the student find the desired mental state of focused concentration, which corresponds to decreased slow wave and increased fast wave activity in the EEG, using the operant conditioning paradigm described above and then *pairing* that mental state with doing an academic task. The fact that the desired mental state is being retained is evidenced by continued auditory feedback even if the visual focus is on text or a math question. If the auditory feedback should stop, the student is instructed to return their attention to the NFB display screen until they again have a steady feedback state. Then they resume the academic task.

We increase the likelihood of the student approaching academic tasks in this focused mental state during everyday life by pairing the mental state with learning metacognitive strategies during their neurofeedback training. (For more on metacognition, see Section XV.) Then, when they are at school or doing homework and consciously think of the strategy, they should immediately go into the desired state of focused concentration, which is the physiological state they were in when they learned the strategy.

Other Relevant Learning Paradigms

Shaping
Shaping refers to conditioning by successive approximations. Animal trainers use shaping extensively and can get animals to do extraordinarily complex behaviors by rewarding little steps in the desired direction. Rewarding a behavior or a sequence of neurophysiological occurrences *shapes* the contributing components of that sequence in a way that results in an increased frequency of that sequence occurring (Sterman, 2000). Shaping is done when you reward a small shift in the microvolt level of a particular frequency band, and then, as the client is successful, you change the threshold to make it a little more difficult. This is part of *operant conditioning*. When working with people with Attention Deficit Disorder, for example, you reward a shift toward more mature patterns, which means reducing the dominance of slow wave activity in the theta range.

Incidental or Associative Learning
Incidental or associative learning occurs when things are unintentionally paired with reinforcers. The red light that indicates muscle activity on some neurofeedback instruments is an example. Although it is necessary to reduce EMG induced artifact in the EEG, you do not want the less important information regarding EMG activity to be more prominent than the reward for the mental state you are training. If the client focuses mostly on the EMG light, your learning curve for EEG changes may take longer. You may want it prominent at first, however, so that the client learns to reduce EMG so that they get quality feedback with less artifact. This *associative learning* can be both a help and a hindrance. We want the associations to be with things such as strategies that they can take with them and transfer to situations outside of the office. In our training we change times of day, materials used, instruments, feedback screens, and trainers. We don't want a transference cure due to a desire to please a particular trainer, we want a change in EEG patterns. Our hope is to reduce to a minimum this type of pairing to stimuli which only occur in the office situation.

Secondary Reinforcers
Secondary reinforcers, such as praise and tokens, can be used to further reinforce learning of brain wave states. The tokens can be exchanged for prizes to help motive a child. This is particularly useful for children with ADD who tend only to be able to focus if something is inherently interesting to them or becomes interesting because there is a tangible payoff for doing it. Skinner would call the tokens a "generalized conditioned reinforcer" since they can be used for a variety of self-selected rewards. Money has the same role for adults who work for that reward. Whatever you use as a reinforcer must be desired by the person learning or they will not be motivated to work for it. Remember that motivation is only important in operant conditioning. Classical conditioning relies on reflexive responses.

Generalization
In its simplest form, this term means that what the client learns in the office doing neurofeedback will also occur at other times, places and with other people and tasks. We know that the ability to generalize is severely impaired in some disorders such as autism.

We have already touched on the importance of taking generalization into account and used the example of having the student use metacognitive strategies during the training session and then use the same strategy when starting a task outside of the center. There are many methods you can use. For example, a young child can turn on the state by looking at the top of their pencil and focusing steadily on it for few seconds then gradually broadening that focus to include the book or the black board. For persons who are tense we recommend breathing techniques that we have already paired with an advantageous mental state in the training sessions. An obvious method used by coaches is the warm up exercises for athletes. Getting the client to use self-cueing of some kind, like a word ("focus") or movement (sitting straight, breathing calmly) that has been paired with the production of the desired state, can promote generalization.

The observation that the results of neurofeedback training generalize is something that sets it apart from other treatments for ADHD. The use of medication does not produce generalization of improved behavior or neater handwriting when the drug wears off. Behavior modification that works in one class does not usually generalize to another class or to the playground where the same contingencies and rewards are not in place.

Extinction

In classical conditioning, *extinction* occurs when the conditioned stimulus is no longer paired with the unconditioned stimulus over a number of trials. In operant conditioning it occurs when a behavior is no longer reinforced (rewarded). Since we want to have lasting effects, we want the response (production of the desired mental state) to be resistant to extinction. This is why secondary reinforcers are important. Pavlov found that even after several years a conditioned response could be restored to full strength with a few trials, so relearning is much more rapid than original learning. Sometimes having a client who has ADD back for a few refresher training sessions is a good idea if things seem to slip in terms of concentration.

When you train a person to do a particular skill, their ability will decrease over time unless the skill is practiced. However, if there is intermittent reinforcement of the skill, the tendency to lose the skill to the point of it being extinguished altogether is markedly decreased. In real life, the student should receive positive reinforcement (praise, better grades) for their new ability to self-regulate attention, which should further reinforce the behavior (mental state).

Note: Learning theory alone does not explain why neurofeedback results appear to last. Most people working in the field hypothesize that structural changes in the brain are also a factor. Changes in the production of neurotransmitters or the way they operate at the synapse may also occur. The mechanisms for immediate change and for lasting change have not yet been established; however, several possible mechanisms appear to exist and are discussed further in Section VII.

Which Conditions Are Appropriate for Neurofeedback Interventions?

Diagnosed Conditions in Which NFB May Be Helpful

The list of conditions for which there is considerable research, including controlled studies published in peer-reviewed journals, is short at the time of writing. As the AAPB/SNR joint guidelines discuss (La Vaque et al., 2002), it is important to distinguish between validated applications, those with some support, and those that are experimental. Seizure disorders and Attention Deficit Disorder are in the first group. The second group includes treatment of depressed mood, treatment of alcoholism and addictions, helping those with closed head injuries (CHI)/traumatic brain injury (TBI), and work with children who have learning disabilities. Applications that look promising due to clinical reports of improvement but which are not yet verified include Tourette syndrome and other movement disorders (Parkinson's disease, dystonia), Asperger's syndrome and high-functioning autism, "brain brightening" in the elderly, obsessive compulsive disorder, and generalized anxiety disorder. When anxiety is part of the symptom picture, it makes sense to include biofeedback.

The work on seizure disorders is well reviewed by Sterman (2000) and in a meta-analysis by Tan (Tan et al., 2009) Joel Lubar has been the leader in research in Attention Deficit Disorder. A multi-site study has established norms for theta/beta ratios (Monastra et al. (1999). In addition, a meta-analysis has been published by Arns (Arns et al., 2009).

Vince Monastra has also published research showing that improvements in ADHD symptoms achieved with neurofeedback continued after training, whereas improvements with stimulant medication alone were not sustained when the drug was withdrawn

(Monastra, 2002). A controlled study by Gani has also demonstrated this lasting effect (Gani et al., 2011).

A Therapeutic Procedure to be Used by an Experienced Psychotherapist

Neurofeedback can be used as an adjunctive procedure in psychotherapy. This use is based on the observation that slow wave activity, particularly in the theta range, can be associated with a *hypnagogic state* (the state we all experience between wakefulness and sleep) that allows for what Sigmund Freud termed primary process thinking. (*Hypnopompic* refers to the partially conscious state preceding wakening.) In the hypnagogic state the client is not consciously evaluating the ideas that drift through his or her mind or float up from the "unconscious." This training comes under the heading of Alpha-Theta training. There has been considerable work done with alcohol dependency starting with Peniston's work (Peniston & Kulkosky, 1990). It will be briefly described in the interventions section of this book.

Optimizing Performance

This type of work is not usually within the purview of healthcare professionals, though a professional such as a psychologist may do the initial evaluation. Training sessions can be carried out by coaches, trainers, and teachers. Training that combines neurofeedback with biofeedback allows the participant to have a flexible brain, for example, being able to produce calm, relaxed yet alert, focused concentration with appropriate reflection before action. A diverse population may benefit from this type of work since difficulties with attention span, concentration or being a bit impulsive interfere with the student using their full intellectual potential or with executives and athletes reaching their top functioning.

Many children currently (incorrectly) diagnosed with ADHD fall into this group. This group would fall into Thom Hartmann's description of the 'Hunter' mind. They can hyperfocus when there is something of interest that they want to pursue, but have difficulty with time management and with concentration for things that are boring or slow-paced. They do not qualify for a formal diagnosis of ADHD as one could not say their functioning is impaired by their ADHD symptoms to a clinically significant degree, but they are underachieving. The underachievement is frustrating to them, their teachers and their parents. Without intervention, there may be "impairment" as they get into high school or university. Neurofeedback training can play a preventive role in such cases, giving the child self-regulation skills so that their behavior and learning improves over time rather than worsening.

A second group that can benefit from optimal performance training is athletes. Sports require both intense concentration and an ability to shift mental states quickly. Golfers, for example, must analyze the shot they need to make, taking myriad variables into consideration (wind, lie of the ball, distance to the green or the hole, etc.). This mental work requires beta activity, but they must shift into alpha to release their shot effortlessly. This training can help an athlete find the zone where performance in their sport seems effortless and automatic. Jim Robbins wrote an article called "The Mental Edge" about athletes and neurofeedback (*Outside*, April 2001).

A third group who make prime candidates for optimal training is comprised of executives. They often work under intense pressure and to tight timelines, so they need to be able to handle stress and work efficiently. This requires good management of their psychophysiology. For example, breathing techniques quickly produce a calm state. They must shift gears mentally to switch from careful observation and/or listening to processing and making decisions. It is a great asset to be able to choose your state: calm and reflective, or energetic and enthusiastic, depending on what kind of interaction you want to have with colleagues. Adults other than business people can also benefit. We worked with a graduate student who improved her concentration and organization in order to complete the thesis requirement for her degree. On another occasion, we assessed a university professor who had an impressive curriculum vitae with over 150 publications (articles) but felt unable to sustain his focus to complete a book.

Another application for optimal performance is with music performance. Rae Tattenbaum has reported her work at meetings and John Gruzelier has done some elegant controlled research with students at London's Royal College of Music (RCM) (AAPB *Proceedings*, 2002). The results over two years were impressive enough that NFB became part of the curriculum at the Royal College in the fall of 2002. It seems that whatever level you are at, you might reach a higher level with some training. Remember that optimal performance is considered an experimental application of neurofeedback as it does not yet have a sufficient, published research base to be considered an established application.

Assessing Efficacy

A joint Efficacy Task Force of the Association for Applied Psychophysiology and Biofeedback (AAPB) and the Society for Neuronal Regulation (SNR), now called International Society for Neurofeedback and Research (ISNR), developed standards for efficacy research methodology, and a template for rating the level of efficacy of each application. Two articles about this important endeavor were published in 2002, appearing in both *Applied Psychophysiology and Biofeedback* and the *Journal of Neurotherapy*. They are entitled "Task Force Report on Methodology and Empirically Supported Treatments: Introduction" (Moss & Gunkelman, 2002), and "Template for Developing Guidelines for the Evaluation of the Clinical Efficacy of Psychophysiological Interventions (La Vaque & Hammond, 2002). Readers are referred to these two important papers. They were meant to be used as the foundation for a series of scientific reviews and practice guidelines to be published by both societies. This was updated by Yucha and Montgomery (2008) and is discussed in Section XXX.

A literature review is beyond the scope of this text. There exists a *The Byers Neurotherapy Reference Library,* published in 1998 and available through the AAPB bookstore. For a more current source, a helpful listing of articles, arranged according to conditions in which neurofeedback has been applied, has been compiled by Hammond and is available on the web at www.isnr.org. It primarily includes outcome studies and case reports. Look for "Comprehensive Neurofeedback Bibliography" under Neurofeedback Archive on the website of the International Society for Neuronal Regulation. (The "International" was added to the SNR name as of 2003 in recognition of the Australian and European chapters of that society.) As of mid-2003, the Hammond list included:

- Epilepsy
- ADD/ADHD, Learning Disabilities and Academic- Cognitive Enhancement
- Anxiety Disorders, PTSD, and Sleep Disorders
- Depression, Hemispheric Asymmetry and Anger
- Addictive Disorders
- Brain Injury, Stroke, Coma and Spasticity
- Chronic Fatigue Syndrome, Fibromyalgia and Autoimmune Dysfunction
- Pain and Headache
- Plus a dozen other conditions with single case studies

SECTION II
Origin of the Electroencephalogram (EEG)

Please remember during all of the following that despite the knowledge contained in textbooks and articles, what is really known at this time about the brain could be likened to what Galileo knew of astronomy. Many breakthroughs have been made in neuroscience, especially during the Decade of the Brain in the 1990s, but the field is still in its infancy. Nonetheless, what we are beginning to understand is fascinating and much of it is relevant to our work in neurofeedback. The discovery of greater *neuroplasticity* is one such relevant finding. It is not just that we start life with billions of neurons that get pruned, but our brains can make new neurons and grow new connections among existing neurons throughout our lives. For an interesting and readable account of brain function in the elderly, see the book *Aging with Grace*, in which David Snowdon provides a popular account of the research known as the Nun Study. Neuroplasticity is also discussed by Norman Doidge (Doidge, 2007, 2015).

Definition
What is an EEG?

The brain's neurons communicate by the conduction of electrical currents along dendrites and axons. Chemical conduction using neurotransmitters occurs at the synaptic junctions between nerves. It is somewhat like a huge complex city that is dependent on its electrical wiring. The analogy immediately fails, however, because the brain is far more complex. Each of the billions of neurons has thousands of connections, though it is postulated that there are only four synapses separating any two neurons in the brain. (That last fact comes from a German neuroscientist by the name of Poppel, who was interviewed for a Lufthansa flight magazine article published in April 2002. The source attests to how popular the world of neuroscience has become.) Perhaps the worldwide telephone network is a better analogy than a city's electrical system, since there are local, regional, and widespread connections. (This is further discussed in Sections VI and VII.) For fast, long-distance communication, the brain uses myelinated fibers (white matter) just like fiber optic cable in telephone systems is faster than regular cable for voice transmission. We have not yet discovered the equivalent of satellite transmission, though perhaps there is an as yet undiscovered brain equivalent that would explain telepathy – the phenomenon that first got Hans Berger interested in brain activity.

The *electroencephalograph* is an instrument that detects and amplifies the electrical activity in the brain. The EEG instrument measures the potential difference between pairs of small electrodes (sensors) placed on the surface of the scalp using a highly conductive medium. This is usually a conductive electrode paste such as 10-20 conductive paste or Elefix, although saline solutions are also used. The electrodes record electrical activity produced by certain neurons (nerve cells) in the brain called pyramidal cells. The resulting recording is called an *electroencephalogram* (EEG), *electro* because you are measuring electricity (the potential difference between the activity at two electrodes), *encephalo*, which refers to the brain, and *gram*, which means *writing*. Many hospital instruments still use pens that actually draw the brain waves on paper that is rolled past them. The instruments used for neurofeedback purposes display the results on a computer monitor. In either case, one has a wavelike line that shows the amplitude of the electrical activity over time. Different frequencies can be seen in the tracing/display. The unit of measurement for the frequencies is cycles per second, or Hertz (Hz), named for Heinrich Hertz, a German physicist who died in 1894. Amplitude is usually measured in microvolts, or millionths of a volt. Different frequency ranges correspond to different mental states; for example, alpha (8-12 Hz activity) is a resting state.

Why Bother with the EEG?

In general, the EEG is helpful as a way of monitoring brain activity because it is noninvasive and has excellent temporal resolution. In these respects it is better than imaging techniques like PET and SPECT, though they have better spatial resolution. You know what the brain is up to from moment to moment when you look at the dynamic EEG; that is, which areas are resting and which are active.

There is interesting research on how brain map patterns can correspond to diagnostic categories. Studies done with American whites, American blacks, Scandinavians and Chinese have all yielded similar results. E. Roy John, in his March 2000 presentation at the annual meeting of the Association for Applied Psychophysiology and Biofeedback (AAPB), gave an overview of the neurometric approach that he has developed at his Brain Research Labs, Department of Psychiatry, New York University. Since 1973 he has used the EEG to produce brain maps, and with his colleagues, notably his wife Leslie Prichep, he has done mathematical transformations of the data to find patterns that correspond to diagnostic categories. This is painstaking work, and they have 2,008 values in their matrix that plots electrode placements by frequencies. In the AAPB presentation, they noted that the utility of their 82 diagnostic classifications, which can discriminate with 85-90 percent accuracy, lies in being able to predict treatment response. For example, in the elderly there is a 94 percent correct discrimination between depression and dementia.

This information has important treatment ramifications, particularly in terms of which drugs the psychiatrist will prescribe. John and Prichep's neurometric approach can distinguish ADD from normals with 90 percent accuracy, and these researchers can build a discriminant function that distinguishes between responders and nonresponders to stimulant drugs even when the symptom picture is the same. John has noted that among the advantages of using the EEG are the findings that it is stable over time and not influenced by culture.

Using a single-channel placement at Cz, the multi-site study led by Vince Monastra and Joel Lubar established theta to beta power ratios that had even higher sensitivity for distinguishing between ADHD subjects and those in the control group (Monastra et al., 1998).

Whereas John and Prichep's findings have mainly been used by psychiatrists to guide drug treatment, the finding that most interests neurofeedback practitioners is that the patterns found by means of the EEG can be changed through neurofeedback. These changes can ameliorate symptoms or optimize performance.

Here is a summary of *six good reasons for using the EEG,* both diagnostically and for changing brain and behavior patterns through the learning process called operant conditioning, which came to be known as neurofeedback.

1. **Mental states can be "defined" by the EEG.** Certain frequency bandwidths correspond to particular mental states. These were described for each commonly used bandwidth (such as theta, 4-8 Hz) in Section III. Examples are: theta – internally-oriented, drowsy, drifting off, memory retrieval and visualizing; alpha – internally-oriented, contemplative, perhaps daydreaming and/or meditative states; SMR – calm states where individuals reflect before acting, beta – alert, problem-solving, often externally oriented states.

2. Certain **brain wave patterns** (and we are referring here to normal brain waves, not abnormal ones as are found in seizure disorders) **correspond to common disorders** or syndromes. A good example is the high theta and low beta pattern (high theta/beta ratio) found in persons diagnosed with Attention Deficit Disorder (Monastra et. al., 1998).

3. Both animals and **humans can learn to alter their brain wave pattern** by means of operant conditioning/EEG biofeedback. The earliest work was done in the 1960s with Sterman demonstrating that cats could increase a particular rhythmic pattern in the 12-15 Hz range that he named *sensorimotor rhythm* or *SMR*. Increasing these frequencies was found to be associated with a reduction in sensory input being relayed to the cortex and a reduction in motor output.

4. Both animals and **humans show changes in behavior when they have learned to change their brain wave frequency patterns** by means of operant conditioning. Again, the earliest work was that of Sterman, now Professor Emeritus at UCLA, with cats. The means of operant conditioning of brain waves used with the cats was that they were rewarded with a milk and chicken broth mixture as they produced SMR. The cats became still, yet alert, as they increased

SMR. When the contingencies were changed and they were rewarded for reducing SMR, they learned that, too, and they became twitchy cats, flicking their ears and tails. Further work demonstrated that this training to increase SMR resulted in the cats being resistant to seizures. Once work was begun with human subjects who had epilepsy it was noted that as they reduced the frequency and severity of seizures, their symptoms of hyperactivity were also reduced. This led to applying the techniques with hyperactive children. Work done principally by Dr. Joel Lubar at the University of Tennessee over the last 35 years has demonstrated that children can learn to decrease slow waves in the theta range and increase fast waves in the beta range, resulting in a marked increase in attention, with decreased impulsivity and hyperactivity. Other variables that changed included statistically significant improvements in their performance on traditional intelligence tests (Wechsler Intelligence Scale for Children), on continuous performance tests (Test of Variables of Attention, TOVA), and in school performance. Work with athletes has demonstrated that differences between experts and intermediate level individuals can be clearly distinguished with the EEG (Landers, 1991). Promising work is now being carried out in the area of improving athletic performance.

5. **Brain maps** using 19 active electrodes can **help distinguish psychiatric syndromes.** Much of this work has been done by E. Roy John at New York University, as noted above. The brain maps help predict response to medication. When excess alpha is found centrally (above the cingulate gyrus), 80 percent of patients diagnosed with obsessive compulsive disorder will respond to SSRIs. In OCD patients with central theta, only 20 percent responded. In a similar vein, Richard Davidson (1998) found that depressed individuals show less activation (higher alpha) in the left frontal lobe. Elsa Baehr and associates have demonstrated that depressed patients can respond positively to operant conditioning using the EEG (Baehr, Rosenfeld, Baehr & Earnst, 1999).

6. **Brain maps** using 19 active electrodes can **demonstrate communication patterns** between different areas of the brain. The terms used for this kind of work are *coherence* and *comodulation*. As mentioned earlier, data from full cap (19-lead) assessments can be compared to normative databases. Databases have been developed by E. Roy John, Frank Duffy, Robert Thatcher, William Hudspeth, M. Barry Sterman, Yuri Kropotov, James Thompson and David Hagedorn. Statistical comparisons may demonstrate too little or too much communication between areas of the cortex. This information can then be used to do training that helps individuals overcome some of the symptoms of various disorders. Coherence training may prove to be particularly useful with mild closed head injuries. The approach is to do training that normalizes the EEG.

In conclusion, this section's main point is that mental states can be changed through neurofeedback. The change can be targeted toward normalization or toward optimal performance.

Target: Normal Patterns

Operant conditioning can move a client *toward normal patterns* when used to produce a mental state that is relaxed, calm, reflective, alert and focused, with an appropriate degree of movement. These changes may overcome symptoms of ADHD or decrease symptoms of some seizure disorders. Indeed, neurofeedback is among the preferred treatments for these two conditions, as noted in the section concerning biofeedback on the National Institutes of Health website for alternative and complementary medicine. NFB may also ameliorate symptoms in anxiety, depression, addiction, movement disorders, and closed head injury. It may improve difficulties in socializing, as found in Asperger's syndrome or in high functioning autism, though there is less published literature for these applications.

Target: Optimal Performance

Operant conditioning of brain wave patterns can move an athlete or a business person *towards optimal performance* for work situations, academic and athletic performance. Again, there needs to be more work published. There are intriguing results from studies such as John Gruzelier's regarding improved music performance, particularly with respect to the interpretive, emotive aspects of performance.

How Is it Possible that the Very Small Voltages Produced by Nerve Cells Can Be Detected?

Electrical activity that we measure comes from the cortex. More precisely, each pyramidal cell acts rather like a little battery to produce a dipole. Dipoles are important because to detect electricity we must have a potential difference between two points. The cortical site, such as Cz, will have electrical activity beneath the sensor due to the dipole created when a pyramidal cell is activated. The site to which it is referred, such as the nose, mastoid bone, or an ear lobe, is usually much less electrically active.

This electrical activity depends on special characteristics of the pyramidal cell. Other cells in the cortex do not have this ability to create dipoles, though they do influence the pyramidal cells. Roberto Pascual-Marqui from Switzerland (*Proceedings*, Society for Neuronal Regulation annual meeting, 2000), who has done brain research in Zurich and who developed LORETA, has given an eloquent explanation, which is summarized below.

The Physiological Basis of the EEG

The EEG is defined as the difference in voltage between two different recording locations plotted over time (Fisch, 1999). The EEG is generated by the synchronous activity of postsynaptic inhibitory and excitatory potentials involving large groups of cortical pyramidal cells. These pyramidal cells' postsynaptic potentials form an *extracellular dipole layer*. "This dipole layer parallels the surface of the cortex projecting opposite electrical polarities towards the cortical surface compared to the innermost layers of the cortex." (Fisch, 1999). The postsynaptic potentials have a long time duration (15-200 milliseconds). These potential changes summate and the EEG records the potential (+ve or –ve) directed towards the electrode on the surface of the scalp.

The charge will differ depending on whether an excitatory postsynaptic potential (EPSP) or inhibitory postsynaptic potential (IPSP) has been generated in the area of the cortex beneath the electrode. The standard electrode used in neurofeedback is called a macroelectrode and it detects the activity of a very large number of neurons beneath it. (Microelectrodes are much smaller, less than two microns in diameter, and are used for measurements of electrical activity within the brain, as, for example, in animal research when the electrodes are implanted beneath the skull.) Each electrode can measure the electrical activity of an area of about six square centimeters. Action potentials which travel down the axons or dendrites of these cortical cells have a very short time duration (1 ms) and that electrical activity does not significantly contribute to the EEG.

If how all of this happens is already clear to you, you can skip the rest of this section. If you wish a bit of review, the next sections will explain about action potentials, postsynaptic potentials, and the current thinking about mechanisms that govern the production of the EEG.

Pyramidal Cells

Terms

Sink – Where positively charged cations entered the cell, leaving a negative charge in the extracellular space. The *sink* may be at the base, middle or apex of a pyramidal cell dendrite.

Source – Where current leaves a cell

Dipole – Electric field between source and sink

Macrocolumn – The neurons in the cortex are arranged in groups called macrocolumns. Each column consists of a group of cells several millimeters in diameter and six layers deep. These groups contain pyramidal cells, stellate cells (excitatory) and basket cells (inhibitory). They also contain glial cells. The glial cells outnumber the pyramidal cells. These cells are important for their role in supporting pyramidal cells: providing nutrition, removing waste products, and giving structural support.

Measurement of Postsynaptic Potentials

In the following diagrams the nerve axon that is connecting with the pyramidal cell is excitatory. If it were inhibitory, then the electrical charges marked on the diagrams in the extracellular space would be opposite to those shown. The positive (+ve) would be negative (-ve).

Example #1, an excitatory postsynaptic potential (EPSP) at the *distal* end of a pyramidal cell dendrite.

Influx of sodium makes for what is known as an active *sink* at the level of the synaptic input from another cell's axon. An active *source*, which is positive, is created outside the pyramidal cell body at the other end of the dendrite. The negative charge (sink) is created outside the cell when sodium, which has a positive charge, rushes into the dendrite due to chemical changes that make the surface more permeable to sodium. This inrush of positive ions into the distal end of the dendrite, as shown in the diagram opposite, leaves a negative charge outside the portion of that dendrite next to the surface of the scalp and just under our electrode. Inside the dendrite the positive charge is towards the surface of the cortex and the negative end of this intracellular dipole is toward the pyramidal cell body.

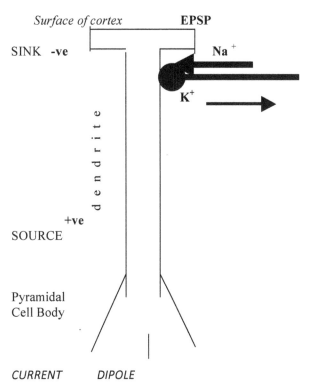

The voltage measured by an electrode on the surface of the scalp above this pyramidal cell's dendrite (and referenced to a point some distance away from it) would be *negative*. It would be measuring an **EPSP** (excitatory postsynaptic potential). An **IPSP** (inhibitory postsynaptic potential) is not shown in this diagram. Inhibitory neurotransmitters make the surface membrane less permeable to sodium, yet potassium (also with a positive charge) continues to be released so the charge outside the dendrite's membrane would be positive. The electrical charge detected by an electrode on the surface of the scalp above this site would then be positive. That is, it would be opposite to the reading on the surface of the scalp produced by an EPSP at the distal end of the pyramidal cell dendrite.

Example #2, an excitatory postsynaptic potential (EPSP) at the *proximal* end of a pyramidal cell dendrite

If the synaptic connection took place near the base (cell body) of the pyramidal cell, then the active sink (-ve) is closer to the soma (body) of the pyramidal cell and the source(+ve) would be at the distal end of the dendrite, closer to the cortex.

The voltage measured by an electrode on the surface of the scalp above this pyramidal cell's dendrite and referenced to a point some distance away from it would be ***positive***. The current dipole is in the opposite direction to the first example.

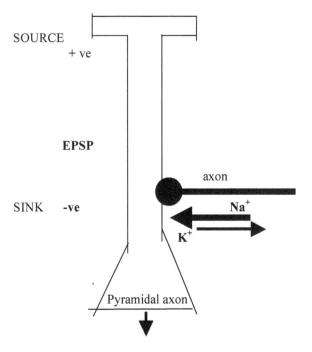

The Neurofeedback Book, Dr. Thompson, (www.addcentre.com).   Published by the AAPB. (www.aapb.org)

Conditions for Current Detection

Having looked at this diagrammatic representation of a pyramidal cell, we must ask ourselves how such small voltages could possibly be detected. The simplest way of understanding this has again been provided by Pascual-Marqui. He explains that four conditions must exist before there is sufficient electrical activity to allow for detection at the surface of the scalp.

Directionality

What would happen if the pyramidal neurons were arranged haphazardly?

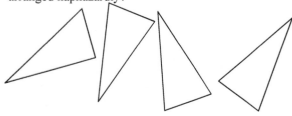

The contribution of these pyramidal cells would sum to zero and no EEG would be detected.

However, pyramidal cells in the cortex are lined up perpendicular to the surface, though not as perfectly as in this schematic diagram because of the convolutions of the cortex.

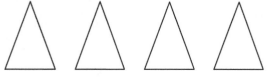

Synchronicity

Cells must be firing together in a synchronized fashion to produce current flow. If all the cells were firing randomly rather than at the same time, then the sum of their potentials would be close to zero at any given point in time.

This simultaneous firing is possible. One clear mechanism that affects timing concerns subcortical structures controlling the rhythm of firing in the cortex. The clearest contributor is the thalamus. We have already noted how it controls theta, alpha and SMR patterns.

Similarity of Position (Proximal or Distal)

The same thing must be happening at precisely the same time for the majority of cells within a cluster or macrocolumn of neurons. We must have the **simultaneous** occurrence of discharge at the synapses of the axons which connect to the dendrites of the pyramidal cells **at the same level on the dendrite**. The postsynaptic potentials (**PSPs**) will then be of the same charge; for example, all will have a sink near the top of the dendrite, leaving a negative charge near the surface of the cortex.

Valence

The valence (+ve or –ve) must also be the same for virtually all of the pyramidal cells within a spatial cluster or they will cancel each other out. Thus it has to be the **SAME TYPE of input** (excitatory or inhibitory) for virtually all the pyramidal cells within a cluster.

These four conditions are well established. They are the basis of the production of an electrical charge (+ve or –ve) that can be measured on the scalp and recorded as the EEG.

Note: The pyramidal cell is the only neuron whose dendrite can produce an action potential (dendrites normally produce excitatory and inhibitory slow potentials that can summate at the axon hillock to produce an action potential.)

How Neurons Communicate

The job of neurons is to communicate with other neurons. They do this through an exquisitely designed system that involves both electrical and chemical transmission of information.

The Nerve Cell Resting Potential

Think of the cell as a castle. The castle is at peace and sits in a resting state. The enemy outside the wall will enter the castle if they can breach the wall. To conduct an impulse, a breach has to be made in the castle wall. The defenders, on the other hand, will repair the wall and force the enemy troops out again in order to return to a resting state. In the case of nerve cell walls, a breach can be affected in two ways: chemical or electrical.

In the resting state, neurons have what is called a *resting potential*. The resting potential is just the potential difference between the inside and the outside of the cell. It measures between –50 and –100 mV. The inside of the cell is negative compared to the outside. When at rest this potential is about minus seventy millivolts (–70 mV). Think again of the membrane as a castle wall. Inside the cell or 'castle' are high concentrations of negatively charged large protein and amino acid anions (A^-), a positive cation, potassium (K^+), and a low concentration of chloride

(Cl-) which is negative. Outside the cell wall is the opposite array: high concentrations of sodium (Na+) and chloride and a low concentration of potassium. There is also another important positive ion outside the cell wall, calcium Ca^{2+}. It will come into our discussion later when we discuss presynaptic neurotransmitter release. The main point to remember is that under special circumstances, sodium, potassium and chloride can all cross the membrane (castle wall); but think of the very large negative protein anions as immobile in the sense that they cannot cross the wall.

The overall resting -ve charge inside the membrane or cell (with the sodium kept out of the castle and the potassium kept in) is maintained by an active process which requires energy. This process, referred to as a *sodium-potassium pump*, transports ions against diffusional and electrical gradients. By means of this process sodium (+ve) is transported out of the cell and potassium (+ve) in. Sodium (Na) is kept at a concentration about 10 times higher outside the cell than inside. Sodium must be transported out against both the +ve electrical state outside and against its own concentration gradient (higher outside). Potassium, on the other hand, is just transported in against a concentration gradient.

Why Doesn't This Electrical Difference Dissipate?

The membrane is about 50 times more permeable to potassium than sodium. Potassium will diffuse slowly out of the cell due to the concentration gradient. This will make the inside of the cell even more negative, which will compete with the concentration gradient for attracting the potassium. If this was the only process that was going on, the membrane would reach an *equilibrium potential for potassium ions* of about –85 mV. However, there is a small steady influx of sodium due to both its concentration gradient and the negative internal electrical charge. This produces a *resting potential* of about –70 mV. This second process will, in turn, encourage more potassium to leave the cell in an attempt, so to speak, to move toward the *potassium resting potential* of –85 mV. This tends to encourage more sodium to enter the cell. If this *leakage* of both sodium and potassium continued, the potassium and sodium gradients would gradually be lost. As mentioned above, an active process called the *sodium-potassium pump*, which requires continual energy input – adenosine triphosphate (ATP) is the source – is necessary to prevent this. It pumps a little more sodium out than it allows potassium in and thus maintains the –70 mV *resting potential* (negative inside the cell) (Campbell, 1996).

The Postsynaptic Potential
How do nerve cells connect, resulting in an EPSP or an IPSP?

The Synapse
In the previous diagrams of the pyramidal cell the line with a bulbous end represents an axon from another cell. The bulbous end represents the *synaptic terminal*. The synaptic terminal releases neurotransmitters, which attach to specific receptor sites on the *postsynaptic membrane* of the pyramidal cell's *dendrite*. These axons may come from *excitatory* neurons such as *stellate* cells, *inhibitory* neurons such as the *basket* cells, or from other kinds of neurons, including other pyramidal cells or neurons at deeper levels, such as those in the thalamus.

The axon ends at a synaptic terminal. It can be at the top or at the bottom of the large dendrite. The arrival of the axon's nerve impulse at this *synaptic terminal* results in an influx of Ca^{2+}. These calcium channels are activated when the membrane potential is held below about –65 mV for approximately 50-100 ms. The resulting rise in Ca^{2+} concentration inside the synaptic terminal results in the small vesicles filled with neurotransmitters to fuse together and, then, to fuse to the inside of the presynaptic membrane. Then the neurotransmitter is released through the presynaptic membrane into the synaptic cleft. The neurotransmitter crosses the synaptic space and attaches to a specific receptor site (protein) on the dendrite and causes the postsynaptic membrane to become temporarily permeable to specific ions. The postsynaptic potentials can be either excitatory (depolarize the membrane) or inhibitory (hyperpolarize the membrane).

Excitatory Postsynaptic Potential (EPSP)
If the neurotransmitter and the postsynaptic receptor site are *excitatory,* then the membrane (the sheath of the dendrite) becomes permeable to sodium. It enters the cell because of the negative charge inside the cell and because of the concentration gradient with more sodium being outside the cell membrane. This results in a drop in the potential difference from a starting point of about **–70 mV inside compared** to outside of the membrane. The membrane begins to **depolarize**. *(The definition, from Dorland's Medical Dictionary (2007), of "depolarization" is: "the reduction of a membrane's resting potential so that it becomes less negative".)* (Remember from the

previous discussion that it is not the electrical activity of the action potential that is measured when recording an EEG, but rather the charge left in the extracellular space as the sodium rushes into the cell.) If more than one EPSP overlaps in time (*temporal summation*) or two or more presynaptic endings release neurotransmitters at the same time (*spatial summation*), then the internal potential of the postsynaptic area may begin to **depolarize and reach about −50 mV**. At this point it will reach what is called the *threshold potential*. The membrane potential will suddenly change to a positive internal potential of about 10 mV. The membrane is said to be *depolarized*, and this will initiate another process called the *propagation* of a nerve impulse. This process is discussed below under "Action Potentials."

Inhibitory Postsynaptic Potential (IPSP)

If the neurotransmitter is *inhibitory,* then the opposite process occurs. The change in membrane permeability allows potassium ions to exit from the cell, and **negatively charged chloride ions,** which are in high concentration outside the cell, **move into the cell** due to the concentration gradient. (This movement happens despite the fact that the electrical gradient is not in favor of their movement in.) This makes the potential difference move in an even more negative direction, inside versus outside the cell. This is called *hyperpolarization*. This makes it less likely that sufficient depolarization will occur to cause an action potential. Two amino acids which are *inhibitory* transmitters are gamma amino butyric acid (GABA) and glycine.

Summation

EPSPs and IPSPs *summate*. This is an important concept both for understanding what is measured in the EEG and for understanding a completely different process, the *action potential*. In the first case, the EEG, we are not able to detect the change when a single pyramidal cell receives input from an axon. As previously noted, we can only detect a positive or negative charge in the extracellular space, compared to a relatively neutral reference point, if the dendrites of a very large number of pyramidal cells receive axon input that is the same (either EPSPs or IPSPs), at the same time. It is the summation of these inputs that may result in an electrode placed on the scalp detecting either a negative or a positive charge.

With respect to the generation of an action potential, perhaps it is just common sense that the same principle applies. A single input might not depolarize a membrane to the point where it will suddenly go through complete depolarization. Several similar inputs along the dendrite may bring it to this point. It is an algebraic sum, so that the inhibitory inputs subtract from the effect of excitatory inputs and can prevent the membrane from reaching its *threshold potential*. Once it reaches its threshold potential, however, it is an all-or-none phenomenon.

Neurotransmitters

In most instances it is the receptor site that governs whether a transmission will be *excitatory or inhibitory*. Receptor sites respond to neurotransmitters, specialized brain chemicals that are needed to conduct nerve signals from one neuron to another.

Acetylcholine

The most common neurotransmitter is acetylcholine. Acetylcholine is the excitatory neurotransmitter at neuromuscular junctions. In the central nervous system (CNS) it may be either excitatory or inhibitory. It is the neurotransmitter in the parasympathetic division of the autonomic nervous system. It is involved in recording memories in the basal forebrain and the hippocampus and is deficient in Alzheimer's disease. In the reticular activating system it has a role in attention and arousal. It is also involved in the control of the stages of sleep.

There are three other groups of neurotransmitters that you will commonly encounter in your reading, but keep in mind that this is a partial list: there are well over two hundred neurotransmitters.

Biogenic Amines

The first group is the ***biogenic amines*** (catecholamines) including: norepinephrine and dopamine. Norepinephrine and dopamine are derived from the amino acid *tyrosine*. There are also a second group called the **indoleamines** and they include serotonin. Serotonin is derived from the amino acid *tryptophan.* Dopamine is generally excitatory, serotonin is usually inhibitory and norepinephrine is both. The common neurotransmitter in the sympathetic portion of the autonomic nervous system is norepinephrine.

Dopamine

Dopamine has been researched with respect to many disorders. LSD and mescaline may produce their hallucinogenic effects by binding to dopamine receptors. Schizophrenia may involve an excess of dopamine, and Parkinson's is related to a reduction in dopamine. It is hypothesized (Malone et al., 1994) that ADHD is associated with reduced dopaminergic activity in the left hemisphere and increased noradrenergic activity in the right hemisphere. It is the principle neurotransmitter in the brain's reward or pleasure circuit. This circuit involves the structures along the medial-forebrain-bundle pathway described below in the section on neuroanatomy.

Too much dopamine has been reported in the following conditions: hallucinations; psychosis, including the *positive symptoms* of schizophrenia such as paranoia; Tourette syndrome; obsessive compulsive disorder (agitation and repetition); and in overly excited states including euphoria and mania.

Amphetamines and cocaine are catecholamine agonists. They block the reuptake of dopamine and noradrenaline from the synaptic cleft and thus increase the availability of these transmitters to the postsynaptic neuron. This effect in the nucleus accumbens may be important in understanding the excitatory effects of these drugs and their ability to create a chemical "high." Alcohol, nicotine and caffeine can also increase dopamine in the nucleus accumbens.

Too little dopamine has been reported in Parkinson's disease with its tremor and inability to start movement, with the *negative symptoms* of schizophrenia including lethargy, misery, catatonia and social withdrawal, in adult Attention Deficit Disorder and in addictions.

Norepinephrine

Norepinephrine arises principally from neurons located in the locus coeruleus. This nucleus is described in the section on neuroanatomy. It has connections through the medial forebrain bundle to the hypothalamus. Its primary excitatory function in the central nervous system (CNS) is related to arousal and attention. It is released during stress and may be a part of the fight or flight response, and it is involved in emotions such as fear, anxiety and possibly mania. It is also thought to have a role in learning and the formation of memories. Too little norepinephrine may be associated with depression and too much with mania. It may be in excess in some anxiety disorders. It may, however, be depleted in patients who have had chronic stress.

Serotonin (5-hydroxy-trypamine, or 5-HT)

Serotonin is produced in the brain stem and released by the Raphe nuclei. It is primarily an inhibitory neurotransmitter. It is involved in the regulation of pain, mood, appetite, sex drive and in falling asleep. It may also be involved in memory. It is a precursor for melatonin, which in turn is important in biological rhythms. Low levels of serotonin are thought to be related to a number of psychiatric disorders including depression, obsessive compulsive disorder (OCD) and aggression. Selective serotonin reuptake inhibitors (SSRIs) are used to treat these conditions.

Amino Acids

The second group of neurotransmitters is the amino acids. This group includes the two inhibitory transmitters: gamma amino butyric acid (GABA) and glycine. It also includes glutamate and aspartate, which are excitatory transmitters. The anxiolytic medications (benzodiazepines), alcohol and barbiturates may exert their effects by potentiating the responses of GABA receptors. GABA will open both potassium and chloride channels and thus hyperpolarize the neuron and make it more difficult for that neuron to be depolarized. The neuron is effectively inhibited.

GABA

GABA is possibly the most important inhibitory neurotransmitter in the CNS. The entire CNS is a system in which, when a neuron is stimulated, a feedback loop is activated to inhibit or stop that neuron from continuously firing. These feedback loops often use the neurotransmitter GABA. This is the braking and stabilizing mechanism of the CNS.

Glycine

Glycine is found in the lower portions of the brain stem and in the spinal cord. In tetanus (lockjaw) the bacteria releases a glycine blocker. The removal of glycine's inhibitory effects is responsible for the muscles contracting continuously.

Glutamate

Glutamate is essential in learning and memory and in an important process called *long-term potentiation (LTP)*. Long-term potentiation is the process whereby a postsynaptic cell changes (is enhanced) in response to episodes of intense activity across the synapses. It is thought to be crucial to memory storage. Further research is needed to show whether long-term potentiation is due to an increase in neurotransmitter receptors, increased synaptic connections, or both.

Whatever the mechanism, the postsynaptic cell can depolarize more in response to a neurotransmitter. How this takes place may be as follows: glutamate activates what is called a *non-N-methyl-D-aspartate receptor*, causing an influx of sodium into the postsynaptic terminal. This depolarization has the effect of displacing magnesium (Mg^{2+}), which is blocking a second N-methyl-D-aspartate receptor site. This site is then activated by glutamate with a resultant influx of calcium ion (Ca^{2+}). This influx of Ca^{2+} results in the activation of other "messenger" pathways and the release by the postsynaptic cell of a *paracrine*. A paracrine is a chemical that is released by a cell and then acts to alter cells in its immediate vicinity. In this case it may act on the presynaptic ending and enhance the release of the neurotransmitter glutamate. The postsynaptic membrane also appears to be changed in this process to become more sensitive to glutamate. It is theorized that the postsynaptic cell may develop more glutamate receptors (Silverthorn, 1998). The importance of this for our work in neurofeedback is that it could be another important theoretical framework for understanding how only a few sessions of neurofeedback might result in sustained changes in the CNS.

Neuropeptides

The third group of neurotransmitters is the *neuropeptides*. These are short chains of amino acids. They are responsible for mediating sensory and emotional responses. Among them is **substance P,** which mediates the perception of pain. Measurements of substance P in the cerebral spinal fluid (CSF) are assisting clinicians in the diagnostic work-up of persons suffering from fibromyalgia. The *endorphins* are also neuropeptides. They function at the same receptors that receive heroin and morphine, and are thought of as naturally occurring analgesics and euphorics. They are found in the limbic system and the midbrain. The ventral tegmental area of the midbrain and the nucleus accumbens in the frontal lobe have opiate receptors (see neuroanatomy section). A third type of neuropeptide is **neuropeptide Y (NPY)/ polypeptide YY (PPYY)**. This substance is found in the hypothalamus and may be related to food intake and eating disorders.

Action Potentials

There are two processes that can lead to membrane depolarization. The first is the response to a neurotransmitter. Traditionally, we thought this occurred just at *synaptic terminals* (electrochemical connections between neurons), but it is now accepted that there are receptors at many sites along an axon and neurotransmitters in the extracellular fluid can float some distance from the site of their release and attach to these receptors. The second process is a *voltage-sensitive* change. This refers to the fact that depolarization in one section of a neuron will activate depolarization in the adjacent portion of the neuron. This raises the question of why nerve impulses are not chaotic, running in both directions. We will discuss the creation of an *action potential*, and then demonstrate why the impulse runs only in one direction.

As mentioned at the beginning of this section, the postsynaptic potentials created in the extracellular space outside the pyramidal cells' dendrites are of a relatively long duration and summate in a manner that can be detected by an electrode on the surface of the scalp. Sustained postsynaptic potentials may cause current to flow along the surface of a cell body or dendrite. The area of the neuron at the base of that neuron's axon is called the *axon hillock*. This is the *integrating* center of the neuron. The depolarizing changes in the cell may summate to the point where the potential at the axon hillock is changed sufficiently (critical level is >10 mV and the critical change is a move from the *resting level* of about –70 mV to the *threshold for excitation* of –55 mV), such that the membrane suddenly loses its charge and an *action potential* is produced and propagated along the axon to the next synapse. The electrical change is a temporary reversal in charge along the cell membrane. It is about 110 mV and lasts about 1 ms. It is all or nothing. This sudden change has the effect of inducing a similar change in the adjacent membrane that is in a *resting state* and so a current is propagated down the axon. The permeability of the adjacent membrane to sodium suddenly increases about one thousandfold over that of the *resting state*. It is, however, unidirectional. It is not propagated in the reverse direction due the structure of the *gateways* for sodium. There are two gates for sodium. The first opens instantly when activated by an appropriate chemical, or by an electrical change. The second is a relatively slow *gate* which closes shortly after sodium enters the cell and will not open again until its resting state is again achieved. While the active sodium pump is removing sodium from the interior of the cell to reestablish the old negative resting potential, the slower gate stays closed. Therefore, a second depolarizing stimulus cannot open this gate. This insensitive time is called the *refractory* period. Thus the current can only be propagated in one direction down the axon. The potassium gates are slow to open in response to depolarization compared to the first sodium gate, and thus, potassium flowing out of the cell is present during, and assists in, the repolarization phase. Indeed, it is these potassium channels that cause a bit of an *undershoot* and hyperpolarization at the end of the repolarization phase. *Action potentials* are very brief local currents. They are not what is being measured in the EEG.

The changes in the cell membrane affecting the permeability of K+ and Na+ are known as the *Hodgkin cycle*. The sodium-potassium pump uses energy to maintain a resting potential and that resting potential allows the neuron to respond quickly to a stimulus; just as a poised arrow is ready for flight, so the energy it requires is well justified.

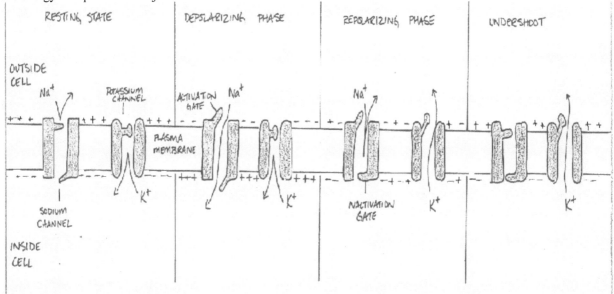

Schematic Diagram of a Receptor Site to Show the Progression of Electrical Changes in the Hodgkin Cycle (After Campbell, 1996)

*A note on two types of synaptic receptor sites: The postsynaptic receptors discussed here are also called **ionotropic** because when a neurotransmitter binds to this type of receptor, an ion channel is opened. Their action is local and is very fast (a millisecond). You may also hear the term* nicotinic *receptor. Nicotine will lock acetylcholine receptor channels in the open state. Nicotinic receptors are a type of ionotropic receptor for acetylcholine and are the type of receptors that open ion gates at neuromuscular junctions of striated muscles and at some neuronal synapses. Also, a slightly different type of nicotinic receptor is found in the autonomic nervous system (ANS).*

*Metabotropic receptors are a different type of receptor. In contrast to the ionotropic receptors, the effect of their action is diffuse and slow (seconds to minutes). Their action involves the production of secondary chemical "messengers," which can influence the metabolism of cells and produce long-lasting changes. You may also hear the term **muscarinic**. These are a type of metabotropic receptor. Muscarinic acetylcholine receptors are found in the smooth muscles of the pupils, glands, blood vessels and so on.*

The reason for mentioning that there are different types of receptors is that, in future, we may discover that some of neurofeedback's lasting effects may come about, in part, due to effects on receptors that cause a change in the metabolic activity of neuronal pathways.

Since action potentials are *all or none,* the strength of a nerve impulse is governed only by the *frequency* of action potentials. The action potential itself begins at a single site. The impulse is propagated by a series of depolarizations and resultant action potentials. The speed of transmission is increased first by axon diameter, and second by a process called *salutatory* (from the Latin word for *to leap*) conduction. This second process is enabled by myelinization of the axon. There are gaps in the myelin sheath called *nodes of Ranvier,* and the action potential *jumps* between these nodes, skipping the myelinated region in between. Think of myelinated axons as being the superhighways with faster speeds. (Campbell, N.A. et al., 1996)

A note on myelinization: Myelinization in the cortex differs from that in the peripheral nervous system. In the latter, Schwann cells form the myelin sheath, whereas in the cortex the myelin is generated by oligodendroglia cells. Myelin contains fat and this produces the white color. The deeper layers of the cortex are thus called white matter, whereas the upper layers are grey matter. In the spinal cord and peripheral nervous system the white matter is on the outside and grey matter on the inside – the reverse of the brain where the grey matter is on the outside. Grey and white matter have different densities, as well as different coloring since the fat holds more water. When there is a head injury, the grey and white matter will move at different speeds due to their different densities. The resultant sheer forces lead to diffuse axonal injury (DAI). This type of injury can be detected in the EEG, though it may not show up using brain imaging techniques such as magnetic resonance imaging (MRI). The EEG also has better temporal resolution than MRI. On the other hand, MRI has better spatial resolution as it can look at deeper structures. The use of mathematical transformations known as LORETA may improve the spatial information that an EEG can yield.

Schematic Diagram of an Axon Showing Schwann Cells
The speed of transmission along a nerve fiber is increased by an increased axon diameter and by a process called *salutatory* (from the Latin word for *to leap*) conduction. This second process is enabled by *myelinization* (cells of Schwann in the peripheral nervous system) of the axon. There are gaps in the myelin sheath called *nodes of Ranvier*. The action potential *jumps* between these nodes, skipping the myelinated region in between.

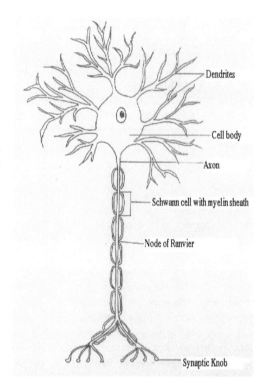

Other Cortical Cells

We have already mentioned *inhibitory cells*. The clearest example of these are the numerous *basket* cells which are found throughout the cortex. *Stellate cells* (so named because they look star shaped) are excitatory. Only the pyramidal cells create dipoles that can be measured on the scalp. They act like little batteries with positive and negative poles and produce current. Other cortical cells **do not contribute** directly to the electricity recorded by the EEG. They do contribute indirectly, however, in so far as they influence the pyramidal cells.

Glial cells are actually more numerous in the cortex than are pyramidal cells. They provide the infrastructure and support the pyramidal cells. Glial cells provide nourishment and remove waste, so they are important for an efficient brain. It is reported that Einstein's brain had a higher-than-expected number of glial cells.

Subcortical Influences

Like the *basket* (inhibitory) and *stellate* (excitatory) cells in the cortex, cells in the subcortical structures do not contribute directly to the electrical potential that we are measuring. These cells are not *dipoles*. They do, however, influence the pyramidal cells. In general terms, one can say that the thalamus is the major pacemaker that influences rhythmic activity – theta, alpha, and SMR.

Thalamic Pacemakers

There are over 100 billion neurons in the brain and 97 percent of neuronal connections are within the cortex. But this activity is modulated by thalamic pacemakers, so the influence of thalamocortical connections (representing only about one percent of the neural connections) far outweighs their number. Thalamic pacemakers produce different cortical rhythms depending upon which cortical loops they activate. The thalamic cells are *relay cells* that may be either in an *active* (relay or *working*) mode, or in a standby (idling) mode. The standby or idling mode is one where these cells are not, for example, receiving cortical input. In this circumstance these thalamic cells hyperpolarize. They then fire in a bursting, oscillating, pattern. Anderson and Anderson (1968) proposed that thalamocortical cells send fibers to the cortex. These fibers give off branches that go to interneurons, which then inhibit the thalamic activity. Then the thalamic neurons recover and go again into excitation. They give off another synchronized volley, which goes both to the cortex and to the thalamic inhibitory interneurons. If this inhibition lasted one-tenth of a second, then the rebound oscillatory (cyclic) excitation would occur 10 times per second and would be recorded as an alpha wave. The nucleus reticularis (one of the many nuclei within the thalamus) has such intrinsic pacemaker properties, and is responsible for the EEG sleep spindle. This type of activity is, in general, responsible for rhythmic wave patterns. Sterman's research with cats demonstrated that if thalamocortical connections are cut, the brain produces no theta, alpha or SMR waves, just delta.

Interruptions to this kind of rhythmical activity (*desynchronization*) can occur as a response to input from ascending neuronal systems that produce arousal in part through "activation of ascending cholinergic projections of the basal forebrain and brainstem and projections from the Raphe nuclei and locus coeruleus" (Fisch, 1999, p 14). Understanding how both synchronous and desynchronized EEG activity is produced is key to understanding EEG biofeedback's usefulness.

Lubar states: "Changes in the cortical loops, as a result of learning, emotion, motivation, or neurofeedback for that matter, can change the firing rate of thalamic pacemakers and hence change their intrinsic firing pattern." He also notes that changes in this intrinsic firing pattern mean changes in mental state. In work with clients who have ADD this usually means that slow waves have been decreased and SMR increased. (Lubar, 1997 after Nunez)

The diagram below is given to emphasize corticothalamic communication.

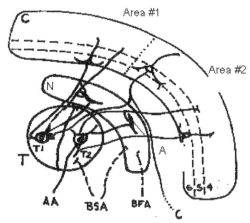

Figure with thanks to Maurice Barry Sterman after Sherman and Guillery

T = Thalamus; AA = Ascending Afferents
C = Cortex; 6,5,4, are layers in the cortex;
BFA Basal Forebrain Afferents to the reticular nucleus of the thalamus. BSA = Brain Stem Afferents

Just above 'A' is an axon from a pyramidal cell in layer 6 to T2 (thalamic sensory relay nucleus). Note how the axon gives off a branch that goes to the reticular nucleus of the thalamus (N) which, in turn, sends an inhibitory axon to the thalamus to stop the firing of the nucleus on which that cortex axon terminated.

This diagram emphasizes corticothalamic interaction. There are two principles schematically represented here. First, "Area #1" of the cortex is shown to send axons down to the sensory relay nucleus of the thalamus. This can influence the thalamus to pay attention to (facilitate) or turn off (inhibit) attention to specific sensory inputs. Second, Area #1 (imagine it to be, say, visual) communicates to T2 (a thalamic association nucleus), which in turn communicates with the cortical "Area #2," which has a different function than cortical Area #1. This is why, if you were to cut cortical connections along the dotted line between Area #1 and Area #2, these areas would still be observed to be communicating with each other.

Cortical Electrical Communication

Macrocolumns

The pyramidal cells and their surrounding support cells (stellate and basket cells) are arranged in groups. Each vertical column contains hundreds of pyramidal cells. The columns are parallel to each other and at right angles to the surface of the cortex. Many adjacent groups may receive the same afferent axonal input and thus fire in unison, allowing for a large enough potential that is measurable at the surface of the scalp. Each pyramidal cell may have more than 100,000 synapses. At least 6 cm^2 of cortex with synchronous activity is necessary to obtain a reliable scalp EEG recording. (Dyro, 1989). The amplitude of the recording will, of course, vary with the amount and type of tissue between the electrode and the cortex. Children, for example, have thinner skulls and higher amplitude EEG. It should be understood that you cannot tell from the EEG whether the potential under your electrode is generated by IPSPs or EPSPs. Research has shown, however, that in a seizure discharge the spike is produced by excitation and the slow wave by inhibition.

Resonant Cortical Loops

Cortical-to-cortical connections are far more abundant than thalamic-to-cortical connections, as has been noted above. EEG rhythmicity occurs due to direct thalamic influence, and may also occur due to these interconnections between groups of cortical cells (Traub et al., 1989).

Understanding the electrical activity of the cortex that is detected by the EEG must, in addition, include some comment on the frequencies produced, which correlate with different distances between columns of cells within the cortex. Joel Lubar has given a lucid overview of this (Lubar, 1999). He explains that distance is one of the factors that dictates the frequency of the waves that are detected in the EEG. Subcortical rhythmic influences, such as those arising from thalamic nuclei, are the other major factor. A simplified way of looking at this is as follows.

The cortex works in terms of three major resonant loops. They are:

1. Local: This loop of electrical activity is between macrocolumns that are close neighbors. It appears to be responsible for high frequency (> 30Hz) gamma activity.

2. Regional: This loop of electrical activity is between macrocolumns which are several centimeters apart. It appears that this activity is in the range of intermediate frequencies: alpha and beta.

3. Global: This loop of electrical activity is between widely separated areas; for example, frontal-parietal and frontal-occipital regions. Areas can be up to 7 cm apart. The activity produced is in the slower frequency range of delta and theta activity.

All three of these resonant loops can operate spontaneously or be driven by thalamic pacemakers.

Clinical Tip

When you work with clients it is often helpful for them to realize that their EEG does change with stimulation. They can close their eyes, then you can have them open their eyes. The client will observe that occipital alpha, and often alpha in central locations, is suppressed by visual stimulation. If you observe the wicket shape of a Mu rhythm at C3 or C4, then this can be suppressed by voluntary movement such as closing their fist on the opposite side of their body from the observed rhythm. It is of real interest to ADHD children and their parents that they can observe theta production, then suppression of theta with a concomitant increase in beta (16-18 Hz) activity either frontally or centrally, when they retrieve information to do a math problem and then do the calculation and produce an answer. The trainer can also run the EEG during a boring task when theta or low alpha predominates (specific frequency is unique to the individual) then pause the EEG and ask the client what was happening. Most clients respond that they had started the task then drifted off. Whereas it is true that theta is sometimes an important active state, crucial for memory retrieval, for example, the ADHD subject tends to go into this state inappropriately and remain there, oblivious to what a teacher or parent is saying. It is the ability to appropriately self-regulate the mental states that is assisted by neurofeedback. In the early stages of working with a client the foregoing exercises may be a helpful means for allowing the client to observe that they can self-regulate their brain activity and that this regulation is reflected in the recorded EEG rhythms.

Communication Linkages

It has already been mentioned that most of the communication (more than 95 percent) going on in our brain appears to occur between cortical areas, much of it in the same hemisphere. Less than 5 percent is due to thalamocortical connections, yet these connections have tremendous influence on what we observe in the EEG. Steriade, in Ottawa, Canada, demonstrated that when the cortex was cut communications remained between distant parts of the cortex (Steriade, 1990). This communication, therefore, is posited to be relayed subcortically through the thalamus. **Synchronous activity** is usually due to thalamic influence. It may also be associated with cortical lesions or seizure activity.

The transmission of a signal through a **volume conductor** is nearly at the speed of light. Examples of volume conductors are cerebral spinal fluid (CSF), brain tissue, skull and scalp. Thus, similarly appearing waves occurring at the same moment in time at nonadjacent sites probably originate from the same generator. If there is a time delay then this must involve a synapse and therefore a different cellular origin. (Fisch, 1999, p 16)

Corticocortical Coupling

Coupling is a term that has relatively recently come into vogue. Coupling means to join two objects, as in coupling between train cars. It is beginning to appear that for any particular mental state there is an optimal coupling between different areas of the brain.

Hypercoupling

To quote Lubar, "Neocortical **states** associated with **strong corticocortical coupling** are called **hypercoupled** and are associated with global or regional resonant modes." (Emphasis is added.) Hypercoupling thus means that large resonant loops are involved. Biochemically, it appears that the dominant neurotransmitter in this type of coupling is serotonin. Hypercoupling is said to be appropriate for states such as **hypnosis and sleep,** and also for **visualization.** Hypercoupling is also correlated with decreased attention.

Hypocoupling

Hypocoupling is associated with small regional and local resonant loops and thus higher frequencies. Biochemically, it appears that the neurotransmitters involved in this type of coupling are acetylcholine, norepinephrine, and dopamine. Hypocoupling is said to be appropriate for **information acquisition, complex mental activity and increased attention.**

We generally **use NFB to train** people to produce activity associated with hypocoupling and **resonance in local and regional areas** in order to be capable of better attention and better learning. Although what we are measuring is cortical activity, regulation of this activity may be primarily controlled from subcortical structures, principally the links from the thalamus to macrocolumns of cortical cells.

SECTION III
The EEG: Frequencies and Normal and Abnormal Waveforms

The electroencephalograph is an instrument that detects the alternating electrical currents that are produced by groups of neurons in the brain. The resulting recording is called an *electroencephalogram (EEG)*.

Definitions and Descriptions of Common Terms

Quantitative EEG

Definition
QEEG refers to Quantitative EEG. This term is used when the software that analyzes the EEG signal quantifies different aspects of the EEG. In this step, something is gained and something is lost. This process does not reflect information about specific morphology (shape) of waves or about the relative abundance of certain types of waves (infrequent bursts of high amplitude waves may, for example, be averaged out), but it makes it easier for you to visualize, in broad terms, what may be going on. The computer is able to display these numerical values in a table as a spectrum of magnitude (average amplitude over a time period) or power (amplitude squared) plotted against frequency, or as a topographic map showing activity at different electrode sites.

Normative Databases
Comparisons can also be made to normative databases. Suffice it to say that a number of different databases are available for purchase, and services exist to interpret 19-lead assessments. Databases/interpretation software currently available include (in alphabetical order) those by Frank Duffy, William Hudspeth, E. Roy John, Yuri Kropotov (WIN-EEG), Sterman and Kaiser (SKIL), Robert Thatcher (NeuroGuide), Thompson and Hagedorn (Evoke Neuroscience).

Amplitude and Power
Typically the QEEG will provide information about amplitudes in microvolts (μV) and power in picowatts (pW) for specific frequencies at specific sites. Calculations concerning ratios, standard deviations, and other statistics are also done. Usually QEEG refers to a 19-lead assessment, but values from one active location can also be quantified. A single active electrode at Cz, for example, was used to generate theta/beta power ratios comparing individuals with ADHD to controls in the multisite study done by Monastra, Lubar, Green and Linden (1999). The term QEEG is therefore independent of the number of sites and channels used in gathering the information.

The interpretation of a 19-channel QEEG is not for the beginner in the field of neurofeedback, and so it will just be touched upon in this text. Examples from 19-lead assessments will be shown.

The EEG Spectrum (The Frequency Domain) – Fast Fourier Transform (FFT)

Transformation from Time to Frequency Domain
The transformation from the *time-related domain* of the raw EEG to the *frequency domain* for statistics is carried out by a mathematical calculation called a *Fast Fourier Transform* (FFT). Jean Baptiste Fourier showed that any repetitive signal could be broken down into a series of sine waves. The breaking apart of a complex wave into its component sine waves may be called a Fourier analysis. Fisch notes that "The FFT function is based on the fact that any signal can be described as a combination of sine and cosine waves of various phases, frequencies and amplitudes" (Fisch, 1999, p 125). Squaring these Fourier

coefficients may be done to give a *power spectrum* that shows the power, measured in picowatts (pW), at different frequencies at a particular point in time. This type of display is used, for example, in summarizing information in many assessment programs designed by different equipment manufacturers. Power is also used in Thatcher's NeuroGuide program. That program uses microvolts squared but not multiplied by a constant, and it calls the units picowatts. The spectrum used in other programs, such as the BioGraph, is in microvolts (an *amplitude spectrum*), which corresponds better to the raw EEG but does not show such dramatic differences between frequency band magnitudes since the units are smaller. (Squaring a microvolt ratio approximates a ratio in picowatts; however, variability in the measurements that make up the averages used in the ratio can alter this approximation.) It should be noted, however, that the EEG is not truly made up of only sine wave signals and the EEG frequencies will sometimes have harmonics. Nevertheless power spectra are reasonably accurate and a valuable tool used in assessment work with the EEG.

Absolute Band Values and Ratios

Absolute band values are calculations based on the area under the spectral curve for that frequency band (for example, 4-8 Hz). Dividing two *absolute band values* to obtain a ratio (such as a power ratio for theta/beta calculated as the value of 4-8 Hz activity divided by the pW value of 13-21 activity) theoretically approximately corrects for skull and tissue thickness differences between individuals. This is the rationale behind doing studies comparing these ratios in individuals who have difficulties with attention span (Monastra et al., 1999; Jansen, 1995; Mann, 1992).

Cautions

Remember: once you have "quantified" the EEG, you have values and pictures (spectra and topographic brain maps) that have no information about wave morphology or about how frequently a particular wave form appeared in the EEG. Spikes and waves are not evident in this information. The topographic maps also do not help with localization of an activity. Topographic maps are a technique for displaying the scalp distribution of EEG activity from a large number of electrode sites. They show clearly what electrode positions have maximal power of a particular frequency band in the EEG. However, one must be very cautious in drawing conclusions from these maps. Remember that a few very high-amplitude waves may give the same value as extensive lower amplitude bursts of the same frequency. The analysis is also prone to artifact effects, so the EEG data must be carefully examined and artifacts removed before the quantitative analysis is done. **It is imperative that you keep referring back to the raw EEG** before coming to any conclusions about the data. Topographic maps are impressive, but keep in mind that there are only 19 "real" values from a 19-site assessment. All the values indicated by the colors between these points are interpolated (estimated). The only way to improve spatial resolution is to increase the number of channels. However, for the purposes of neurofeedback applications, this has very little increased value.

Wave Forms, Frequencies, Phase and Synchrony

Morphology

Morphology (or waveform) refers to the shape of a wave. The following terms are only briefly defined in this basic text because you are probably not going to use them, but you may hear about them in lectures you attend.

Regular and Irregular Waves

Regular waves may be sinusoidal or may be arched (wickets) or saw-toothed (asymmetrical, triangular). *Irregular* waves are constantly changing their duration and shape.

Monophasic/Biphasic/Triphasic, Transients, Complexes and Rhythmic Wave Forms

A wave is *monophasic* if it goes either up or down, *biphasic* if it goes up and down, and *triphasic* if it has three such components. A *transient* wave is one that stands out as different against the background EEG. A *complex* is a sequence of two or more waves that is repeated and recurs with a reasonably consistent shape (Fisch, p 145). Sinusoidal waves like alpha waves or spindle waves (such as sleep spindles or the similar waves called *sensorimotor rhythm*) are described as being *rhythmic*.

Activity: Generalized Versus Lateralized Waves

The activity that is observed may be generalized, lateralized or focal. *Generalized* waves are widespread and diffuse, occurring at the same time in most of the channels being recorded. The origin may be inferred by finding that the wave is maximal in one location as seen in a referential recording, or by observing a phase reversal in a sequential (bipolar) recording. *Lateralized* waves are those more frequently observed on one side of the head. This distribution of waves often may represent an abnormality. *Focal* waves are localized to one site or area.

Phase

Waves may be *in phase*. This would mean that the troughs and peaks of the waves in one area were occurring at the same time in another area of the brain. If they do not coincide, then they are *out of phase*. However, the waves may go up and down in a similar manner consistently, but not quite coincide. There would then be a time delay which is expressed as a *phase angle*. If the angle was 180°, then the peaks would point in opposite directions and it would be called a *phase reversal*.

Note from Fisch: "Phase: (1) Time or polarity relationships between a point on a wave displayed in a derivation and the identical point on the same wave recorded simultaneously in another derivation. (2) Time or angular relationships between a point on a wave and the onset of the cycle of the same wave. Usually expressed in degrees or radians." (Fisch, p 450)

Note: For the term "derivation" you can substitute the word 'channel.' Derivation refers to the process of recording from a pair of electrodes in an EEG channel (Fisch, p 443).

Synchrony

If the same kind of waves occurred simultaneously on both sides of the head, they would be in phase and *bisynchronous*. Waves that occur in different channels without a constant time relationship to each other are called *asynchronous* (Fisch, p 152)

Dominant Frequency and Age

Age is a factor in determining the dominant frequency in the EEG. The dominant frequency, measured with eyes closed in adults, is typically in the alpha range, around 9 to 10 Hz. It is generally reported that higher peak alpha frequencies are found in brighter people; that is, the adult whose eyes-closed peak alpha is at 11 Hz probably has a higher IQ than an individual whose peak alpha is at 9.5 Hz. Higher is not necessarily better, however; Tom Budzynski noted that some of the brilliant Silicon Valley scientists he has worked with have very high alpha but are quite brittle.

In the frontal and central regions below age 3, delta is dominant. From 3 to about age 5, theta is dominant. Low alpha becomes dominant around age 6 to 8, and then it gradually moves to a higher-frequency alpha, around 10 Hz, as the individual reaches adolescence. It is important to keep the developmental aspects of the EEG in mind when working with different age groups. What would be considered excess theta in a 12-year-old client with ADD would be quite normal in a 4-year-old.

Rhythms and Asymmetries

Alpha

Alpha amplitude is normally higher on the right but this difference should not exceed 1.5 times (Gibbs & Knott, 1949). The alpha rhythm should exceed 8 Hz in adults. If it never exceeds 8 Hz, this is probably abnormal. A difference of 1 Hz in frequency of the dominant alpha rhythm between the two hemispheres indicates an abnormality in the hemisphere with the lower frequency (Fisch, p 185, 187). Alpha rhythm is found predominantly in the posterior region (occipital leads), and apart from prefrontal alpha (which according to Fisch should be considered eye movement artifact until proven otherwise), frontal and central predominance of alpha is abnormal.

Alpha rhythm is usually blocked, or at least attenuated, with eye opening. The absence of any reduction is abnormal, as is unilateral blocking (Bancaud's phenomenon). Eyes-closed alpha represents an alert awake state, and alpha attenuates as the person becomes drowsy and theta increases.

Alpha is most strongly associated with the visual system, and seems to correlate with a resting state with decreased visual input.

Beta

Beta activity is found >13 Hz in adults. It is desynchronized rather than being rhythmic. It is almost always a sign of normal cortical functioning. Asymmetry in beta between hemispheres should be no more than 35 percent of the amplitude of the side with the higher amplitude. If the difference is greater than this, then the side with the lower amplitude may be abnormal (Fisch, p 181).

States of Consciousness, Wave Forms and Bandwidths

Historical Context

Different wave forms are seen in different frequency ranges. A very simple analogy can be made to waves on a lake. There are very large waves produced when a ferryboat goes by, smaller, regular waves when an outboard motor boat passes, and little irregular or desynchronized ripples when a gust of wind blows across the lake. Alternating current goes positive, then negative, then positive again. Alternating current will, therefore, produce "waves" when graphed over time. This was first described by Richard Caton in England in 1875 when he used a string galvanometer to show electrical activity from the cortex of rabbits, and displayed the waves by shining a light that made the waves visible as a shadow on the wall. Hans Berger, in the 1920s, was the first to record and report on the human EEG. He used paper recordings with a series of pens linked to the different EEG channels. That continued to be the method until computers offered the possibility of a digitalized signal that could be displayed on a computer screen. Some hospital EEG recordings are still done on paper, but neurologists are also switching over to computer displays rather than paper recordings.

Frequency Bandwidths

Frequency Ranges

Before describing *bandwidths,* it is important to understand that the *frequency* of a wave is just the number of those waves produced in one second. If you imagine that a motorboat produced four waves that passed a dock in one second, each wave lasted for 250 ms. The frequency of this wave would be four cycles per second (cps). This is usually called 4 Hz. *Hz* is an abbreviation from the name of the physicist Heinrich Hertz, who first described waves this way in the late 1800s. Four Hz is in the *theta* range, as we will describe below. The EEG is merely a wavy line. This line consists of waves of many different morphologies (shapes) and frequencies, measured in terms of the number of waves in one second. You may have faster waves riding on slower waves. All the frequencies are mixed together in the wavy lines we call an EEG.

Bandwidths refer to **frequency** ranges. For example, the easily identified, relatively high amplitude, synchronous alpha wave is seen when we close our eyes and it usually runs in a frequency range between 8 and 12 cycles per second (Hz). However, alpha waves may also run more slowly, especially in children, at just 6 or 7 Hz, or they may be faster, 13 or 14 Hz. It is the morphology and not just frequency of the wave that determines whether it is alpha. **Location, amplitude,** and **reactivity,** such as alpha to eyes opening and closing, offer further clues. A *typical bandwidth* or frequency range is not static. The following section discusses some *typical bandwidths* and the age and/or mental state when a particular bandwidth is most frequently observed.

The Spectral Array

To make it easier to see which bandwidths are at high amplitudes and which are at low amplitudes during different mental states and activities, a *spectral array* is used. This is a **histogram showing the amplitudes for each frequency,** usually from 2 to 32 Hz or more. It is helpful to have a spectral array that goes a little above 60 Hz so that you can easily detect electrical interference and artifact produced by the effect of muscle activity on the EEG. High 60 Hz activity can also indicate poor impedance readings, reflecting a poor contact between the electrode and the skin. (In Europe, Asia and Australia it is 50 Hz.) To understand what a spectrum is, imagine that you have 61 containers or bins in a row. Then you ask the computer to pull out of that very complex EEG, which has waves at all the different frequencies, all the waves at 1 Hz and put them in the first bin. You ask it to do the same with all waves at 2 Hz and place them in the second bin, then all those at 3 Hz into the third bin, and so on for each frequency up to 62 Hz. Then you tell the computer to display these bins as a histogram with the height of each bin or column representing the power of that frequency in picowatts (or the amplitude in microvolts, depending on which EEG instrument you are using). The extremely fast modern computers can do all the mathematics (called *Fast Fourier Transform*, FFT) to give you this graphic picture in milliseconds. The amplitudes usually decrease fairly uniformly as you move from 2 Hz to 62 Hz, in part because the skull attenuates the faster frequencies more than slower frequencies. Thus, when you see a sharp rise (or dip) in any particular bandwidth, that pattern is different than you would usually expect. The exception to this general rule is that adults with eyes closed will have a dominant frequency in the alpha ranges. There will be a blip (a sudden increase in amplitude) representing the alpha activity.

Magnitude and Amplitude

We spoke above of the magnitude of the wave being *an average* of the power expressed in picowatts over a defined time period. The amplitude is the height of the EEG wave measured in microvolts (µv), a millionth of a volt. In most equipment, each bin or bar in the histogram represents the average **amplitude** of that frequency of the EEG over one second in time. The relationship between the two measures is that power (pW) is the square of the amplitude (µv) x 6.14. Thus, power measurements will yield larger measures.

Sensitivity and Gain

EEG instruments have been calibrated to give you an accurate estimate of the amplitudes of the EEG waves. This amplitude measurement is based on a comparison of the EEG signal to the height of a wave of a calibrated signal. Your amplifier has its amplification rated in terms of *sensitivity* and *gain*. Thus, amplifiers have known *sensitivities* which are recorded as a number of µv/mm. (A higher sensitivity means lower amplification of the recording.) Thus, if a calibration signal of 50 µV causes a deflection of the wave of 7 mm, then this amplifier has a *sensitivity* of 7 µv/mm. Therefore any signal that is 4 mm in height would be 28 µV (7 mm x 4 µv) in strength or *amplitude* (see Fisch, 1999, p 45, 149). (This discussion assumes that the calibration signal was recorded at the same filter and gain settings.) You will adjust sensitivity when you read an EEG if you work with different age groups because children have much higher amplitude waves than adults. You will decrease sensitivity and thus lower the amplification of the recording for children.

Gain refers to a ratio of the voltage of a signal at the output of the amplifier to the voltage of the signal at the input of that amplifier. Thus, a gain of $10V/10\mu V$ = 1 million. You will see *gain* mentioned in the specifications for your amplifier. It will probably be recorded in *decibels*. A gain of 1 million (or 10^6) is 120 decibels. A simple way to calculate this is to multiply 20 times the power to which 10 is raised. (20 x 6 = 120 decibels)

Correspondence of Mental States to Typical Bandwidths

There is nothing inherently good or bad about any frequency or any frequency range. All frequencies are appropriate at certain times or for certain tasks. The most flexible and efficient brains seem able to change quickly from one frequency range to another depending upon task demands. Think of frequency ranges being like gears in a car; you need them all and you want to be able to shift smoothly and quickly between them.

Delta

0.5-3 Hz (0.5-3 cycles or waves in one second) waves are called *delta*. This is the dominant wave form in infants. These waves appear to originate in the cortex (layer V) and correlate with periods of reduced pyramidal cell neuronal activity. These waves are found during sleep at all ages. Stage four sleep will be more than 50 percent delta waves. They may also be seen in the waking state in infants, in some learning-disabled children and in people with brain damage. Delta activity is the dominant activity in normal infants up to about 6 months of age. Caution: Eye blinks and eye movements can produce waves that look like delta.

Theta

3-7 Hz, 4-7 Hz or 4-8 Hz waves are called *theta*. For the most part, the origin of this rhythm appears to be in the thalamus and in the limbic system (septal area). Hippocampal theta had been recorded in rats and more recently in humans. It appears to be related to memory retrieval and to the ability to control responding or not responding to stimuli. Theta waves dominate the EEG from about 6 months to around 6-7 years of age. In older clients, predominance of these waves in the waking state appears to be associated with being drowsy and tuned out from what others may be discussing or from conscious observations of the environment. It is thought that this kind of theta arises from thalamic nuclei. You might still be walking and not trip over objects in your path or bump into things as these actions are rather automatic. You might even think of some very creative things when in this state since it corresponds to the hypnagogic state before sleep onset. Indeed, Thomas Edison reportedly took naps with a small metal ball in his hand and a metal pie plate on the floor beneath each hand. When he went into a creative mental state just before falling fully asleep his muscles would relax, his hands would open and the ball hitting the plate would wake him. He would immediately write down his ideas. Psychotherapists use this mental state to allow a patient to freely bring up memories, fantasies and associations. Waves at 7 Hz may also be seen when a person is visualizing.

In peak performance states you will observe brief rises in about 6 to 8 Hz during cognitive processing of information. Perhaps this is indicating some visualization techniques are being utilized, or perhaps the theta reflects functions to do with memory and cognition. The point where the power spectra for alpha amplitude intersects with that for theta is called the *transition frequency*. Excessive amounts of theta are typical in individuals with Attention Deficit Disorder.

Alpha

8-12 Hz waves are called *alpha* if they have a regular sinusoidal form. The origin of this rhythm also appears to be in the thalamus. These symmetrical waves are seen in about 90 percent of people when they close their eyes. These waves are the dominant frequency of the EEG (measured with eyes closed) after 9-11 years of age and through adulthood. We like to think of alpha as one kind of "resting" state. When you finish solving a problem (for example, a pilot landing a plane or a child answering a multiplication question) the brain appears to "rest" briefly in alpha. When teaching children about brain waves, we often joke that the brain is a very lazy organ. It will take a rest (in alpha) whenever it can. We often reflect on a problem in alpha, and it may be quite a creative state. Clients who feel anxious and stressed may show decreased alpha. Alpha dominates the EEG measured at a central location when a person is daydreaming and also when a person is meditating (at least those meditation traditions that encourage an inner focus, such as yoga). You also see increased alpha when someone smokes marijuana and the increased alpha (eyes open) may persist for a couple of days.

Low Alpha

8-10 Hz is often referred to as **low alpha**. (Low here refers to the frequency range, not the amplitude of the waves. To be more accurate, what we refer to as *low alpha* is alpha below that individual's peak, eyes-closed alpha frequency. This peak frequency may decrease with age or cognitive deterioration. As noted previously, brighter individuals usually demonstrate a higher *peak alpha frequency*. Meditation often is associated with an alpha state. It is a calm and relaxing state but it is also a state in which we are not attending to the world around us. We are in our own head, so to speak. When students ask, "Isn't alpha really good?" we may answer that it is, but remind them that going into a meditative mental state in the classroom or boardroom won't be appreciated. You want the mental state to match the situation. Both low alpha and theta rhythms demonstrate diurnal variations. Higher amplitudes of these waves are observed around 11 am, 1 pm and 3 pm. The degree of change in amplitude and the exact times vary between individuals and may also vary with fatigue. The peaks are independent of food intake. These diurnal variations are important to the professional who is doing pre- and post-EEG measurements. To compare data, these measurements should be done at the same time of day.

High Alpha

11-12 Hz (or 11-13 Hz) may be referred to as **high alpha.** The bandwidth 11-12 Hz can be associated with a mental state of open awareness. An open awareness implies being capable of responding to a wide range of changes in one's environment. In athletics this state is associated with fast reflexes and accurate responses. The awareness of a professional goalie in hockey or soccer or a black-belt martial artist about to fight several opponents would likely be reflected in 12 Hz activity (though this has not been researched in real life situations). It is also associated with the mental and physical calm required in that readiness state before action: for example, the moment just prior to and during the release of an arrow by an archer (Landers, 1991). Open awareness is part of being in the zone, the ideal mental state achieved by top performers in most any field of endeavor. Encouraging the production of 11-13 Hz activity is probably the most common goal in neurofeedback optimal performance training.

Beta

Beta waves are above **12 Hz.** Except for sensorimotor rhythm, as described below, beta waves are produced in the brain stem and the cortex. In the cortex, beta indicates local activity in a specific area beneath your active electrode. When producing beta we are usually awake, alert, externally focused, logical, problem solving, attentive. It will be seen when we are listening to a speaker or solving a problem. However, we may also be tense and anxious. Beta rhythm is normal, but an asymmetry of more than 35 percent may indicate an abnormality on the lower amplitude side (Fisch, 1999, p 192). Excessive beta, on the other hand, may be due to medications such as benzodiazepines or barbiturates. This broad range of beta can be broken down into smaller frequency ranges that correspond to more precise kinds of cortical functioning, as follows.

Sensorimotor Rhythm

13-15 Hz is called sensorimotor rhythm (SMR) when it is found across the sensor-motor strip. As noted in the history paragraph, this frequency range does not have a Greek letter name as Sterman named it more recently (1967). It is a very specific **spindle-like** waveform produced in the ventral-basal nucleus of the thalamus. It, too, is a kind of "resting" wave state. It occurs when there is a decrease in the activity of the sensory and motor pathways that run through the thalamus; that is, it occurs when there is less attention paid to sensory input and when there is decreased motor output. It is necessary, but not sufficient, to be still for this rhythm to be produced. There is a change in muscle tone such that the person (or cat or monkey in Sterman's early experiments) is mentally alert without muscles being tense. It is measured across the sensorimotor strip of the cortex, and these frequencies found elsewhere would be called beta. The waveform in other areas of the cortex would usually also be different: desynchronized fast waves, not a spindle-like rhythm. Note that the spindle shape of the waves is very apparent when microelectrodes are implanted in the brain as in Sterman's early experimental work, but it is not so easily seen in recordings from the scalp. SMR appears to be associated with a calm mental state with increased reflecting before acting. It is thus important to train up (increase) SMR in those who have problems with hyperactivity and/or impulsivity.

Low Beta

16-20 Hz we often refer to as **low beta.** We refer to it as *problem-solving beta*. Of course this beta is found from 12 to 15 Hz, and sometimes at frequencies higher than 20 Hz. It can be easily demonstrated for children and their parents by having them watch the EEG and a spectral array (a histogram of the amplitudes of each frequency) while you ask the child to multiply, say 7 times 8. This can be quite dramatic. Momentarily, activity around 17 Hz suddenly increases in amplitude, and at precisely the same time, theta waves and low alpha (8-10 Hz) waves decrease in amplitude if the child can do the math.

High Beta
Beta spindling

This refers to bursts of beta waves in a rising and falling spindle-like pattern. Although it can be less than 20 Hz, most spindles involve fast beta, above 20 Hz. They can be associated with epileptic auras. It may be due to a disease process and cortical irritability. It can be seen in ADHD. In Part II of this book, we discuss how training down beta in the area involved may correspond to a decrease in the client's symptoms.

19-21 Hz or 20-23 Hz beta is often observed to be raised (above the levels of the beta between 16-18 Hz) in anxious clients. It may correlate with emotional intensity. You must check with each client to establish if a rise in this area is correlated with productive cognitive work, productive but too-intense work, or unproductive intense or anxious thinking.

Note: It is crucial that the neurofeedback practitioner always check what their client is experiencing. It is important that you check without suggesting what you think a particular frequency band should represent.

24-36 Hz beta is often seen to be at very high amplitudes in clients who are worried and ruminating. These clients often feel stressed out. They may be hypervigilant. We may also note a peak around the mid-20s in individuals if they or others in the family have a problem with alcohol or substance abuse. It may be a marker for a tendency to deal with anxiety by using alcohol or drugs. But caution is emphasized because these high frequency bursts of beta are also observed in persons with very high intelligence who are multi-tasking. Higher frequency ranges of beta (>30 Hz) are sometimes referred to as *gamma*.

Sheer Rhythm

38-42 Hz beta is sometimes referred to as **Sheer** rhythm, after David Sheer who did some work in the 1970s concerning enhancing 40 Hz activity. This particular rhythm has been observed to be important in learning. It may correspond to a type of attention where the subject is bringing together different aspects of an object into a single percept. It is, therefore, referred to by some clinicians as a *binding* rhythm, and is thought to be associated with peak performance. Research at Penn State University, State College Campus, has shown that athletes on a balance board, instructed to lean as far forward as they can, show 40 Hz activity at the moment they correct their balance to prevent falling. An athlete who has suffered a concussion shows poorer balance and does not show the 40 Hz increase. We are, therefore, careful not to include this range of beta in any *inhibit* range we use, to eliminate interference (see under artifacts) from muscle electrical activity (EMG).

Ambient Electrical Activity

50 Hz in Europe, Israel, Asia or Australia, or **60 Hz** in North America, often represents interference from electrical activity in the room. If it is raised, it is a good warning to check all possible sources of unwanted electrical artifact, including your electrode connections. Sometimes, lowering the impedance readings between each of your connections will be sufficient to lower this frequency, as *common-mode rejection* means the preamp will not amplify that frequency. Note, too, that when impedances are different between sites, currents can be induced in the leads leading to electrical artifact. Other possible sources of ambient current include lamps, extension cords and any electrical appliance, such as a pencil sharpener or even a cell phone.

EMG

Fast sharp waves or 'H' forms are seen when there is EMG (electromyogram) interference. The frequency of most muscle activity is higher than 60 Hz, but it is of high voltage. It will overwhelm the instrument's filters and interfere with the accurate reading of brain waves at lower frequencies. In our work with the EEG it is called an *artifact*, that is, electrical activity of noncortical origin. EMG artifact will proportionately inflate the lower amplitude higher frequencies more than high amplitude slower frequencies.

Broad Descriptive Terms Used in Neurofeedback

Slow Waves

This term usually refers to any waves running at a rate less than 12 Hz. These include delta, theta and alpha waves.

Fast Waves

This term usually refers to any waves running at a rate greater than 12 Hz. (12 Hz may be slow or fast depending on whether the waveform you are referring to is alpha or SMR/beta respectively.)

Wave Forms Not Frequently Encountered in Neurofeedback Work

Lambda Waves

Lambda waves are positive sawtooth-shaped waves usually found in the occipital region. They are "evoked" by visual scanning or looking at detailed material. They last about 100-250 ms. They are infrequently observed in typical hospital EEGs because these recordings are done with eyes closed. You may see them quite frequently if you do full cap assessments because you might be recording while the subject is reading. Lambda is only considered abnormal if the waves are markedly asymmetrical. This can suggest an abnormality on the side with a lower amplitude (Fisch, p 193). We will just mention, not focus, on this wave form in the discussions in this basic text.

In the figure below, Gunkelman notes that the lambda waves are sharp downward deflections (positive) that may be seen repeatedly when the subject is reading.

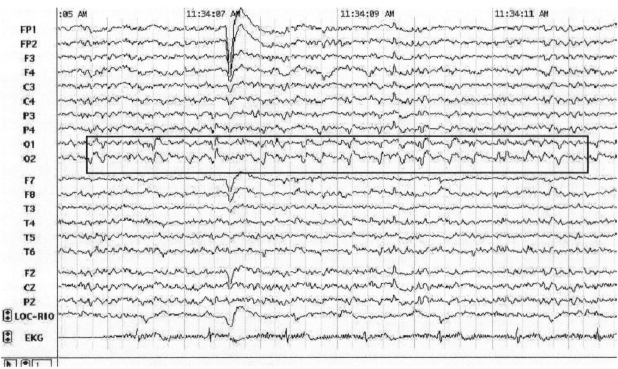

Lambda is the occipital downward (positive) sharp deflection seen repetitively.
Thanks are extended to Jay Gunkelman for sharing this example.

Mu Waves

Mu waves may fool you. They look like alpha waves and are usually found in the 7-11 Hz frequency range. Mu is usually observed at C3 and C4. Because many of you will be working with clients who have difficulty with attention and concentration, it is important that you distinguish this activity from central alpha, which may be a marker for certain types of ADD. Mu is blocked by making a fist. This blocking of Mu is, for the most part, on the *contralateral side* (the side of head opposite the side of the body where the client made the fist). You may also try having the client close their eyes and then open them. When they open their eyes, alpha is blocked but Mu will remain in the central region. The only real indication of an abnormality with Mu is if it is found only on one side of the head. Though most people do not produce Mu, it is considered a normal variant. It may be found in about 7 percent of the population. However, Mu is found in a much higher proportion (over 50 percent) of clients who have ADHD.

Morphology
- Mu waves have a pointed top and rounded bottom (or alternatively, a rounded top and a pointed bottom.) It is, therefore, often referred to as a *wicket* rhythm. This wave shape is monomorphic rather than sinusoidal.
- Contrast this with biphasic waves which have two points like those seen with EMG (muscle activity).
- Contrast it also with sinusoidal waves characteristic of alpha (usually not pointed).

It is reasonably safe to say that, for the most part, you will not normally see alpha centrally without it also being observed in the occipital and parietal regions.

Figure below shows an example of Mu waves seen at F4-C4 in a longitudinal sequential montage.

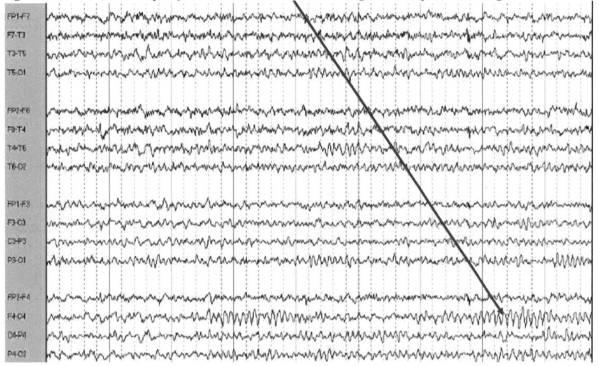

Laplacian Montage, note Mu at C4 – aC4.

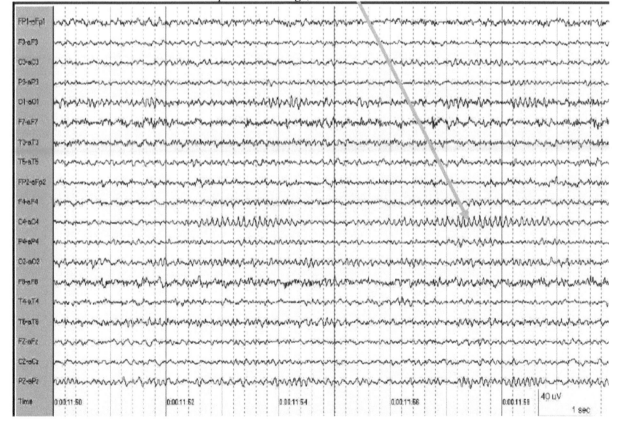

Waveforms Less Commonly Seen in NFB Work

As we stated at the beginning of this section on waves, waves may be described as: *regular*, such as a burst of theta waves, or *irregular*, such as a burst of theta with beta waves riding on top. Waves may be described as *sinusoidal*, such as a burst of alpha waves. They may be said to be *spindle-like* (e.g., SMR), beginning small and building up in amplitude, then falling off in a spindle shape – a shape familiar to those who know the weaving trade. Less often, waves may be described as *sharp* or *spiked*, or as combinations of these, *spike-and-wave* or *poly spikes* or complexes.

Spikes

Spikes: *These waves have a duration of 20-70 ms. This wave looks just like its name says. It is a spike and it would hurt if you sat on it. The spike is a distinct wave form in terms of its rate of climb and descent. A spike's amplitude is usually higher than other background activity, about 40-100 µV.*

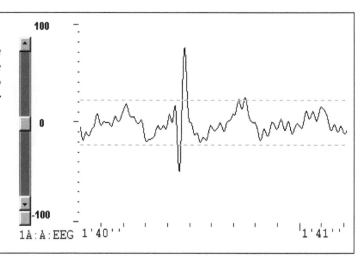

Although this sample of EEG is from a woman who has a seizure disorder and this sample is a spike, the morphology is that of **a myogenic spike biphasic**. In other words, it is muscle-generated and goes in two directions. There is **not** an "after-following" slow wave, the hallmark of many cerebral spikes. A spike with an after-following slow wave is the wave form that is characteristic of seizures. This is an interictal sample from a client who has partial complex seizures. Note also the high-amplitude slow-wave activity. However, since the electrode was placed 1 cm above a point, one-third of the way along a line drawn from external ear canal to the lateral canthus of the eye (appropriate for picking up interictal activity), it is also a placement that is liable to detect a lateral rectus spike, which reflects electrical activity from the muscle involved in lateral eye movement The spike may appear without visible movement of the eye since these spikes are like a single motor unit (SMU), not a contraction. This spike is thus an artifact (electrical activity not of cerebral origin). Thanks to Jay Gunkelman for interpretation.)

Spike and Wave

Uniform three-per-second spike and wave "pairs" are characteristic of an absence (petit mal) seizure. Note very high amplitude (>160 µv) of these waves.

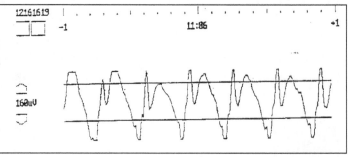

Sharp Waves

These waves have a duration of 70-200 ms. They do not have as sharp a point as a spike and are more like the tip of a pencil rather than the needle-like spike. They are described as having a high velocity segment. They are bi-directional. (See figure below.)

Sharp Transients

These are groups of sharp waves. If they are in infrequent bursts, then they are usually just called *nonspecific*. However, *complexes* containing spikes and sharp waves that are repeated may represent *interictal* (between seizures) epileptiform activity. This kind of activity may occur between seizures in a patient who has epilepsy. It is possible to have complexes that last for a few seconds but do not correspond to any observable clinical manifestations of a seizure. In such cases, the term *subclinical electrographic seizure pattern* may be used.

Paroxysmal Discharge

A *paroxysmal discharge* refers to one or more waves that stand out from the rest of the EEG activity. They usually begin and end abruptly. Although they may be seen in patients who clinically present with a seizure disorder, they may also be seen in people who have never had a seizure, particularly when a person is drowsy.

Epileptiform Paroxysmal Discharge

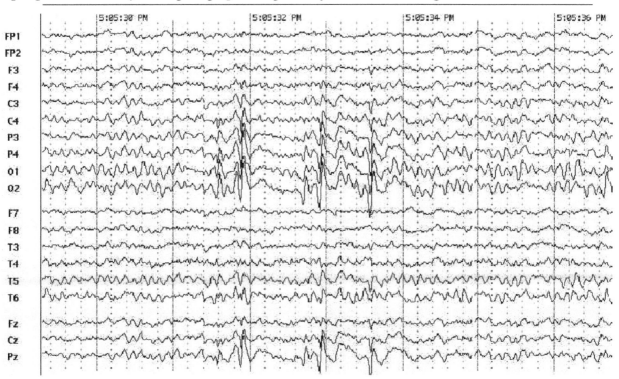

This epileptiform paroxysmal discharge sample has been provided by Jay Gunkelman.

Paroxysmal Hypnagogic Hypersynchrony

This term refers to synchronous, slightly notched, sine waves that are higher in amplitude than surrounding waves and run at about 3-5 Hz. The burst may last for a couple of seconds. You may see this in normal children when they are sleepy.

Sleep Spindles, V Waves and K Complexes

Sleep spindles look like SMR spindles and are in the same frequency range: 12-15 Hz. Like SMR, they are maximal over the central regions, but unlike SMR they are seen throughout the cortex and not just over the sensorimotor strip. Apparently they arise from a different generator, are only present during sleep and are seen in Stage II. V waves occur during sleep and are negative with an amplitude of up to 250 μV and a duration of less than 200 ms. K complexes are seen in stage II sleep and are sharp, negative and of high amplitude followed by a longer duration positive wave. People doing neurofeedback are not concerned with the specialized field of reading sleep EEGs, so these waves are mentioned, but most practitioners will never encounter them in practice.

Square Waves

This term refers to a fundamental frequency plus harmonics. This terminology is familiar to musicians. On the EEG it may look like a slow wave theta or delta with faster, lower amplitude (beta) waves riding on top. A harmonic is a whole number multiple of the fundamental frequency. With a number of harmonics riding on the fundamental wave, the waveform becomes visually squared off. The first harmonic of a theta wave that is seen in the EEG is usually lower in amplitude than the fundamental frequency and three times its frequency. Examples are shown in the intervention section of this book.

Slow Waves

Localized Slow Waves

These waves are under 8 Hz and may indicate a localized lesion (for example, a stroke), or abnormality (for example, migraine, transient ischemic attack, mild head injury, supratentorial brain tumors). The tentorium is a large infolding of dura matter (connective tissue) separating the cerebellum below from the cerebrum above. Child tumors are more frequently inferior, and adult tumors are more frequently superior, to the tentorium. (The other indentation of dura separates the two cerebral hemispheres on the dorsal surface, and is called the falx cerebri.) Delta wave forms are usually surrounded by theta in these instances. The delta doesn't respond to maneuvers such as eye opening or hyperventilation. Deep lesions (such as those in the internal capsule) do not show focal delta, but may result in hemispheric or bilateral delta. (Fisch, p 349)

Bilaterally Synchronous Slow Waves

These may be seen in children when they are drowsy. However, in an alert, resting adult, these delta waves may indicate deep midline structural damage. They may appear intermittently in bursts and be higher in amplitude than surrounding wave forms. This phenomenon may be due to diffuse gray matter damage. Frontal intermittent rhythmical delta activity has the acronym FIRDA. Interictally, 3 Hz waves of this kind may be seen in some clients who have absence seizures.

Generalized Asynchronous Slow Waves

These waves (<8 Hz) occur over both hemispheres but demonstrate no constant time relationship from one side to the other. They are seen in normal individuals who are sleepy and in children who have a fever. However, a marked amount of these slow waves may indicate an abnormality (usually quite nonspecific and seen in such disorders as migraine headaches, head injuries, high fever, encephalopathies, degenerative diseases, dementia, and in some clients with Parkinson's). Generalized asynchronous slow waves are the most common and least specific EEG abnormality. You should always have this checked out by a neurologist. (See Fisch, 1999, p 363-376 for a detailed discussion.)

Continuous Irregular Delta

Polymorphic delta results from lesions affecting white matter.

Abnormal Beta

Neurologists may view an isolated reduction in beta as a fairly reliable sign of local damage. When beta activity is asymmetrical, it is said to be abnormal on the side with less beta if the difference is greater than 35 percent. (Note: An asymmetry in alpha activity needs to be greater than 50 percent to be considered abnormal.) Decreased beta may also be seen during a migraine attack. A reduction in amplitude of all waves occurs for a few seconds after a seizure, and a localized decrease in alpha and beta may be seen for a short time (from seconds to a few minutes) after a focal seizure.

Abnormal Amplitude

There is, generally speaking, no upper limit. The lower limit is 20 microvolts in any channel in any montage during eyes-closed wakefulness, with exceptions that include alerting, mental effort, eyes opening, anxiety, or drowsiness.

A **bilateral decrease in alpha** may be due to anxiety or a disorder, usually toxic or metabolic. A general **increase in beta** is usually due to tranquilizers or sedatives (Fisch, 1999, p 407). **Slowing of alpha**, that is, a lower peak alpha frequency, may be a sign of head injury.

A **unilateral reduction of alpha** may indicate an abnormality. If this occurs while doing mental arithmetic in eyes-closed condition, it may indicate abnormalities in the ipsilateral parietal or temporal lobe (Westmoreland, 1998). Failure on one side to block alpha on eye opening may also indicate a lesion in these areas, and is known as Bancaud's phenomenon.

Sleep

Normal Stages of Sleep

Most readers of this text will not be recording during sleep. We will, therefore, give just a brief overview of the wave patterns seen during normal sleep.

Wakefulness with eyes closed: Alpha dominates and is highest in the occipital leads, while beta may dominate the frontal leads.

Stage I: Alpha attenuates and irregular slow waves dominate the EEG in the frequency range of 1 to 7 Hz.

Stage II: This light sleep phase is characterized by the appearance of V waves, sleep spindles, and posterior occipital sharp transients (POSTs).

Stage III: In this stage, high amplitude delta waves begin to appear. POSTs are still present and sleep spindles may be present as well.

Stage IV: More than half the record will be delta waves at a frequency of 2 Hz or less.

REM: This refers to a rapid eye movement phase in which the subject is usually dreaming and may recall the dreams on awakening. The EEG is characteristically of low voltage and despite prominent theta and the lower frequency alpha, it may look much like an awake recording. There is a distinct lack of muscle tone that is different from the awake state.

Sleep is cyclic through the night beginning with stage I and proceeding to stage II, III, and IV, then going briefly back to stages III and II and into REM before beginning the cycle again with stage I sleep. Each cycle is between 80-120 minutes long. The percentage of time spent in stages III and IV decreases with age.

Sleep Disorders

Many trainers using NFB will see clients who have Attention Deficit Disorder, and one must keep other pathological conditions in mind when making a differential diagnosis. The marker for ADD (increased theta) is also found in individuals who have either narcolepsy or sleep apnea. In clients who have *narcolepsy,* REM sleep will occur at sleep onset. However, people who do not have narcolepsy but take a daytime nap may also show REM onset at the beginning of their nap. Multiple naps with a very short time before falling asleep (less than five minutes), all beginning with REM sleep, may indicate narcolepsy. Sleep deprivation for any reason may result in excessive daytime drowsiness and REM onset naps.

Frequent long periods (>10 seconds) without breathing may indicate sleep apnea. It is most common in people who snore and those who are obese. Essentially what happens in this disorder is that the airway becomes obstructed when the person is asleep. Muscles relax, allowing fat to fold in and close off the airway, or the person may have enlarged tonsils or adenoids that produce the blockage. The result is poor quality sleep, periods of reduced oxygen flow to the brain, and a daytime EEG that looks very drowsy with lots of slow waves. If you have any question about these disorders you should have the client checked out by the appropriate medical professional.

An excellent book that covers all aspects of sleep is *The Promise of Sleep* by William Dement.

Abnormal EEG Patterns

This is not the subject of this book. The only reason for describing and showing a few abnormalities is to **warn you to refer these persons immediately to the appropriate medical practitioner.** You may do NFB training for many years and never see any abnormal pattern, unless of course you are using neurofeedback to treat people who have a seizure disorder or a head injury. When you do see abnormal wave patterns, it will hopefully be in people who have already been the subject of neurological evaluation and so you know what you are dealing with in their EEGs.

It is nevertheless possible that you might see an unusual pattern. What might it look like? The following are just a few selected examples.

Seizures – General Description

Seizures are usually brief periods of motor, sensory, mental, or autonomic disturbances of sudden onset, often accompanied by a change in consciousness and paroxysms of unusual EEG activity. They may be followed by a brief period of paralysis of the function that was most involved in the seizure. Recurring seizures due to a cerebral abnormality are termed *epilepsy*. This is as opposed to a seizure caused by a reaction to a transient circumstance, such as alcohol withdrawal, hypoglycemia, or fever. A single febrile seizure in an infant is thus not an indicator of epilepsy.

Seizures are described by different terms. The first set of terms you may encounter are: symptomatic, idiopathic and cryogenic. ***Symptomatic epilepsy*** is defined as seizures resulting from a cerebral disorder. ***Idiopathic epilepsy*** are seizures without any identifiable cause in a patient who has had a normal neurological examination. ***Cryptogenic epilepsy*** is the term used when the patient has seizures without an identifiable cause but who has a cognitive impairment and/or neurological deficits.

Generalized Seizures

There are more than 100 different kinds of seizures. Although certainly an unwanted abnormality, there are a number of famous people who had seizures, including Charles Dickens, Vincent Van Gogh, and Sir Isaac Newton.

Generalized seizures involve both hemispheres. Examples are a tonic-clonic seizure, myoclonic seizure or an absence seizure. A ***tonic*** seizure involves flexion of muscles. The contraction may last up to a minute and there is loss of consciousness. A ***clonic*** seizure is characterized by rhythmic myoclonic movements lasting for one or more minutes and associated with loss of consciousness. The combination termed a *tonic-clonic seizure (or **Grand Mal** Epilepsy in older texts)* you will not forget once you have seen it. The client loses consciousness, followed by sustained *tonic* contractions. The client turns blue (cyanotic) and his or her heart rate and blood pressure rise. After about 15 seconds, rhythmic *clonic* movements begin. This progresses to violent jerking and the client may bite his or her tongue. This phase lasts for about 30 seconds. Incontinence may occur.

Atonic seizures refer to a sudden loss in muscle tone.

The ***myoclonic*** seizure is characterized mainly by sudden, brief, flexor muscle contractions. It can cause the patient to fall, and can, therefore, be dangerous.

Partial Seizures

Partial seizures involve a single area in just one hemisphere. If consciousness is not impaired it is termed a ***simple partial seizure***. If change in consciousness is involved, it is termed a *complex partial seizure*. A partial seizure may be motor, sensory (burning, tingling or other sensations), autonomic (sweating, flushing, epigastric sensations) or psychological (distortions of cognition, including time; feelings, including fear and anger; hallucinatory; and so on). An example of a simple motor partial seizure would be the sudden loss of expressive or receptive language (aphasic seizure).

Complex partial seizures (temporal lobe seizures) usually involve the inferior and medial part of the temporal lobe. It is generally accepted that Joan of Arc had a seizure disorder that was the source of her visions. Now that a localization for religiosity has been identified in the temporal lobe (see Rita Carter, p 13), it is interesting to speculate about the focus of St. Joan's seizure activity.

The EEG in Seizure Disorders

Spike-and-Wave Pattern

The EEG activity may be *ictal* (during the seizure) or *interictal* (between seizures). It may be *localized* or *generalized*. Thalamic projections appear to be involved, since suppression of thalamic function abolishes spike-and-wave discharges. This early *centrencephalic* theory (Penfield & Jasper, 1954) has been modified to include the primary role of the cortex in seizure propagation and of the reticular formation in modulating cortical excitability (corticoreticular theory, Fisch, p 300).

Focal epileptiform activity often consists of localized spikes and sharp waves seen in a few neighboring electrodes. This activity may be surrounded by irregular slower waves or followed by an after-going slow wave. Focal spike and sharp waves may appear before and after a generalized discharge. (Fisch, 1999, p 271)

There are many disorders that may show epileptiform activity. These are the purview of neurologists, not neurofeedback practitioners. One rare disorder, *Landau-Kleffner syndrome,* is characterized by a progressive disturbance of language comprehension and speech. The spike-and wave-complexes are in the temporal region.

Another pattern that involves speech shows central bisynchronous spikes associated with Rhett's syndrome. This disorder of females is associated with slow decrease in motor and language skills, beginning often in the second year of life. Hand wringing is a prominent symptom. There are generalized complex partial or simple motor seizures that end by about age 10. After this time, the EEG is dominated by delta activity.

Absence Seizures

Generalized 3 Hz high-amplitude spike and slow-wave activity is a pattern you may see, if you evaluate a lot of children for ADHD, Inattentive Type. This pattern is characteristic of *absence seizures* (petit mal epilepsy). Someone with this problem is often mistakenly thought to be daydreaming and inattentive when, in fact, an unrecognized absence seizure disorder is the primary reason for the child's lack of attention in the classroom. There is no loss of muscle tone. Inattentiveness is momentary. If you tell the client something during the attack there will be no memory of it afterwards. They just stare off into space and nothing registers for a few seconds. These children may or may not also have ADD but the seizure disorder must be the first problem to be addressed.

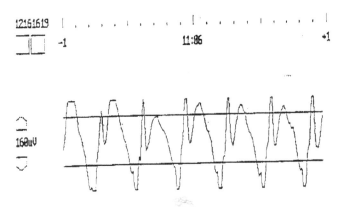

Y axis microvolts; X axis time

The above figure is from a recording made with an 8-year-old girl. It shows a sample of raw EEG 11 minutes into the session using an F1000 instrument. This recording demonstrates the typical 3-per-second spike-and-wave pattern seen in absence seizures (petit mal). This young girl was referred to the ADD Centre by her family physician with a diagnosis of ADHD. She was repeatedly inattentive (many times an hour) in the classroom. Note the very high amplitude, >160 µV, of the spike-and-wave activity.

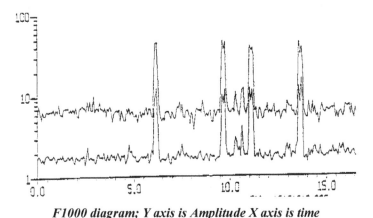

F1000 diagram; Y axis is Amplitude X axis is time

The top line represents slow wave (4-8 Hz), and the bottom, fast wave (16-20 Hz). This diagram is included to demonstrate both the frequency of these seizures (four in the course of 15 minutes) and the very high amplitude of the waves.

Note: Spike-and-wave patterns may sometimes appear with no relationship to a seizure disorder (Fisch, p 333).

Non-Spike and Wave Pattern

Non-spike and wave patterns are also found in some seizure disorders. For example, bursts of rhythmic slow waves may be seen in temporal or fronto-temporal regions in complex partial seizures. Detailed descriptions of the EEG in seizure disorders can be found in any basic neurology textbook.

Illustrations of a Simple Partial Seizure Disorder

In the figure below, the subject is an 11-year-old boy who has simple partial seizure activity in the left frontal lobe. Note the spikes with following slow wave.

This boy was referred to the ADD Centre by his doctor with a diagnosis of ADHD. His teachers had complained to his parents that he was not paying attention in class.

When this activity was observed on the EEG we referred him to a Neurologist who specialized in reading EEGs and working with seizure disorders

Sample of Eyes-Closed Resting EEG – Linked-Ear Montage

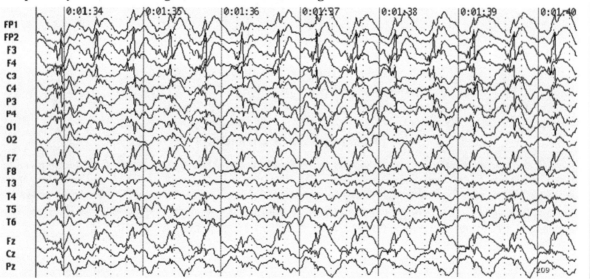

The next figure below shows the same EEG using a longitudinal sequential montage. This shows that the source of the activity is close to F3. We always look at each EEG with a minimum of three different montages. With some EEGs, important aspects of the data are more clearly seen in one montage than in another. The figure below also illustrates the very high amplitude of the spike and wave activity at time 1:41 compared to the regular EEG at time 1:45.

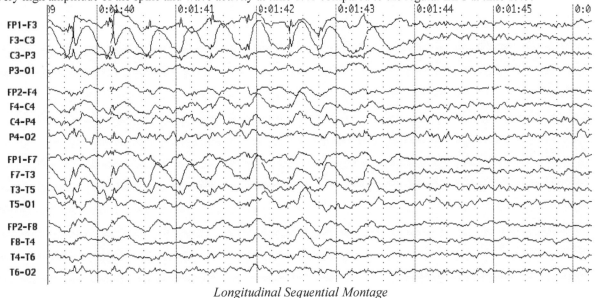

Longitudinal Sequential Montage

The next figure is a topometric display from the SKIL (Sterman Kaiser Imaging Laboratory) program. It compares the subject to the SKIL database for theta, 4-8 Hz.. The light gray lines are 2 standard deviations (SD) above and below the heavy black line (joined squares), which shows the database average. The joined circles (red line) are the subject's values in the 4-8 Hz theta range. Note the high values at F3 and P4. Time of day correction is on.

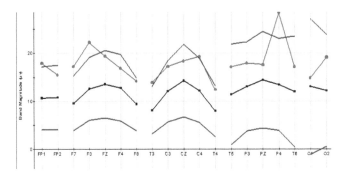

Below is a "brain map" display of the same data.

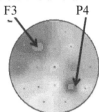

In this brain map, the amplitude range is between 0 and 28.5 μV. High amplitude is the brick-red color. Low amplitude is green. Note high F3 and P4 theta.

In brain maps, the top is anterior, frontal. The bottom is posterior, occipital. There are 19 small dots in each circle. Each dot is a 'site' in the 10-20 electrode system (see detailed diagrams in Section VI) and represents the location of an electrode.

Below, the same data are displayed in 1 Hz frequency bins and compared to the SKIL normative database at 3 SD. Red is above the mean, and dark blue is below the mean. The dull brick color seen in the 1 Hz display is the farthest from the mean.

The figure below is a comparison with the SKIL database at 3 SD. Note the very high theta and beta activity in the left frontal region (Frontal in the brain map is at the top of the circle).

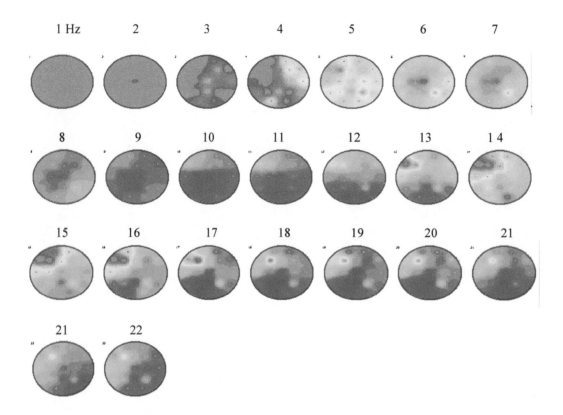

SECTION IV
Measuring the EEG: Instruments and Electronics

When we monitor the EEG using an electroencephalograph, we are measuring the potential difference between pairs of small electrodes. The voltages measured are in millionths of a volt. We can induce voltage differences of thousands of volts just by rubbing our feet on the carpet and approaching another person. The obvious dilemma and question is, how does the EEG instrument detect the tiny voltages produced by groups of neurons in our cortex and eliminate the influence of enormous voltages from extraneous sources? First let's consider what *potential difference*, *current*, and *resistance* mean.

You probably recall from high school physics that for a direct current (DC), such as you have in a flashlight run by batteries, there is a relationship between potential difference measured in volts, current measured in amperes, and resistance measured in ohms. The relationship was described by the German physicist Georg Ohm in 1826 as *Potential Difference (V) = Current (I) x Resistance (R)*. A similar formula describes this relationship for alternating currents (AC). AC is what you get from a wall plug, and also what is measured by an EEG. The formula is: *Voltage (V or E) = Current (I) x Impedance (z)*. Impedance is a more complex construct than resistance because its calculation requires measurements of not only the resistance of the conductor, but other factors such as capacitance, inductance, and the frequency of the alternating current. These terms and impedance measurements will be described in more detail later in this text.

In these formulas, *current* is the rate of electron flow through a conductor. It is measured in *amperes*. *Potential difference* may be thought of as the force or pressure that causes the electron current to flow in one direction. Thus, the current flows due to the potential difference between source (–ve site) and destination (+ve site). The *resistance* (or impedance with AC current) is the opposition to this flow inherent in the material through which the current is flowing. Thus R and z both refer to the opposition to the flow of electrons through the conducting circuit. Opposition to electron flow is high in substances such as rubber where most of the atoms contain outer electron layers that are full. This increases the resistance of the electron in the outer layer to being dislodged. These substances make good insulators and very poor conductors.

Current measures the rate of transfer of electric charge from one point to another. (Your utility company measures the number of electrons passing through the meter in each second. They measure it in *amperes* where an *ampere = 6.28 x 10^{18} electrons* [coulomb]). *Electrical charge* refers to the negative charge carried by electrons. Electrons orbit around their nucleus, but at different orbiting distances that may be thought of as *energy levels*. Each energy level contains a precise number of electrons. Nearest to the nucleus are two electrons. In the next level there are eight, and in the third level there are 16 electrons. It is the electrons in the outermost layer that are responsible for electricity. This electron layer may be incompletely filled. When this is the case, the electrons in that layer are less tightly held in place and collisions may dislodge them. Imagine that a loose electron acts like a billiard ball. It collides with another electron, is captured by the atom it collided with but that atom's electron is sent off on a different course to strike the next atom and so on in a chain reaction. It is this sequence that is responsible for what we know as an electric current.

In our work with neurofeedback, we use the potential difference between a +ve and a –ve electrode as a measure of the *amplitude* of the EEG signal. For brain waves this is expressed in microvolts (µV). A microvolt is a millionth of a volt.

An analogy that is often quoted is the flow of water from a water tower. The height of the tower dictates the pressure that will move the water through a pipe. Pressure is like potential difference (voltage) in an electrical circuit. The rate of flow of the water is the "current." The diameter of the pipe corresponds to resistance in the electrical circuit. A small-diameter pipe will resist the flow of the current. The current can be increased either by increasing the pressure (a higher water tower) or decreasing the resistance (a wider pipe).

The brain produces an alternating current. This current may be thought of as a sine wave. To measure the amplitude of this wave we usually measure from the top of the positive wave to the top of the negative wave and we term this a *peak-to-peak* measurement.

The EEG Instrument
How Does My System *Read* the EEG Signal and Keep Out Unwanted Electrical Activity?

The scalp electrodes used for neurofeedback are macroelectrodes (>5mm), which are detecting microvolt differences between electrodes at two different sites. Merely walking closer to the client can induce currents in the wires, and in virtually any room that you will be using for NFB there will be many other electrical influences that can affect the recording. The simplest and most dramatic example is static electricity. Rub your shoes on the rug and put your hand with a finger pointing forward towards the extended hand from a person seated in front of you. A spark will jump between you and your friend's hand. The potential difference between you and your friend can be as high as 10,000 volts. The current, however, is extremely low. This potential difference is in volts but we are only measuring microvolts! It doesn't shock your friend but it will have a major effect on the EEG recording!

Thus, the first step in measuring the EEG involves an instrument called a *preamplifier*. It amplifies this tiny microvolt difference by more than 100,000 times and does not amplify the other electrical signals. The second step is to change the alternating analog current to the digital form that the computer can work with, by a process called *sampling*. The third step is to take this digital signal and display it in a manner that makes it easier to read by *filtering,* which means showing the portion of the EEG that is of interest to us and filtering out the rest of the signal. The next sections will describe these processes in more detail.

Amplification
Why a *preamp*? The preamp amplifies the EEG current by a huge amount so that other influences in the environment will be small and inconsequential in comparison to the amplified EEG current. It only amplifies the differences in voltage between inputs.

Your electrodes are picking up tiny amounts of electrical current. You are measuring millionths of a volt (microvolts). As in the example given previously, just walking towards your client after rubbing your shoes on a carpet can create a voltage difference between you and your client of thousands of volts. This will induce an electrical current in the wires attaching your client to the amplifier. Long wires may pick up more induced electrical changes. Thus, having short wires that run to a preamp that is placed on the client's shoulder, or even on the headband, should reduce the problem. (There is less wire that can act like an antenna.) Other instruments have the preamp in the same box as the encoder, which means longer wires and more opportunity for other electrical activity to be picked up. Having a well-shielded cable is one way to handle the problem of long wires. Cables for Focused Technology's F1000 equipment, for example, have an extra wire wound inside to pick up extraneous current so that the current does not have much influence on the wires carrying current from the electrodes. Some instruments, such as Thought Technology's ProComp+ and the Infiniti instruments, have a preamp that can be clipped onto the client's collar. Thus, the wires from the electrodes to the preamp can be short.

Whether the instrument has a preamp close to the electrodes, well shielded cables from electrodes to amplifier, or both, the goal is to reduce the amount of induced electrical current affecting the system.

The preamplifier is a small unit that, ideally, would be placed as close to the electrode site as possible because, once amplification has taken place, ambient electrical field effects will have much less influence on the recorded EEG. In Sterman's research with Top Gun U.S. Navy pilots there was a preamp built into the cap worn by the pilot at each electrode site. This is an elegant but very expensive solution to the problem of wires between the electrode and the preamp picking up unwanted current.

Calibration of a full cap EEG instrument is done by applying a standard voltage to all input channels. This ensures that the voltage read is accurate and that all inputs are amplifying and filtering the signal in the same way. Most neurofeedback instruments do not require calibration measurements before each use. You would only calibrate if you suspected a problem. We recommend that practitioners have two EEG instruments. When a problem is suspected, the trainer can rapidly plug the client's electrodes into the second instrument to check the readings.

How does the amplifier work? The amplifier detects and amplifies differences between two inputs. It amplifies changes in the signals to each input to the same degree but in opposite directions, referenced to an electrical reference built into the amplifier. It does this by reversing the polarity of the second input so that the two currents are effectively subtracted. The amplifier only amplifies this difference between the two inputs; hence, it is called a **differential amplifier**.

In simpler, lay-person, language, imagine you have one wire attached to an electrode at Cz and a second wire attached to an ear lobe. All the electrical interference activity in the room, but not electrical activity coming from neurons in the brain, will hit both wires with the same amplitude, at the same frequency, and in the same phase relationship at exactly the same time. Ideally, the only thing different between the two wires is the tiny microvolt current from a site on the scalp such as Cz that is due to activity from brain cortical neurons. When the polarity is changed in one of the wires all the interfering high voltage activity will cancel out. This leaves only the small microvolt activity from brain neurons, which is then greatly amplified in the preamp.

In this way the amplifier is said to *reject* signals that are in common to both inputs. This is called *common-mode* rejection. The machine is wired so that greater negativity at Input 1 than Input 2 causes an upward deflection of the signal.

(Note: Your ground wire, on the client's scalp, does NOT go to the ground in the sense that we call the third wire on an electrical plug a ground wire. Think of the term "ground" when referring to your third electrode as an electrical housekeeping wire. True ground is no longer the reference point for measurements. (Frank Diets, the electrical engineer who designed the F1000 biofeedback/neurofeedback instrument, personal communication.)

The *common-mode rejection ratio* is the ratio of the common-mode input voltage divided by the output voltage (Fisch, p43). This ratio should be greater than 100,000. Failure of this system to eliminate external common-mode artifact is probably due to either a difference in impedance between the two electrodes and/or a poor "ground" connection.

A second amplification is carried out after signals are filtered. This is called *single-ended amplification* because it just compares a single input with the "ground" and amplifies that signal.

Filtering

Your computer amplifier has two filters that help to minimize distortions that could make it quite difficult for you to read the EEG. These are the *high-pass filter* and the *low-pass filter.* In some instruments, this filtering is done in the *preamp* that has input from three electrodes (positive, negative, and ground), and is placed between those wires and the *encoder,* which sits on the desk. In other instruments it is in the same box on the desk with the encoder. Filtering takes place after the differential amplification and before the second, single-ended, amplification. A third type of filter called a *notch* filter is present in some instances to filter out a narrow band of activity such as 60 Hz. These filters do not just cut off waves below or above them. It is a more complex process of attenuating the unwanted frequencies – that is, reducing their amplitude by a set percentage. You may read about this process in a basic EEG textbook such as Fisch (p 46-54). An unavoidable, unwanted effect of the low-pass filter is that the filter can distort artifact potentials (such as muscle artifact) by lowering the amplitude and slowing the observed frequency, with the result that the waves may look as if they are cerebral in origin.

High-Pass Filter

The *high-pass filter* is meant to attenuate waves (reduce the amplitude) that come in at a frequency below its cut off. It lets waves *pass through* it that are higher than its cutoff frequency. It is not an all-or-none type of filtering, but a gradual elimination of frequencies. Most of our instruments have high-pass filters at 1 or 2 Hz because we are usually only looking for wave forms above 3 Hz when doing EEG biofeedback. In hospital work, however, lower frequencies are considered for interpretation of the EEG. Instruments such as the ProComp+ or Infiniti have their high-pass filter set at a lower frequency (0.5 Hz). Delta waves can be clearly seen with this instrument, though you must be careful to distinguish between delta and eye-movement artifact.

In some EEG instruments, such as the Lexicor, this high-pass filter may be turned on or off if they are reading the EEG. You will usually have it on during feedback. A low cutoff gives an EEG that includes delta, which may be useful. However, it means that any interference that is sufficient to overload this high-pass filter, such as the start-up of an air conditioner or a pump, may result in a spurious signal while the amplifier takes time to recover. The resulting EEG may have the appearance of a single square wave several seconds long (with spurious harmonics up through beta). There are pros and cons to every design, and as we discuss instrumentation, you will frequently find that the engineers had to make decisions based on trade-offs. Remember that a high-pass filter is set at a low number such as 0.5 Hz or 2 Hz. It is therefore sometimes called a low-frequency filter.

Low-Pass Filter

The *low-pass filter* is meant to only let through, for our observation, waves below its cutoff point. Many older biofeedback instruments had a low pass at approximately 32 Hz. The instruments we now use more often have a low-pass filter that can be set as high as 62 Hz with the option to set it to a lower value, such as 40 Hz, on one of the display line graphs to make the EEG somewhat cleaner (less electrical and muscle electrical activity) and easier to read. Thus, frequencies higher than the chosen setting are not registered. The F1000 has a digitally tunable low-pass filter. It is set to 61 Hz for an online FFT display that goes out to 63 Hz. During feedback, however, it is set to 45 Hz to reduce the influence of 50/60 Hz interference. The ProComp+ and Infiniti from Thought Technology have low-pass filters that are also above 61 Hz. This higher cutoff point allows us to observe the EEG at higher frequencies. This is important when we are attempting to distinguish cortical activity; for example, ruminative activity around 30 Hz, or cognitive binding activity (Sheer rhythm) around 40 Hz, from activity in these frequency ranges that may be due to muscle (EMG) artifact. Electrical activity from lights, computers, extension cords, etc. is usually very regular and distinct on the EEG and can be seen at 60 Hz in North America or at 50 Hz in Europe, Asia and Australia.

Other sources of interference may be more of a problem to very sensitive instruments than they are to some other less sensitive or older instruments due to a very low *noise level allowance*. The boosters that truckers use for their two-way radio communications, for example, may result in a rise in amplitude starting at high frequencies and moving down to lower ones, like surf in Hawaii.

Band-Pass Filters

A *band pass* is just the frequency range (for example, 4 to 8 Hz) that is chosen by the practitioner to use either for statistical comparisons or for neurofeedback. In neurofeedback, the practitioner chooses frequency bands to inhibit or to enhance. How these are chosen is described in the intervention section of this book. Some software computer systems allow the practitioner to choose both the type of filter (IIR, FIR, FFT) and the width of the frequency band that is desired for statistics or for neurofeedback training. In other systems the type of filter is chosen by the engineer who designed the program and it cannot be changed.

Sampling Rate

The original EEG can be said to be in an 'analog' or continuous wave form. This wave must be broken up into tiny parcels or samples to be used by your computer. Breaking up the continuous wave into small pieces is called *sampling*. This sampling is performed by an analog to digital (A/D) converter. It may be located in the *encoder* of the ProComp+ and in the interface board of the instrument. Modern inputs to the encoder always use female plugs. Female plugs are used because they cannot accidentally be connected to a power source, an error that could be damaging to both equipment and the person connected to that equipment.

A fast sampling rate is crucial for obtaining accurate information. The maximum frequency that can be reconstructed in a filter is based on the *Nyquist principle*, which means that for the digital results to be accurate, the sampling rate needs to be at least twice the maximum frequency of the analog signal. Technically then, 128 samples per second would allow you to view frequencies up to about 64 Hz, although in practice, instruments with this sampling rate usually just handle frequencies up to 32 Hz. This is the basis of the F1000 online spectral display and also approximately the rate read by the Lexicor. Other instruments, such as the ProComp+ biofeedback systems and the Neuronavigator, have a sampling rate of 256 samples per second or greater. Thought Technology's Infiniti can offer choices of sampling rates up to 2500 samples per second. Faster sampling rates allow the practitioner to more accurately observe higher-frequency wave forms. For example, a sampling rate of 256 cycles per second (cps) can accurately show frequencies of one-quarter that rate, or 64 cycles a second (Hz). Although a division by two is acceptable, as a rule of thumb, most manufacturers divide the sampling rate by four to get an approximate maximum frequency for accurate EEG wave frequency analysis. Thus, to get an EEG spectrum that goes accurately to about 64 Hz, we want an instrument that will sample at rate of 256 samples per second, and for 32Hz you can sample at 128 samples per second. High sampling rates are necessary for analytical analysis of single waveforms. It is referred to as *oversampling,* and 8x to 16x the frequency being analyzed is a common standard.

The sampling rate of 64 cycles per second used by some older instruments allowed a faster FFT calculation. This was an important consideration with older, slower computers. The trade-off with respect to sampling rate vs. breadth of frequency range is that higher sampling rates take longer for calculation and this can slow down the feedback. Today's fast computers make this less of a problem.

A sampling rate that is too slow will make the analog signal incorrectly appear to be running at a slower frequency than it actually is. This effect is called **aliasing.**

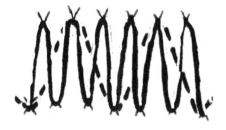

In the above diagram, the actual wave is the solid line and the incorrectly interpreted wave is the dashed line. The true wave is theta at 6 Hz. This is seen if 13 samples ('x' points on the drawing) are taken. If only 5 samples are taken, then the line drawn between the points makes it appear as if the EEG wave form is delta at 2 Hz.

If you draw for yourself a wave sampled at 42 samples per second and then draw a second wave just using every third sample or 14 of those samples, you will see that the first wave is 21 Hz while the second is only 7 Hz.

In addition to having an assigned *sampling* rate, the *analog to digital converter* (ADC) also is assigned a *voltage range* and a *bit number.* The number of "bits" refers to the number of amplitude levels that can be resolved. An 8-bit ADC will have 2^8 or 256 amplitude levels. This would be ±128 discrete voltage levels in the voltage range allowed by that ADC. Too few bits means that small increases in voltage will be overemphasized. Too narrow a voltage range means that a large voltage change would be cut off.

Types of Filters

Three types of digital filtering are: *finite impulse response (FIR), infinite impulse response (IIR)* and *fast Fourier transform (FFT).* The FFT filter can provide a much sharper cutoff than the FIR filter. Both of these filters correct to give reasonably accurate phase relationships. The FIR filter computes a moving average of digital samples. The number of points that are averaged is termed the *order* of the filter. Some instrument programs, such as the original ProComp+/Biograph programs, allowed you to choose both the *order* and the type of filter. Each filter attenuates the same frequency ranges in a slightly different manner; for example, an IIR filter has a much sharper slope than an FIR filter.

The implications for the practice of neurofeedback lie in the realization that when you sample a certain frequency range, say 4-8 Hz, the frequencies outside that range are attenuated but not entirely eliminated. In particular, the frequencies at each end of the range will get through to a certain extent due to the shoulders (degree of slope) on either side of the filter.

The following two diagrams are illustrations taken from an older the ProComp+ Biograph instrument. They compare an FIR Blackburn filter in the first diagram with an IIR Butterworth filter for the same bandwidth 13-15 Hz.

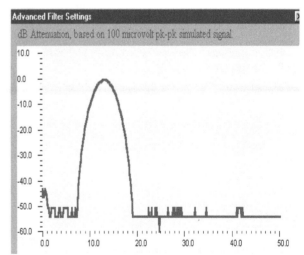

FIR Blackburn Filter for 13-15 Hz

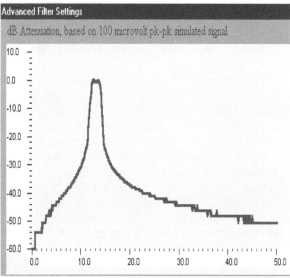

IIR Butterworth filter for 13-15 Hz

We use the IIR filter for doing statistics because we find that we get more consistent results. It is the filter now used on Thought Technology instruments. However, when a fine-tuned analysis is done, it demonstrates that the IIR filter is so narrow and precise that 13-15 Hz may actually turn out to just be 14 Hz activity. The exact range depends on the "order" of the IIR filter. Without getting too complicated, just remember a simple rule of thumb: whatever filter you use, you must use that same filter for all your statistics because different filters give quite different statistical values for each bandwidth.

This diagram shows a three-lead referential placement. The active lead is placed at Fz, and the reference is on the left ear. The ground has been placed on the right ear.

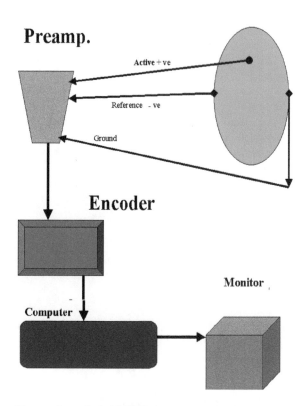

Illustration of an NFB System
This diagram shows the basic functions usually carried out by software programs in the encoder and the computer.

Filtering may be digital (hexagon) or FFT (oval).

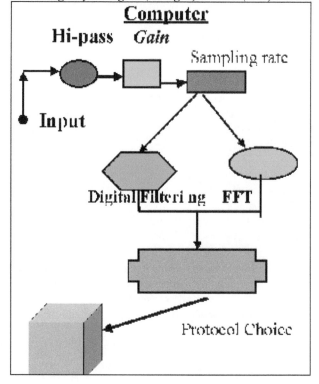

A *Fast Fourier Transform (FFT) filter,* which is a program inside your computer, can take the EEG information and mathematically transform it to give an average voltage for a specific frequency over a specified time frame. This in turn produces a histogram where the x axis is frequency in Hertz and the y axis is amplitude in microvolts or power in picowatts. This type of graphic display can help you show students and parents different mental states in a way they can understand and picture. For example, recall the example we mentioned earlier about asking the student to do a multiplication question in his head. Stop the recording immediately after they answer. Now go back and replay this data to show how the brain activity changed. You may see a sequence such as the following: theta went up in the left frontal area when they were trying to recall the answer from memory. Then when they actually were figuring it out, beta increased. Then theta dropped and beta remained high as they calculated and gave you the answer.

Computations used to be quite slow for FFT filters, but computers now run at >700 megahertz. This has speeded this process considerably so it is now possible to use FFT displays for feedback.

Montage Reformatting

Montage reformatting is a process whereby one can take different views of the same data, referencing one electrode site to any other site or combination of sites. Each montage is a different way of referencing the 'active' electrode site. It is standard practice in 19-channel assessments to use a linked-ears reference to collect data. The montage reformatting is done after the data is collected.

For single-channel assessment and training using NFB we usually use the ear lobe or the scalp over the mastoid bone as a reference site. This is because, compared to other sites on the scalp, these sites have very little electrical activity. We can then assume that changes we are recording are due to the site on the scalp where we have positioned our *active* electrode. In a full-cap assessment, linked ears are often used in what is termed a *common electrode reference montage*. When we do a *sequential* or *bipolar* recording, we are referencing two electrodes to each other. In a full-cap assessment, *sequential (bipolar) montages* may be done using adjacent pairs of electrodes in the 10-20 placement system. In a 19-lead recording, the computer can do a number of different *montages*. It could, for example, reference an active electrode to an average of all the other electrodes (*average reference montage*). It could reference the active electrode to the average of the electrodes immediately surrounding it, which is called a *Laplacian montage*. For a full discussion of Laplacian mathematics that relate to EEG analysis, see Hjorth's 1980 article. Each *montage* is just a mathematical reworking of the data, which can easily be done by the computer software. Examples of the same data shown with sequential and Laplacian montages were used in the examples of measures in the last section.

Each of these different ways of looking at the data will have both advantages and disadvantages. The *sequential* (bipolar) and *Laplacian* montages are good for viewing highly localized activity. This would be of more value to neurologists. The *common reference montage,* on the other hand, is excellent for detection of widely distributed currents and for analyzing asymmetry. It is also very good for the detection of artifacts. It is not so useful for viewing localized activity. A sequential (bipolar) recording may register a lower value for theta and, relatively, a higher value for beta than a referential recording. This is because theta is a more generalized activity than beta and will be canceled out by the differential amplifier in the sequential recording to the extent that it is found at both sites.

Electrical Terms

Electrical Outlets

Electrical wires are color coded. Generally black carries the main current and would then be the dangerous wire. White usually is neutral and carries the current returning from the instrument. Green would then be the ground. However, **don't trust this generalization** – always read the color coding for any wired apparatus that you are using and always have a professional electrician handle the electrical wiring in your center. Colors can be confusing. The EEG wires we use commonly use a different color coding than the one described above that electricians use.

Capacitors

Capacitors are formed by two conductors separated by an insulator such as air. A capacitor will store a charge. Where this concept can be of importance to you is in giving you some understanding of why it is preferable not to use extension cords with your instrument. If there is a gap between the extension cord and the plug, you have created a capacitor. Current can actually flow between the wires (from the black to the green).

Optical Isolation

In our work, steps are taken to protect the patient. Optical isolation is one such step. This term refers usually to a process where we separate the encoder from the computer by a glass connecting cable optical isolator. An optical isolator is a component that transfers electrical signals between two isolated circuits by using light. Thus the opto-isolator prevents high voltages from affecting the system receiving the signal and, in this manner, it protects the patient from any untoward malfunction in the wall electrical circuit, computer, and instruments that are being used to record the EEG.

Thus the digital current from the encoder is changed to an optical form for transmission to the computer. The computer converts it back to a digital current form for computer analysis. Apart from the very high speed of optical transfer, it has the added advantage of protecting (isolating) the patient from the electrical connection of the computer to the wall circuit. Ordinary electrical current cannot be conducted through this glass 'wire' optical connection from the computer to the encoder. The encoder must, therefore, have its own electrical source. This is usually a battery source with very low current produced.

Electrical Artifacts

The manufacturer of your EEG instrument has attempted to minimize artifacts in the EEG recording. There are also precautions you can take when you put on the electrodes. Nevertheless, despite having a good instrument and taking appropriate precautions, you will observe electrical artifacts. You must be able to identify waves that are not due to neuronal activity. This section deals just with electrical artifacts. Other kinds of artifact, such as those due to eye blinks or EMG, are covered in another section.

What May Interfere

Electrical wires act as antennae. They will pick up 60 Hz activity (50 Hz in Europe and Asia). This activity is always present in our offices.

One of the biofeedback trainers didn't like fluorescent lighting in the office. He brought in an old standing lamp. It took several days for us to realize that this was the reason we weren't able to get the EEG equipment in that office space to work properly. Bad wiring in an old lamp can ruin your EEG record!

The electrode wires can also pick up some frequency outputs from a nearby radio station. Truckers with CB boosters can give your machine and you a small ulcer! They are illegal in many jurisdictions because of their effects on hospital equipment. However, this law does not seem to deter the booster users. These boosters may overwhelm the amplifier's high-pass filter and be observed as high-amplitude surges at a variety of frequencies.

Even movement of people in the room may have an effect. Potential differences between objects in the environment and the wires to your electrode, which is on the client's scalp, can induce current in the wires. The simplest example of this, as previously mentioned, is static electricity produced when you rub your leather-soled shoes on the rug and approach another person with hand outstretched. This can produce a spark. What we didn't know when we playfully did this as children was that there could be as much as 3,000 to 10,000 volts difference between the two of us. Think of yourself as a container of negatively charged electrons compared to your client. You know that negative repels negative. As you approach your client, you actually induce a current in the wires. But Ohm's law states that voltage = current times resistance ($V = IR$). You are changing the current (I). Therefore, the voltage will change. It will change at frequencies controlled by the rate of your approach. Electric wiring in the room, lights, and other instruments can all have the effect of inducing unwanted currents. They will induce current at frequencies associated with their source. Electrical wiring in North America, for example, has a frequency of 60 cycles per second. This can be clearly seen at a very high amplitude on the spectrum display unless you do a good job of making sure that impedances are the same at the electrode sites so that your differential amplifier will *reject* this voltage.

What You Can Do to Minimize these Problems

Troubleshooting Pointers for the Office

Many electrical artifacts will be from a source that affects all of your wires equally. Your instrument has a differential amplifier. It will only amplify waves that are different in phase and magnitude between each pair of sites. Effectively it is *rejecting* that which is in common. (Common-mode rejection will be explained in more detail later in this chapter.) However, a 60 Hz source, such as your electric lighting, can be common to both electrodes but appear different when the waves reach the amplifier. This could occur if the connection between the scalp electrode site and the amplifier was different for the two electrodes. The amplifier might then compare the electrical activity from these two electrode inputs, and because the waves appear to be different, amplify this difference in voltage between the two electrodes. The amplifier's common-mode rejection would not come into play. Then this induced interfering signal would be amplified and ruin your EEG recording.

Thus, you want to reduce differences between each site and your amplifier as much as possible. Several steps may help you to do this. First, your electrodes should all be made of the same material. This is usually gold or tin. You should not have two metals at the site, as would occur if some of the gold plate had worn off one of your electrodes. Electrodes should be carefully cleaned after each use. In addition, dead skin has a very high resistance. When you measure potential difference (voltage), it is proportional to current (I) and to the resistance to direct electrical (DC) current flow. You will recall this relationship is described by the equation $V=IR$ (Ohm's law). Impedance (Z) is just the term that replaces R when you are talking about the resistance to the flow of an alternating current (AC). Brain waves that we work with in neurofeedback are AC, not DC. For us the relevant formula is $V=IZ$. Therefore, it follows that if you did an excellent site preparation at Site A and lowered the impedance at this site, but you did no preparation at Site B, although the voltage induced at each site came from a common source, by the time it reached the amplifier it would not appear to be the same. It might not then be rejected.

If you pay careful attention to site preparation as measured by your impedance readings and you clean your electrodes between each use, then you should minimize most electrical interference.

Other hints to minimize the effects of induced changes are as follows:

- Don't let your electrode wires form loops or move during the recording.
- Have specially shielded wires (as with the F1000) and/or the preamplifier as close as possible to the electrode site and have the wires between the electrode and the preamplifier as short as possible.
- Use a headband like those used by tennis players. This keeps the wires stabilized and reduces movement.
- Braid loose wires. This will help equalize the effect on all wires, but specially protected wires are best. In more sophisticated wiring, such as the special shielded wires in the F1000, induced current will flow harmlessly to the amplifier through the wire's special covering and to the ground. With unprotected wires, when the resistance of the amplifier is high, it flows to your client.
- Remove any equipment, such as desk or standing lamps, electric pencil sharpeners, and so on, that may be producing interfering signals.

We run a learning center. The teaching staff regularly use pencils. One enterprising trainer put an electric pencil sharpener in a training room. For days we examined the equipment, carefully redid site preparations, changed electrodes and turned off all electrical lighting, but to no avail. The EEG signal could be read, but the readings for fast waves were too high. By accident, one of us unplugged the pencil sharpener. All electrical interference disappeared. This interference was present even when the sharpener was not in use but was plugged into the wall socket.

Even extension cords can distort the EEG signal.

The authors were presenting at a professional meeting. A member of the audience was attached to the instrument. The EEG signal was virtually unreadable. The connection to the amplifier and the encoder were all changed but to no effect. The author then decided it must be something to do with the hotel's wiring. The first step to check this was to disconnect the extension cord and run on batteries. All interference disappeared. It turned out that the extension cord had been responsible for the artifact.

The building's electrical wiring may induce electrical artifact into the system.

The author was invited to work with the senior vice presidents of a large firm. The EEG feedback system that worked perfectly in his office showed completely overwhelming interference with extremely high amplitudes observed from 56 to 63 Hz. Amplitudes at other frequency bands varied from day to day. The only way to do the neurofeedback sessions was to run them using a laptop on battery power.

We **strongly recommend** that you try your EEG equipment in any new office you intend to rent. On one occasion an associate in a very old building (wall plugs were for two-pronged plugs) had to run copper wires out the window to the ground to obtain good grounding and decrease interference. On another occasion, a colleague couldn't run his equipment from his old office because an engine somewhere in the building was interfering. Every time the engine started in the basement, his EEG amplifier was overwhelmed and the EEG flattened for several seconds. He rented another office.

In an office we were about to rent, I found electrical interference on the EEG. I had plugged the computer in with the wires attached to my head at several times during the day and found that at 5:30 p.m. it was impossible to read the EEG. It turned out that 5:30 p.m. was when the cleaners started their work several floors below ours. We do much of our training in the evening to avoid taking children out of school so it was an important discovery. I demonstrated this problem to the landlord before I signed the lease. The landlord agreed to run special wires to our offices. These wires would not be connected to any other outlets. This avoided our having interference when other offices and cleaners turned on equipment using wiring that was connected to our electrical wiring. When you check a new office site, do it at several different times of day.

Simple Equipment Troubleshooting

In the foregoing discussion we have recommended that you do your best to limit the number of sources of electrical activity that might interfere with your equipment. You have unplugged all unnecessary electrical equipment and turned off other equipment such as cell phones. You should also try increasing the distance between your EEG amplifier and your computer. If you have done all these steps and you still have high 60 Hz activity and/or unexplained EEG activity or inactivity, then begin to check your equipment in a logical, stepwise manner.

Begin by Checking Impedances, Offsets and Electrodes

The EEG amplitude dropped. The readings were lower in amplitude than last day. However, the proportions of one frequency to another seemed reasonably appropriate. The theta/beta ratio appeared close to the ratio that had been observed in the last two sessions. The EEG looked normal to a trainer who had only done a few hundred hours of work with EEGs. However, she intuitively felt it just wasn't right and called in a more experienced trainer. They checked impedances and changed wires, to no avail. Then the EEG suddenly went flat. The client was changed to another EEG instrument to finish the session.

The next day the same machine appeared to be working perfectly. Then, after a couple of hours, it again suddenly dropped in amplitude, and after a few minutes went flat. Humidity and overheating were both considered, but these did not seem to explain the sudden occurrence of this problem. The EEG paste was changed for a tube used in another room where the instrument was working. Finally the author was called. The first question asked was, "What are the offsets?" These had been checked by the more senior trainer. He replied that they seemed high for the scalp electrode. He replaced this electrode. He noted, however, that the high impedances remained for this site even when he put on three other brand new electrodes. He had even moved the child, with these same electrodes still attached, to a different amplifier (different EEG instrument) and the connections all worked perfectly.

The author tried a short wire electrode from a different batch of electrodes. The offset dropped from 85 to 5, and the EEG on the apparently faulty machine has run perfectly ever since.

A second phenomenon had been observed with this same machine when the above problem was investigated. The trainer found that, instead of a flat line when he removed one of the three electrodes at the end of the cable to the amplifier, waves suddenly appeared on the screen. It was as if he was getting an EEG with only two electrodes inserted into the connection to the amplifier. In our Centres, we always insist that the trainers have a way of looking for high frequency artifact. In this case the trainer was able to see a regular spiky high-amplitude, high-frequency artifact. He recognized that this did not represent an EEG, but was a complex electrical artifact.

There are several points that come out of this example. First, unless you have EEG instruments from more than one manufacturer and can compare EEGs on different instruments, you might be fooled into thinking you were getting a normal EEG when what you were recording was either an incorrect (low-amplitude) EEG or artifact. Second, it is possible to get a whole batch of faulty new electrodes. Third, it is helpful to be able to check offset as well as impedance. A high offset tells you that there is a problem with your wire. It may look fine on the outside but be broken on the inside. (However, we are finding that this option to test offset is no longer available with many instruments.) Fourth, to an inexperienced person, complex electrical artifact can mimic a poor EEG. Fifth, it is always helpful to have experienced backup when things don't look quite right. The following corollary to this is: Never underestimate the importance of the technical support department of your equipment manufacturer.

Make Sure the Computer-to-Encoder (or Amplifier) Connections Are Working Properly

When you do not appear to be detecting the EEG on the monitor, there are a series of steps you can take to identify and correct the problem. First make sure that your encoder (encoder/amplifier) is detected by your computer. Put your hand over the end of the place where electrodes are inserted. For example, this could be a cable end in the A620, or a preamp in the ProComp$^+$ or Infinity. Now move your hand and shake the cable. Waves may appear where an EEG should be on the display screen. This is a simple way to demonstrate that there is still a connection from the computer to this point. Some instruments, such as the Biograph Infiniti program from Thought Technology, will show a sign on the display screen that states whether or not the computer is detecting the encoder.

On multichannel encoders, if the computer is not detecting the encoder/amplifier for the EEG, check to see if other channels on the encoder/amplifier work, such as a second EEG channel or an EMG, temperature or EDR channel. If nothing is detected, then check your connections (an optical cable in modern instruments) and check the batteries in the encoder itself. Try replacing the wire or optical cable from the encoder (or amplifier) to the computer. If changing the connection does not work, try putting the wire or the optical cable into the port on a different computer and seeing if the encoder is detected by that computer. If it is, then you may have to reload the program and/or repair the first computer. It could be a hardware fault in the first computer. However, before doing anything dramatic, do try plugging your computer into a different wall outlet. (We have had both a faulty wall outlet and a faulty extension cord.)

If all of these steps fail, try replacing the whole encoder with one that is working on another computer. In 99 percent of cases you will have identified the problem by this point. However, if you haven't, call the technical support for your system.

The EEG was a flat line. "Dr. R" did each of the steps above, but no encoder was detected by the program in the computer. She phoned the manufacturer and a new replacement encoder arrived the next morning by courier. It worked.

It is important to have a good relationship with your supplier. Some companies do respond immediately. However, it is important that you have carried out reasonable troubleshooting before you call. Again, it is very helpful to have more than one of each type of EEG equipment if you are doing a lot of neurofeedback/biofeedback work. Cars, stoves, refrigerators and washing machines break down, and it can happen to computers and EEG equipment, too.

If the Encoder Is Detected But the EEG is Not Being Recorded, Check the Connections from Encoder to Scalp

If the encoder is detected but the EEG is not being recorded, then we usually begin troubleshooting from the head down. There is a saying in clinical medicine that applies here, "Common things are common." Look for simplest errors first. The fault usually lies in a wire. Often it is one of the electrode wires. Try replacing each electrode in turn. If the EEG is still flat or of poor quality, then replace the wire that attaches the preamplifier (which may be clipped to the client's shoulder) to the encoder (on the desk). This is the second most common problem. Wires do break. If the problem persists, then try replacing the cable from the preamplifier to your short electrode wires. If these steps fail to solve the problem, then try replacing the preamplifier itself. When doing workshops we always take two of everything.

All of the difficulties that are mentioned above can be expected to occur over time. It is wise to have backup electrodes and cables. Ideally, one should also have a backup EEG instrument and computer to avoid downtime when there are problems with an instrument. At our Centre we have EEG instruments from 14 different manufacturers, though the equipment in regular use is from only four different manufacturers. All of them will puzzle you with little glitches from time to time.

What the Manufacturer Has Done to Minimize Artifact – The Differential Amplifier

The amplifier *differentially* amplifies the electrical information it receives. This means that signals of the same frequency that are different *in magnitude and phase* at each site are magnified, while signals that are the same are not magnified.

For example, signals that differ can be magnified at the preamp by a factor that is more than 100,000 times greater than the amplification of changes which are in common in *magnitude and phase* between the two electrodes. A number of current instruments have a common-mode rejection ratio much higher than 100,000/1. Every 20 decibels (DB) multiplies the ratio by 10. Thus, a voltage amplification of 10/1 is 20 decibels. 100/1 is 40 DB. Modern instruments may be rated as 120 DB, which is 1,000,000/1. This effectively acts to *reject* the unwanted *common* artifactual currents.

Examples of things that might be in common are jaw or neck muscle tension, heart electrical activity or spurious electrical changes in the environment. The latter might be caused by people moving in the room, electrical lighting, wall outlets, pencil sharpeners, an electric razor, a tape recorder, a CD player, a hair drier, an extension cord, and so on.

Electrical Artifact at Only One Site

Unfortunately not all electrical artifacts (extraneous noise) are in common to all electrodes. An important example is **electrode movement**. Movement of an electrode creates a new waveform with its own frequency, but only in that electrode's connection to the amplifier. It is a DC potential created by galvanic action between the electrode, skin and conductive paste. With movement, there is a change in the geometry of this *galvanic cell*. It will alter what may be termed V_{offset}. Try having your client move his head or wiggle one of his ear lobes. You may observe a large slow-wave artifact in addition to a high-frequency artifact. Getting a very good connection that shows low impedances is one-way to minimize this. Another helpful trick is to put the electrode wire under a head band so that small head movements will not result in wire movement near the electrode site.

If the jar of electrode paste is left open to the air for long periods of time, or if the paste is frozen in transport during the winter, changes may occur that interfere with the paste's adhesive and conductive qualities. This paste should be replaced.

Differential Amplifier

Important Note: This section is purposefully both too brief and too oversimplified from the point of view of anyone experienced in electronics. Nevertheless, we hope it will be helpful to those who have a clinical or teaching background and less knowledge of electronics.

What Are We Measuring?

Briefly, the amplifier receives input from the positive active electrode and the negative reference electrode and it measures the difference between these two inputs: Thus, the EEG that we observe is: $V_{measured}$ (+ve) from Site #1 - $V_{measured}$ (-ve) from Site #2 at frequencies we specify. (V = voltage, +ve = positive, -ve = negative.) It is the potential difference (V in microvolts) between two electrodes. Why do we call one site positive and the other negative? Let's look at this in slightly more detail.

The Differential Amplifier
General Description

The original concept for the differential amplifier came from the work of Thomas Edison. The actual amplifier, however, was not developed until the 1930s. In simplest terms, think of the electrode at one site going to the preamp. There is a potential difference between the site on the scalp and a comparison within the amplifier that involves your third electrode, which you call a *ground*. Many years ago the comparison actually was with "ground." As noted previously, in modern amplifiers there is not any direct connection of your client to the ground. The measurement and calculations are done within the amplifier. Now visualize the wire from the second site entering at another point in the amplifier. Think of the first site being positive and the second negative. Often a +ve sign is seen on the electrode that you use as your *active* electrode and a –ve sign is on the other or reference electrode. In reality, the polarity for the second electrode connection is changed in the amplifier such that the second input is *inverted* and becomes –ve with respect to the first electrode, which is +ve.

As was previously explained, these two potential differences – the active site to the amplifier (+ve) and the reference site to the amplifier (-ve) – are compared. Any voltage that is the same in both will therefore cancel out. Thus, any induced electrical activity from another source such as a nearby lamp (60 Hz) should be the same frequency and amplitude, and should be *in phase* on both wires. The *+ve wire*, so to speak, from the active electrode will, in this case, be the mirror image of the *-ve wire* from the reference site with respect to this 60 Hz current, and the two will cancel out and therefore NOT be amplified. The EEG, on the other hand, will be different as recorded in each of these wires, and therefore, will not cancel out and will be amplified. Thus, the difference is amplified between the two EEG voltages that have simultaneous input to the amplifier. This is the unique function of a *differential* amplifier. This concept is represented in the diagrams below by conventional circuit diagrams. The active (+ve) electrode and the reference (-ve) electrode enter the amplifier on the left, and the output voltage is seen on the right. The ground is represented by parallel horizontal lines.

Conventional Circuit Diagram for an Amplifier (A)

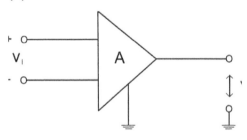

Schematic Representation of a Differential Amplifier

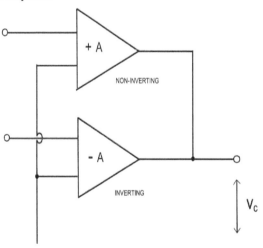

In the next diagram, the 60 Hz electrical interference signal has come equally into both inputs of the differential amplifier. This *common-mode* signal is, therefore, cancelled out and does not appear in the output. The alpha wave is high amplitude on the top, active (+ve) electrode, and much lower amplitude on the bottom, referential (-ve) electrode. When these are subtracted and amplified, a very nice picture of alpha appears at the output.

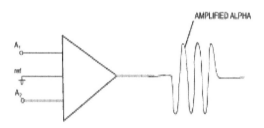

In summary: The amplifier detects and amplifies differences between two inputs. It amplifies changes in the signals referenced to the *ground* of the amplifier (an internal calculation, not a direct ground). It does this by reversing the polarity of the second input so that the two electrical inputs are effectively subtracted. The amplifier only amplifies this difference between the two inputs; hence it is called a *differential amplifier*. In this way, the amplifier is said to *reject* signals that are in common to both inputs. This is called *common-mode* rejection.

Common-Mode Rejection Ratio

The ***common-mode rejection ratio*** is the ratio of the common-mode input voltage divided by the output voltage. This ratio should be >100,000 in newer instruments. Failure of this system to eliminate external common-mode artifact is likely due to either a difference in impedance between the two electrodes and/or a poor *ground* connection.

The importance of having almost the same impedance at all electrode sites can now be better appreciated. If the impedances were very different, then the induced voltage from a common electrical source would not appear the same when the *differential* amplifier compared the active and reference inputs. Therefore, it would not cancel out and might, therefore, be amplified. This would result in a large artifact in the recording. This is, therefore, a good point at which to review impedance in a little more detail.

How You Can Get the Best Quality EEG

Impedance: Rationale

In the foregoing we have recommended having good impedance readings for all electrodes. The most frequent reason put forth for this is to ensure that artifacts in common to all electrodes are seen to be the same by the differential amplifier. A second reason is to ensure that the amplitude of the EEG recorded in different sessions is comparable. High impedance will decrease the amplitude of the EEG record. A third, less often cited reason is that good connections may result in your recording less electrode movement artifact (discussed above). These are different considerations. Below we will discuss the first problem, ensuring that common artifact inputs are rejected.

Impedances should be virtually the same between all pairs of electrodes. As previously noted, if impedances differ, then even if the interference (for example, a 60-cycle *noise* from lights, motors and so on) is the same at each site, it may "appear" to be different, and so the *common-mode rejection* by the amplifier will not eliminate it appropriately. Your first step on seeing a high 60 Hz bar on the spectrum output of your instrument is to immediately stop feedback and check your impedances. You almost invariably will find that one electrode has come loose. This has changed "z" impedance at the electrode site. You must redo your prep and retest impedances. Let's learn a little more detail about this.

What is Impedance?

Definition

Electrode impedance may be defined as the resistance to alternating electrical current flow. You should distinguish this from the electrical term *resistance*. As previously outlined, resistance is the inability of a part of an electrical circuit to allow the passage of a *direct* (constant voltage) current (Fisch p 44). Since the EEG is an alternating current, we must deal with impedance measures. You should always check impedance for each electrode site using a specially constructed impedance meter that passes a weak alternating current that mimics an EEG frequency produced by neuronal alternating potentials. The current flows from the selected electrode, through the scalp, to all other electrodes connected to the meter. The current is at 10 Hz to approximate a common EEG frequency. Ohm's Law for a direct current is V=IR. However, for an alternating current, this becomes V = IZ, where Z is the impedance of the circuit. This is what we are dealing with, because electrical activity we measure in the brain is AC, not DC. Both resistance (R) and impedance (Z) are measured in ohms. In a DC circuit, when a potential difference occurs between two points, then instantly a current flows and will flow as long as that potential difference remains. In an AC circuit, the current will instantly flow but will not remain flowing.

Mathematically (when inductance is not a major factor):

Impedance $(Z) = \sqrt{[R^2 + (10^6/2\pi fC)^2]}$ where f is the frequency of the alternating current in Hz, and C is the capacitance.

Capacitance

You are not expected to be an expert in electronics. However, there are a couple of points to be learned by looking at this formula. First, C (capacitance), which is measured in microfarads (µF), refers to the storage of electrons.

Note that *capacitance* refers to the storage of electrical energy on two parallel *plates* of conductive material separated by an insulating material. **Capacitors block the flow of direct current,** while an **inductor (described below) impedes the flow of alternating current**. The electrons flow into the capacitor and then away from the capacitor as the current alternates. Capacitors are an important element in biofeedback circuits. Capacitance is, therefore, an essential consideration in calculating impedance.

A capacitor consists of two conductors separated by a resistance. It introduces a time factor because current will rise instantly, then the capacitor stores electrons with the result that the current will gradually decrease over time. Thus DC current is stopped and only AC current can pass. Cell walls act as *capacitors*. So do your electric wires that run from the client to the preamp.

If C and f were held constant (as they would be in a DC circuit) then the 'Z' would vary directly with R. However, this is not the case in an AC circuit. The formula shows that 'Z' will go up as C goes down. 'Z' also varies inversely with frequency. Thus, as frequency increases, the measured impedance will rapidly decrease. For this reason a standard measurement must be introduced so that we are all talking in the same language. The international standard is to use a 10 Hz frequency (AC) when measuring impedances for our electrode sites.

Inductance

Note: This is mentioned for completeness, but is not an important consideration for the work of neurofeedback practitioners.

We have **not** considered this factor in our standard equation for impedance. However, if in an instrument an alternating current passed through a wire that lay within a second coil of wire, then the changing magnetic flux that is produced around the first wire will induce a voltage in the second wire. This induced voltage in the coiled wire has the opposite polarity to the original. It will oppose a change in current flow within the original wire. This counter voltage is called *inductance,* and its unit of measurement is the *Henry (L).* This is another type of resistance, but in this case it is to alternating current flow and is called *inductive resistance* (X_L). It is calculated in the following manner:

$$X_L = 2\pi fL$$

In this equation, f is frequency (Hz), L is the inductance of the element under consideration, and its unit of measurement is the "Henry." (Cohen, 1989, p 323-335).

If inductance was a factor in the circuit, then the formula for impedance would be changed (enlarged) to include both capacitance (as discussed previously) and inductance. The formula for impedance becomes:

$$Z = \sqrt{(square\ root\ of)}[R^2 + (2\pi fL - 10^6/2\pi fC)^2]$$

Note: For the majority of readers, electrical formulas are not really important. We include them only when we think that they may be helpful in giving an overview to the reader. An expert in electronics can easily find what we have included to be far too simplistic. These experts should read appropriate textbooks on instrument design and use this textbook only for its clinical explanations.

Can the Client Feel Anything when Impedance is Measured?

Yes, a few young children and the occasional adult have been able to detect a tingling feeling when impedances were checked. The manufacturer of your meter should have met electronic standards that guarantee that the amount of current is entirely within safe limits. Your instrument will be run by batteries, which further assures you that no dangerous connection with a high voltage or current is being applied. Likening the sensation to a cat tickling their ear may be useful. Measuring devices may use either a sine wave or a square wave. Therefore, if one meter gives a sensation, you can try a different type of impedance meter. Some manufacturers build an impedance check into their instruments. You should ask about the criteria to see if they are adequate.

What Are Acceptable Impedance Readings?

The impedance (resistance to the flow of alternating current) at the electrode site should be as low as possible. As a rule of thumb, <5 kohms impedance in all combinations of leads with <1 kohm difference between leads is excellent and meets research standards. If you do this, then differences due to resistance when measuring 'V' at different electrode sites become negligible compared to differences that are proportional to actual current at the source.

The resistance of the instrument's amplifier is a constant. It will be different in different instruments. In the voltage-divider formula given below, you will be able to see that when the amplifier resistance is very high, then the resistance (or impedance, for our purposes, since EEG is an alternating current) at the electrode site will have less effect on measured voltage as compared to systems with an amplifier that has a low input resistance.

What May Occur If You Don't Measure Impedances

Without good consistent connections (low impedances), you may not be giving quality feedback with your amplitude readings, and, therefore, threshold settings will change from day to day. If impedance differs between electrode pairs, then any movement can cause spurious readings. The most common problem is either that the trainer has been pressed for time and has not tested impedance, or something has occurred to alter the electrode connections during the session. Perhaps the impedance readings were good after the hook-up was first done, but something has disturbed the connection (client rubbed their ear or scratched their head, pulled on a wire, etc.). When impedance is retested, it may be found that the impedance between electrodes is not only high, but is quite different between pairs of electrodes. Correcting this will usually mean the raw EEG appears clean and smooth.

John, a neurofeedback trainer, was training a hyperactive child. The child liked to scratch his ears. Partway through the session, there appeared to be excessive fast-wave activity. The amplitude readings for the child's beta and SMR were above the usual readings for that client. The impedances were rechecked and found to be quite different between electrodes.

When the impedances were improved, the high beta activity (24-32 Hz) reading dropped from 10-15 µV to 4 µV. The 45-58 Hz activity dropped to <2 µV. The SMR and beta amplitudes moved from falsely high amplitudes to readings that corresponded to that client's last session. John started a keep-your-hands-still game with this child. This involved the child balancing a token on the back of each hand while doing feedback. John rewarded him at the end of each two-minute time span for having the tokens still there. The child felt very good about it and got rewards. The electrodes stayed in place and quality feedback was received.

If you are calculating a potential difference, you want to ensure that your measurement of electrical activity actually reflects the neuronal activity at the sites being used (or lack of it, in the case of the reference electrode). The above example demonstrates a case where the client would not receive accurate feedback unless care was taken to monitor the EEG and assure that the quality of that EEG was maintained throughout the session.

Why Modern Amplifiers Are More Forgiving – The Voltage Divider Model

You will hear it said that with high input impedance in the amplifier, measuring electrode impedance is not important. Certainly it is far less critical than in older equipment. However, the above example is one from a session on a very high input impedance amplifier. Certainly, the old low impedance amplifiers required a lot of exacting attention to impedances at the electrode sites. Why this is so may be understood (albeit in a superficial way) by a brief introduction to what is termed the *voltage-divider model* as briefly explained below.

Note: Most readers can skip the section in italics below as it is not crucial to understand these fine points in order to do neurofeedback. It is provided for those who want the in-depth information.

Where Does the Term "Voltage Divider" Come From?

This term is commonly used when discussing changing sensitivity on a recording instrument. You are all aware of how an adult EEG may be very low amplitude compared to the very high amplitudes observed in young children. To read the EEG on your display screen, you must change the sensitivity, and therefore, the size of the EEG on your screen. To allow the operator to change the sensitivity of the instrument, a chain of resistors, in series, are attached to the output of a differential amplifier. A current passes through the resistors ($R_1 + R_2 + R_3$). Then, by Ohm's law, $V = (R_1 + R_2 + R_3) I$. (alternatively: $I = V \div (R_1 + R_2 + R_3)$). If all three resistors were the same, then, if the switch was placed after the first resistor, the output would be 1/3 V, and if placed after the second resistor, it would be 2/3 V. This chain of resistors wired to a switch is called a voltage divider. *The total voltage will divide itself across the three resistors in proportion to their values.*

The same voltage divider concept applies to the measurements being made by the amplifier. Initially, imagine the "differential" amplifier as being two amplifiers. (This was diagrammed above.) In this example, for each electrode there are at least two resistances we must consider. Because we are discussing an alternating current, the term impedance *must be substituted for the word resistance. You can think of the voltage being changed by the resistances to current flow. This means that the output voltage is proportionate to each impedance according to the formula:*

$V_{output+} = V_+ \times Z_{amplifier} \div (Z_{site+} + Z_{amplifier})$ *for the +ve electrode site, and*

$V_{output-} = V_- \times Z_{amplifier} \div (Z_{site-} + Z_{amplifier})$ *for the –ve electrode site.*

Thus, the first impedances (Z_{site+} and Z_{site-}) are at the level of the scalp, and the active(+) and reference (-) electrodes. We want these impedances to be very low. The second impedance ($Z_{amplifier}$) is at the input to the amplifier, and we want it to be very high. If this is done, then the voltage measured will be much more related to the input impedance of the amplifier and more forgiving of different impedances at the electrode sites because the voltage will have divided itself across these impedances in proportion to their magnitudes. As the amplifier impedance becomes very large, the voltage measured at the amplifier will very closely approximate the true EEG voltage.

Now let us imagine the whole circuit (not just each electrode going to the amplifier):

When you think of the connections between the scalp and the amplifier, imagine that when you make the final connection, then a current can flow. It flows, in this hypothetical example, in a circle (circuit). The current is generated in the brain and flows out one site, around through the amplifier, and back through the other site and the brain. For schematic purposes, imagine three impedances to the flow of this current. These impedances are in series. The three are: the first electrode site (Z_{site+}), the amplifier, and the second electrode site (Z_{site-}). You will then measure the voltage (potential difference) across the large resistor (impedance to flow) in the amplifier ($Z_{amplifier}$).

By Ohm's law, $I = V/R$. Again, change R to Z (impedance,) because we are dealing with alternating current. Then for the active electrode you would have:

$$I = V_{+(input)}/(Z_{site+} + Z_{amplifier})$$

And for the reference electrode:
$$I = V_{-(input)}/(Z_{site-} + Z_{amplifier})$$

(Strictly speaking, we should use lowercase letters for V (or E) and for I and Z when we are discussing alternating rather than direct (AC not DC) currents. In order to make it a little easier to read, we have not followed this convention in this section.)

For the voltage output of the entire differential amplifier, substitute for I in the Ohm's law equation ($V = Z \times I$) for our hypothetical circuit and get:

$$V_{output+} = Z_{amplifier} \times [V_{(input\ to\ amplifier)} \div (Z_{site-} + Z_{site+} + Z_{amplifier})]$$

Thus, if the impedances at the two sites are very small, and the impedance at the amplifier very large, then the output voltage will be relatively independent of the impedances at the two electrode sites. It will vary with the EEG input voltage, which is then amplified.

Brief Summary

In review, potential differences between objects in the environment and the wires to your electrode on the client's scalp can induce current in the wires. As we approach our client, we actually induce a current in the wires. But $v = iz$ (voltage = current x impedance). You are changing 'i.' Therefore, the voltage will change. It will change at frequencies controlled by the rate of your approach. Electric wiring in the room, lights and other instruments can all have the effect of inducing unwanted currents. They will induce current at frequencies associated with their source. Electrical wiring in North America, for example, has a frequency of 60 cycles per second. This can be clearly seen at a very high amplitude on the spectrum display unless you do a good job of decreasing impedance at the electrode sites. If you do a good job, the common-mode rejection function of the amplifier will eliminate this artifact. You hope that all interfering induced currents will be the same in all the wires, and that they will be rejected by the amplifier, given that you have impedances that are very close between each pair of electrodes.

In more sophisticated wiring, such as the special shielded wires in the F1000 (no longer being manufactured), induced current will flow harmlessly to the amplifier through the wire's special covering and to the ground. With unprotected wires, the resistance of the amplifier is high, so it flows to your client.

You can minimize the effects of these undesirable electrical interferences by using shielded cables or having shorter wires, which is possible if you have a preamp very close to the electrode site. You can reduce differences between electrodes by paying careful attention to site preparation (dead skin has very high resistance, and hair spray acts as an insulator and must not be used before sessions). Measure your impedance and clean your electrodes between each use. Also remove any electrical equipment that seems to produce interfering signals.

Now you know a little about how the manufacturer of your EEG instrument has attempted to minimize artifacts in the EEG recording. Nevertheless, all these mechanisms can be overwhelmed or bypassed. The neurofeedback practitioner must still be able to identify waves that are not due to neuronal activity.

SECTION V
Neuroanatomical Structures, Connections and Neurochemistry

General Orientation

The scope of this section is to give an overview of neuroanatomy and neurophysiology as it pertains to our understanding of the EEG, as well as to behaviors involving attention, impulsivity, memory, learning, executive functions, connation (emotion plus cognition), speech, reading, involuntary motor activity and emotions. A basic knowledge of neuroanatomy is essential in order to understand recent literature, which quotes findings from studies that have used techniques such as MRI, PET scans, SPECT scans and LORETA.

It is important to review a few basic points concerning the cortex and subcortical structures, and to recognize the importance of communication between structures. Keep in mind that neurophysiology is not precisely understood. Only some of the rudiments are beginning to be elucidated. Thus, different ideas will be found in different texts.

It should be understood by the reader that there is quite a bit of redundancy in this section and that this is intentional: first, because learning neuroanatomy is difficult so a bit of repetition is helpful; and second, so that one can consult a subsection and find it comprehensive without reading everything that came before.

Teaching even a short course on neuroanatomy and physiology is beyond the scope of this book.

This section is limited to:
1. A brief overview of Neuroanatomy and the general functions of different parts of the nervous system.
2. A discussion of aspects of the neocortex and of subcortical structures that contribute to functions that can be influenced by NFB training, such as memory, attention span, speech and language, reading and involuntary movement.

Note: You may rightly feel that this section jumps from one theory to another. This is a reflection of the present state of knowledge. It is rather like having a few hundred pieces and attempting a thousand-piece jigsaw puzzle. We are trying to understand where they fit. We don't pretend that it can be done accurately; however, the few pieces we can match do help us understand a little bit about observed behavior and why our EEG biofeedback may have a positive influence.

Getting Your Directions

Basic Terms
The front of the brain behind the forehead is called the *anterior* or *rostral* portion. The back of the head is *posterior*. The top of your head is termed *superior* or *dorsal*. The undersurface is *inferior* or *ventral*. Each side of the brain would be said to be *lateral*. The portion of a lobe of the brain facing into the middle of your head would be termed *medial*. Your back is said to be a *dorsal* surface, whereas your front is *ventral*. Note the overlap in terms: dorsal can mean either top or back; ventral can mean either below or front. As a memory trick, think of the dorsal fin on a fish. It is on the fish's back and on top as he swims.

Sections of the Brain
If you stand at the *lateral* side of a brain and slice that brain with your knife pointing away from you and across the brain from left to right (ear to ear) and you cut from the top dorsal surface down to the base or ventral surface, then you have made a **coronal** or *transverse* section.

If you stand in front of the brain and hold your knife in a horizontal plane from front to back on one side of the head and you slice across so that you could now lift off the top of the brain, then you have made a **horizontal** section.

If you stand in front of the brain with your knife reaching from the front to the back and slice it from its dorsal to its ventral surface, then you have made a **sagittal** section. You will often encounter textbook pictures of a midsagittal section because this neatly slices the brain in half so that you may see the *medial* surfaces of the left and right hemispheres.

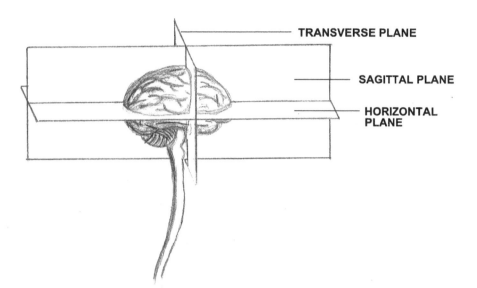

Gyrus, Sulcus and Fissure

A *gyrus* is a broad ridge of cortex and a *sulcus* is a dip or valley in the cortex. A *fissure* refers to a very deep dip or valley. (Don't get too confused. The terminology here can get quite formidable and it may be slightly different in different textbooks.) One good neuroanatomy text (Smith, 1961, p 157, 171) calls the separation or groove bordered above by the frontal and parietal lobes and inferiorly by the temporal lobe and medially by the insula, the "lateral fissure" one time and the "lateral sulcus" another. A basic biology text (Carlson, 1984, p98) refers to the same groove as the "lateral fissure." People generally agree that the large groove between the two hemispheres is a "fissure," the longitudinal (cerebral) fissure.

The two hemispheres are separated by the longitudinal cerebral fissure.

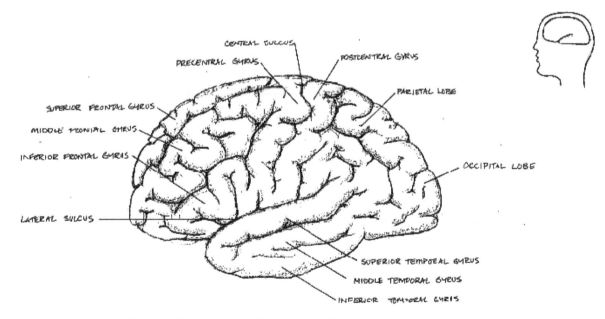

Lateral Aspect of the Cerebral Hemisphere (after Smith, 1962)

The Tentorium and the Falx Cerebri

The *tentorium* is a large infolding of *dura mater* (connective tissue) separating the cerebellum below from the cerebrum above. Child tumors are more frequently inferior, and adult tumors superior, to the tentorium. The other indentation of dura separates the two cerebral hemispheres, along the dorsal surface and is called the falx cerebri. When folded in, the dura matter forms a hard ridge. When there is a blow to the head, the neurons in the cortex can be damaged as they hit this tentorium if the force is from above, or damaged by the falx cerebri if the force is lateral. Thus, a person can sustain medial damage even though the blow is on the side of the head.

Gray and White Matter

Gray matter refers to the cell bodies of the neurons, their unmyelinated axons and dendrites and various types of support cells, especially glial cells (half the cells in the brain, which support pyramidal cells). *Glial* cells include: *astrocytes*, which provide nutrition and structural support, act like phagocytes and form the *blood-brain barrier*; *microglia*, and *oligodendroglia*, which, like the Schwann cells in the peripheral nervous system (PNS), produce myelin for the fibers in the CNS. There are about 3×10^{10} neurons in the human cortex. Gray matter is found in the cerebral cortex, the corpus striatum, and the septal region.

White matter refers to areas that contain mainly myelinated axons. The *myelin sheath* of the nerve fibers is responsible for the white color. The "superhighways" of the brain that allow for high-speed transmission of signals are a part of the white matter. White matter is less dense due to its fat content. *Myelinization* is not complete until the 2nd decade of life. Thus a teenager's brain is very much a work in progress, and executive functions, such as inhibition, are not well developed. Myelinization allows for faster transmission of nerve impulses. There is more white matter in right cerebral hemisphere (Tom Budzynski, *SNR 2001*) than the left cerebral hemisphere since there is more long distance communication there. Budzynski compares the right hemisphere to the functions of an Apple computer that uses more pictures and Gestalt organization, and contrasts this to the left hemisphere that has sequential organization comparable to PC computers. He notes that the abilities tested on a standard IQ test are overwhelmingly left hemisphere functions.

Blood Supply – Overview

The brain weighs only about 1400 grams (2.5-3 pounds) yet it receives more than 20 percent of cardiac output, 25 percent of the oxygen and 25 percent of the glucose available to the body. The *internal carotid arteries* supply the rostral portion, and the *vertebral arteries*, which join to form the basilar artery, supply the posterior or caudal portion of the brain. There are connections between these two great arterial supplies. The anterior cerebral artery runs anterior off the internal carotid artery. The medial cerebral artery runs laterally off the internal carotid artery. The posterior cerebral arteries arise from the basilar artery. (See the end of this section for more detail.)

Meninges and CSF

The entire brain is covered by a layer of connective tissue. The outer layer is called the *dura mater*. The middle layer is termed the *arachnoid* membrane. The inner layer, which carries the smaller blood vessels, is called the *pia* mater. There is a gap between the pia mater and the arachnoid membrane called the *subarachnoid space*. This space is filled with cerebrospinal fluid (CSF). Large blood vessels pass through this space. This arachnoid membrane and the CSF are only present in the central nervous system (CNS). The CNS comprises the brain and spinal cord. The peripheral nerves and the autonomic ganglia are covered by fused pia and dura mater. Large chambers filled with CSF are called the lateral **ventricles** (two of them, one on each side) and the third and fourth ventricles. These ventricles are connected to each other. The fourth ventricle is connected to the **central canal** of the spinal cord.

A bridge of neural tissue called the ***massa intermedia*** passes through the middle of the third ventricle. It connects the two lobes of the thalamus. It may be missing in some normal people. This communication link is of interest since it may help explain why training SMR on one side of the brain will have an equal effect on the opposite side. That finding suggests that there is good communication between the two lobes of the thalamus. This may be one way that we can understand how EEG patterns that depend on thalamocortical connections can spread across both hemispheres.

Approximately half the CSF is replaced in about three hours (half-life). There is about 125 ml of CSF in an adult brain. CSF is produced by the ***choroid plexus***. It flows from the fourth ventricle to the subarachnoid space via the ***foramen of Magendie*** (a single opening) and the ***foramina of Luschka*** (two openings). (*Foramen* means a small opening. The plural is *foramina*.) Blockage of the flow of CSF can result in a condition known as hydrocephalus.

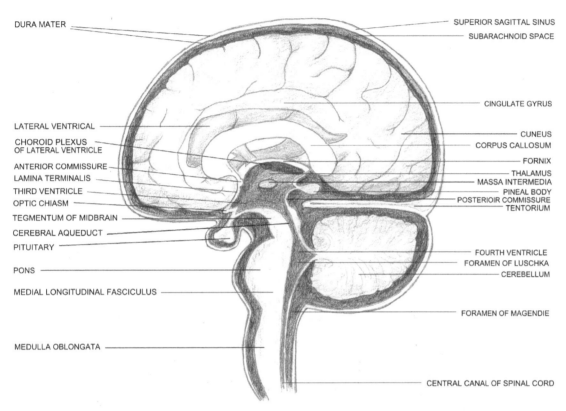

Lateral Aspect of the Brain (after Carlson, 1986)

Neuroanatomical Structures

Introduction and Overview

This section begins with a brief overview of the basic anatomical structures in the brain. This is followed by a description of the development of selected major internal features of the forebrain. Early development leads naturally into an overview of the communication linkages within the hemispheres, between the hemispheres, and between the hemispheres and the diencephalon, midbrain and spinal cord. This basic neuroanatomy overview will be followed by a section on the functions of the principle lobes and nuclei of the brain.

The brain may be thought of in three major parts: *forebrain, midbrain* and *hindbrain*. In broad general terms, the forebrain is responsible for higher reasoning, the midbrain for emotions and motivations, and the hind brain for survival, instincts, and automatic responses. Note that other ways of describing these divisions are also possible; for example, looked at phylogenetically, we have a reptilian brain surrounded by the limbic system, which in turn is wrapped in the neocortex. For an articulate discussion of how this triune (three-in-one) brain structure relates to human emotions, read *A General Theory of Love* by Thomas Lewis, Fari Amini and Richard Lannon, three psychiatrists from three different generations of psychiatric thought. It attempts to blend psychoanalytic concepts with psychobiology/drug effects and more recent neuroscience discoveries.

Forebrain (Telencephalon and Diencephalon)

Cerebral Hemispheres

Two symmetrical cerebral hemispheres are covered by the cerebral cortex and contain the basal ganglia and the limbic system. The cerebral cortex is grooved (*sulci* and *fissures*). The tissues between the sulci are called *gyri*. The EEG that we record comes from the cortex or *gray matter*. The cortex is subdivided into frontal, parietal, temporal and occipital lobes. A *central sulcus* separates the frontal and parietal lobes. The *lateral fissure* separates the frontal and parietal lobes from the temporal lobe. Most of the surface of the hemispheres is covered by what is termed the *neocortex*. The neocortex has six layers. (This is, perhaps, the best definition of a mammal, an animal with a six-layered cortex.) The pyramidal cells constitute the largest part of the third and fifth layers. The granular cells are present in higher numbers in the third and fourth layers, while the fusiform cells are found in the sixth layer.

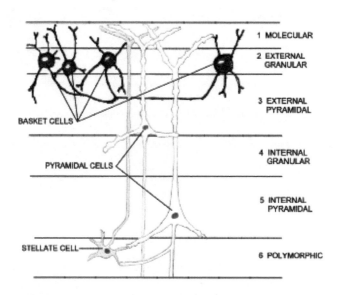

Schematic Representation of the Neocortex *(after Smith, 1962)*

On the medial surface, the edges of the hemispheres have another type of cortex called the *limbic* (limbus means "border") cortex. A portion of this is called the *cingulate* cortex. It lies between the cingulate sulcus and the corpus callosum. The septal region includes a portion of the cingulate gyrus, the superior frontal gyrus, and the cortex anterior to the lamina terminalis. The septal area is important for its role in consciousness.

The hemispheres are connected by large bands of white matter or *commissures*. The largest of these is called the *corpus callosum*. These fibers are myelinated, and therefore, white in color. *Axon* connections unite homotropic (same site) areas in the right and left cerebral hemispheres.

The **central sulcus** divides the motor (rostral) and the sensory (caudal) cortex. This part of the cortex sends motor commands and receives sensory (touch, temperature, pain) information that has been relayed from the periphery by specific areas in the thalamus. The axons come from peripheral sensors (specific information) and ascend, for the most part, in the dorsal column of the spinal cord to cross over to the opposite side in the medial lemniscus to synapse in the ventral posterior nucleus of the thalamus before continuing to the somatosensory cortex of the postcentral gyrus of the parietal lobe.

The visual cortex is located in the occipital area and the auditory cortex in the temporal lobe within the lateral fissure.

The remainder of the cortex is called **association cortex.** In the frontal lobes the association cortex is involved in important cognitive functions including executive functions such as planning.

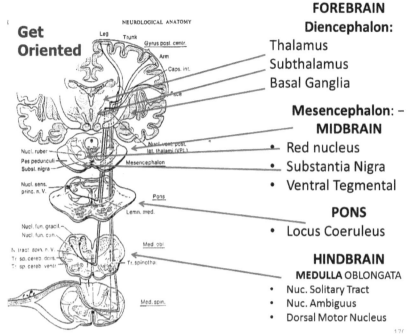

Illustration modified after Smith (1962) modified after Rasmussen (1932)
To show the relative positions of the forebrain, diencephalon, midbrain (mesencephalon), and hindbrain.

Limbic System

The limbic system consists of the *hippocampus, amygdala and septum* in the telencephalon, and in the diencephalon, the *anterior thalamic nuclei and the mammillary body*. The major axon connecting three parts of this system is referred to as the *fornix*. It connects the anterior thalamic nuclei with the hippocampus and the mammillary bodies. These connections will be discussed in the next section.

Amygdala

The *amygdala* is located in the rostral end of the temporal lobe. It has two groups of nuclei: corticomedial and basolateral. The older corticomedial nuclei project axons through the **stria terminalis** to the hypothalamus and forebrain. The newer basolateral nuclei project through the **ventral amygdalofugal pathway** to the hypothalamus, preoptic region, septal nuclei, midbrain tegmentum, and periaqueductal gray matter. The amygdala receives input from thalamus, hypothalamus, midbrain, and the temporal lobe. (See diagram later in this section.)

The *amygdala* is also connected to the tail of the caudate and to the *uncus*. In part, the grey matter of the uncus is continuous with gray matter of the amygdala. It is at the rostral end of the hippocampal cortex. The *fornix* or "arch" connects the hippocampus with the anterior thalamic nuclei and the mammillary body. As previously noted, the hippocampal cortex has connections with the frontal, parietal and temporal lobes and the cingulate via the cingulum. Thus, the amygdala has connections to all the areas concerned with emotions, the autonomic nervous system, and the endocrine system. Animal experiments involving electrical stimulation of the amygdala result in aggression while bilateral removal of the amygdala can result in a tame animal that appears indifferent to danger. It is involved in having affect (like or dislike) towards others. It is an integral part of the system that controls autonomic and endocrine responses to emotional states.

The amygdala is also important in memory. It appears to lay down unconscious memories that bring back an autonomic nervous system – body state information – that accompanied the emotions evoked by the event (Carter, 1995). This is particularly evident with respect to traumatic memories. (Contrast this with the hippocampus, which is involved in laying down conscious memories.)

It is rather as if the thalamus is the "hub" of the wheel with connections radiating like the spokes of a wheel to every corner of the nervous system. The amygdala and hippocampus are rather like the gearshift, governing, in many ways, aspects of thinking, feeling, and behavior.

For a detailed discussion of the functional aspects of these areas of the brain the reader is referred to Sections VI and VII.

Basal Ganglia

The structures that traditionally comprise the basal ganglia include the globus pallidus, caudate nucleus and the putamen. More recently the nucleus accumbens has been included. The basal ganglia-thalamic system is involved in the selection of actions (Kropotov & Etlinger, 1999). For appropriate executive actions this system must flexibly select sensations, cognitions, and motor actions and also inhibit inappropriate sensations, motor actions or irrelevant thoughts (Kropotov, 1997).

The basal ganglia are involved in the motor system and are also important in NFB work because of their importance in movement disorders, including Tourette, Parkinson's and dystonia. They also play a role in learning.

Basal Ganglia Nomenclature

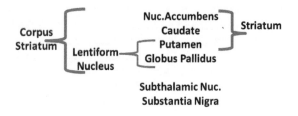

Striatum

The striatum comprises the putamen, caudate, and the nucleus accumbens and is thus a subdivision of the basal ganglia. (Recall that the term *basal ganglia* refers to the globus pallidus, caudate nucleus, putamen, and the nucleus accumbens.) The striatum interconnects with the substantia nigra. It is important for automatic movements. In Parkinson's disease, dopamine-producing cells are depleted in the substantia nigra. The striatum is also involved, however, in higher functions. It appears that it may play an important role in the selection of which "program" will be acted upon. Its connections through the globus pallidus may, in turn, result in a release of inhibition of specific thalamic nuclei. This may facilitate focus both by the cortex and by the superior colliculus (which orients the person in the selected direction).

Normally, the striatum exerts a balanced inhibition of the globus pallidus (GP). The GP normally inhibits parts of the thalamus. Inhibiting the GP thus allows the thalamic neurons to be in a relay mode. These neurons can then facilitate the transfer of sensory and other signals to the cortex. In ADHD it is hypothesized that the striatum exerts too little inhibition on the GP, which in turn will increase its inhibition of the thalamus. When this occurs, it is hypothesized that the thalamus goes into standby mode and the executive functions, such as engagement and disengagement operations, are impaired (DeLong, 1990; Sterman, 2000; Kropotov, 1997). It allows the thalamus to facilitate the cortical elements of a selected program (Kropotov, 1999).

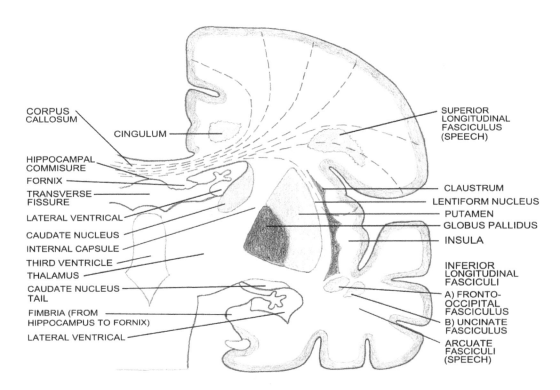

Schematic Diagram of a Transverse Section Through the Right Cerebral Hemisphere (After Smith 1962)

Diencephalon
The diencephalon lies between the telencephalon and the midbrain and contains the thalamus and the hypothalamus.

Thalamus
The name thalamus comes from the Greek word for "inner chamber" according to some, and "antechamber" according to others. Both are appropriate in meaning because it is an inner structure and, like an antechamber, things pass through the thalamus to go elsewhere. All sensory information except smell passes through the thalamus before going on to the cortex. It has two lobes connected by the *massa intermedia*, which goes through the third ventricle. The nuclei within the thalamus project to specific areas of the cerebral cortex. Their connecting fibers are called *projection fibers* or *axons*. **The lateral *geniculate*** nucleus of the metathalamus projects to the visual cortex. The *medial geniculate nucleus* projects to the auditory cortex. The *ventroposterior* nucleus projects to the somatosensory cortex. The *ventrolateral nucleus* projects information from the cerebellum to the motor cortex. The *anterior nuclei* project information from the mammillary bodies (nuclei of the hypothalamus; see Smith, p 68) to the cingulate gyrus. The *midline nuclei* and the *reticular nuclei* partially encapsulate the thalamus and project to other areas of the thalamus and cortex in feedback loops that are important in our understanding of EEG rhythms. (after Smith, 1961, p 96, 185)

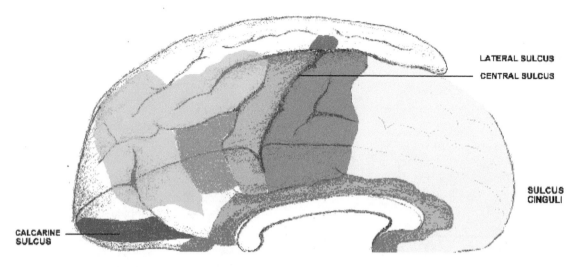

Nuclei of the Thalamus

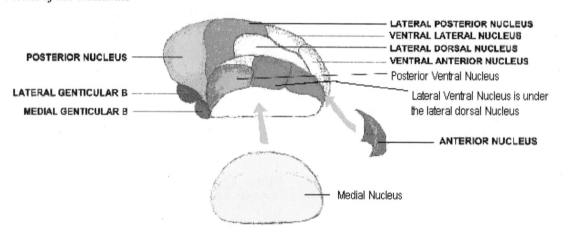

The above diagram is color coded to demonstrate how specific areas (nuclei) in the thalamus link to specific areas in the cortex. **Note:** *The temporal lobe (at top of figure) has been "folded up" to show the auditory area (red). (After Smith, 1962)*

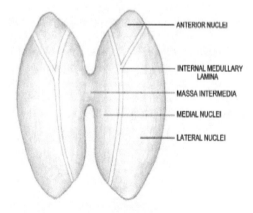

Schematic Diagrams of the Thalamus (after Smith, 1962)

Hypothalamus

The hypothalamus sits under the thalamus and on either side of the third ventricle. The *optic chiasm* is just anterior to the pituitary stalk, which comes out of the base of the hypothalamus. Just anterior to this is an area that will come up in lectures on NFB and brain function, the ***preoptic area.*** It is important in our work with BFB because it is involved in the control of the autonomic nervous system. It is also a keystone in the control of the endocrine system. The pituitary gland is attached to its base and thus, as Carlson humorously puts it, this system is responsible for "the survival of the species – the so called 4 F's: fighting, feeding, fleeing and mating" (Carlson, p 104).

Even maternal behavior appears to be hardwired to a certain degree. The medial preoptic area, which contains estrogen receptors, is important for nurturing the young. It is linked to the hypothalamus.

Midbrain (Mesencephalon)

The midbrain comprises the tectum and the tegmentum. The *tectum,* or roof, of the mesencephalon contains the superior and inferior colliculi. The superior colliculi are part of the visual system and are involved in visual reflexes and reacting to moving stimuli (Carlson, p 107). The inferior colliculi are part of the auditory system. The auditory pathways synapse in the inferior colliculi and proceed on to the medial geniculate nucleus, which is part of the metathalamus of the diencephalon (Smith, p 90), and then the auditory cortex. The *tegmentum* is an extension of the subthalamus and includes the anterior portion of the *reticular* formation, the *red nucleus,* the *substantia nigra, periaqueductal* gray matter, and the *ventral tegmental area.* The substantia nigra projects fibers to the caudate nucleus. Its production of dopamine is reduced in Parkinson's disease. The red nucleus is crucial in our understanding of how increasing SMR may assist clients who are hyperactive and also those who suffer from movement disorders, such as dystonia and Parkinson's disease. The reticular formation contains over 90 nuclei and networks of neurons extending down into the brain stem. It receives all kinds of sensory information and communicates with the thalamus, cortex and spinal cord. It is crucial to our understanding of sleep, arousal and attention. The periaqueductal gray matter is involved in fighting and mating in animals. The ventral tegmental area secretes dopamine and projects to the basal forebrain (median forebrain bundle). It is involved in learning and has been implicated in schizophrenia.

Hindbrain (Metencephalon, Myelencephalon)

The *metencephalon* comprises the cerebellum and the pons. The *myelencephalon* contains the medulla oblongata. This area is also called the brain stem, and it contains the nucleus solitarius and the nucleus ambiguous, which are important in later sections of this book when we discuss the control of variations in heart rate. The *cerebellum,* like the cerebrum, is covered with cortex. It is involved in coordination and motor performance. Damage results in jerky, exaggerated movements on the same side of the body as the lesion. The *pons* lies on the ventral surface of the cerebellum and is involved in sleep and arousal. The myelencephalon consists of the *medulla oblongata.* It is involved in the functioning of the lungs and the heart. It is also implicated in the maintenance of skeletal muscle tone.

An important nucleus in the hindbrain, which has connections to the forebrain, is the *locus coeruleus.* It is important in the production of the neurotransmitter norepinephrine. This is thought to be important in Attention Deficit Disorder, as will be discussed later in this chapter.

Early Development of the Forebrain

Development

The easiest way to learn the anatomy and the interconnections between areas in the brain is to make a mental picture of how the brain developed.

During the early stages of development in the embryo, what eventually becomes the brain is just a tube of neural tissue, an extension of the spinal cord. The rostral end of this tube develops as an *association* mechanism to detect sensory inputs and coordinate and direct motor actions. Four bulges appear along this hollow tube. These become the forebrain *(telencephalon),* the connecting portion of the forebrain to the midbrain *(diencephalon),* the midbrain and the hindbrain. At the rostral end, in the bulges that become the telencephalon, the outer aspect of the walls becomes the cellular gray matter, while more medial aspects contain the myelinated white matter of the interconnecting axons of these nerve cells. (This is the reverse of the rest of the tube that becomes the diencephalon, midbrain, hindbrain and spinal cord where the gray matter is more central and the white matter more peripheral.) The hollow part of this tube becomes enlarged into four spaces containing cerebral spinal fluid. These spaces are called ventricles.

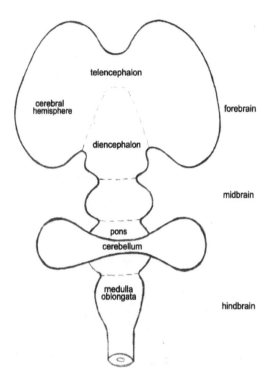

Schematic Diagram of The Developing Brain (after Smith, 1962)

Ventricles

Gradually the bulges in the rostral end of the neural tube take on more definition. Each side of the most rostral end of the tube, the telencephalon, bulges out laterally. These lateral bulges of neural tissue will become the left and right cerebral hemispheres. Imagine blowing out a rather thick-walled balloon on either side of the rostral end of the neural tube. The hollow centers of each of these bulges of neural tissue expands to become the left and right *lateral ventricles*. Instead of air, like a balloon, they are filled with fluid called *cerebral spinal fluid* (CSF). A small tube at the anterior end of these two lateral ventricles remains to connect them to the original hollow center of the tube. This little tube is called the *interventricular foramen (foramen of Monro)*. The hollow portion of this part of the brain, the diencephalon, enlarges and is called the *third ventricle*. It is located in the center of the diencephalon between two masses of neurons in the lateral walls of this tube. This grey matter thickening in the lateral walls of the diencephalon will become the left and right lobes of the thalamus. The sensory pathways ascending up the spinal cord (the caudal end of the tube) will synapse here before going on to the left and right cerebral hemispheres.

Each lateral ventricle lies in a C-shaped form anteroposteriorly from the frontal lobe (anterior horn) back past the interventricular foramen through the parietal lobe (body) and down and forward again into the temporal lobe (inferior horn) in a 'C' shape. The lateral ventricle has a tail into the occipital lobe (posterior horn). The caudate nucleus is in the lateral wall of this ventricle. The caudate nucleus ends as a bulge in the tip of the inferior horn of the lateral ventricle. This bulge of gray matter is the amygdala.

In the body of the ventricle, at its inferior medial corner where the caudate is juxtaposed to the diencephalon, is the choroid plexus. This plexus, which produces CSF, runs medially from its attachment to the diencephalon-caudate juncture to attach to the fornix (described later). The "body" portion of the lateral ventricle has the corpus callosum above it and a medial wall, called the septum pellucidum. The choroid plexus remains attached to the caudate and coextensive (Smith, p 205) with the fibers forming the fornix (and fimbria) all the way around the C-shaped curve to the tip of the inferior horn.

Development of the Corpus Striatum

A mass of neural tissue in the floor of each of these lateral ventricles expands. It becomes very large, filling most of the ventricle. What had been an oval-shaped ventricle space now becomes squeezed into a 'C' shape over (and medial), around the posterior aspect and under this mass of cells. This large cellular outgrowth into the left and right lateral ventricles is now adjacent to the lateral wall of the diencephalon. It is called the *corpus striatum*. Note that this term is somewhat confusing since it is different than the term *striatum* that comprises only the putamen and caudate as described and illustrated above (Smith, p 198-203). The corpus striatum is made up of the *lentiform* nucleus, the *caudate*, the *amygdala* and the *claustrum*. The *lentiform nucleus* is so named because it is shaped rather like a lens between the diencephalon and a relatively thin section of the cortex lateral to it called the *insula*. This lentiform nucleus comprises the *putamen* laterally, the *globus pallidus* medially, and inferiorly at its base, the *innominate substance* containing the *anterior perforated area*.

As you are now surmising, all these nuclei that you keep hearing about came originally from the same mass of tissue that grew out of the original floor of the left and the right lateral ventricles. As it grew, so did its connections to the diencephalon and midbrain. Therefore, the amount of connecting tissue posterior to the interventricular foramen and the lateral aspect of the diencephalon necessarily thickens. A small portion of the medial wall of the lateral ventricle remains paper-thin with no nerve cells invading it. This portion is the choroid membrane that contains capillaries and partly encircles the attachment of the hemisphere to the diencephalon. All the rest of the neuronal tissue that forms the walls of the lateral ventricles also thickens. This outer skin of our imaginary balloon became differentiated into six layers, which later formed the cerebral cortex. The true definition of a mammal, as mentioned before, is a creature with a six-layered cortex.

Development of the Internal Capsule

Naturally the emerging neocortex is not just forming as a disconnected mass. The neurons in this emerging cortex are maintaining their original afferents, those connections going to the new cortex from the caudal end of the tube, the spinal cord, to the rostral end via the diencephalon's thalamic links to the cerebral hemispheres. These neurons also preserved *efferent* connections (from the new cortex) to the rest of the original neural tube. Some of these efferent and afferent fibers went through the bulge of neural tissue forming on the base of the lateral ventricle called the *corpus striatum*, and they eventually formed a thick band called the ***internal capsule***. Thus, this internal capsule pierces the corpus striatum, resulting in the caudate portion of the corpus striatum looping around the anterior part of the internal capsule, then coursing above it in a posterior direction. Below (inferior to) the internal capsule lies the putamen and globus pallidus. Medial to it lies the thalamus in the diencephalon. (See diagram above called *Schematic Diagram of a Transverse Section Through the Right Cerebral Hemisphere.*)

The afferent sensory fibers within the internal capsule came from sensory organs via the thalamus on their way to the sensory cortex. All senses except smell go through the thalamus first. Smell goes directly to the brain (including parts of the cortex and amygdala) and secondarily to the thalamus along its own olfactory tract. The motor cortex sends its efferent axons down through the internal capsule to sweep over the left and right lateral aspects of the diencephalon as the ***crus cerebri***. This becomes the ***basis pedunculi*** on the ventral surface of the midbrain. At the caudal end of the pons segment of the hindbrain, the basis pedunculi becomes the *pyramid*. The pyramid runs along the ventrolateral surface of the medulla. Then the fibers enter the spinal cord and ***cross over*** to the opposite side to become the ***lateral corticospinal tract***, which innervates the skeletal muscle system.

Communication Links
Brief Overview of the Major Communication Pathways

Commissures
These are large bundles of fibers that link the two hemispheres. Think of these bundles as the highway links between countries.

Corpus Callosum
The word ***commissure*** means connection. The largest connection between the hemispheres is the *corpus callosum*. Developmentally, the first commissure to unite the hemispheres is a cellular or gray matter one. The cells are derived from the septal region. (See diagram below called *Schematic diagram, horizontal plane, to show the anterior and hippocampal commissures*.) These cells invaded the anterior wall (***lamina terminalis***) of the foramen (interventricular foramen) between the left and right lateral ventricles. Large bands of fibers connecting, first, the anterior portions of the hemispheres, and then the posterior portions, are added to this commissure, creating the connecting links between these portions of the hemispheres. Fibers of the corpus callosum – going laterally after reaching the lateral border of the ventricle – dive downwards, medial to the insula, and give a thin white matter cover to the lateral aspect of the lentiform nucleus. This white matter cover is called the *external capsule*. The corpus callosum grows thicker, and as it grows in a posterior direction, it forms a canopy over the diencephalon. The space between this canopy and the diencephalons became known as the ***horizontal fissure***. (See diagram above called *Schematic Diagram of a Transverse Section Through the Right Cerebral Hemisphere*)

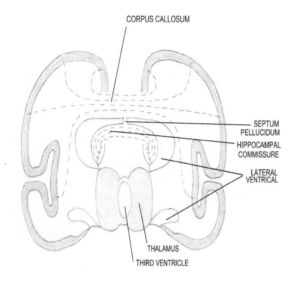

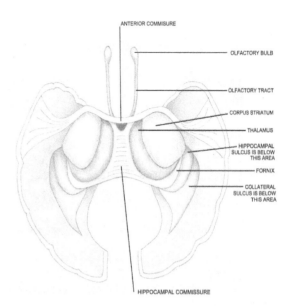

Schematic diagram, horizontal plane, to show the anterior and hippocampal commissures. (After Smith, 1962)

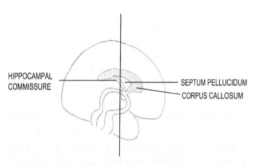

Schematic diagram of a transverse section through the cerebral hemispheres and the diencephalon. (After Smith, 1962)

Anterior Commissure

Two other small commissural bands exist. The first is at the anterior/inferior portion of the corpus callosum. This point is the developmental origin of the corpus callosum, just in front of the interventricular foramen. This band of fibers is the *anterior commissure.* It connects the neocortical areas in the temporal lobes. The anterior commissure pierces the corpus striatum and passes below the internal capsule to fan out in the anterior portion of the temporal lobes.

Hippocampal Commissure

The other commissural band of axonal fibers is the *hippocampal commissure.* It is a thin lamina in the inferior border of the septum pellucidum coursing posteriorly from the interventricular foramen across the top of the diencephalon to become part of the *fornix,* which runs along and connects to the hippocampus on each side. This will be described below. (See diagram below and also the diagram above called *Schematic diagram, horizontal plane, to show the anterior and hippocampal commissures.*)

Association Fibers (within either the left or right cerebral hemisphere)

Short Association Fibers

These U-shaped fiber bands link adjacent gyri. Think of these as the major roads within a city.

Long Association Fibers

These fiber bands link large areas of the cortex that are long distances from each other. Think of them as the super highways. There are four major such bands of association fibers: the cingulum, the superior longitudinal fasciculus band, and the inferior longitudinal fasciculus, which contains two bundles of fibers.

The Cingulum

If you were to look at a *sagittal section* of the cerebral hemisphere just lateral to the midline, you would see the large association communication bundle of fibers for the limbic system. As noted previously, the limbic cortex is a C-shaped region on the medial surface of the cerebral hemisphere. It includes the cingulate gyrus, parahippocampal gyrus, hippocampus, and the parolfactory area. It has connections with the frontal, parietal and temporal lobes. It connects to these lobes through the *cingulum* in the cingulate and parahippocampal gyri. It is a bundle of long association fibers that is at the core of the limbic system. It is involved, therefore, in the conscious perception of emotions.

The Superior Longitudinal Fasciculus
Imagine another sagittal section, but this time cut more laterally. Now, pretend you are standing beside the brain looking the other way at it, that is, looking towards the center of the brain. In this view you would see a large bundle of *association* fibers that course above the insula and between the frontal lobe, the occipital lobe, and after turning around the end of the lateral sulcus, the temporal lobe. One possible important function of this communication system is speech.

The Inferior Longitudinal Fasciculus
In the same kind of view (sagittal section), you will also see two smaller bundles of fibers at the inferior border of the insula. The top bundle of fibers run from the frontal to the occipital poles. Below these are shorter association fibers joining the orbital cortex with the temporal pole called the *uncinate fasciculus*.

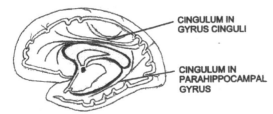

Sagittal section to show cingulum (after Smith, 1962)

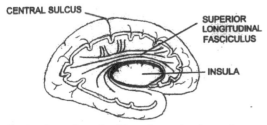

Sagittal section (looking from the lateral aspect of the hemisphere towards the midline) to show the superior longitudinal fasciculus. (After Smith, 1962)

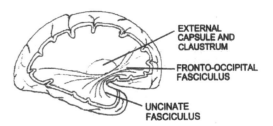

Sagittal section (looking from the lateral aspect of the hemisphere towards the midline) to show the inferior longitudinal fasciculus. (After Smith, 1962) Another diagram titled 'Figure to Show Uncinate Fasciculus from Gray's Anatomy (1918)' is shown in Section VI.

Development of Important Connections Between the Hemispheres
The Role of the Neural Tissue in the Vicinity of the Anterior Wall of the Third Ventricle

Lamina Terminalis – One Keystone
In the foregoing description of the early development of the rostral end of the neural tube, we described the growth of the corpus striatum. We noted that connections to and from the original neural tube extend into each cerebral hemisphere behind the interventricular foramen within that thickened wall of the cerebral hemisphere immediately lateral to the diencephalon.

Now let us consider the wall of the third ventricle anterior to the foramen magnum. This anterior wall of the third ventricle is called the **lamina terminalis.** It remains very thin. At first, the connection here between the two hemispheres consists of grey matter from the **septal region** of the medial-frontal part of the developing cerebral hemispheres. These cells from the septal region had invaded this anterior wall of the third ventricle, which had been the rostral end of the embryonic neural tube. The *lamina terminalis* lies between the optic chiasma below and the anterior commissure (temporal lobes – memory functions) above.

Septal Area and Anterior Perforated Area – More Keystones
The gray matter immediately anterior to the lamina terminalis on each side is the septal region, which plays a role in consciousness. The septal region

connects to the cingulate gyrus and the superior frontal gyrus. Thus, the septum is a small section of the medial surface of the hemisphere adjacent (and connected) to the lamina terminalis. Developmentally, cells from the septum invade the lamina terminalis with the right hemisphere, thus connecting at this point with the left. As this connection thickens, it forms the anterior commissure and the corpus callosum. The anterior commissure contains some fibers from the olfactory tract. Most of its fibers, however, connect the temporal lobes. The septum helps to form the *septum pellucidum,* a thin anterior vertical partition below the cerebral commissure (corpus callosum) and between the right and left lateral ventricles. It is immediately adjacent and connected to the **subcallosal gyrus (part of the septal area)**. The subcallosal gyrus goes along the anterior border of the lamina terminalis to the *anterior perforated area.* The subcallosal gyrus is immediately posterior to the paraolfactory area, which is next to the anterior end of the cingulate gyrus.

Anterior Perforated Area

The *anterior perforated area* (APA) is an area that is perforated by small blood vessels that supply the corpus striatum. This area forms the inferior medial part of the innominate substance of the lenticular nucleus. It borders posteriorly with the *amygdala.* The amygdala is continuous anteriorly with the innominate and with the putamen. It is also connected to the *uncus*. In part, the grey matter of the uncus is continuous with gray matter of the amygdala. Anterior to the anterior perforated area, the lentiform nucleus joins the caudate. The olfactory bulb is quite easy to see in most medial sagittal sections. It attaches by a visible stalk to the anterior border of the anterior perforated area. The APA is beside the anterior half of the ventral (inferior) surface of the diencephalon. At the lateral aspect of the APA is the insular neocortex. (See diagram below called *Midsagittal diagrammatic representation of the diencephalon and surrounding forebrain.*)

Hypothalamus, Septum and Anterior Perforated Area
A Hub of Communication Pathways

Thus far we have seen that the olfactory bulb sent a stalk of fibers to the anterior perforated area. Fibers go on to the septum and the amygdala. Fibers connect the anterior perforated area, the septum, and the amygdala to the **hypothalamus.** The hypothalamus lies below the thalamus and is continuous with the septal area and with the anterior perforated substance.

Thus, the amygdala connects either rather directly to the hypothalamus via the APA, or indirectly by a long circuitous route via the stria terminalis. Complementing this back door route is a septal region "back door" to the dentate and hippocampus via the *longitudinal stria.* The longitudinal stria consists of fibers that arose in the hypothalamus, go to the septal area, then the subcallosal gyrus, penetrate through the corpus callosum, run posterior to the splenium, then go around this curve to descend and run anterior to the dentate (Smith, p 188). (Note figure below called *Midsagittal diagrammatic representation of the diencephalon and surrounding forebrain.*)

The hypothalamus is critical in the regulation of blood pressure, pulse rate, body temperature and perspiration. It is involved in all the body's homeostatic mechanisms. It is important in hunger, thirst, water balance, sexual behavior and lactation. It is important for our biological clock, and thus, our circadian rhythms. It is, with the amygdala, an integral player in our fight or flight responses. The lateral aspects of the hypothalamus appear to be involved in both pleasure and anger. The medial portions seem to be more involved in aversion. The anterior hypothalamus is involved in parasympathetic activity and the posterior portion is part of the sympathetic system. Modulation of affective responses may also involve the hypothalamus. Thus, the hypothalamus, with its influences on the endocrine system (the pituitary gland develops as an extension of the ventral hypothalamus) and the autonomic nervous system, is essential to homeostasis and to our emotional responses to our environment.

How These Structures Relate to the Fear Response

More than 100 years ago, a Swiss psychologist, Édouard Claparède, reported that a woman patient who had severe amnesia could nevertheless acquire a fear response. She unconsciously and automatically associated his handshake with pain after he had pricked her hand with a pin when he was shaking hands with her. Although she could not recall who he was the next day (although she had been introduced to him on a daily basis for some time), she automatically withdrew her hand when he extended his.

Joseph LeDoux, Cornell University, studying conditioned fear in the 1980s, found that rats could not learn when a portion of their thalamus called the auditory thalamus was removed. He concluded that the auditory cortex, which is responsible for integration of sounds into conscious awareness, was not involved in the automatic fear response. He discovered that a second connection went from the auditory thalamus directly to the lateral nucleus of the amygdala and hence to the central nucleus. The **central nucleus of the amygdala** links to brain stem areas that control autonomic responses, such as changes in breathing, heart rate, skin temperature and skin conduction. These changes occur as a response to a fear-producing stimulus. The amygdala also controls the automatic 'freezing' of motion that occurs when a sudden, fear-producing stimulus, like the sound of the rattler of a rattlesnake, is detected. It is an extremely fast automatic response (often to an emotional memory) that occurs long before the conscious brain (cortex) can figure out what is going on (*declarative memory*) and initiate a voluntary response (*procedural memory*). Declarative memory is laid down by the hippocampus. Emotional memory seems to be "marked" by amygdala connections in a way that gives it primary importance before conscious thinking can occur.

To take control of this automatic response initiated by the central nucleus of the amygdala, an animal has to learn to shunt the stimulus from the lateral nucleus to the **basal nucleus of the amygdala.** The basal nucleus links to brain areas that can initiate a new, helpful, action. Beta-blockers, used to help people who have had a traumatic experience, do not block the automatic emergence of the memory but do decrease the emotionally driven autonomic responses. These drugs may gradually make the uncomfortable responses to the memories decrease.

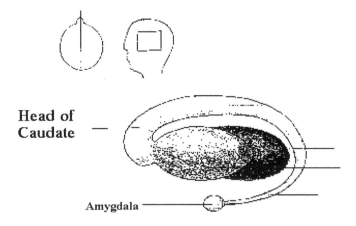

Head of Caudate

Amygdala

Thalamus (darker, and medial to the lentiform nucleus)

Lentiform Nucleus (lateral to thalamus)

Tail of Caudate

Diagrammatic sketch (sagittal section) looking from left lateral aspect to medial, to show some relationships of the basal ganglia to the thalamus.

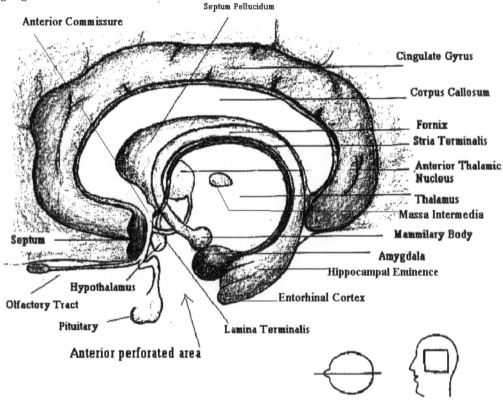

Midsagittal diagrammatic representation of the diencephalon and surrounding forebrain (After Smith, 1962)
Note 1: The stria terminalis joins the fornix at the interventricular foramen and then terminates on the hypothalamus.
Note 2: The fornix joins the hippocampus to the hypothalamus and the mammillary body, and these connect to the subthalamus, tegmentum and the red nucleus (Smith, p 85, 187).
Note 3: The stria terminalis joins the amygdala with the hypothalamus.

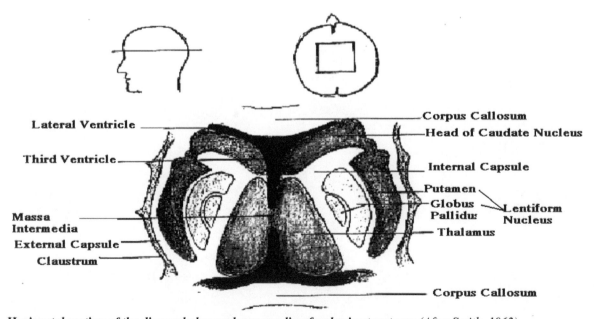

Horizontal section of the diencephalon and surrounding forebrain structures (After Smith, 1962)

The Hippocampus – Another Keystone

If you now visualize the anterior, superior, lateral corner of the diencephalon (thalamus and hypothalamus), you will recall that the balloon-like bulge that grows out laterally still has to remain attached. Think of the skin of this "balloon" at this point of attachment becoming the medial portion of the cortex of the cerebral hemisphere. The inferior portion of this cortex is a primitive cortex with fewer layers than the rest of the *neocortex,* and it is called *archicortex*. It has only three layers, as compared to the neocortex, which has six layers. At its anterior end, this archicortex connects with the amygdala and the anterior perforated area. At the posterior end it is continuous with what is termed *entorhinal cortex* in the wall of the hippocampal sulcus, and this gives rise to the fibers of the fornix that eventually connect with the hypothalamus. The *fornix* or "arch" also connects the hippocampus with the anterior thalamic nuclei and the mammillary body. In addition, the hippocampal cortex has connections with the frontal, parietal and temporal lobes. It is connected with the cingulate via the cingulum, a large band of association fibers. The subcallosal gyrus, the parolfactory area, the cingulate gyrus, the parahippocampal gyrus, and the hippocampal formation together form the limbic system. The limbic system links directly to the amygdala, hypothalamus, and thalamus. Papez suggested as early as 1937 that this system's interconnections with the frontal, parietal, and temporal lobes was at the basis of sensing emotions. The hippocampus, thus, has connections to the areas concerned with emotions, the autonomic nervous system, the endocrine system, and consciousness. It is perhaps not surprising that it plays a key role in understanding the laying down of memory and recall. It is the hippocampus that allows the animal to compare the present situation with past memories of similar situations. This is an important function for survival.

Development of Linkages

There are connections between the hemispheres and complex linkages between nuclei in the hemispheres and the brain stem. In addition, there are very important connections between the limbic system, the extrapyramidal motor system and decision-making portions of the cerebral hemispheres.

The Anterior Commissure

We have already noted that a commissure is a connection, and that, developmentally, the first commissure to unite the hemispheres is a cellular or gray matter one. The cells are derived from the septal region, which invaded the anterior wall of the third ventricle (***lamina terminalis***) anterior to the foramen (interventricular foramen) between the left and right lateral ventricles. The fibers connecting first the anterior portions of the hemispheres and then the posterior portions are then added to this commissure, forming the connecting links between the hemispheres. The commissure grows thicker and forms a canopy over the diencephalon. We also described above the two other, smaller commissural bands. The first, located at the anterior/inferior portion of the corpus callosum (the developmental origin of the corpus callosum just in front of the interventricular foramen), is the ***anterior*** commissure, which connects the neocortical areas in the temporal lobes. It pierces the corpus striatum and passes below the internal capsule to fan out in the anterior portion of the temporal lobe.

Hippocampal Commissure and the Fornix

The other band is the ***hippocampal*** commissure, which goes across the top of the diencephalon to become part of the ***fornix.*** The fornix connects to the hippocampus in each hemisphere. Along the hippocampal sulcus is the hippocampal formation, containing the hippocampus and the dentate nucleus. This hippocampal cortex connects with the *entorhinal cortex* anterior to it and to the bulb of cortex there, called the *uncus,* which is continuous with the amygdala and the anterior perforated area, as noted above. (See figure below.) This entorhinal cortex is the origin of the fornix. The fornix contains ***commissural fibers*** that cross to the other cerebral hemisphere, allowing for connections between the right and left hippocampal formations. It also contains ***projection*** fibers that course anteriorly to the *lamina terminalis.* There, by the *interventricular foramen,* these fibers descend into the diencephalon, enter the superior surface of the hypothalamus and continue on to make connections with the medial part of the mammillary body. (As previously noted and diagrammed above, fibers from the stria terminalis arise in the amygdala and join the fornix as it enters the hypothalamus anterior to the interventricular foramen.)

Mammillothalamic Tract

In the mammillary body, the fibers synapse with connections in the ***mammillothalamic tract,*** which connects to the ***anterior nucleus of the thalamus***. In addition, this **tract connects to the *cingulate* gyrus**. It has **other fibers** descending through the **subthalamus** to the **tegmentum** of the midbrain and the **red nucleus**. The red nucleus connects to the muscle spindles in the skeletal muscles. These connections to the amygdala can be important in understanding the automatic fear response.

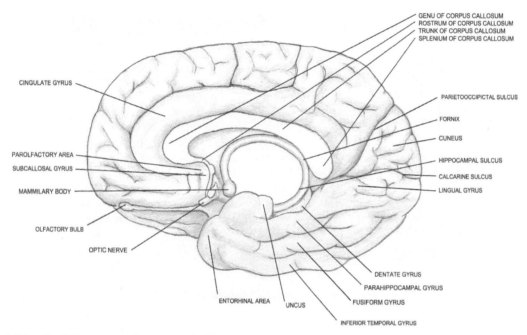

Midsagittal diagrammatic representation of the diencephalon and surrounding forebrain. (After Smith, 1962)

Brief Summary

The foregoing has described a complex network of connections that links areas of the brain that are concerned with emotions, the autonomic nervous system, and the extrapyramidal motor system. These connections imply an interrelationship between the limbic, autonomic, and motor systems. These interconnections may influence skeletal muscle tone through the red nucleus and muscle spindles. This aspect may be of importance to us when we are theorizing about the effects of NFB combined with BFB in improving the quality of life of clients with movement disorders. Reducing hyperactivity and other kinds of unwanted motor activity, such as movements in dystonia or tics in Tourette syndrome, is perhaps also facilitated by these interrelationships. As we increase SMR and practice diaphragmatic breathing, thalamocortical loops are affected, and perhaps there are also other interconnected networks that are affected.

In addition, recall that we mentioned above that fibers of the fornix (which comes from the hippocampus) connect with the hypothalamus and that the *fornix* also connects the hippocampus with the anterior thalamic nuclei and the mammillary body. In addition to this, the hypothalamus connects to the ***medial nucleus of the thalamus***, which connects in turn with the ***prefrontal*** and ***parietal*** cortices. This provides a connection from the cingulate to the prefrontal cortex. This kind of connection is important in our discussion of such disorders as obsessive compulsive disorder (OCD).

Motor Pathways: Cortex to Spinal Cord

Voluntary Control – Pyramidal System

The *pyramidal* motor pathways from the cortex descend as the *corticobulbar* (voluntary movement via cranial nerves) and the *corticospinal* (voluntary movement via the spinal nerves) tracts. A third tract, the *corticopontine* tract, activates the cerebellum to coordinate muscles involved in these movements. These tracts descend through the internal capsule to the *crus cerebri* on the lateral aspect of the diencephalon to become the *basis pedunculi* of the midbrain running in this basal or ventral position through the pons segment to become the *pyramid* on the lateral surface of the medulla. From here they enter the spinal cord and *cross over* to the opposite side to become the *lateral corticospinal tract.* They cross over and, thus, the motor cortex of the left side of the brain controls motor movements on the right side of the body and vice versa. The *internal capsule* is called *internal* because it is medial to the lentiform nucleus and *capsule* because it has the appearance of forming a layer over this nucleus (medial and superior). The *internal capsule* is like a fan. The handle is the crus cerebri (basis pedunculi). This contains the motor pathways. The fan portion opens up between the caudate above and the lentiform nucleus below.

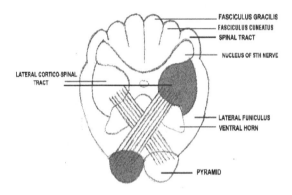

Section of spinal cord at the level of the medulla oblongata (after Smith, 1962)

The above schematic diagram is at the level of the medulla oblongata at the point where the motor fibers of the pyramid cross the midline and descend in the corticospinal tract. The uncrossed sensory fibers in the fasciculus gracilis and fasciculus cuneatus synapse and then cross the midline just above the section of the medulla shown above and become a ventral fiber tract, the medial lemniscus. Here they join the sensory fibers, in the ventral spinal thalamic tract, that had crossed earlier in the spinal cord. This sensory tract, called the *medial lemniscus,* ascends to the diencephalon and the thalamus.

Note: There is also an external capsule. Fibers of the corpus striatum going laterally dive downwards medial to the insula and give a thin white matter cover to the lateral aspect of the lentiform nucleus, and are called the external capsule. In so doing these fibers detach a chunk of grey matter from the corpus striatum, which is called the claustrum. *As previously noted, the anterior fibers of the corpus callosum course into the frontal lobe, and the posterior ones into the occipital lobe.*

Extrapyramidal Motor Control

The cortex also has *extrapyramidal motor* areas (premotor). The pathways from these areas have many intervening synapses in different areas of the motor system. These areas include the nuclei within the *corpus striatum* - including the lenticular nucleus (putamen and globus pallidus), which send fiber bundles to the *subthalamic* nucleus. The first bundle pierces the posterior limb of the internal capsule (lenticular fasciculus) and the second courses around the anterior border of the internal capsule (ansa lenticularis). At this juncture (subthalamus), fibers from the lenticular nuclei cross the midline. They connect in the midbrain with the *red nucleus,* the *substantia nigra,* and the *reticular formation,* and from there descend to influence (extrapyramidal) the striated muscles. This influence differs from that of the pyramidal tract motor pathways in that these extrapyramidal paths are not discrete to individual muscles, but rather involve movement of whole portions of the body such as a limb. Damage within this system results in awkwardness, not paralysis. The many ascending and descending connections with other nuclei in the midbrain, such as the red nucleus, may result in rigidity and/or tremor.

Sensory Pathways: Spinal Cord to Cortex

The sensory pathways in the spinal cord are labeled in the figure below. Note the different pathways for different aspects of sensation: light touch, proprioception (position sense), pain (dual pathways) and temperature. These pathways will ascend to synapse in the thalamus before ascending to the sensory cortex. These will not be discussed in further detail in this textbook.

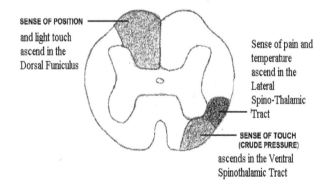

Section of spinal cord to show sensory pathways (after Smith, 1962)

Functional Significance of the Lobes of the Brain

(This is covered in much greater detail in Sections VI, VII, VIII.)

The Cerebral Hemispheres – A Broad Overview

Frontal Lobes

The frontal lobes have an anterior portion termed *prefrontal*. The prefrontal areas are involved in executive functions: the voluntary control of attention, the inhibition of inappropriate and/or unwanted behavior, the planning of actions and executive decision making. They have a role in the maintenance of arousal and the temporal sequencing of complex entities such as the expression of compound sentences. They are not fully developed until the second decade of life, which helps explain why teenagers are more spontaneous and reactive than adults. The symptoms of ADHD are, in large measure, attributable to frontal lobe dysfunction or underactivation. The more posterior portion of the frontal lobes is involved, to a large extent, in motor functions and speech.

The Left Hemisphere

The left hemisphere is involved in speech, articulation and writing. It has a major role in auditory and verbal representation. Object naming and word recall reside in this hemisphere. It is also involved in the representation of visual images evoked by auditory input. The left hemisphere functions also encompass letter and word perception and recognition. Thus, there is representation of abstract verbal forms and perception of complex relationships.

The left frontal lobe regulates speech production and syntax. It can regulate attention, inhibit actions, and switch responses. It is usually involved in analytic, sequential processing. It can utilize speech (inner dialogue) to regulate behavior. It is responsible for operations that require sequencing, such as those one can test by asking the client to recite a series of numbers or count backwards by sevens. As previously noted, Tom Budzynski uses the analogy of the left hemisphere being a PC computer running on DOS, whereas the right hemisphere is like a Macintosh computer.

Left hemisphere dominance is usually thought of as being characterized by a lack of emotion, introversion, goal-directed thinking and action, with an internal locus of control. Processing may tend to be slow and serial in keeping with the tendency to be careful and sequential. The aptitudes tested on IQ tests are largely left hemisphere functions. The dominant neurotransmitter is dopamine.

The Right Hemisphere

The right cerebral hemisphere is also important in the regulation of attention and in the inhibition of old habits. It is more involved, however, in sensing the Gestalt (image of something) and in parallel processing. It attends to spatial relationships. It is more responsible for the representation of geometric forms. It is also responsible for orientation in space and holistic perceptions. Though one typically thinks of language being in the left hemisphere, the emotional aspects of language, such as the information conveyed by intonation, are a right-hemisphere function.

Right-hemisphere dominance is usually thought of as being characterized by distractibility, stimulus-seeking behavior, seeking novelty, and change, being emotionally involved, expressive and extroverted. There may tend to be an external locus of control. There may be more of a tendency for hysterical, impulsive, and even manic behaviors. Processing tends to be fast and simultaneous in keeping with the Gestalt or holistic tendencies. The dominant neurotransmitters are noradrenaline, which is associated with speeding up activity, and serotonin, which may be thought of more as slowing down activity (Tucker, 1984). The cortex under F8 may be related to social appropriateness and the emotional aspects of behavior. Disturbances may result in impulsivity, aggressiveness, and disinhibition.

For a readable book about the temperament and style of children who seem more ruled by this hemisphere, get *Right-Brained Children in a Left-Brained World*.

Right Brained or Left Brained

Sometimes people talk of boys being more right brained. Piaget used the terms, *assimilate* and *accommodate*. The right hemisphere is more responsible for assembling descriptive systems (*accommodation*), making changes that broaden the existing cognitive structure. Dominance in these functions leads to seeking regularities in a new Gestalt through experimentation. This is an inductive reasoning process. It occurs in response to new information, and leads to understanding abstract information and making generalizations. Persons who excel in this kind of thinking can adapt to new things.

They can look at evidence and extend their thinking into something else; that is, generate a hypothesis. Children who have Asperger's syndrome share with autistic children deficiencies in these kinds of functions.

The left hemisphere is responsible for integrating and applying these well-formed descriptive systems; that is, *assimilate* the new information (bring in and make part of) into an existing cognitive structure. This is a process of deductive construction of previously developed schemata. In deductive processes the individual can take evidence and come to a conclusion. Example: The cows are all lying down under the trees, there is distant thunder – I deduce it may rain.

Left-hemisphere dominance should make it easier to follow set rules. The child with Asperger's does better in situations where there are well-set rules. Unlike the autistic child, the child with Asperger's syndrome may do very well verbally, a left-hemisphere function (see also Rourke, 2001).

In Both Hemispheres
The central area of the cortex, which is anterior to the central sulcus, governs skilled movements. Posterior to the central sulcus, it registers sensations.

Temporal Lobes
The temporal lobes are responsible for auditory processing and, on their medial aspect in the hippocampus, for short term and working memory. They receive frontal input (possibly inhibitory via the uncinate fasciculus), limbic input, and parietal sensory inputs. They may play a large role in the integration and comprehension of new information and on the emotional valance of thoughts and behaviors.

Parietal and Occipital Lobes
The occipital and parietal lobes are involved in visual acuity. The visual recognition of simple shapes is registered posteriorly in the occipital lobes. As one moves forward (anterior), there is an ever-increasing complexity of pattern recognition. The parietal lobes process raw sensory information into perceptions. Understanding these perceptions is then carried out by the frontal lobes. Which sensory inputs then make it to conscious awareness may be governed by the dorsolateral frontal cortex. On the left side, this is approximately under the F7 electrode, and disturbances in this area may correlate with distractibility.

Interestingly, parietal high alpha training (encouraging higher amplitude above 10 Hz) may result in a decrease in low alpha (below 10 Hz) frontally (diffuse alpha projection system), and shift the mean frequency of alpha to a higher level (posterior specific projection system). When done in the left hemisphere, this can reduce dysphoria and increase alertness and attention.

Medial and Basal Zones of the Cerebral Hemispheres
As a broad general statement these areas are involved in affective processing. The orbital and medial prefrontal cortices are strongly connected to all of the limbic areas, and in particular to the amygdala. These areas are involved in affect processing. All aspects of emotional processing and attaching emotional significance to events (a function of the amygdala) involve these areas. The medial prefrontal cortex may also be involved in turning off the amygdala's fear systems. If there is damage here, fear would not be properly controlled. Therefore fear and anxiety might persist inappropriately and the autonomic center outputs of the hypothalamus presumably could then lead to observable symptoms of anxiety. The medial prefrontal cortex also gives tonic inhibition to the central nucleus of the amygdala. Thus damage to the medial prefrontal cortex could also be expected to result in an increase in sympathetic drive.

The ventromedial prefrontal region is important in learning concerning social interactions. The lateral nucleus of the amygdala is critical in assigning emotional valence to events. The central nuclei of the amygdala are more involved in emotional expressions, including flight or fight responses. Dysfunction in the interconnections between the orbital and medial wall of the prefrontal cortex and the amygdala and/or dysfunction in the lateral and central nuclei of the amygdala may therefore lie at the root of some of the difficulties encountered in the autistic spectrum disorders (Shultz, 2000).

As previously noted, the limbic cortex is a C-shaped region on the medial surface of the cerebral hemispheres. It includes the cingulate gyrus, parahippocampal gyrus, hippocampus, and the parolfactory area. In addition to the foregoing, the basal-forebrain area is involved in sleep-wake regulation. Concept organization is also a function of this area (Robert Thatcher – personal communication).

Disorders Involving Cortical Functioning

Importance

Increasingly, NFB practitioners may be doing full cap assessments on a wide variety of clients. These assessments can reveal hyper- and hypo-communication links between different cortical areas. Since NFB using coherence and/or comodulation training can perhaps help to normalize these communication links, it is important that we have a general understanding of the cortical areas involved. A basic dictum is that you base your NFB training on an understanding how the client's difficulties correspond to the EEG assessment findings, coupled with your knowledge of brain function and correlations between EEG patterns and mental states. Two case examples may help to clarify this. (These cases will be expanded in Part II, Section VI under Interventions.)

Troy, age 13, was completely oblivious to the meaning of emotions expressed by others. He appeared baffled rather than angry, worried or saddened by loudly expressed anger towards himself or other people. He appeared flat or, alternatively, overly excited with events that would be expected to evoke emotions. He did not express emotions appropriately either verbally or nonverbally. His behavior was impulsive and, as with many autistic children, it could be best understood if one assumed an underlying anxiety. He was in a general learning disability class for very low-functioning children. He had no friends. Troy had been diagnosed as severely autistic at the children's hospital and independently by two other psychiatrists. He did have language skills, but did not always respond to questions. Sometimes he would burst out laughing and flail his arms without any external stimulus being apparent.

The right hemisphere's involvement in the expression (anterior) and understanding (parietal-temporal) of emotional communications has been known for many years. The underlying major symptom of anxiety coupled often with inexplicably impulsive outbursts and/or actions are well known in individuals who have autistic spectrum disorders or Asperger's syndrome. The calming effect of right-sided (C2, C4) SMR training in anxious and impulsive individuals has also been demonstrated. Therefore, the use of 13-15 Hz augmentation and decrease in the "tuning out" dominant slow wave (3-10 Hz range) made good sense both neurologically and from experience. His initial EEG profile showed excessive theta and low alpha (alpha below 10 Hz) that contrasted with a sharp drop in activity above 12 Hz.

The degree of success that Troy reached was not expected, and we wish to be very cautious about making too much of the success with Troy and a small number of other children until large case series replicate this work (see Thompson et al., 2010). Over the course of a year he took 80 NFB sessions. He was placed in a normal classroom when he entered high school at age 14, and by tenth grade, took several advanced courses. He was socializing, and new friends in his class asked him to their birthday parties. Of course, he was still recognizably different, but he had begun a "new" life. A seven-year follow-up was done by telephone, and his father reported that Troy was in university and did not want anyone to ever know he used to have a problem.

Roger, a 32-year-old man, could not read the signs at the race track where he took care of horses, or even the menu at McDonald's. He was a thoughtful, sincere, hard-working man who was engaged to a woman who had gone to university and was a voracious reader. He had been in special education at school and had tried literacy courses as an adult in order to learn how to read, but to no avail. His fiancé learned of neurofeedback and encouraged him to try the training. The active electrode was placed just anterior to Wernicke's area, and he was taught basic reading skills (phonetic awareness plus some sight words) at the same time as he achieved activation of this area (lowered his dominant slow wave and raised beta (16-18 Hz). After 60 sessions he had achieved his objectives and was able to read the cards at his next birthday party and simple novels. He had achieved functional literacy. (This example will be expanded in the next section.)

Damage Within the Cerebral Cortex

Most NFB practitioners are not trained in neurology, so this section gives a few examples of higher cortical functioning that involve the connections between different areas of the cortex. An overview of a small selection of disorders that are related to dysfunction of different areas in the cortex is perhaps the easiest way to make sense of, and remember, the functions of cortical areas.

Right Cerebral Hemisphere Problems

Recognition and Expression of Emotions

Recognition of emotions appears to involve the **right brain, posterior to the central sulcus.** Damage to the right temporal-parietal region has been found to impair an individual's ability to detect differences in tone of voice, which conveys emotional meaning (Tucker, 1977).

Anterior right hemisphere lesions impair the **expression of emotion** both nonverbally and verbally. Electrodermal responses (EDR) also demonstrate reduced reaction to emotionally loaded stimuli in patients with right hemisphere damage (Morrow, 1981).

Recognition that damage to the **right frontal lobe** could result in euphoria and emotional indifference is long standing (Carlson, p 674, after Babinski, 1914). The term *anosognosia* is applied when the person seems to ignore, and even be completely oblivious to, paralysis of their left side. This is due to right-hemisphere parietal lobe damage. Damage to the left hemisphere, frontal lobe, on the other hand, may result in depression.

Although we recognize the meaning of language with the left hemisphere, the meaning of innuendo, and tone is recognized in the right hemisphere, primarily at the right temporal-parietal junction. When we speak, our voices convey emotional information through intonations, melody and rhythm. We put emphasis on different aspects of what we are saying. These aspects of speech are called *prosody*. This aspect of speech can be present even when the individual is making no sense at all, a condition called Wernicke's aphasia. However, the opposite is also true. The person may construct speech accurately and yet not put across the meaning appropriately due to a lack of correct intonation. Prosody is a function of the right hemisphere. Expression is frontal, comprehension is more posterior (Ross, 1981). A person could, for example, express their own emotions but not recognize the emotions expressed by others if the dysfunction is in the right temporal-parietal region.

Left Cerebral Hemisphere Problems

Agnosia
Agnosia means a failure to know. (Agnosia in general may be thought of as a failure to **perceive**.) This is a higher level function, and it therefore implicates a cortical dysfunction, not just a sensory problem. A person who had visual agnosia could see but would not be able to visually *identify* the items that they could see in their immediate environment. Oliver Sachs, a neurologist who writes reflective books about his interesting patients, describes a case of agnosia in *The Man Who Mistook His Wife for a Hat*.

Apperceptive agnosia refers to an inability to recognize objects based on their shape. He would distinguish color, hue and size, but not shape. He could walk along the sidewalk and not trip over objects placed in his path. This client would not be able to copy a drawing, a common test used by psychologists. A more specific type of visual agnosia is *prosopagnosia*, where the client is not able to recognize faces. These higher-level functions appear to reside in the **medial** aspect of the **occipital and posterior temporal cortex bilaterally**. At an even higher level of functioning, the person may be able to perceive an object correctly and even copy a picture correctly. However, they are not able to name the object just from sight (*perception defect*). (They can name it if they feel it, so "naming" is not the key problem.) This is called *associative* **visual agnosia.** In this case, the visual associative cortex may be normal and speech is normal, but the connections (white matter) between these areas and areas used in naming and verbalizing may be damaged.

Anomia
A defect in the **associative cortex involved in speech and language is** both similar and yet different than a defect in the visual associative cortex. A person with a difficulty in speech and language might have what is called *anomia*. Like people with a visual agnosia, people with anomia can both see and recognize an object, such as a tricycle, but be unable to name it. They have a general difficulty with retrieving names. Even without the object in sight, they might be trying to talk about a tricycle but be completely unable to come up with the right name for it. They might, however, describe its function, and they will recognize the name if you remind them.

The **temporal lobe** contains the *auditory association cortex*. The parietal lobe association cortex contains areas responsible for **localization in space and perception of space**. Damage to these areas will result in corresponding deficiencies.

Parietal Lobe Damage
The parietal lobes are involved in the integration of sensory information and perception of the body. They are also involved in the perception of spatial relationships between objects around us. Lesions to

the left and right parietal lobes can have quite profound effects on our functioning.

Lesions to the parietal lobes can result in a loss of awareness, and neglect, of body parts and surrounding space (parietal lobe lesions). *Gerstmann syndrome* occurs when there is damage to the **left** parietal lobe. It comprises left-right confusion, *agraphia, acalculia, aphasia,* and *agnosia*. **Right** parietal lobe damage may include difficulty drawing, *constructional apraxia* (difficulty making things) and *anosognosia* (denial of deficits). Damage to **both sides** causes *Bálint's syndrome,* which includes *ocular apraxia* (inability to control gaze) and *simultagnosia* (inability to integrate components of a visual scene) and *optic ataxia* (inability to accurately reach for an object with visual guidance). *See also the discussion of Brodmann area 7 in Section VIII.*

Clearly, all the association areas are constantly communicating with the frontal lobes in order for us to make meaningful decisions concerning movements and speech.

Apraxia

Damage to the frontal lobe, parietal lobe and corpus callosum can result in motor apraxia.
Apraxia means "without action." It is the inability to execute a learned movement. As noted above, **constructional apraxia** refers to an inability to construct or even to draw an object. Perception of geometrical relations is commonly tested in standard psychological testing before beginning NFB. It is due to lesions in the right parietal cortex.

Limb Apraxia
An Example of Complex Interconnections Involving Limb Movement
A limb apraxia refers to unintended movements of a limb. If you ask a right handed patient to pretend he is screwing in a screw, your command is analyzed in the left parietal lobe (*Wernicke's* area). The interpretations are communicated to posterior areas in the parietal lobe, the left prefrontal cortex, and the left motor association cortex and other areas. These connections evoke memories of how to do this action and begin the process of coordinating the actions. The resulting decisions must be conveyed in the anterior portion of the corpus callosum to the right hemisphere and to the correct portions of the motor association cortex and then to the precentral motor cortex to control the movements of the left arm. Damage at the site of transfer (corpus callosum) to the right hemisphere will interfere with the left arm doing the desired movements. Damage to the *left motor precentral gyrus* will interfere with movements of both arms. The right arm is partially or completely paralyzed, and the left will not perform the task properly. Damage even earlier in the sequence described above might involve the posterior regions of the parietal lobe and would affect the appropriate movements of both limbs.

Complex Interactions Between Hemispheres

Functions: Speech, Language, Reading and Writing
Preschool children who show the symptoms of ADHD have an extraordinarily high incidence of speech and language disorders (Love & Thompson, 1988). It is also known that these children have a higher incidence of reading disorders once they are in school. It has been known for many years that certain areas of the left hemisphere are involved in speech and in reading. Using NFB to encourage the child (or adult) to increase fast-wave and decrease slow-wave activity over particular areas of the left hemisphere can produce improvements in language development, decrease articulation difficulties and improve reading ability.

During neurofeedback training of clients suffering from speech output problems, it makes sense to place the electrode over *Broca's area* and near C3 over the motor cortex when the symptoms are related to speech output. The electrode is usually placed over Wernicke's area and the insula for reading difficulties. Gradually, however, the training will become a little more sophisticated. It will be based on full cap assessments and on data that can demonstrate that certain areas of the brain are communicating either too much or too little with each other. The following is a case example.

As previously described, Roger was 32 years of age. In his school years he was placed in special education to teach him to read, but he remained illiterate. His mother had also been illiterate. He was unable to decipher most signs. He could not read a grade-one level book. He had seasonal employment caring for horses at the track and the Horseman's Benevolent Society kindly funded his training.

Logically, intervention was based on basic neurophysiology and knowledge of brain function. His sight was excellent and his speech good. He could understand verbal communication and things that were read to him. He was having difficulty deciphering the words and could not make sense of sound-symbol correspondence. (For example, he did not link the letter B with the sound "buh.") He had no difficulty understanding when it was read to him. We postulated that the cortical area involved might include the insula on the left side and perhaps the angular gyrus at the occipital-parietal-temporal lobe junction, since activity there affects the visual-spatial-language skills involved in visual word recognition. When he was figuring out a problem, we could observe a clear rise in 17 Hz activity and a corresponding drop in 5 Hz activity. We therefore operantly conditioned an increase in beta (16-19 Hz) and a decrease in theta (4-7 Hz), with the active electrode in a location approximately over Wernicke's area (about the middle of a triangle formed by C3, T7 and P3). We classically conditioned (paired) reading with this mental state. Throughout the course of his neurofeedback training he experienced success. Milestones he related to us included writing his first letter (to his old teacher in Nova Scotia) and reading the cards he received on his birthday. After 60 sessions he was reading short stories and simple novels at about a fifth-grade level. He had achieved functional literacy.

Anatomical Areas Involved
Wernicke's Area
This area is known to be important for **understanding** speech. Understanding involves the inferior parietal lobe and the auditory association area of the superior temporal gyrus. If there is damage to this area, speech will still be fluent. Grammar will be good, but the person will **speak nonsense.** They speak a kind of jargon in which (without understanding what they are saying) they substitute inappropriate or the wrong words to the extent that they are unable to understand their own speech. These people have similar difficulties with reading and writing.

Near Wernicke's
A lesion **near Wernicke's** area at the back of the left temporal lobe leaves the individual with inability to analyze word meanings.

Local Connections
Damage to the connections between the **auditory cortex** and **Wernicke's area results in** the person being able to hear sounds, but they are **word deaf.** They cannot understand the meaning of *spoken* words, yet they are able to read, write, and speak normally.

Inter-Lobe Connections
Damage to connections between **Wernicke's** and **Broca's** areas leave the individual **unable to repeat** what is said to them if it is more complex than a single word or phrase. That is, unless it has meaning that can be recognized in some other way. *This may involve the arcuate fasciculus which runs within the* **superior longitudinal fasciculus (SLF).** *The SLF is a band of very long association fibers running anteroposteriorly in each cerebral hemisphere. These fibers run from the frontal lobe through the parietal lobe above and below the insula (around the end of the lateral sulcus). It has branches to the temporal lobe, and branches that run all the way posterior to the occipital lobe. (Clearly, in the left hemisphere this communication link is essential for speech.)* It is particularly helpful if these individuals can visualize what is being talked about. They are not able to repeat, for example, a short series of unrelated words. They do, however, understand what is said, can speak well and can intelligently answer a question. *See also Wernicke's aphasia in Section VIII of this book.*

In contrast, in some individuals, the opposite problem can occur. An *overactive* connection between **Wernicke's** and **Broca's** areas may result in the involuntary repeating of words or phrases (*echolalia*).

Broca's Area
Damage to **Broca's area** interferes with the individual's ability to instruct the motor cortex, and results in defects in the articulation of speech. Damage to Broca's area and its **association areas** may result in a condition where the person is able to understand what has been said to him and be conscious of what he wants to say, but not be able to say it. This person may substitute staccato nouns and verbs so that the flow of his speech is lost and it sounds like a telegram. Ungrammatical speech and writing errors may be observed. In some instances these persons may also fail to understand spoken or written grammar.

Connections Between Speech Areas and Surrounding Cortex
Damage to the connections to **speech areas** from their **surrounding cortex** may result in the person being unable to **understand speech.** However, they can repeat words or finish well known phrases (e.g., "roses are red...")

Angular Gyrus
This is the region of the parietal lobe that sits at the *occipital-parietal-temporal* junction behind (posterior to) the lateral fissure. It involves *visual-spatial-language* skills, and therefore **visual word recognition.** The left hemisphere angular gyrus is required for reading. Damage to this area can result

in **alexia** (inability to read) and in **agraphia** (inability to write).

Disconnection of the angular gyrus from the left visual cortex

This means that words cannot be recognized in the normal fashion. However, an intact right visual cortex can still recognize letters and convey that information via the corpus callosum to the left hemisphere speech and language areas. The letters may, therefore, be named. As the patient says the sequence of letters out loud, the word can be recognized through auditory input. Subvocalizing can also work. Although the client cannot read silently, he may be able to read slowly out loud and understand what he is reading. He can also write sentences or copy words from a page.

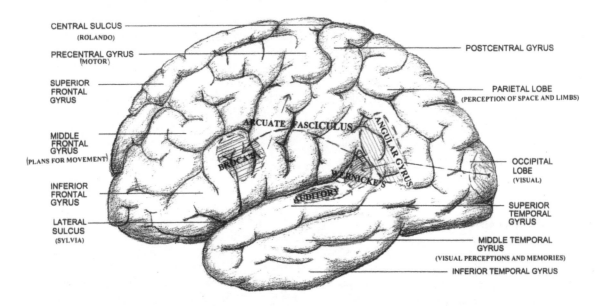

Lateral view of the left cerebral hemisphere to show Broca's and Wernicke's areas (after Smith, 1962; Carlson, 1986)

Neurologically Based Disorders
Reading Disorders

At the most severe extreme, the complete inability to recognize a written word (alexia), may be associated with lesions of the left and right occipital lobes. The damage to the left occipital cortex would disconnect the angular gyrus from all right visual field input. If this damage is accompanied by a lesion in the right visual cortex, then no information from the left visual field is available, and reading is impossible. Even without damage to the right occipital cortex, if the connection between the right visual cortex and the left hemisphere via the posterior portion of the corpus callosum is damaged, then information from the left visual field cannot be transferred, and thus is not available to be analyzed by the posterior left hemisphere.

Dyslexia (*dys* means "poor" or "faulty") – One type of dyslexia appears to be, "a *disconnection* syndrome" in which language areas fail to work in concert. The connections between some of these areas are not functioning properly. Reading subsumes two basic ways of reading. The first is chiefly ***phonological***. Phonology refers to the sounds that letters represent. The second is, in a sense, ***morphological***: we perceive the shape of the word. Most reading combines both processes (sounding it out and recognizing sight words that are memorized).

An individual with *phonological* difficulties would have trouble decoding non-words, whereas they would immediately recognize a meaningful whole word. On the other hand, a person with a *morphological* problem has difficulty recognizing sight words. This person might sound out the word *Peugeot* as pu-gee-ot or read *yacht* as ya-chet, even though they knew both these words orally. Those who have trouble recognizing the whole word may also have difficulties in spelling, since good spellers know when a word looks right.

Two types of dyslexia – *dysphonetic* (trouble sounding out) and *dyseidetic* (trouble with visual processing of the letters) – are approached differently in NFB. Lubar has suggested that for dysphonetic dyslexia you place the active electrode at F3 or F5 (for bipolar, try F3-P3 or F7-P7 (T5). For dyseidetic dyslexia try P3 or P5 (for bipolar, again, try F3-P3 or F7-P7 (T5). In both cases you would be attempting to increase activation of these areas by increasing fast wave activity (low beta 16-20 Hz) and decreasing the dominant slow-wave activity (theta or thalpha, which is a combination of high theta [~ 5-7 Hz] and low alpha [8-10 Hz]). (Reference: Joel Lubar workshop handout of applications, supported by peer-reviewed journal papers.)

Some people who have a reading difficulty have dysfunction in the insular area of the brain. The *insula* is lateral to the corpus striatum. Think of it as the cortex floor of that large indentation in the lateral aspect of the cortex called the lateral sulcus. The overgrowth of cortex during development forms two "lids," like thick heavy eyelids that are the parietal lobe above and the temporal lobe below with the supramarginal gyrus at the posterior end of this lateral sulcus. Immediately posterior to the supramarginal gyrus is the angular gyrus. The lateral fissure goes down the center of this region. Other work using brain imaging techniques has demonstrated that in some students with dyslexia the insula does not activate appropriately, and each language area is activated singly and not in unison (Paulesu, 1996). Incoming words appear to get jumbled.

Agraphia
An individual who has agraphia is unable to write. A number of attributes are important for writing. These include the memory for sounds, and perhaps for symbols which represent sounds. This is located primarily in the left superior temporal gyrus. Visual symbols are in the visual association cortex.

Left-Handedness and Difficulties with Reading and Writing
Left-handed people are significantly more likely to have specific learning disorders, including dyslexia. The literature does not distinguish between "true lefties," "partial lefties," and "acquired lefties." *True lefties* have brain functions anatomically reversed. (Language is in the right hemisphere.) *Partial lefties* demonstrate mixed dominance and may be ambidextrous. Brain functions are not reversed anatomically, and language areas are in the left cerebral hemisphere. This should not be surprising considering that the right brain controls the left hand, but reading and writing are, for the most part, left-brain activities. *Acquired lefties* are those whose left-handedness was not a result of parental genes but was due to some influence during pregnancy or after birth. These children may show mixed dominance. They are more prone to learning disability problems. Their hand, eye, and foot dominance may differ.

Always remember, however, that statistics are easily misunderstood. While many dyslexics may be left-handed, most left-handers are not dyslexic! It is also true that, areas involved in language tend to be larger in women. Men may tend to be more visual and are said to be more 'right-brained.' There are questions as to whether testosterone slows the growth of the left brain. Remember as you read these generalizations that **what is true statistically for the majority may not apply to that single individual with whom you are working.**

Hints for Students Based on Neurophysiology
As a general rule, we tell students that the human brain consciously does one thing at a time. They will find that they cannot effectively be reading their textbook and writing notes while singing along with their favorite rock star all at the same time. They can alternate their focus, but this will waste time and is not the most efficient way to study. On the other hand, it is true that one can do two things at once if they involve separate areas of the cortex. The student could draw a picture and speak or sing at the same time. If this is true, would it sometimes be helpful for a fidgety student to draw a picture, especially if he can make the sketch symbolize what the teacher is saying or what he and others are discussing? For some students this can be a helpful strategy. However, the student must recognize that his brain will be moving off and on what the teacher is saying, so this strategy must be used judiciously.

Each of us has our own dominance: visual, auditory, or kinesthetic. Some people say that they always "hear" in their mind a word that they are trying to spell or write. Others say they are quite the opposite. They virtually always look up and visualize the word. Most of us are somewhere in between. Teachers would do well to avoid pushing their personal preference onto the student. Strategies for teaching that utilize both visual and auditory modalities and also add the kinesthetic modality can work for either extreme and help individual students to use their own dominance to their own benefit. In early grade school years, kinesthetic learners should be encouraged to "feel" letters. Drawing letters in sand or tracing letters that they have cut out of sand paper are techniques that can help hands-on learners master the alphabet.

We feel that it is helpful to incorporate learning strategies into our work with NFB. Therefore, we include Section XV in this book on using metacognitive strategies in training sessions. (See also the "Clinical Corner" comments in the *Journal of Neurotherapy*, 6(4), 2002, by Judy Lubar and Lynda Thompson, on the question of combining on-task activities with neurofeedback.)

Brief Summary of Selected Injuries to the Association Cortex

Injury to the **somatosensory association cortex** can result in difficulties perceiving the shapes of objects that cannot be seen. Drawing or even following a map may become difficult. With injury to the **visual association cortex** the person can still see, but will be unable to recognize objects unless they are able to touch them. Injury to the **auditory association cortex** will result in the person being able to hear, but not being able to perceive the meaning of speech. Damage to the **junction of the visual, auditory, and somatosensory association cortex,** where the three posterior lobes meet, can result in difficulties in reading and writing.

Temporal Lobe Damage

As noted above, the temporal lobe, along with adjacent areas in the occipital and parietal lobes and other connected areas in the frontal lobe, is involved in all aspects of speech, language, reading, and writing. In particular, Wernicke's area is crucial in the understanding of spoken words. This includes the identification and categorization of objects. It has close connections to the amygdala and the entire limbic system, and damage could result in emotional disturbances, including aggressive behavior. It is also crucial in laying down and retrieving memories. Of interest to our work with students is the fact that visual memory decay time is about 500 ms, whereas auditory decay is closer to 10 seconds. This explains how you can repeat back the last few words your wife said in order to refute her complaint that you were not listening to her. The temporal lobes are important in selective attention. Damage to the temporal lobe can result in what is called prosopagnosia, mentioned previously under left-hemisphere damage, and visual agnosias. In normal people, the fusiform gyrus specializes in the recognition of human faces. Studies of children with autistic spectrum disorders indicate different activation of the fusiform gyrus when trying to identify emotions. This fusiform gyrus lies on the medial aspect of the temporal lobe between the parahippocampal gyrus above and the inferior temporal gyrus below.

You may recall that at its anterior end, the parahippocampal gyrus curls around the end of the hippocampal sulcus to form the uncus, which is continuous with the amygdaloid nucleus. At this bend is the olfactory sensory area. At its posterior end, the parahippocampal gyrus forks with the superior end being continuous with the gyrus cinguli, which turns and runs anterior and superior to the corpus callosum. The inferior end is continuous with the lingual gyrus, which runs posterior below the calcarine sulcus to the occipital pole. The visual sensory area is on the medial aspect of the occipital lobe at the posterior tip of the lingual gyrus. The counterpart to the hippocampal gyrus on the medial aspect of the temporal lobe is the superior temporal gyrus, which lies on the lateral aspect of the temporal lobe and contains the auditory sensory area.

Occipital Lobe Damage

Damage to the occipital lobes can result in a variety of problems related to vision. Damage to one side will cause loss of vision to the opposite visual field. Thus, the deficits range from defects in vision; difficulty identifying colors; difficulty recognizing words, drawn objects, or movement of objects; to having illusions and hallucinations. Damage more anteriorly in the left occipital lobe near the parietal temporal junction may result in difficulties in reading and writing. The functions of the visual cortex are anatomically very specific, and a great deal of research has gone into delineating discrete areas and their functions.

The Limbic System

General Description

This system is involved in emotions, learning, and motivation. Large portions of this system lie beneath the corpus callosum. In the three-tiered brain model (reptilian, emotional, and neocortical), this is the middle layer. In simple terms, you can think of the area above the corpus callosum as being the conscious brain, while below it is the unconscious brain. The limbic system consists of a system or network of interconnected neuron groups. In the diagram below you can see the limbic cortex, hippocampus, amygdala, septum, anterior thalamic nuclei, and the mammillary body. As noted in the section on anatomy, the *fornix,* or "arch," connects the hippocampus with the anterior thalamic nuclei and the mammillary body. As noted, when discussing the development of the cerebral hemispheres, the limbic cortex is a C-shaped region on the medial surface of the cerebral hemisphere. It includes: the cingulate gyrus, parahippocampal gyrus, hippocampus, and the parolfactory area. It has connections with the frontal, parietal, and temporal lobes, connecting to them through the cingulum. The cingulum comprises a bundle of long association fibers that lie within the cingulate and parahippocampal gyri. It is at the core of the limbic lobe. It is involved, therefore, in the conscious perception of emotions. A surgical procedure to remove tissue (*cingulotomy*) could tame a wild animal. In humans, cutting certain fibers in this area (*cingulectomy*) has been used to treat depression, anxiety, and obsessive behavior.

Although these groups of neurons are involved in the limbic system, they also have many other functions. Animals with damage to the septal hippocampal area have trouble finding their way around their environment. Mice with this kind of damage do not build nests (see Carlson, 1984, p 416). The hippocampus has a central role in laying down memories. In the amygdala, negative emotions are generated and links to negative memories are stored.

The amygdala is inhibited during hard work on relatively nonemotional tasks. It seems the Romans had it right with their adage: *Labor callum dolori obducit.* (Hard work removes the sting of pain.) Getting dysphoric clients to work on a project is often helpful. This might also tie in with happiness being associated with activation in the left hemisphere.

The anterior nuclei and the dorsomedial nuclei of the thalamus are part of the limbic system. Destruction of these nuclei can leave a person able to perceive sensation and pain without being bothered by the pain (Carlson, p 268). The limbic system has extensive connections to the cortex above it and the hypothalamus and pituitary below it. The hypothalamus is located beneath the thalamus. It is particularly important as it controls the autonomic nervous system (sympathetic and parasympathetic) and the endocrine system as previously noted. It controls the behaviors necessary for survival: fight and flight, feeding, drinking, sleeping, and sexual behaviors.

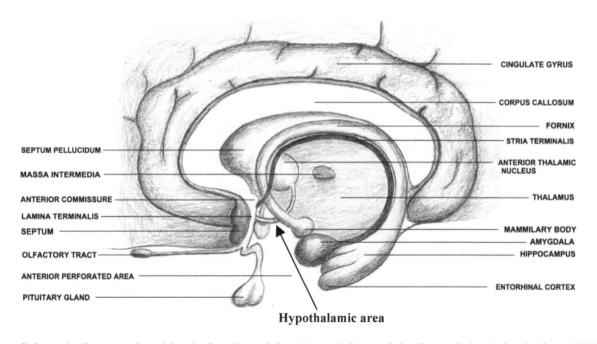

Schematic diagram of a midsagittal section of the telencephalon and the diencephalon (after Carlson, 1986; and Smith, 1962)

Learning – Relationship to the Limbic System

We have described classical and operant conditioning in Section I. We will briefly review it here in order to lead into a discussion of the neurophysiological connections that may be involved in learning.

If you accidentally picked up a hot kettle and burned your hand, then you would be cautious and carefully test the heat of a kettle's handle from that time on. Pain is an *unconditioned stimulus,* and your withdrawal of your hand is the *unconditioned response,* that is, a reflex. From that time on, whenever you approached a boiling kettle (now a *conditioned stimulus*), you would hesitantly touch the handle and withdraw your fingers (*conditioned response*) testing the degree of heat. This is **classical** conditioning. The conditioned stimulus *automatically elicits* a conditioned response once it has been paired with an unconditioned stimulus that automatically elicits an almost identical response.

In **operant** (*instrumental*) conditioning, a stimulus situation is more likely to elicit a particular behavior (a response) if that behavior has been rewarded in the past. In NFB work, we use rewards of visual and auditory feedback concerning success to reinforce the individual for putting their brain in the desired mental state. Then we pair that mental state with doing academic work. Soon the academic work (*conditioned stimulus*) will elicit the appropriate mental state (*conditioned response*).

Neurologically, it is of interest to know what pathway is involved in linking a *discriminative stimulus* with the behavioral response. (If a bar press activates a reward only when a red light comes on, the red light is the discriminative stimulus.) The ***reward*** system appears to involve the *medial forebrain bundle* of axons that run anterior-posterior from the midbrain (ventral tegmental area) to the rostral basal forebrain. This bundle passes through the lateral hypothalamus. The tectum is dorsal to the aqueduct, and the tegmentum is ventral to it. The tegmentum contains the red nucleus. The midbrain is like a "stalk" holding up and entering the forebrain. (The forebrain, remember, consists of the two cerebral hemispheres and the diencephalons – these together are called the *cerebrum.*) The midbrain has a thick layer of *central gray matter* surrounding the aqueduct. This bundle passes through the lateral hypothalamus.

Stimulation of the **dorsal hypothalamus may cause defensive or flight responses**. Stimulation of the **medial hypothalamus can produce irritability and aggression** (Carlson, p 516).

Extensive studies have revealed many areas that appear to **reinforce behavior** if electrically stimulated. These reward areas include the medial forebrain bundle, the prefrontal cortex, caudate, putamen, nucleus accumbens, amygdala, substantia nigra, locus coeruleus, various thalamic nuclei, and the ventral tegmental area (Carlson). It is a reasonable summary to say that connections among the limbic system, the ventral tegmental area of the midbrain, the ***nucleus accumbens*** and the cortex are involved in the experience of **pleasure** (reward). (The nucleus accumbens is a region of the basal forebrain adjacent to the septum and anterior to the preoptic area. It is adjacent to the medial and ventral surface of the caudate. It is important in paying attention to a stimulus and in responding to reward.) A rush of the neurotransmitter dopamine in the reward system is associated with pleasure. Dopamine is produced in the substantia nigra and in the ventral tegmental area. Connections go from the ventral tegmental area to the nucleus accumbens, septum, caudate and the prefrontal cortex. Blocking this pathway to the nucleus accumbens blocks the reinforcing effects of cocaine (Carlson, p 527).

The **medial forebrain bundle (MFB)** is the connecting link (ascending and descending) between the midbrain and the rostral basal forebrain. This *medial forebrain bundle* passes through the hypothalamus. The major catecholaminergic pathways pass through the MFB. (Recall that the locus coeruleus produces norepinephrine.) It also contains serotonergic axons reaching from the brain stem to the diencephalon and telencephalon. Dopaminergic pathways from the substantia nigra and the ventral tegmental area to the nucleus accumbens, septum, prefrontal cortex and the caudate are also involved, and stimulation of these pathways may also be reinforcing. Electrical stimulation of this bundle has been used as an experimental model of reward in animal research for the past fifty years (Olds & Milner, 1954). The release of endorphins is also an important component of the pleasure response. Both dopamine and endorphins work within this system. In addition, neuromodulators including both serotonin and dopamine have important effects on the ventromedial area of the prefrontal cortex. These transmitters are involved in feelings of well-being as well as in euphoria and mania. In broad terms, **activation of the frontal portion of the right hemisphere is more involved with negative emotions, whereas on the left the**

connection is with positive emotions (Davidson, 1995). This latter finding is important in NFB because studies have demonstrated that increasing alpha activity in the right frontal cortex compared to the left, the so-called alpha asymmetry protocol, has a positive effect on depression (Rosenfeld, 1997). In this approach, the alpha is used as an inverse indicator of activation, so the goal is to activate the left frontal area where positive thoughts and approach behavior seems to reside (Davidson, 1995).

Basal Ganglia

As previously noted, the basal ganglia comprise the globus pallidus, caudate nucleus, and the putamen. This system is involved in movement disorders such as Parkinson's disease. Some of these nuclei have also been implicated in ADD. They are, therefore, important in our understanding of how we may be influencing the brain with NFB training. Lesions in the caudate nucleus may cause **abulia.** This is a syndrome that includes apathy and loss of initiative and of emotional responses (Bhatia & Marsden, 1994). Lesions in the basal ganglia may also result in slowing of information processing, inability to sustain attention, memory defects, and impairment in verbal fluency, and motor planning. Difficulty shifting set is observed, which can result in perseveration. The ability to 'shift mental set' is crucial in order to stop one's course of action and take a different direction or approach. This is discussed in Section VIII under Brodmann areas 10 and 11, Like nondominant parietal lobe lesions, there may also be neglect of the side contralateral to the lesion (Kropotov, 1999, after Heilman et al., 1994).

Decreased dopamine production in the substantia nigra is the core problem in **Parkinson's**. The substantia nigra is directly connected to the ventral tegmental area of the midbrain. When the ventral tegmental area (noted above in the discussion of reward systems) is activated, dopaminergic pathways along the MFB inhibit the nucleus accumbens, which reduces the inhibitory effect of that nucleus (GABA mediated) on the globus pallidus, which results in increased activity of these neurons. This is, in turn, associated with increased locomotion. It is postulated that when the opposite occurs, namely decreased dopaminergic activity, it results in decreased voluntary movement. A major symptom of Parkinson's is, not surprisingly, poverty of movement and freezing, that is, being unable to voluntarily move your muscles.

The caudate nucleus, which is a part of the *striatum,* appears to be involved in the suppression of involuntary movements. The major neurotransmitter in this system is, once again, dopamine. Obvious examples of disruption in this system include ADHD, Parkinson's disease and dystonia. In these latter disorders (which are often seen together) the substantia nigra's decreased production of dopamine throws the system of excitation and inhibition between the nuclei and the various neurotransmitters involved in this system out of balance. The relative influence of acetylcholine, for example, becomes too great when there is insufficient dopamine. The result behaviorally is episodes of rigidity, tremor, and involuntary movements typical of dystonia. These unwanted symptoms may be dramatically increased if the patient is mistakenly given adrenaline (such as during a dental procedure) or if they become excited.

Damage just to the substantia nigra produces a decrease in activity but no tremors. The ventral thalamus-cortical loops appear to be involved in tremors.

In **Huntington's chorea** (chorea is from the Greek word for *dance,* and the movements are fast and jerky), the caudate and putamen degenerate. Interactions between these components of the basal ganglia are required for smooth, stable movements. Damage to the globus pallidus or the ventral thalamus causes deficiency of movement *(akinesia)*. It may also result in a kind of mutism due to an inability to use the muscles necessary for speech. Thus, the globus pallidus and the ventral thalamus appear generally to be excitatory, whereas the caudate and the putamen appear to have inhibitory functions (Carlson, p 313).

Blood Supply to the Brain

Overview

As noted in the introduction to this section, the brain receives more than 20 percent of cardiac output. The **internal carotid arteries** supply the rostral portion, and the **vertebral arteries** supply the posterior or caudal portion of the brain.

There are connections between these two great arterial supplies. The anterior cerebral artery courses anterior off the internal carotid artery. The medial cerebral arteries run laterally off the internal carotid artery.

The **internal carotid arteries** enter the skull on the lateral aspects of the pharynx. Each carotid artery enters the skull, takes a rather tortuous route, and finally arrives immediately lateral to the optic chiasma where it ascends to the *anterior perforated area*. Here it divides into the **anterior** and **middle cerebral arteries.**

Cerebral Arteries

Anterior Cerebral

The *anterior cerebral artery* courses anterior and medial above the optic nerve to the front of the *lamina terminalis*. Just anterior to this point, the right and left anterior cerebral arteries are within 1-2 mm of each other. Here they are joined by a short *anterior communicating artery.* The anterior cerebral artery arrives at the genu of the corpus callosum and then courses posterior along the corpus callosum to the splenium. The branches of this artery supply the medial surface of the hemisphere almost as far posterior as the junction of the occipital and parietal lobe (fissure). They course over the margin of the hemisphere to supply a portion of the lateral surface. Through the anterior perforated area, the anterior cerebral artery supplies the head of the caudate, anterior pole of the lentiform nucleus and the anterior limb and genu of the internal capsule. It supplies the septum pellucidum and the septal area. Thus, it supplies a portion of the motor area that controls the lower limbs. Its pathways in the internal capsule affect the motor supply to the head and arm. The septal region governs consciousness; interruptions of blood supply to the corpus callosum may result in apraxia.

Middle Cerebral

The *middle cerebral artery* courses from just lateral to the optic chiasma, posterior and laterally into the *insula* and on to the lateral fissure. It supplies the inferior portion of the lateral aspect of the frontal lobe and much of the temporal lobe, but not the uncus, which is supplied by branches of the anterior cerebral artery. Coursing through the central sulcus, it supplies motor and sensory areas for the upper half of the body. Vessels entering the lateral aspect of the *anterior perforated area* supply the external capsule, putamen, caudate, lateral globus pallidus and much of the internal capsule.

The vertebral arteries enter the skull through the foramen magnum, and ascend along the inferior surface of the medulla to join each other at the junction of the medulla and pons. Here they form one vessel, the **basilar artery.** At the end of the pons segment, the basilar divides into right and left **stems of the posterior cerebral arteries** on the midbrain. These ascend to the diencephalon, where they give off the **posterior communicating artery,** which courses along the base of the diencephalon to join with the internal carotid. This is very important because it means that the two systems, vertebral and carotid, are connected. It means that if, for example, the vertebral blood supply was compromised, some arterial supply could come from the internal carotid.

Posterior Cerebral

The *posterior cerebral artery* lies between the parahippocampal gyrus and the brain stem, coursing to the splenium, where it enters the calcarine sulcus. It therefore supplies the hippocampal formation, including the hippocampus and the dentate nucleus, and the inferior temporal gyrus on the one hand and the visual area of the cortex on the other. It supplies the lateral aspect of the occipital lobe. It supplies the inferior temporal area. Importantly, this artery also supplies the choroids plexus of the midbrain (red nucleus, most cranial nerves), diencephalon, third ventricle and the lateral ventricles. It also supplies the thalamus and subthalamus.

Anterior Choroidal Artery

There is another, smaller, artery that should be mentioned. It is the **anterior choroidal.** It arises from the internal carotid lateral to the optic chiasma, coursing to the choroid plexus of the inferior horn of the ventricle. It supplies the optic tract, the optic radiation in the sublenticular part of the internal capsule, the posterior limb of the internal capsule, the globus pallidus, the subthalamic nucleus, uncus and part of the hippocampal formation.

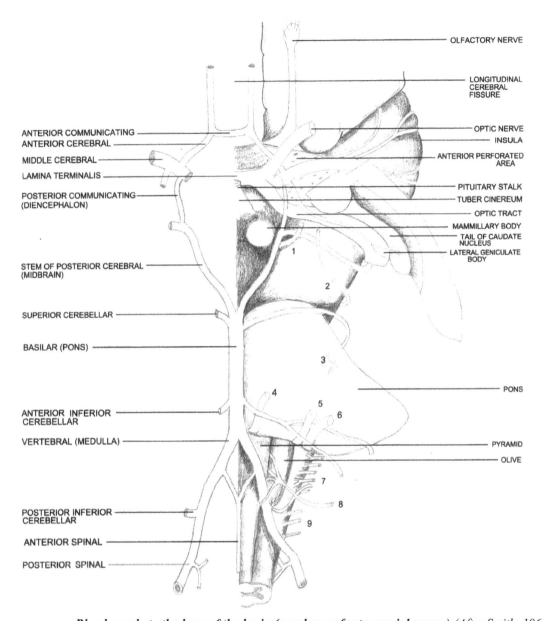

Blood supply to the base of the brain (numbers refer to cranial nerves) (After Smith, 1962)

Brief Summary

If you have followed this description, you may note at this juncture that the arterial supply from the vertebral basilar system is linked with the internal carotid system in a circle at the base of the midbrain and diencephalon: basilar artery – stem of the posterior cerebral – posterior communicating artery – internal carotid – anterior carotid – anterior communicating – anterior carotid – internal carotid – posterior communicating – posterior cerebral – basilar.

The Autonomic Nervous System

Overview

The ANS is part of the motor system. In the viscera, this system is distinguished by the fact that ANS motor nerves are paired: excitatory and inhibitory. You may think of the excitatory portion being, for the most part, sympathetic, while the inhibitory is parasympathetic. The ANS controls the activity of smooth muscle, cardiac muscle, and glands. It controls musculature in the digestive system, gall bladder, urinary bladder, eyes, blood vessels, respiratory system and in hair follicles. All the sphincters in the digestive and urinary systems are controlled by smooth muscles. The glands include the adrenal, salivary glands, lacrimal glands, and sweat glands. The ANS influences the digestive system, including the pancreas, the respiratory system, the urinary system, the cardiovascular system, and aspects of the reproductive system and glandular system.

Sympathetic Division

For the most part, you can think of the sympathetic nervous system (SNS) as being involved in processes that activate and expend energy (catabolic). The cell bodies of the SNS are located in the thoracic and lumbar regions of the spinal cord. The axons exit through the ventral roots of the spinal cord. The majority of these preganglionic fibers synapse in ganglia in the sympathetic chain or trunk, or in ganglia of the prevertibral plexus. This is in contrast to the parasympathetic system, where the ganglia are all close to the organ being innervated. The nerves of the SNS are motor nerves but the fibers do contain some sensory nerves from the organs.

The synapses within the sympathetic ganglia are acetylcholinergic. However, the terminal *buttons* on the target organs are noradrenergic. An exception to this, which you might predict, is the innervation of the sweat glands, which is acetylcholinergic. The endings in the adrenal medulla are acetylcholinergic because it is innervated directly by preganglionic fibers. The output of the medulla is adrenergic (epinephrine and norepinephrine). This is important because these hormones, apart from acting like the sympathetic system (for example, increasing heart rate), also act to stimulate the production of glucose (for energy) from glycogen in muscle cells. The adrenal cortex, on the other hand, is not under direct nerve control, but is stimulated by adrenocorticotropic hormone (ACTH) from the anterior pituitary to secrete cortisone. The anterior pituitary is, in turn, controlled by hormones released by the hypothalamus. These hormones enter arterioles, which feed small veins in a specialized vascular system that feeds the anterior pituitary.

In our work, we are able to measure sympathetic stimulation with respect to its effects on increasing heart rate, peripheral vasoconstriction (which is observed in a decrease in finger - peripheral - skin temperature), and in increasing palm and finger sweating (electrodermal response). It has also been found that there is sympathetic innervation of the muscle spindles in the skeletal muscle system.

Parasympathetic Division

For the most part, you can think of the parasympathetic nervous system (PSNS) as being involved in processes that inhibit and those that produce or conserve energy (anabolic). It increases the secretion of digestive juices and the blood flow to the gastrointestinal system. It affects salivation and intestinal motility. In contrast to the SNS, the nerves of the PSNS leave the CNS in cranial nerves 3, 7, 9 and 10, and in sacral nerves 2, 3 and 4. The cell bodies or ganglia of the PSNS are located close to the organs that they supply. Therefore, compared to the SNS, the postganglionic fibers are relatively short. Although these are motor nerves, they do carry sensory afferents from some organs, such as the chemoreceptors in the mouth, stretch receptors in the lung and receptors in the carotid body. The terminal buttons of both the pre- and postganglionic fibers of the PSNS are acetylcholinergic.

In our work, we are able to observe the effects of the PSNS with respect to its effect on slowing heart rate with respiratory expiration.

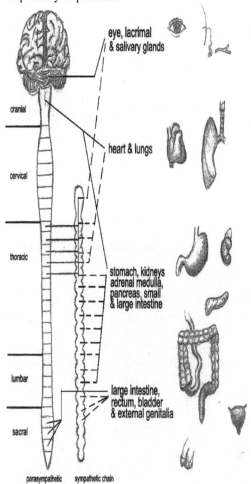

Schematic diagram of the autonomic nervous system

Brief Overview of the Neurophysiology of the Stress Response

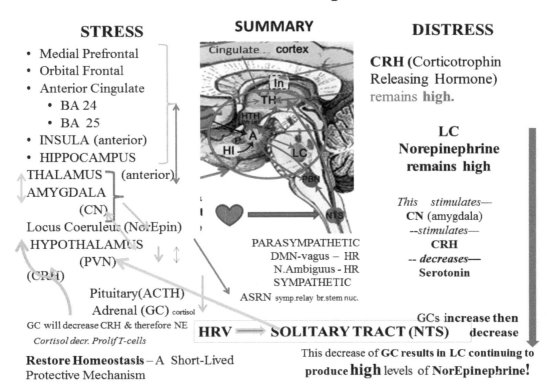

STRESS
- Medial Prefrontal
- Orbital Frontal
- Anterior Cingulate
  - BA 24
  - BA 25
- INSULA (anterior)
- HIPPOCAMPUS

THALAMUS (anterior)
AMYGDALA (CN)
Locus Coeruleus (NorEpin)
HYPOTHALAMUS (PVN)
(CRH)
Pituitary (ACTH)
Adrenal (GC) cortisol

GC will decrease CRH & therefore NE
Cortisol decr. Prolif T-cells

Restore Homeostasis – A Short-Lived Protective Mechanism

SUMMARY

PARASYMPATHETIC
DMN-vagus – HR
N. Ambiguus - HR
SYMPATHETIC
ASRN symp.relay br.stem nuc.

HRV ⟶ SOLITARY TRACT (NTS)

DISTRESS

CRH (Corticotrophin Releasing Hormone) remains **high**.

LC Norepinephrine remains high

This stimulates—
CN (amygdala)
--stimulates—
CRH
-- decreases—
Serotonin

GCs increase then decrease

This decrease of GC results in LC continuing to produce **high** levels of **NorEpinephrine**!

Figure to give a diagramatic overview summary of the normal response to stress (called STRESS) compared to the response to chronic stress (called DISTRESS). Detail is in the text. Similar biochemical findings can be found in post-traumatic stress disorders). This diagram suggests that heart rate variability training results in baroreceptor feedback to the solitary tract nucleus in the medulla and this can communicate with the paraventricular nucleus of the hypothalamus and this can have an effect on the system's response to stress. See also Section XVIII on heart rate variability training.

ASRN is Autonomic Sympathetic Relay Nuclei (in the brain stem); ACTH is Adrenocorticotropic Hormone; BA is Brodmann Area; CN is Central Nucleus; CRH is Corticotrophin Releasing Hormone; DMN is Dorsal Motor Nucleus of the vagus; GC is Glucocorticoid; HR is Heart Rate; HRV is Heart Rate Variability; LC is Locus Coeruleus; PVN is Paraventricular Nucleus (of the hypothalamus)

The stress response is an adaptive biological mechanism that was important for survival of the human species. As previously noted, we explain to children that when our ancestors were in great danger they had to fight or flee to survive. Children enjoy imagining a tiger jumping out of the forest and figuring out how the early caveman would handle this emergency in order to survive. He would need to have maximum energy (oxygen and nutrition) in the large muscles of his shoulders and legs. This would require an immediate increase in blood flow to these areas. He did not need to be digesting food or even have much blood flow to his hands and feet. This is why heart rate increases and the finger tips are cool when someone is anxious. Also, the caveman had to be sure his spear would not slip in his hand, so the hand becomes moist when we are under stress. The sympathetic nervous system is responsible for these reactions. This sympathetic system also has another protective function. It prevents extensive bleeding and tissue damage in the acute phase of stress. *Homeostasis* is maintained despite the stressors.

The stress response is controlled by interaction along a brain stem (locus coeruleus)-amygdala-hypothalamus (sympathetic system)-pituitary-adrenal axis (AHPA), with the frontal lobes and hippocampus having a major influence through connections to the amygdala and hypothalamus. The best overview of this that we have read is a paper entitled "The Hypothalamic-Pituitary Adrenal Axis," written by Carolyn Smith-Pelletier of York University, Toronto, Canada, and presented at the annual meeting of the Canadian Medical Association (2003). Hopefully this will eventually be published. Some of it is briefly summarized here.

When overstressed, this AHPA system can fail. When this AHPA axis becomes dysregulated a number of disorders may be observed. Hypoactivity may result in symptoms of anxiety, fibromyalgia, and chronic pain. A factor in this is that hypothalamic, corticotrophin-releasing-hormone (CRH) has analgesic actions and lower levels could contribute to pain. These are disorders for which we have traditionally used biofeedback (see Section VII, Autonomic Nervous System). Hyperactivity of this AHPA axis, on the other hand, has been implicated in depression.

Biochemically, the normal response to stress involves a sequence of events. A simplified version of what occurs is as follows. The thalamus and amygdala signal the locus coeruleus (LC) in the brain stem to release norepinephrine (NE). This has the effect of increasing selective attention to new stimuli and arousing the sympathetic division of the autonomic system (blood pressure, heart rate, sweating). The effect of increased NE in the AHPA axis is to increase hypothalamic output of adrenocorticotropic hormone (ACTH), which stimulates the adrenal glands to produce glucocorticoids (GC). Cortisol counteracts the effects of increased norepinephrine and epinephrine. GC *down-regulates* the CRH producing cells, which in turn decreases the CRH stimulation of the locus coeruleus to produce NE. The overall effect is to reinstitute homeostasis. Endogenous opiates are also released. These increase the pain threshold.

In addition, in the normal stress response the central nucleus of the amygdala stimulates the hypothalamic production of corticotrophin-releasing hormone (CRH) and arginine vasopressin (AVP). These act to increase AHPA axis activity. Serotonin can also increase the AHPA axis activity. NE and CRH are both involved in the affective responses to stress, including fear, constricted affect, and even stereotypical thinking.

When the brain senses that the full stress response is no longer essential, the AHPA axis is dampened by frontal and hippocampal input. This results in a decrease in ACTH production and a return of the system to its resting state.

The stress response is designed as a short-lived protective mechanism that can improve the chances for survival in a fight or flight situation. Acute improvements in arousal level, memory, focus, and in the immune system occur. However, when stress becomes chronic, abnormal alterations in the system can occur. Corticotrophin releasing hormone (CRH) from the hypothalamus may remain very high despite the high levels of GC. This actually stimulates further output of NE from the locus coeruleus. The increased NE will cause additional increases in the central nucleus of the amygdala activity, and this will cause an increase in CRH production from the hypothalamus. It becomes a vicious circle. In unpredictable chronic stress, high CRH and high NE will decrease serotonin (5-HT) levels. This may result in symptoms of anxiety and depression.

In chronic anxiety and stress, CRH and NE levels are held high and the CRH stimulation from the central nucleus of the amygdala will override the modulating effects of GC feedback and continue to activate the CRH production. The decrease in serotonin seen in chronic stress will initially still allow AVP to stimulate the AHPA axis when a new acute stress occurs, which, albeit perhaps temporary, is a good factor. However, this lowered serotonin level will also have the effect of decreasing the modulating effects of the hippocampus on the AHPA axis. Further anxiety will then begin to result in dysregulation of the AHPA axis. The production of CRH will then decrease under this chronic ongoing stress. GC levels will then fall, with the result that the modulating influence of GC feedback (to lower the production of NE from the locus coeruleus) is impaired. There is an increase in NE levels and a less adaptive reaction to new stressors. This is what is seen in disorders such as fibromyalgia and chronic pain (Pacak, 1995).

Thus hyperactivity of the AHPA axis with resultant increased CRH increases response sensitivity to noxious stimuli and depletes the immune system, while hypoactivity can result in chronic pain. The symptoms of overactivity in the AHPA axis are familiar to anyone in the mental health field. These clients exhibit a narrowing of their perspective, producing a focus on their own preoccupations with ruminative thinking and poor cognitive performance. Their sleep is impaired, which further compounds their difficulties.

Stress interferes with sleep. There are many biochemical substances involved in this complex process. Only a few of them will be mentioned in this very brief overview. GABA is important in the initiation of sleep. GABA is activated by cholinergic pathways from the brain stem to the thalamus and to the forebrain. GABA decreases the blood flow to the thalamus. It is the shutting down of the thalamus that is responsible for the loss of consciousness in sleep. GABA also dampens the NE output of the locus coeruleus. However, as we have noted above, anxiety increases NE production. If this cannot be reduced, then sleep onset will not occur. GABA also acts to decrease basal ganglia functions causing muscle "paralysis" (except for breathing) early in sleep. With high anxiety, this transitory phenomena may still occur momentarily, but sleep does not occur.

In normal sleep, in very general terms, the biochemistry is different in the first and second halves of the process. During the first half of the night, growth hormone (GH) is produced and GC are reduced. Slow-wave sleep is initiated, in part, by a decrease in ACTH and GC caused by hippocampal influences that decrease the AHPA axis activity during this first half of sleep.

An increase in AHPA axis activity is therefore associated with insomnia. Thus, in chronic stress or chronic pain or fibromyalgia the client will be likely also to suffer from insomnia.

During the second half of the night, most people have a more shallow sleep. This is caused by a decrease in GH, increased GC and a rise in CRH. Increased CRH increases ACTH and GC, and these substances will reduce slow-wave sleep. CRH can induce awakening. (CRH in the central nucleus of the amygdala is also involved in the induction of hyperarousal and fear-related behaviors.)

There are male/female differences in this process, but these will not be covered here. In general, the AHPA axis and CRH activity is higher in females. This might make females more susceptible to insomnia with stress. Aging decreases growth hormone releasing hormone (GHRH) and thus GH. This results in a decrease in deep or slow-wave sleep with a concomitant decrease in immune regeneration and a reduction in the pain threshold. The increased influence of CRH results in frequent awakening.

The immune response is dampened by glucocorticoids (GC). As we noted earlier, chronic stress initially increases GC, and will suppress the immune response and even lead to a reduction in the size of the thymus and a reduction in circulating lymphocytes.

Chronic stress is clearly a very dangerous phenomenon that can have a number of untoward effects. Biofeedback has been demonstrated to be somewhat effective in helping to decrease the symptoms caused by stress. We feel that it is likely that future work will demonstrate that the combination of neurofeedback and biofeedback may assist clients to even more effectively self-regulate their responses to stress in their daily lives.

Linking Neurophysiology to Disorders That May Be Amenable to Neurofeedback Interventions

Now that we have had a brief overview of neuroanatomy, we can hypothesize about how the various structures may work together in selected conditions in which NFB is used. These disorders involve links between the cortex, basal ganglia, and the thalamus. These disorders include:

1. Memory problems
2. Obsessive compulsive disorder
3. Tourette syndrome
4. Reward deficiency syndrome
5. Attention Deficit Disorder (ADD)
6. Movement disorders: Dystonia, Parkinson's disease
7. Asperger's syndrome

Disorders Involving Memory

Memory may be understood in terms of a specific group of neurons that have come to fire in the same pattern each time they are activated. Memory is laid down by *long-term potentiation* (LTP). This is a process wherein each time the neuron sequence is fired, the links between the neurons are strengthened. This follows from Hebb's rule of learning, which states: "When an axon of cell A... excites cell B and repeatedly or persistently takes part in firing it, some growth takes place in one or both cells so that A's efficiency as one of the cells firing B is increased (Hebb, 1949). An example is motor memory when a person learns keyboarding; the fingers seem to remember where to find the letters once the person has practiced enough.

When doing NFB, you can teach students to facilitate this process of repetition and LTP. For an example, see the case of Jane in Section XV, Metacognitive Strategies.

Types of Memory

Procedural Memory
Memories for procedures, such as the ability to play a card game like bridge, are stored in the cerebellum and putamen. Instincts, on the other hand, which are genetically encoded memories, are stored in the caudate nucleus. Memories associated with fear, such as phobias and flashbacks, are stored in (or triggered by) the amygdala.

Episodic Memory
Memory of a recent happening is best remembered if associated with emotions. These conscious memories are stored like films, encoded and held for about two to three years in the hippocampus. (The hippocampus is part of the limbic system, as outlined above. It is situated in the medial temporal lobe.) Profound stress has been shown to have a detrimental effect on the hippocampus. Hippocampal memory includes **working memory.** Working memory involves keeping information in mind while performing some operation; for example, doing a two-step math problem in your head. Memories may return to consciousness during dreaming. It is easiest to recall them when the individual is in the same state of mind (emotionally and chemically) that they were in when that memory was laid down. The inability of a person who drank too much one night to recall things he did or said when he awakens the next morning is an example of state-dependant memory. Here are two examples.

Michael, age 21, was a university student. One night he returned home after studying at a friend's house while his parents had a party. The next morning he awoke early to go to his final exam. He drank some of the tomato juice that his parents had left in the refrigerator with his breakfast and left for the exam. He was unable to recall any of the material that he had been able to recall easily with his friends the night before the exam. The tomato juice had really been a Bloody Mary mix. Although not obviously inebriated, his brain was in a different chemical state from when his learning had taken place. Fortunately, the professor sympathized with his situation and he was able to successfully rewrite the exam later in the day and he did very well.

James had thoroughly reviewed material for his final three-hour examination, and had been teaching some of his classmates the material the night before the exam. He was a bright, laid-back student who preferred studying with friends in a relaxed environment. The next day he felt extremely tense at the start of the exam. He said later that he went "blank" when he read the first question. He couldn't remember a thing. After almost an hour of feeling panic and frustration, he decided to give up and then felt very relieved and relaxed as he got ready to leave the exam hall – "the heck with this nonsense," he was saying to himself. Suddenly, he recalled everything. He returned to his desk, started writing furiously and got one of the top marks in the class.

The first example illustrated a chemical change of state; in the second it was an emotional change of state. Memory and recall can be state-dependent. It is helpful for students to have control over their state of mind when studying, going to lectures and taking exams. We use biofeedback to help them learn to relax and self-regulate.

Long-Term Memory
Hippocampal memories gradually become stored as what we call *long-term memory* in the cortex of the temporal lobe. Episodic memories can take as long as two years to be laid down in the cortex so that they become linked together independently of the hippocampus. Recalling formal memories activates areas of the temporal lobe as well as areas in the frontal lobe. It may be that the frontal lobe involvement is necessary to bring the memory to consciousness.

Semantic Memory
The meaning of information, the facts, are registered and stored in the cortex of the temporal lobe independent of the personal memories of where you were or how you felt when you learned the fact. The latter details are part of episodic memory. Facts are encoded in the temporal lobe and retrieved by frontal lobe activity. In *semantic dementia* there is a loss of semantic or factual memory. The patient tends to forget the names of objects and what they are for. A study at the Imperial College of Medicine and reported at the annual meeting of the AAPB 2003 demonstrated improvement in semantic memory with SMR training.

Other

Memory Retrieval
Although the hippocampus and the temporal cortex are most often highlighted in discussions of memory,

it is important to keep in mind that the frontal cortex is always involved in retrieval of any memory.

Memories Are Constantly Changing
Memories are not just recalled, but also reconstructed. Memories are constantly undergoing changes. Errors in and distortions of recall are inevitable. Fragments of events are remembered. When the fragments are recalled, our brain works to make sense of them. Even a little suggestion in one direction may change the "meaning" of the facts compared to a suggestion made in another direction. This "new" or revised memory is then stored, and so the process proceeds (Carter, p 170). This is an important consideration when discussing emotionally laden events with clients. It is particularly important if the therapist is doing alpha theta therapy. Here, the client is encouraged to enter a mental state in which they are supremely susceptible to recalling fragments of both memory and fantasy. This susceptibility extends to linking the two when laying down "new" memories, which may actually be *false memories*.

Memory and Emotion
The association cortex connects with the entorhinal cortex, which impinges on the dentate gyrus. This connects to the hippocampus, which connects via the fornix to the mammillary bodies and so to the limbic system and the thalamus and hence to the cortex. These connections are all two-way. Memory involves this system. Emotions and memory are inextricably entwined in this system, which connects the cortex, involved in the laying down of memories and the limbic system.

Problems with Memory

Korsakoff's Syndrome
Alcoholics affected with this syndrome are unable to form new memories (*anterograde amnesia*). The person can still recall old, long-term memories. It is associated with deterioration of the mammillary bodies, and in the dorsomedial nuclei of the thalamus.

Anterograde Amnesia
The same picture of anterograde amnesia may be observed with bilateral damage or removal of the hippocampus. The patient remains forever in the past. Such a patient can still learn procedures through repetition. However, even as they get better at the procedure, they have no memory of having practiced it before (Carter, p 172).

Alzheimer's Dementia
This illness is due to deterioration of the hippocampus. Recent memory is gone, and these patients often can't find their way around and frequently get lost.

Senile Dementia
This is a similar process to Alzheimer's and has the same results, but it comes on at a later age.

Obsessive Compulsive Disorder (OCD)
This disorder affects about three percent of the population at some time(s) during their life. The pathophysiology involves an overactive loop of neural activity between the **orbital prefrontal cortex,** which is involved in feeling that something is wrong, to the **caudate nucleus,** which gives the urge to act on personal memories or on instincts such as cleaning or grooming, to the **cingulate,** which is important in registering conscious emotion, and which can keep focus or attention fixed on the feeling of unease. The client's error detection system is stuck on alert. The caudate nucleus is involved in automatic thinking. For example, when you check that you closed the fridge, turned off the stove, locked the door. Overactivity of this circuit means constantly checking and rechecking. Brain scans show that when a person with OCD is asked to imagine something related to their compulsion (such as dirt if it is someone who must compulsively clean), their caudate and prefrontal cortex light up. In NFB we tend to work to reduce this overactivity. We have found this overactivity during the EEG assessment at a number of different sites. Although it is usually found frontally (F3, Fz or F4), it may also be found centrally and even at Pz. The assessment guides the placement of the electrode for neurofeedback. The task is to decrease very high amplitude beta where this has been observed in the QEEG assessment. Alternatively, one may work on the right side and decrease beta while increasing high alpha, a suggestion that has been made by Lubar. We have decreased high beta activity (21-34 Hz) at a midpoint between Fz and Cz. Although not published, some clinicians have experienced some success by using regular biofeedback to decrease anxiety (discussed later under Adjunctive Techniques) while also doing neurofeedback, increasing 11-15 Hz activity at C4 and P4. What is in common to these suggestions is decreasing very high amplitude bursts of beta at sites where this has been observed on the QEEG, and

decreasing anxiety and tension using biofeedback and neurofeedback.

Tourette Syndrome

In Tourette syndrome, the **putamen** is overactive. The putamen is related to the *urge to do fragments of preprogrammed motor skills*. The putamen is linked to the **premotor cortex,** which governs the production of the actual movements. Instead of the relatively complete procedures seen with compulsions, the motor and vocal tics appear in Tourette syndrome as fragments of known actions. Oliver Sacks has written a superb description of a surgeon with this disorder in his 1995 book, *An Anthropologist on Mars*. In NFB work, we tend to encourage the increase of SMR (usually between 13-15 Hz), and decrease theta at C4. There have been studies using C3 and Cz placements as well. We also decrease sympathetic drive using biofeedback concerning respiration. The rationale for these procedures is given below under Movement Disorders, and has been published (Thompson & Thompson, 2002).

Reward Deficiency Syndrome

This is similar to OCD, but the compulsive behavior revolves around a pleasurable activity. The person can never get enough of that which they feel they need, such as food, drugs, or gambling (Blum, 1996). In these disorders the cortex is stimulated by something internal (such as hunger) or external (news report on dog races that relates to that person's gambling compulsion). The limbic system is associated with a strong *urge* or desire. When the activity is carried out, the limbic system causes the release of opioid-like substances, which result in a dopamine surge and a feeling of satisfaction. At the time of writing we do not have evidence of the efficacy of neurofeedback for this difficult syndrome. We suggest that you follow the assessment findings and the procedure outlined under OCD above. You must discuss the experimental nature of this work with you client and only proceed with their consent.

Attention Deficit Disorder
(This is expanded in Section XXVI)

Introduction
There are a number of different theories and findings that attempt to delineate the areas of the brain that are primarily involved in attention. Needless to say, one theory doesn't always link cleanly to another. This section will thus present a few different perspectives. Perhaps one of the most elegant formulations was proposed by Molly Malone and her colleagues in 1994. It will be outlined first. Since theta is considered a biological *marker* for the majority of people with Attention Deficit Disorder (ADD), and because this is highly relevant to NFB, we will attempt to paraphrase Sterman's thoughts on its production. This will be followed by a short summary of the essence of a number of other suggestions concerning attention. The reader is also referred to another section in this book on ADHD, Section XXVI. It emphasizes networks and efficacy studies.

Background to Themes About ADD
In general, the left and right cerebral hemispheres are involved in different aspects of attention as follows.

The Left Cerebral Hemisphere
This hemisphere is involved in focused, selective attention that favors contralateral space. It is involved in information processing that requires foveal vision, object identification, and what Malone refers to as the ventral/anterior system. The principle neurotransmitter for sustained attentional activity and information processing in this hemisphere is dopamine.

It is postulated that in ADD there is reduced dopamine in the fronto-mesolimbic system in the left hemisphere. The type of cognitive processing that is affected and deficient is that which requires slow, serial effort. This type of processing is called *tonic*. The left hemisphere is biased in the direction of carrying out routine and repetitive activities. It is this kind of processing that may be improved with stimulant medication. With the EEG one can observe that as the task becomes boring, the person with ADD will drift off the task. There is an increase in slow wave activity. This is similar to the slowing of electrical activity as we become drowsy. In ADD it is really a modulation of functions that is the primary problem. On the one hand, arousal may increase and actions without reflection may occur; on the other hand, the tolerance for routine, boring activities is low in these individuals and their arousal may suddenly drop rapidly. These characteristics may be improved by stimulant medications and also by NFB. Often generalizations about arousal are fuelled by older beliefs that the majority of people with ADD were hyperactive. Arousal is an issue, but a complex one. Most people with ADD have extremes of arousal, in both directions, that are not always appropriate to the situation in which they find themselves. They may be impulsive on the one hand

and literally fall asleep in a classroom on the other hand. Stimulants may help them to better modulate arousal while the drug is in the body at the proper dose. Biofeedback using skin conductance (electrodermal response) combined with NFB can have the same modulating effect – and give a long-term result.

Theoretically, many of the foregoing symptoms may be due to underactivation of dopaminergic activity in the left hemisphere. Dopaminergic overactivity, on the other hand, is thought to be associated with blunting of affect, excessive intellectual ideation, and introversion. Pathologically, it may underlie other disorders including: paranoid states, anxiety, obsessive compulsive disorder (OCD) and schizophrenia.

The Right Cerebral Hemisphere

This hemisphere is involved in the general maintenance of attention and arousal. It regulates information processing that requires peripheral vision, spatial location, and rapid shifts in attention. These aspects of attention appear to involve the noradrenergic system.

The noradrenergic system is thus more involved in initial attention plus arousal and wakefulness. It appears to be related to alertness and to responses to change and to new stimuli.

In Attention Deficit Disorder there might be excessive locus coeruleus norepinephrine production and excess noradrenergic stimulation to the right cerebral hemisphere. This represents minor brain stem involvement in arousal (usually arousal is discussed in terms of the involvement of the pontine reticular system). This excessive norepinephrine production has been suggested because clonidine affects α_2 inhibitory receptors in the locus coeruleus and has been observed to reduce ADD symptoms, possibly because it dampens activity in this area.

People with ADD seem to have automatic processing that is fast and simultaneous. These are called "phasic" attentional abilities. This style of attention is biased towards novelty and change. It has also been noted that overactivation of the noradrenergic system in the right hemisphere is associated with extroversion, histrionic behavior, impulsivity, and manic behaviors (Tucker, 1984).

Brief Summary of Dopaminergic and Noradrenergic Theory (Malone)

According to an excellent paper by Molly Malone and a number of other ADD experts, ADD is characterized by a relative **left hemispheric underactivation.** This involves the left-anterior-ventral-**dopaminergic** system. Combined with this is a **right hemisphere overactivation** or overarousal. This overarousal involves the right-posterior-dorsal-**noradrenergic** system. It could be postulated that this overactivation might be due to a lack of left hemisphere inhibitory control of the right hemisphere. Medications, when appropriately used, may correct this imbalance, but high doses may actually reverse this imbalance and produce a different set of problems.

Also discussed in this formulation is the possibility that ADD might be due to left-sided, dysfunctional, dopamine-rich frontal-striatal connections with decreased glucose metabolism, plus dysfunctional connections (smaller rostral corpus callosum) adversely affecting regulation of attention between two hemispheres (Castellanos, p 246). As an aside, females have a larger corpus callosum (more connections between the two hemispheres). Rates of diagnosis of ADD are much higher in boys – at least 6:1. It may be posited that the right hemisphere functions are being modulated less by the left hemisphere in boys.

In addition, a question is raised as to whether in the ADD inattentive type, there may be right frontal dysfunction. This group demonstrates a tendency for hemispatial neglect of left side. (This is condition is found with brain lesions on the right.) Further, there is the possibility that this right frontal dysfunction might be responsible for a decreased inhibition of right posterior areas, which would explain the high distractibility to external stimuli in these individuals.

This theoretical framework is supported, in part, by the observed actions of stimulant medications. These may be listed as follows (Malone, 1994):

- Stimulants in animal studies block uptake of norepinephrine (NEP) and dopamine in the striatum, hypothalamus and the cortex.
- Stimulants facilitate release of dopamine (but not NEP) from the striatum (which includes the caudate). It has been observed that the left caudate is small in individuals with ADD. However, stimulants will increase activity in the left striatum (increase blood flow).

- Stimulants dampen activity in the locus coeruleus.
- Stimulants increase left hemisphere processing speed.
- Stimulants decrease right hemisphere processing speed.

Sterman's Hypothesis Concerning the Production of Theta

Cortex-Putamen-(Globus Pallidus)-Substantia Nigra-Thalamus

This model is based on M. Barry Sterman's hypothetical framework. Decreased blood flow to, and metabolic activity in, the cells in the frontal/prefrontal areas (including the motor **cortex**) may *lead to* reduced excitation by the motor cortex (layer VI) of the inhibitory cells in the **putamen.** This may result in thalamic relay cells producing bursts of activity at slow frequencies, and then by projection, to the anterior association cortex (layer IV). This results in theta activity being produced that we pick up in the EEG. Two parallel pathways for this origin of theta are proposed: the first involves a direct effect of the putamen on the substantia nigra, while the second is a more indirect route involving an external globus pallidus – subthalamus – (internal globus pallidus) – substantia nigra pathway. Either way, the substantia nigra would increase its inhibition of areas of the thalamus, resulting in the thalamic production of rhythmic bursts at frequencies in the theta range.

To recapitulate this hypothesis, there may be direct reduction in the inhibition by the putamen of the inhibitory cells in the **substantia nigra,** which would result in the substantia nigra *being* released to increase its inhibition of the **thalamus.**

There may also be a longer chain of events. The reduced influence of the cortex on the putamen may result in the putamen producing increased inhibition of the external globus pallidus. (Then the external globus pallidus will not be inhibiting the subthalamus.) This would release the excitatory effects of the subthalamic nucleus on the internal globus pallidus and the substantia nigra. This would also result in an increased inhibition of the thalamus.

The net result of both pathways is increased inhibition of the ventral lateral (VL), ventral anterior (VA) and the centromedian (CM) nuclei of the thalamus. This would then result in hyperpolarization of the thalamic cells. These cells would depolarize then repolarize in a slow rhythmic manner. This would begin an oscillatory process that would be conveyed to the cortex. The end result of this cycle would be **theta in the EEG** (Sterman, 2000; DeLong, 1990).

Other Sources of Theta

Another postulated source of theta comes from animal (rat) studies. Theta is said to be paced by cholinergic pathways that project from the septal nuclei to the hippocampus. This would correspond to the observed involvement of theta in memory retrieval. Recall of words corresponds to synchronization of theta. This is thought to represent hippocampal theta (Klimesch et al., 1999). It has now been shown to occur in humans and is called 'central midline theta.' Future studies will probably help our understanding of a number of different mechanisms involved in theta production. Different kinds of theta, possibly from different *generators*, may be found to correspond to different mental states.

Movement May Help Children With ADD

A theoretical model, such as Sterman's, can lead both to an understanding of observed behavior and to changes in training strategies. Those with ADD are often observed to be much more active than their counterparts in situations where this may not be considered appropriate, such as in lectures or when listening to others in a social context. Most persons who work with children who have ADHD try to have them stop these seemingly nonproductive movements. On the one hand, this need for physical activity makes sense. The brain can only consciously consider one thing at one time. Therefore, we don't want the child to be consciously thinking of twirling his pencil when he is supposed to be listening to a history lecture. On the other hand, Sterman's formulation makes us recognize that if the child doesn't maintain activity in the cortex, then the result will be inhibition of thalamic production of theta activity, less active processing in the frontal cortex and a drop in alertness.

We teach all our students the importance of maintaining focus on the cognitive activity they are supposed to be doing. This requires that they not focus on fidgeting types of activity. On the other hand, we recognize that with some individuals, if they don't move, their alertness may drop off rapidly. Think of Greeks manipulating their worry beads while thinking something through. We suggest they find a means of remaining "active," which will increase (and not decrease) their attention to the problem at hand. Movement during a lecture might involve drawing something that would relate to the lecture content, or it could involve note taking. Drawing would theoretically involve a different area of the brain and not compete and interfere with the left hemisphere problem-solving cognitive work. However, the student must be careful not to let their focus linger on the drawing task.

With the advent of notebook computers, one can type notes as a motor activity. For those ADD types who like to *multitask*, there is the possibility of running two computer programs at once. The first records the lecture while the second incorporates the key points from the lecture into another document for a paper or for teaching.

In Sterman's proposal the sequence may involve:
1. The individual **moves around** to activate the cortex and counteract an increase of theta.
2. The individual feels more **alert**.
3. The individual reduces motor activity in order to read material and **learn**.
4. This experience fits a model of **reciprocity** between motor activity and visual attention – *beta up at one site, down at the other.*
5. Very soon, however, the increased activation of the cortex decreases as the person is sitting still, so theta starts to increase again.
6. The inhibition of the thalamus takes over again and theta increases.
7. The individual moves to activate the cortex again… and so on.

This is just a way of trying to understand how theta may be produced when we are not actively using our cortex. Remember, brain waves don't cause anything. Brain waves merely are like a flag. A flag tells us if there is wind and what direction the wind is coming from, and perhaps a little about the strength of the wind. Brain waves merely signal that something is going on or not going on, and with what intensity. They tell us a little about the person's mental state and how it is changing, or remaining stable, over time.

The Three-Element Theory of Attention: Arousal, Orientation, Focus

This is expanded and networks are described in Section XXVI.

Arousal
Arousal involves reticular activating system (RAS) activity. The nerve fibers from this system stretch as far as frontal cortex. They control consciousness, sleep/wake cycles, and the level of activity in the brain. The axons activating the prefrontal lobe release dopamine and noradrenaline, and beta activity is observed.

Orientation
Orientation involves the superior colliculus (to turn the eyes) and parietal neuronal activity to disengage attention from the current stimulus.

Focus
The lateral pulvinar nucleus in the thalamus operates like a spotlight shining on the stimulus, locking on, shunting information about the target to the frontal lobes, which then lock on and maintain attention.

Francis Crick, the co-winner of the Nobel Prize in 1962 for his work on DNA, is now studying the question of consciousness (the relationship between mind and brain) by looking at the visual process in humans. He notes that if we don't pay attention to some aspect of the visual scene, our memory of it is very transient and can be overwritten by a subsequent visual stimulus (Carter, p 205).

Other Findings Concerning Areas Involved in Attention

Orbitofrontal Cortex
This area of the cortex and the prefrontal cortex are said to be involved in inhibiting inappropriate action. Decreased functional activity seems to be greatest in the left frontal area during tests of intellectual or attentional functions (Sterman, 2000). Amen demonstrated a decrease in blood flow to the prefrontal cortex during a continuous performance task in 65 percent of his ADD subjects compared to 5 percent of the controls (Amen et al., 1997).

Left Frontal and Prefrontal Cortex
Decreased activity in this area is not only implicated in ADD, but also in studies of affective disorders (Davidson, 1999).

Anterior Cingulate Cortex (ACC)
This cortex focuses attention and facilitates tuning inwards to one's own thoughts. The ACC and the insula are the principle components of the Salience network (Section XXVI). Increased blood flow on the **right** side of the ACC suggests that attention is focused on **internal events.** The ACC **distinguishes** between **internal and external** events and is underactive in schizophrenia, where the subject is unable to distinguish their own thoughts from outside voices.

Ventromedial Cortex
In this area of the cortex, **emotions** are experienced and **meaning** is bestowed on perceptions.

Dorsolateral Prefrontal Cortex
This area is involved when you **hold** a thought in mind, **select** thoughts and perceptions to attend to, **inhibit** other thoughts and perceptions, **bind** the perceptions into a unified whole, endow them with **meaning, conceptualize, plan** and **choose**.

In her excellent book, *Mapping the Mind*, Rita Carter describes a ***self-will experiment***. In this experiment, it was found that lifting a finger will only involve activation of the prefrontal cortex when the subject is not told which finger to lift but must decide on their own. It seems that following an order does not bring an individual's prefrontal cortex into play, but making a choice does.

Neurotransmitters, Metabolic and Structural Differences
In ADD there appear to be genetic differences as compared to non-ADD controls with respect to dopamine receptors. These differences may lead to reduced dopaminergic activity (LaHoste et al., 1996). Deficiency in dopamine-related functions in the frontal-striatal system is the basis of some theories as to the emergence of ADHD symptoms (Charcot et al., 1996; Malone et al., 1994; Sterman, 2000).

ADD is primarily a frontal lobe dysfunction. The information from various measuring techniques converges to indicate decreased frontal activation. The EEG shows increased theta, Dan Amen's SPECT studies show decreased perfusion, and Zametkin's PET studies show decreased glucose metabolism.

In addition to these findings in the frontal lobes, in ADD the blood perfusion and metabolic activity in the caudate may be decreased (Hynd, 1993; Zametkin, 1990). Magnetic resonance imaging (MRI) studies have shown decreased tissue volume in the frontal, premotor, sensorimotor cortex and in the caudate of ADD subjects (Hynd, 1993; Filipek, 1997) as reported in Sterman's review (2000).

Reticular Formation
The reticular activating system is only fully myelinated after puberty. This may partially explain why younger children have a short attention span. The RAS plays a major role in maintaining attention. A threat causes **activation** (a rush of adrenaline) that **closes down all unnecessary activity,** leaving an alert brain that appears on a brain scanner as quiet. It also inhibits body activity. Breathing is shallow and quiet, and the heart rate is slow. In this state, however, **activity remains** in the **superior colliculus**, the **lateral pulvinar nucleus** of the thalamus and the **parietal cortex.** These areas are involved in **orienting and focusing.**

Right Hemisphere Deficiencies in ADD
Imaging studies demonstrate decreased activity in the right hemisphere in the following areas: the **anterior cingulate** (fixing attention on a given stimulus), the **prefrontal cortex** (planning actions and controlling impulses, which also involves the orbitofrontal cortex) (Arnsten, p 186; Castellanos, p 251), and the **upper auditory cortex** (integration of stimuli from several different sources). Perhaps a lack of activity here prevents the child from grasping the whole picture, and instead, he views the world in fragments with one stimulus after another vying for attention.

Neuroimaging studies have also demonstrated differences between children who have ADHD and controls. Smaller volume in the right prefrontal brain regions, caudate nucleus, globus pallidus, and a subregion of the cerebellar vermis have all been demonstrated (Castellanos, 2001).

Hippocampus – Septal Nuclei – Prefrontal Cortex Circuit
The diagram below illustrates one set of connections involved in focused attention

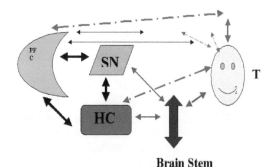

Brain Stem

Fundamental Links to Modulate Attention

BS = Brain Stem; T = Thalamus; SN = Septal Nuclei; HC = Hippocampus; PFC = Prefrontal Cortex

The hippocampus is important in selective inhibition of attention to things that might distract one's focus from the central problem. It also is a part of the brain's system for arousal, alertness, awareness, and orientation in the process of **focusing** attention on one set of environmental signals while excluding others. (In animals, retrieval of memories in this process appears to be associated with the production of theta waves.) The hippocampus is also involved in laying down a conscious memory of events.

The prefrontal cortex processes incoming information about an event or a problem and then organizes a response. In this process it signals septal nuclei. This takes about 300 ms and may be related to the P300 seen in ERP research. (Sieb, 1990). These nuclei then feed back to the hippocampus, thus completing the circuit. Inputs to all parts of this circuit come from the brain stem and the thalamus.

Stimulant Treatment and Neurofeedback Treatment for ADHD

The Stimulant Medications

In adults and in children younger than six, the response to stimulant medication is positive in only about 50 percent of cases. In school age children there is about a 75 percent positive response rate to a trial of a stimulant drug.

Amphetamines raise excitatory neurotransmitters that stimulate the areas that are underactive in ADD. There are numerous studies that demonstrate that both persons who have a diagnosis of ADHD and normal controls experience performance improvements when prescribed stimulants. The reader is referred to the chapter in *Stimulant Drugs and ADHD* titled, "Comparative Psychopharmacology of Methylphenidate and Related Drugs in Human Volunteers, Patients with ADHD and Experimental Animals" (Mehta et al., in Solanto et al., 2001 p 303-331). Stimulant medications have been shown to increase blood flow in the dorsolateral prefrontal cortex and posterior parietal cortex during tests of spatial working memory (Mehta et al., 2000). Complex working memory tasks and sustained attention may also improve with stimulant medication (Mehta et al., 2001 after Elliot, 1997; Koelega, 1993). Tests of cognitive flexibility in normal volunteers have shown that methylphenidate (40mg) can facilitate the shifting of attention toward newly relevant stimuli, but mean response latencies were increased (Mehta et al., 2001). Stimulant medications reduce the rate of both omission and commission errors on continuous performance tests (Losier et al., 1996).

Both neurofeedback and the stimulant medications improve the symptoms of ADHD (inattention, distractibility, hyperactivity, and impulsivity) in a reasonable percentage of children who have this disorder. There may be some similarities in their actions. However, the goal of NFB is to effect a lasting change in the system through learning, while the goal of stimulant medication is to produce a temporary shift while the drug action persists. Theoretically, the cortical activity that stimulants produce has inhibitory effects on the limbic system, and thus helps to substitute thought for action and produce a more controlled and focused behavior. It has been found that some people who have ADD may have a genetic alteration in the D4 dopamine receptors (Solanto, 2001, p 356). Ritalin blocks the re-uptake of dopamine, particularly in the striatum. Frontal- striatal circuits that also involve the globus pallidus and the thalamic nuclei may be implicated in ADHD. The frontal areas appear to be important in selection of appropriate stimuli, suppression of irrelevant stimuli and suppression of inappropriate actions. These are the three cardinal symptoms of ADD: inattention, distractibility, and impulsivity. Theoretically, the same circuits may account for the observed increase in theta seen in the EEG in children with ADD. As explained above, in ADD the striatum may exert too little inhibition on the globus pallidus. The globus pallidus then inhibits certain thalamic nuclei. These neurons become hyperpolarized and fire in a specific rhythm in the theta range. This activity is conveyed to the cortex and detected in the EEG. (Kropotov, 1997; Sterman, 2000).

EEG differences between children with ADHD and normals (higher theta/beta ratios) have been shown in a number of studies (Mann, 1992; Janzen, 1995; Monastra, 1999; Clarke, 2001) and in a meta-analysis by Arns (Arns et al., 2012). Interestingly, stimulants do not appear to lower the theta/beta ratios significantly, but neurofeedback training has been shown to have this effect. This work was originally done by Joel Lubar and has since been replicated by others. EEG studies are beginning to demonstrate different subtypes of ADHD clients according to different patterns found on brain maps (Chabot, 1996; Sterman, 2000). Three groups are clear in Sterman's work: excess theta in the prefrontal cortex, generalized excess of theta in frontal and central leads, an excess of alpha activity at C3, C4, and frontally. Other researchers (Chabot, 2001; Clarke, 2001) have noted a group of ADHD subjects who have high amplitudes of beta. In our experience, many clients with ADHD symptoms who demonstrate high amplitude beta (and a low theta/beta ratio at F5 and F6), or high amplitude spindling beta at F3 and Fz and/or F4, also have high theta/beta ratios at Cz). This high-amplitude beta, in our experience, is most often directly associated with other disorders, including depression with ruminations, obsessive compulsive disorder, and so on. This high beta has also been observed when one client with a diagnosis of schizophrenia kept tuning out and attending to repetitive delusional thoughts. (This observation will have to be replicated in other centers.). Further, we usually find that this high amplitude beta is often between 23 and 34 Hz. If anxiety is a component of their difficulties, then high amplitude bursts of 20-22 Hz activity may also be observed.

One group that is an exception to the above rule of thumb are those individuals who lapse into intense cognitive activity during testing and produce high amplitude bursts of beta between 14 and 19 Hz. Some of these clients also tune out and show bursts of high amplitude theta or alpha. Their overall theta/beta ratio is low. Their problem is, however, not an inability to focus when they want to, but rather variability and tuning out in high thalpha when they feel they should have been attentive to someone speaking or to what they were reading. The lesson to be learned from this description is that power averages and post-hoc EEG analysis do not tell the whole story. The clinician must revert to old-fashioned "know-thy-patient" clinical observation and skills. **In order to correctly interpret the EEG for a particular client, the clinician must, initially, repeatedly pause the recording and inquire about what was going on in the client's mind with each EEG shift. This will allow the clinician and the client to better understand how those shifts in the EEG correspond to shifts in that client's mental state.**

Brief Summary

Research on the fine points of how stimulants may act or on how theta rhythms may be produced does not tell the whole story. For clinicians who have treated thousands of children using medications, behavior modification, educational approaches, and now using NFB, it is apparent that all of these approaches can be beneficial. The medications give the quickest response and are most useful in acute, *urgent* situations. Medications, behavior modification and remedial educational approaches all give good *short-term* outcomes to many children. For a *long-term* change, NFB may prove to the treatment of choice. Neurofeedback is based, for the most part, on two premises: first, increased theta activity does correspond to decreased attention to external stimuli (and less retention of material that the child is either reading or being taught), and second, on the observation that increasing SMR corresponds to a decrease in both hyperactivity and impulsivity. NFB takes time (often 40 or more sessions) to take effect, but the effects appear to be long lasting. Lubar has done one 10-year follow-up (Lubar, 1995) showing that results lasted and Gani and colleagues in Europe did a multicenter, controlled study demonstrating that the effects last (Gani et al., 2008). Preliminary results from a long term, 10 to 20 years post-training, study of ADD Centre clients in Toronto Canada is also showing consistent long lasting effects after NFB training.

We know that stimulants improve the performance of many of these children who have ADHD. On the other hand, the precise reason for these observed improvements, although well researched as noted above, remains somewhat confusing and at times contradictory. The stimulant medications may affect these circuits as previously noted. However, the effect on the EEG is minimal (Lubar, 1999), whereas one would have expected to see marked changes in the EEG. In addition, some observant parents report that their children's eyes *glaze over* (go out of focus) just as frequently on the medications as off them. These same parents will comment that, although their child seems to still go out of focus frequently, they do stay with the task longer and perform better on stimulants than off. Certainly there is reasonable agreement that stimulants, used in children who have ADHD, are excellent chemical restraints and do have a marked effect on impulsivity and hyperactivity, and thus, do fairly uniformly have a secondary positive effect on focus, sustaining concentration and work completion. However, they are drugs and they can have unwanted effects which may prove to be long term, although there are no controlled studies to help with this question. Indeed, as one neuropharmacologist succinctly characterized it in a professional lecture, the child on a high dose of Ritalin may look like a deer caught in the headlights of car. Further research is needed and it would be helpful if this included a neurologically and functionally valid operational definition of attention in the classroom situation.

In summary, stimulants are effective for the short-term management of behavioral symptoms. Neurofeedback produces lasting effects on behavior, academic achievement and even on IQ scores.

A Case Example – Before the Availability of Neurofeedback

Johnny (fictitious name) was a handful as a preschooler and he was referred to a child psychiatrist once he entered Kindergarten. Ritalin (methylphenidate) in a dose of 10 mg twice a day was prescribed at age 6 when he was in the Spring-term in Senior Kindergarten. Although it was somewhat effective in controlling his impulsivity, it was principally effective in decreasing the hyperactivity and decreasing his distractibility to external stimuli. Nevertheless, there remained a "roller coaster" effect due to its short duration of action. Additionally, there was rebound hyperactivity and increased emotionality as the drug effects wore off each day. He also had no appetite while on the drug and was not gaining weight.

Therefore a trial of Dexedrine (dextroamphetamine) was attempted using 5 mg twice a day. It was more effective than Ritalin had been. Nevertheless, on reviewing the total symptom picture in the Fall of the school year, it was clear that Johnny had become very discouraged, frustrated and depressed. He was crying over the slightest occurrence and seemed very irritable; he couldn't be comforted or calmed when he was upset. He demonstrated some regressive behavior, such as hiding under the dining room table when he didn't want to do something. He was very sensitive to any comments that he perceived as being critical, so it was hard to correct or discipline him. He became obsessive in the sense of being fixated upon and unable to let go of any idea he had his mind set on, such as a toy he wanted. He was becoming very, very difficult for his parents to handle.

The next year, in December in Grade One, the diagnosis of depression was considered and the psychiatrist decided to place him on an antidepressant. Desipramine (a tricyclic antidepressant) was chosen as it can be effective in ADHD and it may have less anticholinergic effects as compared to medications such as imipramine. Within three weeks there was a very marked beneficial effect. The temper tantrums stopped and he was able to control his anger and calm himself down. He wasn't nearly as fragile and irritable as he had been. In play therapy sessions it was clear that he could focus and be creative with the materials, and he became much more patient when his parents and the psychiatrist were talking. He would find things to occupy himself at home, which mother describes as an "all time first."

However, by the end of January, some of the beneficial effects were wearing off and he became increasingly impulsive. Dexedrine was reintroduced in order to better control impulsivity and increase his attention span in school. This seemed to be effective at doses of 10 mg at 8 am, and 5 mg at noon. It was necessary to be careful and do regular checks on tricyclic blood levels due to the possibility of an adverse drug interaction (each of the medications could increase the level of the other one).

In March, the good effects noted above for the desipramine seemed to be wearing off. This can happen, and when it does, one normally merely raises the dose by a small increment. One hopes this will only have to be done once or twice. However, Johnny was already at a dose of 150 mg, which can be considered at the high end of an effective adult dose of this medication. A blood level was done and it was found to be in the therapeutic range. Rather than raising the dose further, a different tricyclic was tried in the hope that it might be as effective at a lower dose level.

A trial with amitriptyline (Elavil) was begun. On bringing down the dose of desipramine, Johnny became quite emotionally fragile or brittle and was quite depressed and very difficult to handle. He was completely unable to occupy himself. He seemed to feel badly about his behavior, but unable to control it. On a dose of 100 mg Elavil he stabilized and was behaving as he had on the much higher dose of desipramine. The improvement didn't last.

Every medication used and every combination had initially had dramatically good effects. These initial good effects, however, were not sustained, and the parents and psychiatrist were left wondering if he was worse off now when off medications than he had been before medication treatment was started.

The story of Johnny underscores why it is such a boon to have neurofeedback as an alternative to using drugs. Among the most motivated users of neurofeedback services are the parents of children who have had unacceptable side effects when stimulant drugs were prescribed. Another even larger group are the parents who are reluctant to use drugs in the first place. Now the serotonin selective reuptake inhibitors (SSRI) are available and are being used with children. Some children are being prescribed combinations of these and other medications for which there is no substantial research. No matter what the drug or the combination of drugs, the stories for many children in the twenty-first century are not very different than what we observed with Johnny in the 1970s.

A Case Example – ADD Symptoms Treated with NFB

With his freckles and red hair, love of the outdoors and disinterest in school, Jason was a veritable Huck Finn. Though bright, he was restless, poorly organized, and could not get down to homework. Written assignments were a particular problem. Medications had been effective in keeping him in his seat, but there was still gross underachievement in school and his grade-seven year was in jeopardy. His parents felt that the medications took away his exuberance, and they were also concerned about the appetite suppression and the effects on growth as he approached puberty. Neurofeedback training was focused on reducing 4-8 Hz activity and increasing 12-15 Hz with placement at Cz referenced to the right ear. After 40 one-hour sessions, done twice a week, Jason demonstrated improvements in most areas. He was off stimulant medications, parent questionnaires were no longer in the clinically significant range, though they were still somewhat elevated for task completion and distractibility. His TOVA and IVA scores had improved, but impulsivity/response control scores were still outside the normal range. A further 20 sessions of training were done on a once-a-week basis, and the NFB training was combined with the teaching of metacognitive strategies. Initially these had focused on filling in skill gaps in math, and strategies for reading comprehension. The focus then shifted to time management (rewarding him for using his agenda) and written work (see the "hamburger method" in Section XII). After 60 sessions, retesting demonstrated normalization of attentional variables on both computerized tests and a theta/beta ratio that was no longer extreme. IQ testing revealed a 17-point gain, and basic academic skills on the Wide Range Achievement Test were above grade expectations. Parents remarked that he had "grown up a lot." At graduation, being presented with his official ADD Centre mug, he grinned proudly from ear to ear.

Movement Disorders

Hypothesis Concerning Involuntary Motor Movement

Dystonia is a good example of how a theoretical hypothesis can lead to a training regimen. When asked by a woman with Parkinson's disease and dystonia if neurofeedback might help her regulate certain disabling symptoms, it was necessary to come up with a rationale as to why it might be worth trying. Her EEG showed a marked decrease of activity in the SMR range. This was not surprising, considering the amount of tremor and unwanted dystonic movement.

Shortly after meeting her, she asked if we would do a quick EEG (single-channel at Cz) at a group gathering and explain what she was experimentally attempting with us. Measurements were done on 14 other patients who had Parkinson's. About one-third of the patients were not on medication at the time. There was no obvious difference in the EEG between those on and off medications. All demonstrated very low SMR, 13-15 Hz activity.

It was hypothesized that a two-pronged approach to the neurological control of muscle tone through the muscle spindles might have a beneficial effect. The rational for such a hypothesis is based on the following knowledge. Muscle spindles, which are involved in muscle movement and tone, have double innervations, cholinergic and sympathetic. Both of these systems can be operantly conditioned. The patient can consciously control sympathetic/parasympathetic balance even in difficult situations through diaphragmatic breathing. This could be paired with raising SMR. The combination might have a direct effect on the control of involuntary movements.

The theoretical background to this hypothesis arises from the following knowledge. Gamma efferent fibers from ventral horn of the spinal cord to the contractile ends of the intrafusal muscle fibers receive input from supraspinal efferents that can arise from the red nucleus. The muscle spindle, in turn, sends somatic afferent fibers back to the red nucleus. This results in a feedback pathway from the muscle spindle that influences the activity of the red nucleus in the midbrain. The red nucleus has been shown to decrease firing when the thalamus is producing the spindle rhythm that Sterman named *sensorimotor rhythm* (SMR) (Sterman, 2000). Sterman demonstrated that SMR can be trained by means of operant conditioning. Theoretically, training to increase SMR could decrease muscle tone and, perhaps, unwanted movements. Simultaneously training for calm relaxed sympathetic/parasympathetic balance in autonomic nervous system functioning might also have a direct beneficial effect on muscle tone by means of the sympathetic influence (Grassi, 1986; Banks, 1998) on muscle spindle activity. To attain this balance in the autonomic nervous system the client is taught: (1) diaphragmatic breathing, and (2) to increase the amplitude and quality of RSA (respiratory sinus arrhythmia), also termed heart rate variability entrainment (Thompson & Thompson, 2002). Since breathing is paired with raising SMR, the patient is instructed to consciously begin diaphragmatic breathing at a rate of about six breaths per minute in order to initiate the beneficial effects. The application of this theoretical formulation to the actual case is described later in this text.

Understanding Asperger's Syndrome

See also Section XXV

Introduction: Symptoms and Neuroanatomical Considerations

Autistic spectrum disorders have core symptoms "characterized by the triad of impairments of social interaction, communication, and imagination associated with a narrow range of repetitive activities" (Wing, 2001, p xiv). Relevant DSM-IV diagnostic codes are Pervasive Developmental Disorder (PDD) and Asperger's Syndrome (AS). Incidence is on the rise, and currently estimates are that more than 1 child in 150 is affected. Brain differences include: smaller cells in the limbic system (Bauman, 2001); larger brains due to more growth in grey and white matter during the first three years of life (Courchesne, 2001); fewer Purkinje cells in the cerebellum (Courchesne, 2001); different activation of the fusiform gyrus for facial recognition (Pierce, 2001); and abnormal interaction between frontal and parietal brain areas (Pavlakis, 2001). With respect to Asperger's syndrome, a major difference as compared to PDD is that delayed language is not characteristic of AS. Individuals with AS want to have social interactions, but lack the social graces to do it appropriately. They often present like little professors with extensive knowledge in their area of interest. Symptoms overlap with Attention Deficit Disorder.

EEG brain maps show less activation in the areas of the right hemisphere that process emotional information (Thompson, Citation paper, proceedings, AAPB 2003, Thompson et al., 2010). The amygdala, orbital and medial prefrontal cortex, medial and temporal areas and the thalamus are all involved in the process of attaching emotional significance to stimuli and are most likely of central importance in understanding the autistic spectrum disorders (Schulz, 2000). Temporal lobe dysfunction has been found (Boddaert, 2002).

Neurophysiologically, the medial and basal zones of the cerebral hemispheres are critical in the emotional aspects of social interactions; the ventromedial prefrontal region is important in learning concerning social interactions, and the lateral nucleus of the amygdala is critical in assigning emotional valance to events. The central nuclei of the amygdala are more involved in emotional expressions including flight or fight responses. Dysfunction in the interconnections between the orbital and medial wall of the prefrontal cortex and the amygdala and/or dysfunction in the lateral and central nuclei of the amygdala may therefore lie at the root of some of the difficulties encountered in the autistic spectrum disorders.

Are children with Asperger's more right brained or left brained? This was discussed in Section V: Neuroanatomical Structures, under Functional Significance of the Lobes of the Brain. It was concluded that AS is clearly a disorder where the right hemisphere is not doing its job, whereas left hemisphere functions (language, scoring well on IQ tests, following routines) are not so problematic. Emotional understanding and expression depend on right hemisphere activation, and is deficient.

Early Child Development

A Very Brief Theoretical Review
(Reference: Thompson & Patterson, 1986)
The key principles underlying normal development are:

1. The child's earliest interactions **evoke reactions** from the environment, and from these interactions and feedback the child forms **basic underlying assumptions** about self and self-in-relation-to-the-world.
2. The infant and then toddler will begin to **interact** in a manner which tends to elicit feedback from the environment that **reaffirms** these assumptions and thus **allows** a **psychic homeostasis** or equilibrium to be maintained.
3. In the first three years of life, the infant-toddler normally moves through a **predictable sequence** of behavior. First there is movement **towards** the mothering figure to satisfy **homonomous** (inborn social) needs. This is followed by movement **away** from the mothering figure to satisfy **autonomous** needs.

If the environment flows with the child, welcoming the closeness and clinging (7 to 11 months and again around 18 months) in the first stage, then greeting the curiosity, exploration and needs for autonomy of the second stage (approximately 14 to 16 months and again at about 22 months), then it is likely that the child will **develop assumptions** of being acceptable and will exude a **confidence** that the **world will meet a basic modicum of their needs**.

Temperament and Stages of Development
The normal child goes through a **well-defined progression** of steps or **stages** of development in the first few years of life in each of the following areas: motor, cognitive, speech, and social development. In

addition, each child is born with a **unique *temperament***. Extremes in certain aspects of temperament can lead to disorders such as ADHD.

Behavior disorders, on the other hand, tend to arise from *negative* and/or *conditional messages* during the early years when the mothering figure is, for the most part, the *universe* to the infant/toddler.

Inborn Deficits
Note: Although usually inborn, there are many cases where parents report a sudden onset of these deficits that may be due to a reaction to a toxic substance.

In-born deficits/delays in normal development can occur in each of the *staged developmental lines:* motor, cognitive, speech, and social. Delays in cognitive development result in general learning disabilities (slow learners) or specific learning disabilities. Delays in social development may result in autistic spectrum disorders: autism and Asperger's syndrome. An inborn temperament difference such as is seen in ADHD is highly correlated with other developmental delays in both cognitive and speech development (Love & Thompson, 1988).

In conditions such as ADHD, LD, and Asperger's, the **differences** in these children as compared to the average child would tend to **alter** and even govern the way each of these children initially, and during the first few years of life, began to **view self** and self-in-relation-to-the-world. In response, the **autistic child** or the child with **Asperger's** may have defensively withdrawn (autistic withdrawal) due to primary deficiencies in their:
- Ability to **socialize**
- Ability to sooth themselves and remain calm (**anxiety** is high)
- Connative (cognitive/emotional) understanding of social nuances/**innuendo**
- Ability to *infer meaning* (later called abstraction)

They may later develop ritualistic, **compulsive** behavior patterns that can have the effect of helping them at least maintain a primitive known psychological equilibrium (Thompson & Havelkova, 1983; Thompson & Patterson, 1986). These children are hard to socialize and seem to be stuck at a primitive level of egocentricity and omnipotence usually seen at an earlier stage of the toddler's development.

ADD and LD Children May Need to Use Behavior to Make Their Environment Predictable

The children with ADD or, even more clearly, those with hyperactivity, in many circumstances will **cause the environment to react** in a controlling manner. This reaction **runs counter to the child's autonomous** strivings, and it may be perceived as negative feedback about self. Children who feel negative about themselves tend to demonstrate negative behaviors. These behaviors guarantee that their environment will react predictably, albeit negatively, towards them. This establishes a kind of predictable equilibrium in which the child has a kind of control. It is a vicious cycle.

The child with learning disabilities may unconsciously **need to sustain a more positive** sense of self **by withdrawing effort** from that which appears futile and/or **obtaining attention** from peers through behaviors that receive negative adult feedback – the class clown, the instigator, and so on. Again, these behaviors establish a kind of predictable equilibrium in which the child has a kind of control over his environment.

Asperger's Syndrome Treatment Requires a Multimodal Approach

Neurofeedback has dramatically changed the outcome for our clients with high-functioning autism and for those with Asperger's syndrome, both in terms of speed of change and the quality and amount of change (Thompson, 2010). It is certainly not a quick fix, but at least in many cases it may no longer require a lifetime of careful management. For these clients, NFB is only one part of a multimodal approach. Reports demonstrating the effectiveness of NFB for autism and for Asperger's have been steadily appearing in the literature (Sichel, Fehmi, 1995; Jarusiewicz, 2002; Thompson, 1995, 2003, 2010). However, efficacy would be at Level 3, probably efficacious, and therefore the approaches suggested here should be presented to the client and family as experimental and based on case studies and a case series of 159 subjects by Thompson, Thompson, and Reid-Chung in 2010.

Practical Pointers when Working with a Young Child

If you want the young child who has behaviors which place them on the autistic spectrum to do an activity with you, then it is quite effective if you begin by *'joining'* with the child in whatever activity they are doing. It may be an activity that fits their obsessional interest. Then very gradually *'redirect'* the child into a more appropriate activity / behavior or the activity that you want them to partake in. Initially, if possible, do not speak. Use hand and body posture signals. If you must talk, use as few words as possible. The following example is one showing how the author (M.T.) has handled young, fairly bright, high functioning autistic children.

Jason, age 5, was screeching and running around the waiting room. The words he used were not related to the situation he was in, and were repeated incessantly. He was dancing around the waiting room flailing his arms in the air. His mother was talking at him without stopping, explaining over and over how he must stop what he is doing and go with the therapist. The therapist was also attempting to explain why he should come with her and the reward he would get when he came.

The author came into the room and signaled with his hand to mother and the therapist to stop talking and sit down. The author noted that Jason had come with his favorite cars and blocks. The author took several blocks and a car and sat on the floor on the opposite side of the room from Jason, who had now stopped screeching and running and was rocking back and forth near his toys. He did not look at Jason but sat very still and thought about how Jason would come and join him in play. After a couple of minutes the author moved three blocks slowly to make a crude bridge. Then he held out a block in Jason's direction, still without looking at Jason. Jason took the block and put it near the bridge. The author took another block and placed it on top of Jason's. He then offered Jason a second block. This interaction proceeded for a short time. Then the author got up on one knee, and, still without looking directly at Jason, offered Jason his hand. Jason took it and went quietly to the next room to do his NFB session.

As sessions proceeded, Jason was taught to calm down whenever a downward motion of the trainer's hand was used. When Jason calmed, his SMR (13-15 Hz) rose, and activity began on the screen. The hand signal worked when they went to the waiting room at the end of the session. Mom was asked to say nothing and use this simple hand signal. When Jason calmed she was to give him a simple reward (quiet, short praise is usually sufficient).

The above example is not unique, but is based on over 40 years of experience. If you must intervene verbally, do so using very short, firm, direct commands. One to three words is far more effective than a sentence. An explanation at this juncture will be ignored. Explanations are done later through Socratic questioning that helps the child develop a rationale for appropriate behavior. Playacting the new behavior helps to make it more available to the child the next time this situation occurs. Remember, initially these children will not generalize. Each learned behavior must be taught and be repeated in different environmental settings and with different people. Fortunately, the children who now get NFB training appear to generalize much more easily than children who did not receive this therapeutic modality in past years. Gradually, more verbal communication becomes appropriate. At this juncture, telling stories becomes a very effective way of teaching appropriate social behavior and helping the child to assimilate rules.

Practical Pointers when Working with an Adolescent

The following procedure can be important when you and others are working with adolescents diagnosed with Asperger's syndrome. As a general principle, try using as <u>few</u> words as possible when you are explaining a situation or making a request. Many of these individuals will mentally drift off and stop registering what you are explaining if you make more than one point at one time. It is important that you avoid the use of indefinite terms such as: "probably," "usually," "try," "sometimes," and so on. Remember, they tend to be literal and concrete in their thinking, even when they have very high verbal IQs.

With older adolescents and young adults it can be quite helpful to work out a **memory trick** that fits *their interest area* in order for them to follow the sequence before taking action:

1st STOP 2nd REFLECT
3rd DECIDE on correct behavior
4th Now act

If the adolescent has acted in a manner that is socially inappropriate, make sure that you first define the difficult situation precisely. Second, ask them what the effect was on each of the other persons. Then state simply and clearly your impression of the reactions of others to what they did. Third, ask them what they will do next time this happens. Then state concretely and briefly exactly what you would suggest they should do next time. Now ask them to enact with you the whole situation. You will try to portray the other people and they will act the way that they think might be appropriate. Now change roles with them and playact it again. Be sure that the playacting is concrete. Help them decide what to say. You may actually have to write down and give them a list of phrases that they can use.

Ask the adolescent to think of a person they feel is a good example socially. Then specifically identify the behaviors that make that person a good example. Write these down. Suggest that they try to copy how that person behaves. Ask them if they could listen carefully and watch that individual and try copying a few of their behaviors. These should be behaviors that they have first discussed with you. Always be careful to model appropriate behavior. Jokes are not usually understood. However, as NFB begins to take effect, you will notice that the adolescent begins to smile more often at the right times and begins to demonstrate a sense of humor. Other helpful advice is to have them concretely define their prime area of interest. Then tell them to listen to the other people and not to talk about their own interest area when they first enter any group of people. They may mention it, however, if asked about it.

With the young adolescent you might suggest that the parents use the *set-up-for-success* principle by having another child share an event that doesn't require too much interaction. The parents should supervise the event to make sure it all goes well. Events such as going to a movie or sports event, a place with rides, a science center, museum and so on, where adults can stay between kids, can turn out to be lots of fun for the potential new friend.

No matter what the age or the extreme of the behavior, you must always model being calm and reflective. Do NOT mirror the child's impulsiveness; model reflective thoughtfulness. Anxiety may be at the root of the child's impulsive or unusual behavior. Regardless of how hard you try to cover up tension, these children will always sense it. You must be genuinely calm. If you cannot be, ask a colleague to work with the child.

Finally, remember that this condition **is a developmental deficiency** (inborn), or a deficit possibly caused by exposure to a toxic substance, such as mercury. It has been hypothesized, with considerable documentation, that the incidence of autistic spectrum disorders may be increased in a small group of susceptible individuals due to mercury exposure related to use of the mercury-containing preservative (Thimersol) in multi-dose vaccines. This has not been supported by large epidemiological studies, but that could be due to the statistical analysis used. Most children are able to eliminate toxic substances, including mercury, with no untoward effects. However, some children are not able to eliminate mercury quickly enough, perhaps due to differences in their digestive systems, and it is hypothesized that this makes them more vulnerable. It might correspond to the onset of autistic symptomatology in infants whose development was initially normal. Another theory, from a research group at Harvard, relates the increased incidence to electromagnetic fields (EMF), and radio frequency radiation (RFR). This is a well thought-out proposal and for more information the reader is referred to publications by the BioInitiative Working Group (Herbert, 2012).

Whether inborn or acquired due to toxic exposure, it **is not learned** bad behavior! Initial insistence on appropriate social behavior is like asking an almost blind child to read the blackboard. Kind, understanding but firm assistance and **close supervision** may be needed in social situations for a number of years.

Results of Neurofeedback Training

In the children that we have trained over more than twenty years with neurofeedback, the EEG patterns at Cz initially resembled those of children with ADD. However, the high amplitudes of the slow waves and the dip in the SMR frequency band tended to be more extreme. Excess slow-wave activity in either the delta through theta range or excess alpha activity was found. Peaks at 7 Hz had the morphology of pediatric alpha. Full-cap assessments showed slowing (theta or excessive theta or low alpha, 8-10 Hz) in the right parietal region (P4) and some slowing at T6. There was high-amplitude theta at Fp1, F3, Fz and Cz. There were also differences in coherence and comodulation.

Sufficient training (sometimes more than 100 sessions) consistently produced a decrease in theta/beta ratio with the clearest change being an increase in SMR. IQ increases of about 10 points were found. With respect to the TOVA (Test of Variables of Attention), on initial assessment some of the autistic children could not complete the test. At the other extreme, children with Asperger's syndrome often scored well even prior to training. TOVA data showed that variability was the weakest score, and the majority of subjects with Asperger's syndrome were in the normal range for impulsivity and attention span. Part of the reason for these good scores may have been that the children with Asperger's syndrome were anxious, and in general, wanted to please the examiner and follow the rules. All the client's TOVA scores improved with training. Academic testing using the WRAT (Wide Range Achievement Test) also demonstrated consistent improvements with training. Perhaps the most clear-cut positive change, however, was in the social functioning of all of these clients. Parents noted improved social interactions: children went from having no friends to initiating and maintaining some peer friendships. The largest improvements were in those who received more than 80 sessions. Autistic clients were all difficult to work with. Those with AS were easier to work with once they knew the routines (Thompson, 1994, 2003, 2010).

An example of one severe Asperger's syndrome child's improvement when he was requested to draw a person is shown below. (For a discussion of this case, see "Steven," age 12, under the section on Intervention.)

| *July 1998, on assessment* | *October 20, 1999, after 40 sessions.* |

Discussion

EEG differences observed in autistic spectrum disorders provide a rationale for using neurofeedback. Excess slow-wave activity corresponds to being more in their own world; low SMR is consistent with fidgety and impulsive behavior and also with the tactile sensitivity exhibited by many; high left prefrontal and frontal slow-wave activity is consistent with lack of appropriate inhibition; high slow-wave activity in right parietal-temporal area is consistent with inability to interpret social cues and emotions. Improved social interaction found in conjunction with EEG shifts makes sense: more activation means more alert to the outside world and thus better able to benefit from socialization efforts. The positive results support neurofeedback as an intervention in autistic spectrum disorders, particularly Asperger's syndrome. Further research could build on these observational data.

Glossary of Some Terms that Also Appear in the Foregoing Text

Abulia: loss of interactional spontaneity.
Agraphia: inability to write due to word-finding difficulties
Alexia: inability to read
Anomia: inability to name objects
Apraxia: lack of awareness, and neglect, of body parts and surrounding space
    Constructional apraxia: inability to correctly construct objects
    Gerstmann syndrome: left parietal lobe damage with resulting left/right confusion
    Ocular apraxia: inability to control gaze

Acalculia: inability to calculate
Aphasia: inability to talk
Agnosia: lack of ability to know or understand
    Anosognosia: denial of deficits
        Simultagnosia: inability to integrate components of a visual scene
        Optic ataxia: inability to accurately reach for an object with visual guidance.
Dyscalculia: difficulty with mathematics
Dyslexia: difficulty with reading
Perseveration: loss of flexibility in thinking, persistence of a single thought. It is related to organic brain damage.

PART TWO

Functional Neuroanatomy Organized With Reference to Networks, Lobes of the Brain, 10-20 Sites, and Brodmann Areas

Published in 2015 separately, included here in its entirety. (Questions and References that appeared in the original publication are included in those sections at the end of this volume).

SECTION VI

Introduction/Orientation

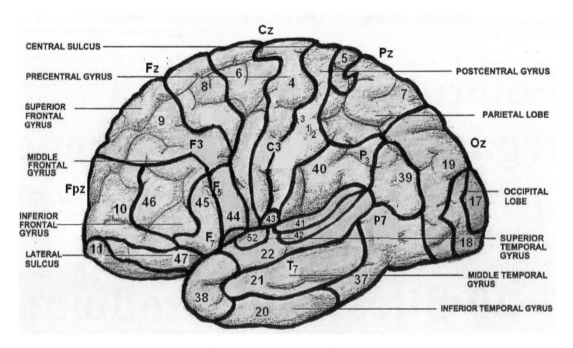

Lateral view to show lobes of the brain, Brodmann areas, and 10-20 electrode sites
(Drawing by Amanda Reeves and Bojana Knezevic)

On the diagrams the numbers go from 1 to 52, but do not frustrate yourself looking for every Brodmann area from 1 to 52 because there are some you will not find since the numbers jump from 12 to 17 and also from 47 to 52. This is because BAs 14, 15, and 16 are cell types found in the insular cortex of nonhuman primates and they are not found in the human cortex. The diagrams thus omit BAs 14, 15, &16. With respect to BA 13, it was identified by Brodmann in monkeys but neuroanatomists today feel it is also found in the human cortex where it acts as a bridge between the lateral and medial areas in the insula. Since the insula is an infolding of cortex, it would not, in any case, be visible in either the lateral (external side view) or the midsagittal view shown in these diagrams. BA 13 is included in the discussion because it is identified as a source using LORETA, and thus is relevant for people doing neurofeedback. With respect to BAs 48 through 51, BA 48 is in the retrosubicular area, a small part of the medial surface of the temporal lobe that is part of the hippocampal region, so it is not visible on the diagrams. BA 49 is found in rodents, and BAs 50 and 51 are found in monkeys. The last number, BA 52, is found in humans. You will find it in the diagrams, in the superior temporal lobe near the junction of the frontal, temporal, and parietal lobes.

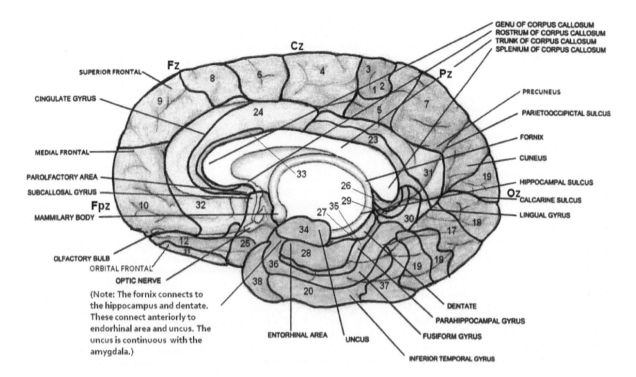

Midsagittal Section to show Lobes of the Brain, Brodmann areas, and 10-20 Electrode Sites
(Drawing by Amanda Reeves, Bojana Knezevic)

NOTE: These first two diagrams, along with brief notes on the main functions associated with each of the Brodmann areas, were first published as a four-page booklet through the International Society for Neurofeedback and Research (see www.isnr.org) in 2007. For quick reference, those booklets are still available on the ISNR website, and proceeds from the sales go to the ISNR Research Foundation.

The collaborative work on the original four-page document that was donated to the ISNR was done by Michael Thompson, M.D. (ADD Centre & Biofeedback Institute of Toronto, Canada), Dr. Wu Wenqing (Friendship Hospital & Capital Medical University of Beijing, China), and James Thompson, Ph.D. (Evoke Neuroscience, New York, NY). These authors developed that document in an attempt to simplify and put in summary form work done by Korbinian Brodmann in the first decade of the twentieth century, combining his original maps of the human cortex with insights concerning functional correlates of the so-called Brodmann areas developed by many others who subsequently studied and added to this knowledge. The work was begun as part of the training of staff at the ADD Centre and was expanded during the time when James Thompson was doing his doctoral work at Pennsylvania State University concerning assessment of concussion in athletes. It was developed further during the six months that neurologist and epileptologist Dr. Wu spent at our ADD Centre while on sabbatical from his positions in Beijing. We hoped the information would be helpful to people who do neurofeedback in a thoughtful way that reflects its neuroanatomical underpinnings. There was a perceived demand for an expanded version of that short booklet and this book has been written to meet that need. We invite you to use the information and give us feedback regarding its utility.

In the Second Edition of *The Neurofeedback Book*, we expand on the discussion of each area and also link the primary functions of each Brodmann area with mention of the networks that involve those functions.

However, it must be kept in mind that each region/lobe of the brain functions in a number of networks that link modules and many Brodmann areas. The individual Brodmann areas do not have unique functions but rather have functions that overlap with adjacent BAs and they may also share functions with other Brodmann areas that are often in distant areas of the brain. In this manner, Brodmann areas are, for the most part, just component parts of complex networks.

Behavioral Interpretations of Intrinsic Connectivity Networks have been investigated in some detail and published by Laird et al. in 2011. Their publication is quite detailed and is based on about 30,000 MRI and PET scans. The reader is encouraged to read Dr. Thatcher's summary of the work by Laird et al. (Thatcher, Biver, & North, 2015). Laird's work defines 18 specific groupings of sites and functions. Their findings do not appear to differ from the functions and networks described in this text but it is a useful addition, providing an addtional perspective, for practitioners who are doing LORETA Z-score neurofeedback.

Primary Functions Related to General Areas of the Cortex

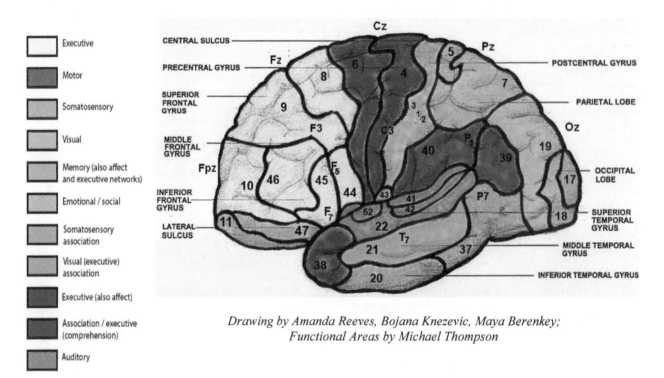

Drawing by Amanda Reeves, Bojana Knezevic, Maya Berenkey;
Functional Areas by Michael Thompson

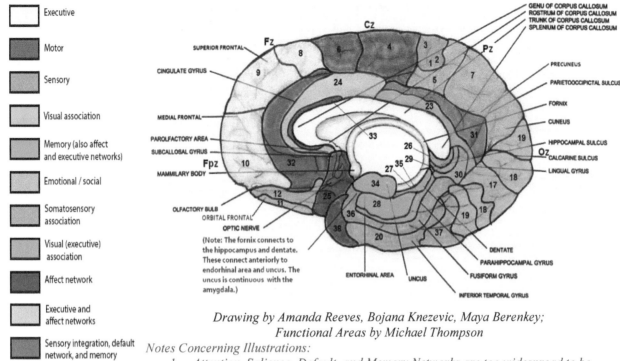

Drawing by Amanda Reeves, Bojana Knezevic, Maya Berenkey;
Functional Areas by Michael Thompson

Notes Concerning Illustrations:
1. Attention, Salience, Default, and Memory Networks are too widespread to be usefully illustrated in this diagram.
2. Brodmann Area 32 is more executive than affect but is purple for color-contrast in diagram.

These two diagrams provide very **broad overviews regarding how Brodmann areas relate to some of the functional networks in the cortex.** It is much too broad an overview and therefore, of course, basically incorrect. For example, memory and emotional regulation clearly must involve many more areas than indicated here. We have not labelled any area as 'attention' because that would include most of the areas that are already labelled! We also made the decision NOT to have four diagrams to show both left and right hemispheres, though there are some differences in functions of the dominant and non-dominant hemispheres that will be mentioned in the text. That kind of detail seemed inappropriate in an overview diagram. We attempted to use general but appropriate terms for areas. For example, the word 'comprehension' for BAs 39 and 40 can be understood, after reading the text, as being primarily language comprehension (Wernicke's area) and other aspects of learning in the dominant hemisphere, while covering nuance, innuendo, emotional, and nonverbal, even spatial reasoning, in the nondominant hemisphere. The finer points will be apparent as you read further: these are just overview diagrams to help the reader get oriented before we move on to more detail.

Thanks Are Extended

Special thanks are extended to Mark Dubin, Dan Lloyd, Richard Soutar, Bob Thatcher, David Kaiser, and Jonathan Walker, who either gave us their direct input concerning Brodmann areas and their functional correlates or whose work on Brodmann areas is available on the Internet. These people were consulted or used as primary references because all have spent considerable time compiling data on this subject. With respect to understanding basic features of neural networks, special thanks are given to Juri Kropotov and his work on event-related potentials, which has helped to delineate networks in the brain. Thanks are given to Jay Gunkelman, Barry Sterman, and David Hagedorn for assistance in the interpretation of EEGs.

Caution

When you read this text, please keep in mind that a Brodmann area (BA) does not (necessarily) correspond precisely to a named gyrus on the cortex. In the following we have tried to give as close a correspondence as possible regarding how specific International 10-20 sites relate to the Brodmann areas on the surface of the cortex. Always keep in mind that there are individual differences and thus variations in the brain's topography across individuals. There are also developmental changes in brain structure, particularly with respect to frontal lobe maturation, which is not complete until about age 25. As more fibers become myelinated and the frontal lobes become larger, there is a slight forward shift in locations. Thus, what is seen at CZ in a child may be better measured at FCZ (half-way between CZ and FZ) in an adult.

Brodmann dissected adult brains, so this section is based on findings in adult brains, though the information can certainly be extrapolated to work with children. Also note that Brodmann himself cautioned about the limitations of his maps due to the fact that labelling the surface gyri does not show the large areas of the cortical surface that are hidden in the infolding of the cortex at each sulcus and fissure.

Lateral view of the Left Cerebral Hemisphere.

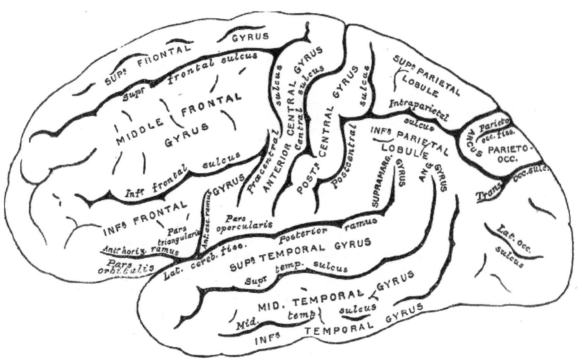

This figure is a Lateral view of the Left Cerebral Hemisphere from Henry Gray, Gray's Anatomy of the Human Body, *1918. This lateral view and the medial view below are shown here to illustrate sulci and gyri that are not labelled in our other diagrams.*
"Gray726." Licensed under Public Domain via Wikimedia Commons -
http://commons.wikimedia.org/wiki/File:Gray726.png#mediaviewer/File:Gray726.png

Midsagittal View of the Left Cerebral Hemisphere

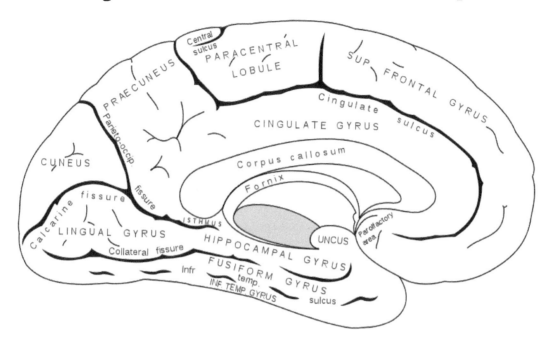

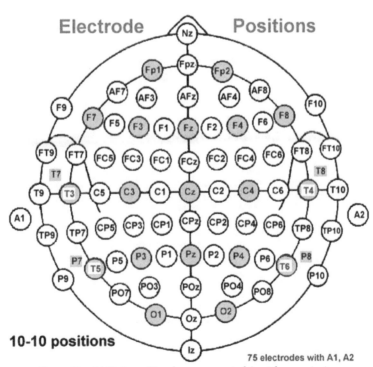

*From David Kaiser (Brodmannarea.info) with permission.
The 19 sites that we will refer to in the following discussion are shaded.*

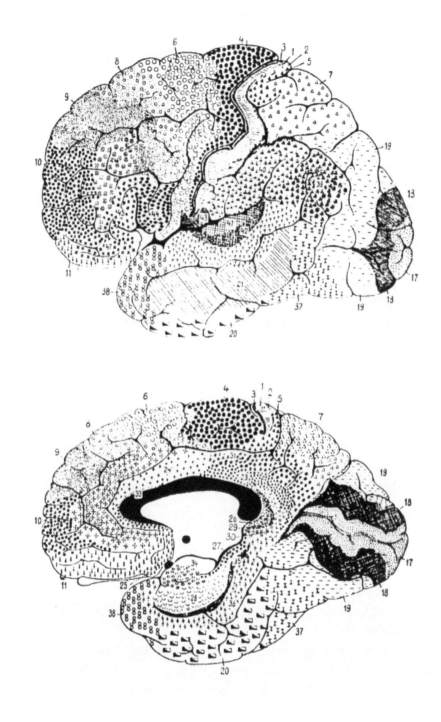

Cytoarchitectonics of human brain according to Brodmann (1909), (public domain)

Korbinian Brodmann said, "One thing must be stressed quite firmly: henceforth functional localization of the cerebral cortex without the lead of anatomy is utterly impossible in man as in animals. In all domains, physiology has its firmest foundations in anatomy."

Dr Korbinian Brodmann, German Neurologist
(November 17, 1868 – August 22, 1918)

Historical Note Concerning Korbinian Brodmann

His work: It is now more than 100 years since Korbinian Brodmann published his work on mapping the mammalian cortex according to different cell types, their spacing and organization. It was while working from 1901-1910 at the Neurobiologishes Laboratorium in Berlin, part of an Institute for Brain Research founded by Oskar and Cecelia Vogt, that he conducted these studies. At that time he also became editor of the *Journal fuer Psychologie und Neurologie,* which survives today as the *Journal of Brain Research.* Between 1903 and 1908 he published his findings delineating areas of the brain according to cell morphology, or cytoarchitecture, in that journal. His best known work from that series is the article about the human cortex published in 1908. In all, he studied 64 different species of mammals. Doubtless inspired by the interest at the time in evolution, he focused on comparative anatomy, developing the idea that there was an old cortex shared across various mammalian species and a more recently evolved neocortex that had developed to a high degree in primates, particularly in humans. In 1909 an expanded version of his work, *Vergleichendes Lokalisationslehre der Grossgehirnrinde*, was published by J.A. Barth in Leipzig. The goal mentioned in the book's preface was to produce a comparative, organic theory of the cerebral cortex based on anatomical features. In 1994 this work was finally translated into English by Laurence Garey with the title *Localisation in the Cerebral Cortex,* but Brodmann maps were being widely used long before the monograph was translated. Brodmann's observations remain relevant today.

His maps were created by doing meticulous dissections and then, using a microscope and a method of staining cells that he learned from Franz Nissl, to examine a slice of cortex. He began with cells in the middle of the lateral side of the brain in the post-central gyrus, calling that area 1, and then proceeded to give a number to each subsequent group of cells that had a different and distinct cytoarchitecture; thus, the morphology, or shape, and organization of the cells differs from one Brodmann area to another. He had become interested in neuroanatomy and its relationship to psychiatric disorders when working for a year (summer 1900 until August 1901) at the Municipal Mental Asylum in Frankfurt, where he met and was greatly inspired by Alois Alzheimer. This mentoring ignited the passion for neuroanatomy that he was to pursue intensively while working as a researcher in Berlin for the next ten years of his life.

His Life: His famous work, which was actually not as well appreciated when he was alive as it came to be later, built upon a steady academic trajectory that began in the local elementary school in Hohenfels, a village north of Lake Constance in southern Germany where he was born. (There is now a Brodmann Museum in the town hall, open the first Monday afternoon each month.) He finished high school at a Gymnasium in Konstanz in 1889 and then pursued medical studies, graduating as a doctor in 1895. He took a position in a clinic in Munich for his final training but, due to coming down with diphtheria and having to convalesce, he did not proceed with his original plan of starting his own medical practice in the Black Forest area. Instead he embarked in 1896 on further studies in Psychiatry, Neurology and Psychology in Berlin and other cities. Final training in Frankfurt that included working with Alzheimer was followed by taking the position in Berlin at the Vogt's brain institute from 1901 – 1910. When the University of Berlin failed to grant him his doctorate in 1910, which was required to become a professor and have a steady income, he left for the University of Tuebingen where he worked from 1910 – 1916 in the anatomical laboratory of the University Clinic. The war disrupted his research and curtailed further publications concerning his work as he began working with neurological patients at the military field hospital in Tuebingen. He was recognized for his services and then, finally, was given a paid position in Halle, where he met and married a much younger woman, Margarite Franz, in April 1917. In 1918, the year their daughter Else was born, he was appointed as the Director of Anatomical Topography at the German Research Institute in Munich. Just as his life was looking happier and more secure, he developed a blood infection, likely from a small cut incurred while doing an autopsy, and he died of sepsis on August 2, 1918. One is left to speculate on what further knowledge he might have added to the field of functional neuroanatomy had he lived into his late eighties like his colleagues Oskar and Cecelia Vogt who died in 1959 and 1962 respectively. Oskar Vogt, in his retirement, wrote a biography of his colleague.

Preliminary Comments about the Organization of this Section

The Brodmann areas (BAs) are discussed in this text by working through **each anatomical area according to the lobes** in which they are located and the related 10-20 electrode placement sites, rather than by listing the Brodmann areas in numerical order. This is because the individual Brodmann areas do not have unique functions but instead have functions that overlap with adjacent BAs and share functions with other Brodmann areas that are often in distant areas of the brain. Each Brodmann area is, for the most part, just one part of one or more complex networks.

Thus, though neuroanatomy is involved, this section is written with the sole goal being to summarize some of the functions within networks that may be expected to respond to neurofeedback (NFB) + biofeedback (BFB) operant conditioning techniques. We decided NOT to merely 'list' functions under each Brodmann area (BA), though there is a summary table at the end. In addition to our previous comments regarding lack of specificity, the lists can be long and there is so much overlap between areas that they lose their clinical usefulness and become overwhelming for anyone without an eidetic memory.

Since this is not an anatomy textbook, but rather a **text written to assist people doing clinical practice where neurofeedback and biofeedback** and, at times, transcranial direct current stimulation (tDCS) are the principle methods of intervention, the organization of the information starts not in the middle, as Brodmann did, but rather with the frontal lobes and electrode sites Fz, F3, and F4. We provide a general overview of the functional significance of cortical areas. Then each Brodmann area for that area of the brain is discussed. The areas are covered in the following order: Frontal Lobes, Central (Cz, Pz), Temporal Lobes, Cingulate Gyrus, Parietal Lobes, and Occipital Lobes.

Things to Keep in Mind
Please remember when reading this explication of functional neuroanatomy with an emphasis on Brodmann areas that each Brodmann area (BA) delineates cortical areas that are differentiated by discrete cell histology and the hypothesis is that each of these areas may have a primary function and also many different functions associated with it. This summary gives rudimentary information associating Brodmann areas with the standard sites of the International 10-20 System of electrode placement. The updated site names are used for the temporal and parietal areas; that is, rather than the old sequence of F7 – T3 – T5, we have used F7 – T7 – P7. In the right hemisphere, the longitudinal sequential placements F8 – T4 – T6 became F8 – T8 – P8. Since most of the databases used in the neurofeedback field were developed before the neurologists changed the nomenclature, you will find both the old and new designations for T3/T7, T4/T8, T5/P7, and T6/P8 used.

The description of Brodmann area functions leads into a discussion concerning neural networks that we posit can be affected using a combination of neurofeedback (NFB) plus biofeedback (BFB).

Neurofeedback Training is Guided by Brodmann Areas and Networks

Brodmann Area Functions Overlap

The primary functions for each Brodmann area (BA) are given in the next section of this monograph. The functions described in the text represent a best guess, based on our clinical observations and on the published and unpublished work of others. Please be advised at the outset that the detail of functions for each Brodmann area is necessarily FALSE. Why? Because all functions emerge from the interaction of many areas and they are not localized to a single area of the brain. This is not a twenty-first century version of phrenology. An expert in Brodmann areas, Dan Lloyd of Trinity College, Hartford, CT, writes, "The **typical BA is differentially engaged in 40% of behavioral (cognitive, perceptual, emotive) domains**" (Lloyd, 2007, personal communication). What lies behind this observation is the fact that **each Brodmann area represents just one area out of many that are involved in one or more neural networks** that involve cortical-subcortical connections; thus each BA becomes involved in a coordinated activity with many other functionally related cortical areas, **depending on what task the brain is doing**.

This may be one reason why practitioners have obtained good results when only doing neurofeedback (NFB) over a single site, such as Cz. That one site lies above BA 4 (primary motor cortex). BA 4 lies above BA 24, the anterior cingulate, which is involved in multiple networks.

Why Use Single-Channel Training?
Training Over Central Midline Structures

Approximately 50% of the EEG amplitude at any single site, such as Cz, arises from neurons directly beneath the electrode, and 95% arises from an area within a 6 cm distance around the site (Thatcher, 2012, quotes Nunez et al., 1981, 1995, 2006). Training at central midline sites such as Cz, Fz, and Pz is hypothesized to **influence key areas**, such as the cingulate gyrus, that are involved in many networks including, but not limited to, the **Executive, Affect, Salience, and Default networks. These** are discussed in more detail in the following section.

A network serves to synchronize the functions of groups of neurons in many different but related sections of the cerebral cortex. The **Attention network,** for example, which is one of the Executive networks, may be affected by training at Cz, and this network **synchronizes the functions of neurons** in the frontal and parietal lobes, anterior cingulate gyrus, hippocampus, the frontal eye fields, and the intraparietal sulcus (Coul and Nobre, 1998). In addition, there appear to be areas that act to switch the brain from one network to another. For example, when the **Attention network is on, the Default network is off** (Sridharan et al., 2008; Fox et al., 2005). The insula is postulated to operate as **a switch** that turns the Default network off and the Attention network on, and vice versa (Sridharan, 2008).

Both the Attention and the Default networks show markedly decreased activity during sleep. The posterior cingulate gyrus shows significant deactivation in both sleep and anesthesia. These examples show how complex the interactions are and how any single BA can be involved in multiple networks at different times.

Doing neurofeedback (NFB) using a referential placement with the active electrode at Cz and the reference electrode on one earlobe, we have seen many case examples where training has apparently had an effect on several networks. This is possibly due to NFB influencing a number of areas within the **anterior cingulate cortex (ACC)**. As noted above, the ACC is a central structure that is involved in a number of neural networks including the Executive (and Attention) network, the Affect (and distress) network, and the Salience network. We have chosen to use the singular for each network but in reality each network is a group of networks.

If the practitioner **combines NFB** for improved attention **with BFB** for relaxation and includes heart rate variability training, then symptoms that relate to anxiety usually decrease. If the practitioner combines the NFB with teaching **metacognitive strategies** ("learning to learn" techniques) and does some training on-task, then an increase in academic performance and in intelligence, as measured by IQ tests, has been found in conjunction with an increase in attention span and concentration (Lubar et al., 1995; Thompson & Thompson, 1998; Thompson & Thompson, 2010).

We hypothesize that we must be affecting complex networks that involve many different cortical and subcortical areas in order to achieve such wide-ranging changes in cognitive and affective functioning with relatively simple single-channel feedback procedures. Indeed, it may be possible that single-channel NFB training, in some patients, could have a **theoretical advantage over training that involves multiple sites**. By having an effect on a network from a single central site, and by only having a direct influence on one site, single-channel NFB may **allow the brain to balance** the degree to which other areas within a network are changed.

NFB at one site avoids making incorrect decisions about the degree to which multiple sites are being 'normalized' when we use Z-score directed LORETA NFB. It could be argued that single-channel NFB might be a more 'balanced' approach to changing the brain, and it does not have the theoretical danger of adversely affecting a site that is outside database norms due to genius or due to compensatory mechanisms. This theoretical dilemma can only be resolved by years of data collection and research.

Why Use LORETA Z-Score NFB?

Nevertheless, with some complex cases, the practitioner may wish to be more precise and try to have an effect on a number of different areas with some of them being distant from surface electrode sites. One may also want to simultaneously train a number of different parameters, such as amplitude, phase and coherence. For these clients we use LORETA Z-score neurofeedback (LNFB).

Importance of Networks

In our work using NFB we are practically always attempting to optimize the performance of neural networks. **Networks are** 'chains' of linked groups of neurons that function together to accomplish a goal. In this context it may be useful to recall what our grandmothers said and our coaches emphasized, "The chain (team, network) is **only as strong as its weakest link**." Thus with cortical dysfunction, our NFB training must either strengthen the performance of that 'link,' or help the network readjust in order to compensate for its dysfunction. The brain has a **plasticity** that should allow this to take place, as has been the case with other interventions, too. (For excellent examples of a range of neuroplastic changes, see Norman Doidge's 2010 book, *The Brain that Changes Itself,* and its 2015 sequel, *The Brain's Way of Healing.*

Cortex-Basal Ganglia-Thalamus: How to Activate One Network – Inhibit Others

In order to have such broad effects, the cortical area to which we are directing our NFB must have a way of connecting to many functionally related, though spatially distant, cortical areas. At the same time, other functionally nonrelevant areas of the cortex must be inhibited. In this manner, many functionally related cortical areas are synchronized and recruited to operate as a specific network to accomplish the task at hand.

The brain is constantly working to reduce uncertainty and make the environment both consistent and predictable. It is finding meaning in information, making patterns and associations. It is continuously analyzing and reanalyzing information in order to understand it in a personally relevant manner. Thus, when we do an assessment, whether it be eyes closed, eyes open, or on a specific task, the brain is always working. There is no true 'resting' state. The brain is continuously activating specific networks and the Default network comes into play when a person is awake but not focused on a particular task. With activation of relevant networks happening, how does inhibition of pyramidal cells and then of systems of whole nonrelevant networks come about?

Remarks Concerning Inhibition in the Cortex

Local inhibition of specific pyramidal cells is done immediately after a pyramidal cell fires by basket cells that are close to the pyramidal cell. Basket cell firing may have other important functions, and it has been noted that gamma synchronization of basket cell inhibition of pyramidal cells may be one key for understanding some aspects of cerebral dynamics. This, however, is not a subject for a basic textbook and it will not be discussed here. What, then, is one possible mechanism that could be responsible for activating a single network while inhibiting often widely distributed nonrelevant cortical areas in order to focus attention and action on a single function to accomplish the task-at-hand? An inhibitory mechanism that is available, which can affect widely distant cortical areas, involves links from the cortex to subcortical structures that are broadly labelled the 'basal ganglia.' These connections, from the cortex to the basal ganglia and then back to the cortex via the thalamus, can focus action while simultaneously inhibiting areas that are not necessary for a particular task.

The Basal Ganglia

The structures that comprise the basal ganglia include the dorsal striatum (putamen and caudate), subthalamic nucleus, the substantia nigra (pars compacta (SNc) that produces dopamine and pars reticulata (SNr) that has functions similar to the pallidum), and a limbic sector that includes the nucleus accumbens (ventral striatum), ventral pallidum, and ventral tegmental area (VTA). VTA efferents provide dopamine to the nucleus accumbens (ventral striatum) in the same way that the substantia nigra provides dopamine to the dorsal striatum and the globus pallidus. You will also hear another term 'lentiform nucleus,' which means shaped like a lens. The lentiform nucleus comprises the putamen laterally, globus pallidus medially and the innominate substance, which contains the anterior perforated area, inferiorly.

Figures to Help Visualize these Structures

Several figures are given below to remind the reader of the locations of some of these structures. Four of these figures have been taken from other parts of *The Neurofeedback Book* to help you see the relationship between the basal ganglia, the thalamus, and the cortex. Note when reading the text and looking at the figures that 'dorsal' is towards the top of the head, but in the brainstem and spine, it is at the back. 'Ventral' is towards the base of the skull but, in the spine, it would be towards the front of the body. Lateral means towards the side and medial is towards the middle of the head. Thus a medial or midsagittal section would slice down the vertical plane in the middle of the head from anterior (front/forehead) to posterior (back/occiput).

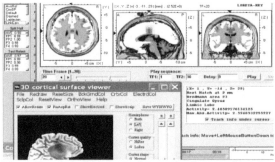

The 'slices' shown along the top in the graphic above are LORETA images: Horizontal, Sagittal and Coronal respectively. In the bottom left hand corner is placed the three dimensional sagittal slice from the LORETA program.

This figure, taken from an analysis done with NeuroGuide, shows the 'slices' of the brain as they are seen using LORETA. It shows a LORETA source correlation in Brodmann area (BA) 23, cingulate gyrus. The activity was 2.5 standard deviations above the database mean using the NeuroGuide program. This finding reflected excessive amplitude of 20 Hz activity in a 42-year-old woman who had anxiety symptoms (affect network).

Note that LORETA images, though they look like MRI scans, are mathematically derived from surface EEG data. For cortical sites, there is correspondence between LORETA source localization and MRI findings. LORETA however, does not reveal subcortical sites.

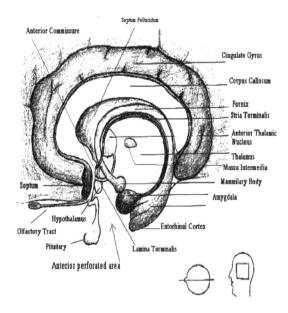

The above figure is a midsagittal section that shows the position of the thalamus below the cingulate gyrus and the corpus callosum.

The figure below is a coronal, also called transverse, section that shows the relationship between the putamen, the globus pallidus, and the thalamus as you move medially from the right lateral aspect of the cortex (an infolding of cortex called the insula) towards the centre of the brain where you see the third ventricle. The same structures are mirrored in the left hemisphere.

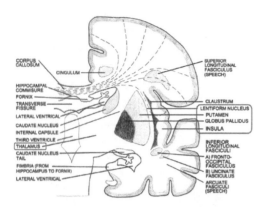

Schematic diagram of a transverse section through the right cerebral hemisphere (Amanda Reeves, after Smith 1962).

The figure below shows these same structures plus nuclei that influence them. The basal ganglia and the cortex are directly influenced by the substantia nigra, which produces dopamine. The following diagram includes the basal ganglia, the thalamus, and the substantia nigra.

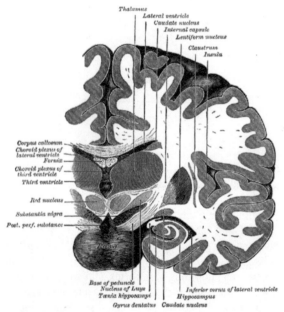

Gray's Anatomy *(public domain). Schematic Diagram of a Transverse Section Through the Right Cerebral Hemisphere and midline structures (after Smith 1962) to show the Red Nucleus and the Basal Ganglia and the Substantia Nigra (Nucleus of Luys is more often called the subthalamus).*

The following figure shows the thalamus and a broad general overview of its projections to different cortical areas. All sensory information except smell passes through the thalamus before going on to the cortex. The thalamus has two lobes (a left thalamus and a right thalamus) connected, in about 85% of people, by the *massa intermedia* (shown in a previous figure), which goes through the third ventricle. The nuclei within the thalamus project to specific areas of the cerebral cortex, as seen in the next diagram.

Note: The temporal lobe (at top of figure) has been 'folded up' to show the auditory area (red).

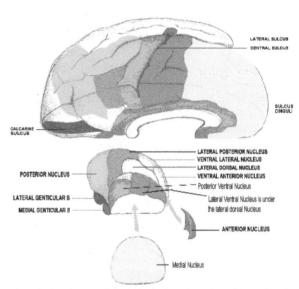

The thalamic nuclei and their functionally related cortical areas (adapted from Smith, 1962)

The above figures are **included to assist the reader, when we outline cortex-basal ganglia-thalamus-cortex loops,** to picture how influencing a single area of the cortex may have the effect of activating one neural network and simultaneously inhibiting others. For example, feedback at Cz might affect BA 24 within the anterior cingulate gyrus, and LORETA NFB might affect this area even more directly. As we will see in greater detail below, this area connects to the ventral striatum in the basal ganglia. Lateral inhibition within the striatum will select one specific network to become active while other networks, and therefore cortical areas that were less relevant to a specific function at that moment in time, are inhibited.

In basic overview terms, the putamen inhibits the globus pallidus (GP) (pallidum). The GP fires at a very high frequency inhibiting the thalamus. Each of these three structures can be thought of as being 'functional maps' of the cortex. Thus if, for example, a motor area of the cortex activates a specific functional area of the putamen then that area will inhibit a specific functionally related area of the GP. This inhibition stops the GP high rate inhibitory firing to a corresponding specific area of the thalamus. Since the thalamus is connected to all the areas of the cortex, this means that only that one functional gate is opened while the others remain inhibited. The result is that areas of the cortex that are functionally related areas to this open gate (a network) are activated while other networks within the cortex remain inhibited.

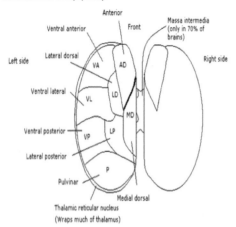
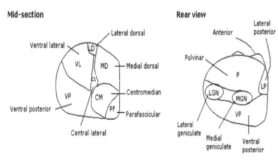

Nuclei of the Thalamus
Permission is granted to copy, distribute, and/or modify this document under the terms of the GNU Free Documentation License, Wikipedia.

Other diagrams can be found on the internet under *Human Neuroanatomy: An Introduction.* James R. Augustine. (2008) Elsevier.

SECTION VII
Functional Networks and Behavior
Frontal-Subcortical Circuits
(Special thanks to Tammy Binder, M.D. for her complete revision of this section)

Five Examples of Cortical – Basal Ganglia Circuits

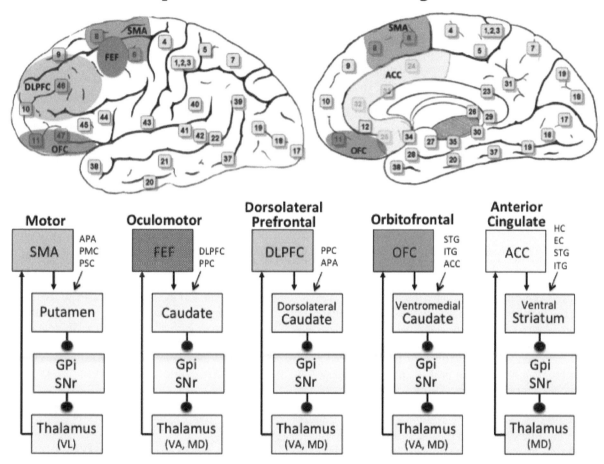

Circuits adapted from Alexander et. al. (1986). Colored boxes at the top of each circuit represent cortical areas from left to right: SMA (supplementary motor area); FEF (frontal eye field); DLPFC (dorsolateral prefrontal cortex); OFC (orbitofrontal cortex); ACC (anterior cingulate cortex). Uncolored boxes represent subcortical structures: GPi (internal segment of globus pallidus); MD (medial dorsal nucleus of thalamus); SNr (substantia nigra, pars reticulata); VA (ventral anterior nucleus of thalamus); VL (ventral lateral nucleus of thalamus). Abbreviations used outside of boxes represent 'open loop' contributions to the circuit and likely provide 'context' to the striatum: APA (arcuate premotor area), EC (entorhinal cortex), HC (hippocampal cortex), ITG (inferior temporal gyrus), PPC (posterior parietal cortex), PMC (primary motor cortex), PSC (primary somatosensory cortex), and STG (superior temporal gyrus). Diagram of Brodmann areas modified from the 20th U.S. edition of Gray's Anatomy of the Human Body, available on Wikipedia.

The striatum is a relatively quiet structure. By contrast, the internal segment of the globus pallidus (GPi) and the substantia nigra pars reticulata (SNr) have high baseline rates of activity that provide tonic inhibitory drives to specific nuclei of the thalamus.

The five circuits shown in the diagram remain remarkably anatomically segregated as they traverse the subcortical structures. They are named according to their cortical area or function.

The frontal lobes have long been considered the centre of 'executive function,' which Kropotov (2009) defines as "the coordination and control of motor and cognitive actions to attain specific goals." Other components of executive function include: voluntary control of attention, inhibition of inappropriate and/or unwanted behavior, planning, decision making, working memory, monitoring, and the use of feedback to improve performance.

As shown above, on the basis of neuroanatomy and function, Alexander et al. (1986) described five parallel frontal-subcortical circuits. Each circuit has the same general structure: a specific area of frontal cortex projects to specific areas of basal ganglia, then thalamus, before returning to the originating region in frontal cortex and to functionally related areas.

In his book *The Frontal Lobes and Voluntary Action,* Richard Passingham (p. 220) proposes that the frontal-basal ganglia system as a whole is involved in the process of deciding "what to do...what response is appropriate."

'Appropriate' behaviors often require that deliberative, planned actions occur rather than more reactive, automatic or over-learned actions that are usually faster to execute. Consider, for example, a quarterback who calls for a short pass play that he and his tight end have practiced many times. Now consider the possibility that just as he is about to make the pass he notices that his star wide receiver has just broken free of the defense and is wide open, halfway down the field. In the current context, the quarterback suddenly needs to inhibit the 'preloaded' motor program for the short pass in order to plan and execute the long pass. How does the quarterback do this?

Our brains often prepare competing plans of action in parallel. In the case described above, the motor plan for the long pass needs to compete with and win over the faster, preloaded and over-learned short pass motor plan. In order for this to occur, the brain must be able to inhibit all potential movement plans until the best plan can be chosen given the current goals and circumstances. The brain achieves this by creating programmable movement gates. These gates help prevent the fastest plan (in this case the short pass) from always 'winning.' Gating of movement until a single winning plan emerges is also important to prevent 'blended' movements in which two or more motor plans are performed at the same time. In the case of the quarterback, this would result in the ball being thrown somewhere in between the two receivers, which is clearly not a desirable outcome. The organization of multiple parallel circuits through the basal ganglia, together with basal ganglia inhibition of the thalamus, serves to create a large number of programmable gates. Note that baseline firing rates of GPi (the internal segment of the globus pallidus) and SNr (substantia nigra, pars reticulata) are high, resulting in tonic inhibition of thalamic neurons. In other words, at baseline, the gates are closed.

Now imagine cortical plans of action sending excitatory projections to the striatum. Each plan activates striatal neurons in its own segregated circuit. These excited striatal neurons inhibit striatal neurons in other circuits (activated by alternative plans of action) by lateral inhibition (either directly or mediated by inhibitory interneurons) and, in this way, different plans of action can compete. At the same time, the active 'winning' plan's striatal neurons also inhibit the tonically active GPi neurons within its own circuit, causing a pause in their activity. This pause in the activity of GPi neurons releases the thalamic cells in the same circuit from inhibition. The transient excitation (i.e. disinhibition) of the thalamic cells provides an excitatory signal back to the area of frontal cortex that generated the winning plan – a kind of 'go' signal – allowing the plan to execute.

Please note that the above discussion is a simplification. For example, neurons that receive 'go' signals from the thalamus to execute a particular plan are in the same cortical area but in a different cortical *layer* than the neurons that generated the plan. (Recall that the cortex of a mammal has six layers.) For more details, refer to Brown et al. (2004).

Let's now consider each circuit in turn:

Motor Circuit

The **motor circuit** is involved in the planning, execution and inhibition of voluntary movements of the body. Disruptions at any level in this circuit lead to disorders of motor control such as those seen in clinical conditions like Parkinson's disease, where basal ganglia dysfunction results in excessive inhibition of movement called bradykinesia (slowed movement), which can be thought of as difficulty in opening movement gates, as well as inappropriate opening of movement gates, which causes tremor.

Oculomotor Circuit

The **oculomotor circuit** is involved in the planning

and execution of voluntary eye movements. Disruptions at any level in this circuit interferes with the ability to voluntarily choose to look at a particular object or location while resisting the natural tendency to make eye movements to look at compelling distracters or competing objects. This circuit is also necessary to make eye movements to remembered locations.

Dorsolateral Prefrontal Cortex (Executive) Circuit

The **dorsolateral prefrontal cortex** is involved in many aspects of executive function including the ability to solve complex problems, plan ahead, focus and sustain attention, regulate actions according to task demands, and shift gears when task demands change. It also plays an important role in working memory: the ability to keep things in mind long enough to direct action, such as remembering a new phone number long enough to dial it. Patients with a disruption at any level of the dorsolateral prefrontal circuitry exhibit a classic 'dysexecutive' syndrome characterized by being concrete, perseverative, inattentive and disorganized, with poor memory search strategies, impaired reasoning and reduced mental flexibility (Tekin and Cummings, 2002). They are particularly impaired on the Stroop task in which a person must override the more automatic skill of reading words in order to name the color of the ink that the word is printed in. This becomes challenging because there is a conflict between the word itself and the color it is printed in. For example, when a subject sees the word 'RED' printed in large block letters in blue ink, the correct response is to say 'blue.' This requires the ability to suppress – or gate – the strong tendency to simply read the word, that is, to say 'red.'

Orbitofrontal Cortex (Social) Circuit

The **orbitofrontal cortex** "is the neocortical representation of the limbic system" (Bonelli and Cummings, 2007). The **orbitofrontal circuit** mediates empathic and socially appropriate behavior (Chow and Cummings, 1999) and may select objects and actions based on their subjective value (Dranias, 2008). Damage to this circuit results in behavioral disinhibition, emotional lability, poor judgment, and irresponsibility toward family and social obligations (Bonelli and Cummings, 2007). For example, a patient with damage to this area may not be able to apply social norms to prevent helping themselves to food from someone else's plate if they are hungry and the food is appealing. That is, the value of socially appropriate behavior is no longer biasing behavior, so more reflexive responses reflecting more basic drives (in this case eating when hungry) drives behavior.

Anterior Cingulate Cortex (Affect) Circuit

The **anterior cingulate cortex (ACC)** has different functional zones (Nee et al., 2011) that can be seen on the diagram of Central Midline Structures that appears at the beginning of the next section.

i) **Pre-** and **subgenual ACC (PACC)** is thought to be linked to emotional networks that activate when an error in performance has occurred. Brodmann area 25 is a part of PACC that has received particular attention by virtue of its consistent overactivity in depression. Remarkable remissions of severe, treatment-resistant depression have occurred when this abnormally high activity is disrupted by deep brain stimulation (DBS) (Mayberg et al., 2005; Holtzheimer and Mayberg, 2011). It is as if DBS to BA 25 'closes the gate' to overwhelming negative mood states. (See Dobbs, 2006, for a very readable history behind this intervention.) Neurofeedback combined with psychotherapy has also been shown to influence BA 25, and also other areas linked to depression (Paquette, 2009).

ii) A more dorsal area of ACC called the rostral cingulate zone (RCZ) or **supragenual ACC (SACC)** has an anterior portion involved in monitoring cognition to predict errors that suggest increased cognitive control is required (Brown and Braver, 2005). Such requisite increased vigilance or arousal might then be achieved through ACC projections to the brainstem **locus coeruleus (LC)** (Aston-Jones and Cohen, 2005) that, in turn, sends noradrenaline (NE) back to many areas of cortex. The increased noradrenaline available to cortical neurons is believed to sharpen their responsiveness to input and the widespread simultaneous changes in cortical neuron activity modulates cognitive performance. There is an inverted U-shaped relationship between tonic LC activity and performance on tasks that require focused attention. This means that poor performance occurs at both low (underaroused) and high (anxious) levels of tonic LC activity. Optimal performance occurs at moderate tonic LC activity, which allows for the largest phasic LC activations in response to goal-relevant stimuli (Aston-Jones and Cohen, 2005; please also see Sara and Bouret, 2012, for a recent review of LC effects on cognition). It is tempting to speculate that ACC/LC interactions play an important role in learning to self-regulate arousal and attention to meet task demands.

Large bilateral lesions to the ACC lead to akinetic mutism, which is seen as an awake but underaroused state of "profound apathy" (Bonelli and Cummings, 2007). Such patients rarely move or talk spontaneously and have limited or no response, even to direct questions and requests. They are indifferent to pain, thirst, or hunger. Thus neurosurgical lesions have occasionally been deliberately placed in the ACC of patients with severe, intractable pain. People with damage to the ACC circuit have also been reported to experience "a diminished ability to conceive new thoughts or participate in creative thought processes" (Chow and Cummings, 1999).

iii) The **posterior part of SACC** is more connected to motor networks and may be involved when uncertainty or conflict make appropriate response selection challenging (Nee et al., 2011). For example, in the go/no-go task, subjects may be asked to push a button (a 'go' trial) when they see the letter 'A' but to inhibit their response (a 'no-go' trial) when they see the letter 'B.' Usually subjects are exposed to more 'go' trials and develop a predisposition to respond (i.e. push the button) that must be inhibited in order to perform correctly on subsequent 'no-go' trials. In addition to the apathy syndrome described above, patients with ACC lesions typically perform poorly on this task.

Go vs. No-Go: Direct, Indirect and Hyperdirect Pathways

Note that the circuit diagrams above and much of the discussion so far has focused on the 'direct' pathway through the basal ganglia and how this generates 'go' signals. There is also an 'indirect' or 'no-go' pathway whose neurons inhibit direct pathway neurons (within the striatum) and also inhibit tonically active neurons of the external segment of the globus pallidus (GPe). The GPe tonically inhibits several structures but only its output to the **subthalamic nucleus (STN)** is illustrated below. Activation of the indirect path can therefore lead to increased activity in the STN (via disinhibition) while the 'hyperdirect' pathway activates the STN directly.

Increased STN activity can generate 'no-go' signals by sending glutaminergic *excitatory* inputs to the neurons of the GPi/SNr, thereby **enhancing thalamic inhibition**. Recall that the 'direct' path has the opposite effect on GPi/SNr: cortical signals transmit through the striatum to transiently *inhibit* GPi/SNr cells, resulting in transient disinhibition of the thalamus. The opposing effects of direct 'go' and indirect and hyperdirect 'no-go' pathways on the GPi/SNr are shown below.

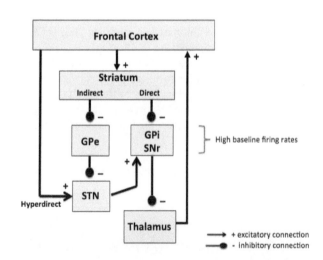

*Simplified frontal-subcortical circuits illustrating selected aspects of the direct, indirect, and hyperdirect paths. Activation of striatal neurons in the '**direct**' pathway leads to inhibition of neurons in the internal segment of the globus pallidus (GPi) and substantia nigra pars reticulata (SNr). Activation of striatal neurons in the '**indirect**' path leads to inhibition of tonic activity in the external segment of the globus pallidus (GPe). This inhibition of GPe disinhibits the subthalamic nucleus (STN), which then provides an excitatory drive to the GPi/SNr. The frontal cortex can also excite STN cells directly via the '**hyperdirect**' pathway.*

By adding extra excitatory drive to the GPi/SNr neurons, high STN activity opposes and sometimes even overcomes the direct pathway's inhibition of the GPi/SNr. In this way, STN activity can slow down, and may even prevent the direct pathway's disinhibition of the thalamus, thereby disrupting the transmission of a 'go' signal back to the cortex.

Indeed, abnormally high STN activity is seen in Parkinson's disease (PD) and contributes to the slowed movement that is seen in this disease. In fact, knowledge of this circuitry led to the development of neurosurgical techniques, including deep brain stimulation (DBS) of the STN, to specifically disrupt these pathologically strong 'no-go' signals in PD.

In normal functioning, 'no-go' signals are important to interrupt behavior when new information, such as a new goal or target, becomes available. Consider once again the example of the quarterback who has called for a short pass and is in the beginning stages of executing the motor program to make this pass. This means that the short pass motor program has already 'won' in the direct path of the striatum. However, if at the last moment he notices a wide receiver with excellent field position, 'no-go' mechanisms can intervene and interrupt the already selected motor program.

One 'no-go' mechanism that may be particularly relevant to behavioral interruption and task switching involves strong projections from the thalamus to the striatum. Neurons of the centromedian-parafascicular complex (CM/Pf) of the thalamus are activated by unexpected salient stimuli, and project to striatal cholinergic interneurons. Activation of these striatal cholinergic interneurons (known as tonically active neurons or TANs) transiently produces increased outflow of the indirect pathway and decreased outflow of direct pathway. This shift in processing to favor the indirect path's output results in the interruption of the current motor program and transient movement suppression (Minamimoto, 2008; Ding, 2010; Tan and Bullock, 2008, Smith, 2011). Following this, a new competition in the direct pathway of the striatum can be initiated. In this new striatal competition, the high reward possibility of the long pass can outcompete the short pass, 'win' and be executed.

In addition to (i) task switching and (ii) interruption of ongoing behavior, indirect and hyperdirect pathways are thought to play important roles in (iii) delaying voluntary action, (iv) locking out competing responses to allow a selected program to fully execute, (v) inhibiting motor programs once they have been executed in order to prevent perseveration, and (vi) shifting to the next motor program in a smooth sequence of action. These last two roles are explicitly incorporated in a computational model of speech production that, when 'lesioned' (by abnormally high levels of dopamine in the striatum or by specific and localized white matter abnormalities), reproduces many features of stuttering (Civier et. al, 2013).

Shifts in the relative activation of direct, indirect, and hyperdirect pathways are also thought to affect speed vs. accuracy trade-offs, a long-observed phenomenon in behavior whereby forcing quicker decisions can lead to a loss of accuracy, whereas emphasizing accuracy in decision-making can lead to slower responses (see Bogacz, 2010, for review).

Note that increased dopamine in the striatum increases activity in the direct path (via excitatory D1 receptors on striatal neurons in the direct path) and decreases activity in the indirect path (via inhibitory D2 receptors on striatal neurons in the indirect path). The net effect favors direct over indirect pathways. In this way, dopamine release into the striatum can speed performance. But too much dopamine can also produce aberrant performance. (See Civier, 2013, for an example.)

Mechanisms that favor the indirect and hyperdirect pathways over the direct pathway have the opposite effect and can slow performance. Consider the example of increased activation of GPi/SNr by the STN. The more active GPi/SNr neurons would then require stronger-than-typical outputs from the direct pathway to become inhibited (in order to disinhibit the thalamus and generate a 'go' signal back to cortex). The extra time it takes to generate this stronger-than-typical direct pathway output results in slower performance. At the same time, this delay gives selection mechanisms within the striatum more time to operate and allows later arriving or initially weaker inputs a better chance to compete and grow in activity than they otherwise would have, if quick responses were demanded.

Consider once again the Stroop task, where faster reading of words competes with the slower process of naming the color of the ink that the word is printed in, especially when there is a conflict between the word itself and the color of its ink. When we view the word 'RED' printed in large block letters in blue ink, two potential verbal responses are prepared in cortex and send competing bids to striatum. Because the pathway for reading words is overlearned, it is faster, and so the verbal plan for 'red' is generated more quickly, going on to quickly activate striatal output neurons in the direct path. If at this point these striatal neurons are active enough to inhibit GPi/SNr activity, then a 'go' signal will be sent back to cortex and an incorrect response to the task will have been made since the person will say 'red.'

On the other hand, if STN excitatory drive increases GPi/SNr activity, then a higher activity level is required of striatal neurons to generate a 'go' signal; the initial activity level of striatal neurons in the 'red' direct pathway won't be sufficient. In this way the higher GPi/SNr activity level has set a higher 'response threshold' that prevents the impulsive response 'red,' and gives the later-arriving striatal bid from the verbal plan for 'blue' a chance to compete.

Note that the task demand – in this case the instruction to report the color – is likely encoded in lateral prefrontal cortex. As long as this task demand is kept 'in mind' and active, it can provide contextual signals to bias cortical and striatal competitions towards the correct 'blue' verbal response. To summarize, correct performance on the Stroop task requires at least three elements: a high enough response threshold to prevent fast but incorrect responses, the ability to keep task demands in mind, and enough time to allow task demands to help grow the slow but correct response to a high enough activity level to win cortical and striatal competitions.

Recall from the previous description of the anterior cingulate cortex (ACC) that it is involved in performance monitoring and predicting when errors are likely to occur. These roles led Frank (2006) to propose that ACC inputs to STN could be a mechanism for increasing response thresholds when errors are more likely, resulting, as we have seen, in slower but more accurate responses. While this hyperdirect pathway hypothesis is currently an influential model, other 'no-go' mechanisms such as the previously described centromedian-parafascicular complex (CM/Pf) thalamostriatal pathway may prove to be even more relevant.

Relevance of These Pathways to SMR Production

At this point, it is interesting to consider the possibility that the sensorimotor rhythm (SMR) arises from a basal ganglia 'no-go' state and that SMR training enhances the ability to voluntarily shift into this 'no-go' state. This is consistent with findings (see Sterman and Thompson, 2014, for review) that SMR is associated with immobility and was originally observed when cats needed to inhibit a previously rewarded response. The finding by Boulay et al. (2011) that reaction time tracked SMR production in a go/no-go task (i.e. longer reaction times during high SMR production and shorter reaction times during low SMR production) are exactly what would be predicted by 'no-go' and 'go' modes of basal ganglia functioning respectively. The proposal that SMR is produced by a 'no-go' state is also consistent with the observation of increased fMRI activity in the striatum during SMR production (Birbaumer, unpublished results reported in Sterman and Thompson, 2014), and after successful SMR training, while performing a Stroop task (Levesque et al., 2005). Because the striatum is known as a 'silent structure' due to the low percentage of neurons active at any time, it is likely that these fMRI findings reflect 'no-go' inhibitory mechanisms, which are more likely to be synchronous and extend over larger areas (Bullock et al., 2009) compared to more focal 'go' pathways.

The view that SMR training might actually be striatal 'no-go' training is also consistent with what is already known about mechanisms underlying SMR generation. The rhythm itself arises from interactions between two populations of neurons within the thalamus: inhibitory neurons in the thalamic reticular nucleus (nRT) and excitatory thalamocortical neurons in the ventrobasal (VB) complex.

A crucial property of thalamic neurons that underlies their role in generating a wide variety of cortical rhythms is that, when they are inhibited strongly enough, they move from tonic activity to a bursting mode of firing. In the case of SMR, strong enough inhibition of neurons in VB causes them to become hyperpolarized. They recover from hyperpolarization with a burst of activity that excites nearby nRT neurons. The excited nRT neurons then reciprocate by inhibiting the VB neurons that excited them. This will once again hyperpolarize the VB neurons and the cycle will repeat. In this way, alternating bursts of activity continue between these two thalamic nuclei. Because VB also sends excitatory projections to primary somatosensory cortex (S1), the oscillatory activity in VB results in oscillatory activity in S1, detectable by EEG.

The thalamocortical oscillations that underlie SMR have been known for some time. What has been less clear is what might initiate the process during voluntary SMR production. Given that behavioral correlates of SMR are consistent with 'no-go' activity in the basal ganglia, it is interesting to note that GPe, the main output of the 'no go' indirect path, projects to nRT (Bullock, 2009). At baseline, GPe provides tonic inhibition to nRT. However, in the 'no-go' state, the indirect pathway's output would inhibit GPe neurons, thereby *disinhibiting* neurons in the nRT. The resulting increased activity in nRT could then initiate the oscillatory process described above (i.e., by inhibiting VB neurons enough to trigger their bursting.)

Same Frontal-Subcortical Circuits, Emphasis on Open-Loop Integration

Understanding how the brain learns to select and enact the most adaptive behavioral plan given the circumstances will certainly involve understanding the flow of activity through frontal-subcortical circuitry. Potential cognitive and other action plans are generated in frontal areas that are part of the closed loops that are emphasized in the original circuit diagrams shown previously. The loop is 'closed' when thalamic projections carry the 'go' signal back to the originating frontal area. The 'open loop' contributions to each circuit can be thought of as carrying contextual information. These circuits are sometimes drawn with equal emphasis on the closed and open loops to highlight how the basal ganglia can coordinate wide areas of cortex in the service of a particular task.

A Motor Network

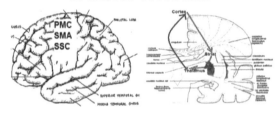

Motor plans from supplementary motor cortex join with motor state information from primary motor and somatosensory cortex in the putamen, then 'winner' outputs to internal GP/substantia nigra, pars reticulata, then to ventrolateral and ventroanterior thalamus, then back to supplementary motor cortex.

For example, in the motor network, note that supplementary motor cortical plans (BAs 6,8) join with information on the current state of the motor system provided by arcuate premotor (BA 8), primary motor (BA 4) and sensorimotor (BAs 3,1,2) cortices to select the best action, taking into account the current motor state. Neurons from all these functionally related cortical areas will project to partially overlapping populations of **striatal medium spiny neurons** (**MSNs**) in the putamen. Here it is important to recognize that there is a clear funneling of inputs from a large number of cortical neurons to a much smaller number of MSNs. In fact, each MSN is estimated to receive approximately 10,000 afferent inputs from different areas of cortex (Lawrence, 1998).

The forced intermingling of inputs from different cortical areas on the same MSN makes MSNs well suited to be pattern classifiers that can come to learn – through previous reinforcement learning – what plan, given the detailed information about the current context (from both closed- and open-loop areas) is most likely to be successful and rewarded in the present moment. The MSN that receives the best *combination* of plan, context, and goals will be the most active, resulting in the execution of the plan that activated it (by the flow in the circuit through GPi and thalamus as described previously). At the same time, its high activity will lead to the inhibition of other MSNs (through lateral inhibition), ensuring that competing plans that are not as well suited to the current situation are not performed.

A Spatial Network

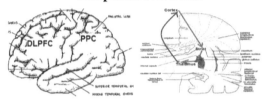

Spatial information goes from posterior parietal and dorsolateral prefrontal to head of caudate to internal GP/substantia nigra, pars reticulata, then to ventro-anterior and mediodorsal thalamus, back to cortex.

Lawrence et al. (1998) modified the original circuits in a way that highlights distinct parts of the lateral prefrontal cortex. The more dorsal part, referred to as **dorsolateral prefrontal cortex** (**DLPFC**) consists of BA 9, and the dorsal aspects of BAs 10 and 46. The DLPFC circuit receives spatial information from **posterior parietal cortex** (**PPC** BA 7) and is known to be involved in spatial working memory, such as the ability to hold a particular location in mind when the cue for that location disappears. Note that PPC is part of the 'where' visual stream. (The 'what' visual stream is more ventral and includes the ventro-lateral prestriate cortex (parts of BAs 18, 19) and the inferior temporal cortex.)

More ventral parts of BAs 10 and 46 are referred to as **ventrolateral prefrontal cortex** (**VLPFC**). This circuit receives object information from **inferior** and **superior temporal gyri** (**IT** BA 20 and **ST** BA 22), and is involved in working memory for objects. Note that this is part of the 'what' visual stream involved in object recognition.

A Visual Network

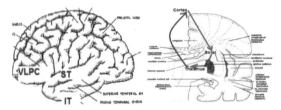

Visual information from inferior and superior temporal joins ventrolateral *prefrontal inputs in* tail *of caudate, then 'winner' outputs to internal GP/substantia nigra, pars reticulata, then to ventroanterior and mediodorsal thalamus, then back to VLPC.*

Lawrence et al. (1998) also modified the original circuits by separating the lateral prefrontal cortex into the VLPFC described above and the **orbitofrontal cortex (OFC)**. They then combined orbitofrontal and **anterior cingulate** cortices into one 'Affect Network' and added known, important contributions from the **amygdala** to the **hippocampal** and **entorhinal** inputs to this network.

Affect Network

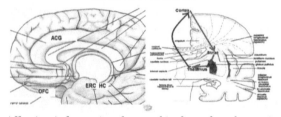

Affective information from orbitofrontal and anterior cingulate joins information from hippocampus, entorhinal and amygdala in nucleus accumbens, then 'winner' outputs to ventral pallidum to medial dorsal thalamus, then back to orbitofrontal and anterior cingulate to regulate mood and emotional reactions.

Drawings by Amanda Reeves, modified from Kropotov, 2009 and Lawrence et al., 1998.

Affect Network: Affect Information Flow

In simplest terms, information related to 'affect' or emotion may traverse from the orbital frontal cortex (OFC), medial frontal cortex, anterior cingulate gyrus (ACG), hippocampus (HC), amygdala, and entorhinal cortex (ERC)/uncus area to the basal ganglia, including the nucleus accumbens and ventral pallidum. From there, signals go to specific functionally related areas of the thalamus, such as the medial dorsal and anterior nuclei of the thalamus. The thalamus then connects back to areas of the cortex including the anterior cingulate, which has functions related to control of the Affect network. The result is regulation of mood and emotional reactions (after Kropotov, 2009).

The orbital and medial prefrontal cortices as well as the amygdala are key areas for understanding anxiety (Davidson, 2002; Thayer, 2012), as is the anterior cingulate cortex (Matthews et al., 2004). The anterior cingulate involvement in depression was described earlier.

Also described earlier in the separate sections on the orbitofrontal and anterior cingulate circuits, other functions of this network include: 1) the selection of goals (objects and actions) on the basis of their subjective value, and 2) the monitoring of error likelihood in order to increase arousal and vigilance to better match challenging task demands.

Concluding Remarks

In summary, when cortical-basal ganglia-thalamocortical circuits are functioning properly, the best cognitive or behavioral plan, given the current context, previous learning, and current goals, will be selected with the aid of the basal ganglia. As we have seen, the basal ganglia coordinate wide areas of cortex into functional networks. When not operating properly, the current context and goals have less influence, and overlearned, reactive responses may dominate. 'Gating' mechanisms slow down prepotent responses to allow potentially better plans a chance to be selected. Complex interactions within the basal ganglia together with the structure of 'go' and 'no-go' pathways create many opportunities to select, interrupt, and otherwise control behavior in real time.

Central Midline Structures (CMS): An Anatomical and Functional Unit

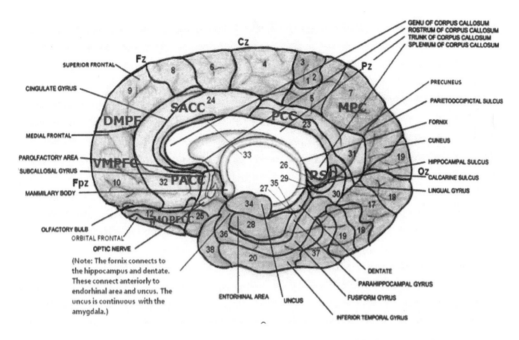

Central midline structures in BOLD (after Northoff, 2006). We superimposed bold letter abbreviations in the above figure to indicate areas that are considered to be important central midline structures. We suggest that one should also include in discussions of CMS, in both hemispheres, the hippocampus, the entorhinal-uncus area, and the insula.

SACC = supragenual anterior cingulate cortex (BAs 24, 32) **DMPFC** = dorsomedial prefrontal cortex (BA 9)
MPC = medial parietal cortex (BAs 7, 31) **PCC** = posterior cingulate cortex (BAs 23, 31)
RSC = retrosplenial cortex (BAs 26, 29, 30)
  MOPFC = medial orbital prefrontal cortex (BAs 11, 12);
  VMPFC = ventromedial prefrontal cortex (BAs 10, 11);
  PACC = pre- and subgenual anterior cingulate cortex (BAs 24, 25, 32);

CMS areas are involved to different degrees in all the major neural network systems: **Affect, Executive, Salience, and Default**. They are vital structures in our understanding of most human behavior.

Neural Foundation of a Sense of 'Self'

Central midline structures are involved in our sensing of who and where we are, both with respect to how we relate to others and with respect to how our body relates to the space around it. It has been stated that patients with lesions in ventral CMS are unable to develop a coherent model of their own self (Damasio, 1999). Damasio noted that lesions that compromise the ability to **experience emotional feelings** and to sense the body are often located on the right side of the somatosensory complex and involve, in particular, the **right insula**. These patients have a compromised **sense of self**. He notes that activity at the level of the parabrachial nucleus (PBN), the **nucleus tractus solitarius (NTS)** and the hypothalamus may be important in establishing a sense of self. (The PBN and NTS structures are discussed later in detail under Heart Rate Variability training.) The NTS and the PBN convey information to the right insula via the ventral-medial nucleus of the thalamus, and some sense of self may emerge at this level. The insula connects to the anterior cingulate cortex and the medial frontal cortex. These areas are also all key structures in the Affect network. Damasio concludes that all of these structures and connections are the **neural foundation for the self,** the grounding of the material 'me' (Damasio, 2003).

Default and Salience Networks

The Executive, Attention, and Affect (Emotional) networks are quite well known, and functions related to these networks will be described when Brodmann areas that relate to their functions are discussed in Part II. Two more recently delineated networks, the Default and the Salience networks, deserve some additional comment here.

Default Network

When nothing is consciously going on, your brain is still active. Your thoughts may include: **insight, imagination, judgment, thoughts about self and self in relation to others**, as well as autobiographical memory. The Default network plays a vital role in forming a sense of identity, self-concept, self-perception, and, more broadly, one's concepts of social and family systems. This **self-reflection** activity is one important aspect of homeostasis within the brain – the set-point to which brain function returns after being engaged in other mental activity.

Aspects of the Default network are laid down in infancy and thus can be significantly **damaged by negative emotional experiences in early childhood.** Its connections with all the key areas involved in the Affect network mean that early negative experiences can be expected to have major effects on self-concept and emotions in later life. This is particularly true due to the fact that, to the infant, the mother (or the mothering-figure) is the universe (Thompson & Patterson, 1986). Any disruption in this relationship will alter the psychosocial development of the child (McCrone, 2002).

Anatomy of the Default Network

The **Default network** comprises a number of central midline structures including the ventral-medial prefrontal cortex, anterior cingulate cortex and, for spatial self-reference, lateral parietal cortex, cuneus and precuneus, and the posterior cingulate. All of these areas are concerned with understanding and constant monitoring of *self and self in relation to others* (Supekar et al., 2010). Thus the Default network is involved in self-referential processing and understanding others' intentions. These functions are important in our work with autism spectrum disorders (ASD). Remember that the Default network becomes active when the brain is not engaged in specific cognitive or motor tasks.

The **Default-mode network** (DMN) includes the dorsal and ventral medial prefrontal cortex (dMPFC and vMPFC; Brodmann areas (BAs) 10, 9, 32, and 24), the posterior cingulate cortex (PCC, BAs 23/31), retrosplenial cortex (RSC, BAs 29/30), and the lateral posterior cortex (LP, BAs 39/40) (Buckner et al., 2008; Fox et al., 2005; Greicius et al., 2003; Gusnard and Raichle, 2001; Raichle et al., 2001).

The Default network also has **connections to Mirror Neuron areas** and these are critical in forming a concept of self and self-in-the-world. Using fMRI, regions can be observed to become active (light up) in the medial orbital and medial prefrontal areas and in medial parietal areas when activities related to the Default network are involved. Thus these areas and functions are subsumed under the term 'Default Network' (Raichle, 2010). The posterior cingulate and subgenual cingulate areas may be involved in depressive ruminating (Berman et al., 2011).

The Default network has been divided into dorsal and ventral areas, but here we are treating it as a single functional unit. In very general terms, one can say that when executive and/or motor networks are active, the Default network is turned off. The opposite is true when executive cognitive functions or motor functions are turned off. In that situation, the Default network is active.

It is thought that the Default network may not be functioning normally in people with autism spectrum disorders.

The Salience Network

The frontal lobes are also important in determining the 'salience' of information; that is, deciding what incoming information is relevant or important. The Salience network (Seeley, 2007) includes the dorsal and anterior cingulate gyrus (BAs 24 and 25), insula, middle superior frontal area, and the para-central area (BAs 4, 5, 6), subcallosal gyrus, entorhinal area (BA 28) and uncus (BA 34), parahippocampal gyrus (BA 27), hippocampus (BA 35) and fusiform gyrus (BA 36). As with other networks, it is apparent that we cannot isolate a function to a single area or lobe of the brain. We must think more in terms of how areas function together in a network that subsumes a particular function. Clearly one would expect overlap with memory areas as well as areas involved in emotions and sense of self when determining relevance.

The right dorsal anterior cingulate cortex (dACC) and the right insula **may together act as a switch** between the frontal-parietal Executive network and the Default network when determining salience (Sridharan et al., 2008). It is also apparent that the ACC will switch between the Affect network and the Executive network when executive tasks are activated, such as when there is a task demanding divided attention (Devinsky et al., 1995).

However, there are always **other networks that subsume similar functions** but in a different context. This is clearly true for determining 'salience.' An example is a salience map (or network) to rank items in the visual field for locus of attention (Bisley, 2006). In this Salience network, the lateral intra-parietal area (LIP) connects to the frontal eye field (FEF) (as does the upper auditory cortex), superior colliculus, pulvinar nucleus, and inferior temporal area (BA 20) (visual pattern recognition). This network determines the significance (salience) of new visual information.

The Salience Network (in more detail)
The following discussion is repeated in the section in the second edition of *The Neurofeedback Book* on **Attention Networks**. Salience combined with parietal, cingulate, and frontal functions, particularly in the right hemisphere for sustained attention, are **major components of attention networking.**

When discussing choosing the 'salience' of new information, regardless of whether it is from internal or external sources, more emphasis should be placed on the role of the anterior portion of the insula and the anterior cingulate gyrus than is done in Bisley's description of visual salience determination. As stated by Menon & Uddin (2010) in their review article "Saliency, switching, attention and control: a network model of insula function." (*Brain Struct Funct*, 214, 655–667): "The anterior **insula is an integral hub** in mediating dynamic interactions between other large-scale brain networks involved in externally oriented attention and internally oriented or self-related cognition." These authors postulate that the **insula is sensitive to salient events,** and that: "Its core function is to mark such events for additional processing and initiate appropriate control signals." These authors reinforce the now-accepted view that the **anterior insula and the anterior cingulate cortex are the major elements in a Salience network.** This network functions to segregate the most relevant among internal and extra-personal stimuli in order to guide behavior. They consider the anterior insula (AI) to be the hub of the salience network. This is particularly important in our work where we combine NFB with HRV and where we have observed that many of our patients who have suffered a traumatic brain injury (TBI), have an anxiety disorder, or have a diagnosis of Autism Spectrum Disorder (ASD) show the left and/or right insula to be outside the database norms. For example, hyperactivity of the right insula might correlate with misinterpretation of the salience of mundane events and lead to anxiety. On the other hand, hypoactivity in ASDs might correlate with a lack of attention to social communication.

In the figure below, the reader can understand that a bottom-up salient event is detected by the anterior insula and ACC. The insula is particularly sensitive to salient events. The insula can then communicate important events to other networks, including those for sustaining attention, executive/cognition, working memory functions and the affect networks. With the posterior insula it will regulate the autonomic nervous system and this will have an effect on HRV and other autonomic nervous system (ANS) functions through connections to the amygdala, locus coeruleus, and the hypothalamus. Thus the **insula is a key link** between incoming bottom-up stimuli and the brain regions such as the ACC that are involved in monitoring this information.

Cognitively demanding tasks have the effect of decreasing activity in areas of the brain that we have previously described as being within the Default network while increasing activity in executive networks. The central executive network (CEN) includes the dorsolateral prefrontal cortex (DLPFC), and the posterior parietal cortex (PPC). The default mode network (DMN) includes the ventromedial prefrontal cortex (VMPFC), posterior cingulate cortex (PCC), medial temporal region, and the angular gyrus. The CEN is important for working memory, and for judgment and decision-making for goal-directed behavior.

The PCC is involved in self-referential thinking and autobiographical memory (Buchner et al., 2008), and the VMPC is important in thinking about self and others (Amodo & Frith, 2006).

The diagram below may assist the reader in picturing Salience network major functions.

This diagram emphasizes the central importance of the Salience network in the dynamic switching between activity generated by the central Executive network (CEN) and the internal self referential activity of the Default mode network (DMN). Menon and Uddin suggest that the Salience network can be conceptualized as having a hierarchy of salience filters that can amplify inputs to different degrees. There are direct connections between the intraparietal sulcus and the insula, and this may be one important link for determining the salience of incoming sensory stimuli.

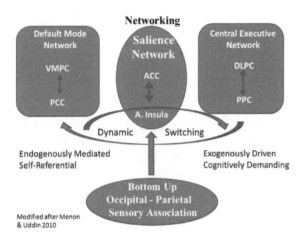

Illustration of major Salience Network connections. VMPC = Ventral Medial Prefrontal Cortex; PCC = Posterior Cingulate Cortex; ACC = Anterior Cingulate Cortex; A. Insula = Anterior Insula; DLPC = Dorsolateral Prefrontal Cortex; PPC = Posterior Parietal Cortex

Why Training at Central Sites Affects 'Networks'

Surface Sites Fz, Cz, Pz, and/or at Deeper Central Midline Structures

The behavioral changes of the client appear to correlate with the client's mental state when doing the NFB with the electrode on a central site; that is, at or between Fz-Cz-Pz. (Cz is the most common single-channel placement for children and FCz the most common site used in adults.) For example, if the patient is doing part of their training on task, and they are using a metacognitive strategy to accomplish the task across a number of sessions, then **the changes that will be reinforced should include** functions associated with the Executive network that is involved in that task. In fact, we do measure marked changes occurring in academics, attention, and IQ measurements. Changes in executive functions have been confirmed in large case series by objective testing done before and after 40 sessions, and can be reviewed in publications by a number of researchers and clinicians. Increases in Full Scale IQ scores of about 10 standard score points, for example, have been reported in four different studies (Linden, 1996; Lubar, 1995; Thompson & Thompson, 1998, 2010).

More about Networks

The thalamo-cortical-basal ganglia loops and connections delineated above underlie our hypothesis regarding how training over the central midline structures (CMS), especially when combined with heart rate variability training, may have such profound effects on complex networks that involve functions related to Affect, Executive, Motor, Salience, and Default networks. When you re-train those loops, it is like re-setting the pacemaker in the brain. Jim Robbin's got the analogy right when he wrote *A Symphony in the Brain*. The thalamus is like the conductor, deciding which sections of the brain are needed for a particular task, how loudly each one should play, and how they interact. It may also be a reason why coherence and phase training, which influence thalamic outputs and cortical communications, may become increasingly important in EEG biofeedback.

Importance of Heart Rate Variability (HRV) Training – Connections to Nucleus of the Solitary Tract in the Brain Stem

The Affect network that we have discussed in the cortex is also connected to important structures in the brainstem, such as the locus coeruleus in the pons and the nucleus of the solitary tract (NST) in the medulla. (You may also see the Latin, *nucleus tractus solitarius* or NTS in some texts.) The NST receives direct afferents from the aortic and carotid baroreceptors. This is one reason why heart rate variability (HRV) training may have such a profound effect on the Affect network and, in particular, on controlling anxiety. HRV, however, is covered in some detail its own section and, therefore, it will not be described here. We mention it to underscore the great importance of combining HRV training with training using NFB. The combination allows you to influence the same structures from two different directions, hitting both afferent and efferent connections. Recall that afferents go *to* the CNS (prefix from the Latin preposition *ad*) and efferents come *from* the CNS (prefix from the Latin preposition *ex*.)

The next two figures are given to help orient the reader to structures that may help the reader to understand why NFB and HRV trainng can act synergistically to improve patient functioning.

The reader can find a description of these important connections in the section of *The Neurofeedback Book* on heart rate variability training.

Brainstem, Pons, Midbrain, and Basal Ganglia-Thalamic Regions.

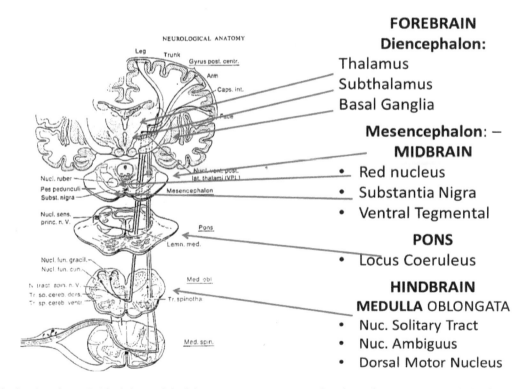

Modified after Smith 1962 (slightly modified from Rasmussen 1932). The above figure is given to help the reader get oriented with respect to the brainstem, pons, midbrain, and the basal ganglia-thalamic regions.

Heart – Brain Connections

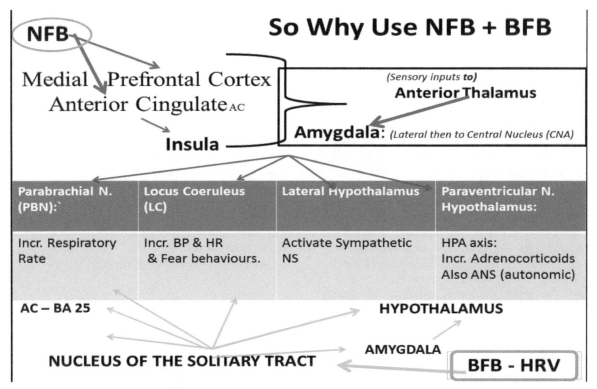

Figure: This figure illustrates how NFB applied as training from the 'top-down,' and HRV given as training from the 'bottom-up' can both affect the same central structures. (Note that our use of the terms 'top-down' and 'bottom-up,' for the past 20 years, is different from the more recent use of these terms in cognitive psychology, economics, and other disciplines. Indeed, in other areas of this book we have used these terms differently when describing parietal sensory bottom-up versus frontal modulation top-down when illustrating the salience network.)

Brodmann Areas and Networks

Brodmann Areas and Networks

Our working definition of a ***network***, for purposes of doing neurofeedback (NFB) and biofeedback (BFB), is **a 'net' of interconnected, functionally related groups of neurons.** In the cortex, these 'groups of neurons,' for convenience of thinking about cortical functions, may be broadly viewed as corresponding to Brodmann areas. Brodmann posited that if the cytoarchitecture of cells is the same, then the function must be the same. This is called a 'structure-function' point of view and is summarized as 'structure dictates function.'

As previously noted, the functions that will be listed in this book for each BA emphasize just the principle functions for that area, but remember that each function also involves many other areas. The reader is asked to understand that our emphasis in training has, accordingly, moved from being primarily concerned with functions at individual Brodmann area sites to an emphasis on training networks that involve a number of Brodmann areas, plus cortical connections to subcortical structures.

Phase-Reset

Thatcher argues that **when different functional systems are working together they are 'phase-locked.' Coherence is a measure of phase-locked activity.** Phase-locked areas produce a relatively high amplitude and synchrony of EEG activity. When there is a shift in task, there is a shift in the sets of neurons involved. This shift involves cortex-thalamus-cortex linkages. The **'shift' will recruit a different set of neurons** that then become phase-locked. During this change-over, called a 'phase-shift' (or **'phase-reset'**), the EEG amplitude drops and there is 'asynchrony.' In this book we use the rather simple overview term 'network' to refer to a set of sometimes distant cortical areas that are functionally connected and interrelated. Thatcher notes that phase shift and phase lock in the coordination of large masses of neurons in functional modules and hubs is what gives rise to the variations in the electroencephalogram (Thatcher, 2012, p 329).

Cortical Connectivity – Modules, Hubs and Nodes Maximize Efficiency

Robert Thatcher has noted that the brain is organized into a relatively small subset of modules or hubs that represent clusters of neurons with **high within-cluster connectivity and sparse long distance connectivity**. This means that there are fewer long distance links for each neuron, which **maximizes the efficiency** of the brain (Buzsaki, 2006; Thatcher, 2012, p305). In routine clinical work, we do not often use the more complex terms such as 'functional modules' (Achard et al., 2006; Hagmann et al., 2008; Thatcher, 2012) or 'node,' though others speak of a **network being a number of nodes connected by links** (Bullmore, Ed & Olaf Sporns, 2012; Thatcher 2012). For our work with NFB, it is sufficient to understand that if one has an effect on a central point within a functional neural network, one will have a reasonable probability of affecting that entire functional network, even though some of the areas within that network are in distant areas of the cortex. We have used this as an argument to explain the positive effects of NFB done over the FCz site on Affect and Executive networks, as seen in the changes in functioning of clients who have symptoms of Asperger's and autism (Thompson and Thompson, 2010).

Nevertheless, because the reader will find the terms 'module,' 'hub,' and 'node' appearing in the literature, we will briefly explain these terms with diagrams. To do this we have used information gleaned from the publication by Patric Hagmann, Leila Cammoun, Xavier Gigandet, Reto Meuli, Christopher J. Honey, Van J. Wedeen, and Olaf Sporns (July, 2008). This was a major cooperative effort between University of Lausanne in Switzerland and the University of Indiana, along with Harvard Medical School, in the USA. We recommend that our readers consult the original paper for more details. These authors made a comprehensive analysis of cortical connectivity. They state: "In the human brain, neural activation patterns are shaped by the underlying structural connections that form a dense network of fiber pathways linking all regions of the cerebral cortex." These authors used diffusion imaging techniques, which allowed the noninvasive mapping of fiber pathways. They **constructed connection maps** covering the entire cortical surface.

Computational analyses of the resulting complex brain network revealed regions of the cortex that were highly connected and highly central. These areas form a structural core of the human brain. They note that key components of this core are portions of posterior medial cortex that are known to be activated when the brain is not engaged in a cognitively demanding task. We have referred to this as the Default network. These authors state that, because they were interested in how **brain structure relates to brain function**, they also recorded **brain activation patterns** from the same participant group. They found that structural connection patterns and functional interactions between regions of cortex were significantly correlated. They concluded that, "Based on our findings, we suggest that the structural core of the brain may have a central role in integrating information across functionally segregated brain regions."

A

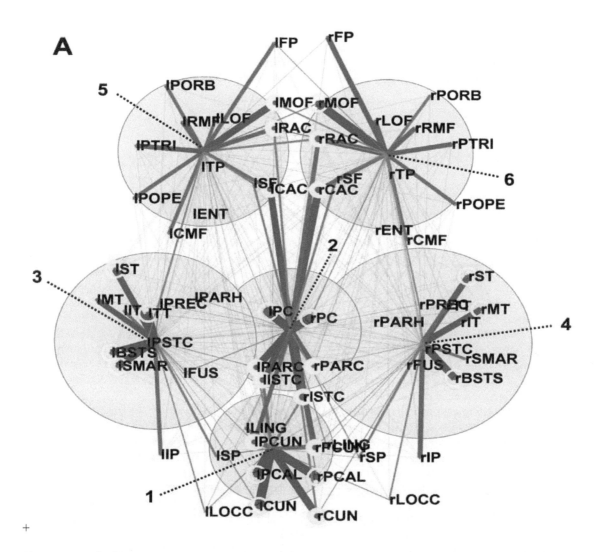

Hagmann et al. (2008). Mapping the Structural Core of Human Cerebral Cortex. (freely available online).

Abbreviations Used Regarding Regions of Interest
The regions shown are **those that demonstrated the highest connection weight, or connectivity**, between each pair of regions of interest. The diagram thus shows a weighted network of structural connectivity across the entire brain. The 66 cortical regions have two-part labels: a prefix for the cortical hemisphere 'r' (right hemisphere), or 'l' (left hemisphere), and one of 33 designators listed below, largely in alphabetical order.

BSTS (bank of the superior temporal sulcus), CAC (caudal anterior cingulate cortex), CMF (caudal middle frontal cortex, CUN (cuneus), ENT (or elsewhere labelled ERC is entorhinal cortex), FP (frontal pole), FUS (fusiform gyrus), IP (inferior parietal cortex), IT (inferior temporal cortex), ISTC (isthmus of the cingulate cortex), LOCC (lateral occipital cortex), LOF (lateral orbitofrontal cortex), LING (lingual gyrus), MOF (medial orbitofrontal cortex), MT (middle temporal cortex), PARC (paracentral lobule), PARH (parahippocampal cortex), POPE (pars opercularis), PORB (pars orbitalis), PTRI (pars triangularis), PCAL (pericalcarine cortex), PSTC (postcentral gyrus), PC (posterior cingulate cortex), PREC (precentral gyrus), PCUN (precuneus), RAC (rostral anterior cingulate cortex), RMF (rostral middle frontal cortex), SF (superior frontal cortex), SP (superior parietal cortex), ST (superior temporal cortex), SMAR (supramarginal gyrus), TP (temporal pole), and TT (transverse temporal cortex).

Hubs

The connector 'hubs' have **above average strength** and a high proportion of cross-module connectivity; these are marked as yellow circles. Provincial hubs have above-average strength; they are marked as unfilled yellow circles. In their analysis of connectivity between areas of the cortex, six areas emerged. Four contra-laterally matched modules were localized to frontal and temporal-parietal areas of a single hemisphere. The two remaining modules comprised regions of bilateral medial cortex, one centered on the posterior cingulate cortex and another centered on the precuneus and pericalcarine cortex.

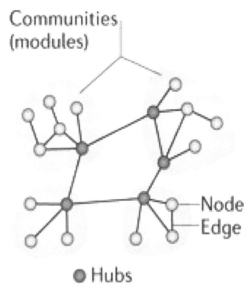

Diagram from: Ed Bullmore & Olaf Sporns (2012)
The economy of brain network organization, Nature Reviews Neuroscience

Functional Module

A **'functional module'** (Achard et al., 2006; Hagmann et al., 2008; Thatcher, 2012) consists of related nodes. A **node** is a large group or **cluster of neurons with high interconnnectivity**. A **Hub** is the key central connecting link for a group of Nodes. A neural **network is** a number of **nodes/modules with functions in common,** connected by links (Sporns, 2011). A neural network may, therefore, also include interconnected modules. It's a bit like travel between airports. There are links to regional airports (nodes) and then there are links to international airports (hubs). To go a long distance, you connect from a local airport to a regional one in order to get to the international airport in the next hub. Interestingly, a German neuroscientist by the name of Poppel posited some years ago that any two neurons can connect with no more than three synapses, which makes sense in terms of connections organized in nodes and hubs. **Phase shift & phase lock** act in the coordination of large masses of neurons in functional modules containing nodes and hubs and, according to Thatcher, this is what gives rise to the electroencephalogram (Thatcher, 2012, p329).

As mentioned above, Thatcher notes that the brain is organized into a relatively small subset of modules with hubs that represent clusters of neurons grouped in nodes with high within-cluster connectivity and sparse long distance connectivity. He notes that this means less in the way of long-distance links for each neuron and maximizes the efficiency of the brain (Thatcher 2012, p305; Buzsaki, 2006).

SECTION VIII
Functions Related to Lobes of the Brain, International 10-20 Sites, and Brodmann Areas

Introduction

Part 2 is organized by lobes of the brain because that is how most practitioners will initially approach their work with NFB. The sections on the Cingulate Gyrus and the Frontal, Temporal and Parietal lobes will each begin with a general introductory overview of important functions and then describe the Brodmann areas (BAs) contained within that region.

A great deal of repetition is to be expected because the functions of a number of different Brodmann areas overlap. The reader can reasonably assume, for the most part, that areas with similar functions are connected in networks. The repetition is intended to help readers find details concerning functions of particular BAs, networks, and lobes of the brain without needing to jump back and forth.

Neurofeedback (NFB) practitioners, when doing training on surface sites with one or more channels of EEG, usually utilize the standard International 10-20 electrode placement sites established for use by neurologists. Therefore this section will initially order the BAs according to their location in lobes of the brain and will also relate them to the standard, 19-channel 10-20 sites, beginning with those over the frontal lobes.

Some practitioners may do LORETA Z-score NFB; they will find the list of BAs correlated to the primary functions of each area particularly helpful. That section is followed by summaries concerning the major functional networks in the brain combined with a listing of the BAs that relate most strongly to the functions of each network.

Suggestion Regarding How to Use This Section
This section has been written to provide NFB practitioners with a useful reference guide. After you do a 19-channel assessment with LORETA source identification, you will know that certain 10-20 sites (and, with LORETA, certain BAs) are outside database norms. We suggest that you can then use this section of the book to look up those 10-20 sites and Brodmann areas that are outside the database norms in order to consider what networks and functions are most likely involved, given your patient's major symptoms. From there you may then make informed decisions regarding the placement of your electrodes or, for LORETA NFB, the most important sites and parameters (amplitudes, coherence, phase, etc.) to target with your feedback.

For purposes of doing either surface site NFB or LORETA NFB, the following functional areas are grouped according to their proximity to the International 10-20 sites wherever possible. This was done because most NFB practitioners use surface sites for training purposes and only a smaller number of practitioners have the knowledge, experience, and equipment necessary for carrying out LORETA Z-score NFB training.

Reminder: BA Functions Overlap!
Keep in mind, as previously noted, that functions of Brodmann areas listed here are only highlighting some aspects of each area's functions and that any given function likely involves a large number of different areas that operate as a network to achieve that function. Moreover, Brodmann areas themselves are not as precise across different individuals as those

depicted in diagrams. One Brodmann area, as drawn below, may overlap another, and a Brodmann area may overlap more than one gyrus.

Summary of Locations of Brodmann Areas

(Remember that BAs can overlap more than one gyrus and a gyrus can have more than one BA associated with it.)

- Areas 1, 2 & 3 – Primary Somatosensory Cortex (frequently referred to as Areas 3, 1, 2)
- Area 4 – Primary motor cortex
- Area 5 – Somatosensory association cortex
- Area 6 – Premotor cortex and supplementary motor cortex (also called secondary motor cortex or supplementary motor area)
- Area 7 – Somatosensory association cortex
- Area 8 – Includes frontal eye fields
- Area 9 – Dorsolateral prefrontal cortex
- Area 10 – Anterior prefrontal cortex (most rostral part of superior and middle frontal gyri)
- Area 11 – Orbitofrontal area (orbital and rectus gyri, plus part of the rostral part of the superior frontal gyrus)
- Area 12 – Orbitofrontal area (used to be part of BA11, refers to the area between the superior frontal gyrus and the inferior rostral sulcus)
- Area 13 (& 14) – Insular cortex
- *Areas 14, 15 (anterior temporal lobe) & 16 – Not found in humans*
- Area 17 – Primary visual cortex (V1)
- Area 18 – Secondary visual cortex (V2)
- Area 19 – Associative visual cortex (V3)
- Area 20 – Inferior temporal gyrus
- Area 21 – Middle temporal gyrus
- Area 22 – Superior temporal gyrus, memory comparison area of auditory cortex, the posterior part is considered to contain a portion of Wernicke's area
- Area 23 – Ventral posterior cingulate cortex
- Area 24 – Ventral anterior cingulate cortex
- Area 25 – Subgenual cortex (part of the ventromedial prefrontal cortex)
- Area 26 – Ectosplenial portion of the retrosplenial region of the cerebral cortex
- Area 27 – Part of piriform cortex
- Area 28 – Posterior entorhinal cortex
- Area 29 – Retrosplenial cingulate cortex
- Area 30 – Part of cingulate cortex
- Area 31 – Dorsal posterior cingulate cortex
- Area 32 – Dorsal anterior cingulate cortex
- Area 33 – Part of anterior cingulate cortex
- Area 34 – Anterior entorhinal cortex (on the parahippocampal gyrus)
- Area 35 – Perirhinal cortex (on the parahippocampal gyrus)
- Area 36 – Parahippocampal cortex (on the parahippocampal gyrus); fusiform gyrus
- Area 37 – Fusiform gyrus, occipitotemporal gyrus
- Area 38 – Temporopolar area (most rostral part of the superior and middle temporal gyri)
- Area 39 – Angular gyrus (part of Wernicke's area in Dominant Hemisphere)
- Area 40 – Supramarginal gyrus (part of Wernicke's area in Dominant Hemisphere)
- Areas 41 & 42 – Primary and Auditory Association Cortex
- Area 43 – Primary gustatory cortex
- Area 44 – Pars opercularis. Contains mirror neurons. (part of Broca's area in Dominant Hemisphere)
- Area 45 – Pars triangularis.(Broca's area in Dominant Hemisphere)
- Area 46 – Dorsolateral prefrontal cortex
- Area 47 – Inferior prefrontal gyrus
- Area 48 – Retrosubicular area (a small part of the medial surface of the temporal lobe)
- *Area 49 – Parasubiculum area in a rodent*
- *Areas 50 & 51 – Found in monkeys*
- Area 52 – Parainsular area (at the junction of the temporal lobe and the insula)

Figures: Lobes, Brodmann Areas, Functions (Networks)

Primary Functions Related to General Areas of the Cortex

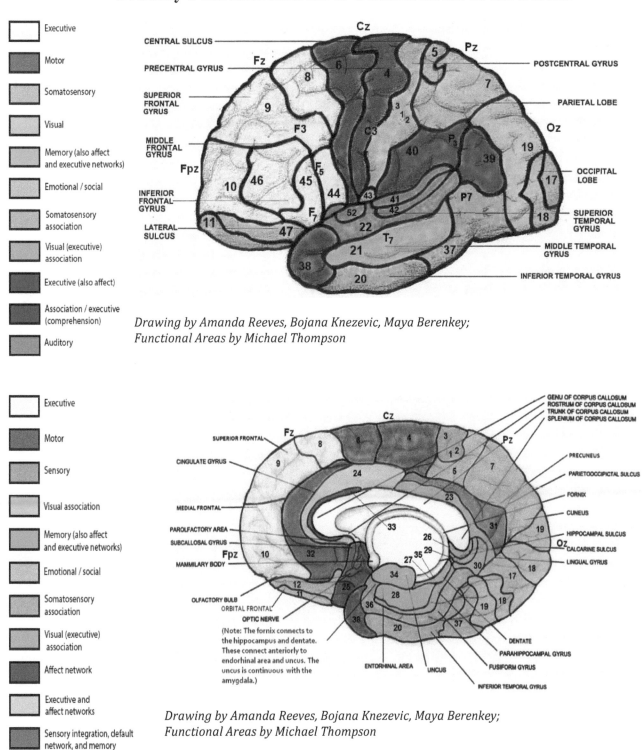

Drawing by Amanda Reeves, Bojana Knezevic, Maya Berenkey;
Functional Areas by Michael Thompson

1. Attention, Salience, Default, and Memory Networks are too widespread to be usefully illustrated in this diagram.
2. Brodmann area 32 is more Executive than Affect but is purple for color contrast in diagram.

Note that the diagrams above, for purposes of consistency, show just the left hemisphere for the lateral view and the right hemisphere for the midsagittal view. The reader can extrapolate to the homologous sites viewed from the other side of the brain. Keep in mind that an electrode location is on the scalp, not on the actual brain tissue, and thus scalp locations will actually be over slightly different aspects of the brain from one individual to the next.

Note that each region/lobe of the brain has a rather long INTRODUCTION before describing any functions under its Brodmann areas. This is because the same region will function in a number of networks. The individual Brodmann areas do NOT have discreet and unique functions but, rather, have a primary function and also functions that overlap with adjacent BAs and, indeed, they may co-ordinate with other Brodmann areas in distant areas of the cortex to perform some functions. In this manner, each Brodmann area is, for the most part, a component part of complex networks rather than an area that functions independently.

Order of Description of Functions

Lobes or Cortical Areas	Associated Networks	Principle Brodmann Areas	10-20 Sites
Frontal (Lateral & Medial)	Executive Networks Medial: Executive and Affect Networks Medial (BA 9, 10): Default Network	8, 9, 10, 46, 45, 44, 6 47 Lateral and Medial discussed together for each BA --Inferior prefrontal 9, 10, 11, 46, 47	Fz F3, F4 F7, F8 FP1, FP2
Frontal Orbital - Ventral/medial	Affect network – Social (Default network)	11, 12	F9, F10 FPz
Central	Somatosensory Cortex Sensory Network Motor Network	BAs 4, 6, & 1, 2, 3, 5 BAs 1, 2, 3 Primary (+ 43) Somatosensory Cortex BA 5 Somatosensory Association Cortex BA 4 Primary Motor Cortex BA 6 Motor Association Cortex	Cz C3, C4 Cz C3, C4
Temporal			
- Lateral Aspect	Executive – Auditory	41, 42, 22; 21, (52)	T7 & T8
- Temporal Pole	Executive, Semantic – also Affect Networks	38	T7 & T8 (F7, F8)
Temporal - Medial Aspect - (+ *Amygdala*)	Executive Network – Memory Affect network (+Default)	26, 27, 28 36, 37, (48) 35, 34	(Pz) (P7, P8)
Temporal - Inferior	Object identification; part of the 'what' pathway.	20, 37	T9 & T10
Insula	Affect & Executive	13	C3-T7
Cingulate			
Anterior (ACC)	Affect - Executive - Default	Affect: 24 (also Executive & Default & Salience networks) Affect: 33, 25 Executive: 32 (also Affect & Default)	Cz FCz Fz
Posterior (PCC)	Default Executive – Sensory	23, 31, 26, 29, (30)	Pz
Parietal	Executive, Sensory, Attention, (& Default) Networks; Spatial Awareness, part of the 'where' pathway.	5, 7	PCz, Pz
Temporal – Parietal - Lateral Aspect	Executive Network (& Default) Comprehension/Language Learning Gesture, Emotion, Nuance	39, 40 DH: 39, 40 NDH: 39, 40	P3 & P4
Inferior-Parietal Lateral	Attention & Salience	37	P7 & P8
Parietal Lobes - Medial Dorsal Aspect	Somatosensory Association	7	Pz
Occipital	Visual	17, 18, 19	O1, O2

Important: *The electrodes used for surface EEG NFB may reflect the activity from a number of Brodmann areas. The functions of Brodmann areas overlap with other Brodmann areas. Thus it is inevitable that under each principle BA being discussed, other BAs will be mentioned.*

Notes on Doing Neurofeedback and Heart Rate Variability Training

Please recall that this book is not written to detail anatomy and connections. Rather, it is written to give an overview of some of the sites and neural connections/networks that the practitioner may influence with NFB + BFB training. In part, the importance for most NFB practitioners is to recognize that **NFB applied over one key site within a neural network may affect the entire network**. Also, that NFB may need to be applied to both hemispheres for different effects. For this reason, and also to (theoretically) directly influence some deep midline structures while also affecting coherence and phase relationships between different areas of the cortex, LORETA NFB may be more efficient than surface NFB.

LORETA gives source localization, naming both the Brodmann area and the cortical area (such as BA 24, Anterior Cingulate) that is related to the EEG measured on the surface. When combined with a database, such as NeuroGuide, one also has information about the degree of deviation from the mean for people in the normative sample of the same age and sex for particular frequencies. However, always remember that you must choose the sites for doing LORETA NFB very carefully so as not to adversely affect sites that are deviant but for a positive reason, either related to a person's strengths or activated to compensate for sites that are not functioning normally. For example, when we found significant deviations in the auditory cortex of a physician who was also a musician with perfect pitch and no symptoms related to functions of the temporal lobe, we certainly did not want to normalize those deviations.

In addition, BFB and, in particular, HRV training, because they influence connections between the solitary nucleus and many higher nuclei, including those in the thalamus, cingulate gyrus, and the hypothalamus, may also have effects on a number of neural and endocrine systems and networks. Understanding connections and networks is helpful for doing careful, effective work with clients.

Frontal Lobe
Introduction – Functions Related to the Frontal Lobes

Due to the important involvement of the frontal lobes in the symptoms of many patients who are being treated with NFB, this introduction to the functions of the frontal lobes is more comprehensive than the introduction to other cortical regions. What follows here is a general overview of some of the functions subsumed by the frontal lobes. Most of these functions can be influenced by neurofeedback training. In addition, the functions related to different BAs in the frontal lobes overlap to such a large degree that this section will discuss **functions in general terms before specific functions of the individual BAs** are discussed.

Rather than always using right and left designations, we will often refer to the dominant hemisphere (DH) and the nondominant hemisphere (NDH). In most people the DH is the left hemisphere and they are usually right-handed. However, it is also true that many left-handed people and most ambidextrous people have their DH on the left. The NDH is the right hemisphere in the majority of people. In most individuals, language and verbal functions primarily involve the DH.

The **prefrontal** cortex makes up about one third of the human neocortex. On the lateral surface it corresponds to the superior, middle, and inferior frontal gyri anterior to the premotor region (BA 6). The prefrontal region includes BAs 8, 9, 10, 11, 44, 45, 46 and 47. On the medial surface it includes the anterior cingulate gyrus and the medial prefrontal area comprising BAs 24, 32 and BAs 8, 9, 10, 11, 12. In the orbital/inferior surface it includes BAs 10, 11, 47, and a portion of BA 13 (Augustine, 2008, pp 356, 357).

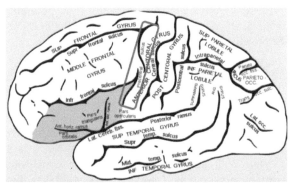

Figure from *Gray's Anatomy* (public domain) to show a lateral view of the prefrontal cortex, inferior, middle, and superior gyri with a red rectangle around the posterior border at the precentral sulcus.

The prefrontal cortex has major connections to the dorsal aspect of the anterior cingulate cortex (ACC). The ACC is a central structure in the ***Executive network,*** and it connects with all functionally related structures, including the dorsolateral prefrontal cortex. The frontal lobes are crucial for all executive functions, such as the evaluation of incoming stimuli, planning an appropriate response, and inhibiting inappropriate responses (Knezevic, 2010).

Definition of Executive Functioning

Executive functioning is one key function of the frontal lobes. It is broadly defined as the overall capacity to manage perception, attention, selection, decision making, inhibition, memory, planning, problem solving, logical thinking (deductive and inductive), sequencing; to monitor, evaluate, and self-correct outputs; and to respond verbally/nonverbally, motorically, and socially for the purpose of the attainment of defined goals.

Although the frontal lobes are crucial for all executive tasks, including the voluntary control of attention, the inhibition of inappropriate and/or unwanted behavior, the planning of actions and executive decision making, these lobes also have many other important functions.

These functions include:
- A role in the maintenance of arousal.
- Temporal sequencing of complex entities such as compound sentences (mainly the dominant hemisphere (DH).
- Hand-writing (Exner's area in the DH).
- Speech articulation (Broca's area: DH BAs 44, 45).
- Frontal mirror neurons (mirror neurons were first observed in monkeys in the Left Frontal Operculum (F5) close to Broca's area).
- Auditory and verbal representation.
- Object naming and word recall (usually DH).
- Representation of visual images evoked by auditory input.
- Letter and word perception and recognition (with input from the occipital and parietal regions).
- Representation of abstract verbal forms.
- Perception of complex relationships.

Some of the Functions of the DH (Left) Frontal Lobe

The **dominant hemisphere** (left) frontal lobe is crucial in the regulation of speech production and syntax. It is important in sustaining and regulating attention, inhibiting actions, and switching responses. It is critical to **analytic, sequential** processing. It is where we utilize inner speech (inner dialogue) to regulate our behavior. The DH's frontal lobe is the location of operations that require sequencing, such as one can test by asking the client to recite a series of numbers or count backwards by 7s. It is involved in regulating routines, **recognizing patterns,** and deciding what to do in different situations.

As a very broad overview, **left hemisphere dominance** is usually thought of as being characterized by a lack of emotion, introversion, goal-directed thinking and action. It is also involved in having an internal locus of control. Mental processing may tend to be relatively slow and serial in keeping with the tendency to be concrete, careful, and sequential. The dominant neurotransmitter is **dopamine.** On the one hand, one may recognize in this description, if taken to the extreme, the pattern displayed by some of your patients who have a diagnosis of Asperger's syndrome. On the other hand, you may also recognize that, **as we grow older**, we move more into this type of functioning and away from the more right-brained functioning that will be described below.

Frontal Lobe Injury

In very general terms, patients who have damage to their frontal lobes show a decrease in spontaneous facial movements and speak fewer words than they did previously. These patients demonstrate **difficulties with flexibility, problem solving, attention span, and memory**. In addition, frontal lobe damage appears to have **more of an impact on divergent than convergent** thinking. This may result in the person appearing to be unmotivated or even lazy, lacking initiative, and appearing rather inflexible and unable to consider alternative options or strategies for solving problems. Despite these deficiencies, in some cases, there is minimal effect on IQ testing, which may be due to the fact that **IQ testing places more emphasis on assessing convergent** rather than divergent thinking. In addition to these general findings, each patient will have specific difficulties that relate to their specific injury.

Some patients with **left (dominant) hemisphere injury** will show more **problems in verbal** areas, be rather apathetic, **depressed**, and have right-sided motor deficiencies particularly with respect to fine motor movements. On the other hand, a patient with more damage to the **right, nondominant, hemisphere** may demonstrate more problems in **nonverbal abilities, lack social graces**, act without tact, lack restraint, and have left-sided motor deficiencies.

In all cases, to varying extents, there may be symptoms related to problem solving, spontaneity, judgment, impulse control, and both social and sexual behavior. People close to the injured patient may comment that the **patient's personality has changed.** These patients show a relative inability to correctly interpret feedback from their environment. They may be noncompliant and show risk-taking behavior (Thatcher 2012). They **may perseverate** (repeat or prolong an action, thought, or utterance after the stimulus that prompted it has ceased). **Orientation** in space may be compromised. Mood can be affected. From the standpoint of general emotional demeanor, as is discussed elsewhere under disorders of affect, activation in the left dominant frontal region has been associated with positive thoughts and approach behavior while right nondominant frontal region activation is more often associated with negative thoughts and avoidance behavior (Davidson, 1995, 1998). So left frontal damage may put one at particular risk for depression.

Reasoning: Deductive & Inductive

Since it is a quintessential frontal lobe activity, reasoning is going to be discussed in detail as an example of how complex and interrelated brain function is within the cortex. You will see how many areas are involved, though the frontal cortex is the main player.

The left/dominant frontal lobe is critical for cognitive work that we generally call reasoning skills. **Reasoning is the cognitive process of drawing inferences from given information.** All arguments involve the claim that one or more propositions (the premises) provide some grounds for accepting another proposition (the conclusion). Arguments may be broadly divided into deduction and induction.

Deductive arguments are arguments that can be evaluated for validity. Validity is a function of the relationship between premises and conclusion and involves the claim that the premises provide absolute grounds for accepting the conclusion: e.g., All men are mortal. Socrates is a man, therefore Socrates is mortal. With **deductive reasoning the conclusion drawn must be true if the premises, whatever they may be, are true.** Thus deductive reasoning takes pieces of evidence and then comes to a valid conclusion from this material.

Inductive arguments are never automatically valid. They must be evaluated for plausibility or reasonableness. The premises provide only limited grounds for accepting the conclusion: e.g., Socrates is a man. Socrates is mortal, therefore, all men are mortal. The conclusion may or may not be true. The premises of an inductive logical argument indicate some degree of support (inductive probability) for the conclusion. **Thus the premises suggest truth but do not ensure it**. A clear example of a false conclusion based on inductive reasoning would be: All crows are birds. All crows are black, therefore all birds are black. Inductive reasoning generally proceeds from detailed facts to general principles. It is commonly construed as a form of reasoning that makes generalizations based on individual instances. In this sense it is often contrasted with deductive reasoning.

Anatomical Relationships in Reasoning

Anatomically, both induction and deduction activate a similar left frontal-temporal system. Induction differs from deduction in greater activation of the medial dorsal prefrontal cortex (BAs 8 and 9). Induction may also have somewhat more right-hemisphere involvement. Deduction is a little more akin to dependence on known patterns and absolute proof, which is a left-hemisphere function. As an example of how complex reasoning networks become, here are some details summarized from research by Goel and his colleagues (2004).

Deductive reasoning (compared to baseline) is associated with activation in the left cerebellum, bilateral inferior occipital gyrus (BA 18), left inferior temporal and occipital gyri (BAs 37/19), left middle temporal gyrus (BA 39), right superior parietal lobule (BA 7), bilateral middle frontal gyrus (BA 6), bilateral putamen, and left inferior frontal gyrus (BA 44, Broca's area) (paraphrased after Goel et al., 2004).

Inductive reasoning (compared to baseline) activates a similar network consisting of the right cerebellum, right lingual gyrus (BA 18), right middle occipital gyrus (BA 18), bilateral superior parietal lobule (BA 7), left inferior parietal lobule (BA 40), right middle frontal gyrus (BA 6), putamen, and left middle frontal gyrus (BAs 8/9/45). This rather complicated 'localization' underscores how a single BA is almost never acting alone!

In the above detailed analysis, **deduction** appears to be associated with activation that is somewhat more left/dominant hemisphere-based, **whereas induction** is somewhat more right/nondominant hemisphere. However, reasoning is definitely a dominant-hemisphere attribute and only the dominant hemisphere is capable of making final inferences (inductive reasoning). There is greater left inferior frontal gyrus (BA 44/Broca's area) activation for deduction than induction. Conversely, the left dorsolateral (BA 9) prefrontal cortex (along with right superior occipital gyrus, BA 19) show greater increases for induction. There are several possible reasons posited for the greater involvement of **Broca's area (BA 44) in deduction** than induction. Broca's Area is part of the phonological loop of working memory, and deductive reasoning has greater working memory requirements than inductive reasoning. Broca's Area is also involved in syntactic (order of words) processing involving logic. Thus enhanced activity in Broca's Area during deduction may be a function of greater engagement of syntactical processing and greater working memory requirements.

Inductive reasoning, on the other hand, is sensitive to background knowledge rather than logical form. The increased activity in the left dorsolateral prefrontal cortex may thus be due to the use of world knowledge in the generation and evaluation of hypotheses (Goel et al., 2004; Grafman, 2002), which is the basis of inductive inference.

In summary, both kinds of cognitive reasoning, deduction and induction, do not just activate left and right prefrontal cortex, respectively. Both forms of reasoning involve the left prefrontal cortex. Consistent with its greater requirements for syntactic processing and working memory, **deduction is characterized by increased activation in Broca's Area while induction involves greater activation in the left dorsolateral prefrontal cortex,** consistent with its need to access and evaluate world knowledge.

The important observation for our work with NFB is that both types of reasoning involve extensive overlapping networks of cortical areas. At the ADD Centre, we will do single channel NFB to activate these networks while the subject is doing cognitive tasks using appropriate strategies. We hypothesize that, by doing some of the training on task, we are reinforcing an optimization of the networks necessary for the particular task, namely, attention, focus, and the specific cognitive functioning necessary for that task. Our combination of NFB with BFB assures that this is carried out in a calm, relaxed state. Published results of large case series of people with ADHD (n=111) and with Autism Spectrum Disorders (n=159), which document improvement on academic and intellectual measures as well as on behavioral ones, support the effectiveness of this approach

(Thompson & Thompson, 1998, 2010). The addition of anodal transcranial direct current stimulation (tDCS) for a few minutes prior to doing regular neurofeedback may prove to give added benefit/more rapid results.

The nondominant (usually right) cerebral hemisphere

The right frontal cortex is involved in the regulation of **attention**. The right hemisphere, frontal, anterior cingulate, and parietal, in conjunction with cholinergic activity from the basal forebrain (nucleus basalis of Maynert and the substantia innominata) are important in **sustained attention** performance (Sarter et al., 2001). This hemisphere is also important in the **inhibition of old habits**. The nondominant frontal lobe is involved in understanding metaphors. **Autobiographical 'Sense of Past Self'** may reside primarily in the right frontal NDH. This is in spite of the fact that memory for most verbal functioning is considered to be in the dominant hemisphere with an emphasis on frontal lobe functioning for remembered item retrieval.

Most item retrieval is carried out in the dorsal medial prefrontal cortex (DMPC and left PFC) with links to the hippocampus (Northrop, 2006) and the anterolateral temporal lobes (we will describe under the temporal lobe, connections that involve the uncinate fasciculus, which links these areas to the orbital frontal cortex [OFC]). Researchers such as Seger (2004) note that there are numerous studies that emphasize predominately the right temporal cortex (RTC) and right prefrontal cortex (RPC) (and right uncinate fasciculus) for both autobiographical memory – self in the past (Damasio, 1999, 2003) and for recognition of one's own face vs. faces of others (medial AC and LPC).

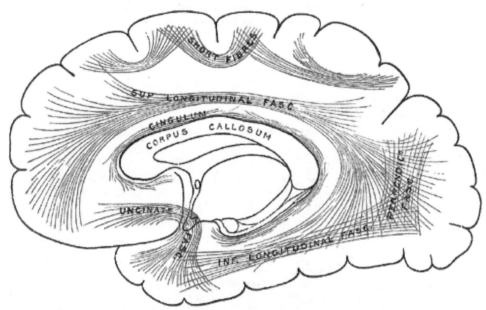

Figure to show Uncinate Fasciculus from Gray's Anatomy *(1918) (public domain). This figure illustrates the Corpus Callosum, which connects the left and right hemispheres, and Fasciculi that connect areas within the hemispheres. These myelinated pathways contain both excitatory and inhibitory pathways, though it has been argued that they are predominantly excitatory.*

Memory

Memories of experienced events have sensory and perceptual components and are therefore, in part, stored in the occipital cortex and related networks. On the other hand, memories **for imagined events are stored frontally**, probably because they contain imagery that is generated in the frontal lobes. As memory declines in some older persons, it is episodic memory with the encoding of new events that deteriorates first. This can leave retrieval of old memories intact, for the most part, including autobiographical memory.

Encoding New Images

It is the networks that involve the right parietal (cuneus, precuneus) with the left temporal regions that encode specific new images, along with the hippocampus, that are most **affected in early memory loss** associated with aging.

In addition, it has been observed that the right prefrontal cortex can show an increase in both alpha and beta activity during anger memory as well as during negative memories concerning self experience (Cannon, 2012).

Other Functions, Nondominant (Right) Frontal

The right frontal areas are **heteromodal** (versus the left, which demonstrates more modality-specific associations). The right frontal lobe integrates sensory information from many sensory channels. It *synthesizes information* from many *distant cortical areas,* like a fleet of airplanes vs. a fleet of taxis (left hemisphere) (Goldberg, 2001). In addition, there is more white matter in the right hemisphere and there are also more spindle cells in the right hemisphere. These spindle cells relay information from distant areas.

The right hemisphere **comprehends novelty**. It may convert novelty to stable familiar patterns or routines that are then part of left hemisphere functioning. Searching for novelty implies dissatisfaction. Seeking newness appears to be **driven more by norepinephrine** than dopamine and may be associated with restless and hyperactive, exploratory behavior. This decreases as we age. As people **grow older, they move from right to left dominance** both cognitively and emotionally.

In very general terms, **verbal IQ** is usually attributed more to left hemisphere (DH) activity while **performance IQ** is attributed more to right hemisphere (NDH) functions. Performance IQ involves some timed tests. Speed on these tests declines with age in adults. This is one of the reasons that norms for IQ are age-adjusted. (In children, speed increases with age, in part because more fibers become myelinated.) **Right hemisphere dominance** is usually thought of as being characterized by distractibility, stimulus-seeking behavior, seeking novelty and change, being emotionally involved, expressive, and extroverted. There may tend to be an external locus of control. There may be more of a tendency for **hysterical, impulsive, and even manic behaviors.**

Processing may tend to be fast and simultaneous in keeping with the Gestalt or holistic tendencies of the NDH. The dominant neurotransmitter is **noradrenalin**, which is associated with speeding up activity, and **serotonin**, which has long been associated with slowing down activity (Tucker, 1984). However, it is also true that decreased serotonin levels may be associated with inflexibility and reliance on known patterns (left hemisphere) as is found in obsessive compulsive disorder (OCD).

In addition, there are a number of other functions that show **predominantly nondominant,** right hemisphere, activation. These include:
- Sensing the Gestalt
- Parallel processing
- Attending to spatial relationships
- Representation of geometric forms
- Speech intonation, musicality, and vowel sounds
- Expressing emotions
- Orientation in space
- Holistic perceptions
- Responding to novel situations
- Intuition
- Eureka-like insight

Lesions in the right prefrontal region may affect any of these functions. The patient may exhibit **constructional apraxia** (inability to copy drawings or to manipulate objects to form patterns or designs). There may be difficulties with learning a maze or with nonverbal visual memory. Psychiatric syndromes, such as **manic behavior**, may be observed.

Comments Regarding Frontal Lobe Lesions

In very severe lesions the patient may not react to environmental instructions or cues and may react to irrelevant stimuli. There may be either echolalia or loss of voluntary speech. Agraphia is common and, in general terms, the patient appears confused.

Orbital Frontal Cortex and Links to the Temporal Lobe

The basal forebrain area is involved in **sleep-wake regulation**. **Concept organization** is also a function of this area of the hemispheres. The orbitofrontal cortex is critical for both **value-based behavior** and learning when value must be inferred. We have already noted that the orbital frontal cortex may be involved in **understanding the effect of one's own behavior on others**. The patient who has a lesion involving this area may be **emotionally labile** with decreased impulse control. **Loss of control** of anger, inappropriate laughter, crying, or inappropriate sexual behavior may be observed. However, with respect to these functions, a slightly different emphasis may be observed in the right and left hemispheres. **Impulsivity**, for example, may be more associated with dysfunction in the NDH orbital prefrontal cortex. Note that both these areas are discussed in further detail under Brodmann area 11 below.

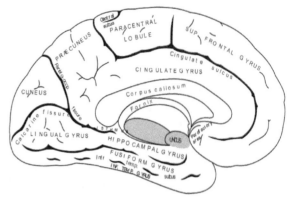

Diagram from Gray's Anatomy *showing uncus of Parahippocampal Gyrus*

Inferior/Basal Areas of the Dominant Hemisphere (Left)

The orbital/basal surface of the frontal lobe is connected to the temporal lobe by the **uncinate fasciculus**. Structures connected by the uncinate fasciculus from the temporal pole to the orbital surface of the frontal lobe have functions, in the left hemisphere, that include associations with general intelligence, verbal and visual memory, and executive performance. **Damage to the uncinate fasciculus on the left** side may be related to social anxiety and to symptoms of schizotypal personality disorder. These functions may be due, in part, to its connections to the limbic system.

Connections of the Uncinate Fasciculus

The uncinate fasciculus connects the entorhinal-uncus area, amygdala, and hippocampal region in the temporal lobes with frontal areas and, in particular, the orbitofrontal cortex and orbital-medial frontal cortex. The left uncinate fasciculus can relate to proficiency in both declarative and auditory-verbal memory (Mabbott, et al., 2009).

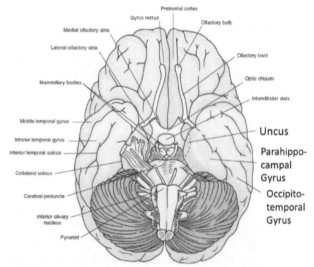

Inferior view adapted From Gray's Anatomy 1918. To illustrate location of Uncus.

Connections of the Uncinate Fasciculus in the Right, Nondominant Hemisphere

Similar areas on the right side may have somewhat different functions. The right temporal cortex (RTC) and the right prefrontal cortex (RPC) are connected by the right uncinate fasciculus. As noted previously, one related function on the right side is **autobiographical memory**, memories of one's self-in-the-past (Damasio, 1999) and recognition of one's own face versus the faces of others. This last function also involves the medial aspect of the ACC and the dominant (left) parietal cortex (LPC).

Deficiencies in the uncinate fasciculus connections in both hemispheres may correlate with symptoms that are found in autism spectrum disorders.

Lateral and Medial Aspect – Executive Network
Fz, F3 and F4: BA 8

BA 8

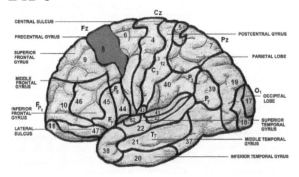

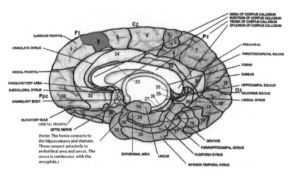

This area is **executive cortex** located in the prefrontal association cortex. As previously outlined, it is involved in **deductive and inductive reasoning**, monitoring, planning, calculating, sequence learning, working memory, speech and language processing, translation, generating sentences, and processing uncertainty. It is important in **memory retrieval**. It contains the frontal **eye-fields** and is involved in horizontal saccadic eye movements. It is a key area in **executive attention** (particularly visuospatial and visuomotor), self monitoring, and executive control of behavior. It is also important in self-image processing. Possibly related to an effect on these areas, it has been shown that LORETA Z-score NFB directed to this area can be effective to improve executive attention (Cannon, Congedo & Lubar, 2007, 2009). It is involved in motor learning and imagery.

Lesions affecting this area can result in tonic deviation of the eyes towards the side of injury. Along with other BAs in the premotor association cortex, lesions can affect motor planning, and coordination of complex sequences of movements, in addition to affecting working memory, executive attention, and some aspects of reasoning and planning. (See previous comments on inductive and deductive reasoning.) Because in the **dominant hemisphere** it overlaps with Broca's area, there may also be an effect on speech including an inability to **initiate speech** if there is damage in this area. It is also said to have some involvement in networks for **generating sentences,** language translation, and lip reading. Less often noted are possible changes in the **anticipation of pain** and in motor aspects of imagery.

FP1-F3, FP2-F4: BAs 9 and 10

Functions related to Brodmann areas in the regions between Fp1 and F3 and between Fp2 and F4: Note that the diagrams, for purposes of consistency, show just the left hemisphere for the lateral view and the right hemisphere for the midsagittal view. The reader can extrapolate to the homologous sites viewed from the right side of the brain. Keep in mind that an electrode location is on the scalp, not on the actual brain tissue, and thus scalp locations will actually be over slightly different aspects of the brain from one individual to the next.

Superior Frontal: BA 9

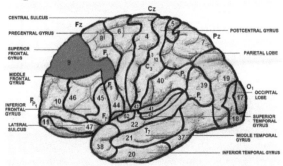

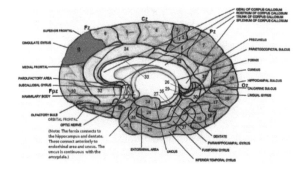

Right, Nondominant Hemisphere

Brodmann area 9 in the right hemisphere may be affected by NFB training that is carried out over the Right Superior Frontal area between Fp2 and F4. It is part of what is often referred to as the dorsal-lateral-prefrontal cortex (DLPFC). Together with BAs 45, 46, and the superior part of 47, it takes in much of the lateral aspect of the superior frontal gyrus and middle frontal gyrus. As will be clear to the reader from the diagrams, this is not a well defined area in the sense that it comprises a number of Brodmann areas. It may be considered to take in parts of areas 8, 10, 44, and, even more broadly, BA 11 and other areas. It is a very important area for most executive functions.

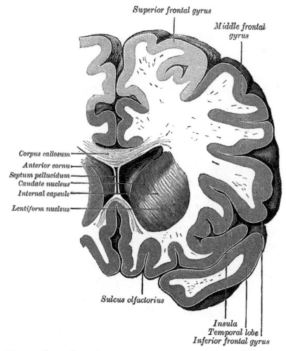

Figure from Gray's Anatomy

It is the executive control area for motor planning and organization. It is involved in the integration of sensory and mnemonic information and in the categorization of data. It is a key area in networks involved with working memory. It is important in networks that are involved in the interpretation of the emotional states of others and in empathy and forgiveness. Inference is a high level executive function that involves this area. Metaphors are liable to be a problem after a lesion in the right DLFC. An example of a metaphor is, "He's a bear in the morning." Understanding this metaphor would be difficult for a person with Asperger's syndrome. Such a patient would be able to interpret more easily a simile such as, "He is *like* a bear in the morning." The right frontal cortex is involved in word recall and, in conjunction with the parietal operculum and precuneus, in the interpretation of metaphors (G. Bottini, R. Corcoran, R. Sterzi, E. Paulesu, P. Schenone, P. Scarpa, R.S.J. Frackowiak, and D. Frith, 1994). Robert Sapolsky (2010) stated that understanding symbols, metaphors, analogies, parables, synecdoche and figures of speech also involves the insula, the angular gyrus, and the anterior cingulate cortex (ACC). The ACC is part of the network for monitoring and subjective evaluation of stimuli (later noted with respect to pain) and it is an essential part of the process in understanding metaphors. The point is that understanding metaphors involves many brain regions, not just a single area.

The right frontal areas are also among several important areas for people who use sign language. These areas are also among many areas involved in the detection of novelty.

BA 9 Both Hemispheres

Lesions in these areas result in difficulties with most executive functions. These functions include abstract thinking and **working memory,** such as encoding and retrieval functions. For the most part, in the dominant hemisphere, this involves verbal and language functions, whereas in the nondominant hemisphere, this involves visual-spatial information.

In general terms, this area in the prefrontal association cortex is part of the **executive cortex.** It is involved in short-term and working memory and in carrying out arithmetic computations. It is important in error detection and in verbal fluency. It is one of the areas involved in **understanding intentions and regulating behavior.** In conjunction with the cingulate gyrus, BA 9 can also play a role in **sexual arousal.**

In the dominant hemisphere it is involved in verbal fluency including generating sentences and word-stem completion. It is involved in **understanding idioms and categorizing** items.

Many aspects of cognitive executive functioning may be tested using **standardized tests**, such as the Wechsler Intelligence Scale (WISC, WAIS), other intelligence tests, the Tower of London (TOL), and the Wisconsin Card Sort (WCS), to name only a few. It is important to do one or more of these tests of cognitive functioning. It is crucial with patients who have a history of learning difficulties and in patients who have experienced a concussion, even in the seemingly distant past, because the results will help guide your treatment and assist in objectively evaluating the patient's progress in the treatment program. We try to make standardized testing part of the pre-training assessment and it is repeated, free-of-charge, after 40 sessions of training. It is very helpful to be able to demonstrate with objective measures that change has taken place, in addition to having anecdotal reports and questionnaire data. Consider online testing if you want to save time, such as that available through CNS Vital Signs division of Pearson, a publisher and distributor of psychological tests.

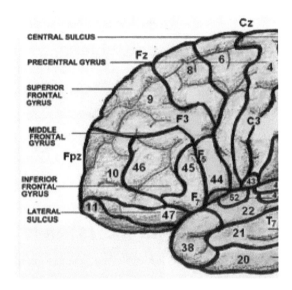

Another important function of this cortical area is governing *attention to and interpretation of emotional and contextual cues*. Although these functions may be more in the nondominant hemisphere, they are found in both hemispheres. In fMRI experiments, Goldberg et al. (2006) found evidence that the superior frontal gyrus when interacting with the sensory system is involved in self-awareness.

The **right, nondominant, prefrontal** cortex is associated with cognitive functions including: working memory, episodic memory, semantic monitoring, and motor imagery (McLeod et al., 1998). Again, however, these functions are not unilateral and both hemispheres are implicated in these functions. This understanding may be important if one is placing the cathode over this area when doing transcranial direct current stimulation (tDCS). The cathode may result in a decrease in activity, and it is often placed over the right orbit in experiments when the anode is placed over an area in the left hemisphere. Although the authors have neither personally experienced nor found any reports of negative effects resulting from this placement of the cathode for tDCS, it is still a possibility that should be recognized.

Left, Dominant Hemisphere, Superior Frontal: BA 9

Brodmann area 9 in the Left Superior Frontal gyrus, between Fp1 and F3, governs **attention to logical information and to detail**. It is also involved in organizing responses. Some liken it to an orchestra conductor. It is also a key area for **phonology** that involves BAs 9, 46, 6, and for **semantics** BAs 9 and 37. It can be involved in the interpretation of **idioms** such as, "It's raining cats and dogs." Patients with Asperger's syndrome often have difficulty with idioms because they take things literally, and idioms, by definition, are turns of phrase that defy literal interpretations because they do not mean what they say.

The **superior frontal gyrus** is also important in producing **laughter** (Fried et al., 1998). Fried found, when exploring to find the source of seizures in a 16-year-old female patient, that stimulation of a two square centimeter area on the left superior frontal gyrus produced laughter consistently. Fried and his colleagues found that increasing the level of stimulation current increased the duration and intensity of the laughter. At low currents only a smile was present, while at higher currents a louder, contagious laughter was induced. The laughter was also accompanied by the stopping of all activities involving speech or hand movements (Fried, 1998). The patient, AK, reported that the laughter was accompanied by a sensation of merriment or mirth. She would attribute it to something in her environment each time, not the stimulation.

Fp1, FP2 (To the left and right of Fpz in the figure)

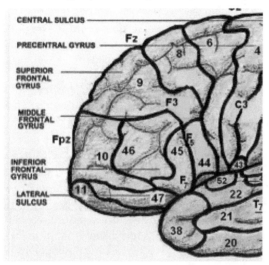
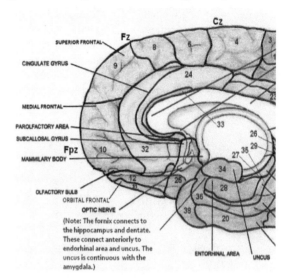

Functions Related to Brodmann Areas Near Fp1 and Fp2: BAs 10 and 11

Fp1, the left prefrontal area: BAs 10 and 11 (see also the Orbital Frontal section for additional discussion of BA 11)

BA 10   BA 11

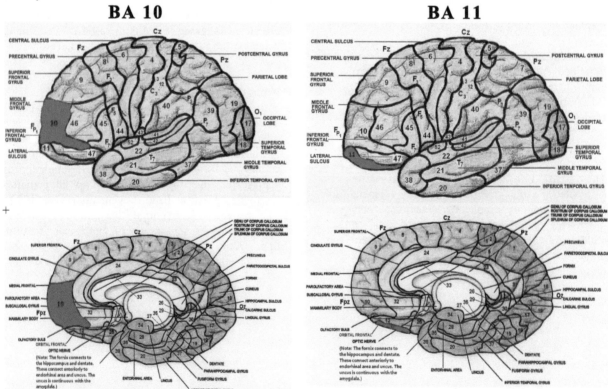

The **middle frontal gyrus** has a fundamental role in **memory** strategies and in controlling memory. Brodmann area 10 is involved in working memory including memory encoding and retrieval. The NFB practitioner understands that these are network functions and even at Cz one may observe an increase around 5.5 to 6.5 Hz when memory encoding and retrieval is taking place. Professionals who are beginning NFB work should watch/record their own spectrum while doing an EEG recording when they are concentrating on mental arithmetic problems or memorizing material. You may be as surprised as the first author was when his 5.5 Hz average amplitude was found to be very high, measured at FCz, compared to the rest of the spectral values at this site when doing math in his head.

BAs 10 and 11 are key areas for **executive functions** including planning and evaluating complicated goal-directed behavior (sometimes called 'executive control of behavior'). These areas are essential for **understanding complexity** in pictures, text, and behaviors. They are involved in high level cognitive reasoning including **making inferences** about other people. These areas are critical to categorizing and, in particular, to **semantic categorization.** They are important in the ability to stop one's course of action and take a different direction or approach, and are thus involved in **shifting mental set**. Deficiencies may result in impulsive judgments.

When this general area is not functioning normally there may be difficulty with abstraction. **Abstraction** is the act of considering something as a general quality or characteristic, as distinct from concrete realities, specific objects, or actual instances. For example, "Mary is laughing." is a concrete observation, whereas describing her as being happy is more abstract. This area is also important for drawing **analogies** (finding a similarity in some respects between things that are otherwise dissimilar). This area is involved in networks related to making decisions that concern emotions.

As previously described in more detail, this dominant (left) frontal area is also important in networks for **language processing**. This includes verbal strategies such as syntactic processing, generating sentences, and, perhaps more in the NDH, metaphor comprehension.

Interestingly, this area may also involve quite diverse functions that include attention to baroreceptor stimulation and to painful thermal stimuli.

Brodmann areas 10 and 11 of the prefrontal cortex are functionally **linked to BA 46**, particularly on the left. They are involved in executive cognitive functions including working memory, and the ability to **plan, reflect, make decisions**, **shift mental set**, initiate response, evaluate response, correct one's response, and to modulate and **inhibit impulses**. These areas are important for the highest levels of cognition involving careful **analysis, judgment**, and decision making involving complex problems or situations and inferential reasoning.

Detection and **processing of errors** involves the cingulate gyrus in addition to these frontal executive areas. The EEG can correlate with the complexity of the context of cognitive activity (Kurova et al., 2007). As previously noted, standard intelligence testing such as with the Wechsler Intelligence Scales, will help to assess many of these cognitive functions.

The **dominant (left) cerebral hemisphere near** Fp1, Brodmann areas 11, 10 and 46, is also involved in verbal analytic cognition and **approach behaviors**. Richard Davidson's work on emotions showed that left frontal activation was associated with positive thoughts and approach behavior. In the dominant hemisphere the anterior cingulate and dorsal-medial PFC is important in **theory of mind** functions (Amodio & Frith, 2006). Calculations and numerical processes are also carried out in these areas.

Fp2, the Right, NDH Prefrontal Area

BAs 11, 10 and 46 on the right side share many of the higher level functions described above. The **nondominant** hemisphere may differ from the dominant hemisphere, however, in that dysfunction may result more in **overactive behavior and this may correlate with irritability, impulsivity, tactless**, and even manic behavior. Other patients may demonstrate **panic, and avoidance** behavior. Davidson found right frontal activation in association with a focus on **negative thoughts and social avoidance**. These areas have important **links to the amygdala**. The right prefrontal area involving BA 10 is also involved in inhibiting action. Elsewhere we have noted that the medial aspect of the prefrontal cortex exerts tonic inhibition of the central nucleus of the amygdala (Thayer, 2012).

BA 10, NDH

Brodmann area 10 in the right prefrontal cortex is involved in aspects of memory recall including spatial memory, **working memory, calculations, and attention.** It is also involved in networks for the recognition of emotions and understanding of **metaphors** and in analyzing **risk versus benefit**. In the nondominant hemisphere, increased activity may be linked to unpleasant emotions, **avoidance behavior** and **anxiety**. Increased activity can correlate with increased activity in the anterior cingulate, posterior cingulate and cuneus, and also increased activity in the insula and in the amygdala-hypothalamic-pituitary-adrenal axis. BA 10 also **responds to baroreceptor stimulation** and may have a role in the regulation of the autonomic nervous system. This finding may be important in understanding the effects of HRV training.

F3: BAs 46 (8, 9)

BA 46

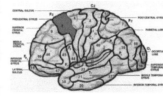

BA 8

BA 9

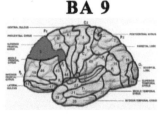

As previously noted under the general discussion of the prefrontal cortex, the functions related to BA 46 (also involving BAs 8 and 9), in the prefrontal, dorsolateral cortex, include making **judgments, planning, monitoring**, revising, **sustaining attention**, working memory, inhibition of responses, and verbal and episodic **memory retrieval. Spatial information is processed** through both the **parietal and dorsolateral frontal cortex** areas. All of these functions are part of networks that involve, as previously outlined in Part 1, specific cortical-basal ganglia-thalamic-cortical circuits.

BAs 46, 8, 9, DH

Functions related to BAs 46, 8, 9 in the DH (left) hemisphere under F3 include problem solving, **sequencing, concrete thinking** (extreme in Asperger's), and **deductive** reasoning. Left frontal lobe activation is also considered to be associated with positive emotions and **approach behavior** (Davidson, 1995).

F4: BAs 46, 8, 9, NDH

As previously noted under the general discussion of the prefrontal cortex, the functions related to BA 46 (8, 9) in the **NDH (right)** hemisphere include evaluating the **Gestalt** and **the context** of information and situations. In the past it was felt that inductive creative thinking was right hemisphere. Although this is important, both deduction and induction have major components in the left (dominant) hemisphere, as previously discussed. Important, however, is the **shift** that takes place from youth where creative thinking and novelty seeking – **right hemisphere functions – shift** more and more **with age** to the more concrete and pattern-dependent thinking of the dominant (left) hemisphere. The right frontal lobe, when more active than the left, has been found in association with negative emotions and **avoidance behavior**. The nondominant (right) hemisphere is involved in **metaphor**ical thinking, short term retrieval of **spatial-object memories**, vigilance, and both selective and **sustained attention**.

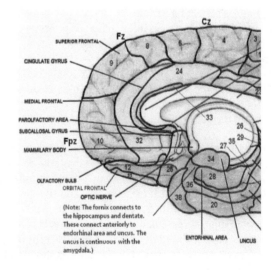

Frontal Cortex

Lateral and Medial views

F5: BAs 44 and 45

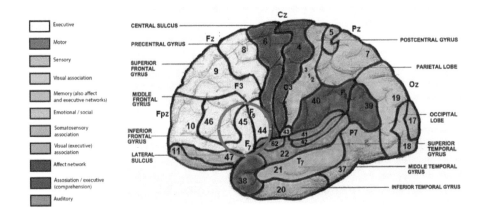

F5 – F7; BAs 44, 45

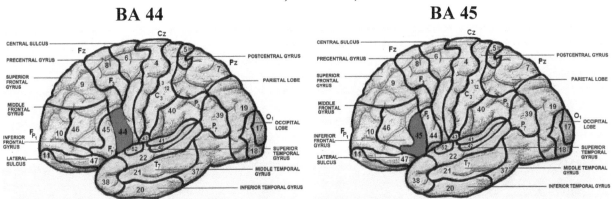

BA 44   BA 45

BA 44, Dominant Hemisphere

Brodmann area 44 is called the **pars opercularis** ("part that covers"). It **covers a portion of the insula**. It was previously mentioned as being important in **deductive reasoning**. Brodmann areas **45 (Pars Triangularis) along with BAs 44 and 46**, in addition to their importance as being part of **Broca's** area in the dominant hemisphere, contain neurons related to the **mirror neuron system** (MNS). The MNS is postulated to be involved in the imitation of movements, and perhaps also in copying appropriate social interactions, as well as being critical to understanding and predicting the behavior of others.

Although **mental rotation** involves the nondominant parietal regions including the angular gyrus, frontal BA 45 is also involved. Important **MNS areas** in the left hemisphere (there are corresponding areas on the right) include frontal near F5, the temporal pole, the temporal-parietal junction, and activities in the anterior insula and the anterior cingulate gyrus. Each area is postulated to have mirror functions that correspond to the functions of that area of the cortex. **Mirror neurons have strong connections to the limbic system including the anterior cingulate (AC)** (Iacoboni & Dapretto, 2006). The cingulate and the insular cortices both contain mirror neuron cells (Ramachandran & Oberman, 2006).

A functional magnetic resonance imaging (fMRI) study demonstrated that activity of the MNS is **correlated with empathic concern** and interpersonal competence (Dapretto, Davies, Pfeifer, Scott, Sigman, Bookheimer & Iacoboni, 2006; Pfeifer, Iacoboni, Mazziatta & Dapretto, 2005). It has also been shown that children with autism spectrum disorders (ASD) have reduced activity in MNS regions during tasks that require the child to mirror facial expressions of different emotions (Dapretto et al., 2006). Findings at the ADD Centre using 19-channel QEEG and LORETA source locations show these areas to be quite consistently outside the database norms (> 2SD) for same age children in clients diagnosed with autism spectrum disorders (Thompson & Thompson, 2010b). There is also a reduction in the volume of gray matter in BAs 44 and 45 in children who have a diagnosis along the autistic spectrum.

BAs 44, 45, DH

The **dominant** frontal cortex under F5 includes Brodmann areas 44 and 45. As previously noted, **Broca's area** is a name that has been applied to this general region, especially when one is discussing verbal language functions including semantics, phonetics, verbal understanding, categorization, speech articulation, selection of information from competing sources, grapheme to phoneme conversion, grammatical processing, and fluency of speech.

Paul Broca

Paul Broca (1824 – 1880) was the French physician and anthropologist who published the first anatomical proof of localization of brain function.

Paul Broca (1824 – 1880)

Broca's evidence, presented in 1861, was based on autopsy findings in a man who had lost the ability to produce speech. He confirmed his initial finding with a dozen more autopsies on the brains of similarly afflicted patients performed in the next two years.

Anterior Insula & Speech

Interestingly, the **anterior insula** may also be involved in some of these **motor-speech functions**. It is essential for processing sequential sounds and for lexical inflection. BA 44 of the dominant hemisphere, as mentioned above, includes frontal **mirror neurons** whose activation is important for empathy and for **understanding the intention of others**. These areas have mirror neuron functions for expressive speech. The mirror neuron areas may also be important in **internal speech** and in the internal rehearsal of information. It is one part of the working memory networks and is important in episodic, syntactic, semantic, and declarative memory encoding. BA 45, the pars triangularis, has a role in the cognitive control of memory. It is a key area in **selective attention to speech**. It is important in sign language, too. When reading aloud, written language needs to be **decoded and correctly pronounced** and this is another function of Broca's area. The role played by these areas in **deductive** reasoning (dominant hemisphere) and inductive reasoning (dominant and nondominant hemispheres) has been previously discussed.

BAs 44, 45, NDH

In the **nondominant** hemisphere, **BAs 44, 45** are involved in expression of **vocal tones and prosody.** Along with the supramarginal and angular gyrus area at the temporal-parietal junction, this area may be involved in the perception of nuance, innuendo, and intonation in speech, though these frontal regions are more involved in language production (putting emotion into speech) and the parietal regions are more involved with language understanding (interpreting tone of voice). It is one part of a network necessary for sentence comprehension and grammar. It is involved in the generation of **melodies** in addition to music enjoyment. It is necessary for arithmetic calculations and sequencing of data. It is involved in **networks for the inhibition of actions**. It is important for word generation and working memory including **syntactic working memory,** episodic and declarative memory, encoding of words and of faces. It is important for arithmetic and **music appreciation**. In conjunction with the dominant hemisphere, it is important for inductive reasoning.

When there is **dysfunction** in these areas **in the nondominant hemisphere,** the patient may present with a **motor aprosodia** where vocal tones are flat or they do not seem to be emotionally appropriate to the situation, although the patient may understand emotions perfectly. Clinicians who work with people with Asperger's syndrome will not be surprised to learn that LORETA analysis for 19-channel EEG data often identifies abnormal findings with NDH BAs 44 and 45 as the source location. Those with Asperger's symptoms often talk in a monotone without any apparent emotional expression. The seemingly flat affect communicated by the monotone voice occurs even when they are experiencing emotions, as shown by other physiological measures. This **lack of emotional expressiveness** in a person's voice is also found in people who have suffered a stroke affecting this area; it is referred to by neurologists as motor aprosodia.

It is important to repeat that the nondominant inferior frontal cortex that includes these areas of BAs 44 and 45 has been found to be important for the **inhibition of behaviors** (Aron, 2004).

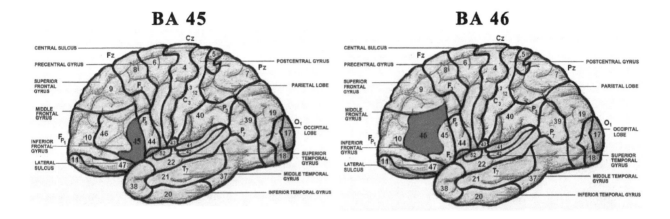

F7 (and between Fp1, F3, and F7): BAs (38), 44, 45, 46, (47), DH

F8 (and between Fp2, F4 and F8): BAs (38), 44, 45, 46, (47), NDH

F7 (and between Fp1, F3, and F7): BAs (38), 44, 45, 46, (47), DH

NFB over the F7 site can influence **Brodmann areas 38, 44, 45, 46, and 47** on the left side. In general terms, think of the functions previously described under the dorsolateral prefrontal cortex and the **executive system** as being influenced by training done at the F7 location. These include visual and auditory **working memory, selective attention**, word retrieval, and functions related to Broca's area including generating words, **verbal fluency, lexical functions, phonemes, grammar, categorization,** syntax. These areas are involved in episodic, declarative, and working memory. As previously mentioned, these areas are a keystone in most aspects of executive functioning.

F8 (and between Fp2, F4 and F8): BAs (38), 44, 45, 46, (47), NDH

Training at the F8 site can influence Brodmann areas 38, 44, 45, 46, and 47 on the right side. In general terms, think of functions previously described concerning the area in the right hemisphere that is homologous to Broca's area. These include spatial & visual working memory, sustained attention, prosody, sensing the Gestalt, and conscious processing of facial recognition information and of information related to emotions. Sensing the **Gestalt** refers to seeing the pattern or the whole picture. **Mood regulation** involving emotion-contextual information can be related to this area, too. The area between F8 and T8 (formerly T4) may be found to be outside database norms in rare cases where **labile mood** with sudden mood changes, such as sudden anger, and impulsivity may be found. The right ventral prefrontal cortex is involved in **anxiety** associated with psychological stress and the ventral lateral prefrontal cortex may be involved in the **activation of cortisol** release (Wang et al., 2005).

This area of the cortex has **strong links to the medial prefrontal cortex, insula, and anterior cingulate**, which are areas involved in increasing cortisol secretion in response to stress. Although the ventral lateral prefrontal cortex does not have strong connections to the hippocampus, these other areas do. Functions related to structures connected by the right uncinate fasciculus may be influenced by NFB over an area between F10 and T8. These can include the capacity for **autonoetic self awareness,** which refers to the ability to **re-experience previous events** from one's past as a continuous entity across time (Levine et al., 1998).

Anger, however, may be primarily related to the Affect network and the stress response system in general, with both left and right ventral prefrontal areas being involved. Some research has demonstrated that **motivational direction** in 'dispositional anger' is an approach-related motivational tendency with a negative valence. It is associated with greater left- than right-anterior activity. This suggests that the anterior asymmetry with emotions may vary as a function of motivational direction rather than affective valence (Harmon-Jones et al., 1998). This is a somewhat different emphasis than that utilized in the NFB alpha-asymmetry protocol developed by Peter Rosenfeld and put into practice by Elsa Bhaer (Rosenfeld, 1996, 1997) that was based on the work of Richard Davidson (Davidson, 1995, 1998) showing left frontal activation was associated with approach behavior and positive mood. The lesson for practitioners is a fundamental one. In clinical practice, do not immediately apply any protocol without doing a careful clinical assessment and EEG assessment. Then follow the findings to design an intervention for the individual patient.

BA 46

Brodmann area 46 roughly corresponds to a portion of the **dorsolateral prefrontal cortex (DLPFC)**. The DLPFC comprises the lateral aspects of BAs 9, 10, 11, 12, 45, 46, 47. The DLPFC plays a key role in executive functioning including sustaining attention and working memory. As previously mentioned, we **define executive functioning** as the overall capacity to manage perception, attention, selection, decision making, inhibition, memory, planning, problem solving, logical thinking (deductive/inductive), sequencing; and to monitor, evaluate and self-correct outputs; and to respond verbally/nonverbally, motorically, and socially for the purpose of the attainment of defined goals.

Lesions in this area have been associated with what is sometimes termed 'dysexecutive syndrome'. Lesions to the DLPFC **impair short-term memory** and cause **difficulty inhibiting responses**. Lesions may also eliminate much of the ability to make judgments about **what is relevant** and what is not. Lesions will also cause problems in the ability to organize. In addition, the DLPFC has been found to be involved in exhibiting **self-control**. The Tower of London Test **(ToL)** is one example of a test that can help the practitioner to **evaluate** many aspects of executive functioning that are primarily related to **BA 46, particularly in the dominant hemisphere**. The ToL involves ten trials of changing the position of colored rings by placing them on three pegs in order to copy a pattern shown by the examiner. It requires the subject to inhibit immediate responses, plan, shift mental-set, use working memory, initiate a thought-out response, and then monitor and evaluate the results of that response. The required cognitive functions all depend on good prefrontal lobe functioning. Part of this functioning involves the dorsolateral prefrontal cortex near F5 and extending downward to F7.

F9 and F10 & Orbital Surface

Neurofeecback using these sites may have an effect on BAs 11, 47, 38. These sites, which are not included in the usual 19-channel 10-20 system sites, may sometimes be used in NFB training due to the need to influence important functions of the orbital surface of the frontal lobes. These functions include understanding the effects of one's own behavior on others.

The orbital surface of the left frontal lobe was the area pierced by the 13 lb. iron tamping rod in the sad case of **Phineas Gage**. This railway worker displayed a total change in personality after the brain injury and, in the words of his physician, became "fitful, irreverent, impatient, vacillating." In general, he became extremely difficult for anyone to get along with and he often got into fights. Though Phineas Gage had a very dramatic injury to his **left orbital frontal** cortex, less extreme but similar difficulties are now being reported in athletes and veterans who have suffered concussions involving this area of the cortex.

The orbital surface of the frontal lobe has connections to the temporal lobe and other structures by way of the uncinate fasciculus. The **left uncinate fasciculus** includes functions relating to proficiency in **auditory-verbal memory** and declarative memory (Mabbott et al., 2009).

As noted above, the **right uncinate fasciculus** is involved in the capacity for **autonoetic self-awareness**; that is, re-experiencing previous events from one's past as part of a continuous entity across time (Levine et al., 1998). (Synonyms for autonoetic consciousness are self knowing and self awareness.)

BA 47

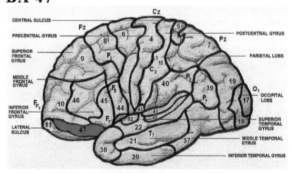

Brodmann area 47 is the part of the frontal cortex in the human brain that curves from the lateral surface of the frontal lobe into the ventral (orbital) frontal cortex and is therefore involved in many functions related to **deductive reasoning, language, and emotions**. It is below **BA 10 and BA 45,** and parallel to **BA 11**.

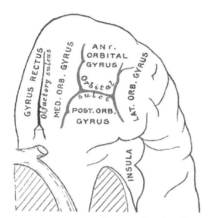

Gray's Anatomy *to show Orbital Sulci*
Public Domain via Wikimedia Commons –

In the human, **on the orbital surface,** it surrounds the caudal portion of the orbital sulcus from which it extends laterally into the orbital part of the inferior frontal gyrus. Like nearby areas previously described, including Brodmann areas 45 and 46, it is part of the **Executive network**. In the dominant hemisphere it is important for **categorization and aspects of language** and speech, and is also important in episodic and **working memory**. It is one of the key areas for **deductive** reasoning. It is important for making inferences. In conjunction with other cortical areas, it helps in **monitoring** actions and incoming information. It is activated when inner speech is used during the performance of tasks. With the insula it may be involved in processing **temporal coherence in music** (Cannon, 2012, p160). It is important in monitoring behavior.

Inferior Frontal Gyrus in the DH

The posterior medial parictal cortex and left prefrontal cortex (PFC) have both been implicated in the recollection of past events. The posterior precuneus and left lateral inferior frontal cortex are activated during episodic source memory retrieval. The **precuneus and the left inferior PFC are important for regeneration of rich episodic contextual** associations. The left ventro-lateral frontal region/frontal operculum is involved in searching for task-relevant information (BA 47) and subsequent monitoring or scrutiny (BAs 44 and 45). Regions in the dorsal inferior frontal gyrus are important for information selection (BAs 45 and 46) (Lundstrom et al., 2005).

Semantic memory comprises general word knowledge, such as word meaning and word use, and it functions to store, retrieve, and associate this information in conjunction with environmental stimuli. On the other hand, **working memory** refers to the short-term maintenance and manipulation of information during processing; keeping something in mind long enough to do something with it, like remembering a telephone number long enough to write it down. Finally, **episodic memory** is employed for long-term storage and recall of previous experiences or episodes, and it allows people to reflect upon their personal past. Entailing more than just event memory and event recall, **episodic memory** includes a special awareness for subjective time known as autonoetic consciousness; it enables people to mentally travel backwards in time and knowingly **retrieve information** from a given personal experience (Lundstrom, 2005). Episodic memory allows current knowledge to be associated with past experiences.

Connections to the Default Network

Several areas of the brain show decreased activity during cognitive tasks and increased activity in the so-called resting state. These include parts of the posterior cingulate cortex (PCC) and medial frontal region incorporating portions of the medial frontal gyrus (MFG) and ventral anterior cingulate cortex (vACC). These areas are part of the **Default mode network** that is engaged during rest and disengaged during cognitive tasks. A decrease in activity is usually inferred when theta increases.

'Hippocampal' Theta and Memory

With the **engagement of working memory, 4 - 7 Hz** (sometimes more specifically around 6 Hz or 5.5 - 6.5 Hz) **synchronized activity may increase and be related to a decrease in hippocampal metabolism** (Hampson et al., 2006; Uecker, 1997). It is detected over the frontal midline. This is sometimes termed **hippocampal theta** and thus the source is different than the excess theta produced in the thalamus that is a marker for Attention-Deficit/Hyperactivity Disorder.

Hampson et al. state that an inverse relationship between activation (as measured using functional imaging) and engagement appears to exist in the hippocampus. They give the example of deactivation in the hippocampus (and adjacent regions) during a transverse patterning task and during virtual navigation of a radial arm maze (Astur & Constable, 2004; Astur et al., 2005). They note that these tasks require engagement of the hippocampus but hippocampal activity decreases during these tasks. They posit that coding of information occurs in terms of neural synchrony rather than rate of neural firing. The **theta rhythm reflects this synchronized activity** and is associated with memory and cognitive function (Gevins et al., 1997; Tesche & Karhu, 2000). They relate this to the observed decreased metabolism (Uecker et al., 1997).

Connectivity between regions is increased during these tasks (Hampson et al., 2006). Clinically, Hampson's observations give support to the notion that we must pay attention to low amplitude findings and increase synchronized slow wave activity when it is well below the database norms. In particular, if amplitudes are low, it may sometimes be appropriate to consider enhancing 6 Hz 'hippocampal theta' because it may be associated with memory functions.

NDH Right Inferior Frontal Region BAs 47, 12

Another function that has been observed for the inferior frontal gyrus (IFG) has come from research concerning the **right lateral inferior frontal** gyrus during **go/no go tasks**. In such a task, the participant has to **inhibit** a prepotent response (for instance, not pressing a button when one signal appears after pressing for another signal). This task involves the general area of BAs 47 and 12. (References that give good descriptions of this include Aron et al., 2004; Kringelbach et al., 2005; Menon et al., 2006; Li et al., 2006.)

It seems that the same area is also implicated in **risk aversion**: higher risk aversion is correlated with higher activity in the IFG (Christopoulos et al., 2009; Knoch, 2006). This might be explained as an inhibition of a signal to accept a risky option. Disruption of activity of this area with transcranial magnetic stimulation (TMS) or transcranial direct current stimulation (tDCS) can lead to a change in risk attitudes, as behaviorally demonstrated by choices involving risky outcomes. Theoretically, dysfunction in this area could lead to dangerous behavior. On a more positive note, it might be possible to decrease such negative behaviors using LORETA NFB.

Orbital Frontal/Inferior Frontal Cortex
Affect Network – Emotional/Social: BAs 11, 12

BA 11

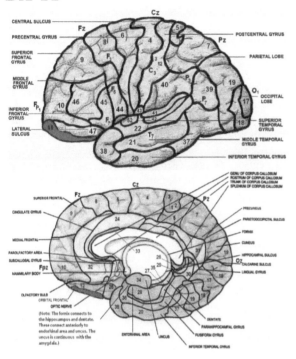

The ventral medial orbital frontal cortex is equivalent to the ventromedial **reward network** (Ongur, 2000). As a reward network, this area is broader than the orbital surface of the frontal lobe and it includes Brodmann areas 10, 11, 12, and areas in the cingulate gyrus 25, 32, and the insula BA 13. However, the term *ventromedial prefrontal cortex* is used differently in different articles. In this section we will just discuss BAs 11 and 12.

Brodmann area 11 includes the **orbital frontal association** cortex. BA 11 covers the medial part of the ventral surface of the frontal lobe. It is bounded on the rostral and lateral aspects of the hemisphere by the frontopolar area 10, the orbital area 47, and the triangular area 45; on the medial surface it is bounded dorsally by the rostral area 12 and caudally by the subgenual area 25. These surrounding areas may influence the activity at BA 11 and 12, the orbital frontal cortex. This area of the cortex is involved in the modulation of affect and is an integral part of the **Affect network**. The ventral lateral prefrontal cortex (VLPFC) in both the nondominant (ND) and dominant (D) hemispheres, in conjunction with the ND insula, amygdala, and anterior temporal pole (BAs 21, 38) may, for example, show increased activity with subjective **sadness**.

BA 11 is also important in social and cultural behavior. It is involved in **empathy** and, along with areas in the right parietal-temporal junction, in **understanding what others may be thinking**. Compromised function in the orbitofrontal lobes, an area linked to social and cultural behaviors, will result in the patient demonstrating difficulty with abstract rules, empathy, reflecting on the **effect of their behavior on others** (consequences of actions) and executive functions including self-expression, problem solving, **willpower and planning.** In addition, reward and conflict are analyzed in this area, as are unexpected outcomes to actions or events. **Lesions can lead to personality changes** including **disinhibited behavior** such as swearing, compulsive gambling, drug and alcohol abuse, and hypersexuality. Lesions can lead to disinhibited and aggressive behavior, as was seen in the famous case of Phineas Gage.

Phineas Gage (1823-1860)
As previously mentioned, Phineas Gage (1823-1860) was a foreman with a railway building crew when in 1848, at age 25, he suffered a severe brain injury. A 13-pound iron bar, 3 ft. 7in. x 1¼ in., that he had been using to tamp an explosive charge, flew upwards when the charge exploded prematurely. The bar went cleanly through his cheekbone and skull, exiting from the top of his head. It pierced the orbital surface of his left frontal hemisphere. He was never unconscious. There was, however, an immediate change in temperament and behavior. He had been a wonderful, thoughtful, kind, individual who was well liked by everyone. He was also said to be the most efficient and capable foreman at his company.

 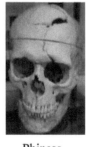 

After Accident   Phineas Gage's skull   Dr J. Harlow

After the accident he became fitful, irreverent, impatient, and vacillating, showing a very negative personality change. As time went by he worked hard as a stagecoach driver in Chile and did make some recovery. He had figured out how to live even with respect to social interactions. His case was followed until he died in San Francisco, a result of a seizure, by his physician, Dr. Harlow, and Dr. Harlow's notes have become the classic account of the effects of brain damage involving the orbital frontal and ventral medial prefrontal cortex.

This area is involved in the analysis of **idioms**, the encoding of words and, in a network that includes the nondominant fusiform gyrus, putting **faces to names**. As can be understood from these functions, this area of the frontal lobe is considered to be a higher order cortical association area. It is an important part of the **Default network,** which is activated when one considers aspects of 'self' and 'self-in-relation-to-others.' It plays a key role in feeling **empathy**, an objective understanding of the emotions of another person. In addition, it has areas from an early time of phylogenetic development that are involved in olfaction. This is doubtless because a good sense of smell confers survival value in many species.

BA 12

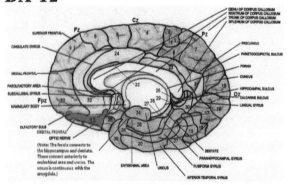

Brodmann area 12 occupies the most rostral portion of the medial aspect of the frontal lobe and includes areas that were originally subsumed under BA 11. This area has been labelled as part of the **paralimbic association** cortex and it is involved in **assessing the affect or emotional value of different reinforcers**. It is also a higher level association cortex that is involved in the **generation of hypotheses**, discrimination of **'same' and 'different,'** making decisions, and in formulating expectations. It has a role in **sensitivity to reward versus punishment**.

Nucleus Basalis of Meynert and Acetylcholine

The nucleus basalis of Meynert (NBM), though it is not located in BA 11 or 12, is **discussed here because of its relation to executive** functions and its connections to executive areas in the dorsal lateral prefrontal cortex, in addition to connections to the basal forebrain and areas within the temporal lobe. The NBM is a diffuse grouping of neurons within the basal forebrain that projects to widespread areas of the neocortex. The NBM is **inferior to the globus pallidus** and within an area known as the substantia innominata. The NBM is immediately inferior to the anterior commissure and superior and lateral to the anterior portion of the hypothalamus. It is the main source of **acetylcholine** (Ach) in the neocortex (Mesulam & Guela, 1988). Neocortical cholinergic innervation principally comes from the Nucleus Basalis of Meynert. It is interesting to note that the **orbitofrontal cortex projects into the nucleus basalis** and thus has the possibility of influencing cholinergic input to the entire cerebral cortex (Mesulam, 1988). This is important in part because Ach is a key neurotransmitter in memory networks.

Anatomical Connections

The other regions **projecting to the nucleus basalis** include the hypothalamus, nucleus accumbens, piriform cortex, entorhinal cortex, temporal pole, anterior insula, septal nuclei, and posterior parahippocampal cortex. We have noted previously that these areas also influence the hypothalamus and the autonomic nervous system. The Nucleus Basalis of Meynert and the cholinergic innervations to the hippocampus and temporal lobes are crucial to laying down new (episodic) memories and are said to be the **first areas to be affected by Alzheimer's** disease. However, there is also an **abnormal connectivity between the posterior cingulate and the hippocampus in early Alzheimer's** disease, which is accompanied by mild cognitive impairment (Zhou et al., 2008).

Alzheimer's and Exercise

We have noted above that the Nucleus Basalis of Meynert and the cholinergic innervations to the hippocampus and temporal lobes are the first to be affected by Alzheimer's disease. However, there are factors that may **influence the degree and speed of progression** of this disorder. Two such factors that should also be considered when discussing deterioration in both physical and cognitive abilities from any cause are diet and exercise. Diet and supplements are of major importance and these will be discussed further in the section on TBI interventions.

Exercise is an important consideration because it has been demonstrated that exercise can **improve neurotrophic factors**, such as the production of brain-derived neurotrophic factor (BDNF) and glial cell neurotrophic factor (GDNF), which are widely distributed throughout the brain and which modulate brain plasticity by **increasing neuritic outgrowth and synaptic transmission.** They are also protective against specific nigral (from the substantia nigra) toxins.

Such toxins include 6-hydroxydopamine (6-OHDA), which destroys neurons and is used to induce experimental Parkinson's in animals. Another such toxin is called MPTP. MPTP (1-methyl-4-phenyl-1,2,3,6-tetrahydropyridine) is a neurotoxin precursor to MPP+, which causes permanent symptoms of Parkinson's disease by destroying dopaminergic neurons in the substantia nigra. It has also been used to study disease models in animal studies related to Parkinson's disease.

In **Parkinson's disease,** nigral BDNF and GDNF expression is diminished (Chauhan et al., 2001). Both these factors cross the blood-brain barrier and levels can therefore be measured in the serum and in saliva. In serum, the level of these neurotrophic factors has been shown to rise proportionally after exercise. It has also been demonstrated that regular **exercise in humans reduces the risk of mild cognitive impairment (MCI)** (Geda et al., 2010), dementia, and Alzheimer's disease (Hamer & Chida, 2009). It has been shown that patients suffering MCI or dementia experienced **significant cognitive improvement with exercise** as compared to subjects randomized to a sedentary intervention (Baker et al., 2010; Kwak et al., 2008).

BDNF is a neurotrophic factor that is critically linked to neuroplastic changes (Lisanby et al., 2000), and might serve to index neuroplastic effects induced by repetitive transcranial magnetic stimulation (rTMS) (Brunoni et al., 2013). In addition, it has been shown that brain-derived neurotrophic factor **(BDNF) may induce the accumulation of AMPA receptors** at synapses previously devoid of these receptors. The α-amino-3-hydroxy-5-methyl-4-isoxazolepropionic acid receptor (AMPA) is a glutamate receptor (Bear, 2004; Cook & Bliss, 2006). These receptors play a **key role in long-term potentiation (LTP)** at the synapse, and are discussed further under the section on LTP in *The Neurofeedback Book*. It is hypothesized that NFB can facilitate LTP through repeated strong activation at synaptic junctions and that this may be one factor responsible for the lasting effects of NFB.

Orbitofrontal and Ventromedial Prefrontal Cortex (VMPC)

At the time of writing this text, the functional differences between the orbitofrontal and ventromedial areas of the prefrontal cortex have not been entirely established. It is known that the areas of the ventromedial cortex **superior to the orbitofrontal cortex** are much less associated with social functions and are **more involved with the regulation of emotions.** Research in developmental neuroscience also has suggested that neural networks in the ventromedial prefrontal cortex are rapidly developing during adolescence and young adulthood. These areas function in conjunction with the amygdala to regulate emotions. This regulation will alter cortisol levels. (See also the section concerning the Stress Response in *The Neurofeedback Book*.) Those with low self-esteem will show high levels of cortisol release in response to tasks that they associate with stress (Cannon, 2012, p70; Holmes & Rahe, 1967; Dedovic et al., 2005, 2009).

The **left lateral and medial orbitofrontal cortex** areas have also been found to be highly active during guessing tasks. An increase in probabilistic scenario complexity is associated with an orbitofrontal cortex activity level increase, suggesting the special role that the ventromedial prefrontal cortex plays in decision making containing uncertainty. In addition, patients with lesions in the ventromedial prefrontal cortex tend to have difficulties when considering the future consequences of their actions when making decisions.

The **right half of the ventromedial prefrontal cortex** is associated with regulating the interaction of cognition and affect in the production of empathic responses. **Hedonic (pleasure) responses** are also associated with orbitofrontal cortex activity level (Kringelbach et al., 2005). These functions suggest that the ventromedial prefrontal cortex is associated with a person's judgment when deciding on preferences. They also suggest that the ventromedial prefrontal cortex has a key role to play in constructing one's **sense of self.** It is one of the links in the Default network.

The ventral medial prefrontal cortex gives **tonic inhibition,** via GABAergic mediated projections, to the central nucleus of the **amygdala.** The amygdala connects to both the autonomic nervous system and to the endocrine system. If there is hypoactivation of the prefrontal cortex, there may be deficient inhibitory control of the amygdala, and it can become active in difficult circumstances, such as with threat and uncertainty (Davidson, 2000; Thayer, 2012, 2006).

Studies with **post traumatic stress disorder** (PTSD) support the idea that the ventromedial prefrontal cortex is an important component for reactivating **past emotional associations** and events, therefore essentially mediating the pathogenesis of PTSD.

Treatments geared to the inhibition of the ventromedial prefrontal cortex may therefore be helpful for those suffering from PTSD. It has been shown that the right half of the ventrolateral prefrontal cortex, which is active during emotion regulation, is activated when people are given an unfair offer in an experimental scenario. In other experiments, specific deficits in reversal learning and decision-making have led to the hypothesis that the ventromedial prefrontal cortex (**VMPFC) is a major locus of dysfunction** in the mild initial stages of the behavioral variant of **frontotemporal dementia**.

These are all areas that require further research, but they may give the NFB practitioner a sense of some of the functions that may be affected with NFB directed to these areas, as can be done using LORETA NFB (LNFB).

One particularly notable theory of VMPFC function is the somatic marker hypothesis, accredited to António Damásio. By this hypothesis, the VMPFC has a central role in adapting somatic markers – **emotional associations**, or associations between mental objects and visceral (bodily) feedback – for use in natural decision making. This account also gives the VMPFC a role in **moderating emotions** and emotional reactions. Ventromedial prefrontal cortex lesions have also been associated with a deficit in processing gender-specific social cues.

When we do NFB using surface electrode sites, the VMPFC and orbital cortex are distant from these surface sites. However, we can influence these areas by having an effect on other cortical areas that are within the same neural network. For example, surface NFB at Fz and FCz (between Fz and Cz) may affect the anterior cingulate cortex and, thus, the Affect network. Electrode placement at either F9 or F10 may affect the lateral aspect of the orbital frontal cortex. However, F9 and F10 may also be affected by both muscle and eye-movement artifacts, making it a difficult area to work; thus, the theoretical value of using LORETA Z-score NFB (LNFB) to directly influence these deep cortical areas becomes apparent.

As previously noted, in the **nondominant hemisphere, the right inferior prefrontal** cortex includes BAs 44, 45, 47, 12 (noted by Aron, 2004), and the superior frontal sulcus BA 10. These areas are involved in **behavioral inhibition**. This is an important consideration when designing a NFB program in a patient who exhibits impulsivity (Aron et al., 2004; Konishi et al., 2005).

Central Regions

Cz, C3, C4: BAs 4, 6, and 1, 2, 3, 5 – The sensorimotor strip

C3, C4
BAs 1, 2, 3 is the Primary Somatosensory Cortex
BA 4 is the Primary Motor Cortex
BA 5 is the Somatosensory Association Cortex
BA 6 is the Premotor Cortex
When reading this section, the reader will recall that, for sensory and motor functions, the right hemisphere relates to the left side of the body and the left hemisphere to the right side.

BA 6

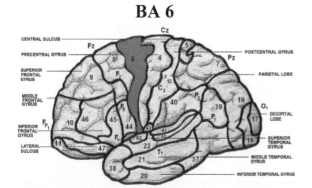

BA 6

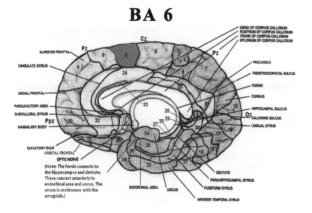

BA 4

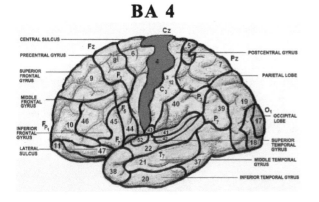

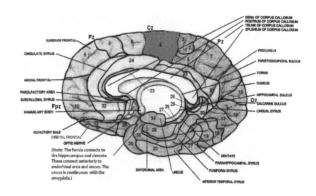

In work with NFB, we use these central sites of Cz, C3, and C4 for enhancing sensorimotor rhythm in the 12-15 or 13-15 Hz frequency range. In seizure disorders, Sterman's approach involves decreasing slow wave activity at the site of the seizure focus, if it is known, combined with increasing sensorimotor rhythm, usually alternating the feedback site between C3 and C4. This training is known to stabilize the cortex and make it more resistant to seizures. We assume when doing this NFB that we are having an effect on the thalamic production of SMR, although we may also, at these frequencies, be having some effect on high-frequency alpha and on cortically generated, low-frequency beta.

The functions of the cortical area immediately beneath our surface electrodes are: **BA 4** Primary Motor (Cz is over the part of the homunculus that connects with the **lower extremities** and C3 and C4 reflect the right and left hands, respectively), **BA 6** Premotor Association and integration, **BAs 3-1-2** Primary Sensory, and sensorimotor integration, **BA 5** Somatosensory Association cortex, involved in **proprioception.**

Functions include control of voluntary movements, swallowing, blinking, and so on. Functions may include aspects of imagery, and even verbal encoding and motor memory. This motor cortex area may play a role in **hyperalgesia**, which is an increased sensitivity to pain.

Mu activity (brain waves in the alpha frequency range but with slightly different morphology, namely, pointed at one end) can sometimes be recorded in some individuals at the C3 and C4 sites. Recall that the areas of BA 4 and BAs 3-1-2 under C3 relate to the right hand and under C4 to the left hand. The Mu activity simply reflects motor quiescence with respect to the hands; for example, Mu at C3 means the right hand is not moving, nor is the person imagining moving the right hand (as might happen if they were doing imagery of something like hitting tennis ball). Mu is a normal variant but does not occur in the majority of people.

We assessed a teenager who received a kick to the left side of his head when sparring in karate who subsequently had trouble in music (playing his trumpet) and in French Language classes at school. A 19-Channel EEG assessment showed high amplitude slow activity at C3 and hypocoherence (disconnect) between C3 and all other sites. NFB training at the C3 site resolved his problems.

Lesions caused by a **stroke or transient ischemic attack (TIA)** affecting BA 4, motor cortex, can result in paralysis of the opposite (contralateral) side of the body in areas such as the face, arm, hand, or leg. The illustration of what is called the homunculus indicates the distribution of motor activity reflected in the cortex. This illustration has been derived from the localization work of Wilder Penfield. Penfield was originally trained in the United States and then went as a Rhodes Scholar to Oxford in Great Britain. He is best known as a neurosurgeon at McGill University in Canada. He worked with Herbert Jasper. Penfield became the first Director of McGill University's world-famous Montreal Neurological Institute.

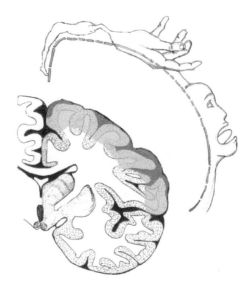

Figure by Maya Berenkey
*The **cortical homunculus** (Latin for 'little man') is a pictorial representation of the anatomical divisions of the primary motor cortex and the primary somatosensory cortex.*

BAs 1, 2, 3: Primary Somatosensory Cortex

BA 1, 2, 3

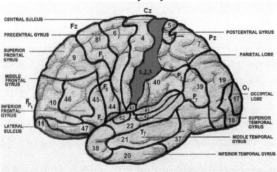

BA 1, 2, 3

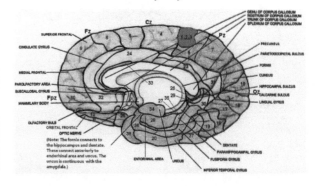

BA 1

Brodmann area 1 is central to a number of functions including **localizing sensations** such as touch, vibration, temperature, proprioception, and pain. The motor components govern hand and finger movements, movements of the tongue and mouth, swallowing. They are part of networks that allow the **anticipation of pain** and other sensations.

BA 2

Brodmann area 2 may be more involved in **agraphesthesia** (difficulty recognizing a written number or letter traced on the palm of one's hand) after parietal damage.

BA 4

Moving **anteriorly to Brodmann area 4**, the functions relate more to **sequences** of muscle movements: hands, fingers, mouth, tongue, eyes, eyelids, limbs. It is involved in **breathing** and complex body actions. The somatosensory cortex is involved in voluntary movements, organization of movements, and it also has mirror neuron functions. These areas contain **mirror neurons that become active when observing the actions of other people**. These mirror neuron functions seem to be impaired in persons with autism; that is, the mirror neurons in the area around C3 may not fire in the person with autism when he watches someone throwing a ball.

In a movement network these areas have important connections that include connections to the thalamus, cerebellum, and premotor cortex. Functions of this network include internal representation of action, and even a role in **anticipating and understanding actions.**

BA 3

Within the primary somatosensory cortex, **near the representation of the face** in the homunculus diagram and at the transition between the somatosensory cortex and the motor cortex, the cells in **Brodmann area 3** respond to proprioceptive and visual stimuli as well as to **vestibular** stimuli. Many of these neurons are activated by moving visual stimuli as well as by rotation of the body (even with the eyes closed), suggesting that these cortical regions are involved in the **perception of body orientation** in extra-personal space (Gray, 2013).

Cerebellum

This area is not being described in this text because it is not a primary site for NFB training. However, it is important in networks that the NFB practitioner is influencing and this brief note may therefore be of assistance:

Cerebellum in Latin means little brain. The cerebellum is the area of the hindbrain that controls motor movement coordination, balance, equilibrium, and muscle tone. Like the cerebral cortex, the cerebellum is comprised of white matter and a thin, outer layer of densely folded gray matter. The folded outer layer of the cerebellum (cerebellar cortex) has smaller and more compact folds than those of the cerebral cortex. The cerebellum contains hundreds of millions of neurons for processing data. It relays information between body muscles and areas of the cerebral cortex that are involved in motor control.

The cerebellum is involved in several functions of the body including:
- Fine Movement Coordination
- Balance and Equilibrium
- Muscle Tone

Of critical importance for some of our patients is the involvement of the cerebellum in **vestibular input** to the cortex. Patients after a concussion often demonstrate difficulties with balance and controlling themselves when walking and doing the tasks of daily living. The vestibular system with its connections to the cortex appears to be dysfunctional in these patients. Assessment of this system has been a major part of the assessment procedures used with concussed athletes that were developed at Penn State University (Thompson & Hagedorn, 2012).

The vestibular network encompasses cognitive and sensorimotor functions including perception of motion, spatial navigation and memory, perception of orientation of oneself (head and body) in the environment, and visual processing related to gravitational cues. The vestibular pathways originate in the vestibular nuclei in the brainstem, and the information goes to the cerebellum, thalamus, and the cerebral cortex. This network provides the brain with signals concerning gravity, such as which direction is up. Vestibular signals are of crucial importance for oculomotor and postural reflexes, and they are also basic to conscious perception and spatial cognition. The **pathways to the cortex apparently go through the ventral posterior nuclei of the thalamus to the Sylvian fissure (usually called the lateral sulcus or lateral fissure), insula, retroinsular cortex, frontoparietal operculum, superior temporal gyrus, and the cingulate cortex** (Lopez, Blanke & Mast, 2012).

Figure to show Operculum: Frontal – Parietal – Temporal – (Insula)

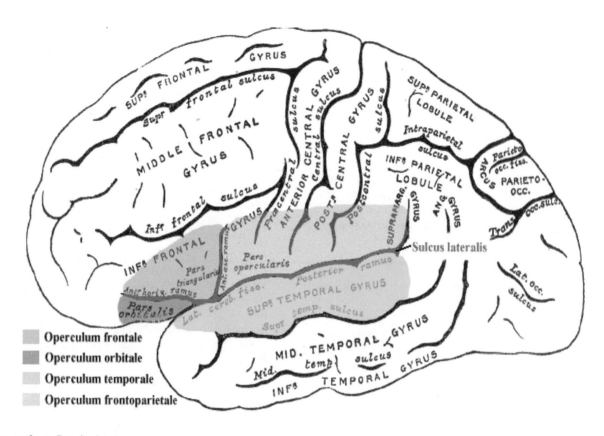

Figure from Gray's Anatomy
Portions of the frontal, temporal and parietal lobes form an 'operculum' (a covering or lid) that covers the insula.

BA 6: Premotor Cortex

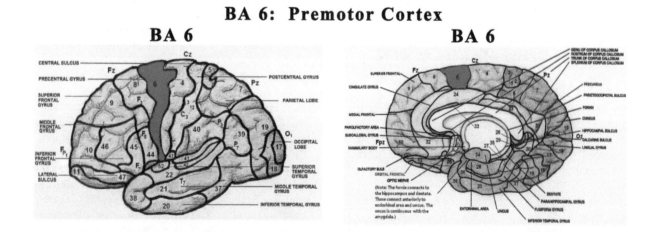

Rostral or anterior to BAs 1, 2 and 3 is **Brodmann area 6**. This region extends forward to the precentral sulcus between BA 6 and BA 8 that forms the **posterior border of the prefrontal cortex**. BA 6 is involved in a wide range of basic functions (motor, language, memory, attention), and in complex functions, such as planning, problem solving, calculations, frequency detection, behavior control, mirror neuron functions, and in the **dominant lobe: deductive** reasoning and the processing of **emotions** related to reflections on 'self' (part of the Default network). It is the key BA for **planning and initiating** movements and motor **sequencing** during movement. Thus a form of **kinetic apraxia** (problems with skilled movements independent of any motor weakness) will result from lesions in this area. It is important for **sensory guidance** of movement. It may be involved in smiling and laughter. In the **nondominant** hemisphere it is important for same-different discrimination and responding to strong odors.

Brodmann area 6 is a **motor association area** that is involved in **sequential movements** and sensorimotor voluntary action. Control of **trunk** and proximal body musculature resides here. In conjunction with BAs 8 and 9, it is involved in the **smooth sequencing** and flow of movements. It is essential for the ability to do rapid alternating movements (**diadochokinesis**), carrying out complex actions, such as threading a needle, and for rhythmic movements such as finger tapping. Near the midline, along with the cingulate cortices, it is crucial in **monitoring errors** (more right hemisphere) and **inhibiting responses** (more left hemisphere). It is crucial for motor learning, sequencing, complex movements, such as laughing and smiling, and the voluntary control of breathing.

In the **dominant (left) hemisphere** it is involved in networks for mnemonics, phonemes, calculations, **speech**, speech motor programming (overlapping with Broca's area), and handwriting (**Exner's** area). In the dominant hemisphere it controls speech perception, language processing, **language switching**, phonological (sounds of letters) and syntactical processing of the arrangement of words to form a phrase or sentence, object naming, lip reading, word retrieval, and lexical decisions concerning word meanings and pseudowords.

It is a key area for both working memory and episodic long-term memory, and is involved in mnemonic rehearsal. In the **nondominant (right) hemisphere** it is more involved in acoustic **rhythms, melodies**, and topographic memories. It is one of the areas that is involved in both **deductive** reasoning and in **working memory**. It has neurons related to **guiding eye movements** with direct links to both temporal and parietal areas.

BA 6 is also important in **attention to visual actions and in the recognition of objects**. It can be involved in **imagining movements**. This region is **vital for attention**, both visual-spatial and visual-motor. The **frontal eyefields** guide visual movement and are a part of Salience networks for attention to visual and auditory stimuli. This region has a role in attention to human voices and to rhythm and sequential sounds.

As previously mentioned, lesions can result in **kinetic apraxia** with movements that appear rough and uncoordinated due to difficulties with motor sequencing and the planning of movements. (For extreme detail, the reader can go to: http://neuro.imm.dtu.dk/services/jerne/brede/WOROI_325.html).

F3-C3: BAs 6, 40, DH

In the left frontal area, approximately between F3 and C3, is an area of the cortex related to **handwriting** called **Exner's** area. C3 may be closer to BAs 6, 40, and 43, and be related to functions that involve attention and execution of action.

BA 5: Somatosensory Association Cortex

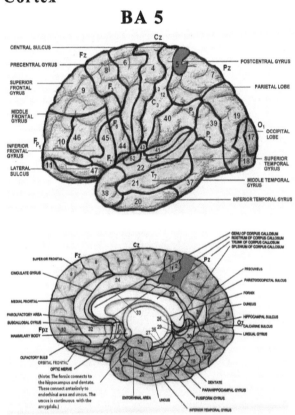

Brodmann areas 5 and 7 are in the superior parietal lobe. The intraparietal sulcus separates them from the inferior parietal lobe, BAs 40 and 39. Between Cz and Pz, in Brodmann area 5, one major function is the **perception of patterns**. This area, particularly in the nondominant hemisphere, is in a network for spatial perception and imagery of movements. It is therefore important in the use of tools and other objects. It may be involved in networks for **mirroring** gestures and movements, and is important in bimanual manipulations. It is involved in networks that are important for **working memory** and the conscious recollection of events. It may be involved in the processing of emotions and **reflections on 'self'** during decision making. It also has some role in spatial mnemonic (memory) processing, semantic categorization, detection of rhymes, temporal context recognition, and saccades (lateral-medial eye movements). It is important in **localizing pain** and in touch. Pain perception could prove to be an important function for NFB work. From all of these roles the reader will understand that it is **critical to the attention networks**.

The **right NDH somatosensory** cortex is more involved in **visual-spatial processing, spatial attention,** spatial awareness, mental rotation, and spatial imagery involved in deductive processing. It can be important in imitation and motor learning. It may also be important in **perception of personal space.**

In the **dominant hemisphere,** it is involved in **temporal context recognition. Tactile** localization is important in the superior parietal lobe whereas tactile recognition is a function of the inferior parietal lobe.

Lesions in this area can result in **ideomotor apraxia,** which is the loss of the ability to produce purposeful, skilled movements that are independent of any motor weakness, paralysis, coordination, or sensory loss. Lesions may also result in **astereognosis** (tactile agnosia), which is the loss of the ability to recognize objects by feeling them without being able to see them. (http://neuro.imm.dtu.dk/services/jerne/brede/WOROI_324.html)

Temporal Lobes

Introduction
General Discussion
Functions related to Connections to Orbital and Medial Frontal Areas

Introductory Overview

Homologous structures in the right and left hemispheres have similar intrahemispheric connections, act in the same general networks (such as the Affect and Executive networks), and often have very similar functions. For the areas under corresponding electrode sites for the dominant and nondominant hemispheres, we will therefore describe general functions that may be influenced by NFB over specific sites and **will only give differences in functions** of the dominant and nondominant hemispheres when there appear to be clear differences documented in the literature. By dominant hemisphere we mean the hemisphere in which the primary language functions reside, which is the left hemisphere for people who are right-handed and also for the majority of people who are left-handed. The right hemisphere may be dominant in a small percentage of people who are 'pure lefties,' that is, people who are left-hand, left-foot, and left-eye dominant, and whose language functioning resides predominantly in the right hemisphere.

For example, as previously described, it is clear that **Wernicke's area** in the dominant (usually left) hemisphere is primarily involved in language functions. Deficient function will be experienced as difficulties in language comprehension. The homologous area in the nondominant (usually right) hemisphere is concerned more with emotional aspects of speech, understanding of feelings conveyed by emotional tone, innuendo and nuance, and reading nonverbal cues. Deficiencies in these right hemisphere, parietal area functions, when they occur in people who have had strokes, are referred to by neurologists as sensory aprosodia (Ross, 1981).

On the other hand, T7 and T8 are close to BAs 40, 41 and 22, which are areas within the auditory cortex. Although one might expect the right ear to relate to the left **auditory cortex** and the left ear to relate to the right cortex, this is only partially true. Hearing from both ears goes to both sides with an emphasis on the left side for verbal/language input.

Influencing CMSs with NFB: For central midline structures (CMSs), the authors suggest that the NFB practitioner may be able to influence the appropriate side by doing the feedback while the client is carrying out task(s) that relate to the functions that one wishes to target. For example, if we are working with a student on mental arithmetic, which involves memory, recall, visualizing, sequencing, and manipulation of numbers, we will do these tasks while encouraging activation of the left cerebral hemisphere. We will teach the student a strategy and then have them carry out tasks while simultaneously doing the NFB plus BFB (such as effortless diaphragmatic breathing). On the other hand, we often are working at a central location (Cz) and we may be influencing a whole network that links memory and visualization and spatial reasoning areas. The task may push the brain into involvement of the correct neural network, in this example, the Executive network. We feel that the NFB practitioner's task often involves much more than just placing electrodes over a site and doing operant conditioning to change the amplitude of particular frequency bands. We recommend combining NFB with BFB and metacognitive strategies to have the biggest impact by influencing broader networks, especially networks that are appropriate to the particular needs of each client. Certainly, operant conditioning of SMR is beneficial for many, many conditions. Combining SMR enhancement with HRV training for managing stress has a role to play, too. However, even here, strategies for achieving a calm relaxed state may be important, and the strategies (which might include cognitive reframing or negotiation strategies or other skills related to stress management) represent added value for the client.

The temporal lobes are involved in Sensory networks and in the primary organization of sensory input (Read, 1981). The **lateral aspect** of the temporal lobes includes part of the auditory cortex. Input from both ears goes to both the left and right auditory cortices. **Verbal encoding** results in left-lateralized activation of the inferior prefrontal cortex and the medial temporal lobe (MTL). **Pattern encoding** activates the right inferior prefrontal cortex and the right MTL. Scenes and faces show approximately symmetrical activation in both regions. Lateralization of encoding processes is determined by the verbalizability of stimuli (Golby et al., 2001).

The **medial aspect** of the temporal lobes includes the **hippocampal region.** This region comprises the hippocampus and adjacent structures such as the parahippocampal gyrus, dentate gyrus, and the entorhinal cortex. This region is part of the limbic system and therefore the Affect network. It is also part of the Executive network and has important roles in the consolidation of information from short-term to long-term memory. This region receives **cholinergic input from the medial septum**. Interference with this input results in severe memory impairment. The hippocampal region is also involved in spatial navigation. It is one of the first regions of the brain to suffer damage in **Alzheimer's** with early memory loss and disorientation. Bilateral damage can result in **anterograde amnesia** (inability to retain new memories). Even **déjà vu** is said to be associated with the temporal lobe. Both left and right side lesions can affect recognition of visual content (e.g., recall of faces). Lesions more of the **middle temporal** area may result in an inability to **retain a series**, or just the correct order of a series, of sounds, syllables, or words in memory.

Animal studies have shown that hippocampal damage can result in difficulty learning to inhibit responses that they have previously been taught. In addition to Alzheimer's, damage to the hippocampus can result from oxygen deprivation, encephalitis, and medial temporal lobe epilepsy. Atrophy of this area can also occur with prolonged stress and with depression.

Seizures in the temporal lobe can have dramatic effects on an individual's personality. **Temporal lobe epilepsy** can cause perseverative speech, paranoia, and aggressive rages (Blumer and Benson, 1975). Severe damage to the temporal lobes can also alter sexual behavior, increasing such activity (Blumer and Walker, 1967). **Personality changes** such as loss of libido, loss of sense of humor or sudden religious conversion have been associated with a damaged temporal lobe as well.

Of interest to NFB practitioners is that long-term potentiation was first researched in the hippocampus. In addition, we have noted previously that midline frontal 5-7 Hz (3-10 Hz possible) theta waves are associated with memory encoding and recall, and these waves reflect a pyramidal cell rhythm from the hippocampus that may be controlled by cholinergic output from the septal area.

Left (Dominant) Temporal Lobe
Language can be affected by temporal lobe damage. Left temporal lesions disturb recognition of words and individuals with temporal lobe lesions can have difficulty placing words or pictures into categories. The medial aspect of the temporal lobes that includes the hippocampal area with the parahippocampal gyrus are highly associated with memory skills as previously described. **Left (dominant) temporal** lesions result in impaired memory for verbal material. These lesions result in decreased recall of verbal and visual content, including speech perception. If the left temporal lobe is damaged, perception of language and memory of words will be impaired.

The auditory cortex in the left hemisphere, temporal lobe, communicates **with the left frontal** lobe and has major control over functions that have to do with speech and language, including **speech articulation.**

NFB Over Dominant Hemisphere Temporal Lobe
In summary, NFB over T7 (T3) may influence functions of the left temporal lobe. It may influence the **integration of auditory, visual, perceptual, and memory** inputs, including those necessary for reading. These inputs can include: **phonemic** (letter to sound), **lexical** (sight words), and **semantic** (meaning) inputs. NFB may also have beneficial effects on other problems that relate to temporal lobe functions. These can include, for example, some people who are suffering from tinnitus (de Ridder, 2010) though it should be noted that tinnitus is complex and not necessarily easy to treat. Training over the left temporal lobe may also have some beneficial effects on visual memory and sequencing.

Right (Nondominant) Temporal Lobe
If the right, nondominant, temporal lobe is damaged, **perception of sounds and shapes** is impaired. Right temporal damage can cause a **loss of inhibition of talking.** Right side lesions may adversely affect the recall of nonverbal material, such as **music** and drawings. Right side lesions can also result in decreased recognition of **tonal sequences** and impair many musical abilities. Right anterior temporal lobe damage can interfere with recognition and recall of patterns (both visual and auditory) that are not coded verbally. Visual analysis is generally impaired and there is an impairment of maze learning. There is an inability to recognize faces (**prosopagnosia**). A patient may have difficulty recognizing common objects when they are viewed from unusual angles.

*The following case demonstrates some of the effects of **nondominant temporal-parietal and frontal residual damage** after experiencing encephalitis. A 44-year-old woman had been a brilliant student and top of her class in university until she suffered encephalitis at age 22. Two decades later she still suffered from profound memory loss for visual experiences, had a mild speech impediment, disorientation (stumbled into objects), and the left side of her face drooped slightly. She had a markedly decreased ability to think in abstract terms. Prior to* the illness she had been superb at visualization, but damage to right parietal area meant that, since the illness, she needs to have everything presented concretely (left frontal function). She had been highly involved in social events in school and university. Post-encephalitis she was unable to understand innuendo, nuance, and meaning of changing emotional tone in conversation (sensory aprosodia, right parietal temporal damage). She was verbally still bright. She is a very capable woman and she has devised means for interpreting what people mean in social situations. Numerous medical specialists told her that there was nothing wrong, because MRI and CT scans showed no abnormalities and she did well in IQ tests, which are quite dependent on left-hemisphere verbal attributes. She was greatly relieved after her QEEG evaluation that someone could show definite findings in areas of the brain that did precisely correspond to her symptoms. She was so happy that someone could finally confirm the source of difficulties she had experienced for over 20 years that she actually cried with relief. Some of the QEEG findings are summarized in the following two figures.

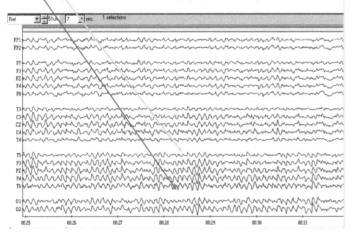

T6 >> T5 LS. was visualizing very poorly – previously excellent. Decreased T6 activity: Great difficulty correctly understanding social innuendo and the non-verbal aspects of social communication (right parietal and right frontal).

BA 20
T9, Inferior Temporal Lobe

The **inferior temporal lobe** in the dominant hemisphere, along with its connections to the frontal lobes, is involved in **representation of visual images evoked by auditory input**.

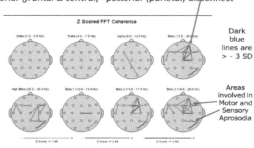

Post-encephalitis - Controlling impulsivity very difficult (right prefrontal?); disconnect FP2-F3 & Cz
SKIL Comodulation (not shown here) demonstrated complete anterior (frontal & central) - posterior (parietal) disconnect

The temporal lobe is a key area in a number of important cognitive functions. The medial aspect of the temporal lobe is an extremely important target for NFB but it may require LORETA NFB to reach some of the structures located deep in the cortex. As suggested above, single channel neurofeedback over T7, T9, and areas between these sites and parietal sites such as P7 may influence a number of different functions. Below we will separate these, discussing first the more superior and lateral portions of the temporal cortex which includes the auditory cortex. This discussion is followed by a connected but somewhat more posterior region in the temporal parietal junction often, in the dominant hemisphere, referred to as **Wernicke's** area. This broad general area includes aspects of the angular gyrus and the supramarginal gyrus which lie in the parietal lobe just superior to the temporal lobe. Dysfunction in these areas may affect a great many functions related to learning. This is followed by a brief overview of functions of the inferior portion of the temporal lobes.

Discussion of Brodmann areas will look at the following areas:
- Lateral Aspect, Superior and Middle temporal lobe
- Lateral Aspect, Inferior temporal lobe
- Lateral Aspect, Temporal – Parietal Junction
- Medial Aspect including BAs 34, 26, 27, 28: uncus, parahippocampal gyrus, hippocampus
- Medial Aspect and Inferior Temporal Lobe, parahippocampal gyrus, fusiform gyrus

Temporal Lobes – Lateral Aspect
T7 (formerly T3); T8 (formerly T4):   BAs 41, 42, 22; 21, 43, 52; 20

Overview Illustration

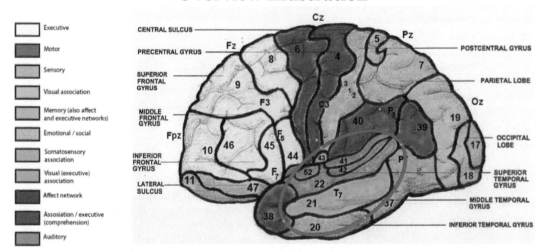

BAs 41, 42, 22 Including the Auditory Cortex

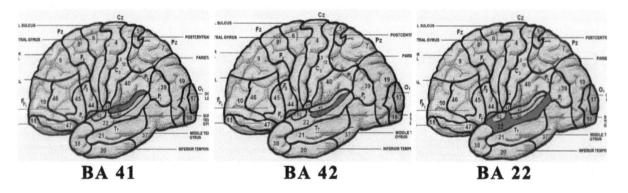

BA 41   BA 42   BA 22

Auditory Cortex

The **auditory cortex** is in the posterior portion of the superior temporal lobe. The auditory cortex comprises: **Brodmann areas 41, 42, 22**. The pathway for auditory information, in general terms, is from the ear to the medial geniculate area of the thalamus, then to the temporal lobe. In the superior temporal gyrus, BA 41 and BA 42 comprise the auditory association cortex while BA 22 in the superior temporal gyrus is auditory memory cortex. Complete **bilateral damage** to these areas will result in a **loss of the conscious awareness** of sound even though the ability to react reflexively to sound remains because that is based on subcortical and brainstem functions.

BA 41 & 42

Brodmann area 41 is the first cortical region of the auditory pathway. BAs 41 and 42 are involved in hearing different frequencies, pitch, harmonic tones, intensity, acoustic patterns, and volume. It is part of the auditory working memory network. Lesions can lead to a loss of awareness of sound.

Transverse Temporal Gyri

Brodmann area 41 includes the **transverse temporal gyri** that are also called **Heschl's gyri**. This is an area of the primary auditory cortex that is within the lateral sulcus with the gyri running mediolaterally towards the center of the brain rather than front to back as is true for all other temporal lobe gyri (see figure below). This is an area that is activated with sounds, but apparently may **also become active during visual word recognition** and with auditory short-term memory. It may also activate when observing the face of another person who is speaking, which may be a mirror neuron function.

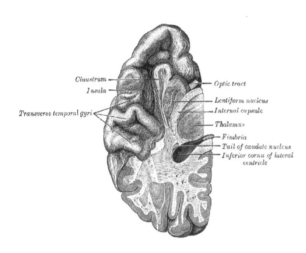

From Gray's Anatomy *(public domain) to show the transverse temporal gyri.*

BA 22

Brodmann area 22 is essential for **auditory working memory**. For example, when measuring evoked responses using an **auditory odd-ball paradigm** (the task might be to count the high tones in a series that is mainly low tones), **a comparison with previously heard signals is reportedly done in this area and it occurs approximately 175 msec after the stimulus**. This is an experimental example of how BA 22 is important in **comparing present input with past experience**. Functioning in the auditory cortex is also concerned with the relationship of **phonetics** (letter-sound relationships) and in the **categorization** and organization of information. It is also important in normal functioning of both **episodic memory** (remembering events) and **declarative** memory (remembering facts) and with grasping the whole picture as a **Gestalt** versus sensing auditory input in fragments. This last function may be dysfunctional in autism, just as it is with sensing a visual Gestalt.

In some patients, lesions in the left dominant temporal lobe can leave some of the functions noted above intact but leave the patient with other functions deficient. An example is **word deafness**. Dysfunction in the auditory cortex may result in the patient having an increased threshold for some sound frequencies. Older patients may complain of hearing **loss for high frequency** sounds and often have an associated problem with **tinnitus**. In these cases there is an association between the nerve deafness, thalamic connections, and the tinnitus (de Ridder, 2010). BA 22 deficits may also interfere with a person's ability to **localize sounds** and to decode speech sounds (phonemes). As this becomes more severe, auditory comprehension of speech may become quite difficult. Lesions involving BA 22 include **paraphrasias** (incorrectly produced words). In conjunction with the frontal lobe, dysfunction in the left temporal area may contribute to perceptual distortions.

The **superior temporal lobe,** auditory cortex, in both the right and left hemispheres is involved in **auditory perception and processing**. The dominant hemisphere is more involved than the nondominant, but both hemispheres have input from both the right and left ears. The auditory cortex is also important for the processing of meaning (semantics) in both speech and vision. The superior temporal gyrus includes an area (within the Sylvian fissure) where auditory signals come from the cochlea via the medial geniculate. Similar to the cochlea, the primary auditory area contains neurons that respond to particular frequencies, a kind of **tonotopic** organization. Adjacent areas in the superior, posterior, and lateral parts of the temporal lobes are involved in high-level auditory processing. The auditory cortex is important in the integration of stimuli that come from different sources. **Like the inferior lateral areas of the parietal lobe, the auditory cortex connects to the frontal eye fields (FEFs)**. It is possible that dysfunction here could prevent the person from grasping the whole picture and instead view the world in fragments with one stimulus after another vying for attention. As noted above, this may be one of the problems experienced by children with autism. In some children who have a learning disability, their visual perception regarding **discrimination of detail** is not functioning normally and this can relate to malfunction of the dominant (left) parietal-temporal region (between T7 and P7).

Conduction aphasia, characterized by good auditory comprehension and fluent but disordered speech production, is classically viewed as a **disconnection syndrome.** Conduction aphasia may result from damage to cortical fields in the dorsal portion of the **left posterior superior temporal gyrus** which participates not only in **speech perception**, but also in **phonemic aspects of speech production** (Hickok, 2001).

BA 22

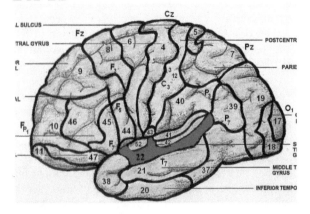

Data from experiments using evoked potentials demonstrates that sound travels from BA 41, then 42, to BA 22. In **Brodmann area 22, the incoming stimulus is compared to memory traces**. If the stimulus does not match memory traces, then one can observe a higher amplitude N1 (negative going wave at about 100 msec after the stimulus) event-related potential, followed by a high-amplitude P2 (positive going wave occurring about 200 msec after the stimulus) in an auditory oddball test, described in more detail under N1-P2 event-related potentials (ERPs). From the BA 22 area of the auditory cortex, auditory information is transferred to the anterior cingulate. Kropotov has given an example of this in his textbook (Kropotov, 2009). Using evoked potentials, he has demonstrated that information travels from the ear to the medial geniculate area in the thalamus in about 116 ms, to BA 42 after about 132 ms, and then to BA 22 within 200 ms. From BA 22, it travels to the anterior cingulate cortex in about 236 ms.

Brodmann area 22, however, is also important in other functions related to language, such as word and **sentence generation in the dominant** hemisphere.

In the **nondominant** hemisphere it is important for detection and **understanding emotional tone** of voice (happy, angry), nonverbal sounds, and affect-prosody. It is important in learning a tone-based language.

Lesions in the dominant hemisphere can result in **Wernicke's aphasia**, which is discussed below in the section **Lateral Aspect, Dominant Lobe, Temporal – Parietal Junction**.

BA 21

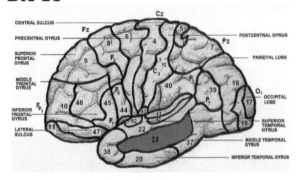

Brodmann area 21, in the middle temporal gyrus, has a role in both auditory and visual **memory and event sequencing**. It is part of the auditory cortex and is involved in processing of complex sounds. It is a key area in monitoring text and speech. It is important in **word and sentence generation.** It is activated during executive functions such as decision making. Lesions here can affect semantic tasks. It is part of the network for **deductive reasoning** and may have a role in categorization.

The **anterior** temporal areas are more involved in **syntactic** (order of words) functions. The **posterior** temporal areas are more concerned with **temporal relationships and naming** functions. The **dorsal** areas of the left posterior temporal-parietal lobe are involved in early integration of the **phonological-lexical-semantic features of printed** words. The **ventral** areas of the left occipital-temporal lobes are involved in fast word identification.

In the **nondominant(right)** hemisphere, BA 21 can play a role, along with the supramarginal (BA 40) and angular gyrus (BA 39) areas, in understanding **prosody** (the emotional aspects of speech conveyed by modulation, timing, and tonal qualities of speech), and in attributing intention to other people.

BA 43

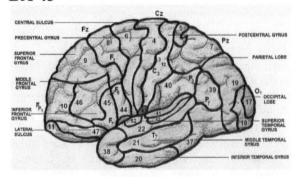

Brodmann area 43 is also referred to as the Gustatory Cortex. BA 43 is at the junction of the insula, frontal, and parietal lobes and occupies the postcentral gyrus and the precentral gyrus between the ventrolateral extreme of the central sulcus and the depth of the lateral sulcus at the insula. It is in a portion of the **sensorimotor cortical area** in the inferior margin of the postcentral and precentral gyri where the frontal-parietal operculum merges with the insula just below the inferior termination of the central sulcus. This area may participate in the sensorimotor representation of the **mouth and taste** reception. It may also overlap with the sensorimotor cortex areas for the **digits**. It can respond to **vibro-tactile** digit stimulation. It can also respond to speech.

The following diagram gives an overview of how the information coming from the medial geniculate to the mid-superior portion of the temporal lobe, auditory cortex, is then conveyed to different areas of the cortex for different purposes. In very general terms, the rostral (anterior) portion of the temporal lobe is related to the **'what'** of auditory information processing, and the superior portion of the parietal cortex is related to the **'where'** (Kropotov, 2009). As noted previously, the time relationships of the movements of information can be investigated using evoked potentials, as has been carried out by Juri Kropotov.

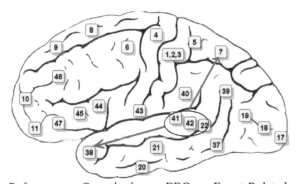

Reference: Quantitative EEG, Event-Related Potentials and Neurotherapy. *Kropotov, 2009. Diagram adapted from* Gray's Anatomy *(public domain).*

The above figure illustrates how the auditory cortex connects anteriorly to the entorhinal cortex (ERC) and BA 38 of the temporal cortex to determine the 'What,' and posterior and superior to BA 7, parietal cortex to determine the 'Where' for incoming auditory information.

BA 52

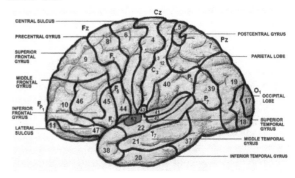

Brodmann area 52 is the parainsular area (at the junction of the temporal lobe and the insula). Functions of this area are assumed to overlap with those of the superior temporal lobe and insula.

BA 20

Inferior to T7 and T8:

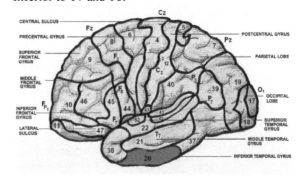

The inferior temporal gyrus, **BA 20**, is an association cortical area that participates in **higher order visual processing,** including the analysis of visual form, the representation of objects, visual memory, visualization, and visual categorization. In the **dominant hemisphere** it may also have functions related to Wernicke's area. It may play a role in **lexical categorization** (nouns and verbs are lexical categories).

In the **nondominant hemisphere,** it may be involved in the **comprehension of semantic ambiguity.** It is important in **recognition memory, semantic memory, self-recognition,** self-reference, and perhaps also in self-regulation. It may also have a role in networks involved in the **empathic understanding of the emotional states of other** people. Although the nondominant lobe may be more involved in **visual fixation and interpretation,** these functions are present in both hemispheres. However, in the nondominant (right) side, for the most part, it is associated with the representation of **complex object features,** such as global shape. It is involved in face perception. A minor role may be in metaphor comprehension.

Interestingly, this area is also involved in **telling the truth, and having suspicion.** It is involved with other temporal areas of the left hemisphere in language comprehension, **lexical categories,** and in a network that possibly includes the right prefrontal cortex for the comprehension of metaphors. It has a role in identifying intention and, in the right hemisphere, in ambiguity of categories, perceptual closure, and working memory. Also of interest, the inferior aspect of the temporal lobe BA 20 is involved in **creativity and introspection**.

Kropotov has also demonstrated that one can follow the **time sequence after a visual stimulus** using evoked potentials and this is shown in the heavy black lines in the figure below. Visual stimuli can travel from the lateral geniculate nucleus of the thalamus to the occipital lobe and from the lateral posterior nucleus of the thalamus to BAs 1, 2, 3.

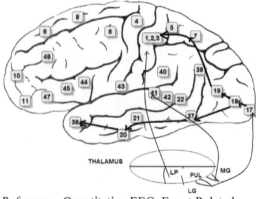

Reference: Quantitative EEG, Event-Related Potentials and Neurotherapy. *Kropotov, 2009. Diagram adapted from* Gray's Anatomy *(public domain).*

To do feedback over this area, the practitioner must be alert to EMG artifact from jaw, neck, or shoulder muscle tension, and be careful not to inadvertently encourage muscle tension with NFB. The practitioner must distinguish between a true increase in beta activity (which is an indicator of activation) and increased beta band amplitude that is due to EMG artifact. It is very easy, given the low amplitude of fast activity, to mistake the effects of EMG for an increase in beta.

Connections

The **inferior temporal lobe** in the dominant hemisphere, along with its connections to the frontal lobes, is involved in **representation of visual images evoked by auditory input**. Patients with frontal lobe lesions have difficulty in the temporal organization of information, such as lists of words. A name can pass through Wernicke's area and then, via the angular gyrus, arouse associations in other areas of the cortex. In this manner, the **angular gyrus is a connecting link** between sensory input and speech production areas. It is an important association cortex that **combines auditory and visual** information that are both necessary for reading and writing. It is involved in the steps between reading and speaking. Damage to the angular gyrus can result in **alexia** (inability to read) with **agraphia** (inability to write).The patient is unable to match spoken letter sounds to written letters. They can only spell the very simplest words.

Right hemisphere lesions in the parietal-temporal-occipital cortex can result in an inability to understand temporal relationships and, thus, future planning. There may also be visual-spatial defects, and problems in retaining a visual image.

Left hemisphere lesions in these general areas result in **aphasia**, inability to evoke a visual image in response to a word, inability to organize sequentially, draw proper angles, and copy designs.

NFB Over the Dominant Temporal Lobe

NFB over T7 (T3) may influence functions of the left temporal lobe. It may influence the integration of auditory, visual, perceptual, and memory inputs, including those necessary for reading. These inputs can include **phonemic** (letter to sound), **lexical** (sight words), and **semantic** (meaning) inputs. NFB may also have beneficial effects on other problems that relate to temporal lobe functions. These can include, as previously noted, some people who are suffering from tinnitus (de Ridder, 2010), though it should be noted that tinnitus is complex and not necessarily easy to treat. It may also have some beneficial effects on visual memory and sequencing.

Amygdala

The amygdala needs to be discussed along with the medial regions of the temporal lobe, and we have chosen to introduce the amygdala before discussing these medial regions.

The Amygdala is not part of the cortex, and thus does not have a Brodmann area number assigned to it. Nevertheless it is a crucial structure when we discuss the Affect network. In addition, the gray matter of the amygdala is continuous with the gray matter of the uncus, and it is therefore an essential structure when we consider the important connections of the uncinate fasciculus. The amygdala lies at the rostral end of the hippocampal cortex. The older corticomedial nuclei of the amygdala are connected to the hypothalamus by the stria terminalis. The newer basolateral nuclei project through the ventral amygdalofugal pathway to the hypothalamus, preoptic region, septal nuclei, midbrain tegmentum, and periaqueductal gray matter.

The dorsal periaqueductal gray matter is related to the sympathetic NS, rostrally for fight and caudally for flight, and the ventral-lateral periaqueductal gray matter is related to functions of the unmyelinated vagus.

Stephen Porges (2007) notes that the unmyelinated vagus originates from the dorsal vagal nucleus in the medulla. The unmyelinated vagus is the key structure that enables some animals to freeze and feign death. Freezing allows a drop in heart rate and blood pressure and a rise in resistance to pain.

The **central nucleus of the amygdala** is related to sympathetic drive, fight and flight, and is inhibited by the medial frontal lobe, the superior temporal cortex, and the fusiform gyrus. It is an integral part of the system that controls autonomic and endocrine responses to emotional states. The amygdala receives inputs from many sites within the cortex, including the medial frontal cortex, the cingulate, the temporal lobe, in addition to important inputs from the thalamus, hypothalamus, and midbrain.

The left and right amygdalae serve somewhat different functions in the processing of emotions: the left may respond to a specific stimulus, whereas the right may respond to any stimulus (Glascher & Adolphs, 2003).

In general terms, the amygdala is involved in processing **unconscious memories** and may be involved in bringing back an autonomic nervous system body state which accompanied the emotions that were linked to an event. This is a form of state-dependent learning. It seems to be designed to activate and remember emotional events (Buchanan et al., pp 289–318, 2009; Kim & Gee et al., 2011; Whalen, 2003). These events may include experiences characterized by feelings of fear, disappointment, and horror, but also of joy and celebration. It is also the key area involved in **traumatic memories**. The amygdala's role in unconscious memories lies in contrast to the role of the hippocampus in laying down conscious memories.

The amygdala is a central key in the **stress response,** which is controlled by interaction along a brain stem (locus coeruleus release of nor-epinephrine) - amygdala-hypothalamus (sympathetic system) - pituitary-adrenal axis (AHPA). More detail can be found in other sections of *The Neurofeedback Book*, particularly in the section on HRV and in the discussion of TBI. Not surprisingly, the amygdala may be hyperactive in people with borderline personality disorder (Donegan, Sanislow, Blumberg, Fulbright, Lacadie, Skudlarski, Gore, Olson, et al., 2003).

Although we are discussing each separate cortical area and some subcortical nuclei, the reader should always keep in mind that these structures have major connections that involve basal ganglia to thalamus links.

For decades we have said in our lectures that it is as if the **thalamus is the hub of the wheel** with connections radiating like the spokes of a wheel to every corner of the nervous system. The anterior cingulate cortex, the amygdala and hippocampus are rather like gearshifts, in many ways governing aspects of thinking, feeling, memory, and behavior.

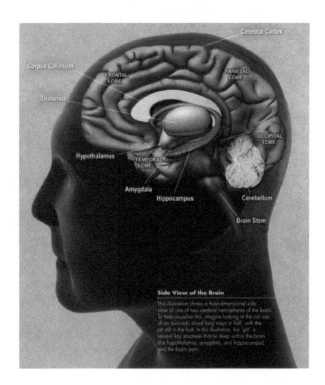

Public domain figure. It is used with great thanks to an unknown US Government employee.

This figure is included to provide practitioners with an overview of the relative locations of the thalamus, hypothalamus, amygdala, and hippocampus. The thalamus location in the brain has been compared to the pit in an avocado. The relationships between these structures can be difficult to visualize using only two-dimensional diagrams.

Figures to Illustrate Amygdala

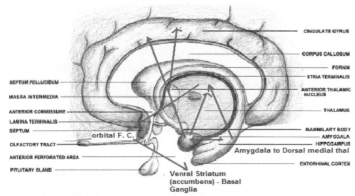

Some serotonin-related connections of the amygdala, hippocampus anterior thalamic nucleus, orbital cortex and the nucleus accumbens Figures adapted from Smith, 1962.

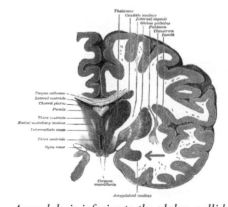

Amygdala is inferior to the globus pallidus

Amygdala

The above figures suggest how the amygdala may be involved in many areas that influence emotional responses. As described previously, the amygdala is involved in far-ranging Affect-related phenomena including appetite, sexual behaviors, fear responses, anger, aggression, reward behaviors and aversion responses. It also has a role in alerting and orienting, and thus in many decision-making processes, in addition to its role in responding to emotionally laden stimuli. Some aspects of its functioning will appear in a number of the following sections of this book.

Temporal Lobe – Medial Aspect
BAs 34, 26, 27, 28
Plus links to uncus, amygdala, parahippocampal gyrus, hippocampus, fornix

Introduction to the Medial Aspect of the Temporal Region

The medial aspect of the temporal region includes the amygdala, hippocampus, medial and inferior temporal lobes, and posterior aspects of the cingulate gyrus. Particularly important in this region are structures within the Executive memory network that are involved in **episodic and declarative memory.**

Deep inside the medial temporal lobes lie the hippocampi, which are essential for memory function – particularly the transference from short- to long-term memory and control of spatial memory and behavior. Damage to this area typically results in anterograde amnesia. **Anterograde amnesia** refers to a loss of the ability to create new memories after the brain trauma that caused the amnesia. With a severe concussion there can also be partial or complete inability to recall the event that caused the damage (retrograde amnesia).

A client was severely injured and rendered unconscious when a truck T-boned his car. This client noted cheerfully that he had no fear of driving because he had no recall whatever of the accident.

Not just structure, but physiology is important, and functions of the hippocampus are controlled, for the most part, by a neurotransmitter, **acetylcholine** (Ach), from the septal area. The septal area and the hippocampus turn out to be only a part of the networks involved in laying down memories.

Papez Circuit

The entire **Papez circuit** is thought to be critical in memory. Thus, in short-term memory, all of the following are involved: the hippocampus, which connects to the dorsomedial nucleus of the thalamus (which projects to the orbitofrontal and temporal cortex) and the mammillary bodies (which project to the anterior nucleus of the thalamus and then to the cingulate gyrus), and the fornix, which connects the hippocampal region with the amygdala. It should also be acknowledged that severe memory disturbance only appears to occur with bilateral lesions of the hippocampus. It makes sense that our brain has some built in redundancy for a function as important as memory.

The **fornix** joins the hippocampus to the mammillary body, which has major connections to the anterior nuclei of the thalamus and this connects to the anterior cingulate. The fornix also becomes the hippocampal commissure that joins left and right medial temporal areas, and is shown in the figure below. The rostral temporal poles are connected by means of the anterior commissure. All of these structures are involved in the Affect networks, as described previously.

To review, affect information goes from the orbital frontal and medial frontal cortex, anterior cingulate, hippocampus, amygdala, and the temporal entorhinal cortex to the basal ganglia, and in particular, to the nucleus accumbens. The nucleus accumbens connects to the ventral pallidum, which is inhibitory. This links to the medial dorsal nucleus of the thalamus and the anterior thalamic nucleus. These thalamic nuclei then connect back to the neocortex to regulate the anterior cingulate and thus all aspects of the Affect network.

The Affect network regulates mood and emotional reactions. It becomes more and more clear that many structures that are primary parts of the Affect network are also critical to laying down new memories. One's earliest memories are nearly always tied to events that were emotionally charged. Thus these two functions of memory and emotional experience are discussed together in this brief overview. It should also be no surprise that the hippocampus is involved in a number of psychiatric disorders (Cannon, 2012, p163).

Figures to Illustrate Commissures
Hippocampal, Anterior, Corpus Callosum

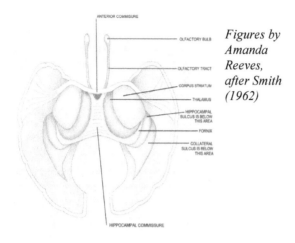

Figures by Amanda Reeves, after Smith (1962)

The hippocampal commissure links the left and right hippocampal areas.

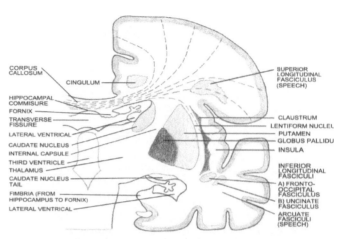

Transverse Section, Right Cerebral Hemisphere. Note the corpus callosum lies over (superior or dorsal to) the hippocampal commissure, which lies above the horizontal fissure. This fissure is above the Diencephalon. Note in cross section the Cingulum and Superior and Inferior Longitudinal Fasciculi.

Cingulum

The **cingulum is a fasciculus** that runs predominantly through the centre of the cingulate gyrus. The cingulum is a large band of association fibers that connects the temporal lobe with the cingulate gyrus, parietal lobes, and the frontal lobes (particularly the orbital surface). It is **connected to all parts of the limbic system.** It is involved in the **conscious perception of emotions**. One can conclude that the hippocampus has connections to the areas concerned with emotions, the autonomic nervous system, the endocrine system, memory, and consciousness.

Seizure Symptoms

Functions of the temporal lobe are complex. One way for the reader to obtain a better understanding of the range of these functions is to **consider symptoms** that may be observed in **simple partial seizures** (SPS) that originate in the temporal lobe. ('Simple' means that consciousness is not altered.) SPS can involve the amygdala and/or small areas of the temporal lobe, such as the hippocampus or other specific areas. A SPS may only cause **sensations** without any overt motor movements. These sensations may be cognitive, such as **déjà vu** (a sense that you have experienced this before), **jamais vu** (the phenomenon of experiencing a situation that one recognizes but that nonetheless seems very unfamiliar), a specific single memory or a set of memories, or amnesia. The sensations may be purely sensory, such as **auditory** sounds or tunes, gustatory, such as a **taste, or olfactory**, such as a smell that is not physically present. Sensations can also be **visual**, or they can involve feelings on the skin or in the internal organs. The latter feelings may seem to move over the body. More unusual possibilities include **psychic** sensations, which can occur as an out-of-body feeling (de Ridder et al., 2007).

Emotions can also occur, such as dysphoric or euphoric feelings, fear, anger, and other emotions. Often, it is hard for persons with SPS of temporal lobe origin to describe the feeling. SPS are sometimes mistaken for the 'aura' that is associated with a grand mal seizure when it is actually an independent event and no generalized seizure occurs.

It is also possible that the temporal lobe may be the origin of complex **visual hallucinations**, including religious ones, such as are attributed to St. Joan of Arc. Hallucinations originating in the temporal lobe become more complex in the more anterior regions, and may include the experience of out–of-world phenomena or even ghosts. The most vivid hallucinations tend to be triggered from the right, not the left, temporal lobe (Penfield & Perot, 1963).

Temporal Pole: A Junction of Inferior Frontal and Temporal Regions through the Uncinate Fasciculus

BA 38
F7, F9-T7; F8, F10-T8;
(T7 in the figure was formerly called T3 in the 10-20 system; T8 was formerly called T4 in the 10-20 system)

Brodmann area 38 is at the anterior end of the temporal lobe, known as the temporal pole. It is directly connected to the **entorhinal cortex** and thus to the uncus. The neurons of the uncus are continuous with the amygdala. These structures are directly connected through the **uncinate fasciculus** with the orbitofrontal cortex.

This area may be involved in moral judgments. These connections mean that BA 38 is involved in both **Affect and Executive** (memory) networks for experiencing emotional stimuli (fear and threat) and for **experiencing emotional states**. It is one area involved in emotional attachment. It is said to be involved in the visual processing of emotional images. Thus it is not a stand-alone area that functions independently of other areas. Although some functions related to BA 38 have been suggested to be hemispherically differentiated, such as female sexual arousal on the left and male arousal on the right, these kinds of distinctions require more research. It must be considered as networked with the amygdala and the orbital frontal cortex, and functionally considered with these areas.

In the **dominant hemisphere,** it is one of the areas involved in **selective attention to speech**, speech comprehension, semantic processing, long-term memories, comprehension of **narrative**, and word retrieval. It is one of the areas involved in the **comprehension** of both **irony and humor,** and is involved in aspects of **inferential reasoning**. It is involved in responding to **music**. It overlaps with the auditory cortex, and is part of the area involved in attention to speech and tone.

It is an important part of networks related to memory and in particular, to **memory retrieval**. Importantly, BA 38 is one of the first areas to be affected in **Alzheimer's** disease, and it is also one of the areas most **frequently damaged in traumatic brain injuries.** It may be an important area responsible for figure-ground difficulties, such as having difficulty **distinguishing speech from background noise**, which is a problem that some patients with TBI experience.

This area can be important in **visceral-emotional responses,** and thus we can postulate that dysfunction may result in symptoms that occur in depressive disorders.

F9 and T7 electrode sites overlap BA 38, which is found in the rostral portion of the temporal lobe, as can be seen in the above figure. It is likely that electrodes placed in this general area may pick up activity from (and thus be able to influence) more medial structures too, including the uncus with its connections to the amygdala. As noted above, the amygdala has activity related to emotions such as fear, anger, and anxiety, and is important in behavioral reactions to these emotions. The uncus is linked by the **uncinate fasciculus to** other structures, too, including the hippocampus and the orbital surface of the frontal lobe, which includes BA 47 and 11. The **hippocampus** has functions related to short-term memory, but it also is a part of the old Papez circuit for emotions, which more recently has been expanded and called the limbic system. It is therefore related to **mood regulation as well as memory.** Memory functions are facilitated in part by its connections via the uncinate fasciculus, which runs between the uncus on the rostral portion of the temporal lobe, past the amygdala and hippocampus, and then to the orbital surface of the inferior frontal gyrus, BA 11. We have noted previously that this pathway and the orbital surface of the frontal lobe may have a role in understanding the effects of one's own behavior on others. It is, therefore, very important in our work with people who have autism spectrum disorders and deficits in that ability. It is difficult to influence these areas with surface electrodes and, though one can do single-channel feedback with placement between F9 and T7, one may be better able to target these areas with the use of LORETA NFB training if that is available. Otherwise, one might try central training, such as the FCz site, and rely on network properties to influence this area.

Medial Aspect, Inferior Temporal Lobe, Parahippocampal Gyrus, Fusiform Gyrus – Memory and Affect Networks

For surface NFB: Inferior to T7 (around T9) and T8 (around T10):   BAs 28, 34; 36, 37, 27, 35, 36

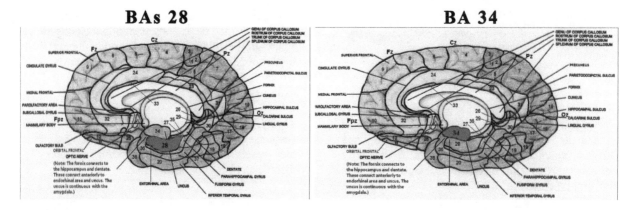

BA 28

Brodmann area 28 includes areas involved in the **paralimbic** cortex on the medial aspect of the temporal lobe, which includes the **entorhinal cortex** (BAs 28 and 34), **uncus** (BA 34, composed of grey matter continuous with the amygdala), and the **parahippocampal** gyrus (BAs 27 and 28). We have noted previously under BA 38 that the **entorhinal cortex (ERC)** is the main interface between the hippocampus and the neocortex.

Particularly in the **nondominant hemisphere**, the ERC-hippocampus system plays an important role in **autobiographical**/declarative/episodic memories and, in particular, **spatial** memories. This includes memory formation, memory consolidation, and **memory** optimization during sleep. The ERC assists in the association of stimuli from the eye and the ear. The ERC receives input from other cortical areas such as the perirhinal and parahippocampal cortices and the prefrontal cortex.

Thus the **ERC receives highly processed input from every sensory modality**, as well as input relating to ongoing cognitive processes. **ERC** neurons process general information, too, such as **directional** activity in the environment, which is in **contrast to functions of the hippocampal** neurons, which usually encode information about **specific places.** This suggests that ERC **encodes general properties about current contexts** that are then used by the hippocampus to create unique representations from combinations of these properties (Jacobs et al., 2010).

As previously noted, the entorhinal cortex is one of the first areas to be affected in **Alzheimer's** disease. In 2012, neuroscientists at UCLA conducted an experiment using a virtual taxi video game connected to seven epilepsy patients with electrodes already implanted in their brains, allowing the researchers to monitor neuronal activity whenever memories were being formed. As the researchers stimulated the nerve fibers in each patient's entorhinal cortex as they were learning, they were then able to navigate better through various routes and recognize landmarks more quickly. This signified an improvement in the patients' spatial memory. With over 30 million people worldwide today suffering from Alzheimer's disease, the study at UCLA was highly relevant. The researchers helped discover that ongoing stimulation of the entorhinal cortex did not necessarily provide a means to boost one's memory. Instead, it was found to be **effective only when trying to learn information** that was **important to that particular individual**. This discovery provides evidence for a possible mechanism that could enhance memory for relevant information. They also observed that entorhinal stimulation resulted in a resetting of the phase of the theta rhythm (Suthana et al., 2012).

In the **dominant-hemisphere BAs 28 and 34**, temporal lobe, the ERC-hippocampus system plays an important role in **declarative and episodic memories** and, in particular, all verbal memories. In **both hemispheres**, these functions include memory formation, memory consolidation, and memory optimization in sleep.

Relation to Dementia
As previously noted, the ERC-hippocampus system is one of the first areas to be affected in **Alzheimer's** disease. Other dementias affect slightly different areas in their early stages: **Picks** disease affects the frontal and temporal poles BAs 8, 12, 38, while **Diffuse Lewy Body** disease affects first the ACC BA 24, Insula BA 13, and parahippocampal gyrus BA 28. Interestingly, cognitive impairment can also occur in older people with no evidence of dementia (Choo et al., 2009, 2011).

Anatomical Connections
With any discussion of memory, one must consider the involvement of a number of important locations. For **long term memory**, these are most often in the central midline structures including the septum, hippocampus, the parahippocampal gyrus and, indeed, the whole Papez circuit. The original **Papez circuit** (hippocampal-mammillo-thalamic tract) comprised reciprocal connections between the hippocampal formation (hippocampus, dentate nucleus), via the fornix, to the mammillary body, anterior thalamic nucleus, cingulate gyrus and, via the cingulum, back to the hippocampal formation.

Other connections have been added to this Papez circuit and the unit is now referred to as the **limbic system,** or emotional brain. These additional areas include the amygdala, hypothalamus, prefrontal cortex, and association cortex. The **amygdala** connects to the entorhinal cortex, which has direct connections to the cingulate cortex. The cingulate projects to the temporal lobe and the orbital cortex.

Now we understand that the **fornix divides**, with some fibers going **to the septal** area (which has direct connections to the hippocampus) and others going to the mammillary body. From the mammillary body fibers go to the anterior nucleus of the thalamus and from there to the anterior cingulate cortex (ACC). This ACC feeds back through the cingulum to the entorhinal area, which connects with the hippocampus from which the fornix emerges. The **septum** turns out to be a key link. It produces **acetylcholine,** which activates the hippocampus and is thus a key in the initial phase of memory production.

Subiculum of Hippocampal Region
*The **subiculum** (Latin for 'support') is the most inferior component of the hippocampal formation. It lies between the entorhinal cortex and the hippocampus. It receives input from the hippocampus and entorhinal area and is the main output area of the hippocampus. The pyramidal neurons of the subiculum send projections to the nucleus accumbens, septal nuclei, prefrontal cortex, lateral hypothalamus, nucleus reuniens, mammillary nuclei, entorhinal cortex, and amygdala. The nucleus **reuniens** within the midline area of the dorsal thalamus links the medial prefrontal cortex with the hippocampus and regulates the amount of neural traffic according to changes in attentiveness* (Vertes et al., 2007). *The pyramidal neurons in the subiculum exhibit transitions between two modes of action potential output: bursting and single spiking. The transition between these two modes is thought to be important for routing information out of the hippocampus.*

Long-Term Memory
The hippocampus, the temporal lobes, and the structures of the limbic system that are connected to them are essential for the consolidation of long-term memory. To be recorded in long term memory, information must pass through the Papez circuit. Any lesion in this circuit, for example, caused by a **stroke or by TBI,** will result in some form of memory impairment.

In **Korsakoff's** syndrome, due to alcoholism with vitamin B1 deficiency, in addition to confabulation, disorientation, and confusion, the patient will have **anterograde amnesia** with an inability to store new information in their long-term memory.

Memory is consolidated and made permanent within the Papez circuit. It is then **permanently stored outside of that circuit** in the area of the cortex from which the sensory information originally came, for example, visual information in the occipital cortex and auditory in the temporal cortex. For this reason, these **long-term memories are not lost** when there is damage to the nuclei within the Papez circuit, such as the hippocampus. Only the ability to form new long-term memories is compromised.

Within the **Papez circuit, the hippocampus** is a key structure for encoding new long-term memories. It has the capacity to facilitate associations among various parts of the cortex. It should be no surprise that the Papez circuit is also central to the Affect network and the limbic system as well as being included in the network for encoding memory. The hippocampus can link, for example, a tune that you heard at a concert with the faces of the people who accompanied you to the theatre. Instead of this memory fading, it may be strengthened and eventually etched into long-term memory due to the strength of your interest in that occasion and the strength of the emotions associated with it.

The **parahippocampal gyrus** is a grey-matter cortical region of the brain that surrounds the hippocampus. This region plays an important role in **memory encoding and retrieval**. The anterior part of the parahippocampal gyrus includes the perirhinal and entorhinal cortices. The posterior part is continuous with the medial aspect of the fusiform gyrus. The parahippocampal gyrus is important in the **recognition of scenes** such as landscapes, cityscapes, or rooms (rather than faces or objects). As was previously noted, damage (for example, due to stroke) to the **parahippocampal place area** (PPA), which is a subregion of the parahippocampal cortex that plays an important role in the encoding and recognition of scenes, often leads to a syndrome in which patients cannot visually recognize scenes even though they can recognize the individual objects in the scenes such as people and furniture. The right, **NDH,** parahippocampal gyrus also enables people to **detect sarcasm**.

Intense activation of the temporal lobe, hippocampus, and amygdala may be associated with sexual, religious, and **spiritual** experiences. **Chronic hyperstimulation** is said to be able to induce an individual to become hyper-religious or visualize and experience ghosts, demons, angels, and even God, as well as claim demonic and angelic possession or the sensation of having left their body. There are many references for these and similar kinds of experiences in the literature, but two older articles serve as an introduction to the effects of experimental stimulation of the cortex, namely Bear & Fedio (1977) and Penfield & Perot (1963). **Fear** is one of the most common feelings that can be generated by stimulation of these structures. It can include feelings of panic, terror, or of something horrible about to happen (Chiesa et al., 2007; Gloor et al., 1982; Penfield & Jasper, 1954).

BA 34: Uncus

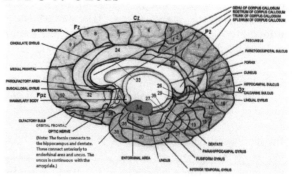

The anterior extremity of the parahippocampal gyrus is recurved in the form of a hook called the uncus. It is a part of the olfactory cortex. The connections of the uncus have been described above in the discussion concerning the temporal pole, entorhinal area, the amygdala and the uncinate fasciculus. Seizures that are preceded by the perception of disagreeable odors may have a source in this area of the temporal lobe. It is also involved in **social and emotional associational processing** functions due to its links to the amygdala, parahippocampal gyrus, and the frontal lobes.

More detail concerning Temporal Lobe Functions
The uncus, in conjunction with the amygdala and hippocampus, is involved in **visual closure**. These structures contain neurons that respond selectively to **faces and complex geometric** and visual stimuli as noted above. The **inferior and middle temporal** lobes are the **recipients of two diverging (dorsal and ventral) streams of visual input** arising from within the occipital lobe and the pulvinar and dorsal medial nucleus of the thalamus. The **dorsal stream** is more concerned with the **detection of motion** and movement, orientation, binocular disparity. The **ventral stream** is concerned with the **discrimination** of shapes, textures, objects, and faces. This information flows from the primary visual to visual association areas and is received and processed in the temporal lobes. It is then shunted to parietal lobe, and to the amygdala and entorhinal cortex (the gateway to the hippocampus), where it may then be 'learned' and stored in memory (Joseph, 2000).

The **anterior temporal** region is more involved in the initial **consolidation and storage** phase of memory, whereas the **posterior region** is more involved in memory retrieval and **recall**. In addition, the temporal lobes directly interact with the frontal lobes in memorization and remembering. Indeed, the greater the activation of the frontal and temporal lobes (and associated tissues), the greater is the likelihood that subjects will remember, whereas reduced activity is associated with forgetting. Hence, these areas **interact to promote memory and retention**. In consequence, if the frontal lobe is injured, even if the temporal lobes are spared, patients may demonstrate significant memory loss due to an inability to correctly search for and find the memory.

Hippocampal stimulation may be associated with memory-like hallucinations, including feelings of familiarity, and dream-like hallucinations. Hallucinations require interactions between the amygdala, hippocampus, and the neocortex. It is only when the hallucination involves the neocortex of the temporal lobe that the individual becomes conscious of it. **In sleep,** the right hemisphere becomes highly active during REM sleep; conversely, the left brain becomes more active during non-REM sleep.

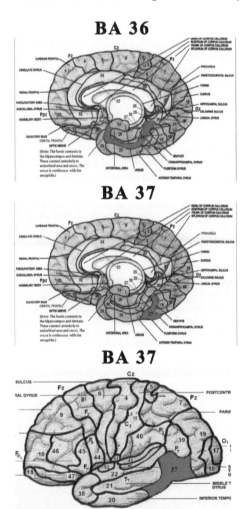

BA 36

BA 37

BA 37

BA 37

Brodmann area 37, which contains part of the fusiform gyrus, is also known as the occipitotemporal gyrus. The underside or **ventral aspect** of the temporal cortices appear to be involved in visual fixation and high-level **visual processing of complex stimuli**. The **visual processing of scenes** involves the **parahippocampal** gyrus. Anterior portions of this ventral stream for visual processing are involved in object perception and recognition.

In the **inferior medial aspect**, this visual processing includes facial recognition in the fusiform gyrus BAs 36 and 37. This includes **face identity** and judgment of familiarity. It is also part of the Executive network related to deductive reasoning.

BA 37, DH

The visual word form area in the dominant (left) hemisphere, the **fusiform** gyrus holds a **prelexical representation** of visual words. The visual word form area responds **only to written stimuli**, not to spoken stimuli, independent of their semantic content (Dehaene et al., 2002). It is an area that is involved in networks for **semantic** relations (meaning of words) and categorization. Face-name associations are carried out in this area, too. It is important in sign language. It is another area that may be involved in the interpretation of metaphors. It is involved in **word retrieval,** and lesions in this area can lead to word finding difficulties. It is involved in **monitoring color and shape** and in drawing. It is part of a network that **relates to intentions**. It is part of networks related to the laying down of new memories and short-term memory, and is involved in episode encoding. It may be involved in true-false memory recognition.

BA 37, NDH

The nondominant (right) hemisphere BA 37, which includes part of the **fusiform gyrus,** is similarly important in the recognition of faces, and lesions can lead to difficulties in face recognition (called **prosopagnosia**), and in face-name recognition. It is important for drawing, too. Lesions can also lead **to constructional apraxia,** which is a visuoconstructive disorder. BA 37 is a part of the right hemisphere network for **sustained attention.** It is also one of the areas that is involved in the attribution of intentions to others.

The fusiform gyrus is involved in inhibiting the central nucleus of the amygdala. When this inhibition is deficient, sympathetic drive may increase. Functioning of this area appears to be deficient in autism spectrum disorders.

BAs 27, 35, and 36
BA 27

BA 35

BA 36

The **parahippocampal** cortex bilaterally is a key area (along with the hippocampus) for **short-term working memory**. As with other cortical areas, the left hemisphere is more related to verbal and the right hemisphere is more related to visual-spatial functions, but there is a great deal of overlap. BA 27 and parts of BA 35 and BA 36 play an important role in the encoding and recognition of **scenes** (as contrasted to recognition of faces or objects). Part of BA 36 overlaps with the fusiform gyrus. fMRI studies indicate that the hippocampal region of the brain becomes highly active when human subjects view **topographical scenes** such as images of landscapes, cityscapes, or rooms; that is, **images of places**. After a stroke affecting this area, the patient might be able to recognize people and objects but not recognize the scene in which they are being viewed.

In addition, the **right, NDH, parahippocampal** gyrus may play a crucial role in identifying **social context** including paralinguistic elements of verbal communication, such as **sarcasm**. It is also an area that is important in memory including faces, pictures, novelty, negative stimuli, olfactory, and gustatory experiences, as well as auditory and visual memories. It is involved in **procedural memory** consolidation and in recognition and recall. It is in a **network for autobiographical** memory and for insight. In addition, it is part of the limbic system and is involved in the Affect network and modulation of emotions. It is important in emotions that include **embarrassment and regret**. It may have a role in hunger and craving. It is part of the **Executive network for both categorization** and for decision making. Interestingly, the right parahippocampal area is activated when a person who is known to be accurate with **telepathy** is engaged in such activity (Persinger & Saroka, 2012). **BA 35** is important in the retrieval of words, acquisition of new memories, short-term memory, and spatial mapping. It is also involved, along with other areas, in the detection of novelty.

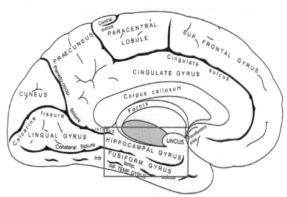

Reproduction of a lithograph plate from Gray's Anatomy, *Medial surface of left cerebral hemisphere. Image is in the public domain.*

NFB to Target Central Midline Ventral Structures

Practitioners with LORETA NFB may prefer to use that program to target activity in the central midline areas in order to be more precise in training for normalization. On the other hand, many practitioners appear to have been successful in altering activity involving Brodmann areas on the medial aspect of the temporal lobes and the temporal-parietal areas using single-channel NFB on the surface of the scalp. In the NeuroGuide Manual, Robert Thatcher has outlined which surface electrode sites can be expected to reflect activity from these deeper structures. Indeed, in this NeuroGuide QEEG assessment program, if you right click on any of the 19-channel sites, the BAs reflected in activity at that site are shown. The next figure illustrates this point. In this figure the arrows point to BA 30. If activity were outside the database norms at this site using LORETA, then this activity might be reflected, for example, in the surface EEG at Pz, P3, P4, T5, and T6.

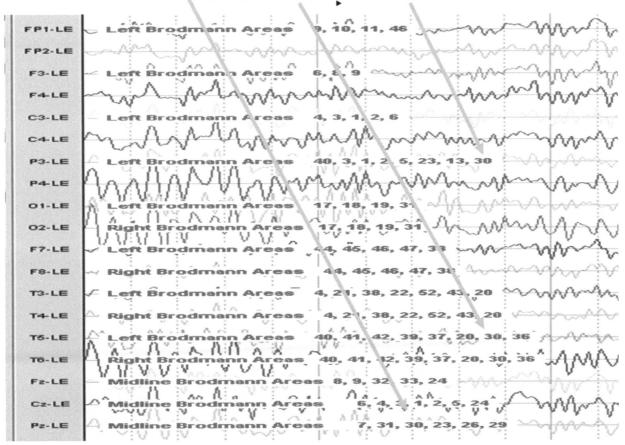

Functions of the area between F7 and T7 and between F8 and T8, and effects of lesions on CMSs that are important for memory.

Temporal Lobe Functions and Epilepsy

The hippocampus and amygdala have the lowest **seizure thresholds** of all brain structures and, when injured, they may develop seizure activity and thus give rise to temporal lobe epilepsy. Automatisms associated with **temporal lobe epilepsy** (TLE) include staring, searching, groping, lip smacking, spitting, laughing, crying, clenching the fist, confused talking, and so on. Common visceral reactions associated with **TLE auras** include feelings of a racing or fluttering heart. Body sensations include numbness, tenseness, pressure, and heaviness. Feelings of strangeness or familiarity are also common and déjà vu occurs in up to 20% of such patients, usually with right cerebral seizures within the temporal lobe. Some claim to have feelings like wanting something but not knowing what. Olfactory hallucinations are not uncommon. There may also be gustatory sensations. Fear is one of the most common feelings, including feelings of panic, or of something horrible about to happen. There may be sexual sensations.

Hence, there is a large range of auras and experiential phenomenon associated with temporal lobe seizures, many of which reflect the possible activation of those structures innervated by the amygdala. Indeed, many of these same feelings and behaviors can be triggered by direct stimulation of the amygdala. In addition, patients may act out on these emotions and auras. There have been reported instances of patients suddenly lashing out and even attempting to attack those close by while in the midst of a seizure. Some patients exhibit behaviors indicative of extreme fear, or conversely, extreme pleasure, including laughter and mirth.

Depression (lasting from hours to weeks) may occur as an immediate sequela to the seizure. Some patients experience confusion. A depressive aura may also precede a seizure by hours or even days. Some patients complain of emotional blocking, and feelings of emptiness. Patients have also described a sense of dissociation and being out of and looking at their own body (Joseph, 2000). The diverse phenomena associated with TLE are a reflection of the diverse functions of the temporal lobe.

Insula

BA 13
C3-T7; C4-T8

Key areas of the Affect network include the **cingulate gyrus and Brodmann areas 24, 25 and 33, plus the anterior portion of the insula, BA 13.** These areas have connections with the amygdala, septum, medial and orbitofrontal cortex, anterior insula, ventral striatum (nucleus accumbens), periaqueductal grey, and the autonomic brain stem motor nuclei. BA 25 has direct connections with the nucleus of the solitary tract in the medulla, as is discussed in a section on HRV training and the stress response in another part of *The NFB Book*.

BA 13 Insula

The insula is an infolding of the cortex between the inferior frontal cortex and the temporal pole. It is a portion of the cerebral cortex that is folded deep within the lateral sulcus. The lateral sulcus may also be called a fissure. It separates the temporal lobe from the parietal and frontal lobes. Thus it is not visible in the diagrams in this book that show the Brodmann areas. It is therefore outlined in the following illustration.

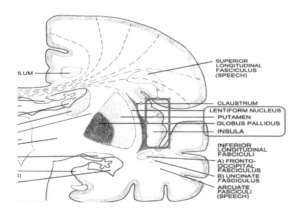

Figure to illustrate the relative position of the insula with respect to the frontal (superior) lobe, temporal (inferior) lobe, and the basal ganglia (medial). In this figure the cingulate cortex is superior/medial with the cingulum (just the label "...LUM" is visible) shown running through its central core. Below the cingulate cortex are dashed lines that indicate the corpus callosum.

Insula (BA 13 and Close to BA 43)

When dissecting the human brain, the insular cortex may be seen by placing your hands on each side of the lateral sulcus and then separating the frontal and temporal lobes and looking between them. The insular cortex is continuous with these two lobes, but it becomes an infolding of the cortex rather than expanding laterally during early development. It may be possible to have an effect on this important area of the cortex using an electrode site between C3 and T7, or C4–T8 on the right. Alternatively, LORETA NFB may have a more direct influence. After 19-channel data collection is done, and before using LORETA NFB, we observe what frequency is (or frequencies are) outside the database norms for BA 13, the insula, using LORETA. If that same frequency is outside database norms near a site between C3 and T7, then we may first try surface NFB to normalize it while we do tasks related to the patient's symptoms.

Due to the importance of the insula for normal balance between the sympathetic nervous system (right insula) and the parasympathetic nervous system (left insula), we always combine NFB with HRV training. Since we cannot specifically target the small area of neurons responsible for a specific network, we use cognitive strategies and tasks to try to have the person's brain activate particular networks while our NFB encourages appropriate EEG activation. Symptoms targeted for improvement would include those symptoms that might be expected to be related to an important aspect of the functioning of the insula and networks connected to the insula. Training to help a client where the insulae are affected, as they might be after a TBI, is a complicated endeavor. These patients will show a decrease in heart rate variability (HRV) measurements so it makes sense to combine NFB with HRV training.

The insula is of import in the **Affect network** and is a key component in some executive functions. It may also be involved in the Default network. The two insulae play a role in diverse functions linked to **emotion** and the regulation of the body's **homeostasis**. These include higher level emotions such as love, joyfulness, appreciating humor, expressing fear or disgust, and evaluating sexual arousal. It has some role in conscious sensing of heat, touch, vibration, taste, and olfaction. Its activity is increased, along with that of the amygdala and anterior cingulate cortices, in emotion-related tasks. These areas have been shown to be linked to heart rate responses and to the presentation of emotional facial expressions (Yang, 2007).

The insula is involved in **Executive networks and in perception and motor** control. It is activated in networks for comprehension, understanding word meanings, and verbal memory. It has a role in cognitive functions including categorization, action planning, being aware of errors, and verbal memory. It is linked to networks that involve perception and motor control. It is important in **self-awareness.** One can immediately see that the insula (possibly different areas in the insula) is thus important in **Affect, Executive, Salience, and Default networks.** As part of the Salience network, it has a role in turning the Default mode off and on (Gotman, 2005). The right insula is a key in switching between central Executive and Default modes during attentional tasks (Sridharan, 2008).

The insulae are important in **autonomic** nervous system control and **affect**. As is noted above, the left insula influences the parasympathetic system and the right insula influences the sympathetic system. The right insula is involved in conscious emotional and physical self-awareness. With the amygdala, the **right insula activity decreases when** there is increased activity of prefrontal regions that regulate responses to negative emotional stimuli. That is, cognitive reappraisal of negative emotional stimuli can down-regulate the insula and amygdala (Lutz et al., 2009). **In contrast, the right insular cortex is activated** with feelings of disgust, which elicits avoidance behavior. You may recall that due to Davidson's work (1990), we tend to associate avoidance with right frontal activation, as contrasted to approach behavior, which is more associated with left frontal activation. Activation of the insula may also occur with fear.

The insular cortex plays a role in the experience of **bodily self-awareness**. There are mirror neuron-like links between external and internal experience that are related to the functioning of the insula. It plays a role in the affect component of pain, depression, anxiety, and feelings of hopelessness (Sheline et al., 2009). The insula is believed to process convergent information in order to produce an **emotionally relevant context for sensory experience**. The insulae send output to a number of other limbic-related structures, such as the amygdala, the ventral striatum, the nucleus accumbens, and the orbitofrontal cortex, as well as to motor cortices. The insular cortex has been suggested to have a role in **anxiety** disorders, and emotion dysregulation. While the amygdala may be more involved in social anxiety, activation of the insula may be more related to trait anxiety and the processing of threat assessment (Shah et al., 2009).

Anatomical Connections

The insula is an important area because of its multiple connections. The insula has major connections to the hippocampus and the hypothalamus. Connections to the hypothalamus influence the autonomic nervous system, as has been described previously. The **right insula** affects the **sympathetic** and the **left** affects the **parasympathetic** nervous system, in part through connections to the nucleus accumbens, which balances the sympathetic and parasympathetic systems (Nagai et al., 2010). Nagai states, "The insular cortex has **reciprocal connections** with the anterior cingulate gyrus, amygdala, entorhinal cortex, medial and orbitofrontal cortex, and temporal pole, and has afferent connections with hippocampus formation."

Additionally, the insular cortex has dense reciprocal connections **with subcortical autonomic** core centers including the lateral hypothalamic area, nucleus tractus solitarius, and parabrachial nucleus, and these centers are also reciprocally connected to each other.

Importance in HRV Training

Of particular importance to the biofeedback practitioner who uses HRV training is the influence of these structures on **vagal efferent** output. These connections, and thus the whole affect network, may be influenced by both NFB and BFB, and thus we suggest that practitioners consider combining NFB with BFB interventions. Changes in heart rate can correlate with activity in the right middle insula because the right insula influences the sympathetic nervous system.

The heart and the **baroreflex** system have direct **vagal afferent** connections to the nucleus of the **solitary tract (NST)**. This was discussed earlier with an illustration under the Frontal Lobes, "Importance of Heart Rate Variability (HRV) Training – Connections to NST in Brain Stem." The NST, in turn, connects to the parabrachial nucleus (respiratory control), the locus coeruleus (norepinephrine production and a key in understanding the stress response), the thalamus, hypothalamus (control of the autonomic nervous system and pituitary outputs), the amygdala, and the cingulate cortex.

These connections mean that this network has roles not only in **mood modulation** but **also** in related physiological **regulation of the autonomic nervous system and of endocrine** functions. In addition, this

Affect network is also important in conditioned emotional learning, vocalizations expressing internal (emotional) state, assessing motivational content, and in assigning **emotional valence** to internal and external stimuli. However, the major function of this network is to modulate autonomic activity and internal emotional responses. If one had to pick one key structure that could regulate the whole network, it would most likely be the anterior cingulate cortex (Devinsky et al., 1995).

BAs 14, 15 and 16 are found in monkeys but not in humans, so they are not discussed in this text.

Cingulate Gyrus

Fz, Cz, Pz
Anterior FCz: Executive BA 32; Affect BAs 24, 25, 33
Central Cz: Affect BA 24
Posterior Pz: Affect – Executive – Default BAs 23, 31; 29, 30, 26
Important connections exist to the insula for all of these areas.

BA 32

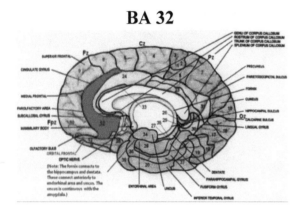

BA 24

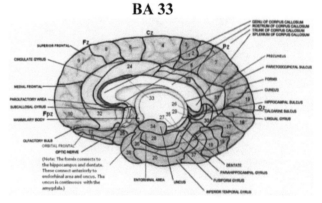

BA 25

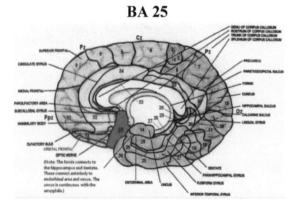

BA 33

Introduction to the Cingulate Gyrus
BAs 24, 25, 33
FCz

Anterior Cingulate Cortex

Many functions are influenced by the anterior cingulate cortex (ACC). The ACC contains a very high density of opioid receptors. It is involved in the emotional experience of **pain**. It is important in possible gating functions for pain between cortical and subcortical structures. It is part of neural networks involved in the identification and the conscious perception of pain. It is involved in **sexual** arousal. It is important in **attention** networks, both auditory and visuospatial.

The medial surface of the ACC is important in motivation (BA 24, 32, 33, 25) in conjunction with mesial frontal areas (BA 6, 8, 9). Amotivational syndrome can be observed as one consequence of marijuana use. Profound amotivational syndrome is described in *The Oxford Textbook of Medicine* (2003) volume 1 page 961. The reader may also reference *Current Drug Abuse Review* (2008) January, 1(1), 99-111.

It may be involved in more complex functions such as having a sense of **balance** and even timekeeping. It is important in **monitoring** stimuli and in response selection. The ACC may detect the need for executive control, and may signal the prefrontal cortex to provide that control. (This latter function is poor in patients who have ADHD.) It is an important component in **reward** networks that involve both Executive networks (the DLPC) and Affect networks (limbic system including medial, orbital, and insular cortices and the nucleus accumbens, amygdala, and the ventral tegmental area). These same areas, plus the left temporal-occipital junction, are involved in networks related to the **appreciation of humor** (Moran et al., 2004).

As noted at the beginning of this section, the ACC has close connections with the insula. **Negative stimuli** will activate the insula and amygdala. Both fear and disgust will activate the insula. Disgust is a powerful emotion that results in avoidance behavior, which we previously noted may involve the non-dominant hemisphere (NDH). Disgust activates the right insular cortex plus the right medial and orbital prefrontal cortex, the hippocampus and amygdala. We have also noted previously that all of these areas are involved in Affect networks.

However, the **prefrontal cortex does exert control over insula and amygdala** activity. It can down-regulate these areas during reappraisal of negative stimuli. There may be problems with this regulation in certain psychiatric syndromes, such as in **borderline personality disorder** (Donegan et al., 2003). The anterior cingulate cortex has direct connections to all of these areas, and it thus influences all of these control mechanisms.

Of interest to the NFB practitioner is evidence from other fields of research, such as **protein analysis**, which shows that the ACC is structurally damaged in major psychiatric disorders. **Disease-specific protein changes within the ACC** have been found in psychiatric disorders such as depression, bipolar disorder, and schizophrenia. This includes cytoskeletal and mitochondrial dysfunction. Thus, protein changes in the ACC have been shown to be important components of the neuropathology of major psychiatric disorders (Beasley et al., 2006). This evidence concerning the wide-ranging involvement of the ACC gives further support to our **emphasis during initial NFB sessions on normalizing** the QEEG abnormalities associated with the ACC. It also helps explain the positive effects that have been observed in hundreds of patients treated with single-channel neurofeedback at Cz and FCz.

The **ventral portion of the Affect network** also involves other structures such as the amygdala and hippocampus. Like the ACC, damage to the **hippocampal formation (BAs 27, 34, 35, 36)** has been implicated in a number of **psychiatric disorders** including depression, bipolar disorder, schizophrenia, and addictive disorders. The Internet or Rex Canon's summary (Cannon, 2012, p163) offer specific references. There is decreased hippocampal volume in persons who have executive dysfunction with depression (Frodl et al., 2006).

When reading about these structures, the reader may encounter the term **'piriform lobe.'** The piriform lobe or piriform complex is a three-layered cortex that consists of the cortical amygdala, uncus, anterior parahippocampal gyrus (Brodmann area 27), entorhinal cortex, lateral olfactory gyrus, and the cingulate cortex just above the corpus callosum. The 'piriform lobe' is considered to be a part of the **paralimbic** system. It is quite clearly involved in the Affect network. Certain parts of this system may be influenced by NFB training on surface sites over the temporal lobe and at Cz, as well as by LORETA Z-score NFB. In addition, the network as a whole may be influenced in a positive manner using HRV training, as might be expected due to the afferent connections of the vagus to the nucleus of the solitary tract in the medulla and the previously outlined connections of this nucleus to important structures that are involved in the Affect network.

BAs 8, 9, 32, 33, 24
Fz:
Fz placement is posited to influence the following Brodmann areas in the central area of the frontal cortex: BA 9 in the prefrontal cortex, BAs 33, 32, 24, related to the ACC, and BA 8 (ventral). It also may influence the frontal eye fields in terms of **motor control, focus, and action observation.**

Anterior Cingulate Gyrus and Medial Prefrontal Cortex
Functions related to Affect networks in the **medial frontal cortex, BA 32, and anterior cingulate cortex** (ACC) include: emotional inhibition, modulation of emotions (sensitivity) and behavioral responses. As previously noted, it also has a role in motivation and attention. BA 32 is the 'dorsal' ACC, and has an important role in Executive networks.

Functions related to **Executive networks in the medial frontal cortex** executive system include: action monitoring, shifting attention, judgment of events (emotional and cognitive), cognitive flexibility or inflexibility, obsessive thoughts, compulsive behaviors, excessive worrying, argumentativeness, oppositional behavior or 'getting stuck' on certain thoughts or actions. **Monitoring** enables the brain to evaluate errors and the quality of action execution and alerts the executive control mechanisms to allocate resources and take corrective action.

Dysfunction in the rostral anterior cingulate gyrus and the medial frontal lobe, as might be observed in a patient with an obsessive compulsive disorder, may be reflected in evoked potentials at Fz and Cz; namely, low amplitude and a long latency for the P400 component as compared to a database.

Dorsal Central Midline Structures (DCMS)

The dorsomedial prefrontal cortex (DMPFC) and the supragenual anterior cingulate cortex (SACC) are connected with the lateral prefrontal cortex. In a meta-analysis of studies on the **cognitive control of emotion**, the dorsal prefrontal regions were characterized by functions that included reappraisal and evaluation of, and explicit reasoning concerning, emotional stimuli (Ochsner and Gross, 2005). This dorsal CMS could also be involved in reappraisal and evaluation of self-related stimuli. It has a strong **evaluative or judgmental** component. It is involved in tasks involving reading other people (**theory of mind**) (Frith and Frith, 2003). In PET studies, in addition to the dorsal medial prefrontal cortex (DMPC) and the temporal pole, it is **the ventral medial prefrontal** cortex that has been shown to be involved in **self-referenced tasks** and to knowledge about one's self (D'Argembeau et al., 2005).

Patients with lesions in **dorsal CMS** show disturbances in social interactions (Damasio, 2003). Northoff and his colleagues make clear statements as to the importance of CMSs. To paraphrase their conclusions, they note that **self-referential processing** is mediated by cortical midline structures. Since the CMS are densely and reciprocally connected to subcortical midline regions, they posit an integrated cortical–subcortical midline system underlying understanding self and self-in-relation-to-others. They conclude that self-referential processing in CMS constitutes the **core of our sense of self** and is critical for elaborating experiential feelings of self (Northoff et al., 2006). This self-referential processing is one aspect of the Default network. These kinds of findings strongly support the importance of combining either single-channel NFB at CZ (or FCZ) or LORETA Z-score NFB with HRV training.

Anterior Cingulate, Other Connections

As previously described under "FCz," the Anterior Cingulate Cortex (ACC) may be thought of as the keystone for both Affect and Executive networks. It is crucial in monitoring errors. As previously stated, the ACC connects with the medial and orbital prefrontal cortex, insula and the entire limbic system. In addition, it receives input from the brain stem that, importantly for work with NFB and BFB, includes vagal afferents from the heart to the nucleus solitarius in the medulla. As previously noted, this nucleus connects to the locus coeruleus (LC) (noradrenaline production) and, both directly and through the LC, it connects to the hypothalamus and to the limbic system including the insula and the ACC. These are important connections for understanding a number of related disorders. Indeed, Porges (2007) has stated that one characteristic of some psychiatric problems, such as anxiety disorders, autism spectrum disorders (ASDs), **reactive attachment** disorder, and even the **hyper-vigilance** seen in people with ADHD, is **an inability to inhibit defensive reaction** responses in what would normally be considered safe environments. The anterior cingulate appears to be a central component in these disorders that involve the Affect network. The **ACC** also has direct links to the hypothalamic-pituitary-adrenal (HPA) axis. These connections play a major role in the human stress response and hence, once again, we stress the importance of treatment that combines HRV training with NFB to control stress (Thompson & Thompson, 2007). Affect can be complex and involves a system of Executive and Affect subnetworks. An example is depression.

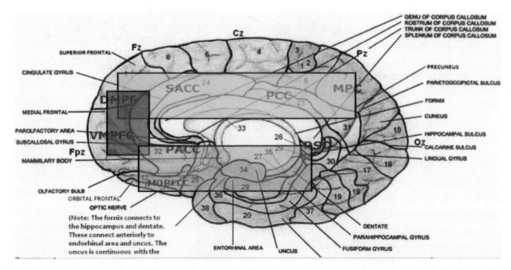

Areas of the Executive and Affect networks involved in depression (adapted after de Ridder, 2010)
- *Executive/Cognitive* Aspects of Network = **GREEN** (BAs 9 and 46, 23, 24, 31, 32, 40)
- *Affect/Vegetative* Aspects of Network = **ORANGE** (BAs 25, 13, anterior insula, hippocampus, hypothalamus)

• ***Integration and connection*** of Cognitive and Vegetative = **RED** *(BAs 24, 9, 10, 11) (red box can be extended down to include BA 11).*

In depression, the **vegetative/autonomic Affect** network symptoms involve a ventral compartment (e.g., sleep disturbance, loss of appetite and libido), which includes the hypothalamic-pituitary-adrenal axis, hippocampus, insula (**BA 13**), subgenual (SG) cingulate (**BA 25**), and the brainstem. It should be noted that the solitary nucleus connects to BA 25 and this could be one important factor in understanding how HRV training can have positive effects in depression.

Sleep disturbance is a major factor in depression. Sleep disturbances **will decrease cortisol, increase inflammatory responses** (related to **increase in cytokine IL6**) and the risk of infectious disease, and **decrease cellular immunity.** For example, studies have demonstrated that shift workers report more colds than nonshift workers. Some studies give a general overview saying that adults are at risk when they have less than five or more than nine hours of sleep. Others suggest that most adults need eight to nine hours and are at risk of flu when they have less than seven hours. Both stress and sleep deprivation can increase pro-inflammatory cytokines. This will increase the severity of depression, and is usually associated with **poor concentration, anhedonia** symptoms, and social withdrawal (Prather, Rabinovitz, Pollock & Lotrich, 2009; Prather, Marsland, Hall, Neumann, Muldoon & Manuck, 2009; Buysse et al., 2011).

The **executive components** of the Affect network in depression are found in the dorsal compartment, which modulates attentional and sensory-cognitive symptoms (e.g., **apathy, attentional and executive deficits**). This includes the dorsolateral prefrontal cortex (**BAs 9 and 46**), the dorsolateral anterior cingulate (**BA 24'**), posterior cingulate (**BA 23, 31**), the inferior parietal lobe (**BA 40**), and the **striatum**. The dorsal prefrontal components, including the ACC, are also involved in the cognitive control of emotion including reappraisal, evaluation, and explicit reasoning concerning emotional stimuli.

Information **concerning cognition and emotion** from the two compartments (dorsal and ventral) is **integrated** by the rostral anterior cingulate (BA 24), medial frontal cortex (**BAs 9 and 10**), orbital frontal cortex (**BA 11**), and frontopolar areas, as is illustrated in the above figure. **Depressive ruminations** may involve the **Default network** with connections involving the posterior cingulate and the subgenual cingulate areas (Berman et al., 2011).

The **central-midline areas shown for depressive disorders are also important in drug abuse and in bipolar** disorders. Robbins has reviewed substance abuse and its relation to these neuroanatomical structures (Robbins et al., 2009). **Bipolar** disorder is a severe and recurrent illness involving periodic cycling of both manic and depressed states. This disorder can result in significant impairments, including high rates of divorce and suicide, high rates of alcohol and substance abuse, and erratic work performance despite, at times, remarkably productive intervals in the manic phase. Bipolar disorder is often diagnosed late in the course of the illness, or may be initially misdiagnosed as other illnesses, such as unipolar depression (Nusslock et al., 2012). Children may be first be given a diagnosis of ADHD, which is later recognized as being bipolar disorder in their adolescence.

In bipolar disorder, there is **manic state-related increased activity in the left dorsal anterior cingulate,** and left head of the caudate (Blumberg, 2000). In QEEG assessments, it has been our experience that high amplitude spindling beta may be seen with LORETA analysis showing a source in both the ACC and the frontal lobes, particularly the right NDH frontal lobe.

Cingulate Cortex

In the **foregoing section, we dealt with the principle functions** of the first division, the **Affect network components**. Now we will review the **second and third components** of the anterior cingulate cortex, **executive and motor** respectively.

Fz and also Cz; Second Division of the Anterior Cingulate Cortex: Cognition Division (Dorsal)

A second, major division of the anterior cingulate cortex is the executive division. In the previous discussion of the functions of the ACC, the Affect network was emphasized. The executive division includes **Brodmann areas 24'and 32'**. (Note that 24', read as 24 prime, refers to the more dorsal layer of BA 24.) We are not able, using LORETA, to make this fine distinction between areas involved with the Affect network (e.g., BAs 24, 33) and areas mainly said to be involved with the Executive network (BAs 24' and 32'). The executive division also has **connections to the DLPFC** and other areas of the frontal cortex involved in **attention, concentration, planning, evaluating,** and all aspects of executive/cognitive functioning.

Other important functions of the executive components of the ACC include initiation, motivation, and **goal-directed behavior**. **Maintaining attention is a key function** and attention to space is to the contralateral side. However, the reader should recognize that **not all attentional processes** are related to ACC functioning. For example, the left dorsolateral cortex is activated during reading and semantic processing, but the ACC is not activated, even though reading requires attention (Peterson, 1990). The ACC is important in the **retrieval of short-term memory**. It is involved in the assessment of the **motivational content** of internal and external stimuli before movement.

The ACC is important in the **regulation of context-dependent behavior patterns** seen in disorders such as **Anxiety, Panic, Obsessive Compulsive (OCD), Sociopathic Personality, and Asperger's syndrome**. The neuroanatomy of panic disorder has been reviewed by Gorman (Gorman et al., 2000).

Third Division of the Anterior Cingulate (Motor)

A third network of functioning that involves modulation by the anterior cingulate cortex is motor control. The locations for this ACC motor regulation are the motor areas in the **cingulate sulcus and nociceptive cortex** that communicate with the **midline and intralaminar thalamic** nuclei (Peterson, 1990, p280). The motor area has projections to the **spinal cord and to the red nucleus**. It has some control over premotor functions.

The cingulate cortical site includes the **nociceptive area,** which is important in **response selection** and in cognitively demanding information processing. The major function of this cortical area is in response selection associated with skeletomotor activity and responses to noxious stimuli.

Anterior Cingulate
BAs 32, 24, 25, 33
FCz
Affect (and Executive) Networks

BA 32

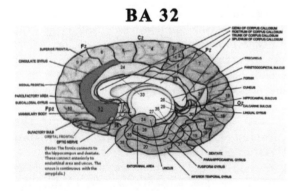

Brodmann area 32 comprises the most anterior (rostral) portion of the anterior cingulate cortex. It is also **called the dorsal anterior cingulate**. It is part of the **Executive network**. It is under Fz and FCz when doing neurofeedback. The dorsal region of anterior cingulate gyrus is associated with rational thought processes and may be seen to be active during the Stroop task. It is also a part of the **Affect** network, and this is involved in regulation of mood and appraising of emotions, including negative emotions, and also rewards. It is an area involved in the anticipation of stimuli and the **detection** of unexpected errors, or when rewards were expected but not received. This area is important in the regulation of mood. It is part of the network involved in **identifying pain**. It is also important in sustaining a response set.

BA 24

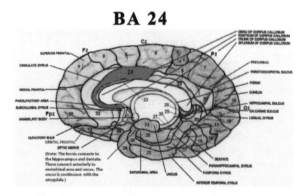

Brodmann area 24 includes a large portion of the anterior cingulate gyrus. The anterior aspect of the cingulate has already been mentioned when discussing the frontal lobes and NFB at FCz. The posterior portion of BA 24 lies under an electrode placed at Cz, which is probably the most frequently used location for single-channel neurofeedback, especially with children. Much of our work with NFB is done to influence the networks associated with the anterior cingulate cortex including BA 24. The anterior cingulate is associated with the Affect, Executive, Salience, and Default networks. It can be described as the **hub of the emotional brain (also called the Affect network or the limbic** system), and we often find that BA 24 is the source location for frequencies that are outside database norms when we do a LORETA analysis of the QEEG in people with anxiety disorders or people with autism spectrum disorders, including Asperger's syndrome.

CZ is a primary site for feedback when we are working with ADHD. In this disorder we are usually decreasing whatever frequency range has excessive slow-wave activity (such as 4–8 Hz or 3–7 Hz in children, 3–10 Hz in some adolescents and adults), enhancing SMR (12–15 Hz or 13–15 Hz), which encourages the client to be calm and relaxed while also focused and concentrating. For part of each session we often change the enhance frequency to 15–18 Hz beta while the client is learning and carrying out cognitive strategies. This can be extraordinarily helpful, as reported in our published case series (Thompson & Thompson, 1998, 2010). The dorsal areas of the ACC are more related to Executive network functions, and the ventral and rostral ACC to the Affect network. Both the rostral-ventral ACC (more under Fz) and the posterior cingulate cortex, BA 31, under Pz, are related to Default network functions.

Brodmann area 24 is also important in a large number of **Executive network** functions, including both inductive and deductive reasoning. It is critical to memory networks, and its functions include aspects of working memory, prospective memory, delayed memory, mental timekeeping, and inhibition. It is important in most aspects of **attention,** including auditory and visual-spatial attention, and divided attention. It is linked to networks responsible for verbal fluency and object naming. It is important in the ability to multitask. It is a key link in Affect network and will influence all aspects of mood and mood swings. It even has some role in such diverse areas as sexual arousal, taste, and pain endurance.

Due to the number of networks that involve the anterior cingulate cortex, when doing NFB, we also use metacognitive training and tasks. We do NFB to normalize frequencies that are outside database norms while the patient is doing tasks related to correcting their symptoms in order to activate particular networks that are relevant to their problems. If working at CZ, the symptoms targeted would be those that might be expected to be related to functioning of the anterior cingulate for a particular patient. In this manner, using metacognitive strategies and related tasks, we postulate we can be a little more specific with respect to involving and constructively optimizing the functions of networks that are important for that person.

BA 25

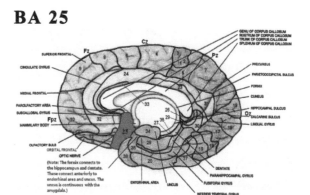

Brodmann area 25 is also called the subgenual area, area subgenualis, or subgenual cingulate because it is "below the knee" of the corpus callosum. It may be influenced by work done with NFB at Fz and Cz by virtue of its involvement in networks that also involve the more superior aspect of the cingulate gyrus under Cz and the medial aspect of the frontal lobe under Fz. It is also involved in moral reasoning.

BA 25 is very deep in the midline cortex. LORETA NFB may have a much more direct effect. This is an area that is very important in the **Affect network.** It is an area that is considered a key to intervention with patients who have an intractable depression. This region is rich in **serotonin** transporters and is considered to influence a vast network. It influences areas such as the hypothalamus and brain stem, which are involved in changes in appetite and sleep; the amygdala and insula, which affect mood and anxiety; the hippocampus, which plays an important role in memory formation; and some parts of the frontal cortex responsible for self-esteem.

A surgical procedure that involves **deep brain stimulation** (DBS) in BA 25 has been used to treat intractable depression (Mayberg et al., 2005). DBS has also been used to treat obsessive compulsive disorder as reviewed by Greenberg et al. (2006). In 2005, Helen S. Mayberg and collaborators described how they successfully treated a number of depressed people — individuals virtually catatonic with depression despite years of talk therapy, drugs, even shock therapy — with pacemaker-like electrodes (deep brain stimulation) in Brodmann area 25 (Mayberg et al., 2005; Kennedy et al., 2011). BA 25 is said to be **metabolically overactive** in treatment-resistant depression, and this appears to correlate with poor therapeutic response to traditional treatments for depression.

However, of importance to the practitioner who uses a combination of NFB and BFB, the **solitary nucleus** has direct connections to BA 25. Thus HRV training will send vagal afferent feedback to the medulla, solitary nucleus, and thus be potentially helpful in influencing BA 25, and thereby, perhaps, the symptoms of depression related to dysfunction in this area of the cortex. Indeed, depressed patients have been shown to be helped by HRV training (Karavidas et al., 2007). In addition, NFB combined with psychotherapy has been shown to affect virtually the same areas as DBS with similar beneficial results and a much less invasive procedure for patients who have depression that does not respond to traditional therapies (Paquette et al., 2009).

BA 33

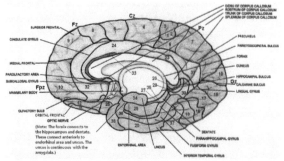

Brodmann area 33 is another part of the rostral anterior cingulate gyrus and thus is a part of the Affect network. It may be particularly important in emotions related to hearing pleasant music. It is important also in networks that involve sexual arousal.

The Caudal Division of the Cingulate Cortex – Posterior Cingulate Cortex

BAs 31, 23
Pz
Posterior Cingulate Gyrus

BA 31

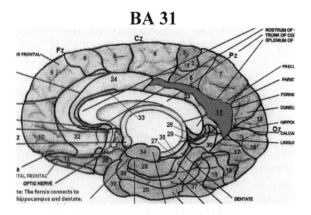

BA 23

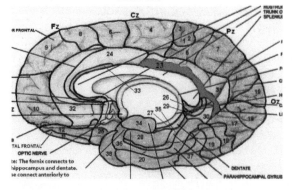

Pz lies above the posterior cingulate gyrus **Brodmann area 31 and below it, Brodmann area 23.**

BA 31

Brodmann area 31 is an association cortex area in the posterior part of the cingulate gyrus. As with other parts of the posterior cingulate, a portion of BA 31 is in the medial parietal lobe. It is a region that is crucial for the **evaluation of stimuli**. It is important in **memory** and has a role in **pain** perception. It is important in **attention**, **comprehension**, and **visual processing** (Choo et al., 2012; Choo et al., 2009). It has a key role in the **integration and synthesis** of almost all sensory input. In our work with university students whose chief complaint is an inability to **assimilate new information during a lecture**, we are finding hypocoherence (sometimes called disconnects) between Pz (over the posterior cingulate) and central temporal, central, and frontal areas. With NFB training directed at changing this, they have improved and become successful in their learning. Without research, this remains a clinical observation only. BA 31 is also important in evaluative judgment and precautionary reasoning. In keeping with part of this Brodmann area being in the medial parietal lobe, it has functions related to parietal **associational integration of sensory inputs**. The cingulate part of this area is a cortical component of the limbic system. It is a key component of the **Default network** and is important in self-evaluation. It is one area involved in the distinction of **'self' from 'other.'**

The posterior cingulate region is also important in the **recall of episodic information,** and it is one of several brain areas that produce anterograde amnesia when damaged. This occurred with the TBI experienced by the 52-year-old woman whose case is described below. Other functions include visuospatial processing and memory. It is part of networks involved in **identification of pain**. This area of the cingulate has connections to the hippocampus, posterior parietal cortex, posterior parahippocampal gyrus, and to the dorsal striatum.

The cingulate gyrus part of this area is a cortical component of the **limbic system** and is thus a part of the Affect network in addition to its roles in the Executive and Default networks.

BA 23

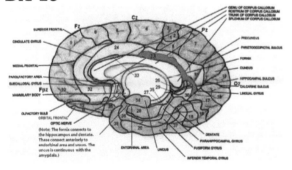

Brodmann area 23 is involved in the **Default network** when the brain is not engaged in active tasks and it is important in aspects of thinking about self versus other(s) and 'self-in-the-world.' Like BA 31, it is also part of the networks for the **synthesis and integration of incoming sensory information**. It is important in learning complex procedures, memory encoding and retrieval, learning complex procedures, visual discrimination, lexical categories, and high order tasks such as **evaluation and judgment**. It is involved in classical **conditioning** and in the conditioning of fear. It is linked to emotions evoked by words. It also has a role in processing temperature.

The following case assessment reveals difficulties that encompass many of the areas described above under the dominant hemisphere temporal – parietal area and cingulate gyrus (described below). Twenty-one years before presenting at our Centre, a 52-year-old woman was hit in the jaw. She fell backwards, striking her head on the hard floor. She was briefly unconscious. When she awoke she was told that she had tried to strangle her girlfriend who found her and was attempting to help her by putting her in a car to transport her to hospital after the assault. She remembers thinking she was outside her own body and that she could see parts of her (forearm, hand) separated from her and floating in space. She had been a brilliant student, perhaps genius level intelligence, able to sort accounting figures and sheets with ease in a part-time job while she was a student. After the injury she had difficulty with even relatively simple accounting, and still cannot order figures in proper columns and rows. Putting numbers and charts together and organizing things — putting all the little pieces together to see the big overall picture — is still, as it has been for the last 21 years since her injury, extremely difficult. There are also major problems with memory. She will pay her phone bill three times, not realizing she already paid it. She will purchase oranges when she goes shopping even though she has far too many already at home. She will sometimes pick up the phone and look at it and say to herself, "What do I do with this?" She has difficulty recalling timelines in terms of what year things happened since the injury. On the other hand, she has an excellent and very detailed and precise memory from long before the injury.

Below are some of the EEG findings for this client that appear to correlate with her symptoms.

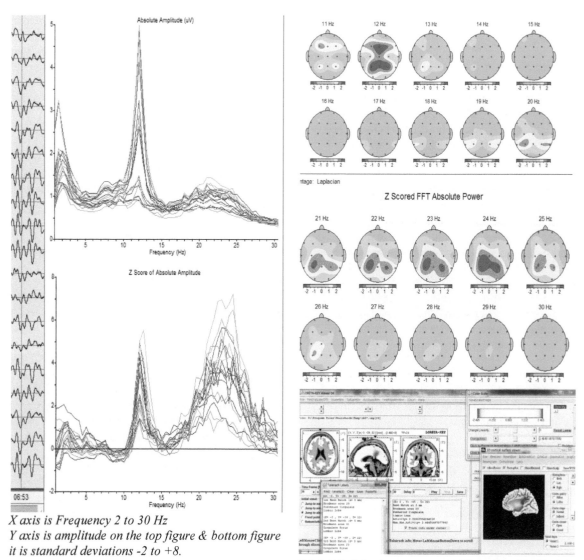

X axis is Frequency 2 to 30 Hz
Y axis is amplitude on the top figure & bottom figure it is standard deviations -2 to +8.

EC 24 Hz 2.5 sd, posterior cingulate BAs 29, 31, 23

LORETA shows high-amplitude spindling beta 24 Hz, which was 2.5 standard deviations above the database mean with a source in the posterior cingulate BAs 29, 31, 23. We postulate that this may contribute to the patient's difficulties with synthesis and integration of new incoming information. She feels this is a key issue. Her particular learning problems with columns and rows of figures could relate to injury to the DH, left, angular gyrus area. Her difficulties with memory may relate to her being well outside database norms in areas that reflect EEG activity influenced by BAs 31, 29, and the left angular gyrus, the supramarginal gyrus, the left entorhinal cortex, parahippocampal area, and the inferior frontal gyrus near and including the rostral ventral cingulate gyrus BA 25. The involvement of BA 25 and the anterior cingulate including BA 24 and right insula may all be related to her easily precipitated extreme anxiety. The spindling beta in the parietal regions may contribute to a major difficulty handling any 'startle' response, whether it be auditory or visual. Despite all these difficulties, she is a very pleasant, hard working individual who has been finding ways to cope with her deficiencies for two decades.

Posterior Cingulate Gyrus and Anterior-Medial Aspects of the Parietal Lobe Overlap

BAs 26, 29, 30, Pz:

BA 26: Medial Parietal Lobe/Posterior Cingulate

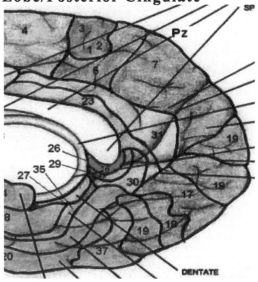

Brodmann area 26 lies in the retrosplenial region of the cerebral cortex. It is a narrow band located in the isthmus of the cingulate gyrus in the medial parietal lobe. It is bounded externally by the granular retrolimbic area, BA 29. It is an area of the association cortex in the transitional region between the posterior cingulate gyrus and the medial temporal lobe; this area is a cortical component of the limbic system.

It has been suggested that in post-traumatic stress disorder (PTSD), triggering traumatic memories may reduce blood flow in this area (as reflected in PET scan data). It is assumed that this represents a decrease in the activation of BA 26. This may correspond to an increase in activity in the motor cortex and represent readying for motor action in response to the stressful stimuli (posted in *Brain* by Dr Justin Marley on April 10, 2012).

BA 29

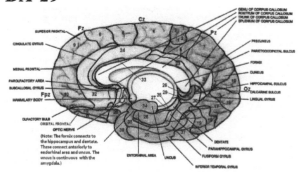

Brodmann area 29 is a medial area of the cortex that may include functions that also relate to the posterior cingulate and the posterior medial aspect of the temporal lobe. It is a part of the **paralimbic cortex**. It may be involved in **semantic associations** (not conscious) and novelty, but its functions are not as well researched as other areas. It lies between the posterior cingulate and the hippocampal area and we posit that it may be an important link between the integration and synthesis of new information in the posterior cingulate area and memory.

BA 30

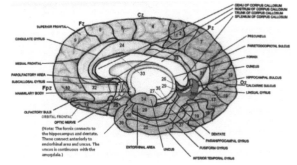

Brodmann area 30 is in the retrosplenial region and is a part of the **cingulate cortex and cuneus**. It is a **transitional area** between the **posterior cingulate cortex and the temporal lobe.** The retrosplenial region includes **Brodmann area 26 and Brodmann area 29,** as well as **Brodmann area 30**. The retrosplenial cortex has dense reciprocal projections with both the anterior thalamic nuclei and the hippocampus. Thus it is likely that it has functions in the **Affect, Memory, and Default networks**. It has a role in attending to speech, listening to sentences and, in particular, to **emotional aspects of words**. It is part of the paralimbic cortex and has direct links to the anterior thalamus and other limbic areas, and thus is important in **mood regulation**. It is in an area that is one of the **central midline structures (CMS)** involved in the **Default** network, and thus has a role in appraisal of one's self (**insight**). All of these functions mean that we are alert to its possible dysfunction in people with Asperger's syndrome.

Parietal Lobes

Introduction
Brodmann areas under Pz that may be influenced by NFB at this site include: BA 7 somatosensory cortex and working memory, BA 31 Posterior Cingulate Gyrus, and BAs 23, 26 and 28, which are medial and can be seen in midsagittal section views of the cerebral hemispheres. Areas 26 and 28 might require LORETA neurofeedback training.

Anatomical Boundaries
The parietal lobe is defined by four anatomical boundaries: the central sulcus (Rolandic fissure) separates the parietal lobe from the frontal lobe; the parieto-occipital sulcus separates the parietal and occipital lobes; the lateral sulcus (Sylvian fissure) is the most lateral boundary separating the parietal lobe from the temporal lobe; and the medial longitudinal fissure divides the two hemispheres, and thus the two halves of the parietal cortex.

Immediately posterior to the central sulcus, the most anterior part of the parietal lobe is the postcentral gyrus (Brodmann area 3), the secondary somatosensory cortical area. Separating this from the posterior parietal cortex is the postcentral sulcus, which can be seen just posterior to the central sulcus in the figure from *Gray's Anatomy*.

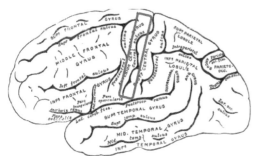

From *Gray's Anatomy (1918). This drawing shows the Central Sulcus.*

Functions
The parietal lobe plays an important role in **integrating sensory information** from exterior sensory inputs and from various parts of the body. It is important in the **knowledge of numbers** and in seeing relationships between numbers. Particularly on the **NDH right side**, it perceives the **manipulation of objects in space**. Portions of the parietal lobe are involved with visuospatial processing. The parietal association cortex contains areas responsible for **perception of space**, and for localization of **self in relation to objects in space**. These areas can encode the location of objects. Although multisensory in nature, the posterior parietal cortex is referred to as the **dorsal stream for vision** (as opposed to the ventral stream in the temporal lobe). This dorsal stream has been called both the **'where'** stream (as in spatial vision) and the **'how'** stream (as in vision for action). The ventral intraparietal (VIP) area receives input from a number of senses (visual, somatosensory, auditory, and vestibular).

The **medial intraparietal (MIP)** area neurons are part of the central midline structures (CMSs) and dorsally may be more involved in the integration and synthesis of incoming sensory information and in communicating this information to the temporal and frontal cortices.

The **anterior intraparietal (AIP)** area contains neurons responsive to **shape, size and orientation** of objects to be grasped, as well as for manipulation of the hands. It responds both to viewed and remembered stimuli. It has the necessary connections to the frontal lobes and the frontal eye fields so that **goal-related activity can be 'remapped'** when the eyes move. It is also claimed that both the left and right parietal neural systems play a role in self-transcendence, a personality trait that may measure predisposition to spirituality.

The parietal lobe is divided into two hemispheres – left and right. The **left hemisphere** plays a more prominent role for right-handers and is involved in symbolic functions in **language and mathematics**. The **right parietal** lobe plays a more prominent role for true left-handers, and is specialized for images and for the understanding of **maps**. These are functions that require the perception and retention of **spatial relationships**. The parietal association cortex enables individuals to **read, write, and solve mathematical** problems. The sensory inputs from the right side go to the left side of the parietal lobe and vice versa.

Damage to the right parietal lobe can result in the loss of imagery and the ability to visualize spatial relationships. This was one of the problems of the 44-

year-old woman who had encephalitis as a young woman in the example given earlier. Extensive damage will result in a neglect of the left side of space and the left side of the body. The person will be unaware of this loss and this is called **anosognosia**. Representing the left side may even be neglected when drawing. A novel treatment for this disorder is to use mirrors to fool the person into thinking that their left is their right so that they will move their left limb.

Damage to the **left parietal** lobe will result in different problems that include problems in mathematics, reading, writing, and understanding symbols. Damage to the left side of the parietal lobe, however, will not result in neglect of the right side of space and body.

Cortical Information Flow

There is considerable evidence that a person's cortex is essentially divided into two functional streams: an occipital-parietal-frontal pathway that processes 'where' information, and an occipital-temporal-frontal pathway that provides the 'what' information to the individual (Ungerleider and Mishkin, 1982, reported in Kim & Robertson, 2001).

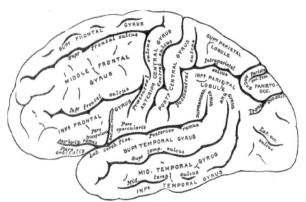

Gray's Anatomy *(1918)*

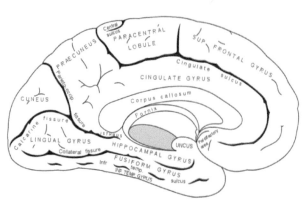

Gray's Anatomy *(1918), midsagittal section*

Parietal Lobes, Dorsal Lateral Aspect

BA 7; BA 7:
Pz:
Somatosensory Association Cortex

BA 7

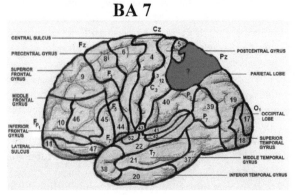

BA 7

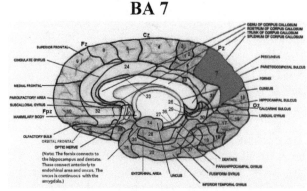

Dorsal and Lateral Aspects of BA 7

This somatosensory association cortex is one key area in the **Default** network, and it has motor, memory, attention, and cognitive and emotional understanding functions. It is involved in the perception of pain. In the **nondominant hemisphere** it is important in the **conscious awareness of visuospatial** events and in **understanding emotions**. It is important in **working memory** and comprehension. It is involved in visuomotor coordination (reaching to grasp an object) and bimanual manipulation.

In the **dominant** (left hemisphere) this area includes various networks that are involved in such functions as: **verbal comprehension**, episode recall, handwriting and spelling, language, imaging, shifting attention, and the use of tools.

In the **nondominant** (right hemisphere) this area participates in understanding emotions, gestures, nuance, and innuendo. It is critical to **visual-spatial comprehension** including attention, mental rotation, stereopsis, and judgments or estimates that involve visual and spatial aspects of drawings and of the environment. This area is also part of the networks that are involved in a perception of personal space. It is important in visuospatial memory.

In both hemispheres, the **superior** parietal lobe is involved in rhythm detection, semantic categorization, temporal context recognition, and tactile localization, while the **inferior** parietal lobe is involved more in tactile recognition. In both hemispheres mirroring functions are present, which are important in learning manipulation of objects and tools. **Lesions** may result in **ideomotor apraxia** (inability to perform skilled purposeful movements unrelated to weakness, paralysis, lack of coordination or sensory loss). Lesions can result in **astereognosis**. This is a form of tactile agnosia where the subject is unable to identify common objects by feeling them if there is a barrier that prevents them from seeing the object. As with so many areas of the brain already discussed, it is one part of the working memory networks for motor, verbal, visual, auditory, and even emotional functions and is involved in the **conscious recollection of previous experiences**. It is also part of the networks (that include the anterior cingulate cortex) for the perception of **pain**.

Important for all of our work with NFB, this area in the nondominant hemisphere, in conjunction with frontal areas, is involved in a **sustained attention network,** particularly with respect to **visual-motor attention.**

Syndromes Associated with Parietal Lobe Functions

The following syndromes are briefly described in this book for the sole purpose of helping the practitioner understand some of the functions that may be affected when NFB is carried out over parietal lobe sites. Many are very rare syndromes so it is not necessary to learn them, just note how the symptoms relate to a particular area of cortical damage.

BAs 40 and 39 DH: Gerstmann's Syndrome

Gerstmann's syndrome is associated with lesions in the dominant (usually left) parietal lobe. This syndrome is characterized by four primary symptoms:

- Dysgraphia/agraphia: deficiency in the ability to write
- Dyscalculia/acalculia: difficulty in learning or comprehending mathematics
- Finger agnosia: inability to distinguish the fingers on the hand
- Left-right disorientation

The lesions causing one or more of these symptoms are attributable to dysfunction in the **angular and supramarginal gyri** near the temporal and parietal lobe junction. Such dysfunction may occur after a stroke or a TBI. In addition to exhibiting the above symptoms, many adults also experience **aphasia**, which is a difficulty in expressing oneself when speaking, and in understanding speech and written communications. Often both reading and writing are also affected.

In children the symptoms that have been associated with this syndrome may be developmental, but most cases are not identified until the child reaches school age and must learn to read and do math. As would be expected, these children exhibit poor handwriting and spelling skills, and difficulty with math functions, including adding, subtracting, multiplying, and dividing. They have an inability to differentiate right from left. They may even have difficulty discriminating among individual fingers. In addition to the four symptoms listed that are part of this syndrome, many children also suffer from constructional apraxia, an inability to copy simple drawings. Frequently, there is also an impairment in reading. Children with a high level of intellectual functioning, as well as those with brain damage, may be affected with the symptoms associated with this syndrome.

BA 7, Lateral Aspect

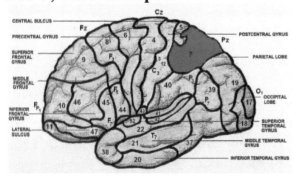

Bálint's Syndrome

Bálint's syndrome was named in 1909 for the Austro-Hungarian neurologist Rezs Bálint. It is associated with bilateral parietal – occipital lesions (Ropper, Allan, Brown, Robert, 2005). People who suffer from this syndrome have an inability to perceive the visual field as a whole (**simultanagnosia**), difficulty in fixating the eyes (**ocular apraxia**), and inability to move the hand to a specific object by using vision (**optic ataxia**). Optic ataxia is associated with difficulties reaching toward objects in the visual field opposite to the side of the parietal damage.

Since simultanagnosia is the inability to perceive simultaneous events or objects in one's visual field, people with Bálint's syndrome perceive the world erratically, as a series of single objects, rather than seeing the wholeness or Gestalt of a scene. It is a specific deficit of attention. Recovery in simultanagnosia may be related to finding meaning to the total scene and thus expanding their very restricted attentional window. Simultanagnosia is a profound visual deficit. It impairs the ability to perceive multiple items in a visual display, while **preserving the ability to recognize single** objects. It may be difficult for the person to disengage from an object once it has been selected and move to other objects or to retain a sense of that first object while viewing a second object. The patient sees their world in a patchy, spotty manner.

Ocular apraxia is exemplified by the inability to carry out familiar movements when asked to do so. As previously described, persons with apraxia understand commands and are willing to carry them out; however, they are physically unable to perform the task. In optic apraxia there is an inability to voluntarily guide eye movements and change to a new location of visual fixation. The most frequent and disabling deficit is the symptom of unilateral spatial neglect. This neglect manifests as a bias of spatial representation and attention on the same side as the lesion, while ignoring the opposite side. This was described above under dysfunction of the right parietal lobe. In addition, there are disturbances of space representation after traumatic damage to the right hemisphere's parietal lobe.

Optic ataxia is the inability to guide the hand toward an object using visual information where the inability cannot be explained by motor, somatosensory, visual field deficits, or acuity deficits. Optic ataxia is seen in Bálint's syndrome, where it is characterized by an impaired visual control of the direction of arm-reaching to a visual target, accompanied by defective hand orientation and grip formation. It is considered a specific visuomotor disorder, independent of visual space misperception. Optic ataxia is also known as misreaching or dysmetria and it refers to a lack of coordination of movement, typified by undershoot or overshoot of intended position with the hand, arm, leg, or eye. It is sometimes described as an **inability to judge distance** or scale. Bálint described a patient who, while cutting a slice of meat that he held with a fork in his left hand, would search for it outside the plate with the knife in his right hand. The ability to grasp an object is also impaired.

Anatomical Lesions Causing Symptoms in Bálint's Syndrome

The visual difficulties in Bálint's syndrome are usually due to damage to the top part of the temporal-occipital lobes on both sides of the brain. The temporal-occipital area refers to the lateral and posterior portions of the brain. In Bálint's syndrome, the superior or dorsal part of the parietal lobes on both sides of the brain may also be affected. The parietal lobes are the middle portion of the dorsal aspect of the brain. There is usually bilateral damage to the posterior parietal cortex. The damage and the syndrome can originate from single or multiple strokes, Alzheimer's disease, intracranial tumor, or brain injury; in short, anything causing damage to the posterior superior areas of the parietal lobe (BA 7).

Parietal Lobes, Medial Dorsal Aspect

BA 7
Pz:

BA 7: Cuneus, Precuneus

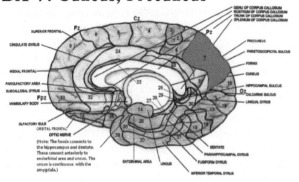

The medial superior areas of the parietal lobe are called the cuneus and precuneus. These areas are responsible for the **organization, integration, and synthesis** of auditory, visual and kinesthetic inputs and for orientation in space. They are some of the last areas to become myelinated. It is important in high-level functions including attention, self-referential and self-recognition tasks, theory of mind, working memory, and many other cognitive processes. The **precuneus** has a role in visual-spatial imagery, episodic memory retrieval, and taking a perspective on 'self.' These posterior medial areas have a very high metabolic rate. There is a decrease in this Default mode activity during non-self-referential, goal-directed tasks. Thus a patient might not feel anxiety when they are absorbed in a cognitive task.

These areas have extensive connections to other medial and lateral parietal areas, the superior temporal cortices, both the medial and dorsolateral frontal cortices, the orbital surfaces of the frontal lobes, and the anterior cingulate (Cavanna, 2006). In other words, they are an important part of the Default network.

BA 5 (Previously discussed in 'Central Region)

The posterior portion of **BA 5** is important in the perception of tactile objects, the body, and in orienting in extrapersonal space. As previously mentioned, damage to these areas may contribute to agnosia and apraxia.

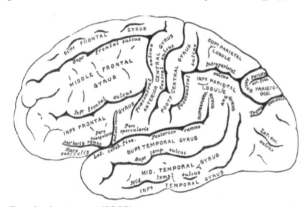

Gray's Anatomy *(1918)*

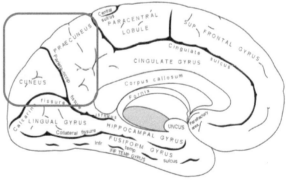

Gray's Anatomy *(1918),* midsagittal section to view medial aspects of parietal lobe

The precuneus, cuneus, and the posterior cingulate may be more active at rest and activation will be attenuated during a motor or cognitive task. These areas subserve core consciousness and represent **self in relationship to the world**. They are involved in monitoring the outside world and are important for psychologically assessing and evaluating 'self-in-the-world.' This is a form of visual and emotional response analysis and involves making decisions about another person as compared to oneself (Seger, Stone & Keenan, 2004). Perhaps importantly, it has direct connections to the inferior right parietal lobe (between P4 and T6/P8), which is part of the **right parietal to frontal Mirror Neuron** network for imitating. Such imitation involves interpreting innuendo and understanding the meaning of communications by others in order to be able to imitate others. Difficulties with all of these functions are observed in patients who have Asperger's or an autism spectrum disorder.

A seven-year-old autistic girl developed an apparent excess of high frequency beta in the P4 area during NFB training. This occurred at the same time as she improved in her social functions and became almost indistinguishable in her behavior from her bright, socially competent, non-autistic, twin sister. Even her mother immediately saw the correlation and said, "Don't train down that high beta. Everything is going so well!" This increase in activation at P4 appeared to represent a compensatory change. We use this case as one of our examples when we warn professionals to be careful when using Z-score training and to not use one-size-fits-all protocols and normalize everything, but rather to carefully assess the patient before determining which frequencies and sites are appropriate for that individual's NFB training regimen.

The precuneus appears to be involved in self-related mental representations during rest—in other words, **self-consciousness**. In our assessments of patients who exhibit high levels of **anxiety**, we may note high amplitude of high-frequency beta in this posterior medial area. We also sometimes see spindling beta more than 2 SD above the database means between P4 and P8 (T6) in the general area of the right angular gyrus. We have postulated that this type of finding might correlate with the observed hyper-vigilant state in these patients and thus relate to their anxiety. This is observed in some **hypervigilant** patients who have **difficulties falling asleep** and going back to sleep when they awaken during the night.

Temporal-Parietal Junction, Lateral Aspect

(As it relates to learning, dominant hemisphere language comprehension, nondominant hemisphere prosody functions)

BAs 39, 40
T7-P3 x C3-P7;

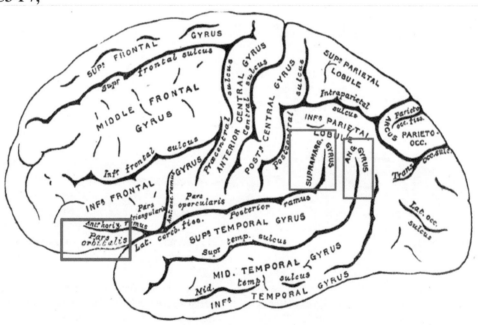

Figure to show the temporal-parietal junction. Adapted from Gray's Anatomy *to show the angular and supramarginal gyri. Also identified is the pars orbitalis*

Introduction

In the **dominant hemisphere** the temporal-parietal junction comprises Wernicke's area. This area encompasses part of BAs 39 and 40 in the supramarginal and angular gyrus areas, and extends inferior into the posterior superior aspect of the temporal lobe, taking a small posterior portion of BAs 41, 42 and 22. Functions of this area and the homologous area in the nondominant hemisphere differ. For the most part, dominant hemisphere, left-sided lesions interfere with verbal comprehension and memory processes, whereas **nondominant** (right-sided) lesions interfere with visuospatial (nonverbal) processes and the interpretation and comprehension of emotional aspects of language and gestures and nuance and innuendo in communications.

The left frontal operculum area (F5, or Broca's area) is the key area for speech articulation. However, the temporal lobe (BAs 21 and 22) and the supramarginal gyrus, angular gyrus, and insula are involved in auditory and verbal representation. These areas (see figure below) and their connections, such as via the **arcuate fasciculus**, are all involved in **letter and word perception and recognition.** These areas with their connections to the frontal lobe have been described in other sections of this monograph. They are important in the **comprehension** of abstract verbal forms, object naming, word recall (involving the hippocampus for auditory information on the left and visual information on the right), and perception of complex verbal relationships.

Patients with frontal lobe lesions have difficulty in the temporal organization of information, such as lists of words. A name can pass through Wernicke's area and then, via the angular gyrus, arouse associations in other areas of the cortex. In this manner the **angular gyrus is a connecting link** between sensory input and speech production areas. It is an important association cortex that **combines auditory and visual** information that are both necessary for reading and writing. It is involved in the steps between reading and speaking. Damage to the angular gyrus can result in **alexia** (inability to read) with **agraphia** (inability to write). The patient is unable to match spoken letter sounds to written letters. They can only spell the very simplest words.

Right hemisphere lesions in the parietal-temporal-occipital cortex can result in an inability to understand temporal relationships (future planning), visual-spatial defects, and problems in retaining a visual image. A study directed at investigating the neuroanatomical correlates of difficulties with the clock drawing test (CDT), which is usually considered to be a parietal lobe test, indicated that impaired performance on the CDT most strongly correlated with damage to right parietal cortices (supramarginal gyrus) and to left inferior frontal-parietal opercular cortices. In the CDT, the person is asked to draw a clock and to set the time to twenty minutes 'til four. Time setting errors were predominant in people with left hemisphere lesions, perhaps because of the interpretation of language involved in the setting of the hands to "twenty minutes 'til four." Visual spatial errors predominated in patients with right hemisphere damage; for example, leaving out numbers from the left side of the clock or bunching up all the numbers on the right side (Tranel, Rudrauf, Vianna & Damasio, 2008).

Left hemisphere lesions in these general areas result in **aphasia**, inability to evoke a visual image in response to a word, inability to organize sequentially, and the inability to draw proper angles and copy designs.

Introduction to Lateral Aspect, Dominant Lobe, Temporal-Parietal Junction, Approximately at the Center of an 'X' drawn between T7-P3 and C3-P7

Wernicke's Area

At approximately the center of an 'X' drawn between T7-P3 and C3-P7 is a site that lies roughly over the angular gyrus and the supramarginal gyrus, which is usually referred to as Wernicke's area. Carl Wernicke (1848 – 1905) was a German psychiatrist and anatomist who identified this area as being related to language comprehension.

Carl Wernicke (1848 – 1905)

Wernicke's area lies mainly in the posterior, superior temporal gyrus, and spans the region between temporal and parietal lobes. It plays a key role (in tandem with Broca's area in the frontal lobe) in language. Both Wernicke's aphasia and Broca's aphasia have been described elsewhere in this text and, for ease of reading, a portion of this is copied below. The functions of the left temporal lobe are not limited to simple perception of what one hears but extend to comprehension, naming, verbal memory, and other language functions.

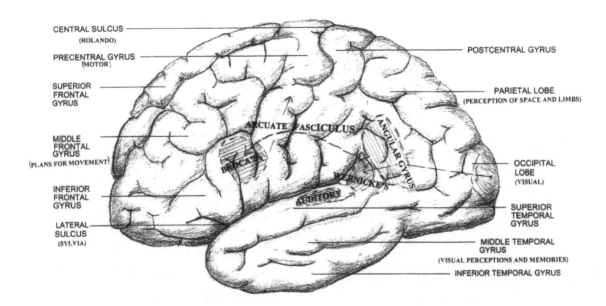

Lateral View of the Left Cerebral Hemisphere to show Broca's and Wernicke's Areas (Amanda Reeves after Smith, 1962; Carlson, 1986). The left frontal operculum area (F5, or Broca's area; BAs 44 and 45), is the key area for speech articulation. However, the temporal lobe (BAs 21 and 22) and the supramarginal gyrus, angular gyrus, and insula are involved in auditory and verbal representation. These areas and their connections, such as via the **arcuate fasciculus** *between Wernicke's and Broca's areas, are all involved in* **letter and word perception and recognition.** *These areas with their connections to the frontal lobe are important in the* **comprehension** *of abstract verbal forms, object naming, word recall (involving the hippocampus for auditory information on the left and visual information on the right), and perception of complex verbal relationships. The proximity of Wernicke's area to the auditory cortex is important for all of these functions.*

Wernicke's area is known to be important for **understanding** speech. Understanding involves the inferior parietal lobe and the auditory association area of the superior temporal gyrus. If there is damage to this area, speech will still be fluent and grammar will be good, but the person will **speak nonsense.** They speak a kind of jargon in which (without understanding what they are saying) they substitute inappropriate words or the wrong words to the extent that they are unable to understand their own speech. These people have similar difficulties with understanding when reading and expressing themselves when writing.

A **lesion near Wernicke's** area in the posterior section of the left temporal lobe leaves the individual with an inability to analyze word meanings.

Damage to the **connections between the auditory cortex** and **Wernicke's area results in** the person being able to hear sounds but they are **word deaf.** They cannot understand the meaning of spoken words, yet they are able to read, write, and speak normally.

Damage to **connections between Wernicke's** and **Broca's** areas leave the individual unable to repeat what is said to them if it is more complex than a single word or phrase, particularly if the sounds are unfamiliar. That is, unless the words have meaning that can be recognized in some other way.

Communication between Wernicke's and Broca's areas involves the superior longitudinal fasciculus and a portion of it called the arcuate fasciculus, a band of very long association fibers running anterior to posterior in each cerebral hemisphere. These fibers run from the frontal lobe through the parietal lobe above and below the insula and around the end of the lateral sulcus. It has branches to the temporal lobe, and branches that run all the way posterior to the occipital lobe. Clearly, in the left hemisphere, this communication link is essential for speech. There are two pathways, direct and indirect, between Wernicke's and Broca's areas (Catani & Jones, 2005). Damage to the direct pathway may produce conduction aphasia, whereas damage to the indirect pathway spares the ability to repeat speech but impairs comprehension. The symptoms of conduction aphasia suggest that the connection between posterior temporal cortex and frontal cortex plays a vital role in short-term memory of words and speech sounds that are new or have just been heard. The arcuate fasciculus connects these two regions and circulates information back and forth, possibly contributing to short-term memory.

In the majority of people with tone deafness, the superior arcuate fasciculus in the right hemisphere is not functioning properly. This suggests a disconnection between the posterior superior temporal gyrus and the posterior inferior frontal gyrus. The posterior superior temporal gyrus may be the origin of the disorder.

Comprehension in these individuals can be improved if they can visualize what is being talked about. They are not, however, able to repeat even a short series of unrelated words. They may, on the other hand, understand what is said, be able to speak well, and intelligently answer a question.

In contrast, in some individuals, the opposite problem can occur. An overactive connection between **Wernicke's** and **Broca's** areas may result in the involuntary repeating of words or phrases (**echolalia**).

Broca's Area (BAs 44, 45)

As previously described in the frontal lobe section, damage to **Broca's area** interferes with the individual's ability to instruct the motor cortex and results in defects in the articulation of speech. Damage to Broca's area and its **association areas** may result in a condition where the person is able to understand what has been said to him and be conscious of what he wants to say, but not be able to say it. This person may substitute staccato nouns and verbs so that the flow of his speech is lost and it sounds like a telegram without connecting words such as 'if,' 'and,' or 'but.' Ungrammatical speech and writing errors may be observed. In some instances these persons may also fail to understand spoken or written grammar. There is difficulty with prepositions. Broca's aphasia is usually accompanied by motor symptoms with right hemiparesis and right central facial paresis.

Damage to the **connections to speech areas** from their **surrounding cortex** may result in the person being unable to **understand speech.** However, they can repeat words or finish well-known phrases (e.g., "roses are red...").

BA 39: Angular Gyrus, Introduction

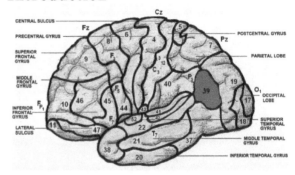

Brodmann area 39 is the region of the parietal lobe that sits at the occipital-parietal-temporal junction behind (posterior to) the lateral fissure. It is immediately posterior to the supramarginal gyrus. It involves visual-spatial-language skills and therefore **visual word recognition.** The left hemisphere angular gyrus is required for reading. Damage to this area can result in **alexia** (inability to read) and in **agraphia** (inability to write). It is important in the dominant hemisphere for generating sentences and verbal creativity. It is involved in networks for higher executive functions such as inferential reasoning. It is necessary for arithmetic calculations and numerical facts. (Note that BA 44 is also important for doing arithmetic.)

Learning Difficulties

Dysfunction of this area can result in a number of other learning difficulties, too. These may include difficulties with calculations (also intraparietal sulcus area), **left-right discrimination,** and technical competence. These persons may show symptoms of **agraphia** including reversing letters, reversing order of letters, crossing lines and difficulties copying material. They may also have **apraxia** (inability to execute a learned movement on demand) and dysphasia, including difficulties with meanings of words (a type of **semantic aphasia**), labeling objects, or interpretation of pictures such as cartoons. Identification of shape, size, and texture may be difficult. In more general terms, some individuals may not have major deficits but merely be clumsy, have a poor sense of direction, poor ability to draw, and they may be poor at spelling. All of these learning problems, though frequently said to stem from the dominant hemisphere, may also involve the nondominant hemisphere. Damage in this area and also in BA 40 is related to **Gerstmann's syndrome,** a condition that was originally described as including problems in four areas: writing (agraphia or dysgraphia), math calculations (dyscalculia), recognizing fingers (finger agnosia), and left-right discrimination. However, more often patients will exhibit only one or two or three but not all four of these symptoms. (For more details see: Mayer, Martory, Pegna, Landis, Delavelle & Annoni, 1999).

Of interest with respect to the autism spectrum disorders, this area is important for functions related to **Theory of Mind,** that is, the ability to theorize about what is in another person's mind and appreciate their point of view. Its functions also include spatial focusing, **action sequences**, and situational context. **Action authorship** – one's judgment concerning whether an action was done by them or by someone else – is also a function of this area. Sight reading of music is another function (particularly in the nondominant hemisphere). It is even involved in reflective self-awareness and seeing things differently from others, which may be functions of the Default network.

When there is a **disconnect between the left angular gyrus and the left visual cortex**, words cannot be recognized in the normal fashion. However, an intact right visual cortex can still recognize letters and convey that information via the corpus callosum to the left hemisphere speech and language areas. The letters may, therefore, be named. As the patient says the sequence of letters out loud the word can be recognized through the auditory input from his/her own voice. Subvocalizing can also work. Although the individual cannot read silently, he may be able to read slowly out loud and understand what he is reading. He can also write sentences or copy words from a page.

BA 40: Supramarginal Gyrus, Introduction

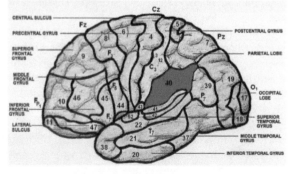

Brodmann area 40 is in the inferior parietal lobe and contains the supramarginal gyrus. In the **dominant** hemisphere it is important in attending to phonemes, verbal creativity, formation of sentences, categorization, and semantic processing. It is crucial for arithmetic calculations with integers. The left (dominant hemisphere) supramarginal gyrus is a general nodal point for **short-term auditory working memory** that is involved in both **music and linguistic** processing. It is important for the short-term memory of linguistic phonology and short-term pitch memory. It is an area that links kinesthetic memories with auditory commands. It may be part of networks for the conscious recollection of previously experienced events.

In **both hemispheres** this area is important in **deductive reasoning** and the ability to sustain attention. These areas are also important in other cognitive activities, such as distinguishing same and different, conflict detection, and for emotional working memory.

Lesions can result in so-called **ideomotor apraxia**, associated with deficits in the ability to plan or complete motor actions that rely on semantic memory. The patient can perform an action automatically, such as tying shoelaces, but the action cannot be performed on request. When there are lesions in BA 40 in the **dominant** hemisphere, the patient will exhibit errors in writing, spelling, word choice, and syntax. Right–left disorientation is common, as is **acalculia** (difficulty with simple arithmetic including addition, subtraction, multiplication and division).

Nondominant Supramarginal Gyrus

In the nondominant hemisphere, the supramarginal gyrus is important for visual grasping, **gesture** imitation, visual-motor planning, **music** performance, and in empathy. Deficits can result in **constructional apraxia**, which involves such problems as an inability to copy designs, difficulty with block construction tasks, inability to locate places on a map.

Mainly in the **nondominant** hemisphere, BA 40 is also important for imitation of movements and same or different discrimination. **Astereognosis** may occur. In this disorder, the patient cannot identify common objects by touch. In the nondominant hemisphere, the supramarginal gyrus is one of the areas near the temporal-parietal junction involved in prosody, which refers to **speech intonation, rhythm**, and emphasis. Thus dysfunction can result in **sensory aprosodia**, which involves difficulties with understanding the emotional content in communication, that is, reading tone of voice, **nuance**, and innuendo, as well as reading body language. Impairments in empathy and understanding the intentions of others may also be observed.

Lateral Aspect, Nondominant Lobe Temporal-Parietal Junction; Approximately at the Center of an 'X' Drawn Between T8-P4 and C4-P8

This is a site that lies approximately over the angular gyrus and the supramarginal gyrus in the non-dominant hemisphere. In most people this is near the temporal-parietal junction in the right hemisphere over **BAs 40, 39 and 22.** It is the homologous site to Wernicke's area in the dominant hemisphere. The functions of this area were noted above under BA 39 and 40. They include: **spatial-emotional-contextual** perception, symbol recognition, ability to understand emotional tones, and the comprehension of **innuendo**, nuance, and the emotional aspects of both gesture and verbal communications (Ross, 1981). It can also be important in nonverbal memory. Low activity in this area, as indicated by high amplitude slow waves (theta or alpha), in our experience, is often seen in patients who have Asperger's syndrome. The same symptoms of sensory aprosodia can be present after a stroke affecting this area, as reported by Ross (1981).

Parietal Lobes – Lateral and Inferior

P3 and P4; P7 and P8; BAs 40, 39, 37

BA 40

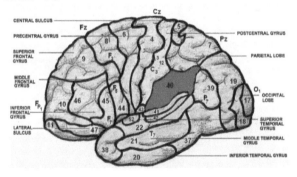

Brodmann area 40 is part of the parietal cortex in the human brain. The functions of the supramarginal gyrus and the angular gyrus as they are involved in language comprehension and learning were covered above in the temporal – parietal junction discussion of Wernicke's area. This will not be repeated here. The **inferior part of BA 40** is in the area of the supramarginal gyrus, which lies at the posterior end of the lateral fissure, in the inferior lateral part of the parietal lobe. This location may underlie its involvement in the **integration of tactile and proprioceptive** information and in somatosensory spatial discrimination. **Same and different** differentiation involves this area in conjunction with the ACC and the prefrontal cortex. It is part of the networks for memory.

The **supramarginal gyrus** part of Brodmann area 40 is the region in the inferior parietal lobe that is involved in reading, both in regards to meaning and phonology. It is important in the network for **deductive** reasoning, as was previously discussed. It has **mirror neuron** functions. In the nondominant hemisphere in particular, it is involved in visual-motor activity including the imitation of gestures.

Dysfunction in this area can have effects on all of the learning functions described above under the angular gyrus. BA 40 lies just superior to Brodmann area 41 and the auditory cortex and lesions in this region of the temporal lobe may affect auditory comprehension as described above under the heading: Temporal Lobe and Frontal Lobe – A Functional Unit. This area is also important, particularly in the non-dominant hemisphere, in a network for sustained attention.

P4; BA 40 – Right Hemisphere

An electrode placed over or near P4 may influence activity emanating from BA 40 and also other Brodmann areas including: 3, 1, 2, 5, 23, 13 and 30. In general terms, the functions of these areas include spatial position, visualization of spatial organization, context, left-side body awareness. As previously mentioned, damage to this area results in anosognosia for left body and space. These and other functions are described above under the nondominant parietal lobe functions.

When this area is overactive, which may correlate with increased beta activity, it may relate to the person being hypervigilant with a **'busy brain.'** As previously mentioned, hyperactivity in this area may correlate with **hypervigilance** and can make it difficult for that individual to fall asleep. Some people may be anxious (even exhibit panic attacks) but at the same time be bright intellectually.

P3; BA 40 – Left Hemisphere

An electrode placed over or near P3 may influence activity emanating from BA 40 and also other Brodmann areas including: 3, 1, 2, 5, 23, 13 and 30. NFB at this site can influence activity in the supramarginal gyrus that may include cognitive reasoning, attention, spelling and verbal short-term memory, imagination (BAs 40, 24, 10). It may also influence activity in the angular gyrus (BA 39) and functions such as the recall of a series of numbers (digit span), with dysfunction resulting in a syndrome called acalculia. These and other functions have been described above under the dominant parietal lobe functions.

Note: Overactivity in the parietal lobe may be accompanied by **parietal spindling beta**. In the parietal region, however, this spindling beta may reflect sensory hypersensitivity or sensory defensiveness (auditory, visual, kinesthetic). It is important for the NFB practitioner to be aware of this because using NFB to decrease the spindling beta can be helpful for some patients.

The supra marginal gyrus and the angular gyrus are shown in the adjacent figure. One can see that they are approximately in Brodmann areas 40 and 39. They lie roughly at the center of an X with lines drawn between C3 - P7 and P3 - T7.

Note the supramarginal gyrus and, just posterior to it, the angular gyrus. They lie just posterior to the lateral cerebral fissure and above the posterior end of the superior temporal gyrus.

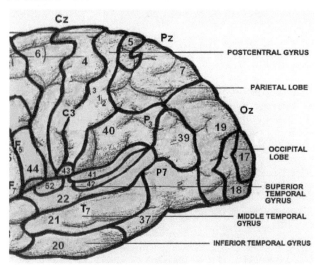
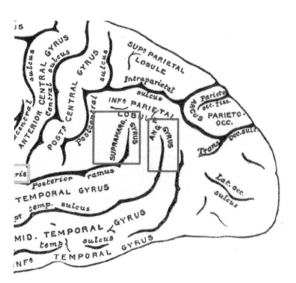

P7; BAs 39, 37

BA 37 (Described under Tempral Lobe)

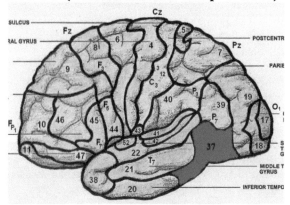

BA 39

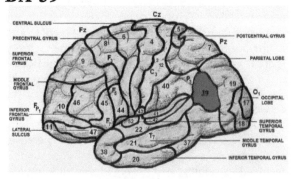

These areas functionally overlap with BA 40 and the posterior portion of BA 22. As described above, BAs 40 and 39 include the supramarginal gyrus and the angular gyrus. In the dominant hemisphere this would include the posterior portion of Wernicke's area and connections to temporal areas. **Sentence generation** and verbal creativity, in addition to right-side **body awareness,** are functions associated with BAs 39 and 37. BA 37 was previously discussed under the medial aspect of the temporal lobes.

Functions include **semantic categorization**, word retrieval, word generation, **face-name** association, attention to semantic relations and associating words with visual percepts. It is important in sign language. It is involved in reading and in making **orthography-phonology links**. Note that orthography refers to a standardized system for using a particular writing system (script) for a language. Children develop links between the sounds of speech (phonology) and how the language is written. BA 37 is involved in this process. It is also part of the networks for deductive reasoning as was described under the introduction to the frontal lobes. NFB carried out over P7 may also influence aspects of other Brodmann areas such as: BAs 40, 41, 42, 20, 30, 36, and possibly BA 21.

P8

In the **nondominant hemisphere, P8** lies over BAs 39 and the posterior superior portion of BA 37. BA 37 is important for drawing. Other functions of these areas include left-side body awareness and fear. As previously described, damage to this area results in **anosognosia** for the left side of the body and space. NFB carried out over P8 may also influence aspects of other Brodmann areas in addition to BA 39 and BA 37, such as BAs 40, 41, 42, 20, 30, 36, and possibly 21.

In the **dominant hemisphere, P7** lies over BAs 39 and the posterior superior portion of BA 37. Apart from reading, functions of these areas include calculations, verbal construction, and meaning. For many areas of the parietal lobes under both P3 and P4, P7 and P8, **focusing and refocusing attention** is another function. Attention, however, also involves large networks of activity that connect the occipital and parietal regions to central midline structures, as well as many areas in the temporal and frontal lobes. We postulate that one way to have appropriate effects on focusing, and concentrating aspects of attention, is to have the client use mental strategies and carry out specific tasks while doing the NFB training. In this manner we posit that the appropriate Attention networks will be activated and trained.

We did **not** show an area for attention in the networks, because, like memory, Attention networks involve such a large number of cortical areas. Frontal, central and parietal ares are all involved. **Focusing and refocusing attention** is one key to efficient cognitive functioning. For visual stimuli, the **frontal eye fields** (approximately between F3 and C3) connect to the superior colliculus in order to turn the eyes towards an object of interest. These areas are **reciprocally connected to the lateral inferior parietal** areas. Lateral inferior parietal (LIP) neuronal activity is necessary in order to disengage attention from a current stimulus in order for focus to change to a different stimulus. Related to this function, lateral intraparietal area (LIP) **salience mapping** is a function that is important in order to pay attention and to rank items in a visual field for locus of attention (Bisley et al., 2006). LIP areas connect to: frontal eye field (FEF), superior colliculus, pulvinar, inferior temporal BA 20 (visual pattern recognition). It determines the relative significance of incoming information, and transfers this salience data to other appropriate cortical areas. However, decisions as to how this salience information is used depend on the function and needs of areas that receive these projections from LIP and are not made in the parietal cortex.

Vestibular Connections to Inferior Parietal-Temporal BA 22, Posterior Insula

Symptoms of **nausea and vertigo** and even choking may be associated with inferior parietal lobe and BA 22 dysfunction. Nausea may also be associated with the posterior area of the insula. On the other hand, persons who suffer from Meniere's disease, a disorder of the inner ear, experience episodes of vertigo, nausea, vomiting, diminished hearing, and tinnitus and may demonstrate pathology in BA 25, 38, and 10. They may be treated with restriction of salt/caffeine/tobacco and may be prescribed diuretics, meclizine, and benzodiazepines (reference, Neurology on the web from the University of Kansas Medical Center).

Further posterior, the parieto-occipital cortex is important in **procedural memory**. This, again, is just an initial link in one aspect of the memory network systems that have been discussed under the cingulate gyrus, medial aspects of the temporal lobes, the Papez circuit, and the Executive areas of the frontal lobes.

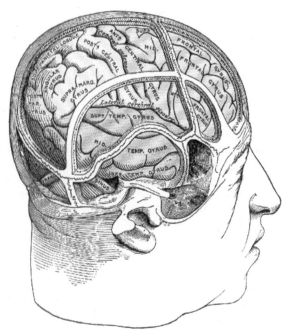

Gray's Anatomy 20th edition *(1918) (public domain)*

This figure from Gray's Anatomy *is included to assist the reader in visualizing the relationship of the angular gyrus and the supramarginal gyrus to the scalp. One can see that these areas would be in a region covered in part by P8 (T6) and P4 on the right and P7 (T5) and P3 on the left.*

Occipital Regions: O1, O2

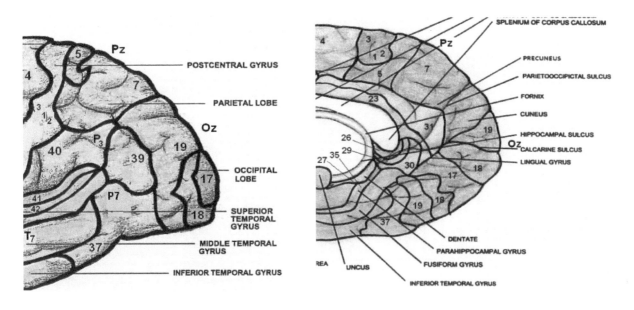

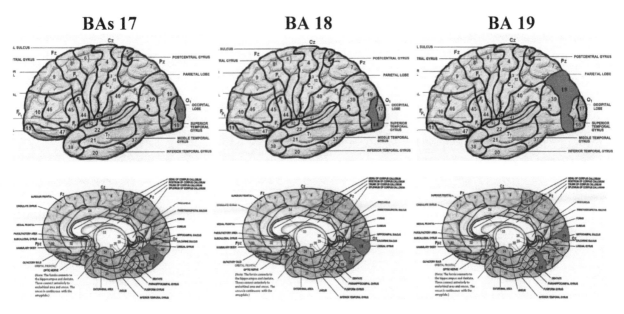

Neurofeedback in the occipital area may have effects on: BAs 17, 18, the visual cortex; BA 19, the visual association cortex; and BA 31, the posterior cingulate gyrus. The area below the calcarine sulcus responds to the upper half of the visual field while above the calcarine suclus responds to the lower half of the visual field. The primary visual cortex (V1) corresponds precisely to the visual field. Complete bilateral lesions of the occipital lobes can result in **Anton's Syndrome**, a condition in which there is cortical blindness with denial of being blind. BA 17 detects intensity of light and recognizes color, contours and patterns. **More complex functions**, such as face encoding, **require networks**. For faces this may include BAs 17, 18, 19 (occipital); 44, 45 (frontal); 37 (fusiform); and the right hippocampal formation.

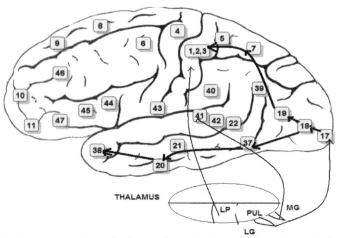

LG = Lateral geniculate
MG = Medial geniculate
PUL = Pulvinar
Visual stimuli from the lateral geniculate go to Brodmann area 17 of the visual cortex. Auditory stimuli from the medial geniculate go to Brodmann area 41 of the auditory cortex.
LP = Lateral posterior thalamic nucleus (to BAs 1, 2, 3 of the somatosensory cortex)
In this diagram the medial and lateral geniculate (MG and LG) are illustrated as loops attached to the posterior portion of the thalamus. Diagram below shows location.

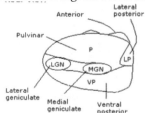

Reference: Adapted from Quantitative EEG, Event-Related Potentials and Neurotherapy. *Juri D. Kropotov, Academic Press, Elsevier: San Diego CA, 2009. Drawing after* Gray's Anatomy.

BA 17

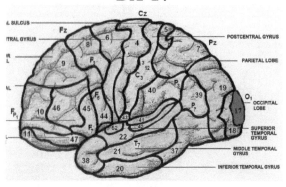

BA 17

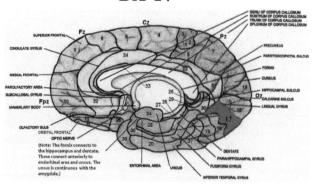

The **primary visual cortex is BA 17.** It is sensory cortex located in and around the calcarine fissure (calcarine 'sulcus' in figure above) in the occipital lobe. BA 17 receives information directly from its ipsilateral lateral geniculate nucleus in the thalamus. It is an area that is important in perception of light intensity, distinguishing visual patterns, contours, colors, and motion. It is also involved in perception of spatial orientation. BA 17 transmits information to two primary pathways, called the dorsal stream and the ventral stream, as noted by Kropotov (2009). The dorsal stream goes to the dorsomedial area and then to the posterior parietal cortex. The **dorsal stream** is referred to as the **'Where Pathway'** or the 'How Pathway.' It is associated with motion, representation of object locations, and control of the eyes and arms, especially when visual information is used to guide reaching or saccades. **Saccades** are a mechanism for fixation. They are rapid intermittent eye movements, as occurs when the eyes fix on one point after another in the visual field. Saccades are quick, simultaneous movements of both eyes in the same direction and are initiated cortically by the frontal eye fields (FEF), or subcortically by the superior colliculus.

The **ventral stream** goes to the inferior and middle temporal cortex and to the entorhinal cortex. The ventral stream is referred to as the **'What Pathway.'** It is associated with form recognition and object representation. It is also associated with storage of long-term memory. The ventral stream is associated with discrimination of shapes, textures, objects, and faces.

BA 18

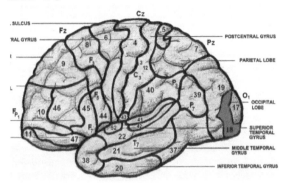

BA 18

Brodmann area 18 is in the secondary visual cortex (V2), which is also called the **prestriate** cortex. It is a visual association area. BA 18 is important in the **perception** of light intensity, motion, and in monitoring **color, shape**, and visual **patterns**. It is also involved in distinguishing whether a stimulus is part of the **figure or the ground**. Object recognition memory is a function of this part of the cortex, as is converting short term object memory to long-term memory. **Face-name association** and word and face encoding are related to this area. However, this and most other functions noted here also involve BA 19.

Lesions can result in **visual agnosia**, a condition in which the patient is unable to recognize patterns, objects, faces, colors. Oliver Sacks' book *The Man Who Mistook His Wife for a Hat* describes a neurological patient with a lesion in this area. In the dominant hemisphere damage/dysfunction may result in an inability to read (alexia). Topographical agnosia may occur. The patient may be unable to track visual motion patterns and discriminate finger gestures. **Orientation and selective attention** may be compromised.

BA 19

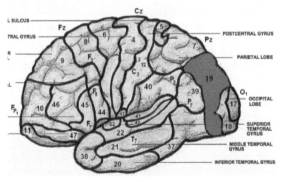

BA 19

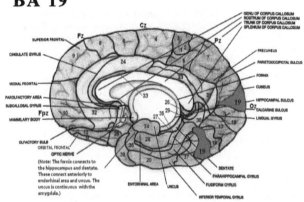

Brodmann area 19 is the associative visual area (V3). It overlaps in functions such as **light intensity** and **pattern detection** with BA 18. Brodmann area 19 refers to a region in the occipital cortex that borders the parietal and temporal lobes. It comprises parts of the lingual gyrus, the cuneus, the middle occipital gyrus, and the superior occipital gyrus of the occipital lobe. It is bounded anteriorly by the parieto-occipital sulcus and rostrally by Brodmann areas 39 (parietal lobe) and 37 (temporal lobe), and posteriorly by BA 18 (occipital lobe). BA 19 along with BA 18 comprise the **extrastriate** (or peristriate) cortex.

Brodmann area 17 is the **striate** cortex. It is the primary visual cortex. In normally sighted humans, **Brodmann areas 18 and 19 are** the **extrastriate cortex. The extrastriate cortex is a visual association** area, with feature-extracting, **shape and pattern** recognition, attentional, and multimodal integrating functions. It is one of the areas that can detect **light intensity**. Although functions of BA 18 and 19 overlap, only area 19 is important in the **spatial orientation of a stimulus (i.e., "Where is it?")**. It is also one of the areas along with the fusiform gyrus for processing **phonological properties of written** words. Brodmann area 19 has been noted to receive inputs from the retina via the superior colliculus and pulvinar nucleus of the thalamus, and may contribute to **focusing attention**. It is involved in visual-spatial orientation and in sustained attention. It is possible it also contributes to the phenomenon of blindsight.

Blindsight refers to being cortically blind due to a lesion in the striate cortex. These persons can respond to visual stimuli that they do not consciously see. If such a patient is asked to detect and localize a visual stimulus or to discriminate amongst visual stimuli on their blind side they may be accurate as to location or type of movement. This is known as Type 1 blindsight.

Type 2 blindsight is when a patient states that there has been a change, such as a movement, within their blind area. Blindsight is said to challenge the common belief that perceptions must be conscious in order to affect our behavior. Blindsight suggests that it may be possible that that human behavior can be guided by sensory information of which the person is not consciously aware. It is **the converse of** the previously mentioned form of anosognosia (**Anton's agnosia). Anton's agnosia is also called** Anton-Babinski syndrome where there is complete cortical blindness and denial of that blindness. This denial may include confabulation concerning a supposed visual experience.

In patients blind from a young age, this area has been found to be activated by somatosensory stimuli. (For more detail and a clinical example, see Doidge, 2010.) Because of these findings, it is thought that area 19 is the differentiation point of the two visual streams, of the 'what' and 'where' visual pathways. The **dorsal** region may contain motion-sensitive neurons, and **ventral** areas may be specialized for object recognition. It discriminates according to light intensity and patterns. It monitors color and shape. It is a key area for visual attention, orientation, selection, and is part of the working memory network for visual and spatial inputs. It is a part of networks for visual imagery, and inferential reasoning. It is an area that may also be involved in theory of mind and in memory tasks.

As previously discussed with respect to the inferior temporal cortex, Kropotov (2009) has demonstrated that one can follow the time sequence after a visual stimulus using evoked potentials. This is shown in the heavy black lines in the above figure. Visual stimuli travel from the lateral geniculate (LG) to the occipital lobe and auditory stimuli go from the medial geniculate (MG) to BA 41 in the auditory cortex.

From front to back, frontal cortex to occipital lobes, the pathways in our brains are finely tuned and organized with awe-inspiring complexity. As we try to influence particular areas and functions, we must always respect these intricate interconnections and forever strive to understand them better.

Appendices to PART TWO

Table of Networks and Brodmann Areas

The lists become too long if one tries to be specific; therefore only highlights are given.

FUNCTION of NETWORK	Brodmann areas (Major sites only)	Comment (Only a few areas noted; networks for attention and for memory, for example, involve many other areas.)
Executive	6, 8, 9, 10, 44, 45, 46, 47 (24, 32, 33), 39, 40	Dorsolateral frontal Cingulate, Parietal (supramarginal, angular)
Memory	34, 35, 36, 37, 31, 29, 20, 21, 28, 35, + 9, 10, 6	Ventral central midline + frontal for working memory and spatial memory + temporal for visual, auditory memory and memories of past events. Other areas for some aspects: 8, 9, 10, 45, 46, 47
Attention	7, 39, 31, 24, (+ right frontal: 10, 6, …)	Sustained attention nondominant frontal-parietal area. Visual: 17, 18, 37; For speech: 38, 47, 22, 23, 24.
Learning and Language	39, 40, 35, 22 posterior, 44, 45, 46, (37, 39, 47)	Mainly dominant hemisphere for language and many other areas are also involved; e.g., 6, 8, 9, 10, 24, 31, 32, 33, 38, 40, 5, 7, even 20, 21, 22)
Affect	23, 24, 32, 33, 25, 34, 13, 9, 10, 11, 12, 38, 47, 34, 35, 36	Central midline structures/cingulate, medial frontal and orbital frontal
Default	31, 23; 29, 30 39, 40; (21) 24, 32, 10, 9 24, 10, 32, (11, 25) 28, 35, (27, 48)	Posterior Cingulate & Retrosplenial cortex Inferior Parietal Lobe (& Temporal) Dorsal Medial Prefrontal Cortex Ventral Medial & (in brackets) functionally related areas: Inferior Prefrontal & Subgenual Cingulate Hippocampal Region (Medial Temporal)
Integration of Information	31, 23, 29, 26, 30	Posterior cingulate, midline cuneus and precuuneus
Somatosensory	1, 2, 3, 5, 40, (7, 40, 31)	Sensorimotor strip
Motor	4, 6, 8, (planning 6, 13, 40, 32, 33)	Sensorimotor strip – It is now known that there is overlap between traditional primary sensory and motor areas.
Visual	17, 18, 19, (integration 20)	Occipital and inferior temporal
Auditory	41, 42, 22, (21, 38)	Temporal (frontal lobes then become involved)
Gustatory	43	
Olfactory	34, 11	

The phrase 'preferential recruitment' may be used. It can refer to functionally related BAs becoming activated when any one of the corresponding group are activated. In work using NFB it is an important concept because it implies that when the practitioner encourages activity at a specific site, that other areas will be influenced if they are linked by networks to that site.

Table of Regions of Interest, Lobules, Brodmann Areas

TABLE II	LOBULES			
REGIONS of INTEREST	Frontal-Brodmann Areas	Temporal-Brodmann Areas	Parietal-Brodmann Areas	Occipital-Brodmann Areas
Anterior Cingulate	25,24,32,33,10			
Extra-Nuclear - 1	13,47			
Inferior Frontal Gyrus - 1	9,10,11,44,45,46,47			
Medial Frontal Gyrus - 2	6,8,9,10,11,25,46,47			
Middle Frontal Gyrus	6,8,9,10,11,46,47			
Orbital Gyrus	11,47			
Paracentral Lobule - 3	5,6,31,4			
Precentral Gyrus - 3	4,6,43,44			
Rectal Gyrus	11,			
Subcallosal Gyrus - 2	25,34,13,			
Superior Frontal Gyrus	6,8,9,10,11			
Fusiform Gyrus		20,36,37		
Inferior Temporal Gyrus		20,21,37		
Insula		13,		
Middle Temporal Gyrus - 1		21,20,38,37,22,39		
Parahippocampal Gyrus		28,30,35,36,34		
Sub-Gyral - 1		20,21,37,40,6		
Superior Temporal Gyrus		13,21,22,38,39,41,42,		
Transverse Temporal Gyrus		41,42,		
Uncus		20,28,36		
Angular Gyrus -1			39,	
Cingulate Gyrus			23,24,31,32	
Inferior Parietal Lobule			7,39,40	
Postcentral Gyrus			1,2,3,5,7,40,43	
Posterior Cingulate			23,29,30,31	
Precuneus			7,19,31,39	
Superior Parietal Lobule - 1			7,	
Supramarginal Gyrus -1			40,	
Cuneus				7,17,18,19,30
Inferior Occipital Gyrus - 1				17,18,19
Lingual Gyrus				17,18,19
Middle Occipital Gyrus - 1				18,19,37,

Groupings of Brodmann areas according to regions of interest and lobes of the brain.
From NeuroGuide.com. Reprinted with permission.

Functional Neuroanatomy
Organized with Reference to Networks, Lobes of the Brain, 10-20 Sites, and Brodmann Areas

Order of Description of Functions
Lobes, Networks, Brodmann Areas, 10-20 Sites

Lobes or Cortical Areas	Associated Networks	Principle Brodmann Areas	10-20 Sites
Frontal (Lateral & Medial)	Executive Networks Medial: Executive and Affect Networks Medial (BA 9, 10): Default Network	8, 9, 10, 46, 45, 44, 6 47 Lateral and Medial discussed together for each BA --Inferior prefrontal 9, 10, 11, 46, 47	Fz F3, F4 F7, F8 FP1, FP2
Frontal Orbital - Ventral/medial	Affect network – Social (Default network)	11, 12	F9, F10 FPz
Central	Somatosensory Cortex Sensory Network Motor Network	BAs 4, 6, & 1, 2, 3, 5 BAs 1, 2, 3 Primary (+ 43) Somatosensory Cortex BA 5 Somatosensory Association Cortex BA 4 Primary Motor Cortex BA 6 Motor Association Cortex	Cz C3, C4 Cz C3, C4
Temporal			
- Lateral Aspect	Executive – Auditory	41, 42, 22; 21, (52)	T7 & T8
- Temporal Pole	Executive, Semantic – also Affect Networks	38	T7 & T8 (F7, F8)
Temporal - Medial Aspect - (+ *Amygdala*)	Executive Network – Memory Affect network (+Default)	26, 27, 28 36, 37, (48) 35, 34	(Pz) (P7, P8)
Temporal - Inferior	Object identification; part of the 'what' pathway.	20, 37	T9 & T10
Insula	Affect & Executive	13	C3-T7
Cingulate			
Anterior (ACC)	Affect - Executive - Default	Affect: 24 (also Executive & Default & Salience networks) Affect: 33, 25 Executive: 32 (also Affect & Default)	Cz FCz Fz
Posterior (PCC)	Default Executive – Sensory	23, 31, 26, 29, (30)	Pz
Parietal	Executive, Sensory, Attention, (& Default) Networks; Spatial Awareness, part of the 'where' pathway.	5, 7	PCz, Pz
Temporal – Parietal - Lateral Aspect	Executive Network (& Default) Comprehension/Language Learning Gesture, Emotion, Nuance	39, 40 DH: 39, 40 NDH: 39, 40	P3 & P4
Inferior-Parietal Lateral	Attention & Salience	37	P7 & P8
Parietal Lobes - Medial Dorsal Aspect	Somatosensory Association	7	Pz
Occipital	Visual	17, 18, 19	O1, O2

Important: *The electrodes used for surface EEG NFB may reflect the activity from a number of Brodmann areas. The functions of Brodmann areas overlap with other Brodmann areas. Thus it is inevitable that under each principle BA being discussed, other BAs will be mentioned.*

Summary of Locations of Brodmann Areas

(Remember that BAs can overlap more than one gyrus and a gyrus can have more than one BA associated with it.)

- Areas 1, 2 & 3 – Primary Somatosensory Cortex (frequently referred to as Areas 3, 1, 2)
- Area 4 – Primary motor cortex
- Area 5 – Somatosensory association cortex
- Area 6 – Premotor cortex and supplementary motor cortex (also called secondary motor cortex or supplementary motor area)
- Area 7 – Somatosensory association cortex
- Area 8 – Includes frontal eye fields
- Area 9 – Dorsolateral prefrontal cortex
- Area 10 – Anterior prefrontal cortex (most rostral part of superior and middle frontal gyri)
- Area 11 – Orbitofrontal area (orbital and rectus gyri, plus part of the rostral part of the superior frontal gyrus)
- Area 12 – Orbitofrontal area (used to be part of BA11, refers to the area between the superior frontal gyrus and the inferior rostral sulcus)
- Area 13 (& 14) – Insular cortex
- *Areas 14, 15 (anterior temporal lobe) & 16 – Not found in humans*
- Area 17 – Primary visual cortex (V1)
- Area 18 – Secondary visual cortex (V2)
- Area 19 – Associative visual cortex (V3)
- Area 20 – Inferior temporal gyrus
- Area 21 – Middle temporal gyrus
- Area 22 – Superior temporal gyrus, memory comparison area of auditory cortex, the posterior part is considered to contain a portion of Wernicke's area in the Dominant Hemisphere
- Area 23 – Ventral posterior cingulate cortex
- Area 24 – Ventral anterior cingulate cortex
- Area 25 – Subgenual cortex (part of the ventromedial prefrontal cortex)
- Area 26 – Ectosplenial portion of the retrosplenial region of the cerebral cortex
- Area 27 – Part of piriform cortex
- Area 28 – Posterior entorhinal cortex
- Area 29 – Retrosplenial cingulate cortex
- Area 30 – Part of cingulate cortex
- Area 31 – Dorsal posterior cingulate cortex
- Area 32 – Dorsal anterior cingulate cortex
- Area 33 – Part of anterior cingulate cortex
- Area 34 – Anterior entorhinal cortex (on the parahippocampal gyrus)
- Area 35 – Perirhinal cortex (on the parahippocampal gyrus)
- Area 36 – Parahippocampal cortex (on the parahippocampal gyrus); fusiform gyrus
- Area 37 – Fusiform gyrus, occipitotemporal gyrus
- Area 38 – Temporopolar area (most rostral part of the superior and middle temporal gyri)
- Area 39 – Angular gyrus (part of Wernicke's area in Dominant Hemisphere)
- Area 40 – Supramarginal gyrus (part of Wernicke's area in Dominant Hemisphere)
- Areas 41 & 42 – Primary and Auditory Association Cortex
- Area 43 – Primary gustatory cortex
- Area 44 – Pars opercularis. Contains mirror neurons and is part of Broca's area in Dominant Hemisphere
- Area 45 – Pars triangularis. Broca's area in Dominant Hemisphere
- Area 46 – Dorsolateral prefrontal cortex
- Area 47 – Inferior prefrontal gyrus
- Area 48 – Retrosubicular area (a small part of the medial surface of the temporal lobe)
- *Area 49 – Parasubiculum area in a rodent*
- *Areas 50 & 51 – Found in monkeys*
- Area 52 – Parainsular area (at the junction of the temporal lobe and the insula)

PART THREE

Assessment and Intervention

This portion of the book begins with case examples showing a variety of clients. In each case, the assessment led to appropriate procedures being applied. It becomes clear from these illustrations that one-size-fits-all does not apply when doing neurofeedback. Even practitioners whose software is set for particular protocols will tell you, if asked, that they make adjustments to individualize the feedback. The rationale for basing intervention on an objective assessment is presented. This includes setting goals and enabling objectives with the client. The text will then discuss how to carry out EEG assessments that lead to decisions regarding neurofeedback (NFB) training. This is followed by a section on the physiological variables used in general biofeedback and a description of how to carry out a brief psychophysiological stress assessment. The next section is on intervention, first neurofeedback and then how to add biofeedback (BFB) to the NFB training. Finally, there is a section on metacognition, those executive strategies that guide our thinking and planning. Our approach to training using NFB is that it is not a stand-alone intervention. Weaving some metacognitive strategies into the training once the client is in a receptive mental state provides added value.

Even by the time this book is published, new and better assessment and neurofeedback/biofeedback training programs will be on the market. Better statistical analysis of assessment data and more immediate feedback of statistics and trends will be one improvement. More versatility in display screens and more ability to adjust the displays to the unique needs of each client will likely also be available in each new generation of programs.

However, the principles outlined here will, for the most part, not change. Most of our present NFB and BFB display screens will still work very well for our clients.

SECTION IX
The Basics of Assessment

Why You Do an Assessment

If you are going to use any biofeedback procedure, it is imperative that you first do an assessment to look at the person's baseline functioning. This assessment should ideally include all of the physiological measures that are relevant and that might be used in training. Careful history-taking, discussion of current functioning plus future goals, as well as objective physiological measures, all have bearing on how you design a training program for your student/client. There is a very basic principle in **systems theory**. You alter any component in a system and all the other elements in the system will shift their functions in order to accomplish the goal or function of that system. In keeping with this principle, we often find on reassessment that functions that were not directly physiologically targeted have improved. A good example is social appropriateness improving in clients with ADD and those with Asperger's syndrome when the main intervention has been to encourage a calm state while maintaining attention to the outside world.

This systems theory principle applies to how we use NFB and BFB with clients. Differences between people and/or new knowledge mean that we must shift how we deal with similar appearing clinical "entities." How a specific individual responds to different types of measurement and to different biofeedback presentations will help determine the best approach for that client. Therefore, we do not use specific protocols in the usual way that that term is used. Instead, we use an overall model for decision making that is based on the premise that our knowledge base is continually changing and that each client is unique in their responses both to our measurements and to our feedback techniques. Assessment measurements are a tool that **help us decide which feedback modalities** to use and **how and where** to do feedback. The model below also allows for change in the way we practice over the years as we learn more about neurophysiology and as our biofeedback instruments improve. The *Decision Pyramid* model we use is diagrammed in the figure below.

The Decision Pyramid

Decisions concerning placement of electrodes and the frequency bands to enhance or inhibit should be based on rational considerations and, where possible, research. One should avoid hearsay or belief unless you are being "intuitive," and then you should discuss with the client/parents the placement as being experimental. Research may be based on clinical experience and consist of a case series. These studies should have a large number of clients, appropriate diagnostic criteria that can be replicated, a good description of what was done, objective pre- and post-treatment testing, and perhaps a discussion of deviations from the general findings for specific cases.

Rational decisions should include consideration of our present knowledge about the functional significance of different areas of the brain (Section VIII) and the correlations of mental states to the activation of different EEG frequency bandwidths.

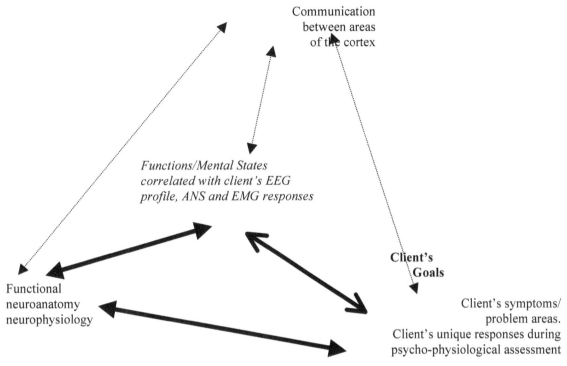

Diagram of **The Decision Pyramid**

The decision pyramid covers the components of an assessment. These include client goals and the assessment of EEG amplitudes at different frequencies. It suggests that with some clients you may additionally wish to do a *psychophysiological stress profile* for information about the autonomic nervous system and electromyogram changes with stress. It requires that we use the assessment information in conjunction with knowledge of neurophysiology and anatomy in order to derive appropriate placements of sensors for training. The top of the pyramid regarding communication between areas of the cortex is relevant when 19-lead assessments are done. The base of the pyramid applies to all neurofeedback work.

Rational decisions concerning what and where to train should include consideration of our present knowledge about the functional significance of different areas of the brain and the correlations of mental states to EEG frequency bandwidths.

Candidates for Assessment and Training

An Overview

Who is a reasonable candidate to assess and train, given the present literature and professional experience? The first group are those who have a known disorder and wish to improve functioning or quality of life. This group can be subdivided according to present experience using NFB. Conditions where there is a solid research literature include ADHD and seizure disorders. Other conditions where research literature is building include learning disabilities, depression (dysphoria), anxiety and tension, alcohol and substance abuse, headaches and migraine, obsessive compulsive disorder, closed head injury and concussion, Asperger's syndrome and autism. (Migraine requires a combination of biofeedback for relaxation and neurofeedback to decrease high beta (21-34 Hz) and increase SMR (13-15 Hz).

Other conditions that are supported by anecdotal experience and presentations at professional meetings include other mood disorders, Tourette syndrome and other movement disorders (dystonia, Parkinson's disease), Pervasive Developmental Disorder (especially high-functioning autism), facilitation of psychotherapy, chronic fatigue, fibromyalgia and stroke rehabilitation.

The second group includes people who wish to optimize their performance. These are usually students, academics, athletes, or professionals in business, law, medicine, or skilled trades. Although there is not a lot of published literature, there is some with respect to athletes, such as Lander's work with archers published in 1991. John Gruzelier did a two-year pilot study at the Royal College of Music in London, England. Results were impressive enough in his study that neurofeedback became part of the curriculum starting in Fall 2002. These elite students showed improvements, in particular with respect to interpretive and emotive aspects of their musical performance, after neurofeedback training that encouraged increased theta. *Alexandre Bilodeau is a wonderful example of an Olympic athlete who publically gave credit to his NFB and BFB training after he won a gold medal for Canada in the men's moguls at the 2010 Winter Olympics in Vancouver.*

There has been a joint publication by the Association of Applied Psychophysiology and Biofeedback (AAPB) and the Society for Neuronal Regulation (SNR) concerning standards for judging research in our field. Other helpful bibliographies are Ralph Byers' *Neurotherapy Reference Handbook,* available from the AAPB, and Cory Hammond's more selective bibliography published on the ISNR website. Efficacy is discussed in Section XXX.

The Common Factors

Given the growing number of disorders where favorable results are being observed, it is of interest to discuss what some of these disorders have in common that neurofeedback is favorably influencing. They are, in the main, conditions for which traditional medical approaches do not yield consistent or complete control.

Attention to External Stimuli

Attention has been found to be improved by decreasing the dominant slow-wave amplitude. In children this is usually in the theta range, 4-8 Hz. In adults it may be in the low alpha range, 9-10 Hz., or the range Lubar calls "thalpha," 6-10 Hz.

Excess theta is a key difficulty in Attention Deficit Disorder (ADD/ADHD) and in perhaps the majority of children with Learning Disabilities (LD). This is also a key difficulty observed in many clients who have had a closed head injury and in some who have suffered a stroke. It is an important component of the profile in both Autistic Spectrum Disorders, both Asperger's syndrome (AS) and Pervasive Developmental Disorders (PDD or autism).

In Asperger's syndrome. the high slow-wave amplitude is found in the central, frontal and prefrontal areas (Cz and Fz), as it may be in ADHD. It may also be seen at P4 and T6 if a 19-lead assessment is done. (One must be careful here and be sure that that high-amplitude alpha is also present in the eyes-open condition.) Both hyper- and hypocoherence and comodulation are observed depending on sites and frequencies. (This will need to be clarified by future research. For your work, it is best to take each case as unique and work to normalize the EEG.) We seem to be finding right hemisphere, high-amplitude, low-frequency, alpha rhythm (eyes open) as a sign that parts of the brain are 'idling' more often in these clients. This right hemisphere 'idling' combined with prefrontal slow-wave activity, correlates with the difficulty these clients have with interpretation and expression of social communication. Deficits in the ability to sustain attention are also found in the elderly and in people who are depressed.

Prevention of the Spread of Epileptiform Activity
See also Section XXVII

Decreasing the frequency, duration and intensity of seizure activity has been observed as a result of increasing SMR (13-15 Hz) training. The literature on this is reviewed by Barry Sterman, the original investigator in this field (Sterman, 2000), in the special issue of *Clinical EEG* devoted to neurofeedback. A meta-analysis was carried out by Gabriel Tan (Tan et al., 2012). It is perhaps not only those with diagnosed epilepsy who can benefit, but also those with undiagnosed subclinical seizure activity. This is thought to occur in almost 10 percent of children diagnosed with ADHD.

Stability

Here there are probably a number of inhibitory mechanisms involved, which will be gradually elucidated by research in the years to come. **Instability** in terms of mood appears to be favorably stabilized by increasing high-alpha and SMR, 11-15 Hz. Both adults and children appear to become calm, less impulsive and able to reflect more before speaking or acting (Thompson &Thompson, 1998). We also find that anxious clients become more self-assured with less fluctuation in mood. A small number of anxious clients may have very high-amplitude, high-frequency alpha, so care must be exercised and a careful initial assessment done in order that one not blindly apply a protocol that raises high frequency alpha with these clients. Some clients with bipolar disorder appear to become less extreme in the manic phases of their mood swings with enhancement of SMR. However, these observations concerning mood still require more case reports and research.

Stabilizing the extrapyramidal system and, specifically, the muscle spindle gamma motor and sympathetic input may be the reason for successes that are beginning to be seen with clients who have movement disorders.

Normalizing of the EEG in Overactive or Underactive Areas and/or Normalizing Communication Between Areas of the Cortex

Activation of an area of the brain that is relatively inactive ("idling") is a logical approach for certain conditions. *Dysphoria* (mild to moderate depression) appears to be improved in a proportion of clients when they practice activating their relatively less active left frontal lobe. Movement disorders may improve when the client is encouraged to control and sustain a higher amplitude of 13-15 Hz, SMR. Sustained activation of areas in the left hemisphere associated with language functions may be what is assisting learning-disabled clients to improve their academic performance. (A caution here: Tom Budzynski has stated that you should not do beta training on the left side with children who have Reactive Attachment Disorder. He says RAD children will show a worsening of behavior with increased left-side activation.)

One has to question whether activation may also be a key factor in the improvement of social functioning in clients where the principle difficulty relates to interpretation and expression of emotion. These are primarily functions of the right cerebral hemisphere that include expressing emotions and interpreting nonverbal social cues. (These problems are found in children with nonverbal learning disabilities and also in Asperger's syndrome.) It may be, in those cases where these symptoms improve, that the improvement is related to activation of the right cerebral hemisphere through decreasing the dominant slow wave and increasing low beta (12-15 Hz).

Encouraging the client to decrease an overactive area may also be helpful in some clients. For example, when extremely high-amplitude high beta or beta spindling is observed (usually in central and right frontal locations), encouraging a decrease in high-beta (24-34 Hz) can correlate with a decrease in depressive ruminations and/or obsessions and compulsions.

The foregoing are all examples of normalizing the EEG. Obviously, the oldest and best example comes from Sterman's work with seizure disorders. He always works from a 19-channel assessment with brain maps and takes a two-pronged approach: reducing slow-wave activity at the site of the epileptiform activity and increasing the protective SMR activity across the sensorimotor strip. He alternates enhancing SMR at C3 and C4.

Improving communication between areas of the brain is an emerging field. Numerous excellent reports are being given at professional meetings regarding coherence training. More work on this area and on comodulation training will come as newer instruments come onto the market that allow practitioners to do this type of training. At our Centre we have used LEXICOR, Neuronavigator, NeuroPulse, Mitsar, and Evoke Neuroscience equipment. Other commonly used instruments are the Discovery, Deymed, and so on. Whatever the instrument, the practitioner then chooses a program that will produce brain maps, and uses LORETA for source correlations and databases to compare their client to norms of the same age and sex. The reader is referred to Section XVII for further information.

Brief Summary

Both NFB and the combination of NFB with traditional BFB are powerful tools that a client may use to overcome a few symptoms, or symptom patterns, that are common to a number of conditions. Some very general and important neural circuits are being changed. It may be that when one or two key symptoms are removed and/or key neural circuits are changed, the individual is able to self-regulate. These people begin to be able to manage their symptoms and then to compensate for their remaining problems. You are training for mental flexibility and for control.

Case Examples

In the following case examples, history and assessment findings were found to correlate. Names have been changed to protect confidentiality. These cases demonstrate that it is possible to set realistic feedback goals. For a few of these cases we have summarized the training program and given the results in order that the reader can refer back to this list for a quick reminder of some typical training parameters. We have not attempted to list all the disorders that may be helped. The underlying difficulties that are being addressed with neurofeedback are often similar for more than one disorder. For example, under Movement Disorders, the example is of dystonia, but Tourette syndrome responds to the same approach. Later in this section we will describe how to set up appropriate feedback screens for the combined NFB and BFB training mentioned in these examples. When electrode placements are mentioned (Cz, P3, etc.), you can check the location in the diagram of the 10-20 electrode placement system found later in this chapter.

Optimizing Performance in Adults

Case #1: Jane, age 39, a bright, successful businesswoman, felt she could be more productive if she could be relaxed and calm when dealing with her employees and her peers. She was somewhat tense and anxious. When assessed, it was found that her peripheral skin temperature was 85 degrees and it dropped rapidly to 79 degrees when she was asked to imagine a difficult business situation. She had practiced yoga for some years and easily demonstrated diaphragmatic breathing with excellent synchrony between respiration and heart rate variability (RSA). However, she had not applied this ability to work situations. She was able to relax her shoulders, but during the stress test she unconsciously tensed her shoulder and neck muscles (trapezius and occipitalus).

Case #2: John, age 47, complained of similar symptoms of tension at work. Compared to Jane, he was less anxious, but he complained more of muscle tension. His peripheral skin temperature was 95 degrees. It changed very little when he was asked to imagine a very difficult situation at work. On the other hand, during his stress test his respiration changed markedly, becoming very shallow, rapid, and irregular. His EMG reading from his trapezius (shoulder) muscles was quite high. John had a second area of concern. He had always been extremely creative and his IQ was in the very superior range. Nevertheless, his school performance had been highly variable. He had gotten over 90 percent in subjects he liked and with teachers he respected, but earned only 60s and 70s when he didn't like the teacher.

You might decide that maintaining a relaxed mental state while under stress, as reflected by an increase in skin temperature, could be a useful exercise for Jane but would be less important for John. On the other hand, learning diaphragmatic breathing while monitoring the pulse rate for respiratory sinus arrhythmia (RSA) might be the most helpful relaxation technique and feedback for John. Jane did not need to learn this technique. However, she did need to use a simple heart rate monitor while doing stressful phone calls in order to apply her ability to attain excellent balance between sympathetic and parasympathetic activity to stressful situations. Both of these clients could benefit from EMG feedback to help them learn to relax their muscles.

Both John and Jane complained of constantly going over and over both business and personal problems in their minds. Both would awake at night and ruminate. John and Jane both demonstrated a dip in SMR at 13-14 Hz when stressed. Both of them also demonstrated high 27-32 Hz activity. Jane also demonstrated relatively high average amplitude 21-23 Hz. Often this corresponds to anxiety.

In addition to his difficulty handling stress, John showed very high 6-10 Hz activity when doing math tasks and when reading boring material.

Given these findings on the EEG, it could be helpful to both of these clients if they learned to raise high alpha and SMR (12-15 Hz) and lower high beta (approximately 22-34 Hz). John would benefit from lowering thalpha (6-10 Hz) activity when doing boring academic-like tasks.

Attention Deficit Hyperactivity Disorder

Case #3: Jason, age 8, was very active, impulsive and could not seem to attend for more than a few minutes during class at school. He was very bright and tremendously creative. His doctor felt he had a problem with short-term memory and, when tested, it was also thought that he had a central auditory processing (CAP) problem. He loved to draw and build. His EEG demonstrated extremely high 3 to 6 Hz activity and slightly lower SMR than beta. His theta/beta power ratio (4 – 8Hz) /(13 – 21 Hz) was high (14).

Jason was a very straightforward and uncomplicated case of ADHD. He had supportive parents and was not a behavior problem though he was underachieving at school. We used the assessment to set the bandwidths and locations for training. We worked at both Cz and C4 locations and encouraged an SMR increase and a decrease in theta. After 50 sessions both bands had altered appropriately on retesting. His impulsivity and hyperactivity had decreased, his attention span had increased, his measured IQ came up by 16 points and he was achieving in the top half of his class academically. He did not have any problem with memory and his functioning was such that there was no reason to re-do the CAP testing.

Learning Disability

Case #4: *Sam, age 11, presented with severe learning disabilities in addition to ADD symptoms. He was entering grade 6, but his reading and math were at late grade 1 and early grade 2 levels despite intensive special tutoring both at school and home. His parents were both teachers. He had optimal support in a wonderful family and extended family. Despite his academic subject difficulties, he gave the appearance of being a bright articulate boy. Delta activity was observed in the EEG and somewhat high theta activity. Training was begun in August. Sam was functioning at a fifth grade level in all subjects within 4½ months (retested by his school), and was in the top half of his grade 6 class at the end of the year. He remained an A-level student into high school. Training included decreasing his dominant slow wave while increasing SMR, 13-15 Hz, at C3 and increasing beta 15-18 Hz at P3 while doing reading exercises. Years later he entered Royal Military College with a scholarship to study engineering. He had said to his parents, "They saved my life!" His parents then wrote to us with his permission to let us know all the good news.*

Sam is used as an example because we have a long follow-up history at the time of writing. When he graduated his father made the observation that the training had saved his son. His assessment was more difficult than more recent cases of children with learning disabilities both because he had a reasonable attention span despite being physically restless and because we did not have full-cap capabilities when he was seen. Now we are able to pinpoint to a greater extent where delta and theta activity are seen. In students like Sam it is often between the T3 and P3 areas or in the middle of the triangle formed by C3, T3 and P3. There may also be coherence differences. These can involve this general area. (Both these observations require more research.) Because reading difficulties are thought to involve Wernicke's area, the insula cortex, and the arcuate fasciculus, we usually decrease slow-wave activity over P3, C3, and F3 and encourage activation with beta activity (15-18 Hz) when reading (*operant conditioning*). We teach strategies for handling academics while the child receives feedback (*classical conditioning* that pairs the calm, alert mental state with academic tasks). These strategies are described in Section XV, Metacognitive Strategies.

Movement Disorders: Example of Parkinson's with Dystonia

Case #5: *Mary, age 47, has both Parkinson's disease and severe dystonia. She has had most medications that are used for these disorders and has even partaken in experimental surgery. Prior to doing NFB and BFB, walking was extremely difficult and she was still unable to control the sudden onset of gross motor movements or the sudden inability to turn when standing or even to get up when lying down ('freezing'). An EEG assessment at Cz showed very low SMR (13 and 14 Hz) activity and high alpha (9 and 10 Hz) activity. A 'stress-test' looking at traditional biofeedback measurements demonstrated low peripheral temperature, high EDR, and shallow, rapid, irregular breathing with no synchrony to heart rate variations (RSA).*

Medications help Mary, and she has used a variety of drug combinations during the 14 years that she has had Parkinson's disease. Increases and decreases in dopamine may be part of a biochemical understanding of the problem and of a medical approach to it, but they do not answer the important question of how to control her sudden, precipitous onset of gross dystonic movements. In dystonia, a noradrenergic predominance has been proposed (Guberman, 1994). It stands to reason that our traditional drug approach is necessary but not sufficient. In Mary's case one obvious factor in precipitating worsening of symptoms is stress. Increased sympathetic drive is a factor. Noradrenalin is also a factor. One iatrogenic precipitator of a major dystonic episode can be the injection of even a very small dose of adrenalin, as might occur in a dental procedure.

If an increase in SMR is associated with decreasing spinothalamic tract activity and a decrease in the firing rate of the red nucleus (which has gamma motor system efferents to the muscle spindles), then it would appear sensible to see whether raising this rhythm would correspond to a decrease in undesirable muscle tension, jerks and spasms. Raising this rhythm has also been observed to be associated with a feeling of *calm*. Muscle spindles are involved in reflexes and in governing muscle tone.

They are innervated by the motor/sensory pathways to the spinal cord, red nucleus and basal ganglia, and by the sympathetic nervous system. We might have some constructive influence on the first half of this innervation by raising SMR. Indeed, Sterman demonstrated in his early research with cats that the red nucleus stopped firing when the thalamus started to fire producing SMR rhythm. Perhaps training Mary to relax would affect the second (sympathetic) part of this spindle in a helpful manner.

We trained Mary to get diaphragmatic breathing at six breaths per minute in synchrony with variations in heart rate (that is, to achieve good quality *respiratory sinus arrhythmia*, or RAS), and to associate this breathing style with increasing her SMR using NFB. She was then able, for the most part, to bring both 'freezing' and gross unwanted dystonic movements under her voluntary control. Daytime alertness is also a problem with Mary and most people with Parkinson's disease. Bringing her alpha (9 and 10 Hz), which had been too high, a little more under her control is helping with this. Mary was formerly an avid reader. For five years she had not been able to read a novel. After training she is again able to finish a book and she enjoys reading. The improved focus for reading was achieved after about 20 training sessions. The improved motor control and a reduction in medications was consolidated after about 40 sessions. She has also resumed her work doing crafts and writing and illustrating small books.

Professional Athletes

Case #6: *Joan, age 28, is a golf professional. Her dream was to play well enough to make the woman's tour. She felt her game could be improved, but had no idea how to do this other than practice. She was also frustrated that some of her golf students did not seem able to improve despite many lessons over an extended period of time. The EEG showed higher than expected 5-9 Hz activity. When specifically asked, it turned out that Joan had been a bright student, but had found it difficult to remain focused in classes. So this ADD pattern was not too surprising. She was outwardly calm and certainly not at all anxious, but, when assessed, her breathing rate was irregular, shallow, and rapid. Her skin temperature was 88 degrees. Her shoulders (trapezius) became tense with the math stress test. When we discussed what occurred mentally during her approach to the ball and in the back-swing, she realized that she was ruminating and worrying about the outcome of the shot. This corresponded to increases in her 23-34 Hz activity and to an increased breathing rate and increased muscle tension in her face, shoulders, and forearm. She then talked about being a very outgoing person, but said that this covered up the fact that she was often quite tense and anxious and a great worrier. Training involved a stepwise progression through different mental states. This will be described later.*

Case #7: *John, age 21, is a professional racing car driver. He was quite aware that he had been inattentive and impulsive as a student. He was happy to be finished with high school and following his passion for driving. His EEG demonstrated numerous bursts of slow-wave activity, 4-7 Hz and 9-10 Hz. His breathing was rapid and shallow. His shoulder muscles became tense with stress.*

The goal for John was mental flexibility since he would not be sustaining intense beta throughout a car race. He would be going in and out of beta problem solving depending on the situation. He is very good at that. He does extremely well during races when he 'locks-on' to the car ahead that he is trying to pass, but does not do as well without other cars around him on the track. His qualifying times are not as good as they should be and he wanted to do training so that he could stay focused and be more consistent during qualifying laps. John does not want to mentally rest in a state where he is at all tuned out. For him, tuning out appears to correspond to EEG frequencies below 10 Hz. He is rightly concerned that if this happened on the track, his reflexes to handle a sudden situation change could be too slow and this could result in a poor standing or, even, in injury. He wants to *rest* in a *readiness* state where his reflexes both mentally and physically are optimal. This requires that he turn on low beta instantly to solve problems, then reflect in a state where he is aware of every aspect of the track conditions and all cars around him. This desirable mental state is generally between 11 to 15 Hz. This state is also associated with a sense of calm, but alert relaxation. This mental state can also be reflected by biofeedback measurements. He wants his muscle tone to reflect relaxation in muscle groups not in use and an appropriate tone (not overly tense) in groups that are necessary for continuous shifting, braking and steering. This state is also reflected in the presence of synchrony between heart rate and respiration, a responsive level of EDR (alert, but not over-aroused), and a peripheral skin temperature in the 90s.

The EEG assessment using both EEG and biofeedback modalities with both Joan and John led to the design of a combined NFB and BFB program that rapidly produced changes in how they performed athletically. In Joan's case, she now taught her golf students how to breathe and relax their shoulders, and in two weeks had greatly improved all of their games. One man who had been with her for two years went from making 50- to 80-yard drives to driving over 200 yards. All her students increased the distance they could hit, usually almost doubling their driving distance. She now wants to set up feedback training at her golf club.

Hunter Mind Traits

Case #8: *Allison, age 12, is a good example of the type of interesting students and professionals who come for NFB training. Allison is used as an example because she also represents an important problem in research studies - choosing a control group. Allison was to be included as a 'normal' control in a study looking at EEG theta/beta ratios in Chinese children with ADHD in Canada, Hong Kong, and mainland China. Allison was the older sister of an ADHD boy who was doing NFB training. Her continuous performance test (TOVA) was normal. Questionnaires done by her parents were well within the normal range. She stood near the top of her class in school. She was an obedient, polite, well-behaved child. She agreed to do an extra EEG assessment to validate the ratio obtained by the researcher doing the study. Using a ProComp$^+$ machine, the researcher had obtained microvolt ratios for 4-8 Hz/16-20 Hz of 2.9 and 2.7 for eyes open and a math task, respectively.* **(As a rough measure, ratios above 2.5 in children are usually associated with ADHD diagnosis.)** *A different machine (A620) yielded ratios for eyes-open baseline and a math task of 3.3 and 2.2 respectively. Allison agreed to do a third EEG on a different A620 instrument. A math task was done and the ratio was 2.8. Allison was asked if anything was different between the tests. She responded, a little embarrassed, saying that she had drifted off in the first test where she was just staring at a dot. She loved math and on the second test she said her mind had not wandered very much. On the third evaluation using a math task she said that she had become a little bored and drifted off. She noted that this was somewhat like her original test with the other doctor. Her EEG ratios reflected her mental states during each test. When asked, she said that she did drift off in school, but it wasn't a problem because school was "so easy." Allison had a superior IQ. She had a positive attitude towards school. She was a very pleasant girl who would strive to do what was asked of her. Allison has Hunter Mind characteristics. Her EEG ratios were a little high and often in the range associated with ADHD. She was very bright, intelligent and energetic. Thom Hartman's analogy of a Hunter Mind would apply because she could both scan and then hyperfocus on something of interest. She survived, indeed thrived, in the Farmer's world of school because of her high IQ and desire to please parents and teachers. She does not have ADHD. Although she does have most of the traits, she is not impaired to a clinically significant degree by them. However, she does have a Hunter Mind and she probably should not be in the normal control group of a study that compares EEGs in ADHD children and normals.*

Note: The Hunter Mind scans and then goes into hyperfocus when the desired target is found. In a classroom the teacher would say such a student was distractible. The student with a Farmer Mind is attentive to the teacher no matter how boring or repetitive the subject matter may be.

The Edison Syndrome

Perhaps the easiest way to resolve the over-diagnosis of ADHD and the confusion of ADHD with oppositional defiant disorder and conduct disorder, and also resolve some of the difficulty in properly 'diagnosing' a control group for studies, would be to remove ADHD from the DSM altogether. It could be replaced with a diagnostic category that recognizes 'incapacitating hyperactivity' and/or inattention. On the other hand, one could leave the ADHD diagnosis in the manual but insist that the present diagnostic category be properly used. If this was done, ADHD would include those persons whose symptoms (and not those of associated disorders) significantly impaired functioning. If this were the case, then a new term could be introduced for those without impairment but with a degree of underachievement.

Such a new term would not be a diagnostic category in the manual of medical disorders. It would give a positive frame of reference and include all the very bright, capable, creative individuals who have that cognitive style so well exemplified by historical figures such as Thomas Edison and Mozart. The term "Edison Syndrome" has been proposed for this group. They fit the diagnostic criteria of inattention, distractibility, impulsivity (spontaneity) and, at times, hyperactivity, and they do get bored in school. Their performance is often quite variable. Academically, they may hyperfocus and do extremely well on subjects that catch their interest, but underachieve on subjects where they don't find the teacher interesting. These children, using neurofeedback, can learn to sustain focus even in boring situations. They decrease the variability in their performance. This is a group where NFB can be extremely effective.

Diagram: The Hunters and the Farmers

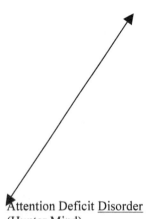

Edison Syndrome (Hunter Mind)
These are the bright, capable, creative individuals who fit the diagnostic criteria of inattention, distractibility, impulsivity (spontaneity) and, at times, hyperactivity but who do not fit the category of being "impaired to a clinically significant degree. They are often very good at developing new organizations but tend to move on to new challenges rather than staying on to maintain and administrate them.

Attention Deficit <u>Disorder</u> (Hunter Mind).
These are the individuals who exhibit all the symptoms of ADHD and are significantly impaired by these symptoms. (Not just by the symptoms of associated disorders such as oppositional defiant disorder.)

Farmer Mind (normal in the sense of most common)
This style is found in the majority of the population. They do not exhibit the symptoms of ADHD. They tend to be comfortable with daily routines and maintaining organizations rather than creating them.

A caution using ADD as an example: Control group studies (Ritalin versus NFB or cognitive behavior therapy versus NFB) should demonstrate a reasonably objective diagnostic workup to prove a clear difference, with respect to the primary targeted symptoms (attention span, distractibility and impulsivity), between the control group and the subjects. Studies on ADD are a clear example of problems in this area. Control groups should not contain successful, nonproblematic (to others) "Hunters" (term introduced by Thom Hartmann). The control group should have persons who fit Tom Hartmann's Farmer Mind classification (non-ADD and non Hunter mind). The A.D.D. group should be the 'Hunter' mind subjects whose functioning is impaired by the symptoms. If some individuals have behavior problems then this should be part of the matching process between groups.

Case #9: Mike, age 29, wanted to return to university to do an MBA. He had an engineering degree and said he had had to work much harder than his fellow students even to get just passing grades at university. He was extremely anxious, particularly in performance situations. He would wake up at night worrying about his work and his social relationships. He had to succeed in the GMATs (graduate achievement tests) yet he could not get-down to studying. They are an incredibly stress producing exam. When asked during the assessment to do even simple math, he stumbled over it and began to make silly errors. His respiration rate rose to 40 breaths per minute (BrPM) and was shallow and irregular. His heart rate rose and was completely out of synchrony with his breathing. His skin temperature dropped, electrodermal response rose markedly and his muscles tensed. The electroencephalogram with placement at Cz showed high thalpha, 6-10 Hz, low SMR, 13-15 Hz, and high beta with peaks at 21 Hz and 27 Hz.

Training, therefore, emphasized learning how to remain relaxed (breathing at 6 breaths per minute, skin temperature up, forehead EMG down, and SMR up), but alert (EDR remaining high but not extreme), focused (decreased thalpha), not anxious or ruminating (decreased 20-23 Hz and 24-32 Hz). We taught him metacognitive strategies that focused on test taking skills and reading comprehension. He sent us an e-mail immediately after getting his results. He had attained an almost unbelievable result – 94th percentile rank. He attributed his success to the accuracy of the training in pinpointing his problems (assessment) and helping him to overcome them (training). Best of all, being able to produce a relaxed, yet alert state even during a stressful situation like an exam, is a skill he should retain for the rest of this life.

Depression/Ruminations

Case #10: Lorraine was 63 years of age when she started NFB training. She was extremely bright and active. She had been under unremitting stress for some time. She complained of depression and was constantly ruminating (having negative thoughts). For more than 35 years she had been treated for severe bipolar disorder. She had been on virtually all appropriate and available medications during the course of this time. Her childhood history demonstrated clear ADHD symptoms, but she was an exceptionally bright student. Her EEG and stress test are discussed as a detailed example in the intervention section of this text. Neurofeedback combined with biofeedback was carried out. In a subsequent follow-up session three years later, it was found that she had remained off medications and has been functioning reasonably normally without medication for the entire time.

Note: This is a single case. Other cases have also improved. For case examples of depression the reader is referred to work by Elsa Baehr in her discussion of NFB techniques and results (Baehr et al., 1999).

Autism

Case #11: (Also cited in the previous section on Neuroanatomy) Troy, age 13, was completely oblivious to the meaning of expressed emotions of others. He appeared baffled, rather than angry, worried or saddened by loudly expressed anger towards himself or other people. He appeared flat, or merely overly excited, even with events that would be expected to evoke emotions. He did not express emotions appropriately, either verbally or nonverbally. His behavior was impulsive and, as with many autistic children, it could be best understood if one assumed an underlying anxiety. During his sessions he would sometimes start to flap his hands or laugh suddenly with no external stimulation producing those behaviors. At school he was in a general learning disability class for very low functioning children. He had no friends and no interest in other people. Troy had been diagnosed as autistic both at the local children's hospital and independently by two other senior psychiatrists. Parents were told by a psychiatrist that the only help available was counseling to help them accept their son's permanent disability.

The right hemisphere's involvement in the expression of emotion (anterior area) and understanding of emotional communications (parietal-temporal areas) has been known for many years. The underlying major symptom of anxiety, coupled often with inexplicable impulsive outbursts and/or actions, are well known in individuals who have autistic spectrum disorders or Asperger's syndrome. The calming effect of right-sided (C2, C4) SMR training in anxious and impulsive individuals has also been demonstrated. Therefore, augmenting 13-15 Hz and decreasing in the 'tuning out' dominant slow-wave activity (3-10 Hz) made good sense, both neurologically and from experience.

During his Grade 8 year and the summer thereafter Troy took 80 NFB sessions. He was placed in a regular classroom for high school and took advanced courses in math in Grade 10 after having been in special education since kindergarten. He was socializing, had made friends in his class and ate lunch with them. Of course, he was still recognizably different. He did not go in for sports, but otherwise he was able to fit into the regular activities of the school to a reasonable degree.

Eight years later in a telephone follow-up, Troy's father indicated that Troy was too busy to take time to come in to do the retesting at this time. (He lives a long distance from the center.) His parents said that they thought his reluctance was also because he didn't want to recall that anything was ever wrong with him. He is doing well both socially and academically and he is in a university.

The degree of the positive result with Troy, however, was not expected. We wish to be very cautious in making too much of his success (and the successful outcomes with a number of other children), at least until there is a large case series to replicate our case series (Thompson et al., 2010, Reid et al. 2010).

We have previously noted that in autism and in Asperger's syndrome, high slow-wave amplitudes may be found in the central, frontal and prefrontal areas (Cz and Fz), similar to ADHD patterns. However, slowing may also be seen at T6 and, occasionally, very high amplitude of theta or low frequency alpha may be observed at P4. Right hemisphere, high amplitude, low frequency (7-8 Hz), activity that has the morphology of sinusoidal alpha waves is a kind of 'idling.' This right hemisphere 'idling,' combined with prefrontal slow-wave activity, may be a reflection of the difficulty these clients have with expressing themselves appropriately socially, expressing affect, and interpreting social communication. In particular, they have an inability to interpret abstract references, inferences and innuendo and to understand nonverbal communication. We are finding improvements in social functioning in our Asperger's clients when we use P4, T6 low beta (13-15 Hz) training, in addition to decreasing the high amplitude low frequency (7-8 Hz) alpha (or theta when this is elevated) in this region. We combine this with decreasing left prefrontal and frontal high amplitude, slow-wave activity and increasing SMR (12-15 Hz) in the right central region (C4). With some clients we use LORETA Z-score NFB, which will be discussed in Sections XVII and XXV.

Asperger's Syndrome

Mild Asperger's Symptoms

Case # 12: Brad, age 20. Brad had failed in his first year in a community college. He came for assessment and training largely because of difficulties with attention span and concentration in class and when doing assignments. However, his parents had another concern. Brad had never had friends in his own peer group. Exploring his history, symptom picture and especially his style of social interacting suggested mild Asperger's syndrome, rather that ADHD, Inattentive Type. His expression of affect was different. He would laugh suddenly when he thought he should, but it was often at an inappropriate time. The laugh did not sound genuine. His emotions and expressions were flat. He had limited comprehension of nonverbal communication and innuendo. Despite the obviously flat emotional appearance, Brad demonstrated physiological signs of anxiety that were confirmed with a psychophysiological stress assessment. This type of assessment involves measurement of pulse, heart rate variability, respiration, peripheral skin temperature, electrodermal response, and electromyography activity in different states; for example, under the stress of doing difficult mental math and then while learning to relax.

The 19-channel EEG demonstrated high theta frontally and centrally, and high amplitude low frequency activity (7 Hz and 9 Hz) that had the morphology of alpha at P4 and T6. This P4 and T6 slow-wave activity can be seen in the eyes-open segment of the EEG shown below.

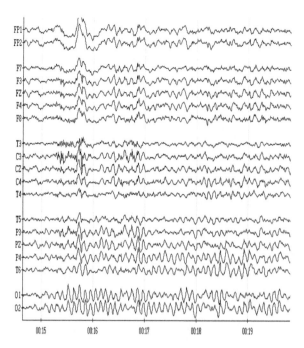

In addition, high frequency (23-30 Hz) spindling beta was observed frontally. Initially, training was to enhance 12-15 Hz activity and decrease theta (3-7 Hz) at Cz. Then Brad suppressed high beta (23-30 Hz) activity at F3 and Fz. Later, training also included enhancing low beta (13-15 Hz) and decreasing the high amplitude low frequency (7-9 Hz) at P4 and T6. This was followed by two-channel training that combined decreasing left frontal high amplitude slow-wave activity and increasing SMR in the right central region (C4) where there was a dip in SMR at 12-13 Hz. Concurrently, Brad practiced rhythmic diaphragmatic breathing at six breaths per minute in order to relax while remaining alert and learning cognitive strategies for school. Brad was easy to work with as he liked routines and was anxious to please.

After 40 sessions his parents found him more sociable. His listening improved and he managed much better at his part-time job at a fast-food outlet. He could take orders and do cash, whereas before he was only in the back doing routine food preparation. He returned to college and passed his courses except in the case of not doing a group project because he did not have a group. Even more remarkable given his initial presentation and history, he was able to understand jokes and appropriately carry on conversations with good eye contact.

Severe Asperger's Symptoms with a Low IQ

Case #13: Ben, a 15-year-old boy, had well over 100 training sessions. Ben had hydrocephalus as an *infant and, though treated, he remained a child with some cognitive deficits and a lower IQ (1st percentile rank). When he presented at the Centre, he demonstrated extreme hyperactivity and impulsivity. His attention span was very short. He would burst into one office after another regardless of what the practitioner in that office was doing. He asked incessant questions in a loud voice but did not always wait for an answer before his next question. He irritated everyone. Often his questions were personal and inappropriate and they could be about the person you were working with at the time. His conversations were entirely one-sided. He was overly formal, pedantic, precise, and his vocal tone was rather flat. He was completely oblivious to his effect on others – he couldn't "read" their reactions to him. Intellectual assessment showed him to be developmentally delayed but his problems went far beyond being a slow learner.*

Socially, he would only play with much younger children or just want to interact with adults. He had to control the interactions. His special interest was cars. He knew all the makes and liked to ask people what kind of car they drive.

He demonstrated very high theta and low SMR (13-14 Hz) activity at Cz. After 40 NFB training sessions (to decrease theta and increase SMR) he would walk calmly into the center and wait for his trainer. If he wanted to ask a question, he would wait outside the room and not speak until you indicated he might. He could ask a sensible question and wait for an answer. He has learned that some questions are not OK to ask someone outside of his family. For example, he asked one of the staff if it is OK to ask people how old they are or how much money they earn. He is learning through observation and copying how to behave appropriately. One of the most dramatic examples of change in Ben was his response, upon retesting, to being asked to draw a person. His pictures at the time of initial assessment and after 40 sessions were shown earlier at the end of the last section on Neuroanatomy. His brain maps are shown below, derived from the SKIL program (Sterman Kaiser Imaging Labs).

In 2002 Ben was continuing to improve, so he continued training once week. He was behaving in a more mature manner and was far less egocentric. He was still below average intelligence but he communicated well now with both staff and students. When the school psychologist did a reassessment, the report stated she would not support the previous diagnosis of Asperger's syndrome because he made good eye contact and was appropriate in conversation with adults. He was entirely bilingual, Cantonese and English, and translated for his parents. Ben has become a model for what neurofeedback can do in terms of increased calmness and social awareness. Parents in his community have noticed the improvement in Ben and have brought their own children for assessment and training.

This case is given not only because of the complex symptom picture, but also to demonstrate that children in a lower IQ range, in the first percentile, may benefit from NFB training, though it will take longer and goals will be more modest. Also, due to the length of time necessary for training a child with severe Asperger's, some of these children are given a partial scholarship.

Brain maps done with Ben using SKIL *(severe Asperger's syndrome symptoms with low IQ).*
Eyes-Closed Data:

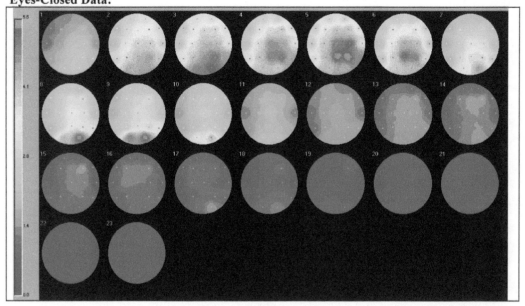

Eyes Open   *Red = 4.4 µV   Light Blue = 1.1 µv*

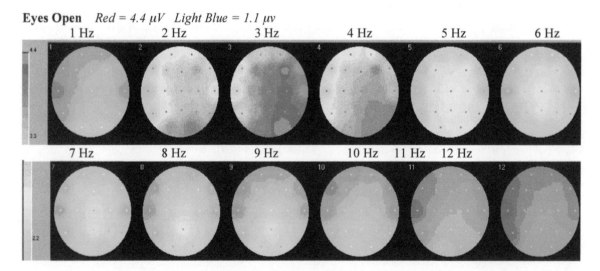

Note that eyes-open data show slowing in the right hemisphere. Note also that they are similar but not quite the same as the eyes-closed data shown earlier. The brick-red color in the right frontal (top of map circle) for 3 Hz is at F4 and may correlate with motor aprosodia symptoms. The brick-red color in the right posterior (bottom of map circle) for 3 Hz is between T6 (P8) and O2, and may correlate with sensory aprosodia symptoms.

Figure Below: **Brain Map,** eyes open, comparison with **database** with a time-of-day correction on. *Red = 1.5 SD above the database mean. Note the slowing in the right temporal region.*

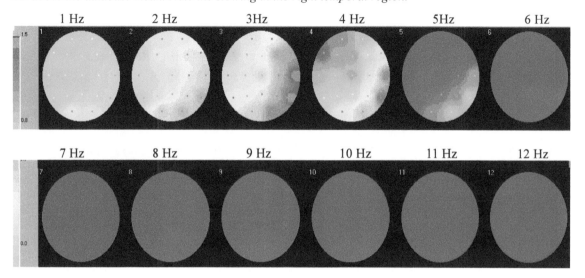

Figure Below: Brain Map, eyes open, showing comodulation of the dominant frequency range, 3-5 Hz, compared to the statistical database with a time of day correction on.

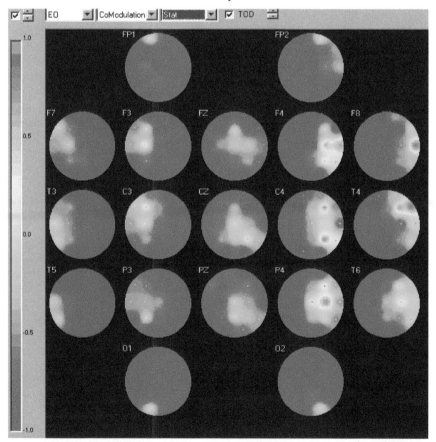

In the brain map above, this adolescent with Asperger's syndrome demonstrates a right-to-left disconnect (presumably not communicating at an expected level compared to the SKIL database) and some evidence of hyper comodulation ('spectral correlation') within the right hemisphere, particularly between F4, C4 and P4. Social contextual functions would be expected to be degraded by this local disturbance in function.

These EEG findings and results of NFB in children who exhibit the symptoms of either high functioning autism or Asperger's syndrome are particularly interesting in light of imaging study findings and the findings on autopsy of individuals diagnosed with autism. For example, Dr. John Sweeney, a University of Illinois psychologist, found that activity in the prefrontal and parietal cortex is far below normal in autistic adults when they perform tasks that involve spatial memory. Functional imaging has demonstrated temporal lobe abnormalities and abnormal interaction between frontal and parietal brain areas (Pavlakis, 2001). As noted in the section on Functional Neuroanatomy, these areas of the brain are involved in planning and problem solving. They are important for keeping an ever-changing spatial map in our working memory. Sweeney used an experiment in which the subject must keep track of the locations of a blinking light. Nancy Minshew, a neurologist at the University of Pittsburgh, has collaborated with Sweeney and has suggested that connections between these areas of the brain are not functioning properly in autistic children (Minshew, Luna, Sweeney, 1999). This kind of finding suggests that in our work with the EEG, we should look carefully at *coherence* and *comodulation* between these and other areas of the brain in children who are diagnosed with autistic spectrum disorders. Perhaps training to improve deficiencies in communication and/or lack of appropriate discrimination in function between areas of the brain might be helpful.

Also of interest to NFB practitioners are the findings of Margaret Bauman, a pediatric neurologist at Harvard Medical School. She found abnormalities in the limbic system on postmortem examination of subjects who had been diagnosed autistic. The cells in the area that includes the amygdala and the hippocampus were found to be atypically small and tightly packed together compared with the cells in normal subjects. Edwin Cook, University of Chicago, suggested that they look immature.

Neuroimaging studies carried out by Karen Pierce, a neuroscientist at the University of California at San Diego, demonstrated that in people with autistic spectrum disorders, the fusiform gyrus did not react when they were presented with photographs of strangers, but lit up when shown photographs of parents. Persons who have symptoms in the autistic spectrum seem to register emotional responses in their brains even if they do not express the emotion outwardly (Pierce, 2001). This finding of decreased activity extended to other areas of the brain that normally are expected to respond to emotional events. In normal people, the fusiform gyrus specializes in the recognition of human faces. This fusiform gyrus lies on the medial aspect of the temporal lobe between the parahippocampal gyrus above and the inferior temporal gyrus below.

[You may recall that at its anterior end the parahippocampal gyrus curls around the end of the hippocampal sulcus to form the uncus, which is continuous with the amygdaloid nucleus. At this bend is the olfactory sensory area. At its posterior end the parahippocampal gyrus forks, with the superior end being continuous with the gyrus cinguli, which turns around the splenium (posterior end of the corpus callosum) and runs anterior and superior to the corpus callosum. The inferior end is continuous with the lingual gyrus, which runs posterior below the calcarine sulcus to the occipital pole. The visual sensory area is on the medial aspect of the occipital lobe at the posterior tip of the lingual gyrus. The counterpart to the hippocampal gyrus on the medial aspect of the temporal lobe is the superior temporal gyrus, which lies on the lateral aspect of the temporal lobe and contains the auditory sensory area.]

Another finding, one which we cannot directly examine with the EEG, is that of Eric Courchesne. He has found that there are a smaller number of Purkinje cells in the cerebellum of autistic persons. Purkinje cells are an integral part of the brain's system for integrating sensory information with other areas of the brain. One could speculate that the work done by NFB practitioners to decrease slow-wave activity and increase low beta and SMR in the 12-15 Hz range may have some beneficial effect on this system. This might, in part, account for some of the good results observed with clients who have autistic spectrum disorders.

Courchesne also has reported from brain imaging studies that the brain in autistic subjects is normal at birth but is larger than normal by age three (Courchesne, 2001). MRI-imaging demonstrated increases in both gray and white matter of the cerebral cortex. These differences are most likely genetic. As previously discussed, other theories abound, including concerns about environmental pollutants, Wi-Fi, and vaccinations for measles-mumps-rubella (MMR). One hypothesis regarding MMR is that it is not the vaccine itself but the fact that some batches have been preserved with thimerosal, a substance containing mercury, which could be toxic to the immature immune system. There are suggestions that there may be an increased genetic sensitivity to mercury due to other factors that might be encountered in the earliest years of life, such as the infant body's ability to eliminate such toxins.

We have found that the children and adults with Asperger's symptoms have characteristics that overlap with other disorders including: ADHD, obsessive compulsive disorder (OCD), anxiety and depression, subclinical seizure disorder, allergies, sensory disturbances. The symptom picture and treatments in the majority of traditional centers working with these clients differs little from that which has been carried out for decades (Thompson & Havelkova, 1983; Attwood, 1997; Wing, 2001).

We have found that constructive change can be considerably accelerated in some children and adults by the addition of NFB to the traditional approaches. It was encouraging that at scientific meetings in early 2003 there were independent reports of improvement in clients with Asperger's syndrome (AS) from three different practices: Michael Linden in California, Betty Jarusiewicz in Massachusetts, and the authors in Toronto, Canada. Going beyond clinical anecdotes and case series to controlled research will require research money and clinical trials.

High Functioning Asperger's Syndrome

Case #14: Peter was an extremely bright but rather eccentric man in his early 50s. He came for neurofeedback training because he wanted to improve his attention when teaching bridge and when playing bridge competitively. Taking his history, it became obvious that Peter's symptom picture was indicative of Asperger's syndrome rather than attentional problems. He was highly intelligent and had an IQ in the gifted range around 130 but, despite a university education, he had never had regular employment. He gave a detailed clinical history in a detached manner with a monotone voice. He described being seen by a number of child psychiatrists and psychologists in his youth and noted, "Nobody knew what to make of me, but they said I was not schizophrenic." His best adjustment was at university where he says being eccentric was all right. For a time he worked in a stamp and coin business (collecting was among his intense interests). In his 30s he had decided that he needed to change his behavior, and so he began to copy his stepfather. He even got married to a woman older than he, whom he chose for a wife because she had no faults and he could relate to her as his stepfather had related to his mother. Peter's EEG was characterized by an extremely high alpha amplitude at 11 Hz. He reported that he did often tune out; indeed, he did not drive because he said it would be unsafe due to what he termed his "white-outs."

With training he learned how to control his excess alpha, and he also became calmer (less anxious) and more organized, with SMR up-training. His IQ tested at 140 after training. He had increased the number of bridge classes he was teaching. About two years after finishing training he dropped in, looking very well turned out in blazer, tie and flannels. This was in contrast to his previous style of dressing, which had been based on comfort ahead of fashion. (In his first interview, for example, he had come with his trousers belted around his chest.) He had been in the area and wanted to share a copy of a glossy bridge magazine which featured an article he had written, "The Psychology of Bridge." Not only had his ability to organize his thoughts and get them down on paper improved, he now had something to say about how to navigate the social scene and get invited back to play again. He reflected that NFB training had had its greatest effect in the social domain.

Similar EEG and Symptom Findings Post-Encephalitis

The brain map below shows 9-10 Hz eyes-closed comodulation comparison with the SKIL database. The subject is a very bright 44-year-old woman who suffered from encephalitis when she was 22. At that time she had profound memory loss, speech impediment, disorientation. She stumbled into objects and the left side of her face drooped. She lost all ability to think in abstract terms. Before becoming ill she was superb at visualizing things. Even now she still must have everything presented concretely. The anterior/posterior disconnect seen in the brain map below may correspond to her post-encephalitis loss of ability to visualize or even to interpret visual information or infer the abstract meaning of written material. She still has great difficulty correctly understanding social innuendo and the nonverbal aspects of social communication. Below the brain map is a sample of the eyes-closed EEG from the same recording. (The same findings were present at 2 SD from the mean, but the figure is presented here at 1 SD in order to be distinct for printing purposes.) Similar findings were found in the eyes-open conditions doing reading and math.

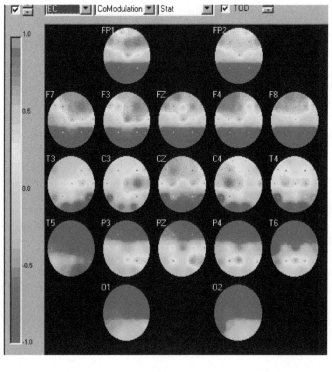

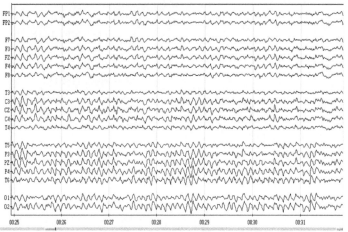

Brief Summary

The EEG assessment, which is often combined with a stress assessment for autonomic nervous system and EMG measures, leads to the design of an individualized, and therefore appropriate, combined NFB and BFB program. This careful assessment allows the practitioner to use these same measures to **assess physiological progress as training proceeds**. Physiological shifts appear to correspond to long-term changes in both productivity and quality of life. In addition to doing NFB and BFB, it is a good idea to complement this work with other stress reduction strategies, cognitive strategies, appropriate educational or work settings, and social support.

Careful assessment in each of the foregoing case examples led to more precise biofeedback. The assessment also enabled the client and the practitioner to evaluate physiological changes and correlate these with progress experienced subjectively by the client as training proceeded. Although it is a diverse group of clients, the four *common factors* suggested at the beginning of this section (attention, epileptiform activity, stability, and normalization of EEG patterns) encompass all of the difficulties presented by these clients. Depending on the client's symptom picture, the goal is either normalization or just self-regulation. Normalization is the goal if there is a disorder. Self-regulation, that is, increased control of mental states, is the primary goal if there is no disorder or if the client has not come to correct a disorder. This was true in the examples of an engineer applying to do postgraduate studies and the professional golfer who wanted to improve her golf.

How Assessment Relates to Goal Setting

Rationale for Doing an Assessment

EEG Assessment

An EEG assessment is useful if you agree that:

1. There are differences in the EEG patterns between different mental states.
2. These differences can be detected by the EEG.
3. These EEG findings relate directly to what we should encourage with feedback: that is, training the right parameters allows a person to self regulate their mental state and, over time, normalize or shift their EEG pattern.

ANS and EMG Assessment

An assessment of the ANS (autonomic nervous system functions) and EMG (muscle tension) is useful if you agree that:

1. There are differences in the ANS patterns between relaxed and tense mental states.
2. These differences can be detected by the psychophysiological measurements used.
3. The findings relate directly to what we should encourage with biofeedback.

Before You Commence Training

In our experience, the EEG and ANS/EMG findings accurately reflect the histories given by students and clients. However, if you do encounter a discrepancy between the client's history and your findings, then this must be clarified and understood before commencing neurofeedback (NFB) and/or biofeedback (BFB). You must also set realistic goals with your client and then delineate enabling objectives with clear steps for reaching each of these goals. In this manner objectives are linked to procedures that you can do with your NFB and BFB equipment.

The case examples presented here clearly demonstrate that history and EEG findings do correlate and that it is possible to set realistic feedback goals.

Goal Setting

Most Common Client Goal: Optimize Performance

Overview

Whatever the specifics of their original reason for coming, most clients want to optimize conscious, task-oriented performance. The goal to optimize performance can be expanded as follows:

To help a student/client achieve self-regulation of their mental state; in particular, the ability to have mental flexibility and to be able to produce a self-defined optimal mental state. This is usually a state of relaxed, alert, aware, calm, focused, problem-solving concentration, though different aspects will be emphasized with different clients and at different times with individual clients.

Each of these states, such as being relaxed or alert and so on, can be reworded as an *enabling objective.* To achieve each *enabling* objective, you can use one or more biofeedback procedures that are designed to assist the student/client in learning self-regulation. The following is a brief overview of these enabling objectives and some of the biofeedback methods that can be used to improve them.

To Achieve a *Relaxed* Mental and Physical State

To achieve a relaxed state, we will most often have the client increase EEG activity between 11-15 Hz and teach them to breathe diaphragmatically at a rate of about six breaths per minute (BrPM). For most adult clients, this rate, 6 BrPM, will result in their breathing being in synchrony with their heart rate. This is covered in more detail in Section XVIII on heart rate variability training. In children the respiration rate will be a little faster than in adults. The heart rate will increase with inspiration and decrease with expiration (RSA). At the same time the client is taught to warm their hands and relax their muscles. The client is taught to gradually relax his forehead, followed by their neck, shoulders, arms, and hands. Relaxing often corresponds to a decrease in anxiety when this pattern is present. Anxiety may correlate with a high or labile electrodermal response (EDR). Chronic stress, on the other hand, may result in an abnormal response to a new stress. The EDR will go flat rather than showing an increase, and finger temperature may actually increase rather than decrease.

Anxiety may also be reflected in the EEG. In the EEG it usually corresponds to an increase in 19-23 Hz activity relative to 15-18 Hz beta. A relaxed frame of mind implies that the individual is not negatively ruminating and worrying about aspects of their life. This type of unproductive mental activity may be found in conjunction with an increase in the amplitude of the EEG somewhere between 24 and 33z as compared to beta activity immediately above and below that bandwidth.

Note that these general guidelines must be adjusted for individual clients. Depending on the EEG findings in assessment, one may, for example, train 12-15 or 13-15 Hz.

To Achieve an *Alert* Mental State

To accomplish this objective we encourage the student to maintain their electrodermal response (EDR) at a level that corresponds to alertness; It should, however, not be as extreme as it could be with anxiety (too high) or with a drop in mental alertness (too low). EDR is measured between two electrodes placed across the palm or on two fingers. The EDR corresponds to sweat gland activity, which is controlled by the sympathetic nervous system. For most young children, being alert corresponds to a level between 12-16 micromho (µMhos), adolescents 10-15 µMhos, middle-aged adults 8-12 µMhos, and as low as 3-7 µMhos in older clients. (As always, there are exceptions to this general outline.) It is usually desirable to see the student's/client's EDR constantly shifting but not fluctuating wildly between extremes.

*Note: You may encounter the term galvanic skin response (GSR) in your reading. This is the old unit of measurement of skin conductivity or '**resistance**.' It is measured in **ohms**. This was changed to its inverse (1/GSR) and called '**conductivity**' or electrodermal response (EDR). EDR is more intuitively obvious because it rises with alertness and excitement. The unit of measurement is just ohm spelled backwards, '**mho**.'*

To Be *Aware* of the Environment

In athletics this corresponds to the ability to respond rapidly to changing conditions. For example, a goalie in hockey must be aware of every player's position around the net. In this mental state the EEG will show a clear increase in 11 to 13 Hz. The athlete moves rapidly in and out of this state depending on the task. For example, an archer will be in *narrow focus* (beta, 16-18 Hz activity) when judging the wind and the distance. The archer will then move to this state of calm, *open awareness* (11-13 Hz) just prior to the release of the arrow (Landers, 1991).

Flexibility is important in terms of moving appropriately between different states (small changes between different ranges within alpha appears to be important in performance). As implied above, the alpha rhythm appears to be associated with a number of functional states. Some authors have said that low alpha (approximately 8-10 Hz) is associated with general or global attention. We find this is more often global in the sense of internal rather than external attention. People with good (versus poor) memories desynchronize (show desynchronized beta activity) and low alpha drops in amplitude during the encoding and retrieval of memory (Klimesch, 1999). A high alpha band (approximately 11-13 Hz) is associated with task-specific attention. (Klimesch, Pfurtscheller & Schimke, 1992; Cory Hammond and Barry Sterman, personal communication). A rise in alpha amplitude is seen in this range in high-performing athletes (Landers, 1991). In this case, the words 'task-specific' may actually refer to what we term *broad total awareness*, which is a state of intense focus and readiness, without tension that is seen in the athlete.

To Be Appropriately *Reflective*

This mental state usually corresponds to a temporary increase in the amplitude of high alpha, around 10-13 Hz. A task such as scanning text, organizing the important facts and registering them in one's memory, requires that a student move from narrow external focus on the material to internal reflection. Moving flexibly and continually between states is characteristic of effective learning.

To Remain *Calm* Even Under Stress

This mental state is also associated with being both *aware* and *reflective*. It may be attained by increasing 11-15 Hz and by decreasing 24-35 Hz. The 11-15 Hz band corresponds to the production of high alpha for reflection and sensorimotor rhythm for a sense of *calm*. As discussed above, increases in 21-35 Hz are often found to correspond to moments when a person is anxious and worrying or ruminating (usually negatively). This kind of ruminating mental state is the opposite of being calm. It is not conducive to efficient action or to the breadth of associations that would be necessary for creative thinking. In a calm mental state a person may also be physically relaxed. This state may, therefore, be associated with diaphragmatic breathing, good RSA, warm hands, and relaxed musculature. One can also be calm mentally while carrying out a physically strenuous task. In this instance, breathing and muscle tension would vary with the task.

To Be Capable of Sustaining a *Narrow Focus*

Being capable of sustaining a *narrow focus* for the required length of time for a specific task is a key to efficient academics and to optimal athletic performance. The student who is continually distracted by either internal or external stimuli will not achieve at the level of their intellectual potential. The athlete who loses focus in the middle of a golf swing will not win that round. Most of us produce bursts of waves either in the theta or the alpha range (somewhere between 3-10 Hz) when we are *internally* distracted.

To Be Able to Turn On and Remain in a State of *Problem-Solving Concentration*

Clients need to be able to turn on and remain in a state of ***problem-solving concentration*** until a task is complete or a problem is solved. This sustaining of focused concentration is a key to success both in school and in business. This state is usually associated with 16-18 Hz activity. It is also said to be associated with activity in the 39-41 Hz range known as Sheer rhythm.

However, one should also note that cognitive tasks require both memory retrieval (episodic memories) and semantic processing. Theta may be associated with tuning out from the external environment but it is also apparently necessary for *memory recall*. By asking the student to recall information, you can observe the frequency of theta and the wave-form produced for that individual. It will be different than the theta produced when that student is in a rather dreamy, *tuned out* state. Often more synchronous theta is evident in the tuned out state. Alpha-theta training in very accomplished music students (people who probably quite readily produced a lot of beta) was found to enhance aspects of their musical performance (Gruzelier, 2003). In addition, semantic processing with eyes open has been observed to be associated with a decrease in synchronous high alpha, (Klimesch, 1999). Clearly, one does not just shift into high-amplitude beta and stay there to solve problems. There is a constant shifting between frequencies as one actively receives and processes information.

Brief Summary

When working, studying or being involved in an athletic event, the efficient individual is constantly fluctuating between the beta of narrow problem-solving focus, high-amplitude alpha for reflection and theta memory retrieval. Sterman has also noted that there is a brief burst of alpha activity after successful completion of a task that required sustained concentration. In the research he conducted with top gun pilots, an alpha burst was observed when the wheels touched down during landing. These bursts may be called *event-related-synchronization*. This seems to be a way in which the brain rewards itself.

When one is working on optimizing performance it is usually beneficial to use various combinations of neurofeedback and traditional biofeedback. If you have an instrument that does only one of these, you may wish to expand sometime in the future. Most clients require a combined approach. You can get good results with either, but our view is that the combination is more powerful. (A good thesis topic would be to do efficacy comparisons.)

Normalization of the EEG

Some clients have specific disorders and needs, where the goal of training is related to reducing a defined symptom or problem such as a learning disability, seizure disorder, movement disorder, stroke, speech disorder, depression, autism, Asperger's and so on. In these clients it is appropriate to refer to *normalization of the EEG*. Normalization includes such factors as bringing the amplitudes of frequency bands and the communication (coherence or comodulation depending on what instrument and database you are using) between areas of the brain into the ranges found in normal databases. This kind of work is facilitated by a full-cap ($\geq$ 19 leads) assessment and, for complex cases, LORETA Z-score NFB. This is more advanced work. Usually one is working on variations of normal brain waves, such as excess theta in the majority of children with ADHD, rather than abnormal waves. Epilepsy is the obvious example where normalization, in the sense of having less epileptiform activity, is clearly the goal.

The EEG Assessment (QEEG)
An Overview from Assessment to Training
Steps to Optimize Performance

1. Decide on which mental state the client/student wishes to encourage and/or discourage. This investigation is facilitated by the intake interview

and history taking. Goals may be clarified by psychological and performance testing as appropriate for your student/client. Goals should be realistic and put in writing. After the EEG, ANS (autonomic nervous system) and EMG (electromyogram) assessments, these goals will be linked to specific objectives for neurofeedback combined with biofeedback.

2. Assess the client's central nervous system (CNS) state by means of the EEG. Delineate how the client's quantitative electroencephalogram (QEEG) differs from expected age values. If you are not using a database, you can look at the pattern, both watching the dynamic EEG and looking at average values in different frequency ranges for the time period. Some programs produce a histogram and statistics, others provide values from which you can create your own graph. *Recall how Jane and John demonstrated differences from normally expected values in their EEG assessments, and how these differences did correspond to the problems they were experiencing at work.* (Case examples #1 and #2, above.)

3. Place electrodes at the sites you decide on based on EEG data, client's goals and history, and based on your knowledge of neuroanatomy and brain function (see the Decision Making Pyramid near the beginning of this section.) You want to inhibit the frequencies prone to EMG artifact in order to provide accurate feedback. Encourage the client to decrease the appropriate slow-wave frequency band(s) while simultaneously increasing the appropriate fast-wave frequency band. (Appropriate, here, means appropriate for that client's EEG pattern, symptom picture, and goals.) Remember, **the EEG acts like a 'flag' that reflects brain functioning. You can infer from a flag's activity the wind's velocity (amplitude) and direction. You make inferences about the brain's activity by reading the EEG.** You are now ready to set up display-screen instruments to assist the client to self-regulate their mental state using EEG biofeedback or neurofeedback.

4. Decide on the training screen you wish to use. The selection will depend on the EEG equipment you are using. Some instruments allow you to customize the visual feedback and audio feedback, and nearly all instruments offer a range of feedback screens. It is helpful if there are methods for measuring achievement. These may include percentage-of-time over or under threshold and a measurement that gives the client a sense of how often they are able to hold a mental state for one or more seconds (*decrease variability*). It is also helpful if the raw EEG is available for viewing. Children, in particular, become very good at *pattern recognition* quite quickly.

5. Assess the *flags* that reflect information concerning the client's autonomic nervous system (ANS) such as: peripheral skin temperature, electrodermal responses, respiration, pulse rate, heart rate variability with inspiration and expiration (termed *respiratory sinus arrhythmia* or RSA), and musculoskeletal nervous system (MSNS) activity in terms of muscle tension (EMG).

6. Assist those clients who demonstrate anxiety and tension to relax in order to facilitate their work on self-regulation of their mental state. Both you and the client can see how they are doing if you set up feedback for autonomic nervous system/electromyogram (ANS/EMG) feedback. If you do not have general biofeedback capabilities, you can still work on diaphragmatic breathing, imagery, and other relaxation techniques. Remember, however, that you are usually not trying to achieve the kind of relaxation associated with being in a hammock. **You want to maintain mental alertness but be free of tension.** Use the example of the top gun pilots to epitomize being alert yet physically relaxed (not tense).

7. Do the training until the client has mastered an appropriate degree of self-regulation and achieved the goals that were set out initially.

Note: In the following, though the specific examples will often involve applications to ADHD and/or to optimizing performance, the principles apply to the general use of neurofeedback and biofeedback. At the time of writing, applications of EEG biofeedback other than for ADHD and seizure disorders would be considered experimental. This may rapidly change as more papers are published.

Doing the EEG Assessment (QEEG)

Overview
When we speak of assessment, remember that we are assessing the client's suitability for neurofeedback.

We are not assessing this EEG in order to make a diagnosis.

This section outlines how to make decisions concerning the type of EEG assessment to carry out, the electrode sites to use, the preparation of the sites and the importance of measuring impedances between each pair of electrodes. Different *filtering* (FIR, IIR, FFT) will affect your data. This will be noted, though much of the equipment currently available on the market does not offer a choice: you use the filter that the manufacturer decided was best. This is followed by a consideration of *conditions* under which the recording is made (eyes open or closed, with or without tasks), types of artifacts that may affect the recording and how these artifacts are recognized and removed. This section will conclude with a short summary of some of the different EEG patterns you are likely to encounter.

A Note about Disorders Appropriate for Neurofeedback Intervention

It is important to communicate clearly to your client that you are only interested in variations in the distribution and amplitudes of NORMAL brain waves. (The exceptions would be work with epilepsy or closed-head injury.) Doing QEEG assessment as a preparatory step before working with clients to help them learn self-regulation is a totally different process than that carried out by the neurologist when he does an EEG assessment. Let us underscore again that the neurologist is looking for abnormalities, whereas we are looking, by and large, at variations on normal. As well, we are not doing a diagnosis. We are looking for patterns that indicate whether a person is an appropriate candidate for neurofeedback intervention.

An EEG is defined as 'normal' if it does not contain 'abnormal' waves or patterns "which are known to be associated with clinical disorders" *(Fisch, p 141). In this definition, clinical disorders are understood to refer to medical illnesses, such as: seizure disorders, dementias, conditions causing encephalopathies and space occupying lesions such as tumors, subdural hematomas, and aneurysms. We do not usually interpret the term "clinical disorders" (when it is used referring to neurological conditions) to include mild to moderate changes in mood, movement, impulsivity, attention or focus. These are not, in themselves, the usual reasons a physician would refer a patient to a neurologist for an EEG assessment. In this regard, a traditional EEG read by a neurologist is not helpful when diagnosing ADD, but a QEEG is invaluable when planning a neurofeedback intervention for this disorder. Norms for theta/beta ratio measured at Cz have been published and they discriminate ADHD from controls with 98 percent accuracy (Monastra et al., 1999).*

*Where symptoms are due to a serious, significantly incapacitating medical illness, then **that individual must first be assessed by a physician before you begin your work**. A child with difficulties in school but no physical symptoms would usually not fit into this category. This being said, some people use the term disorder (as in Attention-Deficit/Hyperactivity Disorder) incorrectly for children who are having difficulties, such as being restless or daydreaming or underachieving. ADHD should be reserved for those who are **"impaired to a clinically significantly degree"(DSM) by difficulties with attention span, distractibility and impulsivity.**(Note, this does **not** refer to symptoms from other, comorbid disorders such as oppositional defiant disorder.)*

*It is important in ADHD to be clear concerning severity of the primary symptoms. There are a small percentage of children who do, for example, fit the diagnostic label of Attention Deficit **Disorder**. These are the children who usually cannot attend regular school programs due to the severity of the impulsive and hyperactive symptoms. These children require medications and should be first seen by a physician. They may then attend NFB sessions and it is likely that over time their need for medication will lesson. Many of these children may later be able to come off medications altogether as they learn to self-regulate and manage their symptoms. In one clinical study, 80 percent of the children who were taking Ritalin when they started neurofeedback training were off the stimulant drug by the time they had done 40 sessions of training (Thompson & Thompson, 1998).*

The Principle of Parsimony

In general we should follow the medical motto, "do no harm," and also follow the principal of parsimony. Being parsimonious suggests that we first try the least invasive, least disruptive intervention that has a chance of being effective. (The late Naomi Rae-Grant, a very wise and practical professor of child psychiatry, always stressed this principle, especially when she was the Director of Children's Mental Health Services for the Province of Ontario in Canada.) We should all keep these basic tenets in mind as we work with clients. If a child is impaired to a significant degree, then that child may have to be referred to a physician for a treatment that will allow the child to attend school, such as stimulant medication for a very hyperactive child. This, however, is *invasive* and ideally should only be used for a short time until the child can learn self-regulation. Self-regulation to manage ADHD symptoms can be accomplished using neurofeedback. Joel Lubar has worked in this field for more than 35 years and has tracked his clients' outcomes. He reports a 90 percent success rate using neurofeedback with children who have ADHD.

On the other hand, the term "clinical disorder" is appropriately applied to some conditions that should always be seen first by a physician, but which can also be helped with neurofeedback training as an adjunct to medical management. This list includes disorders such as seizure disorders (Sterman, 2000) where there is good research (i.e., good controlled studies that have been replicated), depressive disorders where there is just a little research (Rosenfeld, 1997; Baehr, 1999), and Parkinson's disease where work is just beginning (Thompson & Thompson, 2002). There is also a small base of literature on closed-head injury, and the State of Texas passed legislation in 2002 that ruled that neurofeedback as an intervention should be reimbursed when used in the treatment of brain injured patients (Lynda Kirk, personal communication).

Goal of the Assessment

Goal
One goal for assessment is to ascertain what is different from usual patterns in your client's EEG so that you may decide on an EEG normalization training program. You must first compare a number of variables, such as comparing the power of the EEG in different frequency bands to expected age norms. Database norms for children doing active tasks, such as reading and math, are very limited. But for eyes open and eyes closed the NeuroGuide database has a group of over 400 children divided into age groupings because the norms for children change rapidly as they grow older. Norms for persons doing activities with their eyes open and with their eyes closed have been gathered by Sterman and are contained in the SKIL program for analyzing 19-lead EEG data. Large databases for EEGs with eyes open and closed are available from several sources, such as those from Frank Duffy, E. Roy John, Robert Thatcher, and William Hudspeth (Duffy et al., 1989), and from Thompson/Hagedorn Evoke Neuroscience. However, you may have difficulty comparing your data to a particular database (see Congedo et al., 2002).

For a one- or two-channel assessment, we suggest the following brief list of objectives. Usually one is trying to assist the student to achieve self-regulation of mental states that are reflected in the EEG, so keep that in mind as the purpose of collecting data.

Objectives
1. To be able to graph amplitude (either microvolts or picowatts) versus frequency (1 to 61 Hz) to show a pattern typical for the individual client.

2. To ascertain the dominant slow-wave frequencies within the 3-10 Hz frequency range since these are the frequencies you will likely wish to inhibit.

3. To be able to compare the power in different fast-wave frequency bands from 12 to 61 Hz. These can be relatively wide bands such as: high alpha 11-13 Hz, SMR 13-15 Hz, low beta 16-20 Hz, high beta 21-24 Hz, 25-36 Hz (particularly 27-32 Hz for some clients), 39-41 Hz (Sheer Rhythm), 45-58 Hz (which can reflect muscle activity) and 49-51 Hz in Asia, Australia and Europe (which may warn you that your electrode connections should be redone or that there is external ambient electrical activity that you may need to reduce). Rather than the foregoing wide bandwidths, it is preferable for some purposes to do 1 Hz band topographs, as can be seen, for example, in 19-channel assessments using the SKIL program. However, overlapping 2 Hz bands as can be used in single-channel assessments with instruments such as the ProComp+/BioGraph. The Autogen A620 program also does a very well thought-out assessment that is particularly useful for ADHD when used in conjunction with the norms for theta/beta power ratios (4-8/13-21 Hz) from a multisite study (Monastra et al., 1999). The Biograph program with the Infiniti instrument from Thought Technology can give data for assessments. The graphic displays can be very helpful. Graphs of amplitude (y-axis) and frequency (x-axis) are shown later in this section and in Section XI.

4. To be able to calculate the ratios of slow (theta and alpha) frequency bands to fast-wave (SMR and low beta) frequency bands.

EEG Assessment – Outline
To fulfill objectives for either optimizing performance or normalizing an EEG pattern, it is helpful to have an *EEG assessment profile presented as a histogram or graph.* Some instruments do this

for you. With others you can collect the necessary data to plot amplitude (in microvolts) on the vertical or 'y' axis against frequency (in Hertz) on the horizontal or 'x' axis. This will enable you to distinguish EEG frequency patterns and design an EEG feedback program. In the future we hope that most EEG instruments will do this for you as part of their software. But all instruments should provide you with accurate statistics that can be readily copied and pasted into Excel in order that a graphic display can be made. At the time of writing only three out of seven manufacturers' instruments at our Centre either do this or allow you to collect the necessary data to do this for yourself on a summary graph sheet.

We therefore recommend that, regardless of which instrument you choose to use for neurofeedback, you should also have one instrument that will enable you to collect and graph data from a careful assessment. There are some instruments that do both data collection for assessment purposes and feedback.

EEG Assessment – Method

Overview of Electrode Site Locations

An electrode or *sensor* is usually a tiny gold or tin cup. Whatever material you use, all your sensors must be of the same metal. The first thing you want to know is where to place these electrodes on the scalp. This is done in a worldwide standardized and systematic way. It is called the international 10-20 system (Jasper, 1958). This system is reviewed in the diagram below. The name is descriptive. You will note on the diagram moving from the front of the head (face), which is called *anterior,* to the back of the head, called *posterior,* that the 10 and the 20 of '10-20' refer to 10 percent and 20 percent of the distance between the *nasion* (anterior) and the *inion* (posterior). You can easily feel both of these bony protuberances. The nasion is the little notch above (*superior to*) your nose and below your forehead. The inion is the little bump or ridge at the base of your *occiput* (back of your skull) and above your neck. You measure the distance between these two points with a measuring tape, preferably one in centimeters to give more accurate readings. A typical measurement is 36 cm. Ten percent of this distance above the nasion gives a midpoint between two prefrontal lobe sites called Fp1 and Fp2. Twenty percent of the distance between nasion and inion takes you to the next point on the diagram, Fz. The next 20 percent, moving from front to back, takes you to Cz, which is the halfway point between nasion and inion.

Odd numbers always refer to electrode placements on the left side of the head. Even numbers refer to electrode placements on the right side of the head. The letter 'z' is used to denote any point along the central or midline between nasion (Nz) and inion (Iz).

The letters F, C, P, O and T refer to areas of the brain: Frontal, Central, Parietal, Occipital and Temporal.

In a similar fashion, you measure transversely from the *preauricular notch* on the left side to the same notch on the right side of the head. The preauricular notch is an indentation you can feel just in front of the ear canal. It is most easily felt if you open and close your jaw. Measuring from the preauricular notch on the left side to the preauricular notch on the right side, 10 percent of the total distance brings you to a point above the left preauricular notch called T3. The next 20 percent measured from T3 brings you to C3, and the next 20 percent brings you to the central point Cz. Frontal, central and parietal sites used in the 19-channel full-cap assessments are the main locations for most of our work with neurofeedback.

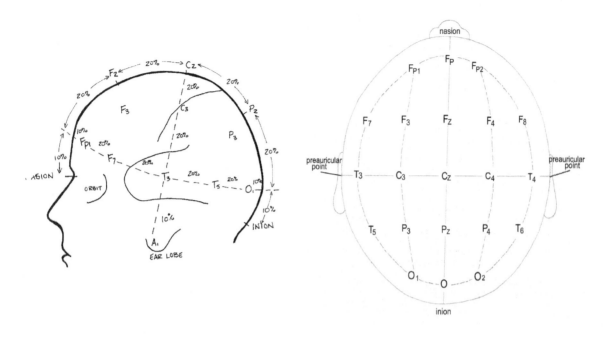

Side view   *Bird's eye view*

Diagram of the International 10-20 Electrode Placement System

The neurologist is often looking for seizure activity. We are not usually doing that, but you should be aware that interictal epileptiform activity (that which occurs *between seizures*, as contrasted to activity during a seizure) is most often detected in the anterior temporal region. An electrode to detect this activity may be placed 1 cm above a point on the line between the lateral canthus of the eye and the external ear canal that is a third of the distance from the ear canal.

A Styrofoam 'head' from a hairdressing supply house, though not anatomically correct and lacking a proper inion, is useful to help you practice marking these points on a model head so you can visualize them more easily. It is not as easy as it sounds if you want to be precise about placements. Fortunately a cap is available, in different sizes, that has the electrodes built in for the 19 sites on the scalp that are used for standard EEG recordings. It looks rather like a bathing cap. A neurologist or an EEG technician, however, may measure for each location rather than using a cap. To obtain sites encircling the head, they measure the circumference. This is done by beginning the measurement from the midpoint between O1 and O2, ('O' for occiput), which is 10 percent above the inion on the anterior-posterior midline of the head. Alternatively, one can begin from a point that we call Fpz that is 10 percent of the anterior-posterior distance, but above the nasion. The measurement for the circumference then goes through Fp1 (frontal pole on the left) approximately above the middle of the eyebrow, which is five percent of the distance around the circumference to the left from Fpz, then F7, T3, T5, O1, O2, T6, T4, F8, Fp2, which are all 10 percent of the total circumference from each other. The diagram above indicates measurements for just one side, from Fpz to midway between O1 and O2, so the percents are 10 percent and 20 percent of half circumference.

Don't be confused if you see designations that differ from the above! Renaming of sites has occurred in the last decade and in some literature you may find the Expanded 10-20 System is being used. In that case, T3 and T4 are now T7 and T8 while T5 and T6 are now P7 and P8. This is according to a modified and expanded 10-20 system proposed by the American Clinical Neurophysiology Society. (See Appendix II in Fisch, 1999, for the details and a diagram, and also Jasper, 1958.) Some labs are using 128 leads and there is even research being conducted using more than 200 leads so that maximum spatial resolution is achieved. The highest number of channels used by neurofeedback practitioners is typically 22 leads: 19 on the scalp, a ground, and the two ears for reference.

Choosing Electrode Placements for Assessment

Now you must decide how many channels you will use for your initial assessment. Think of a *channel* as the gateway where you plug the wire from your electrodes into the encoder that sits on the desk and is attached to the computer. Three leads go into a preamplifier that is closer to the electrodes (on the scalp in the very sophisticated research done with pilots, or clipped to the person's shirt near the shoulder). These three leads are the *active* and *reference* electrodes and the *ground* wire. One lead comes out of the preamplifier and goes to a single channel or entrance point on an encoder that sits on the desk or is attached to the client's belt. In some equipment the preamplifier and the encoder are in a single box on the desk. The encoder may have more than one entrance or channel for sets of wires. For most people, the choice is between one and two channels. For more complex cases, a full-cap (19 channels) is preferable. The encoder may have other channel inputs for other biosignals used in regular biofeedback, such as temperature, EDR, EMG, pulse and respiration.

The most common placement for a single-channel recording is Cz. This site is less influenced by artifact because it is far from the eyes (eye-blink and eye-movement artifacts) and from the jaws (a common source of muscle artifact). Cz is also far from the ear reference so that there is less common mode rejection and you record a higher amplitude activity than you would at sites closer to the ear lobe. It provides information about activation (or lack of it) in the central region and across the sensorimotor strip. The most common type of ADHD characterized by theta in frontal and central regions can be readily seen at the Cz location. For adults, move the electrode slightly forward to FCz (halfway between Fz and Cz) for an ADD assessment.

Types of Single-Channel QEEG Assessment

A minimum of three leads are necessary to record the EEG. It is analogous to getting electricity from a three-prong wall electrical outlet. These leads are: an active electrode (+ve), a reference electrode (-ve) and a *ground*. Electrical activity is measured as the potential difference between two sites, that is between your positive and negative leads. In what is called a *referential* recording, you will usually initially place an active electrode at Cz with a reference electrode attached to the left ear or placed over the mastoid. A sequential placement (in older terminology *bipolar*) also uses three electrodes. In this case, however, you are measuring the potential difference between two electrodes that are both "active" in the sense of being over an area of the brain where cortical electrical activity is produced. More on this shortly.

Types of Single-Channel EEG Assessments

The Three-Electrode Referential Placement

For most clients we use a referential placement. As explained earlier, *referential placement* means that you place an *active* (+ve) electrode at one site, for example, Cz, and a *reference* (-ve) electrode on the left (or right) ear lobe or over the mastoid bone. The word *active* means that you presume most of the EEG electrical activity that you are recording comes from this site. *Reference* means that you presume that this site is relatively inactive. The ear lobe has been found to have less than 15 percent of the electrical activity of other sites on the scalp. A third electrode is called a *ground*. This wire is not actually connected to ground. It goes to the amplifier and does what could be termed *electrical housekeeping*. In some instruments, such as the F1000, you can demonstrate a good quality EEG without this wire if the client remains motionless and there is no movement in the environment. In practice, however, you must be as careful about the attachment of the ground electrode as you are about the two that produce the voltage difference you are measuring.

Where Should I Place the Electrodes When the Results at Cz Do Not Parallel the Clinical History?

Using ADHD as the example, if you look at the assessment of a person who has all the symptoms of ADHD and you do not see any deviation in the EEG from what you would expect in an individual who does not exhibit this cognitive style, then you should look a little further. Try placing your active (+ve) electrode on Fz. You may then measure the high-amplitude slow-wave activity (theta or alpha) that was not seen at Cz. There are different EEG patterns for ADD, and you will pick up most of them by placing your electrode at Cz and then trying Fz to get a clear picture of high slow-wave activity relative to beta. In some clients you may find that theta is higher at C3 or C4 than at Cz. Thus, if you are not able to do a full-cap and your client's EEG does not appear to match the clinical history, move the active electrode to different sites. *High-beta ADD* may show more in the frontal sites. (In our experience, *high-beta ADD* is usually associated with other difficulties. The clients do drift off focus on what is being said to them or what they are reading, but they note that this is because their mind is 'racing' on about other matters. Usually these other matters are personal problems that they ruminate about repeatedly. Most often the high beta is above 23 Hz, but some clients also have very high amplitude bursts between 15 and 20 Hz and even occasionally between 13 and 15 Hz. If you take the time to watch the EEG and pause it at these points of increased beta amplitude, your client may be able to tell you exactly how their mental state changed and what they were thinking about. The majority of these clients will complain of dysphoria if not depression.

For some clients, where the symptom picture is clear but the Cz site and other sites referenced to the left ear do not show ratios significantly different from expected values, we will do one or more of the following:
- A bipolar assessment
- A two-channel assessment
- A full-cap assessment

Sequential (Bipolar) QEEG Assessment

When you use a sequential placement, you have two active electrodes in sequence on the surface of the scalp. Since all measurements are *bipolar,* because you are always measuring the potential difference between two sites, use of the term bipolar to refer to measurements where there was cortical activity at both sites has fallen out of favor and been replaced with the term *sequential*. This is in contrast to a referential placement where one electrode is considered to be relatively inactive (ear lobe or mastoid bone). The sequential montage may measure activity between electrodes placed either longitudinally (anterior-posterior *lateral*) or *transversely*, (right-left *lateral*). Thus, both these placements may be referred to as *lateral*. The activity measured is coming from dipoles between the electrodes. It is a difference between two points in the cortex. This is a different measurement than the referential placement that measures almost vertically to the surface of the cortex and may be referred to as a *radial* measurement.

For an initial assessment of a child thought to have ADHD, the sequential placement will place the positive (+ve) electrode at FCz (10 percent anterior to Cz, which is halfway between Fz and Cz), and the negative (-ve) at PCz (10 percent posterior to Cz – where percent refers to the percentage of the distance between the nasion and inion). To assess an older adolescent (age ≥16) or adult, we will place the electrodes slightly more anterior at Fz and Cz. Usually a sequential placement will give lower amplitude readings due to *common mode rejection.* (There will be more activity in common between Fz and Cz than between Cz and the ear.) Occasionally, we will observe a much higher theta/beta ratio with the sequential placement than we do with a referential placement. Lubar suggests that you train using the placement that gives the highest ratio.

With the *referential* placement, you can reasonably assume that any electrical changes being measured are from the site under the electrode on the scalp.

With the *bipolar* placement, you cannot be sure what the electrical measurements represent. For example, a Fz–Cz sequential placement indicating that theta was high amplitude could mean it was high at Fz (relative to Cz) or that it was high at Cz (compared to Fz). It could also be a phase difference, that the wave at Cz was *out of phase* with the wave at Fz. With waves 180° *out of phase,* you would register a higher amplitude of that frequency compared to waves that were *in phase*.

To make it more complicated, if you were trying to measure changes over time, it is conceivable that you could have a situation where you might conclude, after training using a sequential placement, that no electrical change had taken place using a referential montage, and yet see a significant change using a sequential montage. This is diagramed below. In this figure, the beginning higher amplitude (amplitude is the y axis) is theta, and the lower amplitude at the starting point (time is the x axis) is beta or SMR. **This is a hypothetical example.**

Inhibit and Reward Frequencies *Changes over time with training*
(Adapted from Discussion With Professor Joel Lubar)

*Below is a diagram representing sequential measurements done at **Fz-Cz***
Bipolar Measurement Done.

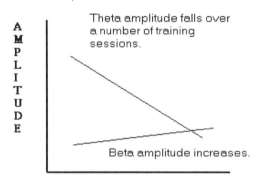

Measurements are with sequential placement

*Below is a diagram representing referential measurements done at **Fz***

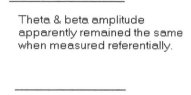

Theta & beta amplitude apparently remained the same when measured referentially.

Measurements with referential placement

In both the above diagrams and the diagram shown below, the y axis is amplitude in microvolts, and the x axis is the number of sessions from pretraining assessment to post-training reassessment.

<u>Explanation</u>
Y axis = amplitude (μv)
X axis = time (treatment)
Referential: apparently **no** change
Bipolar: large change due either to **phase** alteration or to do with the fact that a different axis of activity (lateral versus radial) is being measured.

*Below is a diagram representing referential measurements done at **Cz***

Theta & beta amplitude apparently remained the same when measured referentially.

One other possible explanation of this is that the rate of transmission of information between the two sites (Fz and Cz) may have changed (not the true amplitude). This could change the phase relationship which would appear as a change in amplitude. Lubar has remarked that using a sequential placement, though it has the disadvantage of not knowing at which site change has taken place, may have the advantage that it gives the brain more possible ways to learn the task of changing the theta/beta ratio.

Interpretation of Findings from Your One-Channel Assessment Bandwidths

ADD is a good example for EEG assessment. It is relatively simple to do but it demonstrates the care that must be taken in interpreting the EEG findings. For example, the clinician must consider other factors that may affect the ratio of theta/beta activity. These include anxiety and/or alcohol problems that may increases beta, especially above 20 Hz. This may lower the theta/beta ratio. If one doesn't carefully remove artifacts, the data may be distorted in a way that makes theta appear high. For example, 4-8 Hz activity may be high due to electrode movement, eye movement or eye blinks. This may give a false impression of a high theta/beta ratio. Beta activity may appear high due to muscle tension, giving a false impression of a low theta/beta ratio.

The ratio, itself, may not reflect what is going on due to measurement factors. For example, if you use 4-8 Hz for theta but the client's high activity is 7-9 Hz, then your theta/beta ratio may look fine as it does not reflect this client's tuning out in what we call *thalpha*. If a power ratio 4-8 /13-21 Hz is used, but alpha is very high spilling over into 13 and 14 Hz range, then a person with clear ADD symptoms might be reported as having a ratio within the normal range. Another problem is that not just artifact, but real cortical activity, such as a brief absence seizure, will affect the ratio by producing a short burst of extraordinarily high slow-wave activity. There is simply no substitute for a careful analysis of the raw, dynamic EEG. Being able to see an FFT-derived spectrum of 1 Hz bins of activity for each one-second epoch of EEG activity is extremely helpful as well. This is shown in our assessment screen below under the heading Additional Single-Channel Assessment Procedures..

The diagrams below represent a range of clients measured on different instruments. They demonstrate how EEG patterns differ among clients with different symptom pictures. They also demonstrate different ways of looking at the data from a one-channel assessment.

Case Example: Philip, Age 10, ADD without Hyperactivity

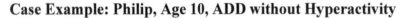
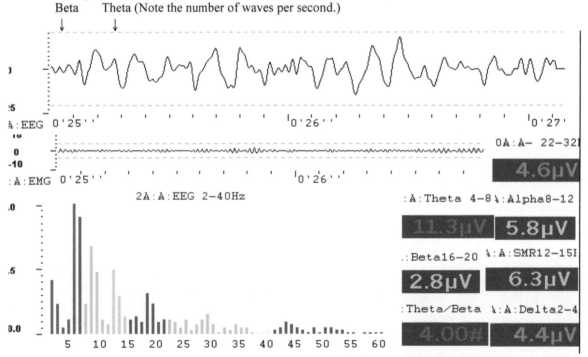

The above figure is an example of a one-channel EEG assessment screen created using ProComp+/ Biograph from Thought Technology. The screen shows a two-second EEG sample and a spectrum that relates information about the first second shown at the top. The student, Philip M., is a 10-year-old boy. He demonstrates all the symptoms associated with Attention Deficit Disorder. When Philip's mind wanders off topic, his EEG demonstrates an increase in amplitude at 6-7 Hz. In the figure above. the theta/beta ratio (4-8/16-20) is four. Microvolt ratios above three are fairly typical of young children who have ADD symptoms. Children without ADD symptoms are usually below 2.5. Philip is not an impulsive child. He has been diagnosed with ADHD Inattentive Type (DSM-IV 314.00) and his SMR activity is consistently higher than beta. There is virtually no effect of muscle activity on this EEG (45-58 Hz is very low) as shown in the "EMG" line below the EEG line. This two-second sample was consistent with the statistics for an artifacted three-minute sample done eyes open.

Case Example: Mike, Age 29, ADHD and Anxiety
In the figure below, Mike, case example #9, said he had tuned out and could not recall what I was saying to him.

Alpha wave is seen here. Compare its repeated sinusoidal nature to most theta and beta waves.

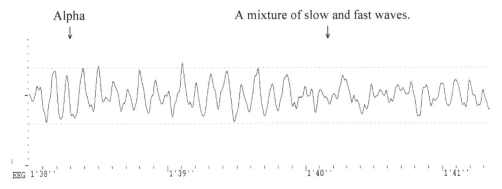

Case Example: Barry, Teenager, Non-ADD
In the figure below, a Fast Fourier Transform (FFT) power spectral array from the A620 (Stoelting Autogenics) program shows the EEG from a referential assessment (Cz to the left ear lobe). This EEG was taken from a client who reported **no** ADD symptoms. Note the reasonably high level of activity through the beta range to 18 Hz.

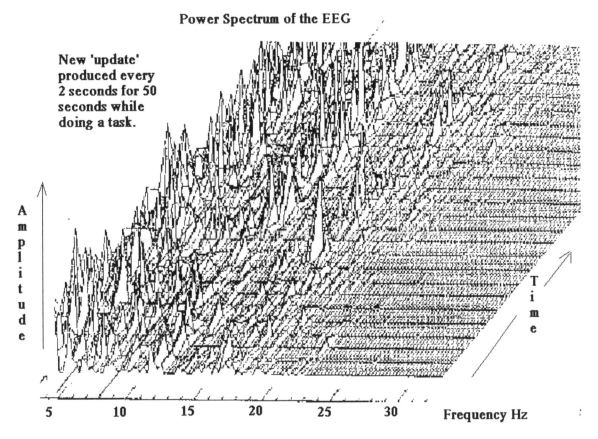

Case Example: John, Age 10, ADHD, Combined Type
The figure below is another Fast Fourier Transform (FFT) power spectral array from the A620 (Stoelting Autogenics) program showing the EEG from a referential Cz to the left ear lobe assessment. This EEG was taken from a client who demonstrated moderately severe symptoms of ADHD. Note the striking lack of activity through the beta range compared to relatively high activity in theta.

Although the power spectrum showing the ADHD pattern is at a slightly lower magnification than the one in the foregoing diagram, the differences are clear. (The magnification is usually increased for older clients since amplitude decreases with age, in part due to the adult's thicker skull.) Numerically, in the non-ADD example above, the theta/beta (4-8/16-20) µV ratio was 1.2 whereas the same ratio in the second, ADD example, was 3.2. These two samples demonstrate a very clear difference in the amount of slow-wave as compared to fast-wave activity. While teaching in Basel Switzerland, we used the analogy that the activity between 2 and 20 Hz is like Canadian geography for the person with ADHD, with the western mountains (lower frequencies) and the plains of the prairie provinces (higher frequencies), whereas the non-ADHD pattern is like Swiss geography with mountains and valleys across the whole range.

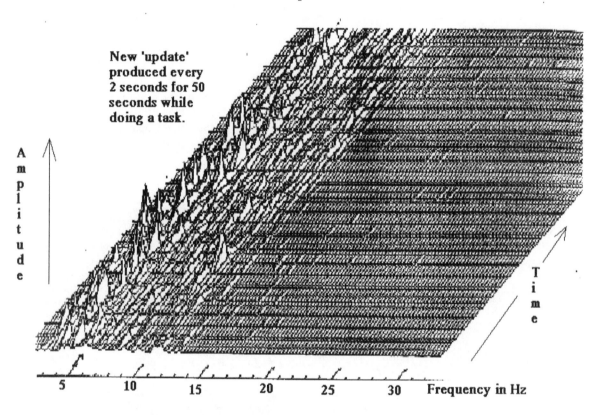

Power Spectrum of the EEG

Case Example: Mary, Age 47, Parkinson's Disease plus Dystonia

*Note: We find the same pattern (though less extreme) in **Tourette** and the same treatment regimen has been successful in the clients who present with Tourette syndrome.*

This is Mary, case example #5. She is a 47-year-old female who has severe Parkinson's and dystonia with daytime sleepiness. Note the extreme depression of SMR and elevation of low alpha. Note, this pattern was the same with sequential Fz-Cz and with referential Cz-Left ear placement. (y-axis is in microvolts.)

The SMR amplitude rose with training and dystonic episodes decreased.

The next figure constitutes a histogram. It was drawn from a three-minute artifacted collection of EEG data that had been taken using a sequential recording at Fz-Cz. A sequential placement ("bipolar" using older terminology) was used to try to reduce the movement artifact. Unwanted movement, both tremor and dystonic movements, were not within this client's control at the time of assessment.

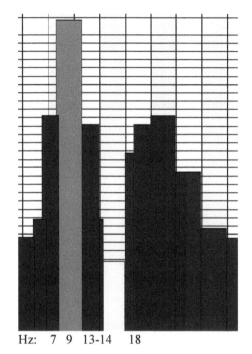

As you can see, it is very helpful to graph your data. It also allows comparisons of frequency band amplitudes over time as training progresses.

Case Example: Joan, Age 28, a Professional Golfer with ADHD

Joan came to the ADD Centre to add to her concentration skills and to improve her performance in golf. However, on taking her history it became evident that she had had some difficulty staying focused in school and had many symptoms of ADHD. In addition, she worried about her performance constantly while playing in competition. When she worried, 29-32 Hz beta activity increased and SMR (13-15 Hz) activity decreased. This kind of pattern is also seen in clients who are depressed and tend to ruminate. The client's mind is likened to a car engine. The accelerator has been pressed to the floor and the engine is on maximum revolutions. However, the car is in first gear and effectively going nowhere!

Below is another method of graphing the results of an assessment. In this example Joan's data has been exported to a spreadsheet. The spreadsheet below is a relatively simple one from Microsoft Works. This kind of spreadsheet allows one to overlay reassessments over a period of time in order to follow changes in frequency amplitudes with training.

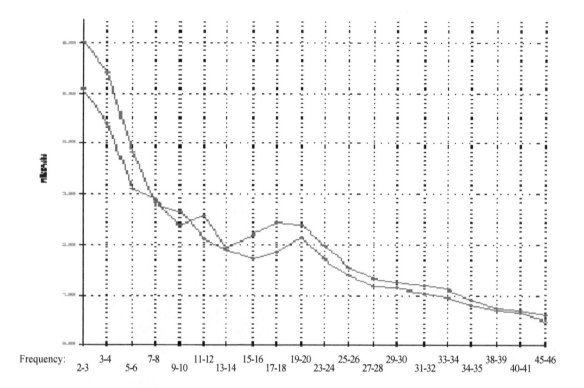

Frequency: 2-3, 3-4, 5-6, 7-8, 9-10, 11-12, 13-14, 15-16, 17-18, 19-20, 23-24, 25-26, 27-28, 29-30, 31-32, 33-34, 34-35, 38-39, 40-41, 45-46

The above graph is enlarged below so as to better look at the 4 to 28 Hz range.

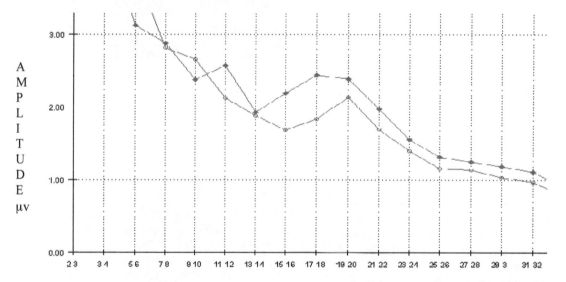

In this example, a referential Cz – left ear lobe recording from a three-minute artifacted sample is diagrammed. Open circles (red) show the pretraining record. Note the high theta compared to SMR and beta activity. The line of filled circles (blue) shows that after five training sessions slow waves (<7 Hz) and low alpha (9-10 Hz) have dropped, while high alpha (11-12 Hz), which she had been training up, has increased. Low beta (15-18 Hz) has also increased as she focuses more on analyzing the terrain for her next putt.

The x-axis is in two Hz bins. After being shown this progress report, she then worked on being very calm and raising 13-15 Hz (SMR) activity and not being anxious and worrying when performing. When she learned to do this, higher beta frequencies (20-34 Hz) dropped well below her initial assessment levels and her golf performance improved further.

Case Example: Lorraine, Age 63, (case #10 above) Bipolar Disorder and a History of ADHD

One second of EEG activity and its corresponding spectral analysis is represented in the figure opposite. The average spectral analysis of a three-minute artifacted specimen of eyes-open EEG showed much the same pattern.

Example of high amplitude, high-beta

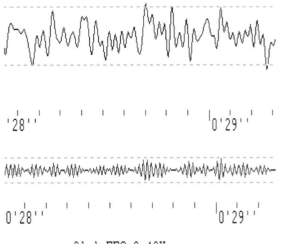

Lorraine is 63 years of age. She is extremely bright and active. She has been under unremitting stress for some time. She complains of depression and is constantly ruminating (negative thoughts). A stress profile was done during which she was asked to imagine a very stressful situation. When she did so her breathing became rapid, irregular, and shallow and her heart rate increased. Her skin temperature increased (rather than the expected decrease) and her EDR was flat. This corresponds to other clients who have experienced long-term chronic stress. It was in keeping with the chronic nature of the stress she was re-experiencing. The EEG at FCz (halfway between Fz and Cz) looked like the pattern opposite, with very low SMR and high 29-32 Hz activity. The same pattern could be seen at Cz and frontally with F4 > F3 for the 31 Hz activity.

In contrast, when she began a mental math task, which usually puts subjects under tremendous stress, her breathing became regular and its rate decreased, as did her heart rate. Her temperature and EDR both decreased slightly. Her 29-32 Hz activity dropped off to a very low level and 16-18 Hz activity increased. She explained that doing even extremely difficult math was a tremendous relief compared to just sitting and trying to relax. She said that while doing math she didn't have circular negative thoughts going through her head.

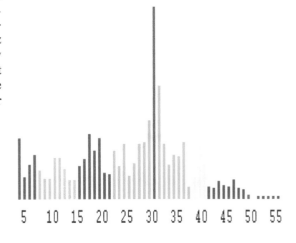

(*Remember neurophysiology: The amygdala holds unconscious negative memories and relates them to the physiological state which corresponds to those thoughts or that incident. This aspect of the amygdala's functioning seems to (relatively) turn off when you are intensely involved in a cognitive problem solving task or creative operation.*)

In the sessions, Lorraine was encouraged to relax, externally focus and 'empty her mind' while remaining very calm. To help her do this, Lorraine chose a complex screen with multiple feedbacks. She said that she could then choose what she would focus on and the steps she would take to meet her objectives. The training program for Lorraine included simultaneous feedback of respiration and pulse for RSA, together with forehead and/or trapezius EMG. Two-channel neurofeedback was used: a Cz channel to encourage a decrease in 28-33 Hz/13-15 Hz ratio, and an F3 channel to encourage a decrease in 6-10 Hz activity and an increase in 15-18 Hz to activate the left frontal lobe. *When follow-up was carried out it was found that she had remained off medications and has been functioning reasonably normally for the past three years.*

Case Example: Brad, Age 19, Asperger's Syndrome (Case #12 Above)
High Beta – Beta Spindling

Brad demonstrated a lack of ability to interpret innuendo and to make inferences about the meaning of written communication. He was inappropriate in social communication. He would smile or laugh at incorrect times during a conversation. He had very poor attention span except in his own area of intense special interest. He demonstrated extremely high Cz theta, T6 idling in low alpha and high-amplitude bursts of beta spindling in the frontal region. This kind of beta activity may be seen in some cases of ADD. The lack of activation in the right temporal area is a pattern we have observed with Asperger's syndrome though there is not yet any controlled research of this finding. The figure below shows an example of this pattern. It is a relatively high amplitude beta.

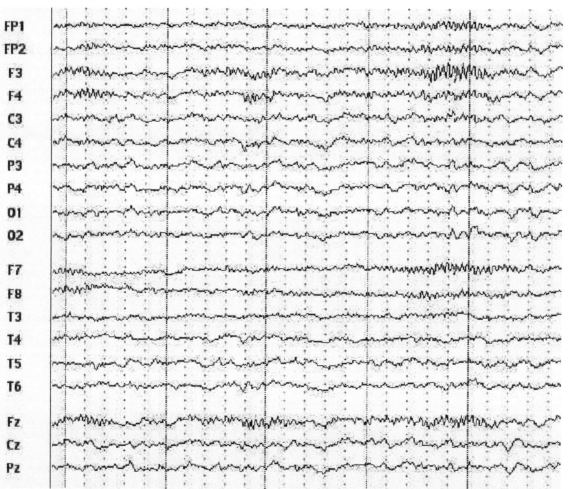

Frontal Beta Spindling in ADD with thanks to Jay Gunkelman for assistance in interpretation.

Case Example of Asperger's Syndrome, Age 62

High Slow-Wave Activity at T6

An eyes-closed example of a segment of EEG in a 62-year-old male who has mild to moderate Asperger's syndrome is shown below to give an example of high slow-wave activity in the T6 area. T6 is oriented for spatial and emotional contextual comprehension, and non-verbal memory.

This man has never been able to visualize. His emotions have always been flat. He has experienced high anxiety throughout his life and has only recently stopped taking anxiolytics.

When talking to an old friend, he has a sense that he is just talking to himself – not really connecting. He doesn't have the feeling or the sense that the other individual (e.g., giving feedback to a client) is hearing or understanding what he is trying to put across. He has done psychological testing with clients and given feedback on their results for about 20 years but he feels that this has become more difficult to do in the last two or three years. In addition, he feels his ability to sustain attention and complete tasks has deteriorated. He has had a full medical and neurological work-up but nothing abnormal has been detected.

He has always 'craved isolation' and feels this has also become more extreme. He dreads going out and being with people. He says that it is more peaceful just being by himself.

Recently, he has been feeling mildly depressed and this appears to correspond to high slow-wave activity at F3, which was more easily seen in eyes-open recordings. In the eyes-closed recording shown in the figures below (also observed with eyes open), note in particular the high amplitude slow-wave activity at T6.

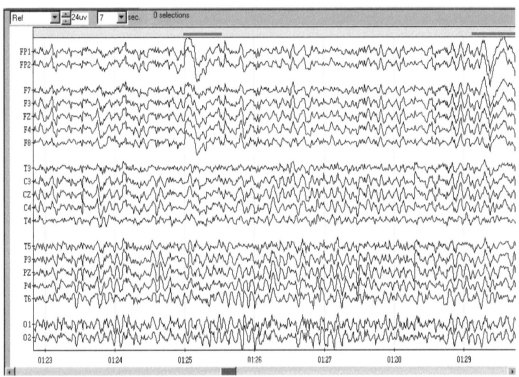

Eyes closed referenced to linked ears.

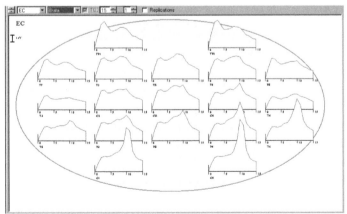

Eyes closed spectrum from the SKIL program.

Intervention

When the total record was reviewed, it was found that the eyes-open math task demonstrated, in addition to the P4 and T6 findings shown above, high slow-wave activity at F3 and Fz. Therefore, in the initial sessions, this man was encouraged to increase frontal beta (15-18) Hz and decrease 4-10 Hz activity at F3 referenced to linked ears or to the right ear. Activation of this site tends to counter depressive affect.

He also demonstrated high thalpha (6-10 Hz) at Fz, perhaps associated with inattention. High amplitude beta spindling at 30 Hz was also noted at Fz. This might represent cingulate activity, corresponding to mild compulsive tendencies and severe ruminating. Training, therefore, encouraged a decrease in thalpha and in high beta frequency (28-32 Hz) activity at Fz.

For calming and stabilization and to help him control anxiety, respiration and EMG biofeedback was combined with neurofeedback. CPz (halfway between Cz and Pz) and Pz training was done to enhance 11-14 Hz and to suppress 4-10 Hz.

To improve his ability to visualize objects and events T6 training was done. He has never been able to use visual imagery. Work at T6 might also improve his nonverbal memory. In addition, this area is involved in emotional contextual comprehension. The hope is that activating this area and 'normalizing the EEG' will result in a better ability to understand and respond to social nuances, decrease the flatness of his affect and decrease his anxiety during social interactions.

To achieve these goals, T6 was referenced to the left ear (referencing to the contralateral ear reduces common mode rejection) and he was instructed to enhance 13-16 Hz and suppress 4-10 Hz. This was coupled with respiration and EMG biofeedback and having him attempt to visualize a pleasant object (a flower) or a scene.

Units of Measurement

In most of the foregoing discussion, the units used for relating the amplitude of the waves have been microvolts. However, some of the literature is reported in power units called picowatts. A *picowatt* $\approx 6.14 \times \mu v^2$. It is important that the reader know what units are being used when reading the literature.

Units: Picowatts or Microvolts

Not only do the frequency bands used in various research studies often differ from the ones most of us use when doing neurofeedback, but the units are also different. Take for example the landmark multisite study regarding theta/beta ratios in ADHD (Monastra et al., 1999) entitled "Assessing Attention Deficit Hyperactivity Disorder via Quantitative Electro-encephalography: An Initial Validation Study" (Neuropsychology, Vol. 13, No. 3, 424-433). The study reported mean theta/beta ratios 4-8/13-21 Hz in picowatts. The A620 assessment program will do this calculation either for amplitude (measured in µv) or for power (measured in picowatts). However, most NFB instruments record only amplitudes in microvolts. This presents a small challenge if you want to compare your data with the Monastra norms and you are not using an A620. (It is also built into Biograph Infiniti Thompson Setting Up for Clinical Success assessment screens.)

What is the relationship between microvolts and picowatts? Recall that In mathematical terms the relationship between amplitude and power is: $(microvolts)^2 \times 6.14 = picowatts$. However, we are usually just trying to relate the published literature findings for ratios in picowatts to the microvolts that we are measuring on a daily basis with our students. A simple approximation used by some practitioners is that the $(ratio\ in\ microvolts)^2$ is similar to the ratio in picowatts. However, this simpler calculation mainly holds true if there is little variability in the sample amplitudes.

Jon Breslaw, an economist and a statistician in Montreal Canada, kindly gave us the following example to consider:

Two data sets:
Data Set 1: Theta in microvolts: 20 20 20 20 20
Beta in microvolts: 10 10 10 10 10

Data set 2: Theta in microvolts: 40 5 5 40 10
Beta in microvolts 10 10 10 10 10

Means: Data set 1: theta 20, beta 10. theta/beta = 2
Data set 2: theta 20, beta 10 theta/beta = 2

If this had been done in picowatts, and leaving out the factor of 6.14:
Data Set 1: Theta in picowatts: 400 400 400 400 400
Beta in picowatts: 100 100 100 100 100

Data Set 2: Theta in picowatts: 1600 25 25 1600 100
Beta in picowatts: 100 100 100 100 100

Means: Data set 1: theta 400, beta 100, theta/beta = 4
Data set 2: theta 670, beta 100, theta/beta = 6.7

*He then noted that, "Squaring the theta/beta ratio to make comparisons to the Lubar studies reported in picowatts is not valid if there is any degree of variability. The **mean of the squares is not the square of the mean**."*

He suggests that, "To replicate the Lubar results, one can take the microvolt reading, and create a data channel equal to the square of the microvolt reading using compute, and multiply the original channel with itself."

Theta/Beta Ratios in ADHD

Where this discussion is important at the present time is in working with ADHD. The most comprehensive published work cited above (Monastra et al., 1999) reports ADHD ratios 4-8/13-21 Hz in picowatts. Many people, however, will be using microvolt ratios, and often programs will calculate a theta/beta ratio that is 4-8/16-20 Hz since these frequency bands are often used in training. Our clinical observations, using microvolts, provide a less precise guideline: In general, most children between 7 and 11 will have a microvolt ratio (4-8/16-20) less than 2.4 unless they are observed to be having problems with attention span in school. The majority of children we see in this age group who are definitely having attention difficulties in school will have µV ratios >2.8 and often >3. The ratio decreases in older age groups. In general, however, our conservative estimate is that the majority of adolescents and adults have ratios below 2.1 unless they are reporting difficulties with attention span. (We gave an SNR presentation of an unpublished study comparing adults with and without an ADD history that found a cut off at 1.8) The majority of our adolescent population who have difficulty with attention span at school and when doing homework have ratios above 2.4 before training. As a rough guideline, one should suspect ADD if the 4-8/16-20 theta/beta microvolt ratio is >2.5 in children above age 7 and if it is >2 in older adolescents and adults.

Two-Channel QEEG Assessments

Rationale

As mentioned earlier, some equipment allows two channels of EEG biofeedback where 'two channels' refers to two sites being monitored with the use of two separate preamplifiers (and, in some equipment, two input jacks into the encoder). Each of these channels have three electrodes: active, reference and ground. This is a different set-up than that used when doing a 19 or more channel full-cap assessment. In the *full-cap assessment* you have a common ground lead and a common reference (usually linked ears). You can also get electrodes for two channels that have a common reference and a common ground. This helps ensure more accurate comparisons between activity at the two active sites.

Why Use a Two-Channel EEG Assessment

A two-channel assessment is carried out in order to compare the relative amplitudes of different frequency bands at two different sites. This can be very useful. We suggested above that if a client had all the expected symptoms of ADD, but your Cz placement did not show any frequency bands that were outside of the expected range given the overall amplitude of that client's EEG, then you should redo the assessment with the active electrode placed at a different site such as Fz. Similarly, you could assess two sites individually to compare different bandwidth amplitudes in the two hemispheres. In both instances the two single-channel assessments could be done at different times. (You use the same 'active' electrode and move it to a new site.) However, in this instance the results may be reasonable for planning training but they are not truly comparable because you have done two separate single-channel assessments at two different times.

A two-channel instrument allows you to get around this problem since you can compare, for example, F3 with F4 at the same moment in time. It is also more efficient than having to move the active electrode and take additional EEG samples. This is an important consideration since fatigue will affect the EEG. (Drowsiness = increased theta.)

How to Set Up Two Channels

Electrode Placement

With the one-channel referential assessment described above, the first assessment is typically done with the active electrode at Cz referenced either to the left or to the right ear or mastoid. Cz is chosen as it is less prone to eye movement and muscle artifact and gives the highest amplitudes for most frequencies since it is farther from the ear reference. (Less common mode rejection means higher amplitudes.) In some EEG instruments there are two or more channels available for EEG. This opens up a number of possibilities for electrode placements and referencing.

Comparing Two Central Locations

A client with ADD may be assessed with Channel 1 (A) active electrode at Fz (or F3) and Channel 2 (B) active electrode at Cz. This usually allows you quickly to identify the appropriate area for your inhibit frequency. Some people with ADD symptoms of inattention and distractibility have a very high amplitude at a particular slow-wave frequency when they 'tune out' that appears frontally and not centrally. In these cases it may be better to place your reward frequency active electrode at a different site than this frontal electrode that is being used to inhibit the dominant slow wave. An electrode to reward the production of SMR, for example, should be placed over the sensorimotor strip. A reward in beta for improving speech, verbal memory or reading would be placed at an appropriate site on the left hemisphere. This type of feedback **requires two channels** if you are to do it simultaneously, though it could be done with one channel if you train one aspect first (for example, decrease 7-9 Hz at F3) and then the other (increase 12-15 Hz at Cz) later. The sites could be alternated within each session or every other session. Alternatively, you could train at one site until the associated symptoms for one site are under control and then change to the second site. If being hyperactive and/or impulsive was the biggest problem, one would do SMR training first. If tuning out was the main complaint, you would decrease the frontal slow-wave activity first.

Assessing Differences Between Hemispheres

In some clients you may wish to compare activity in the two hemispheres. We find this useful for clients who complain of ruminating and becoming internally distracted by concerns over matters not relevant to the task at hand – in other words, the worriers. In these clients you can compare a site in the left hemisphere with a similarly positioned site in the right hemisphere, such as F3 and F4 both referenced to Cz (or reference them to their ipsilateral earlobe or mastoid). You can purchase electrodes where the two ear electrodes are attached to each other. Two leads come off this common wire between the two ear electrodes and these go to Channel A and Channel B, respectively. This results in a common reference for the two channels of EEG. You can also purchase a ground electrode that has two leads that go to Channels A and B, respectively, giving a common ground for both channels. (See IMA in the Resources section to order these electrodes. They can also be purchased from Thought Technology.) This type of assessment may be particularly useful for a client with dysphoria or mild to moderate depression. (Clients with severe depression should of course also be under medical care).

When you are comparing the EEG in two channels, it is essential that you have good impedance readings that are virtually the same for both channels.

Now that you have your electrodes in place, compare the activity at the two sites for different frequency ranges. For catching the worriers, one of these should be within the 22-34 Hz range. The precise bandwidth(s) for comparisons will depend on your equipment and your hypotheses about your client and the EEG patterns that would mirror their symptoms. In the assessment you might observe single Hz or overlapping two Hz bins depending on the constraints of your programs. We are finding that when clients think negative thoughts, 23-34 Hz activity measured either in microvolts or as percent-time-over-threshold (which you can put on your screen with some instruments) tends to be higher in the right frontal region. When the same client thinks positive thoughts it may tend to be higher on the left. (This observation requires replication in other centers and is not true in every case.) We also find that when the client relaxes and becomes externally, openly aware, activity in this frequency band (23-34 Hz) drops in both hemispheres, and the 11-15 Hz range shows an increase. We are not finding large overall differences in 9-10 Hz or in 11-15 Hz comparing right and left hemispheres, although there are reports in the literature concerning higher frontal alpha in the left hemisphere than the right in depressed persons (Baehr, 1999). The higher left hemisphere alpha is taken as an indicator that the left frontal area is resting; that is, there is a lack of activation in the area. This may be associated with depression. Davidson has found left hemisphere activation to be associated with approach tendencies (being engaged in the world, not withdrawing) and positive affect (Davidson, 1998). We emphasize the necessity of respecting individual differences and doing careful assessments and progress testing to ascertain the precise frequency bands (and locations) that correspond to the mental states that concern each individual client.

Clinical Tip

If you have symptoms of both tuning out and dysphoria, here is a suggested method to make electrode placement relatively quick and simple.

With Electrodes That Are Not Linked

In your first assessment, do: Channel 1, F3-Left ear and Channel 2, Cz-Right ear. Then you only need to move the Channel 2 active electrode to F4 in order to do a second assessment comparing the activity at similar sites over the two frontal lobes.

With Linked Electrodes (Preferable)

In the diagram below, note that there are linked-ear reference electrodes. In addition, the ground is linked and goes to both preamps **(Channel A and Channel B)**. This arrangement makes it very simple to check impedances. It assures that the impedances are the same for reference and ground electrodes at the two sites so that comparisons between the two active sites are valid. The ground can go anywhere. It is shown here in a frontal location.

Diagram of Linked-Ear Reference and Common Ground Set-Up

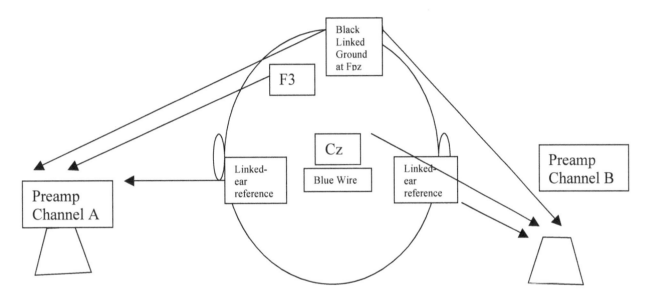

An Example of a Two-Channel Assessment

John was a client who complained of dysphoria (mild depression) and anxiety.

John is a 25-year-old university student who complained of lack of focus and a disorganized approach to his academic work. He finds himself ruminating about small problems in life when he should be attending in class or studying. When a two-channel assessment was done with active electrodes at F3-left ear and F4-right ear, it demonstrated that, when he thought about negative things in his life, percent-time-over-threshold for 23-34 Hz rose much higher on the right than the left. The reverse was true when he thought about pleasant things in his life. Both percent-over-threshold readings dropped dramatically when he was calm, relaxed, and highly focused on the trainer describing a metacognitive reading strategy for his university science subjects.

The most important finding here is not so much that the high beta is on one side or the other, but that the client's mental activity related to worrying and ruminating seems to correspond to high amplitude, high beta activity. This may be a subtype of ADD in that the client is internally distracted by these thoughts and not attending to external stimulation. When the client becomes calm and focused, this excess high beta activity decreases.

John's Two-Channel EEG Assessment

The figure below is a four-second sample of John's EEG at F3 (Channel A) and F4 (Channel B). Both are referenced to their ipsilateral ear. Impedances were below 5 Kohms and within 1 Kohm for each pairing of electrodes. Note the high high-beta activity on the right side. This pattern was consistent with findings from a three-minute artifacted sample of his EEG.

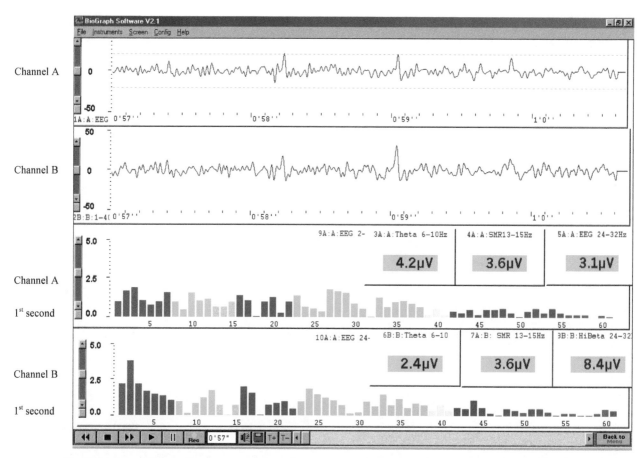

The histograms shown above are one-second averages of the magnitude at each frequency from 1 to 61 hz.

When this assessment was done, John was feeling somewhat discouraged and dysphoric. The spectrums and the figures refer only to the first second (57'-58') for each EEG channel. Note the high slow-wave activity on the right side (Channel A) and the high high-beta, 24-32 Hz, in the right frontal area. Note artifacts at 58.2 and 59.0 seconds. These would be excluded from a statistical analysis.

When you do these two-channel assessments, you should do the statistics in one or two Hz bandwidths. Then graph it using a spreadsheet program that allows you to superimpose the two graphs or histograms (one for each channel). This will make it easy to visually compare the amplitude of different frequency bands in the two hemispheres. This comparison will make it relatively simple to choose appropriate inhibit and reward frequency bandwidths for each hemisphere for training. The most common meaningful comparison deals with activity in the frontal lobes at F3 and F4. The spreadsheet allows one to automatically graph and superimpose progress assessment results in different colors on to the original assessment graph in order to assess progress.

To decrease dysphoric ruminating and increase a sense of calm, John was successfully trained using a two-channel placement (Channel A at F3 and Channel B at F4) to decrease 23-34 Hz at F4 (compared to F3). Then he was trained to decrease 23-34 Hz overall with Channel A at Fz, while increasing 11-15 Hz with Channel B at C4. When only one channel was used, the active electrode was placed at FCz.

The focus of training for John was then changed to emphasize a decrease in distractibility and an increase in concentration when carrying out a cognitive task. He was asked to maintain his calm mental state while keeping his dominant slow wave very low (5Hz and 9Hz peaks were evident initially in John's case) and raise 15-18 Hz when problem solving or analyzing a chapter in a textbook. (On assessment it had been found that he increased 17 Hz when problem solving.) This training was combined with teaching metacognitive strategies and practicing holding the appropriate mental state while carrying out academic tasks.

To do this training a different two-channel placement was used. Channel A was used to decrease his thalpha (5-9Hz)/low-beta (15-18 Hz) ratio of activity with the active electrode between F3 and C3 referenced to the right earlobe. Channel B training had the active electrode just anterior to C4. It was used for decreasing the relatively high 23-34 Hz activity in the right hemisphere while at the same time encouraging an increase in 13-15 Hz activity. John was also receiving biofeedback concerning respiration and forehead EMG during these sessions. At the end of training he felt he was able to control his mental states and stop his negative ruminations.

Conclusion

The real point in doing a careful assessment is to find out what is true on the EEG for your client. Then respond to these findings by placing your active electrode over the relevant area and increase and/or decrease the appropriate frequency bandwidths. In the future, coherence and/or comodulation training may make this a more efficient process, though we do not yet have much published data regarding outcomes for different groups/different coherence patterns using coherence training. With similar patients one might alternate these sessions with LORETA Z-score NFB. But the point being made in this section is that very good results can be achieved with single-channel and two-channel assessment and training.

Importance of Impedance in the Two-Channel Approach

When you are comparing EEG frequency bands in two channels the impedances must be virtually the same. If they are not almost identical, then any apparent differences between the hemispheres may just be due to the different electrode connections (Hammond & Gunkelman, 2001). You might then come to incorrect conclusions about the activity in the two hemispheres.

Theoretically, using high resistance amplifiers, impedance differences should not present a problem. However, try experimenting and you will find (as we have) that it does make a difference. It makes a huge difference if the client should move their head even a small amount! Impedance should be between 100 and 5000 Ω. Measures for each pair of electrodes should be within 1 Kohm of each other. Electrodes with much higher impedance may cause 60 Hz artifact and attenuate the recording (Fisch, 1999; *Fisch and Spehlmann's EEG Primer*).

Impedance will be further discussed under "Site Preparation."

A 19-Channel (Full-Cap) QEEG Assessment

Advantages

Nineteen active leads plus a ground and reference electrodes on the ears are typically used in what is called a ***full-cap*** assessment. (In research labs the number of leads can multiply to over 200 to attempt to get better spatial resolution.) This type of quantitative EEG allows one to collect data at standard 10-20 sites around the scalp. Each of these sites is referenced to linked ears. The data collected, however, can be analyzed in various ways using different *montages* other than linked ears (see below). The information can be used to produce brain maps that plot the activity at the 19 sites and interpolate the data for the spaces between the electrode locations. As noted previously, the term 'full-cap' is apt since this type of assessment usually utilizes a soft, thin, cloth cap that looks very like a swimmer's cap. (If the client is a child who does not like the idea of putting a cap on, compare the cap to what water polo players wear.) The cap has 19 tiny electrodes built into it. The wires from these are sewn into the cap and are joined into a cord that hangs like a queue at the nape of the client's neck. This cord has a plug at the end that fits into the EEG machine.

In our field it was traditionally an EEG recording system developed by LEXICOR, Inc. (Golden, Colorado). This company has also contributed to teaching full-cap EEG recording methods. In 2003, the NeuroNavigator became available that provided the option for both QEEG and neurofeedback. It is no longer available.

Available options include the DeyMed from Czechoslovakia, the Mitsar from Juri Kropotov's group in St. Petersburg, Russia, the Discovery from Brainmaster, NeuroPulse, and eVox from Evoke Neuroscience. eVox does 19-channel EEG assessment with LORETA analysis + Evoked Potentials + Heart Rate Variability (HRV) + Continuous Performance test in a single 20-minute assessment session. Neuropsych testing battery is optional. This amplifier can also be used for single-channel NFB or LORETA NFB plus HRV training. They provide almost immediate reporting after the assessment as the data is instantly sent to their New York office over the internet. The practitioner has the option to also do their own analysis with their own program and at our Centre we do this using NeuroGuide.

Montages

In a full-cap assessment various *montages* may be used. This does not affect the collection of data as it is something that is done at the interpretation stage using computer software. The data that are collected using a linked-ear reference may be *remontaged* using computer programs designed for analyzing the results of the EEG recording. A 'standard-of-practice' is to always use three different montages to look at your data.

Linked-Ear Referential Montage

Recordings are usually carried out using a linked-ear reference and a forehead ground. In the computer programs for interpreting the results of the EEG assessment you may use this montage to interpret the data. Neurofeedback practitioners often look at this montage first when artifacting the data. Other montages may then be used to help bring out any abnormalities in the record or to help clarify what might be artifact.

Sequential (Bipolar) Montages

Sequential (bipolar) montages compare adjacent pairs of electrodes in the 10-20 placement system. These may be either *longitudinal* pairs (such as F3-C3) or *transverse* pairs going across the head (such as C3-Cz). The term 'bipolar' refers to the amplifier. There used to be both bipolar (also called *differential* or *push-pull* amplifiers, using an active, a reference and a ground electrode) and monopolar (single-ended, using active and ground electrodes) amplifiers. Nowadays there are only differential amplifiers (Notes from correspondence with Jay Gunkelman).

Recording should always be done referentially so that the remontaging can be done. It would be best, according to some electroencephalographers, if the recording were done using a single ear montage. A single ear reference avoids what is called *shunting*. Shunting occurs as a result of an impedance imbalance between the ear electrodes. Shunting produces a skewing of the topographic maps towards one side, due to the impedance mismatch. The shifting follows the path of least resistance. In practice most of us are using instruments where the linked-ear reference is built into the hardware and there is no choice concerning how the recording is carried out. We must, therefore, be careful to check that the impedances are the same in the two ear electrodes..

'Global Average' Laplacian Montage

An active reference montage using common average reference (CAR) will reference an active electrode to an average of all the other electrodes. This means that all of the 10-20 electrodes are added together in Input 2 of every amplifier. This then serves as a reference for each of the electrodes in Input 1 (the 'active' electrode). Thus, each of the 19 will contribute 1/19th of the total activity in input 2. Strictly speaking, this should be called a *'global average' Laplacian technique*, where weighting factors are added to correct for inter-electrode distances. Pierre-Simon Laplace was a mathematician and physicist. The name 'Laplacian' refers to an application of his mathematics to the EEG by Hjorth (1980).

'Local Average' Laplacian or Hjorth Montage

A *'local average' Laplacian or Hjorth montage* will reference to the average of the electrodes immediately surrounding each active electrode (Fisch, 1999). Most practitioners when they say 'Laplacian' they are really leaving out the name 'Hjorth' yet they mean this 'local' average rather than the global average.

Brief Discussion Concerning Full-Cap Assessments and Montages

Each *montage* is a mathematical reworking of the data that can be done rapidly by the computer. Each of these different ways of looking at the data will have advantages and disadvantages. As with many aspects of instrumentation, there are trade-offs when it comes to choosing a montage. As noted earlier in this text, the *sequential (bipolar)* and *Laplacian (Hjorth)* montages are good for viewing highly localized activity. This would be of great value to neurologists looking for abnormalities. The *common average reference montage (global average Laplacian technique)*, on the other hand, is excellent for detection of widely distributed currents and for analyzing asymmetry. It is also very good for the detection of artifacts. It is not so useful for viewing localized activity.

Regardless of the montage, one must always pay attention to strict removal of artifacts from the EEG before analyzing it to produce brain maps. This process will remove *transients*. (As previously mentioned, a *transient* wave is one that stands out as different against the background EEG.) These transients may include clinically important data (e.g., an epileptiform burst). Always remember, the morphology of the EEG is diagnostically important. You can only see waveforms (*morphology*) by looking at the raw EEG. You do not want to miss waves such as triphasic waveforms that could indicate a toxic/metabolic encephalopathy, or periodic discharges such as may be seen in other encephalopathies, or epileptiform bursts. These discharges would alert you immediately to the need to have the record read by experts and to refer your client to the appropriate medical specialist, usually a neurologist.

Instructions for placing the 'cap' on the client's head are given in detail with the cap when you purchase it and are not repeated here. You will have to have individual supervision in order to learn how to properly put a cap onto a client and make sure that the cap is giving accurate information to the computer. This will require that you learn how to bring the impedances for all the electrode pairs to <5 KΩ and within 1 KΩ of each other.

The full-cap assessment has the advantage of doing all sites at the same instant in time. EEG has wonderful temporal resolution. You can record under different conditions such as: eyes closed, eyes open (looking at a single point to minimize eye movement artifact), reading and math. This type of assessment allows for a conventional investigation of amplitude differences between sites at different frequencies under different conditions. Additionally, it allows for an analysis of communication between different sites. This is traditionally called *coherence,* which is calculated using a cross-correlation co-efficient concerning the waveforms at two different sites.

Definitions of Common Terms

Terms that arise in discussion about how different areas of the cortex are communicating with one another include: *phase, coherence, synchrony,* and *comodulation.*

Phase

Covariance in time is called **phase.** When waves that are morphologically the same occur at the same time at two different sites they are said to be *in phase. Conducted phase* refers to time delays between two sites. With a time delay the waves will be said to be *out of phase.* This is measured in terms of degrees, 0-180°.

Propagated phase refers to a *focal phase reversal.* This type of phase reversal is best seen in sequential recordings between adjacent pairs of electrodes. A phase reversal in this case may point to the source of an EEG phenomenon.

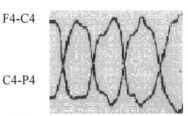

The type of phase reversal shown in the figure above could appear in the EEG of a person who has seizures. The origin of the seizures in this instance is likely at C4.

Coherence

(Special thanks to Dr. M.B. Sterman for his help with the following three sections)
Coherence refers to a cross correlation between frequencies. It indicates the degree of shared activity between sites. Coherent waves have the same morphology. The amplitude of the waves is not a factor. Being *in phase* is not a factor. There could be shifting in phase or out of phase as long as the morphology of the waves is the same. Time is also not taken into account as it is when calculating comodulation.

Calculations for coherence compare the waveform at two sites and then do an FFT spectral analysis. Comodulation calculations (described below) do the spectral analysis first and then compare the rise and fall of the amplitude of a specified waveform.

Coherence is expressed as the square of the correlation coefficient ($[r]^2$). Correlation coefficients range from –1 to +1. Since coherence is the square of the correlation coefficient, it will range from 0 to 1. It is a measure of the linear association between two variables, in our case the EEG waveforms recorded from two different scalp locations, independent of the EEG amplitude at either location.

Coherence is calculated as the *Fast Fourier Transform* of the cross correlation between sites across the frequency spectrum. It discloses the frequencies with the most shared activity between any given pair of sites across a sample of data. Thus, the specific timing of coherent activity is not a factor. The waveform morphology and phase are compared between sites, independently of amplitude. Coherent waves have the same morphology and therefore frequency. There can be a shifting in phase as long as the morphology/frequency of the waves is the same. As long as the morphology is the same, you have the same kind of activity (say theta or alpha) at the two sites, and there is coherence. Phase relationships can be derived from this measure separately.

Clinically, coherence represents the degree to which two sites have similar waveforms (vary together) and may imply that the two sites are getting input from a common generator. As a child matures, coherence will increase between central and temporal and parietal regions in all bands. However, with maturation, coherence will decrease between the frontal lobes. This implies that with maturation there is greater cortical differentiation in the frontal regions of the brain (Thatcher, 1986). Learning disabled children, on the other hand, demonstrate increasing frontal coherence in theta, alpha, and beta bands (Marosi, 1992).

Why Use a Coherence Measure?

The EEG measures summations of extracellular electrical changes produce by radially generated activity within pyramidal cells that lie perpendicular to the surface of the cortex. It measures the extracellular summation of EPSPs and IPSPs as discussed earlier. The EEG is *blind* to about 2/3 of the cortex's electrical activity. It doesn't measure *lateral* current flow or intracellular activity. A magnetoencephalogram (MEG) measures magnetic activity in the brain. This may be thought of as *lateral* current flow or intracellular activity. The MEG, however, is blind to the extracellular potentials measured by the EEG.

The EEG amplitudes are also *blind* to the longitudinal myelinated fiber tracts. These tracts connect different areas of the brain. These are the intrahemispheric fasciculi and the interhemispheric commissures discussed in the neuroanatomy section of this book.

Coherence and phase measurements allow us to infer the *connectivity* between areas of the brain. When differences from databases of normals are observed, this can lead to a NFB program that attempts to 'normalize' our measurements of these connections with the objective of seeing a corresponding 'normalization' of cognition and behavior.

Comodulation

Dr. M. Barry Sterman has often been referred to as the "grandfather" of the NFB field because he was the first researcher to establish that operant conditioning of brain waves was possible. He has produced a database that is included with the SKIL software. Using SKIL (from Sterman Kaiser Imaging Labs), one can produce brain maps and topographic displays. This software has been helpful in introducing many clinicians to the complex world of 19-channel EEG assessments. In this software, SKIL has introduced a new measurement of the integration between different areas in the cerebral hemispheres, termed *comodulation*. Simply put, comodulation assesses the degree to which a given frequency is increased and decreased in its spectral magnitude simultaneously at two different sites over time. Comodulation refers to a spectral correlation; that is, a correlation between shared spectral content across time. In contrast to coherence, comodulation assess the degree to which waves of the same frequency are moving together over time. It is a measure of collective integration rather than connectivity. As previously noted, these calculations require that the FFT be done first for a given frequency band, and then the cross-correlation between sites is compared over time. Any frequency band can be selected for this comparison, and it can reflect either the subject's own unique integration or be compared statistically with a normative database. Comodulation, accordingly, stresses changes over time, a focus not provided in coherence analysis. However, like coherence, amplitude at the two sites does not matter.

Thus. comodulation estimates are time locked. Comodulation is a stepwise calculation; for example, it may be done each quarter second. In this case the waves must stay lock-stepped together for that ¼ sec to be registered as comodulating. Comodulation is a new term that will not be familiar to neurologists (they may refer to it as spectral correlation) but people doing neurofeedback may hear more about it in the future. Remember that sensorimotor rhythm (SMR) was also a term coined by Sterman, and it has become part of the basic jargon in our field.

Why Are Coherence and Comodulation Important in NFB?

Measures of coherence or comodulation lead directly into a different way of training. To normalize the EEG the client may be trained to either increase or decrease communication or functional coordination between two or more sites on the scalp.

Synchrony

Synchrony is a term used in Lexicor programs. It refers to a mathematical calculation that considers phase and coherence in a single calculation. The waves must be, therefore, the same shape (morphology) and are in phase (going up and down at the same time) at the two sites to be considered synchronous.

Spectral Density

It is important to remember that all frequency analysis methods must estimate amplitude over time and thus include *incidence*. Simple amplitude cannot be assessed. Thus, *spectral density* refers to the incidence and amplitude of a particular frequency band. For example, 50 µV of 6 Hz activity for ½ second would have the same spectral density as 25 µV for 1 second. This metric is also sometimes called *spectral magnitude*.

19-Channel Assessment – Case Examples

Note: In later sections of this book all the assessments are shown using more commonly used programs such as WinEEG, NeuroGuide, and eVox. Each of these programs has its own database.

In this section we show illustrations from an older program called SKIL. The following 19-channel EEG data samples are interpreted using the SKIL program from Sterman Kaiser Imaging Labs. Similar analyses may be carried out with other programs and other databases. A number of these examples have been provided by Dr. Sterman, and we gratefully acknowledge their provision. These examples show:

1. Amplitude in 1 Hz frequency bands
2. Amplitude of particular frequency bands compared to the SKIL normative database
3. Comodulation, for a chosen frequency range (usually the dominant frequency), of each of the 19 sites with all other sites.

Goal of this Section

In this book on the fundamentals of neurofeedback we will not describe how to do a full-cap assessment, since that involves hands-on training. It is an art to inject the gel and abrade the surface of the scalp enough to get good impedance readings but without breaking the skin, and do all this quickly. Our staff generally have the cap on with excellent impedances in less than eight minutes. The basic principles regarding preparation and impedance checking are analogous to what is done with three electrode assessments. The objective here is to demonstrate, first, how information from a full-cap assessment may extend your understanding of a client's difficulties and, second, how this information may suggest additional training interventions.

A full-cap assessment can elucidate many features that cannot be seen in a single-channel assessment. However, nothing is perfect. (Remember those tradeoffs). Some full-cap assessments do not, for example, give accurate information concerning frequencies above the mid-20s (SKIL and NeuroGuide databases) using older amplifiers. This will change with modern amplifiers such as eVox and modern databases. It also gives no information about the autonomic nervous system or about muscle tension (EMG) unless these recordings are done simultaneously with the EEG such as is done with the eVox system.

Case Examples

This section has been left in from the first edition. Most practitioners are not now using SKIL, but it may still be of interest. NeuroGuide analyses are very commonly used and are given for the case examples in later sections of this book.

The first case is Sean, a man with ADD and the co-morbidities of learning differences and mild depression. He was initially assessed and trained based on a single-channel EEG and a stress test. He was showing significant improvement concerning management of ADD symptoms. The full-cap was done to provide a more detailed assessment that might guide training to address the LD and depressive problems. A second client, Betty, had a closed head injury. Her training was not initiated until a full-cap assessment had been carried out.

Case Example #1 - Sean, a 34-Year-Old Man with ADD, LD and Depressive Problems

Sean was married with two young children. He was employed as manager of a large commercial greenhouse. He had dropped out of high school due to frustration associated with undiagnosed learning difficulties and inattention. He came to the Centre after testing by his company identified personality variables that were affecting his job performance; in particular, his communication style with staff (which could be abrupt) and his weak organizational skills. He also reported tensions in his marriage, anger management problems, and symptoms of depression. At work he was having difficulty maintaining his focus on the task at hand. He "had a short fuse" and could become rapidly upset and angry.

Cognitively, he had some difficulty explaining things in a sequenced, organized fashion. He might say things impulsively. He had poor listening skills. These characteristics were frustrating for his employees. Nevertheless, his creative problem solving, high energy, excellent hands-on skills and ability to hyperfocus and get a job done meant that the greenhouse he managed was the top producer among the many sites owned by a large corporation.

IQ testing revealed that he was bright, with an overall IQ just above the 75th percentile rank. However, his performance IQ was lower than his verbal IQ. Attention to visual detail was far below his other abilities and he was very slow at reproducing visual symbols and had difficulty constructing whole objects from their components. Being impatient, having a disorganized approach rather than applying strategies, together with difficulty sustaining attention, were associated with these weak scores.

The initial assessment demonstrated high amplitude 4-9 Hz activity and a dip in the SMR range at 13-15 Hz measured at Cz referenced to the left ear lobe. His 27-31 Hz activity was high. The stress test showed how he became rapidly frustrated with math. During this part of the test peripheral skin temperature dropped, and EDR, respiration rate, heart rate and EMG forehead readings all rose. Training was initially focused on improving attention span (decrease 4-9 Hz activity), becoming calm and reflective (increase 13-15 Hz activity) and focusing his thoughts on a task at hand rather than ruminating about work and home (decrease 27-31 Hz activity). He was also taught to relax and breathe at six breaths per minute. Excellent progress was made in each of these areas during the first 25 sessions.

We had several hypotheses when doing the full-cap assessment. First, that his mild depression might be reflected in *hypercomodulation* frontally. Second, that his learning difficulties might correlate with inappropriate 'idling' in areas of the brain that are involved in synthesis and integration of visual, perceptual, verbal and reading tasks. These patterns would be expected to emerge because he was engaged in a cognitive task for part of the assessment, namely, the silent reading and mental math conditions. Third, that the learning difficulties might also be associated with a lack of differentiation (hypercomodulation between areas of the brain) and/or *disconnects* (hypocomodulation) between areas of the cerebral cortex. Note that we like to have a rationale to justify the extra cost of a full-cap assessment. With simple ADD, a single-channel assessment will often suffice since problems are evident at the central location and training is also done there. Here are Sean's results:

1st – Amplitudes Using 1 Hz Frequency Bins

In the following brain maps, red is high amplitude and blue is low amplitude, green and yellow are in the middle range. These data are from Sean, age 34, a three-minute sample of EEG done with eyes open doing math.

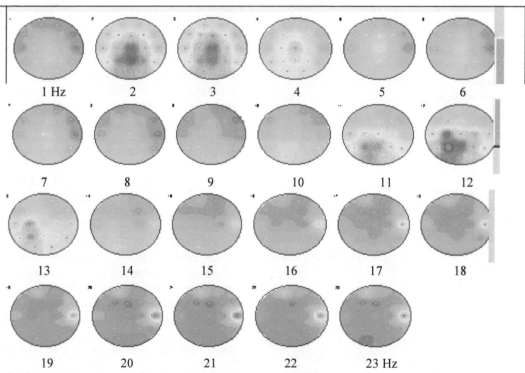

Brain maps showing amplitudes of each frequency while doing math

*The above display is called a **topographic map**. In each circle the front of the head is towards the top of the page and the back of the head (occiput) is towards the bottom of the page. This brain map displays information about activity in terms of raw data concerning amplitude of single hertz frequencies. It was generated from the SKIL program. EMG activity was high throughout the record at T4. This accounts for the higher amplitude seen at T4 on the maps from 14 Hz through 23 Hz.*

Sean had unusually high amplitude low frequency activity (delta through low theta) activity at Cz and Pz compared to other sites. In addition, he had high amplitude alpha at 11 and 12 Hz at P3, and alpha was somewhat elevated at C3 and Pz compared to alpha at other sites. These findings relate to the second and third hypotheses outlined above concerning problems with synthesis and integration of information. Although we had picked up the high amplitude slow-wave activity at Cz on the initial one-channel assessment, the full-cap allowed us to observe the alpha at P3 and plan an intervention to address the learning problems.

(Note: The above image was done after 25 sessions of training at Cz that was aimed at decreasing 4-9 Hz and increasing 13-15 Hz activity.)

Sean's early data had appeared to fit a *beta minima* pattern at F8 in eyes closed recordings, but it was not so evident in this recording while doing mental math. This low amplitude right frontal beta may sometimes be seen in persons who lack impulse control, and this was a problem that manifested itself as sudden angry outbursts in Sean's case.

2nd – Amplitude of Particular Frequency Bands Compared to the SKIL Database

Sean demonstrated an unusual pattern in his dominant frequency, the 10-11 Hz range, when doing mental math. (His dominant alpha frequency had been determined from eyes closed data, which is not shown.) This could be seen on a *topometric analysis* that allows a comparison of the client's data with a normal database. It displays the amplitude of a chosen frequency band at each of the electrode sites. Sean's results are in red, the line joining solid circles. The average for the normal database is the dark black line, the line joining squares. Light gray lines above and below represent +2 SD (standard deviations) and –2 SD respectively. Ninety-five percent of the population is within two SD of the mean. Sean is outside the normal limits at C3 and P3.

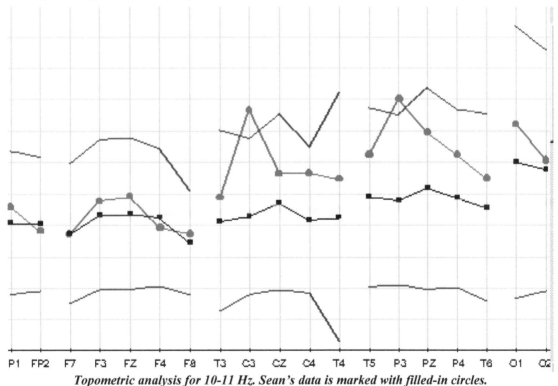

Topometric analysis for 10-11 Hz. Sean's data is marked with filled-in circles.

3rd – Comodulation of Each of the 19 Sites with All Other Sites

The 19-lead assessment also permits you to observe communication links among different areas of the brain. The electrode site will, of course, communicate (comodulate) with itself perfectly (+1). The display below is a statistical comparison of each site with all other sites compared to the SKIL normal database for similarly aged men doing a math task with eyes open. This program also allows one to compare the client with a normal database for eyes closed, eyes open, and a reading task.

The color at the active site is usually yellow surrounded by light green. If that site comodulates too much (≥2 SD above the mean) with another site, then the other area will be an orange to red color. If it comodulates too little with another site (≥ 2SD below the mean), then that area will be a blue color.

IMPORTANT: All of the following is discussed only to show the kind of decision making processes that may evolve in the next few years. At the time of writing, relating 19-lead assessment findings with respect to comodulation to clinical observations must be considered only as a way to generate interesting hypotheses. There is no solid published research base at this time as the technique is quite new.

However, there is a great deal of published data using NeuroGuide, and this will be shown in later sections

Comodulation with Eyes Open Doing Math

A comodulation analysis, according to Sterman, is best done using the client's dominant frequency. For most adult clients this is in the alpha range. It can be a very narrow band, even a single Hz, or a two- or three-Hz range; for example, 9 Hz or 9-10 Hz. If there is hyper- or hypo-comodulation, it will stand out best using the dominant frequencies, but might not show up if a broad range such as 8-12 Hz was used. In Sean's case, there was a bimodal distribution with two peak frequency bands. The first figure shows the dominant alpha between 10-11 Hz. The second shows in the 3-5 Hz range.

The figure below shows comodulation in the dominant alpha frequencies while doing a math task (statistical comparison with normal SKIL database at 2 SD).

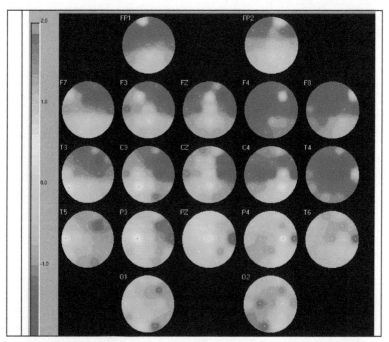

Comodulation in the dominant alpha frequencies while doing at math task. *(Statistical comparison with normal SKIL database at 2 SD.) (On the raw EEG for this case example, there was some muscle artifact at T4. Therefore, NO comodulation interpretations are made concerning the comodulation of T4 with other sites.)*

In the foregoing figure, 10-11 Hz activity is used as it is Sean's dominant frequency. He was doing a math task. Here comodulation between sites is compared against the SKIL database. Dark blue is –2 SD from the mean of the database. In this figure, we see a suggestion of hyper-comodulation between central and parietal regions, particularly between C3 and P3 (possibly indicating a lack of discrimination of function). We also note hypo-comodulation – disconnect – between sites on the left and the right frontal and central regions. Clinically, we hypothesized that there was a possibility that this disconnect could relate to some of Sean's difficulties with synthesis and integration in learning. Training could include encouraging comodulation between C3 and F4, C4, and also between F4 and Fp1, F3.

In the figure below, a comodulation comparison with the SKIL database shows hyper-comodulation between the central and parietal leads in the 3-5 Hz frequency range. The data are with eyes open, doing a math task. The prefrontal area and F7 and F8 appear to be hypo-comodulating with other frontal and temporal areas. This could be artifactual and due to eye movement, although careful artifacting had been done. It might also be postulated that, since the prefrontal areas are involved in impulse control, this finding might relate to his difficulties with anger management.

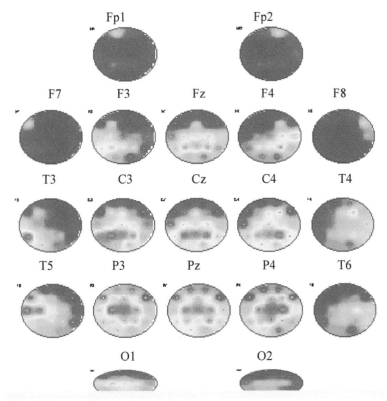

Brain map showing comodulation during a math task. (Statistical comparison with normal SKIL database, 3-5 Hz band.)

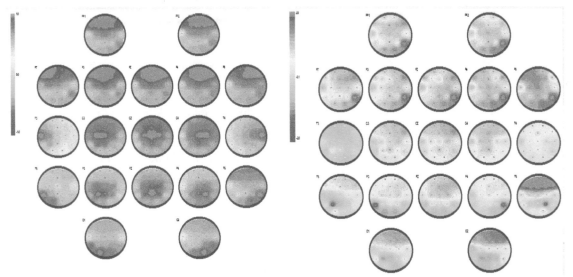

Comodulation data, 10-11 Hz. The range is +1 to −1, as it would be for any correlation coefficient. Red is high, blue is low.

*Comodulation, 10-11 Hz, compared to SKIL normal database (+2 SD from the mean is red and −2 SD is blue). For **a sense of what would be 'normal,' look at T3**. The green color indicates*

Comodulation With Eyes Closed
Concerning the above figures: We wondered if eyes closed might give a different picture than the math above. Indeed, when this was analyzed, it gave two clear pictures that had not been seen when he was struggling with the math task. **Comodulation With Eyes Closed** used the dominant frequency range, which for Sean was in alpha at 10-11 Hz. Thanks are given to M. Barry Sterman for his assistance in interpreting these comodulation figures.

Sean, as noted above, was age 34 and was mildly to moderately depressed. The figure on the left, above, shows raw data. It demonstrates hyper-comodulation in the frontal area with eyes closed. This pattern has been frequently observed by Sterman in depressed persons.

The figure on the right above shows a comparison of cross-correlation coefficients with the SKIL comodulation database. It indicates clear disturbance in communication/coordination between the frontal and prefrontal cortex and the right posterior temporal cortex. This is an example of 'hyper-comodulation.' One could postulate that discrimination of functioning was impaired.

We questioned whether this might be related to Sean's difficulties with anger control and with understanding and responding appropriately to nonverbal communication at home and with his subordinates at work. It might also be another finding that related to his learning difficulties.

Suggestions for Training Arising From Comodulation Findings
The eyes closed assessment led to two objectives for training. The first objective was to decrease comodulation in the frontal area. The second objective was to decrease comodulation between F3, Fz, F4 and T6. Since direct training of comodulation is not yet possible, another strategy would be training to activate the left frontal area (F3). This area's functioning would then become more independent from activity at the other sites. However, comodulation findings, in our experience, do correlate with coherence findings, and coherence training is easily done with virtually any commonly used NFB equipment.

Important Cautions
As is true of all cutting-edge science, little is known about comodulation. The comodulation differences that are two standard deviations from a normal database may be taken as a concrete finding, a differ-

that it comodulates in the average range with all other sites.
ence from the norms. However, all interpretations of what this may signify, and suggestions as to which of the client's symptoms these differences from the normal database might relate to, **are only hypotheses.** At the time of writing there is no published research concerning whether these deviations from the norm relate to specific symptom pictures. Nor is there research to show that training to increase or decrease comodulation has an effect on the symptom picture. As equipment that allows this kind of training is developed, we can look forward to its clinical application and await the results.

A second caution is that no comodulation conclusions can be drawn in any channel in a frequency range affected by an artifact such as muscle activity. In Sean's case, this means that we cannot look at comodulation related to T4 in the beta frequency range.

For Sean, a One-Channel or 19-Channel Assessment: Which Should We Have Done?
In Sean's case, the single-channel (Cz), three-lead assessment using the decision pyramid led to informative results. It demonstrated very high slow-wave activity at 3-5 Hz and high 27-31 Hz activity. Because he had high 11-13 Hz alpha, it was difficult to say definitely if SMR was low. However, given his history of impulsivity and the fact that we found 14-15 Hz activity was approximately the same as 15-16 Hz, we could conclude that encouraging him to decrease slow wave activity 3-9 Hz, and also decreasing 27-31 Hz, while increasing SMR 14-15 Hz, might be an appropriate starting point. The initial results, based on this single-channel assessment, were good. Nevertheless, without the full-cap we would have missed the parietal location of very high 12 Hz activity. We would also not have been aware of a possible frontal-central disconnect and a possible relative lack of left to right frontal communication when doing a math task and the hypercomodulation noted in the eyes closed condition. Acting on these additional findings might help his mild learning difficulties. This additional work would be based both on a knowledge of neurophysiology and brain function, and on a comparison of his EEG with a normal database. The recommendations for further training were based on 'normalizing' the EEG. However, Colin was quite satisfied with his results following our initial training based on the single-channel assessment and he opted not to do further training. Since his job did not require new learning to any appreciable extent, the learning difficulties were not posing a problem.

Note Regarding Equipment

The NeuroNavigator will have software for training comodulation. Lexicor equipment and all modern equipment can be used for coherence training. Remember that coherence and comodulation are related – both involve the calculation of cross-correlation coefficients; but comodulation adds in a time dimension. With comodulation, the question is whether you have the same kind of activity (same morphology of the waves) going together over time. With comodulation the correlation is done after the FFT analysis. (Comodulation is also called *spectral correlation*.)

Importance of this Example

It is important for the reader to understand that this example was chosen, in part, because it was not a medical illness that the client demonstrated. A full-cap was not necessary. Perfectly adequate learning did take place before the full-cap was done. In most cases where the client presents with just ADD, the full-cap may **not give additional information that changes your intervention**. However, Colin's case was one where we did not understand the unevenness of his subtest scores on the IQ testing. He could afford the full-cap and it did give additional information. Nevertheless, he decided that he had accomplished his objectives without acting on this further information. What to recommend concerning both assessment and training is a judgment call that must be made with the client in the light of all factors, including the client's main objectives and finances. This concept will be developed further in examples that follow.

Case Example #2 – Dawn, 48-Year-Old Woman after Several Closed-Head Injuries.

Use a Full-Cap Assessment if There Is a Medical Problem

Clients who have more complex histories (for example, head injury, seizure disorder, depression, and so on) can be expected to show differences from the general population. They will often have major difficulties with attention span, concentration, and recall. You must use a full-cap assessment to adequately assess these individuals and plan a neurofeedback intervention to complement their total treatment program.

Having done the full-cap, Dawn's EEG, below, is an example of one instance when you would decide that a full-cap should be interpreted by an expert. Although this may be interpreted as within normal limits, you can see paroxysms of high-amplitude waves through much of the record. This may represent more than just a slight variation from expected normal frequency distributions.

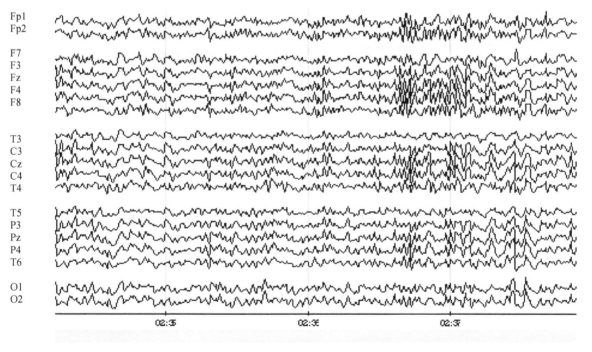

Four seconds of eyes-open baseline data (no task)

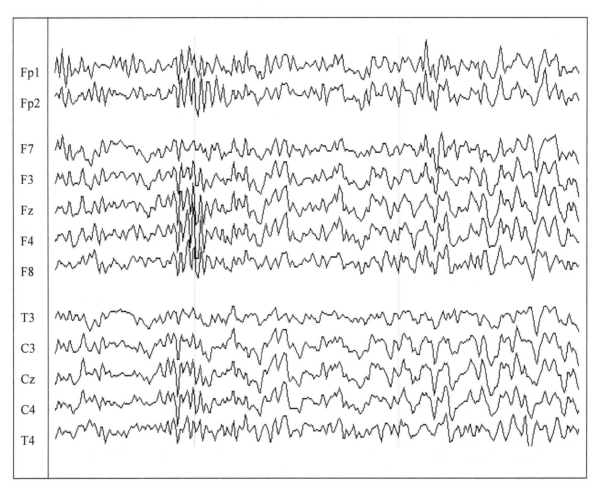

Magnification of anterior sites, two seconds, eyes-open data, showing sharp waves.

Dawn had been in three minor car accidents without hospitalization. In one of these there was a whiplash type of injury. She was previously an articulate, intelligent lady. Now she was having extreme problems with memory. She often had to close her eyes to attempt to recall a series of things in order to compose a sentence. She used to be able to rapidly read and synthesize information. She could no longer do this. She was not even able to help her teenage children with their homework, though she tried. Previously, she was an excellent source when they wrote essays. Focusing and concentrating was difficult. Standard neuropsychological testing at a large hospital center did not show her deficits. When a three-lead EEG was done at Cz referenced to the left ear, it demonstrated repeated bursts of what appeared to be sharp waves. Because of this finding, her history of injuries and her symptom picture, it was recommended that a full-cap assessment be done and sent to another professional (QMetrx) for specialized evaluation and second opinion. A sample from the full-cap is shown above, with an enlargement of the anterior leads shown in the second figure. These bursts were seen throughout the record. Her brain map showed a clear disconnect between right and left frontal lobes when comodulation statistical comparisons to the normal database were done for 4-7 Hz and for 13-18 Hz.

This analysis also demonstrates a mild (above 1SD) hypercomodulation between central (C3, Cz, C4) areas and parietal and posterior temporal areas (P3, Pz, P4 and T5, T6). This may point to a lack of discriminatory functioning; that is, more of the brain is being recruited to process information than would usually be required. The second objective for comodulation training might be to encourage her to decrease comodulation between these areas.

As noted above, this 48-year-old woman had three car accidents. At the time this EEG was done, she was having significant difficulty with memory recall. She would close her eyes and write with one hand on the palm of the other to help her recall information.

4-7 Hz comodulation is compared to the SKIL database. The result demonstrates a right to left frontal lobe disconnect. The objective for comodulation training would be to increase left to right frontal communication.

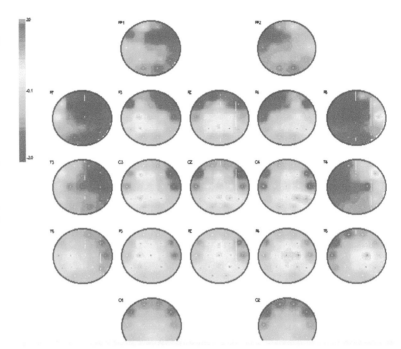

Case Example # 3 – Christian, 18 Years Old, Severe Learning Disability, Social Awkwardness, Dysphoria

This example is used only to show how a QEEG with topographic displays can reflect the symptom picture and suggest a focus for neurofeedback intervention.

Christian has been in special education throughout his school years. Reading is at a Grade 5 level. He is well motivated and parents are supportive. The first topometric display is eyes closed raw data. The posterior, occipital alpha is appropriately the dominant frequency. However, the peak alpha at 9 Hz at Fz and Cz is not the expected pattern. It corresponds to his ADD inattentive symptoms.

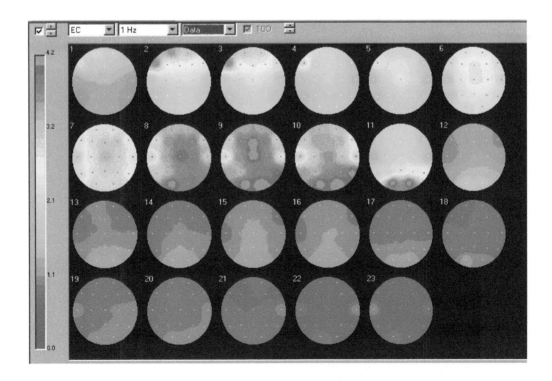

The next topometric display (shown below) indicates comodulation at his dominant frequency range of 8-10 Hz (eyes closed). The lack of comodulation T5 to the left central and frontal areas and to P4 corresponds to his difficulties with language and expressing himself. It also corresponds to his language-based learning disabilities. The hypercomodulation observed between C4 and Fz and C4 and F2 may correspond to difficulties in emotional expression.

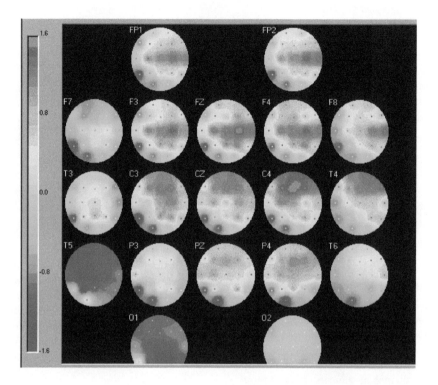

On a topometric map for comodulation data at 8-11 Hz (not shown), Christian demonstrated hyper-comodulation in the frontal areas corresponding to his dysphoric symptoms.

In addition, he displayed abnormal jerking movements and twitching. SMR demonstrated a dip in the 13-15 Hz range at C3, Cz, and C4.

The neurofeedback training program for Christian should initially emphasize increasing SMR at Cz in Channel A. At the same time, he will be encouraged to increase activity in Channel B (decreasing the dominant slow-wave activity and increasing 14-17 Hz beta) at T5 then at F3 while reading. Later, comodulation training may be attempted.

Case Example #4 – Jane, Age 8, Reading Disability

Jane was an 8-year-old girl who was not even reading at a beginning Grade 1 level. This is a specific learning disability in a girl with normal intelligence. She also exhibited all the symptoms of ADHD, some difficulties with geometric shapes and mild social inappropriateness (slightly behind age level). In the record below, taken while she was reading, the raw EEG shows some slow activity on the left side. Note the large burst of slow activity at P3 and Pz and at C3. The slow activity seen at Pz, P4 and T6 may correspond to her not always being appropriate with older professional people and acting younger than her age (this can only be a hypothesis). This slowing could also correspond to visual spatial comprehension difficulties and her difficulty understanding geometric forms. Her ophthalmologist has also noted that she has problems with geometric shapes and trying to reverse them.

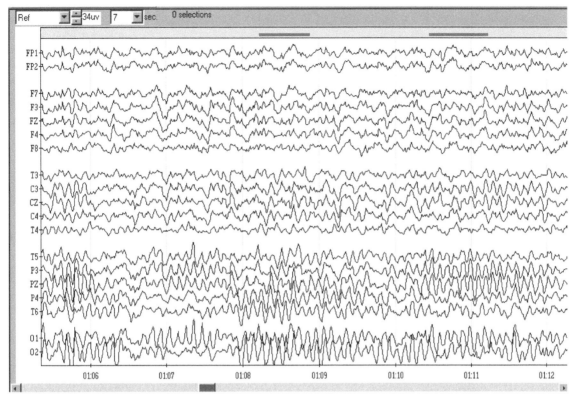

EEG, linked-ears reference, while reading.

In the brain map, with eyes closed, there is also some slowing. The amplitude of 3-5 Hz activity is compared with the SKIL database. This database does not yet include children; therefore, the only usefulness of the information is the comparison of the left and right side in the temporal-parietal region and seeing if this corresponds to what is observed on the raw EEG.

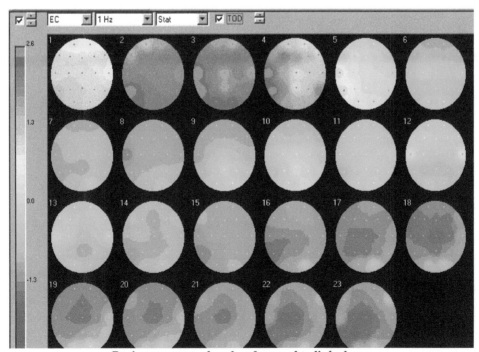

Brain map, eyes closed, referenced to linked ears.

We now use NeuroGuide, which is a good database for children and has publications on findings with learning disabilities. With the SKIL database one could only cautiously make hypotheses about areas that are affected based nonneuroanatomical knowledge, and see if these correspond to expected findings on the EEG. In this case, we would expect slowing in areas that correspond to known brain areas for reading on the left and for visual spatial manipulation on the right. The above EEG findings do correspond, and therefore lead to suggestions for neurofeedback training. Jane was asked to decrease dominant slow-wave activity in areas where it was observed. This is discussed further under interventions.

Artifacts in 19-Channel Assessment

Example of Muscle Artifact and Left-Right Asymmetry

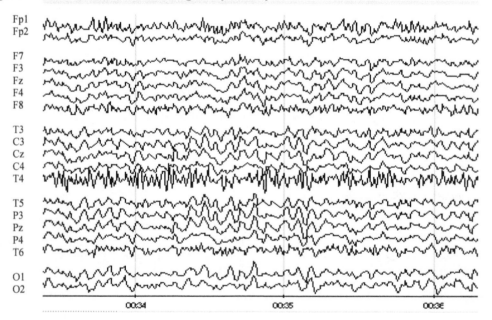

The above example of a 19-lead EEG (eyes open) assessment demonstrates several points. First, it is sometimes difficult to obtain a clean record without artifact. This client had a mild injury to the right side of his jaw. The muscle artifact at T4 and a bit at F8 and T6 is consistent throughout the record. There is also muscle artifact in the left prefrontal lead. An additional observation is that there is a consistent difference throughout the record between the left and right cerebral hemispheres that is independent of the differences due to EMG artifact. The left side demonstrates more slow-wave activity in the central and parietal regions.

19-Channel Assessment Examples
Provided by M.B. Sterman (SKIL Program) to Illustrate How a 19-Channel Assessment May Give Valuable Additional Information

There were a number of databases in use when these examples were done, such as those by Frank Duffy, Robert Thatcher, William Hudspeth and E. Roy John's group at N.Y.U. The figures for the following examples came from the training manual for the SKIL program. The comments are a précis of Sterman's remarks, plus some additional observations. SKIL provides a method of artifacting and interpreting 19-lead assessment data that has been collected with a 2 Hz high-pass filter on for the data collection. Comparisons with Sterman's database are also utilized by the program. Numbers, such as #14, refer to the figures in the SKIL manual published in 1999.

ADHD – Comparison of Data: EC, EO, Reading and Math – *only the SKIL program does this kind of comparison.*

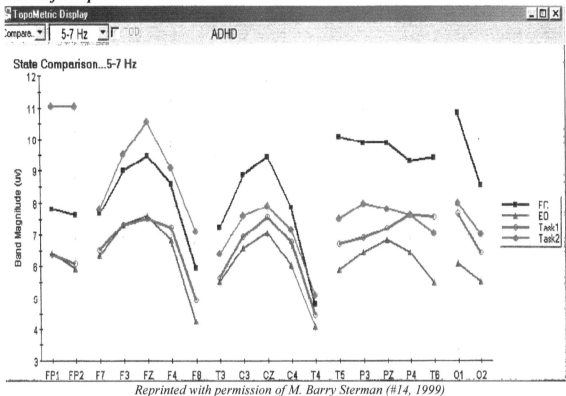

Reprinted with permission of M. Barry Sterman (#14, 1999)

The above figure is a state comparison plot from the SKIL program. The client is 16 years of age and has ADHD and affective symptoms. This pattern of high amplitude frontal 5-7 Hz activity was only seen during a math task and not on the eyes closed, eyes open and reading conditions. Black with squares is eyes closed. Blue with stars is eyes open. Pink, Task 1, with open circles, is reading. Red, Task 2, with filled in circles, is doing math. This state comparison also illustrates that the state change from eyes closed to eyes open produces the biggest change in theta activity.

ADHD and Math Task – Comparison to Database

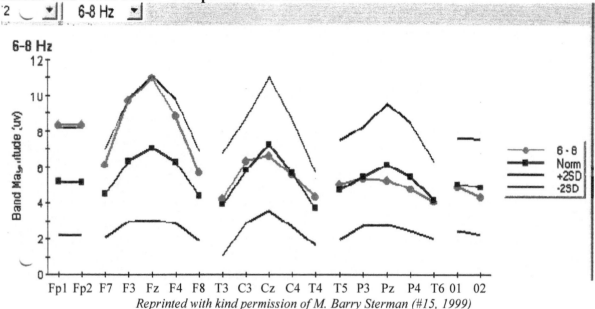

Reprinted with kind permission of M. Barry Sterman (#15, 1999)

The client is 23 years of age and has ADHD symptoms. This pattern of high prefrontal and frontal 6-8 Hz activity was recorded during a math task. The red line with filled in circles is this client's data, heavy black is the database, and the gray lines represent +/–2 SD from the norms. Note that in both of these examples, Fz is a better site for seeing the excess theta than Cz. This is in line with Lubar's suggestion that Cz should be used for children with ADHD, and the leads should be moved more frontally with older adolescents and adults.

Head Injury

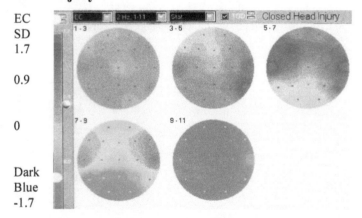

Brain maps (comparative data) after head injury. Reprinted with kind permission of M. Barry Sterman (#19, 1999)

Topographic map (SKIL program) comparison with normal database: This 28-year-old female client had suffered a blow to the left frontal-temporal region in a car accident. Note the high amplitude 5-7 Hz activity at T3 and F3. Note also the 'contra-coup' focus of high amplitude 5-7 Hz activity at T4 and T6.

Since the accident, her IQ dropped (126 TO 96) and she had symptoms that included episodes of confusion, word aphasia, memory deficits, and short attention span. NFB training would suppress 5-7 Hz at F3, T3, T4 and T6.

Head Injury With and Without Time of Day Corrections

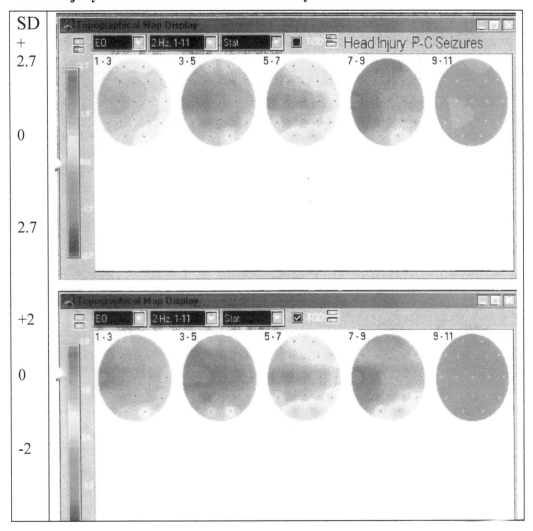

The above figure shows brain maps after head injury with and without time of day correction, and is reprinted with permission of M. Barry Sterman (#20, 1999). It shows a topographic map after head injury in a 42-year-old male. This figure demonstrates the importance of a time-of-day correction, which is only done in the SKIL program. The practitioner using other programs can and **should take time of day into consideration** when booking a patient for a follow-up appointment. Both figures demonstrate a significant increase in amplitude of 7-9 Hz activity at T3. The bottom figure is time-of-day corrected, however, and shows a clear 3-5 Hz focus at T3 and C3 and a 1-3 Hz band focus at T3 that are not seen in the top figure. Training would decrease these high amplitude bands. The reference electrodes would be placed on the contralateral (right) ear to decrease suppression of signal strength due to common mode rejection.

ADHD (Child)
Owl or Monkey pattern in the brain map of a child with ADHD.

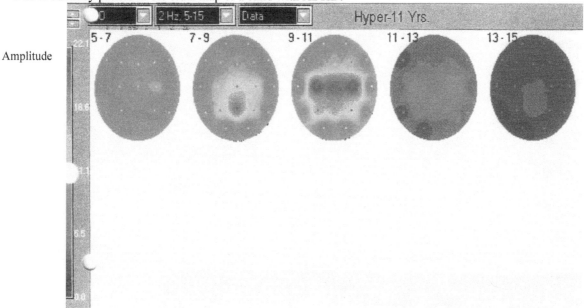

Reprinted with permission of M. Barry Sterman (#21, 1999)

Topographic map, 11-year-old male diagnosed with ADHD: This figure demonstrates Sterman's *owl* or *monkey* pattern. Sterman suggests that this is an EEG subtype of ADHD. Note the high 9-11 Hz activity at C3 and C4. The child is too young to compare to the SKIL normal database. On the raw EEG, the waveform was not the wicket pattern seen with a Mu rhythm. This must be distinguished from alpha, as previously explained. NFB would suppress 7-9 or 7-10 Hz activity and increase SMR 13-15 Hz. You should avoid suppressing 11 Hz activity in this case, due to its proximity to the SMR activity that you would be attempting to increase.

Seizure Disorder

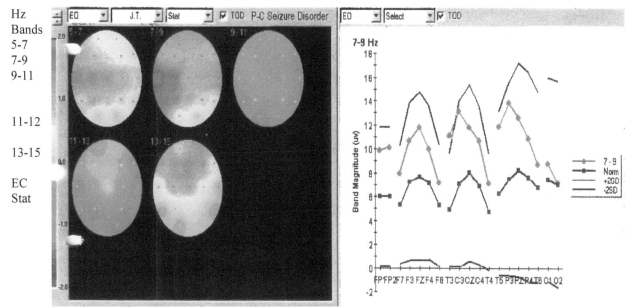

Brain map and topographic display (comparative data) in a client with seizures. Reprinted with permission of M. Barry Sterman (#24, 1999)

Convergent analysis, using the SKIL program, in a client who has partial-complex seizures. The 7-9 Hz activity is raised at the anterior temporal seizure focus. The topographic maps compare amplitudes of different frequency bands to the normal database. The topometric plot on the right gives details concerning the spectral distribution of the 7-9 Hz band.

Comodulation in Depression – Comparison with Database
Hypercomodulation in the frontal regions in depression (comparative data)

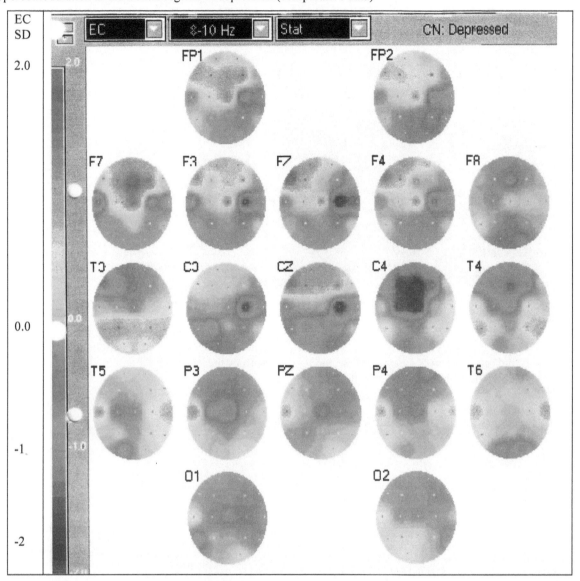

Reprinted with permission of M. Barry Sterman (#27, 1999)

This is a statistical comparison, using the normal database on the SKIL program, in the eyes closed condition, of the comodulation of 8-10 Hz activity in a depressed female in her early forties. Note the hyper-comodulation between F7 and Fz at 2 SD from the mean for the normal database. Sterman suggests that this hypercomodulation may represent a lack of differentiation between these sites and in thalamo-cortical regulation. He notes that this hyper-comodulation is more likely to be due to a change in thalamocortical than intracortical connections, because severing intracortical connections has been shown not to abolish covariation between cortical sites (Contreras, 1997; Sterman, 1998). Note also the hypo-comodulation between C4 and Cz and Fz.

Frontal – Central Disconnect Pattern

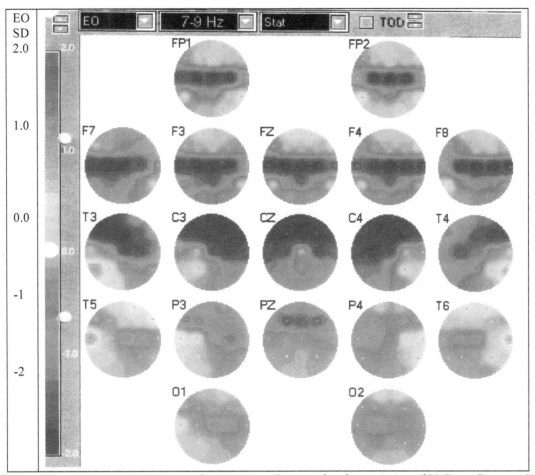

Frontal central disconnect pattern: statistical comparison. Reprinted with permission of M. Barry Sterman (#30, 1999).

Comodulation comparison with normal database for a 23-year-old male: Note the hypo-comodulation in the 7-9 Hz band between frontal and central cortical sites. Sterman has termed this type of pattern a *frontal-central disconnect* or *dissociation*. In this example, Sterman postulates that a lack of shared activity between these areas might result in impairments in reciprocal interactions and the individual's ability to regulate sensory input and motor output. In this particular patient, Sterman observed facial tics, stuttering, insomnia and attentional problems.

A Note on Pre- and Post-EEG Comparisons

It is important for clients to be shown clear pre- and post-NFB treatment comparisons of the frequency ranges that they have been working on. This can be easily done in single-channel work (Thompson & Thompson, 1998). It is also possible to plot this type of comparison data for 19-channel assessments. It is always gratifying for a client to see that their behavioral changes correspond to electrical changes.

A caution, however, is needed. Training, for client who has a specific disorder such as ADHD involves 'normalizing' the EEG. In this type of work you would expect to see an overall EEG change. For ADHD, this often includes a decrease in theta and/or low alpha and an increase in SMR. For an executive or professional athlete, training does not usually involve normalization of the EEG; rather it emphasizes self-regulation where the client learns to control mental states. Thus, with these clients, post-testing should demonstrate their ability to consciously enter a specific state. Measurements would, therefore, have to be made when they were in, and then out of, the desired state. The EEG should demonstrate a shift. For example, a professional golfer (see later example) may want to block out all inner thoughts, concerns, and worries. When she does this, the amplitude of a 27-31 Hz band decreases. Post-testing should demonstrate her ability to do this. You are training for a more flexible brain.

Single-Channel or 19-Channel Assessments – When Is Each Appropriate?

Here are three important considerations to open the discussion on EEG assessment. First, who are you in terms of professional qualifications and experience in reading the EEG? Are you an educator, coach, trainer, therapist, or neurologist? Second, what is the purpose of your EEG? Third, what do you plan to do with the results of the EEG assessment?

The Advantages of Doing a Full-cap Assessment

The 'gold standard' is the 19-channel EEG assessment. This is also frequently referred to as a full-cap assessment. It allows you to view what is happening at each frequency at all 19 of the standard sites in the 10-20 electrode system. As noted above, when done before and after neurofeedback training, you can see what has changed in terms of brain wave activity not only at the site(s) where training was done, but at all sites. Certainly you would use this for any client who had a history of a head injury or a seizure disorder since it would indicate the best site(s) for training. You must use it if you are going to do coherence or comodulation feedback. You would also want to use it for any research study in which the EEG brain map would be helpful for differentiating between different diagnostic categories; for example, differentiating between different EEG subtypes of ADD.

The Drawbacks of a Full-Cap Assessment

This ideal kind of assessment has some drawbacks. It is expensive, takes a great deal more background and training to analyze, and, with many commonly used programs, it does not look at the spectrum up to 62 Hz. It is important for our work with NFB to be able to assess and monitor the frequencies not only up to 23 Hz but also above the mid-20s up to 62 Hz. In some cases, particularly if anxiety and ruminations are a concern, you want to look at the high-beta frequencies. At the time of writing, this kind of assessment was not generally available using the 19-channel instruments commonly used by neurofeedback practitioners.

It is expensive because of the initial cost of acquiring an instrument capable of 19-lead assessments. The 'caps' do break down and must be fixed or replaced. Also be aware that it takes a little more time to set up and collect information. In addition, it takes a great deal more time and expertise to read and interpret the data once it is collected. Families and individuals usually only have a certain amount of money that they are willing or able to spend. You must decide whether it is absolutely necessary to advise the family that this must be done when it may mean that they would only be able to afford an inadequate number of sessions of training. In many cases this may be greatly interesting academically, but not influence what and where you train. Your interest in having data must not override common sense considerations with respect to helping the family.

If you are likely to be able to obtain clear information as to the probability of a condition and good-enough training parameters to help that condition from a simple single-channel, three-lead assessment, then you should probably follow the Principle of Parsimony: do the least intrusive and least expensive intervention that has a probability of success. In the case of neurofeedback interventions, this would become: do the simplest and least invasive and disruptive intervention first. Therefore, if a three-lead assessment can give sufficient information to set parameters and carry out a successful training program, you may be justified in doing that rather than the full 19 channels. Always ask yourself, "How can I provide the best service?" and, "What is realistic for this family in this situation?"

A practical consideration is rapport with your client. On the one hand, it is impressive if you do 19 channels. On the other hand, a single-channel assessment can be a quick and painless hookup, whereas a full-cap assessment takes more time and can be uncomfortable for some clients.

Another practical consideration is experience and training. Most biofeedback practitioners have one- or two-channel instruments. Most of these individuals can do an adequate job of reading the EEG produced with their equipment. These people are not electroencephalographers and do not have the training or experience to read a full-cap assessment. They should not pretend that they do. If they decide to do a full-cap, they must either outsource the job to a trained practitioner or undertake extensive training so that they can interpret the full-cap assessment themselves. Interpreting the data from a 19-lead assessment involves both art and science. Different databases and different montages may give different results, and, as shown at early SNR annual meetings when a panel was given the same EEG data on a selected case to interpret, even the experts may disagree on some points of interpretation. It is the gold standard for neurofeedback assessments, but it is not entirely objective. We need more validity and reliability data concerning this procedure. Of course, neurologists reading EEGs are also not entirely objective. Sterman quotes a statistic that the correlation coefficient for different neurologists reading the same record is only .56, and the correlation for the same neurologist doing it at two different points in time is about the same. Here is an excellent thesis topic for an enterprising student: Investigate the correlation in findings from 19-lead assessments done for the purpose of designing neurofeedback interventions both across different practitioners and with same the practitioner over time.

In summary, full-cap assessment is excellent and necessary for certain purposes; however, most software programs for doing the full-cap, at the time of writing, are not looking at the spectrum of activity to 62 Hz. For this reason **we always also do a single channel assessment up to 62 Hz** using the Infiniti at sites that were found to be outside the database norms using the full-cap.

Secondly, people getting started in neurofeedback will not yet be doing full-cap assessments on a regular basis and are not yet comfortable with reading the output. Thirdly, a full-cap assessment is much more expensive, particularly if the information is sent away to be read elsewhere. It may use up funds that clients might otherwise have been able to use for training. Finally, the cost may reduce the number of clients who are able or willing to pay for post-training EEG assessment.

General Discussion: When Do We Recommend a Full-Cap Assessment?

In this section we are stating a personal bias. There are centers that do a full-cap assessment on every client both pre- and post-training, and they might say it is the only way to know what changes you have made in the brain's functioning. Our view is that it is hard to justify the extra cost of $500-$1,500 in every case, particularly if training can proceed with the information from a single-channel assessment or from repeated single-channel data collection. Collect data at Cz, move the electrode to Fz, collect data again, and repeat for as many locations as are appropriate to provide the data needed in order to proceed with a particular client.

We recommend 19-lead full-cap assessment for clients who have more complex symptom pictures. This includes all clients who have a history of head injury or a seizure disorder. It includes clients who have a complex array of symptoms such as may be found in Asperger's syndrome or in clients who have a combination of problems, such as depression and anger management. At the ADD Centre/Biofeedback Institute of Toronto, this has been done using Lexicor, NeuroNavigator, NeuroPulse, Mitsar, and now eVox instruments. To analyze the data we originally used the SKIL and WinEEG programs, but now are using mainly NeuroGuide and the eVox programs.

Data can be sent by CD, memory stick, or via the internet to companies or senior professionals that provide interpretations. Some of these companies have a neurologist who specializes in EEG write a report on the data. This is in addition to a report or summary written by an EEG specialist (usually not a neurologist) who is capable of advising on the best neurofeedback training program to carry out. Experienced individuals can both read the EEG and give valuable advice on appropriate neurofeedback parameters for training that are specific to the client's EEG findings. Unless one is doing large numbers of QEEG assessments and has invested in purchasing the normative databases, it makes economic sense to outsource the interpretation.

However, it is always best to do the data collection and interpretation oneself. Data interpretations by persons who did not sit with the patient during the recording, and do not know the patient, **can be grossly misleading!** We have seen reports by very well respected practitioners where findings well outside database norms, for example, were interpreted as abnormal and recommendations were made to normalize the EEG at those sites. However the data really demonstrated genius level functioning of that area of the brain. Clearly this can lead to very poor outcomes. There is no substitute for learning neuroanatomy and the functions of different Brodmann areas and doing your own interpretations. Then sending the EEG to another practitioner is just a useful second opinion. You won't be making gross errors.

Do use the services of those who have looked at thousands of EEGs, who have purchased or created normative databases for comparison purposes, who know the intricacies of remontaging the data to yield the most helpful information, and who have knowledge of neurofeedback **for a second opinion**

It is a wonderful learning experience to get intimate with the EEG data yourself and come up with hypotheses as to what it all means, but it takes time to become an expert and a second opinion is very helpful to your learning process. Our view, as noted above, is that there is a great deal of art as well as science in quantitative EEG analysis. As in much of our field, the adage "a little knowledge is a dangerous thing" should be kept in mind by those starting to do 19-channel QEEG work. People who have just taken a few weekend workshops are not ready to do a 19-lead QEEG interpretation solo any more than an aspiring pilot is ready to take over the controls without his flight instructor beside him.

There are exceptions to every rule. People with extensive training in neurophysiology, anatomy and pathology who have had previous clinical training and experience in this area may learn more rapidly than others how to run this kind of equipment and interpret the results.

If there is any concern about a primary medical condition, such as a seizure disorder, head injury, and so on, your client MUST be seen and assessed by the appropriate medical practitioners before undertaking EEG biofeedback. The QEEG that you are doing is being done to look at data concerning *normal* brain waves. **You make no claim regarding assessing, training or even recognition of abnormal waveforms or patterns.** This is the task of a specially trained neurologist in conjunction with an EEG technician.

Effect on NFB Results

At the time of writing, we are not aware of any studies that investigate the question of whether there would be a difference in outcome if a full-cap had been done as compared to a single-channel, three-lead assessment. Many of our students want to improve focus and attention span and, at times, decrease tension. There is good evidence in the literature that a single-channel assessment at Cz works very well for recognizing patterns seen in ADD (Monastra; Linden; Lubar; Thompson & Thompson). We do obtain a great deal of useful information concerning the higher range of beta (above 28 Hz) activity using a single- or two-channel instrument. For this reason, as previously emphasized, we always have to do this whether or not a full-cap was also done. The multisite study headed by Monastra and Lubar showed good reliability and validity data using the single-channel assessment at Cz. Nevertheless, as pointed out above, there are cases with additional problems where a full-cap may be very helpful and the cost is justified.

Overview of Interpretation of Findings Using Different Types of Assessment

Single-Channel Findings

With a *referential* montage with the active electrode placed at Cz you can be reasonably certain that the electrical activity recorded reflects the cortical activity under the active electrode. It is unlikely that you will have a false positive result. On the other hand, you may have a false negative. We have already discussed a good example regarding ADD. A Cz assessment may not show high slow-wave relative to fast-wave activity when the client is tuning out, but a second assessment using a frontal placement may pick up the problem and show the ADD symptom picture.

With a *bipolar* montage you will not know which of the two electrodes has the higher amplitude electrical activity under it. Perhaps there is no difference between the two waves at the two sites in terms of type of activity, but a potential difference (voltage) is being recorded because they are *out of phase*.

Two-Channel Findings

Two-channel results may clarify differences between two different areas of the brain. However, special caution must be exerted to assure that impedances were the same for all electrode placements before findings at different sites are compared. Again, many areas of the brain are ignored, and findings are therefore still quite limited. Another caution is to be sure that the influence of muscle tension artifact is approximately the same for both sites. This can be very subtle and can make beta activity for the site where there is more EMG influence appear (falsely) high compared to the other site. It is easier and better to compare two sites using two channels with a common reference or a 19-channel assessment where the whole picture can be seen.

Full-Cap Findings

These findings are obviously more complete. Different sites may be compared and communication between areas of the brain elucidated. However, as previously noted, it takes many years of experience to accurately interpret the findings. Each way of referencing the electrodes will yield different information. In the standard, linked-ears reference, the electrode sites close to the ears will show low amplitudes compared to more central sites due to amplifier common mode rejection. The topometric graphs from Sterman's SKIL program allow the clinician to visualize this. False impressions of frontal and/or prefrontal high slow-wave activity (due to high temporal slow-wave amplitudes that have contaminated the ear references) may be quickly clarified by switching to a sequential montage. This is just a keystroke during an analysis of the data, and does not require a different type of electrode placement as it would when only using a one- or two-channel instrument. Unusual sites of high- or low-amplitude activity at different frequencies can be distinguished. Hyper- or hypocommunication between areas of the brain can be evaluated and compared to normal databases. All of these findings will provide additional suggestions for NFB training.

Advantages of Also Doing Single-Channel Assessments to 62 Hz

For some clients, doing an additional assessment on an instrument that shows the spectrum to 62 Hz may have some advantages. For example, we find that negative ruminating may be reflected in high-amplitude activity between 23 and 27 Hz or 28 and 36 Hz. This high frequency, high amplitude beta is sometimes called a type of ADD. Certainly it is a form of tuning out, and these individuals do have difficulty sustaining attention to other things. However, we feel it is more often a clear reflection of internal ruminations, which are often associated with depressive affect. (Of course, ADD plus depression and or anxiety is a common comorbidity in adults.) Constructive cognitive activity and attention, on the other hand, may be reflected in activity in the 38 to 42 Hz range, the so-called Sheer Rhythm. The influence of muscle can usually be distinguished from these high frequency beta (also called gamma) bursts by examining the effect of EMG on the activity above 43 Hz. If activity above 43 Hz is consistently low compared to frequencies below 42 Hz, then it is reasonably likely that the activity you are observing below 42 Hz is not a reflection of muscle tension.

Ambient electrical activity may raise 50 Hz activity in Europe and Australia, or 60 Hz in North America. Observing a higher amplitude of these frequencies compared to 48 and 52 Hz in Europe, Asia and Australia, or 58 and 62 Hz in North America, may be a reflection of a difference in impedances between the electrode sites. You should attempt to correct this by re-prepping the electrode sites and checking impedances again. All of these observations are useful not only during an assessment, but also during the actual NFB sessions. Not only will clients change mental activity and muscle tension during their sessions, but they may move and loosen an electrode, thus changing its impedance.

Quality Data Collection

The following tips apply to data acquisition whether it is one channel or 19 channels. They also apply whether one is doing an assessment or running a training session. One is naturally going to be very fussy about things like low and equal impedance readings when collecting data for a 19-channel assessment because you are basing important decisions on the interpretation of this data. Yet training outcomes are also going to be influenced by the quality of the feedback, which depends first and foremost on the quality of the EEG information. The saying among computer users of "garbage in – garbage out" is also very much true when dealing with EEG data. If you are going to be spending the next hour with a client, giving them support while they try to make EEG changes and exercise their brain, you want to be using the best feedback information you can obtain. So make it a habit to be careful about the following things all the time.

Site Preparation

Having decided on which sites you are going to use for assessment, you now must prepare the site for accurate recording. The first step is to clean the site and test for *impedance*. Impedance is the resistance to the flow of an alternating current (AC). However, to understand impedance, we should first review two terms: *voltage* and *differential amplifier*.

Measurements – Voltage and Amplification

When measuring the EEG, we are measuring tiny amounts of electricity. Remember that a microvolt is a millionth of a volt. You are always measuring the potential differences between two sites for the different EEG frequencies. EEG is an alternating current, and most instruments measure *peak-to-peak* voltage differences. Occasionally you will use equipment or read a report where *root-mean-square* (RMS) is the measurement used. The values are smaller because the relationship is that RMS = 0.707 x peak-to-peak values. Occasionally you might see the value expressed as *Peak* value. Peak-to-peak = 2x peak value.

Differential Amplifier

(Also covered in the earlier section on 'electrical artifact')

The original concept for this type of amplifier came from the work of Thomas Edison. The actual amplifier, however, was not developed until the 1930s. In simplest terms, think of the electrode at one site going to the preamp. There is a potential difference between the site on the scalp and a 'comparison' within the amplifier that involves your third electrode, which is called a *ground*. Many years ago, the comparison actually was with ground. Now there is not a direct connection of your client to ground. The measurement and calculations are done differently within the amplifier. The wire from the second site enters at another point in the preamplifier. As previously explained (in Section IV), think of the first site being positive and the second negative. Often a positive (+ve) sign is seen on the electrode that you use as your *active* electrode and a negative (–ve) sign is on the other *(reference)* electrode. Now these two potential differences, the active site to the amplifier (+ve) and the reference site to the amplifier (–ve) are compared. Any induced current from another source, such as a nearby lamp (60 Hz), should be the same and *in phase* on both inputs. The +ve wire from the active electrode will, in this case, be the mirror image of the –ve wire from the reference site with respect to 60 Hz current; the two will cancel out and therefore NOT be amplified. The EEG voltages, on the other hand, will be different as recorded from each of these wires, and therefore will not cancel out and will be amplified. Thus, the difference between the two EEG voltages that have simultaneous input to the amplifier will be amplified. This is the unique function of a *differential amplifier*.

The importance of having almost the same impedance at all electrode sites can now be better understood. If the impedances were very different, then the induced voltage from a common electrical source would not appear the same when the *differential* amplifier compared the active and reference inputs. Therefore, it would not cancel out and might be amplified. This would result in a large artifact in the recording. The *common mode rejection* feature only works well if you have low and equal impedance readings.

Preparing the Electrode Site

The first step before preparing the electrode site is to prepare your clients by telling them exactly what you are going to do. You thus get permission to attach electrodes to their head and earlobes. Reassure them that the paste used will be easily removed with alcohol at the end of the session, and that they will leave with their hair looking as it did when they arrived. After you have measured the head to accurately find the site that you want to use (such as Cz) according to the 10-20 electrode system, mark it with a felt tip pen. (A practical suggestion: Do not mark it with a red pen as that may look like blood when you rub it off.) Then part the hair so that you have a flat base and scalp showing. To obtain a good connection between the scalp and the electrode (good connection = a low impedance), you may begin by cleaning each site with an alcohol swab to remove sweat and skin oils. It is then essential to rub the site vigorously with a nonallergenic prep such as Omniprep or Nuprep to remove dead skin cells and anything else that might act as an insulator. (Hairspray, for example, is a good insulator). Nuprep is slightly less abrasive and is now used at our Centres. It is imperative, however, that you do **not** break the surface of the skin. Now carefully put some 10-20 EEG paste on the site and press it into the skin. Cotton swabs are handy for applying both Nuprep and 10-20 paste. Put more EEG conductive paste in the electrode cup, leaving a small mound where it will touch the skin. Place the electrodes on to the prepared sites. Do **not** allow any bare electrode metal to touch the skin, as metal to skin would have different conductance than metal-electrode paste-skin. Jan Hoover, whose company (J&J) is a manufacturer of equipment and electrodes, says, "A good daub of conductive paste is the best way to ensure good electrical conductivity." This is, of course, why many electrodes are designed with little cups that hold gel. Flat gold electrodes, available from IMA in Florida, are easier to clean, but you have to be more careful about ensuring a cushion of paste between electrode and skin.

At this juncture, you must check the impedance between each pair of electrodes. For a one-channel referential assessment, this means three checks: active to reference, active to ground, and reference to ground.

Impedance – Definition and Measurement

As explained briefly under Section IV, Measuring the EEG, electrode *impedance* may be defined as the resistance to the flow of an alternating electrical current. You must distinguish this from the electrical term *resistance*. Resistance is the inability of a part of an electrical circuit to allow the passage of a *direct* (constant voltage) current (Fisch, p 44). You should always check impedance for each electrode site using a specially constructed impedance meter that passes a weak alternating current from the selected electrode through the scalp to all other electrodes connected to the meter. The mild current should alternate at 10 Hz to approximate a common EEG frequency. Ohm's law for a direct current is $V=IR$ (voltage = current x resistance). However, for an alternating current this becomes $v=iz$ where z is the impedance of the circuit. This is what we are dealing with because electrical activity in the brain is AC, not DC. Both resistance (R) and impedance (z) are measured in ohms. In a

DC circuit, when a potential difference occurs between two points, then instantly a current flows and will flow as long as that potential difference remains. In an *AC circuit,* the current will instantly flow but will not remain flowing.

Mathematically:

$$\text{Impedance}(z) = \sqrt{R^2 + (10^6/2\pi f C)^2}$$

where π is 3.14, f is the frequency of the alternating current in Hz, and C is the capacitance. You are not expected to be an expert in electronics. However, there are a couple of points to be learned by looking at this formula. First, C(capacitance), which is measured in microfarads (μF), refers to the storage of electrons. A capacitor consists of two conductors separated by a resistance. It introduces a time factor because current will rise instantly, then the capacitor stores electrons with the result that the current will gradually decrease over time. Thus DC current is stopped and only AC current can pass. Cell walls act as capacitors. So do the electric wires that run from the client to the preamp.

If C and f were held constant (as they would be in a DC circuit), then the z would vary directly with R. However, this is not the case in an AC circuit. The formula shows that z will go up as C goes down. Also, z varies inversely with frequency. Thus, as frequency increases, the measured impedance will decrease. For this reason, a standard measurement was introduced so that we are all talking in equivalent terms. **The standard is to use a 10 Hz frequency (AC) when measuring impedances for our electrode sites.** *This frequency was chosen as it is the most common dominant frequency found when adult EEG activity is measured with eyes closed.*

Notes Regarding Measuring Impedance

You must have an impedance meter unless you have equipment that either checks impedance (like the EEG-Z preamp available for the ProComp+ and Infiniti) or has such high input impedance that you can just check your raw EEG signal and 60 Hz in the spectral display to ensure you have good connections (F1000 equipment). But a caution: this is not good enough if your client moves. Good clinical practice requires that you always check impedances. An *impedance meter* is not an ordinary meter for measuring resistance purchased at an electrical outlet store. As noted above, it is an instrument specifically constructed for EEG work. The specially constructed impedance meter passes a weak alternating current (that mimics an EEG frequency) from the selected electrode through the scalp to all other electrodes connected to the meter. The current is at a 10 Hz frequency to approximate a common EEG frequency. It is essential that you have good impedances for all channels. Otherwise, the EEG will be recording differences that may be, in part, due to differences in your connections. This is particularly important when you do a two-channel assessment where impedances must be virtually exactly the same in order to compare frequency bands in two hemispheres. As noted above, theoretically, with high resistance amplifiers this should not be a great problem. However, try experimenting and you may find that it does make a difference. (It can make a large difference if the client should move their head even a small amount). Impedances should be between 100 and 5000 Ω. Electrodes with much higher impedance may produce 60 Hz artifact and attenuate the recording. Impedance less than 100 Ω is an indication of an abnormal *shunt,* or a short circuit, between electrodes (Fisch, 1999). This could be caused, for example, by electrode paste connecting and creating a bridge between two adjacent electrodes.

Electrode polarization may occur if a direct current is used to measure impedance. This would then encourage current flow in one direction and resist flow in the other direction. This would distort recordings and is another reason for using an alternating current for checking impedances. Direct current is appropriate for measuring resistance but not impedance, as explained above. So you cannot use an inexpensive meter from your local hardware store. You must purchase an impedance meter or use equipment that has a built-in impedance check. (Note: Electrode connections that use saline rather than conductive paste are prone to poor impedance.)

The most common indicator of unequal impedances is the appearance on the output recording of 60 Hz electrical artifact from alternating current sources in the recording area (or 50 Hz in countries where that is the standard electrical current). The amplitude of this artifact will be proportional to the inequality in impedances and the strength of the signal. In full-cap assessments the ground is usually placed at Fpz. (The result when an electrode detaches from the scalp is that the other input is compared to the ground, giving a clear reading of eye movements that are easily recognized.)

You are doing an excellent job if impedance is <5 Kohms and the pairs are within 1 Kohm of each other. As previously noted, impedance should be between 100 and 5000 Ω. Electrodes with much higher impedance may cause 60 Hz artifact and attenuate the recording (Fisch, 1999).

Note: For most of our neurofeedback work, the most important factor is that the impedance readings for each pair of electrodes all be very close to one another. If all three impedance readings were between eight and nine, this would be more acceptable than two being at eight and one at two. If you have equal impedance readings, movement artifact and externally induced electrical artifact will be minimized. Induced artifact will be discussed further under artifacts and filters.

What If the Impedance Readings are Higher than 5 Kohms?

The answer depends to a great extent on which instrument you are using. Instruments differ in characteristics such as input impedance of the amplifier (which is an independent issue from the question of the impedance readings you get from your electrodes.) In our experience, all commonly used instruments require a strict adherence to the above recommendations for impedance to ensure good EEG readings. We redo the preparation of the electrode site with alcohol cleansing, gentle abrading with Nuprep, and then spreading 10-20 conductive EEG paste until we get those impedances down. We may let the client rub their own ear lobes as they know what the goal is and often tend to be more vigorous in the rubbing than the trainer would be. Children love to hold the impedance meter, and are quite willing to tolerate a little squeeze on the ear electrode if they can watch the result of the reading getting down to the magic <5 Kohm range. With good impedances one will see a good consistent EEG on these instruments.

Instruments such as the F1000 (Focused Technology) or the ProComp+ or Infiniti (Thought Technology), where the input impedance of the amplifier is high and work has gone into protecting the wires from the electrode site to the preamp, may be more forgiving of impedance readings that do not meet research criteria. (Research criteria, as you know by now, are that all pairs should be below 5000 ohms and within 1000 ohms of each other.)

You should also always check the EEG spectral display before proceeding with feedback to ensure that it looks like good EEG and that 60 Hz. is not too high. (It also indicates clearly if you forgot to turn the instrument on, which will happen occasionally!) We have experimented with the Thought Technology instrument and could see very little difference in the EEG tracings with higher impedances *if they were all close to one another* (for example, all three readings around 14 Kohms). There were two exceptions to this. The first was found when we moved the electrode a little bit by pulling on the skin or wiggling the ear such as might occur during feedback sessions. When there were good impedances, it took a great deal more movement of the skin to affect the EEG readings. The second exception was when the client turned his head. With poor impedance and a difference in impedances between the electrode pairs, even a small movement of the head (30 degrees) caused a large artifact in the EEG. With reasonable but not great impedance readings (<18 Kohms) but almost no difference between pairs of electrodes, there was little discernible difference in the EEG. These observations need to be replicated. Poor impedances do lead to a lower amplitude EEG. This will affect comparisons between sessions and pre-post neurofeedback training assessments.

Impedance: Client's Perspective

Consistent impedance readings from session to session mean that data comparisons are more meaningful; that is, more likely to represent changes the client is making in their EEG rather than differences in electrode contacts with the scalp that affect the measurement of the EEG. We have observed that the total amplitude of the EEG is lower when there are high impedance readings. This makes a big difference to clients when they are trying to equal, or improve on, yesterday's percent-over-threshold scores or point score when doing feedback training.

This might also be of interest from the standpoint that lowering total amplitude is being put forth, in some quarters, as being useful for optimizing performance. This is sometimes referred to as a frontal 'squash' protocol. Certainly, an overall decrease in amplitude of the EEG would have one assume that theta and alpha were lower and that there were no major artifacts in either slow frequencies (due to eye movement) or faster frequencies (due to muscle activity). It could imply that greater desynchronized activity was taking place. The normal adult, eyes-open EEG displays low voltage and is fairly flat. There are also some good anecdotal reports of its usefulness.

However, for optimal performance we emphasize a rise in certain bandwidths such as 11-13 Hz. We also emphasize mental flexibility, that is, being able to change quickly from one state to another as appropriate for task conditions. Thus lowering of the total EEG amplitude would not usually be the sole objective.

Last, but by no means least, equal impedances between each of the electrode pairs may help to minimize cardiac artifact during NFB training. The heart produces about ten times the electrical output of the brain, and EKG artifact due to volume conduction is sometimes a problem. (An example is given in the section on artifacts.)

If you are going to sit with a client for an hour, you want them to have valid, reliable feedback. Taking a few minutes to get good impedance readings is the first step in providing quality neurofeedback.

Offset
The handheld impedance meter also allows you to check the offset of the wires. High offset (above 50, and certainly above 100) may be your first warning that a wire is breaking or that your gold electrode cup is deteriorating. Deterioration will occur rapidly if you do not immediately clean your electrodes after each session. It will cause a little 'battery-like' effect at the site. Another indicator of a faulty wire may be the occasional flattening of the EEG.

Site Preparation – A Caution
You must be careful that you do not allow the Nuprep or the 10-20 EEG paste (or other electrode gel) to cover the space between any two electrodes. If it does, this bridge will cause a short circuit. If you aren't careful, it is possible that you might miss this when reading the EEG quickly before you record data. This could mean that you would have to redo your recording at another time. An example of this problem is given below.

A salt bridge is not the only possible cause when you see two sites with identical recordings. On one occasion we sent back a new EEG instrument because it had an internal fault that linked the two electrodes.

With Full-Cap Be Careful That Gel Doesn't "Bridge" Between Electrodes

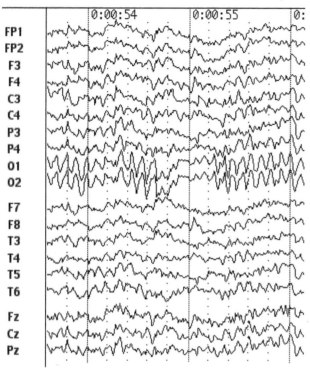

Referential montage (linked ears)
Note how O1 and O2 are exactly alike. This may be due to what is termed a "bridge." In the above example it is a **salt bridge** due to excessive electrode gel that bridged between the two electrodes.

Gunkelman notes that a cap requires 3.5 to 4.0 cc's of gel but many people use over 5. Movement of the cap (or even having it pressed against a recliner without a neck support to elevate the cap) can smear gel too.

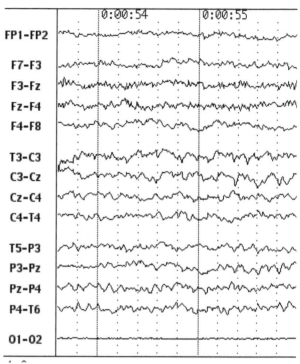

Sequential Montage (old term "bipolar")
Common mode rejection makes it appear as if there is no signal at O1-O2 because the signal was identical at the two sites. Gunkelman points out that this is a good example of how a different **montage** would instantly clarify the problem. Note the flat line O1-O2. The 'salt' bridge is caused by the salts in the gel (stannous chloride). This causes a short.

Thanks extended to Jay Gunkelman for this example.

Collecting Data

Introduction
It is helpful to collect data using a screen made for the specific purpose of doing an assessment. This screen should have a means for artifacting the data once it is collected. Shading segments, for later inclusion or exclusion when calculating statistics, is the method used both in the BioGraph program and the SKIL program. In the older A620 or the Lexicor programs, screens that show two seconds of data are either accepted or rejected. In NeuroGuide you shade the segments that you accept. It is also helpful if this assessment display screen has some secondary means for clarifying the frequency spectrum and amplitudes of the waves for each frequency from 2 to above 60 Hz, and for showing when EMG activity is particularly high. An example of this type of screen is shown under "Assessment, Artifacting and Teaching Screen," shown below, with permission of Thought Technology.

Become Familiar with Your Client's Typical Artifacts
Your first task is to look at your client's raw EEG to differentiate different types of artifact before you begin data collection. Artifacts may have different appearances from one student/client to the next. Eye blinks, for example, though they all have a similar triangular form, will look different and affect slightly different frequency ranges depending on things like the individual's eye shape and how quickly they blink. You must, therefore, have the student/client do things that produce artifacts while you observe and make a mental note of what that person's artifacts look like. In this way you will recognize them if they

occur later in the EEG sample you use for your assessment statistics.

Ask your client to blink, roll their eyes, look left then right, up then down. Have your client clench their jaws vigorously then very gently. Ask them to wiggle their ears (people who can do this like to show off their talent), furrow their brow, tense their neck and shoulder muscles, lean their head forward, then left, right and backwards. Watch to see which movements produce artifacts and what the artifacts look like. You will need to recognize these artifacts when you review the assessment EEG. If you notice a regular, approximately once-per-second waveform, it may be the electrical activity of the heart, referred to as a *cardioballistic* or *EKG artifact*. In some clients, moving your electrodes to a position where they are not over a blood vessel may be helpful. It is important to make absolutely sure that the impedance readings between electrode pairs is the same for each pair. In the worst-case scenario, if you cannot get rid of a consistent artifact when you are using a referential placement, you may have to try a sequential placement in order that the amplifier's *common mode rejection* lessens the artifact. The sequential placement will decrease the artifact at least to the extent that it is affecting the two sites used for this *bipolar* placement equally; that is, the waveform produced by a noncortical source is *in phase* and at about the same amplitude and frequency at the two sites.

You should also watch the client and the EEG during each of the conditions as you record the EEG. You watch the client to observe eyes glazing over and any movements. You glance back and forth to the EEG and get a general impression of how it is changing with the client's changes during the actual recording. (At least, on those instruments where you can see the EEG while it is being recorded, you can look back and forth, if the raw EEG is not visible then note the time when an event occurs, such as "eye blink at 38 seconds.")

Conditions for Data Collection

Conditions are usually eyes closed (ec), eyes open (eo), reading (r) and math (m). You do different conditions because clients may show patterns that are outside the range of usual values more, or even exclusively, in one condition and not in others. *Eyes closed* is the condition used by neurologists, and some databases have been collected only in this condition. The NeuroGuide, eVox, and WinEEG databases are eye open and eyes closed. Eyes closed may make it easier to differentiate high delta and low theta activity in a client who has frequent and high amplitude eye-blink artifact. (Though you still get eye movements with eyes closed.) On the other hand, keeping one's eyes closed in business or school is frowned upon, so it is usually of major interest to us to see what is happening in the EEG when the client must just look to the front and remain quiet for a few minutes (so-called eyes open resting state, though the brain is never totally resting and probably the default network is activated) or when they silently read a passage or do mental math calculations. Do they tune out? Do they become tense, anxious and start ruminating? We therefore recommend that you do the EEG assessment using more than one condition. It is essential in our work with ADD to do a minimum of at least one *active* condition – a reading or mental math task since that academic work will elicit the tuning out in most clients who have ADD.

If you are doing a 19-lead assessment, be aware of what high-pass filter to use. The SKIL program and Sterman's database require the use of a 2-Hz high-pass filter. Most other databases use no high-pass filter so you just let pass the frequencies selected by the equipment manufacturer. These might start at 0.5 Hz. (For feedback, as contrasted with assessment, the frequencies sampled usually start at 2 or 3 Hz. to avoid the very slow frequencies that make your EEG recording appear to sway.) It is more expensive to produce instruments that sample down to 0.1 or 0.5 Hz. so that is another factor that manufacturers keep in mind. As an aside, you must be able to sample those very slow frequencies when you work with *event-related potentials*. For ERPs the same electrodes are used and some of the same EEG hardware, but the data collection procedures and the software for analysis are entirely different.

Remember to Minimize Artifact During Recording

Some researchers (for example, Sterman) have reported that with clients who have ADD, the math task is the best condition to clearly bring out the QEEG differences as compared to age norms. Lubar, who designed the assessment program for the Autogen A620 over a decade ago, has always advocated at least a three-minute sample that includes one minute eyes-open baseline and two minutes of silent reading. If it is possible, you should do all four tasks: eyes closed, eyes open, reading and mental math. Each of these conditions should be done for about three minutes. It is always a good procedure to ask your client to blink several times before recording in order to minimize eye blink artifact during data collection. During the *eyes-open* condition, cover the screen and have the client look at a single dot on a piece of paper that is placed in front of the monitor screen. This will help to minimize eye movement artifact. During the reading task, place the material to be read on a stand so that the client does not have to turn their head or bend their neck to read. The best data is usually recorded between 30 seconds and 3 minutes. The first 30 seconds can reflect EEG activity related to a change of state (such as from eyes open to eyes closed) and after 2 ½ to 3 minutes the client may become drowsy.

Math Tasks

Sterman has standard math tasks that can be downloaded from the SKIL website (www.skiltopo.com). Alternatively, use *serial sevens*. In this task the client keeps subtracting seven (in his head, not aloud) from a large number, say 200 for children and 900 for adults. The sequence goes 200…193… 186…179, and so on. When the data collection is finished he must tell you the number he got to. This instruction ensures that they try to stay at the task. The advantage of this task is that the subject can do the task silently without moving or looking around. In the serial sevens task, we usually ask the client to look at the paper in front of them or at the dot in front of the monitor screen while doing the math task. This will minimize eye movement artifact. A third possible task requires that they write down the number that is the sum of the last two numbers spoken by the tester. This they can usually do with very little movement if they sit in a comfortable position and just look at their pen and the paper they are writing on. This minimizes eye movement artifact. The examiner starts by giving two numbers (say 2, 3). The client writes 5. The examiner gives another number, perhaps 7. The client must add this number to the last number the examiner said, which in this example was 3. The client writes down 10. The examiner might then say 9. The client would have to recall that 7 was the last number the examiner said, add to it 9 and write down 16. The examiner continues saying numbers, making it difficult enough to challenge the client. The advantage of this task, for the population we work with, is that it involves working memory, which tends to be a weak area in clients with ADD. Therefore, it is likely to show up the person's tendency to drift off and forget what they heard a moment ago.

Time of Day Effects

When saving data, note the type of task used and the time of day. Time of day will affect the EEG. Sleep latency studies indicate that there are certain times of day when people are naturally sleepier, independent of other factors such as whether they have just eaten a meal. When sleepy, there is higher theta. This occurs usually around 11 am and again, to an even greater extent, in the early afternoon around 1 to 2 pm. This EEG finding confirms what some societies already respect through their customs, such as having a siesta after lunch or having school go from 8 am to 12:30 or 1:00 pm. When we are working with people at this time of day, we often use an EDR monitor and train them to control their alertness level. (Good posture helps.)

Purpose of a Special Assessment and Artifacting Screen

We do the initial data collection on a specific screen for three purposes. First, it is done to remove artifacts that could distort the data and thus the interpretations you make from the data. Second, this screen is used to help you interpret the EEG data by assessing the amplitude of various frequency bands. Third, this display screen is used to teach the client/student (and parents in the case of children) about the EEG and the manner in which it corresponds to mental states.

The example used below is from ProComp+/ Biograph. It shows an assessment done at Cz. We used to use the Autogenics (A620) assessment with clients but now use the more modern Infiniti from Thought Technology. In selected cases we will do a 19-channel assessment using Mitsar and eVox equipment. We are not endorsing a particular brand of instrument, but we are recommending that you do some form of assessment with whatever instrument you have. It does not matter which instrument you use as long as you are consistent and can do pre-post evaluations.

Assessment, Artifacting and Teaching Screen

Introduction

It is important that one of your EEG instruments have a screen that shows about two or three seconds of the raw EEG. Since the EEG of an older person will be of a much lower amplitude than the EEG of a child, it is important that you are able to adjust the display range on this screen to make it easy to read. On the screen shown below, the spectrum and the boxes that display the amplitude, in microvolts, for a particular bandwidth refer to the first second of the raw EEG. You may notice that the display items that require mathematical calculations by your computer may show a slight lag behind your 'raw' EEG.

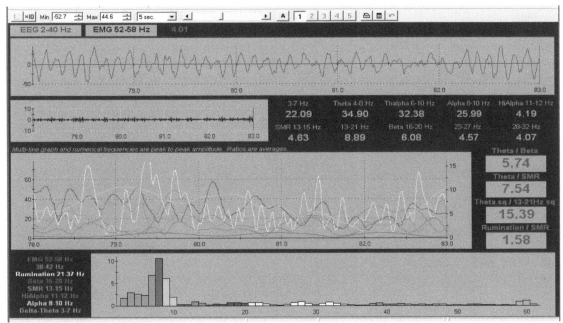

A single-channel assessment screen created using the BFE Thompson Setting Up for Clinical Success suite with Thought Technology Infiniti instrument

The Screen

It is essential that you be able to view the raw EEG signal. It is helpful if the screen automatically adjusts the y-axis scale or if you can manually change the amplitude of the display because adults will usually have a lower amplitude EEG record as compared with children. (It is most likely that this is due, for the most part, to the tissue between the electrode and the brain and not the actual neuronal activity; that is, factors like a thicker skull in adults and more fluid as the brain shrinks a bit with age.)

We also find that a display showing more than four seconds of EEG sometimes obscures the waveforms to a degree that makes it a little harder to recognize the morphology (shape) when we are only artifacting one or two channels. (On the other hand, when we artifact 19 leads we use 5- to 7-second segments.) It is helpful to be able to view a line graph under the EEG that reflects muscle and ambient electrical activity. In North America we use 45-58 Hz to show the effect of muscle activity on the EEG. In Europe and Australia we use 52-62 Hz. Alternatively, we may use 45-70 Hz. In the latter case, the combination of this EMG indicator plus the EEG spectrum help us to differentiate activity that may be due to electrical interference (60 Hz in North America, 50 Hz in Europe and Australia) and/or muscle as compared to actual brain wave activity.

It is very useful to simultaneously see the raw EEG and the spectrum. (The spectrum in the above example corresponds to the first second of the EEG above it.) It can be quite a challenge to distinguish true delta wave activity (which is not artifact due to movement or eye blink that imitates delta) on a one-channel assessment from low frequency theta when you just observe the raw EEG. The spectrum will show you more accurately where the highest amplitudes are. The pink bar indicates the highest amplitude when you use Biograph software as is done in the above figure. In this example, the spectrum shows that the one second on the right hand side of the screen has predominantly 8 Hz alpha.

Another helpful feature is to have frequency ranges (theta, alpha, etc.) that are color-coded. In addition, it is useful to be able to see the relationship between commonly used frequency bandwidths as you move through the EEG. We put a theta/beta ratio box as one of the boxes that you can see on the right-hand side of the figure above, because this ratio is helpful in distinguishing the most common type of ADHD in children. In the foregoing example the theta/beta ratios are very high. This patient also has long bursts of alpha as is shown above in the raw EEG.

Note regarding equipment:
The ProComp+/Biograph is used for illustrative purposes as it is equipment that is currently in common use and is still available. There are other

instruments that also allow for quantitative assessments that are not full-cap, 19-channel. The Autogen A620 assessment program developed by Lubar is excellent and is the one that we used with nearly every client for the first 10 years at the ADD Centre, to generate pre- and post-training statistics and profiles. We have done more than 5,000 such new client assessments and closer to 20,000 if reassessments are included. The assessments are then repeated after 40 sessions of neurofeedback to assess progress in training. Monastra, Lubar et al. have published norms from a multisite study for mean theta/beta ratios for ADHD and control subjects using the A620 instrument. These norms are very helpful in working with clients who present with attentional problems. We have compared assessment data with the A620 and the Biograph Infiniti and found that the results are virtually the same.

It is also possible to select a portion of good EEG data and calculate your own theta/beta ratios (or any other ratios you wish) using the F1000 equipment from Focused Technology. The F1000 regrettably went out of production in 2002 though there is still support for equipment already sold. The F1000 has always been a favorite of ours because the electronics are so good. There are analog as well as digital filters for some screens. You also have the option of simultaneously monitoring temperature and EDR. We wish Frank Dietz a happy and healthy retirement from the equipment manufacturing side of the neurofeedback business, but look forward to his continuing contributions to the field.

Some other instruments are designed for doing feedback and cannot be easily adapted to do an assessment. On-the-other-hand, software such as that used with the popular Brainmaster and with EEG Spectrum's Neurocybernetics system can be very useful for neurofeedback training. On some instruments that are designed primarily for training you may want to do an assessment on another instrument first to ensure that you are using appropriate parameters for your client. In addition, you may wish to have other equipment available to do biofeedback at the same time as you are doing neurofeedback with instruments that do not have the combination. The Roshi system works on different principles and does some entrainment as well as feedback. It has shown promise for some applications, particularly optimal performance, for example, Dan Chartier's research with golfers (ISNR presentations and personal communication). Clinicians report high client satisfaction with the Roshi. The Peak Achievement Trainer (PAT) has also been marketed for use with athletes. It uses a sensor on a headband, and the protocol calls for reducing the amplitude for a very wide frequency range from delta through beta. This set-up would seem to leave the EEG data prone to artifact, so you would certainly want to ensure good connections (low impedance readings) and try to minimize eye movements. The goal of reducing the wide range of EEG activity would, in practice, reduce slower frequencies more since they have the higher amplitudes to begin with. (Recall that the skull attenuates the higher frequencies more than the lower ones.) This might be helpful for improving attention. However, you cannot do basic feedback work such as raising SMR amplitude or working on coherence; so although we have it we do not use it. Since we have not used the Roshi we cannot comment further.

Important: The foregoing discussion is meant to show that there are a broad range of instruments available. Our discussion is not inclusive. We are only able to give examples in this book from equipment and software that we presently use. There are many other instruments and software systems. You may find it helpful to discuss the advantages of each instrument and software with manufacturers who exhibit at conferences such as those given by AAPB (Association of Applied Psychophysiology), BFE (Biofeedback Foundation of Europe), and ISNR (International Society for Neurofeedback and Research).

Purposes of the Above Assessment Screen

This screen is used both when the client is first seen and at the beginning of each neurofeedback training session done with this equipment. It is even used when a patient has a 19-channel assessment. Initially it is used to obtain an EEG profile. This profile helps in decision making concerning where electrodes should be placed and the frequency bands that should be enhanced or decreased. It is also used at the beginning of each training session to check that you have a good EEG signal and to remind the client of the purpose(s) of the NFB training session. In each instance the clinician is using the screen for *three* purposes: assessment, artifacting, and teaching.

Assessment

To assure that sufficient data without artifact is collected during the initial assessment, the trainer/clinician can watch this screen while the client is carrying out each of the data collection conditions. The trainer can also observe what the artifact looks like on the EEG when the client blinks, swallows, or moves. The trainer can also mark down, or place a marker (by hitting the space bar), for each change of condition (EO, EC, reading, math) so that statistics can be done separately for each condition. Alternatively, the clinician may pause or restart for each new condition. At the ADD Centre we prefer to do each condition separately and copy the statistics into a specially prepared Excel sheet that automatically graphs the data and produces ratios for us. This will be shown in a later section in this book. At the beginning of each training session the clinician will usually only do a brief reassessment using whichever task seems to give the best profile. (Use the same condition each time.) As noted previously, the best data is usually collected after the first 30 seconds, which represents a change of state (for example from eyes open to eyes closed), and before three minutes has elapsed as a subject may start to get drowsy at that point. For daily session recordings, you may find it helpful to start the recording after the client has been doing the task for 30 seconds in order to avoid this problem with change of state. Then record for 30 seconds to a minute, artifact the resulting EEG with your client, print the statistics and discuss the objectives for the session.

Artifacting

While reviewing the data, the clinician should mark artifacts that are to be removed when doing statistics. Mark data for artifacts and *exclude* these data when calculating statistics on the report screens. Of course, you should attempt to minimize artifacts in the first place, not just get rid of them later. You do this by asking the client to blink before recording begins (so they need not blink as much during the recording) and to sit comfortably and still while you are recording. Remind them to keep their muscles relaxed, with the mouth slightly open, and the tongue just suspended, not touching teeth or roof of mouth. (These instructions are important because the eye has its own polarity, think of the cornea as positive and the retina as negative. The tongue is also a dipole with the tip negative relative to the base.) Ask the client if he has any metal in his scalp or jaw and teeth. Make sure the client has turned off his cell phone and tablet.

If you are doing a recording with the eyes open, have the client look at a dot on a piece of paper covering the screen (if not reading). But make sure she isn't looking up and wrinkling her forehead. Most of the record, under these circumstances, should be relatively artifact free. There are circumstances when you would choose to mark data for inclusion rather than exclusion. Examples are when you have a very poor record and you just select the good bits, or when you wish to have statistical information on particular kinds of activity, such as paroxysms. You must decide to either include or to exclude data before artifacting so that you get a consistent record when you do your statistics.

The 45-58 Hz EMG line below the EEG helps you distinguish muscle artifact from high beta activity. (Note: Be careful not to include, either here or in the 'inhibits' on your feedback screens, 39-41 Hz activity. This bandwidth is sometimes referred to as the Sheer rhythm. It is named after David Sheer who reported in the 70s that the frequencies around 40 Hz were related to certain attentional processes. In addition, Steriade has noted that this rhythm reflects attention and this will be discussed in a later section of this book.)

As suggested above, at the beginning of each training session it takes only a few seconds to artifact one or two minutes of data. This time can have a second function of reminding the client about the functional significance of different frequencies and what their training session will emphasize. Clients, especially children, often become very good at pattern recognition and may play an active role in artifacting.

You should make sure that one of your single-channel instruments has a special assessment screen that is designed specifically for removing artifacts. This screen should allow you to highlight an artifact using the mouse. This is the method used in the Biograph Infiniti. An alternative method is to simply reject any two-second segment of the EEG that contains an artifact. This method is used in the A620 assessment program (Stoelting Autogenics).

Teaching

One of the most important objectives of an assessment screen is to help your clients understand the EEG pattern and its possible meaning in terms of mental state changes. It is a useful tool when one explains to parents or to the client how frequency bands correspond to mental states. Having a spectral array and boxes showing the microvolt reading for particular frequency ranges on the assessment screen can help you do this. For example, the clinician might discuss the following in an assessment screen:

1. In the figure shown above, dominant slow-wave frequency is at 8 Hz. This client may be internally focused and not fully aware of the external environment at this point in time. If we had observed and marked the times when this client appeared to drift off (eyes may seem to glaze over), then this pattern might correlate with these observations. To help the client become aware of this change of state the trainer may pause the EEG when this waveform is observed and then ask the

client what their brain was doing a moment ago, just before the screen was paused.

2. If the amplitude of the EEG is lower in the SMR range 13-15 Hz than in the beta range 16-20 Hz, then the client might attempt to increase the SMR amplitude. If the low SMR activity is present throughout much of the recording, then it may correspond to a child's hyperactivity and impulsivity. If both SMR and beta were much lower than theta 4-8 Hz activity, which is typical of those with ADD, then you might encourage the child to increase these beta frequencies. We have also repeatedly observed extremely low SMR readings in clients who have Parkinson's disease. They have difficulty controlling unwanted movements, such as tremor.

3. The beta amplitudes between 23 and 36 Hz are usually lower than in the beta range between 15 and 19 Hz. This is a normal pattern. When they are higher, check with the client what kind of thoughts were going through his mind during the recording. You may find that this activity corresponds to a combination of anxiety, tension, and worrisome ruminations.

You can show your client a section with slow waves and ask them to estimate the number of such waves that could fit into one second. Let them actually count the waves, if they want to do so. Then show a second segment of EEG where they are really tuned in and, perhaps, problem solving. It will contain a large proportion of low voltage, desynchronized fast waves. Have the student estimate how many waves of this sort would fit into one second of time. Usually they estimate between 16 and 20. Some students become so good at pattern recognition that they actually prefer to watch the EEG in feedback sessions rather than a game screen. Motivation for biofeedback training increases profoundly when students understand what it is they are trying to do in terms of learning the task of *self-regulation* of brain wave activity. It also helps recruit their cooperation in keeping EMG artifact to a minimum when they understand that it affects the quality of the feedback they are receiving.

Common Artifacts

Introduction
Artifacting, as previously discussed, is the process in which you review the EEG recorded from each condition (eyes closed, eyes open, etc.) and look for activity that is likely not of cortical origin. In other words, it is not true EEG that reflects neuronal activity. You usually look at segments lasting from two to four seconds at a time. As mentioned above, your assessment program should have a simple means for marking or rejecting data that contains an artifact. Artifacting should be rigorous. If it is not well done, you may have statistics that distort the real picture; for example, thinking you have beta activity when it is really the result of muscle artifact, or interpreting eye blinks as if they were delta or theta activity. Distortion may lead to incorrect conclusions about the EEG patterns, which might then lead to incorrectly deciding on what frequencies to inhibit or encourage during training. Your training result would then quite possibly be due to that the person learning to inhibit eye blinks when you thought you were actually reducing delta activity.

You should endeavor to exclude from the statistical analysis segments that are contaminated by eye movements, blinking, muscle contractions or cardiovascular activity. Electrode movement and eye movement or blinks usually produce artifact in the low frequency range (1-3 Hz, delta). However, occasionally it may affect the amplitude of waves in the theta range up to 7 Hz. Muscle artifact usually produces changes (increased amplitudes) in the higher frequency range (above 11 Hz, SMR and beta) although it can influence even the alpha range if it is high enough in voltage, as in teeth clenching.

What are Artifacts?
An artifact in EEG parlance refers to a segment of unwanted data. When doing an EEG assessment, 'unwanted' refers to data that is not generated by neurons in the brain. Hammond and Gunkelman (2002) state, "An artifact refers to a modification of the EEG tracing that is due to an extracerebral source." This is a good definition as it includes both interfering biosignals produced by the person and also external sources of electrical interference. These segments of unwanted data can be caused by a number of different mechanisms. The most common are:

- Movements of the electrodes on the scalp
- Movement of wires
- Eye movement
- Tongue movement
- EKG (heart) activity
- Facial, scalp or neck and shoulder muscle tension
- Electrical activity near the equipment (lamp, extension cord)
- Waves from a strong radio transmitter or other electrical source

Why Is Artifacting Important?

A gross artifact such as that caused by a sudden electrical surge or a movement of the client can make it impossible to read anything in the EEG. More subtle artifacts may be more damaging to our neurofeedback work precisely because they are subtle and may not be detected. Undetected movement artifacts or eye blinks may make it appear that slow-wave activity is higher than it really is. Undetected muscle tension artifact may make it appear that there is higher beta or SMR activity than is really taking place. Cardiac artifact is much more difficult to analyze as it affects a number of different frequency bands. Fortunately, it is fairly easy to see in the raw EEG, and you can often move the electrodes and minimize or eliminate it. In addition, it is usually constant and, therefore, *changes* in bandwidth amplitudes are more consistent than with other, more transient artifacts. This will be discussed later in more detail.

General Rules for Detecting Most Artifacts

The first rule is simple: a relatively high amplitude potential that occurs only in one channel (at one site) is probably an artifact.

The second rule is simple, too: repetitive waveforms that appear simultaneously in unrelated head regions are usually artifact. The reason for these two common sense guidelines is that true neuronal activity usually will have a predictable physiological distribution with a maximum potential where the activity originates. Then, there will be a gradual decrease, with increasing distance, at other sites. Abrupt neurological changes do not just 'jump' to a distant site on the head.

Artifact – Electrical Sources
This is discussed extensively under Instrumentation.

Artifact – Muscle (Reflecting EMG)
EMG stands for electromyography. We use it as a short form when we are talking about the activity in the EEG recording that is not neuronal, but is a reflection of the electrical activity from skeletal muscles. True EMG activity is usually above 60 Hz, and thus at a much higher frequency than we are measuring with the EEG. However, EMG activity is also usually of much higher voltage. It also may have harmonics at lower frequencies. It may even be of a high enough amplitude to *overwhelm* the instrument's low-pass filter. You can have the student demonstrate this overwhelming of the low-pass filter by asking them to clench their teeth. They will observe with a very light clenching of their jaw that only the higher frequencies on the spectrum increase. Slightly more tension will overwhelm all frequency bands. In explaining this to a client, a useful analogy is to say it is like a loud noise drowning out a whisper. The EEG signal is like a whisper.

When doing EEG biofeedback, have your client understand that they want to learn control of their brain, therefore they must keep muscle tension to a minimum so that they receive quality feedback that truly reflects brain activity. With young children you demonstrate how EMG activity will shut off the point counter. They don't want to do that! Another fact you can share that will encourage the client to keep the frequencies affected by EMG as low as possible is to explain that work with athletes shows that reaction times are faster if muscles are relaxed rather than tense. You can mimic an elderly person holding a steering wheel with shoulders all hunched up and tense. Contrast this with a top gun pilot who is physically relaxed, yet mentally alert while flying. They want to emulate the pilot, which corresponds to low EMG readings.

During feedback it is very helpful to have a muscle inhibit (or a 45-58 Hz percent-over-threshold) on every feedback screen. Otherwise you will not know if an increase in the student's fast-wave reward frequency, (for example, SMR 12-15 Hz or beta 16-20 Hz), is due to muscle tension or actual brain activity in the 12-15 Hz or 16 to 20 Hz range.

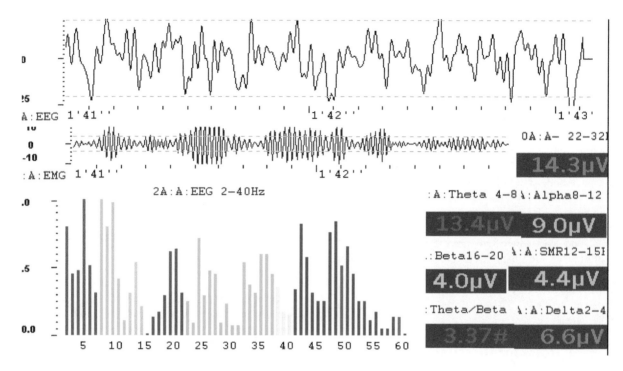

The above figure shows muscle artifact. Compare this with the figure shown in the section above on one-channel EEG assessment. It is the same child in the same session. (A slightly different assessment screen with a box for theta/beta ratio was used for this recording.) You can see that the amplitudes of most bandwidths are grossly increased. In particular, note the differences in the EEG wave (top), and the "EMG" (44-58 Hz) band or 52-60 Hz band in Europe and Asia (center line graph). For NFB sessions these frequencies are used as an "EMG" inhibit. (On the other hand, you may decide to use an inhibit on 22-32 Hz activity during feedback sessions. Using this range as an inhibit discourages EMG as well as anxiety and ruminative activity.) Note the very large increase on the right side of the spectrum (gray, above 43 Hz). This is the area of the spectrum that first shows an increase in amplitude when a child tenses his jaw or his neck muscles.

The raw EEG in this example has used an FIR, Hamming, 2-40 Hz band-pass filter.

EMG activity is fairly easy to distinguish with 19 channels. Unlike the EEG seen in the single-channel recording, it is usually seen affecting only one or two channels. The channels affected are usually seen in the periphery at T3, T4 (jaw tension) and sometimes at O1, O2 (neck tension), or at Fp1, Fp2. (forehead tension)

The figure opposite shows how muscle artifact may appear in one temporal region (T4). More often it is seen bilaterally at T3 and T4.

Helping the client to be aware of tensing their jaw or neck is a first step. Then the trainer will assist the client to find ways to relax these muscle groups. This may involve the use of biofeedback, which will be discussed later.

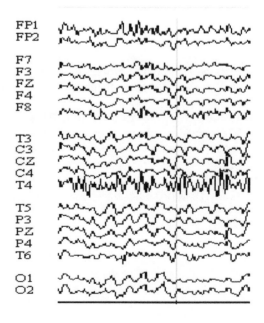

As previously mentioned, when doing NFB to inhibit the influence of muscle, you may choose to inhibit 45-58 Hz (in Europe, 44-48 Hz or 52-60 Hz) frequencies that mainly reflect muscle tension. In addition, you may wish to be warned if there is a sudden increase in electrical interference. You would then use a 45-70 Hz inhibit. This will catch the 60 Hz electrical activity. These inhibits avoid affecting good Sheer rhythm activity in the 38-41 Hz range. Research suggests that Sheer rhythm is associated with attention. (Some clinicians have recommended reinforcing 38-41 Hz with neurofeedback. Sheer's research suggested that this may be effective when working with those who have learning disabilities. However, it is a difficult range to reinforce because amplitudes tend to be very low and muscle artifact may interfere a great deal.)

It is important to be able to distinguish high amplitude beta activity that is due to muscle tension from activity due to worry and ruminations. A client who is tense and anxious may demonstrate excessive activity somewhere in the 21-35 Hz. range. We find that the anxiety component appears to correspond to a rise in the 19-23 Hz range, whereas the constant worrying and ruminating seems to more often correspond to a peak somewhere between 24 and 36 Hz. The peak differs between clients but is consistent for a particular client. When this EEG picture is seen and found to correspond to the client's complaints, then you may choose to train down one or more frequency bands in the 20-35 Hz range. In addition, you would place a severe inhibit on activity that corresponds to muscle activity in the 45-58 Hz range. In this manner you allow your client to know which kind of activity is causing an increase in high beta and assure that muscle tension does not influence the EEG feedback.

Despite the foregoing discussion, it is true that some 19-channel EEG biofeedback instruments do not register activity above 32 Hz. These instruments, like older instruments that are no longer used, tend to treat all increases in the 24-32 Hz range as muscle artifact. In working with children with ADD this may, for the most part, be true. With anxious and/or worried adult clients it is frequently not true.

On those instruments that did not allow you to differentiate between muscle activity and high beta neuronal activity, the practitioner just inhibited 24-30 Hz and thus inhibited muscle influences while at the same time discouraging tense ruminative thinking. As previously noted, we feel quite strongly that it is important that you know which factor is responsible for the activity in this range. Then you can make an *informed* decision on how to design your feedback screen and how to advise your client. (If you only know that 24-30 Hz keeps rising and you keep telling a client to relax their jaw and neck, when these muscles are perfectly relaxed, you are not being very helpful, and you may, in fact, increase their already elevated anxiety.)

Artifact – Electrode Movement

V (voltage) = I (current) x R (resistance) according to Ohm's Law. As noted above, since the EEG is alternating current, it is actually impedance, not resistance, that concerns us. However, the principle is the same. Impedance is, in simple terms, resistance to the flow of AC current. So here we will keep it basic and use just Ohm's basic equation. Because the current (I) is so small, any change in resistance (R) or impedance (z) will have a major effect on voltage (V). Even a very small movement of an electrode will massively alter R (or z) and thus V. Further, it may do so in the frequency range that we are measuring, such as 3-6 Hz. This will mean that you might interpret a rise in theta as the client tuning out when, in fact, it is just due to a small movement of the electrode on the skin or a poor connection. This can be very frustrating for your client, who cannot relate their genuine attempts to remain highly focused to a decrease in theta.

Electrode movement will usually be different at each electrode site and therefore not removed by the amplifier's *common mode rejection* feature. It usually results in a high amplitude slow wave in the delta or low theta range.

There are a few simple procedures that we routinely employ to minimize movement artifacts. We always have the client wear a headband. We purchased hundreds of these at a discount from the local sporting goods store. They can be thrown in a bag after use and washed. Regular clients have their own headband tucked into their training file. It makes the file bulky but follows the principle that nothing that touches one client's head touches another client's head without being cleaned first. You put the headband on and tuck the electrode wires under it. The kids can have flashy colors or their favorite sports logo and we tell them they look like a tennis star. Headbands have the added advantage of making it easier to put the electrodes on a child's head because children often do not sit perfectly still. Be consistent and no one is likely to object.

Artifacts from EMG and Electrode Sway

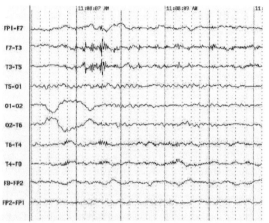

EMG artifact in the temporal regions and electrode sway movement artifact in the right occipital region. With thanks to Jay Gunkelman

Artifact – Eye Blink

The cornea is 100 mV electropositive with respect to the retina (Fisch, 1999, p 108). There is over a 100-millivolt difference between the aqueous and vitreous humor (Hammond, 2001). When the client blinks, the eyeball rolls upward (*Bell's phenomenon*). When the eyelid touches the cornea, it is like touching the positive end of a little battery. This increases the positivity in Fp1 and Fp2, whereas F3 and F4 may become relatively more negative (Dyro, 1989). The result is a sudden large-amplitude slow wave that is seen on the EEG. Movements of the eye will also result in currents that produce artifact on the EEG, particularly in the frontal leads. It is always higher in amplitude anteriorly and decreases as you view sites that are more posterior. Sections of the EEG with this type of artifact must be rejected during the assessment or you will have a false impression of high slow-wave activity when looking at the statistics. The eye blinks and eye movements mimic delta (and sometimes theta) activity. As previously noted, during feedback you may also decide to reject this type of artifact by placing threshold lines above and below the EEG line graph with an *inhibit-over-threshold* command. By doing this you assure that there will be no auditory or visual feedback during an eye blink. For example, the EEG Spectrum and Biograph programs both allow this. The amplitude of these artifacts is usually much higher than the regular EEG waves, so they extend beyond the threshold lines. Below is an example of an eye blink artifact in a single-channel EEG with the active electrode at Cz referenced to the left ear. In this example, white threshold lines are seen above and below the EEG. The large-amplitude eye blink artifact has gone outside these threshold lines.

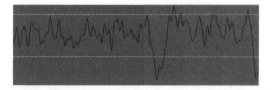

Having threshold lines placed like this is a doubly effective method as the trainer can see both artifact and the EEG. Many clinicians feel you must watch the raw EEG during feedback sessions both to assure that the EEG remains of good quality and to observe the EEG for artifacts.

Note, however, that if your instrument's program still calculates an average amplitude for frequency bins during the time that that wave is being produced, then the threshold lines would only cut out the excess amplitude of that wave but the wave itself would remain in the calculations. As a result, your data would be skewed to give a falsely high average power for that frequency bin. Some programs deal with this difficulty by stopping all feedback instruments, sound, animations, percent-over-threshold calculations and points, for the time period that any waveform goes outside of threshold boundaries (With thanks to Frederick Arndt, engineer, personal communication).

Dealing with Artifacts

For assessments you must carefully remove artifacts before calculating statistics. For statistics recorded and graphed during your training sessions this would require too much time. You have to just assume your eye-blink artifacts, and, with steady diaphragmatic breathing, your cardiac artifacts are fairly consistent. For muscle artifact you must be consistent in having your patient sit still and not talk during the recording. If they move around or talk, do not use that three-minute segment in your statistical comparison graphing.

Eye movement Artifact in a 19-Lead Recording

Eye movement is one of the most common artifacts.

Two eye-blink artifacts are demonstrated in this recording. Note how the primary effect is seen in the frontal electrodes.

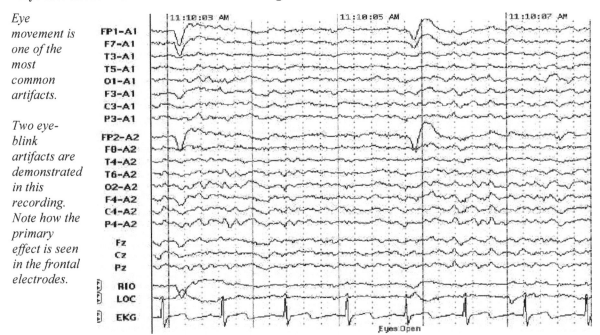

Eye Movement Artifact: With thanks to Jay Gunkelman

Artifact – Eye Movement

A vertical eye movement may give a downward deflection (positive) of the EEG because the positively charged cornea is moving towards Fp1. A downward movement of the eye would give the opposite deflection. This is most easily seen in a longitudinal sequential montage. It is usually seen frontally (Fp1-F3 and Fp2-F4). A movement to the left may show a positive deflection (downward) in F7 and the opposite deflection in F8. We usually just think of this as the whole EEG seems to move up in one case and down in the other so that, for example, the two EEG lines appear to come closer to each other. A spike preceding this slow wave may just represent EMG activity and should not be confused with a spike and wave. The 19-channel referential montage shown below demonstrates lateral eye movement.

Lateral eye movement while reading

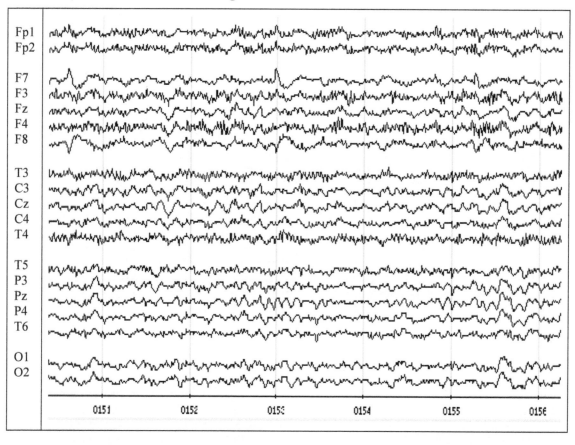

*In the above recording, the subject was reading. Note the **lateral eye movement is** seen best in leads F7 and F8 preceding time: 0151 and at 0153.*

On this record it is also possible to see a small **evoked potential** at 0155 and an artifact (**transient**), maximal in the occipital leads, just preceding 0156 at the end of this segment. There also appears to be **muscle artifact** at F3, F4, T3 and T4.

A *transient* is best thought of as "a change in kind" from the ongoing activity. Transients may have little effect on averaged bandwidth activity, but variability is increased (Hammond, 2001). You should typically remove transients, treating them as artifacts, following the maxim: "When in doubt, leave it out."

Know the Effect of Artifacts on Frequency Bands that are Involved in NFB Training.
The diagram below is a topographic display taken from the SKIL program and thanks are extended to Barry Sterman for artifacting this EEG sample for this figure. This example shows the amplitude of delta wave activity before (blue lines) and after (red lines) artifacting the EEG for eye blinks. Note how the removal of artifact activity does make a visible difference in the average amplitude of the delta wave activity in the prefrontal and frontal leads. No difference can be observed, in this example, in the central or parietal leads. Thus, with this client, eye blink would have an effect when the active lead was placed frontally, but probably not when a central placement was used. When tested in this way, there are a small number of clients where the eye-blink artifact activity affects the amplitude of frequencies as high as 5 or 6 Hz even at Cz. It is crucial that the neurofeedback practitioner who is setting up a training protocol know the extent to which this kind of activity is affecting the recordings.

Effects of manual artifact correction are seen in differences between levels of frontal and prefrontal slow activity

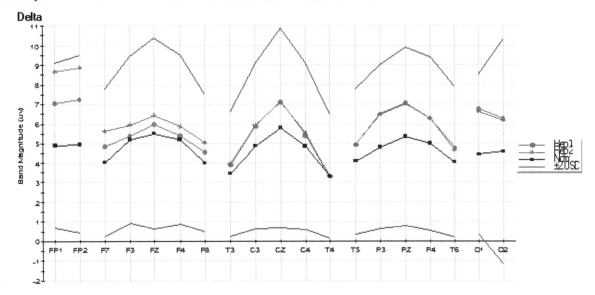

Movement and Muscle in Review

Eye blinks, rolling eyes, and other eye movements can give **a false impression of high delta and theta.** Theta is also affected by heartbeat (a consistent artifact) or any electrode movement over the skin. Muscle tension can give a **false impression of increased SMR or beta amplitudes**.

Artifact – Heart

A regular, high-amplitude artifact at a rate of approximately one per second is usually not due to movement, but rather to electrical activity related to the regular contractions of heart muscle. This electrical signal may be carried through arterial connections. It would make sense that this activity may be more pronounced if an electrode is placed over or close to an artery. However, Jay Gunkelman tells us that this "electrical artifact also is picked up as a 'field effect' that has nothing to do with arterial closeness." He went on to note, "A *pulse artifact* (slow sway due to electrode movement with each heart beat) is from arterial or venous pulsation, unlike the cardioballistic sharp waves." He feels that, "the linking of the ears makes the cardioballistic artifact much less of a problem, since it has reversed polarities in each ear, and tends to cancel... though not in all individuals." We have used linked-ear electrodes in the past when doing NFB with the F1000 equipment and found this helpful in eliminating such artifact.

We have seen this phenomenon in a client who was awaiting valve surgery. However, we most often have this problem with people who have very large, short necks. The EEG segment below is from such a client. The recording is from a single channel at Cz referenced to the left ear lobe.

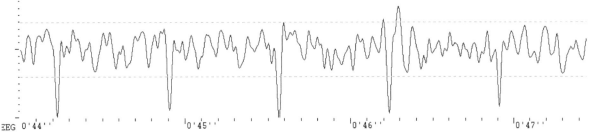

This artifact is usually seen as a regular wave approximately once per second. It does affect your bandwidth calculations. It is quite a frustrating wave to try to analyze. We have found it can differ in different people as to which frequencies it increases. It is a complex waveform and, as such, it always increases frequencies in more than one bandwidth. Quite often it will influence not only delta, but also alpha and beta (Hammond, 2001). Because it is in common to many electrodes, **it will artificially**

increase coherence or comodulation (Hammond & Gunkelman, 2001). It is much more prominent in some individuals than in others, and is more often seen in adults compared to children. Being a relative constant, you can take some consolation in assuming that changes in band wave average amplitudes will be fairly consistently affected during feedback sessions. Since we are usually looking for changes, this is helpful. However, prevention is the easiest solution. During an assessment, the first step for prevention is to make sure that impedances are low and exactly the same between the ear electrodes. If doing a 19-channel assessment you may just have to analyze the EEG using a different montage (Laplacian or sequential). For NFB sessions you can often eliminate the problem by moving each of the electrodes in turn until it disappears. If this doesn't work, you may find that a sequential (bipolar) montage will eliminate it by means of the amplifier's common mode rejection.

In the figure below the electrodes were moved from the above referential placement to a sequential (bipolar), Fz-Cz, placement. There is some decrease in the cardiac artifact amplitude.

0'7''   0'8''   0'9''

The cardiac artifact is still definitely present, but at a lower amplitude relative to the rest of the EEG. This is a good example for another reason: it demonstrates, in the second EEG, how difficult it can be to see the effect of the cardiac influence on the EEG when the waves are not large. It might have been difficult to pick out the cardiac artifact if we had been rapidly moving through just the sample from the sequential placement.

The more consistent the wave, the less may be its overall effect when averaging and comparing amplitudes in different bandwidths for each segment of the training session. The heart rate effects may be a little more consistent if you train the client to relax and breathe diaphragmatically at six breaths per minute (see the section on RSA – respiratory sinus arrhythmia – later in this book and the section on heart rate variability training, or HRV). Because the changes that cardiac artifact will make in your recording are fairly consistent over time, you can still do successful feedback. Absolute values will be affected, but *average changes* in fast- and slow-wave amplitudes will be reasonably correct, and your client can still learn self-regulation.

Artifact – Tongue and Swallowing
The base and the tip of the tongue form a dipole, with the tip being negative. Movement will mimic delta activity. It is fairly easy to see this artifact as the entire EEG line will move gently up or down in the frontal and temporal areas. This mimics frontal intermittent rhythmic delta activity (FIRDA) that neurologists may associate with a structural lesion of the brain (Fisch, p 354). Clients with dystonia may demonstrate this kind of artifact and be unable to control muscle spasm.

Artifact – Other Eye Movements
Eyelid Flutter
Watch your client's eyes during the recording. Unfortunately, an anxious client may flutter their eyelids. This may mimic frontal delta or theta waves in the 2-4 Hz range (Dyro, 1989) or even "thalpha" activity in the 5-10 Hz range. Only careful observation of the client while also observing the EEG will distinguish this kind of artifact from true frontal slow-wave activity. The figure below shows this activity at Fp1 and Fp2.

Eye Lid Flutter Artifact in Fp1 and Fp2 in an eyes-open, linked-ear reference recording

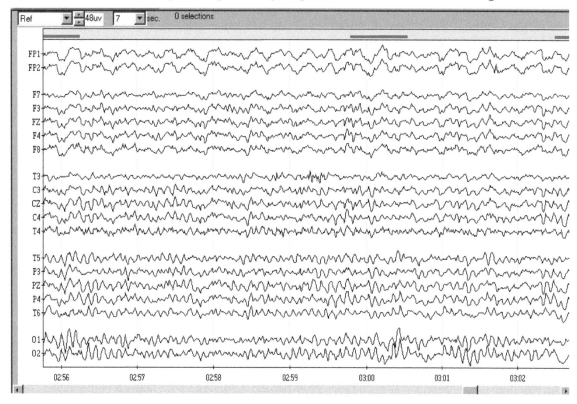

Electrode Pop Artifact at P4

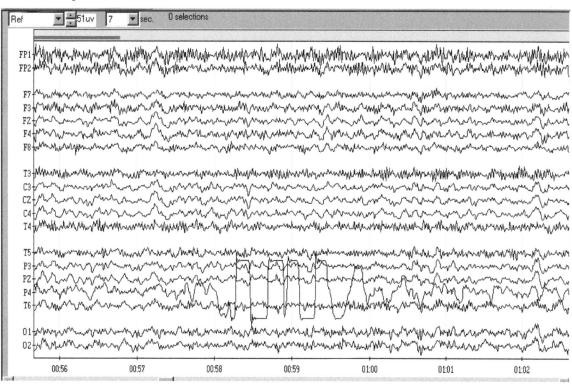

Electrode Pop Artifact
In the above figure the electrode at P4 lost its connection. The EEG was paused and more electrode gel was put into the site. The artifact disappeared.

Note also the muscle artifact. It is particularly evident at Fp1, Fp2, F3, T3 and T4. This 12-year-old autistic child was grimacing and furrowing his brow.

Drowsiness
This may produce spike-like activity. More often it just looks like excess theta or thalpha activity in the frontal regions. In the eyes-closed condition as a client becomes drowsy, you may observe slow eye movements under their closed eyelids. Sometimes you will observe a fairly sudden drop in occipital alpha activity. You may also observe an increase in theta waves. You should watch for these signs when doing an eyes-closed recording. It is a signal for you to pause the recording and alert the client. Slow eye movements can mimic theta activity on the EEG. If you just notice it when reading the EEG, then you should reject those epochs. Good examples of this kind of phenomenon are given in the book *The Art of Artifacting* by Hammond and Gunkelman.

Additions to Artifacts

Earlier in this section, coming we defined an artifact as electrical activity in the EEG recording that is not due to neuronal activity. A more precise definition is **"electrical activity of noncerebral origin."** Artifacts can be of biological origin, such as tics, or eye blinks, or of nonbiological origin, such as electricity coming from an old lamp or an extension cord. Illustrations were shown of the most common artifacts, including eye blinks and reflections of muscle activity. The instrument and program you use should have a means for observing the raw EEG and of removing these artifacts. In this edition of the book we are only giving a small number of additional illustrations to assist the beginners in this field to identify artifacts.

Some of these, such as a figure showing a metal bar inserted during dental work in one young patient and the effect of a hearing aid in another patient, are unusual artifacts and are included here only to reinforce the practice of asking each patient if they have any metal objects in their scalp, mouth, teeth and gums, or their neck. The patient should also be asked to turn off any electrical devices that can be turned off such as a cell phone, pager, or tablet. They should also be asked about medical devices such as hearing aids and asked to turn off anything that is not essential for their health while you do the recording.

Figure to demonstrate the effect of muscle tension on the EEG

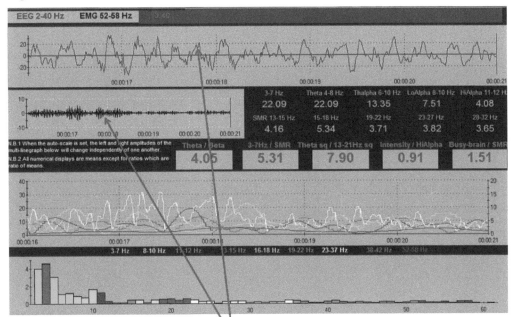

Child with diagnosis of ADHD. Note high amplitude theta waves on right side and the **theta frequency waves with muscle Artifact** "riding on top" of the EEG on the left hand side. Some of this slow wave activitry may be related to eye movement.

In these assessment screen figures, as previously explained, the top line is filtered to show 2-40 Hz activity. The dark red line below it on the left is showing 52-58 Hz, which is considered to reflect the effect of EMG activity on the EEG. We therefore refer to it as "EMG noise." True EMG training is typically done in the 100-200 Hz range and can go as high as 500 Hz.

Figure to show a difference between spindling beta and an artifact caused by muscle activity

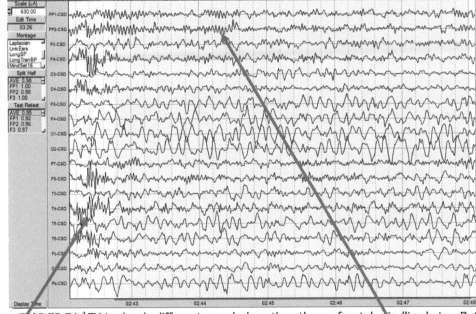

EMG F8, F4, `T4 is clearly different morphology than the prefrontal spindling beta. But often the prefrontal muscle artifact does look very like spindling beta!

If you are unsure, follow the adage, "When in doubt, leave it out."
Figure taken from the raw EEG of an autistic child.

The Neurofeedback Book, Dr. Thompson, (www.addcentre.com).   Published by the AAPB. (www.aapb.org)

Figure to show a rare artifact caused by a single motor unit

Below is a specific type of EMG Artifact. This is a "single motor unit" (SMU) muscle fasciculation at T3. It is a series of bi-phasic EMG spikes with a "series of H-H-H" morphology.

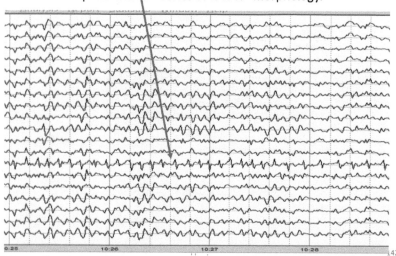

This artifact is not often seen.

The figures below demonstrate how cardiac artifact can have a large effect on the amplitudes of a number of different frequencies. This effect can make an abnormal power ratio appear falsely normal.

Cardiac Artifact

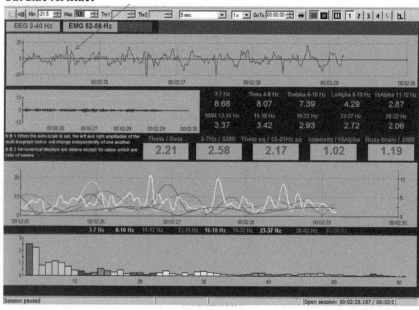

Blue = Only Eye movement & EMG taken out.
Red = Cardiac Artifact Removed.
Same up to 6 Hz and again at 9 Hz
Biggest difference is at 14 Hz
Becomes very close below 5 Hz & at and above 20 Hz

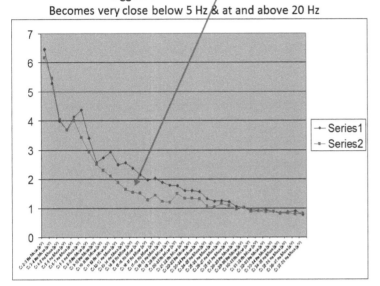

Note that there is virtually no effect on amplitudes below 6 Hz and above 25 Hz.

EKG REMOVAL	Before	After
C: 4-8Hz MeanSq/13-21Hz MeanSq	2.22	**3.78**
C: 3-7Hz Mean/13-15Hz Mean	2.58	3.76
C: 4-8Hz Mean/13-15 Hz Mean	2.39	3.45
C: 4-8 Hz Mean/16-20 Hz Mean	2.25	3.04
C: 6-10 Hz Mean/16-20 Hz Mean	2.08	2.57
C: 19-22 Hz Mean/11-12 Hz Mean	0.97	1.09
C: 23-35 Hz Mean/13-15 Hz Mean	1.13	**1.54**

Removing heart artifact suddenly reveals abnormal ratios. This fit his description of tuning out (high theta/beta) in classes, and of having a very busy brain, which appears to corelate with high 23-35/13-15 Hz.

This artifact must be carefully removed for assessment and progress testing reassessments. However, it may be considered a relative constant when doing training, particularly if the patient is breathing in a regular manner such as would be the case if HRV training were combined with NFB. Thus during training, if theta decreases and beta increases in a session, it is probable that those changes are real. Your absolute values are influenced by the cardiac artifact, but you can still make reasonable interpretations of the relative changes that are observed during and across training sessions.

Figure to show how poor impedance with a loose electrode can result in a single channel showing cardiac artifact (thanks are extended to Jay Gunkelmen for his assistance in the interpretation of this artifact)

EKG artifact only at T6, but there is a reason that the EKG is seen there. It is due to a bad T6 electrode contact. (Always look at other montages and it will be clear.)
The EKG is seen due to the **high impedance at T6, and the electrode is loose**. Put a little more gel in the electrode and it disappears. (Thank you to Jay Gunkelman for suggestions.)

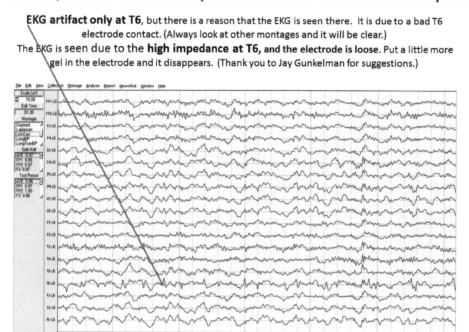

EEG recordings showing artifacts were generated using the NeuroGuide program.

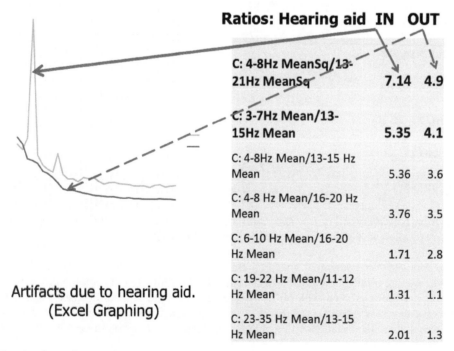

Artifacts due to hearing aid. (Excel Graphing)

Ratios: Hearing aid	IN	OUT
C: 4-8Hz MeanSq/13-21Hz MeanSq	7.14	4.9
C: 3-7Hz Mean/13-15Hz Mean	5.35	4.1
C: 4-8Hz Mean/13-15 Hz Mean	5.36	3.6
C: 4-8 Hz Mean/16-20 Hz Mean	3.76	3.5
C: 6-10 Hz Mean/16-20 Hz Mean	1.71	2.8
C: 19-22 Hz Mean/11-12 Hz Mean	1.31	1.1
C: 23-35 Hz Mean/13-15 Hz Mean	2.01	1.3

Single-channel recording at Cz with the reference to the left ear. The graph showing amplitude in microvolts (y-axis) versus frequency (x-axis) was done using Excel.

Artifact at F8 Boy age 13. This patient suggested that this might be related to **metal pole** placed into a front tooth after an injury – He is correct about this – we experimented and whenever he wiggles it this Artifact appears.

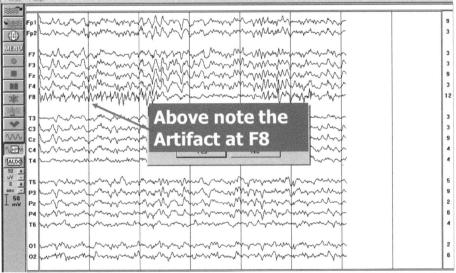

This figure is from the SKIL program.

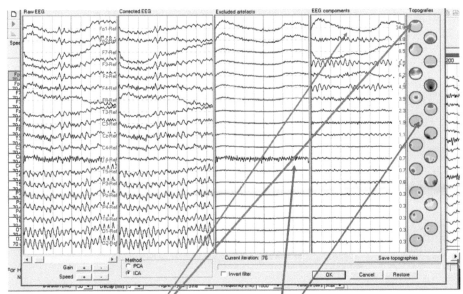

RAW EEG   CORRECTED EEG   EXCLUDED Artifacts   EEG Components
ICA: Independent Component Analysis.
We eliminated only FP1 FP2 Slow sway and T4 EMG Artifact. However, this very considerably lowers the amplitude of all components at T4.

This figure is from the WinEEG program. Indepenent Component analysis can be very helpful in removing artifacts, particularly with a persistent EMG artifact. However, it can lower amplitudes at those sites, and Thatcher has said that it will interfere with doing an accurate coherence and phase analysis.

Independent Component Analysis (ICA)

Definition

Independent Component Analysis finds independent cortical generators of potentials recorded from the scalp. In formal terms, Dr. Kropotov defines it as "a method of solving the blind source separation problem" (Kropotov, 2009). It can also be helpful for artifact removal. At the ADD Centre, we use the Mitsar (one of our 19-channel EEG instruments) along with the WinEEG program for some of our recordings. Using the WinEEG program for the quantitative analysis of the EEG, you are able to use independent component analysis to remove suspected high-amplitude artifact waves at individual 10-20 sites. Because at the same 10-20 electrode site there are many different EEG frequencies that all have different generators, the practitioner can theoretically remove one of these while leaving most of the other frequencies and waves from other generators intact at that site. Thus, a very high-amplitude slow-wave frequency due to eye blinks can be removed, as shown below. The high-amplitude slow-wave artifact is removed without removing other true EEG waves at the same site that come from other generators. In this manner, you retain the real EEG while removing an artifact. While this may not be entirely true for all artifacts, it is considered valid and is a 'good-enough' assumption when removing a small number of very high-amplitude artifacts.

Juri Kropotov, with his team in St. Petersburg, Russia, developed WinEEG.

The key assumption used in ICA is that the sources are statistically *independent*. This assumption seems to fit the EEG data quite well. Kropotov has noted in his workshops that statistical independence means that measuring a potential generated in a separate source, say the left sensorimotor strip that generates mu-rhythm, at a given moment does not allow any prediction regarding the source in another brain region, such as the right sensorimotor strip or the occipital area, at the same moment. Kropotov in his ERP workshop also notes that the assumption of statistical independence seems to be also true for components observed in ERPs.

This information is expanded in a 2009 textbook by Kropotov; the interested reader can refer to his books for more detail.

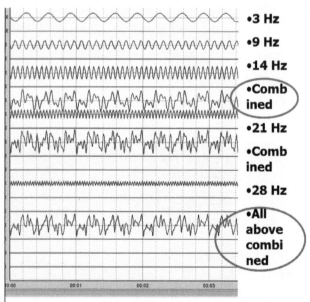

Independent components may reflect different generators for different frequencies. The amplitude of 3, 9 and 14 Hz were doubled (lower frequencies have higher amplitude).

Imagine a large ship's waves + a small boat's waves + ripples from the wind.

*The **wave you observe has many independent components** from different generators contributing to it. Here, 3 + 9 + 14 + 21 + 28 Hz from a sine wave generator make the final wave form, which resembles spontaneous EEG.*

The figure shown above is derived from the NeuroGuide program in order to illustrate how adding each of even a small number of frequencies together makes a complex wave that looks more and more like the wave form that the practitioner sees in a raw EEG.

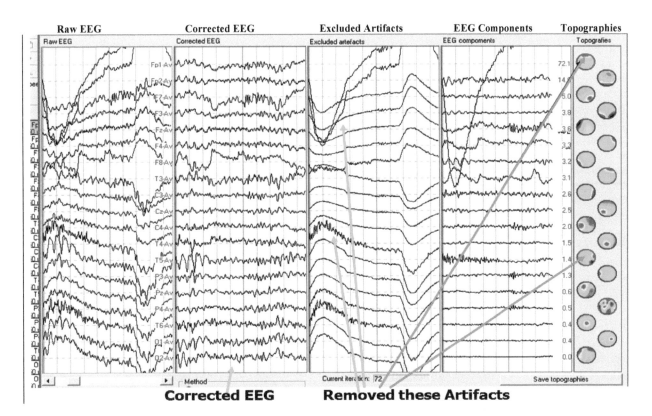

ICA: Here we removed artifact related to eye movement.
Note that the sites from the top down are: Fp1, Fp2, F7, F3, Fz, F4, F8, T3, C3, Cz, C4, T4, T5, P3, Pz, P4, T6, O1, and O2, and each site is referenced to the average of all sites.

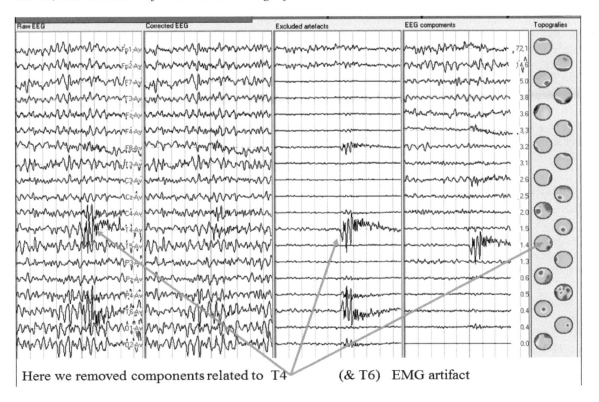

The Neurofeedback Book, Dr. Thompson, (www.addcentre.com).   Published by the AAPB. (www.aapb.org)

The above two figures were generated using the WinEEG program. Note that not all the artifacts are removed. Only removed were the artifacts that had extremely high amplitude.

When using this program, the practitioner will quickly notice that removal of more than one or two large artifacts results in a reduced amplitude at the sites where artifacts were removed. To remove an artifact, the practitioner left-clicks the red high-amplitude site in one of the topographies shown on the right hand side of the figure. In the top figure we clicked on the Fp1 and the T4 10-20 EEG sites. In the second (lower) figure we clicked on T4 and T6.

For a more complete description, the reader is referred to the textbook by Juri Kropotov (2009).

Effects of Medications on the EEG

A detailed account of the effects of medication is beyond the scope of this text. We will just give a general overview. For a more complete understanding, the reader should consult Bauer (1999), and the chapter by Bauer and Bauer in Niedermeyer and da Silva's book, *Electroencephalography* (1999).

The changes and updates in this section on medication and toxic substance effects are largely thanks to Jay Gunkelman. They are taken from his workshops, handouts, and his online article, "QEEGT, Drug exposure and EEG/qEEG findings: A technical guide by Jay Gunkelman, QEEGT." In the following, the reader will note that there is a **generally reciprocal effect between alpha and beta**: as brain stem stimulation desynchronizes the alpha generators, beta is seen. During states of under-arousal, this relationship is not seen. When the subject is alerted, initially both alpha and beta increase.

Benzodiazepines, Barbiturates and Tranquilizers

These medications can significantly increase beta activity – particularly beta over 20 Hz. There may also be a slight decrease in alpha. They also increase sleep spindles (Fisch, p 417). **Neuroleptics,** however, are found to generally increase alpha power and reduce beta power (Hughes & John, 1999)

Barbiturates (Phenobarbital, etc.)

Rhythmic 18 to 26 Hz activity is initially observed frontally, but can spread to the entire cortex. Increased dose will result in an increase in slowing, then the beta activity will decrease and the slowing predominates. The voltage of the EEG at this point will decrease and can even go to an isoelectric (absence of electrical activity) pattern when the patient enters a barbiturate-induced coma.

Benzodiazepines (Valium, Ativan, etc.)

These medications can decrease alpha and increase beta activity, particularly over 20 Hz. The beta may be a spindling beta. Intravenous Valium is used for its antiepileptic qualities in cases of *status epilepticus.*

Neuroleptics

Tranquilizers, such as *chlorpromazine* or its equivalent, increase the coherence of the EEG and decrease beta. They can increase temporal and frontal sharp theta transients. There is a reduced alpha blocking with sensory stimulation. This may correlate with the memory disturbance that can occur with these medications.

In addition, Gunkelman notes that in cases of *dopamine receptor hypersensitivity* (which produce symptoms of tardive dyskinesia such as restlessness, muscle stiffness with pain, tremors, purposeless movements, and other extrapyramidal side effects) and there are prolonged bursts of mixed fast/sharp transients and slowing. There is a potentiation of latent epileptiform activity, even with lower doses.

Thioridazine also increases faster activity, accounting perhaps for its commonly reported antidepressant effects.

Marijuana/Hashish/THC

This drug will increase frontal alpha. With chronic exposure there can be increased frontal interhemispheric hypercoherence and phase synchrony. Effects on the evoked potentials have been noted as well. Unfortunately there can be other serious side effects, including a twofold increase in the incidence of schizophrenia among young people who started using marijuana in their teens. Even more common is the risk of amotivational syndrome. In the clinic, the increase in alpha can easily be seen the next day. If you know that your patient uses marijuana, ask them to refrain for a few days before the recording in order to get a true baseline. Most of the young people who use marijuana have been shocked that there could be such a marked effect on their brain activity, and that knowledge changes their recreational use of marijuana. This, coupled with Dan Amen's pictures of SPECT scans of the brains of chronic marijuana users, which show "holes" where blood flow has been decreased, have been quite effective in encouraging teens to re-evaluate their behavior. Most quit the habit early in training.

We also had a parent phone confidentially after he sat through his daughter's assessment to reveal that he had been smoking marijuana regularly for 14 years. After learning about the effects on the brain, he went through the Employee Assistance Plan at his company and did a six month treatment program. It was an unexpected example of the power of educating clients and their families about brain functions during the initial assessment.

Antidepressants

The tricyclic antidepressants, such as amitriptyline and imipramine, will produce an increase in slow (theta) activity, and can also decrease alpha and increase fast beta frequencies in the mid- to upper-20 Hz range. Amitriptyline in particular will produce more slowing, and very small doses may be used to help a patient fall asleep. Iproniazid produces a smaller increase in slower activity but a larger increase in fast (beta) activity. In addition, tricyclics can produce generalized asynchronous slow waves and spike and wave discharges. They may also increase sleep spindles.

The monoamine oxidase (MAO) inhibitors, such as isocarboxazid, can increase 20-30 Hz activity but decrease slower and higher frequencies, which is rather similar to the effects of stimulant medication.

The selective serotonin reuptake inhibitors (SSRIs), such as Prozac, Paxil and Zoloft, will produce a mild frontocentral beta increase in the general range of 18-25 Hz with a decrease in alpha anteriorly.

Lithium

Lithium can produce generalized asynchronous slowing, and some slowing (decrease in peak frequency) of alpha. It may also increase theta and increase beta. Unfortunately, there is a strong potentiation of latent epileptiform activity. The reader will note that this mimics the tricyclic antidepressant profile, but it may show more fast activity.

Lithium overdoses produce a marked slowing of the EEG, with triphasic discharges. This may be due to liver toxicity and the associated metabolic disturbances, similar to the findings in *hepatic encephalopathies*. Unfortunately, post-hospitalization, these slower findings (down to even as slow as 4 Hz) may be noted many weeks following discharge. In a child, normal 9 Hz alpha may return only after many months.

Phenothiazines, Haloperidol and Rauwolfia Derivatives

These medications may slow alpha (decrease the peak frequency) and produce asynchronous slow waves even at nontoxic doses. There may also be increased synchrony.

Alcohol

The immediate effect of consuming (acute use) an alcoholic beverage is an increase in the amplitude of theta and low-frequency alpha. This is not a surprise as most people feel somewhat sleepy after consuming alcohol. It also decreases beta frequencies above 20 Hz. The increase in alpha and decrease in beta could reduce feelings of anxiety or a tendency to ruminate. Drinking to "forget your worries" is thus a real phenomenon. People who are alcoholic, and perhaps people without a current drinking problem who might be at risk for alcoholism, often show an increase in beta (usually above 20 Hz and around 24-26 Hz) and decreased thalpha (6-10 Hz) and alpha. These differences from the profile found in most people may also be seen in family members who have no history of alcoholism but may be prone to developing a problem. This pattern has been noted in children whose parents or grandparents had alcohol-related problems. An interesting hypothesis is that people prone to becoming alcoholics may 'normalize' their EEG by decreasing fast beta and increasing alpha with the consumption of a small amount of alcohol. Continuing to consume more alcohol at one sitting, however, results in severe problems, both in the acute phase of being drunk and, long term, due to developing dependency and an addiction that leads to increasing dysfunction at home and work.

Stimulants

These medications can produce some increase in beta with a decrease in theta, but the effect is small and difficult to detect. Methylphenidate (*Ritalin, Concerta*) produces a decrease in delta and theta, with a more pronounced posterior alpha increase and an increase in low beta, with effects delayed up to six hours, compared to the more rapid effects of the amphetamines. The theta decrease may, in part, be secondary due to increasing alertness. Note that stimulant drugs have major effects on brain neurotransmitter activity. Cocaine and Ritalin, for example, are taken up by the basal ganglia (Amen, 1998, p 86) and enhance dopamine availability. With cocaine the onset is rapid, giving a high feeling. Ritalin prescribed in the usual therapeutic doses is slow to produce effects and thus is not considered to be addictive. The cocaine 'reward' effect may be due to its stimulation of the ventral tegmental area

(Bozarth, M.A., 1987, Neuroanatomical boundaries of the reward-relevant opiate-receptor field in the ventral tegmental area as mapped by the conditioned place preference method in rats. *Brain Research, 414*, 77-84.)

Note that amphetamines require just half the dose for the same effect: 5 mg of amphetamine is equivalent to 10 mg of methylphenidate. In people with ADHD, people with excess theta are usually positive responders to stimulants, whereas the inattentive presentation that often has high alpha, or the subtype with high-amplitude, high-frequency beta who get distracted by their worries, usually are poor responders to stimulants and either derive little benefit or suffer more side effects. People with the high-beta subtype often experience increased anxiety.

Research in the 1970s at the Hospital for Sick Children (HSC) in Toronto looked at the dose response curve using a paired associate learning task. They found that low doses improved learning performance, but that higher doses required to get a behavioral response of decreased impulsivity and hyperactivity actually interfered with learning. There were also state-dependent effects on learning; that is, material learned in one state (either on or off meds) was better recalled in the same state. Other studies have demonstrated that growth hormone, which is normally secreted at night, is secreted during the day when a child is taking Ritalin. The result, especially when combined with the loss of appetite that is a very common side effect, will be an inhibition of growth. Growth suppression of about 1 cm per year was found in the three-year follow-up of The Multimodal Treatment of Attention Deficit Hyperactivity Disorder (Jenson et al., The MTA Cooperative Group, 2004; The NIMH MTA follow-up: changes in effectiveness and growth after the end of treatment. *Pediatr113*:762-769). This study was the largest study ever undertaken on the treatment of ADHD, but it had a serious omission in that it did not include NFB treatments.

Gunkelman notes that the arousal level changes the EEG responses. When a stimulant is given to an underaroused subject it may increase alpha, whereas in a normally aroused subject, stimulants decrease alpha, and in an anxious subject (low-voltage fast EEG variant) alpha will not be changed by a stimulant.

Caffeine and Nicotine
These substances will suppress alpha and theta. Withdrawal may result in an increase in alpha and theta frontally. One reason that some people give for being reluctant to quit smoking is that they do not wish to lose the mental sharpness that smoking a cigarette provides due to the effects of nicotine, which has similar effects to caffeine but is much more addictive. Indeed, nicotine is one of the most addictive substances, and research in rats has shown that a single exposure can create craving. Morning coffee is a way to wake up your brain due to its caffeine content. Caffeine is a moderate stimulant that has a moderate length of effect, about 6-8 hours. There can be alpha and theta increases with withdrawal, maximal the second day, resolving with resumption of caffeine consumption. The degree of change in both frequencies corresponds well to the subjective withdrawal severity.

LSD, Cocaine, and PCP (Phencyclidine or Angel Dust)
LSD and cocaine both increase fast activity. LSD, however, will decrease alpha, whereas cocaine tends to increase it. Phencyclidine (PCP) increases slow activity. With Lysergic Acid Diethylamide (LSD-25), the baseline EEG seems to determine the effect, with decreased alpha and increased beta from a normal background. With slower EEGs, however, there is an increase in both alpha and fast activity. The low-voltage fast EEG shows little change in spectral profile with exposure. The stimulant effects of this powerful drug may cause convulsions at higher doses.

PCP, phencyclidine, or angel dust shows a marked increase in slow activity, with paroxysmal activity and extreme voltages noted with increased dosage. Convulsions have been reported.

Heroin and Morphine
Shortly following administration, there is increased alpha, with slowing of alpha peak frequency during the euphoric high. With increased dose there is increased slowing, and like barbiturates, the EEG may go isoelectric. There may also be an increase in theta and delta. There is an increase in REM sleep noted with opioids.

Antibiotics
Antibiotics can result in a marked increase in the amplitude of slow waves including theta. This can be important to note in a clinical setting where the theta/beta ratio is one of the assessment tools used in deciding on a NFB intervention for ADHD. A parent may bring their child when they assume their child is no longer infectious and they are just finishing off the last days of a course of antibiotics.

Antihistamines

Antihistamines, and even so called nonsedating antihistamines, can result in a marked increase in the amplitude of slow waves including theta. As is true with antibiotics, this can be very important in a clinical setting where the theta/beta ratio is one of assessment tools used in deciding on a NFB intervention for ADHD. When a child has a chronic allergy, such as hay fever, a parent may not think of mentioning it because their child has always been on antihistamines and they no longer really think of these over-the-counter drugs as "medications."

Thyroid Hormone Levels

Post-thyroidectomy, beta is increased and alpha decreased. Alpha peak frequency is decreased. After treatment with thyroxine the EEG returns to normal range (Pohunková et al., 1989).

Methanol

The EEG shows marked slowing, which correlates with the extent of *acidosis* more than the blood levels of methanol. This has been shown to be quite *neurotoxic*, with optic nerve blindness noted commonly in chronic abuse/exposure.

Solvents

The EEG shows slowing; that is, the dominant frequency, alpha, becomes lower. Since higher peak alpha has been found in association with higher IQ and better problem solving, and a slowed peak alpha frequency is seen after head injuries and especially in coma, this is not desirable.

Hormones

Vasopressin, usually in the form of DDAVP (desmopressin acetate), increases amplitudes in the high-alpha band.

Cyproterone acetate is an antiandrogen with clinical effects on premenstrual complaints. A decrease in frontal alpha and increased beta may be observed.

Tuberculostatics

INH, *isonicotinic acid hydrazide*, is an irritant to the CNS. Large doses can hypersensitize the CNS with the EEG showing bursts of paroxysmal activity when exposed to photic stimulation.

Mercury

With initial exposure to this neurotoxin (and many other heavy metals) there is an increase of faster activity. With higher concentrations there is an increase in both fast and slow activity, with eventual paroxysmal activity of an epileptiform nature.

Organophosphates

The insecticides are known to form *peripheral neuropathies*, though they also have central actions. The EEG shows slowing and paroxysmal bursts. However, if the exposure is high enough to induce coma, there can be paradoxical spindling fast activity.

Chlorinated Hydrocarbons

Also insecticidal, these chemical compounds are fat soluble. Stored and accumulating to a toxic level, they are known to cause convulsions. Neurologically, there are bi-temporal sharp discharges and anterior slowing; that is, increased theta or even delta activity. Rarely, spikes may be observed, with or without convulsions.

Lead (organic)

Cerebrotoxic effects can be strong. IQ points may drop significantly even with trace measurable exposure. Dementia progresses with increased exposure, with eventual convulsions. The EEG shows diffuse slowing in subacute exposure, with increased exposure leading to paroxysmal discharges. Inorganic lead has weak cerebrotoxicity.

Aluminum

Increased aluminum is commonly seen in *dialysis encephalopathies*, with *myoclonic* activity seen behaviorally. Though not well documented, the EEG shows slowing and also excessive fast activity. At autopsy, the aluminum is found concentrated anteriorly in the brain.

Other Rare Events

Toxic Materials

This subject is beyond the scope of this book. Poisons such as lead, aluminum, mercury, various insecticides and chlorinated hydrocarbons will all cause major clinical and EEG changes.

Withdrawal

Withdrawal from medications may have an effect on the EEG. This is especially true if the withdrawal is sudden and after a long period of time on that medication. Generalized epileptiform activity may be seen.

Encephalopathies

These are beyond the scope of this text. You should always ask your client about medical illness. The EEG may be quite confusing to you, showing bisynchronous and asynchronous slow waves and generalized epileptiform discharges. If you do an

EEG on a client with any chronic illness, we would strongly advise that you get a second opinion from an expert in reading EEGs.

Artifact – Other

Hyperventilation may cause an increase in frontal theta and delta activity.

Sweating produces large, slow, up and down movements of the EEG line, like low rollers on the ocean surface. This slow undulation of the EEG may in part be due to a loosening of an electrode's connection to the surface of the scalp. This artifact is most often seen frontally (not maximal at F7, F8 like eye roll), and it does not reverse phase like horizontal eye movements do. `It is usually seen in more than one channel. You may also be recording electrodermal responses (EDR) when doing feedback. These sympathetic nervous system cholinergic responses are also seen in the EEG. Usually they give an isolated slow-wave pattern, 1 or 2 Hz, lasting only one or two seconds. They are usually frontal and central. They are most often seen in response to a sudden unexpected stimulus such as a loud noise or a visual stimulus. The distance from the brain to the finger tips is relatively long and the EDR reaction seen in the hand or fingers will, therefore, occur slightly after it is seen in the EEG.

Bridging between electrodes occurs if you use too much electrode paste or if the client is sweating or has a wet head. It is an electrical short circuit between two electrodes. An example of this is given above under Site Preparation. (There can be rare instances when the bridging is due to a problem with the cap or hardware.)

Electrode Pop, as noted above, refers to an electrode suddenly losing its connection with the skin surface. It is usually seen as a very large abrupt deflection in one or more channels. A more subtle 'popping' of an electrode may be seen as an irregular series of spikes in a single channel (Dyro, 1989, p 14). If you observe unusual activity of this nature, check your electrode connections and check your impedances. It might also be that you have a faulty wire.

Evoked Potentials may also be called a 'transient.' These are seen as a single abrupt change in the recording that is usually observed in several channels. As a general rule you will only reject that epoch if it is 50 percent greater than the background activity. These changes increase the variability and reduce the reliability of the recording, but have little effect on averaged data (Hammond, 2001). A visual evoked potential can occur between 80-150 ms after a flash of light and is positive.

Parkinson's (PD) patients may produce a slow rhythmic theta in the occipital regions due to head tremor (Westmoreland et al, 1973). It is abrupt, occurs at a single frequency, and presents as a change isolated from the rest of the background activity. We have observed low SMR activity and increased alpha in those with PD.

Preventing Artifacts

Before clients come for their assessment, they should be asked to shampoo their hair twice or even three times. They should not use any hair conditioner or hair spray. If it is raining, they must keep their hair dry. You should also request that they go to bed early and get nine hours of sleep if they can. They should refrain from drinking beverages containing caffeine on the day of their assessment. When recording, make sure that no electrical appliances such as table lamps or portable telephones are turned on. The impedances must be below 5 Kohms and within 1 Kohm of each other. It is particularly important that impedances for the two ears compared to the ground be the same for linked-ear references (Hammond, 2001). It is also very important that homologous (interhemispheric) electrodes be balanced as perfectly as possible, because, as you become more sophisticated in your work, you are likely to do types of neurofeedback training that require you to accurately discern differences between homologous sites on the left and right side of the head (Hammond, 2001; Weidmann, 1999; Davidson, 1998; Heller, 1997; Rosenfeld, 1997). Differences in impedances between electrodes can make it falsely appear as if there are differences between cerebral activity in the two hemispheres.

When doing a 19-channel assessment, have the client sit in a comfortable chair. A small towel rolled up behind the neck may assist some clients to relax their neck muscles. This will also assure that the back of the cap is not touching the chair. Infrequently, it may be necessary to gently place a finger over closed eyelids to discourage eye movement. It is helpful for most clients if you present as calm, quiet, and reassuring. Ask the client to relax their shoulders, jaw, neck, and tongue. Some clients tend to hold tension in their jaw muscles. Suggesting that they allow their teeth and lips to rest slightly apart, with their tongue floating, may help them relax their jaw musculature.

Clients appreciate knowing as much as possible about the procedure you are doing. They also appreciate the care you are taking to assure an accurate recording of their brain waves. Tell them you need to get familiar with how artifacts look in their EEG. With the equipment running and the EEG visible on the screen ask them to move their eyes to the left, right, up and down, both with eyes open and eyes closed. With eyes open ask them to blink, move the tongue up, down, and sideways and push it against their palate, teeth and cheeks. Next, ask the client to say "lift" and tense their jaw and wrinkle their brow. A short form for entering your observations when the client does each of these actions can be copied from *The Art of Artifacting* (Hammond & Gunkelman, 2001). This book will also give you practice in identifying artifacts. Make sure your client is not chewing gum. You may need to ask some clients to remove their dentures or dental appliances. We ask clients who wear contact lenses to also bring their glasses since they may blink more while wearing lenses. Pause the recording after each artifact is produced and discuss it with the client.

Now that you and your client have observed various possible artifacts, you can record and play back a short segment of EEG. Look at this segment of EEG to see if there are any major artifacts which you should try to reduce with the client. With many clients you will find EMG activity in one or other temporal leads. With these clients you will have to find some way to help them relax their jaw so that the EMG activity decreases. Sometimes you will observe occipital EMG due to neck tension. Do the best you can but don't be discouraged; there will always be the occasional person who is just not able to decrease muscle tension. Even with this careful preparation you may have to pause the recording if you begin to see artifact reappearing. A few clients will begin to fall asleep when you are recording the EEG. As mentioned previously in this section on artifacts, you must watch for this. If it happens, you can have the patient wiggle their fingers, arms and legs from time to time in order to maintain alertness.

As previously noted, to avoid 'change in state' data affecting your overall statistics, you should have your client be in the required state (eyes closed (ec) or eyes open (eo) or activity (reading or math) for at least 30 seconds before you begin recording.

Last, and very rarely encountered, if you see spike and wave activity frontally, ask your client about fillings and possible jaw or facial surgery. Different metals in the oral cavity may cause this kind of artifact.

Always ask your client about medical conditions and medications and record this in your notes. Send EEGs from these clients to a professional electroencephalographer and neurologist to be interpreted.

False Impression of Frontal Slow-Wave Activity

The following illustration is not an example of an artifact. Rather, it is an example of how the apparent EEG waveforms can be misinterpreted. This example is given as a caution to warn you to make sure that what you see is really what you've got, and to demonstrate the value of looking at the raw EEG and being able to remontage it. In the case shown below this was necessary in order to clarify the source of the apparent frontal slow wave (alpha). In this case the ears, which are usually electrically inactive, were contaminated by the alpha activity in the temporal regions. We are always measuring the potential difference between two electrodes. We usually are safe in making the assumption that the linked-ear reference electrodes have very little electrical activity compared to the scalp electrode. The example below is a relatively infrequent occurrence where the high-amplitude electrical activity in the alpha range was located in the ear electrode as compared to the usual state of affairs where the scalp electrode has the higher electrical activity.

The figures below show the same data using two different montages.

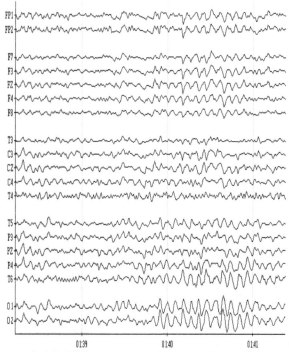

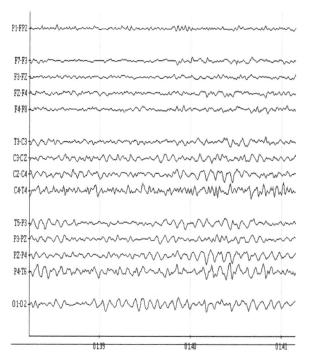

Sample of EEG with eyes closed using a referential linked-ears montage. Note the suggestion of frontal dominant slow wave activity.

This is the same sample of eyes-closed EEG but viewed with a transverse sequential montage. Note that the frontal dominant activity is eliminated. This demonstrates that the origin of this apparent frontal slow-wave activity is really slow-wave activity near the ear reference electrodes. This is termed "reference corruption." It could be quite misleading both in terms of diagnosis and with respect to prescribed intervention.

Looking at the EEG Recording

Some programs, such as SKIL, will automatically reject the first 30 seconds of recording. This helps avoid what is called a *state change*. The clinician should then remove all other visible artifacts before accepting the remainder of the record for analysis. This must be done carefully and severely. With possible artifacts remember the two important adages: "Garbage in means garbage out," and "When in doubt, throw it out."

Some assessment programs have what is termed *automatic artifacting*. This is mainly helpful in removing frontal eye blink artifact. It is never a substitute for careful artifacting on the part of the practitioner. Automatic artifacting will sometimes eliminate good EEG data. It will also miss artifacts that do not reach a required amplitude. In addition, it may not pick up artifacts that occur outside of the frontal and central areas.

EMG activity that is consistent throughout the record, and therefore cannot be eliminated, is usually readily identified in 1 Hz topographic maps and in power spectrum diagrams. Identification is relatively easy because this EMG activity will usually be morphologically different from the EEG waves, at a high frequency, and be isolated to specific locations, such as T3 and/or T4.

Pathological Activity

Absence Seizures

Absence seizures are reasonably easy to distinguish in the figure below. You are not trained to recognize abnormal EEG activity. Nevertheless, it is important that you be able to spot this high-amplitude, three-per-second spike and wave activity. It is important because these children may be misdiagnosed as being ADD (ADHD, Inattentive Type) and subsequently come to you for NFB treatment due to classroom behavior that is inattentive. They should be seen as soon as possible by a neurologist. There is **no** published research evidence at this time that this type of seizure activity can be helped by using neurofeedback. The research on using neurofeedback to reduce seizures has been done with patients whose seizures had a motor component. We have worked with children who had absence seizures combined with ADD. They were under the care of a neurologist and taking medication. Some have demonstrated a decrease in seizure frequency as well as in the ADD symptoms, but this may just have been coincidence. Some children with absence seizures develop grand mal or partial complex seizures as they grow older. It is unknown at this time whether SMR training might have a protective influence with respect to this complication.

Absence seizure with spike and wave activity seen at Cz

Uniform three-per-second spike and wave pairs, which are characteristic of an absence (petit mal) seizure. Note the very high amplitude (>160 µv) of these waves.

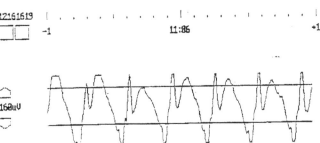

Closed Head Injury

Severe closed head injury may show rhythmic delta activity. These clients should always be assessed and treated medically first. Hughes notes that: "There is a broad consensus that increased focal or diffuse theta, decreased alpha, decreased coherence, and increased asymmetry are common EEG indicators of the postconcussion syndrome" (Hughes & John, 1999, p 198). If you are going to do NFB you should consider doing a full-cap assessment first. It is likely that the EEG assessment would demonstrate decreased communication between some areas in the brain. This may be helped using coherence or comodulation training.

Hypoglycemia

Hypoglycemia may be accompanied by an increase in theta and delta wave activity. This should normalize when blood sugar levels are restored to normal.

Grand Mal and Partial Seizures

Grand mal and partial complex seizures both produce bursts of sharp wave activity. These patients must always be first assessed and treated medically. It would be very unusual for you to be the first person to notice this activity. However, you may see clients with subclinical EEG seizure activity in the form of frequent paroxysms of waves as is illustrated in the figure below. You are not qualified to make this kind of diagnosis. If you see bursts of waves you do not understand then immediately have the client obtain a neurologist's opinion before you agree to do NFB work with them. The example below is of a subclinical seizure in an eyes-closed recording. It shows spike and wave activity in the posterior left frontal lobe from a seizure focus near F3. This client's parents had never observed a seizure that showed any motor or sensory activity. He was brought to the center with a diagnosis of ADHD inattentive type due to difficulties with his attention span in school. He was sent for neurological evaluation and the EEG findings were confirmed.

Subclinical seizure activity: Sequential (bipolar) longitudinal montage of a sample of the resting EEG with eyes closed

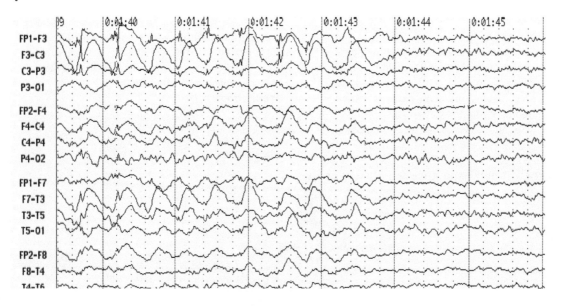

The same patient, but here a linked-ear montage is used. Subclinical seizure activity: Linked-ear referential montage of a sample of the resting EEG with eyes closed

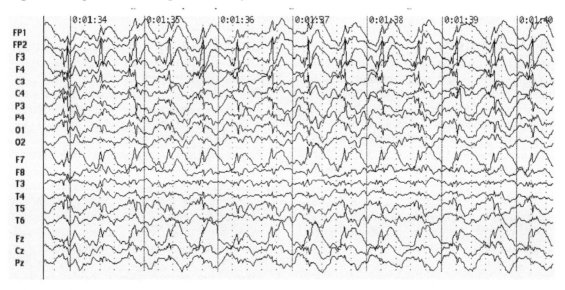

Unusual Analyses

A separate analysis of transients

Occasionally you may see bursts of unusual activity. These clients should always be under the care of a neurologist in addition to doing neurofeedback. It would be unusual for you to be the first person to see this kind of activity in a client but it has happened at our Centre. In one case the paroxysms were bursts of high-amplitude slow waves and sharp waves. This teenager turned out to be having myoclonic seizures but he had not mentioned to anyone his occasional sudden collapses in the shower and at home. He tried medications prescribed by a neurologist after he was referred for medical assessment but debated whether the side effects were worse than the symptoms. We insisted that he make all decisions about medications with his neurologist. He also had symptoms of Attention Deficit Disorder. We therefore agreed to work with him to decrease his dominant slow wave and learn metacognitive strategies to use in school. In addition, however, we followed the published recommendations of Sterman and put a special emphasis on also increasing his SMR.

The figure below is a sample of the EEG of a 19-year-old man with a history of frequent myoclonic seizures. He came to us because he tuned out frequently and, therefore, couldn't follow arguments the teacher was developing in classes. These bursts were frequent throughout the entire record with eyes closed. This is an example of the type of case where neurological consultation and treatment is essential.

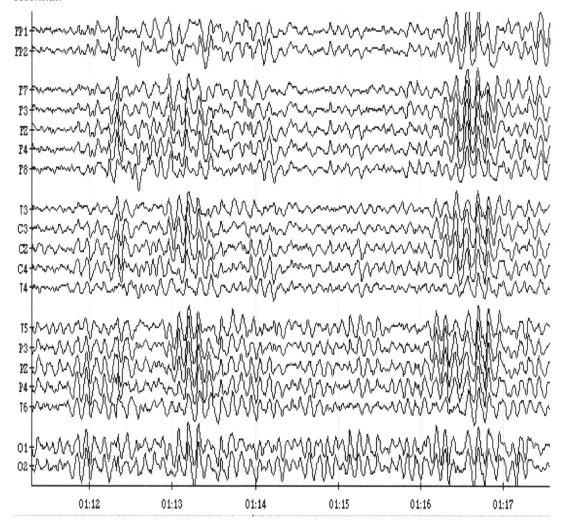

When you do the EEG and see paroxysms of unusual waves, it may be helpful to do two separate statistical analyses. First, mark each of these paroxysms as if it were an artifact, omit those segments and do your statistics. If you are doing a full-cap, look at the topographics with these segments eliminated. Second, analyze only these segments. The location and the characteristics of these waves may then clearly stand out.

'Unusual activity' can be waves that represent pathology. On a 19-lead assessment using a sequential montage, you might occasionally see what is called a *phase reversal*. In these instances, the EEG wave on one line (from one pair of electrodes on the scalp) seems to point down to a rise in a wave in the EEG from another pair of electrodes. These waves may, in fact, be pointing to a site of pathology that is the origin of seizure activity. Presumably such a client is already under the care of a neurologist. If not, require that the client see the appropriate medical practitioners as soon as possible.

Can the EEG Distinguish Subtypes of Some Disorders?

Distinguishing subtypes of seizure disorders has been done by neurologists for many years. The late E. Roy John and Leslie Prichip at New York University (NYU) have specialized in correlating brain mapping data with psychiatric disorders. Their sophisticated neurometrics can, for example, distinguish subtypes of depression and predict medication response. Now it is becoming possible to distinguish a few different EEG subtypes of ADD. In the example of Peter, we saw a distinct subtype of ADD where slow-wave activity was not picked up in the central leads but was in the frontal region. A full-cap allows for an even more precise analysis of EEG subtypes and this is outlined in Section X under ADD.

Research Using Full-Cap Assessments to Distinguish Psychiatric and Neurological Disorders

Introduction

This discussion is for a more advanced book concerning research findings and the EEG so it will only be touched upon here. Most of these findings are not yet being utilized by NFB practitioners. Sometimes brain maps are being used to guide psychiatrists as to which medications a patient would most likely respond. This work has been done extensively by the late E. Roy John and Leslie Prichep at New York University in Manhattan. We refer the interested reader to articles by E. Roy John, Leslie Prichep, Frank Duffy and Bob Thatcher (Hughes, 1999; John, 1989; John, 1988; Duffy, 1994; Prichep, 1993; Thatcher, 1989).

A brief summary of some recent findings has been reviewed in the article "EEG Findings in Selected Neurological and Psychiatric Conditions" (Hughes, 1999), and what follows is a very brief summary of the highlights.

Various brain imagining techniques have given information concerning structural and functional pathology in mental disorders. These include: QEEG (quantitative electroencephalogram), MRI (magnetic resonance imaging), PET (positron emission tomography indicating regional cerebral metabolic rate), SPECT (single photon emission computed tomography indicating regional blood flow), MEEG (magneto-encephalography), ERP (event-related potentials), LORETA (low-resolution electromagnetic tomography). Hughes and John state that these techniques have "unequivocally established that mental illness has definite correlates with brain dysfunction." They go on to state that "QEEG and ERP methods afford the psychiatric practitioner a set of noninvasive tools that are capable of quantitatively assessing resting and evoked activity of the brain with sensitivity and temporal resolution superior to those of any other imaging method." A few of the findings in various conditions are summarized below.

Dementia

In *organic delusional states,* increased slow-wave activity may be observed over both temporal lobes. In delirium there is slowing of the EEG with increased theta and delta activity.

In *Alzheimer's* there is a diffuse increase in slow-wave activity and a decreased mean frequency of the dominant alpha activity. The slowing is proportional to the severity of the dementia. The dominant eyes-closed alpha peak in normals is 10 Hz and it may be higher and be seen at 11 or even 12 Hz. In patients whose *mental capacities*, including memory, are *decreasing,* the peak frequency is lower and this may be calling 'slowing' of the peak frequency or 'slow alpha.' In dementia, in addition to the increased theta, and in more severe cases, increased delta activity, there is a decrease in beta. There is also a decrease in occipital alpha power. In *frontotemporal dementia* (*Pick's* disease), the changes are more localized. These EEG changes are focal in *multi-infarct dementia*. These EEG changes, however, are not present in *depression,* which is an important factor in making an accurate differential diagnosis and distinguishing dementia from depression in the elderly.

Cerebrovascular Disease

Decreased regional cerebral blood flow correlates with EEG slowing.

Mild Head Injury

Increased theta, decreased alpha power and/or decreased coherence and asymmetry are characteristic; examples are given elsewhere in the text.

Schizophrenia

There is a relatively low mean alpha frequency and alpha power and an increase in beta activity in patients with schizophrenia. An EEG assessment, however, may not demonstrate this if the patient is on neuroleptic medications, due to the fact that these medications typically increase alpha power and decrease beta activity. There may also be an increase in frontal theta and delta activity. Not all characteristics are found in all patients. There appear to be subgroups that respond differently to different medications. Other changes include a decrease in Stage III, IV, and REM sleep. There is also increased delta in the left anterior temporal region. There is increased interhemispheric coherence, which distinguishes patterns in schizophrenia from the findings of decreased coherence in depression. This may be helpful in the differential diagnosis of schizophrenia from bipolar illness.

In work at the ADD Centre we have noted that when the client is 'hearing voices' and/or having intense delusional thinking that the EEG shows peaks in the high beta range, often somewhere between 21 and 32 Hz. These peaks disappear when the client is assisted to externalize and normalize their thoughts.

Rod, age 20, had been a client of the Centre when he was 12 years old. He was diagnosed with ADHD. He did very well, became a straight-A student, and went into higher mathematics at the university. At age 19 he began having delusional thoughts. He was diagnosed with schizophrenia. He was no longer able to maintain his focus and concentration on his work and failed his year. Medications helped him stay in touch with reality, but he couldn't get motivated and couldn't concentrate. With his family's help, he returned to the ADD Centre to see if he could improve sufficiently to return to school. He demonstrated high-amplitude beta peaks at 21 and 25 Hz. When these occurred, a dip was observed in 13-15 Hz activity. The author gave him an academic task and asked for his help in critically analyzing material the author had written in order to give suggestions for improvement. Rod did this and as he worked the high beta dropped sharply and 13-15 Hz activity increased. However, at times there would be a sudden reversal, 13-15 would drop and 21 Hz and 25 Hz rose. The trainer intervened and asked what had happened. In every instance the delusional thinking had returned.

Depression

Decreased activation of the left frontal area compared to the right has been used to guide NFB treatment. Studies have shown increased theta and/or alpha power. This may not be observed if the patient is on antidepressants as these medications generally reduce alpha activity. Coherence is decreased between the frontal lobes. Six-per-second spike and wave complexes have been reported, particularly in the right hemisphere.

More recent work has found anxiety plus depression results in high-frequency beta spindling, and a small number of patients with 14 Hz beta spindling. Dysfunction in the region of Brodmann area 25 has become a major finding. These findings are discussed in Section XXIV.

Bipolar Disorder

In contrast to the unipolar depressions described above, alpha appears to be reduced and beta increased.

Anxiety and Panic

There is usually a decrease in alpha activity. In panic disorder, paroxysmal activity may be observed. Temporal lobe abnormalities are also found. (Note, however, in adults with ADHD, anxiety is a common comorbidity. These individuals may have high amplitude alpha.) We have also observed increased beta activity between 19 and 22 Hz, though this is not reported in Hughes' paper.

Obsessive Compulsive Disorder (OCD)

There are two groups of patients with OCD. In the first group there is an increase in alpha relative power. These patients usually respond (82 percent) to selective serotonin reuptake inhibitor (SSRI) interventions. In a second group of OCD patients there is an increase in theta activity, and about 80 percent of these patients do *not* improve with SSRI medications.

Learning Disabilities

Diffuse slowing (theta and/or delta) and decreased alpha and/ or beta activity have been reported. Since there are many different kinds of LD, it is not surprising that EEG findings are less consistent than in other disorders. Examples including dyslexia are described elsewhere in this text.

ADHD

As noted in other areas of this text, increased theta and/or low alpha may be seen centrally and/or in frontal locations. In addition, there may be hypercoherence and interhemispheric asymmetry. See also Section XXVI.

Alcoholism

Increased beta activity is found in people with alcoholism. There is a high incidence of comorbidity with ADHD in adults and the typical EEG findings of ADHD may therefore be somewhat masked. The increased beta makes the theta/beta a less accurate guide when there are also problems with addictions.

Additional Data Interpretation – LORETA

(Low Resolution Electromagnetic Tomography) *With thanks to Joel Lubar for this description*

Almost 150 years ago, two physicists, Green and Gauss, independently developed a mathematical theorem showing that if there is a distribution of electrical activity or charges on the surface of a three-dimensional hollow object such as a sphere, then by means of a complex vector analysis it is possible to localize inside the object the generators of the surface distribution of electrical activity. This procedure is known as the inverse solution. In 1994, Dr. Roberto Pascual Marqui and his team in Zurich, Switzerland, implemented this inverse solution with some appropriate modifications and updating to provide a methodology known as *low-resolution electromagnetic tomography,* or LORETA. Actually, there are an infinite number of inverse solutions for determining the generators of activity inside of a three-dimensional solid with a surface charge distribution. Pascual Marqui wrote several papers in which he described the various inverse solutions and showed that LORETA was the most accurate.

In terms of practical application, the three-dimensional object that interests us – the head – consists of the skin, skull and the cerebral cortex. These three components provide what is called a three concentric shell model, and provides more accurate localization of the internal generators than a single shell model does. LORETA includes the gray matter of the brain, which encompasses the cerebral cortex and the hippocampal formation; it also localizes activity in nearby structures such as the parahippocampal gyrus, insular cortex and other internal temporal lobe structures. All of the gray matter that contains dipole generators is part of the LORETA solution space. In the original publications the gray matter is broken down into 2394, 7x7x7 mm cubes called *voxels*. For each voxel there is a three-dimensional vector, which is part of the solution in terms of localizing the internal generators of the surface EEG activity. The LORETA technique cannot be used to localize monopole activity such as found in thalamic regions, brain stem and other subcortical nuclear structures.

There are a number of studies that have shown the relationship between LORETA tomography, functional MRI, and PET scans. One of the newest developments that Dr. Lubar's laboratory has been developing is the potential for using LORETA for neurofeedback. For example, teaching an individual to change activity in the cingulate gyrus, which is known to be a generator of theta activity in the brain. This might be particularly helpful since the anterior cingulate plays a very important role in a number of complex cognitive and emotional functions. LORETA neurofeedback has now become a useful clinical tool. We have found it to be extremely effective in clients who have had a concussion or who suffer from complex disorders such as autism. For a more detailed description of LORETA and its use in both assessment and treatment the reader is referred to Section XVII on LORETA and LORETA Z-score neurofeedback.

The figure below has been copied with the kind permission of Joel Lubar.

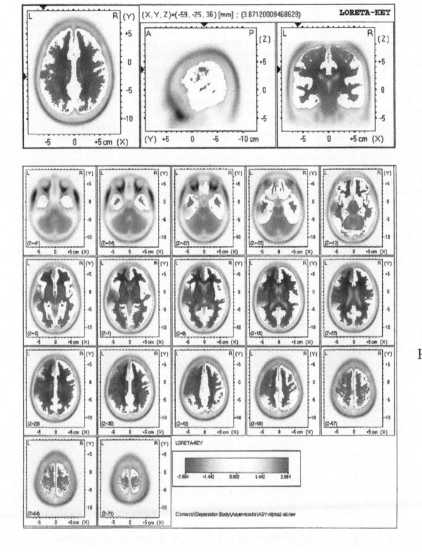

The diagram above represents the difference between 15 chronically depressed females and age-matched controls. The left-sided alpha activity is very clear. This diagram will also be found in an article by Joel Lubar and published by the International Journal of Psychophysiology, *Elsevier Science Publications. Acknowledgements are given to Elsevier Science Publications and thanks to Dr. Lubar for permission to use this figure.*

Training in Assessment Procedures

To learn how to do QEEG with one or two channels, you may attend workshops at the annual meetings of AAPB (Association for Applied Psychophysiology and Biofeedback – chapters in North America and Australia), ISNR (International Society for Neurofeedback and Research – chapters in North America, Australia and Europe), BFE (Biofeedback Federation of Europe), and workshops sponsored by manufacturers such as Thought Technology or brokers such as Stens Corporation and American Biotech. Intensive training for beginners and intermediate level practitioners may be obtained from a small number of large clinical centers such as the ADD Centre (www.addcentre.com) in Canada. You should then obtain personal mentoring from an experienced practitioner. For QEEG assessment with 19 or more channels, you must attend specific courses and obtain personal instruction.

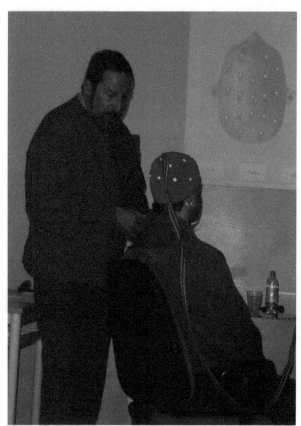

Above, Barry Sterman is demonstrating to a group of psychologists how to put on a full-cap correctly. He will then demonstrate how to collect data from 19 channels

At the time of this writing, workshops and/or training concerning 19-channel QEEG were been given by a small number of individual practitioners and by a large clinical center, The ADD Centre, in Toronto, Canada. The beginner can look for a good clinician who works in their area for assistance.

An invaluable resource for artifacting is the already-mentioned SNR publication entitled *The Art of Artifacting* (2001) by D. Corydon Hammond and Jay Gunkelman. The purchase enriches your knowledge and also the field, since the authors have donated all proceeds to the Society for Neuronal Regulation. Check out www.snr-jnt.org. For a basic textbook, we recommend *Fisch and Spehlmann's EEG Primer, Basic Principles of Digital and Analog EEG* (Fisch, 1999). For educational opportunities, check www.aapb.org, www.bfe.org, www.isnr.org www.thoughttechnology.org, and for certification information, www.bcia.org.

SECTION X
Autonomic Nervous System (ANS) and Skeletal Muscle Tone (EMG) Assessment

Introduction
In this section we will discuss the purpose of measuring physiological responses to stress and tension. The areas that can be easily measured will be listed and the methods for measuring them discussed. This is not a book on autonomic nervous system (ANS) and electromyogram (EMG) biofeedback. There are many excellent texts on this subject. George Fuller wrote a text on the basics that is very easy to read (Fuller, 1984) and still relevant. Mark Schwartz's *Biofeedback: A Practitioner's Guide* is a classic reference text. In 2003, Don Moss edited a comprehensive tome on mind-body medicine. Our overview will be brief and specific to combining biofeedback (BFB) with neurofeedback (NFB).

What Can Be Measured?
We are able to measure autonomic nervous system (ANS) activity, sympathetic and parasympathetic branches, and the skeletal muscle tone with an electromyogram. The ANS governs a large part of our daily lives. All our internal organs are regulated through this branch of the nervous system. It functions automatically and unconsciously. However, it is possible to influence this part of our nervous system consciously through the use of biofeedback. The common measurements that are used are listed below.

For the Autonomic Nervous System
- Peripheral skin temperature
- Electrodermal responses (EDR)
- Heart rate (pulse)
- Respiration
- Respiratory Sinus Arrhythmia (RSA)

For the Skeletal Muscle System
Electromyogram (EMG)

Notes Regarding the Autonomic Nervous System
This has been previously described in the neuroanatomy section of the book. It is reviewed here to give an example of how it may be explained to the client.

The ANS includes the sympathetic and the parasympathetic systems. The main neurotransmitter for the parasympathetic system is acetylcholine. The main neurotransmitter for the sympathetic nervous system is noradrenaline. The parasympathetic system is generally thought of as acting to restore the body. Its major functions are related to rest and relaxation. The sympathetic system, on the other hand, is involved in expenditure of energy with resulting increases in blood pressure, heart rate and the utilization of energy (glucose metabolism, oxygen). Sympathetic drive is associated with the fight or flight response that was related to survival for our ancestors. This type of response was first described in physiological terms by Walter Cannon in 1915. As previously noted, the release of adrenaline when a person is stressed appears to be related to improved blood clotting, increased EDR and alertness, and a concentration of blood flow to the brain and to large muscle groups that are necessary for fight or flight.

It is important to recognize that every client is different in terms of their physiology. For example, one anxious client may show a peripheral skin temperature of 70 degrees Fahrenheit, or 21 degrees Celsius, while a second, very tense client, has a peripheral skin temperature of 93 degrees Fahrenheit. With the second client you may find the tension reflected in other modalities, such as respiration, RSA (respiratory sinus arrhythmia), pulse, or the EMG. Despite these individual differences, it is nearly always very helpful to look carefully during the assessment at the synchrony between respiration and cardiac variability. With most clients it is a good starting point to teach them diaphragmatic breathing and relaxation techniques. The exception is the client who has grand mal or partial complex seizures. With these clients you must be very cautious when teaching diaphragmatic breathing because accidental hyperventilation could precipitate a seizure.

Goal of the Assessment

The goal of the assessment is to discover how a particular client responds physiologically to mental stress. These findings may then be used to set up a biofeedback program to help that client self-regulate, that is, control their own physiological responses even under stressful circumstances. In addition, practicing this control may produce an automatic, unconscious, beneficial change in that client's response to stress in the future.

This goal is really the same as our usual goal when doing work with neurofeedback; namely, to produce an optimal state of mental and physiological functioning. In this state the client is both relaxed and alert. This will broaden associative capabilities and perspective, decrease fatigue, allow calm reflection on alternative approaches to tasks, and when combined with high levels of alertness, improve reaction time and increase response accuracy. The individual will be flexible in terms of mental state and resilient in terms of their physiology.

In the first or second interview after EEG parameters have been set, we use a structured interview process with the client to fill out a single-page questionnaire. This questionnaire, an assessment profile called the TOPS Evaluation, also outlines the variables that we can measure and use to give feedback. (This form is found below under Method.)

It is crucial to remember that seldom are there simple miracle cures. You must gently but firmly help your client to generalize what they learn. The techniques for doing this include attaching a new habit, such as relaxing with diaphragmatic breathing, to old habits. These should be regular daily activities, such as driving or answering the phone. The new learning that combines both NFB and BFB must become an integral part of daily living.

Brief Overview of ANS and EMG Measures

Peripheral Skin Temperature

This is perhaps the simplest and yet one of the most useful measurements in biofeedback. A *thermistor* is placed on the pulp of the distal portion of the little finger. When the subject becomes tense, the sympathetic nervous system will cause the arterial blood vessels in the finger to constrict. The result is a cooling of the finger. Most people in a relaxed state demonstrate a temperature of 93 to 95 degrees Fahrenheit (33.9 to 35 degrees Celsius).

Even children should be able to learn to raise and lower their peripheral skin temperature. Children enjoy understanding the advantages of the hands becoming cold and wet when a person is tense or afraid. To make it interesting and fun we ask them to imagine what a caveman would have to do when a saber-toothed tiger jumped out of the woods. To save his own life he had to fight or flee. If his hand was dry, his spear would slip out of it. To fight or flee he needed all his energy conserved for his brain and his large muscle groups. Therefore. arteries to these areas dilated while other arteries, to areas which were not immediately so necessary (like the finger tips), would constrict under the influence of the sympathetic nervous system.

In addition to helping a client learn to relax, skin temperature feedback has been shown to be useful in decreasing migraine headaches (Andreassi, 1995, p 327, after Blanchard, 1978). It is also the first step in the Peniston protocol used in the treatment of alcoholic patients.

Electrodermal Response (EDR)

EDR is a measure of skin conduction (SC). It is most easily measured by placing a silver/silver chloride sensor on the distal, ventral (palm) surface of the index and ring fingers. The middle finger prevents the two sensors touching, which would cause a short circuit. The conductance in µMhos of a very small current of electricity is measured between these two sensors. (This measurement is the inverse of GSR, galvanic skin response, which is measured in Ohms for skin resistance.) Alternatively, the sensors may be placed at two sites on the palm of the hand. Skin conductivity increases when the sweat glands open. Like skin temperature, this is regulated by the sympathetic nervous system. We use the same imaginary story and ask the child to imagine the caveman fighting the saber-toothed tiger. Early man needed his hands to be moist (a bit sweaty) in order to hold his spear without it slipping. Children enjoy, understand and remember this example.

Extremely high arousal may correlate with high anxiety and tension. This may be seen in PTSD. A flat, unchanging EDR may also correlate with tension (often chronic) in some individuals. You may find that your adolescent and adult ADD clients demonstrate low arousal a short time into a training session. Their alertness level is dropping as they become bored with the activity. In these ADD clients, raising their alertness correlates with increasing their EDR. Teaching these clients to maintain alertness is important if you hope to succeed with neurofeedback training. Research has demonstrated that children with higher levels of arousal and more labile EDR are better able to sustain attention and can perform faster on assigned tasks. Higher EDR is associated with better learning of novel material and improved memory recall (Andreassi, 1995, after Sakai et al., 1992).

Optimal performance may be associated with increases in heart rate (HR) and skin conduction (SC) that may, in turn, be related to improved attention and short term memory (Andreassi, 1995, after Yuille, 1980).

Heart Rate (HR)

We measure HR and blood volume (BV) using a *plethysmograph* with a photoelectric transducer. For BV measurements, the difference in magnitude between the lowest point of a pulse and its peak is expressed as a percentage of the average. A sensor on the thumb is used and it is held in place comfortably with either an elasticized band or skin-sensitive tape.

Heart rate increases are greater for stress associated with anger, fear or sadness than increases noted with surprise, happiness and disgust (Andreassi, 1995, after Ekman, Levenson & Friesen, 1983). We have all experienced heart rate increases with frustration and with a defensive response to, and/or rejection of, a stimulus. HR decreases with the orienting response, stimulus acceptance, or being given a reward. One would also expect differences in Type A versus Type B personalities (Friedman & Rosenman, 1974). It has also been found that HR responses are greater when the person is capable of exerting some control over the event (Andreassi, 1995).

Respiration

There is a vast literature on the importance of proper breathing; for those interested in this topic, we refer you to Robert Fried's work. Breath work is emphasized in all the martial arts and in Eastern traditions of meditation and yoga. For our purposes, the main application of breathing is for relaxation and tension reduction. We tell clients, "If your breathing is relaxed, it is likely that you are relaxed."

In a training program using BFB, most clients can be rapidly taught to breathe deeply, more slowly and regularly. Older children enjoy pretending they are blowing up a balloon in their tummy while balancing a small ball on their shoulder. In this manner they breathe deeply while keeping their shoulders relaxed and motionless. They easily understand that rate of breathing is controlled by pCO_2 (partial pressure of CO_2) in the bloodstream, and that this will be decreased if they blow out all the CO_2 from the bottom of their lungs. In this way they lower their breathing rate to a comfortable five to eight breaths per minute. An ideal rate for most adults is six breaths per minute. Children breathe at a slightly faster rate.

Correct breathing appears to be a particularly helpful adjunct when you are doing SMR training. Producing 12-15 Hz activity across the sensorimotor cortex is associated with being physically calm while maintaining mental alertness. We encourage this in children who are hyperactive and impulsive. We always teach them diaphragmatic breathing as well. The old advice to take a deep breath before responding, especially if you feel angry, is an invaluable technique for the child with ADHD and for their parents, too. The combination of SMR up-training and slow diaphragmatic breathing (about 6 BrPM) has been used successfully in a single case report of treating Tourette syndrome (Daly, 2002) and in Parkinson's Disease plus dystonia (Thompson & Thompson, 2002).

RSA (Respiratory Sinus Arrhythmia)

When a student/client breathes diaphragmatically and effortlessly at a rate of about six breaths per minute, the heart rate will follow a sinusoidal pattern. This pattern correlates with the respiration pattern and is called respiratory sinus arrhythmia (RSA) (Basmajian, 1989, after Fried, 1987.)

Measurement for respiration is carried out by monitoring the degree of stretch in at least one respiratory band (a kind of lightweight belt with Velcro closure). Ideally, you would place one sensor band around the chest just below the armpits to measure thoracic breathing. The second sensor would be placed around the abdomen at a level of maximum change during inspiration and expiration. If only one sensor is used, it should be placed around the abdomen. When a single sensor is used, thoracic movement may be detected by using EMG sensors over the trapezius muscle. A second EMG sensor could be placed over the occipitalis muscle group.

As noted previously, the photoplethysmograph (PPG) is usually attached to the thumb or put on a finger (ventral surface). Its light source and photodetector register changes in blood flow and thus heart rate.

With stress the heart rate will increase. Stress may result in the client's breathing becoming shallow and irregular. Some clients may reverse the normal process by moving their diaphragm up with inhalation and down with exhalation. With stress one does not usually observe synchrony between heart rate variations and inspiration and expiration.

At six breaths per minute, most adults will demonstrate synchrony between their breathing and their heart rate. Children breathe at a slightly faster rate. When graphed, a beautiful synchrony is observed. There is an increase in heart rate with inspiration when vagal inhibition of HR is decreased (also some sympathetic nervous system effect), and a decrease in heart rate with expiration due to vagal parasympathetic inhibition of HR. This is said to be the only measure that allows the clinician to influence the activity of the parasympathetic nervous system and thereby affect the balance of sympathetic to parasympathetic activity. One of the best instruments to follow these variations in heart rate is made by Thought Technology. Another instrument made specifically for this without other measures is called the Freeze Framer from a company called Heartmath. However, the Biograph Infiniti from Thought Technology allows you to look at RSA while simultaneously monitoring other modalities, including EEG. It has the advantage that all your recordings are instantly entered into their program (called Cardio-Pro) that will correct for artifacts and give you all the international statistical measurements, including the standard deviation of the artifact corrected (normalized) R-R EKG intervals called SDNN and other measures. See Section XVIII for more detailed information.

All of us tend to link the way we breathe to different activities in our lives. The objective in training is to have the client take over control of how they breathe even in stressful situations. When breathing is predominantly thoracic, there may be an associated hyperarousal. This is a catabolic state which is thought to predispose the body to pathology. The catabolic state is associated with a decrease in the production of white cells for the immune system, increased salt and water retention, decreased repair and replacement of cells, and a decrease in the synthesis of protein, fat, and carbohydrate. Cardiac output and blood pressure may also increase. However, a positive result of being in a "stressed state" has been identified by research that suggests that the release of adrenaline with stress may improve blood clotting and, thus, be a factor in improving survival in some situations (Andreassi, p 29, 1995).

Pepper notes that: "Effortless diaphragmatic breathing reduces sympathetic arousal and promotes an anabolic state, which encourages regeneration." This has been shown to improve a variety of disorders including asthma, coronary heart disease, pain, panic, and hypertension. It has also been shown to improve athletic performance (Pepper, 1997). The clinical result of achieving RSA synchrony, which is found in conjunction with effortless breathing, is usually a sense of a release in tension and an increase in physical and mental relaxation.

Note regarding measurement: If your instrument uses a '%' measurement, then this may be confusing at first. These *measurements* **cannot be used to compare one session to another session**. For the respiration sensor, '%' means percent of possible tension of the sensor itself. It is not a physiologic measure: 100% means the sensor is stretched to its maximum length, 0% means the cable around the abdomen is too slack. With the PPG sensor for pulse, '%' is a measure of the percentage of light being reflected. This is useful to see changes during a session, but it may be highly sensitive to a number of artifacts such as the angle of the instrument on the skin or ambient light around the subject's finger. Movement of the finger may grossly disrupt the readings. Thus, percent measurements are used for monitoring changes within a session. They are not comparable between sessions or between individuals.

Electromyography (EMG)

EMG readings reflect the depolarization and repolarization of muscle fibers. There has been some research showing that in clients with ADHD, behavior improves with muscle relaxation, as do their scores on reading and language tests. *Locus of control* is the best predictor of lasting success, and it has been shown to move from external to internal with successful EMG feedback (Andreassi, 1995, after Denkowski et al., 1984). We postulate that future research will demonstrate this shift in locus of control correlates with success in many modalities of biofeedback and neurofeedback. Logically, locus of control should become more internal as the client learns to self-regulate, since internal locus of control means the person feels they or their attributes are responsible for outcomes. (I got a poor grade because I did not study enough.) Those with an external locus of control believe outcomes are dependent on luck, chance, or the behavior of others who have more power. ("I got a poor mark because the teacher does not like boys.")

It is important to recognize that research has shown that relaxation of one targeted muscle group does not necessarily generalize to adjacent muscles (Andreassi, 1995, after Fridlund et al., 1984). Nevertheless, training a client to relax a muscle, such as the frontalis (forehead) or trapezius (shoulder), does help that client to recognize the difference between tension and relaxation. This recognition helps most clients to generalize learning how to relax to other muscle groups.

The raw EMG has both positive and negative signals. This is usually amplified by a factor of >1000 and mathematically changed to root-mean-square (RMS). RMS allows the EMG to be seen as a positive signal. Normal resting muscle activity is usually <4μv. High RMS amplitude indicates muscle tension. To detect muscle tension, +ve and –ve electrodes are usually placed two centimeters apart and parallel to the muscle fibers. A third electrode, the ground, is placed equidistant from the other two electrodes (Sella, 1997). The ground electrode does the electrical housekeeping including filtering out electrical noise. If you want a general measure of total upper body tension, try putting an EMG electrode on each forearm (with the ground further up the forearm on one side). Heart (EKG) artifact may be seen as sharp spikes at one-second intervals. Absolute values are not very important: the goal is a downward trend in the EMG recording (Stephen Sideroff, personal communication).

Artifacts may be dealt with in several ways. The EMG amplifier should have a preset *notch* filter, to filter out the prevailing environmental electrical *noise* (50 Hz in Europe and 60 Hz in North America). In addition, one may use a narrow 100-200 Hz filter to filter out the heart muscle signals when using a chest placement near the heart. A constant high signal may indicate that one electrode is not properly attached to the sensor or one electrode is not making contact with the skin. A short burst of high signal may indicate a movement artifact or a low battery.

To learn self-regulation of EMG, the client is asked to tense and then relax the muscle being measured several times. During this exercise the examiner and the client observe the changes in the EMG recording. The client then attempts self-regulation by varying the tension in the muscle group and observing the changes on the meter. To do this, the client changes the tension in the muscle being recorded by varying amounts (e.g., 100%, 50%, 25%, 10%, 5% of maximum contraction holding each state for five seconds, then relaxing). To improve muscle control (as might be desired post-stroke), this would be done on a flexor muscle while the client attempted not to contract the opposing extensor muscle group. Thus, a client who is having difficulty flexing a limb would first work on relaxing the opposing extensor muscle group. When the client achieves some success in consciously relaxing the extensor group, then work would begin on a combination of relaxation of the extensor while contracting the flexor (the late Bernie Brooker, Lynda Kirk, personal communications). EMG biofeedback has been applied to rehabilitation with great success by Bernard Brooker's group at the Miami School of Medicine Rehabilitation Department. Indeed, they have been setting up similar centers worldwide (Brazil, Germany, India). They are sometimes literally able to make the lame walk. In a dramatic case that illustrates both the success and limitations of training specific muscle groups, Brooker was able to work with a concert pianist to restore function to his hands after he suffered a severe head injury when he was attacked and robbed when performing overseas. After EMG biofeedback training he was able to return to the piano and the concert stage. He was, however, still not able to button his shirt, a motor act that involved a different muscle group than playing piano.

Biofeedback using EMG combined with thermal feedback and relaxation has been shown to decrease both systolic and diastolic pressure (Andreassi, 1995, after Cohen & Sedlacek, 1983). Chronic tension headache may also be treated using EMG feedback. For headache, electrodes are often placed on the forehead. In this case the electrodes may be placed slightly further apart. This can be done in a consistent manner if you use the same 'strip' each time (for example, a Velcro band with active electrodes 7 cm apart, as is available for use with equipment from

Thought Technology). In this placement the ground electrode is between the two active electrodes. In a relaxed state the forehead EMG is <2 μv. Alternatively, electrodes may be placed over the left and right trapezius muscles and/or on the neck (occipitalis) region (Arena, 1997).

Associated Therapeutic Techniques

During all of the above work the client is assisted to learn ways of relaxing and to generalize what they learn to their everyday lives. Below you will find a brief discussion of some of the main therapeutic techniques used with these clients.

Imagery

Imagery is often encouraged during biofeedback training sessions, especially if stress management is the goal. The client is asked to release all muscle tension while producing images of scenes or events that are personally relaxing. The client may then alternate these relaxing images with images that have the opposite effect on them, that is, images that simulate a crisis. A crisis may be defined as "a personally perceived potential adaptive incompetency" (Thompson, 1979). The client may be asked to imagine an event that fits this definition and then alternate this image with relaxing images while monitoring and gradually gaining control over their physiological responses. Whenever possible, the imagery used by the client should involve all the senses. In other words, imagery is not just about picturing things but also using your imagination to hear, smell, feel and even taste. Dr. Vietta (Sue) Wilson, a sports psychologist at York University in Toronto, Canada, who has worked with elite athletes for decades, gave us an example of working with swimmers. When they practice a race using imagery, she encourages them to see the sequence of scenes in the pool, feel the water on their skin, hear the splashing and the sounds of spectators, smell and taste the chlorine, and finally, visualize themselves finishing ahead of other swimmers. The sequence using imagery should take the same time as the actual race. An interesting aside is that Wilson found when she was doing brain maps with her swimmers that there was a male-female difference in activation during imagery. In females the areas associated with language lit up more, indicating that they did more self-talk, whereas the male swimmers showed more activation in areas associated with visualization.

The name of Jacobson has been associated with progressive muscle relaxation, although similar procedures had been taught by others for many years. Imagery of warmth, heaviness, and pleasant, tranquil situations may help the client to relax. *Autogenic training* refers to a specific series of exercises and is discussed below.

Autogenic Training

Autogenic training was done prior to 1930 in Germany (Stoyva, 1986). There are six standard exercises: limb heaviness, limb warmth, cardiac, respiration (diaphragmatic breathing), solar plexus warmth, and forehead cooling. More often, only four modifications of these exercises are done: heaviness, warmth, respiration, and solar plexus warmth (Basmajian, 1989, p 170). During this type of relaxation training, the trainer may use suggestive phrases such as: "Your arms and legs feel heavy and warm." In practice, most people who talk about doing autogenic training are not doing the full regimen of exercises.

A simplified version is to teach effortless diaphragmatic breathing that is regular, without chest or shoulder muscle tension. The student/client may practice these exercises at home, beginning with having each limb first feel heavy, then feel both heavy and warm. Then, while maintaining this state of limb relaxation, the student/client pretends that they are gently blowing up, then deflating, a little balloon in their abdomen. After each exhale they briefly 'rest' and increase their feelings of having relaxed muscles and a sense of warmth (hands, arms, legs) before beginning their next inspiration. Soon they may also begin to feel warm in their solar plexus area. Some students/clients may need to precede the relaxation exercises with tensing-then-releasing exercises for the hand, arm, shoulders, neck and jaw muscles in sequence. This helps them feel the difference between tension and relaxation. Some students/clients may augment this with the use of imagery. They may imagine an image that represents, for them, peace and calm and good feelings. Other students/clients find that they can use either a cue word or scene to bring on this state of calm. It is important to note that up to 40 percent of clients may experience discomfort while doing relaxation and/or autogenic training (Strifel, workshop on autogenic training, AAPB Annual Meeting 2000).

Systematic Desensitization

Systematic desensitization is the next step for some clients. It is a good method to help *transfer* the student's/client's ability to induce a calm mental and physiological state into the real world of stressful situations. To do this, the students/clients imagine themselves in progressively more stressful situations. This technique has been shown in controlled studies to be very effective in migraine control (Basmajian, 1989, after Mitchell & Mitchell, 1971)

Shaping

Shaping involves giving the student/client a fairly simple task, recording their success, and then gradually making it more difficult until you reach the desired behavior. It entails successive approximations that are recorded. This technique is used during the sessions and in home assignments, as appropriate to the client's present situation and goals.

Generalization

Generalization of what they have learned in training sessions to their everyday lives is the key to successful training. We use the adage, "Attach a habit to a habit." It is very difficult for any of us to just form a new habit, no matter how good we know it would be for us. Thus, if you want a habit, such as diaphragmatic breathing at 6 BrPM, to generalize and become a new habit, ask the student/client to list activities that occur every day. These may include: getting out of bed, sitting down to eat, brushing their teeth, answering the telephone, opening the mail, reading email, driving to and from work, and so on. The student/client must then, without exception, relax for two or more deep, diaphragmatic breaths before beginning each of these tasks throughout the day. We say, "without exception," because we know that any exception will lead to more exceptions and to an eventual failure of this attempt to form a new habit and generalize their relaxation exercises to everyday living. Most students/clients can master this assignment in a very short time. Then we request that they maintain the relaxed breathing exercise that they do in sessions, and at the beginning of each of these daily routines, throughout these regular activities. They are encouraged to make notes in their SMIRB (see below) pocket book on how successful they have been, preferably at a time and place that is already routine, such as when they have dessert at dinner each night or coffee in the morning.

Compartmentalizing

Compartmentalizing is another key technique to assist in generalizing relaxed self-regulation to everyday living. Compartmentalizing means that each of your problem areas has its own compartment and does not flow over into other areas of your life. For example, when you are at home playing with your children you should not be thinking about work. Doctors have to learn to severely compartmentalize. They cannot be worrying about one patient when they are treating another. Their family life will be a disaster if they bring home the tragedies of the emergency room to the family dinner table.

To help the student/client compartmentalize, we suggest they purchase a small, 3.5 x 7 inch, book that will easily fit in a shirt, vest or trouser pocket or purse. Some of us just use a pocket daytimer because it has the added advantage of being used as a personal reminder book and it has a section for telephone numbers. The Quo Vadis agenda planning diary has a telephone number section that slips into the flexible cover. Buy an extra diary or just an extra telephone number refill and use the blank pages, which are perfect for making a few notes. High-tech people may utilize the note-making capabilities of their electronic organizer, but in our experience this does **not** work as well. We call the back few blank pages of this refill our SMIRB. (The front pages are used for short-term and long-term goals.)

SMIRB stands for: Stop My Irritating Ruminations Book. These pages are used for recording, organizing, and controlling one's intrusive worrying thoughts. We tell the client that we all tend to ruminate at times. If there were a solution to the problem they are worried about, they would have acted on it. The reason they ruminate is because, at this time, there is no acceptable solution. This recognition is often, in itself, helpful for a client who is worrying. We suggest that they give each major area of worrisome thoughts (spouse, children, finances, exams, job, future career, and so on) two pages. The repetitious thoughts are to be organized and listed on the left-hand page. The right-hand page contains any ideas for rectifying the situation or for *reframing* the way the client thinks about it. Most of these right-hand pages will initially be blank.

Once they have their SMIRB set up, the client is told to establish a time each day (they can write it in the daily agenda section) when they could sit alone and worry for 15 minutes to half an hour with a cup of tea or coffee without interruptions. The client is told that they must make an absolute decision – no exceptions

– to reserve worrying for this time. If a new worrisome thought comes into the client's mind, then it must be entered onto the appropriate list. Otherwise, when a worry comes up, the client is to reassure himself that his worry will not be forgotten because it is in his book and will be reviewed during his next 'worry-time.' In reality, most people have the same worries again and again, (finances, in-laws, weight) without many entirely new worries cropping up.

One of the nice outcomes of this SMIRB technique is that our clients get a better sleep. They make an absolute rule that no worrying is allowed in bed. If they awake worrying, they are to keep their eyes closed and decide if that worry is listed in their book. If it is, they reassure themselves that they will not forget to think about it during 'worry-time' the next day. If they decide that it is not in the book, then they immediately turn on a light and carefully write it into their SMIRB. Then they think of some repetitive pleasant activity such as wind-surfing or paddling their canoe or walking through a flower garden – no people, just a pleasant scene or activity.

The author devised this method and began using it with patients in the 1960s. Amazingly, this very old-fashioned, simple technique has been more helpful than medications and other treatments for countless patients over the decades.

Light-Sound Stimulation and Subliminal Alpha

The field of audiovisual stimulation (AVS) has been rapidly expanding. There are many products in the market that combine photic stimulation with audio beats at particular EEG frequencies. The combined audio and visual stimulation will entrain the EEG in the same rhythm during the time of stimulation. The auditory stimulation is given through headphones, and the visual is done with flashing lights that are built into goggles that resemble dark sunglasses. The manufacturer may offer goggles and programs that allow each visual field to be stimulated by a different frequency. For example, with an ADHD client, the left visual field (right brain) might be stimulated in an SMR frequency between 13 and 15 Hz for calming, while the right visual field (left brain) is stimulated in a beta frequency between 16 and 20 Hz for improving attention and learning. The client sits with his eyes closed in a relaxed position while the equipment runs through a preset program. Each program has one or more preset frequencies and time lengths set for each frequency to be used. There are programs for relaxation, for alertness and for a range of states in between. It has been suggested (Siebert, 1999) that delta sessions be used to assist sleep onset, alpha sessions for relaxation and meditation, low frequency beta (SMR) for calming with ADHD students/clients and beta sessions to "perk up cognitive functioning." Research is starting to accumulate regarding beneficial outcomes with entrainment but controlled studies are few. The late Tom Budzynski worked with AVS, and Dave Sievert, a manufacturer of AVS equipment based in Edmonton, and has done workshops at many conferences.

Because AVS is doing something to the client (providing stimulation to the brain), it is more invasive than neurofeedback. NFB, when all is said and done, is just providing information that, hopefully, assists the client to learn self-regulation. AVS tries to actively change a mental state. The results of the stimulation may differ according to the individual's baseline EEG activity, so one cannot predict with certainty how a given person may react. AVS is certainly a field worth investigating and monitoring.

Some experienced practitioners, such as Paul Swingle, report shorter NFB training times when AVS is added to the NFB program. Dr. Swingle also adds other things to the training program such as the use of "subliminal alpha" – a tape or CD of pink noise (it sounds like running water) with subliminal beats imbedded into the noise that are designed to decrease slow wave (theta) activity. This subliminal alpha can be played using headphones during times when the student/client who has ADD is doing homework. Dr. Swingle suggests that you try this technique while monitoring the EEG to see if it lowers theta. Every client is different. If it lowers theta, then give the client the opportunity to try it at home, and with the cooperation of the teacher, at school. It will work for some clients and not for others. Tapes and CDs can be ordered from Dr. Paul Swingle in Vancouver, B.C., Canada.

SAMONAS Sound Therapy
(Spectrally Activated Music of Optimal Natural Structure)

The goal of SAMONAS sound therapy is to improve listening skills and, thereby, cognitive functioning by means of increasing, first, relaxation and emotional calming, and second, acoustic perception. Relaxation of tension will increase breadth of associations, memory, recall, and efficiency of learning. Psychologically, this may be experienced as both a centering and an energizing effect. Throughout history, many different therapeutic approaches (yoga, relaxation training, meditation and so on) have their roots in this basic understanding. SAMONAS sound therapy appears to offer a simple, pleasant and effective way to promote relaxation.

Improvement in acoustic perception also appears to be a result of SAMONAS. At birth, the human ear has the capacity to perceive distance (depth), direction and the specific character (structure) of sounds with the accuracy we marvel at in many species of animals. Much of this ability appears to be lost as we grow up in our modern environments. In addition, these components are for the most part lost in most recordings, but are preserved by the specialized techniques developed for SAMONAS. Practitioners at the highest levels of the martial arts appear to have relearned some of these abilities. They can perceive and interpret movements and sounds that others cannot. SAMONAS sound therapy may help us to train to improve our listening skills. Improvements in this realm may in turn increase our ability to receive information detail, and combined with the relaxed reflective cognitive style that is engendered, to assimilate and more effectively and efficiently use the information we hear. This may help clients with ADD and with Asperger's syndrome; however, there are no controlled studies at the time of this writing.

How Does SAMONAS Work?

The biological faculty of hearing is innate, but the ability to listen has to be learned. Learning involves the reorganization of neural pathways. It results from emphasizing some neural pathways and de-emphasizing others. New connections among neurons in the brain are formed throughout life. Learning requires that we exercise our brains, just as we exercise our bodies, to remain active and healthy. This is the basic premise underlying NFB. Similarly, SAMONAS sound therapy is built on the premise that appropriate exercise of the most fundamental organ in our bodies related to learning, the human ear, can improve the total functioning of the human organism. The ear has neural connections within the brain to areas that are linked to virtually all other organs in the body. It has direct links to speech centers.

SAMONAS uses advances in recording technology to develop CD recordings that promote exercise of the human listening/hearing pathways. It is based on the premise that improvement in the quality of sound, combined with an increase in the information the sound carries, will enable the student/client to not resist listening unconsciously or automatically. The student/client, thereby, gains more information through auditory channels.

We learn to tune in to meaningful sounds and tune out the surrounding *noise*. All babies babble the same, but as they grow, sounds not heard in their native tongue drop out. Similarly, native speakers of different languages show different audiograms. Italians, for example, have peak hearing in higher frequencies (4000 Hz) and French have great sensitivity around 1500 Hz, which corresponds to a nasal sound. Our hearing is flexible enough to adjust to the demands placed upon it. Humans hear in a range from as low as 16 Hz up to about 20,000 Hz. As we age, the higher frequencies drop out so the elderly may only be able to hear up to 10,000 Hz. We discriminate best in the 1000 to 3000 Hz range, which is the key range for speech.

As we develop, there is a psychological conditioning taking place that determines what we listen to and for how long. We tune out not only uncomfortable noises, but also sounds that are emotionally undesirable. Sounds that are tuned out at one point in our lives for psychological reasons may continue to be inappropriately and unnecessarily tuned out as we mature. Theoretically, SAMONAS recordings may enable us to rejuvenate this system to achieve more complete listening.

Today we are familiar with the concept of stress. With life-threatening stress, our ancestors had to fight or to flee. Those of our ancestors who survived developed nervous reactions, which, through the sympathetic nervous system, cut down blood flow to the skin (reduced bleeding when cut) and decreased digestive activities while simultaneously increasing blood flow to the large muscle groups necessary for fight or flight. Today these body reactions to chronic stress at home, school and in the office have no survival value and can result in a number of psychophysiological disorders. Tension-producing circumstances early in our lives may lead to the inappropriate exercise of the sympathetic nervous system in this same manner. Even when these early stressors are no longer present, this *fight or flight* system is activated by cues in our environment. We become tense, anxious, fatigued and even somewhat depressed without due cause. In a similar vein, we may automatically block out listening to aspects of our environment. In this area of our lives SAMONAS can also be helpful. The relaxation effects of the SAMONAS recordings are said to act directly to train us to habitually counteract these inappropriate stress-related reflexes.

The SAMONAS training may also activate or reactivate unused or unexercised neural pathways, which in turn may allow for improved reception and assimilation of information. The hope is that the student/client will continue to improve both listening and cognitive abilities because once started, improved listening becomes habitual and automatic (unconscious).

How Is "Information" Conveyed by Sound?

Particular tones seem to universally evoke certain response patterns. Low tones, for example, carry little information compared to high tones, but do tend to evoke movement and even flight, which may have had survival value. Deep tones and drumming rhythms have been used to evoke group responses since primitive times. High tones, on the other hand, may be used to evoke joyous activity. High tones are also used to convey information about distance, direction and what is producing the sound.

The character of sound is contained within the harmonics or overtones. In the most basic sense, a flute, piano, violin, horn and the human voice would all sound the same if one only heard the same fundamental tone from each. However, there is no such possibility in nature. All natural sounds have complex harmonics that give the character to the source of that sound. If a sound was at a frequency of 100 Hz (cycles per second), the harmonics or overtones would be whole number multiples of the fundamental 100 Hz frequency, or 200, 300, 400, 500 Hz, and so on. with each frequency multiple being lower in amplitude than the last. The decrease in amplitude is not simple and linear. It depends upon the source of the sound. The pattern of the amplitudes of the harmonics of a particular fundamental tone is different for each source. The pattern of the amplitudes of the harmonics is called the *formant*, and is different for each instrument. It is the *formant* that makes a note played on a flute so easily distinguished from the same tone played on a violin. Thus, it is these higher-amplitude harmonics of each tone that carry and convey the information, the color or timbre, that tells us the unique character of the sound's source. It is the overtones that allow us to identify the tiny nuances of verbal tones that convey mood and subtle meanings. These are analyzed in part at the temporal-parietal junction of the nondominant hemisphere.

In our modern urban world our ears are continuously bombarded with both high- and low-frequency noise (e.g., cars, electrical appliances, music and so on), which for our own sanity and in order to listen to important sounds (e.g., speech) we must habitually tune out. We automatically train ourselves to unconsciously narrow our acoustic perception. Both in our response to stress and to sound we have to some degree become automatons, and narrowed our capability for choice in our response patterns.

SAMONAS may offer clients an opportunity to relearn how to relax and to listen. It offers an opportunity for renewed choice.

How Does SAMONAS Promote Relaxation and Listening Skills?

SONUS: The first step was the production of recordings that differed from those made on tape and even those found on most CDs. These differed in that they reproduced the depth, direction and true natural character of the original sounds. This applied to sounds heard in nature or in the concert hall. This music did more than just sound nice for relaxing: it actually exercised listening skills in terms of giving far more information than past recordings of the same music or sounds had been able to do using conventional recording equipment and techniques. The name *SONUS* was used to describe these recordings. *SONUS* means System of Optimal Natural Structure.

SAMONAS: The next technological advance involved taking this system one step further, to exercise our ability to listen for the information contained in the high-frequency harmonics or overtones of the recorded sounds. This was accomplished through highly specialized developments in recording technology and techniques in the Klangstudio Lambdoma Studio in Germany under the direction of Ingo Steinbach. In this step, researchers, figuratively speaking, placed an envelope around the *formant* of tones and slightly increased its amplitude. They then chose certain music to stimulate the left (cognitive processing) or the right (calming of emotions) hemispheres of the brain and increased the amplitude of these higher, information-carrying frequencies to the right and left ears respectively. Then they went one step further, and in some recordings would exercise laterality by changing the ear to which the increased volume was directed. SAMONAS means Spectrally Activated Music of Optimal Natural Structure. It is *SONUS* with the spectrum of harmonics for the fundamental tones activated to increase contrast and the informational content of the sounds.

In these recordings the overall volume of a piece will decrease at certain points in the music. This has the effect of attracting the attention of the listener. In the SAMONAS process there is still an increase in the information contained in the output of the recording. This is achieved by increasing the relative amplitude of the higher frequency harmonics or partial tones, and increasing contrasts within the music. Steinbach's work in Germany has built upon the earlier work of the French audiologist, Tomatis. Using modern technology, Steinbach has been able to raise the high-frequency formants, which is particularly important because the human ear is most sensitive to sounds in the frequency range important for speech (2 to 4 KHz). Volume for frequencies higher or lower than this must be 10 to 100 times greater to enable these frequencies to be perceived.

Bone Conduction
It has been reported that clients may achieve faster improvement using bone rather than air conduction. This may be due to the fact that humans are quite capable of unconsciously turning off middle ear conduction through tension of the stapedius and tympanic muscles. This tuning out is bypassed by bone conduction.

Evidence
There are no controlled studies using SAMONAS Sound Therapy. There is a great deal of anecdotal evidence, and the theoretical rational appears reasonable. In addition, we do not know of any untoward effects. Much of the recorded music is by Mozart, and there is increasing evidence that Mozart's music has rhythms that reinforce natural brain rhythms. John Hughes, a neurologist and epileptologist, has reported decreased seizure activity when patients listened to Mozart.

Biofeedback Assessment – Method

Introduction
Biofeedback of autonomic nervous system parameters and muscle tension indicators helps the client gain a relaxed, yet highly alert mental state. When combined with neurofeedback (NFB), it assists the client to improve mental flexibility and combine this with physical readiness to achieve optimal mind-body functioning. The first step in this process is to help your client make their goals for doing BFB clear. We find the easiest way to do this is to use Socratic questioning to bring out a logical and achievable set of objectives for a particular client.

It is helpful to understand the normal psychophysiological baseline pattern for the autonomic nervous system and the usual changes to this pattern when a person is subject to a stressor. A diagrammatic representation of this is given below with thanks to Marjorie Toomin. In this diagram, the y axis is amplitude and the x axis is time. (Note: There are exceptions to the usual response to a stressor; for example, a person who has been under unremitting, chronic stress may show an atypical response – temperature rises and EDR goes flat with a new stressor.)

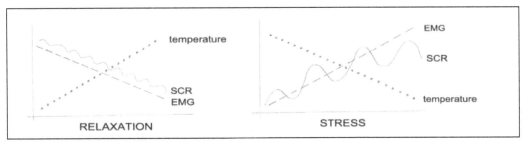

With thanks to Marjorie Toomin.

Goal Setting With the Client
Each client should understand a little about the functions and the physiology of the ANS. The essential areas of the body served by the sympathetic and parasympathetic nervous systems have been outlined and illustrated in Section V, Neuroanatomy.

Tools for Optimal Performance States (TOPS)

Before starting training with adults we discuss objectives for training. Their objectives are clarified as work proceeds with them using a structured interview and filling in a one-page questionnaire. This questionnaire correlates typical client objectives to the parameters that can be measured with combined BFB/NFB equipment. A score from 1 to 10 (high level achievement is a 10) is assigned to each area of our Biofeedback Institute's TOPS Evaluation. TOPS is our acronym of **Tools for Optimal Performance States**. Note that we use the plural, states, to reflect the reality that there is more than one optimal performance state. The state varies with task demands. This form is shown below.

TOPS Evaluation
(A structured interview.)

Achievement
Rate your ability to reach and sustain each of the following desirable states.
Score 1 (cannot reach this state)
to 10 (reach and sustain this state easily)

(<u>Describe</u> beneath or in margin <u>times when you are exactly the opposite</u> to the problem area you are working on during your training sessions. (e.g., I'm a very calm leader in catastrophic situations but I worry a lot about little things in daily life.)

A ____

A. State of Physiological Readiness

1. **Relaxed**   1 ____
   - **Objective:** To broaden associative capabilities and perspective, increase reaction time and decrease fatigue, tension and stress.
   - **Measurement:** ↑Peripheral temperature, ↓Pulse rate, ↓Respiration rate, ↓EMG, moderate EDR, ☐ 20-23 Hz activity.

2. **Alert**   2 ____
   - **Objective:** To efficiently respond to new information. (State of eustress)
   - **Measurement:** Moderate increase in EDR (arousal and performance relate in an inverted U shaped curve), plus increases in 12 Hz and low beta (15-18 Hz).

B ____

B. State of Mental Readiness

3. **Calm**   3 ____
   - **Objective:** To allow reflection and consideration of alternative approaches.
   - **Measurement:** ↑SMR (13-15 Hz), ↓high beta (22-34 Hz); ↑peripheral skin temperature, Respiration.(diaphragmatic), ↓EMG, EDR (control), ↓Pulse, ↑RSA (synchrony) at approximately 6 BrPM.

4. **Aware**   4 ____
   - **Objective:** To broaden input range – a state of calm readiness (as in Sports: Goalie, Martial Arts, Golf, Archery…) and increase creativity.
   - **Measurement:** 11-13 Hz↑.

5. **Reflective**   5 ____
   - **Objective:** To increase accuracy, breadth and completeness of responses.
   - **Measurement:** Bursts of alpha 11-13 Hz and beta 15-18 Hz and brief bursts of theta (with memory retrieval).

6. **Optimistic in Attitude**   6 ____

C ____

C. State of Active Mental Work

7. **Focused**   7 ____
   - **Objective:** To maintain attention to work area.
   - **Measurement:** Decreased slow waves (ϕ and low α) – Increased fast waves (β)

8. **Concentrating and Creative**
   Objective:
   a. To problem solve, elaborate on ideas and make decisions (↑β 16-18 Hz). 8a _____
   b. To have 'fluency of ideas' and be creative (↑left hemisphere bursts of 11-13 Hz α activity; sometimes, a semi-hypnagogic state with ☐4-8Hz 8b _____
   c. **Measurement:** Beta shows controlled increase 15-20 Hz (39-42 Hz) + 'shifts' to high Alpha. Theta states can also be produced at will when internally and creatively oriented.

9. a) **Strategic Goal Oriented Approach with** b) **Openness to New Innovation and a Commitment to Objectives and Time Management** 9a _____
   - **Objective:** To work efficiently and effectively without constricting response possibilities. Apply Active Learning Techniques (see Section XV, Metacognitive Strategies) and SMIRB (Stop My Irritating Ruminations Book). 9b _____
   - **Measurement:** Cognitive and metacognitive strategies, goals are written down, techniques for handling stress are utilized (especially breathing)

10. **Flexible yet Decisive** 10 _____
    - **Objective:** Respond with openness and thoughtfulness and appropriate flexibility to new ideas while remaining able to make decisions and commit to a goal.
    - **Measurement:** Find and weigh the positives in each new situation and decide on direction and actions.

ULTIMATE GOAL: Self-regulate to achieve 'flow'
  Flow _____
  ⇒ **Flow** ⇐   **Flow** is as easy as ABC:
  A and B – setting the stage: **C** – Performing and Producing
  Put it all together, automatically, to be Efficient - Effective - Productive

This form is just a tool. It is used initially to structure a discussion with the client concerning their goals for combined NFB and BFB training. Some clients use it only at the beginning of their training to help them understand more clearly their goals and the kind of work they are going to do in sessions to achieve these goals. Other clients may find it helpful to use the form, and their written comments around each section, on a regular basis to guide them through the training and to decide when they have met their training goals and are ready to graduate from training.

The reverse side of this form is the "Psychophysiological Profile" (illustrated below). It lists scores for measurements in each of these areas. These scores should be put into the appropriate areas of the form with the client. This is done after completing the EEG assessment and the stress test. This is discussed again with the client in the first few sessions in order to clarify objectives for each area of feedback. It is reviewed from time to time as training proceeds.

The Psychophysiological Stress Profile

Stress Assessment

A rapid stress test can be done during the EEG assessment session or at the beginning of the first training session. This is not a research assessment protocol. It is a practical, clinical approach to clarify what general biofeedback modalities, if any, would be important for training. We want to observe changes in respiration including rate, depth, and regularity. We like to be able to see the same types of changes in heart rate: regularity, rate, and extent of variability in rate. To do this requires a screen which shows respiration and the heart rate variations that occur with inspiration and expiration as line graphs. It is desirable, with the client, to be able to correlate these changes visually with changes in peripheral skin temperature, EDR, and forehead EMG. In our Centre we find that the "Assessment Summary Screen" is helpful. It is illustrated below. This screen allows for a quick statistical analysis and graphic display. Using this kind of display, it takes no more than ten to twenty minutes for the majority of our clients to see the effects of stress, contrast these with relaxation, and connect this information to their goals for training. The process begins in the first interview with a quick stress test.

A Quick Stress Test

Versions of this are found in www.BFE.org Suites by Thompsons and by Vietta Sue Wilson. These suites use Thought Technology Biograph Infiniti.

In the stress assessment we want the client to begin in as relaxed a manner as possible. They sit in a comfortable chair and close their eyes. We then explain that we are going to ask them to carry out two tasks that are meant to be emotionally uncomfortable. The first will be to think of and, if possible, mentally experience a very depressing, stressful event. After about three minutes you change the task and ask them to do mental math that will be challenging. After those two stressors they open their eyes and work on some relaxation techniques with guidance regarding how to breathe. Then we will review with them how their body's physiology responded during the test.

The steps for doing a quick assessment are as follows:

a. **Put on all the sensors,** explaining as you do so what each sensor is going to measure. Tell the client that there will be a *baseline* with eyes closed and then two stressful tasks with their eyes closed: Stroop Color Test (we used to ask the client to imagine, or re-live, a stressful event, but some clients could not do this so the technique was replaced with the Stroop test), and doing challenging mental math. This will take about 6 minutes. Then they will open their eyes and watch the screen while you teach them some relaxation techniques. When you have finished putting on the sensors and explaining what you are going to do, ask the client to relax and close their eyes.

b. **Collect data without stress** for 1 to 3 minutes to obtain a baseline for that individual. After 1-3 minutes, place a marker on the data (*marker #1*) if your instrument allows you to do this. Otherwise note the exact time. (In the example given below you will note that we only used two markers – as long as you write down what you are doing and the times when you change tasks, you may vary the type of stress, timing, and markers.)

c. Now ask the client to **imagine a stress-producing situation** (*a personally perceived potential adaptive incompetency [Thompson, 1979]*) for 3 to 5 minutes. Reassure them beforehand that they will not be telling you what it was that they were imagining. Tell them to really put themselves fully, emotionally, into this stressful image. There must not be any distracting sounds during this phase. After 3 to 5 minutes, again place a marker on the data collection (*marker #2*). Alternatively, use the Stroop Color Test.

d. At this point, have the client carry out a **cognitive task** that is designed to progressively increase in difficulty until it is impossible to do. (In most instances you don't let the client know that this is what you are doing.) Two examples are as follows:

Use one of the following tests (or an equivalent), and after about three minutes again place a marker on the data collection (*marker #3*).

- **The Stroop Color Test**: This test is done with eyes open watching a special display on the computer screen. This test uses words that name a color. However, the name of the color does not match the color of the ink used to print the word. For example, the word *green* might be printed in red ink. The client must say the color the word is printed in. On one program, sold by *Thought Technology* and called the *Biograph Stress Protocol*, is a test done with a very fast rate of display so that it becomes quite stressful.

- **A mental math test** for adolescents and adults: The assessor says two digits (3, 5) and the client must add them (3 + 5 = 8). The assessor then states one further digit (7). The client must now add this digit to the last digit spoken previously by the examiner (7 + 5 = 12). The series might progress as follows: 3, 5, (8); 7,(12); 2,(9); 39,(41) 63(102)…

Other, not so stressful math tests are as follows:

- Some clinicians use **serial sevens**. In this test you tell clients to begin at 200 (900 for teens and adults) and subtract 7. They should continue subtracting 7 until they are close to 0. They then start over again and see if they end at the same number. The advantage of this is that they do not speak during the assessment so you have less EMG artifact if you are recording the EEG. They feel stressed because you have said that they will tell you the final number they reach when the time ends or when they are close to zero.

- Another mental math test is the one Sterman recommends when doing full cap assessments: the Paced Auditory Serial Addition Task (PASAT). It can be found on the SKIL website (skiltopo.com).

e. At this juncture, teach the client **methods for relaxing**. Ask clients to open their eyes and tense their shoulders hard for about 10 seconds. Then completely relax and feel their muscles relaxing from their forehead through their jaw, neck, shoulder, upper arm, forearm, to their hands. They should now feel their hands getting heavier and warmer. After about a minute, as they are doing this, again place a marker on the data collection (*marker #4*).

f. Now ask clients to continue with this relaxation and feeling of warmth while they follow your breathing. You demonstrate diaphragmatic breathing and ask if they see your shoulders move at all. They don't. Now they are to do it. They must only breathe by moving their abdomen in and out. They can pretend that they have a balloon in their tummy and are inflating and deflating the balloon. They also can pretend they have a glass balanced on each of their shoulders and they do not want their shoulders or chest to move or the glass would fall off and break. (You must judge if a particular client would find this second image stressful. If they would, omit it.)

Tell your client that you will purposely make a sound as you inhale and exhale and that you will indicate *breathing in* with your hand moving up and *breathing out* with your hand moving down. The client is to follow this paced breathing as you give soft encouragement. You breathe evenly, regularly, at 6 breaths per minute. 'Thsssssssssssss' = in, and 'phhhhhhhh' = out, then say "rest" before you begin to inhale again ('Thsssssss') and so on. Your client can watch their respiration and their cardiac variability on the line graphs on the display screen.

Now place marker #5 on the data, and do this breathing and relaxation exercise with the client for 3 to 8 minutes. With most clients you will observe the finger temperature rising, EMG, EDR and heart rate dropping and a synchrony between heart variability and their breathing. Once they are achieving an appropriate relaxed state, as indicated by the physiological variable being measured, stop the recording and save the data.

Case Example

John, age 39, was asked to think of a stressful life event after about 30 seconds of relaxing. In the example screen shown below, we did not use a marker at that point. At 3 ½ minutes (marker #1) we began the math task. At 5 ½ minutes (marker #2) we ended the math task and began the relaxation. The relaxation instructions for this took less than a minute and he began to breathe evenly at a rate of about 5 to 6 BrPM. The EMG recording on the forehead and back of the neck in this client did not demonstrate major changes in this short time period and were left out of the graph below. Each client will demonstrate a unique combination of physiological responses to stress and relaxation. In training you will want to use those measurements that demonstrate change clearly to the client. Most often, we find that respiratory rate corresponds to the other changes in EMG, EDR and temperature, and we often use it on feedback screens. If alertness is a problem (as in clients with ADD), then EDR is also used. The client should be relaxed <u>but</u> also alert.

With John, when the most difficult math questions were asked, his heart rate rose as high as 96, averaging about 90. It fell to about 70 with relaxation. During the math task John's skin temperature fell from 94.5 to 91.6 degrees Fahrenheit and his EDR rose from 4.6 to 9 µMhos. Respiration became irregular, shallow and rapid, rising to as high as 74 breaths per minute (BrPM). Marker #2 was placed when the math test ended and he was told that he would now learn a method for relaxing. He was then instructed on how to breathe diaphragmatically at 6 BrPM. As he did this, his peripheral skin temperature rose and his EDR fell. His pulse rate decreased. His breathing became deep and regular at 6 BrPM. There began to be symmetry between respiration (inspiration and expiration) and heart rate (increasing and decreasing). He felt subjectively calm, relaxed and alert.

Assessment Summary – Graphic Representation and Statistics

(See also Section XVIII)

Now you can review the data with your client. The assessment summary graph shows a display screen from the *ProComp+/Biograph* that can be used. (Modern programs use the Infiniti/Biograph) It graphs time against amplitude or rate for each of the parameters in a manner that allows the client to see clearly how all the parameters changed with each task or condition.

Summary Graph
Assessment Summary Screen for ANS Variables

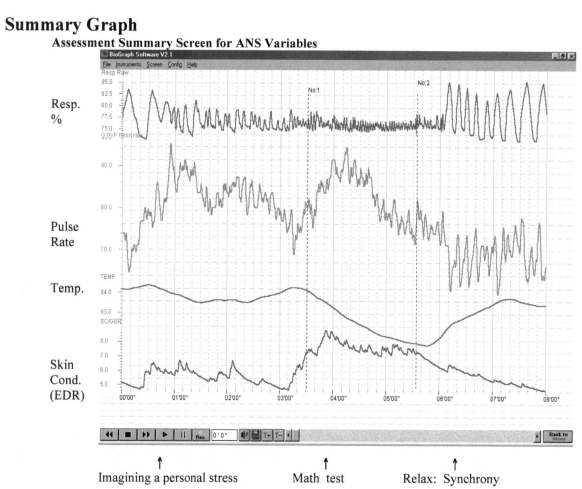

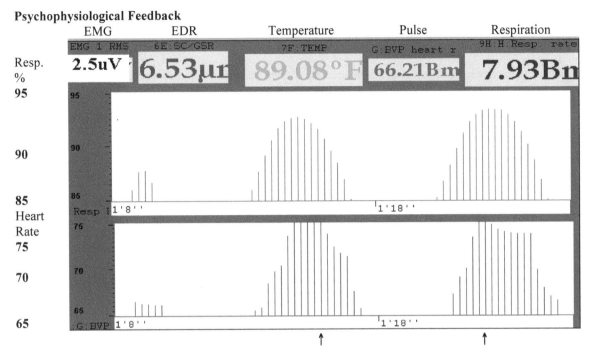

Note the synchrony (arrows) between increasing and decreasing heart rate (bottom) with respiratory inspiration and expiration (top).

The psychophysiological feedback is the screen that John, a 39 year old man, was looking at as he learned to relax his forehead and shoulders, warm his hands and breathe diaphragmatically and regularly at a rate of about 6 breaths per minute. His heart rate was increasing with inspiration and decreasing with expiration. In the above 20 second period he has decreased his breathing rate to 7.9 breaths per minute. He is attempting to raise his peripheral skin temperature as he relaxes. His electrodermal response (EDR) rose to 18 with stress. It should not decrease much more than the current 6.5 mhos as he relaxes or he will become less alert. His forehead EMG had risen with stress to >9 μv, but is now dropping and has reached 2.5 μV as he becomes calmer. With further training he did begin to hold it between 0.9-1.6 μV even with a mental math stress. Many clients show EMG readings on their forehead of 9 to 15 μV when they enter training and these will drop to between 1 and 2 μV with biofeedback training.

During Regular Training Sessions

In his training sessions, John had a respiration line graph and a numerical counter (breaths per minute) on the feedback screen while he was using brain wave feedback to help him decrease nonproductive ruminating and increase a calm state of focused, problem-solving concentration.

As training progressed, John was able to think of stressful things in his life and even do a math stress test while remaining calm, as evidenced both by the measured variables and by his subjective experience. John practiced the diaphragmatic breathing, muscle relaxation, and hand warming when answering stressful calls at work and while listening to others in social situations.

We suggest that you carry out some type of stress assessment when you commence work with an adult client, at intervals during training, and at the end of training. The above example gives a quick psychophysiological stress profile. A more comprehensive assessment, where you give time for recovery to baseline after each stress, is described later in this chapter and under Heart Rate Variability in Section XVIII of this book.

Recording your Data – Statistics

When you have shown your client the graphic profile (above) and they have understood how even a mild, artificial stress situation can affect so many physiological variables, then you should record the data with your client. The form displayed below is one way of doing this. You can write in your client's ratios if your equipment generates this data.

Psychophysiological Profile
Biofeedback Institute of Toronto

MEASUREMENT & Objective	PRE	POST	Associated with	COMMENTS
Neurofeedback (EEG Biofeedback) May be combined with learning metacognitive strategies.				
Theta 3-5 Hz 4-8 Hz (⇓ amplitude and variability)			Tuned out	Variability is reduced with control of tuning out. (Example: The goal on the F1000 is ≤35 units on 20 overlapping 30" screens at smoothing factor of 5) Variability is a very sensitive measure but it fluctuates with time frame and averaging. Therefore, you must have a standard format for measuring this variable.
7 and 8 Hz			Visualizing?	Test this – ask the client to visualize a flower while you observe EEG.
8.5-9 Hz			Dissociation?	
Low Alpha 9-10 Hz (↓ for external focus) (↓ Left frontal in depression)			Internal orientation, reflection, meditation	Ratio α/β = , Relaxation
High Alpha 11-13 Hz (Shifts to high alpha states in the left hemisphere of right handed individuals)			Creativity, Broad awareness, Flow of ideas, Peak performance	Observe internal mental manipulation of ideas, concepts
SMR 13-15 Hz (↑ if impulsive and if emotionally labile) (↑ if anxiety, Asperger's) (↑ if seizures, movement disorders – Tourette, dystonia and Parkinson's)			Calm self control, Reflection before action, Control movement	Ratio φ/SMR =
Low Beta 15-18 Hz (↑ in ADD and LD) (↑ left frontal in depression)			Focused concentration, Decision making, problem solving.	Ratio φ/β = , External input (but genius on easy task needs little time in beta - low amplitude.)
Mid Range Beta 19-23 Hz			Anxiety, Emotionally Intense	
High Beta 24-34 Hz (↓ in tense, overly intense subjects)			Ruminations, Too Intense – "spinning your wheels"	Like driving a car – foot pressed to floor on gas pedal but car left in first gear.
Sheer Rhythm 38-42 Hz (↑ focused creativity)			Attention and Problem solving	May be hard to measure due to EMG artifact.

BIOFEEDBACK			
May be combined with SAMONAS sound therapy and relaxation exercises			
EDR (Electrodermal Response) (↓ if Overarousal/↑ if Underarousal) Attain "Eustress" state		**Arousal level** -- often low in ADD - may be labile and high if anxious, flat with chronic stress	↑ **for alertness** ↓ **if tense, Stabilize if labile**
Peripheral Skin Temperature (Vasodilatation/ vasoconstriction - try for 94°-96°F.)		Low if anxious, fearful.	↑ **to relax** - particularly during tasks
Pulse (also **Blood Volume** measurements)		High if anxious	↓ **rate** and ↑ **variation in synchrony** with inspiration and expiration
Respiration - Diaphragmatic		Anxiety: Rate ↑, irregular, shallow	↓ **rate** to ≈ 6 and more **regular** with increased depth.
RSA (respiratory sinus arrhythmia) Sympathetic/parasympathetic balance		Anxiety: irregular breathing and heart rate which are not synchronous	**For Relaxation** ⇑ synchrony and amplitude of heart rate variation with inspiration and expiration

Note: RSA is the self regulation of breathing for maximum heart rate variability.

Alternative ANS and EMG Assessment Technique

Putting on the Sensors

Though doing a stress profile may be routine for the person providing neurofeedback and biofeedback sessions, remember that it is all new to most clients. This necessitates being careful and relaxed in your approach. Explain everything before you do it, going step by step. There are a large number of sensors when doing an ANS and EMG initial assessment. Some patients may feel a bit overwhelmed and even frightened by this. Patients attending a pain clinic may show distress if any strap feels too tight. Some patients may demonstrate psychological and/or physiological hypersensitivity even to light touch. The trainer should show each sensor to the client and briefly describe **what** and **how** it measures, and **why** this could be important to the client whose psychophysiology is being assessed. In addition, giving the client a simple printed sheet with an illustration showing the sensors on a person with a brief succinct description of the foregoing "what, how and why" can be very helpful. The trainer should continuously check with the client that they are comfortable with how the sensors are being put on. The trainer should obtain the client's permission for each thing they do.

The sensors are placed as described under the section on a brief ANS/EMG assessment above. An additional EMG placement may be used, however, to monitor changes in upper body tension. The ground and +ve electrodes are placed on one forearm and the –ve electrode is placed on the other forearm. Psychologist Stephen Sideroff at UCLA teaches this placement in workshops as a good way to get an overview of total upper body tension.

The purpose of this assessment is slightly different from the foregoing brief stress assessment where the purpose was, first, to allow clients an opportunity to observe how the various parameters changed with stress, and second, to observe how rapidly they could bring these variables under control by following the instructor's example and instructions for relaxation for three to five minutes.

There is a third, additional objective for the more formal and lengthy assessment outlined below: namely, to measure how long it takes the client to recover to their baseline values after being given a variety of stresses. Three stressors and three recovery periods are used.

A Sequence of Steps for this Stress Assessment

This assessment is done in a series of steps. After each step a three-minute period is allowed for recovery. A suggested series of steps is as follows:

1. **Baseline for 3 minutes**. The client sits comfortably. They may look at the screen if they wish to.

2. **Breathe rapidly for 1 minute**. This is a physiological stress. Breathing is shallow and rapid. Most clients will demonstrate a shift in baseline for a number of the parameters being measured.

3. **Baseline/rest period for 3 minutes**. Clients will vary in the length of time required to recover. For some clients, some variables may not fully return to baseline values.

4. **Math test.** Use one of the math tests described previously, or ask the client to subtract serial sevens from 900 (893, 886, 879, etc.) out loud after being told that they are being checked for speed and accuracy. The math tests are a real-life stress for most clients as was noted in the previous description of a stress assessment.

5. **Baseline/rest period** for 3 minutes. The recovery period may be quite long for some clients. This may be particularly **true** for clients who demonstrate performance anxiety.

6. **Talk about a very stressful time** in their life for 3 to 5 minutes. This test is quite useful, but it should only be carried out with a person whom the client accepts as a therapist. If the examiner is not the client's therapist, then the client can be asked to close their eyes and try to visualize and emotionally relive a very stressful event without talking about it. Alternatively, the client can imagine an event that would be extremely stressful if it should occur in the future.

7. **Baseline/rest period** for 3 minutes. With anxious clients, the ability to compartmentalize may be compromised, and recovery after this task may be incomplete and lengthy.

8. See if the client can **learn to relax** even further than their usual baseline over a 3- to 5- minute period. This final phase is the same as the previous, quick ANS assessment. The trainer models being calm and relaxed, and encourages the client to breathe diaphragmatically at a rate of 6 breaths per minute. Many clients will be able to demonstrate for themselves that they are capable of changing most of the variables being monitored as they follow the trainer's instructions.

Note: In Section XVIII a method for determining the client's 'resonant' frequency is outlined. The best breathing rate for your client is at their own resonant frequency and this will be a lower rate for large persons and men. This is the frequency at which there are maximum heart rate variations and it is usually the frequency at which there is synchrony between the breathing and the heart rate as was shown in the above illustration.

Examine the Data

One of the easiest variables for the client to understand is upper body tension. Most clients have experienced a feeling of tension in their neck and shoulders. *Muscle bracing*, in addition to autonomic reactivity, is a common finding. Although changes with stress may vary from one client to another, most clients will have one or two variables that show a marked shift with stress. The variable showing the largest shift, however, may not be the same one that demonstrates a slow recovery to baseline values. How well the client returns to baseline is one of the most important variables for the biofeedback therapist. If you do not return to the baseline, then you will accumulate stress over the course of a day. In addition, clients will react differently to different stressors. All of these variables should be noted and discussed with the client.

John's skin temperature only varied by one to three degrees. However, it was the one variable that did not return to the original baseline during the three-minute rest period. It therefore qualified as a sensitive indicator of John's ability to adjust, self-regulate, and recover. The NFB trainer monitored John's skin temperature during the NFB training sessions. John found ways to associate relaxing with raising his skin temperature rapidly after stressors. John used a small portable monitor at home and work and found he could rapidly have his skin temperature recover to 94.5°F even during stressful phone calls by relaxing and using diaphragmatic breathing, both before picking up the phone and during the phone calls. He began to make better business responses to clients when he was in a calm, relaxed state. These home-use temperature monitors are readily available and very cheap.

During the assessment the easiest variable for John to self-regulate was breathing. Obtaining good RSA was observed in the last step of the initial assessment sequence when diaphragmatic breathing at 6 BrPM was taught. Good RSA was observed to correlate with other variables returning to baseline status (except for temperature). This was the reason that John used diaphragmatic breathing to help him relax quickly during stressful events while he monitored his skin temperature. During training sessions John used both temperature and respiration feedback instruments, in addition to the NFB instruments that fed back information concerning levels of SMR and high beta. The SMR frequencies were reinforced and the high beta, in the range indicative of anxiety, was inhibited.

John was like many other clients in that he would become quite sleepy as he began to relax. This did not benefit his ability to focus, concentrate and problem solve. The trainer, therefore, had John do the feedback sessions in an upright sitting position with excellent posture. Skin conduction was monitored during each session. He had to make sure that it did not drop into a range that, for him, represented a loss of alertness. John's work performance improved and his family life became relaxed and fun as he learned self-regulation techniques.

Theory and Objectives

Recovery to baseline levels is an important indicator of good physiological and psychological health. It is important to retest this variable as BFB training proceeds. Even in normal daily living there are many periods of time between stress and no stress. Autonomic balance is best in those persons who are able to return to a calm and relaxed baseline between stressors. If you do not return to baseline, the stress becomes cumulative and you may reach the point of overload, or become "stressed-out."

Some clients remain in a physiologically stressed state because they know that there will be another stress coming in a couple of hours. It is much better if your client can learn to compartmentalize, so that they can relax between known stressors. Persons who are able to rapidly return to a calm, relaxed physiological state are said to be resilient. Thus, a major goal of BFB treatment is to learn how to return to baseline, even though you know there will be another stress. This newly learned baseline state should be one that is even more relaxed, with less sympathetic arousal (drive) than was observed in their original baseline.

Our goal is not total relaxation, such as reclining in a chaise lounge on a beach. We want the client to maintain motivation and alertness. The client should be able to enter a *eustress* state, which Hans Selye discussed (Selye, 1976). Eustress is that level of stress which is optimal for a person. The eustress curve is an inverted "U" where the y axis is performance and the x axis is stress. Performance is poor both with low stress and high stress, and is optimal in between; thus the inverted U-shaped curve.

Office Environment

There should always be a relaxed time period before you begin a BFB session. If a client rushed to get to your office, their baseline measurements will reflect this physiological state. In the office you should maintain the environmental conditions to be as constant as possible as you do biofeedback. Room temperature must be comfortable. When it is cold in the winter, a time period is needed to acclimatize before a client begins their biofeedback session. If at all possible, the client should train at the same time of day. As discussed previously under NFB, circadian rhythms and our responses to stress and relaxation will change with the time of day. Some BFB therapists will train their clients to relax and then hope to have this training generalize to outside situations. In our Centre we have clientele who are attempting to optimize their performance in school, business, or athletics. We want the office situation to vary from silent and conducive to relaxing, to being more like a normal classroom or office, busy and noisy. We therefore leave doors open and, as the client improves, even allow interruptions such as the telephone ringing or staff coming in to ask questions. We also ask the client to use the "attach-a-new-habit-to-an-old-habit" technique, as described previously, to assist in this generalization process.

Beginning Training Sessions

Start BFB sessions using muscle tension and relaxation exercises because the client can see the changes in the EMG easily. Then do diaphragmatic breathing combined with pulse rate variability feedback to follow RSA. Breathing at 6 BrPM, the client can learn to rapidly turn on a calm relaxed state. Remember the phrase, "When your breathing is relaxed, you are relaxed." When doing the breathing, the client practices feeling heaviness in their shoulders and warmth in their hands.

It is important always to remember that during the early stage of training, the trainer's verbal feedback is more important than the instrument feedback. The trainer should set the instrument thresholds and the goals so that rewards are achieved readily while still having it difficult enough to make them shift in the right direction. It is important for the client to experience success in the initial stage of training.

Change, not absolute values, is the important factor in biofeedback training. For example, the same self-regulation learning is taking place if the patient moves from 24 to 9 as from 12 to 2 in the EMG values. No matter how tense the client may be, they can learn to lower muscle tension. It is very important, however, that the client sees success and that they not become discouraged at the beginning of treatment. The client must understand that training proceeds in small, but truly significant steps, and that, little by little, they will acquire the skill of self-regulation.

Monitoring BFB Variables During a Session

It can be quite valuable to review how some of the variables you are measuring changed with different events during a regular training session. The example given below is taken from the statistics screen of an F1000 instrument. The client in this case was receiving EDR, peripheral skin temperature, and EEG feedback. In the diagram below only the EDR (green line with sharp peaks) and skin temperature (red line with handwritten words along it) recordings are shown.

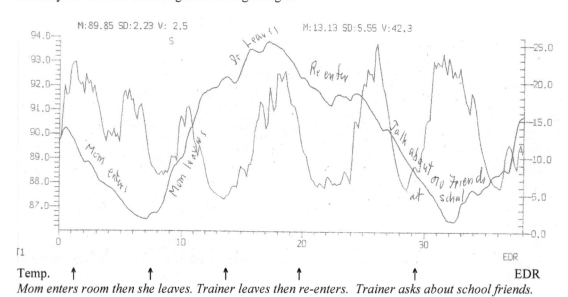

Mom enters room then she leaves. Trainer leaves then re-enters. Trainer asks about school friends.

Daniel was a bright 14-year-old student/client. He was training to improve his attention span and concentration and increase his ability to remain calm and reflect before acting. Daniel had always been a loner. It wasn't that he didn't want to socialize with the other children, it was just that every time he tried he seemed to say or do the wrong thing and he would end up being teased and rejected by his peers. Daniel had learned that it was easier to stay with much younger children who would follow his interests. Daniel was quite anxious and his rather odd behavior could be better understood if one took this into account. Daniel couldn't understand innuendo. He was very concrete in his interpretation of what others were saying and often drew the wrong meaning from expressions that were used in the school yard because he took them literally. Other children took advantage of this to encourage him to do things that he shouldn't and then laugh at him. Daniel was diagnosed with Asperger's syndrome.

In the example above, Daniel became very anxious when his mother came in to watch the session. In the above recording, skin temperature fell and EDR rose. He was greatly relieved when she left and you can see how his skin temperature rose and his EDR tended to fall as he settled back to focus just on feedback (EEG, EDR and finger temperature). The trainer then left the room for a moment to speak to his mother and his physiology indicates that he felt even calmer. When the trainer re-entered the room Daniel's EDR rose and skin temperature decreased. When the trainer then introduced the topic of how he was socializing at school his temperature dropped precipitously and his EDR again rose. At the 32-minute mark this discussion was finished, and Daniel talked about his favorite hobby. His skin temperature began to rise and his EDR fell.

In the sessions, Daniel learned to associate diaphragmatic breathing with increasing his EEG sensorimotor rhythm (SMR) 13-15 Hz. He learned that when he relaxed, his skin temperature and EDR were less variable. As Daniel progressed he began to have less fluctuation in skin temperature and EDR and he was able to apply suggestions for socializing with his peers. By the time he stopped sessions he was being invited to parties given by peers. His special interest in electronics and music allowed him to help set up the music systems for these parties. In this way he turned his special interest into a socially useful attribute.

Biofeedback Has Been Found to Be Effective

The reader is also *referred to* Biofeedback: A Practitioner's Guide, *Editors: Mark Schwartz & Frank Andrasik (in press for 2016) and to a new section in this edition of the* Neurofeedback Book *on Heart Rate Variability,* **Section XVIII.**

It isn't the purpose of this text to teach biofeedback. This has been very well described in other texts such as those by Andreassi (1984), Basmajian (1989), Fuller (1984), and Schwartz (1995). Nevertheless, it is important for the reader to recognize that there are a reasonable number of studies demonstrating, for the most part, significant improvement in a wide variety of disorders using biofeedback. An excellent **review of biofeedback research** has been done by **Carolyn B. Yucha**, Associate Dean for Research at the University of Florida, with the assistance of Donald Moss and Pam Sherwill. We have only made a very brief overview of areas where BFB may be helpful in the outline below.

Anxiety may be decreased using EMG frontal biofeedback or EEG alpha biofeedback (Rice, 1994; Sarkar, 1999). Thermal biofeedback (Hawkins, 1980) and GSR (Fehring, 1983) biofeedback have also been shown to reduce anxiety. Astin et al. have done a review of studies using mainly thermal and EMG biofeedback for arthritis (Astin, 2002). Flor demonstrated that a reduction in pain could be maintained (Flor, 1986). Beneficial effects in asthma have been demonstrated using self-regulation of breathing for maximum heart rate variability (RSA) (Lehrer, 2000). We also recommend that the reader learn this valuable technique and combine it with neurofeedback. Richard Gevirtz gives excellent workshops on HRV at the annual meetings of the Association of Applied Psychophysiology and Biofeedback (AAPB). There are questions concerning improvement of immune function that require further study (Taylor, 1995). Chronic pain is an important area for research. Reduction in abdominal pain with thermal biofeedback has been demonstrated (Humphreys, 2000). It would be entirely logical that immune function, chronic pain and fibromyalgia could be improved by biofeedback (see Section V, under The Neurophysiology of the Stress Response). Back pain has been helped by EMG biofeedback (Vlaeyen, 1995). Thermal biofeedback to improve blood flow has proven useful in improving the healing of nonhealing foot ulcers in diabetics (Rice, 2001).

Urinary incontinence is one of the best investigated areas for traditional BFB. There are well controlled studies demonstrating its effectiveness (Dougherty, 2001; Sung, 2000; Sherman, 1997; van Kampen, 2000). A review of fecal incontinence demonstrates success using biofeedback in a proportion of cases (Heymen et al., 2001). Even vulvar vestibulitis has been effectively treated with BFB (Bergeron, 2001).

Biofeedback may be a useful treatment modality for fibromyalgia (Hadhazy et al., 2000). EEG-driven stimulation may also be helpful (Mueller, 2001). EMG feedback for hand dystonia (writer's cramp) requires further investigation (Deepak, 1999).

The efficacy of BFB for headache has been investigated for years. Thermal biofeedback for childhood migraine has been found to be particularly successful (Hermann, 2002). Frontal and trapezius EMG biofeedback has also been used for headaches (Arena, 1995). Although the degree of response to biofeedback training for hypertension does vary, BFB (EMG, thermal, blood pressure, heart rate) has been shown to be an effective treatment modality, particularly if clinic training is followed by training at home (Yucha, 2001; Henderson, 1998). Both BFB (Morin, 1998) and NFB (Hauri, 1982) have been used with some success to treat insomnia. There are indications that BFB may be useful in reducing risk for persons who have suffered a myocardial infarction (Cowan, 2001). Post-traumatic stress disorder has only had limited investigation, but it may prove promising. (Carlson, 1998). Raynaud's disease has been helped using thermal biofeedback (Peterson, 1983; Yokum, 1985). However, results are dependent on technique (Middaugh et al., 2001, The Raynaud's Treatment Study). Thermal biofeedback has been successfully used to treat repetitive strain injuries (Moore, 1996).

Very promising work using EMG biofeedback with spinal cord injuries requires further replication (Brucker, 1996; Petrofsky, 2001). Some hemiplegic stroke patients have also benefited from EMG biofeedback (Schleenbaker, 1993; Moreland, 1998).

NFB may add a very useful dimension to the rehabilitation of some stroke patients (Rozelle, 1995; Sherin, 2003). EEG biofeedback is helpful in decreasing depression and improving memory and cognitive functioning in patients who have suffered a traumatic brain injury (Shoenberger, 2001; Thornton, 2000).

Temporomandibular pain and mandibular functioning may be helped by BFB (Gardea, 2001). Tinnitus may show some improvements with either EMG BFB (Erlandsson, 1991) or NFB to increase alpha and decrease beta (Gosepath, 2001).

SECTION XI
Fundamentals of Intervention

Overview: Goal Setting, Bandwidths, Electrode Placement, and How to Begin

NFB and BFB are learning procedures. Although the brain is capable of extremely fast ('one trial') learning, this is not the type of learning we are trying to achieve with NFB. Usually, one-trial learning is experienced in traumatic or highly emotional circumstances. Most learning takes time and practice. Beware of exaggerated claims of quick efficacy of some new placement or protocol. You might effect a transference 'cure' in minutes with a client, yet it might not last. Certainly, anyone can experience a sudden change in their neurophysiology. Think of how you might react to news that you had won a lottery. To make matters even more confusing, a high proportion of clients with ADHD are suggestible and have been shown to have higher hypnotizability. (Wickramasekara, personal communication) This is not surprising since excess theta is the marker for ADHD.

On the other side of the coin, a consistent finding by various experienced professionals is that it is difficult to effect a significant and lasting change in children when one or other of the parents is not enthusiastically supportive or, at worst, subtly (or not so subtly) undermining the process. The undermining can be done unwittingly by an anxious, well-intentioned parent who examines in detail every school report and questions why the child, on any test, has not performed as well as they had hoped. Often these parents are quite unaware that nonverbal communication is responsible for almost 80 percent of human communication, and that their child is acutely aware of their concerns even though they may have been very careful "not to say anything."

Biofeedback and neurofeedback are not stand-alone interventions. They are almost always combined with other interventions that correspond to the objectives of, and for, that client. For example: diet, sleep, and exercise are very important variables that need to be discussed and managed. Two quite different examples will be used as illustrations. Each example has an overview of goal setting in terms of relating the client's EEG to their personal goals.

Stages in Training a Child

The first example is Jason. Jason was case #3 in our introduction to Section IX, The Basics of Assessment and Intervention.

Jason, age eight, was very active, impulsive and could not attend for more than a few minutes during class at school. He was very bright and tremendously creative. His doctor felt he had a problem with short-term memory, and when tested, it was also thought that he had a central auditory processing (CAP) problem. He loved to draw and build. His EEG demonstrated extremely high 3-6 Hz activity and slightly lower SMR than beta. His theta (4-8Hz)/beta (13-21 Hz) power ratio was exceedingly high (14). His mother was concerned that his self-esteem was being affected by the negative feedback he was receiving at school.

The **objectives and steps** for working with Jason were:
1. Enhance self-esteem
2. Improve social interactions/behaviors
3. Improve learning and performance

These goals correlate with a sequence of three groups of interventions. We encourage people to group interventions this way, as we do when giving lectures on interventions for ADHD.

Urgent (Short-Term)	Ongoing	Long-Term
Behavior modification Stimulant medication for hyperactivity	+ve Family environment Nutrition Sleep Exercise Tutoring School Extracurricular activities (sports, drama, music, art and so on)	Neurofeedback plus Biofeedback plus Metacognitive Strategies

Jason had a supportive family. All of the first two sets of suggestions had already been set in place well before we saw Jason. Parents had read *The A.D.D. Book* (Sears & Thompson, 1998), and had worked on the positive reinforcement suggestions contained therein and heeded the dietary recommendations. Nevertheless, Jason still displayed ADHD symptoms, and they wanted to do something that could make a long-term change and perhaps even decrease the amount of stimulant medication he required to function in school.

Jason was a reasonably straightforward case. We used the EEG assessment to set the bandwidths and locations for training. We worked at both Cz and C4 locations and encouraged an increase in SMR (12-15 Hz) and a decrease in theta (3-6 Hz in Jason's case). His behavior had begun to change in a positive direction by 30 sessions. Progress testing done after 40 sessions showed improvements on parent questionnaires, TOVA and IVA scores, and theta/beta ratios, but these measures were still indicating significant inattention. He still required medication for school though it had been stopped after school and on weekends. After 60 sessions both theta and SMR had altered appropriately on retesting. His impulsivity and inappropriate high level of activity had decreased, his attention span had increased, his measured IQ came up by 16 points, and he went to the top half of his class academically. He no longer had any discernible problem with memory, and his functioning was such that there was no reason to re-do the central auditory processing (CAP) testing. His parents and their family practitioner gradually lowered his medications and he eventually came off the stimulant drug that he had been taking. He was very proud.

Stages in Training an Adult

Our second example is an adult with quite a different agenda.

Case # 9: Mike, age 29, wanted to return to university to do a postgraduate degree. He had completed an engineering degree, albeit with difficulty, and was steadily employed. He was extremely anxious, particularly in performance situations. Fortunately, he usually exhibited this in a positive way with laughter. He would wake up at night worrying about his work and his social relationships. To study in his chosen field he had to succeed in the GMAT examination, one of the graduate aptitude tests taken as a prerequisite for obtaining admission to some of the better graduate programs in North American universities. This is an incredibly stress-producing exam. When he was asked to do even simple math during the psychophysiological stress profiling, he stumbled over the questions and made silly errors despite having a strong math background from his engineering years. His respiration rate rose to 40 breaths per minute (BrPM) and was shallow and irregular. His heart rate rose and was completely out of synchrony with his breathing. His skin temperature dropped, electrodermal response rose markedly and his muscles tensed as an electromyogram showed. In the electroencephalogram measured at Cz referenced to the left ear, he showed high thalpha at 6-10 Hz, low SMR at 13-14 Hz, and high beta peaks at 21 Hz and 29 Hz. His high-amplitude thalpha and bursts of high-amplitude, high-frequency beta were also observed on a second assessment at F3.

The objectives and steps for working with Mike were:
1. Decrease tension
2. Decrease his negative, nonproductive ruminations
3. Increase his focus and concentration while remaining calm
4. Improve strategies for learning and performing well in a test situation

Training initially emphasized relaxation. He learned to breathe at six breaths per minute (BrPM), raise his peripheral skin temperature and bring his forehead EMG down. As he mastered these biofeedback tasks, he began to feel more confident that he could make both physiological and behavioral changes. At this juncture he began to work on maintaining a sense of calm. He raised his high alpha (11-13 Hz) and his sensorimotor rhythm (SMR 13-15 Hz). To decrease anxiety and ruminating, he learned to decrease 21-32 Hz. He said that to do this he really had to clear his mind and either think only of the task at hand or think of positive experiences while relaxing.

In addition to biofeedback, he actively worked on other suggestions we had made to him. He created a section at the back of his pocket day-planner for recording and organizing his worries. This SMIRB (Stop My Irritating Ruminations) Book was a technique to gain control of intrusive worrying thoughts. Each major area of worry (wife, finances, GMAT exams, university future), was given two pages. The repetitious thoughts were organized and listed on the left-hand page. The right-hand page contained any ideas for rectifying the situation or for reframing the way he thought about it. He established a time each day (noted in his day planner) when he could sit alone and worry for 15 minutes with a cup of tea and without interruptions. He made an absolute decision to reserve worrying for this time. If a new worrisome thought came into his mind, he would open the pocket book and enter it onto the appropriate list. Otherwise he would reassure himself that that worry would not be forgotten because it was in his book and would be reviewed during his next 'worry-time.' He said he felt reassured that to have worries was normal. But to worry all day long when he had to accomplish other things was nonproductive.

However, being relaxed was necessary but not sufficient. He needed to get rid of muscle tension and yet remain extremely alert. To do this he learned to control his EDR, keeping it high enough but not extreme. He had always tuned out in school and while doing homework unless it was a subject he was really interested in. When he tuned out, he would lose his alertness and become sleepy. To increase his ability to focus while studying relatively tedious information, he decreased thalpha, 6-10 Hz. the range that increased when he tuned out. We taught him metacognitive strategies (described in Section XV) and had him do GMAT reading exercises using these strategies while he decreased thalpha and increased beta 17-18 Hz. When he was cognitively problem solving, 17 Hz would rise. He practiced becoming relaxed, with his mind free of worry, then focusing on the task. The auditory feedback reinforced his maintaining this state while he did the reading.

He sent us an e-mail immediately after getting his results on the GMAT exam. He had achieved an almost unbelievable result, a ranking in the ninety-third percentile. He attributed his success to the program, beginning with pinpointing his problems during the assessment and then receiving help to overcome them during training. He said that in the exam he used the sequence of relaxed breathing, calm and focused mental state, and the other strategies he had learned in training and when studying.

As in our work with Jason, other techniques (relaxation techniques, SMIRB and metacognitive strategies) were important adjuncts to the neurofeedback and biofeedback done with Mike. Our clinical experience suggests that neurofeedback is probably crucial for a long-term change to take place in most of the children and many of the adults that we see. **The results achieved with the addition of NFB + BFB are definitely more consistent than the results achieved in previous years with patients using psychotherapy and medications.**

Beginning a Training Session

The first step in every training session with a client, if your EEG instrument has this capability, is to record a raw EEG sample for approximately 1 to 3 minutes. Together with the client, you review the EEG sample and remove artifacts. This assures you that you are obtaining a clean recording without an undue number of artifacts. It allows you to review and discuss your clients' objectives for training, and how their objectives relate to what they will be working on with respect to the EEG in today's session.

A Child

What is a simple EEG goal for young Jason? For an eight-year-old it must be expressed initially in simple terms. Sometimes we put up a simple diagram near the monitor. For a child with ADD, such as Jason, it looks like this. It serves as a visual reminder to him of his NFB objectives:

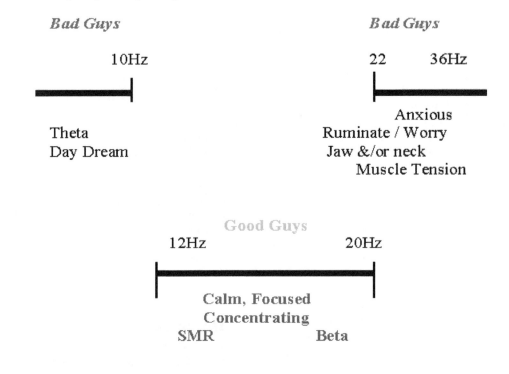

When Jason is achieving his initial NFB goals, we will combine the NFB with learning metacognitive strategies. He needs to become aware of how he best learns and remembers things. Even an eight-year-old can learn to plan an approach to a task and evaluate how he is doing. (Should I subtract using my fingers or by making tally marks? Maybe I should use those flashcards with Mom every night and memorize my number facts.) Then he will practice this combination of NFB and strategic learning while doing academic exercises.

When working with children who have ADD we do not routinely use general biofeedback sensors. A practical reason is that they tend to touch and fidget with them. Relaxed breathing can be taught without a respiration strap being attached. EDR is monitored more often because arousal is a very important in managing ADHD symptoms. The child earns extra tokens for keeping the hand with the EDR sensor still.

An Adult

The goal for an older student or adult client is more often expressed in terms of optimizing functioning. This may mean a decrease in symptoms and/or an increase in positive attributes.

Three seconds of Mike's EEG from his first training session is shown below. The EEG had been paused and Mike was asked what was happening in the previous few seconds. He said that he could not recall what I had been telling him. His mind had wandered and he was thinking of something completely unrelated to the session. With this kind of review, Mike started to learn about the correlation between his EEG and his mental state. We tried to find a mental state that corresponded to his goal of focusing on one task without going off-topic. Using the spectral display of the first one second of this EEG sample, he correctly pointed out that the desirable, externally focused, mental state he wanted to achieve would correspond to a decrease in amplitude and variability of his dominant slow-wave activity at 9 Hz.

Looking at the EEG before the main training part of the session begins is also helpful to the trainer. The trainer might, for example, observe that even though impedance readings had been good initially, when Mike moved around in the chair, settling into a comfortable position, he had moved an electrode such that it now had a poor connection, resulting in 60 Hz activity being a little high. He would then readjust the electrodes before beginning the training. If a cardiac artifact was noticed, then the trainer would try moving one or more electrodes to see if it could be reduced.

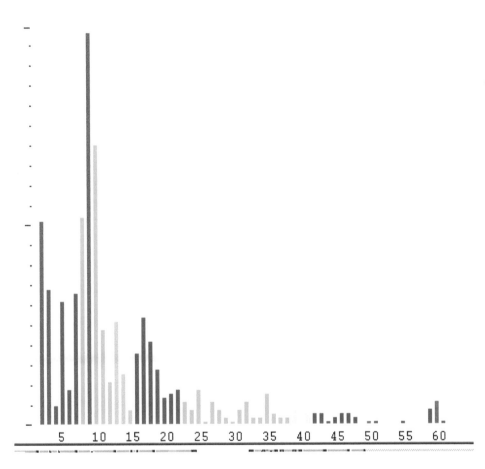

This spectral display of one second (1:38-1:39 on the EEG below) shows how Mike tunes out in low alpha.

He told us that at this moment he did not register what was told to him. He was thinking about something else (also note the high 17-18 Hz activity).

His active electrode was adjusted and the 60 Hz activity seen here dropped to almost zero.

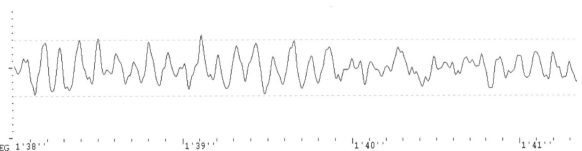

Mike tended to tune out in low alpha as is seen in the left portion of this EEG.

Reviewing the Client's Goals in Light of their EEG

After looking at the brief EEG profile at the beginning of each session, the trainer has the student restate their purpose for being in that session. The student states this purpose in terms of their goal for training. Then they relate this goal to what they wish to change on the EEG and in the biofeedback modalities being used for feedback, such as respiration.

For adults such as Mike, we use a more complex diagram than the one used for children. It has a greater emphasis on optimizing performance and mood using a combination of NFB and BFB. It spells out a state that is relaxed, calm, with broad awareness yet focused concentration. We often put the following reminder sheet near the monitor.

Summary Sheet – Bandwidths and Mental States

Goal: Relaxed yet alert, calm reflective state

20-23 Hz ↓	**Decrease anxiety/tension**
High Beta 24-34 Hz ↓	**Decrease 'racing' mind/ruminating**
High Alpha 11-13 Hz ↑	**Relax and open/broaden awareness**
SMR 12 or 13 – 15 Hz ↑	**Calm, reflecting before acting, 'readiness state'**
Theta and/or Low Alpha (<10 Hz) ↓	**Remain focused**
Finger Temperature ↑	**Relaxed**
Forehead, Jaw, Shoulders and Neck Muscle tension ↓	**Relaxed**
Respiration @ 6/minute	**Diaphragmatic, (Abdominal) Breathing**
Pulse ↓	**Calm**
Respiration synchronous with pulse rate. (RSA)	**Feeling calm**
Skin conduction (EDR) fluctuating between *approx.* 12-18 child; 6-12 adult	**Remain alert**

Consider an 18-year-old student who is not focusing well in school and who is impulsive. The impulsivity can extend to an impulsive approach when answering questions on exams. The goals and an approach to achieving them with such a client are outlined below:

Summary of Goals

1. **To be calm, not impulsive or irritable** and able to tolerate frustration; increase 13-15 Hz

2. **To be relaxed mentally and not ruminating**; keep 24-33 Hz at a low amplitude; as previously described, use a 'stop-my-irritating-ruminations-book' (SMIRB)

3. **To remain free from anxiety**; low amplitude of 21-24 Hz and physiologically not tense – breathing diaphragmatically at 6 BrPM (good RSA), skin temperature 94-96 degrees, skin conduction (EDR) not flat and not extreme amplitude or extreme fluctuations, forehead EMG <4µV

4. **To remain alert**; maintain EDR at a level that reflects alertness (usually above 8, though it will be lower with older clients

5. **To be broadly, externally focused and calm**; high 11-13 Hz (high alpha)

6. **To be able to shift to a narrow focus**; decrease dominant slow-wave activity (below 10 Hz). Although reflection and memory recall and associations requires shifting in and out of theta and low alpha, these frequencies should not be dominant when a student is sitting with eyes open in class or when studying

7. **To be capable of remaining in a problem solving mental state**; increase beta in the 16-20 Hz range (often about 17 Hz)

8. Most importantly, **to be able to self-regulate and to move in a flexible manner between mental states** according to task demands. (One possible feedback screen for practicing going in and out of focus at will is the sailboat screen which will be shown later in this section.).

For any individual client, you must emphasize components from the foregoing that are most important for that client. The same procedures may be used for episodic conditions such as headaches, but the overall goal is stated somewhat differently. Your objective then is to decrease frequency, intensity, and duration of the undesirable events.

The Trainer's Task During an NFB Session

The trainer should remain with younger clients for the entire NFB training session. The exception to this rule is at a stage during training when we are working on 'generalizing.' At this point the trainer and the student may agree to experiment and see if the client can maintain her good results without the trainer sitting beside her. The trainer might sit behind the client or stand at the door so they can still observe the computer but without the child receiving the trainer's verbal feedback and encouragement. The trainer may also decide to have no visual or auditory feedback though still recording data while the client carries out a task. After each of these experiments the recorded data is reviewed and compared with previous time segments when there was active coaching.

For most sessions, however, the trainer is actively involved. The trainer's task will vary with each client. It will be quite different with young children compared to adults. As a general rule, the trainer will have to be more actively involved with young children. The trainer's task is to *facilitate, enable, assist and model and always remain positive!* **Trainers are coaches and, at times, educators**. Trainers are acting as therapists only when professionally trained in psychotherapy and carrying out a psychotherapeutic intervention. This type of intervention, for example, is appropriate when using an alpha-theta protocol for patients who have addiction problems.

We want our trainers to believe in our ADD Centre motto, "You can't change the wind, but you can adjust the sails.." Trainers are not trying to change the basic personality, but rather to facilitate the student learning a new skill, the skill of self-regulation. In students with ADD the new skill allows for sustained, focused concentration on learning tasks.

The trainer's task is to encourage his or her learners. The trainer is coaching but cannot tell the person what to do to attain and maintain a particular mental state. The task is about being, rather than doing. The analogy of learning to ride a bicycle is a helpful one that most people doing neurofeedback use. You cannot tell a child how to balance when they first get on a two-wheeler. Similarly, you cannot tell a child how to concentrate. With practice and the direct "neurofeedback" of the vestibular system in the middle ear the child learns to ride his bike. With practice and the computerized neurofeedback that encourages the calm, paying-attention waves and discourages the tuning-out waves, the child with ADD learns what it feels like to concentrate. When the older students get discouraged or impatient for results, the trainer can remind them that brain change takes time – it is a learning process. If they consider how long it took to learn how to hit a good drive in golf, serve in tennis, skate and turn rapidly for hockey, play basketball, and so on, they will perhaps have more patience. It certainly took more than the equivalent of 40 sessions to master these athletic skills.

With adults or parents the trainer can use analogies that remind them that it takes time and, with enough training, things should finally fall into place. Learning a new skill takes time and practice. Second, be patient, in the early stages it may seem that very little is changing. Some parents enjoy sayings such as: there are orchids that can take 9 years to bloom, or 90 percent of the growth of the Chinese Bamboo tree is in the fifth year! Third, don't give up. If you stop pumping the old hand water pump when you need water, you must re-prime it and start all over again.

The trainer and the parents must always remember that many of their young clients are going to become discouraged or complain that it is boring and they may even want to give up. This is the same phenomenon that every good coach is familiar with – the child is all keen to play on their first hockey team. Then they find they cannot skate and stick handle like Wayne Gretzky after one or two practices and they get discouraged and give up. The trainer, as their coach, has the task of helping their young clients stick with it, facilitating and encouraging them. For the children, a reward system can be used. At the ADD Centre, children and adolescents earn tokens over the course of a 50-minute session when they are working hard and achieving goals set for them. There is an ADD Centre Store (a large corner bookshelf) with books and toys; they can exchange tokens for prizes when they have saved enough tokens. The children can earn about 10 blue tokens during their session. For children who need frequent reinforcement, white tokens can be used, and 10 white tokens equal 1 blue token. There are also red tokens (special bonus tokens worth 2 blues) that can be used for an extra good job, or for using calm focus outside of the center as evidenced by a good test result at school, for example. Tokens earned and redeemed for prizes are tracked on the student's bank account sheet. This level of extrinsic reward system is an important part of the motivational aspect of this program. As children with ADD shift towards more mature (age-appropriate) brain wave patterns, they are usually more able to delay gratification and save their tokens over a number of sessions in order to earn the larger prizes.

Dr. Lynda Thompson at the ADD Centre Store

Note: Throughout the remainder of this section, discussion of means for lessening the effects of artifacts is followed by a description of how to decide on the appropriate placement for the active electrode(s) and how to choose the bandwidths to be enhanced or inhibited. Simple one-channel and two-channel feedback combined with metacognitive strategy training (operant plus classical conditioning) is then outlined. This is followed by pointers on handling more complex cases where neurofeedback is combined with biofeedback.

A Review of Artifacts During Feedback

Importance of Artifact Recognition

It is important to minimize false feedback and to be able to distinguish electrical activity of cortical origin from muscle activity and non-EEG electrical activity. Non-neuronal activity that appears in the EEG may also be caused by eye blinks or movement of an electrode. Preventing artifact from affecting feedback is an important consideration when building or evaluating a feedback screen.

As previously noted, at the beginning of every training session you always want to look at the raw EEG. If your instrument has the capability to do so, always record, artifact, and print out data concerning at least one minute of EEG. Apart from having a consistent record of changes taking place over time, this assures you that you are obtaining a quality EEG signal. You may note cardiac artifact and have to change your electrode sites. If 60 Hz is apparent but impedances are excellent, then one must investigate the wiring and the possibility of an unusual electrical source of interference. This could be things like a table lamp or an extension cord. Check the *offset* as well as the *impedance* as this will give you information about the integrity of your wires. The following examples of troubleshooting come from our clinic.

The trainer asked for some back-up, explaining, "There seems to be an EMG (45-58 Hz) reading that is too high." It was extremely high (over 25 μv). When we looked at the raw EEG, it consisted of irregular, extremely high fast waves. The spectrum showed high 60 Hz activity. If the client made even a small movement, the entire spectrum went so high as to be off the scale. The first step was to check the impedance readings. They were all over 25 Kohms. When the trainer reattached the sensors with more 10-20 conductive paste, he brought them down close to about 3 Kohms for each pairing of electrodes. Then the EEG was excellent, 45-58 Hz activity was <2 μV and 60 Hz activity was at the baseline on the *spectrum. The trainer learned **two important lessons.** First, always do impedance checks both before starting and during the session if something looks awry. (If you work with youngsters who have ADD, electrodes sometimes do shift position.) Second, always look at the raw EEG and the spectrum before beginning feedback.*

The same week, another trainer announced, "I don't understand it, the EEG machine worked with the last client. Now I get a flat line. If I shake a wire I see a sudden change in the EEG line so there is a connection between where I plug in the electrodes and the amplifier and the computer. All my impedances were below 5 Kohms. Why am I getting a flat-line EEG?"

She had done the correct thing and begun with a screen that showed the raw EEG. She had excellent impedances. Asked what her offset readings were, however, she realized she had not done them. When she did, she found the offset between the ear electrodes was 20 but the offset between either ear and the active scalp electrode was above 120. There was a broken wire. Replacing the scalp electrode and lead resulted in a normal EEG. She has learned an important lesson. Check the offsets. Offset can still be done if you are using a separate instrument for checking impedances. However, these are expensive and not in common use because most modern programs have their own means for checking and rechecking impedances. The rule of thumb now is – if the impedances are good and your instrument is detecting activity but not a normal EEG, then just change each electrode, one at a time, until you discover the faulty wire.

These anecdotes reveal some of the technical challenges in trying to give quality feedback. Consider the complex question of impedance. Ideally, you should check impedances and offsets every time you connect a client. Unfortunately, not all instruments are set up to allow for this, and not all practitioners have invested in an impedance meter. Some manufacturers do not sell their equipment without an impedance meter. An occasional manufacturer may say the input impedance of the EEG amplifier is so high that you can still have a good EEG signal even without careful prepping and low impedance. If you hear that, understand that the high input impedance of the amplifier is necessary but **not** sufficient. Still others have a built-in impedance check. The most conservative approach is to verify readings with an external meter such as the Checktrode.

Next, consider the question of the integrity of your wires. An electrode and its attached wire may not appear bent or broken, but a wire inside may be loose. If you suspect a problem, just replace the lead with a new one and see if that solves the problem. The length of time that an electrode lasts is dependent on how carefully it is handled. The connection between the wire and the electrode is a particularly vulnerable spot, so remove electrodes without your client trying to help and do not pull them off using the wire.

You will always have some degree of eye movement artifact. At times the eye-blink artifact may be worse than in previous sessions. It may be that the client has worn contact lenses or allergies are bothering them. It may be that you must change from a referential to a sequential (bipolar) recording to see if the preamplifier's common mode rejection will eliminate, or at least reduce, this problem. If it does not, then you will probably just have to use an inhibit bandwidth that is not affected by eye blinks. In some clients an eye blink may only increase delta range frequencies, but in others it may increase 4 Hz or 5Hz as well.

Joan, age 32, had very high-amplitude eye-blink artifact. It affected 2-5 Hz, but usually not 6 Hz and definitely not 7 Hz. Sequential (bipolar) placement gave slightly lower-amplitude readings in the 2-5 Hz range when she blinked, but it was still significant. We could inhibit feedback every time she blinked by placing two threshold lines above and below the EEG line graph and putting an inhibit-over-threshold into the system so that all feedback would cease when the large eye-blink artifact occurred. This is further described below. (This was using ProComp+/Biograph equipment and software.) However, given Joan's very frequent high amplitude blinks, this would have resulted in a very interrupted feedback session. We, therefore, decided to inhibit 7-10 Hz and reward 15-18 Hz when doing academic work. This worked well and avoided false readings due to her eye blinks, while reinforcing appropriate mental states.

What can you do about artifacts produced during feedback? This is an important question because these artifacts can give your client an incorrect impression of their brain states as they are reflected in the EEG.

First, place inhibits to stop sound or visual rewards when your client is producing unacceptable artifacts. You are not actually artifacting the EEG during feedback, you are merely stopping false information from being conveyed to the client. If you save some of the data from the NFB session, remember that this data **is not** artifacted. The EEG instrument and computer were still recording the EEG, artifact and all, even though the client was not receiving feedback. Before you look at statistics, you must return to a screen that shows the raw EEG and remove artifacts. We use the assessment screen shown above, in the section on Assessment, to do this artifacting with the ProComp+/Biograph. We then print out or graph statistics for 1 Hz 'bins' using a very precise IIR Butterworth filter.

Because you are **not** artifacting data during feedback, you must know what major artifacts may affect NFB for each of your clients. Each client is unique. We will sometimes do the statistics both with and without the artifacted segments. In this way we can obtain an impression of the extent to which that particular client's blinking or muscle tension may affect their EEG during feedback sessions.

Jason, the client in the case example at the beginning of this section on Intervention (and example #3 at the beginning of Section VI, assessment of the EEG), had shown rather large excursions of the EEG with eye blinks. However, when the EEG was artifacted, only 3-7 Hz and not 4-8 Hz activity changed with artifacting. Thus it appears that although 3 Hz activity was increased with eye blink, 4 Hz was not significantly increased. We could thus do 4-8Hz training without worrying that we were just teaching him not to blink.

Minimize the Effect of Eye-Blink Artifact

Whenever possible, eliminate the effects **of eye blinks.** Eye blinks can falsify the amplitude of slow-wave activity, particularly in the 3-5 Hz range. As previously explained, the eye can be likened to a small battery: the retina negative, the cornea positive. When you blink, the eye rolls up and the eyelid touches the cornea, or positive end, of this little battery and a wave is observed on the EEG tracing. It usually looks like a large V. In setups where there is a separate clinician screen (one monitor for feedback, a second monitor to show the raw EEG and statistics) the trainer can observe the raw EEG for artifact. In most children the wave is in the delta range and does not appear to have any major influence above 1 or 2 Hz. In adolescents and adults you may find that it goes up to 3 Hz or higher. Frequent blinks may have a major effect on average theta amplitude calculations. You can reduce the influence on feedback caused by eye blinks and movements reasonably efficiently by putting threshold lines on an EEG line graph on your feedback screens as noted above.

In programs that allow you to modify the display screen, you should consider placing an EEG line graph (or a 3-62 Hz spectral display) on some of the display screens that you typically use during a session or that appear on a monitoring display screen viewed by the clinician. This is important for two reasons:

- First, you make sure through the session that you have a normal EEG (an electrode has not popped, for example).
- Second, by means of threshold lines, you can stop two types of false feedback. First, you put an inhibit on movement artifact and thereby stop feedback to the client when this occurs. Second, you stop feedback when large amplitude eye-blink artifact occurs. By so doing you prevent falsely high percent-over-threshold theta readings on the feedback screen. These falsely high readings can be unduly discouraging to your client who is trying to lower theta.

When recording your assessment data and during artifacting, it is important to note the frequency of blinking and the degree to which the client's eye blinks change frequencies that you hope to be inhibiting (for example in the 3-5 Hz range). However, despite our best efforts, we do meet clients, mostly adults, to whom we are not able to give any feedback in the lower range of theta.

It may be possible on your program to create an inhibit instrument, set thresholds, and then eliminate it from view during most of the session so as not to clutter the screen. This can be done with the Infiniti / Biograph from Thought Technology at this time and, hopefully, with other instruments in the future.

Minimize the Effect of EMG and Electrical Artifacts

You can minimize false feedback by putting an inhibit-over-threshold on 45-70 Hz to prevent most **muscle activity and electrical** activity from giving a false impression of fast-wave activity. Alternatively, you can put an inhibit-over-threshold on 45-58 Hz in North America to show mainly the effect of muscle activity on the EEG without concern for 60 Hz. Have your client grit his teeth to demonstrate the massive changes that muscle contraction can make on beta. Now ask your client to very gently tense his jaw, scalp, and neck muscles respectively and observe the changes in SMR and beta that appear to correspond to a rise in 45-58 Hz activity. In Europe, Australia, and Asia, you might use a 53-61 Hz range for muscle inhibit since their current is at 50 Hz.

In the example given below, note that the bar graph has gone above the threshold that we set for this client. The client had some tension in his jaw. This could give a false impression of high SMR or beta, which might otherwise have been rewarded with sound or game feedback. We usually set the feedback options on the program to stop all feedback, both sounds and movement in a game display, when the client goes above threshold on this bar graph. We called it EMG, though strictly speaking, it is only a reflection of EMG activity. True EMG usually occurs at higher frequencies, well above 60 Hz. Our interest is in how EMG affects the amplitude of EEG activity in the SMR and beta ranges, causing it to appear (falsely) high.

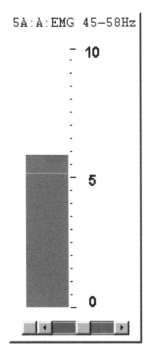

A Bar Graph for 45-58 Hz Activity (in North America)

Even a small amount of jaw muscle tension can shift the beta and SMR readings upwards. We often use a 45-58 Hz inhibit in North America and a 43-48 Hz or a 53-61 Hz inhibit in Europe, Asia, and Australia. We do this so that we have a better idea of the extent to which muscle activity is responsible for a rise in beta frequencies.

*Note that **we do not inhibit 38-42 Hz.** This is because an increase in the EEG amplitude in this range may be associated with useful attention and cognitive activity. Activity around 40 Hz is sometimes called Sheer Rhythm after David Sheer who published in the 70s on 40 Hz being associated with some aspects of attention.*

The bar graph turns red when over threshold.

Minimize the Effect of Electrode Movement Artifact

Electrode movement will usually be different at each electrode and therefore this artifact will not be removed by the amplifier's *common mode rejection*. It is also more significant with even very small head movements if impedance readings between electrodes are high (>5 Kohms).

The large effect of even a slight movement of the electrode can be understood if we recall Ohm's law: V (voltage) = I (current) x R (resistance) (Resistance is used for direct current. We are actually interested in resistance to the flow of alternating current which is referred to as z– (impedance). Because the current (I) is so small, any change in impedance (z) will have a major effect on voltage (V). Even a very small movement of an electrode will massively alter z and thus V, and it may do so in the frequency range we are measuring, say 3-6 Hz, since electrode movement produces a high-amplitude slow wave. As previously noted, this will mean that you may interpret the rise in theta as the client tuning out, when in fact, the increased amplitude is just due to an eye blink or a poor connection. This can be very frustrating for your client!

To minimize electrode movement try the following:

- Tuck the electrodes under a headband. It really will minimize movement artifact if you stabilize the wires. Tell children they look like a tennis star. Between sessions we keep each client's headband in their chart. (The important rule here is that nothing that touches the head of one client touches the head of another client.)

- Keep the impedances between all the electrodes within 1 Kohm of each other.

Place an Inhibit on Electrode Movement and Eye Blink Artifact

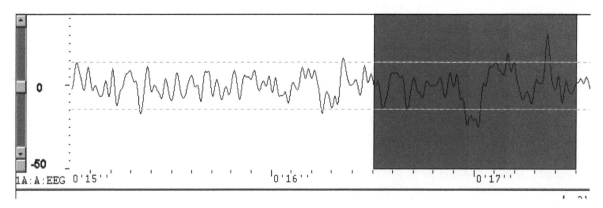

The horizontal dotted threshold lines above and below the EEG can be set so that when the EEG goes outside of these lines it will stop rewarding with feedback. In the above example, the client both blinked and moved. This example also demonstrated how, at the end of a few minutes of feedback during the session, areas with artifact can be marked on the EEG for exclusion from statistical analysis. The gray area has been marked, in this program, by drawing the mouse across the time ruler below the segment in question. The slightly lower amplitude movement artifact at 16.2 seconds was not eliminated. (Note that +ve is a downward deflection on the graph despite the scale reading on the right in the above figure. This is because we have reversed the reference and active electrodes in order to have +ve, be a conventional downward deflection.) Remember Bell's phenomenon: the eyeball rolls upwards when you blink. The positive cornea moves up and the consequent EEG deflection is +ve; that is, by convention, represented by a downward direction on the electroencephalogram.)

Minimize the Effect of Cardiac Artifact

It is also important to watch for **cardiac waveform**s. These are regular waves at a frequency of approximately one per second. They are usually consistent waves, the same general form each time. These complex waveforms can alter different frequency band amplitudes. This problem is rare. It is usually only seen in adults. As noted in the artifact section, it is more often encountered in persons who have a very short, thick neck. When it occurs, your best solution is to redo the assessment, changing your electrode site until it is minimized. Moving to the top of the ear or using the mastoid placement for reference sometimes works. In some cases we have had to use a sequential (bipolar) placement to use the amplifier *common mode rejection* feature to minimize the effect. This also does not always work, but it may decrease the amplitude of this artifact somewhat as shown below. (This is the same example used previously in the section on artifacting.)

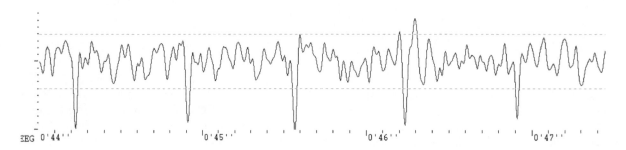

The electrodes for this client, a large man with a very short and extraordinarily wide, thick neck, were moved from the above referential placement to a sequential, Fz-Cz, placement (below). There is some decrease in the cardiac artifact amplitude.

The cardiac artifact is still definitely present, but at a lower amplitude relative to the rest of the EEG. This is also a good example for another reason: the second EEG demonstrates how difficult it can be to see the effect of the cardiac influence on the EEG when the waves are not large. It might have been difficult to pick out the cardiac artifact if one had been rapidly moving through just the second, sequential, sample.

In the section on assessment it was noted that the more consistent the wave, the less may be its overall effect when averaging amplitudes in different bandwidths. The heart rate effects may be a little more consistent if you train the client to relax and breathe diaphragmatically at six breaths per minute (see the section on Respiratory Sinus Arrhythmia later in this book). Because the changes that cardiac artifact will make in your recording are fairly consistent over time, you can still do successful feedback. Absolute values will be affected, but *relative changes* in fast and slow wave amplitudes will be reasonably correct and your client can learn self-regulation.

Brief Summary

Having the raw EEG on the screen is helpful during feedback as it enables you to detect interference, for example, from an electrode that has lifted slightly midway through a session. Without the EEG on the screen you would miss certain kinds of artifacts and the client might be receiving incorrect feedback.

In summary:
- Eye blinks, rolling eyes, and other eye movements can give **a false impression of high theta,** as can heartbeat (a consistent artifact) or any electrode movement over the skin.
- Muscle tension can give a **false impression of increased SMR or beta amplitudes**.
- Heartbeat (a consistent artifact) **can influence a number of frequencies** and every effort should be made to minimize the effects of this source of artifact. (Move the electrode or try a sequential placement.)

Decisions Regarding Frequency Bands and Setting Thresholds

Basic Principles

Your next task is to decide on which bandwidths to reward and inhibit. Then you must decide on the threshold levels for each. Bar graphs are good for setting thresholds because they are easy for a client to understand. Bar-graph screens are available on many systems such as the F1000, the Autogen A620, the ProComp+/Biograph, the Infiniti/Biograph, and other manufacturers. Both Biograph/Infiniti and EEG spectrum programs may also display the same information with boxes. BrainMaster can illustrate it with thermometers.

How you determine appropriate thresholds will depend on the software you are using. The basic principle is that the client must receive enough information to learn the task. If reward rates are very high (threshold too easy), the client is not going to shift his EEG pattern because the computer is indicating he is already in the right zone. If reward rate is very low, there is not enough information about success and the client may become frustrated. You must, as always, respect individual differences. Some people need more frequent reward. This is generally the case with young children and with clients who have ADD. These two groups need a high frequency rate of reward.

The other decision with respect to thresholds is whether to set them initially with great care and leave them unchanged for the duration of training, or to employ a shaping strategy and change thresholds according to the client's performance over time. One cannot say that either approach is better. As with other decisions, it is a matter of trade-offs and depends on your goals and knowledge of you client with respect to such things as frustration tolerance. The set-and-leave-alone approach has the advantage of being able to make session-to-session comparisons that are meaningful using data such as microvolt amplitudes, percent-over-threshold for particular frequency bands, or number of points earned in a time period. The task for the client will be relatively hard initially, and should become easier as training proceeds. With this approach the trainer will need to provide a lot of encouragement and coaching initially to augment the computer feedback, which will be harder to achieve. With the shaping procedure the amount of reward can remain at about the same level across sessions and the client's improvement will be reflected in the thresholds changing: threshold for inhibit frequencies declining and for reward frequencies increasing over time. The Infiniti allows you to do both effectively by shaping with the on-screen threshold line on the bar graph while at the same time entering a threshold figure into the program that does NOT CHANGE when you change the on-screen threshold line. (In Biograph-Infiniti this is under Edit – Edit Virtual Channels, but it should be an available option with other programs, albeit under a different heading.) This is displayed in Biograph as "% >" – a constant (threshold) beside the relevant bar graph. We show this later in this section. Whatever method(s) you choose we recommend that you do **Not** choose automatic thresholding (as previously explained).

An analogy from training a golfer puts this in perspective. If the objective is to be able to consistently sink a 20-foot putt, you could take two different approaches. The analogy to a constant threshold would be to start 20 feet from the hole and keep trying from that distance until the person can do it reliably. This will be very difficult at first and you will have to give lots of praise and encouragement for coming close to the hole since successful shots will be relatively rare. The other approach would be to start one foot from the hole and, when that is mastered, go to 18 inches away, then two feet away, and so on until you get out to the 20-foot mark. The latter method uses shaping, or what psychologists term *successive approximations*.

Both approaches should work whether we are dealing with golf or learning to change brain waves. It would make a good dissertation topic to compare the two approaches with two groups of similar subjects receiving neurofeedback.

Setting an Inhibit Frequency Band

Setting Thresholds

Consider the case example of the eight-year-old boy with ADHD described at the beginning of Section VIII. On assessment, Jason had demonstrated very high 3-7 Hz activity when he was not paying attention. His eye blinks affected 2-3 Hz activity but not 4 Hz. Therefore, a bandwidth of 4-7 Hz or 4-8 Hz could be inhibited in order to help him gain control of focusing and learn to focus externally when this was appropriate. The initial assessment was referential, Cz to the left ear, so this placement would be good for feedback sessions. Further assessment demonstrated that the high theta activity was over the entire frontal and central area, so that placement of the electrode for feedback to inhibit slow-wave activity could be helpful anywhere in the frontal or central regions. The final placement decision could, therefore, be based on where it could be appropriate to reward his fast-wave activity.

With the EEG Spectrum equipment, you use lines above and below the EEG tracings and adjust them until the client is meeting the criteria a certain percentage of the time. Thresholds can be changed without stopping feedback, and the thresholds are stored in the client's disk for the next session. This is a handy feature.

Bar graph showing theta 4-8 Hz activity.

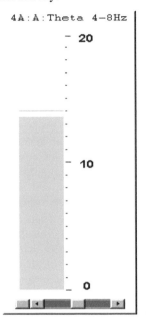

We have decided to reward our client, Jason, for keeping his theta 4-8 Hz activity below a threshold that we have set for him. We based the threshold level on his being able, initially, to keep theta below threshold between 50 and 60 percent of the time.

His threshold for theta had initially been at 20 µv. After 42 training sessions, Jason had progressively lowered that threshold. It was now at 14 µv.

To the left is a bar graph from a ProComp+/Biograph or Infiniti/Biograph screen. This is one of the simplest and most useful instruments used in NFB work. We initially set most bar graph instruments so that the client can get rewarding feedback about 50 percent of the time. This should provide enough feedback for learning to take place without the task being too easy. We always have the color of the bar green when the client is meeting the objective (focused and concentrating) and red when he is not meeting preset threshold criteria. The bar graph shows that Jason has met the criteria; he is below the threshold of 14 µv. Therefore the bar is green. If he goes over the threshold line, it will instantly turn red. Early in training we slow the movements of the bar somewhat so it is easier to follow. As training proceeds we decrease this smoothing or averaging and the movement becomes much faster than conscious thought.

If you are working with a client who has ADD, you could have both a theta inhibit bar graph and an alpha inhibit bar graph on the same screen if both were a problem. With some equipment you are able to use a different sound for below threshold (reward) for each of them. Or, you can use a *sound reward* for below threshold for one of them and use an *inhibit all feedback if above threshold* setting for the other one.

Using Percent-Over-Threshold to Follow Progress

Alternatively, and the method we most frequently employ, you can link a percent-over-threshold counter to each bar graph. This is DIFFERENT THAN THE % > CONSTANT mentioned above because this % follows the on-screen threshold that you may be adjusting as you "shape" training. You might have three bar graphs, perhaps one for theta, one for beta and a third for the ratio of theta/beta. You and the student follow changes in all of them in this manner. One of them may also be linked to a point (reward) counter and a tone can sound for each point earned. An example of having percent-over-threshold linked to a bar graph is shown below. (*Display screens that do this will be shown and described more fully below.*)

The A620 software has a feedback screen that shows three bar graphs (EEG reward, EEG inhibit, and EMG inhibit) with an amplitude scale, percent-over-threshold, a timer and points.

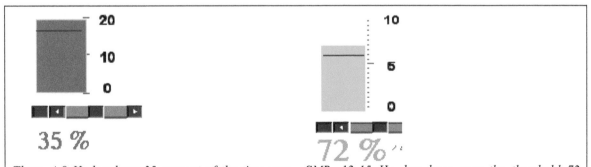

Theta, 4-8 Hz has been 35 percent of the time over the threshold. It is red when over threshold to signal that the client is not meeting the criteria and should try to make it green by bringing it down.

SMR, 12-15 Hz, has been over the threshold 72 percent of the time. It is green when it is over the threshold as a signal that this is good. The client attempts to keep all the bar graphs green (not red).

Once you know what threshold level works well for a client, then you can use the same threshold level for any of your feedback screens. Some equipment, such as the Infiniti/Biograph, will do this automatically for you. In other programs you may have to enter thresholds for each new screen you put up.

By keeping the thresholds constant by means of entering a threshold into the program to give a % > constant for that bandwidth (usually your mean microvolt value found in your initial assessment before artifacts were removed because artifacts are not removed during the training session), you and your client can follow progress during a session and between sessions. You can do this even though during the sessions you may change the on-screen-threshold lines to shape the client's behavior, and change the display screens according to the client's needs. You can also compare their progress from week to week. You can develop and use a tracking client progress sheet to help with this. It is useful to track a client's progress over time. We find that Excel graphs do this beautifully. However, program-generated graphs, trend reports, or even hand-drawn diagrams are useful here.

When dealing with ADD, each client will have one, or perhaps two, dominant slow-wave frequency bands. In young children, this is usually in the theta range. In adults, tuning out may be in low alpha or in the range Lubar has called "thalpha" (6-10 Hz).

Setting a Reward Frequency Band

Choosing an Initial Site and Reward Frequency Band

Again using Jason as the example, two factors dictated the initial reward frequency for him. First, the assessment EEG showed that 13-15 Hz activity was generally lower than both 10-12 Hz and 16-18 Hz. Second, he was a very active and impulsive boy. This was confirmed by his history, clinical observations, and results on continuous performance tests (TOVA and IVA). Rewarding a child for increasing SMR, 13-15 Hz across the sensorimotor strip, in most cases, will result in a decrease in hyperactivity and impulsivity and an increase in reflecting before acting. Although some practitioners have suggested that a C4 placement is preferable, others have obtained good results at other locations across the central strip including C3 and at Cz. With Jason, we decided to use a C4 placement initially.

Since you are training rhythmic activity that is influenced by thalamocortical loops, the change in the sensorimotor (12-15 Hz) activity is not localized to the area under your electrode. (Remember that the activity you record is from a scalp area of about 6 cm^2.) It should be found all across the sensorimotor strip: training at one site should also produce an increase in those frequencies at other sites. This would not be the case if you were training to increase beta, which is a desynchronized waveform largely reflective of corticocortical communication rather than rhythmic activity influenced by thalamic pace-makers.

(Here is another good thesis topic: measure at C3, Cz and C4, but just train at one of those sites in each of three groups to learn to what extent an increase in SMR generalizes.)

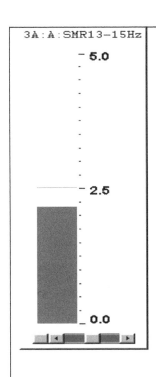

Jason is to raise SMR 13-15 Hz activity. Early on he had trouble with this and a very low threshold had to be set to give him encouragement. When it went below threshold he did not receive sound or game movement rewards.

The bar graph on the left is below threshold. Jason is not meeting the preset criterion (threshold). Therefore, the color of the bar is red. However, the blue 62% shows that overall he has been over threshold 62 percent of the time in this segment of his session. To give encouragement, we often place both a percent-above-threshold and points on to the screen. In the square on the right-hand side, below, points are designated as reward. Points have been linked to the SMR bar graph. This bar graph must be held above threshold for one second to score a point. Points are different than percent-above-threshold because they have a time factor built in. On other screens, Jason had the points linked to theta, and he had to hold theta below the threshold line for a full second to score a point. On still other screens, points were linked to a ratio of theta/beta. On these screens, Jason had to hold the bar below threshold (green) for one second to get a point and hear a beep. In addition to the bar graph and numbers, he would see an animated display – maze, car morphing, etc. This combination of display screens allowed for some variety during training and encouraged Jason to find and sustain a calm, focused, problem-solving mental state.

Changing Electrode Site to Meet a New Objective

When his inappropriately high activity level and his impulsivity decreased (evidenced by progress testing after 40 sessions that collected EEG, TOVA, IVA and questionnaire data), we moved the electrode site and changed the reward frequency. This was done to address the second objective of improving his academics. To achieve this, he must learn to sustain activation of the area of the brain that is involved in problem solving, organizing, sequencing, and analyzing material. F3 was the site we used. Sustaining activation required that he decrease slow-wave amplitude (4-8 Hz) while he increased fast-wave activity (15-18 Hz). The result was sustained attention while solving problems. On the EEG we could observe an increase in 17 Hz activity while he was doing a math problem or reading.

We then paired his sustaining low 4-8 Hz amplitude and high 15-18 Hz activity with having him carry out academic tasks including reading, math and listening to his trainer teaching him learning strategies. If the feedback sound rewards stopped for more than a few seconds, the academic task was discontinued until the feedback indicated he was again maintaining focus.

With a two-channel instrument we might decide to have one channel primarily working on increasing SMR with an active electrode at C4, and a second channel increasing 15-18 Hz (or 16-20 Hz) with its active electrode placed over the lateral aspect of the left hemisphere. This could be, for example, at F3 or at other sites if there were more specific learning disabilities. An example would be training over Wernicke's area for a child who had reading problems involving decoding. At the same time the second channel would also reward a decrease in 4-8 Hz activity. Academic exercises would be done during the feedback sessions, but only while concentration was being maintained.

Below is a display example showing beta, 16-20 Hz, that has been 82 percent of the time above threshold.

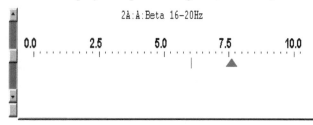

82 %   Above Threshold

This figure shows an alternative method of displaying µV levels. Again, sound feedback, points, percent of time over threshold and games can be linked to the gauge so that Jason will be rewarded for scoring above (to the right of) the threshold line with various forms of feedback.

How Your Assessment Data Are Used to Set Bandwidths

In the section on assessments it was recommended that you be able to look at 1 Hz bandwidths or overlapping 2 Hz bands. The reasons for this suggestion are threefold.

Wide versus Narrow Bandwidths

First, in assessments you can miss marked differences from normal by using wide bandwidths. For example, a large increase in 9 Hz activity may not show up if you are only looking at 6-10 Hz. It will definitely be missed if you look at a standard bandwidth for theta, such as 4-8 Hz. Looking at narrow bandwidths, your assessment will pick out the high amplitude at 9 Hz. You can then be quite specific and inhibit 9-10 Hz or 8-10 Hz activity. In addition, if you are using an assessment program for full cap assessments, such as SKIL, then the precise, narrow bandwidth may show important comodulation differences from a normal database. These may not show up with a wider bandwidth.

Second, when doing neurofeedback you may incorrectly assume you are encouraging a client to raise a specific type of brain wave when, in fact, you are encouraging quite a different waveform.

For example, a client with Parkinson's may have very high 12 Hz alpha activity and a marked dip at 14-15 Hz. Your objective is to increase SMR. In this case the assessment would indicate that rewarding 12-15 Hz would not be appropriate since that could be achieved just by increasing 12 Hz alpha activity. While this may not do any harm, it is not your objective. Given the assessment, you would narrow your training band to 14-15 Hz in order to make it more likely that training would encourage an increase in true SMR activity. You are thus being more specific and targeting the deficiency observed in your assessment. Some instruments, such as the F1000 (unfortunately no longer on the market, but perhaps available secondhand) and Biograph-Infiniti, will allow you to reward a single Hz band such as 14 Hz. (Remember in this discussion that the actual frequencies measured also depend on the filters used. Other frequencies are attenuated but not eliminated when you select 14 Hz.)

Third, in attempting to guide your client you can be misled, and even make the situation worse. For example, if you observe high 24-30 Hz activity, you might assume it reflects EMG activity when the client is actually anxious and worried. On the other hand, if your assessment looks at a wide frequency range up to 61 Hz, you may observe that there is no significant electrical interference (59-61 Hz) and no significant reflection in the EEG of muscle activity (45-58 Hz). In these circumstances, instead of telling your client to relax their jaw, forehead, neck or shoulders (which can make them more mentally anxious and tense because they really have tried to do that), you would recognize that they had a problem with anxious ruminations as reflected by activity in those frequencies. You would not get sidetracked into working on muscle relaxation (though low EMG is always important for quality feedback).

Selecting Bandwidths – Discussion of Case Examples

Bandwidths Correspond to Mental States

In the first part of this book we gave brief descriptions of a number of different bandwidths found between 2 and 60 Hz. Each bandwidth had a specific mental state or mental activities associated with it. This will help you to choose the appropriate bandwidth to reinforce or inhibit, based on the triad of the client's symptoms and goals, the EEG pattern, and the correlation between particular frequencies and mental states.

As a very general statement, we could say that most clients who want to optimize their performance can benefit by learning to achieve a relaxed, yet alert, calm, openly aware state with efficient problem-solving capacity and a minimum of anxiety and ruminations. We want the EEG to give us information related to these mental attributes with minimum artifact.

While it is true that low alpha (8-10 Hz) and high theta (5-7 Hz) states can be useful in meditation, memory recall, relaxation and in psychotherapy, remaining in these states for long periods of time is not normally considered useful in the board-room, classroom, or in an athletic endeavor. They are internal mental states while optimal performance rests on mental flexibility that includes the ability to maintain controlled external orientation. The word 'controlled' refers to the major goal of NFB and BFB: self-regulation of mental states to optimize performance during a task. Except in *alpha-theta* psychotherapeutic work, we are usually training the student to reduce activity in the slower frequencies for reasonable periods of time. We, therefore, place an inhibit on them.

However, as with most general rules, there are well documented exceptions. Psychologist John Gruzelier, working with elite musical performing artists from the Royal College of Music in London, England, demonstrated that increasing theta corresponded to improvements in some aspects of performing (Gruzelier, 2002). This does not seem to contradict the work done in decreasing theta in students with ADD who underachieve academically. Such students are remaining internal-oriented in a non-productive fashion when they should be spending more of their time with an external focus, listening, or reading. The high-performing musical artists, on the other hand, were perhaps getting more in touch with their feelings and enhancing the creative aspects of their performance, which is an internally oriented phenomenon. It is also likely that these students, in contrast to those with ADHD, had no difficulty in producing beta. Also, quite likely their initial theta levels were not high and thus they demonstrated very low theta/beta ratios.

Most adult clients can identify with changing mental states while driving a car. It is quite normal and reasonable to go 'internal' in order to recall the directions to your destination. It is also a common experience to hyperfocus while trying to read a road sign as you approach an intersection. In both cases, however, one's broad attention to all the other traffic around your car is momentarily gone. That period of time, in order to drive safely, must be very brief. Longer time frames are neither safe nor reasonable while driving. These would include going internal and figuring out a problem concerning home or work, or focusing on a cell phone conversation. These longer distractions may result in the driver missing changes in traffic patterns. An accident may be the serious consequence of a longer loss of focus during the important task of driving. For a student, loss of focus means that they miss what the teacher is saying and later fail to recall material that they read.

Clients who demonstrate a relaxed state while remaining alert, externally aware, and capable of quick, accurate reactions to the environment demonstrate *high alpha* (11-13 Hz) activity. With athletes we may reward this bandwidth to help them more easily get in the zone for sports performance. High performing clients may also show some momentary activity in theta that is associated with memory retrieval, creativity, and visualizing. Flitting in and out of theta can, therefore, be useful, but remaining in this state results in being tuned out and not being consciously aware of what is happening in one's external environment. These optimal performance clients also demonstrate frequent bursts of beta, usually around 17 Hz. This is associated with analyzing information and problem solving. It can be interesting for a client to try testing out whether their mental states really do correspond to specific bandwidths. You might suggest that your client check out these hypotheses by creating some image in their mind, visualizing it as they do so. You pause the spectral display of the EEG when they tap the table to indicate that they really have the image. You may see a sudden rise between 7 and 8.5 Hz. Try asking them an oral math problem, pausing just as you can see that they are about to answer. We have found that asking a child to multiply 7x8 results in some alpha or theta activity, then a significant dip in their dominant theta frequency and a rise around 17 Hz. This is so consistent and dramatic that we often let the child demonstrate it to their parents. Perhaps we now have a way of measuring concretely the admonition to put on your thinking cap.

Comments on Bandwidths to Enhance

Rewarding 11-13 Hz

The bandwidth 11-13 Hz near Cz can be associated with a mental state of *open awareness*. An open awareness implies being capable of responding to a wide range of changes in one's environment. In athletics this state is associated with fast reflexes and accurate responses. The awareness of a professional goalie in hockey or soccer, or a black-belt martial artist fighting several opponents are examples. It is also associated with the mental and physical calm required in that readiness state before action. The moment prior to and during the release of an arrow is an archery example that has been studied (Landers, 1991). Open awareness appears to be an ideal mental state achieved by top performers in most any field of endeavor. It can be helpful to any of us. When clients are ruminating over problems in their lives, you may find low 11-15 Hz and high activity somewhere in the 21-32 Hz range. In these clients we have found it useful to reward decreasing the ratio 21-32 Hz/11-14 Hz or 11-15 Hz. By doing this we are discouraging ruminative activity. We are also encouraging both a calm high alpha and SMR state. Because SMR is lower in amplitude compared to alpha we will alternate this with SMR, 13-15 Hz training, and/or training to decrease the ratio 21-32 Hz/13-15 Hz. The precise range used depends on assessment findings. We have clients with whom the dip is very specific, for example at 14-15 Hz. We also have clients with whom the rise in high beta is quite specific when ruminating and then a narrower high beta frequency range is used.

On the other hand, high alpha amplitudes in the 9-12 Hz range can carry over to 13 Hz and even, in rare instances, as high as 14-15 Hz, and mask a low SMR. In this client you would not reinforce in the "sensorimotor" range 13-15 Hz (because from the waveform you can see that it is really alpha, albeit at a higher frequency than usual). This kind of client may show symptoms, such as tuning out and having their mind wander from what they are reading; this clearly corresponds to the slow-wave component of the EEG. The same client may also act impulsively. Since a low SMR in the assessment may have been masked by the high amplitude high alpha, then it is reasonable to set rewards assuming the true sensorimotor rhythm activity may be low. For example, a 16-year-old boy with mixed symptoms of Asperger's syndrome and severe ADHD demonstrated a very high peak at 10 Hz with typical alpha waves seen to be flowing over into the 11-15 Hz band, but not above 15 Hz. He had an average 14-15 Hz amplitude over three minutes that was very close to the average amplitude of 16-17 Hz beta. His symptoms included impulsivity and anxiety. We feel it is reasonable in his case to reward decreasing 9-10 Hz activity and reward increasing 14 Hz or 14-15 Hz.

The bandwidths used should be derived from your assessment. They are also changed as the client changes. For example, as a child with ADD begins to concentrate better and to mature, the dominant slow-wave bandwidth may not just decrease but it may also shift to a higher frequency. For example, it may move from 5 Hz to 7 Hz and then cross into the alpha range. A shift of the dominant frequency upwards is the natural progression with age so it seems that neurofeedback training in some clients with ADD may be encouraging a more mature pattern.

Rewarding SMR (12-15 Hz or 13-15 Hz)

In the ADD Centre, our most frequent starting point is to reward 12-15 Hz or 13-15 Hz activity with the active electrode placed at Cz or C4. The initial EEG profile typically shows very low amplitude (eyes open) at 12 Hz in most people though it may be high in some clients who have ADD or in elite athletes who have put themselves into a state of "open readiness." In clients who show high amplitude 12 Hz but a dip at 13-15 or 14-15 Hz we chose the appropriate frequency outside of the alpha range when we do the settings for NFB. The rationale for SMR training also includes the following considerations:

- For a broad, relaxed awareness we reinforce synchronous high alpha sine waves at a frequency around 12 Hz (11.5-13 Hz). The mental state that corresponds to this band appears to work well in athletics as discussed above. This range is used in peak-performance training.

- In addition, high alpha may appear through the entire 12-15 Hz range, and this waveform appears to be associated with a mental state that embodies inner reflection and creative thought. (Remember, the term *alpha* refers to the morphology of the waves and not just a frequency range. Alpha waves may also appear at slower frequencies, such as 7 Hz, especially in young children. It is sometimes referred to as pediatric alpha.)

- Also in this range, perhaps more between 13 and 15 Hz, is the synchronous SMR spindle rhythm, which is regulated by the thalamus. This sensorimotor rhythm appears across the top of the head along the sensorimotor strip and is associated with decreased sensory input as well as decreased motor output. (In other locations, such as frontal or occipital, 12-15 Hz activity would be low beta, not SMR.) It appears to be associated with decreased red nucleus firing, decreased muscle spindle activity, and decreased muscle tension. When it is increased, the clients tend to become less hyperactive, less fidgety, and less impulsive. They tend to be calm and reflect more before responding. Sleep also improves in some cases. In some clients who have emotional lability or hypomanic behaviors SMR may also be of some benefit since higher amplitudes of SMR are associated with a calming effect. SMR training is being used for pain management and in cases of fibromyalgia.

- Desynchronous beta waves associated with problem solving activity are also found at the upper end of this range, though perhaps more in the 14-17 Hz area.

These components or waveforms come from different origins (or "generators"). The brain waves observed in the EEG represent cortical electrical changes which may be influenced by both thalamocortical and corticocortical activity. The feedback loops within the brain are complex. Each of these waveforms (synchronous alpha, SMR spindles, desynchronous low beta) represents a different mental state though their frequency may be similar. All of the mental states mentioned above are desirable in the process of learning self-regulation. This is particularly true when dealing with clients who present with difficulties in attention, concentration, reflection before action, and calmness.

Caution #1: Some of your clients tune out in alpha. This alpha may be the predominant component of the 12 to 15 Hz band you are rewarding. If this is true, you may be rewarding your client for tuning out.

Suggestion: Look carefully at the raw EEG. When you see clear alpha waveforms, ask yourself if they go into the 12, 13 Hz range on the spectrum. Is the microvolt level of 12-15 Hz bandwidth markedly raised? If it is, consider rewarding SMR at a slightly higher frequency, say 14 to 16 Hz, especially if there is a dip in this range when you generate your histogram summary of your assessment data. In addition, inhibit 7-9 or 7-10 Hz rather vigorously if this low alpha is excessively high.

Caution #2: If you inhibit 3 to 5 Hz, the predominant wave you are inhibiting may be that which is caused by eye blink. In effect you are rewarding your client for not blinking or moving their eyes. There is a large literature on eye blink rates and it is an interesting phenomenon but it is not typically what we are trying to train. On the other hand, this may be the most important frequency to inhibit in some young children. You must watch the child and decide if this frequency reflects eye movement or tuning out.

Suggestion: Look at the raw EEG carefully and note the frequencies that are increased when the client blinks. If blinking is frequent and does raise the amplitude of the 3 to 5 Hz range, then you must be careful about how you interpret the EEG. In addition, it is difficult to actually 'feel' or 'sense' 3 to 5 Hz. If 7 to 9 Hz is also raised in your assessment profile, you can choose to train this down. One advantage of this appears to be that even young children can 'sense' what it feels like to be in the 7 to 9 Hz range or out of this range. The lower frequencies may normalize (decrease) as the client learns to control the 7 to 9 Hz frequency range. You may have to down-train 6 to 10 Hz if your equipment does not allow you to customize the bandwidths.

Rewarding Beta 16-20 Hz (or 13-15 Hz)

The electrode placement is usually at Cz, C3, or F3 when training up 16-20 Hz, as previously described. Rewarding beta is recommended in clients who demonstrate ADD without hyperactivity, impulsivity, or labile mood. It is often the range of choice for the client who is not impulsive and who **is lethargic and has a low alertness level.** (Also consider *sleep apnea* in the differential diagnosis if the person is lethargic to the point of excessive daytime sleepiness. Other symptoms usually include snoring. It is more common in those who are obese and have short, thick necks.) It is appropriate to encourage this range in the left hemisphere in a client who is doing academic work, and in most clients with learning disabilities, particularly language-based LD. It may also be desirable to reinforce this range in the left frontal region for clients with depressed mood. Of course, you always consider the client's EEG pattern plus knowledge of neuroanatomy and physiology in making decisions about which frequencies to reinforce and where to place the active electrode. In children, 15-18 Hz may be more appropriate than 16-20 Hz. Remember that the dominant EEG frequencies shift higher with age.

As noted above, when you reward 13-15 Hz over the sensorimotor strip, you are usually attempting to increase SMR activity. SMR rhythm originates in the ventrobasal nucleus of the thalamus. Anywhere else over the cortex, 13-15 Hz is called low beta. When compared to SMR, desynchronous beta at 13-15 Hz has a different waveform and a different origin. It is the result of localized neuronal activity. It may correspond to cognitive work. There are occasions when you may wish to *activate* an area of the brain in the right hemisphere but not want to encourage higher beta over 16 Hz. A *soft* encouragement of activation is used. Examples are cases of autistic spectrum disorders, including Asperger's syndrome, where the full-cap assessment has demonstrated a localized area of slow activity at, or close to, T6. Although at the time of writing there were no case series in the literature, there have been a number of anecdotal reports of side effects (such as being temporarily cognitively somewhat out of touch with reality) in clients where 16-20 Hz activity was encouraged in this area (Jay Gunkelman, personal communications). In these children we have observed positive effects of increasing 13-15 Hz activity while concomitantly decreasing the dominant slow-wave activity in the right hemisphere.

During a course he taught at the 2001 meeting of the SNR in Monterey, CA, Tom Budzynski mentioned that one should not train to increase 16-20 Hz on the left side in children who have *reactive attachment disorders* (RAS). RAS children reportedly have a worsening of symptoms with this training.

At the ADD Centre we have never seen adverse results with 16-20 Hz training in the dominant hemisphere.

Brief Review of 'Reward' Frequency Ranges

What is the most important frequency to work on? To make that decision, look at the EEG profile and also keep these suggestions in mind:

- To decrease impulsivity and raise 'reflection before responding,' the client should usually attempt to increase SMR 12-15 Hz (or 13-15 Hz) activity across the sensorimotor cortex.

- To increase focused concentration, decrease the client's dominant slow wave. Make decisions regarding location and frequencies based on whichever location is higher than expected. Database norms may be consulted for this purpose. Usually the dominant, eyes-open, Cz frequency will be somewhere between 3-10 Hz in people who have problems with attention.

- To emphasize peak performance, broad awareness, and a calm mental and physical state, the client should be encouraged to increase 11.5-13.5 Hz activity. However, the ultimate objective is for the client to have mental flexibility, the ability to shift mental set appropriately and thus rapidly shift frequencies. The objective is not merely to increase a particular frequency range.

- To emphasize problem solving, use beta 15-18, 16-19 or 16-20 Hz. You could consider attempting to increase 39-41 Hz activity, but as previously noted, this Sheer rhythm is more difficult to work with both because you must severely inhibit any muscle influence (43-58 Hz) and because the amplitude of this frequency band is usually quite low.

- Anxiety and mind *racing* with tension may be associated with increased amplitude in waves anywhere from 20-35 Hz. Looking at frequencies above 36 Hz is helpful in distinguishing muscle tension from mental tension. These high beta frequencies will usually decrease if the client enters a calm state and increases SMR while concomitantly working on relaxing using biofeedback: respiration, RSA and HRV, finger temperature, EDR and muscle tension. (As previously mentioned, the clinician should be cautious about encouraging diaphragmatic breathing in a client who has a seizure disorder. If, in the learning process, such a client hyperventilates, a seizure may be precipitated.)

Electrode Placement for Feedback

Introduction

Decision making regarding placement is most easily done after doing a full-cap 19-channel assessment because you can then compare the relative amplitudes at each site for each frequency. You can also observe how different areas of the brain are communicating with one another by measurements of coherence or comodulation.

However, many readers will not have either the training or the equipment to do this. We will, therefore, look at some general principles and then at two common and relatively simple examples. If you are doing a one-channel assessment, Cz is often the site chosen. It tends to have the highest amplitudes (it is furthest from the ear reference points so there is less common mode rejection) and it is also less influenced by artifact (either eye movement or jaw tension). In addition, it can reflect activity in the cingulate gyrus and correlate with functions in the Salience, Affect, and Executive networks.

Shaping and Rewarding Responses at One Site may Affect Many Sites

Dr. Joel Lubar has experimented with placing electrodes at different sites for training. He found that after training along the central cortex at Fz, Cz, Pz or at FCz-CPz, post-training changes in theta/beta ratios were also noted at the other 10-20 electrode locations (Lubar in Evans & Abarbanel, 1999; Lubar, 1997).

Case Examples

Electrode Placement in ADD without Hyperactivity

Gains from NFB may be seen more rapidly with persons who have ADD without hyperactivity than with those who are quite hyperactive. As an aside, this parallels the finding that there is a different dose-response curve for learning as compared to sitting still for stimulant medications. It takes a higher dose to reduce hyperactivity than to improve performance in a paired-associates learning task. (See Chapter 9 on medications in *The A.D.D. Book,* Sears, 1998.)

Jane, age 17, was a 'couch potato,' according to her mother. She was described as being lazy and unmotivated. When Jane was questioned, however, it turned out that she felt very discouraged. She said she knew she was bright enough but it took her so much more time studying than her girlfriends to get even close to their marks. She admitted she had just given up. She said that she would start reading a paragraph, get to the end, and realize she couldn't recall anything. Her mind had just drifted off even though her eyes may have followed the words.

We began training with the active electrode at Cz. The reference electrode was placed on the left ear with the ground on the right to emphasize the left hemisphere. After a few sessions we decided that activation in the left frontal area was important due to her discouraged and somewhat dysphoric mood. We wanted to pair activation with doing academic tasks. Placement was changed to having the active electrode at F3 and the reference electrode on the right ear. Placement of the reference electrode on the right ear gave higher amplitudes of the waves being measured. If it were placed on the left ear lobe, there would be greater common mode rejection by the amplifier and thus lower amplitudes.

Note: Occasionally you will place the active electrode either more anterior or more posterior either for a clinical reason, such as specific learning disabilities, or for an EEG reason because of findings on a full-cap, 19-channel assessment. You may also choose to reference to the left ear if you decide you mainly want to influence activity in the left hemisphere between your active electrode and that ear. This suggestion must be understood as purely theoretical. It has not been empirically tested. It is based on an equipment manufacturer, electrical engineer's model of the EEG. (Frank Diets, personal communications.) The theory is that you are mainly training dipoles between the active electrode and the reference electrode.

Electrode Placement in ADD with Hyperactivity

Symptoms of Brent, a Nine-Year-Old Boy

Assume for this example that a student has come to you with a combination of ADHD, Combined Type, and dyslexia. It is important to proceed through the decision making process for each client using the decision pyramid (see the Decision Pyramid diagram in Section IX under Assessment). This example is discussed in detail as ADHD, Combined Type. It is the most common application of NFB at the time of writing.

Brent was an extremely hyperactive and impulsive nine-year-old boy. He was in Grade 3, but his reading was at beginning Grade 1 level. He was a child who definitely needed either to have entirely one-on-one teaching or he had to be on a stimulant medication when at school due to his disruptive behavior. Stimulant medication did work well for Brent as a chemical restraint. However, his mother commented that, "His eyes glaze over just as much on the stimulant as off [it]." She was not convinced it directly affected his attention span, although he would spend more time working and finish more school work when, as she put it, he was "glued to his seat." Certainly it had not helped him learn to read. She wanted Brent to learn to control his own behavior and concentration. She also wanted him to catch up in his reading.

The assessment first used an active electrode placement at Cz referenced to the left ear with the ground on the right ear. Brent displayed very high theta at 5-6 Hz. His theta 4-8 Hz/beta 16-20 Hz microvolt ratio was 4.

Now take Brent's case and use the decision pyramid to decide on electrode placements and bandwidths for enhance and inhibit instruments.

Theoretical Approach

1. The decision pyramid requires that we first delineate our immediate objective from the client's history and objectives.

2. We consider the wave pattern that corresponds to that mental state and that, therefore, must be enhanced and/or inhibited to normalize the EEG and hopefully the symptom(s).

3. We consider the client's symptoms in the context of what is known about functional neuroanatomy/neurophysiology. This should assist in delineating the appropriate theoretical site for enhancement and/or inhibition of frequency bands (or sites for coherence or comodulation training if 19-lead assessment data are available).

4. Fourth, we see if our theoretical formulation corresponds to our findings on the EEG.

5. At this juncture we make a final decision on the following parameters: frequencies to be enhanced/inhibited, the site for our active electrode (or electrodes in a sequential placement), the condition(s) that should be incorporated into the feedback sessions, such as reading, math, listening (or combinations of these activities in a work-rest sequence).

Practical Application with Brent

First: We must decide what is the most important symptom to deal with first. Most of us would feel we must deal with the **hyperactivity and impulsivity** before we can adequately focus on his reading. Clinically, experience has demonstrated that raising SMR along the sensorimotor strip will correspond to a decrease in hyperactivity and impulsivity in most children. The red nucleus will decrease firing when the thalamus has increased SMR output. The red nucleus is part of the extrapyramidal system with gamma motor efferents that go to the muscle spindles. Anatomical knowledge and neurophysiological theory suggests that in ADD there may be overactivity related to norepinephrine activity in the right cerebral hemisphere (Malone et al., 1994). This would support a right-sided placement. Therefore, we most often place the electrode at Cz or C4. Cz would be less prone to jaw muscle artifact, so C4 could be used once the frequencies reflecting EMG activity are consistently low. (As an aside, it is known from experience that a C3 placement can also bring good results. This is probably because SMR rhythm is generated in the thalamus and there is communication between the right and left lobes of the thalamus so that training SMR on one side of the brain can have a similar effect on the other side.)

Second: While we enhance SMR, we should increase Brent's ability to **sustain focus**. Studies have demonstrated that decreasing slow-wave activity will assist with this objective (Lubar, 1995). Neuroanatomically there are studies done with subjects who have ADD demonstrating a decrease in frontal blood flow (Amen, 1997) and glucose metabolism (Zametkin, 1990). Slow-wave activity is generally associated with decreased external focus and decreased problem solving activity. In addition, it is hypothesized that there is deficient dopaminergic activation of the left hemisphere in ADD which corresponds to a decrease in focused attention (Malone et al., 1994). Therefore, we will decide to reinforce a decrease in dominant slow-wave activity. Studies are showing that there may be different locations for maximum theta and/or low alpha wave activity in ADD, however, high-amplitude slow-wave activity is usually found over a relatively wide region. Most often it is frontal or frontal and central. Since we found high theta centrally it is reasonable to lower theta in the central region (C1, Cz or C2).

We could do a second assessment and evaluate the amplitude of theta at F3 and/or Fz. If you have a two-channel instrument, then it is quite simple to place a second active electrode (for a second channel) at F3 and use a *linked-ears reference* and a *linked-ground* anterior to Fpz with wires to both Channel A and Channel B in the Procomp2 (or C and D in the Infiniti). This allows you to assess Cz and F3 at the same time. This gives data for comparison that is collected at the same time and the procedure takes less time than moving your electrode from one site to another and taking two EEG samples.

Third: We want to ameliorate his **reading** difficulty. This will require that his brain be both focused and active, in the problem solving sense, while he is learning to read. For this we combine a theta inhibit (sustain focus), and a beta enhance (actively analyzing the reading material). Neuroanatomical understanding of dyslexia points to the optimal placement of the electrodes being over areas of the brain that correspond to reading functions which are in the left cerebral hemisphere. Reading may involve the insula area and the angular gyrus at the occipital-parietal-temporal lobe junction. The latter area could be postulated to affect the **visual-spatial-language** skills involved in visual word recognition. To affect these areas the electrode placement, for most people, would be approximately in the middle of a triangle with angles at C3-P3-T5 or where the two diagonals would cross in a four sided box with corners at C3-P3-T5-T3.

Brent had single-channel training. His first 40 sessions were at C4 referenced to the left ear to raise SMR and decrease theta. His impulsivity and inappropriate hyperactivity decreased markedly and his attention improved. His reading improved somewhat but was still only at a Grade 1 level. The active electrode was moved to Wernicke's area between and at the center of an X formed by lines joining C3-P7-T7-P3 and referenced to the right ear. He increased beta 14-17 Hz and decreased theta while doing appropriate level reading tasks. After 40 more sessions his reading had greatly improved to an early Grade 4 level and he was just entering Grade 4.

Alternatively, for dyslexia, you could do as Lubar has suggested (noted earlier). For dysphonetic dyslexia, place the active electrode at F3 or F5 (or for a sequential placement, try F3-P3 or F7-P7 (T5), and for dyseidetic dyslexia, place electrodes at P3 or P5). In each instance the objective is to have the student increase activity in that area while carrying out the appropriate academic task.

Further Case Examples

Go back and review the earlier descriptions of cases such as Brad and Ben (severe **Asperger's** syndrome) from the standpoint of how the 'decision pyramid' was used to define both electrode sites and bandwidths for neurofeedback training. In particular, note how the 19-channel assessment finding of slowing around T6 with Brad and P4 with Ben corresponded to our neurophysiological understanding of the right brain's role in interpreting and expressing innuendo and emotions. This led to electrode placement at a posterior location in the right hemisphere. Brad and Ben both attempted to activate this area.

Brief Summary

At this juncture you have chosen electrode placement for a one-channel instrument. We will look at more complex cases and two-channel training later. You have decided on the inhibit and enhance bandwidths and the thresholds for beginning training. Now you may consider the speed of feedback (termed 'smoothing' or 'averaging' in most instruments), the types of feedback (auditory and visual), and the screens that you may use. All this is done in order to give clients the most appropriate information to facilitate the task of changing their EEGs and, thus, improving self-regulation skills. It is likely in a nine-year-old like Brent that the highest theta activity will be across the central region (C3, Cz, C4). In adolescents and adults, the slow-wave activity is often found more frontally (closer to Fz).

Speed of Feedback

Some instruments, such as the F1000 (Focused Technology), ProComp+/Biograph or the newer Infiniti (Thought Technology), allow the operator to alter the rate of feedback; for example, how fast a bar graph moves up and down. This is variously referred to as *averaging* or a *smoothing* factor. It relates to the time period of data collection used before the feedback is updated. We find that, in the early stage of work with a client, slower changes allow the client to follow changes more easily. Jumpy movement may even irritate some clients. However, some clinicians believe that faster feedback, often beyond what one can follow consciously, may have a more dramatic effect. (Here is another empirical question for a graduate student to investigate.) Averaging is a bit like the suspension in a car. You can have it very responsive like a sports car and feel every bump, or you can have it very smooth like a Lincoln Town car or a big Citroen. Jumpy will be a little more accurate (faster feedback) but some clients do not like it. You can let your client choose – do they want a smooth ride or a more responsive 'feel the bumps' sporty ride? Better still, in an instrument like the F1000, you can begin the session with a challenge to reach a certain score using a high averaging setting, then change the setting to a faster feedback with a new goal, and then to even faster feedback (less averaging). This can be very challenging and we have found that it will engage even some of the most unmotivated children.

Important: 'Averaging' puts a slight delay on information

This can be confusing when you review data. Say, for example, that you are using sound feedback for a professional golfer as they putt. They wish to review the data after they make the putt to see the precise mental state they were in at specific points during a series of actions such as preparing to putt and putting. You must turn off all averaging to accurately see what was occurring at specific points in time, such as when the client made contact with the ball. (Not all equipment allows this kind of review.) If averaging were being done, then there would be a delay and the raw EEG data would not correspond precisely to the spectrum if both were on the feedback screen.

Similarly, if you want to view a few seconds or minutes of data during an assessment and look at precise points in time, you don't want to use the spectrum because it will average the information. You want bar graphs without averaging so that they will precisely reflect what the EEG at that instant is showing.

Choosing Sound Feedback

Use of Sound Feedback

Sound feedback is often used as a reward signal for meeting the criteria such as being below threshold for the theta inhibit. A sound reward at the instant the client achieves the preset threshold can be used in most instruments. On a bar graph, for example, the sound reward for theta is set to occur when the column moves below the threshold.

Type of Sound Feedback

Discrete Sounds

In some instruments you have further options such as setting a distinct sound. (Digitized sounds such as a *chord* or *ding* can be found in Windows Media). The sound occurs only after the client has held the criteria for a specific length of time (for example, 0.5, 1, 2 or 3 seconds). In some instruments you can set it so that the reward sounds upon moving above or below threshold, and then, if held in that state, it sounds again at specified time intervals (such as 1, 2 or 3 seconds).

Continuous Sound

On the other hand, you may choose to have a continuous sound, a pleasing tone or even music from the client's favorite CD, which remains on only if they continue to meet all the preset criteria. You should see what the client likes and tolerates best and also fit the feedback to the task you are asking your client to do, such as reading or math. Music stopping tends to alert the client to when they *lose* rather than *achieve* the threshold criteria. Music should be without words. You can explain (again) to the client that the brain can only consciously focus on one thing at one time.

Timing of Sound Feedback

There are several other important considerations. First, all feedback must be <500 ms after a client's response. (The response is meeting the criteria.) Virtually any visual or auditory feedback on a NFB instrument can meet this primary criterion. In addition, you must decide what you are attempting to accomplish: 1) feedback for discrete events meeting preset thresholds <u>or</u> 2) prolonged feedback for sustaining threshold levels. Let us expand this, to answer the question of what type of feedback for what purpose.

Discrete Feedback

Do you want discrete feedback corresponding to each time the client attains the desired mental state? The Biofeedback Federation of Europe, using Thought Technology Biograph Infiniti equipment, has designed, with Dr. Sterman's guidance, a suite for training to enhance SMR that follows this principle.

Gail Peterson, a psychologist at the University of Minnesota who teaches learning theory, outlined the criteria for adequate feedback for operant conditioning of a desired response in a guest lecture at the annual meeting of the Society for Neuronal Regulation, October 2000, in Minneapolis. These criteria may be given the acronym SURA. This stands for being **sharp** or immediate, **unique** in that it doesn't otherwise occur, **reliable** in that the equipment produces it reliably, and **ambient,** meaning that it stands out distinctly from the ambient environment. No icon that we use should be identified with any previous learning. In addition, Sterman's early animal research suggests that there should be a two- to five-second recovery period between rewards. The BFE Sterman Suite meets these criteria.

Ideally we would like to meet Sterman's and Gail Peterson's criteria for successful operant conditioning when we are enhancing SMR. To do this we should first pay attention to our office setting. We want to have the least amount of clutter, not only on the screen but also around the monitor. This is done to minimize associative learning (anything associated with the stimulus may become necessary for obtaining the response). Second, when training to enhance SMR, which comes in bursts, we would require a sharp, unique sound and a visual feedback with a delay between rewards. For this type of feedback for SMR enhancement, the visual feedback displays would give a discrete, distinct movement each time the criteria are reached and allow a defined time length between movements when the criteria are maintained would theoretically be best.

Thus for just training to enhance SMR which comes in bursts we would use discrete, clear, unique feedback. However, when training to sustain attention we are combining enhancement of low beta ranges with inhibition and control of high-frequency beta and low frequencies such as theta and low alpha as appropriate for the particular individual. With that type of training for sustaining attention, rests between rewards would be quite inappropriate. School children and university students, for example, might miss half of every equation in math classes if they only learned to sustain attention for short discrete periods of time.

Sterman's recommendations for feedback are based on research concerning synaptic transmissions of impulses. More recent work has demonstrated that the postsynaptic membrane is not the only site of ionotropic receptors that receive a neurotransmitter and open an ion channel. There are presynaptic *autoreceptors* along the length of axons, for example, that bind extracellular circulating neurotransmitters. This system, involving many neurons, may be involved in *sustaining* attention. (Personal discussion with Fred Shaffer, based on material covered in his workshop about neurons that was given at the AAPB 2001 annual meeting.) Thus we postulate a second way to do operant conditioning may be acceptable.

Sustained Feedback

You may decide to reward **maintaining the desired mental state**. Sustaining attention is essential in academics and at work. To do this, set the threshold on your instrument so that it is somewhat easier to reach. You may also decide to set *averaging* to a higher figure. This will ensure that sounds or music will not cut on and off too frequently, which could be quite irritating to some clients. Try to give a pleasant sound for achievement of the desired mental state. The music chosen may be either relaxing or invigorating. For example, the William Tell Overture may be quite invigorating. The slow movement from the work of baroque and classical composers may, on the other hand, be quite relaxing. Your choice depends on your purpose at the time.

As mentioned previously, it is arguable that this type of feedback emphasizes *losing the reward*. Thus, when the music stops suddenly, this loss of rewarding feedback may alert the client to the fact that they have lost the desired state and precipitate them working to actively re-enter the desired mental state. Students report to us, "Now I notice when I tune out."

When to Use Music

Some equipment, as mentioned above, may allow you to use a CD for feedback. This option can allow teenagers to bring in their own CD with music they find rewarding. This provides a further opportunity for you to discuss with the teenager the effect of music on different types of study tasks. As previously stated, the brain can consciously work on one task at one time. We show them visual illusions, such as one where you see either two faces or a vase, and ask them if they can see both images at precisely the same instant in time. They can't. We then joke that on a canoe trip, it is possible to walk across a narrow log over a fast moving river carrying a canoe and chewing gum – just don't taste the gum. If you allow your focus to move to the gum you are likely to find yourself swimming. Music with words may be fine if the subject is one where you learn best by moving in and out of focus (like memorizing a vocabulary list). It is not so good if you are trying to solve a complex math problem or write an essay. Above all, listen to their experiences and beliefs. Try an experiment with diffcrent types of music or no music while they do a series of math problems if they would find this genuinely interesting. It is amazing how the most oppositional students will work with you in the spirit of genuine discovery about what works best for them if you approach it in a truly interested (in them) and scientific manner.

Sample Feedback Screens
Keeping Score – Points and Percent-Over-Threshold

Some EEG instruments allow the user to *link* one feedback display to another. This allows the client to keep score at the same time as they try to meet a challenge, such as keeping the ball to the right on the balance beam in the figure below.

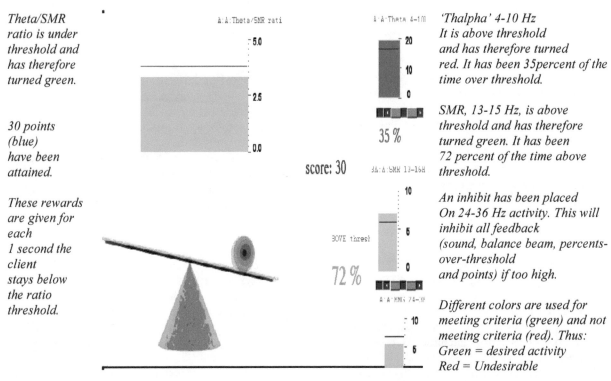

Theta/SMR ratio is under threshold and has therefore turned green.

30 points (blue) have been attained.

These rewards are given for each 1 second the client stays below the ratio threshold.

'Thalpha' 4-10 Hz It is above threshold and has therefore turned red. It has been 35 percent of the time over threshold.

SMR, 13-15 Hz, is above threshold and has therefore turned green. It has been 72 percent of the time above threshold.

An inhibit has been placed On 24-36 Hz activity. This will inhibit all feedback (sound, balance beam, percents-over-threshold and points) if too high.

Different colors are used for meeting criteria (green) and not meeting criteria (red). Thus: Green = desired activity Red = Undesirable

The ball moves to the right when the client remains below the ratio threshold (desired result) and goes to the left when the client goes above the ratio.

Purpose of Screen

The objective is to roll the ball to the far right hand side of the balance beam and hold it there as long as possible. This screen is set so that the balance beam animation is linked to a theta/SMR ratio. It therefore requires an externally focused, calm mental state. If, instead, a theta/beta ratio were to be used, it would require a focused and actively problem-solving mental state. The beta option is used (beta substituted for SMR in the above screen) during the part of a session when the student is listening to a mock academic lecture given by the trainer or when the trainer is teaching the student metacognitive strategies such as active reading, note taking, underlining or math strategies. It makes sense to train for increased beta when on task.

Screen Details

The sound feedback can be discrete, digitized sound, a chord or a ding. The sound could be set so that it requires that the mental state be held, for example, for one or more seconds. The points are often set so that the student is required to keep the ratio below threshold for an entire second. This makes attaining points very different than changing the percent-above-threshold. The *green* percent-over-threshold is linked to the SMR (or beta). The *red* percent-over-threshold is linked to the slow wave (thalpha, 4-10Hz) in this case. The student is told that green and blue are the 'good-guy' colors – increase them. The red is a 'bad-guy' color – decrease it. The 4-10 range is an unusually wide range that was chosen for this client because there were two slow-wave peaks at 5 Hz and 9 Hz. The *dark blue* points are linked to the balance beam animation. We linked a percent-over-threshold to the points and thus to the animation and the ratio, but we decided we had sufficient information on the screen for this client to watch so the instrument was hidden while feedback took place. Percents-over-threshold and points are recorded manually at three- to five-minute intervals on a Client Tracking Sheet. The length of time, two minutes, three minutes, etc., is determined by what seems doable for a particular client. Two minutes of sustained focus is quite a challenge for young children. We like the clients to write down their own scores. We find this gets them more interested in shifting the scores in a desirable direction.

Note that the 24-36 Hz bar graph is being used on this screen as a muscle inhibit as well as an inhibit for nonproductive ruminative thinking. An EEG with threshold lines has been set and hidden in the background in order to reduce clutter on the screen. The threshold lines above and below the EEG are set to inhibit feedback if there is significant eye-blink or movement artifact. This EEG line graph can be brought on to the screen if the trainer wishes to view the EEG during the session.

A Clinical Hint

With some display screens, such as the one above, try counting out seconds, in a barely audible voice, while the student maintains the desired mental state. For example, count: 1,001, 1,002, 1,003… while the ball remains on the right side. Mark a horizontal line in seconds; say 1 to 20. Place a checkmark on how far the student gets on each count. Perhaps give one token for achieving 7 seconds and two for reaching 10 seconds, 3 for 12 seconds and so on. Tokens earned are recorded in the client's bank account at the end of each session and they can use them to purchase prizes.

Empower the Client to Produce Specific Frequencies at Will

A major goal in training is for a client to gain control or self-regulation. It is very empowering for a student to discover that they are able to decrease theta or thalpha on purpose or to raise SMR or beta through conscious control. The balance beam above can be used in this exercise. **If your instrument has a "work-rest" function** then you can set it so that "work" lasts for about 45 seconds and "rest" lasts for about 15 seconds. During the "work" phase the client must roll the ball to the right. During the "rest" phase the client must roll the ball to the left and keep it there. They must do this without changing any aspect of their posture and without moving their eyes or their jaw. (With special thanks to Jim Stieben, Toronto, for his work with clients refining this technique.) This technique follows an exercise model. You work on producing a particular state, then briefly rest, then try to produce the state again. Alternatively, try to turn the mental state on and then deliberately turn it off. It can be likened, for the client, to doing a number of sit ups or chin ups and then resting.

The sailboat display screen (below under H2) available in the Infiniti/Biograph program is also particularly good for this purpose. This screen allows the client to build or destroy the sailboat (which fragments and goes totally black when the client loses the desired state of focus). You make it difficult to build the boat (threshold for theta somewhat challenging for that client). The client must then find and hold a focused mental state (if the boat picture is linked to theta) that "builds" the boat. This is not easy at first, but attaining control can be tremendously gratifying and self-empowering.

We begin the session with the client attempting to hold theta down and thereby obtain the lowest possible percent-above-threshold (in red) and the highest number of points (in blue) possible during a three-minute period. The student will compete with himself five or more times. He tries to improve his scores each time. When he gains some real control over theta the trainer suggests he try a different challenge. He is told to hold the boat for a count of five. Then he is to destroy the boat by letting his focus wander. He is to hold this lack-of-focus mental state with the boat in its blackened form for a count of five and then purposefully build it again and hold it built for a count of five. He is to repeat this process several times until conscious control over theta is possible. Older students may experiment retrieving memories and see if this increases theta.

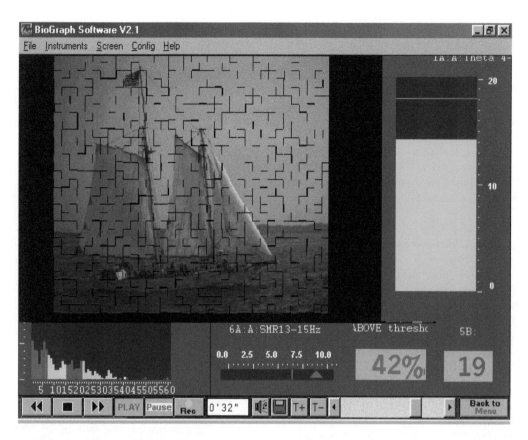

Two inhibits may be used. The first is 45-58 Hz to stop the influence of muscle tension, and the second is on 13-15 Hz and stops feedback if the client goes below threshold. These and the spectrum are sometimes hidden to make the screen less cluttered (the data are being recorded but the information does not appear on the screen). Clients learn to hold all movement (EMG) at a minimum and remain calm and concentrating (to produce SMR) while they ONLY vary their focus and thereby build the sailboat and hold it built (theta below threshold) for a defined period of time. Then they let their focus go on purpose and the boat is destroyed. Then they rebuild it again. They repeat the exercise until it is entirely under their conscious control. This is an exercise for all ages. We have watched clients from age 4 to 64 enjoying it. When the client has control with theta or thalpha, they try it with SMR and then beta 16-18 Hz. Adults may try control of high beta (20-23 Hz and 24-32 Hz) activity. The sailboat is linked to an appropriate bar graph in each case. With young children, after they become proficient at this form of self regulation, we ask them if they would like to show the other staff (and then their parents) how they have control of their own brain waves.

The bottom left corner of this screen has a spectrum. One could substitute the raw EEG. This allows the trainer to see the broader picture of brain wave production and monitor whether artifacts are interfering with the feedback.

Gaining Control of Slow- and Fast-Wave Production

During a session, it may be helpful to use a variety of screens. Each screen should have a clear purpose. In the screen shown below, the client gets a real sense of making the box (which corresponds to the amplitude of the slow-wave activity, *theta or thalpha*, smaller. They try to make it recede into the back of the screen while simultaneously making the fast-wave activity, *SMR or beta*, box large so that it seems to come out of the screen towards them. This kind of screen is simple but very effective. The percents-over-threshold can be recorded for each five-minute period.

Jason demonstrated that he could hold theta below threshold while maintaining SMR above threshold. In his first session, both percent-below-threshold figures were close to 50 percent. Now, in a five-minute time frame, he is able to reach 79 percent above threshold for SMR and 36 percent for theta. Skin Conduction (SC) was placed on his screen because he had a tendency to drop to a low level of alertness when things weren't moving quickly.

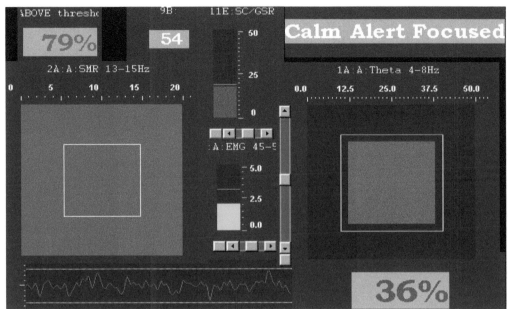
Simple screens like this get better results than games and flashy videos.

Brief Summary

Purpose

1. For the client to attain, through visual and sound feedback, a distinct, dynamic sense of:
   a. The reciprocal relationship between *decreasing* slow-wave and *increasing* fast-wave activity
   b. Being able to find and hold the desired mental state for longer and longer periods of time

2. For the client and trainer to follow progress by tracking brief time periods (two to five minutes) for changes in *percent over threshold* of the slow and fast waves. (Changes in the percent-over-threshold values change rapidly at first and then slow down over time because the calculation is an average over the period of the recording. Even after ten minutes it is difficult to shift the percent.) Points are also scored (white number at the top of the screen) for each one second that theta is held below threshold.

3. For the client and trainer to follow the EEG, and, in particular, to note the changes in the EEG with different mental states, it is highly motivating for a client to be able to observe firsthand that they are truly learning to control their brain's electrical activity.

4. To ensure quality feedback.
   a. **Minimize muscle artifact** – 'Muscle inhibit' in this screen is 45-58 Hz. It would be changed to 52-58 Hz in Asia and Europe. Sound feedback will stop if there is muscle activity in the head and neck region. Because EMG can be very high in amplitude, it may overwhelm filters and give a false impression that beta and/or SMR is raised. You do not want to be rewarding tense jaw muscles and think you are rewarding SMR or beta!
   b. **Minimize movement artifact** from electrode movement or excessive eye movement. The movement artifact inhibit in this screen is achieved by setting the EEG threshold parallel lines at a level that allows normal EEG activity, but that will stop sound feedback should a large amplitude wave appear in the EEG. These dotted lines can be seen above and below the raw EEG signal.

5. **Maintain an alert mental state**. EDR is recorded. If it drops, the client's alertness level may also be decreasing. The client may want to adjust his posture or take some other action to maintain arousal at an appropriate level.

Choice of Games for Feedback Training

Purpose Dictates Type of Game

On the whole, children with ADD have excellent focus for games and things that are novel. Our job as neurofeedback trainers is to have them create their own internal "game" (a method for remaining alert and focused) in order to achieve *self-regulation* and be in control in boring situations. The more interesting, dynamic or three-dimensional the feedback screen, the less useful it may be in the long run. Although using relatively boring feedback theoretically seems logical to us, it requires research studies to elucidate the variables. For example, we may use an interesting feedback display to initially catch the attention and effort of an extremely hyperactive young child or an autistic child. Certainly neurofeedback is never one-size-fits-all, so you want to be able to reward learning the task of self-regulation in a number of different ways. Ultimately, however, our clients must be able to sustain their concentration during boring activities such as doing homework or listening in some classes.

'Resting Waves' Review

Think of the brain as *turning off or resting from external world involvement* when it is in theta or low alpha. These synchronous waveforms (they look like sine waves), in an oversimplified way, are thought to come mainly from thalamic generators (central area of brain) and project up to the surface of the cortex. We record their rhythmic influence on the pyramidal cells in different areas of the neocortex with sensors on the scalp.

Calmly resting in a very positive sense is associated with increased SMR. These synchronous *spindle-like* waveforms are thought to come from other thalamic generators (ventrobasal nucleus of the thalamus) and project up to the surface where cortical neurons are recruited to produce the same rhythmic activity. This cortical activity is what we record. We see higher amplitudes of this waveform when people are in a calm state, with reduced sensory input and motor output (fidgeting) and not tending to be impulsive. Remember the original research did with cats. It is the cat waiting but alert and ready to pounce when conditions are right. SMR should be considered only to be projected to, and therefore recorded over, the sensorimotor strip, immediately anterior (motor) and posterior (sensory) to the central sulcus in the left and right hemispheres. We usually record it from C3 on the left across to C4 on the right

Combining Operant and Classical Conditioning with NFB

(With special thanks to Dr. Joel Lubar for the clarity of his teaching about this kind of training.)

Operant and classical conditioning are more fully discussed in the first section of this book. This section briefly reviews these two learning paradigms and clarifies how these principles are applied during training sessions.

Objective: Transfer to Home and School

The objective is to assist the transfer of NFB training to work, home and school settings. To accomplish this objective, we want the student to unconsciously and automatically turn on a mental state of calm, focused concentration when appropriate. Training for this is carried out in two steps that involve operant conditioning followed by classical conditioning.

To briefly summarize this, recall from Section I that you first use **operant conditioning.** The basic principle underlying this kind of conditioning is that when you reward behavior you increase the likelihood of its recurrence. This is done by having the student find a mental state that results in his meeting the thresholds that have been set for the slow and fast waves in the absence of EMG-induced artifact. This will cause visual and auditory feedback to be emitted. This information acts as a reward and increases the likelihood of that brain state being achieved and maintained again.

Secondly, use **classical conditioning.** The basic principle here is to pair two stimuli. In this case, pair the desired mental state with carrying out an academic task. This is done by having the student find the desired mental state using the operant conditioning paradigm above, and then introducing the academic task while retaining that state, as evidenced by continued auditory feedback. (That state has been trained, through operant conditioning, to turn on when the feedback program starts. It has become a reflex. You are pairing a metacognitive strategy for an academic task to that mental state.) If the auditory feedback stops, the student is instructed to return his attention to the NFB display screen. The student works at getting the appropriate feedback until he again is able to hold it steady for a reasonable period of time. This steady feedback signifies that he has the desired mental state for returning to the academic task.

For this procedure we want a nonintrusive screen such as a fractal display. Alternatively, the screen is covered and the student only receives sound feedback. The counter is usually linked to a soft *chord* or *ding* feedback that occurs at a rate of one every one to three seconds while the state is produced. You can also use a pleasant tone or music. This may sound continuously as long as the desired mental state is produced. The client, in this case, is alerted by the sudden cessation of sound.

The Why and When of Combining Academic Tasks with Neurofeedback

(A version of this section appeared first in the *Journal of Neurotherapy* 6(4) Clinical Corner, with our permission. Judith Lubar's comments on the efficacy of training while on task also appear in that volume.)

As we have stressed throughout this text, the practice of neurofeedback has its roots in research labs. It draws on both learning theory and empirical observations concerning outcomes. Each practitioner also brings his own background and knowledge into play. If you have a background in educational psychology, it feels natural to add the teaching of metacognitive strategies to neurofeedback training. Combining neurofeedback with the teaching of strategies and academic tasks is also supported by learning theory principles and outcome studies.

Neurofeedback is a type of learning, since it involves the operant conditioning of brain wave activity. As Sterman points out in his writings (Sterman, 2000), Thorndike's Law of Effect, which states that behaviors that are rewarded have a higher likelihood of recurrence, is at the core of what we do. Operant conditioning, carefully developed by B.F. Skinner, grew out of Thorndike's trial and error learning experiments. When we reward the production of certain EEG patterns with information about success, using visual displays and auditory feedback from the computer, we increase the probability that the client will produce that pattern again. We do not know the precise mechanisms (perhaps a change in neurotransmitter release, or in receptor sites at the synapse, or structural changes involving greater dendritic arborization over time), but we do observe EEG changes in people who learn the task (Lubar, 1997; Thompson & Thompson, 1998).

When you pair an academic task with the state of being relaxed yet focused, you are adding classical conditioning to operant conditioning. Classical conditioning, whose principles were elucidated by Ivan Pavlov through his experiments with dogs and their digestive systems, involves presenting a neutral stimulus (the conditioned stimulus) just prior to presenting a stimulus that elicits a reflexive response. Pavlov rang a bell before giving meat powder and, after a few pairings of bell and meat powder, the bell elicited salivation even if no food was given. In our work with clients who have ADD, where the goal is to improve concentration, we first use operant conditioning to train the state of being relaxed while sustaining alertness and focus. The feedback comes to reliably elicit this state so that it is like a reflex. (Remember, only reflexive, autonomic responses can be trained with classical conditioning. You use operant conditioning to train voluntary responses.) The feedback now acts like an unconditioned stimulus that produces the response of the desired physiological state (relaxed yet focused). If you now present metacognitive strategies and an academic task along with the feedback, this academic work is the conditioned stimulus that, after enough pairings, will also elicit the relaxed, yet alert and focused, state.

The academic tasks are done with an emphasis on metacognition; that is, executive thinking skills that monitor and guide how we learn and remember things. Examples include active reading strategies, techniques for organizing written work, and mnemonic devices, such as tricks for remembering multiplication facts. (For a fuller discussion, see Section XII of this text and the chapter on "Strategies for School Success" in *The A.D.D. Book* by Sears & Thompson, 1998.) Metacognition is particularly important for students with ADD because they are not naturally reflective: they do not plan their approach to tasks, are not good at time management, fail to make neat and organized study notes, and they always underachieve relative to their intellectual potential. Good students, on the other hand, seem to just naturally apply metacognitive strategies (Palincsar & Brown, 1987). To have the greatest impact, you cannot just do tutoring along with neurofeedback because there is not enough time to cover much content when you see a person for two one-hour sessions a week and the main focus is on getting focused. But there is time to teach one strategy in a session and then try to apply it to an academic task. In the next session you can review that strategy and either reinforce it with more practice or move on to a new one. It is a great advantage to be teaching something when you know (from the neurofeedback) that the person is paying attention. Thus you want the feedback to continue both when you are coaching people concerning a strategy and when they are trying to apply it. If they are reading it will be the auditory feedback that is giving the information. If the feedback indicates they have tuned out, you simply stop the task and let them return to focusing on the feedback until they get back in the zone. Learning principles tell you that you do not want to pair new learning with a tuned-out state of mind.

What are the logistics of fitting in the strategies? First, obviously, you want the client to be in the right mental state before you start. Thus you pair the task with neurofeedback once the feedback is reliably eliciting the desired mental state. The timing will differ from client to client, both in terms of which session first includes strategies and in terms of when during each subsequent session they are introduced. If the client is not performing well one day, maybe due to having an infection and being on antibiotics, you might just do feedback for that session. Another day the same client might be very much in the zone and more strategies and academic work would be covered. Typically, in practice, the first 20 minutes or so are usually spent doing pure feedback – paying attention to paying attention. (This is not 20 minutes uninterrupted; indeed, it may be ten two-minute segments with a client who is struggling to maintain focus. Always respect individual differences and tailor the feedback, and the strategies, to that client and how they are performing that day.) Once the client is reliably producing the desired mental state, the metacognitive strategies and their application to an academic task begins. Now they must think about thinking; that is, be aware of how they learn and remember things and apply it to an actual task.

The reason we bother with strategies/academics in the first place has to do with generalization. Generalization of a response is another concept from learning theory. It means that similar situations (or stimuli) will elicit the same response that was learned during training. In Psych. 101 you perhaps learned that when John Watson conditioned fear of a white rat in little Albert in his (in)famous experiment early in the last century, Albert also came to fear cotton wool and even Watson's white hair. With Pavlov's dogs, a bell with a different tone could still elicit salivation. When we have paired being focused with doing an academic task while receiving neurofeedback, the expectation is that the student will also get focused when they pick up a book to read at home. The student may also use metacognition, recalling the active reading strategies taught in the session. As he thinks about applying them, they should also trigger the relaxation and sustained concentration that was his physiological state when the strategies were learned. Although some of the learning that occurs with neurofeedback is clearly unconscious, we also want to encourage generalization (some of which is also unconscious) with the conscious application of strategies. Parents are impressed if they see their child calmly reading and doing schoolwork during a training session and they are even more impressed when their child starts doing this at home.

Parental notice of changes brings up the second underpinning of neurofeedback, namely, empirical observations. A further reason for pairing academic coaching with neurofeedback is that is has been observed to work. The results at our Centre (Thompson & Thompson, 1998) showed statistically significant gains not only in behavior (measured by TOVA and parents' questionnaire data) but also in academic performance and IQ scores. These results parallel those of Lubar (1995; also reviewed in Lubar & Lubar, 1999), which is not surprising since the Lubars are mentors to us and many others in this field. They advocate a session structure that has five conditions: feedback alone, feedback plus reading, feedback alone, feedback plus listening, feedback alone. This approach follows the principle of getting the person focusing before you introduce the academic task.

In summary, adding metacognitive strategies and applying them to academic tasks makes sense when working with clients who want to improve their concentration, organization, and academic or work performance. The reasons for doing so are derived from learning theory principles involving operant conditioning, classical conditioning, and generalization of behavior. The combined intervention is also supported by empirical observations of favorable outcomes using this approach. Furthermore, there is perceived value to the client because metacognitive strategies can be applied immediately in other learning situations even before the EEG changes are consolidated.

It remains an open question, however, whether adding the academic component improves outcomes, and which measured outcomes are affected. It could be argued that time spent on strategies detracts from learning the EEG task. It would make a good doctoral dissertation to compare neurofeedback alone, metacognitive strategies alone, and the two in combination. Until we have such data, clinicians can justify the combined approach since it is based on established learning principles and published empirical research.

Two-Channel Assessment and Training

Using Two Channels

With two separate channels of EEG one can see what is happening at two sites simultaneously. A common use of this is with a mind mirror display, a term used by Anna Wise in her book, *Peak Performance Mind.* A mind mirror shows a frequency display for left and right hemispheres and it is placed vertically (see figure titled Mind-Mirror below). It was the basic screen used in Val Brown's original software for the ProComp+ equipment.

For the reading to be truly comparable, it is ideal to use linked electrodes for reference and a common ground as shown in the diagram below. This way you know that impedance at the ground and reference sites and most artifacts will be affecting the two active sites equally. Thus, differences in EEG readings at the two sites can be assumed to reflect what is happening at those sites. Of course, eye movement artifact will have more influence on frontal sites and jaw tension will affect temporal sites the most but you have at least eliminated some variance due to impedances and electrical artifacts.

Note that two-channel training is not the same as sequential (bipolar) placement. Sequential placements have two electrodes at active sites plus a ground, and those three sensors comprise one channel of EEG. With a sequential placement you know there is a difference, but you cannot say with certainty which site has higher activity in a particular frequency range.

Two-Channel Montage with Linked Ears and Common Ground

Diagram of Linked-Ear Reference and Common Ground Setup

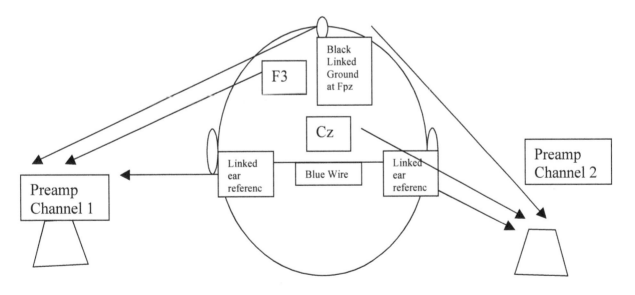

Examples of Two-Channel Training

Two-channel training allows options that may be useful for some clients. There is not yet published research on outcomes, but these suggestions make sense in terms of what we know about symptoms, neurophysiology, and mental states. (See the *Decision Pyramid.*) A two-channel placement may be the option of choice for a learning disabled child who is also impulsive: Channel A at F3 or C3 rewarding beta, inhibiting theta to increase left frontal activation, and Channel B at C4 rewarding SMR to decrease impulsivity.

A two-channel training may also be considered for dysphoria combined with either anxiety or emotional lability. In these circumstances one may encourage the client to increase activity at F3 by decreasing their dominant slow-wave activity and increasing beta 15-18 Hz at this site. Concomitantly they would attempt to increase SMR using Channel B at C4. Three examples for the use of two-channel feedback are described below.

Note that these *are **not** protocols.* They are suggestions for training if certain symptom pictures and EEG profiles are found in a client. This type of training should only be carried out by persons who have training and expertise in these areas. Though the suggestions are not researched, they do logically follow from what is known about brain function (Section VIII).

ADHD Combined with Learning Disabilities

Learning disabilities are the most common co-morbidity found with attentional problems. If you are using a single channel for training you will address the main symptoms that are distressing to others, which usually will be hyperactivity and impulsive behavior. This involves reducing the dominant slow-wave activity and, even more important, increasing 12-15 Hz across the sensorimotor strip. Placement at Cz or on the right side at C4 makes sense in terms of calming effects corresponding to an increase in SMR activity. This is mediated by thalamocortical loops and can be expected to generalize to sensorimotor areas and not be limited to the training site. With two-channel training you can simultaneously address those problems while also activating areas associated with the cognitive functions that are impaired. Most often these students have language-based learning disabilities, so you want to activate areas in the left hemisphere. (A good text that discusses the typical learning problems and offers teaching strategies to deal with them is *Right-Brained Children in a Left-Brained World* by Freed, 1997.) Since beta activity is associated with cortico-cortical coupling and short distance loops, you will choose the site according to which cognitive processes you want to activate. You do not expect beta to generalize. Activation equates to increasing 15-18 Hz (or 16-20 Hz) and reducing the dominant slow-wave activity. Thus, as previously noted, for the child who has trouble decoding words, placement might be tried over Wernicke's Area in the vicinity of P3 to P5. If he has trouble reading aloud it might be near Broca's area approximately in the vicinity of F3 to F5.

Nonverbal learning disabilities may be associated with right side dysfunction. These students have problems with spatial reasoning and thus will usually have weaknesses in some aspects of mathematics, especially geometry. Some also have problems reading social cues and with emotional responsiveness, both reading emotions and expressing them with the subtle modulations you would expect. The *nonverbal LD* style has some communalities with Asperger's syndrome. The parietal-temporal regions seem to show more slow-wave activity (less activation) on brain maps done on people with AS. (This finding has been noted with a rapidly increasing number of clients by Jay Gunkelman (personal communication), plus our own findings). Thus activation of the area around P4 to T4 in the low beta range could be tried. Frequencies of 13-16 Hz may be appropriate.

Decreased activity at F3 and F4 has also been observed. However, you would not likely want to activate right frontal activity alone, since that can be associated with a focus on negative thoughts and *avoidance behaviors*. However, activation of both right- and left-frontal activity may be tried. Increasing spectral correlation (comodulation) between the left frontal lobe and the right parietal and frontal areas may also be attempted in future.

To summarize the above:

- Channel 1 on left side of head: Usually for purposes listed above to help with learning. The site will vary depending on the type of learning disability being addressed.

- Channel 2 on right side of head: Usually for purposes listed above, namely to decrease the hyperactive and/or impulsive symptoms if placed over the sensorimotor strip at Cz or C4. It might also be used to increase activity in the right parietal-temporal region in Asperger's syndrome.

Depressed Clients

Depressed clients may show less activation of the left frontal lobe compared to the right frontal area (Robinson, 1984; Davidson, 1995, 1998; Rosenfeld, 1996, 1997). In EEG terms, it may be observed that beta is lower and theta and alpha wave amplitudes are higher on the left as compared to the right frontal areas. To remember this asymmetry, recall that one's heart (happiness) is on the left and the left prefrontal cortex is associated with processing positive thoughts and *approach behavior*. Robinson's original work noted that damage to the left frontal lobe was associated with *depression*. while damage to the right frontal lobe was associated with *manic symptoms*. The conclusions drawn from these observations were that the left frontal area mediated positive thoughts and approach behavior while the right frontal area mediated negative thoughts and withdrawal behavior. Taking this a step further, a low activation of the left frontal area compared to the right may not only differentiate between depressed patients and normals (state marker), but it may also be a trait marker for vulnerability to depression (Rosenfeld, 1996).

In treatment with neurofeedback, practitioners have observed that increased activation means increased fast-wave activity relative to slow-wave activity. The hypothesis is that activation of the left frontal area may decrease depressed feelings. Jay Gunkelman has suggested that, along with increasing beta over the anterior portion of the left frontal lobe, one may use a second channel to encourage high alpha (11-13 Hz) in posterior regions (parietal). This posterior alpha is thought to compete with the production of alpha in the frontal region. In other words, you have the alpha in a more appropriate location. The reason for paying attention to alpha in this case is that it is an inverse indicator of activation. (In adults, increased frontal alpha suggests the brain is resting, whereas less alpha suggests activation.) Alpha is used because EMG activity can produce artifact that affects the beta range, but it has less effect on alpha. Beta is a much lower amplitude wave than alpha so small changes due to EMG would have a relatively larger effect on recorded amplitudes of beta compared to their effect on the much higher amplitude alpha waves.

Another way to approach the treatment of depression, developed by Peter Rosenfeld at Northwestern University, is called the *alpha asymmetry protocol*. It uses two channels with the active electrodes over the left (L) and right (R) frontal regions. Both channels are referenced to Cz. You can purchase Rosenfeld's protocol from him for a nominal fee.

On some instruments that allow you to set parameters, you could utilize the alpha asymmetry idea put forth by Rosenfeld and put in the formula (R-L) ÷ (R+L). Set up in this way means that a positive number will indicate that the client is happier, because you are measuring alpha – the inverse of activation. You want right-sided alpha (R) to be higher than left sided alpha (L). You let the right side, associated with negative thoughts, rest. and you keep the left side's focus on positive thoughts activated. The most important factor, according to Elsa Baehr, who has done clinical work using Rosenfeld's protocol, is the percent of time during the recording session in which (R-L) ÷ (R+L) is greater than zero. This is a better discriminator of depressed vs. control subjects than the (R-L) ÷ (R+L) score itself. Baehr notes that <55 percent of the time with left frontal activation suggests depression, and >60 percent of the time with left higher than right suggests no depression (Baehr, 1999).

It should be noted by the reader that using a complex formula such as (R-L) ÷ (R+L), the results do not inform you as to whether changes have taken place in R or in L. A positive value only indicates a shift in a ratio, but does not say if alpha on the right increased or if alpha on the left decreased. This is another question worthy of research.

Two-Channel Training to Improve Mood

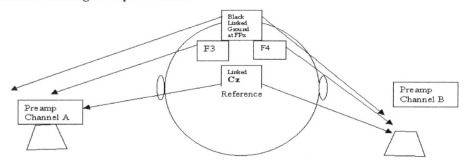

Note in the above diagram how the ground reference electrode has two wires, one for each channel. If a second electrode with two wires attached is used at Cz, it could go into reference inputs on each of the two preamps for Channels A and B (Channel 1 and 2). This would be an easy way to compare activity under the two active electrodes at F3 and F4 because the reference and ground are the same, so comparisons are meaningful. Channel A would be 'L' and Channel B would be 'R' in the formula (R-L) ÷ (R+L) given above. Electrodes linked in this fashion must be specially purchased from a manufacturer such as IMA in Florida.

Although a Cz reference is used by Rosenfeld, others have found that clients differ on which reference point is best, and some clinicians find an ear reference can be used with equally good results. Here is another good thesis topic, to compare outcomes with different reference sites and see if it makes a difference in training outcomes. Still others have suggested that it is more effective to just activate the left frontal area as suggested by Jay Gunkelman.

Whatever method you use, try using as your goal having a greater percent of time that left activation exceeds the level of activation on the right. Of course, you must look at your client's EEG data first. Perhaps, for example, some people who are 'true lefties' with hemispheric functions reversed will show the opposite pattern. Always respect individual differences rather than just following a protocol.

Other approaches to the treatment of depression – and particularly depression combined with anxiety – are discussed in Section XXIV. At the ADD Centre, the majority of patients who are depressed do not demonstrate a clear alpha asymmetry. Their difficulties appear to involve the cingulate gyrus and high frequency beta spindling as is discussed in Section XXIV. For these patients, LORETA Z-score NFB may be the treatment of choice.

Remember that anyone with depression should first be assessed and perhaps treated by a physician. For a patient with serious depression, neurofeedback would be an adjunct, not a substitute, for medical treatment. As is also true of clients with ADD, they may be on medication when neurofeedback is begun. As progress in self-regulation is made, the individual may, in consultation with the prescribing physician, be able to reduce their medication use.

Seizure Disorders

At the seizure site there is often observed to be increased slow-wave activity. Decreasing this slow-wave activity at or near the seizure site appears to be a constructive, normalizing approach. In addition, past research has demonstrated that increasing SMR activity may have a positive effect for those clients who exhibit grand mal seizures, partial complex seizures, or any type of seizures with a motor component. The analogy to setting up a 'firewall' by increasing SMR has been used by experts in this area, notably Sterman (Sterman, 2000). Thus, you are both directly trying to decrease epileptiform activity and encourage competing activity so that neurons near those firing in the wrong way resist being influenced and the kindling of a full blown seizure does not occur.

To this end, the clinician might consider placing one active electrode (Channel 1) near the seizure site to inhibit the slow-wave component. A second active electrode (Channel 2) would be placed along the central motor strip (possibly alternating sessions at C3 and C4, the sites where Sterman worked), and 12-15 Hz SMR activity would be encouraged.

Case Example of Two-Channel Training

Electrode Placement in a Complex Adult Case

Each adult will be different in terms of their symptoms and it follows that there may be corresponding differences in the placement of electrodes both for assessment and for training. The following case is sufficiently complex to allow readers to follow the kind of decision pathway they may wish to follow with their clients.

Peter, age 36, complained of a poor school history, depression with ruminating, impulsivity, and falling asleep when he tried to read textbook material. With a Cz single-lead assessment, the EEG demonstrated a dip in SMR at 13-14 Hz, but no significant rise in slow-wave activity. This did correspond to his impulsivity but not to his constant worrying or inattention.

A second, two-channel, assessment was carried out using F3 and Cz placements for the active electrodes. This demonstrated very high amplitude 7-9 Hz activity, when reading or doing math, at F3 but not at Cz. The theta/beta µV ratio at F3 was 3. Clearly, we now had two tasks. First, we would do SMR training centrally. Second, we wanted to decrease thalpha (7 to 9 Hz in this case; in others it might be 6 to 10) and increase beta (16-20 Hz) at F3. This would be done first while paying attention to feedback, then later while doing an academic task such as reading. (We chose the F3 site for a comparative assessment because we knew we would not be encouraging an increase in beta in the right frontal region as it seems to be related to undesirable states such as depression in some individuals, although this supposition requires further research.)

However, there was still another challenge. Peter was dysphoric and ruminated. We had seen increased 28-32 Hz activity at F3 and at Cz.

We wondered if there would be a difference between activation of the two frontal lobes. Research, as noted above under depression, has found asymmetry between the left and right frontal areas with greater activation on the right. It was concluded that we tend to focus on negative thoughts in the right frontal areas and on positive thoughts on the left (Davidson, 1998). This has led to the hypothesis that, in work with NFB in depressed clients, we should increase beta (15-20 Hz) and/or decrease alpha (an inverse index of activation) in the left frontal area in order to increase activation of the left frontal lobe.

The investigation of the dysphoric mood and ruminating required that we again use two channels. This time we explored the question of differences between the two frontal lobes by doing a two-channel assessment using F3 and F4. Full-cap assessments would be excellent, but it is more time consuming and expensive. As well, most programs for doing the full-cap, at this time of writing, are not looking at the spectrum of activity to 60 Hz. If we do not look beyond frequencies in the mid 20s, you may miss some of the ruminating that is reflected in activity in the 28-34 Hz range. Finally, many readers of this volume will not yet be doing full-cap assessments on a regular basis.

We find it useful for clients who complain of ruminating and becoming internally distracted to put the active electrode for Channel 1 at F3 or slightly anterior to this point. The active electrode for Channel 2 is placed at the equivalent point (F4) over the right hemisphere. Both electrodes are referenced to their ipsilateral ear lobe or mastoid, or to linked ears. Although theoretically we should only require one ground we have found that the results are more consistent if we use a ground for each channel. Better still is to purchase a lead for a common ground as shown in the diagram above. When not using a common ground, we usually ground to the opposite ear from the reference, placing the electrode on the top of the ear. In this type of assessment the **impedance** must be almost **exactly the same** between each pair of electrodes. Otherwise (we again remind you), any difference observed might in part be due to the different electrode connections.

Now that you have your electrodes in place, compare the two sites for different frequency ranges. You should graph in one or two Hz-band segments. For training, use the exact bandwidth that the assessment shows to be problematic: that is, the bandwidth that has an amplitude outside the range that you would expect when compared with other bandwidths for that client and for normal subjects. (Having an idea of what normal looks like is always helpful.) For ruminating and worrying we usually find a marked increase somewhere between 23 and 34 Hz. We are finding that when most (but not all) of our clients think negative thoughts (ruminate) 23-34 Hz activity, measured either in microvolts or as percent-time-over-threshold (which you can put on your screen with some instruments), is higher in the right frontal region. When the same client thinks positive thoughts it is either higher on the left, or this activity disappears altogether. (This requires replication, so please check it out with your own clients.) We also find that when the client relaxes and becomes externally openly aware, this frequency band (23-34 Hz) drops in both hemispheres and the 11-15 Hz bandwidth amplitude rises. With intense, obsessive compulsive clients and clients who are anxious and ruminate, you may observe very high amplitudes of beta between 20 and 34 Hz along the midline Fz, Cz. (You may, less frequently, see this high-amplitude beta activity at lower frequencies.) It would appear that this corresponds to a 'hot' cingulate and it does respond to neurofeedback training with a decrease in the symptoms. (Bob Gurnee has also reported this in meetings of the SNR, 2002, and FutureHealth, 2003). We are not finding large overall differences in alpha 8-12 Hz or in 11-15 Hz, or even 16-20 Hz comparing right and left hemispheres. It will take further work to define precisely which frequencies for different age ranges change. The differences between hemispheres for the high beta between 24-34 Hz, however, are remarkably large and consistent. It is not that the whole range is high but, for individual clients, the increase is somewhere in that range. It might just be 24-26 Hz or 30-31 Hz. Once again (it is worth repeating) check the individual client's pattern and customize the feedback accordingly.

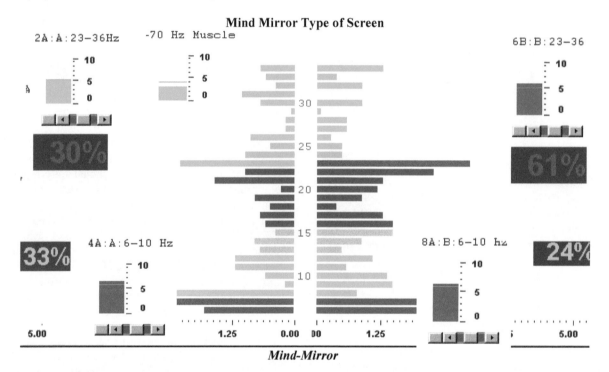

Mind Mirror Type of Screen

Mind-Mirror

The above screen can be used to help the client see the difference between activity in the two frontal lobes. Percent-of-time-above-threshold numbers are linked to each bar graph. In this display, Peter shows higher average frontal high-frequency beta (23-36 Hz) on the right and higher frontal thalpha (6-10 Hz) on the left. The percent-over-threshold numbers show an average over several minutes. If this screen is used for feedback, the bar graphs are set, then placed in the background so that they do not clutter the screen. Respiration, heart rate variability, skin temperature, and/or EDR may be added as further checks on anxiety.

Peter demonstrated a combination of difficulties. Therefore, after using the above screen for a short time for feedback, another screen was made.

Given his history of tuning out when doing school-type work, we would want to decrease 6-10 Hz and increase 16-18 Hz while carrying out academic tasks. Simultaneously, considering his rather marked impulsivity, we would use a second channel and encourage him to increase SMR at C4. We would also, due to his negative ruminations that were associated with high high-beta frequencies in the right hemisphere, want to have him decrease high-frequency beta activity at this site on the right side. We did find that the high amplitude 28-32 Hz activity was easily detected at C4 as well as at F4, making it easier to choose a site for the second channel.

At the same time, we used biofeedback emphasizing increasing EDR for alertness (but not too high because that, in Peter's case, corresponded to his becoming emotionally labile and excited). Peter was to try to maintain his breathing at a steady 6 BrPM even when stressed. This corresponded to good RSA.

Simultaneously, on some screens, we had Peter decrease his forehead EMG and increase his hand temperature. All the biofeedback modalities were based on a stress test assessment.

Two-Channel Feedback to Reduce Ruminations and Impulsivity

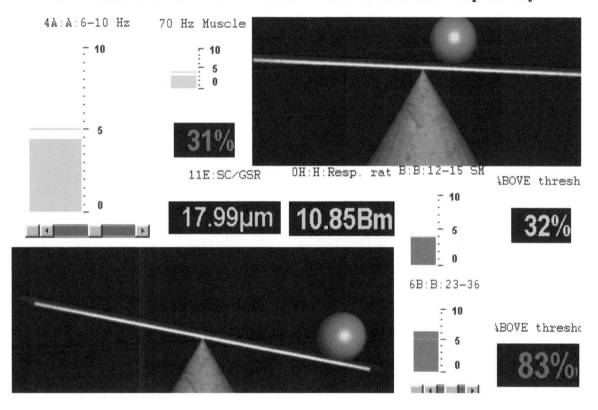

In this display, Peter tried to keep both balls to the right on the two balance beams. The bottom balance beam was linked to Channel 1 or A, the left-side thalpha (6-10 Hz) bar graph. He is to keep the ball to the right, representing low thalpha. The red "31%" represents the time he was over the threshold for thalpha. This is good for Peter, but he is trying to lower it still further. The "Muscle" bar graph at the top is actually 45-70 Hz, representing both electrical and EMG effects on the EEG. It must remain very low.

The top balance beam is linked to Channel 2 or B, right-sided SMR. He is to keep the ball to the right representing high SMR. The green "32%" represents the percent of time he has kept SMR above threshold. Peter is encouraged to become very calm and raise this percent.

The bottom right-hand corner "83%" in red represents the amount of time he is over the Channel 2, right-sided, 23-36 Hz threshold. (For most clients, this is a narrower band that corresponds to a single peak found during their assessment.) He wants to drastically lower this percentage. He is to keep respiration at an even 6 BrPM. It is to be diaphragmatic without shoulder tension (EMG had been monitored on other training screens). SC (skin conduction) or EDR (misnamed GSR on this program's display and the company has since corrected this error) must come down to around 10-12. This is the range it is usually in when Peter is both calm and alert. With training Peter was able to control all these variables. Three years later, Peter telephoned to say that the training has "changed his life" and he would like to see training available in the city where he lives.

Additional Single-Channel Assessment Procedures
(Not detailed in the first edition of The Neurofeedback Book*)*

Hints for Careful Clinical Practice

In our workshops and in our staff training we suggest that the clinician should, initially, when the EEG changes, pause the EEG recording and inquire about what was going on in the client's mind. We recommend that our trainers never assume that they know what client mental state correlates with changes in the EEG spectrum. Earlier in this book we listed different frequency bands, measured at Cz, and gave possible mental states that commonly correlate with each frequency band. For example, 3-5 or 3-7 Hz in the low theta range can correlate with tuning out and even becoming drowsy. However, common correlations do not necessarily apply to your particular patient. With respect to this example, the first author has observed extremely high-amplitude 5.5 Hz when he was doing difficult mathematics in his head such as multiplying 376 x 75. He was certainly not drowsy and not tuned out. (It may have been due to a burst of hippocampal theta that often is central mid-line around 6 Hz and relates to memory encoding and recall.)

We recommend that our trainers suggest to their clients that the clients explore for themselves how different frequency bands correspond to their own mental states. All of our trainers will have done this with themselves when doing self-training. One method is for the trainer to ask the client to try and recall and describe their mental state whenever the trainer says "Now," and pauses the EEG screen. The trainer will do this when he sees a burst at a particular frequency, such as a burst of alpha or a burst of beta spindling.

Interesting correlations may occur. For example, if there is an increase in alpha, the person may report that they had drifted off, or that they were alert but just feeling calm. If the change were increased amplitude of higher beta (>20 Hz) frequencies, which has been termed "busy brain" (Thompson & Thompson, 2006), then the person might report that she was ruminating. This will allow the clinician and the client to better understand how those shifts in the EEG correspond to shifts in that client's mental state. We hypothesize that the mental state will correlate with the network involved (Affect, Executive, Default....) and be a key to guiding the effective use of cognitive strategies during neurofeedback sessions.

We also have the staff graph results during each session and, from time to time, between sessions. This is done for different conditions in order to show a client how their brain waves really do change across time and with different tasks. The following figures show examples of this graphing using Excel.

Assessment – Single-Channel, Cz, Four Conditions

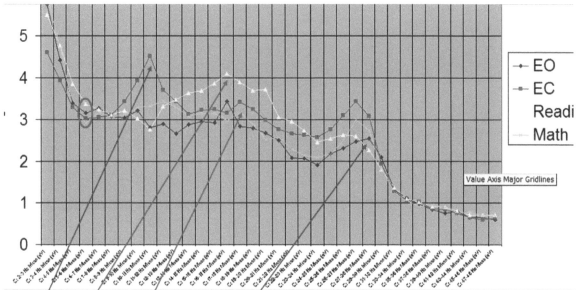

Excel graph of **1 Hz bins Assessment Script across 4 conditions:** Note that **red 'EC'** 10 Hz very high (intelligent, NORMAL, adult), 16 & 26 Hz high (Thinking about the program). **Blue 'EO'** 16 & 27 Hz high (working out how script works), **Yellow 'Reading'** 16 Hz very high & alpha desynchronizes (cognitive activity to remember material), **green 'Math'** 17 & 26 Hz (thinking of strategy for working memory – "that's a multiple of 7 so I'm on track"). Note 6 Hz (red circle) is a little higher in a reading working memory task than it is in EC.

Note how alpha is high at rest (eyes closed) and attenuates on task.

This is a copy of a PowerPoint slide that we use when teaching EEG fundamentals. It is an Excel graph using statistics from assessments using the Biograph Infiniti to compare four mental states: eyes open, eyes closed, reading, and math. It illustrates how amplitudes change at various frequencies with different mental states. The subject was a bright 62-year-old professional with no clinical symptoms. We suggest that our staff try this on themselves during their own training.

Referential vs. Sequential: An Amplitude Comparison

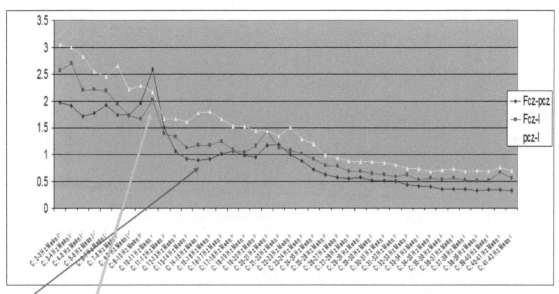

SMR is in phase between the 2 sites because SMR comes from the thalamus to both sites. The amplitude is lower in the sequential placement due to more common mode rejection. Frontal and parietal alpha come from different generators so it may be out of phase giving a high amplitude in sequential.

This PowerPoint illustration is given to demonstrate how two different montages during eyes-open 'resting' state recordings can show different amplitudes at different frequencies. The sequential FCz-CPz recording (blue line) has lower amplitudes for nearly all frequencies except 9 and 10 Hz in the alpha range. Note that sequential montage was previously referred to as "bipolar." The two referential (or "monopolar") montage recordings for FCz referred to the left ear and CPz referred to the left ear have higher amplitudes because there is less common mode rejection when an active site anywhere on the cortex is referred to an ear electrode. Amplitudes in the parietal region tend to be higher than frontal for both regular EEG and ERPs (event-related potentials).

It is postulated that one reason for the relatively high sequential amplitude of alpha and low sequential amplitude of SMR may be the phase relationship between brain waves measured at the two locations. This graph was done in Excel.

Graph of a Training Session Done Without Formatting the Axis

C: Delta-Theta Mean (µV)	27.06	28.89	29.07	29.41	28.03	31.61
C: SMR Mean (µV)	8.25	9.17	9.01	8.54	9.33	8.04
C: Busy-brain Mean (µV)	7.16	6.75	6.47	6.47	6.56	10.8
C: EMG noise Mean (µV)	1.81	1.98	1.81	1.56	1.52	5.38
C:Delta-Theta/SMR ratio Mean	3.28	3.15	3.23	3.45	3	3.93

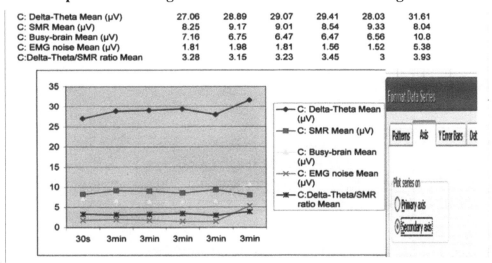

You should click over one of the blue dots and then right-click. When you do that, you will see a menu (a small section of it is shown above). Click "Secondary axis" and then "OK." RESULT: Delta-theta (3-7 Hz), a much higher-amplitude axis on the right, leaving all the low-amplitude beta and ratio lines on a smaller scale axis on the LEFT-hand side. Now your client has a much better view of the changes she made, both in high-amplitude theta and low-amplitude beta waves.

The figure below is an example of reformatting the graph to have a different scale for high-amplitude slow waves compared to the relatively low-amplitude fast waves in a child. In your clinical practice you can make an Excel template to do this automatically when you enter a column of statistics from your session.

After Reformatting the Axis for Delta-Theta

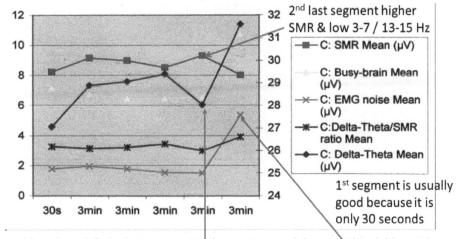

I would not have left the last segment in the session graph because this child must have been moving and /or talking and the last segment is just reflecting EMG artifact. That is **not** a helpful segment of the training review for either the child or the parent.

I would have just discussed how **focused he was for the 3 minutes** in the final (actually here the second last) segment of the real session.

Leave on a positive note.

Note: This graphing is done AUTOMATICALLY by the Excel program when the trainer copies the statistics into a column in an Excel template designed by Andrea Reid-Chung at the ADD Centre.

The above two figures show how we recommend that results in every session be graphed using Excel. These graphs are appreciated by the patients, both adults and little children. By the end of training, clients like to see that, with both the monitor and the sound turned off, they can still get as good or better results by controling their own mental state. This is called a "Transfer Trial" in centers using slow cortical potential feedback (SCP), such as Ute Strehl's group at the University of Tuebingen. These graphs take only a few seconds to produce. They are done automatically at our Centres by using Excel when we enter statistics for each 3- to 5-minute or 10-minute segment of the training session.

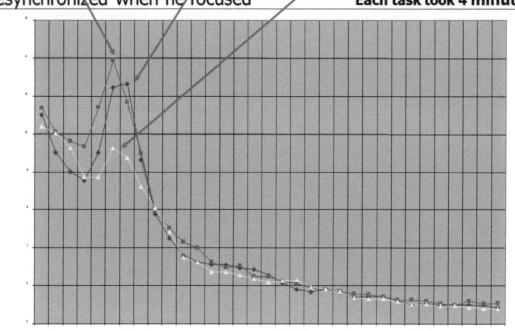

Every 8-12 sessions record (at Cz) client doing different tasks. Below Reading (RED), Math (BLUE), **then highly focused listening to me teaching him math tricks (YELLOW).** The 8- 9 Hz low alpha desynchronized when he focused   **Each task took 4 minutes.**

The above graph was done with a 19-year-old boy who said that the brain waves did not reflect anything that was happening in his brain. He was going to stop training. The trainer told him that that was a reasonable hypothesis and asked him if he would do a little experiment to either prove it true or not true. He agreed.

He then did three recordings, reading (red), attempting two difficult math problems in his head (13 x 17, then 376 x 75 – shown in blue), then listening to the trainer explain the tricks/strategies for doing these mental math problems (yellow). He could not do the second multiplication and got frustrated. He really enjoyed learning the writer's strategies for rapidly doing complex math problems in his head. He was fascinated to see how the amplitude of thalpha (6-10 Hz) changed so dramatically when he was really focused and concentrating. He readily agreed that he had convincingly disproved his original hypothesis and he continued in training.

He also wished to see the actual ratios. The table below comes up automatically in our Excel program. He was impressed at the degree of change that occurred just with a change in his mental state.

C:4-8HzMean squared/13-21HzMean squared	5.96	7.8	5.31
C: 3-7Hz/13-15Hz ratio Mean	3.85	4.02	4.08
C:4-8Hz/13-15Hz ratio Mean	4.28	4.38	3.95
C: 4-8Hz/16-20Hz ratio Mean	3.62	4.44	3.53
C: 6-10Hz/16-20Hz ratio Mean	4.09	4.49	3.5
C: 19-22Hz/11-12Hz ratio Mean	0.74	0.73	0.78
C: 23-35Hz/13-15Hz ratio Mean	0.95	0.88	0.98
Ratios	**RED**	**BLUE**	**YELLOW**

Note that his **ratios decreased** when he was focused and **concentrating (YELLOW). Very attentive** while learning **strategies** (tricks) **for solving the math** problems in his head.   I drew out the grid for 13x17 and then showed how to rapidly do 376 x 75 in his head.
   He was fascinated.   Lesson – make learning a 'game'.

(Trick: Since 75 is ¾ of 100, just do ¾ x 376 then multiply the answer by 100.)

Table shows marked decrease in important ratios when he was focused and concentrating.

Single-Channel Assessment Screen

Assessment is the key to careful, appropriate, and successful intervention. When doing single-channel or two-channel NFB, the staff have the assessment screen as the first screen (of five) so that it can be referred to at the beginning, and any time during, the training session. This is the only way a trainer can be sure that the client is always receiving feedback based on a good EEG and not on artifact. We have seen that kind of error all too often when training people from all corners of the world. We have even seen it occur in professional demonstrations at meetings. The example screen we will show below is from Thought Technology (TT) equipment, and is available from the Biofeedback Federation of Europe, Thompson Setting-up-for-Clinical-Success Suite, but any company's assessment screen can be successfully used if it provides the same items: a raw EEG, some method of clearly observing EMG artifact, a constantly updating 2-62 Hz single Hz bins spectrum so that the clinician can pause and show their patients how the raw EEG waves correspond to changes in the frequency that has maximum amplitude and discover how these patterns correlate with their mental state.

Our patients also like to have available running scores on their important ratios, but that is optional. Of course, for initial and repeated assessments, this screen must have a 1 Hz bin, bandwidth, and ratio statistical scores that pop up at the end of each training segment so they can be graphed. It must also allow artifacts to be removed from the raw EEG. Many errors can result from merely using automatic artifacting – nothing can replace looking at the whole raw EEG when doing assessments, though this would take too much time after each training segment.

Single-Channel Assessment Screen

Boy, nine years old, Note theta. ASD with delayed language.

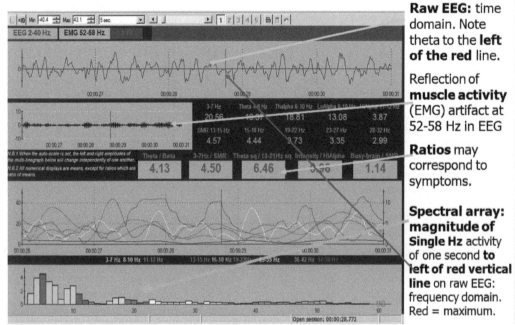

The 'Thompson' Assessment Screen (available www.BFE.org).

Ratios are running averages (with time). Theta/beta is amplitude of 4-8/16-20 Hz bands. We have noted in our publications that all children with a ratio >3 in our clinic have had the symptoms of ADHD. The staff at the ADD Centre find that 3-7 Hz/SMR (13-15 Hz) is the most useful ratio to follow during the training of most children and young adolescents who have a diagnosis of ADHD. Theta (4-8 Hz) squared/13-21 Hz squared is the 'power' ratio used in research studies on ADHD by Lubar. Norms for this theta-beta power ratio were established in a multisite study and are widely used (Monastra et al., 1999, 2002.) Intensity (19-22 Hz)/HiAlpha (11-12 Hz) was a ratio that was found at the ADD Centre to be >1 in many of the clients who presented with anxiety. Busy-Brain (23-35 Hz)/SMR (13-15 Hz) was found to be >1.55 in many clients who presented with a major complaint of negative ruminations. However, this ratio was also very high in intelligent, multi-tasking, busy professionals and academics. The last two ratios were discussed in a publication by Thompson and Thompson (2007).

The bandwidth figures above the ratios can be set as running averages of amplitudes in microvolts. For the one power ratio, the unit is picowatts. The line graph below the ratios shows amplitudes of various bandwidths that are shown in the same colors as the spectrum at the bottom of this figure; this allows for the examination of changes across time, whereas the spectral array at the bottom updates every second for the last second of data recorded. Note that the "raw EEG" at the top of the figure is based on a band-pass filter of 2-40 Hz (high-pass filter at 2 Hz and low-pass filter at 40 Hz). These filters make the EEG easier to read because it does not roll up and down with the underlying shifts in slow cortical potentials, nor does it contain a great deal of electrical and EMG artifact that influences frequencies above 40 Hz to a greater degree. At the ADD Centre, we also have assessment screens that have an additional line graph showing frequencies from 0.5 to 60 Hz that is used when artifacting the raw EEG.

This screen makes it easy to correlate the EEG at the top, shown in the time domain (graphed as amplitude across time) with the spectral array at the bottom, which shows the data in the frequency domain (graphed as magnitude across the spectrum from 3 to 60 Hz) for the EEG observed one second to the left of the red vertical line on the raw EEG. It is a helpful way for a clinician to train his eye to recognize different wave forms; for example, when reviewing the data, you can look at the spontaneous EEG, decide that it looks like alpha, then check on the spectral array to see if the alpha frequencies are the ones with the highest amplitude.

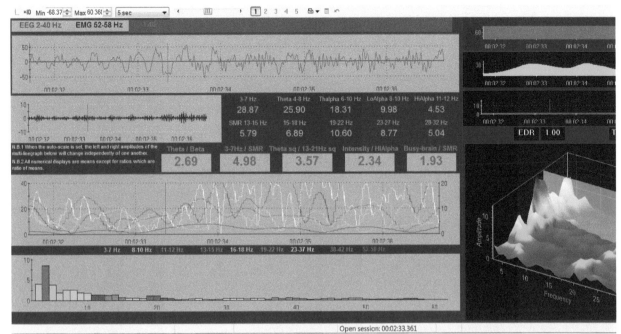

Burst of theta to left of red verticle line. If the red vertical line was moved to the right of the spindling beta immediately to the right of the time indicator of 00.02:35, the red histogram column on the spectrum at the base of the screen would shift to 20 Hz, as can be seen in the 3D spectrum on the trainer's monitor.

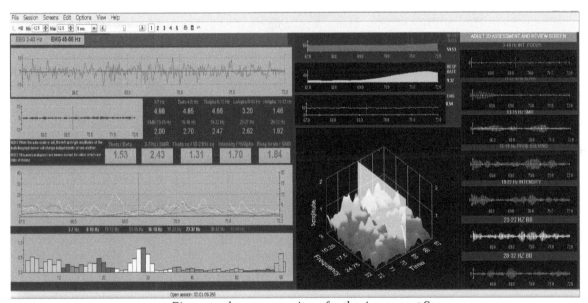

Figure to show two monitors for the Assessment Screen

The above figure shows the patient's monitor on the left-hand side. On the right-hand side is a second monitor, the trainer's monitor. The bandwidth EEG line graphs can be used to follow what the client is producing. Sterman uses this type of line graph to show that bursts of SMR may increase in amplitude with training, and each burst may be followed by a burst of alpha (post-reinforcement-synchronization or PRS). In the above example, a spindle of beta activity during the previous second is seen at 29 Hz in the 2D spectrum. If we move the cursor to the left of this burst on the raw EEG, then it can be seen in the 3D spectrum, as shown on the therapist's monitor to the right of the patient screen. The red line on the 2D spectrum shows as a white transparent 'wall' on the 3D figure on the therapist's monitor. (Note that there is **not** high amplitude at 53-58 Hz, as would have been the case if this was just due to artifact from muscle activity. It is also higher amplitude than the 60 Hz electrical artifact, and therefore not likely to be a harmonic of 60 Hz artifact, which would be lower amplitude. Note: When the EEG is enlarged, the morphology of the waves is sinusoidal beta spindling, not EMG artifact, even though setting the time range to show pictures with five seconds of data results in EEG waves at that frequency that look sharp rather than sinusoidal.)

- *The 3D spectrum really highlights brief bursts of high amplitude waves.*
- *The trainer may ask her **client to click the space bar** while doing training to mark (appears as a vertical dotted line when they review the EEG) when their mental state really shifts. Alternatively, the client can use a hand signal or make a little mark with a pencil. When they signal, you write down the time or hit the space bar. Then you can review that EEG record with the client and see what EEG changes actually occurred when the client's mental state changed. In the above example, note the high amplitude 60 Hz. This 60 Hz could signal poor impedance, but here it was electrical interference.*

Note: Our adult clients like to use a 3D spectrum for self-regulation training. They produce a "clear blue lake" to the right of about 20 Hz, indicating that they have emptied their minds of all intruding thoughts. They did not have a busy-brain for that period of time. In the example figure, the client had a very busy brain and just could not stay focused on the task at hand due to competing inner thoughts.

Dr. Vietta Sue Wilson has worked for decades with athletes including Olympic athletes. She finds that any shift out of the "smooth blue lake" will correlate with poor performance, such as a missed shot in archery.

Conclusion
Clients appreciate being able to instantly see how their EEG changes with mental state and training. Graphing is an important part of this process. It is a reasonable hypothesis, as previously described, that training will tend to alter whatever network is dominant when the training is being carried out. We have postulated that this may be one reason why single-channel training at Fz and Cz can be so effective. It may influence the anterior cingulate cortex with respect to appropriate activation of the Executive, Affect, Salience, and/or Default network(s) depending on what the trainer has the client do during the feedback. Most importantly, over the last more than 20 years of combining NFB with BFB and metacognitive strategies, we have come to the firm conclusion that a key and crucial component to successful outcomes is how the trainer works with the client and what activities they are doing together. See Part III, Section XV on Metacognition for ideas about activities that activate Executive networks.

The Electroencephalogram: Notes on the Interpretation of Low-Power EEGs

The emphasis in this document is on areas of the cortex that we observe outside database norms using a 19-channel EEG assessment followed by Low Resolution Electromagnetic Tomography Analysis (LORETA) for source localization. We then use NFB combined with BFB to change the EEG and, possibly, thereby change functioning in neural networks that involve these sites. However, there are, infrequently, situations where the entire power of the EEG is low. It could be due to a thick skull or increased fluid attenuating the signal as occurs in the elderly. Or it could be that too few neurons are firing in synchrony to produce a higher amplitude signal. This will often be observed after a concussion and amplitude will improve as the patient begins to recover with neurofeedback treatment. When this low-amplitude EEG problem occurs, we must look at the power spectrum profile and the Z-score graph profile. Then we can imagine this profile being above the zero Z-score line. This will help us decide on what frequency bands should be enhanced. In addition we always look at the power ratios as they give a clear indication of where frequency bands are relatively high or low. When necessary, we compare relative power to absolute power Z-score statistics to help us decide on sites and enhance and inhibit frequency bands. Whatever we decide with respect to sites and amplitudes must always correlate with the patient's symptoms in order to make a final decision on the correct enhance and inhibit frequencies for NFB.

Power ratios can be very helpful! These ratios can be several SDs above and outside database norms even when the entire EEG is at or below 0 SD. In the NeuroGuide program one can also use LORETA Z-score coherence analysis, which, even with a very low-amplitude EEG, may show dramatic coherence abnormalities between specific Brodmann areas.

The authors follow Thatcher's insistence that one should generally not use relative power. If you do ever have to look at relative power you must only do it in the context of referring back to absolute power in order to interpret what the relative power values mean. The reasons for this warning are explained below.

Use of Absolute Power versus Relative Power

The simplest rule was stated clearly by Robert Thatcher, "Please avoid relative power." Thatcher (NeuroGuide Forum, July 2012) goes on to say, "Relative power is computed by summing the absolute power and then dividing the absolute power into each frequency in order to create proportions of the total power. Therefore, if there is increased beta you will often see "apparent" reduced theta even though absolute power in theta is normal. Or, if there is reduced alpha, you may see elevated theta or beta even though theta and beta are not elevated in absolute power. One must examine absolute power in order to interpret relative power and therefore, why bother with relative power when one knows that relative power distorts the spectrum?"

On the other hand, relative power can yield significant differences between groups and conditions that may not show up using absolute power. At times, differences between experimental and control groups is larger with relative power than absolute power. In addition, relative power eliminates differences in absolute power that are due to differences in skull thickness. Thus for group data there may be reason to use relative power. However, it is not usually wise to use it when looking at an individual. A problem arises when one ignores absolute power and/or when one makes no attempt to interpret relative power by examining absolute power.

The limitation of relative power is ambiguity, because you are doing mathematics that take you a long way from what is seen in the raw EEG; it may even result in distortions that do not occur with absolute power. As noted above, relative power is calculated by summing absolute power at a specific site and frequency and then dividing that by the sum of absolute power of all frequencies at that site. This transforms power at each frequency into a proportion of the total power at that frequency. Because a proportion or percentage (i.e., multiply by 100) is created, this means that if there is an increase in power in one frequency (e.g., beta, perhaps due to EMG artifact), then there will be a decrease in the proportion of the total at some other frequency, e.g., theta or alpha, etc. Therefore with relative power there is always ambiguity about whether there was a real increase or decrease in power at a given frequency. Also, the distribution of relative power across frequencies is always a distortion of the absolute power distribution. Thatcher concludes, stating that one must examine absolute power in order to interpret relative power and this is "an indisputable matter of simple mathematics" (Thatcher, NeuroGuide Forum, July 6, 2012).

Example: ec Laplacian Montage Brain Maps

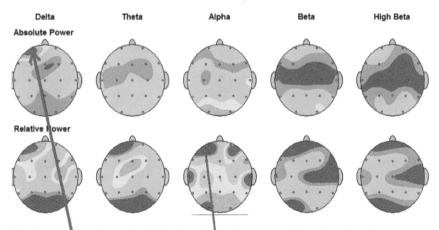

This 22 year old girl actually had high 8 & 9 hz alpha in the right prefrontal region. However, relative power shows alpha to be - 3 SD. This is because delta absolute power is extremely high at FP1 thus alpha is low power *relative to it.*
(Beta & theta also appear low SD in Relative power)

More Complex Example

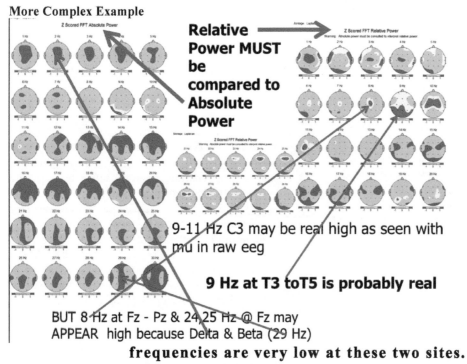

Figure: Example to compare absolute and relative power.

Negative Z-Scores and Relative Power

At times, with relative power, one frequency may appear as if it is too high an amplitude. However, it may be that the real problem is that another frequency is too low. We often observe that either delta or theta is low amplitude in a number of areas. When these low-amplitude (negative Z-score) frequencies normalize with NFB intervention, then the comparatively high relative power Z-score frequencies in some locations may also normalize. The clinician must observe whether the patient's symptoms also improve. The lesson seems to be that **when one observes extreme values in relative power, look for the inverse**. There may be negative z scores in other frequencies observed in absolute power so that by comparison this frequency appears very high. The converse may also be true. (Thanks also to Nancy Wigton who has made similar observations on the NeuroGuide Forum, July, 2012.)

A negative Z-score in absolute power may be interpreted as insufficient neurons firing together to produce sufficient power at that frequency at that site. Thus it is important to think in terms of normalizing both too-high and too-low Z-score sites **when** the known functions of these sites (Brodmann areas) **correlate to** the patient's symptoms. This has become an easier target now that LORETA Z-score NFB training has become available.

For advanced readers, a note from Robert Thatcher: The amplitude of the EEG is equal to the proportion of synchronous generators divided by the square root of the proportion of asynchronous generators. This means that a 10 percent change in the number of synchronous generators can produce a 300 percent change in EEG amplitude (Thatcher, 2012). Hence a change in only a small number of generators that are not representative of the relevant system can produce changes in the surface EEG, but this does not mean that the brain has changed much. Coherence and phase are amplitude-independent and reflect the degree of coupling or synchrony between nodes in a network. Changes in these metrics are more representative of changes in the network than is power.

Additional Notes Concerning Low-Voltage EEG

The low-voltage EEG is somewhat more difficult to interpret than a normal or high voltage EEG. We have already noted that low voltage is one sign of a concussion. Low voltage is found, however, in about 11 percent of normal EEGs. The voltage in these persons can be from less than five microvolts to as much as 10 microvolts with little rhythmic content.

The so-called "Low Voltage Fast" activity is defined as low voltage EEG that is never more than 10 microvolts (peak-to-peak) in the raw EEG amplitudes. The low voltage fast (LVF) pattern is a normal variant and may be associated with anxiety, hypervigilance, alcoholism, or a family history of alcohol problems. It may be seen in post-traumatic stress disorder (PTSD). When you have a low-voltage slow EEG (LVS), be sure that you are not seeing "stage one" sleep and drowsiness from a poorly artifacted QEEG. When you apply the Fourier analysis to the EEG data, a stable state is assumed, not one of a mixture of alert and drowsy states. If there is a true LVS pattern and not drowsiness, a physician may look for metabolic and/or toxic disturbances, or for one of the diffuse encephalopathies (e.g., infectious or posthypoxic encephalopathies, hypometabolic states from hypothyroidism, or Hashimoto's thyroiditis, or early dementia. However, drowsiness is not necessarily characterized by low voltages, but rather by reduced alpha in posterior regions and elevated delta and theta power in midline structures. This is Niedermeyer and Lopes da Silva's description of drowsiness.

Conclusion

To interpret the Fourier analysis spectrum of these EEGs, Thatcher has suggested that one just imagine the whole profile being raised to be at and above the zero standard deviation line for comparison to the normal database. Relative power is used by some practitioners, but as noted previously, one must be extremely careful to only interpret this in relation to absolute power and amplitude values. In our Centre we have found bandwidth absolute power ratios to be very helpful in interpreting these EEGs.

Notes Concerning Objective Measurements with Training Screens

As a general principle, the simpler the feedback screen, the better. It is true that with some very young children one must have interesting screens in order to initially engage them. However, as training proceeds the task is to attend even when the screens are boring. Children with a diagnosis of ADHD, for example, can be incredibly proficient at computer games. The task of NFB is not to entertain, but to learn how to sustain focus and concentration when tasks are relatively boring and repetitive from a child's point of view.

Measurements of Progress During Training Sessions

At the ADD Centre, the client is given feedback concerning their progress during each session. This feedback uses various types of measurements that indicate their progress. As with most NFB programs, the microvolt amplitudes of the bandwidths that are being trained, such as theta 3-7 Hz, SMR 13-15 Hz, beta 15-18 Hz, and high-frequency beta 23-35 Hz (and other bandwidths as appropriate for the particular client), are all summarized at the end of each segment of training (often three minutes or five minutes). These values are graphed as shown above.

Other measurements are on the screen **while** the client is doing the training. Some games, such as bowling, have their own rewards. Bowling has 'strikes.' In addition, our screens have 'points' for sustaining meeting the threshold criteria on the screen. For example, the trainer might set the time for sustaining theta below threshold to two seconds. Then every time this is achieved the client receives one point. This screen is shown below as an example.

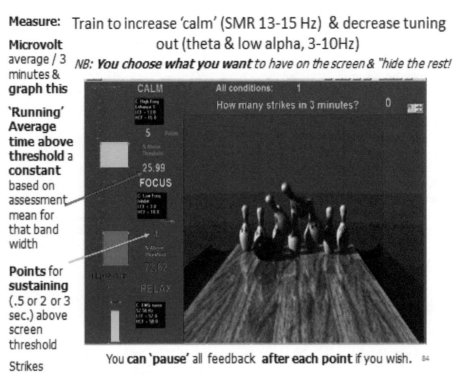

Example of a Training Screen with thanks to Thought Technology Biograph Program and Biofeedback Federation of Europe Thompson Suite, "Setting up for Clinical Success."

The screen in the figure above shows 'points' for sustaining a desired mental state. The period of time (e.g., 2 seconds) is set by the trainer. This screen also shows a 'running average' of percent of time >C (a constant in microvolts set by the trainer). This percentage reflects, over several sessions, how the client is doing relative to their 'average' microvolts for a bandwidth in their original or progress assessment.

On the ADD Centre screens, in addition to points for sustaining attention, the trainer also has a percent number that instantly gives feedback as to how the client is doing relative to their last assessment. This number is called bandwidth (e.g., 13-15 Hz) percent of time >C where 'C' is a constant. The constant (C) is entered into the actual program (under "edit virtual channel settings" in the Thought Technology Biograph program) by the trainer. This number in the first few sessions of training is the microvolt average amplitude for that bandwidth for that client during the initial assessment. It is the average before artifacts were removed because artifacts are not usually automatically removed in this program during the training session. In this manner the on-screen thresholds for each bandwidth can be changed during "shaping" procedures without altering this constant (C). This mean the client can always observe how they are doing compared to their assessment. Their theta percent >C should drop from 50 percent down to 45 percent then gradually to 40 percent, and so on. Their 'SMR' 13-15 Hz percent >C should increase from 50 percent to 55 percent then 60 percent, and so on.

The trainer and client adjust the screen bar graph or animation threshold values to meet the immediate goals of that particular training session or exercise. Importantly, these threshold adjustments do not affect the value that the trainer has previously entered for C.

Authors Recommend: Do Not Use Automatic Thresholding

There are professionals who advocate automatic thresholding, and even worse, some who do not have a trainer working with the client. It is true that at the ADD Centre we may, late in training, have a client work for short periods of time on their own in order to prove to themselves that they can accomplish the desired states and use metacognitive strategies without coaching from their trainer. However, we do not feel that those articles in the literature that have utilized automatic thresholding are setting appropriate standards for carrying out NFB training.

In sessions at the ADD Centre, the learning principle of shaping is used. This principle is completely violated when auto-thresholding is employed. For example, if the patient begins to fatigue, lose interest, or even stop actively participating in the training, the reward signals continue even if they are not producing the desired behavior and in the desired mental state. The patient is being rewarded for changing the EEG based on the previous averaged time period. This may not be an actual change from the starting EEG point. Sherlin has noted that, even worse, it may actually be in the opposite direction than the desired training parameter (Sherlin et al., 2011).

Graphing Training Sessions

The trainers always paste statistics into an Excel sheet at the end of each 3- to 5-minute segment of training. This sheet has been designed to automatically give important ratios and to graph all the microvolt values for that segment of the session. An example of a graph is shown below. In addition, from time to time, the trainer will also graph changes to compare earlier with later sessions. This is shown in the second graph below. Note that the y axes have different scales. This is because with children beta is very small amplitude compared to theta and alpha. For example, changes in SMR amplitude would not be clear, and this is important. When a different scale is used, small microvolt changes in beta can be seen. Then the graph has clear meaning for the patient.

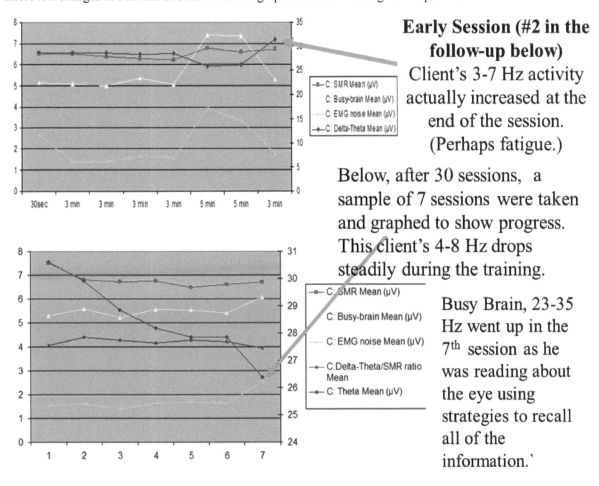

This figure demonstrates how the ADD Centre staff use Excel graphs to show clients how they are improving with training. It also helps to point out areas that may be focused on in future sessions. In this case the trainer would focus less on decreasing theta and more on increasing SMR.

Questions Trainers Might ask Their Clients during Training Sessions

At the ADD Centre/Biofeedback Institute of Toronto, we want our trainers to be continually checking with their clients to learn what the relationship is between their EEG and their different brain states. We suggest that they not assume anything. This introspection should help awareness of mental state and encourage transfer of control of one's thinking from the clinic to everyday life. The following are a few examples of the kinds of questions we like our trainers to be asking themselves and their clients as they are doing the training.

With the electrode at FCz in adolescents and adults, when particular frequencies increase in amplitude we suggest the following.

Increased Theta

Is your client really "tuned out"? Is he inside his own head and not doing the assigned task: reading, math, or looking at the feedback screen? Is she feeling sleepy?

On the other hand, when theta rises in amplitude, one has to also be aware, as in our first example above, that both delta and theta power and particularly 6 Hz theta may increase with working memory tasks – encoding and recall (Aguirre-Perez et al., 2007). In addition, short-term memory tasks may increase both frontal and parietal region gamma (Babiloni et al., 2004). Left temporal theta may increase with lexical retrieval tasks.

With high 6 Hz, is your client encoding or recalling memories OR just tuning out? Both are possible, so check it out with some questioning.

Increased low alpha (8-10 Hz)

Is your client "tuned-out" or dissociated from external stimuli? Is she highly focused on something internal (in her own head)? Lower alpha is related to both attention and response readiness (Klimesch, 1992). Has your client used marijuana in the last couple of days?

With high alpha (10-12 Hz)

Is your client in a broad, open awareness, readiness, state? This may be quite readily observed in people who do high-level martial arts. Upper alpha may also be related to long-term memory with high amplitude at rest and rapid desynchronization (lower amplitude alpha with a shift to beta production) during semantic memory tasks. This can correlate with efficient memory performance (Klimesch, 1998, 1999, 2003, 2005; Sauseng et al., 2003, 2007). In addition Klimesch (2003) reported that an increase in the upper alpha range produced by central frontal and right parietal rTMS could improve cognitive performance. In a small number of patients, very high-amplitude, high-frequency alpha can correlate with anxiety and brittle affect.

Increased SMR (12-15 or 13-15 Hz)

Is your client relaxed and calm? Is she breathing evenly, diaphragmatically at about 6 BrPM? Or is your client an exception to the usual pattern and producing high amplitude spindling beta and not SMR? In this person, does this spindling beta correlate with depression, a hyperactive/restless state, anxiety, or some other undesirable clinical state? If the increase is the opposite of being calm, then the increased 13-15 Hz at Cz or FCz is likely not SMR from the thalamus to cortex, but rather spindling beta that is cortically generated.

With high frequency beta (19-22 Hz and 23-36 Hz)

What is your client *feeling?* Is your client feeling anxiety, depression, or are they in a thoughtful, intense and productive, creative state? How are their thought processes working? Is it a mental state of useful multitasking? Or is this a mental state in which they are negatively ruminating with their mind racing but stuck in a manner that is not productive? Patients can compare the ruminating state to being in a car with their foot jammed down on the accelerator, but the car is in neutral. You are using up energy without moving forward.

The trainer should also be aware that if there is a sudden short burst of high-amplitude activity, but these bursts are relatively short compared to the overall EEG recording time, then the average amplitude of the EEG for that bandwidth may not show much change. This may be particularly true with small but possibly significant increases in the amplitude of an inhibit band when that bandwidth is averaged over a 3- to 5-minute training time segment. These short bursts may correlate with an important mental state (positive or negative) in your particular client, and this could be of significance. We will show an example in the TBI section of a man after a stroke who had bursts, sometimes <1 second, of extremely high amplitude delta in all leads. In that precise timeframe he was completely confused, disoriented, frightened, and had no recall of anything that occurred. In some of these cases **the variability or SD may show** large change. Reducing variability may be important if we are trying to decrease bursts of beta spindling or bursts of delta/theta after a stroke or TBI.

In the case example given below, the average delta activity was not substantially outside the normal database. However, the amplitude of the bursts of 3 Hz activity were extreme. Fortunately we observed the patient during the recording and we did **not** do automatic artifacting after the recording was completed. Automatic artifacting would have eliminated all of these bursts along with the eye blink artifacts! At our Centre we always watch the patient and the EEG as it is being recorded! We saw the bursts and his "reaction." After the EEG was recorded we looked at every second of the EEG, and, as is always done at our Centres, manually removed artifacts. In this process we observed the bursts and set up our training program to target these bursts of 3 Hz activity. The results speak for themselves.

Sam was a 74-year-old man who had suffered a left frontal area stroke with a loss of speech and movement of his limbs on the right side. A year later, after completion of his hospital rehabilitation programs and being told that he would not have any further recovery, he came for assessment to the ADD Centre. He was now able to walk with a cane and move his right arm and hand. Speech was difficult. He had been an avid reader but could not read even a simple page. He had to have constant supervision due to disorientation. He could not remember two simple commands. He would become lost even in his own house. All improvements had plateaued well before he presented for assessment. In the 19-channel EEG he showed sudden bursts of extremely high-amplitude 3 Hz activity at all sites. This correlated with him being totally confused. When these bursts of delta occurred he could not answer any questions and he had to hang on to something to remain balanced in the chair. He was conscious. These bursts lasted from <1 to 4 seconds. When the burst stopped it would take a few seconds for him to realize where he was and what he was doing.

Just 24 sessions of single-channel NFB focused on increasing activity in Broca's and Wernicke's areas combined with HRV training resulted in a return of his ability to read an entire book in a few days and to recall the material. He re-read a book by Eckart Tolle he had enjoyed before his stroke and was able to discuss the ideas with his wife.

Next he began LORETA Z-score NFB and after a number of sessions he was able to find his way using public transportation with a number of changes from subway to bus to come for training. The 3 Hz bursts were no longer observed. He remained disabled, walked with a shuffle, and he was not back to his old, very intelligent professional self, but he had regained his independence and could enjoy his life again. The improvements were substantial, and he and his wife (who was able to return to work rather than staying home to provide care) were very grateful for the neurofeedback intervention.

Additional Terms You May Encounter

- **Phase Shift and Phase Lock**
- **Bicoherence**
- **Chaos Theory and Nonlinear Mathematics**

The purpose of this short section is to very briefly define a few terms that the beginning NFB practitioner may encounter in the literature. In this 'fundamentals' textbook we do not wish to go into detail on these areas, and we refer the reader to other material if they require more in-depth information.

Phase Shift and Phase Lock

Phase: Theoretically, if you imagine a sine wave generator in the center of a sphere, the waves detected anywhere on the surface of that sphere would be in phase with each other. However, the human head is not a perfect sphere, and the thalamus, the major generator of rhythmic EEG activity, is not necessarily absolutely in the center of the skull. Nuclei within the thalamus can generate theta, alpha, and SMR rhythms that are often a focus of training using NFB. There will be different distances for any wave to travel to various areas in the cortex where it is reflected in the EEG measured at surface sites. The waves at any two points will arrive at slightly different times, and thus will likely be out of phase with each other. However, since they were produced at the same moment from the same source, they would maintain the same phase delay across time and thus be coherent (phase locked).

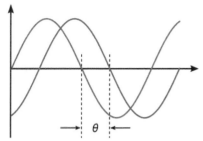

Reference for figure: Wikipedia

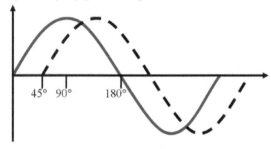

In the above figures two sine waves are out of phase by a time constant (between the two dotted lines) or, in the second figure, by a number of degrees from 0 to 360. In the second figure, the time difference along the horizonal line is calculated as a phase angle and therefore given as 'degrees.' Imagine that one sine wave, from its start at the horizontal line to its return to that line after traversing a full cycle above and below the line, has completed a full sweep around a split clock face or a compass. It has swept through 360 degrees. The amount by which these oscillations are out of step with each other is therefore in degrees from 0° to 360. Thus, two oscillators that have the same frequency and different phases have a phase difference, and the oscillators are said to be **out of phase** with each other. In the figure, the second (dotted line) sine wave lags behind the first (red line) sine wave by 45 degrees. Peaks pointing in opposite directions are out of phase by 180 degrees, as shown below. Phase lock is independent of phase difference. Therefore, for example, two waves can be phase locked even though they have a phase difference of 180 degrees.

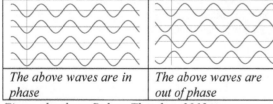

| The above waves are in phase | The above waves are out of phase |

Figure thanks to Robert Thatcher, 2012.

The middle two waves in the second column are out of phase by 180 degrees. However, they are phase locked. (Note that waves of a different frequency also may have a consistent relation to each other and also be out of phase and be phase locked.) Waves that are phase locked are said to be *coherent*. Thus, coherence is a measurement of phase-locked activity. Coherence therefore suggests a common origin for waves.

When two sites in the brain have EEG waves phase locked, we make the assumption that these waves have a common source or generator. Phase locked, in our work with NFB, means coherent, and coherence is one of the EEG measurements that we use for training. NFB practitioners, and particularly those who now use LORETA NFB, may consider recording and giving feedback based on measurements of phase-lock duration and phase shifts. To understand how these measurements are carried out, we refer the reader to papers by Robert Thatcher. We will not attempt to give this very advanced detail in a basic textbook. However, the reader may be interested to know that, even at the time of writing, there have been significant findings published in two areas.

Subjects with a higher IQ have been demonstrated to show an association with shorter phase-lock duration and longer phase shift times. On the other hand, children who suffer from autism (Asperger's and high functioning autism are not included in the experimental group in studies to date), compared to normal controls, demonstrate the opposite, that is, longer phase-lock duration and shorter phase shift. In this research by Robert Thatcher, phase locking occurs only in the alpha 2 (10-13 Hz) band. Phase shift occurs in the alpha 1 (8-10 Hz) band for the most part and to a lesser extent in all other bandwidths.

Phase shifting occurs continuously in the brain. Thatcher points out that the distances between sites in the cortex are too long to account for the short times recorded for the phenomenon of phase shifts and thus the cortex could not be the primary source of this phenomenon. He suggests that the same distance constriction does not exist in the thalamus, where distances between different nuclei are very small. He comes to the conclusion that the thalamus is the only place where this phase shift phenomenon could be taking place (Thatcher, 2009). He notes also that theta and low alpha (8-10 Hz) are involved in resource recruitment by rapid phase shifting, while high-frequency alpha (10-12 Hz) involves resource allocation and binding by phase locking.

As an aside, we wonder if Thatcher's findings could relate in any way to common amplitude findings in autism. At the ADD Centre, we have collected data demonstrating that autistic children have very low amplitude in the alpha 1 low frequency range and higher amplitude in the alpha 2 higher frequency range of alpha (Thompson and Thompson, 2010), as well as higher amplitudes of slow waves, delta and theta, and 13-18 Hz beta. In other words, there is a dip in low frequency alpha (alpha 1) range when you look at the spectral array for an autistic child that is quite different from what is observed in the Asperger's patients that we have assessed. These findings, plus clinical observations, make it difficult to understand the DSM-5 classification deleting Asperger's as a diagnostic category and lumping it together with Autism Spectrum Disorders. Perhaps a reasonable hypothesis might be that a number of symptoms observed in autistic subjects, such as a poor ability to shift mental set and a tendency to lock on to a set of repetitive behaviors, might relate to these EEG findings.

Detail on Phase Reset

In the autistic subject the shorter phase shift durations probably relate in part to deficient GABA. GABA is the brain's neuroinhibitory transmitter. It is the primary neurotransmitter responsible for rapid synchronization of neurons and the rhythmic discharges in the thalamo-cortical circuits. It is these rhythmic discharges that are the source, for the most part, of the EEG that we are measuring. In the autistic subject, deficient GABA results in a short phase shift and consequently a low degree of neural resource recruitment. The longer phase lock duration suggests that the reduced number of neurons that are synchronized at each moment in time are now locked for a longer period than normal, resulting in reduced mental flexibility.

Note that there is a causal linkage between GABA mediated inhibitory postsynaptic potentials and the frequency of the EEG. Cortical pyramidal cell activity is modulated and synchronized by thalamo-cortical circuits (Steriade 2005). We have previously described the thalamus as being like the hub of the wheel. It is the switching and coordinating system, the orchestra leader, the traffic control central 'brain' that sets rhythmic activity. Thatcher likens the phase shift problem in those with ASD to a train station where, in the autistic, the train (information) arrives but the shift time is too short so the next train, so to speak, leaves the station and there is a reduction in information flow. One might hypothesize that this could correlate with findings that, in people with autism, there is local hyperconnectivity and long range hypoconnectivity. This is in contrast to findings in people with ADHD who have local hypoconnectivity and long range hyperconnectivity (Sokhadzi et al., 2012).

Dr. Thatcher has incorporated phase shift and phase lock and Z-scores for these measurements into his LORETA NFB program.

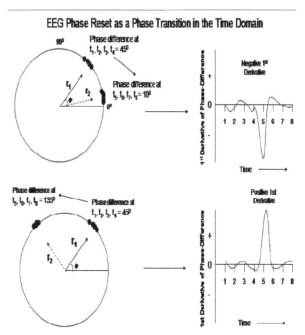

Diagram from Lecture / Workshop by Robert Thatcher. Used with Permission.

Thatcher illustrates **'phase reset'** as above. We previously noted that a sine wave goes through 360^0 and comes back to its starting point. The wave can be illustrated as a circle. Then phase shifting can be illustrated as points on this circle as is seen in the diagrams above. In this illustration Thatcher shows amplitude as the length of the 'r' and the angle of phase at different time points is illustrated by times on the perimeter of the circle such as t_1 t_2 t_3 t_4 (at about 45^0). In the first illustration this locking suddenly shifts in what he calls a 'phase shift' to times t_5 t_6 t_7 t_8 (at 10^0). This illustrates a 'resetting' of phase.

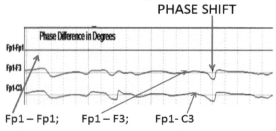

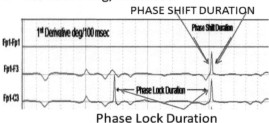

Reference: Illustration from Thatcher (2012). Used with permission. For the full diagram and explanation it is recommended that the reader consult Thatcher's textbook.

Thatcher in his textbook and lectures shows this example of phase differences between Fp1 - F3 and Fp1-C3. The second figure is the 1st derivative which in this example can be thought of as the rate of change of phase in degrees / centiseconds (cs is a unit of time = 0.01 seconds). Phase locking represents synchronization.

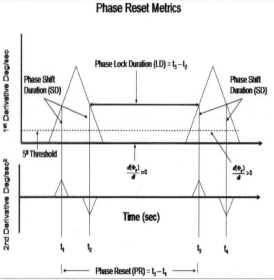

Reference: Illustration from Thatcher (2012). Used with permission. (Times T1,2,3,4 are *not* the same as in the other diagrams.)

Dr Thatcher has always noted that phase lag in brain maps is not easily interpreted in terms of the patient's clinical functioning, but that, despite this, these scores are one of the most important contributions to his NeuroGuide discriminative analyzes for Learning Disabilities and for Traumatic Brain Injury. Phase shift and phase lock, however, give new insights that do appear to correlate with clinical presentations.

In our experience, his Z-score brain maps when combined with LORETA are very easily correlated with the clinical presentation of the patient. Again, for more detail on his work we refer the reader to Thatcher's 2012 text, *Handbook of Quantitative Electroencephalography and EEG Biofeedback*.

Summary

Phase shift identifies and recruits available neurons within a network. Then the phase locking binds together synchronous neurons in interconnected loops to mediate momentary functions. These are released by a subsequent phase shift when a different cluster of neurons becomes phase locked. Thatcher finds a positive correlation between intelligence and phase shift duration and a negative correlation with phase lock duration. His explanation is that long phase shift durations represent an expanded neural recruitment process in which larger populations of neurons are recruited. This would correlate with longer inhibitory bursts in the thalamocortical circuits (Thatcher et al., 2008). Phase lock represents periods of phase synchrony of selected clusters of neurons. If there is too long a phase-lock period, then Thatcher postulates that there would be less cognitive flexibility, fewer neural resources available to be allocated, and a reduced intelligence.

In the autistic subject the shorter phase shift durations probably relate in part to deficient GABA. GABA is the brain's neuroinhibitory transmitter. It is the primary neurotransmitter responsible for rapid synchronization of neurons and the rhythmic discharges in the thalamocortical circuits. It is these rhythmic discharges that are the source, for the most part, of the EEG that we are measuring. In the autistic subject, deficient GABA results in a short phase shift and consequently a low degree of neural resource recruitment. The longer phase lock duration suggests that the reduced number of neurons that are synchronized at each moment in time are now locked for a longer period than normal, resulting in reduced mental flexibility.

Bicoherence

Bicoherence refers to two different frequencies of a single channel or signal (unlike coherence, which takes two channels) being phase locked. This is unlikely to be something that the beginning practitioner will be concerned with. However, it is a term that is seen in the literature. Bicoherence is a squared normalized version of the bispectrum. Bicoherence takes values bounded between 0 and 1, which make it a convenient measure for quantifying the extent of phase coupling in a signal. It is also known as bispectral coherency. The prefix *bi-* in *bispectrum* and *bicoherence* refers not to two time series x_t, y_t, but rather to **two frequencies of a single signal**. Thus, unlike coherence, it can be calculated from a single signal.

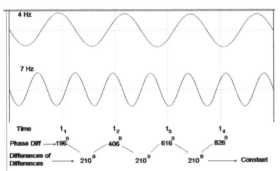

Illustration of the constant phase differences as a function of time when two different frequencies are phase locked. Illustration from Dr. Thatcher's workshop and from Thatcher (2012). Used with permission from Dr. Thatcher.

Summary

Phase locked is a definition of coherence, and this is an important component of our work using neurofeedback. For example, standard deviations that show lower coherence values than the database are sometimes called 'disconnects,' and these can be quite common in traumatic brain injuries (TBI). We also frequently find disconnects in learning disabilities. In contrast, hypercoherence is often a dominant finding in intellectually challenged children. Both hyper- and hypocoherence problems are found in many of our clients and in a variety of disorders. Both coherence and phase values that are outside the database norms appear to be important factors to target when we do LORETA NFB. In our experience with LORETA NFB, amplitude Z-scores improve rapidly, but it requires more sessions to change the coherence and phase Z-score values. However, our patients improve with this work. This impression at our Centre requires research to establish how important it is to normalize phase and coherence values for different patient populations.

Chaos and Nonlinear Mathematics

Given that mathematical terms like 'nonlinear' and 'chaos theory' are not in everyday use by clinicians, a brief explanation of this jargon is needed. Though 'chaos' implies, in common English usage, random happenings that occur purely by chance, this is **not what is meant** when discussing 'chaos theory.' Without going into detail, here is a simple explanation of these terms. Linear implies a line with a beginning point A, then straightforward steps that must, given the outlined conditions, lead to point B. In a simple, straightforward linear model, the output at B should be proportional to the input at A. This does not accurately describe the EEG.

Given that an EEG rhythm may be produced by an area within the thalamus, then it should arrive at its destination in the cortex in a precisely predictable manner. However, **EEG signals are actually the result of the activity of many brain cells,** and the existence of a particular frequency in the FFT does not necessarily mean that there is an oscillator in the brain at that frequency. **Linear analysis has to assume one generator and one measurable output,** but given that the production of electrical activity measured at the scalp is more complex, the mathematics become **nonlinear, which means that output may not be proportional to input from the main source.**

This leads us to the second difficult word, 'chaotic.' Certainly the EEG frequency wave has a "determined" origin or starting point. It is definitely **not just a random** occurrence. It is not changing its origin constantly in a completely unpredictable manner. **Chaotic is therefore quite different than random.** The starting point may be determined as a particular group of neurons in the thalamus in our example. But then almost an infinite number of variations of chemistry, electrical activity, and alterations in blood flow are found in the immediate environment of the signal as it travels. This environment will act upon this wave and change it in various ways. This, then, is a system in which the **output is not in a uniform relationship to the input**. The behavior of a system that contains nonlinear components is harder to model and to predict, but these nonlinear systems are, as we said, not by chance or random. Nonlinear systems can be transformed to a linear system of multiple variables, but it is difficult to do precisely.

The most familiar nonlinear and chaotic system is the weather. It is not a by-chance, random phenomenon. It is 'determined' at its origin; however, simple changes in one part of the system produce complex effects throughout, and its behavior is so unpredictable as to sometimes appear random, owing to great sensitivity to small changes in conditions. Meteorologists usually can only accurately predict weather for the next few days.

To summarize, small differences in initial conditions can yield widely diverging outcomes for such systems as weather and the EEG. This happens even though these systems are deterministic, meaning that their future behavior is fully determined by their initial conditions, with no random elements involved. The deterministic nature of these systems does not make them totally predictable. **Deterministic chaos, or simply chaos,** are the terms used to refer to such symptoms.

Neuromodulation

Neurotransmitters have been described earlier in the book in Section II. Neuromodulation was mentioned in Section I, but will also be briefly discussed in this section. A neuromodulator is any compound that affects the excitability of nerve cells. Neuromodulators can affect the responsiveness of nerve cells to transsynaptic actions of presynaptic neurons. Thus neuromodulation occurs when a substance released from one neuron alters the cellular or synaptic properties of another neuron. Dopamine, for example, can increase both the size and the number of synapses.

Neuromodulation is thus the physiological process by which specific neurotransmitters, such as dopamine, noradrenaline, serotonin, and acetylcholine, regulate diverse populations of neurons. This is in contrast to the classical synaptic transmission from the thalamus in which glutamate, the transmitter to the cortex from one presynaptic neuron, directly influences a single postsynaptic partner.

Neuromodulators, on the other hand, are secreted by small groups of neurons that reside in subcortical and brain stem areas, as was outlined in Section II. These neuromodulators will affect specific areas mainly in, but not limited to, the cortex. Dopamine from the substantia nigra, for example, will affect the firing of striatal neurons, whereas dopamine from the ventral tegmental area will affect the anterior cingulate gyrus and prefrontal areas. A broad difference between neurotransmitters such as glutamate and GABA, also described in Section II of this book, is that neuromodulators go directly to only approximately 15 percent of the cortex compared to the much wider distribution of neurotransmitters such as glutamate. However, the stimulation of the neuromodulator is slow (lasts longer) compared to the very fast action of the neurotransmitter glutamate. The neuromodulators have longer action and their action will diffuse through large areas of the nervous system and have an effect on multiple neurons.

Neuromodulators can also end up spending a significant amount of time in the cerebrospinal fluid (CSF), influencing (or "modulating") the activity of a number of neurons in the brain. In this manner, some neurotransmitters, such as serotonin and acetylcholine, are also considered to be neuromodulators.

Neuromodulation may be contrasted with classical fast glutamate synaptic transmission. In both cases the transmitter acts on local postsynaptic receptors, but in neuromodulation, the receptors are typically G-protein coupled receptors (slow action), while in classical chemical neurotransmission they are ligand-gated ion channels (fast action). These ionotropic receptors are ion channels with a binding site for the respective transmitter, and cause fast changes in the membrane potential of the postsynaptic neuron (in the range of milliseconds).

In contrast, neurotransmission that involves metabotropic receptors (like G-protein linked receptors) often also involves voltage-gated ion channels, and is relatively slow. Metabotropic receptors activate a signaling cascade in the postsynaptic cell that leads to slow changes in the electrical (and also biochemical) properties of the postsynaptic cell – in the range of hundreds of milliseconds, or seconds or even much longer (Synapses, Neurotransmitters, Neuromodulators, Lecture series model systems in Neurobiology; From Molecules to Behaviour. Free University of Berlin, Winter semester, 2007/2008).

Both classes of synapses summate and produce the human EEG. The longer-duration neuromodulators shape the synapses through long-term potentiation (LTP). Thus, neuromodulators can increase the size of the synapse (Thatcher's workshops and Thatcher, 2012, p 40, pdf version). The importance for practitioners is that it appears that EEG biofeedback can affect these changes in the synapses.

SECTION XII
Combined NFB + BFB Intervention Fundamentals

Adding Biofeedback to Neurofeedback

Introduction
Most of a neurofeedback practitioner's time is spent working with a student/client to improve specific areas of functioning. The task is to help the student/client learn how to self-regulate. Increasing attention and concentration and decreasing tension and anxiety are often central in this process. The EEG Profile and Stress Profile assist in decision making. For neurofeedback, the assessment dictates where electrodes should be placed, and the frequency bands to inhibit or enhance. For biofeedback, the assessment will help you to decide, with your client, which instruments best reflect physiological changes that occur with stress. It will show both the modalities that demonstrate the most change and which modalities show a slow recovery after stress. It will also help you decide on the usefulness of EDR for reflecting alertness levels.

On the basis of the assessment, the practitioner discusses with clients what they may reasonably expect from NFB combined with BFB. The practitioner also discusses with clients the things that will be done in addition to NFB and BFB. For example, you will be training your clients in procedures that will help them generalize what they are learning in your office to home, school, athletic and work situations. With students you will be training them in the use of metacognitive and time management strategies for their academic work. You may be training them in quick procedures to turn on a relaxed but alert state. You always want to include a discussion of diet, sleep and exercise, since these all have a direct effect on performance. We call them the common sense variables: Eat well and take a balanced vitamin-mineral supplement. Get enough sleep. Exercise at least three times a week. It seems to be stating the obvious, but remember what Covey has said in his books about successful habits: common sense is not always common practice.

Self-Regulation: Four Steps
There is a logical sequence of steps (the following four headings), which overlap and eventually merge as the client becomes more flexible in his mental functioning and is able to shift to the appropriate mental state for the task at hand. There are further activities that help the client to generalize so that the steps he has learned in training become largely automatic. The client learns to control his own mental and physiological states. This is called self-regulation.

External Open Awareness
For the athlete, executive, or a tense and/or anxious student, the **first step** is to learn to be externally and broadly focused, aware of everything in their environment (open awareness), yet often focused on no one thing in particular. In the martial arts this may be called "no-mindedness" or *mushin*. (It is an application of Zen in the training of these practitioners. At a high level of training, the practitioner may be instantly aware of any change in the environment surrounding him.) The professional athlete, such as a goalie in soccer or hockey, must retain both the open awareness of all players at the same time as having a singular focus on the player with the ball or the puck. This may be more accurately described as a *flicker-fusion-focus,* where the mind is actually moving rapidly through the entire scene. On the other hand, it may be a *soft-eyes* state (Sue Wilson's term) for really perceiving everything in a state of mental and physical *readiness*. In the latter hypothesis, it is almost as if the person's attention is to the *space* around her. Psychologically, when the objective is relaxation, it is important for these clients to imagine a pleasant scene or event. Physiologically the most helpful technique is to breathe diaphragmatically at a rate of 6 breaths per minute. In this state, a balance occurs between sympathetic and parasympathetic systems and in natural biological oscillators (heart rate, respiration, blood pressure). There will be

synchrony (coherence) between oscillators, which is most easily seen by monitoring respiration and heart rate variability (RSA). At the same time the client is to relax their forehead, neck and shoulders and warm their hands. Les Fehmi has used the term "open focus" for a similar mental state. He has also produced audiotapes to help people achieve open focus.

Flexible Shifting of Mental State: Open Awareness to Focused Concentration

The **second step** is to use operant conditioning of brain waves so they can train themselves to consciously move from the open awareness high alpha (11-13 Hz) and SMR (13-15 Hz) to a focused concentrating state where they lower their dominant slow-wave activity and increase beta (16-18 Hz). This may require that the student shift in and out of open awareness. When they shift out of the calm, openly aware state, they move into an external, narrowly focused state where they consciously produce a problem solving type of activation. In this mental state the student will decrease slow waves and increase fast waves (15-18 Hz activity). This shift toward faster, desynchronized activity is reinforced by a visual and/or auditory signal that indicates they have met the preset criteria. Using this traditional operant conditioning they will train themselves to control, and perhaps to permanently change, their brain wave pattern.

Pair Desired Mental State with Learning

The **third step** is to link a desired activity, such as studying, to this efficient mental state using classical conditioning. Associative learning may also be occurring, so the trainer must be alert to what is in the immediate environment. To pair the focused mental state with learning, the feedback is altered so that the student can receive continuous auditory reinforcement for sustaining the mental state of externally oriented narrow focus while performing a task, such as scanning a textbook chapter or doing a math problem. If the feedback stops, the student immediately moves back to the operant conditioning mode, watching the display screen until the desired mental state is again achieved and sustained.

Generalize Techniques to Home, School and Work Environments

The **fourth step** is to train the student in procedures that will help them to generalize this ability to produce a calm, relaxed, alert, flexible, focused, concentrating mental state to home, school and work situations. Breathing techniques linked to daily habits and metacognitive strategies linked to learning new material are examples of effective methods of assisting this generalization.

Heart Rate Variations with Respiration – A Key Variable for Combined NFB + BFB Training

See also detail in Section XVIII

As previously noted, heart rate variability is emerging as one of the most important feedback modalities for rapidly assisting individuals to self-regulate their autonomic nervous system. Heart rate continuously changes or oscillates with healthy functioning. Controlling breathing is the simplest and most powerful means of controlling this system in a helpful manner. Heart rate is normally primarily generated by the sinoatrial node in the heart. Its normal rate is a little over 100 beats per minute. This is faster than a normal heart rate. The 'control' of this fast rate to slow it down is the parasympathetic vagal input through efferents from the brain stem, as is explained in Section XVIII on heart rate variability. Breathing in (inspire) can also be partially linked with sympathetic outflow, which increases heart rate. Breathing out (expire) releases the heart rhythm from the sympathetic nervous system's control and the parasympathetic system's inhibition takes over to slow the heart rate.

Instrument

It is helpful early in training to have a biofeedback instrument that allows your client to observe synchrony between their breathing rate and the increases and decreases in their heart rate. This is called *respiratory sinus arrhythmia* or RSA. Heart rate can be detected using a photoplethysmograph taped to the palmer surface of a finger tip. If your biofeedback instrument does not have this capability, you can purchase a heart rate monitor and even an app that will show a very smooth high-amplitude wave form when you are producing good quality RSA. The Freeze-Framer device from HeartMath in Boulder Creek, Colorado, is an example of an instrument that can be used by your client at home using their own computer. A simple $20 skin temperature apparatus can be added for monitoring finger temperature as a reflection of relaxation.

Begin with Biofeedback

The first step for most of our clients is to learn diaphragmatic breathing. A breathing rate of about six *breaths per minute* (BrPM) achieves a good quality RSA for most adult clients. Children breathe at a slightly faster rate. Next, they learn to raise their peripheral skin temperature, maintain a moderate skin conduction and a relaxed musculature (frontalis is a good indicator of EMG tension and a simple muscle to train, but the shoulder trapezius muscle or the forearms can also be helpful as previously described). This is done as they maintain a good quality RSA. Ask them to practice this breathing, relaxation and hand warming while entering a mental state of awareness of everything in their environment. Opening and broadening their focus from an inwardly directed, narrowed, ruminative focus allows a quieting of the mind and a state of mental alertness.

Heart rate variability or change is greatest in periods of relaxation. In our early example, Jane's heart rate varied regularly from around 88 *beats per minute* (BPM) when she inspired, to about 71 BPM when she expired when she was calm and really relaxed. On the other hand, when Jane thought of a stressful event or discussed an anxiety-producing event, this variation in heart rate almost immediately was reduced from the previous 17 (88 minus 71) BPM to a difference of only 1 to 2 BPM. Heart rate should show approximately a 10-beat difference or variation as the patient inspires and expires. However, our young healthy patients usually show a larger variation. This variation is lost when the client becomes overanxious. This loss in variability corresponds to an increase in overall cardiac rate with stress. It is also much lower after a concussion.

Purposefully changing one's breathing to diaphragmatic breathing at a rate of about 6 BrPM is often associated with a sense of calm control. This can be very helpful in stressful situations. It can also improve efficiency of learning at school and work. Almost needless to say, it is an essential skill for a top athlete.

Now that the client is beginning to control her physiological state using BFB, introduce the NFB variables according to that client's EEG assessment.

Display Options for Combining BFB with NFB

You may have two, three, or four BFB variables on your screen when doing neurofeedback with adolescent and adult clients. These will vary depending on the stress assessment. The ones we most commonly use are: *breaths per minute (BrPM), heart rate (HR), temperature, forehead, trapezius or masseter EMG,* and *EDR. The appropriate variables for a particular client are placed on display screens with the neurofeedback instruments.* Although this is highly recommended for adolescents and adults, it may also be important for younger children who demonstrate anxiety and/or low alertness levels. The trade off is that having sensors on their hands may be distracting to some young children. They may fidget and thus distort the readings. For most of our older clients we use respiration plus one other modality. The reason we use respiration as the primary instrument is because it can change rapidly with stress and, for most clients, it corresponds to changes in other modalities. A respiration rate that is regular, deep, diaphragmatic, at a rate of about 6 BrPM corresponds to increasing peripheral skin temperature and relaxing muscles in most people. However, some clients can breathe in this manner and yet remain tense. Others can relax their muscles and breathe diaphragmatically yet have trouble raising skin temperature. Most people need to be able to change all modalities in order to genuinely self-regulate and sense that they are really beginning to relax.

To adequately follow respiration, clients can be helped with feedback using both a counter (for rate) and a line graph. With a belt sensor around the abdomen, the line graph allows them to see that respiration is regular, diaphragmatic and deep. Rate alone could be brought to six per minute and breathing still be largely thoracic and even irregular. In order to initially help your client to stop shallow thoracic breathing you may use a second respiration belt around the chest (under the armpits) or measure EMG from one forearm to the other. Both methods will demonstrate a rise in amplitude if the client expands their upper chest rather than moving the diaphragm. At the ADD Centre, a finger sensor is used to track variations in heart rate. This line graph overlaps the respiration line graph on the feedback screen. The client quickly learns to obtain synchrony between these two lines as is explained in Section XVIII.

Temperature changes slowly and can be adequately followed, in most people, using a simple box that recalculates every second. In some individuals, however, significant changes are very, very small. Recent observations with ADHD adults showed erratic little finger temperature changes and research is underway to see if this could be used as a diagnostic test for ADHD. In these people an instrument that registers minute shifts can be quite useful. The F1000 had a temperature circle with this capability. When we are just measuring gross shifts in EMG, a simple microvolt box that recalculates each second may do. However, we prefer much faster updating that can be displayed on a speedometer-like line or a bar graph where the calibration can be shifted to display small changes. This kind of instrument is shown on Joan's screen below.

Combining Biofeedback with Neurofeedback – Case Examples

The examples below are further to ones earlier in the text that illustrated the importance of combining biofeedback (BFB) with neurofeedback (NFB). In the section describing one-channel EEG, there was the example of a university student with ADD and performance anxiety. In the section using two-channel training, an example of an older woman with ADD and agitated depression was used. Below, three further examples are given. The first category involves training to optimize performance and the cases are a professional golfer and a racing car driver. The second category involves training procedures and the theoretical rationale for the use of biofeedback plus neurofeedback for problems such as movement disorders (dystonia, Parkinson's, Tourette), fibromyalgia, hypertension, and headache. The third category discusses how we are training business executives to optimize their performance while improving their ability to shift mental states between work and home.

A Professional Golfer

Deciding on NFB and BFB Feedback Parameters

This example demonstrates a decision-making process for using a combination of BFB + NFB to optimize performance. This combined approach was also mentioned in the example of two-channel-training. In that case, it was important to produce a calm, relaxed–yet-alert mental state. The example below is one of optimizing performance in a professional golfer. What may initially appear to be a simple case of improving athletic performance often ultimately involves a complex array of factors that are interfering with the athlete maximizing their potential. For this example we have used Case #6 which is reiterated here:

Case #6: *Joan, age 28, is a golf professional. She felt her game could be improved but had no idea how to do this other than practice. She was also frustrated that some of her golf students did not seem able to improve despite many lessons over an extended period of time. The EEG showed higher than expected 5-9 Hz activity. When specifically asked, it turned out that Joan had been a bright student but had found it difficult to remain focused in classes. She appeared calm and certainly not at all anxious, but when assessed her breathing rate was irregular, shallow, and rapid. Her skin temperature was 88 degrees. Her shoulders (trapezius) became tense with the math stress test. When we discussed what occurred mentally during her approach to the ball and in the back-swing, she realized that she was ruminating and worrying about the outcome of the shot. This corresponded to increases in her 23-34 Hz activity and to an increased breathing rate and increased muscle tension in her face, shoulders, and forearm. She then talked about being a very outgoing person, but said that this covered up the fact that she was often quite tense and anxious and a great worrier.*

Joan used the biofeedback stress assessment screen for 10 minutes in each of her first three sessions. She learned to produce good quality RSA with heart rate increasing with inspiration and decreasing with expiration. When she did this, her peripheral finger temperature rose and her forehead EMG dropped from 5 µV to 1.1 µV. After 10 sessions she was able to lower her thalpha (5-9 Hz in Joan's case) and sustain an external focus while maintaining diaphragmatic breathing at 6 BrPM.

Joan then began to practice putting and chip shots while she received feedback. The encoder was clipped to her belt and she set up the waiting room as a putting green. A screen was customized to meet her needs. She wanted the screen to be complex. This is not unusual, we have found that most of our optimal performance clients ask us to make a complex screen with a lot of information. They can choose for themselves what parameter to focus on at different times. She needed to:

1. Stop anxiety and worrying (decrease 20-23 and 24-32 Hz activity respectively)

2. Be very calm and externally, broadly focused on the green, the hole and the ball (increase 11-13 and 13-15 Hz activity and maintain diaphragmatic breathing at 6 BrPM)

3. Eliminate undue and inappropriate tension in her lead forearm and her shoulder muscles. (EMG on forearm extensors and trapezius)

4. Be consistent in her shot with respect to her breathing cycle. Make the putt during the pause after exhaling and before the next inhalation.

An Initial Display Screen

Golfer's Training Screen

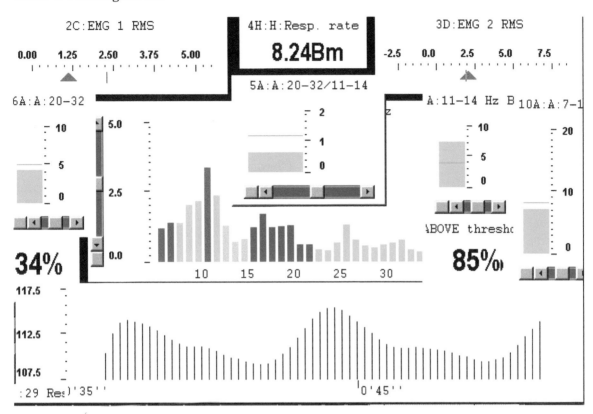

Use of the Above Screen

Joan used the screen above to practice changing mental states before putting. Then it was used by her coach (with sound feedback to her) while she practiced putting in the waiting room. She chose the waiting room because it provided distractions similar to the audience in competition. We also felt she was a good role model for the children.

Screen Design

The display screen shown above was made to provide the feedback while she was putting. At the moment shown above, the spectral array indicates she was in a calm (high 11-12 Hz), externally-focused (low 4-8 Hz) mental state but she was still worrying about her shot (25-32 Hz). Before putting she used the *spectrum* on the screen for feedback while she visualized herself putting and went into the desired mental state for each stage of the putt. While putting she received auditory feedback for keeping the (20-32)/(11-14) Hz ratio below threshold with a second sound to signal whenever EMG was higher than desired. (It went silent when she relaxed.) It had become clear that Joan's 20-22 Hz activity increased when she became anxious about her performance. When she thought about the people watching her or worried about her score or next shot, the amplitude of 26-31 Hz rose markedly and 13-15 Hz decreased. Therefore, the main sound feedback was contingent on her keeping the ratio 20-32/11-14 Hz below threshold. The secondary sound feedback was based on her keeping a chosen muscle group (usually: forearm, shoulder or forehead) below threshold. The trainer would sometimes call out the reading of variables such as EMG if they went outside of the desirable range as part of the coaching.

Our new screens all show HRV in a line-graph display by putting a heart rate (red line) and respiration rate (blue line) at the bottom of the display screen instead of the pure respiration graph.

Physiological and Brain State Changes Occur in a Sequence

Joan would mentally prepare to putt in the following sequence: She stands relaxing her face, shoulders and arms – respiration goes to 6 BrPM, trapezius and forearm EMG go as low as 0.8 μv, 20-32 Hz and 45-

58 Hz (not shown) falls. Beta at 17 Hz increases as she judges the distance, grass conditions, wind, slope and angle. At that point she bends to the correct position to putt with her eyes vertically over the ball, arms hanging totally relaxed – EMG at 0.9 to 1.1 μv, and SMR rises. When she is totally mentally relaxed but focused, she grips her putter a little more firmly and visualizes the putt: 7-8 Hz rises briefly. Then she sees only the hole and the path the ball will take. She completely enters a mentally relaxed state at 11-12 Hz, shuts out all other thoughts (20-32 Hz drops and 11-15 Hz rises), and putts during the pause in her breathing after expiration. At the completion of her putt she relaxes and 9 Hz rises. Because there is mental flexibility and a sequence of EEG patterns associated with a successful putt, the record must be reviewed and analyzed over the total time period. Averages of various frequency amplitudes do not tell the story and can be misleading. For example, a raised 7-8 Hz at the wrong time could mean she was internally distracted. Produced at the correct time in the sequence it was good evidence of visualizing the shot.

Joan's Changes with Training
When she increased 11-15 Hz activity and breathed effortlessly and diaphragmatically at a rate of 6 BrPM, she felt calm, relaxed, and aware of all aspects of terrain, wind and distance. When she visualized the shot, activity around 8 Hz increased. When she evaluated the range and wind, 17 Hz activity increased. Joan found that, just as in her high school years, her mind sometimes wandered off what she was doing, even in competition. She, therefore, did additional training to gain control of theta activity.

Joan continued to practice diaphragmatic breathing, which she knew from the stress assessment screen is associated with relaxed muscles, warm hands and a calm mental state. However, breathing was important for another reason. She had been hitting her putts at different points during the breathing cycle without realizing it. When she learned to putt always at the same time in her breathing cycle (after expiration, in a rest phase), her putts became more consistent.

Professional athletes cannot afford to be anxious, worried or ruminating; hence the emphasis on decreasing 20-32 Hz activity in the above screen. Note also that scores are on the screen so that she can evaluate performance. There are percent-over-threshold scores for both fast and slow waves. Scores may also be used for the slow/fast-wave ratio.

In the display screen shown above, the left EMG is on her leading forearm extensors and the right EMG is on her left trapezius. She found, initially, that her forearm and shoulders were quite tense as she addressed the ball. When she learned to relax and correct her posture, she discovered that she required a new putter with a shorter shaft. The length and angle of her old putter were incorrect for her height with arms relaxed. She learned to control muscle tension. Twenty training sessions resulted in a significant improvement in Joan's game.

As Joan progressed we wrote out the steps with her as follows.

Review of a Sequence of Mind-Body Changes for Optimal Performance in Golf

1st Stand 3 feet away from the ball, relaxing your face, shoulders and arms – *respiration should go to 6 BrPM, trapezius and forearm EMG to as low as 0.8 μv, 20-32 Hz and 45-58 Hz (not shown) should fall.*

2nd *Beta at about 17 Hz will increase* as you judge the distance, grass conditions, wind, slope and angle.

3rd Now you will step up to the ball and bend to the correct position to putt with your eyes vertically over the ball, arms hanging totally relaxed – *EMG low (around 0.9 to 1.1 μv), as SMR rises.*

4th When you are totally mentally relaxed but focused, you will grip the putter a little more firmly and **visualize** the putt: *7-8 Hz will rise briefly.*

5th At this point you are to see only the hole and the course the ball will take. You will completely enter a mentally relaxed state at 11-12 Hz and you will have shut out all other thoughts or awareness of noise and people around you (20-32 Hz will drop and 11-15 Hz will rise). At this point you will exhale (breathing has to be diaphragmatic at about 6 BrPM) and **you will putt.**

6th At the completion of your putt you will not alter this calm mental state. For two to four seconds you will mentally follow through and calmly sense the ball going into the hole. At this point 9 Hz may rise.
At the end of this process the record is reviewed and analyzed over the time period to see if any of the stages require further practice.

The trainer could distinguish good from poor performance and predict whether the shot would go into the cup, with his back to Joan, by watching the display screen. If he saw the correct sequence of brain wave, muscle, and breathing changes, he knew it would be a successful shot.

A Racing Car Driver

Deciding on NFB and BFB Feedback Parameters

Dave, a professional racing car driver, received training that was a variation on Joan's. Dave had had ADHD symptoms in school and fit the Hunter Mind style described so well by Thom Hartmann. He was excellent in competition because he entered a state of hyperfocus. His mind wandered, however, when he was alone on the track during the qualifying laps that determined pole position. His goal was to better manage his attention when there was no other car to chase and to improve his qualifying times. With Dave we had to use mental imagery since it was not feasible to physically simulate the race course. As with the golf pro, the goal was mental flexibility and control of physiological measures reflecting alertness, muscle tension, and breathing. The first emphasis in training was to be fully externally focused and problem solving even when the situation was less stimulating. He started his training by decreasing and controlling 4-9 Hz activity while keeping 17-18 Hz high. Then he practiced remaining highly alert (EDR) and relaxed (diaphragmatic breathing at 6 BrPM with forehead EMG below 1.5 µv) while he imagined scanning the track and his instrument panel (11-15 Hz increased).

An Initial Display Screen
Racing Car Driver's Training Screen

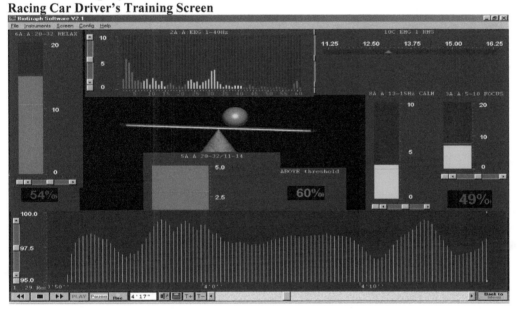

A Business Executive

Goal: Optimize Performance of a Company Executive

Harry was a senior vice president of a large corporation. Many executives lack the flexibility to be in different mental states depending on where they are and what they are doing. They tend to stay in high gear. This high intensity, take-charge-rapidly state is advantageous in some work situations. In other situations, however, they would do better with a calm, receptive mental state. In this receptive state they would appear to have nothing else happening in the world other than paying attention to the person to whom they are listening. When they arrive home to greet their wives and children, they need a different state again. They need to be attentive to their children's interests and participate actively in family life in a relaxed way without thinking about business.

Training emphasizes awareness of different emotional, cognitive, and physiological states. The executive is trained, using NFB + BFB + metacognitive strategies, to adapt to various work and home situations. They learn to rapidly switch mental/physiological states.

Harry's QEEG during a math task, demonstrated very high 21 Hz at Fz and Cz, and very high 11 Hz activity at T5. High 3 Hz and 9 Hz activity and a dip in 13-15 Hz activity at Cz was also noted. In a stress test, during a math task, he demonstrated shallow, irregular, fast respiration with no synchrony with heart rate, raised pulse rate, high forehead and trapezius EMG activity.

The figure below demonstrates an eyes-closed comparison with the SKIL database. It shows at 1.5 SD above the database mean, bright red, high-amplitude, 11 Hz alpha activity at T5 and 21 Hz, beta activity at Fz.

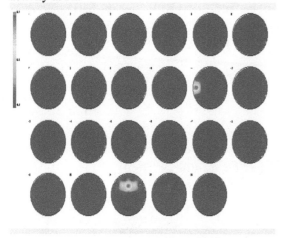

Display Screens and Training Parameters

The display screens and training parameters are determined for each individual according to their profile on EEG and psychophysiological stress assessments, their responses to lifestyle questionnaires (designed by Vietta Wilson, Ph.D., professor, York University, Toronto), and their goals regarding their work and home situations.

Important Caution: Nothing stated below should be considered a 'protocol.' The electrode placements and bandwidths corresponding to mental states are always **hypotheses or assumptions** that must be checked out with every client **before** you decide how to proceed. You check out electrode placements by looking at QEEG data, as previously explained. You check out the correspondence of mental states with bandwidths at each of the chosen sites by asking the client to consciously sense differences in their mental state when they raise versus when they lower the amplitude of the chosen bandwidth.

Training Screens

For feedback, Harry requested that there be a lot of information on any display screen that we used. He would decide on his point of focus from moment to moment and ignore the rest.

Screen #1: Initial Screen

This is a typical screen used for two 3- to 5-minute periods at the beginning and end of each session. Harry was encouraged to attempt to control a number of variables simultaneously and have numerical values (percent of time over threshold) to compare with previous sessions. The bandwidths used are unique to each individual and depend on his initial assessment.

Raising 13-15 Hz controls rolling the ball to the right. "70%" represents the percent of time over threshold for this 13-15 Hz activity. It is colored green to signal that the objective is to increase it.

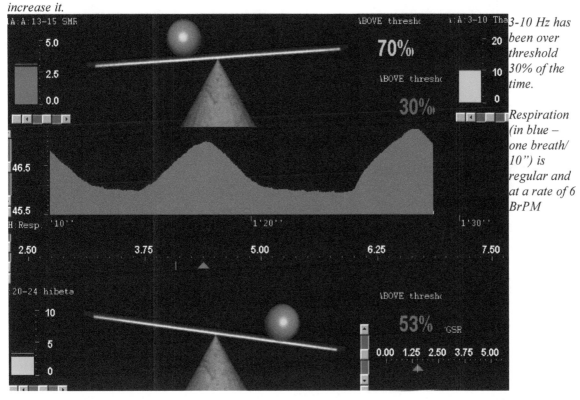

3-10 Hz has been over threshold 30% of the time.

Respiration (in blue – one breath/10") is regular and at a rate of 6 BrPM

Lowering 20-24 Hz controls rolling the ball to the right. "53%" represents the percent of time over threshold for this 20-24 Hz activity. It is colored red to signal that the objective is to decrease it.

Forehead EMG is at 4 µv.

We add points later. *Points are dependent on decreasing variability. The client must hold the required state for three seconds to receive a point.* **Decreasing variability** *is an* **extremely powerful feedback technique** *to improve self-regulation capabilities.*

EDR at 1.3

Note: Lowering 20-24 Hz. would not be a common exercise, but it matched this particular individual's profile. Most of the other executives had high 27-32 Hz activity that corresponded to worrying and ruminating. In those cases, the lower bar would encourage the client to find a relaxed mental state without worries and ruminating that corresponded to a decrease in 27-32 Hz activity. We quickly learned that his high 21 Hz activity was related to extremely emotionally and cognitively intense mental activity that was necessary for his rapid analysis of complex corporation problems. It did not relate to anxiety. This was important for him in business, but this same mental state was ruining his marriage and was inappropriate for some business activities. This screen was therefore DISCARDED for this particular executive.

This is an important lesson. You must know what the assessment shows but you do not *normalize* the EEG until you learn what a particular deviation from the database represents for that client.

Screen # 2: Control 9-10 Hz

The objective is to be able to consciously sense activity in this range. This executive found that the activity in this range increased, and the boat disappeared, when he was internally distracted. When he focused externally and was figuring out a problem, the activity decreased markedly and the boat reappeared. He became capable of doing this at will.

The sailboat jigsaw puzzle is fully formed when 9-10 Hz activity is low (as in this display).

The spectrum shows where maximum activity is in the frequency range from 6 to 24 Hz. Harry was attempting to increase the 11-13 Hz bar graph, next to the spectrum, while lowering the 9-10 Hz bar graph beside, and linked to, the sailboat. He found that this would occur when he had a relaxed, calm, external, open awareness.

With training, he was able to consciously turn this state off and on. If he raised the 24-36 Hz bar graph in the lower right corner, it would go red, and all feedback would be inhibited. This controlled both for EMG artifact and ruminations.

Screen # 3: Control 11-15 Hz

In this step, the executive attempts to find a mental state that would raise 11-15 Hz, and then to find a different mental state that would lower 11-15 Hz.

The objective was to be able to sense both states and be able to control them consciously. Harry found that he could build the sailboat puzzle (increase 11-15 Hz) by breathing at six breaths per minute, relaxing his muscles, warming his hands, and mentally going into a calm, yet alert, open awareness state.

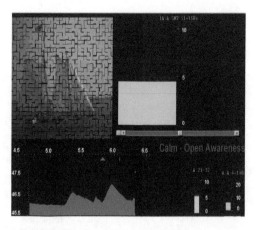

Recall that Harry had demonstrated extremely high Fz, 21 Hz and T5, 11 Hz activity.

Immediately prior to this point, Harry had been breathing in a tense, shallow, irregular manner (blue respiration line graph), and his forehead EMG had been very high (>7 μv). At this moment his forehead EMG was decreasing. (Note that the red indicator is now below the automatic threshold, which is at 6.2 μv.) The 11-15 Hz activity has risen well above threshold, and the sailboat (which had been totally black and gone from view) is now rebuilding. Harry became able, with training, to turn this state off and on. A bar graph placed on the bottom right of the screen is for the EEG inhibits on 23-32 Hz activity.

Screen # 4: Control 21 Hz
20-22 Hz

- The spectrum shows activity between 10 and 26 Hz. In the lower left corner is a respiration line graph.

- Forehead EMG activity is high at 3.5 μv. He was capable of decreasing it to 1.5 μv.

- 24-36 Hz activity (blue in the bar graph on the lower right corner) is below threshold.

- 12-15 Hz activity (red in the bar graph on the lower right corner) is too low at this juncture.
- 20-22 Hz activity is high. Harry said that emotional intensity raised this area. As previously noted, he felt this intensity was valuable when chairing meetings. On the other hand, it was not appropriate in other situations at work and it was ruining his home life. In training **he learned to turn it on and off.**

Screen # 5: The Mind Mirror Screen with EMG and Respiration

This screen was developed because this executive had *dissociated* in his first session when he had worked too intensely on decreasing high beta and increasing 11-15 Hz. Other factors may have contributed to this, such as not eating anything all day (it was 1:30 PM) and having a viral infection. Nevertheless, it was felt that a wise precaution would be to activate the left frontal lobe with Channel A at F3 and use Channel B at C4 for 13-15 SMR or at Pz to raise high alpha, 11-13 Hz.

(When the EEG was carefully reviewed it was noted that 9 Hz had risen near the end of the session and this may reflect the dissociation.)

This screen is complex, but it may be used for training with some individuals. The specific bandwidths would be modified according to the EEG assessment, but the principles are the same for most individuals:

The client moves in steps. These are summarized in the Sequence for Success chart below. The first step is to control physiological variables, respiration and EMG. To do this the client will:

First, relax: Respiration diaphragmatic and regular at about 6 BrPM. Respiration is chosen because, for most executives, changes in respiration correspond closely to changes in other biofeedback modalities including the EMG, EDR, peripheral skin temperature and heart rate measurements, as previously explained. These other modalities may be used on other screens.

Second, attain an alert, calm, focused state that can move between thoughtful reflection and memory retrieval with visualization (6-7 Hz) and high alpha (11-12 Hz) and calm (13-15 Hz) problem solving, focused concentration (16-18 Hz) without undue anxiety (21 Hz). (Some anxiety is a positive driving force, but too much may almost paralyze the system with the complaint of *going blank* or being confused or 'disrupted.') The client must do this without unproductive ruminating (23-35 Hz). Increased *binding rhythm* or *Sheer rhythm* (38-42 Hz) may be helpful (though muscle artifact makes it a difficult rhythm to accurately assess and reward).

The bandwidths on the following screen are altered according to the goal for the executive.

The "Make a Christmas Tree" Screen
Respiration (blue line graph)
'Mind Mirror'

Channel A
Left Hemisphere

Channel B
Right Hemisphere

45-56 Hz inhibit.

Increase 16-18 Hz activity (may alternate with 14-18 Hz)

Decrease 3-5 Hz activity. – the specific frequency band depends on where the individual client is found to tune out.)

Decrease 23-36 Hz (May alternate with decreasing 19-22 Hz (intense emotional state and/or anxiety.)

Increase 13-15 Hz at C4 (But may alternate with increasing 11-14 Hz at Pz for a calm open awareness state.)

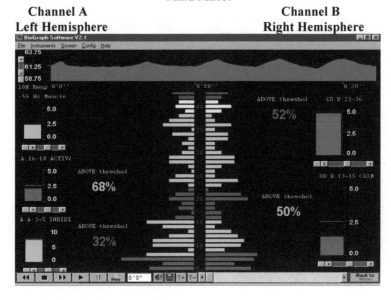

BRING   Red % ↓   Green or Blue % ↑

At this juncture, Harry, our executive, had learned how different frequency bands corresponded to different mental states that he experienced. He wanted to be more active and see if he could control these states in a sequence that might correspond to his activities outside of the sessions. Like many executives he played golf, so putting during the session was both fun and meaningful. More than that, putting could be directly related to the sequence we had recommended to reduce stress and tension at home after work.

The following table shows the sequence of mental steps done with Joan, the professional golfer described previously. The last column shows how this same sequence for an executive might correspond to going home after work. The reader can see how it might apply also to the sequence the executive might go through just before the start of a business meeting. It is a sequence of steps that can be practiced outside of the biofeedback office to help generalize the executive's ability to enter different mental states.

Sequence for Success

	GOAL	*ATHLETE* ACTION	PHYSIOLOGY	EEG	*EXECUTIVE* Coming HOME
1.	Relax (IPS)	Stand 3' away from the hole. Relax forehead, face, shoulders	6 BrPM, RSA EMG ↓ Skin temp. □	45-58 Hz □	Park the car. Talk to yourself, say: "Stop," "Breathe," "Act." Relax muscles and do diaphragmatic breathing at about 6 BrPM.
2.	Solve problem	Judge the distance, grass conditions, wind, slope and angle	Retain relaxed state	17 Hz	Fill out day-planner and SMIRB
3.	a.) Empty-your-mind b.) Lose all anxiety	Step to ball <u>then</u> bend, hang arms loosely, mentally have nothing in the world but ball, green and hole	Totally relaxed – loose, calm, EMG □, Skin temp. □	13-15 SMR □, 24-32 Hz □ 20-23 Hz □	Think of being pleasant with the family and feeling good.
4.	Visualize or 'feel' the next step you will do	See or feel the ball going into hole. Keep eyes open	Remain relaxed. Any movement or tension is seen in EMG	7 Hz □	Visualize yourself approaching family with warmth, openness and then positive interaction with kids
5.	Focus	Shut out all other thoughts or awareness of noise and people Become openly aware of the Gestalt: hole, green, ball – soft eyes.	Exhale and putt	11-13 Hz □ SMR □ Thalpha □ 20-32 Hz □ and 'binding' rhythm 39-42 Hz □	Go into house and focus 100% on smiling with a relaxed posture; <u>then</u> listen, listen, listen and be 100% positive
6.	Follow through	Follow ball in same calm, openly aware state	Remain relaxed with 6 BrPM	9-11 Hz □ (possibly *Post Reinforcement Synchronization*)	Active play (not passive TV or video) until the kids are asleep.

In the above, single-channel feedback is often at Fz or Cz. Two-channel feedback may be at F3 or Fz for the high beta and thalpha feedback, combined with Pz for 11-13 Hz feedback.

In the feedback screen above, the client forms a 'Christmas tree' by decreasing 3-10 Hz, which makes a thin trunk to the tree. Then the largest lower branches reach out at 11-13 Hz and gradually decrease in amplitude through to 18 Hz, becoming narrow above 20 Hz.

Once Harry was able to consciously regulate his psychophysiological profile, it was just a matter of generalizing this ability to his home and work environments. He was asked to list his daily habits. The list included: getting out of bed, showering, eating breakfast, brushing his teeth, driving to work, answering the telephone and so on. None of these activities required high emotional intensity. Therefore he would take two deep diaphragmatic breaths while relaxing his forehead and feeling his hands warm while he entered a mental state where he was aware of most things around him, relaxed, calm and still very alert. He would only enter the high intensity state for certain meetings and for studying materials when this seemed to assist his efficiency and his memory. Harry was now beginning to control his life with less stress, which had been his objective when he entered the training program. Nothing is simple, however. This takes time and other interventions such as family therapy may also be necessary.

Can the Client Have Any Untoward Experiences?

This possibility is not called a side effect, because the experiences do not appear to be caused solely and directly by neurofeedback. However, when you work with extremely bright, intense, fast-moving executives who have 'Type A' personalities, then occasionally one of them may put himself into the initial session with such intensity that he feels 'different' later that day. The previously described executive is a good example of a tense client who only focused on bringing high beta down and SMR, 13-15 Hz, up. An 'overdose' of 13-15 Hz without activation (beta) of the left hemisphere may make one feel a little 'spacey.' The result can be a single session *after-effect* where the individual has practiced, not being externally alert and problem solving, but going into a state of relaxation and practicing a type of meditative total awareness. In addition, if in this effort they ignore the inhibit on 9-10 Hz and raise instead of lower their alpha, they may dissociate. The temporary lowering of their peak alpha frequency, from 11-12 Hz or higher to 10-11 Hz or lower, may result in some loss of semantic retrieval. Remember, this could be for a number of reasons, one being that memory retrieval is state dependent. These effects will wear off in a few hours.

Example: As previously described, in his first session, one senior executive insisted that no one talk while he worked to become relaxed and decrease 20-28 Hz. On assessment he had a high peak at 21 Hz. He wanted to work at lowering this. and did so for 1½ hours (longer than the usual 50-minute session because he was determined to 'get-it'). He preferred trying it alone so he could really concentrate, so the trainer for the most part refrained from coaching or commenting. The client stared at the screen and really worked at decreasing the high beta and increasing 11-15 Hz. Although there was a bar for 6-10 Hz on the screen, he reported later that he did not pay attention to that. After 90 minutes he suddenly looked at his watch, realized he had to be chairing a meeting of senior staff, had the sensors removed and raced out to the meeting. It was learned later that he had not had either breakfast or lunch and it was 1:30 in the afternoon when he finished the session. His administrative assistant, a very observant woman, noted that he did not appear his usual self. He had trouble chairing the meeting in his usual efficient fashion and was a bit 'spaced out.' He described feeling intense, perhaps anxious, later in the afternoon and said he was a bit "out of it." In some ways it was like waking up with a mild hangover with a slight dull headache. Reviewing his reaction at the next session, we thought it possible that he had hypnotized himself and perhaps he had experienced a dissociative state. (A similar response to a brief neurofeedback demonstration was described by Dan Chartier, personal communication, and later noted in the SNR journal.) The executive said later that he had gone into the session believing this was a very powerful tool and he was going to change quickly. This was despite being told it was a learning phenomenon that would take time, and being given instructions to relax and 'let it be' and to spend the first session working on breathing and muscle relaxation.

*In his second session, the trainer sat with him and discussed his experience during the session. The **emphasis was placed on control**, going into and out of mental states that changed the variables as described above under screens 1 to 4. There was also a discussion of hypoglycemia and the importance of breakfast and regular meals. At the end of the second session, his assistant brought him lunch. He ate this while discussing with his trainer ways to generalize relaxation techniques in appropriate situations. There was no repeat of the 'spaced out' feeling after subsequent training sessions.*

A second executive had a very mild but similar experience. However, it was a week later that he had his first session, and he had discussed the first executive's experiences with him. He did not have much time for his first session, and after instructions and discussion of his objectives, he only had a sample of about five minutes with feedback. He was in perfect shape when he left the session. It was later in the day that he reported bloodshot eyes, headache and a 'spaced out' feeling. This raises a remote possibility that perhaps both workers had a mild viral infection. We feel certain that diaphragmatic breathing and muscle relaxation, which was all we had time for, did not cause his symptoms. It is of interest that he knew of the first executive's symptoms, so one wonders if there was an unconscious suggestive factor involved.

Side Effects – An Overview

In thousands of sessions over the past decade we have not had other clients have any changes they would call negative side effects. Very occasionally, if you ask your adult client, they may say that after the first or second session, they felt a little different or 'spacey' for a few minutes. But this rare phenomenon seems to occur, if at all, only after the first session or very early in training and not after that. Nevertheless, the example given was a valuable learning experience. If it is due to the intensity with which the person applied themselves to the feedback, it is worth noting and is easily avoided.

Sometimes you will hear from some practitioners how powerful a few minutes of feedback can be. In our Centre this has not been our experience, possibly because we avoid using verbal or nonverbal suggestions about what they might feel. Anyone experienced in the psychotherapies knows how powerful (and yet temporary) suggestion can be. Some people are much more suggestible than others. People with ADHD, for example, have higher hypnotizability (Wickramasekara, personal communication). Neurofeedback and biofeedback are learning experiences. Learning usually takes time and reinforcement. The amount of time and repetition will be different with different people, but instant, dramatic change, or change within a few minutes, is not likely to last.

Since the foregoing example is a rare, but possible, phenomenon, it is a good idea to have a set training time and to have time to chat with your client for a few minutes at the end as you are removing sensors to ensure that they are not in a spacey state of mind. It is also important that your clients follow good nutritional habits and sleep habits.

Disorders Involving Muscle Spindles

Variables in Common to Movement Disorders and Fibromyalgia

Movement disorders and disorders such as fibromyalgia involve muscle spindles. Muscle spindle activity is influenced by SMR and by sympathetic drive. It may thus be that increasing SMR and decreasing sympathetic drive will have benefit in terms of symptom relief in a number of disorders. Results in clinical settings are preliminary but promising. Movement disorders such as dystonia, Parkinson's and Tourette have a number of variables in common. A few of these variables are as follows:

- Dystonia, Parkinson's, and Tourette all involve involuntary muscle movements. In every case the basal ganglia (caudate, putamen and globus pallidus) are involved.

- Interactions between the basal ganglia are probably responsible for smooth, stable movements.

- Damage to the globus pallidus or the ventral thalamus may cause deficiency of movement (akinesia). Thus, interference with the globus pallidus inhibition activity and with the ventral thalamus functioning appear to result in excitation. The caudate and the putamen have inhibitory functions (Carlson, 1986, p 313). Different combinations of malfunctioning in these nuclei appear to be important in movement disorders.

- In each of these movement disorders an imbalance of neurotransmitters is postulated to be at the core of the difficulty, and in each case, increases and/or decreases in dopamine are implicated.

- In each condition control of muscle spindle activity may be important. This control involves both gamma motor efferent input to the spindles from the red nucleus and sympathetic nervous system control.

Working with ADHD one also sees clients who demonstrate the symptoms of Tourette syndrome since the majority of those with Tourette also have problems with attention. In our experience, the training program described below in the example of Mary has been **just as successful for clients affected with Tourette** as it was for this case of dystonia and Parkinson's. (See also Section XXVIII.)

Theoretically, the same combination of BFB with NFB should decrease the negative activity of trigger points in **fibromyalgia** (Thompson, 2003). Work with fibromyalgia patients has been pioneered by Stuart Donaldson in Calgary, Alberta.

In addition, the calming result seen with increased SMR and decreased high beta (24-32 Hz) combined with decreasing sympathetic drive should be helpful in patients with **hypertension** and in patients who have severe **headaches** (Basmajian, 1989).

Parkinson's with Dystonia, a Case Example

Mary, Age 47, Parkinson's and Dystonia

Mary's example also appears as Case #5 in the case examples. *Her case has been reported in the* Journal of Neurotherapy *(Thompson & Thompson, 2002). To recap, Mary, age 47, has both Parkinson's disease and severe dystonia. She has been treated with multiple medications, has even taken part in experimental surgery. Prior to doing NFB and BFB, walking was extremely difficult and she was unable to control either dyskinesia (the sudden onset of gross motor movements) or the inability to move (freezing) when standing or even sometimes being unable to get up when lying down.*

Assessment and Deciding on NFB and BFB Feedback Parameters

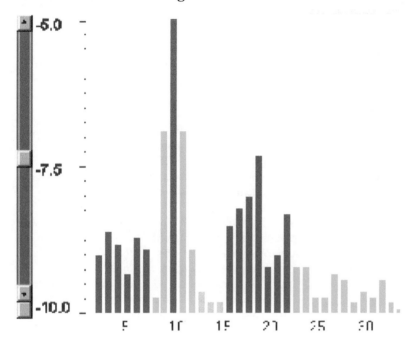

A typical one-second sample above

EEG assessments at Cz referenced to the left ear and bipolar assessments, Fz-Pz, showed very low SMR (13 and 14 Hz) activity and high alpha (9 and 10 Hz) activity. This finding was consistent in all conditions, eyes closed, eyes open, and doing a task.

When averaged over three minutes, there was also a peak of activity at 27-29 Hz.

In addition, a 'stress-test' looking at traditional biofeedback measurements demonstrated low peripheral temperature, high EDR, shallow, rapid, irregular breathing with no synchrony between breathing and heart rate variability (RSA)

Medications helped Mary, but not all symptoms were under control. Quality of life also suffered due to poor concentration, so she had trouble completing tasks, such as reading a novel. Increases and decreases in dopamine may be part of a biochemical understanding of the problem and of a medical approach to it, but they do not answer the important question of how to control her sudden, precipitous onset of gross movements. In dystonia, a noradrenergic predominance has been proposed (Guberman, 1994). It stands to reason that our traditional drug approach is necessary but not sufficient. In Mary's case, one obvious factor in precipitating undesirable events is stress. Increased sympathetic drive that occurs with stress is a factor. Noradrenalin is also a factor. One iatrogenic precipitator of a major dystonic episode is the injection of even a very small dose of adrenalin, as might occur in a dental procedure.

If an increase in SMR is associated with decreasing spinothalamic tract activity and a decrease in the firing rate of the red nucleus (whose activity links the gamma motor system to the muscle spindles), then it would appear sensible to see if raising this rhythm would correspond to a decrease in undesirable muscle tension, jerks and spasms. Raising this rhythm has also been observed to be associated with a feeling of *calm*. Muscle spindles are involved in reflexes and in governing muscle tone. They are innervated by the motor/sensory pathways to the spinal cord, the red nucleus, and basal ganglia, and by the sympathetic nervous system. We might have some constructive influence on the first half of this innervation by raising SMR. Sterman demonstrated that the red nucleus stopped firing when the thalamus started to fire in association with the production of SMR rhythm. Training Mary to relax would theoretically affect the second (sympathetic) input to the muscle spindle in a helpful manner.

NFB + BFB Training

We trained Mary to breathe diaphragmatically at six breaths per minute. At this rate, breathing became synchronous with heart rate variability. She then associated (paired) this breathing style with increasing her SMR using NFB.

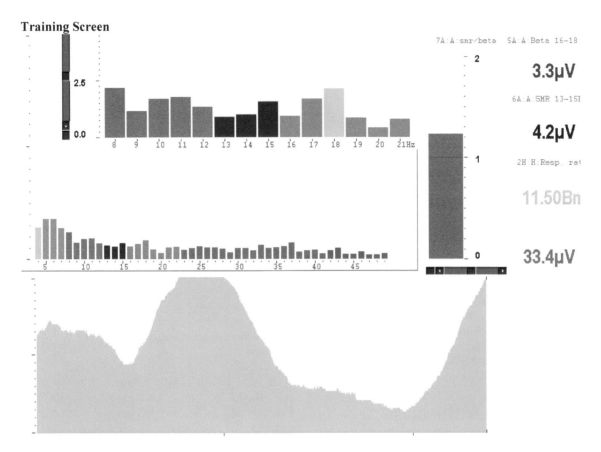

Job number one is to pay attention to respiration shown at the bottom of the screen. At the moment in time reflected here, she must relax and decrease the breathing rate from 11.5 BrPM (the number appears on the right-hand side) to around 6 BrPM. When she has accomplished this, she will then continue to breathe in this relaxed manner and focus on increasing 13-15 Hz and keeping it above 16-18 Hz. Given her initial extremely low 13-15 Hz, this was a difficult task. However, within 25 sessions she had control of this variable. The double spectrum allowed her to see the whole picture and assure herself that she was controlling the effects of EMG on higher frequencies and that she was maintaining a low level of high-beta activity in the 20-32 Hz range. The higher magnification spectrum from 8 to 21 Hz allowed her to focus on her primary goal of increasing SMR. Her goal was to move the green bar (representing the maximum amplitude at that instant in the 8-21 Hz range shown) to 14 Hz. Gradually she was able to do this for an increasing percentage of the time. In the figure it is at 18 Hz, and she is attempting to find a mental state where it will move to 14 Hz. She was there in the frame before this, slipped out and then got back into it in the next frame. (*Note: the spectrum feedback is faster than the figures showing μV levels on the right because it does not require averaging over a one second time period.*)

Mary's Changes with Training

After 40 sessions of training, Mary was able to bring both 'freezing' and gross unwanted dystonic movements under her voluntary control. She did this by focusing on her breathing and putting herself into the same calm state that was associated with increased SMR during training.

Daytime alertness had been a problem for Mary, and this symptom improved as excess alpha (9 and 10 Hz) came a little more under her control. Mary was formerly an avid reader, but had not been able to sustain her focus in order to read a novel for four or five years. Controlling her dominant slow-wave activity has resulted in her being able to concentrate well enough to finish a book and enjoy reading again.

Mary's quality of life was improved by training and she was even able to reduce the amount of L-dopa medication. Long-term effects are not known (follow-up has been three years at this point). There is no pretense that one is treating the underlying disease process. But, symptom management is, in and of itself, a desirable outcome. It is a joy to see this woman doing crafts again, and, she had a book published that she wrote and illustrated.

There has been a case report of a boy with Tourette being treated using a similar approach (Daley, 2001). By raising SMR and teaching relaxed, diaphragmatic breathing, the boy's tics disappeared over the course of a few months. We have also successfully treated a number of patients who had Tourette syndrome (Section XXVIII).

Biofeedback + Two-Channel Neurofeedback

Rationale

With adult clients, one is often faced with complex objectives that require multiple feedbacks. The client's objectives may correspond to EEG differences in different areas of the two hemispheres. In these cases, two-channel training becomes important. To achieve good results quickly requires that the client be able to relax and yet remain alert. This is best achieved by combining biofeedback with neurofeedback. David is an example of such a client.

Case Example – A Business Executive

Deciding on NFB and BFB Feedback Parameters

David is a brilliant young executive who has developed a very dynamic company. He is doing well in business and family life. Unfortunately, he has a bipolar disorder that is unresponsive to medication, and the cycles do affect his functioning. As with most persons who have periodic mood swings, depressive affect underlies both phases of the disorder. His objective in coming to the center was to gain a greater degree of self-regulation. He ruminated about things constantly. He was impulsive. He had always been easily distracted except when he was developing something himself. He wanted to be able to relax and focus on his business and home life without always thinking about immediate problems at work. He wanted to be able to remain alert even when the situation was boring.

His stress test demonstrated rapid, irregular, shallow, thoracic breathing, tense chest, neck and forehead musculature, and a tendency to fall asleep if the session became the least bit boring for him. The two-channel EEG assessment in the F4 right frontal channel and F3 left frontal channel both showed very high amplitude high-beta activity with peaks at 22-24 Hz and 27-29 Hz. The beta activity was slightly higher in amplitude on the right side. High-amplitude slow-wave activity with a peak at 9 Hz was observed in the left hemisphere. When the second channel was put at Cz and then at C4 during the assessment, a dip was seen in the SMR frequency range at 13-14 Hz (compared to 11-12 Hz and to 15-16 Hz) and 17-18 Hz was also low. High amplitude high-beta and thalpha activity was not present at the central sites. These findings correspond to literature findings of low activation (high alpha) in the left frontal lobe in depression.

The display screen shown below was designed to allow David to follow his respiratory rate and forehead EMG as he attempted to relax. Relaxation increased his tendency to become bored and fall asleep. The EDR monitor on the screen forced him to find a way to remain alert. Sitting straight with good posture and on the edge of the chair, not leaning against the back, in addition to being aware of his arousal level with feedback on the screen, helped him to remain alert. His EDR rose from 1.8 in the initial two sessions to ranging between 7 and 9 after a few sessions. Learning to breathe diaphragmatically and relax his forehead EMG (from 15 to <2 µV as training proceeded) took longer. He had difficulty learning to empty his mind of all the worries and insolvable problems in his life. Success in this was reflected in being able to bring down the high amplitude 22-32 Hz activity in Channel A and increase SMR in Channel B. As he succeeded in this self-regulation, his thalpha activity in the left frontal area decreased and he felt subjectively brighter and less dysphoric.

An Initial Display Screen

Example of a display screen at the end of a five-minute segment of training.

Channel A at F3
Thalpha 5-10 Hz →
(Green below threshold).
The red 45% shows that it has been below threshold 45% of the cumulative time.

Channel B at C4
SMR 13-15 Hz →

It has been over threshold 53% of the time.

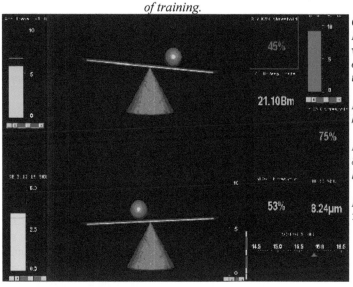

Channel A @ F3
High-beta 22-32 Hz is very high. It has been over threshold 75% of the cumulative time.

His breathing rate is high at 21 BrPM.

His EDR has risen to 8.24 μm and he has become more alert.

Forehead EMG is high, >15 μv.

The client, David, attempts to roll both balls to the right and hold them there. To accomplish this he must decrease slow wave activity in the left frontal area (appropriate activation) and increase SMR at C4 (a calm mental state).

David's Changes with Training

David feels that his work performance has improved and that he is better able to focus on his family life when away from the office. As with the Parkinson's client, one does **not** cure a bipolar disorder, but the ability to self regulate does give more control over the symptoms and improve quality of life.

Tracking Client Progress

Why Track Progress?

Students are more highly motivated when they are challenged to improve and can track objective data to demonstrate the changes they are making. Within a session they can compare a few minutes of data with preceding time periods of the same length to see how they are improving. One also wants to see if there has been a demonstrable shift over a number of training sessions.

Within-Session Tracking

Types of Measures – Points versus Percent-Over-Threshold

During each session we suggest that you record the student's scores every few minutes. With children this is usually 3- to 5-minute time frames, though with very young or very hyperactive clients it may be shorter. We use a *Tracking Client Progress* sheet and may have the student fill in their own scores. This can be done by hand or automatically by pasting the statistics into an Excel tracking sheet. We reward children for beating their last scores or their last day's scores. On some instruments, such as the Autogenics A620 or the Thought Technology Infiniti, a chart with statistics will appear on the screen after each time period. It is important to note that μV average, a point score and a percent-over-threshold score are all different types of calculations. Thus one could conceivably improve their percent-over-threshold score and yet get no points. They could improve their percent-over threshold, get the same points as last time and yet not obtain as good a μV score for a particular parameter (fast or slow wave). The reason for this is that these variables measure slightly different things.

Percent-over-threshold depends on where you set the threshold. If it was for theta, then the student could be just below threshold for much of a five-minute sample, and thus get a reasonable percent-below–threshold score, but seldom stay below threshold for the required time (0.5 seconds or 1, 2 or 3 seconds depending on your settings) to score a point. Thus the **point score** would be very low. The average µV level could be the same in two 3-minute samples, but the student might score a higher number of points in the second sample just because he held theta below the threshold for more than one second more times (more often) on his second try. Higher point scores are associated with better performance in terms of less *variability* in slow-wave activity.

Test Ability to Hold Mental State Without Feedback

When the student has reasonably good control of his brain wave pattern, challenge him to find and hold the same mental state without feedback. You turn off the sound and the monitor and run the program for the same time period (3 or 5 minutes) recording data but not giving feedback. Compare the scores with those obtained with the feedback turned on. This is excellent practice for transferring what has been learned in sessions to home, school, and work. Just as you cannot tell a person how to balance but they can learn what it feels like through feedback from the vestibular system, so you cannot tell someone how to achieve a particular mental state. They must learn what it feels like through neurofeedback.

Obtain Consistent and Accurate Data

To obtain consistent data within each session, we suggest, first, that you record for the same time period each time and under similar conditions, with feedback alone and then on task. We suggested above that this be between three and five minutes for children. Adults may want to record for longer periods. One reason we use the shorter times for children is that it becomes increasingly difficult to shift a percent-over-threshold score as time progresses. Since it is calculated over the total time, it changes extremely quickly in the first few seconds and then slows down.

Pause the program whenever your client wants to move or to talk. Most programs have some simple method that allows a pause. With mature clients you can support their feeling in control by letting them use the mouse or keyboard to pause the program when they want to rest, stretch, ask a question and so on. These pauses result in the data remaining accurate for whatever time period you are using with minimal muscle and movement artifact. It also, even more importantly, emphasizes to the client that they should only do one thing at one time and they must be in control of what they decide to do: work on a task, or, conversely, fidget.

Comparing Different Sessions – Trend Reports

Some programs have built-in mechanisms for computing a trend report. This allows the practitioner to graph trends over a number of sessions for each of the variables (theta, SMR, beta, ratios). If the equipment does not automatically generate such reports, you could still do it manually if you collect comparable data during part of each session.

Our preferred method is to use the excellent graphing provided by Excel. The statistics pop up at the end of a segment of training and the trainer pastes them into a column on our standard report sheet in Excel. A graph is drawn and the line(s) is(are) extended automatically and the ratios are calculated and placed in a list.

Compare Similar Conditions

Remember to compare data that has been collected in the same way *or* to identify and compare different 'conditions' using a graph. 'Collected in the same way' means that the impedances were low between electrodes in each session, that the *active* and *reference* electrodes were on the same sites, and that the client was in the same condition (watching the display screen, reading or doing math). 'Comparing different conditions' is also possible if you have chosen to do a graph and compare each frequency or frequency band (3Hz to 60 Hz) for the two or more conditions. Trends are important, so you should try to ensure that you clearly record the following information for each segment of data saved: the site, reference, time of day, activity (condition) and whether the data was artifacted. On the A620 one generally uses a standard progression of conditions, for example, watching the feedback screen, reading, watching the feedback screen and listening. On the Biograph program we use short preprinted forms to write in this information as we save each data segment or we type it into the Excel report. For example: Cz-L, 11A, R, ART means that the active electrode was at Cz, the reference was to the left ear lobe, the recording was done at 11 am, the student was reading and the data was artifacted. Comparing data that was artifacted with data that was not artifacted may be misleading. Eye blinks, for example, might raise theta and give the impression that the student was more tuned out in that session compared to one which had been artifacted. SMR might appear to be higher in the unartifacted session due to muscle activity.

Time-of-Day Effects

Include time of day because this may affect theta amplitudes in particular. Sleep latency studies show that we all produce more theta in the early afternoon. People with ADD often show more theta when asked to do a task, like reading or mental math, than when not challenged with a task. You can check for yourself how time of day and tasks affect your client.

Compare Artifacted Samples

If possible, do one or two artifacted segments each day, perhaps at the beginning and end of the session. This allows for a consistent trend report to be done after a number of sessions have been completed. There can be large differences in artifacted versus non-artifacted samples. Obey the simple rule: Be consistent.

Summary: Areas to Target for Feedback

One-Channel Feedback

Right-Brain Feedback

Feedback to encourage SMR appears to assist the client to stabilize emotions and reduce anxiety and tension, decrease impulsivity and increase calmness as well as the ability to reflect before responding. For this, we usually place our active electrode either centrally or between Cz and C4. To increase social appropriateness in Asperger's syndrome we use an experimental approach, encouraging an increase in low beta, 13-15 Hz, and a decrease in the dominant slow wave at sites identified in the full cap assessment. This has been between P4 and T6 in a number of clients. (P4 and T6 have shown less activation in the Asperger's clients we have seen, but this needs to be replicated in controlled studies before anything definitive can be said.)

Left-Brain Feedback

Focus and Concentration

We place our electrode on or between Cz and C3 on the left side when working primarily on focus, concentration, and problem solving. We encourage a decrease in slow-wave activity, 3 to 10 Hz (the exact range depends on the client's profile) and an increase in beta activity in the 15-18 Hz or 16-20 Hz range.

As previously noted: A note of caution from the late Tom Budzynski is that training up beta frequencies on the left side in children or adolescents with reactive attachment disorders may result in angry behavior.

Dysphoria/Depression

When dysphoria or mild depression is an additional factor, we may place the electrodes more anterior around F3 at a site determined by the assessment QEEG. The active electrode is placed over the frontal or prefrontal area on the left side to *activate* the brain. The client is encouraged to decrease the dominant slow wave and increase low beta, 15-18 Hz. In addition, if there is a peak between 21 Hz and 34 Hz, they will be encouraged to "empty" their mind (of all ruminative activity) and lower this peak frequency. This has been discussed above under Combining NFB and BFB, and also under Two-Channel Training.

Learning Disability – Dyslexia

Language representation is in the dominant (left) hemisphere for almost all right-handed people and for approximately 75 percent of left-handed people (Panu & Wong, 2002). When doing NFB for dyslexia, we may choose a site slightly towards the posterior third of a line drawn between C3 and P3 and somewhat inferior to it for a few sessions, and then do a number of sessions between C3 and F3. In these instances we decrease slow-wave and increase fast-wave (beta) activity. This method is used in certain kinds of learning disabilities, particularly in training clients with reading, writing, and speech problems. The specific difficulties the client is having are considered in terms of what is presently known about the neurophysiology and neuroanatomy for these brain functions. Wernicke's area (posterior superior temporal and inferior parietal) is used for the comprehension of spoken language and for the initiation of a reply or action. Visual stimuli reach Wernicke's area through the angular gyrus. This is important for the comprehension of written language. Wernicke's area is thus involved in *phonemic knowledge* (decoding written word and sound symbol relationships). It is connected to Broca's area by the *arcuate fasciculus*. Broca's area and Wernicke's area are, together, known as the *perisylvian language zone*. Broca's area processes meaning and sends signals to the motor cortex if one is reading aloud.

NFB training to activate these areas is still considered *experimental* at the time of writing as they are based on present knowledge and clinical experience and not on a great deal of controlled, multicenter, published research.

As noted above, for some sessions we may move the electrode more posterior (closer to Wernicke's area) in some speech and language disorders and in dyslexia. For dyslexia, we reference to the right ear, reasoning that the insula, which has been implicated in some forms of dyslexia, is well below the surface of the scalp and we want to try to influence more than just the dipoles between C3 or P3 and the left ear.

Sequential (Bipolar) Training

In general it is recommended that bipolar training be carried out between different areas (frontal, central, parietal, occipital) of the brain within the same hemisphere or along the central strip. It is not usually done transversely across the hemispheres at homologous sites, such as C3 and C4. However, opinions differ. If observations of beneficial changes are seen with a particular protocol, the next step is research to replicate the observations. This requires defining the population and the symptoms that respond. A rationale should be provided as to why such training should work. One needs to address the question of why creating a potential difference in particular frequencies between two sites would affect particular symptoms.

SECTION XIII
Intervention Summary
Basic Stages of Training and Tracking Progress

Four Basic Components of Training Sessions

Goals and Overview

The learning technique based on neurofeedback can be used to normalize the EEG in cases where it differs from established databases OR it can be used to modify the EEG towards patterns that indicate mental states that are associated with improved functioning. The parameters for training can be adjusted according to goals and the area of interest (work, school, relaxation, performance arts, athletic achievement, etc.). In all cases one is trying to improve mental flexibility.

The overall objective is to help the clients manage symptoms and optimize their performance in areas they have defined as important at this point in their lives. To do this with a student or business person, we take into consideration that person's entire physiological state and carry out procedures in a logical sequence. However, the sequence or components of training **overlap and eventually merge** as the person's mental fundtioning becomes more flexible. Shifts to the appropriate mental state for the task at hand then become largely automatic.

The Four Component Parts of a Typical Training Program

Note: These four components overlap. They are not to be thought of as strictly sequential 'stages.'
Component #1 is used if appropriate to your client's assessment profile and goals. With children you typically start with Component #2. A complete history will precede training, as will a discussion with parents of other appropriate interventions (diet, medications, sleep, extracurricular activities, educational support).

Component #1: Setting Up for Success

As noted in the previous section, general biofeedback modalities may be combined with neurofeedback. This is particularly helpful with optimal performance clients. In the initial stage of training the student/client learns to be externally and broadly focused, aware of everything in their environment (open awareness) yet focused on no one thing in particular. It is almost as if their attention is to the *space* around them. This state closely resembles the *open focus* so well described by Les Fehmi. In its extreme form it approaches the *cosmic experience* described by E. Bucke, the Director of London Psychiatric Hospital near the turn of the last century, or to *Satori* in Zen Buddhism. This mental state is a kind of dissociation. It is psychologically important for the student to be experiencing a pleasant scene in their imagination and associate this image with a self-empowering feeling when entering this mental state. A positive inner dialogue can also be quite helpful.

Physiologically, the most useful technique is to breathe diaphragmatically at a rate of about six breaths per minute. In this state a **balance occurs between sympathetic and parasympathetic systems** and in natural biological oscillators (e.g., heart rate, respiration, blood pressure). In this state it is postulated that there will be synchrony between oscillators. This synchrony can be seen by monitoring respiration and heart rate variability (respiratory sinus arrhythmia or RSA). Of course, one can teach diaphragmatic breathing without using biofeedback sensors, but the use of feedback will usually make learning faster and more effective. Heart rate variability training is the current method used for achieving this calm state.

Component #2: Operant Conditioning

The second stage is to use *operant conditioning* of brain waves so that students/clients can train themselves to move consciously from open awareness, high alpha (11 and 12 Hz), and SMR (12-15 Hz) to a focused, concentrating state in which they lower their dominant slow-wave activity and increase beta. This may require that they shift in and out of open awareness. When they shift out of this open or total awareness state they move in to a narrowly focused state where they consciously produce activation. What one sees in the EEG is a decrease in slow waves and an increase in fast-wave (15-18 Hz) activity. This narrow focus is reinforced by a visual and/or auditory signal that indicates they have met the preset thresholds. In this traditional operant conditioning they will train themselves to control, and perhaps eventually to permanently change, their brain wave pattern.

Two approaches for operant conditioning of brain wave patterns have evolved. The methodology used initially, exemplified by Barry Sterman's work, involves rewarding the person (or, for that matter, cat or monkey) each time they produce a burst of the desired brain activity. The second approach involves continuous feedback, a tone or song, whenever the activity is maintained. These are described below.

Technique # 1: Use of Feedback to Reinforce Finding a Desired Mental State

One method of *knowing* when a desired state is being achieved is by getting almost instantaneous feedback, such as a discreet tone paired with a change in a visual display. Modern filters have a reasonably short latency time and it is assumed that the brain is receiving the feedback faster than the student can move on to their next thought. Feedback should be well within the outside limit of <500 milliseconds that is necessary for operant conditioning.

On the other hand, the sudden absence of a sound can also be effective feedback. If the client's mind wanders, the absence of auditory feedback makes them aware that they have drifted. It is felt that this kind of feedback *catches* them before they move off too far and they can refocus. Some practitioners believe this method may be the most efficient way to effect change.

Operant conditioning involves rewarding a behavior that you want to increase (such as the production of SMR activity in the EEG) and it can be done either by observing a single instrument, such as a bar graph, or by simultaneously observing several instruments. (It is not really simultaneous since the brain consciously focuses on only one thing at a time; rather, it is dividing attention among various feedback compo-nents, paying attention to one and then another in a rapid sequence.) The bar graphs show amplitudes for fast and slow waves and for higher frequency activity that mainly reflects muscle artifact. Knowledge of possible artifacts permits one to distinguish changes in brain wave amplitudes that are due to *real* alterations in the subject's mental state (as opposed to changes that are merely a reflection of other processes, such as eye blinks or muscle tension). Inhibits can be set that will stop all feedback, including sound and animations, if the client moves, blinks, tenses muscles, ruminates or lets his mind wander.

Technique # 2: Using Averaging and Continuous Feedback to Reinforce Remaining in a Desired Mental State Over Time

Smoothing and averaging can be done by a number of EEG instruments. Slowing down the speed of feedback and using visual displays plus continuous sound, such as classical music, can encourage the student to find the desired mental state and hold it for longer and longer periods of time. The student is encouraged to **continuously track** how much of the time they are managing to *remain in* the desired mental state. Percentage-of-time-above-threshold (or below threshold), combined with points that require the client to sustain feedback for two or more seconds, are useful ways to score this kind of activity.

In this type of feedback you may decide not to have any *feedback inhibits,* since it could be irritating to have the music starting and stopping frequently. With respect to the use of an inhibit, the following is true:

a. *The advantage of NOT using an inhibit is:* Clients may become quite keen to find ways to increase their scores and better their threshold levels (for example, shifting their position to increase alertness in order to decrease slow waves and increase fast waves if you are training a person with ADD).

b. *The disadvantage of NOT using an inhibit is that the client may consciously or unconsciously tense muscles. This would artificially raise SMR and beta and reward them for EMG artifact rather than their mental state.*

On the other hand, with adults who remain very still and are not clenching their jaws, muscle tension artifact is usually not a problem. With children, if you don't want to have an EMG inhibit, then *deal with the problem of muscle movement by setting the* EMG (really EEG at 45-58 Hz – or 52-58 Hz in Europe and Asia – that 'reflects' muscle activity) instrument to a very difficult setting, link a percent-over-threshold counter to it, and insist that if they remain over, say, six percent for more than 30 seconds then you will stop feedback and start over. Children do not like stopping and starting. (They aren't getting tokens for the store in our Centre when this occurs.) It is amazing how even children with extreme fidgeting will settle down and work on feedback using this technique. It is helpful if at first you go somewhat overboard in rewarding young children every few seconds for remaining so still. With some young children it may also be helpful to be animated, silly, fun and extremely positive. At the end of the time, say three minutes, you record their scores, either automatically with statistics popping up, or, if your instrument does not do that, on a client progress tracking sheet, and give a token for each score that beats their last three-minute period score. You can create your own sheets and graphs for tracking progress, depending on which EEG instrument you use. Some instruments automatically track relevant statistics that you can print out, or, as is done at the ADD Centre, graph in the Microsoft program Excel.

c. In strict operant conditioning terms, the student's/client's behavior is being rewarded with feedback. However, positive response may also be linked to the secondary reinforcement of your actions as a trainer. If it becomes necessary psychologically (unconsciously) to the client to have you to trigger positive responses, you may have only achieved what is sometimes, in psychotherapy, called a *transference cure*. This is not permanent, and is thus not the mechanism of change we are after. Therefore, you should vary your techniques for reward. At the ADD Centre, we change rooms, EEG instruments, feedback screens, and trainers during the course of training. We want the gains to be within the student/client and independent of specific cues found only in the training environment. This works.

d. If you get electrical changes in the EEG pattern, results should last, according to Lubar, who has reported on 10-year follow-ups with one series of children with ADD. (Lubar, 1995)

If you see positive behavioral changes in a child with ADHD but without the EEG changes, it might be due in part to the therapeutic relationship between the child and the EEG practitioner. There may be heightened self-esteem due to the positive attention and the child may behave better to please the trainer. The therapeutic alliance is important in order to make gains as rapidly as possible, but the gains must be independent of that relationship by the time the child leaves training.

Here is an example of a serendipitous follow-up at our Centre. This young man had been a difficult client to engage, but he respected his trainer; he would work for him but not others. On the other hand, he did make electrical changes and took pride in acquiring this control.

Case Example: A strong-looking, upright, impeccably dressed young man walked into our Centre. I heard another student's mother audibly gasp as he introduced himself in the waiting room and explained politely that we might not recognize him but he had been a client during the summer four years ago. Failing grades (30s and 40s) had been the reason for seeking help, and he had responded well to training. He left training after only 20 sessions because he went away to a military school at the end of the summer. He said he was pleased to report that he had graduated with top marks and as a commander. He attributed his success to his training at the ADD Centre. He wanted to return that summer and finish off a further 20 sessions with us because he was entering a science program at a top university and he recognized that the training would further optimize his functioning. He arranged his sessions, shook hands and left.

The mother who had gasped in the waiting room came forward and quietly explained her astonished behavior. He had, she said, been her eldest son's friend before he went away and he was one of the worst kids in their high school, always in trouble. She had not seen him in years and was astounded at the transformation. Obviously the military academy had been the right school environment for this young man. He himself felt the neurofeedback training before he began there had prepared him to benefit from the experience.

Component #3: Classical Conditioning

The third stage is to link a desired activity, such as studying, to this efficient mental state using *classical conditioning*. To do this, the feedback is altered so the student can receive continuous, auditory reinforcement for sustaining the mental state of externally oriented, narrow focus while performing a task such as scanning a textbook chapter, doing a math problem, organizing a business proposal or hitting a golf ball. If the feedback stops, the student immediately moves back to the operant conditioning mode (watching the display screen and/or listening to the auditory feedback) until the desired mental state is again achieved. One may use a series of discreet tones or, perhaps more effective, a continuous tone while the desired mental state is maintained.

Classical conditioning is when you pair a new stimulus (*conditioned stimulus*) with a stimulus (*unconditioned stimulus)* that already reliably elicits a response. A dog will reliably salivate when given meat *(unconditioned stimulus)*. Pavlov paired a bell *(conditioned stimulus)* with the presentation of meat powder *(unconditioned stimulus)*. Soon the dog reliably salivated when he heard the bell (*conditioned response)*.

We use operant conditioning until we have reliably paired the desired mental state of focused concentration with auditory feedback.

Once the mental state (think of it as a physiological state, just as salivation is a physiological state) can be reliably produced in the training session using operant conditioning, then the client is ready for the second stage using classical conditioning. In this second stage, we *pair* this mental state (now an unconditioned response) with doing an academic task (or a business or an athletic task) (conditioned stimulus). After a few sessions of training the new *conditioned stimulus* of doing this task will automatically and reliably elicit the appropriate mental state. An example is a child coming to automatically go into a focused state when opening a book and beginning to read.

During the task, if the auditory feedback stops (reflecting that the desired mental state has been lost) the client is instructed to return to doing operant conditioning (look at the monitor) until it is regained. Then they can return to doing the task. You do not want to pair being tuned out with the task.

Summary of Learning Paradigms

Operant Conditioning:

Behavior	Reward (behavior repeats)
Mental State →	*Computerized Feedback +*
	Trainer's Feedback (mental state maintained)
15-18 Hz → *tone* → *more 15-18 Hz*	

Classical Conditioning:

Unconditioned Stimulus	Unconditioned Response
Meat →	*Salivation*
Feedback →	*Mental State*

Conditioned Stimulus	Conditioned Response
Bell →	*Salivation*
Academic Task →	*Mental State*

Component #4: Generalizing

The fourth stage is to train the student in procedures that will help them to *generalize* this ability to produce calm, relaxed, alert, flexible focusing and sustained concentration to home, school and work situations.

The first step in doing this is to have the student achieve all feedback objectives without active coaching from the trainer. The second step is to have them achieve all the training objectives for neurofeedback and biofeedback both while sitting looking at a blank screen and then while doing a cognitive exercise without any visual or sound feedback. The monitor and speaker are turned off, but the program is still recording data. This is referred to as a "transfer trial." You measure how well the skill of producing the mental state while receiving feedback is transferring to a situation that doesn't include feedback. If you are doing a series of seven or eight feedback trials within a 50-minute training session, you can make one of the later ones a transfer trial. Obviously this is only done when you feel the person has mastered the task and can be successful.

The final step is also an ongoing step. The client is instructed early in training to practice going into the psychophysiological state, which they are learning in sessions, when they are at home and at school or work. Breathing techniques, imaging, and metacognitive strategies are examples of effective methods of triggering this mental state. They act both as behavioral cues for eliciting the desired mental state and as useful techniques for efficient functioning. Attaching this state to regular habits in daily life, such as answering the phone, reading, listening to others, and driving, is an effective way to have it become a new routine component of one's daily living. Specific methods to help students, professionals, and business people generalize this mental state to their academic and work tasks are given in the chapter on metacognition. For details about metacognition, see Section XV.

Summary of Steps in a Typical Session

Screens and Screen Sequences

Objectives
1. To have clients compete with themselves to remain for longer and longer periods of time in the desired mental state (focused concentration when working with clients who have ADD). This means they get better scores (and children get more tokens if you are using that type of secondary reward system at your office).
2. To learn to *self-regulate* and flexibly control transitions between mental states (EEG patterns, ANS, and EMG variables) and achieve rapid recovery after stress.
3. To have the client end the session feeling empowered and feeling they had some fun as well.
4. To have the client generalize their learning to daily life.

Principle: The screens used and the sequence will vary with the age, stage of training and interests of the client. However, they must also meet the objectives of the training, which means changing the pattern of brain wave activity in the desired direction. They should be coupled with doing cognitive tasks appropriate to the goals of the individual client. Thus a whole session on Pac-Man or Pong or puzzles or car racing or some other animation may not be what the trainer feels is giving the best feedback in the early part of the session, but may be useful for enjoyable "breaks" during the training time. Perhaps favorite (game-like) screens like these are best used with young children as a reward. Less entertaining screens that clearly indicate the parameters for training should be the mainstay for conscious learning. In this manner the client will learn to consciously hold a particular mental state. In real life they must maintain their concentration even with boring tasks, so this must be practiced during their session. Doing brief periods of training, two to five minutes, and using secondary rewards (praise, tokens) maintains motivation for the mental exercises. Even executives will say that they tune out rather quickly when reviewing documents. Doing this kind of cognitive activity while receiving feedback that is informative can be very useful. While receiving feedback, the student learns mental strategies for sustaining interest and attention.

Suggested Sequence
1. **Baseline:** Start with a brief baseline recording without feedback.

2. **Pure Feedback:** Do three to five training segments, each lasting two to five minutes, using a screen that shows percent-over-thresholds and scores (points). The first screens used should give simple, discrete auditory and visual feedback spaced one to three seconds apart.

Adolescent and adult screens may include appropriate biofeedback measures such as EDR, respiration and EMG.

We usually do not save all the data recorded during each segment of feedback. We just record relevant scores (such as percent-over-threshold and points) and compare them to previous scores. After each segment we discuss the client's own observations about their *mental state* during that segment compared to other segments where different scores were attained. Statistics for each segment can be saved, even though the raw data is not. Transferring the statistics of interest to Excel means you can print out a learning curve at the end of each session.

1. **Task:** After three to five short periods (one to five minutes) of pure feedback, we may introduce an academic task. The academic task is only done *when the client is maintaining steady auditory feedback*. We may have to make the thresholds slightly easier in order to achieve steady feedback. This is the *classical conditioning* step in the training session. We may increase the time of an on-task segment. It might be from 5 to 10 minutes, depending on the age of the client and their attention span. We ask questions about the work when we stop feedback (to avoid muscle artifact that may result from talking during feedback). We may *pause* the feedback, if we notice a change in the wave pattern, and ask the client to identify what was going on in their brain at that time. The tasks can be reading, writing, mathematics, or listening to the instructor. They will have to use metacognitive strategies to remember the information for a question period that will follow the listening task. We train the students in the use of strategies for the task so that using strategies is paired with the correct mental state for cognitive work. In time, the strategies will also unconsciously cue the student to turn on the desired mental state.

2. **Pure Feedback and/or** *Work–Rest* **(Self-Regulation):** Then we return to pure operant conditioning for one or two segments. At this step, if the instrument allows for it, we might have the client turn on a focused mental state for 30 seconds (or longer). This is called a *work* state. Then, the client is told to deliberately turn it off this mental *work* state and let his mind wander for 30 seconds. This second condition is called a *rest* state. At the end of several trials, compare the scores. Clients feel quite empowered when they see that the values of the different frequency bands actually change and that they have control over this. Be careful to make sure that there is no difference in 45-58 Hz activity between *work* and *rest* periods. A higher reading in one state could indicate that muscle activity was raising beta.

3. **Task:** At this juncture we may do another academic task with auditory feedback on. Once the client is showing mastery we will have them do some segments of the session with no visual or auditory feedback. The student will compare the results of these segments to similar time frames where they were receiving both feedback and secondary rewards.

4. **Positive Ending:** To finish the 50-minute session we will let the client choose their favorite screen for feedback. You want to end on a positive note and reinforce the idea that they are in control.

Other Pointers

Some children will produce a great deal of muscle artifact. With these children you simply must have an EMG inhibit. Without such a strict EMG inhibit you would not be able to accurately reward the child for producing SMR or beta due to the very large effect that muscle activity can have on beta and SMR frequencies. These active children are often the same children who need to have you emphasize increasing SMR (12-15 or 13-15) at C4 or at Cz in some clients. Remember that SMR is only measured and trained over the sensorimotor strip: C3, Cz, C4.

If you want them to also remain externally focused while they learn to increase SMR, then you might put an *inhibit-over-threshold* on your dominant slow-wave (often 3-7 or 4-8 Hz) instrument. Alternatively, you might relate your primary reward to decreasing a theta/SMR ratio (if training a ratio is possible in the program you are using).

Tracking Trends and Motivating Children

For some feedback segments it is a good idea to have the student attempt to *feel* (or 'sense') the desired **mental state and consciously change it** back and forth. This has been described previously using the example of a sailboat screen. This gives the student a real sense of self-regulatory control and empowerment. At our Centre one of the ways to earn tokens is to demonstrate this control. Some students like it if we use a stopwatch to time how long they can maintain the picture or hold the theta in the *green zone* below the threshold line. With younger students, the trainer counts out loud and we see how high he can get before he 'loses it.' We make up games with the students. For example, we might have little piles of tokens. One white token each time we reach five seconds; two tokens if we reach eight seconds, three tokens for 10 seconds and four for 12 seconds. We will record the numbers on a special score sheet after a session segment of three to five minutes. The child may want to do it again to better their score. Making up a special game for the student is more personal and interactive and we feel this is even better than automatic, machine-controlled games.

Note: There has been discussion, even in the popular press, of using **interactive video games** for feedback. This can be done with attachments to some types of equipment. However, Olaf Palsson and Alan Pope have done the best-known work in this area. They pioneered this approach, feeling it might improve motivation for doing the training (Palsson, 2001). Some instruments do this as their primary mode of feedback. Other instruments have this as a secondary capability. Students with ADD/ADHD typically are past masters at video games. The can usually beat their peers in this type of activity. They go into *hyperfocus*. **We see no need to train them to focus on these items.** There is a place, however, for this kind of activity. We find it very useful as a reward for the young, hyperactive, noncompliant child to get them interested in working during other parts of the NFB session. It is also a useful way to attract the attention of an autistic child. It is used as a reward and as an opportunity to maintain a state of reduced theta and increased SMR, the criteria that must be met to play the games successfully. Presumably this is a different brain wave pattern than is found when playing other computer games. The main question is whether the student is able to transfer this state to an academic setting. Hopefully some follow-up studies can be done to investigate this question.

At the time of writing, however, we **do not** feel this is the best way to train a student to constructively handle boring classroom tasks. Until there is evidence that feedback with a videogame format generalizes to classroom performance, we will, for the most part, use relatively boring feedback screens that give direct feedback about brain wave activity, as the mainstay of our feedback program.

Other Types of Feedback and Cueing

1. Teaching a student to be able to *self-regulate* going in and out of high **alpha synchrony** is said to result in improved cognitive performance (Anna Wise). This kind of peak performance training can be attempted using the mind-mirror type of display, which requires the use of two channels of feedback with sensors on homologous sites, such as C3 and C4. Percent-of-time-over-threshold data for the right and left channels help the client to quantify the feedback. The mind-mirror display places the spectral array derived from FFT calculations for each hemisphere in a vertical position (see executive training, Screen #5) so that one can see how symmetrical (or asymmetrical) the activity is on the right and left by seeing if the bars for the particular frequencies are going in and out to the same degree on both sides.

2. We encourage all students to sit up straight, rather than lean back in the chair. Good posture **increases alertness**. The increased alertness correlates with an increase in EDR, and this can be recorded and shown to the student if you have equipment that does both neurofeedback and biofeedback simultaneously. We also teach them diaphragmatic breathing at a rate of approximately **six breaths per minute.** The younger child will have a faster rate than the six BPM that is best for most adults. There are individual differences at all ages with respect to what rate is best in order to achieve synchrony with heart rate variability. This, combined with learning to raise their peripheral skin temperature and decrease shoulder and/or forehead muscle tension, is done before starting on an academic exercise. We encourage them to use these techniques as a method of triggering ('cueing') getting into the desired, highly focused, problem-solving mental state. This is the zone which they are learning to turn on and maintain for efficient learning and work production.

As noted previously, when they are in the physiological state described above, then using a metacognitive strategy that we have taught them during training sessions will trigger the student to produce the appropriate mental state. The strategy, apart from being useful in itself, acts as an unconscious and automatic 'cue' to precipitate the appropriate relaxed, alert, focused and concentrating mental state.

As mentioned in the preface to the Second Edition, the reader will have observed that in our description of good clinical practice, NFB is never a stand-alone procedure. We recognize that researchers feel a need to do research on individual items such as NFB without other therapeutic or educational inputs but we feel they should clearly state in their reporting that they are not doing research on NFB as it is done in standard clinical practice. In practice it is always combined with the work of a good therapist/teacher/coach/trainer who adds specific metacognitive strategies and tasks, plus appropriate other biofeedback techniques such as heart rate variability training. There is also consideration of other factors such as diet, sleep, and exercise. The implications of positive or negative "research" findings should be evaluated in this context.

On the other hand, large case series that include both subjective data (such as questionnaires and school reports) and objective data (such as standardized testing like the WISC or WAIS or a computerized test battery) may better reflect the value of these combined approaches to the clients.

SECTION XIV
Alpha-Theta Therapy
Combining NFB and BFB with Psychotherapy and Relaxation Training

Introduction

When William Peniston's study showing successful treatment of chronic alcoholics was published in the early 1990s, it created considerable interest. He used a combination of temperature training, imagery, and neurofeedback that encouraged slow-wave activity. Due to the neurofeedback component, it came to be called alpha-theta training. Traditional alpha-theta therapy using the Peniston protocol is described in other texts (White, 1999).

The newest application of alpha-theta therapy has been to enhance the performance capabilities of music students. This is designed by John Gruzelier and his colleagues. They told the students only that they were to receive EEG biofeedback in order to induce a deep relaxation state. The feedback consisted of a pleasant tone sounding when theta activity exceeded alpha activity measured at Pz. Just twenty half-hour sessions produced measurable improvements in the interpretative aspects of musical performance.

Alpha-theta training is not really new; indeed the basic underlying techniques are very old. It is the use of computers and modern high-speed feedback instrumentation to produce the changes in the patient's attitudes and behavior that is relatively new. The addition of biofeedback and neurofeedback has speeded the learning process tremendously. It has made learning the techniques possible for people who might never have been able to change using just teaching or psychotherapy.

At the outset, let us again emphasize that EEG biofeedback for improving attention span and for optimal performance is a completely different intervention than alpha-theta therapy. The first is a learning process that can be carried out by trainers with backgrounds in education, coaching and so on, or in some cases, by the individuals themselves. The second is a psychotherapeutic process, and, because of the state of extreme suggestibility that the client may enter, it should probably only be used by very well-trained and experienced psychotherapists.

Tom Budzynski discusses the use of inducing the theta state for the purpose of having the client in a state where they are not able to resist rescripting suggestions. Hypnosis also utilizes this shift. (Hypnosis as an adjunctive technique is too broad a topic to be covered in this text. The reader is referred to experts in the field, including Eric Wilmarth in Grand Rapids, Michigan, and Cory Hammond in Salt Lake City, Utah.) Even a cough or a movement by the therapist may reinforce a client's recall of a real event or a fantasized one with the possibility of solidifying in their mind a false memory. We will not have our training staff use alpha/theta training in our Centre unless they have extensive psychotherapy qualifications since we have an educational orientation. Nevertheless, we do feel that it should perhaps become part of the curriculum in Psychiatry and in other professions where the practitioner has supervised psychotherapy training for four or more years. The therapist must have training that equips them to deal with memories or imagery or emotions that may be relived when the patient is in the hypnagogic state associated with theta being higher than alpha with eyes closed.

In the mid 1960s, a relaxation psychotherapy technique was being taught to psychiatry residents that will be described here. Not surprisingly, the stages parallel the technique used by Peniston with alcoholics. The stages are a variation of the old *sandwich technique:* Think/say something positive – get to the 'meat' of the issue – think/say something positive. You use this technique continually when working with children. To demonstrate the shift from these very old techniques to modern coupling of them with what we now understand concerning brain waves, we suggest you keep three stages in mind.

Preliminary Caution

Anyone who has worked extensively with autistic children will probably have come to realize how essential it is to think only that which you would like your client to 'hear.' Put simply, you will communicate your thoughts and feelings nonverbally, whether you want to or not. It may even be the case that we are capable of picking up far more telepathically than scientific research has yet been able to elucidate. Whatever the mechanisms, it is important that you be aware of your own thoughts and feelings with each client, and that you not work with clients when you are in a negative frame of mind. The telepathic possibilities are not a well accepted area to talk about in professional circles, but this phenomenon has been reported by a number of responsible professionals. Julian Isaacs, in discussing these phenomena, quotes Nancy White and also references a chapter by Ullman (1975) in Freedman, Kaplan & Sadock, Comprehensive Textbook of Psychiatry, *2nd Ed., Vol. 2, 2552-2561. Baltimore, Williams and Wilkins.*

Here we suggest you keep in mind three processes that merge and run simultaneously. The suggested format is to do the NFB and BFB assessments first, then begin regular feedback and neurofeedback as required to promote relaxation, and then, when the client is relaxed (as indicated by an ability to warm their hands and breathe diaphragmatically at about 6 breaths per minute), begin the imagery. This is followed by neurofeedback that promotes a hypnagogic state; that is, a mental state with eyes closed in which theta activity reaches a higher amplitude than alpha in the occipital region. This is the normal state of affairs as one falls asleep, but the goal in this therapeutic technique is to retain sufficient alertness that one can be conscious of the material that comes to mind and can process this material without falling asleep.

A. Start with BIOFEEDBACK/ NEUROFEEDBACK Assessments/ Training	B. RELAXATION (with attempted 'SATORI') TECHNIQUE	C. Addition of ALPHA-THETA THERAPY

All three processes (A, B and C) should be carried out when doing alpha-theta therapy. They are not 'stages' in the true sense of the term, because they are layered and A and B are still in process as C is added.

EEG and Stress Assessments and Teaching Basic Components of Alpha-Theta Therapy

Stress Assessment

A stress assessment is carried out as previously described. In this assessment we record respiration, pulse rate, synchrony between pulse rate and respiration (RSA), peripheral skin temperature, skin conduction (EDR), and forehead and upper body muscle tension (EMG). We suggest placing the electrodes for the first EMG measurement on the forehead, and placing electrodes on the trapezius or masseter for the second EMG measurement may be helpful. (Note, it has also been suggested that placing the positive electrode on the ventral surface of the left forearm and the negative electrode in the corresponding position on the right forearm may give an impression of overall upper-body tension.)

EEG Assessment

Electrode Placement

With a one-channel instrument the assessment may be done with the active electrode placed at Fz referenced to the left ear or to linked ears.

With a two-channel instrument use a linked-ear reference and a single ground electrode attached to preamplifiers for both Channels A and B. The active electrodes will be F3 and F4 initially, with a second assessment comparing Fz and Cz. These sites will show whether there is a pattern suggestive of anxiety, and may show high amplitude beta spindling that may be associated with obsessive-compulsive symptoms. A third assessment may compare F3 and T3. (T3 is based on the hypothesis that theta activity in this area is associated with memory retrieval.)

EEG recordings may be done at the same time as the stress assessment. However, the EEG assessment is usually done separately to minimize the effects of muscle tension artifact. This is especially important in that we are primarily looking for peaks and dips in the beta range, 13-34 Hz. This beta range we are interested in (13-34 Hz) is very susceptible to the effects of EMG artifact, which can artificially elevate the amplitude of the waves and give a false impression of brain activity.

Set Objectives for NFB Training

With individuals who are anxious, we may see a peak between 20-24 Hz. If they are negatively ruminating, a peak may emerge between 25-32 Hz. In both cases there may be a concomitant dip in the amplitude somewhere between 13 and 15 Hz (SMR) when the site is along the sensorimotor strip.

We are also interested in the degree to which the client can go into high alpha, 11-13 Hz, when calm and externally focused. People with addiction problems tend to have less alpha and more beta than is found in adults without addiction concerns.

Set the EEG parameters according to assessment results.

Training Overview

Step 1: Learn Diaphragmatic Breathing Combined with Relaxation

The client is asked to close her eyes, breathe deeply diaphragmatically, pretending that they have a balloon in their abdomen and imagining that they are inflating it while counting to three, and deflating it while continuing the count to six, then resting while they finish counting to 10. Next, they are asked to tense then release the muscles in their forehead, neck, shoulders, arms, forearms and hands, and to feel the difference in the two states (tension and relaxation).

Now the client is instructed to again focus on their breathing. Once this is going smoothly (RSA recording will tell you how successful they are), they are instructed to continue their diaphragmatic breathing, but in the expiration phase and the resting phase, to sense/feel heat, tingling, and any other sensations in their muscles. They are to start with their forehead and then pretend that the feelings are 'energy' and they are allowing the 'energy' (tension) to move from that muscle to the next in a series moving from their face to neck to shoulder to arm to forearm to hand and then to dissipate out the end of their fingers. They may liken the energy sensations to a light that grows brighter and brighter. As the energy/tension/light moves down, the muscle it leaves is to be completely relaxed. They are to allow the warmth to remain in their hands and fingers.

Step 2: Couple These Exercises with BFB + NFB

Feedback is done that encourages respiration at about 6 BrPM with good synchrony between breathing and heart rate. The client is encouraged to increase peripheral skin temperature. Clients are to maintain EDR in a middle range without volatility. The client is to find a way to decrease EMG readings. Much of the foregoing formed a part of Tom Budzynski's twilight learning state (Budzynski, 1979).

During the feedback stage, the student/client is encouraged to decrease the appropriate bandwidth, which usually lies somewhere between 21 and 34 Hz. The band 45-58 Hz must be held down to ensure that muscle artifact is not affecting the lower frequency beta bands. Concomitantly the client will increase 11-12 or 13 Hz and 13-15 Hz. These are done separately because 11-12 Hz contains high amplitude alpha. The 13-15 Hz range contains much lower amplitude SMR and beta.

Psychologically, in this stage, the student/client is encouraged to find a calm mental state of **'open awareness.'** This is the state athletes enter when they are well trained. For example, the goalie who needs to be aware of every player around the net and yet who can instantly shift to narrow focus on the puck, the black-belt martial artist who is being attacked by several opponents at the same time, the archer just prior to and during the release of the arrow, and so on. Clients were encouraged to find this mental state in the 1960s, but only in the last decade have we understood that it corresponds to an increase in 11 and/or 12 Hz activity. Previously, we could not objectively measure how well the client was doing, but now clients can have instant feedback. It is a similar state to that which Les Fehmi teaches using the term 'open focus' (Fritz & Fehmi, 1982). This state allows a wider context to be sensed and defuses the narrowly focused negative ruminations of the client.

Step 3: Begin Using Imagery and Desensitization

In this stage, the client maintains their relaxed state from Stage #1 and begins the imagery suggested below. In this stage of imagery more alpha will be seen on the EEG and it may be lower in frequency (thalpha in the 6-10 Hz range) than the alpha seen initially. The client is moving to an even more internal, free-association state. One may see a peak of alpha at either 7.5, 8 or 9 Hz. This step is done with the eyes closed.

Remember that for the vast majority of our clients we are not doing psychotherapy (it is, rather, an educational intervention to learn self-regulation), and we prefer **not** to have lower frequencies; indeed we discourage theta production and a big drop in EDR. We do not want the client in a state of high suggestibility. The work with alpha-theta with the most common application being with alcoholics, or those with drug addiction problems, is a different intervention than that discussed in the rest of this book. Another application of alpha-theta training would be if a highly trained psychotherapist was working with their client in long-term psychotherapy and they wanted to facilitate a hypnagogic state so that there would be material to work with in therapy. In these instances all aspects of alpha-theta work would be discussed, including the possibility of abreaction and an appropriate signed release would be obtained. With respect to the utility of increasing theta during a therapy session, there is no published data. On an anecdotal level, a Zurich trained analyst who was using the F1000 equipment mentioned that his classic Jungian techniques, when augmented with increasing theta, speeded up the process of psychoanalysis so that the work could be done in 9-12 months rather than 3-5 years. Instead of waiting for the subject to have a dream that he could recall, the free associations while in the hypnagogic state of high theta (which is akin to a presleep mental state) provided the material for discussion in the therapy session. This work needs to be replicated and published.

Relaxation-Desensitization Technique

The client has begun to achieve a physiologically and mentally relaxed state, using the combination of biofeedback and neurofeedback described above. While the client maintains this relaxed state with eyes closed, diaphragmatic breathing at 6 BrPM, increased amplitude of 11-12 and 13-15 Hz, and decreased amplitude of frequencies between 21-32 Hz (they are receiving soft, pleasant, auditory feedback), begin the following relaxation techniques.

The method described below is as easy as A, B, C. Other techniques can be substituted for the ones described here. The stages described below are the ones carried out and taught by the elder author for the last 25 years. However, most therapists have their own favorite techniques to aid clients with relaxation, and you should use whatever you are comfortable using.

Deal with Major Negative Ruminations

The client is asked to state, and the therapist writes down, a list of experiences, thoughts, or concerns about which they ruminate. Next, the client is asked to order this list from least to most personally traumatic. This is followed by the client putting each of the items on this list as a title on a separate page of their **SMIRB** (stop my irritating ruminations book), which they picture in their mind. The therapist writes these titles on separate pages. (The SMIRB has been described previously and will therefore only be summarized here.) The clients are asked to state the phrases that continually reverberate through their minds when they ruminate about each area in turn, and the therapist lists these phrases below their appropriate heading.

The page on the right side of this notebook is left blank for future ideas concerning actions to rectify the situation or to list different, more positive, ways of reframing it (this is similar to what is now called cognitive behavior therapy or CBT). Later, the clients will rewrite the foregoing into their own SMIRB. Then they are told that this book should always be with them. They are instructed that, for example, if they awake in the night with a negative thought they must ask themselves if the negative thought they are having is in the book. If it is, then they remind themselves that they have a worry time (perhaps a half hour with tea each afternoon) and they will not forget to worry about that thought because it is in their book. If it is not in the book, then they must turn on the light and put it in. It is relatively rare, however, to have a new worry. Most people ruminate about the same things repeatedly.

Note that the SMIRB is a very small book that can fit into a small pocket or purse. The rumination section starts at the back of the book. The desirable goals and enabling objectives and actions section is in the front of the book. Writing positive goals with timelines is an old and well-documented technique for achieving success. Stephen Covey's books are a resource that can be used to expand on goal setting, as is time-management training such as that offered by Charles Hobbs using Day-Timer products.

Imagine an 'Ideal Me'

The client is asked to make, in a very brief form, a statement describing an "Ideal Me." The therapist will carefully write this down. Then the client is to describe succinctly the sensations and feelings she would experience in that ideal state, followed by how others would react to her if they were actually like that. The therapist writes this down carefully for the client to have at the end of the session. Finally the client must describe how they would envisage their environment being (home, school, work, social) if they were in this ideal state. The therapist also notes this.

The client is told that when he is relaxing, this is the state we want him to have (be living in) for himself. They are to take home their notes (that the therapist has made) and add, modify, or delete items. They are to repeat the imagery at specific times during the day. The times are chosen to be paired with habits that are already in place each day. This is also a technique that has been previously described – attach a new habit to an old habit, and it is used with the SMIRB technique described in Section XII (and in *The ADD Book*).

The recovering alcoholic, for example, would typically describe an "ideal me" who was an abstainer. He would picture family and friends enjoying his company without any apprehension that he would drink and spoil the occasion.

Desensitize Yourself to Stress

Step 1: Imagine a Relaxing Scene

In this stage, the client develops a strong positive, calm, relaxing image or scene. The client may develop their own scene. However, experience suggests that most clients want and require assistance in this phase. Therefore, the therapist may need to help them imagine themselves in the most peaceful and relaxing scene possible. You may assist this process by playing music. There are many opinions about what to play, but the main thing is that the patient find it relaxing and pleasant. Piano music by a contemporary artist, such as Michael Jones' Pianoscapes or Seascapes is one option. Classical music may work for others. SAMONAS sound therapy CDs are another option (especially the ones using compositions by Mozart). Mozart's music has been studied by John Hughes, an epileptologist, and he found that it corresponds best to the brain's natural rhythmicity. He has even witnessed patients stop seizuring when Mozart's music was played (Hughes, 2002).

Similarly, there are any number of dialogs on tape, though a personal approach is often better. The following is an example.

Begin by telling the client, in a soft voice, that she is on a beach. The beach stretches for miles in both directions. There are no other people. The ocean is a beautiful light blue and the waves are softly lapping on to the sand. They see a lounge chair and go to lie down on it. It is like no other chair. When they are seated they cannot feel any pressure from the chair, it is like being suspended in warm, body-temperature, salt water. We go on for a bit about the total comfort they are feeling (we use a very comfortable chair). Then we tell them that the only things they can see are a single white fluffy cloud and a seagull. They gull is beautiful and graceful. Without even moving its wings it circles down slowly towards the ocean only to dip its wings slightly to a different angle and rise again in great sweeping circles.

Step 2: Imagine Themselves Being "One with the Scene"

When they are ready, clients arc to allow themselves to imagine that they are floating out of their body and rising to become one with the sky, the soft white cloud, the gull in its gentle circular drifting. They are to feel totally relaxed, strong, aware of everything, aware without narrowing focus, a sense of complete power, a sense of total mastery and oneness.

Step 3: Enter a Personally Perceived Stress

At this point it is quietly suggested to the client, in their imagined scene, that they are to come back into their body on the lounge chair and, while remaining relaxed, rise and go behind the chair, off the beach and up a slight grassy hill. When they reach the top they will look down on a scene that represents for them a problem from their own life. They remain relaxed and detached and watch it unfolding. Then, if they are still able to be relaxed, they enter and correct the problem and then return to the beach and their chair. They again relax.

This is a type of progressive desensitization. The client is instructed to initially choose a scene that is only mildly traumatic and only to move progressively through more traumatic scenes as they are able to feel comfortable, calm and relaxed in so doing.

Step 4: Return to a Totally Physiologically and Psychologically Relaxed State

In this stage, the client has returned to the beach and his chair, and again attempts to mentally become 'one with the sky' and the peaceful scene as a whole.

Remember to end the relaxation session with a gradual return to awareness of the here and now. You do not want your client leaving the session in a dissociated state.

Addition of Alpha-Theta Therapy

Review of Assessment and of Training Procedures

Stage 1: Review EEG, Stress Assessment and Autogenic Training Procedures, and Begin Relaxation Training

Although the original Peniston protocol used autogenic training (Green, 1977) and handwarming as the first stages in training, you could also use the steps described above, that is, combining BFB + NFB with traditional relaxation training. Use of finger-temperature feedback alone may not be as effective as using a number of sensors for different modalities (HRV, temperature, EDR, breathing), as described above. Some practitioners do not do an EEG assessment. If you are beginning in this field, we would suggest you take the time to do careful assessments of both EEG and a psychophysiological profile.

If you are going to do alpha-theta therapy and do not have a full cap EEG assessment, then you may want to do at least a two-channel assessment with electrodes placed at Fz and T3, since these may be the sites you will be using. Alternatively, if you are doing the traditional *Peniston protocol,* you would do your assessment using an occipital site (Peniston, 1993). Then you can reassess after training to see what effect there has been on EEG parameters.

We would suggest that you begin relaxation training using traditional biofeedback and neurofeedback as outlined above, and combine this with the modified autogenic training and imagery described above. Handwarming has received particular emphasis as being important before starting EEG biofeedback. Remember that autogenic training and other relaxation techniques can have negative side effects. Seb Streifel has stated in his AAPB workshops on autogenic training that about 30 percent of patients will experience discomfort. Begin the alpha-theta therapy after the client has mastered the relaxation techniques.

Stage 2: Alpha-Theta Enhancement Introduction

This has three components. First, you must decide on electrode placements and whether to do one- or two-channel training. The theoretical considerations that may affect your decisions in this area are outlined below. Then you will set your inhibit frequencies, your enhance frequencies, and thresholds.

Theoretical Considerations

Theoretical Framework – Alpha and Theta Enhancement

The essence of the alpha-theta process is the premise that alpha represents a bridge between internal and external attention. Theta is a bridge between wakefulness and sleep. Alpha and theta states are also important for and involved in memory and reflection. Training is done with eyes closed, so the expected dominant frequency (particularly in the occipital region) is alpha. When the client hovers in a mental state of higher amplitudes of theta as compared with alpha (this is referred to as the alpha-theta crossover), they are capable of recovering memories with associated emotions and/or insights and/or creative thoughts. Early memories may be evoked harkening back to childhood, perhaps because, in a child, the dominant wave form is theta. This may represent a form of state-dependant learning and state-dependent retrieval of memories.

At the same time as this state is one of creativity and insight, it is also an unstable state in which the patient is prone to suggestion. Wisely used, this opens the door to new learning and the replacement of old and inappropriate 'scripts' laid down in early childhood. Inappropriately used by a therapist, this state could lead to false memories. The key principle for the therapist is to be cautious, well-trained in handling abreactions, and do no harm.

Theoretical Considerations for EEG Feedback Electrode Placements

The traditional Peniston protocol called for an O1 placement for the active electrode, with a linked-ear reference and a forehead ground. However, given the theoretical considerations outlined below, there is a rationale for the use of two-channel training with electrodes at Fz and T3. (Julian Isaacs has also suggested T3-Fz sites, but he uses a bipolar placement. With a sequential electrode placement one is less certain of what might be occurring at each site.) In this manner, the alpha that is encouraged at Fz will relate to the client relaxing while the slowing at T3 may reflect a decrease in cognitive activity related to speech and language processing and correspond to memory retrieval and primary process insights.

Recently, work has been done with music students at the Royal College of Music in London, England, with placement at Pz (Gruzelier & Egner, 2001, 2002, 2003). Researchers found improved musicality (emotional interpretation of the music) in a group of students trained to increase slow-wave activity as compared to groups who received mental strategies, Alexander techniques (posture and movement exercises), aerobic training, or training to increase 15-18 Hz activity.

Theoretical Rationale for Using Fz and T3 Sites

Alpha projection systems are from the thalamus. Julian Isaacs has pointed out (EEG spectrum workshop handouts, 2000) that the 'specific' thalamic projections to the occipital lobe do not necessarily correspond to relaxation. On the other hand, the nonspecific thalamic projections to the entire cortex do reflect relaxation. The Fz site reflects this second type of projection system. (Isaacs notes that the medial forebrain bundle projects theta from the limbic system to the frontal lobe.) In addition, alpha at the T3 site may reflect the speech/language center being inactive, which may be desirable as it may mean that there is less self-talk going on. These suggestions by Isaacs appear to be well thought out, and the concept of more than one alpha generator is well accepted (personal communications with Jay Gunkelman).

Since sequential (bipolar) placement can only reflect differences between the two active sites, a two-channel feedback system might more clearly reflect the activity at each site and thereby better monitor the objectives for training at Fz and T3. However, this means that the program used must allow for computations based on Channel A and Channel B governing the feedback to the client. Both sites must demonstrate the desired alpha-theta state.

A practical consideration for not using O1 site is the effect on the active electrode of the client's position in the chair. For relaxation and for carrying out alpha-theta therapy we often use a comfortable, cushioned high-back chair. Relaxing your head back in these chairs with an O1 active electrode site may result in electrode movement or electrode pop. This will interfere with accurate recording and feedback. If you are using O1, then try using a rolled towel or a small cushion under the client's neck so that the electrode is not touching the chair's headrest.

Training

Step 1: Use Inhibits

While doing alpha-theta therapy, the therapist should inhibit activity in the 2-5 Hz bandwidth. This is not only to inhibit movement artifact, but also, and perhaps more importantly, to prevent the client from falling asleep. It is also important to maintain inhibits on fast beta wave activity. These inhibits include, first, 19-24 Hz that may correlate with anxiety; second, 24-36 Hz that may correlate with worry and ruminations; and third, 45-58 Hz that may correlate with scalp, jaw and neck muscle tension (EMG artifact).

Step 2: Enhance Alpha and Theta

With all the above being carried out, one will set the reward tones to encourage alpha 8-11 Hz initially, and later 8-9 Hz. The reward for alpha should be a pleasant high-pitched sound when alpha is over threshold. The threshold is set so that the client can achieve it about 55 to 70 percent of the time. This will ensure that the client does not become discouraged, and will help the client increase the production of alpha for longer periods of time. In addition. one will reward theta, 5-8 Hz, with a lower tone feedback.

It would be especially helpful if, for the two enhancing signals, the pitch of the two tones blended appropriately and were proportional to the amplitude of the respective waveforms being monitored. However, these options are not yet available on most instruments.

In addition to these enhancement feedback sounds, most programs provide an auditory signal, such as a pleasant ding, when the theta band amplitude exceeds the alpha amplitude.

It is helpful if your program can graph changes in alpha, theta, and beta. Otherwise, you may choose to hand-draw the changes on graph paper. In the graph, make the x axis be time, and the y axis amplitude. With this arrangement high beta will be observed to decrease, and alpha will increase then slightly decrease as high theta increases and a crossover takes place.

Step 3: Imaging

In this stage in most alpha-theta work, the client is to imagine themselves performing a task or being a part of something in the area they would most like to change. They are then asked to change this scene in a way that makes it what they wish it to be rather than what it typically is. This imagery of the situation they want to change, followed by imagery of the scene after the change has been made, is done while they continue with the alpha-theta feedback.

Conclusion

In concluding this section, we can reasonably expect that adding biofeedback and neurofeedback to standard relaxation methodologies should make the process more effective and decrease the time necessary for learning effective self-regulation considerably.

It is a good idea to obey the principle of parsimony. Being parsimonious means you first do the simplest, least invasive, least expensive, and least time-consuming interventions that have a reasonable chance of success. The first thing to try for intervention is to use biofeedback and verbally instruct the student/client in relaxation techniques to use at home, and also get them to implement the use of a SMIRB (Stop My Irritating Ruminations Book), as has been previously described. Psychotherapy and alpha-theta neurofeedback are there as a back-up for those clients who require a long-term therapy. There may also be a developing market for using alpha-theta training alone as a form of deep relaxation to further enhance performance in those who are already at a very high level.

SECTION XV
Metacognitive Strategies

With special thanks to James, Aaron and Katie Thompson for their help in the development of this section.

Note: This chapter may be photocopied and used as a monograph to assist students.

An Introduction

Training to decrease slow-wave electroencephalographic (EEG) activity and increase fast-wave activity is necessary, but not sufficient to maximize beneficial behavioral changes in clients who wish to improve their attentional processes. To work efficiently, the graduates of a training program should ideally be able, at will, to put themselves into a mental state that is relaxed, alert and focused. In this state they can demonstrate concentration and engage in organized problem solving. In addition, graduates should have techniques – metacognitive strategies – that improve their ability to listen, learn, organize, and remember material in a manner that allows them to efficiently and effectively accomplish tasks. These executive thinking skills can be applied with equal efficacy in academics, work, and social situations.

Important Note

This chapter is reasonably detailed and reviews some of the neurofeedback and biofeedback principles and examples given in previous sections. This has been done in order that this section can be read and used independently by learning center/clinic staff, senior students, and parents.

Defining Metacognition

The term metacognition came into use in the mid-1970s at the same time as the first studies of neurofeedback for ADD were published. It was a new term for thinking skills that have always been with us. Metacognition may be understood as thinking about thinking and learning about learning. Metacognitive strategies are those executive functions of the brain that go beyond ("meta" in Greek) regular thinking (cognition) and allow one to be consciously aware of thinking processes, such as how to learn and remember things. Strategies for listening, reading, organizing and remembering are included in this training. The effective learner is able to select appropriate strategies and monitor their use.

In a nutshell, metacognitive strategies boil down to a three-step approach. You ask yourself: 1) What is the job here? 2) How can I best accomplish this job? and 3) How did I do? Metacognition thus involves the executive functions of the frontal lobe, including planning and evaluating.

This chapter will describe an approach that combines self-regulation training with direct instruction in metacognitive strategies. Self-regulation training includes neuronal regulation (EEG biofeedback) and autonomic nervous system regulation using biofeedback of other modalities, such as respiration, temperature, and skin conduction. The examples used in this chapter mainly deal with training students who have Attention Deficit Disorder. However, the principles and strategies described apply equally well to top students, professionals and athletes who wish to further their performance. We emphasize metacognition, because using strategies is what separates superior students from weak ones. Awareness and teaching of strategies varies a great deal from teacher to teacher. We feel that it is an essential component of our NFB work with students. There are good articles in the literature concerning the relationship between metacognitive strategies and school performance (Chen et al., 1993; Palincsar, A.S., 1987; Weins et al., 1983).

Combining Neurofeedback and Metacognitive Strategies

Neurofeedback training for attentional difficulties is based on observed links between brain wave patterns and particular mental states and behavior. It trains the student in self-regulation of brain wave activity. For example, students who typically would produce an excessive amount of high-amplitude slow waves are trained to decrease the amplitude of these waves while attending, reading, writing, and listening. Less fluctuation in slow-wave activity means attention is steadier, since bursts of theta are associated with flickering attention to external stimuli. (It is important to note, however, that theta and/or alpha waves will be briefly observed when the student is recalling information.) Simultaneously, these students increase the production of faster waves associated with activation of the brain for problem solving. When this balance is achieved, students report that they feel focused, attentive, and that they organize and recall material better than they have ever done previously. It is also entirely logical to use this ideal learning state during the neurofeedback sessions and to simultaneously train these students in metacognitive strategies to further increase their learning effectiveness and efficiency. The pairing may also stimulate more rapid transfer of their training to classroom and study situations.

Teaching strategies while you are monitoring brain wave activity means you teach only when you know the student is paying attention. This greatly enhances learning.

In addition to improving attentional processes and reducing impulsivity, goals for training include reducing anxiety and increasing alertness. The overall objective is to improve mental flexibility so that a person can produce a mental state appropriate to situational requirements.

Dealing with Anxiety

In addition to needing help with focusing and using strategies, many students demonstrate performance anxiety in classroom, athletic and social situations. It is associated with fear of failure when some performance is demanded; for example, answering a question, reading a passage, speaking in public, or skiing down a steep slope. This type of anxiety markedly inhibits performance. Often the same task could be performed easily if it were not a demand situation; for example, these children may spontaneously answer a question that is directed to the whole class, but they cannot produce that same answer when singled out without warning. Going blank on a test is another example of anxiety interfering with performance. People with ADD are particularly prone to this problem, and it is brain-based: SPECT scans show decreased blood flow when the child is given a stressor (Amen, 1997). Dan Amen used a math task as the stressor and Sterman also feels that mental math is the best task for eliciting slow waves (tuning-out) in those with ADD. Decreased blood flow parallels increased theta activity. Parents often report frustration that their child knew the material when they reviewed it at home and then forgot it during the exam.

Skin temperature is one physiological measure that reflects anxiety and tension. It can be easily monitored using a tiny heat sensor placed on a finger. The fingertip skin temperatures may initially be low, sometimes in the low 70s, rather than the low to mid-90s on the Fahrenheit scale. Most clients quite quickly learn to self-regulate their temperature. They are able eventually to increase hand temperature at will, and this correlates with a more relaxed state. One client, for example, began to use hand warming prior to ice-skating performances and her coach noted her remarkable improvement (two senior levels in a month).

Many candidates for training also demonstrate very low or labile levels of alertness. This is monitored by measuring skin conduction (EDR). Electrodermal response (EDR) is an autonomic nervous system measure that reflects arousal and alertness. Clients learn to recognize and regulate their arousal level. Instead of drifting off towards sleep when they perceive the teacher as boring, they can choose to stay alert. Both the self-regulation of the EDR and the application of metacognitive strategies help in such situations.

With modern equipment you can now monitor and give feedback on peripheral skin temperature, EDR, EMG and respiratory sinus arrhythmia while at the same time giving EEG feedback and carrying out learning exercises involving the use of metacognitive strategies. The trainer can choose the appropriate combination of modalities for each client. Students are taught specific techniques for attaining a *eustress* physiological state wherein they remain highly alert yet relaxed. (Eustress is the term coined by Hans Selye for a level of stress that is optimal for the individual (Selye, 1976).

Dealing with Daydreaming

Appropriate candidates for combined NFB, BFB, and the teaching of strategies are, in particular, those who have a propensity to slip into daydreaming or drowsiness at inappropriate times. They may tune out in class, when they are doing homework, when the coach is giving instructions or even in everyday conversations. These states are associated with excessively high levels of slow brain wave activity. In most children, this slow-wave activity is in the theta bandwidth, approximately 4 to 8 Hz (Hertz, shortened to Hz, means cycles per second). In adolescents and adults it may also be associated with increases in "thalpha" (6-10 Hz) or in alpha brain wave activity (8 to 12 Hz). These EEG patterns appear to correspond to drifting off topic and brief or prolonged cessation of concentrated work on the subject matter at hand. Clients learn through the feedback of brain wave activity to self-regulate their attention, increase their concentration and maintain their focus until a task is completed. When they start to drift off, self-regulation training allows the student to recognize rapidly that they are no longer tuned into working on the subject matter, and to redirect focus back to the primary task.

Dealing with Learning Disabilities

Many, but by no means all, of the students with ADD also have specific learning deficiencies. They underachieve with academic performance that is low relative to their overall intelligence. All of the candidates demonstrate difficulties in working efficiently at academic tasks, although sometimes they will produce at the last minute. In the majority (exceptions often being those with high anxiety), organization and timely completion of work tasks is a major difficulty. All of the candidates benefit from the combined training using feedback and teaching of cognitive strategies, but those with specific learning disabilities need them most of all. The key in working with LD students is to develop strategies that use their strengths to help them compensate for weaknesses. Thus, the child with weak spatial reasoning skills can be taught to use their stronger verbal abilities to "talk their way" through tasks where other students may just "see" the solution. One example would be learning to read maps. The visual child may just picture the compass points to orient himself. The verbal child will have an internal dialogue, "North is at the top so, if I turn the map upside down, north will be at the bottom and south is at the top." The trainer must be perceptive and figure out how the child best encodes information, or they may offer a number of ways; for example, using visual, auditory, and kinesthetic approaches to learning the letters of the alphabet.

Client Example

"C." is an 11-year-old boy with severe learning disabilities and ADHD. His family is extremely supportive and both parents are teachers. In his history it was related that he had been called "the most profound learning disability ever seen" at a major Canadian hospital's child development clinic.

Despite intensive special education his reading and arithmetic scores on standardized tests were only at an early Grade 2 level when he started the program on August 8, 1994. Re-testing on November 20, 1994, after 40 sessions demonstrated that reading and math were both up to a Grade 5 level. His TOVA (Test of Variables of Attention) profile had also shifted towards a normal pattern. He sits calmly, is no longer restless and fidgety, and can listen attentively. This boy could not multiply even 2x2, and now he takes great pleasure in being able to do easily and quickly all the multiplication tables and has been learning the 13 times table on his own "just for fun." Most importantly, he no longer says, "I can't," but eagerly jumps into each new challenge and really enjoys learning.

*More than **a decade later** we received an email from his mother. She gave us permission to use her email in any publication or talk. She said, "Here is a short history of C... - A Miraculous Success Story."*

During C's elementary school years, his sister would take him to her university to see plays. Afterwards, the students would naturally discuss the particular play. They were astounded by the in-depth analysis that her little brother would contribute. C. is extremely perceptive and can naturally perceive and understand the subtle undercurrents of human drama. Her cohorts thought C. was some kind of genius or child prodigy. They found it hard to believe not only how young he was, but also that he had a severe learning disability.

C. was routinely elected class president of his high school classes. He was heavily involved in sports and was the captain of the school's rugger team. He graduated from his Collegiate as a Provincial scholar with first-class honors. He was also valedictorian for his final year in high school.

After graduation C. took and graduated from a College program in Computer Engineering (we thought university would be too difficult and stressful for him with his disability). However, after graduation he decided he would like to entertain the idea of becoming a teacher so he enrolled at his home-town university in the four-year Computer Science program. In the summers during his university career, C. worked at the Military College. His assigned task was to program their SLOWPOKE reactor to analyze samples automatically. This turned out to be an extremely frustrating and virtually impossible task as the computer program was completely inept, there was no program manual, and the program company refused to help him. Despite almost insurmountable road blocks, he ended up creating a user manual himself and getting the reactor running properly.

C. was the only person to achieve this for the government's slow poke reactors. Subsequently, the College has offered C. a Master's program with them and he is being actively pursued by an up and coming computer software company."

Which Children Benefit Most from Training?

Parents often ask if intelligence is the key factor involved in successfully dealing with ADD. It is true that, as with other kinds of learning, learning self-regulation is generally easier for children who test at a high level on standard intelligence tests. However, IQ scores do not reflect other variables that are important in achievement, such as perseverance or creativity. Intelligence tests originated with Simon Binet, who was working for the school system in Paris, and wanted to develop a tool to help predict which children would benefit from the school system. Standardized intelligence scales like the Stanford-Binet and the Wechsler scales are still used as predictors of school success. They measure skills associated with doing well in academic settings. Intelligence, in the broader sense, is made up of a number of components that include:

- Areas predicting school performance (as tested on standard IQ tests)
- Memory (short-term and longer-term and types of memory such as visual or auditory)
- Motivation
- Persistence (and the factors that seem to stimulate it for a particular child).
- Creativity
- Goal-setting ability
- Self-confidence, "street smarts" and the ability to read social cues
- Approaches to learning and remembering
- Attention span and ability to concentrate

Goleman has termed some of these attributes "emotional intelligence" (Goleman, 1995). Each of the foregoing factors are interrelated. Given a basic modicum of natural ability with respect to the above areas, the effective student also requires the ability to turn on:

- A relaxed (not tense) state of mind and approach to learning
- A high level of alertness
- Flexibility and control of mental states (not in a meditative alpha state or a drowsy theta state when attempting to problem solve complex material)
- Focus and attention with the ability to exclude irrelevant material
- Concentration and a problem-solving state of mind (associated with beta wave production)
- A thoughtful, reflective, considered style (not impulsive)

It is these latter factors that can be most directly affected by neurofeedback training, and these in turn affect each of the factors listed under "intelligence" above. Outcome studies that have included pre-post testing of intelligence have reported increases in overall IQ scores of at least 10 points. ADD Centre retesting on the Wechsler subscales most affected by attentional factors (general information, arithmetic, digit span, coding) consistently demonstrate gains, often of 3 to 4 scale score points (Linden, 1996; Lubar, 1995; Othmer, 1991; Thompson & Thompson, 1998). Data for 103 subjects seen at the ADD Centre are shown below. These data represent an update done late in 1998 of the case series reported in the literature.

	PRE	POST
Verbal (+ 9)	104	113
Perf. (+ 12)	105	117
Full (+ 11)	105	116

The problem with the majority of students who have ADD is that their memory, at least for subject matter covered at school, is poor. Some of these students have given up trying to remember material and use both conscious and unconscious defenses to excuse their shortcomings. One of the more common defenses is some variation on an "I don't care; it's all useless and irrelevant anyway, I want to quit school as soon as I'm old enough and earn money" attitude. This may be directly and confrontationally, or passive aggressively and indirectly, expressed. The passive-aggressive stance is perhaps the most difficult to deal with. In this stance the student may begrudgingly agree to do the work or even smile and be quite pleasant and agree that the work will be done. Then, despite many reminders, it is not completed, completed far below their ability, or completed but just not handed in on time or at all. The terms 'lazy' and 'unmotivated' are often applied to these students, but it really is a neurologically based problem with focusing and sustaining attention for things that are, frankly, boring for them. Over the years these students have become discouraged. Their self-confidence in many academic areas is low. Many have given up.

What we do about this problem is give them hope, because once they start using strategies, life becomes a little easier. Our initial job is to catch their interest in a nonthreatening fashion. The initial interview, where they see their own brain waves and can even change things on the screen by focusing and concentrating, is extremely helpful in stimulating their interest. Virtually all, even the initially reluctant students, want to come back.

But neurofeedback training takes time. Some of the students, though initially fascinated, become quickly discouraged as the novelty of training wears off and the real work of exercising the brain to make a lasting shift toward more mature patterns begins. Metacognitive strategies, apart from being an integral part of the program for maximizing the student's potential, are also a good means for catching interest and producing immediate changes in their academic endeavors. When taught without the feedback, in our experience, the majority of students would use only a few strategies and would return rapidly to their old patterns. When taught during neurofeedback, the students appear to apply many of the strategies on an ongoing basis. Their initial and continuing interest in working on strategies during sessions is also completely different than when strategies were taught without neurofeedback. Again, research on any single component (metacognitive strategies, NFB, HRV) does NOT reflect the true effect(s) of the interventions combined.

Overview of How Metacognition Fits into Training

Here is how metacognition can fit into the training program. We use three to five overlapping stages or steps. Some students have developed habitual counterproductive styles of coping with perceived stress. These coping styles or "bracing" techniques may include becoming very tense and anxious, displaying decreased arousal, opting out, and "distress" autonomic states. These coping patterns are usually learned very early in a child's life and are automatic and usually outside of cognitive awareness. The first two steps in training help counteract these negative coping styles. They are not necessary for all students. Most younger students go directly to step number three and do neurofeedback plus strategies with some work on diaphragmatic breathing added as needed. Adolescents and adults are more likely to receive the total biofeedback, neurofeedback and metacognition combination.

The **first** step, learning diaphragmatic breathing with RSA feedback and raising fingertip temperature, is used with candidates who report performance anxiety. The **second** step, regulating the EDR, is emphasized in candidates who have low arousal (alertness) levels in class or work meetings and who exhibit low or labile EDR. Students report that they feel more relaxed yet alert, awake and energetic when they learn to control these parameters. These first two steps create a eustress state and are relatively quick to learn. They give the client a real sense of empowerment, since self-regulation of temperature and arousal is easier to learn than self-regulation of brain waves. Mastering these first two steps gives the student confidence that they will master the steps involving brain waves, too. The third, fourth and fifth steps run concurrently with the first two, and are done by all clients.

In the **third** step, the student is trained to hold the slow-wave activity at a low microvolt level with decreased variability. Decreasing theta standard deviation and variability results in the student maintaining a steady focus on a topic for increasingly longer periods of time. Parents and students often initially ask what they should be doing in order to decrease both the amplitude and the variability of the theta wave. We use the analogy of learning to ride a bicycle to help them understand that, just as one cannot explain how to "balance," one cannot put into words how to control brain waves. In learning to ride a bicycle the brain receives direct and immediate feedback from the inner ear concerning going off-balance. In training using brain waves, the student is receiving direct and immediate feedback (less than 50 milliseconds delay) concerning going off-focus. Given this directness and immediacy, the student trains to self-regulate and, just as in riding a bicycle, the new learned behavior remains accessible and becomes, for the most part, automatic over time.

The **fourth** step emphasizes the student's ability to find the mental state in which they can continue to hold down the slow-wave activity and increase the fast-wave activity. When in this state, students report that they remain acutely aware of their surroundings but remain totally "absorbed" by a single train of thought and mental activity. We often liken this alertness and focus to the mental state of a high-level expert in the martial arts. The students at the ADD Centre, when highly focused, sometimes report a concurrent sensation in their abdomen and occasionally a mild headache if they come out of this state too rapidly. The fast-wave activity that is being trained may be in the SMR range (sensory motor rhythm, 13 to 15 Hz) or the beta range (15 to 18 Hz or 16 to 20 Hz), depending both on the presenting difficulties of the child, their EEG pattern on assessment, and their initial response to whichever range is initially used. The more impulsive children are usually started in the SMR range. The electrode placement is referential ("monopolar"), and it is usually placed in the central Cz position with the reference and ground electrodes being placed on the ear lobes. In some children who are not impulsive and who have specific left-brain academic difficulties (particularly in reading or language), the electrode may be positioned on C3 initially and then over Wernicke's area. Other electrode positions may be considered for special cases, but are not as frequently used. Electrode placement and bandwidths enhanced or inhibited vary according to the objectives for training and the EEG assessment.

The **fifth step** is to continue doing the first four steps while reading, listening, doing math or written work in a manner that is extremely well organized and that utilizes metacognitive strategies. There is coaching done regarding strategies to increase the student's ability to assimilate, organize, and recall information.

Note that the strategies are taught, modeled, and practiced while the student continues to receive feedback. If they are looking at a page of math, reading or written work, it is the auditory feedback that will give them information about their focus and concentration. The trainer can monitor the relevant EEG parameters and have them stop the academic work and return to pure neurofeedback whenever they tune out. Thus, the metacognitive work is always paired with an alert, focused state. The initial part of a training session is always pure feedback and then strategies are begun after the client has demonstrated she is remaining focused.

The rest of this chapter is a mini-catalog of strategies. Regard them as samples, but recognize that the number of strategies is limitless. The trainer's job is often to do their own task analysis concerning a topic their client is having difficulty with, figure out which part of the task is presenting difficulties, and then come up with appropriate strategies for handling that part of the task.

Sample Metacognitive Strategies

Advantages of Using Metacognition

The strategies presented here are not always new to the client. Good teachers and parents may have tried to share them previously. Unfortunately, the ability to enter and remain in a mental state wherein these strategies are actively and continuously used over a period of time has been a missing component. Being able to apply a strategic approach to academic tasks is almost always an entirely new experience for the students.

At the ADD Centre, we typically train the students in strategies for listening, reading, math, and organizing written work. We train the older students and adults in time management and any other area they desire. Most often it is reading skills. Our first adult client at the ADD Centre was a 56-year old lawyer who reported having to read things three or four times because his attention wandered and he could not recall what he had read when he got to the bottom of a page. Learning to maintain his concentration while reading increased his efficiency and shortened his working day.

Introducing Learning Strategies

When using metacognition, you think before you jump into a task. A simple method may be used to introduce reading comprehension exercises. The student is taught to use the **W-W-H-W paradigm**. That is, "Why am I doing this?" and, "What is it I wish to learn or accomplish?" "How am I going to approach this task?" strategies. These questions, explored in a simple, enthusiastic manner, will engage even early primary grade students, and all ages enjoy the positive reinforcement that follows the final question; "**What have I learned in this session?**" At the completion of each step when working with children, tokens are awarded. Tokens can later be exchanged for prizes, including gift certificates for the local bookstore and music store, when enough have been accumulated in the student's 'bank account.' The bank account sheet is the first page of the student's training file, and tokens are an integral part of the motivation during training.

Memory

The first step toward teaching strategies to students age 12 and over begins with an exercise that is carried out to underscore the need for strategies. Memory is key for learning, so the demonstration relates to the ability to remember simple material. This, in turn, has the effect of catalyzing students' motivation to learn techniques that might improve their memory. The trainer then uses this opportunity to teach the student time management, study organization and a basic learning-to-remember strategy.

Establishing the Need for Memory Strategies

To recall facts for a test or examination, the facts must have been processed through immediate, short-term, intermediate, and long-term memory. Students often feel that their memory is less than it should be. Early in training. students are therefore presented with a short challenge in a fun, game-like manner. They are asked if they include regular, daily review of material presented in their classes as part of a routine study time. The majority do not. They are then asked whether, if they added this to their routine, would such a habit require more or less study time. These questions usually lead to an interesting interchange.

Students are then challenged to a memory task involving telephone numbers. Most agree that they do not find it difficult to recall a telephone number. The student is then given a number to remember. It is easily recalled. Then they are given a second phone number but before they can reply they are asked to look at a picture on the wall and name three colors that are in it. Virtually none can then recall either the number they were just given or the number they had recalled correctly just a minute before.

They are reassured that the failure to recall the numbers when there is interfering material (the distraction of another question in this case) means they are no different than everyone else. They are then asked how this might relate to listening to a teacher talk and then being asked, even five minutes after the beginning of a lecture, what has been said. The analogy is a powerful one, and usually leads to a productive discussion and motivation to improve their memory.

The telephone number demonstration has shown that they have a good immediate memory but they have also just demonstrated to themselves two other facts. First, their immediate memory will fade almost instantly unless they do something else with the information. Second, any simple, quick distraction will interfere with their memory. The trainer, almost in passing, notes that when a person does even a few seconds of work on material (for example, attempting to associate it with something familiar or amusing), they usually will be able to recall it an hour later. The student is then asked to read a few lines of material, make some amusing association and recall it at the end of the training hour. They are then left with the suggestion that they may be able to learn how to recall even a large amount of material if they first learn ways of organizing it. Students are usually shocked that immediate memory fades so quickly, but upon reflection, they agree that "in-one-ear-and-out-the-other" is typical of their memory for new information. Yet they have noticed how easily they can remember amusing things that remind them of something else. This leads to a better understanding of why they must review new material at least once again the same day it is taught in order to lay it down in intermediate memory. This may last longer than a minute or two, or even longer than a couple of hours or days, but it will usually fade within the week. More work must be done with the material for it to be placed in long-term memory. The trainer reassures the student that working together on memory strategies will make school life a little easier. We will ask him to do a simple three-key-facts experiment in class during the course of the next week and report back the results. Then, during our next few sessions, we will coach him in methods for reading and remembering.

Three-Key-Facts Technique

This task requires that the student experiment at school by writing down three "key" facts in each class. (Of course, it may be one fact in one class and 10 in another.) They are told that teachers almost always give away all the questions they are going to ask in the next exam during their lectures. Teachers cannot resist emphasizing what they think are the most important things that they are teaching, and will give away their feelings about important subject matter in various ways such as: saying it more than once, changing their tone of voice, gesturing, writing it on the blackboard and so on. Students may even notice that the teacher gets excited and their pupils dilate when it is something they find particularly interesting. Some students get really involved in playing a "beat-the-teacher" game by predicting the questions on the next test and thereby scoring a top mark in their class.

Case Example, Jane

Jane, age 19, complained that she couldn't recall anything from her classes. She had failed her last set of exams. She noted that when reading and trying to study, she couldn't even recall the material in the first paragraph by the time she reached the end of the second paragraph in a text book.

Her NFB trainer asked her to remember a telephone number. She repeated it back easily. Then a second number was given but she was interrupted before she could repeat it back by being asked to look at a painting and count the children in it. When asked to repeat the second telephone number, she couldn't. The trainer remarked that she had easily memorized the first number, so would she please repeat it. She couldn't. Putting this in the context of trying to remember things during an hour-long lecture when her immediate memory was five to six seconds long convinced her she needed some mnemonic strategies.

One suggested technique was the key facts method. She would look up tomorrow's lecture topic in her textbook and in the notes she had from a student who took the course last year. She would only spend a few minutes doing this before going to bed. She would scan the headings and then the subheadings and look at any picture or table that caught her interest. She would try and talk to herself about what the lecture tomorrow should cover and how it might be organized. When she got to the classroom she would sit near the front and put a small pocket notebook beside her main notepad where she took lecture notes. Every time the lecturer made a gesture or altered his tone of voice in a manner that she detected as meaning that he thought this was a key fact or point, she would write it in her notes but also scribble it in short form in the small notebook beside her. At the end of the lecture she would glance at this small list of facts before walking to her next class. During the class, she would mentally compare the teacher's organization with the textbook and if she felt sleepy or if she was falling behind, she could ask a simple question such as, "Are you going to cover _____ today?" She knew the topic from her last evening's scanning of headings, so it was relevant and not embarrassing to ask, and it would serve to raise her level of arousal and get her emotionally involved in the class. In fact, she might just think about adding a question without actually doing so. This, too, would raise her alertness level and get her feeling more involved in the class.

When she got home, she would put these facts neatly onto a list at the end of her notebook. When she took the same subject two days later, she would add the next few key facts to her list at the back of her notebook.

Somewhat to our surprise, she tried this technique and reported to us that she could recall the key facts when she got home each evening, and after copying them again neatly, she could recall them a week later. At the end of the term, she shared with her trainer a straight A report card. She was near the top of the class. Her succinct conclusion was that she had every exam question in every subject covered in her list of key facts. How could she lose!

It is true in Jane's case that she was ADD and was also doing NFB training to help her focus, so the key facts strategy was not the only intervention. Nevertheless, without a technique to help her improve her recall of relevant facts, we do not think the results would have been this remarkable.

Day Planner and Three-Facts Scribbler

For some students, we suggest that they purchase a day-planning book that we call their ADD-PADD (ADD Centre Planner and Distraction Dissipater). Planners are commercially available. They can also use their school's student agenda. They use the day page for scribbling down the key facts in short form. Then they transfer the facts neatly to a summary sheet for that subject in a section in the front of the book. Alternatively, other students may choose to just write the three key facts from each lecture in a small pocket book, and then transfer them neatly to a blank sheet in their binder for that subject. These should be facts that the student feels are likely to be asked in tests and are facts that should be committed to memory. An example in math could be the formula for the surface area of a sphere. In history class it might be several dates, names, and events. In biology it could be the definitions and critical characteristics of DNA, RNA, and ribosomes. The facts from each class normally take three to five lines, and the total for the day may take less than a page. The student is asked to do one more step. On the way home or while having a snack after reaching home, for five minutes they are to really concentrate and transcribe carefully each of the facts that they feel absolutely must be memorized for exams to another section of the ADD-PADD. This section is divided into school subject areas with three or four pages per subject. When the experiment is completed, the full use of the ADD-PADD is outlined and worked on with each student for keeping track of their school and personal dates and deadlines, and for maintaining sections for every major area in their life to record internal ruminations and distractions (hence the label "distraction dissipater").

Case Example, Vince

Vince was an 18-year-old high school senior. He came to the ADD Centre in September, at the beginning of the school year. He had almost failed the previous year and was in the bottom 10 percent of his class. Testing demonstrated that he was a bright student. He found it very difficult to concentrate in class and literally fell asleep doing his homework in the evenings. Vince was at first reluctant to come for NFB training. His twin sister felt the same way. She dropped out after four sessions. She failed her final high school year. Vince seemed to be somewhat interested in observing his own brain waves. He remained in training. The author introduced Vince to the three-fact method in his second NFB training session. Four months later, Vince came into his first session of the new year and threw an envelope on to the author's desk. When asked what it was, he just said, "Have a look." The envelope contained his report card. All his subjects: algebra, calculus, physics, chemistry and English, were in the 90s. Vince stood first in his class.

I inquired as to how he accomplished this turn-around in such a short period of time. He responded, "Look in my back-pack." I did so and found a rather thick day planner. I was not able to read very much of the scribbles that filled each day, but it was clear they were abbreviated notes, facts and formulas. At the front of the day planner, however, were neatly demarcated sections, one section for each subject. In each section there were about 6 to 12 pages. On these pages were printed, with extreme care, lists of phrases, facts and formulas.

Vince put it quite simply, "I couldn't help but stand first. The facts listed accurately predicted nearly all the exam questions and I was ready with the answers."

Vince had made this exercise a game, a contest. He wanted to outwit the teacher. He had decided to see what percentage of the term exam questions he could guess. He guessed correctly for 100 percent of the questions on some of the exams. This is admittedly unusual, but it can happen.

Vince's story is exceptional. He was very bright and could use that strength once he had improved his ability to sustain focus during class and while doing his homework. He had made use of the other strategies outlined in the remainder of this chapter. However, he did conclude that the three-fact-method had made a significant contribution to his scholastic success.

Vince Used SMIRB

The author noted that there was another set of section markers at the back of his day planner. Vince said this was his version of SMIRB. SMIRB was another suggestion the author had made about halfway through the fall after he noticed that Vince was worried about certain things that were happening in his life. He was worried for his sister, but she wouldn't talk to him. He was concerned about other things that were happening in his family. He couldn't make up his mind about what to do next year. He had brought this up because he felt he was wasting a great deal of time in unproductive thinking. It was also interfering with sleep.

*The suggested technique was that he make a section of his day planner into a **S**top-**M**y-**I**rritating-**R**uminations-**B**ook (SMIRB). This would have two pages devoted to each major worry area. He would list his concerns on the left-hand page, and on the right-hand page he would list possible options for future action. He would put aside one time of day for a 20-minute worry. At that time he would open this section of his book and think about things. If he began to ruminate at any other time of day or night, he was to make sure the 'worry' was listed in his book. If it was, he would forget about it and get on with his work (or sleep) knowing he could think about it later. If it wasn't listed, he would enter it in his SMIRB.*

Vince said that this little technique had also been a positive factor in his being able to achieve high marks. It freed his mind of worries so he could focus on his studies and social life. (We should also note that SMR enhancement can have a significant effect on improving sleep. This may also have made a positive contribution to his success.)

Reading

Four Steps for Reading

Overview

In each step outlined below, the student carries out an internal dialog in which they generate questions concerning what they wish to learn from reading a passage. Then they predict answers to their own questions. Thus, reading becomes a search for the answers.

The four steps or stages for reading a textbook chapter occur:
1. Before opening the book
2. Upon first looking at the chapter
3. While reading the chapter
4. When reviewing what was read

In this process the student is always an active participant and emotionally involved. The student should never be a passive recipient in the learning situation. The students learn that the responsibility for learning being interesting resides with them and not the teacher.

Early in training, the student is introduced to the approach outlined above for reading textbook material. The following exercise has been found to be a useful starting point. Note that strategies are integrated into a neurofeedback training session only after the feedback indicates that the student is concentrating.

The "Nelson versus Napoleon" Exercise

A short example can demonstrate reading (scanning, reading for detail and chunking) and memory strategies (associations and visualization). One example that is frequently used at the ADD Centre for students at or above a Grade 8 level of reading is three pages taken from a history textbook, *The British Epic*. (Grade 3-6 level students choose an animal or bird described in a National Geographic book series and carry out a similar but simplified series of steps.) The student is told that we are giving them just 2 ½ pages of reading with large pictures on each page, that it is about a war that took place in the 1790s, and that we will ask them when they finish it what they had learned. The student receives the information that the reading is about a war, and the approximate time is given, so a good student could use learning strategies for the reading if he knew how to do this. The first question we will ask after the student has read the passage is deliberately open-ended, and no instruction is given as to how they should go about organizing their thinking before reading or about how they should scan, read and learn, or how to review at the end of paragraphs, pages, sections and at the end of the passage. Instead, it provides a baseline for how they currently read and recall information.

This initial reading exercise has been used on hundreds of students, many of whom were in the senior years of high school, university or in postgraduate studies. Nearly all used basically the same procedure. They opened the text, went directly to the assigned page and began reading, starting with the first word of the first paragraph. Only a few of them bothered to look at the titles or the author or observe the time line giving dates and events at the top of the first page of the chapter. Those with extremely high intelligence remembered a sprinkling of facts, and one, a lawyer, recalled the basic strategies of the British and the French. Even though little is recalled, the trainer must remain very positive. The idea is to be a coach and the strategies are presented within the framework of a game and an enjoyable challenge. The student is told that they will now try a few tricks with the trainer to see if they can increase their recall and they are going to do this in four stages.

Four Steps in Detail

Step 1: Before Opening the Book

First, before reopening the book, they are to play a game with the trainer in which they think of all the things they might expect to read about in these pages, given that all they knew initially was that the passage is about a war in the 1790s. They are asked, "What would you like to learn?" Often the trainer must model speaking aloud the thoughts that the student will later internalize. Here and throughout the following steps, you are training them to use what cognitive psychologists call *verbal mediation techniques*. In other words, they use an internal dialog to talk their way through a task.

Generate ideas, headings, and questions and organize these into a grid: Within the framework of generating questions and organizing ideas, the student is guided into forming a simple mental grid. They draw the grid on a piece of paper to provide a visual reminder. When using the grid technique, the rows represent headings for grouping the information that the student wants to learn concerning each of the areas represented by the columns. In this example there may be just two columns representing the two countries: England and France. The rows would be generated by questions commonly referred to as the reporter's questions: **Why** – there should be at least one clear reason for why each country enters a war; **Who** – the Generals, Admirals, Prime Ministers, soldiers, sailors, merchants etc.; including characteristics or important facts about each; **When** – the dates for events and time frames; **Where** – the countries, oceans, cities, battles etc.; **What** – the strategies developed by each of the countries and events that occurred such as the battles, voyages and so on; **How** – the types of weapons, ships, other transportation, clothing, foods, and so on; and **With What Effect**? both on the countries at the time and on future history.

The student and the trainer attempt to predict answers for some of the questions they have generated. When their ideas are exhausted, they then agree to open the book and begin scanning the author's headings.

Step 2: Immediately After Opening the Book
At this juncture the trainer will ask the student if he thinks that the author will cover all the topics that the student has thought of and put in his grid. Then the student checks this for himself.

Scan the author's headings: Before starting to read, the student scans the table of contents of the history textbook and notes what seemed to be happening immediately before and after this time in history. They then open to the chapter, read the headings, subheadings, introduction, and conclusion of the chapter, review the questions at the end of the chapter, if there are any, and glance at the pictures and their captions. This step takes only a few minutes. During it they enjoy seeing what ideas they had generated that the author had forgotten to mention, and what they had missed that the author had included. They remain actively involved because they are checking out their own ideas.

Step 3: Reading and Remembering
The student has already read the required pages once in their usual fashion. However, they recalled very little information. Therefore, the trainer now reads each paragraph with the student and models techniques for memorizing the important facts. (These mnemonic devices are described and illustrated in more detail in *The A.D.D. Book* by Sears and Thompson in the chapter "Setting Up for School Success." Some will also be described later in this chapter.)

Pictures Help Recall of Characters: The trainer may develop a picture such as the Mapping Techniques. Create a mental map (or sketch one) showing Great Britain, Europe, the Mediterranean and North Africa with the West Indies in the distance on the left and India on the right. England has a large hole in its center, representing the Prime Minister "Pitt." Europe has dollar signs (or English pound signs) in each of the countries surrounding France, and there are ships surrounding France in the English Channel, the Atlantic and the Mediterranean Sea. This represents the British strategy. The French have guns pointing at Belgium, Holland and India, representing both the reason for the English becoming involved in a war and the French strategy. Students with ADD tend to be creative with good visual skills (remember Hartmann's Hunter Mind analogy) and creating mental pictures, or even sketching them, provides opportunity to use those positive attributes.

Associate Names with Known People, Places, Things, or Amusing Pictures: Horatio Nelson is represented in a Neilson's chocolate bar wrapper standing on deck looking at a book (battle at Aboukir Bay) with a sphinx in the background (Egypt), Santa Claus on a cruise ship is by one empty sleeve (lost an arm at Santa Cruz) and an eye patch with an apple core or a small car on one eye (lost an eye at Corsica). Napoleon Bonaparte is depicted as rowing away in a small rowboat with his ships being "blown-a-part" in the background.

Students find this exercise amusing and informative.

Step 4: After Reading the Chapter
Speed-read review of the material: The trainer models a quick scanning (rereading) review of the material using the first few paragraphs. In this demonstration the trainer uses a speed-reading technique that involves 'chunking,' grouping important words and phrases together and scanning through unimportant details to find the key words and phrases. The facts are then related to a revised grid or set of mental pictures that will assist future recall of the material.

When the student is asked what they remember after using these techniques they are usually surprised at themselves in that they not only recall more detail, but they also do it in a more organized fashion. They are even more amazed, when they come into their next session and are asked to do a quick review of the passage, at the immense amount of detail they effortlessly recall in an organized fashion.

During the final review, the trainer will ask for much more detail. Taking the passage about the Napoleonic wars, for example, a question might be, "What did Nelson look like?" (He is missing an eye and an arm.) If they know then they are asked, "Where did he lose the use of one eye? (They may have pictured an apple core on the eye patch and recall Corsica.) and where did he lose his arm?" (They may have pictured Santa Claus on a cruise ship and recall Santa Cruz.) At this juncture that trainer may demonstrate, model, or develop through Socratic questions, more advanced methods for laying material down in memory using mnemonic devices that include associations and visualization. These are discussed in the following sections.

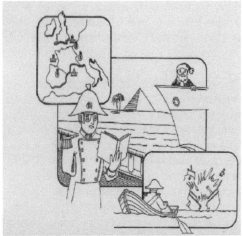

Illustration done for the ADD Centre by the late Dennis Dunn.

Summary of the Four Steps for Reading a Textbook Chapter

1. **BEFORE I OPEN THE BOOK**
   A. Why? What is my PURPOSE?
   B. Why" What is author's purpose in writing this chapter? (Set tone, get in the zone – relaxed/alert/focused/concentrating/steady
   C. MAKE ASSOCIATIONS (develop questions)
   the tree and its branches
   D. ORGANIZE AND SYNTHESIZE
   • Headings/grid and the organizing principle/the red thread
   • Scaffolding and linkages

2. **Right After I Open the Book**
   A. SEARCH AND SCAN
   • Headings/subheadings/pictures/abstract/conclusions
   B. SKIM – key words/phrases – 1st line of each paragraph
   C. ORGANIZE/SYNTHESIZE

3. **As I Read or Listen**
   A. Make notes/underline/reorganize
   B. Use MEMORY TRICKS
   • Simultaneous visualizing
   • Sequential – the Roman room/ mnemonics/rhymes/acronyms/first letter sentences/silly sequenced scenes
   • Make associations

4. Review
   A. Chunk/Key Word Review/Reorganize
   B. What have I Learned?
   C. What is the PRODUCT!

Value of This Approach

Working through these initial discussions and practical experiments allows the students to discover for themselves that applying strategies really does work to make their academic lives easier. The initial three facts and memory exercise leads to the students understanding why they should quickly review last week's material in order to further lay it down in long-term memory. The reading of three relatively simple pages of history gives them basic methods for learning that do not take a lot of extra time. The seed is sown and students become willing to try learning how to learn.

The exceptions are predictable. Occasionally a student from a family where there is disruption, separation or divorce, will resist strategies just as they are resisting other learning experiences. They may be angry and resentful, and not have either the motivation or the emotional energy for learning. Students with low self-esteem comprise another group who may resist trying something new: if they do not try, they cannot fail. Being passive entails less risk. These students need to experience success in learning the self-regulation aspects of training before they can be introduced to strategies. Working on modalities that can be quickly mastered, raising finger temperature, and breathing such that the heart rate follows in synchrony with respiration (RSA), and controlling EDR are the first places to start building a sense of self-sufficiency. To sell the student on strategies, the trainer can comment that these methods, if faithfully carried out, make studying for final exams a faster and more enjoyable process. The student's initial success in recalling material more easily makes this statement believable. These initial steps also lead into further discussion of memory strategies. (See below under Memory)

Basic Principles that Underlie the Four Strategic Reading/Listening Steps

As sessions proceed, the students are challenged with further listening, reading, and presentation-preparation tasks. The same strategies apply when listening to a lecture as when reading a chapter. The tasks are at appropriate levels for their intellectual and academic abilities. (Psychoeducational testing carried out prior to beginning training can be used to establish these levels. Failing the availability of such data, use grade placement and report card information.) These tasks allow the trainers to expand on the strategies and the principles that underlie them. The students are always encouraged to use the neurofeedback in the first half of the session to help them achieve a highly focused, relaxed, alert state before this desirable mental state is paired with an academic task. The auditory feedback is continued while the student learns the strategies and works on the tasks. Any time their focus is lost they can return to watching the screen until it is regained..

Generating Alertness and Motivation

Students must be attempting to figure out something that is important or interesting to them in order to maintain an optimal level of alertness, focus and concentration. This is the first of the four strategic steps: forming a personal challenge at the outset of each lecture or assignment. That the teacher or one's parents will invoke a consequence if one does not do the work is sufficient motivation to sit in a seat, face the front and pretend to listen or work. It is a good reason for falling asleep, because if a student is forced by someone else to do something, the student may unconsciously sabotage the activity and sleepiness is a classic method of so doing. To be motivated to learn, the student must formulate a personally meaningful reason to pay attention and concentrate – which is particularly essential for students who have ADD. This is relatively simple if some aspect of the subject matter is interesting for the student, or if the subject itself is important for a secondary reason. An example of the latter is learning the cardio-pulmonary system for a student who is advancing in the Red Cross and Life Saving levels in swimming. A future air force pilot may fail a high school math credit, yet later do exceptionally well on the math required for navigation when he is a cadet. Knowing he would be dropped from flight training without a good grade made that math a relevant challenge. Older students may motivate themselves with a challenge such as, "By the end of this lecture I will have produced a superior and more organized outline of this topic than the presenter is about to give." This stance requires that students become exceptionally alert and form their own personal organization and grid of the material as it is being presented.

Generating Questions

When reading mysteries or thrillers there is not a problem staying engaged in the process. The nature of the stories makes one automatically generate questions, including the classic "Who dunnit?" The trick for the student is to somehow transform their textbook into the equivalent of a page-turner mystery. In the Nelson vs. Napoleon example, the student was encouraged to generate questions and predict answers to their own questions concerning the material to be learned. The trainer then helped the student reorganize and redesign the scaffolding organization that the student initially made in order to logically order the information. First, the student "free associates," scribbling down, in point form, ideas that come to mind about the topic area. Generating ideas also identifies large areas where the student has only questions. This increases the effectiveness of reading and listening since they are searching for something. Students are taught to continue the process of generating questions and predicting answers throughout the scanning, reading, and reviewing stages.

Generating Organized Thinking

The basic thesis is that **recall is dependent on continuously organizing and reorganizing data** that comes into one's mind. Use the analogy of the garbage-bag mind.

The Garbage-Bag Mind versus the Filing Cabinet Mind

In order to emphasize the need for organization in one's life (and in one's thinking), begin with a dramatic example. The student is challenged to visualize a possible scenario in their social life such as the following: Your alarm didn't go off, you awake to find that two friends just telephoned saying their father is already in the car driving them to the ski hill charter bus. There is just barely enough time to catch the bus, and you must be ready in five minutes so they can pick you up en route to the bus. Otherwise you miss the planned ski day.

The student is then asked how long it will take to find their ski things if all their clothes are contained in a huge garbage bag in the middle of their room. They usually agree it would take much longer than the required five minutes. They are then asked for a solution. They normally suggest labelled, organized drawers or shelves with gloves in one, socks in another, and so on.

A comparison is then drawn. Equate entering a classroom or opening a textbook without mental preparation with the garbage-bag (disorganized) mind. Compare that to a mind that is ready to take in and 'file' new information. Information is easier to recall if it is logically organized. Even the initially most unmotivated students agree they would like to make their study time more efficient, that is, spend less time for better marks. They acknowledge that they have to sit through the class anyway and they might just as well learn the material there as have the hassle of trying to retrieve it and work on it later when they would rather be doing something else.

Using a Grid

Introduction
A grid technique can be used for organizing thinking, discussing, and writing. It is a major help in creating the Filing Cabinet Mind.

In the Nelson versus Napoleon example, after generating as many ideas and questions as possible, the trainer helped the student to organize the data into headings and a grid. Gridding is only one of several organizing methods. Some students prefer to take the central idea (the topic heading), and draw highways or branches coming out from it and place the most important headings on different branches close to the central trunk. Other students will make a list and gradually add new headings. In Nelson vs. Napoleon, the student was encouraged to use the reporter's headings: Who, When, Where, What, Why, How and with What Effect, to get ideas flowing. More advanced students prefer to immediately make a grid with columns and rows. One of the headings may form the columns and the others the rows.

When:	Before	During (battle)	After (effects)
Who	Nelson Napoleon		
What	Weapons Ships		
Where	Map	Aboukir Bay, Egypt	
How	Strategies Map		

Habitual 'Bracing'
In the Nelson vs. Napoleon example, the student created a simple grid. In later sessions more complex examples are used. However, regardless of age and past education, the first time most students are given such a task they become anxious and resist. The students with ADD may freeze and begin their habitual mental turn-off as a part of their dysfunctional "bracing" when put under perceived academic stress. (Dan Amen's SPECT scans show decreased blood flow when students with ADD are given a math stress.) Biofeedback techniques are used to help them learn to self-regulate and relax when under stress. Even if you are not doing any general biofeedback, you can teach slow diaphragmatic breathing to counteract those old, maladaptive responses to an academic task.

Examples of Organizing and Gridding Problems for Students or Business Professionals

Gridding Task – Design a School
Designing a school is a good, nonthreatening introduction to the grid technique. The trainer begins by asking, "What does a school have to do?" If the student "freezes" (goes blank) the trainer jokes with them and makes fun, ridiculous suggestions such as, "I guess the best thing would be to put the 1500 students in one big enclosed space with wire mesh for a roof and scatter a few teachers and blackboards through the area." This is usually sufficient to provoke even the most reticent participant into suggesting that that would not be appropriate. The trainer then rapidly changes vocal tones and challenges the student, saying this is a perfectly sensible idea and would save lots of money and why shouldn't we design our school this way? The challenge becomes a game, with the trainer playing the naïve straight man. Soon the student has outlined functions that must be carried out by a school that are incompatible with all ages being grouped in a single room with no effective roof. Two key areas that must be addressed emerge: education and health or safety. Soon the structures in the design are being dictated by the needs or functions each part of the structure must achieve. Even the younger students come up with excellent needs, including the need for small classes to decrease distractions and group students of equal competence together for more interesting discussions, the need for books and therefore for a library, the need for healthy exercise and learning teamwork and therefore for a gym, and so on.

In this process the light soon flashes on and the student discovers for himself the equivalent of "needs/functions dictate structure." This kind of quick exercise gives an enjoyable introduction to the idea of generating ideas and organizing them into meaningful groups.

A 'Gridding' Task – Design an Introductory Lecture on the Cardiopulmonary System

The trainer may starts with a question such as: "You must give an introductory lecture to your class tomorrow on the cardiopulmonary system." We have had students as young as Grades 6 and 7 really get interested in this challenge. Having done the first few examples, they know to first ask (and we look up with them) the meaning of each of the words: cardio, pulmonary and system. They then use the same procedure of answering the questions of: "What are the needs (to stay alive)?" This question will help them generate a list: take in oxygen and give off carbon dioxide (or blow up like a balloon!); take in nutrients and give off waste material; the basic structures essential to these processes including tubes for transport of gas, tubes and a medium (blood) for transport through the body, a means for moving (exchange) the gas from the lung to the blood, pumps for the respiratory process and for pumping blood, and so on. Students who have never taken this subject in school draw an amazingly accurate diagram of both systems with little prompting! They can move from their idea to examining a textbook illustration and will have some motivation for doing so. Most children and adolescents find medical things interesting. The following two grids are ones the author would develop with students during sessions.

Medical Illnesses
For advanced students, any illness/disorder can be organized using the following grid

	Contributing	Precipitating	Sustaining
Biological			
Psychological			
Soc+ial			
Cultural			
Environmental			

Describe Blood Cells. This is a more difficult task:

GRIDDING Example of A More Complex Issue:
Describe Blood Cells

Why Function	Who	What	How	When	Where	↑ Incr.	↓ Decr.	X
Oxygen	RBC	Shape	Hb	?	B.M.			
Bacteria	WBC Leucocytes							
Viral Infection	WBC Lymph-							
Clotting	Platelets							
Garbage	Macrophage							
Allergies	Eosinophils							

On the right of the dark line are columns for too much, too little, and dysfunctional cells.

A Gridding Task – Discuss the Great Depression
In a history assignment, another student used a grid and the five W's to organize information about the Great Depression. This student developed a grid where the column headings went under the general heading of WHEN, that is, a time frame: before, during and after. To help generate questions and organize information for these three time periods, the rows were labelled: WHERE (countries, urban vs. rural etc.); WHAT (the economics, production, consumption etc.) and WHO (the classes of people and how they were differentially affected, etc.). This student then used the questions of WHY and HOW to help expand information and discussion for every row in every column.

When:	Before	During	After (effects)
Why			
Where			
What			
Who			
How			

A Gridding Task – Design a Living Environment for Another World
Gridding can be a real challenge for even postgraduate university students. It is an essential method to ensure that important areas are not being omitted before beginning an investigation. Once the grid is done, the student makes decisions concerning which areas to include and which to exclude, which to emphasize and which just to touch upon. Without a grid, the student would approach the task in a relatively disorganized manner, and at worst, in a comparative vacuum with little motivation.

Case Example: Susan
Susan, a senior in high school, was asked to pretend that she would have to give a lecture for 30 minutes to her class the day after tomorrow. She was told to imagine that the earth's atmosphere was deteriorating and that she had unlimited funds and one year to create something to sustain life for herself and a few chosen others for the rest of time. Her topic, therefore, was "Ecology and the Biosphere." She had never heard of a biosphere, but guessed it must be some sort of container that sustained life. She was shown the cover of a book on Biosphere2 and told she could use this and the encyclopedia. She was then asked how she would like to proceed.

Susan had previously learned how to formulate **grids** *and the importance of having an* **organizing** principle or 'red thread,' and she had practiced the **three-step procedure for giving a speech** (say what you are going to say, say it, then say what you have said). With guidance and encouragement, after her first exclamation that it was an impossible assignment, Susan was able to help herself go from utter confusion and a giving-up attitude to using the four strategic steps and a methodical step-by-step approach to the problem. She looked up 'ecology' and 'biosphere,' then began a **free association and grouping procedure.** This led to her asking herself what she and her friends would require if they were to have to begin living one year from now in a completely enclosed container with nothing being allowed out and only sunlight being allowed in. She quickly grouped together essential elements such as, air, food, water, energy, waste recycling systems, and so on. This led her to defining an underlying general principle that she would use to tie together the entire talk, "Functions Dictate Structure," or "Form Follows Function." This led to exploration of **How** to build the enclosure and its components. With each step she was asking herself **Where** component parts should be, and **Why and When** each part should be introduced into the system. As she thought this through - without ever opening a book - she was **generating question after question, predicting** a few answers and generating some ideas as to where she might be able to find the information. Susan decided that she would use her general principle to demonstrate to the class that if one followed this principle and fulfilled all the basic needs of the humans and other creatures in this closed system, one could create a system with multiple subsystems that would constantly regenerate its own equilibrium and sustain life.

As she thought about it, she drew small rectangles in the top left and the bottom right corners of a page to represent her **introduction and thesis** and her **summary and conclusion** (proof of thesis), respectively. In the remainder of the page she drew a rough **grid.** As she free-associated, she first created **rows - areas that she would need to discuss for each of her columns** - and filled in major areas mainly under What - Needs. She also had rows for Where and When, but these were not developed initially. Beside each of the subheadings under What (the needs of the humans who were to live in the biosphere), she filled in ideas in her second column, titled "Supporting Ecological Systems and their interconnections." Her third column was "How – Structures." Her fourth column was initially titled "Why and With What Effect." She explained to the trainer that she had been **overinclusive only as a starting** point. She said that, once she had given a general overview in the beginning of her speech, she was going to tell the class that for ease of understanding and time constraints she would take one need, such as the need for fresh water, and expand the biological cycle related to it. She would then list the structures that would be necessary in order to create it in a closed environment. She would use this example as a means to demonstrate how the thinking behind creating a biosphere would evolve.

First step – Look up Key Words then Grid it
DICTIONARY: ecology, Systems, biosphere, ...
The 'Red-Threads' - Goal to sustain life & 'structure dictates function'

What – Needs	How Sources	Supporting Ecological Systems 'Regenerate'	Structure To Accomplish & Support	Failure Safety Mechanisms
Biological •O2 •CO2	Green – algae, trees			
FOOD: •protein •CHO •Fat •Vitamins •Minerals	Fish, animals Fruits			
WATER				
ENERGY	Solar			
Waste Recycling				
Medicines	Herbs – insects			
Psychological	> 10 people			
Social				
Multiple Environments	Forest / Farm / Desert/ Ocean/		Natural recycling	

All of the above was developed over a 20-minute period, and Susan correctly observed that even without opening a textbook she had enough in the way of ideas to deliver a very well-organized and interesting introduction to this topic without any further study. However, she also noted that she had gone from virtually no interest and some negative feelings about this assignment to wanting to continue it on her own, just for her own interest. Susan **had established her own personal motivation** for completing an assignment. Sometimes a student must begin without personal motivation and plunge into the four strategic steps in order to discover for themselves a personal reason for carrying out the assignment! It is part of their job to make the task personally relevant and interesting.

Broader Applications of the Grid Technique
This student created a mental image that she was continually modifying and to which she was steadily adding new information. Continuously reworking one's initial 'grid' is a procedure that is also carried out when reading. The student should review the material he is reading at the end of paragraphs, pages, chapters and so on. To review, they quickly 'scan,' 'chunk,' and make visual images of key information. This reviewing technique is taught and emphasized as one of the key metacognitive strategies with all students. Why? Because it is the step good students naturally do and the step poor students always skip. Students with ADD, for example, read to get to the end and seldom want to take time for review. No wonder they feel discouraged. They were never actively involved and thus cannot remember what they read.

At times, the trainer will just go through the second and third strategies as outlined above. When the student is very interested in a topic and/or has that topic as a project in school, then the trainer may begin reading a textbook chapter that the student has brought in. With older students the trainer may elect to assign a first reading of such a chapter to be done at home. At the ADD Centre, however, we very rarely assign homework. An exception might occur if the student was keen to try a new learning technique on material at home and would like us to go over it in the next session.

Planning a Project Timeline – a Flow Sheet

Projects and independent study assignments are being assigned starting in the junior grades (4, 5 and 6), and high school students are expected to do these without much teacher guidance. Most students with ADD do not have any idea where to begin, so they procrastinate. The old adage that failing to plan is planning to fail is often borne out. Coaching in how to develop a timeline is like preventive medicine.

When project planning is done, the trainer will develop with the student a flowsheet outlining how the student will approach the task. In this process, the trainer works with the student on how to set down a timeline including each step of the project. On the sheet would go the exact time(s) the student says that he will do his rough work plan, first draft of a grid (organization of headings for the paper), scanning of the text(s), reading with note taking, reformulating the draft organization for the written paper, initial draft of the paper, and revisions leading to the final copy. The trainer wants to make sure the student puts down realistic times and goals for each stage of this activity in their day planner. The rule of thumb is that you spend one-third of your time researching and planning, one-third writing the first draft, and one-third editing, revising and doing a final copy. Most students do not allow for that final third, and those with ADD often want to call it done when they have finished the first draft. In the sessions that follow, the trainer will review what the student did and any stumbling blocks encountered.

Brief Summary
Initially we prefer to use materials that we have in the ADD Centre. Most students want us to not only be independent of the school authority network, but also to appear to be completely independent. As some students proceed with training, they become more comfortable that we can be independent and yet make their school assignments easier. At this stage the student may decide to bring in their own textbooks and projects.

In the first stage of looking at material, we emphasize that the students always pretend that they are the author of that text. The student is then asked to begin by imagining that their assignment is to give a lecture on that topic the next day, and that they have therefore far too little time to prepare or even to begin at the first line and read the full text of the material. With this thought in mind, the student works through the first step, 'before opening the book,' and then compares her ideas and organization with that of the author's work that she must read for school. Then the student and trainer may go through the steps for reading, organizing and reorganizing a draft for the project.

The Red Thread

The concept of the 'red thread' is often elucidated while working on the foregoing gridding exercises. Older students quickly carry out the entire process on their own. Before the lecture begins or before opening a book, they generate questions, predict answers, mentally register unknown areas/definitions to look up, decide on **an overriding question** - usually a 'Why' question or **an underlying principle** around which the entire area can be organized. This is the 'red thread' that will tie all the factual information together into a logical sequence or Gestalt.

In the foregoing examples of designing a school, figuring out the cardiopulmonary system and creating a biosphere, function/needs dictate structure was the universal principle. In reading short stories and novels, the author's purpose in writing the story (usually a universal and timeless message for mankind) may be a thread that ties together what might otherwise be seemingly unrelated sections of the story. The students learn that when they present material this linking thread must be very clear to whomever is reading their work or listening to them.

Memory and Recall

Among countless visual techniques to assist memory and recall, the following three will be discussed: The Single Picture Technique, The Roman Room and Mapping Technique, and The Cartoon Technique.

These related techniques were briefly discussed in the original example above, in which the student was asked to read 2½ pages of British history concerning a naval battle between Admiral Nelson and Napoleon Bonaparte. They will be expanded below.

The Single Picture Technique– The Titanic Exercise

Like the British history example, this exercise also **emphasizes the inclusion of as many essential facts in a single picture as possible.** Either a mental picture or a rough sketch can be used. **This picture** is built as the story unfolds about the sinking of the Titanic.

Jacob imagined the name of the Titanic's shipping line's President: "Ismay" as being similar to "is May," then thought of adding a D to make it read "dismay" because the Titanic sank. Jacob made a picture in order to remember the names of the shipping line (White Star), the ship building firm (Harland & Wolf) and the place where it was built (Belfast). He imagined a scene with a hard piece of land and a wolf sitting on this rock, in the middle of the night, surrounded by white stars, while a bell rang quickly in the wolf's ear. He had added a D not only to Ismay to make it "dismay," but also to Harland to make it "hard land." As he read, he added to the picture and retained both the old and the new information despite the fact that new facts usually rapidly replace previously read or listened-to facts.

Jacob's retention and ease of recall was due to his use of an **active process** in which all of the material was altered slightly and placed into a single mental image. He was reducing the material from **multiple units of data to a single unit of related material.** The process also entailed his own creativity, something that students with ADD have in abundance, but which they seldom recruit when trying to memorize facts.

The Roman Room Technique

This technique requires that the student visualize a room where items are placed in a logically organized fashion. The Roman Room technique actually goes back to techniques used by Roman orators. You picture a familiar room and mentally go through it in a particular order, but familiar items have things hanging or sitting on them that trigger recall of the items to be remembered. Using this technique, the student could memorize anything. Let them practice with a list of groceries. To do this, the student might group the groceries into fruits, vegetables, meats, and so on, and place the groups of items in different sections of a familiar room. The individual items in each group might be arranged in their area of the room in the form of statues and paintings (bananas hanging on a picture frame, etc.). The bigger and bolder, the better for the mental images.

The **Mapping Technique** described earlier and the Roman Room technique are very similar. In the original British History example, ships blocked the French ports, ships in battle were placed near the West Indies and Egypt, and dollar or English pound) signs were placed over each of the countries England attempted to use to surround and fight France as part of Britain's strategy in the war.

The Cartoon Technique

This technique is a sequential extension of the above Single Picture, mapping and Roman Room techniques that creates a series of related pictures that allow a **time sequence** of changes in the data to take place. In the British history example above, the next picture in the sequence would include all the data around the battle at Trafalgar where Nelson died, and the next pictures might include events leading up to the battle at Waterloo and the battle itself.

Verbal Techniques to Assist Memory and Recall

Visual imaging has been emphasized up to this point because many of the students who have ADD find this particularly easy and fun to do. Some students, however, find it easier to use verbal mnemonic devices, such as rhymes or acronyms (using the initial letters of words). Another verbal technique is to **invent sentences out of the first letters of words or phrases** in order to help with later recall of information or names. Examples in music are "FACE" for the notes between the lines in treble clef score, and "Every good boy deserves fudge" for the notes on the lines of music. Some acronyms are in such common usage that they have virtually replaced the original word, like SCUBA for self-contained underwater breathing device, or IBM for International Business Machines. Another common example of a mnemonic trick that has been used for many years to help people learn French grammar is the phrase "Dr. (and) Mrs. Vandertamp." This acronym is used to remind the student of the French verbs that are conjugated with être rather than avoir: Devenir, Revenir, Mourir, Rendre, Sortir, Venir, Aller, Naitre, Descendre, Entrer, Retourner, Tenir, Arriver, Montre, Partir.

Sometimes the acronym is a real word, such as in the FACE example for music given above, or using the word HOMES to help recall the names of the five Great Lakes: Huron, Ontario, Michigan, Erie, Superior. To remember where Lake Superior is, you can remember that it is above the others. (Here a visual technique and a word meaning technique are used together.)

Seeing words and images within an unfamiliar name may help later recall.

Case Example, Benny
Benny found names from different cultural groups very hard to distinguish, recall, and spell. He felt this might be due to a mild hearing problem, which meant that distinguishing different and unfamiliar sounds was hard for him. In one of his classes in comparative religion, he had to learn the name of the person who developed Kung Fu. Benny could make no linkages or associations for this name to people or things in his everyday life. The monk's name was Bohidarama. Benny was able to commit to memory the phrase, "Bo-hid-a-ram," and he visualized his friend "Bo" hiding an old goat.

Reading Comprehension
The active reading strategies outlined above improve reading comprehension for a wide range of materials. There are also graded series that can be purchased to practice specific comprehension skills. These include finding the main idea, getting the facts, drawing conclusions, inferring, and using the context to understand vocabulary. Examples include: the Barnell Loft Multiple Skills Series or a similar series of texts such as the Sullivan Readers, which cover Grades 1 to 9. *Timed Readings in Literature,* edited by Edward Spargo, takes passages from short stories and novels graded from Grade 5 to college level. These exercises are a useful beginning to encourage students to read passages carefully. The format is a short passage followed by multiple-choice questions, some of which require factual recall and some of which require more inferential thinking. The advantage of the multiple choice format when doing neurofeedback is that the student will not produce as much movement artifact as they would when printing or writing. It also provides an opportunity to work on strategies for answering multiple-choice questions that are a crucial part of test taking skills. However, this multiple-choice format does not require the student to synthesize and organize information. Therefore, instead of doing the questions, the neurofeedback is paused and the student is taught how to underline the information, jot brief notes, critically analyze the facts, make inferences and summarize the information for the trainer. This will be discussed further under "Self Stimulated Recall," below.

Another excellent resource is the series by Gary Gruber called *Essential Guide to Test Taking for Kids*. There is an edition for Grades 3, 4 and 5, and another for Grades 6, 7 and 8. Gruber is excellent at teaching strategies, and he has also published books on preparing for the Scholastic Achievement Test (SAT).

Using these texts for practice is helpful, but you do not just assign a passage for the student to do. The trainer is essential to provide coaching in strategies. If the child gets a main idea question wrong, perhaps because they focused on a detail mentioned in the passage, it is the trainer who will ask reflectively, "Hmm, is that what it is mostly about?"

Some children have very special difficulties. The child with Asperger's syndrome is a good example. This child, who will take things literally and have trouble understanding social innuendo, may read stories several years ahead of their grade level and answer correctly all the factual questions, yet be completely unable to answer any question that required thinking abstractly and inferring something even from stories several grades behind their age and grade level. The trainer will have to help this child devise methods for answering these questions.

Usually the trainer will use short, simple questions that "lead" the child. Trainers should practice using **Socratic questions**. You never put two questions in a single sentence. Only ask one question at a time. (A useful example of this type of questioning is the kind of questioning used on the old TV show, "Columbo." Columbo, the detective, would appear to be baffled by what the criminal had just said. Then, as he is leaving the room shaking his head, he turns and asks the critical single question that pinpoints the flaws in the criminal's alibi.) It is important to use this type of teaching technique with all students. It is crucial that the trainer severely limit what they "teach or tell" the student. The trainer must not lecture, but should use a short, carefully thought-out series of questions to help the student learn to ask similar questions of themselves (inner dialog) when reading.

When practicing a new strategy expect to work through four similar questions. With the first question, the trainer models how to get the answers. On the second question, the trainer works together with the student and they share the task. The third question is done by the student with guidance from the trainer, and the fourth question is done independently.

Self-Stimulated-Recall/ Organization of Ideas/Synthesis of Data

Three essential comprehension strategies are: self-stimulated recall (as opposed to mere recognition of facts - the emphasis of the popular multiple-choice format), organization of ideas and material, and synthesis of data into a logical, meaningful Gestalt. The trainers will therefore rapidly move from using the questions given at the end of passages in the type of learning material mentioned above to asking the student to recall more from their own memory. Another way to do the questions is for the trainer to read the question and the answers: a, b, c, d. The student must identify, not the answer to the question, but the letter (a to d) that preceded the correct answer. This requires exercising auditory memory.

The next step is a different type of exercise. The student is allowed to read the first sentence of the story and then carry out the four steps/strategies for reading that were outlined above. Even younger children quickly catch on to generating questions that they think the story should answer. After the passage has been read, the trainer asks the student to silently summarize the data in an organized fashion. Shortly, all of our students give far more data in a succinct fashion than even the most difficult text questions ask. This is pointed out to the students who find this complement very gratifying. You are always looking for ways to reinforce the attitude that the student is a good learner. Self-concept is an important component of successful learning. As Henry Ford reportedly said, "Whether you think you can, or you think you can't – you're right."

When it is appropriate, the final step may be to ask the student to analyze how the author has written the passage. Has the author given her thesis in the first paragraph? Has she followed this with three paragraphs, each stating an argument or point and supporting that point with data? Has she summarized, discussed and given a conclusion?

Special Techniques for Short Stories and Novels

The first of these techniques concerns character analysis and is labelled the **billiard game technique**. Most of the students have played pool or have heard of the game of billiards, which is played on the same kind of table but only uses three balls: red, white, and an off-white 'spot' ball. The student is asked to imagine that the pool or billiard table has many sides instead of just four sides. Each of these sides represents a "flat" character or a setting. A flat character is one who never seems to change from one situation to another. It could be, for example, a man who is always angry and boasting, or the opposite, always anxious. A setting might be a school or a frozen lake, and so on. These ideas are likened to the flat sides of the table. In a story there are usually only two or three "round" characters. These are likened to the three balls in the game of billiards. These characters are multifaceted. The student is told to imagine that each time one of the balls (round characters) bounces off the side of the table (a flat character or setting) or caroms off another round ball (which is rewarded with points in billiards) that something important is learned, through that interaction, about an aspect (for example, a fault) of the round characters. (In the game of billiards the player would learn something about angle and spin.) For ease of portraying and understanding the round characters, the student is asked to only consider two important sides of each of the main characters, and develop how the character's thinking and emotions govern their behavior.

Here is an example using "A Kind of Murder," a short story by Hugh Pentecost (Scholastic Scope

Literature, 1991). The student's first task is to find and recall the major facts and issues in the story.

Mary, a high school student, while reading this story, made a visual picture of an emotional scene. In her scene (not an actual scene in the story), Pentecost, the author and also the principle character and narrator of the story, is waving a fist at fellow students defending Mr. Warren. Mr. Warren looks weak and extremely anxious. In the actual story he is a very nervous and rather hapless new teacher at Morgan Military Academy, a boy's private school. However, in Mary's mental image, although Pentacost is waving a fist apparently in defiance and warning at the other boys in order to stand up for Mr. Warren, he is also, at the same time, anxiously turning away from Mr. Warren toward the jeering students. She has caught in her own image the two sides of Pentacost that are developed in the story. He is compassionate, thoughtful and brave, but he is also anxious and concerned for his own popularity, and is not immune to peer pressure. In this manner Mary developed a mental image that portrayed the double bind that Pentecost found himself in and the opposing, internally conflicting aspects of Pentecost's character.

For older students who have symptoms related to Asperger's syndrome, this story provides an excellent background for discussing emotions and emotional relationships. Again, it is important to use simple, single, questions in a Socratic questioning format. For young students we use age-level picture books with expressions and gestures that convey emotions. A young girl with her pet dog can be a useful background picture for discovering how to interpret facial expression and gestures.

The Reporter's Questions Technique for the Elementary School Student

Most of the foregoing techniques are equally effective for all ages of students. However, it goes without saying that students must be taught using a degree of complexity and language that suits their mental age and reasoning skills. Young children love to read about animals, birds and insects. The Nature's Children series is an excellent resource. Before reading about an animal, the trainer may ask, "Why is this animal able to survive?" Then the trainer, through Socratic questioning, helps the child to organize headings they think the author should include, and questions they will be trying to answer as they read. With the younger child or with an anxious student, the trainer can model by saying, "I want to know…," then it is the student's turn to say what they want to know and so on. In this manner the trainer and the student develop questions that they would like to have answered. Together, the trainer and the child think of 10 to 20 questions and then check their ideas against the author's table of contents.

The more advanced students may organize their thinking about an animal, bird, fish or insect using the reporter's questions: Who, When, Where, What, Why and How. In this example, What equals structure. This stimulates the child to think of the animal, bird, fish or insect from the head and working down the body to the limbs. This stimulates thoughts on various aspects of the creature: intelligence, five senses, eating/elimination, reproduction, locomotion and any special capabilities (e.g., venom, web making, etc.). They must then discuss how this creature's structure dictates their functions, such as escaping enemies and finding or capturing food. The student then considers further questions including: Who = (social and family), When = (life span) and Where = (countries and terrain).

An Organized Approach to Reading Complex Scientific Material - the Boxing Technique

Students in computer science, chemistry, physics, and other complex areas that involve formulas rather than literature may use some of the techniques described in the foregoing. Another technique called the "blocking or boxing" technique can also be utilized to maintain concentration and remember material. In this technique the student groups together sections of scientific material into "blocks" that contain **one connecting principle or unit of study**. In a calculus, physics or chemistry text each section must, or at least should, relate to the previous section. In calculus, for example, the 'box' the student makes might contain only half a page of material. For the student to define it as a 'box,' it must contain nearly all of the data necessary to derive the equation or the principle being taught and understand it. The student then endeavors to be clear concerning the linkages of the 'box' under study to 'boxes' previously studied and areas that come after it. The student may use the four strategic steps in shortened form on the material in the 'box,' and then take a break. Before beginning the next 'box,' the student should always do a quick review of the area previously studied. The student then considers how that box *links* to the next unit of study.

Writing Technique – The Hamburger or '3-3-3' Method

One general approach can be used to accomplish many writing and speaking tasks including writing a paragraph, an essay, preparing a debate or answering an exam question. It is commonly called The Hamburger Method. This name is apt because the student organizes the material into three sections: the top bun, the meat and veggies, and the bottom bun. These sections correspond to the Introduction, the Main Body and the Conclusion of the piece. In these three sections the student follows the general principle: *Tell them what you are going to tell them, tell them, tell them what you told them.* In its simplest form, if the student is writing a paragraph the top bun is the introductory sentence, the next three sentences are the meat and condiments (you want some details to add interest, not just a plain burger) and the last sentence is the bottom bun that holds it together. The top and bottom buns are the same material (topic) since you do not want to introduce something new at the end.

The three simple sections noted above become further elaborated for students in Grade 7 and above as they must write essays. Each section is further subdivided into three parts in the following manner.

Introduction: Divide the introduction into three sections: the introductory statement that catches an audience's attention, a thesis statement that says what you are going to discuss or prove, and third, mention the three arguments or points that you will develop in the body of the dissertation.

Main Body: The body of the dissertation is also divided into three sections. These sections are for the three major arguments (points) that will be made. Each of these three arguments must be supported by at least three facts (data). The student is told that two arguments are insufficient and 10 arguments would be too many and quite confusing. Three to five arguments would probably be all right. If there are more points or arguments, the student should sub-group them into three major headings and then each major heading could have subheadings or arguments.

Conclusion: The third section of the dissertation is also divided into three sections. The first section is the summary, stating in brief the three arguments. In the second section the student discusses how the thesis (or hypothesis) was supported or proved. The final section is very short. In it the student makes a conclusion that wraps things up.

Note that this structure works in simple form in grade-school and high school. In its more detailed form on the left hand side of the diagram it is the same structure used by University students and in writing a professional paper.

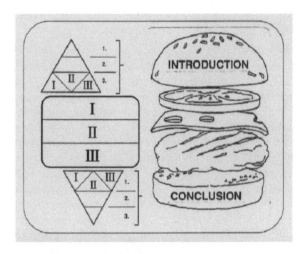

Case Example, Judy

Judy was given 10 minutes to develop a debate. She could either imagine that she would have to do this tomorrow morning in front of her class or that she and the trainer were the local university debating team competing against other universities. The subject was handed to her as it would be in the short topic section of a University debate, on a folded slip of paper or in an envelope. (It is more fun this way.) The subject is: "Panda bears will become extinct."

Judy decided to introduce the topic by holding up an imaginary little panda bear which she would introduce to the audience as "Charlie." I, as her partner, was to hold up a little cap gun as if I were going to shoot Charlie. She would pull Charlie away, turn to the audience and, speaking to Charlie and the audience, say, "I'm so sorry Charlie, but I don't think I can protect you forever. You and all your friends and relatives are going to become extinct." She hoped this would startle the audience, catch their attention, and state the thesis for our side of the debate. She would then say, pointing at me, that we were going to prove today that panda bears would be killed by human actions, that they couldn't escape, and they could not reproduce quickly enough to replace the lost numbers.

In the body of the debate Judy would support the first argument (pandas would be killed) by noting that they live in a region of the world where people needed meat to live, fur to keep warm and more land on which to grow crops. Destruction of the panda's bamboo habitat would necessarily lessen their numbers. Her second argument (panda bears couldn't escape) she would support by noting that, unlike small burrowing animals, they nest on the ground. They aren't particularly fast or fierce, so they are easy prey. Her third argument (panda bears do not reproduce fast enough to fill the gap) she would support by noting that they only have one or two babies at a time, the babies nurse with their mother for about nine months and the offspring take at least six years to grow to maturity and begin to have their own babies.

In the conclusion of the debate Judy was going to summarize her three arguments in one sentence. She would then discuss the situation. First, she would acknowledge the argument she was sure the other team would make concerning the success of conservation measures in various parts of the world, by saying that we, along with all those present, would hope that the world would and could convince the responsible government of the area that they should enact measures to prevent this terrible outcome. She would note, however, that the government in question had too many other pressing issues to deal with. It was highly unlikely they would act rapidly enough to avert this disaster. She would end by stating that, "Our conclusion has to be that panda bears will become extinct."

The trainer can make up amusing topics such as: "Cheese is the most important food in the world," and have the student put their organization of a dissertation on this subject into the hamburger design. If they use the 3-3-3 approach they should have little trouble organizing their ideas, and the essay should have a good flow. They will discover that time spent planning ends up saving time because they are never twiddling their thumbs, wondering what to write next. The harder sell is getting them to spend time on editing. The rewrite typically should take as long as the first draft did. If they understand this in advance they can allocate time appropriately and not resent the editing stage. The analogy is that you do not leave your diamond in the rough. You find your piece of rock, cut your gemstone, and then polish it.

Studying Technique Using Packaging

Packaging is a method for staying up-to-date and maximizing recall. It is a technique for older students. We consider it an essential technique for anyone taking sciences in college or university. In this technique the student makes precise, concise, very neat notes for each section of the subject matter being taught. The student writes **definitions** for each new term. They **organize** (logical sequences of material) **and synthesize** (combine material from different sections/books/notes) to make one clear statement of the facts and conclusions for that section). Each section's 'package' should be less than 10 handwritten (or well-spaced, typed) pages. A section usually corresponds to a chapter in their textbook. The student will make the 'package' using information from class notes, handouts, labs, a last year's student's notes ('cookbook') and relevant chapter(s) from the text book(s). It must be very short, yet contain all the key facts in a logical order that the student can find easy to recall. The student then staples these pages together to make a 'package.' These become their short notes for final exam studying.

Case Example
'A' was busy at his desk. It was a large desk surface but he had very little on it. He had his computer, an open, underlined textbook, lecture and lab notes, and his 'cookbook.' To one side were neatly laid-out series of packages with each title showing. I asked him why he was spending all this time every night doing this. His answer was short and simple, **"I spend time to save time."**

He said that, before exams, he could review the same material in a quarter the time it took his friends. It dawned on me, sometime later, what he had meant. He was the elected Head of his college and participated in every possible activity. He didn't have time to cram before exams, and in science the volume was far too large for that to be effective anyway. He was expanding the three-fact method discussed earlier. His method was to scan each evening the subject matter to be taught the next day. He would ask questions in class to keep up his alertness. He would scan the material covered in class when he got back to his desk that evening, and then summarize everything in 'packages' as the course went along. Keeping up and even keeping a little ahead actually did save time, as evidenced by his participation and leadership in so many other activities.

I asked him what technique he and his top-performing friends used for reading passages. Two techniques emerged. The first is the scanning technique used every day for reading a textbook. However, for these exams, another technique was also used. First, they did practice exams so that they would be familiar with the format and would not have to spend a lot of the time reading the examination instructions. Second, they would read the first paragraph of a passage twice in order to have an extremely clear picture of what that passage was about. Then they would memorize where arguments and facts could be found in the remaining paragraphs. In this way, detailed questions, demanding small factual differences, could be rapidly found in the original passage. The text would hopefully only be consulted once because the information was then distilled into a 'package' that could be reviewed in a fraction of the time it would take to reread the textbook.

A's techniques are exemplary and efficient. He went on to graduate from Medicine. In post graduate studies he was top of his class at Harvard doing a Masters degree. He feels that strategies made a real difference. We try, through Socratic questions, to help the students at the ADD Centre formulate and experiment with similar techniques in their own studies.

Strategies for Mathematics

Introduction

In this section only a few very basic examples are used. As is true for metacognitive strategies for reading and remembering, there is no comprehensive catalog of strategies. Rather, there are as many strategies as there are stumbling blocks. You must continually create new strategies relevant to the task. The purpose here is to illustrate some of the most common difficulties experienced by students. The mathematically gifted students usually do not have these difficulties. Students with attention-span problems who are talented in math reasoning, however, will often have tuned out in class in former years when some of the very basic math concepts and principles were being taught. Now they are well beyond these levels, and their teachers expect the basics of math to have been mastered previously. To help them, you must address their skill gaps, which are usually a failure to learn boring things like multiplication tables, the steps for long division or fractions.

Many of the teenagers who come to the ADD Centre are very bright intellectually, and yet they have never learned simple, basic approaches to written math problems and to handling fractions and equations. The neurofeedback trainers teach the students a few very simple concepts, rules and examples. The purpose is not to teach the child math, for that is the job of the school, but rather to stop the giving-up, "Oh, I can't do that," all-or-nothing reaction of some students. When this attitude is observed, our first objective is to help these children get themselves into a relaxed alert state. Then we assist the child to focus on each small section of the word problem. At this point we have the student ask him- or herself: "What am I being asked in this problem? What facts have they given me? How can I draw the facts so that I can visualize the problem?" It is the approach to the question that can be the key to success. We attempt to help create a positive attitude toward this subject by increasing the student's self-confidence through training. Students must learn how to learn and how to think about problems in mathematics in order to genuinely understand them.

Grade School Math Tips

These tips and tricks are generally only needed for those students who have a specific difficulty with math. The child who can generate excitement at the challenge of solving the problem, just because it feels so good to figure it out for himself, is the child who will progress. The following is a sampling of the kinds of tips that can be used with grade school students while working on having the student feel competent, interested, and then fascinated by the challenges that math offers.

In teaching younger students math, you begin with a concrete math manipulative. This does not have to be a fancy purchased set of objects. It is much better to use things in the child's natural environment like fingers, pencils, or blocks. Later, you use pictorial representations like tally marks or drawings. Drawings can be chocolate bars or a pie or other familiar, 'fun' things that can be divided into fractions. Good mathematicians see the relationships. This is why the spatial reasoning tasks on IQ tests correlate so strongly with math skills. Only after something is illustrated with a tangible object (fingers are the easiest) and the concept is grasped do you move to a pictorial representation such as tally marks. After the student is comfortable with the pictorial representation you can move on to the abstract, that is, a numeric symbol. A deck of cards is very handy for both learning and practicing math facts because it has the number in two corners, but also the back-up of the illustration (five hearts or clubs, etc.). If you take the face cards and put the remaining 40 cards into two stacks and flip one over one card from each stack, you have a great way to randomly practice addition and subtraction facts up to 20, or multiplication facts up to 10x10 = 100.

Addition and Subtraction

In grade school the student must **understand** simple addition and subtraction. We see students as old as 12 years of age who have never grasped what it means to be adding 9 + 8, and 8 + 8 is completely out of their range. To teach the concept we have them use their own fingers and the examiner's, and fully understand 10 +. Fingers are a wonderful math manipulative. When this is achieved through understanding and not rote memory, then we have them remove one thumb after doing 9 + 10. In this manner the child gets rid of, subtracts, one digit so that the same question becomes 9 + 9. They then do a series where they do 10 + and change each to 9 +. For example, 10 + 7 = 17 becomes 9 + 7 = 16. Then they can do 8 + by changing each 10 + answer by subtracting two thumbs. In training trainers for the ADD Centre, we emphasize that they must **never assume a child has grasped a concept just because they give a rote memory correct answer!** They must see the number fact to understand it.

These students require a familiar, concrete representation of number facts. Initially, fingers - yours and theirs - work best. You can check that they are understanding what they are doing by first asking them to tell you if the answer is going to be "more or less" than what they had. Then they must tell you what they have done. When they are adept at using concrete representation and explaining what they are doing, then you can move to using math manipulatives such as counters where numbers are represented by colored plastic discs or cubes. We encourage parents to do things at home to reinforce addition facts. Games that use dice give lots of practice for addition facts up to 6 + 6. The dots on each die give the pictorial representation of the number.

For those who have grasped the foregoing, the concepts of grouping into ones, tens, and hundreds, borrowing and carrying are rapidly taught when necessary.

Multiplication and Division

When addition and subtraction are mastered, the trainer will check that the child has grasped a basic concept in multiplication, namely, that we are multiplying the number of groups times the number of things in that group. As noted above, children who have difficulty in paying attention very often miss the teaching of math concepts when they are taught in the classroom. Unfortunately, we find one teenager after another who, even if they have memorized the multiplication tables, have no really clear understanding of what is meant when they multiply one number times another. The older students should, for example, be able to rapidly multiply virtually any number in their heads. Often we ask the teenagers to multiply such numbers as 17 times 13 or 376 times 75 and 2/3 times 3/4. To the reader it will be clear that these examples are deliberately simple if you visualize them, but they are used in order to emphasize the concept of multiplying and some simple tricks for visualizing figures and fractions. (For example, to multiply 75 by 25, you multiply 75 by 100 and divide by 4.) With the younger children we use a simple grid with houses for the columns and people for the rows. We like to use groups of objects, animals, or people that the child suggests from their interests. Often the children will choose groups of horses in their stalls, teams of hockey or basketball players and so on. We can then have the columns in a grid be the team (stall for horses and so on), and the rows can be the number of players (each position) on each team. The child multiplies the number of teams (groups) times the number of players (rows) on each team to find the number of players on the field if every team is lining up before a big contest like the Olympics.

Division follows the same concept using the same grid diagrams. This basic understanding of multiplication and its reciprocal, division, is necessary in order to begin fractions and algebra, and it is frequently deficient even in high school students. It is therefore expanded below.

The Teams (Groups) Times the Number of People on Each Team Rule

Although you can use commercially available, brightly colored modern math manipulatives, the concepts are often more easily, thoroughly and lastingly assimilated using the time-honored and quite old-fashioned basic rules and hand-drawn diagrams. With the younger children we start by **drawing two (then three and four and so on) little houses** that are attached to each other like row houses. **Each house is at the top of a column** and each is labelled 1, 2, 3 and so on. **We then draw a row** below the houses and put a little person or happy face into each of the resulting squares and ask how many people there are. The answer, of course, is the same as the number of houses. This is the 1x table. We note that this is the same as saying the number of houses times the number of people in each house. We **then draw a second row of people, a third, and so on**. With each addition the child notes the same general rule. They realize that multiplication is simply repeated addition.

The Reciprocal Rule
When the child clearly understands the rule, we introduce the question as to whether three houses times four people in each house holds the same number of people as four houses with three in each (the reciprocal). For children with a specific difficulty in math this is a very difficult question. However, using the grid, they catch on to it quite rapidly, noting that the number of people in the squares is the same. An egg carton is another way to visually display this principle. The student can **discover** that there are not only three groups of four and four groups of three, but that there are six groups of two, two groups of six and one group of 12. The trainer may mention the word **'factors'** at this juncture. (A factor will divide evenly into another number. Two and three are factors of the number six.)

More Difficult Multiplication
Some simple tricks are used to help accelerate remembering tables and thus increase the child's self-confidence. We start with the seven and nine times tables because they are usually considered difficult, and the child's confidence soars when they are mastered.. First we teach the **counting trick,** 7 x 8 = 56 is learned by simply counting from 5 to 8: 5, 6, 7, 8; 5 6 = 7 x 8. Then we teach 3 x 7 as three sevens, and we hold up three fingers (the trainer uses two on the right hand and one on the left and the student mimics it, to make 21. Then we do the same using six fingers, four on one hand and two on the other, to make 42 for six sevens. We call these the **finger tricks**. Seven sevens is one less than 50, i.e., 49. Alternatively, if they like football, say lucky number seven twice (7x7) gives you the 49ers football team!. Four sevens is taught as three sevens (using their finger trick) plus an extra seven, making 28. The latter emphasizes the concept of one extra seven is just seven more. Five times seven is not a problem since they can count by fives using seven fingers. And two times seven is just 7 + 7. This just leaves 9 x 7. For the nine times table, they "discover" that the tens column is always one less than the number they are multiplying times 9, which is logical because they know that if they had, for example, 10 sevens, it would be 70, thus 9 sevens <u>has</u> to be less, that is, sixty-something. They then notice that the two figures in the answer always add up to 9 (e.g., 6+3). Then we decide that maybe we might have discovered a new rule. Subtract one from the number we are multiplying by to get the first number, and then the second number plus the first will always equal nine. We try it out and find that 9 x 9 = 81 and 8 + 1 = 9, 9 x 8 = 72 and 7 + 2 = 9, and so on. By this time most children discover for themselves that the numbers that we have conveniently placed in a column go from 9 down to 1 and from 1 up to 9 in sequence.

We then show them that they have the nine times table at their finger tips. You show the two hands on the table minus one finger trick for the nine times table. For those who don't know this little maneuver, try putting both your hands on the table, palms down, and label the fingers one to ten starting at your far left with the little finger of the left hand. Now fold down one finger, say the fourth finger, which would be the index finger of the left hand. The finger you put down (the fourth finger) stands for the number of nines. The nine fingers left on the table represents the fact you are in the nine times table. The answer to 4 x 9 is represented by the three fingers to the left of the one folded down finger and the 6 fingers to the right of it, that is 36. Younger children really enjoy this maneuver. Make sure you have your fingers held up beside theirs, not opposite them, so that left to right is the same.

Do not assume the student knows that zero times anything is zero and one times anything is that number. Don't assume that they know that two times is just two of them added together (doubling), or that the 10 times table just requires one to add a zero. Check each of these concepts.

The 11 times table merely requires one to place the numbers side by side, e.g., 6 x 11 = 66. However, ask the teenager what 11 x 11 is and you will often get the incorrect answer 111. You then help him use the logic you have been teaching. What are 10 elevensies? He answers, "110". Ask him if there are 11 players on each (hockey or basketball) team how many players would be on the field if nine teams lined up (99). How many if 10 teams lined up (110 by adding a 0 to 11). Then how many are on the field if the 11th team marched out and lined up beside the 10th team? The impulsive student will still blurt out the answer as 111. You ask incredulously how he went from 110 to 111, just one more, when the whole team just marched on to the field. Once he slows down, stops guessing, and thinks before speaking, he calculates 121, and then you take him all the way up to 11x15 and congratulate him on really understanding what he is doing.

When we are doing times tables, most of the children are able to easily count by fives. Therefore, whenever we get to five times a number, we teach them to reverse the question so it becomes a five times table question. They then use their fingers to count the number of fives as they say the series: 5, 10, 15 etc., to themselves. Very soon the children have mastered the 11, 10, 9 and 7 times tables. 8 x 8 = the number of squares on a chess board. The series: Six 4s is 24, six 6s is 36, and six 8s are 48, has a rhythm to it and it kind of sounds right. At this juncture students are surprised to discover that there is very little they haven't learned. In fact, there are only six more facts to learn (3 x 3, 3 x 4, 3 x 6, 3 x 8, 4 x 4 and 4 x 8). These are quite easy, and the trainer can guide the student to figure out a strategy for each. For example, for 3x you can double the number and add one more: 3 x 3 = 6 + 3 = 9; 3 x 8 = 16 + 8 = 24. For 4x you double and then double again; 4 x 4 = 8 x 2=16; 4 x 8 = 16 x 2 = 32. Most of this work can be done while they practice getting themselves into a focused and concentrating state while we put the electrodes on their head and ears. They are rewarded with lots of upbeat praise and sometimes extra points. Parents are told what they have accomplished at the end of the session and the child often will demonstrate it.

Case Example, Jason

Jason, age 11, had gone to a learning center for two years to learn math, especially the times tables. It was a system where the children did drill and practice, but were not taught the concepts underlying math skills. They did hundreds of exercises. (It takes about 800 repetitions for the average child to learn a single math fact by rote). He knew the two times and the five times tables, but little else. He arrived for his first session with a very discouraged mother. In the session the author got him focusing with neurofeedback and then taught him the tricks for the seven times, nine times and 10 times tables. After the one-hour session he answered questions on each of these tables accurately and rapidly for his mother. She was flabbergasted. In the second session he learned the concept of multiplication (he loved calculating using groups of ninjas). He then did the 11 times table, up to 11 x 15, for all the parents in the waiting room. Jason was very proud. He never forgot the times tables, and went on to learn how to add, multiply and divide fractions in the next few sessions.

In this manner, you continue to have children **play with numbers** in order to really be able to manipulate them and understand them. No amount of repetition of math facts will have the same beneficial effects. Also, the beautiful pictures in most school texts that are meant to teach these concepts turn out not to really be very interesting once they have been seen once by these children. We like the children to have mental pictures, but we prefer that these are pictures they have developed themselves through playing with numbers. We do use repetition, however, because for a number of sessions we will go through exactly the same sequence of **"discovering"** all of the foregoing tricks and logic. In our experience students who learn their tables in this manner really become very confident about their mastery and understanding of multiplication. Older students are then willing to be challenged with otherwise seemingly impossible questions such as 13 times 17.

The Concept of Multiplication

Give a student a tough problem to develop self confidence, such as 17x13. To get the answer, the trainer takes them through a logical sequence: 13 is just (10 + 3), 13 x 17 is the same as writing (13)(17) or (10 + 3)(10 + 7). The trainer asks the student to draw the houses and people grid. (Use soldiers with rocket launchers and rockets grid for some of the more reticent boys). For this example, say there are 13 basketball teams (or columns). There are 17 players on each team. The trainer then suggests that since it is so simple to multiply 10 x 10, why not just call the first 10 players the first string? The next seven players are the substitutes. Now the child uses a red pen to outline the 10 x 10 box and write 100 in it. He then multiplies the 7 substitutes times the 10 teams and gets 70. He draws a box on the grid around players 11 to 17 for the first 10 teams in green ink and puts 70 in this box (at the bottom of the grid). Then on the right-hand side of the grid, the student draws a blue box around the squares representing three columns (or teams 11 to 13) x 10 (rows or first-string players for those teams), and puts 30 on the page. The only squares on the grid not enclosed in the red, green and blue boxes are in the lower right hand corner, and they represent 3 columns x 7 rows. The student encloses this area with a pencil and writes 21 in it. The student then **sees** simply that the answer to the problem is 100 + (70 + 30) + 21 = 221. It is not much of a step then to see that they could have done this in their heads using the bracketed figures (10 + 3)(10 + 7). The student is then encouraged to relate this method to the grid (10 x 10) + (10 x 7) + (3 x 10) + (3 x 7).

The 13 Teams (Groups) x 17 People on Each Team Grid

	1	2	3	4	5	6	7	8	9	10	11	12	13
1													
2													
3													
4				100								30	
5													
6													
7													
8													
9													
10													
11													
12													
13				70								21	
14													
15													
16													
17													

Long Division: Use the Family Rule

For long division, the hard part is remembering the steps. Unless the steps have become an automatic checklist the student is inclined to get stuck and then tune out. Memorizing a simple mnemonic can be very useful. Try using the Family Ruler: **D**ad, **M**om, **S**ister, **B**rother, and **R**over (the dog), to help them with the five steps: **D**ivide then **M**ultiply, **S**ubtract, **B**ring down and **R**epeat. Tell them this family will help them remember the steps. Some children learn better using a really silly phrase they make up. One boy came up with "**D**ead **m**onkeys **s**till **b**reathe **r**egularly." (Always ask the student's parent if it is OK to use the Family Rule example before you say it to the child).

Help the child attain a real pride in keeping their number work in neat columns so that they do not become confused. This is particularly important when working with children who have ADD. Children with ADD may be wonderfully artistic, may do script incredibly artistically and may have excellent fine motor coordination. At the same time, however, they may have a very specific handwriting difficulty where, as they attempt to put down a great deal of material, the numbers and letters begin to change in size, shape and spacing, making their work appear messy and even illegible. Cursive writing is particularly difficult for many of them. The same difficulty may be seen when they are doing a full page or more of math. It may also be due to an impulsive carelessness. If uncorrected, this leads to negative feedback from teachers and a downward cycle with respect to motivation and effort. Using grid paper is another helpful aid for keeping math calculations neat. In some European countries math notebooks come with grid lines rather than just horizontal lines.

Division – the Concept

Many of the children have memorized how to carry out simple division, but have never thought about what it was that they were doing. We use the **same grid that they used for multiplication to help them.** They can quickly understand that if they have a total of 12 players altogether (total for all the teams) and four teams, then there will be three players on each team. As noted at the outset of the section on math, begin with something concrete, like dividing 12 pencils into groups of three, then move to a pictorial representation using the grid, and only after these two steps are consolidated do you write 12 ÷ 3 = 4. The student must get to the point where they understand, and truly see, that multiplication and division are reciprocal operations. This is necessary in order to figure out what operation to use in word problems.

Fractions
Analog Time – Simple Fractions

Many children have never learned to tell the time from an analog clock, though most are familiar with digital time pieces. The trainer and the child begin by discussing how to divide a blueberry pie into four equal pieces. The concept of quarters and halves comes rapidly, and the analogy drawn to saying quarter and half past and a quarter to the hour is easily drawn. Many of our trainers have diver's watches. Stories about scuba diving and how important it is to know how many minutes you have been underwater naturally follow a study of the clock face and hands. Kids enjoy the "real thing." Playing down to them with large school supply cardboard clock faces is not only boring, but is actually irritating to some of these children. They can see the dial on a real diving watch very well and they love turning the bezel. The bezel gives them the chance to mark how far 15, 30 and 45 minutes are around the face of the clock. They then rapidly grasp that ¼ hour + ¼ hour = ½ hour. This leads, often within only one or two sessions, to addition of fractions and simplifying the answers (e.g., 5/4 = 4/4+1/4 = 1 1/4).

Another way to reinforce the concept of quarters is with coins: four quarters equal one dollar; or two quarters are half a dollar. Money is also a helpful way to help them understand that decimals are just another kind of fraction and you can convert between decimals, fractions, and percent with all of them being equivalent. So, 25 cents can be written 0.25 and it equals ¼ or 25 percent. Explain that '%' means percent, which is per 100 ('cent' is French for 100 and 'per' means divided by, as in miles per hour). You must make the new learning relevant and link it to things they already know. As a general rule, **if more than one-third of the facts or information is new, then no learning takes place.** Make sure **two-thirds of what you say involves material or concepts they already know.**

Adding Fractions

Once the student has mastered the clock face and pie analogy, we can move to adding and subtracting fractions. However, it is easier for most children to visualize a rectangle (chocolate bar) than a circle (pie) when we later get to multiplication and division of fractions. We therefore move to the chocolate bar as an example at this juncture. It is not always easy for some students to perceive that if they had two chocolate bars (we always initially have them draw the chocolate bars under each of the fractions) and one was divided in two thirds and the other into three quarters, which one would have the bigger pieces. Always ask the student if he likes chocolate. Then, if he does, point to the chocolate bar from which he would like to have a piece. Many students choose to have a piece from the chocolate bar that is divided into four because they look bigger. Help these students to see that dividing it into four makes smaller pieces than dividing it into three. Now they understand that if the chocolate were to be shared with friends on his soccer team, his teammates who received the pieces from the three-quarter chocolate bar might be upset because they were smaller.

The trainer can carry the illustration further. The team and the coaches mean that there are between 15 and 17 people, and we have only got 5 pieces. We need more (and therefore smaller) pieces, and it is essential that everyone be happy that the chocolate is shared fairly, that is, equally. Therefore, the pieces must all be the same size. It is at this juncture that we look at how we would divide a chocolate bar. The chocolate bar, or a rectangular cake of brownies, could be divided into three sections horizontally. What if, however, we then divided it into four sections vertically. Now each of us who had one section of the bar that was cut into three would have our slice divided into four. Our whole chocolate bar has how many pieces now? (Twelve)

We take a second chocolate bar and divide it into four vertical slices. Then we cut it horizontally into three and each of the people who formerly had one slice of the bar cut in four now has their piece cut into three. The total number of pieces is 12, the same as our other chocolate bar! Have the student 'discover' that now everybody will get the same size piece no matter which chocolate bar they choose from.

The **first** chocolate bar.	The 1st chocolate bar is divided into 3 pieces. One piece is missing and therefore there are 2 pieces of chocolate left. (2/3)
The **second** chocolate bar.	The 2nd chocolate bar is divided into 4 pieces. One piece is missing and therefore there are 3 pieces of chocolate left. (3/4)
The 1st chocolate bar is further sub-Divided.	The 1st chocolate bar is cut vertically three more times. There are 8 shaded boxes out of a total of 12 (8/12). Thus there are 8 pieces of chocolate (8/12) in the bar where 2/3 of the bar had been left uneaten.
The 2nd chocolate bar is further subdivided.	The 2nd chocolate bar is cut horizontally two more times. There are 9 shaded boxes out of a total of 12 (9/12). Thus there are 9 pieces of chocolate in the bar where ¾ of the bar had been left uneaten.

Case Example, John

John, age 11, was finding math very hard. He had, in fact, given up. However, John loved chocolate, and after discussing what his favorite chocolate bars were, he was open to talking about a problem where he was pretending he was giving out a chocolate reward after his soccer game. In the above problem he was asked if the pieces were all going to be of equal size. He said, "That's obvious, we just cut them so that there are 12 pieces in each bar." I then asked how many pieces were left since he and I had each had a piece from the original two chocolate bars. He said, "That's easy, there are a total of 17 equal pieces of chocolate left over in the two chocolate bars." I said, "Each piece is one 'twelvesy' (1/12)." He agreed.

I then asked him to write that addition as a fraction. I asked what would go on the bottom (denominator). He said that it would be the number of pieces originally in the chocolate bar before we ate any of it = 12. I then asked what number would go on top (numerator). He said, "Eight." I noted that he had just said (implied that) we had eight twelvesies = 8/12. He agreed. We did the same process for the second chocolate bar and got 9/12. He added these and made 17/24. I asked him what we had nicknamed each piece. He correctly answered, "One twelvesy." I then asked how many twelvesies were left for the team. He said, "17 twelvesies. I asked him what that would look like as a fraction and he wrote: "17/12." His eyes lit up and he crossed out his former answer of 17/24.

We then looked at how many pieces made up one whole chocolate bar and he said, "12." I asked if he could use that answer and simplify 17/12. He correctly wrote: 1 5/12. He then showed me how he figured it out with the picture by putting pieces of chocolate from the second bar to fill up the gaps in the first and finding that five pieces were left over.

It was a simple matter for John to repeat this exercise a couple of times and **derive basic rules for working with fractions for himself.** After working with twelfths, thirds and quarters seemed easy for him.

In summary, the general principles you want to stress are the following:

- The denominator represents the total number of pieces the object is divided into.
- The numerator represents the number of pieces you have left.

- You must make sure the objects are divided into the same number of pieces before you add or subtract to find out how many pieces there are in total.

- Whatever you do to the bottom (denominator) of the fraction (e.g., multiply it by a number such as three or four in the above example), you must do to the top (numerator). The logic of this is quickly apparent to the student when they look at each of their chocolate bars or pies. They multiply the four pieces (in the bar or pie that is divided into quarters), by three to get 12, and they multiply the three pieces they have shaded by three and get nine.

You cannot add apples and oranges, but you can group them as fruit; similarly, you cannot add thirds and quarters until you change both to twelfths.

Don't Let Children with ADHD Fall Behind

These very simple examples are, of course, usually worked through in the classroom. The children we are working with, however, were not mentally present. Most of them were tuned out much of the time when the concepts were being taught. The rest of their classes have long since gone on to higher levels of math. Higher math levels unfortunately assume that the basic principles are second nature to all of the children. **Once children with ADD get behind, the degree to which they tune out rises exponentially!** These children could do very well in picking up some of the math facts and skills that they missed by attending a good supplementary learning center. We often recommend that the children who do not have a significant problem with ADD do just that. Many of the children with ADD, however, have tried that or had tutoring. These children continue to fall behind in class. They have not made a **fundamental shift in the way they learn.** The problem that remains is that they have never learned how to deal with ADD in terms of increasing their attention span and concentration. and in addition, they have never learned how to learn. The neurofeedback trains them in how to pay attention, focus, concentrate, and act in a reflective and less impulsive fashion. When in this state the children learn remarkably quickly. The children can feel this, and a brief experience of this during their neurofeedback sessions usually leads to a positive attitude toward learning in their classrooms. Then they are taught **strategies that work for their kind of cognitive style.** Skill gaps, such as learning multiplication tables or steps for long division, can be taught. The combination means that in the future they will be able to learn in the classroom.

The Be-Fair, Do–the-Same-Thing-to Rule for Fractions and Equations

With selected students, the trainer helps the student discover that the problem of multiplying 376 x 75 in your head is nothing more than a very simple example of the general rule: If you do the same thing to the top of a fraction as to the bottom the value stays the same. You are fair – you treat numerator and denominator the same. Many of the children can see that since it is easy to multiply times 100, they could multiply the 376 by 100 as long as they divided the 75 by 100. But 75/100 is 3/4; therefore, they could just divide the 37,600 by 4 and multiply by 3 to obtain 28,200. A simpler concept that can be taught first is that to multiply by 25, you can multiply by 100 and then divide by 4 (since 25 = 100/4). Manipulations like these with the older students are used to decrease their resistance to thinking and puzzling over math problems, and to engender a feeling that there really can be a very pleasant sense of accomplishment when they take up these challenges.

Multiplication of Fractions – The "Of" Rule

Many of our high school students have been taking fractions for a number of years. However, they often do not really understand what they are doing when they are multiplying or dividing fractions. To grasp what is meant by **multiplying** fractions, the trainer asks the student to multiply 2/3 x 3/4 and then to explain in words, or by a diagram, what they have done. Most students cannot do this. The trainer then makes the problem into a little story. The trainer says, "This (shows a sheet of paper) is a tray full of brownies. One quarter of the tray must be saved for Dad, leaving three quarters available. Your mother says that you and your brother may have 2/3 of 3/4 of this tray if you can figure out how much that is. The trainer then has the student fold a piece of paper into four quarters and remove ¼ (by folding it behind). The trainer asks how many quarters are left. The student answers "Three." The trainer then has the student follow what the mother had said by taking 2/3 of this 3/4 section of pretend brownies. The student refolds the paper into thirds and folds one third behind. Now they have 2/3 of just 3 of those quarters. When they open the sheet and count all the little squares (brownies) they find that they have 6 of 12 sections or half of the brownies. Many students are quite intrigued by the fact that they are really taking 2/3 of 3/4. Whenever they encounter a word problem with 'of' in it, they will remember that multiplication will be the operation needed.

Division of Fractions – The "How many "___sies" are there in ___ Rule

Some of the students have actually been told that there is no way to understand division of fractions, just learn the rule, invert and multiply. They are quite confused by the fact that, with fractions, multiplication results in a smaller number and division results in a larger number for the answer. This is the opposite result to what you get working with whole numbers. If we have been able to pique their curiosity, then we tell them that we can prove that inverting the second fraction and multiplying really does give a sensible (valid) answer. We note they will never have to do this, but it is kind of interesting and fun. We ask the student to take a simple example, such as dividing 2/3 by 3/4. They are told to use the same drawing that they used for addition. They draw two chocolate bars and make all the pieces equal by changing it to 8/12 (eight twelvesies in this bar) and 9/12 (nine twelvesies left in this bar).

Now, off to the side, ask the student to tell you how many apples each student would get if there were 20 apples and five students. He will immediately answer "4." You then circle the denominator in the fraction 20/(5) and tell him that he just asked himself the question, **"How many fivesies are there in 20?"** Tell him that he is to do exactly the same procedure with the fraction division, i.e., 8/12 ÷ 9/12. He is to ask himself, **"How many 9 twelvesies are there in (or can I fit into) 8 twelvesies?"** At this juncture he takes a highlighter and highlights the eight squares in the 8/12 chocolate bar and the nine squares of chocolate in the 9/12 chocolate bar. Now he draws lines to represent moving pieces of chocolate, one at a time, from the 9/12 chocolate bar, into the 8/12 chocolate bar. Only 8 pieces will fit in. He therefore concludes that the answer is that 8 out of 9 pieces, 8/9 of the chocolate, will fit into the other bar. The answer to, **"How many 9 twelvesies are there in (or can I fit into) 8 twelvesies?"** is that he can fit in 8/9ths. In school he has learned to invert one fraction and then multiply. This would become, 8/12 x 12/9 = or 8/9. He then does the same thing to the original and gets: 2/3 x 4/3 =8/9.

Now have the student see that "inverting" is merely an example of the "do-the-same-thing-to the numerator as you did to the denominator" rule. You multiply the denominator (the fraction 3/4) by the same fraction inverted (4/3) to get "1." Then you multiply the numerator fraction (2/3) by the same number (4/3) to get the answer (8/9).

The objective is to have the student begin playing with math in order to see it and thus understand it!

Word Problems

A logical stepwise strategy takes <u>extreme focus and concentration</u> and is crucial to solving most math problems. The first rule is to read carefully – then read it again, putting lines into the problem to divide it into sections. Use a logical four-step approach. Not all the steps are necessary in every question. The steps are analogous to a detective investigating a crime. The detective, on coming to the scene of the crime, first wants the facts. Second, he draws the scene. Third, he wants the truth – a solution. Fourth, he double-checks it to make sure he hasn't made a mistake. The method is therefore as follows:

1. **Facts + Question.** Section the question into separate statements using slashes (/).
    - What are the facts?– List them!
    - What is the question? – Circle the key word.

2. **Draw it + Label it.**
    - Can I draw the facts? – Draw them!
    - Label the drawing(s) with all the facts.
    - Label the unknowns with a letter such as x or y.

3. **The Truth? And Solve it.** Put the facts into a "truth statement," which, in math, is an equation.
    - Translate words into mathematical signs; e.g., "is" becomes an equal (=) sign, "more than" is a plus (+) sign, and so on. There are two types of truth statements:
    - What "truths" can I derive from the facts?
    - What "truth(s)" does the question give me that relate some or all of the facts to each other?

To solve it, first recall with the student the basic rule: whatever you do to one side of an equation, you must do the same thing to the other side of the equation. (This is analogous to the rule with fractions: whatever you multiply the denominator of a fraction by, you must multiply the numerator of the fraction by the same number.) It may be helpful to draw an analogy to a simple balance scale. In order to keep a scale in balance, if you take something away from one side you must take away the same amount from the other side.

4. **Check your answer**

Most high school problems can be thought through logically. Particularly when working with the student who has ADD, one uses **math problems to reinforce a nonimpulsive, thoughtful, reflective approach.** The following problems may be solved by a formula or by a shortcut, but many students do not know that. To these students, some of these challenges may at first appear insurmountable. In their early stages of learning we emphasize the logical four-step approach given above. **Following a defined sequence of steps helps to provide an opportunity to overcome negative "bracing."** The student knows how to get started using metacognitive strategies rather than avoiding tasks.

Example #1
Step I. Facts and the Question

a. Section the question into separate statements using slashes (\).
   - A piece of string is cut \ into two pieces.\ The second piece is five cm more than twice the length of the first piece. \ If the original string is 245 cm long,\ how long is the longer piece when cut?
b. "What are the facts?"   - List them!
   - Fact #1: A piece of string is cut \ into two pieces.\
   - Fact #2: The second piece is five cm more than twice the length of the first piece.
   - Fact #3: The original string is 245 cm long.

Step II. Draw and Label the Facts

a. "Can I draw the facts?"   - Draw them! Label them.

I <----------- 245 cm ----------------------------> I
_____

I ←- x cm -→ I I ←---- y = 2x + 5 cm ---------→ I
_____   _____

b. "What is the question being asked?"
- How long is the longer piece when cut?
c. Label the unknowns with a letter such as x or y.
   - Let the first piece be x and the second piece be y.
d. Put the facts into "truth statements" or equations. Translate words into signs.
e. What "truths" can I derive from the facts?
   - y is 5 cm longer than two times x ('longer than' is the same as a plus sign)
   - Therefore: $y = 5 + 2x$

Step III. Truth Statement or Equation and Solve it

a. What "truth" does the question give me that relate some or all of the facts to each other?
   - $x + y$ = the whole string = 245 cm
   - Therefore, when you *substitute* for y: $x + (5 + 2x) = 245$ cm
b. Solve it for one unknown at a time. (Do the same thing to both sides of the equation – just like on a scale for weighing gold, if you add or subtract from one side you *must do the same thing to* the other side in order to have the scale balance again.)
   - $3x + 5 (-5) = 245 (-5)$ cm, then divide both sides by 3 and get $x = 80$ cm Thus, $2x + 5 = 165$ cm

Step IV. Check your answer

In this case, the student would add $x + y$ to see if they got the length of the original string (80cm +165cm = 245cm).

It is crucial that the trainer always use the same four-step approach and that the trainer not assume the student has the basic concepts that would allow steps to be skipped.

Example #2

Often the wording of a problem confuses the student. Some students give up rather than try to logically think through the problem. The following is an example that the reader can try.

- A line (cord) is drawn between two points, A and B, on the circumference of a circle. The radius of the circle is 3 cm. The angle formed when two lines are drawn from the center of the circle to points A and B on the circumference is 90 degrees. What is the area of the segment between the line AB and the shortest arc of the circle between A and B?

This is a very easy problem providing the student doesn't panic. The student must mark the three key facts in the question, then draw a large clear diagram. Then the student must not waste time. They should recognize and accept that they do not know any formula to calculate an area between an arc and a cord of a circle. Therefore they must think logically and **ask themselves what area formulas they do know.** They know the formula for getting the area of a circle. They also know how to calculate the area of a triangle. The student must figure out from the drawing that he/she must subtract knowns (area of the triangle from the area of 1/4 of the circle) to find the unknown (the area of the segment).

One quarter of the area of the circle (πr^2) /4 minus the area of the triangle ABC ([r x r] / 2) = area of the segment between the cord AB and the arc AB. Pi or π is 3.14 and the radius r = 3cm..

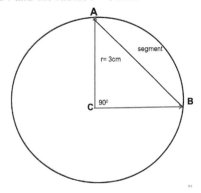

Note: The number of strategies for different aspects of math is large. There are good texts and courses that cover this material in detail. A good example is any of the SAT (Scholastic Achievement Test) preparation books by Dr. Gary Gruber. The key phrase the student must keep in mind is, "When the question looks too complicated, find a way to simplify it."

Brief Summary

The importance of the foregoing processes to help children with basic math concepts does **not** lie mainly in learning how to read a clock, add and subtract fractions, multiply large numbers, solve word problems and so on. These skills should be taught in school. The value of the foregoing process is to train the child to go into a high state of focused concentration and recognize that they have, within themselves, the capability to **figure out things from first principles.** Our objective is to have the students change their outlook and **apply strategies rather than avoiding work or giving up** on something that seems hard. We don't want them to grudgingly memorize formulas for exams. We want them to feel that they can take abstract academic subject matter and make it interesting and exciting. They can use their creativity and develop their own tricks for learning and remembering things that augment the strategies taught during neurofeedback training sessions.

Study Hall

The student must have a determined attitude (and a concrete method) for entering every period of study, **and must never deviate from the basic dictum that every unit of work or of time must produce a "product."** In other words, they do not just flip open a book and start; they set a goal.

We liken study hall time at home or in the library with the production of widgets in a factory. The student must decide before beginning on an evening of study, for example, exactly what the product should be for each unit of this study time. This may be a unit of time, but preferably it would be a unit of work to be accomplished.

With a very young child the parent should ask Socratic questions to help the child develop these goals. Using an egg timer can help awareness of time and allow the child to develop his judgment about how long it will take to complete a task.

To help with setting up a profitable evening study time, the parent may use a three-step study hall procedure: First, the parent might ask, "What would you like to learn about?" Second, they might ask, "Where can we find this information (e.g., the child's notes or school text, the encyclopedia, the dictionary, a magazine or book etc.)?" Third, the parent will find and read the needed material with the child. The idea is for the parent to *facilitate.* The parent must not add to the burden. Rather, the **parent must be perceived as a helper, not a critic.** If possible it is good if the parent can generate excitement about finding answers and generating new interesting questions as they read. Fourth, after the parent has read the material, partly with and partly to a young child, the parent may then reward the child. The parent perhaps gives points for achievement and bonus points for new questions generated during the reading process. The points gradually accumulate towards earning something. Or the reward may be a game the child enjoys or "recreational electricity" (TV or computer time). A desired activity is a strong reward for doing a less desirable activity.

It is **important that parents model** for their child. The parent(s) should read, do bills or write letters, but **not** watch TV or play games on the computer while their child is doing homework. Then, if helpful, parent and child can take pleasant breaks together.

With older students, the parent will demonstrate very genuine interest in the student teaching them how to organize the material for study and reporting to them the essence of the units studied at the completion of each time period. It is **crucial, however, that the student develop the schedule, the units to be studied, and the time** frames. The parent is only a useful tool in the process. The moment the parent begins to dictate the "what, why, when, where and how" of study hall is often the moment when passive aggressive noncompliance begins. One has to sell the parents as well as students on the idea of study hall since the parent is in a position to reinforce this good habit. Tell them that if they were investing $35,000 in a good boarding school, one of the things that they would be paying for would be study hall – a period of time each weekday evening (usually 7 to 9 pm for high school students) that is set aside for studying. Students who need the support of a teacher resource may study in a special hall that is supervised. Stronger students usually study in the library or in their room. Students with an A+ standing in all subjects are excused from formal study time since they have demonstrated that they know how to use effective study habits. Since this is part of the formula for success at boarding schools. Why not develop a similar procedure at home? Friends soon learn to call after study time (especially if only an answering machine gets calls during study time). A call to clarify a homework problem, however, is allowed. Television shows can be taped for later viewing. A family may decide that 4 to 6 is a better time than 7 to 9. Be flexible about when and where, but rigid about the expectation that there is a study hall time. No homework? The time can be spent in review, reading, and test preparation.

Conclusion

Many students are not very efficient learners in standard classroom settings. Students with ADD, for example, are the *'hunters'* so aptly described by Thom Hartmann in his books. They are often quite bright, though they tend to underachieve in school. They are capable of hyperfocus and they are often extremely creative. Simply put, they learn differently than the *'farmers'* do. Our job is to help such students to use metacognitive strategies and **develop learning mechanisms or techniques that fit their cognitive style**. Neurofeedback plus metacognitive strategies can help these students to learn more efficiently and effectively and it empowers the students to remember material and improve their school performance. They finally learn how to learn.

Metacognition – Quick Reference

<u>Cognition</u> = Thinking
- Perceiving
- Learning
- Remembering

<u>Metacognition</u> = Thinking About Thinking
- Guiding and monitoring your cognition
- Being conscious of how you learn and remember things
- Develop strategies to organize, plan, learn, and remember more effectively
- Games for young children *(start techniques early - learn that learning is fun)*

Read Actively, not passively, using Four Steps:
1. **Before You Open** the book or start on a chapter:
   - **Predict what should be covered:** Headings and subheadings the author should include. Brainstorming, Gridding: Rows are questions like who, what, when, where, how, with what effect, <u>or</u> biological, psychological and social. Columns are main areas to be discussed like before, during and after, <u>or</u> contributing, precipitating, and sustaining factors.
   - **Generate questions** about things you would like to learn.

2. **When You First Open** the book or chapter:
   - **Scan** - abstract, conclusion, headings, questions, graphs, tables, pictures.
   - **Skim** – a few seconds a page for first-line nouns (key words) in each paragraph.
   - **Framework** - build the foundations, the girders, the framework of the knowledge, then fill in as you read.

3. **While Reading**
   - **Start** with something that catches your interest, e.g., the caption under a picture or a table that summarizes the main points.
   - **Highlighting and underlining: Use two pens (red and blue) and a highlighter to mark the essentials:** highlight very, very little and keep underlining to a minimum. Circle the major items. The red pen is only used to mark the most important items when you review the chapter. Write summaries in the margins and/or draw summary diagrams.
   - **Only highlight** the thesis (purpose of the material) - the key word for each of the main arguments or each new topic (new section or new paragraph).

- **Circle** the key word for each new thought within the section or argument.
- **Underline** key facts which support each area.
- **Boxing** - Divide reading time and material into small units.
- **Connect** each box to the next as you read. Connections must be logical to you.
- **Number** (sections/concepts) and mark (important points)
- Keep **Generating** questions and **Predicting** answers
- **Chunk** key words and phrases into logical and/or amusing groups.
- **Pictures, Rhymes, etc.** - Use tricks to assist memory and recall.
- **Review** as you read, after each page or paragraph
- **Widgets** - Decide on a product-to-be-accomplished for each unit of time in your study schedule.

4. Review
   - **Organizing, Synthesizing:** Make a **flow sheet,** an overview linking the main concepts.
   - **Packaging:** Make extremely neat and well-organized but brief notes - *only the exam essentials. One package per chapter. The numbering gave you your sections for each package and your marks (or highlighting) gave you the main points to include.*
   - **Lecturing**: To parents or a friend or pretend the walls are your teachers. Teach/Teach/Teach - it's the best way to learn!

Writing an Essay, a Debate or an Exam Question. The Hamburger or '3-3-3' Method

Follow the general principle: Tell them what you are going to tell them, tell them, tell them what you told them.

1. Make three sections – the bottom bun, the meat and veggies, the top bun (introduction, body, conclusion)
2. Have three subsections for each of these sections.
   - **Introduction**: Divide the Introduction into three sections: introductory statement that catches an audience's attention, thesis statement, then mention the three arguments you will develop in the body.
   - **Body**: Make three major arguments (points) in the body of the essay with three supporting facts (data) for each argument.
   - **Conclusion:** Summarize the three arguments, discuss them in the light of your thesis or hypothesis, and then make a wrap-up comment.

Listen Actively

1. Find out from the teacher, if possible, what will be covered in each class for the next term. Buy the textbook. Obtain good cookbook notes from a high-ranking student who is in the year ahead of you.

2. Brief before-class preparation gives you a **Filing Cabinet Mind** rather than a Garbage Bag Mind. Briefly (5-10 minutes) the night before each class do the Predicting, Organizing, then Skimming, Scanning, and Frameworking of the textbook (see reading section above) and/or look at last year's notes from a student in the year ahead. The next day you will recall more information than your fellow students at the end of each class because your mind has been ready to assimilate the information.

3. Set up for success for in-class alertness and information recall by using the *3-fact- method.* This is a method for connecting what is learned at school to reviewing the main points at home. It is a quick way to predict the questions that will be asked on the next test or exam.

4. Scribbling each main idea/fact, or marking it specifically in the notes you are making in class will move that fact from a six-second recall time to about a four-hour recall. Try for three facts from each lecture.

5. Writing each of these key facts neatly in a list that is kept at the back of that subject's binder moves those key facts from a four-hour recall to a seven-day recall.

6. Glancing at, and/or looking up, information on those key facts a few days later moves that information to long-term memory and recall.

7. At the end of the month or term, this list of facts may correspond quite well to the questions that the teacher will ask on the exam.

Organize Your Time

1. **Study Hall** - Your desk has files, calendar, planner and silent phone (with message machine to say you will call back at the end of the hour or at the end of study-time).

    - **Planner** - Keep a calendar to help your recall of things to be done: one calendar that you carry, plus a large, three-month wall calendar for assignment due dates and test dates.
    - **Short-Term** work to be done must be listed for each day and seen as a *week-at-a-glance* overview. Daily activities and homework are listed for each day.
    - **Long-Term** projects should be kept on a separate page (*inserts and/or month-at-a-glance pages*). This section includes projects with a plan for completing library searches, internet searches, outline, rough draft, and final product, with deadlines for each section of the work.

Attach a new habit (looking at your planner) to an old habit (eating breakfast, eating dessert, having a snack after school, sitting in the car).

Keep in mind the adage that failing to plan is planning to fail. With neurofeedback you self-regulate your mental state, and with metacognition you self-regulate your approach to learning.

Metacognition is a powerful adjunct to neurofeedback because they both empower the person to take control.

PART FOUR

Assessment and Intervention: Advanced Procedures

This part of the text describes the use of evoked potentials and LORETA in assessment and LORETA Z-score neurofeedback as an intervention

SECTION XVI
Evoked Potentials (ERPs)

Definition: An event-related potential (ERP) is a brain response that shows a stable time relationship to actual or anticipated stimuli. Event-related potentials are also called *evoked potentials* because they are evoked by an "event". The event can be any visual, auditory, or kinesthetic stimulus. It could also be a psychological event, such as a thought. The most common use of evoked potentials has traditionally been audiologists using auditory stimuli to evoke a brain response to test hearing in people who cannot give a verbal response, such as infants or people with autism. Simpler methods have now been devised to test hearing, but event-related potentials (ERPs) continue to be an active area of research in psychology and are of interest to psychologists and neuroscientists who want to measure and investigate various aspects of brain functioning.

ERPs measure, to the millisecond, the *brain's speed of processing information* (visual, auditory, physical, or even electrical, such as thoughts arising in the brain). This fast processing of information, measured in milliseconds, is what makes humans able to take in billions of bits of information at a time and filter and process this information to make decisions. ERP testing can detect disruptions in this process before it is noticeable in everyday life. ERP measures can indicate slowed physical reactions, slowed decision making, memory impairments, or stress disorders, or they may reflect other known neurologic disorders. These measures are also important in determining brain speed in concussed athletes to assess whether their brain reaction time has returned to baseline (Evoke Neuroscience, 2012).

Measurement

ERPs are measured from the scalp using standard EEG electrodes or by using a standard 19-channel cap. Standard electrodes means the use of *macro-electrodes* (>100 microns) and filters that pass frequencies <50 Hz. ERPs are usually measured along the midline at the central locations, Fz, Cz, and Pz, but may be measured at all 19 sites, as is done with Mitsar equipment and the WinEEG (Kropotov, 2009), or with eVox and the Evoke Neuroscience programs. Of course, ERPs require special software for analysis. Measurement with 19 channels may be done with the standard electrocap that is used for recording the EEG. The evoked waves are time-locked to repeated occurrences of sensory, cognitive, or motor events, so there has to be a task that involves a series of stimuli to evoke the responses. Recording and seeing these ERPs is not as simple as looking at a raw EEG. Evoked potentials are contained within the raw EEG but the waves are small in amplitude and cannot be easily seen when they are embedded in the spontaneous EEG. Therefore, an averaging technique from a large sample of responses to the same stimulus must be used. ERPs are extracted from scalp recordings by a computer that averages epochs of EEG. The epochs analyzed are the one second of data that follow each presentation of the stimulus. At least 20 presentations of the stimulus are necessary for the averaging, and in practice, usually more than 100 presentations of the stimulus are used. The spontaneous background EEG fluctuations, which are random relative to when the stimuli occurred, are averaged out. This leaves only the "time-locked" event-related brain potentials. In this manner, the ERP reflects, with high temporal resolution, the patterns of neuronal activity evoked by an event.

ERPs are named according to whether the waves are positive (P) or negative (N) followed by the appropriate time after the stimulus. For example, N100, also called N1, is a negative-going wave that occurs about 100 ms after each presentation of the stimulus. A typical ERP wave form is shown below.

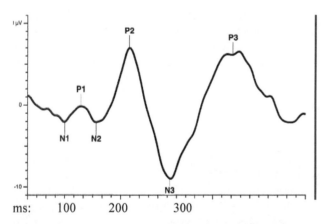

Figure shows a typical visual evoked potential. Note that this graph shows negative down and positive up. In some texts the opposite is shown with negative up. Source: Adapted from Odom et al., Visual Evoked Potentials Standard, Documenta Ophthalmologica 108: 115-123 (2004).

Auditory ERP

An oddball paradigm is commonly used to measure auditory ERPs. If a regular sound such as a beep in a low tone is interspersed with a less frequent high-tone beep, then each of the "oddball" higher pitched tones will evoke the same ERP wave, which becomes visible when they are averaged. In the auditory cortex, followed in a time sequence by other cortical areas, the brain will produce a series of evoked potentials, evoked in this case by an infrequent but meaningful sound). The subject is told to count the high tones, or press a button when they occur, so the person pays attention to them. This particular design is called an 'oddball' paradigm, because the less frequent, meaningful, high-pitched beep causes a different reaction in the brain's electrical activity than is elicited by the more frequently presented low-tone beeps.

Visual ERP

The above figure is from the Evoke Neuroscience Assessment Program using the eVox amplifier. The large circle is the target. The client is instructed to press a button on the end of a small handheld rod whenever the large circle appears. They are to ignore (make no response to) the small circle (a frequent visual distracter), a checkerboard (an infrequent visual distracter), and a noise (an infrequent auditory distracter). The evoked responses are later extracted from the EEG that was recorded during the administration of this continuous performance test. A visual infrequent distractor ERP, for example, is determined by averaging all the responses to the checkerboard that were recorded at O2. Note that the large blue circle is a meaningful stimulus because the person has been told that they must respond to it by pressing the button. ERPs that are earlier than 250 ms are not recognized consciously. Thus, the ERP that best reflects conscious information processing is the P300 response evoked by the large circle as recorded at Pz where the amplitude should be highest. This continuous performance test, also called a Go/NoGo task, additionally yields scores for correct responses (vigilance), incorrect responses (impulsivity), response time, and response time variability, so it makes efficient use of the recording time, providing a great deal of data in just 12 minutes. This is a relatively difficult visual continuous performance test designed originally by John Polich (2004). The patient must decide, each time a circle appears, if the circle is large or small. It is difficult for some people because the difference in the size of the big and the small circles is quite small. There is a practice before the recording begins. The person must meet the criterion for being able to distinguish large from small before the test recording is started.

The program knows the exact time of presentation for each type of stimulus: sound, checkerboard, small circle, large (target) circle. Therefore it can extract the epoch times for, say, checkerboard, and put each epoch for that stimulus in a bin. It can then average JUST those 1-second epochs. You therefore have the ERP evoked uniquely by each type of stimulus.

When a stimulus is detected it will engage a memory-checking system in the patient's brain. If the new stimulus is the same as the previous stimulus, such as having just one more of the small circles in the visual evoked potential test that we use, or another of the regular sounds in the auditory oddball test, then just the sensory evoked responses (N100, P200, N200) will stand out in the averaged signals for that stimulus. If, on the other hand, it is the target stimulus (the large circle in the Evoke Neuroscience testing, or the odd-ball sound in an auditory odd-ball test) then attentional resources are allocated to this target stimulus and the neural representation of stimulus is updated. When a person becomes consciously aware of the signal, the ERP is called a P300. This positive-going wave occurs after the purely sensory evoked potentials and indicates that conscious processing is occurring. The sensory evoked potentials that are automatic and occur without conscious awareness, as shown in the diagram of a typical visual evoked potential, were called N100, P200, N200 (John Polich, 2004, 2007).

Note that automatic processing is done within the first 250 ms in a normal efficient brain, and the conscious evaluation of the stimulus begins after about 300 ms. Hence the P300 (also called P3) is an important measure. It does not always occur right at the 300 ms mark. Up to 450 ms can be normal for an adult's conscious processing to begin. Children will have a slower P300 than adults because there is less myelination and thus slower brain speed. When there has been brain damage there is also a slower P300. In fact, the P300 can occur anywhere from 300 milliseconds to almost 1,000 milliseconds after the event. The amount of time before the appearance of this wave is called its **latency**. A longer latency than expected for a person's age indicates problems with brain speed. As will be discussed later in this section, problems such as ADHD, dementia and concussion may result in longer latencies for the P300 evoked potential.

Example from Assessment Data
(Note: +ve is up and –ve is down in these illustrations)

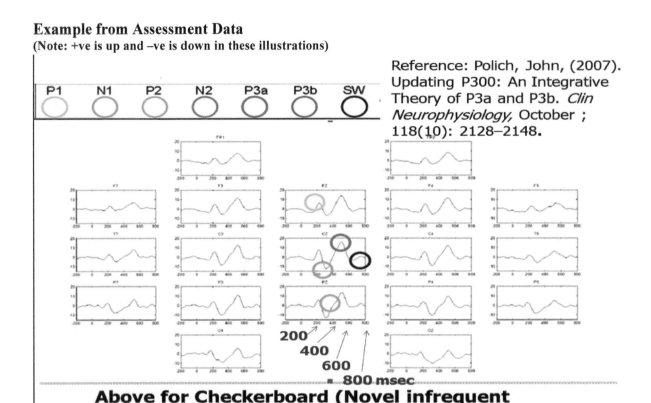

Above for Checkerboard (Novel infrequent Distractor) shows good amplitude (red) P3a

The above figure shows ERPs at 19 standard sites. Hand-drawn colored circles have been used to highlight the evoked potentials in response to the **novel infrequent visual distracter, a checkerboard** figure in this test from Evoke Neuroscience. Here the client demonstrates a good amplitude (red circle) for P3a at Cz. Diagram used with permission and thanks to David Hagedorn and James Thompson, Evoke Neuroscience, after work by Dr. John Pollock. Note that the waves occur at all sites. The maximum amplitude of P3a is usually best measured at Fz or Cz. It is a reflection of attention. The maximum amplitude for P3b, on the other hand, is usually at Pz. P3b response (not shown here) is measured as the P300 response to the target stimulus and is hypothesized to reflect memory processing.

Sensory ERPs (<250ms)

Quick Overview – Types of ERPs

In very broad general terms, when ERPS are evoked by an external stimulus they are termed *exogenous,* and when evoked internally they are called *endogenous.* At about 300 ms one is conscious of a stimulus (exogenous) or of one's own expectation that a stimulus (endogenous) should occur even if it doesn't. In both cases a P300 will occur. An example of an endogenous ERP would be the brain response if you were waiting for your name to be called in an alphabetical list and it was not called: your brain would respond to that omission. This is why ERPs are defined as being a brain response to actual or anticipated stimuli.

It can also be helpful to think in very broad general terms of four types of ERPs (Vaughn, 1969).

Exogenous ERPs
1. Sensory ERPs: N1, P1, N2, P2. There is no conscious awareness of the stimulus at these short time intervals that are all less than 300 ms
2. Motor ERPs
3. Long Latency ERPs: P3a, P3b, P400
4. Contingent Negative Variation (CNV): (Not described here.) This is a brain activation that occurs after a signal alerts a person to prepare to make a response.

Details Regarding Types of ERPs
First Type of ERP
The earliest components of ERPs correspond to perceptual signal detection. These are **sensory ERPs.** They are *exogenous ERPs* **because the stimulus comes from the external environment.** Negative 100 (N1) occurs at about 60-80 ms. N1 is thought to correlate with the signal's arrival at the thalamus. P1 occurs at about 100 ms and may represent signal transmission to perceptual cortical areas such as the auditory cortex for sound, and the cuneus and occipital cortical areas for visual stimuli. Initial sensory processing within the sensory association cortex is noted from 90 to 130 ms after the stimulus.

Sensory awareness corresponds to P2, but does not occur until approximately 200 milliseconds after the stimulus (Railo et al., 2011). P2 correlates with sensory awareness, but not full consciousness. The detection of the stimulus in posterior cortical areas is followed by projection to prefrontal and parietal areas. This occurs about 130 to 280 ms after the stimulus.

In addition, phase-locked gamma activity appears between the parietal and prefrontal cortex at approximately 180-230 ms during a cognitive task (Varela et al., 2000). Posterior temporal, phase-locked gamma is seen at approximately 200 ms (Gurtubay et al., 2004).

Sensory ERPs in More Detail
N1 and P1 occur between 70 and 120 ms after an external stimulus and, as noted above, the N1 component is generally thought to represent the initial extraction of information by sensory analysis of the stimulus, or by the excitation associated with allocation of a channel for information processing out of the primary cortex (Barry, 2003). Note that negative correlates with excitation. These sensory ERPs are evoked by sight, sound, smell, and touch. They are termed *exogenous* ERPs because they are evoked by something outside the individual. These sensory ERPs are labeled according to their millisecond latency time after a stimulus and whether they are positive (P) or negative (N). The P100 and N100 are primarily recorded in the vicinity of the primary sensory areas. The dominant site will be dependent on the specific sensory modality that is involved. Those evoked by visual stimuli, for example, will occur in the cuneus and in the occipital lobe.

P2, in the context of the oddball task, may represent inhibition of sensory input from further processing via automatic stimulus identification and discrimination/classification processes. It also may represent inhibition of other channels of information that are competing for attention and further processing (Barry, 2003).

N1 and P2
ERPs occur with a negative peak at about 80-90 ms and a positive peak at about 170 ms after the stimulus. This is called the **N1-P2 complex**. For sound, the maximum amplitude for P2 occurs in the auditory cortex in the temporal lobe. For visual stimuli there may be an N1 response in the cuneus. However, the combination, called the **N1P2 complex,** is thought to be primarily located in the cingulate cortex. It may reflect aspects of the sensory system such as auditory processing and perceptual processing.

N2
Under active attend conditions, the N2 component is thought to represent an endogenous **mismatch detection** process related to stimulus discrimination. In the Evoke Neuroscience testing, this difference detection would occur after the infrequent target stimulus (large circle) as compared to the frequent non-target distracter (small circle). However, with visual ERP testing this is sometimes difficult to discern. It is more easily observed in the temporal auditory cortex using an auditory oddball paradigm. An example of an auditory oddball paradigm would be the task of counting high tone beeps in a series of mainly low tone beeps. It is associated with attention focusing, categorizing, perceptual closure, and conflict detecting. Its source is the anterior cingulate and the prefrontal cortex (Sokhadze, 2012).

There may be N170 left sensory stimuli mismatch (left angular gyrus) and N170 right emotional mismatch in the region of the right angular gyrus.

Mismatch Negativity (MMN)
(Thanks are extended to Sara Ottonello, Ph.D., Milan, Italy, for her assistance with this section.)

MMN is the difference-waves obtained by subtracting the average of the standard stimulus ERP from the average of the deviant stimulus ERP. MMN magnitude increases with the stimulus deviance but MMN latency decreases when the discrepancy is greater and thus more easily perceived. The latency of the response to the deviant stimulus is longer than the latency of the responses to the standard stimulus.

The existence of enhanced responses to the deviant stimuli is usually interpreted as an indication that the brain effectively stores a "memory trace" of the standard stimuli, so that incoming stimuli are compared to this memory trace, and the mismatching deviants elicit stronger responses (Näätänen, 1992; Näätänen et al., 2001).

Thus MMN, when using the auditory oddball paradigm, provides an index of preattentive auditory memory functioning. It results from a mismatch detection process involving the comparison of incoming stimuli with a memory trace formed by preceding stimuli. For example, the frequent distracter sound (low tone) is distinctly different from the infrequent target sound (high tone) in the test situation and this elicits the MMN response in the brain. A tendency towards a smaller MMN amplitude in ADHD has been reported (Kropotov, 2009). This could be interpreted as paying less attention to incoming stimuli at the level of automatic processing done by the brain. (The MMN precedes the conscious processing. Conscious processing is done after about 300 milliseconds.) This smaller MMN suggests detection abnormalities (Barry, 2003).

Mismatch negativity (MMN) is a term you may hear or read about in research even if you are not measuring ERPs. Kropotov used intracranial electrodes and demonstrated that MMN could be seen in the auditory cortex in the "comparing area," BA 22, of the superior temporal cortex. It reflects a DC potential difference, and Kropotov considers it a measurement of attention. As noted above, it is smaller (lower amplitude) in children diagnosed with ADHD. It can be thought of as reflecting a difference in amplitude between the N2 wave recorded for deviant (target stimulus) and standard (nontarget frequent distracter) stimuli in the oddball paradigm. MMN may be a measure of decay of the auditory trace (Kropotov et al., 1995).

Example

In the figure below from Kropotov, note that the temporal lobe, BA 22, is involved in an automatic change detection process involving sensory input. This area detects a difference between the standard and the deviant auditory stimuli in an auditory oddball task. The deviant stimulus in the oddball task generates an enhanced negative potential, which is observed in BA 22 of the left (dominant) superior temporal gyrus. Frontal-central areas are also involved in the detection of this difference. The figure below is taken from a study done by Dr. Kropotov in St. Petersburg using intracranial electrodes implanted in the temporal region of the cortex. In the example below he shows that MMN is seen in the auditory "comparing area" BA 22, superior temporal cortex. Note that a strong P2 follows the negative wave. Kropotov has noted that it is more difficult to see MMN in other cortical areas.

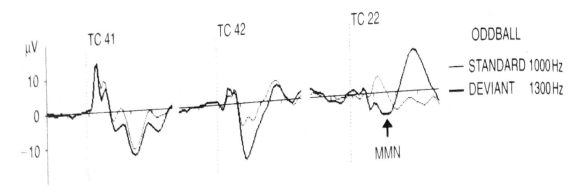

*Diagram shown with permission, from PowerPoint presentations used in workshops by Juri Kropotov. Note in the figure above that **negative is down**. However, neurologists usually view tracings in which negative is up. Negative shifts are typically associated with brain activation. Note also that TC 22 refers to Brodmann area 22.*

At the time of writing MMN is being studied with central measurements at Cz in excellent research on intellectually challenged patients in Italy. MMN was chosen as a measure because it is an automatic process that **detects a difference between an incoming stimulus and the sensory memory trace of preceding stimuli**. This automatic processing is thought to be a good objective measure for research being done with these individuals who are intellectually challenged. The work is being carried out in Milan and Florence, Italy. A preliminary report was presented during at the Biofeedback Federation of Europe (BFE) meeting in Venice, February 2014, by Sara Ottonello. Gualtiero Reali was the second researcher on the project and it was overseen by Marco Bertelli. These researchers have been able to demonstrate that MMN amplitude is decreased and latency increased in people with severe developmental delays (mental retardation). They are now investigating whether NFB training can result in more normal MMN responses and concomitant clinical improvements in these patients. The preliminary results in people with intellectual developmental delays look promising, as seen in the figures below, which are from the initial analysis of their data. The graph compares the ERPs of the subjects prior to training and after eight months of training with the ERPs of a control group who do not have intellectual challenges. One can see that there is a shift toward "normalcy" after training in the disabled group.

Remember when looking at the graph shown below that it is NOT the raw ERP data. Rather it is a graph of a calculation of the MMN, which is the difference between an incoming stimulus and the sensory memory trace of preceding stimuli for a particular group of subjects.

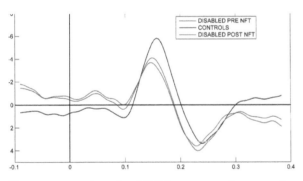

Figure comparing the MMN response in normals with that of people with intellectual disabilities before and after neurofeedback training. (Note that negative is up on the y axis in this diagram. Note also that each of the MMN waves is calculated by subtracting: the average response to the infrequent (deviant) stimuli minus the average response to the frequent (standard) stimuli.)

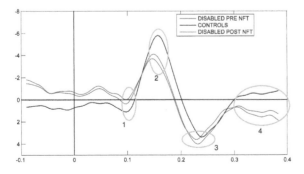

Figure to provide further comment on the components of the MMN response. (Note that negative is up.)

Remember when viewing this graph that MMN is calculated as the **difference between the response to frequent and infrequent stimuli**. Note that MMN is not an N1P2 phenomenon. At #2 note that negative is up on the y-axis. Negative indicates activation. This **negative activation has increased post-NFB** (NFT is neurofeedback therapy in the diagram). **It is still less than the negativity shown by the control** group. At #2, the amplitude difference to the two stimuli in a disabled group is posited to be less than in the control because their neural systems elaborate with less accurancy, so there is less activation. The deviance between the two auditory stimuli at #3 reflects slightly greater inactivation (more positive) in the disabled subjects for the same reason. The MMN latency (the difference in latency, **not** the actual latency) is shorter for the disabled subjects because these subjects discriminate the two stimuli with LESS time deviance. It is as if they were unable to consistently discriminate the difference between the frequent and the infrequent stimuli. It is as if their auditory system analyzed them at times as if they were the same sound.(See also Norman Doidge's discussion of auditory processing in his writing concerning Tomatis Sound Therapy, 2015).

The timed response at #4 is past the MMN stage and is more in the range of the P300, which is a conscious response (not the automatic sensory response) that is involved in being aware of and evaluating the stimulus. We would expect the P300 to be a longer latency in the experimental subjects as compared to control subjects, but that was not being measured in this study.

The auditory MMN can occur in response to deviance in pitch, intensity, or duration. The auditory MMN is a fronto-central negative potential with sources in the primary and non-primary auditory cortex and a typical latency of 150-250 ms after the onset of the deviant stimulus. Sources could also include one from the right opercular part of the inferior frontal gyrus (see Section VIII, BA 44). The amplitude and latency of the MMN is related to how different the deviant stimulus is from the standard. Large deviances elicit MMN at earlier latencies. For very large deviances, the MMN can even overlap the N100 (e.g., Campbell et al., 2007). The MMN data can be understood as providing evidence that stimulus features are separately analyzed and stored in the vicinity of the auditory cortex.

The experimental evidence suggests that the auditory sensory memory index MMN provides sensory data for attentional processes, and, in essence, governs certain aspects of attentive information processing. This is evident in the finding that the latency of the MMN determines the timing of behavioral responses to changes in the auditory environment. Furthermore, even individual differences in discrimination ability can be probed with the MMN. The MMN is also a likely component of the chain of brain events that relates to attention switching due to changes in the environment.

The mainstream "memory trace" interpretation of MMN is that it is elicited in response to violations of simple rules governing the properties of information. It is thought to arise from violation of an automatically formed, short-term neural model or memory trace of physical or abstract environmental regularities (Näätänen, 2004, 2007, 2011). Integral to this memory trace view is an assumption that there are:

1. A population of sensory afferent neuronal elements that respond to sound, and
2. A separate population of memory neuronal elements that build a neural model of standard stimulation and respond more vigorously when the incoming stimulation violates that neural model, eliciting an MMN.

An alternative "fresh afferent" interpretation (Jääskeläinen et al., 2004) is that there are no memory neuronal elements, but the sensory afferent neuronal elements that are tuned to properties of the standard stimulation respond less vigorously upon repeated stimulation. Thus when a deviant activates a distinct new population of neuronal elements that is tuned to the different properties of the deviant rather than the standard, these fresh afferents respond more vigorously, eliciting an MMN.

A third view is that the sensory afferents are the memory neurons (Jääskeläinen et al., 2007).

We cannot comment on these differing views, but they do reflect the kinds of questions being addressed in ERP research. Interested readers should pursue the research. The primary reason for being aware of ERP measures in our field is that they can be helpful in assessment and also helpful in measuring changes achieved with training.

Motor ERPs

The second type of ERP, is the **motor ERP**. Motor ERPs precede and accompany motor movement. They are proportional in amplitude to the strength and speed of muscle contraction. These ERPS are seen predominantly in the precentral (motor) cortex. They are often studied in kinesiology and in work with athletes. The *Bereitschaftspotential*, for example, is studied in kinesiology research and reflects motor readiness.

Long Latency ERPs (>250 ms; e.g., P300)

Long-Latency Potential ERPs
The third type of ERP is called **Long-latency Potentials.** The long-latency ERPs include P300, sometimes subdivided into P3a and P3b, and P400. These reflect subjective responses to stimuli. They occur approximately 250-750 ms after the stimulus, depending on brain processing speed. Most often it will be about 300 ms, and not slower than 450 ms in adults. The P300 can be as slow as 900 ms in some conditions, but this is very unusual and has not been seen in our clinical work. P300s reflect a psychological response rather than just the brain registering the presence of a stimulus.

P300 Components
After unconscious signal detection there is a phase of conscious recognition. Thus, while the earlier components - N1, P1, N2, P2 - are not in the realm of conscious cognitive processing, the P300 components reflect conscious processing of the stimulus. They represent context updating when stimulus events require that an individual's model of the environment be revised. The extent to which this updating process is activated may depend upon the value, significance, or relevance of the stimulus.

The P300 indicates the brain has noticed something and is paying attention. P300 latencies can distinguish between groups, for example, people with ADHD having longer latencies than non-ADHD subjects, elderly subjects have longer latencies than younger adults. Latency of the P300 will also be greater in dementia and after a head injury. More recently it has been found that P300 measurements hold promise for lie detection. Farwell has shown people show longer latencies when telling a lie, presumably because they must hold both the truth and the lie in mind, which takes more processing time (Farwell & Donchin, 2000). Rosenfeld has presented on different scalp distributions with lies than with the truth. He notes a crooked line between amplitude of responses at Fz-Cz-Pz rather than the usual smooth line with amplitudes increasing from front to back (frontal to parietal).

The P300 (P3) component is sometimes divided further into P3a and P3b components.

P3a and P3b

P3a is an attention-orienting frontal component, while P3b reflects sustained attention and memory and is more central-parietal, with the largest amplitude being observed at Pz. P3a, in the Go/NoGo task described above, is primarily evoked by the infrequent, novel, visual distracter (a checkerboard).

P3b, on the other hand, is primarily evoked by the target stimulus, which is a large blue circle in the test described. In ADHD the P3b latency is longer and the amplitude less than is seen with a normal control group (Barry et al., 2003). This 'low and slow' P3b ERP may also be observed after a head injury.

The P3a is usually seen to have its highest amplitude frontally in the premotor cortex, and can be as fast as 250-300 ms in a healthy young adult. The P3b has its highest amplitude in the parietal region at 300-350 ms. Neurotransmitters differ, with **P3a involving frontal/dopaminergic** activity, while **P3b involves parietal/norepinephrine** activity. It is hypothesized that neuro-inhibition is an important mechanism for the P300. The P3b is elicited when stimulus detection engages memory operations. Then rapid neural **inhibition of ongoing but nonrelevant** activity will facilitate transmission of stimulus/task information from frontal (P3a) to temporal-parietal (P3b) locations.

One hypothesis is that P300 signals could occur from the initial need to enhance focal attention during stimulus detection relative to the contents of working memory (Soltani & Knight, 2000b). Hence, a minimization of extraneous stimulus processing would facilitate the transference of incoming stimulus information from frontal to temporal-parietal areas to sharpen memory operations. Whatever the process turns out to be, the primary functional importance of the P300 is that it is able to reflect aspects of the executive network. In addition, P300 suppression at prefrontal sites can relate to impulsivity. In general, **P3a relates to attention and vigilance** and reflects the processing speed for novel stimuli, while **P3b appears to relate more to working memory, stimulus interpretation, and engagement.**

Case Examples

The best way to come to understand the utility of ERPs for assessment is to look at data from a few case examples. Three cases are shown in detail below. The first two show a table with the most helpful ERPs measured at the O2, Cz, and Pz sites followed by the data from all 19 sites for the four different stimuli (large blue circle, small blue circle, checkerboard, and sound). Studying all 76 ERP responses (19 sites x 4 stimuli) may be of research interest, but for clinical purposes, the four ERPs that most clearly reflect brain speed in the majority of patients are the N1 response at O2 for visual processing, the N1 response at Cz for auditory processing, the P3a response at Cz for executive functioning (attention and vigilance), and the P3b response at Pz for executive functioning (information processing and working memory). To avoid being overwhelmed, only these are usually reported.

Case Example #1

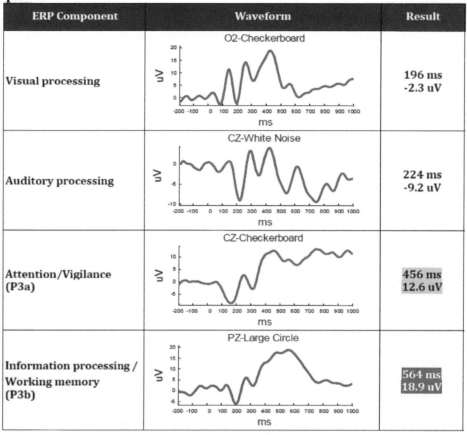

Male ("BS"), age 14. Client appropriate for NFB training to reduce the traits of severe Asperger's syndrome plus ADHD symptoms and learning problems. The visual and auditory ERPs are within normal range. However, the P3a and P3b values are normally <450 ms. The P3a at 456 ms is slow, indicated by yellow, and **the P3b latency (speed) is colored red because it is very slow and considered abnormal.**

The sensory and cognitive ERPs were calculated across the 19 channels for all four stimuli (large blue circle, small blue circle, checkerboard, and sound) used in the task. The four most relevant findings are the four shown in the table, and in particular, the P3a and P3b latencies that were longer than would be expected in a boy of his age. The P3a was recorded at Cz in this example. It is evoked by the infrequent checkerboard stimulus. The P3b was recorded at Pz and is evoked by the large blue circle target stimulus. It is called 'target' because the subject presses a button during a continuous performance test whenever they see the large blue circle but not with any other stimulus. The visual and auditory sensory N1 components were within normal limits. In the figure the visual infrequent stimulus (checkerboard), at O2, and the auditory infrequent noise stimulus in the ear buds (in-ear headphones) called Cz white noise, are the sensory stimuli. As is also the case with quantitative EEG analyses, more data are available than one has time to review, and so just the most relevant information is routinely reported and displayed in the above format using the vertex and occipital locations.

Here are the interpretive comments from an ERP expert, John Polich, who designed the test used to elicit these ERPs while the subject is doing a continuous performance test (CPT):

Special thanks is extended to Dr. John Polich for his help in the interpretation of the following cases at the request of Evoke Neuroscience.

The auditory white noise stimulus condition produced strikingly small amplitudes for P3a and P3b. The data almost seem as if they were "shrunk" across the waveform from all electrodes. The P3b from the target circle is small, but not unreasonably so, with scalp topography that is relatively normal. Note that the N1 is rather late at 200+ ms, but quite large at -9+ μV. There is a well-defined P1 potential before the N1, which reflects initial attentional sensory processing, but again it is delayed and small. The overall morphology of the initial sensory potentials is typical, however, as the amplitude distributions are the usual Cz maximum.

The P3b from the target stimulus condition produced an overall nice looking waveform set, although rather small frontally.

Polich also noted that it might be worth examining the amplitude laterality differences for the frontal and central electrodes, as they are small but present for the P3b and not P3a.

For the White Noise Auditory Distracter

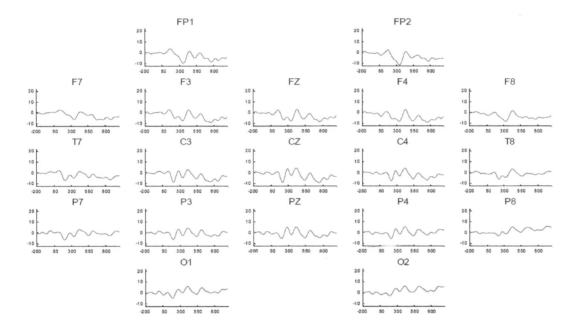

For the Visual Infrequent Distracter
Visual Checkerboard

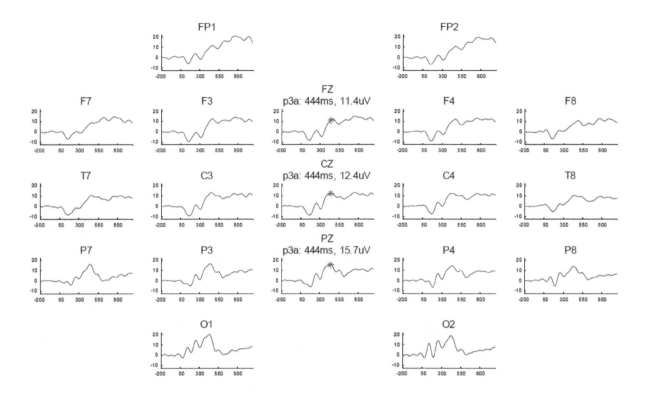

Note P3a marked with a red star.

The visual checkerboard stimulus condition produced an overall typical waveform set, with appropriate latency, amplitude, and distribution properties, albeit somewhat small for a young adolescent. The N1 at about 200 ms (see target visual stimulus) is appropriate as well.

For the Visual Frequent Distracter
Common Stimulus (small circle)

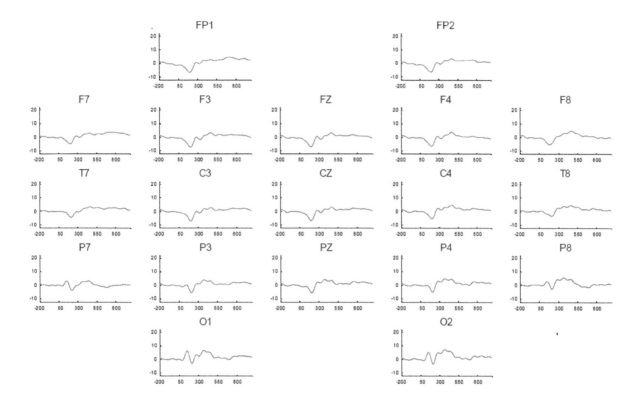

For the Visual Target Stimulus
Target (large circle)

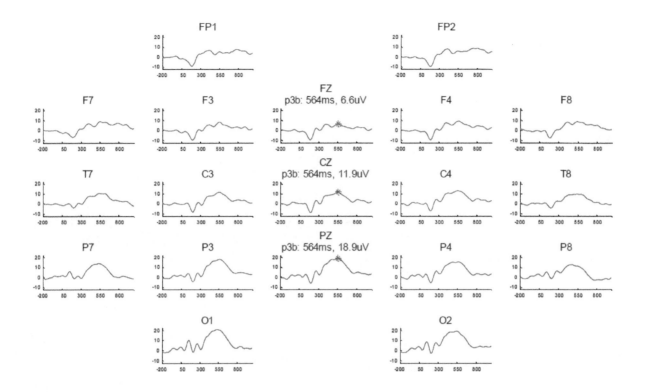

Note the red star marking the P3B at 564 ms.

Case Example #2, Two Years After the Accident ("BR")

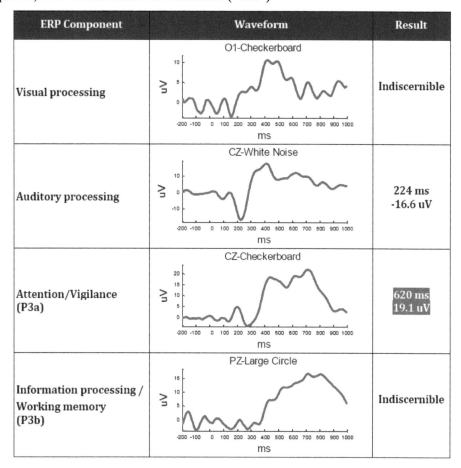

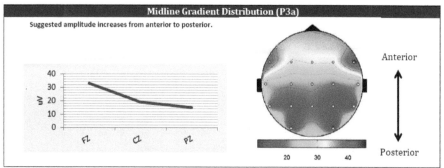

Thanks are extended to Dr. John Polich, who did a second interpretation of this post-TBI ERP and was able to discern the P3b. He noted that, *"The degree of abnormality in the ERP data set appears indicative of major deficits. Both the P3a and especially P3b component exhibit long latencies. The P3b scalp topography is consistent with dementia patients and likely indicates a parietal deficit. If ERPs are viewed as a reflection of the brain inhibitory strength, then frontal areas are quite low and facilitate large P3a potentials (which contribute to P3b amplitude), whereas the damaged brain (almost wherever it is) does not work well at all to yield very small central and parietal components."*

This case is shown to demonstrate that even a couple of years following a very severe head injury, the ERPs may indicate problems.

The P3a and P3b ERPs suggest long latency in this 22–year-old male with a traumatic brain injury. This young man (Br) was riding a bicycle when he was struck by a truck on a country road, and his injuries were severe. He was in a coma for one month. Now, two years postinjury, he continues to demonstrate functional deficits in the executive networks involving attention and impulsivity. He also has severe difficulties with working memory. Verbal functioning, however, was excellent. The morphology and long latency of the P3b was such that a definitive measure was marked as questionable in the original interpretation (in above figure). Visual processing N2 component was likewise not well defined by the morphology and categorized as indiscernible. Auditory processing N2 was found to be within normal limits. The measure referred to by Evoke Neuroscience as the midline gradient shows greater P3a amplitude in the frontal region compared to the central and parietal regions. It is hypothesized that this pattern of **highest amplitude of P3a at Fz is consistent with cortical inefficiency in the frontal** regions. The midline gradient demonstrated an abnormal slope down from anterior to posterior. These ERP findings correlated with the patient's symptoms. The clinician wants to see improvements in these values after NFB.

White Noise

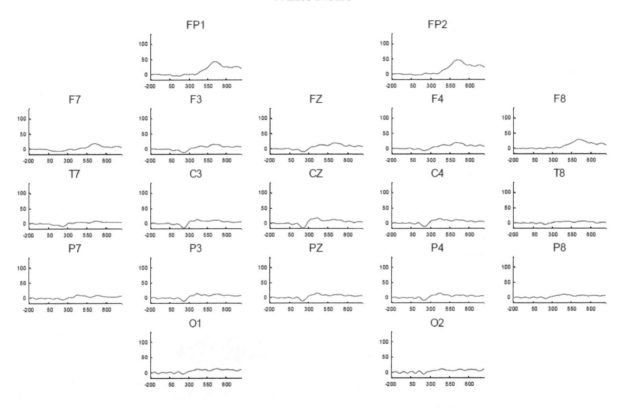

Visual Checkerboard

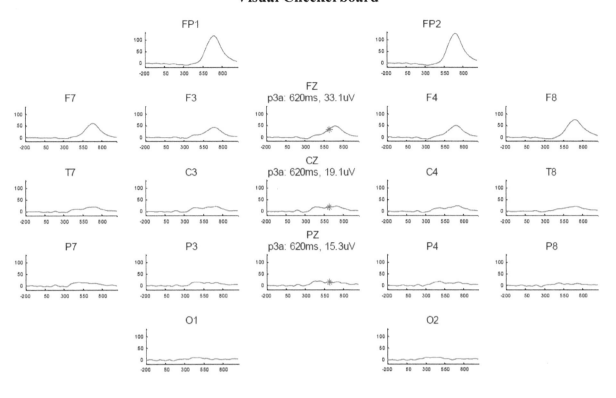

Common Stiumuls (small circle)

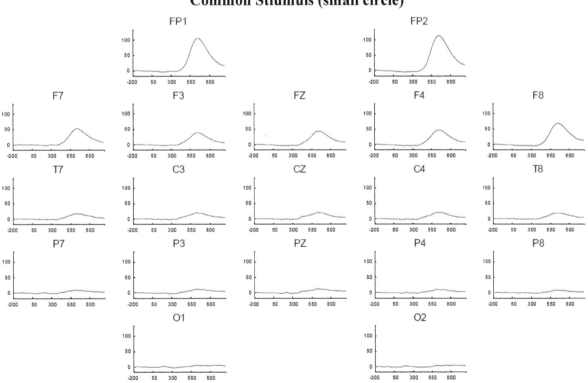

The Neurofeedback Book, Dr. Thompson, (www.addcentre.com). Published by the AAPB. (www.aapb.org)

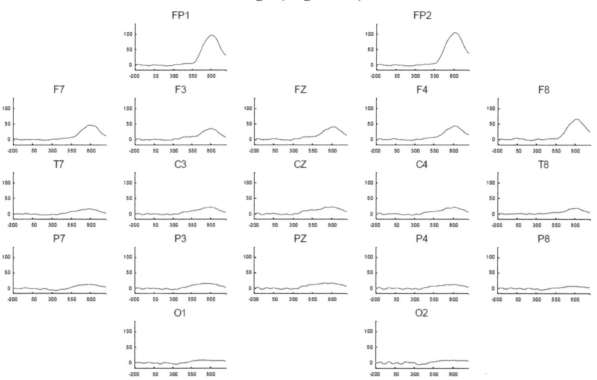

Case Example #3; Hockey Player

This 21-year-old man played hockey at a high level and had a number of concussions. He wanted to be a professional hockey player. He came for assessment and training as an optimal-performance client in order to prepare to play varsity hockey at university on a hockey scholarship. Although he came to optimize performance, assessment findings suggested there was first some normalizing to do with respect to the speed of processing incoming information and heart rate variability measures also indicated problems. In addition, he had a short attention span, memory problems, and learning difficulties and needed to improve his academic standing in order to maintain a hockey scholarship at college.

The sensory ERPs show long latency for a young athlete, though they are not abnormal. The P3b, however, is 633 ms, and is too slow for a young athletic man. His HRV statistics are also well outside the norms, particularly for his young age. His SDNN is 31 ms and total power is 402 ms^2. These measurements are below what is expected for a competitive athlete, and would even be poor for a middle-aged man.

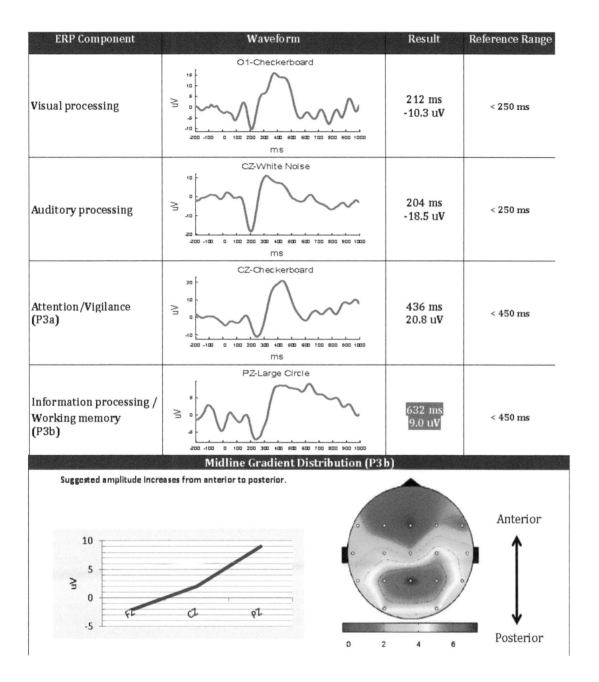

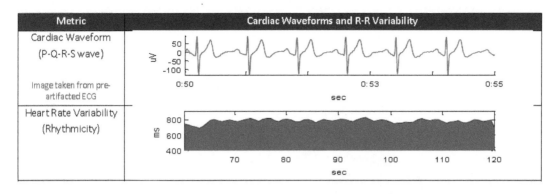

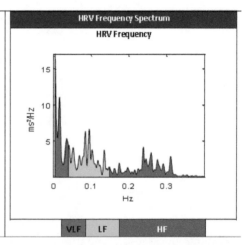

Metric	Result	Reference Range
Heart rate	76 bpm	50 - 80 bpm
QRS duration	0.096 sec	0.06 - 0.12 sec
SDNN (HR variability)	31 ms	75 - 150 ms
Total power	402 ms²	>= 800 ms²
Very Low Frequency (VLF)	119 ms²	low
Low Frequency (LF)	157 ms²	highest
High Frequency (HF)	126 ms²	low

Explanation to Patient

In discussing with a client the reason for recording EEG while doing the continuous performance task, keep the following in mind but tailor what you share to the interest level of the person. In someone recovering from a serious head injury who has severe memory problems, it may be sufficient to say that the test will measure brain speed. For the very sophisticated client, more detail is given. The clinician may explain that when any of the visual images appear on the monitor the patient will not, initially, be consciously aware that it is on the screen. It will take his brain 300 to 400 ms to become consciously aware. However, the patient's eyes do see the image and the brain recognizes that the picture has been presented on the monitor long before there is conscious awareness. In addition, the brain responds with definite electrical changes that we can later observe when we analyze the recorded EEG, looking for the pattern that follows each presentation of a particular stimulus (i.e., target, visual distracters, and auditory distracter). There will be precise peaks seen in the waveforms. As noted in the measurement section above, peaks that are easily identified include a negative N100 (N1), followed by a positive P100 (P1), a negative N200 (N2), a positive P200 (P2), a negative N300 (N3), a positive P300 (P3a and P3b) and a positive P450 (P4). The latencies recorded for each of these peaks will vary according to unique factors that are influencing that patient's brain, and will therefore change if the patient has a condition such as Attention Deficit Disorder (ADHD), learning disability, Alzheimer's disorder, or has suffered a stroke or other kind of brain trauma, such as a concussion. The millisecond accuracy of ERPs yields a temporal resolution superior to any other brain imaging technique available to date. This enables the measurement of basic information processing systems that constitute the building blocks for all cognitive functions, from preconscious attention to later occurring memory processes. Each component reflects neural processes involved in a particular stage of processing.

P300 has been researched and found to be abnormal in a number of clinical disorders. In schizophrenic patients, and even in unaffected relatives of those with schizophrenia, ERPs show decreased amplitude and an increased latency (Blackwood et al., 1991; Sham et al., 1994). Changes in the latency and the amplitude of P300 are also found in patients with bipolar disorder (Muir et al., 1991).

A very helpful study is reviewed in Kropotov's early workshops and later in his textbook (2009).

It demonstrates clear differences between the group with ADHD and the control group, as is shown in our figure below that was adapted with Dr. Kropotov's kind permission, from his PowerPoint illustrations.

ERP Controlled Study by Juri Kropotov

Decrease of Amplitude of NOGO Component in ADHD

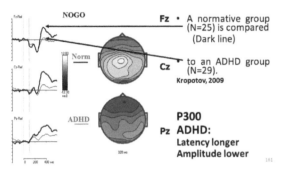

Kropotov has also demonstrated that NFB, using beta training, can result in latency and amplitude of the P300 in children diagnosed with ADHD to shift into the normal range. This is illustrated in the following diagram (modified after Kropotov's PowerPoint illustrations).

Beta NFB Training Effects on Event-Related Potentials in GO/NOGO Task

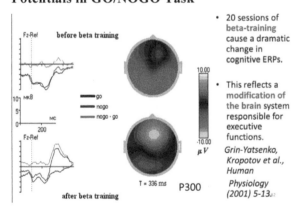

These findings by Kropotov are a real landmark in the studies of the effect of NFB on ADHD.

Differences in ERPs may be used to distinguish Lewy body dementia (LDB) from Alzheimer's. LBD includes both Parkinson's disease dementia and dementia with Lewy bodies. Lewy bodies are abnormal protein deposits that disrupt the brain's normal functioning. These Lewy body proteins are found in an area of the brain stem where they deplete the neurotransmitter dopamine, causing Parkinsonian symptoms. In Lewy body dementia, these abnormal proteins are found in many areas of the brain, including the cerebral cortex. Acetylcholine is depleted. This can result in a disruption of perception, thinking, and behavior. Lewy body dementia exists either **in pure form, or in conjunction with other** brain changes, including those typically seen in Alzheimer's disease and Parkinson's disease. We remind the reader that the **Nucleus Basalis of Meynert** and the **cholinergic** innervations to the **hippocampus** and temporal lobes are crucial to laying down new (episodic) memories and are said to be the first areas to be affected by **Alzheimer's** disease. However, there is also an abnormal connectivity between the posterior cingulate and the hippocampus in early Alzheimer's disease, which is accompanied by mild cognitive impairment.

Lewy Body disease affects first the **ACC BA 24, Insula BA 13, and parahippocampal** gyrus BA 28 (see the section on Brodmann area functions). In contrast to the memory problems seen early in Alzheimer's, Lewy Body dementia is a **frontal-type** dementia with little in the way of memory deficits early in the course of the illness.

Lewy body dementia comprises between 25 to 43 percent of patients diagnosed with dementia (McKeith et al., 1996). Bonanni and colleagues (2010) used an auditory oddball paradigm and measured the auditory P300 responses in 50 controls, 36 patients who exhibited dementia with Lewy body (LBD) pathology, and in 40 patients with Alzheimer's dementia (AD). **P300 latency was longer and the P300 had lower amplitude in LBD compared to AD.** In addition, the anterior-to-posterior scalp **amplitude gradient (Fz-Cz-Pz) was reversed in LBD**, with Fz at higher amplitude than Pz. P300 latency correlated with neuropsychological test scores that indicated **frontal cognitive** deficiencies. These authors also noted what we would term alpha slowing with the dominant frequency found in the high-frequency theta range (they call this pre-alpha) in LBD but not AD. Their study demonstrated that gradient inversion and delayed P300 responses in frontal derivations evidenced differences between LBD and AD patients with a sensitivity of 70 percent and a specificity of 97 percent.

Heart Evoked Potentials (HEP)

Another brain response that is evoked by a stimulus is the wave form that occurs after each heartbeat.

This will be very briefly described at the end of the section on heart rate variability training.

SECTION XVII
Using LORETA Mathematics in Assessment

LORETA in Assessment
Definition and Historical Overview

The acronym LORETA stands for low resolution electromagnetic tomography. LORETA is a mathematical procedure that provides the inverse solution, as is explained below. It takes the quantitative data from 19 electrode sites and finds discrete linear solutions for the sources within the cortex of activity at specific frequencies. LORETA was created by Roberto D. Pascual-Marqui of The KEY Institute for Brain-Mind Research at the University Hospital, Zurich, Switzerland. The first publication concerning LORETA appeared in the *International Journal of Psychophysiology* in 1994. It was titled, "Low resolution brain electromagnetic tomography: a new method for localizing electrical activity in the brain." 'Tomography' refers to the creation of two-dimensional slices of the three-dimensional brain. LORETA images typically show horizontal, sagittal, and coronal slices that look like MRI images (magnetic resonance imaging). Indeed, LORETA has been validated against MRI and found to be accurate for cortical locations. It does not calculate subcortical sources of EEG frequencies. Roberto Pascual-Marqui made LORETA available for all researchers and practitioners at no charge. You can download it online.

Pascual-Marqui's mathematical solution takes into account the fact that the EEG recorded from a given surface electrode is a sum of numerous elemental dipoles located not only under the electrode but also in remote areas of the cortex. The problem of finding these dipoles by knowing only potentials recorded by multiple scalp electrodes is called, in mathematics, the inverse problem; when you measure something on the scalp, what is the source deeper in the cortex? An inverse solution converts observed measurements into information about a physical object or system, such as the brain, and tells us about a physical parameter that we cannot directly observe. In regular clinical work, we cannot put our electrodes inside the patient's brain. **LORETA is a mathematical inverse solution that allows us to take information from external sites on the scalp and convert this into information about many different source locations, some of them deep in the cortex.** Note that LORETA mathematics does not, at this time, go subcortical. Thus the thalamus, for example, will not be identified as a source because it is not within the cortex.

The sites are 'voxels.' Each voxel is three-dimensional and has specific coordinates x, y, z, that precisely give its location in a three-sphere model of the head. 'Three-sphere' because skin, skull and cortex, all have different conductivity. For example, a site with coordinates x46 y33 z40 is in the frontal lobe, BA 8. This is called the Talairach coordinate system of the human brain. It is used to describe the location of brain structures in a manner that is relatively independent of individual differences in the size and overall shape of the brain. The Talairach coordinate system is defined by making two points, the anterior commissure and posterior commissure, lie on a straight horizontal line. Since these two points lie on the midsagittal plane, the coordinate system is completely defined by requiring this plane to be vertical. **Distances in Talairach coordinates are measured from the anterior commissure as origin.** Talairach coordinates are sometimes also known as stereotaxic coordinates. These coordinates are used for magnetic resonance imaging (MRI), positron emission tomography (PET), and other imaging methods. The Talairach atlas is based upon postmortem sections of a 60-year-old French female who had a smaller-than-average brain size. This does introduce some error that was recognized by the authors of the atlas (Talairach & Tournoux, 1988).

Nonetheless, the Talairach atlas is an invaluable tool in modern neuroimaging, and has paved the way for more representative brain atlases, including the Montreal Neurological Institute (MNI) atlas. This MNI atlas is based on 3D statistical neuroanatomical models from 305 MRI volumes (Evans et al., 1993; Collins et al., 1994).

Thus, as previously stated, LORETA is a mathematical inverse solution that allows us to take information from external sites on the scalp and convert this into information about many different tiny source sites deep in the cortex. In the solution to this problem, the brain is modeled as a dense grid of three-dimensional volumes called voxels that lie within a sphere (this is somewhat modified in some versions of this process) representing the head. The original version had 2394 grey-matter voxels of 7 mm x 7 mm x 7 mm. Updated versions have smaller voxels that are 5 x 5 x 5 mm to provide greater spatial resolution. In the LORETA analysis these sources are organized with reference to Brodmann areas. The Talairach and Montreal Neurological Institute atlases are the primary ones used in medicine. The voxels based on these atlases that are used in the original LORETA have a spatial resolution of 7 mm. This is the version of LORETA that we use when we examine the EEG and compare findings for an individual with the NeuroGuide database in clinical practice. A newer version of LORETA, standardized LORETA, called sLORETA, is used in the eVox system developed by Evoke Neuroscience. Both work well in clinical practice. These voxels have a maximum spatial resolution of about one cubic centimeter. The reader is referred to Pascual-Marqui's papers for a detailed description of the mathematics that are involved in this program, or to Juri Kropotov's book (2009) for a short overview of these mathematics. There are three iterations of LORETA: original LORETA, sLORETA and most recently eLORETA – exact low resolution brain electromagnetic tomography), which is said to have zero localization error. Pascual-Marqui has established that this version is a genuine inverse solution with exact, zero error localization even in the presence of measurement and biological noise.

LORETA and sLORETA

*The LORETA method was originally devised by Pascual-Marqui in 1994. Almost a decade later he developed the sLORETA method (Pascual-Marqui, 2002). The sLORETA method differs from regular LORETA by a transform into a relative metric. This transform is **not** the same as relative power. The 's' in sLORETA stands for 'standardization,' in which current sources in each voxel are divided by the variance of the source space. The complex mathematics is a type of normalization. It is not the same as relative power, since total current is not divided into each voxel at each frequency. Instead, the variance is divided into each voxel independent of frequency. The sLORETA normalization helps minimize 'ghost' images and distortions that are due to the use of a spherical head model and improves the accuracy of source localization, especially when there is a great deal of noise. However, the price paid by this normalization is the loss of a physics unit of current as is available in regular LORETA. For example, in regular **LORETA the units are amperes per centimeter cube.** For sLORETA and eLORETA this is not a serious problem (in contrast to the use of relative power calculations of surface EEG, which can be difficult to interpret), since the standardization in sLORETA is not a proportion or a distortion of total energy. The sLORETA calculations accurately estimate the electrical neuronal activity at each voxel based on geometry.*

*The **units in sLORETA are in the form of an "F" statistic, whereas** amperes per cubic centimeter is the unit of measurement in regular LORETA. The maximum "F" tells you the geometric location of current in a volume with good accuracy that is **less influenced by noise** than regular LORETA. The localization accuracy is more important in LORETA than the absolute physics value of current density, so the price of losing units of amperes is not high because location, location, location is what is important.*

(Thanks are extended to Robert Thatcher for this explanation, which we have paraphrased.)

Examples of LORETA Images Generated by the NeuroGuide Program
Male with Symptoms of Depression, Age 35

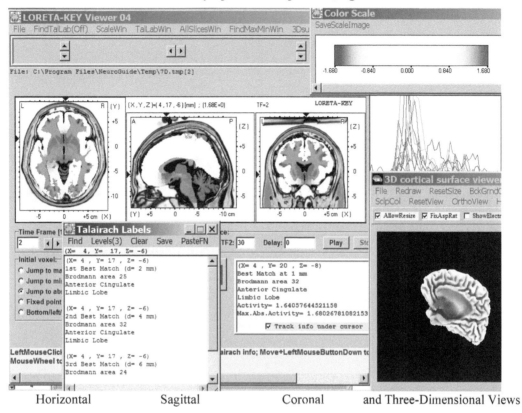

Horizontal   Sagittal   Coronal   and Three-Dimensional Views

This depressed male, age 35, showed excess slow waves in the surface EEG. In the LORETA images (eyes open) using the NeuroGuide database, 1 Hz was >4 SD and 2 Hz (shown above) was >1.6 SD above the database means with a "best match" source in BA 25 of the cingulate cortex.

It has been well established that dysfunction in BA 25 can be associated with intractable depression. Although most of our depressed patients show >2 SD in high-frequency beta (above 20 Hz), with source at BA 25 when they are seriously depressed, this man had abnormal findings related to slow-wave activity in the delta range. Thatcher's dictum has been location-location-location and not frequency, and we keep this in mind when prescribing NFB interventions. The NFB has been successful with patients like this when it is based on the QEEG and LORETA findings.

Male with Depression, Age 32

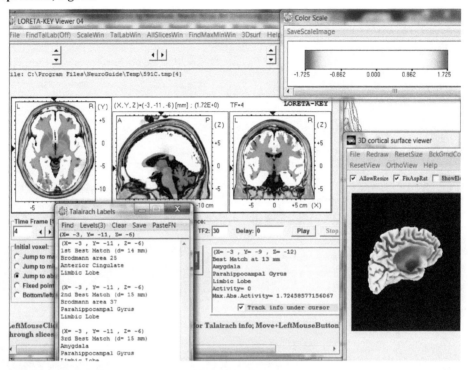

In the LORETA images from a depressed male, age 32, we observed that 4 Hz activity in the eyes-open EEG was >1.7 SD above the database mean. The source was identified as BA 25 in the anterior cingulate, with BA 37 in the parahippocampal gyrus as the second best fit for his data.

Using the eVox equipment from Evoke Neuroscience, the sLORETA image below shows BA 25 to be 3 SD above their 'elite' database mean in the eyes-closed recording.

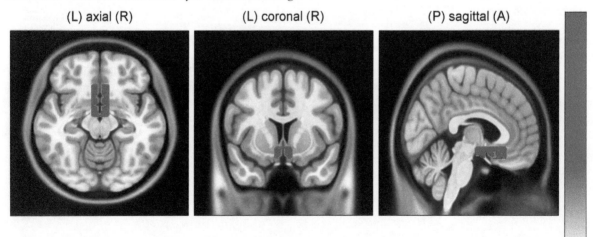

BA 25 identified as source (images from an Evoke Neuroscience report)

Female with Anxiety, Age 30

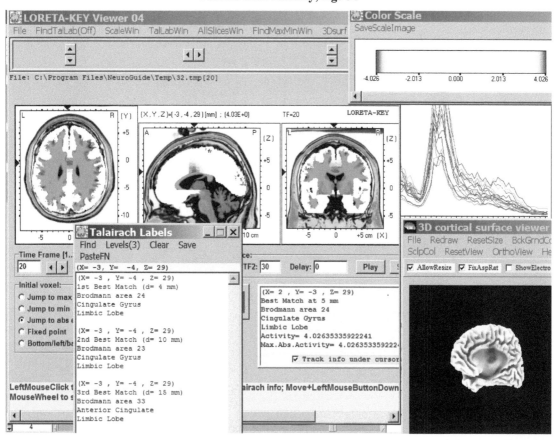

Female (ND), age 30, complained of excessive anxiety. The LORETA images using data from the eyes-open condition for frequencies between 16 and 23 Hz all showed a source derivation within the anterior cingulate (BAs 24, 23 and 33). At 20 Hz the source of the activity in the anterior cingulate is >4 SD above the database mean as shown in the above figure for 20 Hz activity.

Each of the above cases recovered from their symptoms with single-channel NFB training at FCz. These clients did training when LORETA analysis was being used to guide training, but before LORETA Z-score neurofeedback had been developed. Training at one central site, you rely on network properties in the brain to influence deeper structures. We now may choose to use LORETA Z-score NFB (LNFB) with similar patients. As Dr. Thatcher has said innumerable times, it is not the frequency that is key, it is location, location, location: i.e., the location that is outside the database norms is the area of interest for training when the known functions of that BA correspond to the patient's symptoms. Identifying location and applying NFB accordingly should be the route to efficient and effective intervention. If doing single-channel training, the FCz site can influence the anterior cingulate that lies beneath it and thus have an effect on the whole Affect network as well as the Executive network. Combining neurofeedback with cognitive behavior therapy (CBT) for depression would make sense as that therapy would activate both Affect and Executive networks and ensure that they were being influenced. If not training "on task" (in this case, the task is CBT), then one would likely be training the Default network, which could also be helpful.

The case example below is given to show single-point LORETA analysis. This technique can be useful when there are isolated, extremely high-amplitude bursts that might correlate with sudden changes in behavior or mental state. If these **bursts occur relatively infrequently, they may not show up as deviations from the database when the whole EEG time frame (after artifacts are removed) is averaged**, despite their very high amplitude. Single-point analysis allows the clinician to identify the source of these bursts using LORETA. The teenage girl whose LORETA image is shown below had sudden, unprovoked rages in the evenings. These rages were considered to be quite different from her normal behavior. No medications had had any good effect, and family therapy (tried since she was 7) also had no effect on her rages, which occurred only at home in the evening. The raw EEG showed occasional spikes at FP2 and F8. Neurofeedback training decreased the amplitude and frequency of these bursts of high-frequency beta and this change was associated with a marked decrease in her episodes of rage.

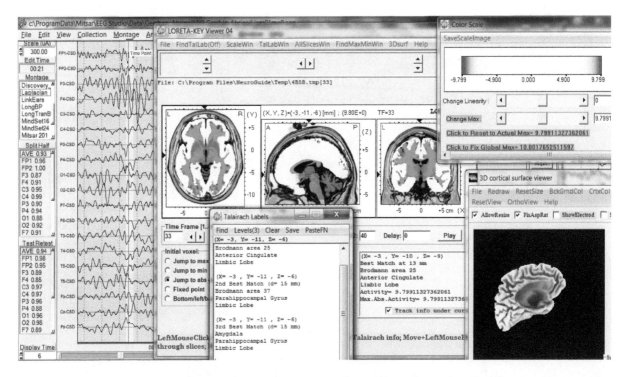

Eyes open, Laplacian montage (AG) single-point LORETA analysis of beta shows that these bursts of beta are from BA 25 and the amygdala. In the surface EEG the beta was seen at F3 and F4.

The great advantage of EEG measures and LORETA analysis for our work with NFB is that this methodology has good temporal resolution compared to traditional methods of neuroimaging with tools such as PET, SPECT and fMRI. EEG and MEG both give excellent time resolution. Adding LORETA analysis accurately provides the spatial component to the already excellent temporal resolution.

Magnetoencephalography (MEG)

Magnetoencephalography (MEG) is a functional neuroimaging technique for mapping brain activity by recording magnetic fields produced by electrical currents occurring naturally in the brain, using very sensitive magnetometers. A major difference between EEG and MEG is that the EEG reflects, for the most part, the activity at the crest of the gyri where the pyramidal neurons are perpendicular to the surface of the scalp. The activity in the walls (tangential) and sulcus (radial) areas is not well detected.

Dipole

*This term can refer to a **magnetic field** that can be measured at the surface of the scalp. Each pyramidal cell can create a dipole: if one end of the dendrite receives input and the extracellular space becomes positive, then at the other end of the dendrite the extracellular space will be negative. Any electrical current will produce an orthogonally (at right angle) oriented magnetic field. It is this field that can be measured with an electrode on the scalp. A current dipole gives rise to a magnetic field that flows around the axis of its vector component (see figure). It takes a very large number of neurons (more than 50,000) to generate a detectable signal. Since current dipoles* *must have similar orientations to generate magnetic fields that reinforce each other, it is mainly the layer of pyramidal cells that are situated perpendicular to the cortical surface in the walls of a sulcus that give rise to measurable magnetic fields. Bundles of these neurons that are orientated tangentially to the scalp surface also project measurable portions of their magnetic fields outside of the head.*

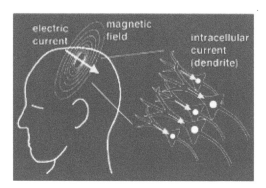

Figure by Tom Holroyd, in public domain.

LORETA Z-Score Neurofeedback (LNFB)

At the ADD Centre/Biofeedback Institute of Toronto LORETA neurofeedback (LNFB) **is done when there are multiple midline and ventral** structures found to be the sources of surface EEG findings and shown to be outside database norms. In addition, there may be a substantial number of pairs of sites showing coherence and phase lag outside the database norms. LORETA helps us to identify sources of electrical activity, deeper in the cortex, that is observed in the surface EEG. When looking at brain maps it is often not clear as to which structures in the cortex we are actually dealing with. Surface sites that show deviations from database norms may be quite distant from the principle Brodmann area (BA) that is the source of the deviation from the database seen in the brain map. For example, BA 25 is deep ventral midline and may be the source for observed deviations in the brain map at F3 and/or F4 and/or Fz. On the other hand, we feel compelled to note that, though one may be tempted to jump in with LNFB, we have often observed the standard deviation (SD) and the affect symptoms in such cases can be normalized using single-channel NFB at FCz combined with heart rate variability (HRV) training. We hope that LNFB may produce a faster result and therefore a decrease in the necessary number of sessions, but whether that is the case remains an empirical question.

In a clinical setting we apply every method that may help the patient. We therefore always combine HRV training with neurofeedback whether it be single-channel or LNFB. One rationale for this is based on many ventral midline structures such as BA 25 having direct connections with the solitary nucleus in the medulla. In the section on heart rate variability, the solitary nucleus (*nucleus tractus solitarius*) in the medulla is described as the key nucleus receiving afferents from the aortic and carotid baroreceptors that are of central importance in understanding the effects of heart rate variability (HRV) training. Therefore, when we note success with single-channel NFB and/or LNFB, it could be postulated that HRV training may have been an important aspect of this success.

For LORETA neurofeedback at the ADD Centre, we are using two different systems. The first system is from Applied Neuroscience. It uses the NeuroGuide program and database. It is widely used for both clinical work and research. A Mitsar amplifier is used at the ADD Centre to collect data for this program, though many different 19-channel systems are compatible with NeuroGuide. The second system now also used at the ADD Centre is a newer system from Evoke Neuroscience. It uses their new and powerful amplifier (eVox) and their newly developed database. At the time of writing in 2015 this database is comprised of adults who were above average in both physical and intellectual functioning. The two systems have different attributes because they are based on different populations. Both work very well in our clinical setting and complement one another. The NeuroGuide database has large numbers of children as well as a good representative sample of adults at different ages, including the elderly. Other databases, such as WinEEG developed by Juri Kropotov's group, also have a wide age range for comparisons. For this section the authors have chosen to describe the NeuroGuide LORETA Z-score neurofeedback program because it is the best known and most widely used program at the time of writing.

In depressed patients, we often observe deviations from the normal database with the source of these deviations being the anterior cingulate (BA 24), the medial frontal area (BA 9 and 10), the rostral-ventral cingulate gyrus (BA 25), uncus, parahippocampal gyrus, along with other central midline structures and the insulae. These areas are all involved in the Affect network and are found in the medial aspect of the cerebral hemispheres and not directly under any particular surface electrode. LORETA Z-score neurofeedback (LNFB) allows us to choose up to 24 areas in the cortex for feedback and to give feedback concerning activity at various frequencies (amplitude, phase, coherence) at all or a selection of these BA sites at the same time.

LORETA as Used in Z-Score NFB

The first publication concerning doing neurofeedback with ongoing LORETA measurements linked to database norms came from the work of Joel Lubar and Marco Congedo (Congedo et al., 2004). Based on LORETA computations, LORETA Z-score NFB can be carried out using the NeuroGuide (NG) program developed by Robert Thatcher and his group at Applied Neuroscience in Florida. Feedback is provided to the patient that allows the patient to bring up to 24 BA sites to within Z-score limits as set by the practitioner. In this program the Z-score variations are calculated using a comparison of the patient's EEG data with Thatcher's NG normative database. As an aside, we have used the WinEEG and SKIL (Sterman-Kaiser Imaging Labs) databases for comparison with the NG database and found the results to be very similar.

For LNFB done using the NeuroGuide database, the Brodmann areas of interest for a particular patient are chosen by matching that individual's symptoms (rated for severity) to Brodmann areas that are both outside the database norms and are known from the literature to relate to the patient's symptoms. It thus follows our decision making pyramid that was described earlier in this text. The continuous neurofeedback display usually comprises bar graphs, animations or movies, plus line graphs of time plotted against Z-scores at each selected site. The display reflects changes in the Z-score deviations from the database at all (or a percentage of) the chosen Brodmann area locations. In this manner, the patient is rewarded for maintaining brain wave activity within a Z-score range or window (say −1.5 to +1.5 SD) set by the therapist at all the designated Brodmann area (BA) sites, also called regions of interest (ROI), that are important for that patient. *(For specific directions on how to use this program, the reader should attend appropriate workshops and read the manuals produced for NeuroGuide by Robert Thatcher. At the time of writing, in North America,*

introductory workshops using NeuroGuide are given by Andrea Reid-Chung (ADD Centre) and advanced NeuroGuide workshops are given by Joel Lubar and Robert Thatcher. Workshops on the eVox system are provided through Evoke Neuroscience by James Thompson and David Hagedorn. The eVox system is sold primarily to physicians but other licensed health care professionals can also purchase it.)

We occasionally observe that an important activity, such as infrequent but high-amplitude bursts at a particular frequency, may appear to correlate with a patient's symptoms. In the case example shown above, the EEG bursts were of 20 Hz spindling beta. LORETA showed the source of this beta to be in BA 25. These infrequent bursts can be called an 'outlier.' The symptom that correlated with the high-amplitude bursts of beta, in this case example from a teenage girl, was an outburst of rage. We used single point LORETA analysis to find the source of the burst of high-amplitude beta. In these cases we definitely use 100 percent criteria, that is, all areas must be within the preset Z-score limit for the patient to receive a reward. Often one sets the criterion to be that 80 percent of the selected areas must be within the Z-score range. By setting the criterion to 100 percent, however, even an infrequent but important outlier must be within the range that you have set. We also narrow the range of deviations that will be targeted in order to really focus on a particular problem area.

Advantage

The advantage over traditional neurofeedback is increased specificity of the training and the possibility to train activity at locations not otherwise identifiable by conventional surface EEG. Examples of these areas might include the fusiform gyrus and the parahippocampal gyrus (involved in memory and affect) and/or Brodmann area 25 (involved in affect symptoms including depression) and the orbital frontal cortex (involved in autistic spectrum and affect disorders). It also allows the patient to train up to 24 targeted Brodmann areas at the same time although we usually prefer a smaller range of targets.

Disadvantage

This procedure takes more time for set-up at each session since it involves continuous 19-channel data collection and the equipment and equipment maintenance costs (amplifiers, databases, electrode caps, gel, syringes, etc.) are relatively high. In addition, we feel it should only be done by experienced clinicians who have a good working knowledge of neuroanatomy and Brodmann area functions, because that is required to make reasonable judgments as to which Z-score deviations correlate with problems and which ones might be outside database norms for good, healthy reasons. These reasons might include compensatory phenomena or high functioning. In addition, it is limited by the original NeuroGuide database only going to 30 Hz. Although this range does cover almost all our routine requirements, it still leaves us using surface NFB for patients who require training at higher frequencies. For example, some patients show beta spindling associated with negative ruminations in the 31 to 37 Hz range. In addition, some patients benefit from careful enhancement of 39-41 Hz in the gamma range. We also prefer to use single channel feedback to specifically enhance SMR (13-15 Hz) at C3 and C4. Often in those cases we may alternate these sessions with LNFB sessions.

Explanation Given to the Patient
How Does it Work?

Low resolution electromagnetic tomography (LORETA) finds the principle source in the cortex of EEG waves that are of interest. In simplest terms, imagine 19 microphones hung in four rows on the ceiling of a large room. If there is a conversation taking place in one area of that room, then the different amplitudes of the voices from that conversation at each of 19 amplifier locations on the ceiling would allow you to deduce where in the room the conversation was taking place. In a similar manner, LORETA mathematics enables the practitioner to infer the source within the cortex of a particular frequency of EEG activity.

As an example, you might be able to show the patient that the slow waves or even spindling beta that was observed at one or more cortical surface sites could, with LORETA, be seen to have a primary source deep in the cortex. It could, for example, be primarily from BA 25 in the ventral rostral cingulate cortex, and the clinician might hypothesize that it could correlate with the patient's depression. Because that site is ventral and medial, and thus far from any surface electrode activity, it might be postulated that LORETA NFB would be a more efficient way to give the patient the opportunity to have positive feedback when they decreased this activity specifically at that cortical site. Similarly, the orbital surface of the frontal and temporal lobes, the insula, and areas in the hippocampal region may be difficult to train with surface EEG. Patients and parents appreciate careful explanations of what they are training and why they are training it.

Clinically, at the ADD Centre / Biofeedback Institute of Toronto, we have used LORETA assessment information with the NeuroGuide, WinEEG, and Evoke Neuroscience programs. We are collecting 19-channel data with our Mitsar, NeuroPulse, and eVox (EvokeNeuroscience) systems. We provide approximately 150 NFB training sessions a week and, at the time of writing, more than 40 of these are LNFB sessions using the NeuroGuide system. In future we may decide to move to adding more sessions with Evoke Neuroscience eVox equipment since it has the most modern amplifiers and it also has an additional three-lead EKG embedded in it, which may allow for simultaneous heart rate variability training with one instrument. The NeuroGuide LORETA Z-score program can be used with nearly all of the instruments that are used by NFB practitioners to collect 19-channel data, including: Mitsar, NeuroPulse, Discovery by Brain Master, and Deymed. The practitioner has choices. We look at the EEG with LORETA using Z-score deviations because we may decide to use either the NeuroGuide LORETA NFB (LNFB) program or the eVox program from Evoke Neuroscience that utilizes sLORETA. In the future there will undoubtedly be other programs developed that will do similar types of feedback.

In the NeuroGuide program, the LNFB correlates the physiological signal with a continuous feedback signal. In this case the physiological signal is defined as the current density in a specified region(s) for training, called a region(s) of interest (ROI). This allows the continuous feedback signal to become a function of the intracranial current density and to covary with it (Congedo, 2006). As previously noted, the advantage over traditional neurofeedback is increased specificity of the training and the possibility to train activity in structures not otherwise identifiable by conventional EEG using scalp measurements alone.

Summary

Thus low-resolution electromagnetic tomographic (LORETA) neurofeedback (LNFB) provides the opportunity to train individuals to influence the electrical activity in regions not easily influenced by surface EEG training. When using these programs care must be exercised to only direct training at areas of the cortex that correlate with the patient's symptom(s) and NOT with other areas of the cortex even though they may be outside of database so-called norms. You do not want to normalize everything by bringing it to the database mean. Normal for a genius is not necessarily at the average score for the group that was measured when developing the database.

Choosing Sites (BAs)

The practitioner will want to examine which BAs their particular LNFB program 'matches' to each of the symptoms on that program's symptom checklist. Then the practioner will know what additional areas they wish to add or delete before beginning feedback for a particular patient. How to do this depends on what program you are using and can be found in the manual for that program.

However, the practitioner must know the main functions that can relate to each Brodmann area (BA) as is outlined in Section VIII of this book. Using this knowledge, the practioner will adjust the program's 'matches' of BAs to symptoms and target the areas that they feel will give maximum benefit to a particular client. The next figure is given to remind the reader of central midline and ventral structures and Brodmann areas.

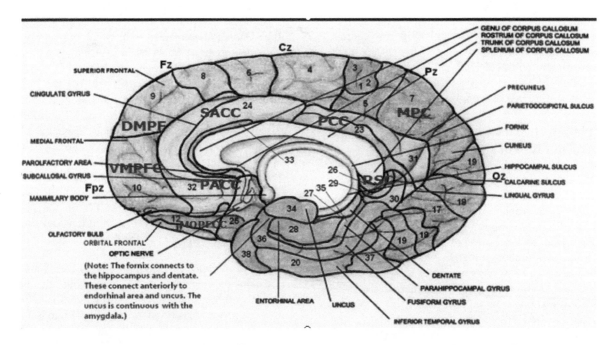

*Figure. **Central Midline Structures (CMS): An Anatomical and Functional Unit** (after Northoff, 2006).*
Drawing from The Neurofeedback Book (2003) and ADD Centre Brodmann Areas Booklet (Thompson, 2007b).
Artwork by Amanda Reeves and Bojana Knezevic.
The bold letter abbreviations in this figure indicate areas that are considered to be important central midline structures when we are discussing neural networks (NN). These abbreviations are as follows:
***SACC** = supragenual anterior cingulate cortex (BA 24, 32)*
***DMPFC** = dorsomedial prefrontal cortex (BA 9)*
***MPC** = medial parietal cortex (BA 7, 31)*
***PCC** = posterior cingulate cortex (BA 23, 31)*
***RSC** = retrosplenial cortex (BA 26, 29, 30)*
***MOPFC** = medial orbital prefrontal cortex (BA 11, 12)*
***VMPFC** = ventromedial prefrontal cortex (BA 10, 11)*
***PACC** = pre- and subgenual anterior cingulate cortex (BA 24, 25, 32)*

The authors suggest that one should also include in any discussion of CMSs, the hippocampus, the entorhinal-uncus area, and the insula in both hemispheres.

In this figure, the septal area and the nucleus accumbens (now considered to be part of the basal ganglia) could have been shown inferior to the subcallosal gyrus. Also important and not shown is the fact that the gray matter of the amygdala is continuous with the uncus. The central nucleus of the amygdala is tonically inhibited by the medial prefrontal cortex (Thayer et al., 2012) and by the fusiform gyrus and the superior temporal cortex (Porges, 2007). The septal region governs consciousness and memory (acetylcholine to hippocampus).

We should note again that, possibly due to our having an effect on neural networks, we have observed changes in these deeper areas of the cortex using conventional surface NFB on central sites such as Fz, Cz, and Pz. In addition, one could reasonably speculate that, **in some instances, it might be advantageous to allow the brain neural networks to adjust themselves** with this simpler form of NFB combined with HRV training. We do not know the answer to these questions. They could be empirically tested.

When we use the Z-score feedback we must first do an EEG assessment and remove the artifacts from the data. Second, with the patient, we rank the patient's symptoms from 0 (least) to 10 in severity. As previously noted, the NeuroGuide program will calculate which BAs are outside database norms in any of the standard NeuroGuide frequency bands. The ranges used in NeuroGuide for amplitude in microvolts and coherence and phase are delta (1-4 Hz), theta (4-8 Hz), alpha 1 (8-10 Hz), alpha 2 (10-12 Hz), and several beta bands including 12-15 Hz, 15-18 Hz, 18-25 Hz, 12-25 Hz, 25-30 Hz. The practitioner will choose a Z-score feedback value that is usually between one and two standard deviations from the database mean for people of the same age and sex. Then the program will match the sites from our artifacted 19-channel EEG that are outside our chosen Z-score feedback value, to the Brodmann areas from the literature that are known to involve the functions that correspond to the patient's symptoms. The practitioner can now see a list of the Brodmann areas that the program has hypothesized as being important to focus on for feedback. The **practitioner can make changes in this**. For example, one of our patients did not check off depression (only anxiety) as a problem. However, in his assessment, BA 25 was well outside database norms. It was not included in the "match" by the program, since depression was not mentioned as a symptom. We therefore entered it, because BA 25 would not automatically be selected, in the NeuroGuide program, without depression being a symptom. (Note: BA 25 is important in many symptoms, particularly in the Affect network, other than depression.) In a similar manner, a very young boy (4.5 years old) who could easily read and spell very long and complex words had Z-score values well outside the database norms for his age in Wernicke's area. We made sure none of these areas were included in the 'match' for training. We also saw a doctor who was a musician. He had perfect pitch and had areas in the left auditory cortex well outside database norms. If we were to do LNFB with him we would have **removed these areas if** they had appeared in the match. You do not want "normalization" of areas of high functioning, just as you would not want to normalize an IQ of 130 to be within 1 SD of the mean (85-115).

Once the training parameters are set, the patient then sits in front of the computer monitor and receives visual and auditory feedback. Often the feedback is a movie or documentary on a DVD. Remember that this is learning based on operant conditioning. The client is rewarded by the movie playing when he meets the criteria that have been set. It is possible to train 168 measurements at one **time**, although that is not necessarily a good way to use this program. It is better to focus on a few key sites and measurements.

During training the clinician can see the BAs and their Z-scores on a second clinician's screen. Those Z-scores that go outside the clinician's chosen Z-score range limits are shown in red for as long as they remain outside these criteria. These values for each BA(s) for amplitude, coherence, and phase are rapidly changing on a list. Thus those showing as red in the moment-by-moment Z-score list on the clinician's monitor, can be identified as the ones that are turning the positive feedback off. In addition, there is a graphic display of Z-score as y axis (time as x axis) that shows each Brodmann area (BA) site as a different colored line. The therapist can put the mouse arrow over a line and see which BA site it represents. The patient's feedback is on a second monitor. If the patient achieves one-eighth of a second (all or none response) below the chosen Z-score criteria line for, initially, 85 percent of the items, then the patient receives both a sound and a visual reward. The visual reward can be as simple as a green dot or as complex as a DVD image fading or growing larger or smaller and then disappearing. The client can, for example, bring in a favorite movie to use as feedback though we usually prefer scenes without words.

Options: As previously noted, in the example of a patient with rage attacks, the clinician has the option of changing the feedback **criteria from the aforementioned 85 percent, which is the default setting, to 100 percent.** That would mean that all targeted variables at all targeted BAs would have to be met in order to get positive feedback. **This is an important option that we use** because bursts from some areas may be the key to a patient's symptoms and can be well outside the SD window that the clinician has set, but they might stay outside the window and thus not be targeted if 85 percent of the other targets were being met. The clinician **also has the option** of changing from using many symptoms to **selecting just a very small number of symptoms or a small number of target areas.** This also allows one to target one key area/symptom at a time.

This discussion can only give the reader a general overview of the NeuroGuide LNFB program at the time of writing. This program is constantly being updated by Thatcher and his team at Applied Neuroscience to make it easier to use and more precise. Further information on this topic may be found in the textbook: *Z-Score Neurofeedback: Clinical Applications*, Robert W. Thatcher, Ph.D., and Joel F. Lubar, Ph.D. (Eds.), Academic Press, 2015.

Examples from a Case where LNFB is being used at the ADD Centre/Biofeedback Institute of Toronto

The EO LORETA image above (as seen in NeuroGuide program) at 6 Hz deviates by 7.5 SD in the inferior orbital-medial frontal cortex. This area communicates via the uncinate fasciculus with the temporal pole, amygdala, hippocampal region, and other areas that are important in the Affect network. The patient was severely depressed, and LORETA analysis implicated ventral central midline structures as the source of activity that was outside database norms. Thus, LORETA NFB was a reasonable option for training. The results of training were good with resolution of the depression symptoms as measured by the Burns Depression Inventory.

The following figure illustrates one way of summarizing central midline structures (CMSs) that may underlie some of the symptoms found in depressed patients.

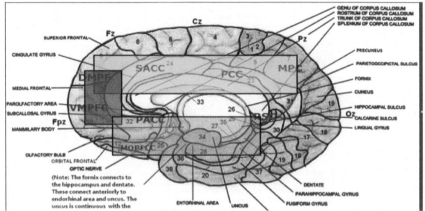

Figure: Overview of CMS involved in Affect Networks

The three colored boxes in this figure highlight aspects of the Executive *and* Affect networks *involved in **depression** as put forth by De Ridder (2009, 2010):*

*Symptoms related primarily to **Executive/Cognitive** Network areas are **GREEN** (BA 9 and 46, 24, 31, 40)*
*Symptoms related primarily to **Affect/Vegetative** Network are **ORANGE** (BA 25, anterior insula, hippocampal region, hypothalamus)*
***Integration and connection** of the above Cognitive and Vegetative areas are **RED** (BA 24, 9, 10, 11)*

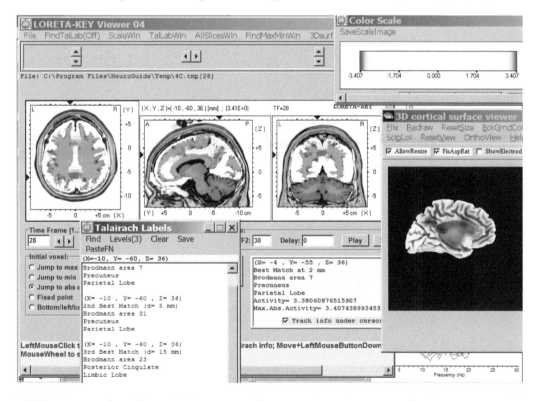

Figure: LORETA images from NeuroGuide program from a patient who presented with symptoms of depression. The maximum deviations from the database were at 28 Hz >3.4 SD. The source location was in the precuneus and posterior cingulate cortex (BA 31, 23). These areas are involved in the Default network (Fair et al., 2008). These regions are also important for the integration and synthesis of sensory inputs. Note that the anterior cingulate is also above the database norms (red) to a lesser degree.

With multiple midline areas outside the database norms, we decided to use LNFB + HRV training. The results were good with this patient. First the insomnia associated with the excessive ruminations resolved, and within 40 sessions the depression had also lifted.

LORETA Training Examples

The diagram below shows how one client responded to LORETA NFB + HRV training in five sessions and then benefited from a further 12 sessions. He is a very bright university student with a diagnosis of anxiety and Asperger's syndrome. His difficulties included anxiety, depression, obsessive compulsive (OCD) symptoms, and inability to pass his courses at the university. All of these problems are now in remission after 20 sessions of training using LORETA Z-score NFB combined with heart rate variability (HRV) training and learning metacognitive strategies. We note that this is much faster than our past experience with patients who had similar symptoms and who did single-channel NFB plus HRV training. Many more cases are necessary, however, before we can come to any reasonable decision as to which method is preferable. We also note that we have seen regression when patients stop training earlier than 40 sessions. Thus we are cautious and advise clients to consider taking more sessions to solidify the changes. (Occasionally, with a small number of patients, a few of these 'extra' sessions may have to be done at a price that does not meet our costs. However, in the final analysis for a clinical/ educational center, a successful outcome for every patient is the important factor.)

Below we show images from his initial assessment that indicate that central midline structures and the insula were involved. These structures are in the Affect network, which correlated with this patient's difficulties.

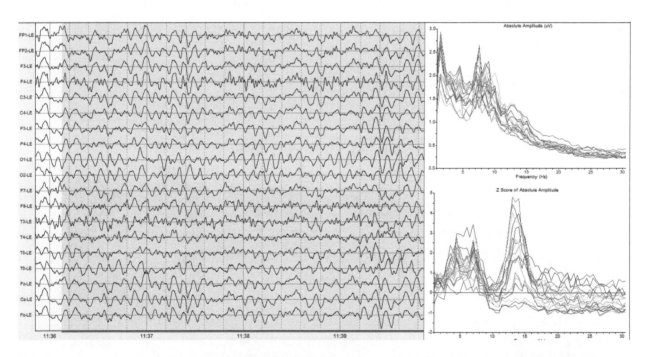

This linked-ears montage profile shows Fz, F4, F3 and F7 >3.9 SD at 12.5-15 Hz. Alpha is relatively low amplitude in the 8-10 Hz range. (Note that we have left some EMG-related activity in the recording, visible at T3 and T4. We will do this when removal of all the artifact caused by EMG will leave us with less than one minute of good data.)

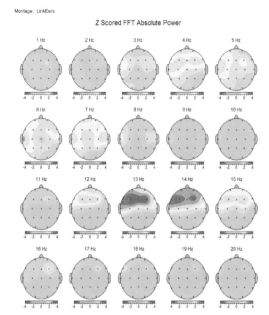

Brain map linked ears shows 3 SD for 4-7 Hz between T3 and F7 and >4 SD 13 and 14 Hz at F3 and F4.

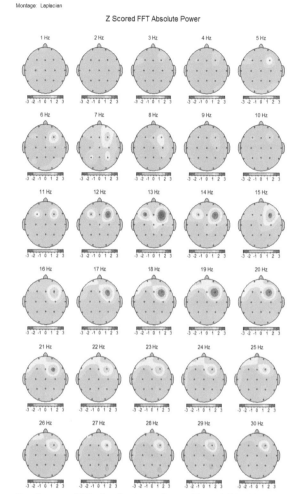

Brain map EC, Laplacian, shows F4 >3 SD 12-15 Hz and about 2 SD, 17-30 Hz. Low power (blue) in beta frequencies is found in the right occipital-parietal region.

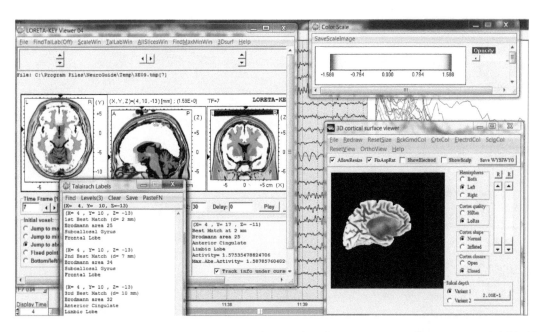

4-7 Hz is 1.59 SD in the subcallosal gyrus at BAs 25 and 34, and in the anterior cingulate cortex at BA 32. This is not a high SD, however <u>bursts</u> of waves at this frequency were very high amplitude and the functions of these areas do correlate with his symptoms.

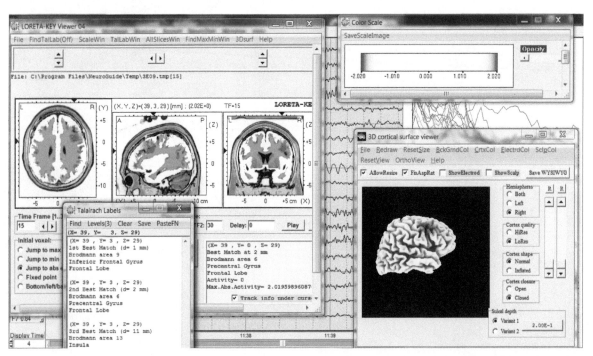

14 Hz and here 15 Hz BA 9, BA 6, and the right insula BA 13 are >2.8 SD in this LORETA image. Abnormalities in the right insula when outside database norms may suggest negative effects on HRV and can reflect Affect problems.

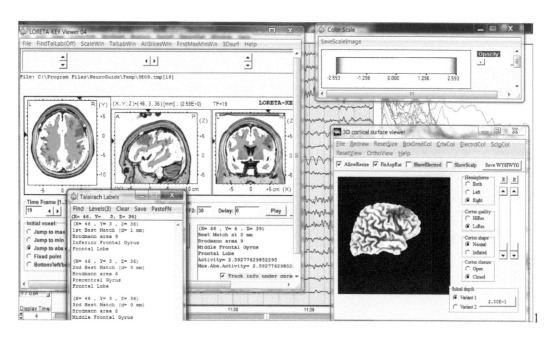

The frequencies from 6-19 Hz were all high amplitude. Here we see 19 Hz is 2.6 SD with source matches for non-dominant hemisphere BAs 9, 6, 8. Spindling beta observed in the raw EEG in this area. We have had similar findings in patients who presented with extreme anxiety.

Hypoactivation, reflected by high-amplitude delta and theta activity, was found in the dorsal ACC and dorsomedial PFC as well as in ventral portions of the ACC and the ventromedial PFC. The amygdala is inhibited by the medial prefrontal cortex (Thayer, Åhs, Fredrikson, Sollers III & Wager, 2012). Dysfunction in this region may correlate with his high anxiety. In addition, the literature does note hypoactivation in these areas in depressed patients (Mayberg et al, 2005; Kennedy et al, 2011). These hypoactivations are said to be significantly more common in post-traumatic stress disorder (PTSD) than in social anxiety and phobia disorders. Chronic anxiety is a major problem for this client.

Note: In PTSD the EEG most often shows one of three distinct patterns: high arousal reflected in fast peak alpha frequency equal to or greater than 11 Hz (thalamic origin), high frequency (fast) spindling beta (cortical origin), low amplitude (voltage) EEG with high fast (beta) activity. There are, of course, many other findings depending on the patient and the reader is referred to Section XXIX on PTSD.

It was recommended to this young man that we have a trial of LNFB due to the number of areas involved and the fact that several areas were ventral midline and distant from surface sites.

In the figure below, the colored lines indicate the standard deviations (SD) from the NeuroGuide database at each of the Brodmann areas that were **matched by the program** with his major symptoms. Each of these colored lines represents a different BA site that is being trained. LORETA NFB is like traditional surface site neurofeedback training sessions in that we expect to see a learning curve over time. These graphs are showing intersession change and we are aiming to bring the client's score towards ±1 SD for most sites or to less than 2 SD for values that were initially much greater than 2 SD.

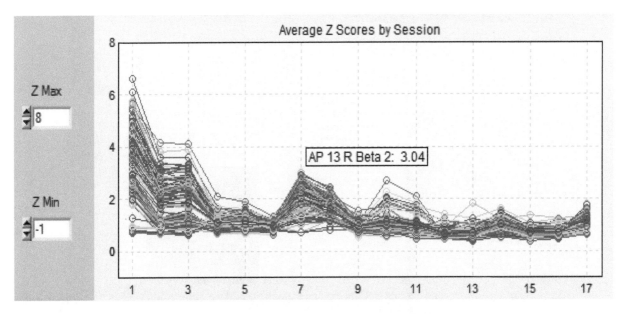

In this figure, the y axis shows Z-scores from 1 to 6.5 SD. The x axis shows the client's results from Session 1 through Session 17. In this display the cursor has been placed over the highest Brodmann area (BA) SD line from Session 7. The pop-up display shows that the SD for amplitude (AP) at BA 13 (insula) for beta 2 (15-18 Hz) was 3.04. This client was chosen to illustrate how a practitioner can be deceived in thinking a client is 'better' after a small number of sessions, when in reality the amelioration of symptoms and a more lasting change in the EEG parameters can take a much longer period of time.

This client had to leave training after Session 6 to return to university in another city and write exams. When he returned he came back into training because he had regressed, with respect to his symptoms, while away. In his first session after returning it was clear that his EEG had also regressed. His statistics improved over the next few sessions, and then he had to leave training because he was travelling for two months. His scores regressed slightly, but he did not notice a regression in symptoms. After a few more training sessions his scores stabilized. The figure above shows that with further training, and by 17 sessions, his Z-scores had decreased from >4-6 SD to <2 SD at all BA sites. We have found that **regression can occur if clients do not do a sufficient number of sessions.** This client remained in training beyond 17 sessions both to solidify the changes and to further optimize his performance.

He also did HRV training at each session. He is an athletic young man and had a very high standard deviation of the artifact-corrected R-R values (SDNN) ranging from 112 to 122 ms, overall power in the range of 2200 ms^2/Hz, and a ratio of low to high spectral frequency (LF/HF) of 0.6 to 1 on his initial assessment. (For an explanation of these terms please see HRV, Section XVIII of this book.) These values are unusually high for our clinical clients, but are within the usual range of the athletes. Despite his difficulties, he was a cyclist and had continued to do extensive daily cardiovascular exercise routines that doubtless contributed to his healthy HRV statistics. With training, his SDNN is now ranging from 143 to 151 ms with power in the range of 7000 ms^2/Hz and a ratio of LF/HF in the range of 2 to 4. Thus, HRV statistics improved a great deal even though they were already good to begin with. The HRV pattern is also better. The peak frequency shifted towards 0.1 Hz. In a clinical setting the focus is completely on the patient, thus, although this patient no longer is hampered by any of his original symptoms, we cannot come to any conclusion as to the relative importance of LNFB and/or HRV in his recovery.

Graphing results is also important during a session to give the patient immediate feedback as to how they are doing. The graph below is from the NeuroGuide system. It demonstrates how rewards do reflect what has been occurring during different segments of the session.

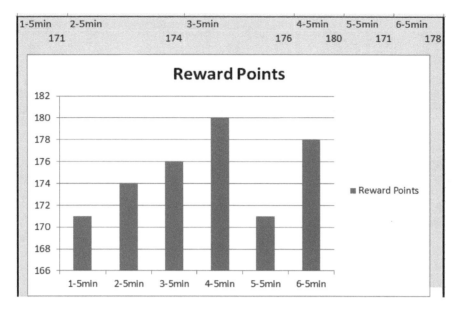

Reward Points at different time segments during a single session.
1. Time segment 1 - door open, the patient was distracted by sounds in the office.
2. Time segment 2 - door closed, the patient became more focused
3. Time segment 5 - screen stopped working (froze, still giving visual feedback but no sound); this resulted in the patient becoming frustrated.
4. Time segment 6 – everything working properly and the patient relaxed

Threshold for this patient was set at 1.9 SD. After each five-minute interval of training (training screen), a score pops up. These scores are plotted on the graph for within-session points or scores. This represents a method for giving meaningful information to a patient about their program during a session.

Case to Show Use of Only a Small Number of Sites

This patient demonstrates the rationale for using ALL-OR-NONE training. We have been successful doing training using the z-tunes-style feedback available on the LORETA Z-score program. Z-tunes enables the client to be rewarded for achieving a certain percent of brain wave patterns below the selected threshold. This makes for an easier starting point for some clients who are easily frustrated. However, in some cases, a client may be receiving rewards during the training yet not be having any significant change in some high amplitude (but perhaps short or infrequent bursting) brain wave patterns that have the highest standard deviation from the database mean. These areas that have the highest standard deviations are also often the areas of greatest concern for the client because they often relate to the most severe symptoms. We must alter our methodology, because with the z-tunes method, the client can still be rewarded with feedback even if these important areas remain high as long as a certain percentage of other brain wave patterns are below the threshold. To resolve this problem we have been **using a combination of the all-or-none function** provided in the NeuroGuide LORETA Z-score neurofeedback program and the program's option to choose how many sites are worked on at any one time. If we are working on many sites, as in the example below, in addition to using an all-or-none setting, we will pick and use only the top three to six matches (the 3, 4, 5, or 6 areas with the highest standard deviations that match the client's most severe symptoms). In the example below we reduced the number of sites (Brodmann areas) trained in Session 11 and then from Session 12 to 16 we focused on training only the top six. During the twelfth session we had an audience (student), and the client performed extremely well under pressure. In the following session the values rose and then gradually decreased over the remaining sessions. All of this client's presenting symptoms are now in remission.

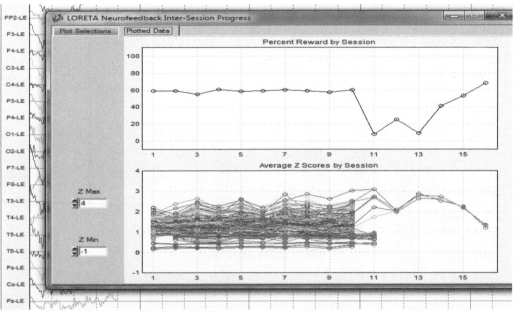

Figure. Progress over 16 LNFB (+ HRV) sessions.

At Session 11, only the six Brodmann areas that included the three having the most extreme Z-score deviations (matching the patient's most severe symptoms) were targeted during the training, and "all-or-none" was used. This meant that the client had to maintain all of the six target areas within the threshold window that was set.

Conclusion Concerning LORETA NFB

LORETA is very helpful in pinpointing areas of dysfunction in the cortex. When the following criteria are met, we will consider designing a LNFB program for a client:

1. The images observed with LORETA during our QEEG assessment support our observations of frequencies that are outside the database norms for surface EEG sites.

2. The LORETA source correlations agree with our knowledge of the functions of those Brodmann areas that are found in the literature and these correspond to our patient's symptoms.

3. The neural networks that are likely to involve those Brodmann areas are ones that would reasonably be predicted to be important to our client and correspond to our client's difficulties.

4. There are a large number of Brodmann areas outside 2 SD and/or a number of the Brodmann areas that are outside 2 SD are medial or deep within the cortex and unlikely to be easily influenced by surface EEG feedback.

5. There are a number of significant coherence and phase differences from the normal database that are identified using the QEEG.

When these criteria are met, then we will design a LNFB program to bring the EEG measurements in relevant areas to within a preset Z-score range, which is usually ±1.5 SDs.

During LNFB training we will observe the changes in the patient's thinking, feeling, and behavior to determine if the patient is moving in the desired direction (without undesirable effects occurring) with the LNFB combined with HRV training. We also **test the client after each session,** alternating between digit span, numbers forward and backward, and language tests, such as word recall and reading comprehension. In addition we do a goal-attainment ranking. This is done in order to be sure that no negative cognitive effects are occurring and that the patient is continuing to advance with this procedure.

To this date of writing, our clients and/or their parents have reported much faster positive changes using LNFB than when they were using conventional surface NFB. However, though this represents experience with more than a hundred patients, this is still a relatively small selected group of clients who have a complex array of symptoms. In addition, we have no way of knowing if the changes seemed faster because some clients had had some initial surface NFB combined with HRV training prior to starting LNFB. In many cases at the ADD Centre, we alternate LORETA NFB sessions with single-channel NFB sessions. In both we combine the NFB with HRV training.

Could We Just Do Single-Channel NFB and Get the Same Results?

The answer is both yes and no. For more than 20 years the ADD Centre staff have obtained consistently good results using single-channel NFB combined with biofeedback and coaching in metacognitive strategies. Although each Brodmann area (BA) does appear to have a primary function, it is also true that each BA can have a number of other functions and be involved in many different functional networks. Also, each BA represents just one area out of many that are involved in one or more neural networks that involve cortical-subcortical connections and a coordinated activity involving many other functionally related cortical areas (Thompson, 2007b). This may be one reason why practitioners have obtained good results when doing neurofeedback over a single site such as Cz (Thompson & Thompson, 1998, 2010).

Approximately 50 percent of the EEG amplitude at any single site, such as Cz, arises from neurons directly beneath the electrode (with electrode diameter of 0.5 to 1.0 cm), and 95 percent arises from an area with a radius of 6 cm measured from the center of the recording electrode (Thatcher, 2012, p 35, 37, 51, after Nunez, 1981, 1995). Training at central midline sites such as Cz, Fz, and Pz likely influences not just that small area of the surface cortex but also key areas lying beneath that area of the cortex, such as the cingulate gyrus, that are involved in many networks. A network serves to synchronize the functions of groups of neurons in many different but functionally related areas of the cerebral cortex. The cingulate gyrus is one central area that is involved in the functioning of the Executive, Affect, Salience, and Default networks (Seeley, 2007; Supekar, 2010; Thompson & Thompson, 2011). Influencing the activity of the cingulate gyrus when the patient's Affect network is actively engaged may be assumed to have indirect effects on many deeper structures that cannot be directly affected by NFB operant conditioning at a site on the surface of the scalp. A group of these deeper areas has been referred to above as Central Midline Structures (Northoff, 2006). Thus, EEG biofeedback training to influence these areas can be done either by FCz single-channel NFB combined with appropriate exercises to activate and influence appropriate neural networks that the client wishes to alter, and/or by LORETA NFB to influence these areas more directly. Both methods are used at our Centres. Either method can be effective. Both methods assume that the targeted site or sites influence a particular network of neurons while inhibiting, or not activating, other networks that are not involved with the targeted functions. Long-distance connections in the cortex along fasciculi and commissures can be both excitatory and inhibitory, but not activating non-involved areas/networks can be achieved through basal ganglia to thalamus connections. Our assumption is that the activated or inhibited cortical area will connect to the basal ganglia-thalamic system to establish the emphasis necessary so that inhibition of non-involved areas of the cortex will remain, while allowing appropriate excitation of all functionally related areas of the cortex. This is discussed under Functional Networks and Behavior in Part Two, Section VII of this book.

Summary

In summary, the practitioner does have a real choice. With many patients, single-channel NFB will influence the site that requires the training and/or the neural networks that can influence changes in a positive direction. On the other hand, LORETA NFB may be a more efficient route to follow if there are many involved Brodmann areas, particularly if these areas are far from surface electrode sites such as in the midline and ventrally. It may also be expected to be more efficient if a number of measurements such as amplitude, coherence, and phase are all outside database norms. Finally, in some instances, such as in patients who have experienced a TBI, the damage usually has effects on so many areas that we postulate that LNFB will be a faster intervention methodology. Thus far, for example, TBI patients at the ADD Centre have been recovering remarkably quickly using LNFB, often with just 40 sessions of training, whereas postconcussion syndrome with regular NFB typically can take double that number of sessions. Nevertheless, it will take hundreds of cases and controlled, multisite studies before we can come to any definitive conclusions.

In our Centre LORETA NFB is being found to be effective in a small number of patients who have suffered from a stroke but there is insufficient research at this time to make further comment. In the future studies will clarify whether or not, or to what degree, this technique may be helpful in a wide range of other disorders, such as Alzheimer's disease. It is theoretically possible that we are at the beginning of a new era in terms of interventions for debilitating brain disorders.

This section has used Dr. Robert Thatcher's NeuroGuide and LORETA Z-score NFB program as the basis for giving examples. However, other programs are becoming available. Evoke Neuroscience (eVox) is one such instrument and program, and its database has been collected with modern amplifiers that go to over 100 Hz. It goes beyond the 28 Hz used in some of the other currently available database programs. The field is expanding rapidly with respect to new tools. Keep in mind that the basic knowledge of neuroanatomy, neurophysiology, and operant conditioning is still needed to apply these tools effectively.

PART FIVE

Other Treatment Modalities

This part describes other treatment modalities that may be combined with neurofeedback and biofeedback. The authors have selected modalities that are evidenced based and do not appear to have major side effects.

SECTION XVIII
Heart Rate Variability (HRV) Training

Heart rate variability training makes sense in light of our Systems Theory of Neural Synergy (Thompson & Thompson, 2008). An emphasis on neural networks is really a systems theory approach to understanding our work with NFB and BFB. The human nervous system is a dynamic network of interconnecting elements. It works to maintain homeostasis and equilibrium. Input to any element within the nervous system will produce change in the other elements. These elements are **synergistic,** they work together producing correlated action where the product is greater than the simple sum of the parts primarily involved. What does this mean for our work? In our work we do neurofeedback (NFB) to influence EEG patterns in areas of the cortex when these patterns correlate with the patient's symptoms. The reinforcement of a relaxed, calm-yet-alert-and-concentrating mental state is one key to optimizing cognitive processing. To achieve this goal we have found that it is more efficient and effective to combine the NFB with peripheral biofeedback (BFB). BFB provides sensory feedback to the brainstem from organs such as the heart.

Peripheral Psychophysiological (Biofeedback) Measurements

At the ADD Centre/Biofeedback Institute of Toronto, we combine peripheral biofeedback with neurofeedback during the same training session. Both components are sometimes carried out at the same time during a session using feedback screens that reflect both EEG and peripheral measures. One can also do segments of just one or the other to get a feel for a particular mental state or for the biofeedback modality, such as diaphragmatic breathing or finger temperature. Most often we are giving feedback concerning heart rate (by way of a blood volume/pulse monitor on a thumb or finger) and respiration (using a breathing belt around the abdomen).

In order to decide which variables would be most useful for a particular patient, we do a psychophysiological stress assessment with many of our adolescent and adult patients. The stress assessment only takes about 20 minutes. In that short time, however, it is a very effective method of allowing the client to observe that even a minor stress will change their physiology. In addition, in the final section of the procedure, they can observe how relaxing and breathing diaphragmatically at about 6 breaths per minute (BrPM) can improve these measurements and facilitate recovery from stress. The client is reassured that none of us can escape stress. We all will react to stress, but what we want to learn is how to recover rapidly both mentally and physiologically.

Method

The figure below shows one of our staff demonstrating this process of a psychophysiological stress test. She and the physician (a family practice resident) have put several sensors on her hands. The hand sensors include: a blood volume photoplethysmograph (BVP) on one finger of her left hand to detect her pulse (alternatively we use the thumb since it has a strong pulse). Heart rate (HR) is being measured using this plethysmograph, which measures red light reflected from blood vessels in a finger or thumb. Instead of measuring R-R intervals with EKG leads, we are measuring pulse to pulse, which gives the same figures at rest (though may not do so during tasks) for the standard deviation of the R-R intervals. With artifacts removed, this is called SDNN (Giardino, 2002). At our Centre the SDNN is calculated after the session using the CardioPro program from Thought Technology. This program was developed to provide automatic (or manual) correction of artifacts in the recording, and to calculate the statistics used internationally for heart rate variability measures. There is a temperature sensor on her right little finger, skin conduction/electrodermal (EDR) sensors on her right index and ring fingers. She has also placed a belt with a stretch receptor around her waist to detect diaphragmatic breathing. On her forehead she has put a headband that contains electromyogram (EMG) sensors to detect changes in muscle tension. Often, a second triad of EMG sensors are placed over the trapezius muscle mass on the shoulder. This second placement of EMG sensors will show us when a patient is using shoulder and thoracic tension to inhale as well as reflect any muscle tension in the shoulder. This shoulder placement of an EMG sensor allows the patient to relax muscle tension and learn diaphragmatic breathing at the same time. In addition, our staff member in this demonstration has one EEG electrode at Cz, a reference electrode on her left ear, and a ground electrode on her right ear.

Heart rate variability (HRV) is one of the peripheral biofeedback variables used at the Biofeedback Institute of Toronto, and it is being measured in this process. These peripheral biofeedback measures and training to improve them, including respiratory sinus arrhythmia, are all described elsewhere in this text. However, HRV training is not dealt with as thoroughly as other variables and will be expanded in this section.

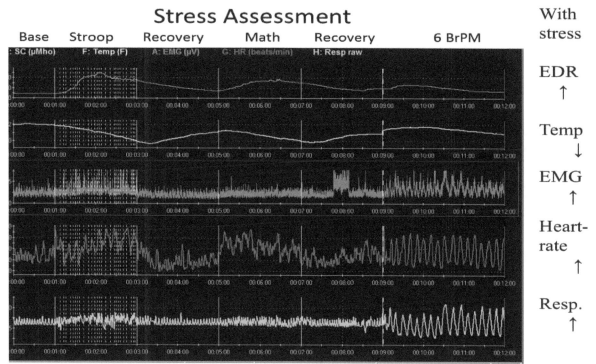

Figure: Stress Assessment Profile
Stress assessments are described earlier in the NFB Book.
Note that with the two stressors, the Stroop color test and the arithmetic task, skin conduction (EDR) increased, peripheral skin temperature decreased, muscle tension (EMG) increased, heart rate increased, respiration became irregular, fast, and shallow. Observe in this example that in the final segment, when breathing at six BPM, this client used her shoulder (and presumably thoracic) muscles (EMG) when breathing. This was noted and she soon learned to relax her muscles and breathe diaphragmatically. This screen is from Thompson & Thompson "Setting-up-for-clinical-success suite," available through BFE.org. It is used with the Thought Technology Biograph Infiniti eight-channel instrumentation. With the Infiniti equipment, different configurations of the channels can be used, but in this text we are using two channels of EMG (2500 samples/second), and two channels of EEG, plus channels for electrodermal, temperature, heart rate, and respiration each recording at 250 samples per second.

The screen in the above figure shows changes in each of the measurements with stress, followed by some recovery when the first stress (Stroop color test) stopped, but somewhat less ability to recover after the second (mental math) stressor. In the last three-minute section the client breathes slowly with the trainer at six breaths per minute with expiration longer than inspiration. The client usually feels a little stressed (performance anxiety) when they are told to follow the trainer's breathing, and this may be reflected by a small rise in skin conduction and a slight drop in skin temperature as shown in this figure. It is possible to also look at EEG changes (usually at Cz) for each section, but this is not shown in this summary.

Clinically, for the control of stress and anxiety, and to attain a calm, focused concentration during the session, the most effective biofeedback training appears, at our Centre, to be HRV. It is also very easy to have all our clients do this training while the EEG sensors are being applied at the beginning of each session and during the EEG NFB sessions. However, the trainers always emphasize that patients *empty their minds of all their worries and unnecessary thoughts and* relax their shoulder, neck, and jaw muscles, and feel their hands getting warm, while doing their diaphragmatic breathing at about 6 BrPM to increase their HRV.

Heart Rate Variability Training

Training Method
The most commonly used screen by our clients is shown in the next figure.

Note that on this 30-second display, we tell the client to make three little *mountains,* all the same shape and height. We also tell them NOT to take their next breath until they really have to breathe. In this manner we attempt to avoid the patient 'over-breathing.' We also educate the patient in the signs of overbreathing, such as feeling light-headed.

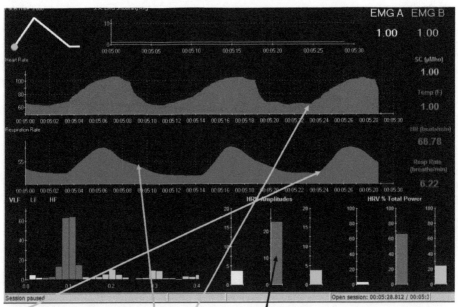

Breathing a little behind her **heart rate** resonant frequency (5.8 BrPM) at about 6.2 BrPM, 28 year old athletic doctor has **HRV of 24 x2 = 48** on her first try. Note short inhale, **long exhale**, breathing peaks the same each time.

This doctor, during the demonstration, was not using other biofeedback variables such as skin conduction, skin temperature, and EMG. These variables are on the right-hand side of the screen should they be required. In the spectrum in the lower left corner, the x-axis is frequency. She is breathing at 6 BrPM and her heart rate is following this and showing six variations per minute. The international statistics for HRV (Task Force, 1996) define very low (VLF), low (LF), and high (HF) frequency for heart rate. Low frequency (LF) 0.04-0.15 Hz is the *purple* range in the feedback screen. This is the range that usually should have the highest power, and it is what we are encouraging in HRV training. It is said to reflect a balance of sympathetic and parasympathetic activity. VLF is *yellow* <0.04 Hz and may reflect sympathetic activity and other variables, but this is not agreed upon at this time. HF is *green,* 0.15-0.4 Hz. HF is said to reflect parasympathetic activity. The ratio of LF/HF should normally be about 1.5-2. In the figure it can be seen that this young athletic doctor has very high LF activity.

At the ADD Centre/Biofeedback Institute of Toronto, clients are asked to find a breathing rate, usually between 5.5 and 6.5 breaths per minute (BrPM), that results in an even amplitude of both breathing and heart rate (HR) changes with synchrony of their HR changes and their BrPM. To accomplish this, they are asked to make inspiration shorter than expiration and to have each of the blue and red *'mountains'* all exactly the same height as is shown in the figure. Some clients initially count to themselves saying: inhale, 1, 2, 3, exhale, 4, 5, 6, 7, 8, hold, 9, 10, inhale... When this synchrony is difficult to achieve rapidly, the staff will use the methodology described by Lehrer (2007), beginning at 7.5 BrPM and decreasing in 0.5 second steps to 4.5 BrPM to find the frequency at which the amplitude of HR variations are maximal in order to establish the resonant heart rate frequency for that particular client. Breathing at one's resonant frequency produces the highest variability in heart rate. Note, however, that if HRV is used with children they will have a much higher (8-12 Hz) resonant frequency than adolescents and adults.

The single- and two-channel NFB training screens at the ADD Centre have a double line graph (red for heart rate, overlapping blue for respiration rate) on many screens so that the client can follow their attainment of synchrony between HR and respiration while they do their NFB (Thompson, Thompson, Thompson, Reid & Hagedorn, 2013). With LNFB, a separate computer and monitor are used for the HRV feedback during some of the sessions while the patient is doing the LNFB. For most sessions, HRV training is done during the time when the cap is being put on and impedance checked. The next screen gives an example of the way in which single-channel NFB can be combined with HRV training.

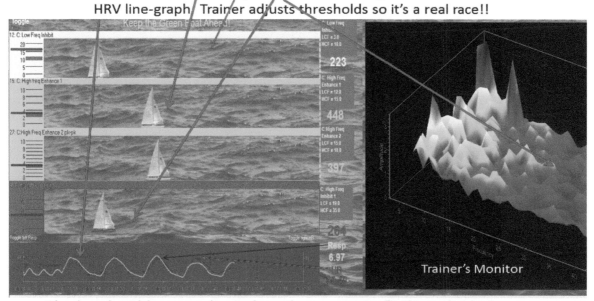

The band widths are adjusted on the screen to fit the individual client's needs as identified in their assessment

In this figure, breathing is shown by the light blue line in the bottom graph. It is measured using a belt around the abdomen that stretches when the client inhales. The heart rate is shown by the red line, and it is measured using a finger plethysmograph. Observe that this client quickly adjusted his breathing and heart rate and breathing changes then became synchronous.

This tall, adult male client presented with anxiety and depression. Being tall and male may be a contributing factor in his HR resonant frequency being unusually low (about 4.5), and he moved into that frequency as training proceeded. He has moved the boat that responds to the production of SMR into the lead position while maintaining synchrony between his heart rate variations (red line) and his breathing (blue line). He did not have high-amplitude, high-frequency alpha at 12 Hz, so his SMR range included that frequency. (If he had, then the SMR frequency range selected for training would have been 13-15 or even 14-16 Hz (14.0-16.0 Hz). *With Biograph Program 15 Hz is 15.0 Hz. To include 15 Hz (15.0-15.9 Hz), set the program to 16 Hz.* With some of our other instruments 15 Hz would actually measure 14.5-15.5 Hz. Thus the practitioner should always check with the manufacturer to establish what the bandwidths are for their instrument.

Peripheral skin temperature, electromyogram, and electrodermal information is displayed if these are needed for the client. In that manner a screen can be customized to the needs of a client. A trainer is always working one-to-one with the client, and during a portion of every session the trainer is teaching the client to use metacognitive strategies. (See Section XV in this book for more details on what these strategies look like.) In addition, the trainer is working with the client to help the client move into mental states that engage different networks, such as the Executive and the Affect network. In any discussion of the training environment, one must not forget the importance of engaging the placebo network in the therapeutic process. The placebo network has distinct neural pathways that go through the hippocampus to the frontal lobes (Benedetti et al., 2005).

Goal of Training HRV

Our goal in HRV training is to increase the heart rate variations (HRV). When this is achieved we will see an increase in the amplitude of these variations and an increase in the standard deviation of the artifact-corrected RR interval measurements (SDNN), as will be explained below. The maximum change will occur when the client breathes at the resonant frequency of their own heart. **When any resonant system (such as the heart) is rhythmically stimulated at its resonant frequency, the external stimulation magnifies the persistent oscillations,** thus greatly increasing total variability. Think of pushing a child on a swing. You know that if you push the swing just as the child has reached the height of the swing's oscillation, then the swing will go higher and higher as you push until it reaches its maximum arc. In this case, the maximum depends on the length of the rope holding the seat of the swing. In the case of the heart, the HRV will increase with training to a maximum for that patient. Breathing diaphragmatically at this resonant frequency will usually be observed to result in the highest amplitude of HRV for that person. In very general terms, higher heart rate variations (for example, SDNN >70) are usually associated with good health. Low HRV (for example SDNN <50 ms) is not considered a good sign. Low values may be found after a concussion. However, also remember that HRV will decrease with age.

Measurements Used in Heart Rate Variability (HRV)

Definitions

- **HRV (heart rate variability)** depends on the variation in interval (R-R) between heartbeats. It is a normal cyclic variation in heart rate. For example, a young healthy person might say that his heart rate was 70 beats per minute (BPM). In reality this number is just an average. In this healthy young person, the heart rate might be varying from 55 to 85 BPM, and the heart rate variation is therefore about 30. We want to see it greater than 10 in all our patients.
- **RSA (respiratory sinus arrhythmia)** is the naturally occurring variation in HR that occurs during the breathing cycle. It is directly proportional to HRV.
- **HRV training** includes monitoring RSA and adjusting breathing patterns to increase HRV.

Heart rate variability refers to the constant variations in heart rate that may be observed in healthy individuals. These variations may be measured in terms of the interbeat intervals (IBIs), These are the intervals between heart beats. This interval is measured between the R waves in the adjacent electrocardiogram (EKG) QRS complexes. The standard deviation of these IBIs is called SDRR and, when computed after all artifacts have been "corrected", is called SDNN. 'N' stands for 'normal'. The interbeat, R1 - R2 and R2 - R3 (IBI 1) intervals are shown as examples in the figure below.

Correction of Artifacts

As with EEG recordings, artifacts will result in inaccurate assessment values. One cannot have artifacts in assessment data (Peper, Shaffer, I-Mei, 2010). The reader will note that, unlike artifact removal from an EEG, in an EKG record time sections must not be removed. Experts can 'correct' the artifacts manually, but fortunately there are programs that will do this automatically. One such program is produced by Thought Technology and is called CardioPro. This program can rapidly do this for regular practitioners. Practitioners using equipment from other manufacturers should inquire from their supplier as to what program they can use to carry out similar procedures. Briefly, the CardioPro program normalizes the IBI data by adding small IBI values together and splitting large ones into two or more IBIs. Pairs of IBI values where one is short and the other long are averaged. It is important to understand that the beat time information (i.e., the moment of each beat) needs to be maintained for the HRV analysis to be valid, and your program should do this for you. (Thanks are given to Thought Technology for this clarification.)

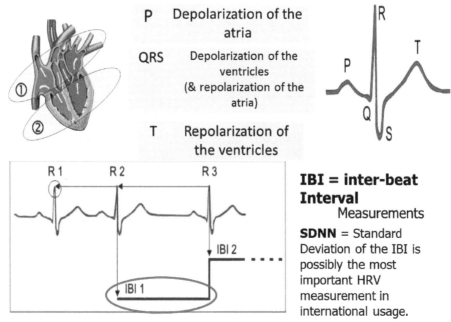

Electrocardiogram (EKG) origin and the standard measurement SDNN. Figure with permission from Thought Technology, Montreal, Canada. This illustration shows the QRS wave of the EKG and the interbeat intervals, R1-R2 and R2-R3. Interbeat (R-R) interval (IBI) is the basis for most HRV statistics. RR interval becomes NN after artifacts are 'corrected' in the record.

Heart-Brain Connections Produce Oscillations

The reader will note that most of our biofeedback measurements, such as skin temperature, EDR, and EMG, involve only the sympathetic portion of the autonomic nervous system (ANS). They are **'tonic'** measurements. HRV is a completely different phenomenon. It is an **oscillating** and not a tonic measurement. It is said to be the only means we have to obtain a simple measurement of the parasympathetic portion of the ANS.

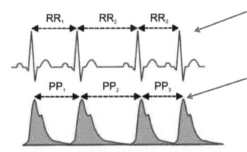

At the ADD Centre we use EKG with BIAT 19 channel EEG assessment (top line RR intervals).

During training we use a finger plethysmograph to measure Pulse to Pulse times.

The CardioPro Program will eliminate artifacts & calculate statistics.

EKG electrocardiogram is measured when we do our 19 channel EEG assessment.

Pulse wave measurement is how we measure the heart beats when doing BFB. (This is called BVP (blood-volume-pulse) and it uses a **photoplethysmograph** which is usually placed over a finger. It **reflects changes in blood flow** detected by an infrared light reflected from the vessels in the pulp of the finger.

The **interbeat interval 'RR'** measured with the EKG turns out to be **virtually the same** time frame as the **pulse to pluse 'PP' interval** measured using the plethysmograph.

Figure to show relation of EKG R-R intervals as is done with eVox (was BIAT) compared to Pulse Transit Time as is done with the Infiniti HRV training programs used at the ADD Centre. (Note, Evoke Neuroscience has discontinued the use of the name 'BIAT.' It is now a program that uses an instrument with a very advanced amplifier and is called 'eVox'). The above figure from Thought Technology manuals shows the relationship between the R-R interval measured with EKG compared to the P-P interval measured with the plethysmograph. Under resting conditions these will give the same statistical values.

The Heart as an Oscillating System

The healthy beating heart is the principle component of an oscillating system called the cardiovascular system (CVS). A healthy heart is constantly changing its rate. Every person's heart has its own unique resonant frequency. This is often close to six variations per minute. However, it will be less for men than women and less for tall persons than those who are shorter because it is partly dependent on total blood volume. It will therefore also be higher for children than adults. When the resonant frequency is achieved, the clinician will note almost perfect synchrony between breathing and heart rate changes. The patient can inhale and observe an increase in the heart rate, exhale and observe a decrease in heart rate (Lehrer, 2007). This was seen in the HRV training screen.

Example of Statistics Obtained with CardioPro

Source	SDNN (ms)	NN50	PNN50	RMSSD (ms)	VLF ms2/Hz	LF	HF	LF/HF	Power ms2/Hz
HR/BVP-Pro/Flex - 1G	87.64	9	0.127	85.309	555.61	1447.7	1063.1	1.36	3066.4

Source	Min.	Max.	Average	St. Dev.	Variability	Mode	Area
BVP HR	57.96	100.7	71.751	8.873	0.124	70.790	-
BVP HR Max-Min	6.124	42.75	14.187	14.28	1.007	7.610	-
Resp. Rate	5.191	8.562	6.540	1.395	0.213	5.612	-
SC-Pro/Flex - 1E	3.376	4.283	3.606	0.183	0.051	3.512	216.3
Temp-Pro/Flex - 1F	93.61	93.81	93.732	0.064	0.001	93.780	5623.73
MyoScan-Pro 400 - 1A	1.541	5.425	2.113	0.363	0.172	2.269	126.769
MyoScan-Pro 400 - 1B	1.000	1.000	1.000	0.000	0.000	0.000	60.000

The second MyoScan sensor (1B) is not being used in this example. Under source ignore the letters 1E, 1F, 1A, 1B as they just refer to the input link of the instrument being used to take the measurements.

Definitions

SDNN
SDNN is the standard deviation (the square root of the variance) of the NN interval. Variance is mathematically equal to the total power of the spectral analysis. SDNN reflects all the cyclic components responsible for variability in the period of recording; therefore it represents total variability. In hospitals this may be calculated over a 24-hour period.

SDANN
SDANN is the standard deviation of the average NN intervals calculated over short periods. SDANN is therefore a measure of changes in heart rate due to cycles equal to or longer than five minutes.

PNN50
PNN50 is the mean number of **times per hour** in which the change in consecutive normal sinus (NN) intervals (IBI values after artifacts corrected) exceeds 50 milliseconds. Ewing and Travis (1984) proposed this measure to help assess parasympathetic (vagal) activity from 24-hour ECG recordings.

pNN50
The pNN50 statistic is defined as the NN50 count/total NN count. This is the percentage of absolute differences in successive NN values >50 ms. The pNN50 statistic has proved to be a useful HRV measure (Bigger et al., 1988). **pNN50 = (NN50/n-1) x 100%**.

This is somewhat correlated to **HR max-min**, in the sense that epochs with higher HR max-min values will probably show higher NN50 values, but there is no mathematical relationship.

RMSSD
RMSSD is the root mean square of the standard deviation of the RR interval (i.e., the square root of the mean of the sum of the squares of the successive differences between adjacent NNs.)

These measurements are further defined in *The European Heart Journal* (1996), and described in the manuals for programs that calculate these statistics such as the CardioPro program from Thought Technology.

The table figure above shows a summary of HRV training variables taken during a NFB + BFB session using Thought Technology (TT) Infiniti equipment. A BVP sensor was used for heart rate recording and the data was analyzed using the TT CardioPro program with automatic correction of artifacts. The calculations follow: Task Force of The European Society of Cardiology and The North American Society of Pacing and Electrophysiology (*European Heart Journal*, 1996).

The next figure need not be studied in detail. However, the Wiggers diagram does help some biofeedback practitioners understand the relationship between the QRS complex, and, in particular, the interbeat R-R interval and ventricular filling and emptying.

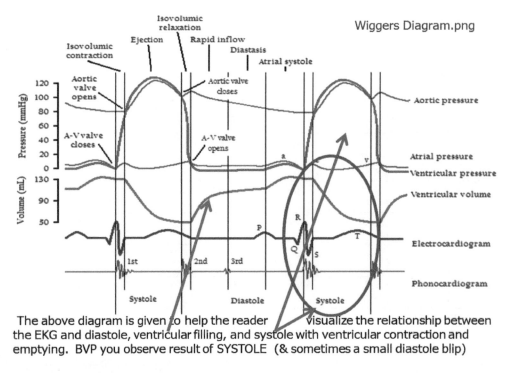

The above diagram is given to help the reader visualize the relationship between the EKG and diastole, ventricular filling, and systole with ventricular contraction and emptying. BVP you observe result of SYSTOLE (& sometimes a small diastole blip)

Wiggers_Diagram.png; Reference Wikipedia.
*A **Wiggers diagram** is a standard diagram used in cardiac physiology. It is named after Carl J. Wiggers, M.D. (May 28, 1883 – April 28, 1963). From 1918 to 1953, he was professor and chairman of the Dept. of Physiology at Western Reserve University Medical School. Wiggers established and was the first editor of the medical journal Circulation Research. In 1952 he received the Gold Heart Award from the American Heart Association.*

Vagal Control of HRV

Without parasympathetic control, the innate rhythm of the sinoatrial node (about 100 BPM) would drive heart rate. Health is tightly related to oscillatory systems (Lehrer & Eddie, 2013). Oscillations in vagal input are the primary system for controlling oscillations in the heart rate. The right vagus nerve primarily innervates the sinoatrial (SA) node. The left vagus primarily innervates the atrioventricular (AV) node. However, there is significant overlap in this anatomical distribution. Atrial muscle is also innervated by vagal efferents, whereas the ventricular myocardium is only sparsely innervated by vagal efferents. Sympathetic efferent nerves are found throughout the atria (especially in the SA node) and ventricles (Klabunde, 2013).

The reader is referred to papers by Stephen Porges (2007), who has developed the polyvagal theory. The vagus contains specialized subsystems that control competing adaptive responses. **Porges' *polyvagal theory*** describes the different roles of the unmyelinated vagus (dorsal vagal complex) and newer myelinated vagus (ventral vagal complex). The ***unmyelinated vagus*** responds to threats through immobilization, feigning death, passive avoidance, and shutdown. The unmyelinated vagus is activated in PTSD when, for example, a patient believes that he/she will die. A different mechanism involves the ***sympathetic nervous system*** that, in concert with the endocrine system, responds to threats to a person's perception of being safe and secure. The sympathetic system responses result in mobilization, fight or flight, and active avoidance. The sympathetic nervous system inhibits the unmyelinated vagus and will mobilize the individual for action. A third system depends on the highest evolutionary level and is called the ***myelinated vagus system.*** The myelinated vagus rapidly adjusts cardiac output and promotes social engagement. Fred Shaffer in his lectures gives an excellent summary grid, which we have shown below:

ANS Component	Activated/Inhibited By	Responses
Unmyelinated vagus (dorsal vagal complex)	Survival threats activate	**Feigning death, immobilization, passive avoidance, shutdown (PTSD)**
Sympathetic-adrenal-system	Survival threats activate	Mobilization, fight-or-flight, active avoidance
Myelinated vagus (ventral vagal complex)	Social engagement, facial muscle relaxation activate; daily stressors inhibit	Bonding, calming, inhibition of sympathetic-adrenal activation, respiratory sinus arrhythmia

Figure to show autonomic nervous system (ANS) components that are discussed in the polyvagal theory. Figure from workshops given by Fred Shaffer who discusses work by Dick Gevirtz (2011), and Porges, S. W. (2007).

In most persons, daily stressors will result in a decrease in the myelinated vagal response and a corresponding increase in sympathetic drive. Relaxing muscle tension, emptying the mind of negative ruminations, and practicing diaphragmatic breathing at one's resonant frequency (about six BrPM) to increase HRV can decrease this sympathetic drive and restore the myelinated vagal system state of calm.

Summary

In a normal individual in a **safe environment**, the **primitive unmyelinated vagal and the sympathetic system defensive activation is actively inhibited,** allowing the phylogenetically newer, myelinated, medullary **vagal system to dominate** (Porges' polyvagal theory). This **results in a decrease in HR and blood pressure** (BP) and an activation of striatal muscle systems in the face, head, middle ear (stapedius muscles), larynx and pharynx so they can function and **appropriately respond to safe social** interactions (Porges, 2004, 2007).

The figure below shows the carotid and the aortic arch baroreceptors. These **baroreceptors are mechanoreceptors that respond to blood pressure** (BP). When the blood pressure rises, the resulting expansion of these baroreceptors sends afferent signals to the nucleus of the solitary tract in the medulla. This nucleus sends signals to many areas in the brain. These connections result in modulation of the activity of the **paraventricular nucleus** of the hypothalamus, which can affect sympathetic and parasympathetic activity, in addition to the hypothalamic-pituitary-adrenal axis. The final result for heart rate is that, in the brain stem, the dorsal nucleus of the vagus and the nucleus ambiguous send parasympathetic vagal efferents to the heart that slow the heart rate. At the same time, sympathetic efferents decrease their activity in the vascular system allowing vasodilation to occur. The result is a fall in BP. Then, when BP falls, the opposite will occur: parasympathetic activity decreases and heart rate will increase. Sympathetic vasoconstriction occurs, but this is not a major factor. The inherent frequency of the sinoatrial node is greater than 100 beats per minute (BPM). Thus, the most important controlling factor of heart rate is the parasympathetic vagal slowing of heart. This is an oscillating system due, in large part, to the feedback to the brainstem from the baroreceptors. This is illustrated in the figure below.

The Central Autonomic Network (CAN) and the Nucleus of the Tractus Solitarius (NTS)

The output of the CAN has connections to the sinoatrial node of the heart via the stellate ganglia (sympathetic at level of the seventh cervical vertibrae) and the vagus nerve. CAN output is under tonic inhibitory control via GABAergic neurons in the nucleus of the solitary tract (NTS). In addition to the already mentioned connections to central midline structures in the brain and to the paraventricular nucleus of the hypothalamus in particular, the NTS has direct connections to the nucleus ambiguus (NA) and the dorsal vagal motor nucleus (DVN) (see Thayer and Lane, 2009, for a complete description of these pathways). These connections are via interneurons between the NTS, NA, and DVN traversing the intermediate reticular zone, and provide input to the cardiovagal motor neurons. In addition, the NTS is a site where the afferent and efferent vagus meet (Thayer et al., 2012).

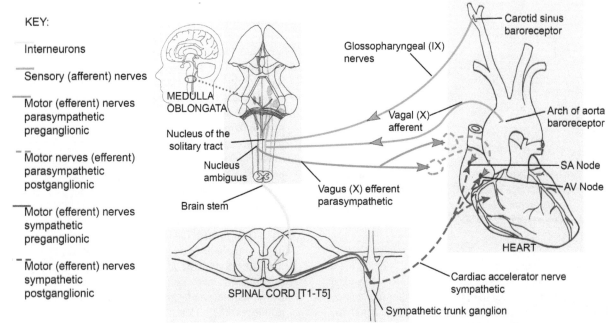

Schematic diagram of afferent and efferent ANS connections between the brain stem and the cardiovascular system. Drawing by Maya Berenkey. The parasympathetic vagus efferent neurotransmitter is acetylcholine (Ach). The sympathetic neurotransmitter from medulla to the sympathetic trunk or chain ganglion is Ach, but the efferent from that ganglion to the heart has a different neurotransmitter, norepinephrine.

HRV Plus Peripheral Biofeedback

For some patients, a portion of their session may benefit from a greater focus on HRV plus other biofeedback measurements. These patients are most often persons for whom anxiety is a major presenting symptom. The screen in the next figure is just one example of a complex biofeedback screen. This screen might be on the trainer's monitor while a pleasant animation is being played on the patient feedback monitor. As with NFB feedback, the visual animation and sound will be dependent on whatever measurements the trainer and patient agree to use.

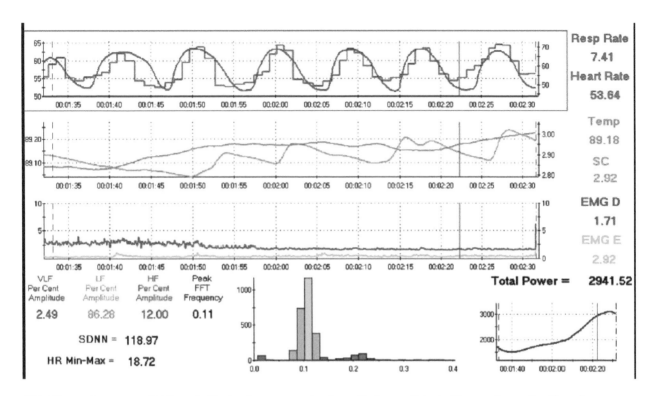

This illustration on an old Thought Technology screen is given, kindness of Evoke Neuroscience, Dr. Hagedorn. (Note: SDNN should read SDRR since the interbeat interval has not yet been corrected for artifacts. This has since been changed to SDRR on the screens in this program.) The steps shown in the HR changes will be explained using a hypothetical model in the next figure.

In this example, six sensors have been placed on the patient to measure psychophysiological variables. Heart rate changes are seen as steps of the red line, while respiration is seen as a blue line. The heart rate is slower than 60 BPM, and there are only about nine steps in each, approximately, 10-second period. (The steps just reflect stepwise decrease in the R-R interval as heart rate increases with inspiration, and a stepwise increase in R-R as HR decreases with expiration.) The respiration average rate is at 7.41, and the heart rate and respiration appear to almost be in synchrony. The heart rate average at this point in the session was HR min-max = 18.72, which is a healthy range. The frequency of the overall HR changes (increase in rate as the subject breathes in, and decreases in rate as the subject exhales) is shown in the FFT spectrum as being close to 0.1 per second (6 changes/60 seconds = 0.1 Hz). The Low Frequency (LF) is 86.28, which is considerably greater that the Very Low Frequency (VLF) or the High Frequency (HF). The standard deviation of the R-R interbeat interval is above 100 at 118.97. (This is labelled incorrectly on this screen as SDNN; it should be SDRR as it has not been artifacted; this label has since been changed by the manufacturer on these screens.) These figures are considered to be in a healthy range.

The second line graph shows skin temperature and skin conduction. The peripheral skin temperature is 89.18° F, which is a bit low but in the normal range. The third line graph shows electromyogram (EMG). We like to see the EMG below 2 and here it is 1.71 and 2.92. The skin conduction (SC) is 2.92, which is somewhat low in North America but is not unusual in some other countries such as Taiwan. Also, older patients often show lower values for SC. In North America this low a value in younger people could indicate that the patient was losing his alertness. Reference: Dr. D. Hagedorn, Evoke Neuroscience, DJ Technologies, Inc.

HR Displays: Smooth Line or Steps

We have been asked why the heart rate goes up in steps in some feedback screens (such as Dr Hagedorn's screen above), while in others it is displayed as a smooth line, as in the red *"mountains"* for heart rate feedback with the blue "mountains' for breathing in the feedback screen shown earlier. In either type of display the line rises with inhalation and falls with exhalation. The smooth line display is shown in our first example screen and in the example of the feedback screen for the young athletic doctor.

The smooth line is just a computer function that can be done to change the stepwise display to a smooth line. Clients at the ADD Centre can choose their display. In general, clients at the ADD Centre prefer the smooth line for feedback.

The steps shown in Dr. Hagedorn's TT screen reflect how the heart rate is changing. The *rise* of each step represents the *change* in IBI. This change, we have said, should ideally be *averaging* more than 70 ms. (We would hope that, with HRV training, this would increase.) The *run* of the step in this staircase would be *averaging* about 1000 ms if the person were breathing at six BPM and the heart rate changes were in synchrony with her breathing such that there were six variations (*mountains*) each minute. The rationale for saying this is that in one minute, 60 seconds, there are 60,000 ms. When this person's heart rate (HR) was 60 BPM, then 60,000/60 = 1000 ms/heart beat or per IBI. At that instant, she would theoretically see one variation or mountain on the screen every 10 seconds, and in our 30-second example display screen shown at the beginning of this discussion there would be three little mountains.

However, as we keep pointing out, as long as we remain healthy, the **heart is not beating at one frequency**. It is an oscillating system and the rate is constantly rising and falling, giving us the beautiful blue mountains in our feedback screen. **As heart rate increases when we inhale, the IBI is decreasing. As HR falls when we exhale, the IBI steadily increases.** The lowest HR in the figure shown below of a 'model' (hypothetical) example client might, for example, be 50 BPM, making the IBI at that point higher, i.e., 60,000/50 = 1,200 ms. But as our model of a client inspires, the heart rate increases and the IBI time falls on *average* by, in the model example, approximately 70 ms each step. At 60 BPM, the IBI would be 1000 ms. If the HR rose to 75 BPM as the model person inspired, then at the top of the blue line (*mountain*) in our display screen, the IBI would be approximately 60,000/75 = 800 ms.

The reader can observe, in Dr. Hagedorn's TT screen above, that the steps are not even. The actual change in the IBI and in the length of the IBI is constantly shifting. We usually observe more steps when we are inhaling and fewer steps as we exhale.

Hypothetical Model to Show HR Changing

In the next figure, a stepwise display is superimposed on a smooth rise and fall line. This is purely a hypothetical example to let the reader visualize the process of steady decreases then increases in the R-R interbeat interval (IBI). Assume that the heart rate (HR) interbeat interval (R-R IBI) changes by an average of 100 ms each beat (as in a young athletic person). Imagine a 10-second time frame and assume that the IBI decreased by 100 ms each beat so that the HR increased for five beats. Then the HR decreased in a similar stepwise fashion by increasing the R-R IBI by 100 ms for each beat for the next five heartbeats. The steps down hypothetically stop, and then, with the next inspiration, the heart rate begins to increase again and the whole cycle repeats itself.

*The result of one such hypothetical cycle is 10 heartbeats in 10 seconds (HR 60 beats/minute). We have drawn this hypothetical example in the next figure. This figure shows a purely hypothetical representation of HR changes over a 10-second time frame. The average IBI is 1000 ms, so in the example we had the IBI start at 1250 ms and decrease in 100 ms **steps** as the HR increased from 48 BPM (60,000 ms/1250 ms = 48 BPM) to an IBI of 750 ms with an HR of 80 BPM. Then the IBI increased in 100 ms **steps** and the HR decreased from 80 BPM to the hypothetical starting point of 48 BPM.*

In practice is it NEVER even, as it has been drawn in this sketch.

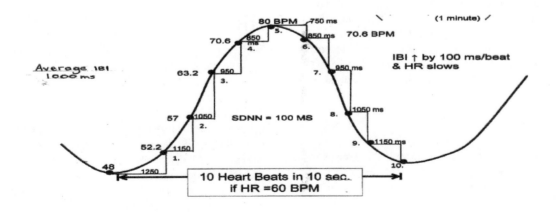

Why We Combine HRV with NFB

Literature Demonstrates Improvements Using HRV Training

Training HRV is important for the majority of clients seen at the ADD Centre. HRV training is known to reduce anxiety and stress (Mikosch, Hadrawa, Laubreiter et al., 2010; Reiner, 2008). It can also improve cognitive functioning in stressful conditions (Prinsloo, et al., 2011). As noted previously, in very general terms, higher heart rate variations – for example, SDNN >70 - are usually associated with good health. Low HRV - for example SDNN < 50 ms - is not considered a good sign. We remind the reader that low values may be found after a concussion and that HRV will decrease with age.

As a result of HRV training alone, researchers have demonstrated improvements in mental processing (Thayer, Hansen, Saus-Rose & Johnson, 2009) and emotional stability (Applehans & Luecken, 2006). HRV training has been investigated in the treatment of emotional disorders, such as depression (Grippo & Johnson, 2009; Hassett, Radvanski et al., 2007) and anxiety (Karavidas; Lehrer, Vaschillo et al, 2007). HRV training has also been shown to benefit physical conditions, like asthma (Lehrer, Vaschillo, Lu et al., 2006; Lehrer, Karavidas, Lu et al., 2008), hypertension (Lin, Xiang, Fu et al., 2012) and cardiac functioning (Lehrer & Vaschillo, 2008).

There are many heart-brain connections, and at our ADD Centre and the Biofeedback Institute of Toronto we have documented evidence of improved brain wave patterns being found in conjunction with HRV training, namely, an increase in the amplitude of sensorimotor rhythm (12-15 Hz) activity (Reid, Nihon, Thompson & Thompson, 2013). Given the multitude of conditions helped when a person increases the variations in the time between their heartbeats, it made sense to add this intervention when working with clients who were appropriate for neurofeedback training.

Can Injury Such as Concussion Affect HRV?

Addition of Heart Rate Variability (HRV) to Assessment Measures

All of our 19-channel QEEG assessments include EKG. The observation that traumatic brain injury (TBI) impairs cardiac function has become increasingly apparent through work with clients who have had vehicular accidents (car, motorcycle, bicycle) and concussions in sports (Baguley, Heriseanu, Felmingham & Cameron, 2006; Thompson & Hagedorn, 2012). As previously noted, at the ADD Centre/Biofeedback Institute of Toronto we have observed large deviations from expected normal values for different measurements of heart rate variability (HRV) in relatively young and otherwise apparently healthy athletes who have suffered a head injury. This has been documented by others investigating the effects of concussion in athletes (Thompson & Hagedorn, 2012). Thompson and Hagedorn demonstrated that, after a concussion, even an otherwise healthy athlete can show major cardiac changes. The amplitude of the variations in heart rate and the standard deviations of the heart rate variations after the effects of artifacts have been removed (SDNN) are both lower than would be expected in a person of similar age and health.

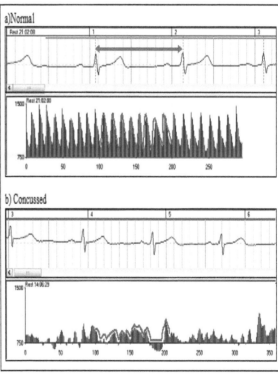

Figure. HRV profile from a healthy adult and from a concussed athlete; from Thompson & Hagedorn (2012).

The figure shown above contrasts heart rate variability in a healthy adult with that of an athlete who has suffered a concussion. It demonstrates that, after a concussion, even a healthy athlete can show a significant irregularity and low HRV values. The top EKG shows a three-second sample and below it is a tracing of variations in heart rate over a 300-second time period, both from the same healthy 36-year-old male. These demonstrate what one would expect in a healthy person: heart rate variations are even with high amplitude and steady rhythmicity. The figure below it shows data from a highly trained male athlete two weeks postconcussion. It demonstrates low amplitude and poor rhythmicity of heart rate variations. The amplitude of the variations in heart rate and the standard deviations of the heart rate variations after artifacts have been removed (SDNN) were also both lower than would be expected in a young athlete. These data provide an illustration of how brain injury affects not just the brain but also the heart. Specifically, it demonstrates that concussion has a negative effect on heart rate variability (HRV).

Similar observations are found postconcussion in clients at the ADD Centre. The following example demonstrates changes observed after 40 sessions of LORETA Z-score NFB combined with HRV training at the ADD Centre. This client had sustained a head injury with loss of consciousness at age five and a second less serious concussion in his teenage years. Memory and attention were both affected and both improved with training.

	SDNN (ms)	Power (ms^2)
Initial Assessment	59.90	1095.42
Following 40 sessions	159.30	7277.85

Thompson, M., Thompson, L., Reid-Chung, A. (2015).

In addition to clinical observations, there are data published by researchers. One example is a study by Baguley and his colleagues (2006). They compared HRV data on 16 TBI age-matched subjects with and without dysautonomia and 16 non-injured controls. The mean SDNN (ms) for the control group was 60.2 with a standard deviation (SD) of 26.9, while the mean SDNN (ms) for the TBI group was 40.6 with a SD of 16.4. The total power (ms^2/Hz) of the controls was 4065 (SD 4082) and in the TBI group only 1656 (SD 1237). The LF power (ms^2/Hz) of the control group was 1304 (SD 1484) and in the TBI group only 189 (SD 152). The researchers demonstrated that the TBI group revealed significant differences in HRV parameters compared to controls.

	Control	TBI
SDNN ms	60.2 sd26.9	40.6 sd16.4
POWER ms^2	4065 sd4082	1656 sd1237
LF power ms^2	1304 sd1484	189 sd152

Heart rate variability statistics from 16 controls and 16 age-matched postconcussion individuals (after Baguley et al., 2006).

Affect Disorders

We have also noted lower-than-expected HRV measurements, such as SDNN and total power, in patients who present with affect disorders. This observation is supported in the literature (Grippo et al., 2009). Grippo's group has reviewed the effects of stress and depression on HRV. They reported that depression can impair cardiac functioning and, in particular, HRV variables (Grippo et al., 2009). Thus, there were good clinical reasons for adding HRV to our work with NFB. In addition, HRV training seems to link neuroanatomically/neurophysiologically with NFB and act in a synergistic manner to improve results, as compared to using either technique alone. To prove this hypothesis will require further research.

Neuroanatomy: What are the links that underlie these clinical observations?
The literature reveals that some of the connections between HRV and brain functioning are being elucidated by investigating the effects of brain damage on cardiac functions.

The Insulae

In many of our TBI assessments LORETA indicates involvement of the insulae (BA 13), and these areas (right insula, left insula) have been among the targets in our work using LNFB. Interestingly, research has demonstrated that injury, as from a stroke, to the right insula has dramatic effects on autonomic balance and heart functioning (Nagai, Hoshide & Kario, 2010; Shah, Klumpp, Angstadt, Nathan & Phan, 2009). Stroke affecting the right insula can result in low values for standard deviation of the artifacted R-R interval (SDNN). It can also alter values for the ratio of the amplitude of low to high frequency (LF/HF) spectral power, in addition to resulting in symptoms of nonsustained ventricular tachycardia and premature ventricular contractions. There may also be an increase in nocturnal blood pressure (BP) and higher norepinephrine levels. Patients with left insular cortex involvement have decreased baroreceptor sensitivity (BRS).

The insulae are important regulators of autonomic balance. The right insula predominately affects the sympathetic nervous system and the left insula primarily affects the parasympathetic nervous system, in part through connections to the nucleus accumbens, which has a role in balancing the sympathetic and parasympathetic systems (Nagai et al., 2010). The insular cortex has reciprocal connections with the anterior cingulate gyrus, amygdala, entorhinal cortex, orbitofrontal cortex and the temporal pole, which are all involved in the Affect and Default networks and may also be implicated in executive functioning. It has afferent connections with hippocampus formation. Additionally, the insular cortex has dense reciprocal connections with subcortical autonomic core centers, including the hypothalamic paraventricular nuclei. As previously noted in the section on functional neuroanatomy, and as will be diagramed below, the paraventricular nuclei can control the hypothalamic-pituitary-adrenal axis (HPA) and both parasympathetic and sympathetic aspects of the ANS. These nuclei also have important connections to the lateral hypothalamic area (sympathetic system), nucleus tractus solitarius (HRV control), and parabrachial nucleus (respiratory control), and all of these centers are also reciprocally connected to each other.

This review goes on to state that using amobarbitol to inactivate either right or left hemispheres has shown that the **right hemispheric inactivation** can induce a significant decrease of BP and an increase in HF amplitude, and that **left inactivation** can induce an increase in HR, BP and LF amplitude. It can also result in a decrease of baroreceptor sensitivity (BRS) by nearly 30 percent. The assessment of baroreflex gain (BRS) as an index of baroreflex function is based on the quantification of RR interval changes related to blood pressure changes, and is measured in ms/mm Hg (Zöllei et al., 2003). For a more complete description of techniques for the measurement of baroreceptor sensitivity the reader is referred to a review by La Rovere, Pinna & Raczak (2008).

Depression

There have been HRV findings in depressed patients similar to those that we have been observing in TBI. Depressed patients show changes in autonomic regulation of the heart. In an extensive review by Grippo et al. (2009) it is noted that there is activation of sympathetic drive and a decrease in vagal tone with a resulting increase in heart rate (HR), decrease in heart rate variability (HRV), and reduced baroreceptor reflex function. This review goes on to note that similar autonomic changes can also be associated with hypertension, increased body mass index, and increased blood glucose level and that these autonomic changes are observed in both acute and chronic cardiovascular conditions. These changes can be predisposing factors for ventricular fibrillation. Important for our work is that HRV has been shown to be reduced in depressed patients both with and without cardiovascular disease, and HRV may be lower in more severely depressed patients (Grippo et al., 2009).

Exercise

As an aside, it should also be noted that exercise training can significantly improve HRV, baroreflex sensitivity, blood pressure and vascular function (Zoeller, 2007), and may also result in improvements in depression (Beniamini, Rubenstein, Zaichkowsky & Crim, 1997). For reasons such as this, we always tell any new client that they will experience a much better outcome if they integrate appropriate diet, sleep, and exercise habits into their training program.

Neuroanatomical Links Between HRV Training and NFB Training

We have already noted that the healthy beating heart is the principle component of an oscillating system called the cardiovascular system (Lehrer & Eddie, 2013). When the resonant frequency is achieved during HRV training, the clinician will note almost perfect synchrony between inspiration and an increase in the heart rate, and between expiration and a decrease in heart rate.

At the ADD Centre some years ago, the question arose as to whether we could go beyond clinical observation and explain why it was likely that NFB combined with BFB could be expected to help patients who had complex presentations more than either modality alone. In simple terms, could we show that the areas of the brain that subsume these different systems and networks are neuroanatomically linked, and that the manner in which they are linked might reasonably be expected to respond to a combination of NFB plus HRV training? Could training to simultaneously improve HRV *and* EEG variables improve the patient's thinking, feeling, and behavior? We concluded that the answer is yes, in theory, and yes, in practice.

The field of combined EEG neurofeedback and peripheral biofeedback has its foundations in functional neuroanatomy and neurophysiology. As is described in the section on low resolution electromagnetic tomography (LORETA), LORETA allows the practitioner to find the source of different frequencies deep within the cortex. This information, when combined with the practitioner's knowledge of the major functions of different Brodmann areas and their relationship to neural networks, can assist the practitioner in deciding whether a specific frequency at that site may correlate with a patient's symptoms. This can help the clinician decide whether or not a pattern correlates with a cognitive, motor, sensory, or affect problem for a specific patient.

Using QEEG and LORETA, clients who present with affect disorders, learning disabilities, autistic spectrum disorders, or TBI (even a number of years after they have suffered from one or more concussions) are found to have a large number of symptoms that correlate with deviations from database norms involving different areas of the brain and different neural networks. Brain areas that may be directly influenced by HRV training and, in particular, by the influence of this training on the solitary nucleus, do have connections to the same regions that we are training in the cortex. The solitary nucleus has connections to cortical areas that are central to Affect, Executive, and Default networks that, in turn, are central to our work with NFB.

Thus, in simple terms, the areas of the brain that are central to these systems and networks are neuro-anatomically linked and have connections to areas that are directly affected by HRV training. These links are illustrated in very broad general terms in the next two figures. Anxiety is used as the example.

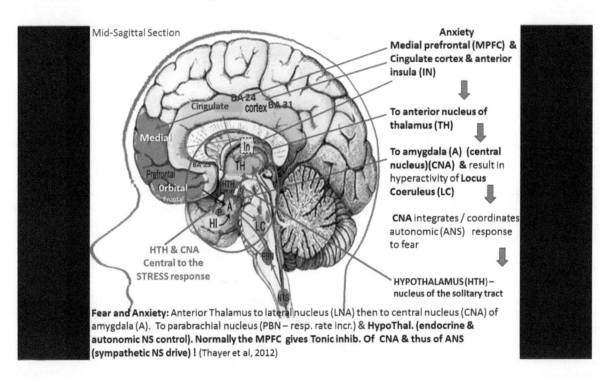

PAGE 582: Central Cortical and subcortical structures involved in response to stress and anxiety – Diagram by Maya Berenkey

Central cortical and subcortical structures involved in response to stress and anxiety *Diagram by Maya Berenkey after figures by Amanda Reeves in* The Neurofeedback Book, *in* Gray's Anatomy *1918, Netter, 1997, and Smith,* Neuroanatomy, *1962.* **Importantly,** *the paraventricular nucleus projects directly to the dorsal motor nucleus of the vagus and the nucleus ambiguous, and thus directly affects HRV. It also projects directly onto sympathetic autonomic relay nuclei in the brainstem.*

TH = Thalamus	*CNA = Central Nucleus Amygdala*	*HI = Hippocampus*
IN = Insula	*A = Amygdala*	*PBN = Parabrachial Nucleus*
HTH = Hypothalamus	*ANS = Autonomic Nervous System*	*NTS = Nucleus Tractus Solitarius*
P = Pituitary	*PVN = Paraventricular Nucleus*	

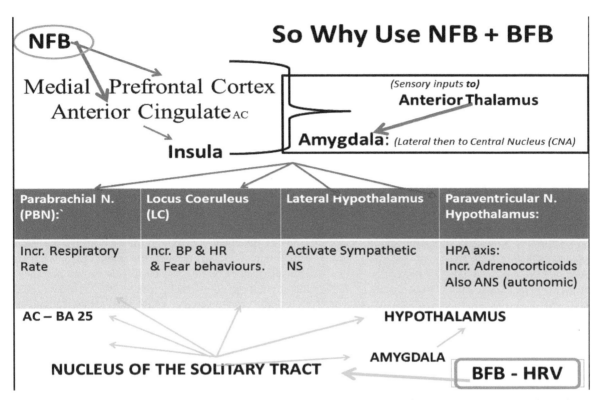

This figure illustrates how NFB applied as training from the 'top down,' and HRV given as training from the 'bottom up' can both affect the same central structures. (Note that our use of the terms 'top down' and 'bottom up', for the past more than 20 years, is different from the more recent use of these terms in cognitive psychology, economics, and other disciplines. Indeed, in other areas of this book we have used these terms differently when describing parietal sensory bottom-up versus frontal modulation top-down when discussing the Salience network.)

Note in the figure shown above that the amygdala innervates the hypothalamic nuclei. It also innervates the parabrachial nucleus, which controls respiratory rate, and the locus coeruleus, which produces norepinephrine and thus can increase blood pressure. This may correlate with fear and anxiety-related behaviors. The central nucleus of the amygdala (CNA) can be important in responses to anxiety and fear. As previously mentioned, this may be particularly true when it is not inhibited by the medial prefrontal cortex (MPFC) (Thayer, Åhs, Fredrikson, Sollers III, Wager, 2012) and the fusiform and superior temporal areas (Porges, 2004, 2012).

The work of Steven Porges gives us other reflections on the importance of HRV training. HRV training increases vagal tone. The vagus has sensory branches to the tympanic membrane. The vagus also communicates with the facial nerve (cranial nerve VII) and thus may influnce facial expression. Due to the facial nerve connetion to the stapedius in the ear, it may influence hearing. These innervations are important in our treatment of patients diagnosed within the autistic spectrum. Porges has said, "The neural pathways involved in raising the eyelids also tense the stapedius muscle in the middle ear, which functionally dampens the transmission of low frequency background sounds and facilitates the ability to hear the acoustic frequencies associated with human speech" (Porges, 2013). This may provide one explanation as to why our patients who have a diagnosis of autistic spectrum disorder become so much calmer and able to socially communicate when we combine NFB with HRV training. In the future we intend to add specialized *listening* training designed by Stephen Porges to the training regimen.

Sound therapies are not a new type of intervention but they have not usually been combined with HRV training. Earlier in this book we discussed the use of Samonas Sound Therapy. Tomatis Sound Therapy is older and better known. It is very well described by Norman Doidge in his 2015 book, *The Brain's Way of Healing,* published in New York by Viking Press.

How Do You Combine HRV Training with NFB and LNFB?

There are a number of instruments that allow one to simultaneously do biofeedback and neurofeedback. Our experience has been with the focused Technology F1000 in past years and now with the Biograph program and the Infiniti instrument (Thought Technology). More recently we have added the eVox from Evoke Neuroscience. The Biograph program has screens that give feedback for EEG parameters, such as decreasing the amplitude of slow waves (theta) and some high-frequency beta waves, while increasing the amplitude of lower frequency fast waves (SMR or beta). All of these programs can both assess EEG and HRV variables and give the EEG feedback at the same time as they also show BFB variables including HRV training. This is particularly important for our feedback demonstrating to the patient when they have synchrony between their breathing and the changes in their heart rate.

When doing LNFB with the NeuroGuide program we use two computers and two different instruments and programs at the same time. For the LNFB we use the Mitsar with the NeuroGuide LNFB program and for the HRV we use the Biograph Infiniti from Thought Technology. The HRV training is done while the trainer is putting on the 19-channel electrocap as well as during the session. It makes good use of the time it takes to put the cap on and achieve proper impedance values for all the electrodes. Clients appreciate being able to use this time productively and to learn a technique they can utilize during their session and also in everyday life whenever they need to feel calm. They can also practice effortless diaphragmatic breathing at the rate that achieves the highest variability in their heart rate, called their resonant frequency (or sometimes resonance frequency), on a daily basis. With practice it becomes easier to utilize this diaphragmatic breathing when under stress such as when driving or when, for example, there is a deadline and one needs to be highly focused during a stressful period at work or school.

For that matter, getting the cap on can be a bit uncomfortable and constitutes a minor stress, so it helps to give a task to focus on during this procedure, which can take 5 to 10 minutes. What could be better than a task that recruits the vagal response, which is associated with a person feeling calm and safe (Porges, 2007). It also helps prepare the client mentally and physiologically to get the most out of his LORETA Z-score session. As the clients use the set-up time to achieve synchrony between their breathing and heart rate, they have reported that it helps get them get into the right mindset for training. They will often continue using the breathing as they strive to change their brain wave patterns during their LORETA Z-score neurofeedback session (LNFB).

We have noted above that some patients will also do the HRV training while they are doing an LNFB session. Note that the LNFB session is done using one computer and a second computer is used for the HRV training because it is not advisable to run both programs on a single computer due to the large amount of data processing that is required in these programs. The clinician can be checking impedances on the computer screen that will also give the LNFB while the client is watching the monitor for the HRV training. It also allows for monitoring what is happening regarding the client's breathing and heart rate during the LNFB, since the clinician could monitor that screen while the client/patient is watching and listening to the feedback concerning her brain waves. For example, is your patient getting anxious or frustrated? Switch back to focusing on the HRV screen until they regain good synchrony between respiration and pulse parameters.

Research needs to be carried out on the effects of combining HRV training and LNFB in different populations, such as those with TBI, learning difficulties, anxiety disorders, or depression symptoms. This would provide groundbreaking and clinically relevant topics for graduate students because, though it has been applied clinically, the benefits have yet to be researched in a rigorous way.

Is There a Relationship Between EEG Training and HRV Training?

In this section we will give examples that illustrate how training HRV may directly affect the EEG and our work with patients.

ADD Centre HRV and SMR Study

In a pilot study done at the ADD Centre, Reid and Nihon showed that training to improve HRV had a positive effect on the EEG with respect to increasing rhythms associated with a calm and alert mental state. This clinical research was carried out to examine the effect of HRV training on the EEG with 20 athletes and 20 clients (diagnoses of ADHD and/or anxiety and/or depression), 23 male and 17 female, ages 16 to 61. Data collection for this study included a three-minute baseline EEG recording at Cz for each client, followed by three minutes of EEG collected while the clients did HRV training and sustained synchronous rhythmic sinus arrhythmia (RSA). Statistics were selected to measure mean microvolt values for sensorimotor rhythm (SMR, 12-15 Hz) under each condition.

Results showed a statistically significant rise in SMR amplitude during HRV training as compared to a baseline resting state. Statistical analysis found no significant differences between the groups when they were compared for age, sex, and diagnosis. The SMR increase held across both clinical and athlete groups for both males and females, older or younger (Reid & Nihon, 2013). This pilot data confirmed the observation made by trainers at the ADD Centre over the past few years, that it seemed easier to raise SMR when HRV training was combined with neurofeedback training. This pilot study requires replication.

Although not known by Reid and her coworkers at the time of the study, Sterman had made similar observations when working with animals in the late 1960s, noting that SMR was produced on the exhale. This was never published. We asked him to go back through his notes and see if he still had any recordings to show the relationship of SMR production to variations in HR. He was able to find the following illustration.

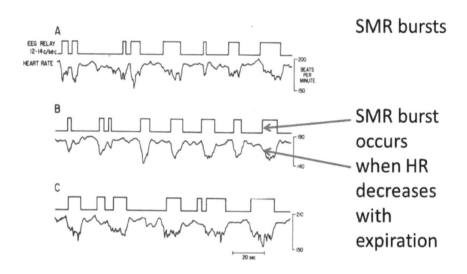

Sterman made a second discovery which we had to rediscover without knowing he had found it 40 years ago. At that time he did not quite comprehend the importance of his own brilliant observations!!! He discovered the correlation of HRV with SMR production.

The above figure, kindness of Barry Sterman, is from Sterman's early research on SMR. Sterman observed in his animal research that, as SMR bursts occurred, heart rate decreased. This is similar to our observations that SMR amplitude appears to increase as the patient exhales and heart rate decreases.

SMR: Origin of the Term SMR and Its Relation to HRV

As noted above, a pilot study done at the ADD Centre showed that training to improve HRV had a positive effect on the EEG with respect to increasing SMR rhythms that were associated with a calm and alert mental state. There was a statistically significant rise in SMR amplitude during HRV training as compared to baseline resting state.

Brief Notes on SMR

In work with HRV to encourage a relaxed calm but alert state, we always include enhancement of SMR. It is therefore appropriate to include a brief overview of this special frequency band.

In the 1960s, Dr. Maurice Barry Sterman observed a rhythm that was similar to Stage 2 sleep spindles in awake, alert, focused, immobile cats who were waiting for a signal indicating that a bar press would be rewarded with food. Dr. Sterman called this 12-18 Hz spindling beta rhythm 'sensorimotor rhythm

(SMR),' because it was recorded over the sensorimotor strip in the surface EEG (Sterman, 1967). Sterman also observed, in some of his early work in which respiration was monitored, that SMR was produced during the exhale. When one does HRV training the exhale is longer than the inhale, so that is one factor that may explain the correlation between increased SMR and improved HRV in that the person is immobile, doing slow diaphragmatic breathing and relaxing with a long exhalation. Sterman noted that this SMR was produced by the ventral posterior lateral nuclei of the thalamus. (Note that NFB practitioners, including Dr. Sterman, usually use bandwidths of 12-15 Hz or 13-15 Hz over the sensorimotor cortex as SMR.)

Later work demonstrated that the amplitude of SMR fluctuated with the individual's degree of vigilance (Steriade & McCarley, 2005, after Rougeul-Buser, Bouyer, Montaron & Buser, 1983). Steriade called the 14 Hz spindles observed over the sensorimotor cortex an **'expectancy rhythm.'** This expectancy rhythm occurs when the awake animal is immobile. He gives the example of a cat watching a hole where a mouse may appear. Steriade noted, as did Rougeul-Buser, that another rhythm occurs along with the SMR when a cat is immobile and watching a mouse that it can see but cannot attack due to a transparent plastic shield. Steriade called this 35-45 Hz spindling a **'focused attention'** rhythm. He noted that this gamma frequency arose from the medial part of the posterior thalamus.

This range around 40 Hz contains the same frequency referred to as 'Sheer rhythm' by neurofeedback practitioners, naming it after Daniel Sheer, who trained the 40 Hz response and found it was associated with improvements in children with learning disabilities (Sheer, 1976). There is also a 40 Hz response when retaining/regaining balance if a person leans over on a balance board as far as they can without tipping (Slobounov et al., 2008). The highest amplitudes of gamma activity have been recorded in experienced meditators (Tibetan monks) when asked to meditate on feeling compassion (Lutz, Greischar, Rawlings, Ricard & Davidson, 2004).

Training can be done for this frequency in the low gamma range, 39-41 Hz, if it is done with extreme care so that EMG artifact from muscle tension is not mistaken for true 40 Hz activity of cortical origin. This has been done for a small number of clients at the ADD Centre. A recent client said that she finds it is associated with feeling extremely clear headed. Another client remarked that it increased her feeling of being one with the people and the world around her and that she better understood the perspective of Tibetan monks.

HRV and P300

From a different perspective, Kaufmann, Vögele, Sütterlin, Lukito & Kübler (2012) demonstrated a link between resting HRV and performance in a Go/NoGo task in 34 healthy subjects. They found that resting HRV accounted for almost 26 percent of the variance of P300 during a brain computer interface (BCI) performance task. These authors note that that success in a P300 oddball based BCI task requires the capability to focus attention and inhibit interference by distracting irrelevant stimuli, and that such inhibitory control is linked to peripheral physiological parameters such as heart rate variability (Kaufmann et al., 2012).

We postulate that future research may be able to show that part of this linkage between HRV and ERP performance is due to direct feedback from the heart. This feedback goes through the connections from the solitary nucleus in the medulla to areas such as the thalamus. It could be that this feedback influences the production of SMR. It may also influence the anterior cingulate gyrus, which is directly involved in executive-attention networks. Increasing SMR by training HRV could become an important factor in optimizing a client's executive functioning. Supporting this postulate is the work of Vernon, who found that an increase in SMR is related to significant improvement in cued recall performance, using a semantic working memory task, and to the accuracy of focused attentional processing using a two-sequence continuous performance task (Vernon et al., 2003).

Heart Evoked Potentials (HEP)

At the ADD Centre, we are not doing heart evoked potentials. At this time of writing there is very little research concerning HEP in neurofeedback centers, though one recent study will be mentioned. Therefore we will not comment on how HEP might be useful in neurofeedback and biofeedback practices in the future. However, we will briefly mention HEP here for the sake of completeness and because excellent work is being done on HEP by Richard Gevirtz and his colleagues at the Alliant University, California School of Professional Psychology.

In 2005, Perez et al. noted that HEP is time-locked with the electrical activity of the heart. These authors state that the existence of a specific response of the central nervous system to the heart action was first reported, simultaneously, by Schandry et al. and by Jones et al. in 1986. They extracted this heart-evoked response from the EEG using the ERP approach by taking the R-wave of the ECG as a synchronous event that 'evokes' a response that, when the random EEG is averaged out, can be observed, as a very low amplitude 2 μV response between 150 to 200 ms post-R wave. They also note that other investigators have suggested that it can be 200-350 or more ms post-R wave.

In a more recent study that is related to our work, MacKinnon et al. (2013) induced patients into emotional states using an autobiographical script of their happiest and saddest memory. The researchers then recorded the amplitude and latency of the HEPs while the patient was diaphragmatic breathing at six breaths per minute.

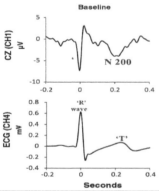

Figure adapted from Mackinnon et al. 2013. The bottom figure shows the electrocardiogram (ECG) from chest leads. The figure above it shows the corresponding HEP recorded at Cz in the EEG. The evoked potential is about 200 msec after the R wave and is negative (N200).

They found that the HEPs during the emotional conditions were most attenuated. This was particularly true of negative emotions. They hypothesized that the signal from heart to brain was filtered in the limbic system. They conclude that their findings support the speculation of many authors who believe that vagal afferents play a role in brain function.

In summary, the R wave may be considered as the stimulus and the evoked response occurs approximately 200 ms to 250 ms after the R wave. The HEP is very low amplitude. The reader will note that the T wave in the EKG occurs within the range of the likely latency of the HEP, and it is said that this can lead to misinterpretation.

TBI Case Example

An athletic 25-year-old female presented after a concussion she had suffered in a motor vehicle accident one year prior to assessment at the ADD Centre. She had postconcussion symptoms that included headaches and memory problems. She was unable to work or even keep house for herself and her fiancé. Intervention involved HRV training combined with LORETA NFB for 40 sessions.

Post HRV training combined with LNFB her SDNN has changed from a pretreatment value of 32 ms to a post-treatment value of 74 ms. Total power had been 454 ms^2 and it had risen after treatment to 2073 ms^2, and her LF power went from 149 ms^2 to 714 ms^2. Her ERPs showed the same morphology, but shorter latencies. For example, her P3a, which reflects attention, dropped from 433 ms to 376 ms. In other words, she had a faster brain speed. The continuous performance test (CPT) also demonstrated improvements: for example, motor speed (reaction time) improved from 544 ms to 439 ms, and her omission errors fell from 11.4 percent to 5.7 percent. Her CPT scores had moved into the normal range.

Her EEG dominant frequency rose from 9.6 to 10 Hz. With most head injuries there is a decrease in EEG power. The power of delta may rise but the power of high alpha and of beta, particularly high-frequency beta, has been found to decrease significantly (Haneef, Levin, Frost & Mizrahi, 2013). Her posttreatment EEG demonstrated a global increase in power. We would postulate that this may indicate increased neural recruitment, suggesting that more neurons in neuronal pools are firing synchronously. This positive change may be the result of the return of long-term potentiation (LTP) and a reduction in long-term depression (LTD). TBI will usually result in a decrease in blood flow and therefore a deprivation of needed oxygen and nutrients to the neurons in affected areas.

In a TBI there is also diffuse axonal injury (DAI). DAI can result in a 'flooding' of glutamate into the synaptic junction. This may result in an abnormal increase in sodium and calcium entry into the post-synaptic ending. This will result in a depression of mitochondrial, Krebs Cycle, production of adenosine triphosphate (ATP), which is the energy source for the neurons and in particular for active transport across neuronal membranes (Thompson, M., Thompson, L., Reid-Chung & Thompson, J., 2013). This series of events may be part of the reason there is a decrease in EEG amplitude post-injury. The increase in total power correlated with her improvement in symptoms. Her overwhelming anxiety is completely gone at this time. She had been a brilliant student before the accident and had competed a masters degree. Prior to her injury she had been placed on propranolol for hypertension. Post-treatment she no longer requires this medication. Another physical change was that before treatment she had required bifocals after her head injury, and shortly before completing 40 LORETA NFB sessions, her ophthalmologist confirmed that she no longer required bifocals. Additionally, her glasses prescription for distance vision was not as strong as it had been before the accident. He remarked that he had never previously seen such a "spontaneous" improvement in vision. Six months after her training she said that her full-time job was not challenging enough and she is considering doing a Ph.D. She was also busy with wedding plans. Clearly her life was back on track.

Side Effects of Over-Breathing

Practitioners should always be careful **when doing HRV training** to have the patient identify and avoid over-breathing (hyperventilation). The normal breathing reflex is triggered by serum partial pressure of carbon dioxide (PP-CO2). **Hypocapnia** (CO2 depletion) with over-breathing can occur rapidly. The **blood pH becomes alkaline**, and the partial pressure of CO2 will drop. Partial pressure (PP) refers to the relative pressure of a gas contributing to the total atmospheric pressure of 760 millimeters of mercury (mmHg). The PP of oxygen at rest is 19 percent, or 144 mm Hg, and of exhaled carbon dioxide (ETCO2) - where ET means 'end tide' - is **5 percent** or 38 mmHg (35-45 mmHg). The PP of exhaled CO2 increases with exercise. This PP of CO2 can be measured using a capnometer in the office while the patient is doing HRV training. This is considered by some to be a gold standard for practice, but most practitioners do not have this instrument, just as most centers teaching yoga, meditation or the martial arts teach diaphragmatic breathing but do not use a capnometer. This instrument is expensive. In addition, some adult patients and many children find anything on their face/nose distracting and even anxiety producing.

Most practitioners find that they do not run into complications if they are careful to appropriately instruct and observe the patient. Effortless diaphragmatic breathing done with a complete exhale feels comfortable. In contrast to comfortable breathing, early signs of over-breathing include one or more of the following: light-headedness, tingling skin, tightness of the chest, sweaty hands, breathlessness, restless mind, memory loss, lights seem brighter and sounds louder, startling more easily, muscle twitching. These symptoms may, in themselves, arouse anxiety and fear of losing control, which exacerbates the condition that one is treating. (Of course, it should almost go without saying: don't psychoanalytically interpret these signs, they are caused by physiological changes and are real!) Other symptoms are: apprehension, anxiety, fear and panic, emotional lability (disinhibition), anger, and even the precipitation of depression. Lack of careful instruction to the patient concerning breathing could conceivably result in an increase in the symptoms the practitioner was attempting to treat.

Why is Over-breathing so Deleterious to Wellbeing?

A decrease in PCO2 will cause a decrease in the ability of hemoglobin to release oxygen. This is called the **Bohr effect.** In addition, these biochemical changes will result in vasoconstriction of the blood vessels and this is very important in the brain and heart. The result of these processes is a marked decrease in oxygen available for metabolism. The increase in blood alkalinity (from pH 7.4 to 7.5 (normal is 7.38-7.4) can decrease blood flow by 50 percent in minutes. Alkalosis results in hyper-excitability and a compromise of the blood buffering system. When this occurs, the ability to regulate acidosis is compromised. Blood pH is an important consideration in training athletes. With over-breathing there is both cerebral hypoxia and a cerebral glucose deficit.

Longer-term symptoms that can alert the practitioner and the client to over-breathing include: impairment of perceptual motor skills, fatigue, and increased theta activity in the EEG. Rarely, a client might experience seizure or migraine. These more drastic outcomes may be due to changes in membrane permeability. Calcium, potassium, and magnesium move across membranes that are normally relatively impermeable. These minerals move into the cells, resulting in irritability of nerves and muscles. Due to effects on the heart the patient may experience angina and even arrhythmias. Unfortunately, the client may be initially unaware of the changes that are taking place if these changes develop slowly over a period of time. Fortunately, these side effects are a rare complication in good clinical practice. Diaphragmatic breathing can have a very good outcome with the clinical result of achieving RSA synchrony being a sense of a release in tension and an increase in physical and mental relaxation.

To avoid these undesirable effects of over-breathing, the practitioner can teach the client about the physiology and psychology of respiration so that they can consciously identify the sensations of over-breathing. We find that over-breathing is uncommon if the patient is comfortable and calm and does rhythmic, restful, effortless, diaphragmatic - *but not dramatic* - breathing at their heart's resonant frequency (approximately 6 BrPM), and if they focus initially on just **effortlessly** taking small breaths and only taking their next breath when they need to take a breath.

It may help your client to focus on a long, relaxing exhale while they are emptying their mind of all circular, repetitive worries and thoughts about all the things they have to do. **Emptying the mind** is essential. Relaxing their muscles as they exhale is also healthy. It can be helpful to have the client put her hands on her thighs (not on the arm of the chair, which can increase muscle tension in the shoulders) and use EMG feedback from the shoulder trapezius muscle to increase relaxing and to counter the client using her shoulders and chest to inhale. Breathing in this relaxed manner makes it likely that your patient will not have symptoms of over-breathing.

Clinical Conclusion for Practitioners

For biofeedback practitioners, the significance of the above observations linking healthy brain function with healthy heart function, and TBI and affect difficulties with disrupted cardiac functions, is that it suggests that heart rate variability training has effects on many of the same basic neural structures that are also directly influenced by EEG biofeedback training. It implies that one can achieve a synergistic effect by combining the two modalities when treating a range of clients, from those with head injuries or problems like depression through to athletes who wish to optimize their performance.

SECTION XIX
Transcranial Direct Current Stimulation (tDCS)

See also Other Cranial Electrical Stimulation (CES) Methods
Also called Cranial Electrotherapy Stimulation

Overview

Transcranial direct current stimulation (tDCS) refers to the application of a very small direct electrical current (DC) between an anode (+) and a cathode (-). When this current is applied to the scalp, it is typically below two milliamps (mA), which means that it is so gentle that it is not felt by the client. The electrodes are commonly connected to a conductive pad that distributes the current over the surface of a saline-impregnated sponge or similar absorbent and conductive material. The size of each pad may be the same or in some cases the cathode may be larger (example: 25 cm^2 and 36 cm^2 respectively). The larger cathode will have less current density beneath it relative to the anode. The anode and cathode are commonly placed either at two sites on the scalp or one site on the scalp and another site that is not over the cortex, such as the mastoid or shoulder. Alternatively, multiple specially designed smaller electrodes with conductive material and various combinations of anode and cathode sensors may be used with some instruments. The cost of instruments varies widely with FDA-approved versions for research and hospital use that can cost upwards of $20,000. There are also basic instruments available for clinical use that cost only a few hundred dollars. The basic equipment is a six-volt battery with connections to two pads/sponges that become anode and cathode when the current flows.

Dosage

Dosage is generally considered to be related to the combination of the number of electrodes, size of electrodes, placement, milliamps, and duration of current. It is also understood that a large percentage of the variance of the effect of stimulation is due to the state of the brain at the time of stimulation. Perhaps this is similar to using metacognitive strategies to put the patient's brain into a particular mental state to activate a particular network when carrying out neurofeedback. Medications can also have an inhibitory or excitatory effect. Carbamazepine (calcium channel blocker), for example, appears to block the beneficial effects of tDCS, while citrulline (a naturally occurring amino acid that can reduce blood pressure) seems to have a positive synergistic effect in cases of depression. transcranial direct current stimulation (tDCS) may be enhanced by glutamate or protein synthesis. Cathodal stimulation appears to reduce glutamate levels, and anodal stimulation reduces GABA (the main inhibitory neurotransmitter). This can be thought of in terms of a balance between excitation and inhibition that tDCS can artificially modulate. While this may appear to be an oversimplification, it is still reasonably accurate. However, the DC effect on neurotransmitters requires further research. There also remains much to learn about dose response curves and mechanisms of action. In our Centre we use a short "dose" of tDCS (5-8 minutes) with current setting between 0.5 mA and 2 mA. We view it as 'priming' the cortex before the NFB training begins.

Anode and Cathode

The **anode (+ve)** stimulation causes neurons proximal to it to depolarize. This results in excitation of the underlying superficial cortical neurons (but the opposite for neurons deep in the sulci). The cathode stimulation (-ve) has the effect of hyperpolarizing neurons proximal to it, resulting in cortical inhibition. (When hyperpolarized, the cell membrane becomes more negative and the extracellular space more positive.) The tDCS is not a neurostimulation technique like transcranial magnetic stimulation (rTMS) that is used in some hospital settings. Rather, it is more of a "brain polarization" technique, the term used in the early tDCS literature. This constant gradient of applied voltage changes local charge of ions, changes pH, effects glial cell changes, and therefore **facilitates rather than** directly causes an action potential. Surface EEG recording may show that the anodal stimulation **increases beta and reduces delta** while the cathode stimulation increases delta. It has also been shown that tDCS has an effect on event-related potentials.

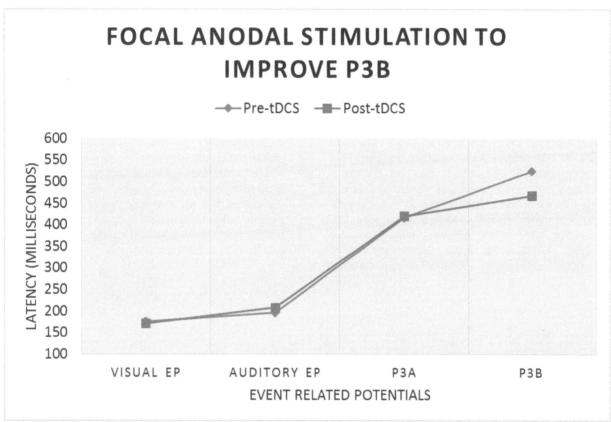

Figure to illustrate the effect of a high density single anode placement at Pz with single cathode placement at Fp2, five minutes of 0.5 mA followed by a 20-minute delay. This stimulation can result in an 11 percent P3b latency improvement with no change to visual evoked potential (VEP), auditory evoked potential (AEP), or P3a (used by permission of David Hagedorn). This carefully done single subject 'experiment' requires research and replication, but it is an interesting observation and the change is real, being about 50 ms faster.

Electrodes and Effects

The sponge or pad type of electrode used in many centers for tDCS is wet. It can be soaked in water or saline. Some authors note that water, unfortunately, may give a minor skin lesion side effect, and saline has not been shown, at low currents, to have this side effect. These authors can be quite firm in stating that saline must always be used. The skin reaction is not a thermal burn but is a pH type electrolysis reaction that has been called "burn" in the literature because the skin is red where the electrode pad had been. Research findings suggest that even minor skin reactions are unusual, while the sensation of itching is much more common. While not a side effect, it is worth mentioning that a postulated contraindication is the presence of a tumor or suspected tumor where increased cancer cell replication would be a possibility.

In clinical practice, the practitioner places an anode or a cathode over the site that is of interest and applies a very small current of about 0.5 to 2.0 milliamps (mA). This is about 400 times less current than is used in electroconvulsive therapy (ECT). Only a very small fraction (said to be less than 25 percent) of this current crosses the bone and tissue separating the cortex from the scalp surface. *Wagner et al. (2007) talk about the percentage of current that gets to the white matter being in some cases only 1 percent, but we typically think of it being about 25 percent.* In the literature one will find a range of pad sizes, usually between 25 and 36 cm^2, and in some instances smaller. The current used at the ADD Centre is generally about 1 mA and always <2 mA. The cathode pad used is about 36 cm^2, and the anode pad is 16 cm^2 *(1 inch = 2.54 cm)*.

The current used in tDCS, unlike electroconvulsive therapy (ECT) or direct brain stimulation (DBS), does not induce neuronal action potentials. However, when we apparently activate neurons with anode electrical stimulation they tend to modulate toward a more normal pattern. In fact, the descriptive term

transcranial direct current modulation may be a more accurate and better name for this process than tDCS. Repetitive transcranial magnetic stimulation (rTMS) is used in hospitals and private practice, mostly for atypical depression. However, it is also showing positive results in other conditions such as Parkinson's disease. In addition, it has been hypothesized (Hagedorn, personal communication) that the use of low frequency (10 Hz) rTMS over locations with excess beta spindling fatigues beta spindling (see image), which may be due to a modulatory process not yet fully understood. In our work with tDCS, the anode stimulation is as close an approximation as we can get to transcranial magnetic stimulation (TMS) and cathodal impact appears minimal. However, there are some reports of significant therapeutic benefit using cathodal placement. The neuromodulatory effect of tDCS only refers to an effect on the resting potential of the neuronal membrane. However, the clinical effects of tDCS appear to be very similar to those reported for TMS and it is much more gentle.

Summary: As noted above, the anode is positive and the cathode is negative. Anode stimulation will result in excitation of the underlying superficial cortical neurons (but the opposite for neurons deep in the sulci). Cathode stimulation will result in hyperpolarization and cortical inhibition. The anode stimulation will increase beta and reduce delta. The cathode stimulation may increase delta. We are finding tDCS to be a helpful adjunct when used just prior to regular NFB training.

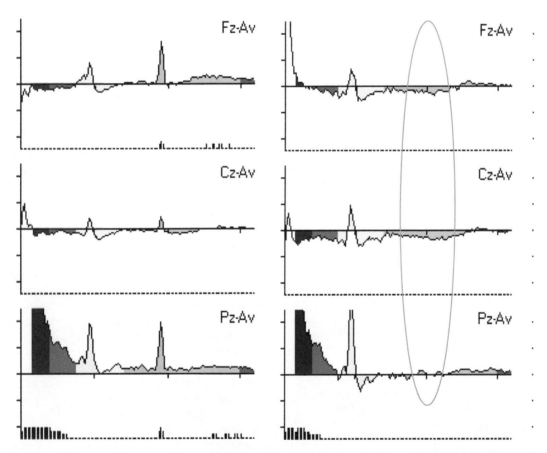

Figure to illustrate that vertex beta spindling evident prior to a single session of 10 Hz rTMS (left column of diagrams) is not seen after rTMS (right column of diagrams). There is an apparent suppression of the generator's ability to easily kindle the cortical area following a single session of rTMS (used by permission of David Hagedorn).

Estimated Effects on Brain Activity

Brain activity is estimated to increase by 20 to 40 percent when the anode current density (concentration of amperage under the electrode) exceeds 40 µa/cm^2 (260 µa/inch^2). On the other hand, cathode 'stimulation' reduces brain activity under the electrode site by 10 to 30 percent at the aforementioned current density (Siever, 2013).

Measurements of Brain Activity

In studies, estimates of brain activity can be measured using scans. A very rough estimate is that brain activity can be measured by a PET scan (Positron Emission Tomography). PET provides a visual display of brain activity by detecting where a radioactive form of glucose goes while the brain performs a given task. Brain activity can also be estimated by Functional Magnetic Resonance Imaging (fMRI) which uses blood-oxygen-level dependent (BOLD) contrast and gives a response that varies with oxygen uptake. SPECT scans measure perfusion (blood flow) in the brain.

Current Control

Battery current can vary, and skin resistance may affect the current path from anode to cathode. Most of the current will pass through the skin rather than going through the cortex. In clinical work, control of current is essential. The practitioner must be sure that the instrument that is used has current controls with fade capabilities, true current control, and contact quality. The instrument used must be able to automatically adjust the current up or down in order to keep it at a constant level (for example, <1 mA), and automatically turn off if it goes above a set level. The instrument used must also be set to automatically turn off if the surface connection is too poor to allow sufficient current flow.

At the ADD Centre/Biofeedback Institute, the interventions in the past have always been educational/learning in nature. Introduction of a technique that involves neuromodulation, no matter how minimal, has been done only with careful prior discussion with the patient, and for children, with their parents. While the tDCS safety literature is very positive and consistent (Poreisz et al., 2007; Nitsche et al., 2008), it remains, at the time of writing, an off-label intervention or experimental in most countries.

Neuronal Membrane Changes

The current under the **anode** electrode induces a lack of positive ions (resulting in negativity) at the basal part of the neuronal membrane. (The neuronal membrane serves as a barrier to enclose the cytoplasm inside the neuron, and to exclude certain substances that float in the fluid that bathes the neuron. It is responsible for establishing an electrical potential inside the neuron.) The anode induces **depolarization** of this part of the membrane. The excitability of the neurons increases and the frequency of the background activity increases (increased beta). The net effect is **anodal activation of neurons**.

Conversely, the current under the **cathode** electrode induces an excess of positive ions near the external part of the basal membrane. This induces **hyperpolarization** (more negative than the normal resting -70 mV potential inside the neurons) of this part of the membrane, making the pyramidal cells less likely to fire. The excitability of the neurons will decrease and the frequency of the background activity will decrease (more slow waves and less beta). The net effect is **cathodal suppression of neurons**. Hyperpolarization inactivates Ca^{++} and Na$^+$ channels. On the other hand, anodal stimulation with depolarization *activates these channels.*

In the section on The Origin of the Electroencephalogram, the authors noted that the changes in the cell membrane affecting the permeability of K+ and Na+ are known as the Hodgkin Cycle. Here only a brief review of the neuronal membrane is given. In the case of the resting membrane potential across the membrane, potassium (and sodium) gradients are established by the Na$^+$/K$^+$-ATPase (sodium-potassium pump), which transports two potassium ions inside and three sodium ions outside at the cost of one ATP molecule. This is called an 'active transport' system. Active transport requires ATP as energy. Anions (An$^-$) in the cell will account for most of the negative charge. These anions are chiefly contributed by proteins in the cell, and they do not cross the cell membrane. Potassium is at a higher concentration in the cell than outside of the cell. The concentration gradient for potassium with a high concentration of potassium ions in the cell should result in the exit of potassium across the cell membrane into the extracellular space. However, the high negative charge (-70 mV) in the cell, due in a large part to the protein anions, keeps the positive potassium in the cell. When these forces (which are applied to K$^+$) are the same strength and oriented in opposite directions, the system is said to be in equilibrium. Put another way, the tendency of potassium to leave the cell by running down its concentration gradient is now matched by the tendency of the membrane voltage to pull potassium ions back into the cell. K^+ continues to move across the membrane, but the rates at which it enters and leaves the cell are the same, thus, there is no net potassium current. Because the K^+ is at equilibrium, membrane potential is stable, or 'resting,' which is at -70 mV inside the cell (neuron). The resting potential of animal cells is determined by predominant high permeability to potassium and adjusted to the required value by modulating sodium and chloride permeabilities and gradients.

Although the cell membrane is permeable to potassium, it is for the most part not permeable to sodium in the resting state. In a healthy animal cell Na^+ permeability is about five percent of the potassium ions' permeability or even less. The resting membrane potential is dominated by the ionic species in the system that has the greatest conductance across the membrane. For most cells this is potassium. The outward movement of positively charged potassium ions is due to random molecular motion (diffusion), and continues until enough excess negative charge accumulates inside the cell to form a membrane potential that can balance the difference in concentration of potassium between inside and outside the cell. 'Balance' means that the electrical force (potential) that results from the build-up of ionic charge, and which impedes outward diffusion, increases until it is equal in magnitude but opposite in direction to the tendency for outward diffusive movement of potassium. This balance point is an equilibrium potential as the net transmembrane flux (or current) of K^+ is zero. For most animal cells, potassium ions (K^+) are the most important for controlling the resting potential.

When the postsynaptic membrane receptor sites are opened by the axon, releasing neurotransmitters into the synaptic space, then Na^+, which is in high concentration outside the cell, will flow into the cell due to the concentration gradient for Na^+. With Na^+ flowing across the membrane into the intracellular space, the cell is said to 'depolarize,' and the -70 mV resting potential drops. Active transport, requiring ATP energy produced by the mitochondria in the cell, is required to move sodium out of the cell against its concentration gradient. This will restore the resting potential, which will return to -70 mV. This need for ATP energy will become important in our discussion of diffuse axonal injury (DAI). DAI results in a decrease in mitochondrial function after a TBI.

Time Length of Effects
Effects of tDCS are not long-lasting, but the time frame for positive change appears to lengthen as more sessions are experienced. At our Centres, when tDCS is used, we will do one short 10- to 20-minute session followed by 35 minutes of NFB every day for the first week (when this is possible for the patient), and then follow this with two sessions a week. It is purely a hypothesis at our Centre, but we postulate that following the tDCS with NFB may encourage the long-term changes that are known to occur with NFB. Our hope is that the tDCS may accelerate the process.

Active Electrodes and Effects on Neurons
It is hypothesized that tDCS has an impact at nodes along the networks in the brain, and thus the effects are on networks rather than just the site of application of the electrode.

It is important to remember that both electrodes are active; there is no true "reference" (inactive) electrode. Observed effects may be due to both electrode placements when both electrodes have been placed on the scalp.

Alterations in regional cortical blood flow (rCBF) are also said to occur. Usually the effect of anode stimulation is to increase rCBF in the area near the anode (Zheng et al., 2011; Lang et al., 2005). Indeed, Zheng and his colleagues demonstrated that the strength of anode current was proportional to this increase in rCBF. The effects on the neurons appear to be due to modulation of sodium and calcium channels. The lasting aftereffects are most likely due to N-methyl-D-aspartate (NMDA) receptor-dependent neuroplasticity (Nitsche, 2003). This is related to what is called long-term potentiation (LTP). The effects can be blocked by giving an acetylcholine agonist (rivastigmine) (Kuo, 2007). *(Note that an **agonist** is a chemical that binds to a receptor of a cell and triggers a response by that cell. Agonists often mimic the action of a naturally occurring substance.)* The reader should go to specific articles on biochemical substance effects for further information on this subject.

Long Term Potentiation (LTP)
Definition: Long-term potentiation is a process wherein a postsynaptic cell is enhanced so that it can depolarize more in response to a neurotransmitter. In essence, it becomes more sensitive.

Importance: LTP is important to NFB practitioners because it is possible that this process may account, in part, for the long-term effects of NFB. LTP will only be briefly described in this text.

Neuroanatomical/Biochemical Changes

A synapse is the communication link between two neurons, and the synaptic cleft is the small gap between the axon of one neuron to the dendrite of another neuron. In response to an electrical message arriving at the presynaptic membrane, a neurotransmitter is released. The neurotransmitter travels through the synaptic space to the postsynaptic receptor membrane. Long-term potentiation involves a biochemical change that takes place on the postsynaptic receptor membrane of the dendrite. What follows here is just one of a number of possible mechanisms.

Receptor Sites

In the figure below, note that there are two major types of receptors on the postsynaptic receptor membrane. These are the AMPA (alpha-amino-3-hydroxy-5-methyl-4-isoxazolepropionic acid) and the NMDA (N-methyl-D-aspartate) receptors. The first, the AMPA receptors, are glutamate receptors. The name 'AMPA' is derived from this site's ability to be activated by the artificial glutamate analog (agonist) AMPA. The APMA receptor is also called a non-N-methyl-D-aspartate receptor. The second type of receptor site is called an NMDA receptor site. NMDA (N-methyl-D-aspartate) is the name of a selective agonist that binds to NMDA receptors but not to other glutamate receptors.

Glutamate Activation

The transmitter glutamate initially activates only the AMPA or non-N-methyl-D-aspartate receptors, as is shown in the diagram. This is because the NMDA receptors are blocked by extracellular magnesium (Mg^{2+}) ions. A unique property of the NMDA receptor is its **voltage-dependent activation**. This means that when glutamate causes an influx of sodium into the AMPA postsynaptic terminal, an electrical voltage depolarization change takes place that allows the Mg^{2+} block of the NMDA receptor site to be released, as is shown in the diagram. This newly opened NDMA receptor site is now able to transfer Ca^{2+} through the postsynaptic membrane into the dendrite. Calcium influx through NMDA is thought to be critical in synaptic plasticity, which is a cellular mechanism, and thus in enhancing LTP and, as a consequence, cognitive processes such as learning and memory. In summary, it results in the postsynaptic cell now being able to depolarize more in response to a neurotransmitter.

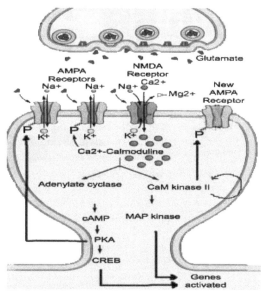

With thanks to: Wikipedia Images: Bruno Dubuc, Website of Canadian Institutes of Health Research; Institute of Neurosciences; Mental Health and Addiction(NMHA)

☺Copyleft; The content of the site The Brain from Top to Bottom is under copyleft. also from Nature Reviews Neuroscience, Nature Publishing, 2005.

When Na^+ enters the postsynaptic membrane on the next neuron's dendrite, the depolarization displaces magnesium (Mg^{+2}) that is blocking the N-methyl-D-aspartate receptor sites. When Mg^2 blockage is removed from these sites, they can be activated by glutamate, allowing an influx of calcium ion (Ca^{+2}).

Result of Calcium Influx

The influx of Ca^{+2} results in a number of different processes. It results in the activation of other 'messenger' pathways. It is said to cause the release by the postsynaptic cell of a substance that is called a 'paracrine.' A paracrine is a chemical that is released by a cell and then acts to alter cells in its immediate vicinity. In this case, it may act on the presynaptic ending and enhance the release of the neurotransmitter glutamate. In addition, the influx of calcium appears to change the postsynaptic membrane, making it more sensitive to glutamate. It is also thought that the postsynaptic cell may develop more glutamate receptors (Silverthorn, 1998). In the diagram, a new AMPA receptor is shown. In addition to these processes, these calcium ions are extremely important intracellular messengers that activate many enzymes by altering their conformation. An important product of these biochemical changes is the production and activation of a protein kinase called PKA.

One of these enzymes is calmodulin, which becomes active when four calcium ions bind to it. It then becomes Ca^{2+}/calmodulin, the main second messenger for LTP. Ca^{2+}/calmodulin then, in turn, activates other enzymes that play key roles in this process, such as adenylate cyclase and Ca^{2+}/calmodulin-dependent protein kinase II (CaM kinase II). These enzymes may then modify the spatial conformation of other molecules, usually by adding a phosphate ion to them. This common catalytic process is called phosphorylation. Thus, the activated adenylate cyclase manufactures cyclic adenosine monophosphate (cAMP), which in turn catalyzes the activity of another protein, kinase A (or PKA).

PKA phosphorylates the AMPA receptors, allowing them to **remain open longer** *after glutamate binds to them. As a result, the postsynaptic neuron becomes further depolarized, thus contributing to LTP. In summary, there is* **a typical cascade of biochemical reactions** *that can have many different effects, but* **most of these reactions appear to contribute to LTP.** *(Reference web: aboutmind.com Memory, Brain, Neurons: Random Facts)*

Summary
In conclusion, glutamate is essential in learning and memory and in an important process called long-term potentiation (LTP). **Long-term potentiation is the process whereby a postsynaptic cell changes and becomes enhanced in response to episodes of intense activity across the synapses.** Glutamate and LTP are thought to be crucial to memory storage. Further research is needed to show the relative importance in long-term potentiation of changes such as a change in the postsynaptic receptor sites, an increase in neurotransmitter receptors, and/or increased synaptic connections. However, **whatever the mechanism(s), the postsynaptic cell can depolarize more in response to a neurotransmitter** when there has been repeated stimulation. The importance of LTP for our work using neurofeedback, and perhaps also for tDCS, is that it could be an important theoretical framework for understanding how only a few sessions of neurofeedback may result in sustained changes in the CNS.

Side Effects of tDCS
Poreisz (2007) studied side effects and only found mild itching, tingling, moderate fatigue, nausea, and seldom=occurring headache. At our Centre, despite using tDCS on a wide range of clients, we have not yet seen any side effects. Do remember, however, that the **electrode sponge must remain thoroughly damp.** A dry area could cause a burn. In addition, the person applying this intervention must be knowledgeable concerning the functions of the Brodmann areas under the electrodes and the likely effect of stimulation and/or inhibition of these sites.

One example of a theoretically possible error would be applying a cathode to the right central prefrontal area with the intention of decreasing pain and having it result in the patient demonstrating anxiety and/or impulsivity. (See our discussion of BAs and medial prefrontal tonic inhibition of the central nucleus of the amygdala for further explanation.)

Comment from Dr. Hagedorn: Side effects can include burn and itching. Both are known in the tDCS community and have been observed by such authors as Felipe Fregni, Marom Bikson, Alex Da Silva, and Nitsche. However, they have also written on the safety of repeated transcranial direct current stimulation. For our purposes the conclusion should be that one must be particularly careful using tDCS if the patient has any skin injury or skin condition and seek medical advice before applying this technology in such cases.

Clinical Effects

Anodal Stimulation
Most research, until the last few years, had been done to produce motor improvements in patients who had experienced a stroke. However, there are also reports of improvements in a number of functions in normal volunteers and in subjects who have learning difficulties and psychiatric disorders. The majority of these studies have used anodal stimulation over a functionally appropriate area of the cortex with the cathode over the right orbit.

Improvements in Cognitive Functions
In normal volunteers, cognitive functions such as **verbal fluency** and learning improved with anodal stimulation over the dominant (left) temporal-parietal junction (Wernicke's area) and cathode over the right orbit (Flöel et al., 2008). **Working memory** improves with anodal stimulation of the left dorsolateral prefrontal cortex (DLPFC) and with the cathode over the right orbit (Fregni et al., 2005; Ohn et al., 2008) while both **phonemic and semantic verbal fluency** improves with the same setting. Verbal fluency did not improve when the anode was placed over the right DLPFC (Cattaneo et al., 2011). **Lexical retrieval** with improvement of naming performance and verbal reaction times occurs with anodal stimulation of the left DLPFC (Fertonani et al., 2010).

Improvements in Disorders

There have also been shown to be improvements in **Parkinson's** disease (Boggio et al., 2006) and in mood in **depressed** patients (Brunoni et al., 2013; Loo et al., 2012; Fregni et al., 2006; Arul-Anandam & Loo, 2009). Some patients with **tinnitus** may improve with anodal stimulation of the left temporal-parietal cortex (Garin et al., 2011). Children with **autism** may show improvement in language syntax acquisition with anodal stimulation of the left DLPFC (Schneider & Hopp, 2011). Anode stimulation of the right motor cortex with cathode stimulation of the left motor cortex will improve functions of the left hand, whereas simple anode stimulation of the right motor cortex alone will not (Vines, 2008).

Other findings have been reported in a range of disorders, and for specific criteria the reader is encouraged to refer to the original papers. Relief in **fibromyalgia** has been shown with anodal stimulation (Fregni et al., 2006). Craving for either food or alcohol has been decreased (Boggio et al., 2008). **Word recognition** in patients with Alzheimer's has been improved (Ferrucci et al., 2008). Symptoms of **obsessive compulsive** disorder can be decreased (Greenberg, 2006) as can symptoms of **Tourette** syndrome (Neuner et al., 2009). Changes in motor functions have also been found; as noted above, there have been reports of improved motor function after strokes.

Depressive Disorders

For our clinical work the recent studies with depressed patients are particularly interesting. In 2009 Nitsche and his colleagues published a review. Loo's group (2012) carried out a double-blind, sham-controlled trial of tDCS with 64 participants who were diagnosed with depression. They note that tDCS shifts the resting membrane potential, with anodal stimulation depolarizing the soma of pyramidal cells, whereas cathodal stimulation results in hyperpolarization. These authors state that the effects of tDCS on neuronal excitability have now been demonstrated in numerous neuroimaging and physiological studies. The anode in their study was placed on the left prefrontal cortex (2 mA, 15 sessions over three weeks). Mood and neuropsychological effects were assessed, demonstrating that there was significantly greater improvement in mood after active than after sham treatment. **Attention and working memory** improved after a single session of active but not sham tDCS. They found no decline in neuropsychological functioning after three to six weeks of active stimulation.

However, in this study, one participant with bipolar disorder became hypomanic after active tDCS. These authors concluded that there is antidepressant efficacy and safety for tDCS. They advise, however, that practitioners **watch carefully for mood switching** with bipolar disorder patients. In another study (Brunoni et al., 2013), tDCS was compared with an SSRI medication in a double-blind, controlled trial with 120 subjects. This study did anodal stimulation over the left dorsolateral prefrontal cortex (DLPFC) and cathodal stimulation over the right DLPFC. This placement was based on the hypothesis that tDCS antidepressant effects would occur due to activation (anode) in the left dorsolateral prefrontal cortex (DLPFC) - a region that may be hypoactive in depression - and it would restore prefrontal activity by increasing activity in this region. They found that tDCS alone, vs. sertraline alone, presented comparable efficacies, and that tDCS alone (but not sertraline alone) was superior to placebo/sham tDCS. This study also **reported mania or hypomania** in seven subjects, five of whom were in the combined treatment group. They conclude that in major depressive disorder (MDD), the combination of tDCS and sertraline increases the efficacy of each treatment. The efficacy and safety of tDCS and sertraline did not differ.

Tinnitus

Transcranial direct current stimulation (tDCS) has been used for tinnitus. The prevalence of this disorder is about 10 percent of the population. There are two general types. The first, objective tinnitus, refers to sound that can be heard with a stethoscope and may reflect either a vascular or a muscular origin. The second type is subjective tinnitus. This has been called a phantom auditory sensation. There may be an associated lesion, as in Meniere's syndrome, in which balance can also be affected. It may occur in conjunction with an age-related hearing loss termed **presbycusis.** This is most often a high-frequency loss. It may be associated with deafness, or with ototoxicity, which is damage to the ear, specifically the cochlea or auditory nerve and sometimes the vestibular system, by a toxin.

The key factor is a **distorted sensory input**. In some cases it is associated with depression, anxiety, and/or insomnia. There is often a **reduction in alpha and an increase in slow waves and gamma** in the auditory cortex. The cause is thought to be a breakdown in thalamic auditory connections including reduced auditory input due to hearing loss. The result is slow thalamic-relay oscillations (delta-theta). Gamma may predominate in the left temporal area when the patient has had tinnitus for less than four years, but if the history covers more than four years, gamma can be more widely distributed. (Note that rTMS may be used to interfere with neural activity in an area to see if it is relevant in tinnitus.) Treatments may include NFB (Dohrmann, Weisz, Schlee, Hartmann & Elbert, 2007) over the left temporal area in some patients. tDCS may also demonstrate a good response with anodal stimulation over the temporal area (Fregni, 2006). Dirk de Ridder has shared the following figure from his lectures. This figure summarizes results with patients at his clinic who suffered from tinnitus.

tDCS for tinnitus,
adapted from Dirk de Ridder

tDCS
- 448 patients (male: 160, female: 288) with chronic tinnitus (> 1 year)
- **29.9% responders** (n=134) 70.1 % non-responders (n=314)
- For responders VAS intensity ($t = 8.62, p < .001$) and VAS distress improve ($t = 8.40, p < .001$)
- **EEG (6 responders** vs 6 non-responders) Findings in these:
  - **Increase in gamma at anode**
  - **Increase in Alpha** (8-12.5 Hz) in BA 21/22

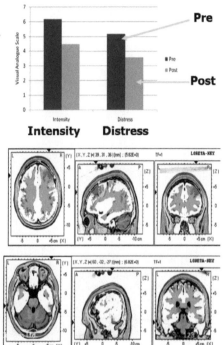

The figure shows results obtained by Dr. De Ridder, a neurosurgeon working in Belgium at that time. On the other hand, Fregni et al. have reported a higher proportion of positive responders to tDCS (transcranial direct current stimulation), but not on as large a number of cases (Fregni, 2006). Thanks are extended to Dr. De Ridder. (VAS refers to a Visual Analog Scale)

Cathodal Stimulation
Most of the successful work with tDCS has been done with anodal stimulation. However, cathodal stimulation has also been demonstrated to be effective in selected conditions. Cathodal stimulation has been shown to decrease pain in patients with fibromyalgia (Angelakis, 2011) and pain perception in healthy volunteers (Antal et al., 2008).

Conclusion
In the foregoing brief overview of tDCS we have noted that a number of clinical conditions have been improved using this technique. In addition, there are a number of controlled studies. However, this is not work that should be done blindly without assessing your client. It is up to the clinician to decide on where the anode and cathode should be placed to help a particular patient. The clinician will use their knowledge of Brodmann area functions and results of a 19-channel QEEG with Z-score database comparisons to help them to decide on appropriate placements. We also suggest that you consider using tDCS to prime an area before doing regular neurofeedback. The tDCS will depolarize, with anodal stimulation, or hyperpolarize, with cathodal stimulation, the pyramidal cells, making them more, or less, likely to fire.

Reminder: Despite our emphasis in this book on NFB guided by the EEG that is a reflection of pyramidal cell activity in the cortex, the practitioner of NFB and BFB should also be aware of slow cortical potentials that are generated by global changes in glial cells. We hypothesize that this change associated with glial activity may more closely correlate with the effects of tDCS.

SECTION XX
Cranial Electrical Stimulation (CES)

With Major Input from **Dave Siever**

Introduction

To augment our own experience with CES equipment, the authors asked for input from Dave Siever (Mind Alive Ltd.), and he provided most of the content of this section, including the figures and examples.

Cranial electrotherapy stimulation, also called cranio-electro Stimulation or simply CES, is primarily a brain-calming technique that passes small pulses of alternating electrical current (AC) through the brain. Note that direct current (DC) stimulation is a different technique and is discussed in a separate section in this book (Section XIX). CES is a subset of TENS or transcutaneous electric nerve stimulation. TENS covers any form of electrical stimulation that is delivered through the skin. Typically, TENS consists of a fairly strong pulse that is used to produce muscle contraction. CES, on the other hand, involves a much weaker pulse that is usually applied bilaterally across the cranium via the placement of two small electrodes, one on each side of the head. Most studies have used electrode placements on the mastoid process, the earlobes, or over the temporal lobes. Over the years, a few devices have also used electrode placements involving one electrode placed on the scalp and the other on the opposite shoulder or arm. CES employs alternating currents (AC) mainly from 0.5-15 Hz, but some instruments use up to 100 Hz. It is hypothesized that the effects of CES are mediated through a direct action on the brain involving the brain stem, the limbic system, the reticular activating system and/or the hypothalamus, although the means of action is not yet entirely understood (Gibson & O'Hair, 1987; Brotman, 1989).

Background Information

The first recorded use of brain electrostimulation was in 129 AD, when the philosopher and physician Galen used electric shocks to treat a variety of ailments, including melancholia, depression, and epilepsy. Scribonius Largus, a Roman physician, was cited as using torpedo fish (a type of electric eel that can deliver a strong electric shock) to treat various ailments including headache and gout (Kneeland & Warren, 1994). Modern research was begun by Leduc and Rouxeau in France in 1902. CES was initially studied for insomnia and called electrosleep therapy (Gilula & Kirsch, 2005). In 1949, researchers in the Soviet Union expanded the use of CES procedures to include the treatment of anxiety as well as sleeping disorders.

Currently more than 40 CES devices have been marketed in the USA, Canada, and Europe. About 200 studies have been published covering a wide variety of clinical applications including improved drug abstinence and cognitive functioning in recovering alcoholics and drug users. On the regulatory side, the Food and Drug Administration (FDA) recognizes CES as a treatment for anxiety, depression, and sleep (serotonin effect). Somewhat surprisingly, it has not been recognized by the FDA for relieving pain (endorphin effect) even though the endorphin effect and efficacy for pain treatment have been well documented.

Physiological Basis for CES

In its normal resting polarized state, neurons are positively charged on their outer membrane and negatively charged inside the cell. The charge inside the neuron in this resting state is approximately –65 to -70 millivolts compared to the extracellular space. If the outer membrane of the neuron is exposed to a negative (-ve) stimulus it will temporarily change the permeability of the membrane, allowing positive ions to flow in, and this will cause it to depolarize (move towards a positive charge), thus reversing the neuron's resting state polarity. This will increase the likelihood that the neuron will activate; that is, depolarize sufficiently to cause an action potential to fire.

When the stimulus comes from a CES device, rather than from the axon of another neuron making contact with a dendrite, it is only the amperage (known as current) that can produce stimulation. Voltage does not. The skin resists (impedes) the flow of current (electrons). The resistance to the flow of an alternating current is called impedance (z), and is measured in Kohms, as is familiar to anyone who does EEG work because impedance between electrodes must always be measured before neurofeedback training is done. Current is measured in milliamps (mA). A voltage potential difference between the two electrodes is necessary for a current flow to occur. For direct current, Ohm's Law applies: voltage = current x resistance, V=IR (sometimes written E=IR). For alternating current, the potential difference between the electrode sites varies directly with the product of the current times the impedance (v=iz, sometimes written V=IZ). Actually, Z is a more complex measurement than R because it also depends on the frequency of the alternating current and other factors, like capacitance. (See Section I on instruments and electronics.) When ear clips or small electrodes are used, the resistance to the flow of the alternating current, called impedance, ranges from 10,000 to 40,000 ohms (10-40 Kohms) from one electrode to the other electrode. This electrical resistance varies widely, as it depends on skin thickness, dryness of the skin, mineral content in the water (sweat), and use of conductive gel. (Recall that good impedance measures between electrode pairs when doing neurofeedback is <5 Kohms.)

It's fairly common for CES devices to stimulate at 1 to 3 milliamps (mA). One mA is 1/1000 of an ampere - a rather small number. To put this in perspective, a typical wristwatch uses a few microamps (1/1,000,000 of an ampere) of current, while a typical LED indicator light on an electronic device uses 2 to 5 mA of current.

Activation of a neuron depends on both the *amplitude* of the current and the *length of time* (pulse width) that the current has been flowing. This relationship gives rise to what is known as a strength-duration curve (McClintic, 1978). A strong (tall) short pulse will activate neurons just as well as a less intense (smaller) pulse of longer duration. By exciting neurons with varying pulse widths and intensities, scientists have developed a strength-duration curve for firing of the neurons. In the figure below, as long as the strength-duration is above the curve, neurons will be able to activate and fire. Thus a very long pulse can exert an effect with very low voltage (and low current combined with low impedance/a good connection). This, in part, is why a 9-volt battery tingles the tongue. It has a relatively low voltage but a good (wet) connection and the current flows for a long time. (A battery, of course, provides direct current or DC).

Strength-Duration Curve for Nerve Stimulation at an Impedance of Approximately 10 Kohms

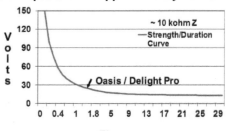

Oasis/Delight Pro refers to a commercial CES instrument available from Mind Alive, Alberta, Canada.

Some devices can deliver pulses above 150 volts with a very high current, but for only 0.1 milliseconds (ms), which for CES is a very short period of time. This approach has a tendency to utilize the capacitance of the body, which is the ability of a living body to store a charge, and it can reduce the risk of blistering an earlobe. However, high voltages can push current through high impedances and might pose a risk in certain circumstances; for example, it could trigger a heart contraction if the electrodes accidentally fell onto the user's chest.

The alternate approach is to use a low voltage and current, but for a longer period of time, on the order of a few milliseconds. The plus side to this approach is that it is relatively safe. The down side is that there is a larger flow of electrons and it could burn/blister the skin if the intensity were turned up too high. Most CES devices use lower voltage and current and longer duration stimulation for the sake of safety. (One can compare this to a modern electric welding machine. These machines operate at 35 volts, as this voltage is considered safe. It is the same voltage used in CES instruments.) The Delight Pro and the Oasis Pro (two commercial instruments produced by Mind Alive (Siever, 2012, 2013) are CES instruments that stimulate in the 1.5 to 2 ms range with a maximum voltage of 30 to 40 volts and a maximum current of about 4 mA.

In studies that are *double-blind*, neither the participant nor the researcher knows when the real treatment or placebo (fake) treatment are being applied. To accomplish this using CES, the real stimulation must be low enough that the participant cannot feel it, and therefore cannot differentiate it from the fake stimulation. Double-blind studies concerning the use of CES have shown excellent clinical results, even though the current is so mild that the subjects could not feel the stimulation. On the other hand, for clinical work, it is usually appropriate for the user to feel the stimulation very mildly.

The Electronics Behind CES

CES exerts its effects on the brain by delivering an oscillating current (AC a sine wave). This alternating current (AC) is similar to the current flow from an office or house wall socket. Because the outside of a neuron is positively charged, a negative pulse will flip its resting state and may cause an action potential to occur, as explained above and in more detail in Section I of this book.

This AC stimulation (tACS) is NOT to be confused with transcranial direct current (DC) stimulation (tDCS), which was described in Section XIX of this book. In tDCS a **direct** current from a battery is applied through a damp sponge (or electrode in some instruments) that is placed on the scalp and travels to another sponge (or electrode) at another location on the scalp or shoulder. The principles governing DC stimulation and AC stimulation are not the same. With tDCS, the anode (+ve) causes the cortex below it to be more negative, and negativity is associated with activation; that is, neurons are more likely to fire.

On the other hand, when applying tACS using CES instruments (and also TENS), the current alternates as a sine wave. In tACS it's only the negative-going (also called cathodic) pulse of the AC that can trigger activation (the interior of the neuron may drop from over -65 mV to -55 mV). Thus, whereas CES flips the resting state of a neuron in an oscillating fashion, tDCS is unidirectional such that under the anode (+ve sponge) the extracellular space becomes more negative. With tDCS the probability is increased that neurons under the anode will fire when their dendrites receive excitatory stimuli from the axons of other neurons.

The predominant hypothesis of tACS action is that alternating fields can increase or decrease the power of oscillatory rhythms in the brain in a frequency-dependent manner by synchronizing or desynchronizing neuronal networks. This can result in 'entraining' certain frequencies, though the lasting effects are not fully explained by this transient entrainment.

The polarity of the two electrodes alternates back and forth. When one electrode provides stimulation (-ve pulse), the other electrode provides the return path (+ve), and vice versa when the polarity reverses (alternates). This is shown in the figure below. So long as both electrodes are not –ve or +ve at the same time, there will always be a current. Figure 2 shows an AC device that alternates uniformly such that it makes pulses at what is called a 50 percent duty cycle, meaning that each electrode is -ve or +ve 50 percent of the time. Some people respond very well to this 50/50 alternating of the current. However, some other clients may experience nausea and would do better with pulses that are of a shorter duration. Note that the current flows only when the electrodes **are of opposing polarity.** The switching (oscillations) can be altered (made shorter or longer). In the figure below, current flows ONLY during the red zones when there are opposing polarities on the electrodes. When the polarities are the same (both +ve or both -ve), there is no current flow and no stimulation (no red line).

Pulses at a 50 Percent Duty Cycle

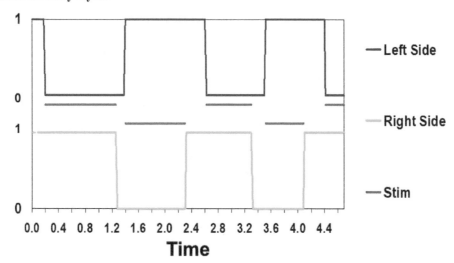

Household Power

Household power alternates from positive to negative 60 times per second (in North America), or 60 Hz AC (alternating current). If you should accidentally grab a "hot" power source, you will get a much more severe shock than from an electrostimulator with the same voltage. This is because at 60 Hz, the stimulus is on for about eight ms (a very long time), then flips polarity for another eight ms, and a lot of continuous current can pass through a person. Due to the long stimulus time, a 60 Hz line shock can really excite nerves and make muscles contract harder than anyone could ever willfully contract their own muscles. As a result, torn muscles, heart attacks, and burnt tissue are possible, depending on where the hazardous wire contacts the body, how well the connection is made, and the reference or ground. For this reason, all electricians must, by law, use nonmetal ladders (such as fiberglass or plastic) to prevent shock through their body and into a grounded ladder (especially if placed on wet ground) if they were to accidentally grab a hot wire. The following table shows the shock hazard at varying currents of 60 AC.

60 Hz AC Currents and Human Experience of Them

1 mA	Perception level
5 mA	Mild shock felt, not painful but startling
6-30 mA	Painful shock, but able to let go if through the hand
50-150 mA	Extreme pain, unable to breathe, severe muscle contractions
1000-4000 mA	Mild skin and internal tissue burns. Heart goes into ventricular fibrillation (heart flutters but does not pump blood)
5000 mA and up	Cardiac arrest

It is not the voltage that does harm, but rather the amperage and the length of time that the amperage is present. With a low voltage, a better connection is needed to get the current (amperage) through. On the other hand, a high voltage can push an excessive current through the body even if the connection is poor.

CES Physiological Studies

Methodology

CES manufacturers generally use frequencies somewhere in the 0.1-40 Hz range. However, there are other ranges on the market including 100 Hz, and one device uses 15 KHz, modulated with 15 Hz and 500 Hz. However, it's mainly the electrical nature of CES itself that produces most of the benefits. Frequency has some influence, but only a minor amount.

As a general guideline, it is believed that CES increases availability of neurotransmitters at any frequency, but it is believed that 100 Hz mainly increases serotonin and therefore that frequency is used to reduce anxiety and depression and to improve sleep. Sub-delta and delta frequencies in the 0.5-3 Hz range mainly increase endorphins, and those frequencies are used to reduce pain and promote feelings of well-being.

Frequency Effects on Brain Waves

A Quantitative EEG (QEEG) study by Kennerly (2006) had 72 subjects. The subjects were divided into two groups. One group (n=38) received 0.5 Hz, and the other group (n=34) received 100 Hz. The results demonstrated that both groups showed increases in alpha with decreased delta and beta activity. The difference was that the **0.5 Hz group showed decreases in a wider range of delta in the EEG, whereas the 100 Hz group showed decreases in a wider range of beta frequencies (serotonin effect)**. This is consistent with the observations in the literature that CES at 100 Hz is more effective at increasing relaxation while reducing anxiety.

CES is thought to primarily exert its effects on the brain via neurotransmitter production. In 1988, Shealy et al. performed a study using the Liss Pain Suppressor (n=5), comparing serum drawn from blood with cerebral spinal fluid (CSF) drawn from lumber subarachnoid space. His study was designed to show that CSF measurements were considerably more accurate than blood serum measures. But in the process he found that the average increases in beta endorphin and serotonin production following CES were quite dramatic, as shown in the following figure.

Blood Serum vs. CSF Measures of Neurotransmitters Following CES

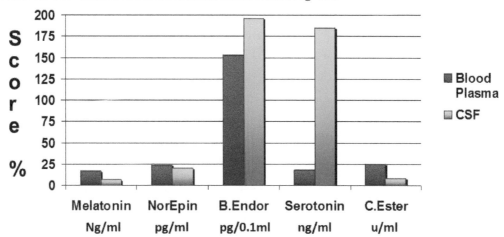

Another Shealy study (1989) using the same device on depressed (n=11) and chronic pain (n=23) subjects showed lesser results, but the greatest increases involved the neurotransmitters that the participants most needed (as shown in the following table). For patients who were depressed, serotonin and norepinephrine were needed most, and these showed the largest increases. Those with chronic pain showed the largest increase in endorphins, which can be helpful with pain.

Neurotransmitter Changes Following CES Among Those With Depression and Chronic Pain.

	Depressed (n=11)			Chronic Pain (n=23)		
	Pre CES	Post CES	% Change	Pre CES	Post CES	% Change
Serotonin	28	44	**57**	40	42	5
B. Endorphin	10.5	7.5	-29	8.9	10.2	**15**
Norepinephrine	212	239	**13**	214	224	4.6

Clinical Studies

By 2002, there were about 200 studies using CES spanning 50 years and using more than 25 devices (Kirsch, 2002). Ray Smith (2006) completed an extensive analysis of the more credible studies of CES and cross-referenced them according to *effect-size*, a type of statistical analysis, so they could be compared with other treatment modalities. The results are shown in the table below.

Studies Using CES Published Before 2005

ndrome	Total # of Studies	# Double Blind	# Single Blind	Other	Total # of Participants	Avg Improve (Effect-Size)
somnia					8	%
epression					3	%
nxiety					95	%
ug Abstinence					5	%
ognitive Dysfunction					8	%

Depression is a debilitating condition. Functional neuroanatomy studies of depression have shown a direct relationship with *hyper*activation of the ventromedial prefrontal cortex (vmPFC) and *hypo*activation of the dorsal lateral prefrontal cortex (dlPFC). As depression subsides, the vmPFC becomes less active, while the dlPFC becomes more active (Koenigs & Grafman, 2009). Shealy et al. (1992) studied blood-serum levels of five neurochemicals (melatonin, norepinephrine, beta-endorphin, serotonin, cholinesterase) in patients with depression. Shealy and his colleagues found that 92 percent of patients with depression had abnormal levels in at least one of the five neurochemicals tested, and 60 percent showed three or more abnormalities. In over half of the patients with depression, they found either elevated or low levels of norepinephrine/cholinesterase ratios. They also found magnesium deficiencies in 80 percent of the depressed patients while 100 percent were deficient in taurine. Neurotransmitter levels play a role in depression. Depletions in serotonin, norepinephrine and dopamine are well documented with major depressive disorder (Nutt, 2008).

The widely varying factors involved in depression may well contribute to the failure of drug treatments in some patients. On the other hand, studies using CES, with a total of over 1000 patients with depression across the various studies, have shown relatively successful outcomes across the board. A meta-analysis study by Gilula & Kirsch (2005) of 290 patients with depression showed a direct comparison of CES against various medications that are used for depression. The following table shows the treatment-effect improvement in depression compared to placebo, obtained from freedom-of-information data as provided to the FDA from pharmaceutical companies when seeking FDA approval. The CES data came from eight studies submitted to the FDA from Electromedical Products International, Inc., to reclassify CES from Class III to Class II for the treatment of depression, anxiety, and insomnia. The CES studies had no reported negative side effects, whereas the drug studies indicated side effects ranging from interruption of liver metabolism to reactions with other medications, and even an increase in thoughts and behaviors related to suicide.

Medication Treatments vs. CES for Treating Depression

Medication	Trials	N	Placebo Effect	Medication Effect	Proportion Placebo	% Meds Over Placebo
Prozac	5	1132	7.3	8.3	0.89	11
Paxil	12	1289	6.7	9.9	0.68	32
Zoloft	3	779	7.9	9.9	0.80	20
Effexor	6	1148	8.4	11.5	0.73	27
Serzone	8	1428	8.9	10.7	0.83	17
Celexa	4	1168	7.7	9.7	0.80	20
CES	8	290	---	---	0.37	63

Although data in a table illustrates relationships, a graph can emphasize differences to a greater extent. The figure below shows the relationship between drug treatments for depression compared to the percentage improvements obtained with CES. This graph shows depression reduction compared to a placebo response that, in the case of Prozac, is only 11 percent better than the placebo.

Drug Treatment vs. CES in the Treatment of Depression.

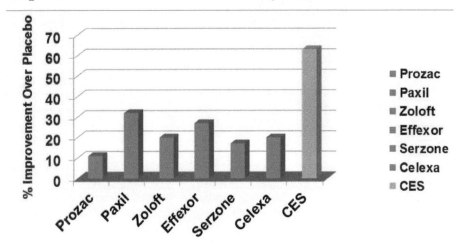

Several studies point to the success of using CES to help with cognitive dysfunction. One such study was carried out by Ray Smith (1998). He examined the emotional and IQ outcomes in 23 children with ADD/ADHD following three weeks of 45 minutes of CES stimulation, five times per week. The children were retested in a follow-up after 18 months. Smith compared three different CES devices (CES Labs, Liss Stimulator, and Alpha Stim). These devices have different stimulation parameters with respect to frequency, pulse width, and intensity. Despite the various device manufacturers claiming to have the best unit on the market, no statistical differences in treatment efficacy were found, though all three had positive outcomes. The figure below shows the reduction in both state and trait anxiety and depression, as measured by the Institute for Personality and Ability Testing Depression Scale Questionnaire (IPAT Depression Index), and the State-Trait Anxiety Inventory (STAI). The figure following that shows the improvements in IQ as found using the WAIS-R in adolescents over age 16 and the WISC-R in children.

Behavioral Outcomes Following CES Treatment at Three Weeks and at 18 Months

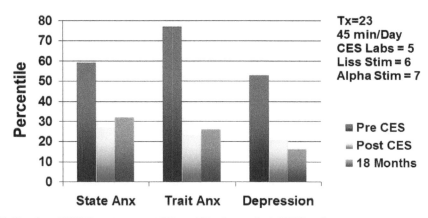

IQ Outcomes Following CES Treatment at Three Weeks and at 18 Months

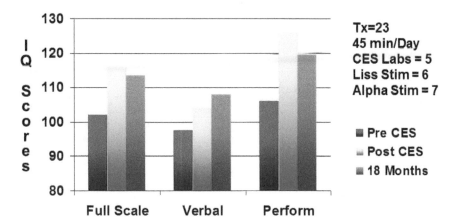

Current Developments Regarding CES

Frequency
Research investigating frequency has demonstrated that CES can entrain brain waves. Although brain wave entrainment is not done with most CES devices, it has been investigated with transcranial AC stimulation (tACS) instruments. The predominant hypothesis regarding the action of tACS is that alternating fields can increase or decrease the power of oscillatory rhythms in the brain in a frequency-dependent manner by synchronizing or desynchronizing neuronal networks. This can result in **entraining the brain in particular frequencies.**

Waveforms
Some manufacturers state that they use a unique "special" waveform, which is better than that used by other manufacturers. Most CES research has used the simplest waveform to generate the pulse. This is a square wave (on and off). Although simple and inexpensive to generate, the downside to a square wave is that the stimulus can sometimes sting the ears, especially when the connection is not a good one. People with naturally thicker and/or drier skin may experience more pain directly from the stimulus than a person with thinner, moister skin. Siever finds that by rounding the front end of the pulse, so it doesn't turn on so quickly (Siever, 2013b), he can reduce the stinging/burning sensation by about 80 percent without loss of effectiveness. This increases user compliance as people obviously prefer not to experience discomfort. The following figure shows an oscilloscope tracing of a rounded waveform versus a traditional square wave.

Oscilloscope Tracing of a Square Waveform vs. a Rounded Wave

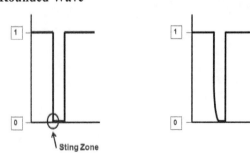

Use of tDCS Compared to tACS
Like repetitive transcranial magnetic stimulation (TMS) that is used in hospitals, tACS may be better suited than tDCS to modulate those cognitive functions that are closely related to brain oscillations at specific frequencies. One approach that has shown clinical effectiveness is to randomize the stimulus frequency, as done by Heffernen (1997) who demonstrated EEG smoothing and pain reduction following CES treatment. In addition to the clinical therapeutic effects, tACS may prove to be a good tool for investigating human brain oscillations and therapeutic applications (Zaehle, Rach & Herrmann, 2010).

The tDCS can up- or down-regulate the stimulated brain areas irrespective of the frequency at which neurons in this area oscillate. Thus, **tDCS might be more effective in regulating those brain functions that are clearly related to specific brain areas** rather than certain EEG frequencies, for example, syntax processing in Broca's area, face processing in the fusiform gyrus, or motor processes in the primary motor cortex. Modulating either a brain area or an oscillation may help to normalize a disturbed function. One can postulate that the effect could also be negative, but at the time of writing, major side effects were not being reported. This is discussed more under the section on tDCS.

Though the approaches and methods are different concerning tACS and tDCS, modulating either a brain area or an oscillation may help to normalize a disturbed function.

Conclusion
There is a reasonable physiological rationale for using CES and considerable evidence regarding the positive effects of CES on the brain. CES is easy to use, the efficacy is quite high, the equipment is fairly inexpensive, and negative side effects are minimal when it is properly done. CES can be a helpful part of the therapeutic tool chest of clinicians who work with cognitive challenges, affective disorders, and pain.

Electromagnetic Stimulation Approaches Light Therapy (Low Intensity Laser), NeuroField and LENS

Laser Therapy

A *laser* is a device that emits light through a process of optical amplification based on the stimulated emission of *electromagnetic* radiation. Lasers produce electromagnetic radiation that is polarized (vibrates in a single plane) and monochromatic. Low Energy Laser therapy is primarily used to decrease inflammation and improve healing. Its use to improve brain function is an entirely new field where research is in its infancy. Nevertheless, it is such a powerful technique that a few points concerning its scientific basis and use should be mentioned in this textbook. For a readable introduction to the basic fundamentals underlying its use, the reader is referred to Chapter 4, "Rewiring a Brain with Light," in Norman Doidge's 2015 book, *The Brain's Way of Healing,* published in New York by Viking Press.

The following information is based, in part, on Dr. Doidge's wonderful summary and on conversations with the innovative surgeon, Dr. Fred Kahn, who has pioneered laser rehabilitation therapy for the last thirty years. He has a brilliant assistant, Dr. Slava Kim. Dr. Kim, while a general surgeon in Kazakhstan, had used lasers to promote healing and learned from the work of Russian researchers Meshalkin and Sergievskii, who investigated low-intensity laser treatment of the blood and then also applied it to cardiovascular problems. Dr. Kim thus had extensive experience with lasers before coming to Canada where he works at Meditech in Toronto with Dr. Kahn. For more information see the website www.meditech.com.

Laser stands for Light Amplification by Stimulated Emission of Radiation. The inner electron orbits in an atom are low-energy when compared to the high energy of electrons in the outer orbits. When an electron falls from an outer orbit to an inner orbit around the nucleus of an atom, a photon of light is given off. This is a *spontaneous* emission of light radiation. Bombarding atoms using an outside energy source, such as a battery in the laser pointers used when giving a lecture, can result in more electrons moving to the high energy outer orbit state. This is called 'inversion.' If the atom continues to be bombarded with energy, it will release a large additional number of photons. These photons will stimulate neighboring atoms to release more photons. If this area contains mirrors, the photons will hit the mirror and be reflected back and hit more atoms and produce more photons. In this manner the production of photons is further stimulated, hence the designation of Light Amplification by Stimulated Emission of Radiation or, more simply, laser.

In its most basic form, a laser is produced using electrical stimulation of a semiconductor diode. The diode can be thought of as a sandwich: the top layer is a substance with a surplus of electrons, then comes a space, then the bottom layer, which is a substance with a dearth of electrons. When electromagnetic energy of a specific frequency is passed through the diode, it stimulates the production of photons and the resulting light is projected in the form of a laser beam. The color or wavelength of light being emitted depends on the type of material being used and the optical configuration of the laser. In laser pointers, for example, the frequencies most often produced are red (635 nm), green (520 nm) and blue (445 nm). The lasers used for healing tissue and in laser pointers are very low power. They are quite different than the medical lasers used in ophthalmology, and completely different from the high energy lasers that may be used in industry and by the military. These high energy lasers would burn tissue. The lasers used therapeutically to heal tissue are therefore called **low energy lasers.**

For further readable information, the reader is referred to "Laser Fundamentals" by Robert Aldrich, Naval Surface Warfare Center, Dahlgren Division, available on the internet.

Photons absorbed by living tissue can trigger chemical reactions in light-sensitive molecules. Different molecules will absorb different frequencies of light. Cytochrome is a light-sensitive molecule that exists in the mitochondria of most cells. Cytochrome, when stimulated by light, will activate the production of ATP (energy in the cell produced by the Krebs cycle). In this way, cytochrome is responsible for converting light energy from the sun into energy for the cells. In this manner a **laser light can increase the energy production in cells by causing the cytochrome molecules to trigger the production of ATP** within the mitochondria.

Cytokines, which can be pro-inflammatory or anti-inflammatory, can down-regulate cytochrome molecules. The **correct type of laser light can activate the anti-inflammatory cytokines.** In this way, laser light can bring an end to chronic inflammation. Cytokines were mentioned earlier in this book in the section on heart rate variability (HRV) and in the section on concussion that discussed how an energy crisis can be created in cells when a traumatic brain injury interferes with cell metabolism. In addition to anti-inflammatory action and increasing the production of ATP, **laser light can stimulate the synthesis of both DNA and RNA.**

In summary, mitochondria are the source of energy production in the cells. Laser light can enter the mitochondria and stimulate the production of adenosine triphosphate (ATP), which, as explained in the section on concussion, is the cell's form of energy in all body tissues. ATP provides the energy for the immune system and for cell repair. This may be the key for our understanding of the almost miraculous healing observed in some patients who have major tissue injury who are treated with low energy laser light.

The author (MT) can personally testify to laser treatment's efficacy. He suffered a complete tear of a supraspinatus tendon in the rotator cuff of his right shoulder while moving a heavy anchor. He was unable to lift his arm more than a few inches from his side without pain. The pain was continuous, and for almost 18 months he could not get a decent night's sleep. After failure of conservative methods including physiotherapy, rest, and gradual exercise, he was booked for surgery. Four weeks prior to having major surgery to repair the tendon, he was introduced to Dr. Fred Kahn by Dr. Norman Doidge. After five laser treatments he canceled the surgery. He had recovered full range of motion for his right arm and gradually began lifting weights again. Rapidly he recovered the complete use of his arm. No treatment works 100 percent of the time, but here was a noninvasive procedure that restored function, even if it could not reattach the torn tendon. More stories about the success of Dr. Kahn's light and laser treatments for a range of patients are shared in Doidge's highly readable book.

Dr. Kahn's method begins each session with about 20 minutes using red light from 180 light-emitting diodes (LED) on a band that covers the area. It penetrates about 1 to 2 cm. We can see light with wavelengths from 400 to about 700 nanometers: 400 is violet and it contains the most energy; 700 is red light and it contains the least energy. For laser healing, red at 660 nanometers is frequently used. The red light is followed by another 20-25 minutes using a second band of infrared lights. This penetrates tissue to a depth of about 5 cm. Infrared has a wavelength of about 840 nanometers, which is not visible to the naked eye. The red and infrared light is said to prepare the tissue for healing and to improve circulation. This is followed by the patient and therapist putting on special glasses to protect their eyes from the laser that will be used next. Then the therapist uses what I like to refer to as a laser gun because it looks like a pistol and is aimed at the target area. This is a powerful focused laser treatment. It will result in a cascade of photons deep in the tissues. Dr. Kahn uses red laser followed by infrared laser. The laser probe can reach as deep as 22 cm. A laser is a very precise narrow frequency range (one nanometer) of extremely high intensity light. I found the probe most effective. Each session with it lasted only a short time, from five to seven minutes.

Dr. Fred Kahn, originator of Meditech Laser Therapy, shared the following with us (personal communication, March 2, 2015). "For head injury patients, we initiate treatment over the cervical spine, which in most cases includes the brain stem and cerebellum. In this area, we utilize both large surface arrays and laser probes extending generally in the midline from the cerebellum to T1 or T2. If symptoms are not relieved by this approach, we add the large surface arrays over the cerebral hemispheres. When tinnitus and symptoms relating to the middle and inner ear are also involved, we treat those areas in addition. We generally avoid using the laser probes over the hemispheres, but do use them liberally around the ears and the cervical spine including the brain stem. We individualize treatment for each patient."

Chronic inflammation is a major problem in many disease processes. Laser light will increase the anti-inflammatory cytokines, lower the number of neutrophils (these are high in chronic inflammation), and increase the number of macrophages that remove damaged cells and other substances. Laser light is also said to activate both DNA and RNA synthesis and thus assist in regeneration of normal tissue. In addition, lasers are said to increase production of neurotransmitters. Serotonin, acetylcholine, and endorphins are among those said to be increased. Interestingly, depression has been treated with light throughout history. All this sounds hopeful for our work with the brain, but research is needed because it is not one-size-fits-all. Lasers are very precise and a tiny difference in the frequency of the light waves can markedly change the effect. Low energy laser is clearly an extraordinarily powerful tool. While its use for dramatically decreasing inflammation has been demonstrated over many decades in Europe and Asia, and benefits for wound healing and for problems like arthritis and joint pain are well established, in our opinion, aiming the laser at the brain will require much more research. There are, nevertheless, some claims for its beneficial use in cases of head trauma, and currently some animal research is being done in Israel. It may become an important adjunctive technique for management of traumatic brain injury.

Inflammation is an important factor not only in concussion, but in other kinds of brain dysfunction, including dementias such as Alzheimer's. In Alzheimer's, the mitochondria show oxidative stress. It is possible, even probable, that we will soon be combining light treatments with neurofeedback relearning methods in the treatment of a number of brain disorders. As with our work with neurofeedback, tDCS, and CES instruments, trainers will have to learn how to properly use this new type of rather complex equipment. Low energy laser treatment can be applied effectively or ineffectively, and both the equipment used and the training of the therapist are important in the outcome for patients. As with any new methodology, especially one that is invasive, caution is warranted until we have more data and research.

LENS and NeuroField

Electromagnetic Stimulation Approaches:

Other adjunctive techniques involve devices that use electromagnetic field (EMF) or pulsed electromagnetic field (pEMF) stimulation. Current systems include the Low Energy Neurofeedback System, called LENS, developed by Len Ochs, NeuroField, developed by Nicholas Dogris, and High Performance Neurofeedback (HPN) founded by Fred Willis, which is being marketed for post-concussion syndrome in particular and is being used with former NFL players. These are all very different systems.

These approaches differ considerably but all stimulate the brain for at least part of the intervention. They are thus invasive and passive, providing stimulation without the client making effort, in contrast to the non-invasive approach of regular neurofeedback that provides information (feedback) that facilitates learning of particular mental states.

At the ADD Centre we have had senior staff take the LENS workshop and we have purchased NeuroField equipment but we are not yet using either of these approaches with clients, mainly because they do not yet meet the criterion of being evidence based. At the time of writing, randomized, controlled trials are lacking. However, for NeuroField, one sham controlled trial was reported by Perminder and colleagues (2012).

LENS

With respect to LENS, we have informally surveyed people using this equipment and have heard many positive and some negative anecdotes. We personally observed negative effects on mood, cognition, and sense of well-being that gradually subsided after a few days in a very bright psychologist and friend who was the volunteer during a LENS workshop. On the other hand, there are anecdotal reports from well-respected clinicians of good outcomes and faster resolution of symptoms when LENS is combined with regular neurofeedback.

There has been one large case series published concerning the use of LENS (Larsen, Harrington, & Hicks, 2006). Stephen Larsen has written a book about therapy that incorporates LENS. Clinicians using these techniques could perhaps collaborate to publish papers involving case series for particular disorders, or for particular symptoms, that respond to this training.

NeuroField

With NeuroField we heard positive anecdotal reports. NeuroField has a method for measuring the EEG and heart rate variability changes that occur after a pulse is delivered and it does use a normative database when looking at the client's response. This ability to measure the response led to our purchase of a unit, which we will try out ourselves for some time while getting further training before making a decision about using it with clients. In a similar manner, we did not introduce LORETA NFB or tDCS until there was sufficient scientific explanation as to why these techniques could be effective and research that supported their use; that is, positive outcomes without undue negative side effects.

Nicholas Dogris Ph.D. and Brad Witala created NeuroField in 2008 and it is now in its third generation of hardware products. Dr. Dogris (personal communication) explains that NeuroField is based on stimulation technology and utilizes pulsed electromagnetic stimulation (pEMF), transcranial direct current stimulation (tDCS), transcranial alternating current stimulation (tACS) and transcranial random current stimulation (tRNS) so as to entrain and activate specific regions of the brain for the purpose of achieving regulated brain states to enhance cognitive functionality.

There are studies conducted on the NeuroField system, most of them conducted by the system's founder. The initial study by Dogris (2009) on the NeuroField system utilized 22 subjects and showed that these subjects responded to a three milligauss stimulation for 100 seconds. This study demonstrated that very low intensity pEMF can impact the QEEG and change power, coherence and phase metrics. It also showed that the brain will replicate or copy pEMF of low intensity ranging from 1-1000 Hz. Another study by Dogris (2011) examined three case studies that demonstrated positive treatment responses for PMDD, ADHD, and PTSD. A study by Moore & Kube (2013) showed a reduction of symptoms on a three year old with Stickler syndrome. Dogris notes that the literature shows that pEMF has been utilized for a wide variety of other problems. Dr. Dogris notes that NeuroField now incorporates pEMF and tDCS/tACS synchronized with EEG operant conditioning and HRV. He states that this methodology continues to be tested by the NeuroField practitioners, but has yet to undergo double blind randomized control trials.

At the time of writing there are various theories attempting to postulate why some patients might get a good effect using some form of electromagnetic stimulation. The interested reader is referred to the manufacturers of these devices for more details.

Dogris, N.J. (2011). NeuroField: Three case studies. *Journal of Neurotherapy,* 15, 75–83.

Dogris, N.J. (2012). The effect of NeuroField pulsed EMF on Parkinson's disease symptoms and qEEG. *Journal of Neurotherapy,* 16, 53–59

Moore, J.D. & Kube, E.L. (2013). Evidence for the Effectiveness of NeuroField in a Three-Year-Old Boy with Insomnia. *Neuroconnections*, Summer Edition, 16-22.

SECTION XXI
Hemoencephalography: pIR HEG and NIRS HEG

Introduction

The following section on passive infrared hemoencephalography (pIR HEG) and near infrared spectrophotometry hemoencephalography (NIRS HEG) is written after years of use of both methodologies at the ADD Centre as adjunctive techniques. Our original work was based on the scientific innovation of the late Hershel Toomim, who is rightly credited with being the 'godfather' of this field. At the ADD Centre, HEG is always used in combination with NFB and HRV training; thus, the contribution of pIR to outcomes cannot be reported. This short description of this methodology relies to a large extent on: Jeffrey Carmen's publications and his website, and to personal communications with the late Hershel Toomim, with Frank Diets of Focused Technology, and with Marc Saab at Thought Technology. We currently use the headgear from TT as our preferred instrument.

Definition and Description

Passive infrared hemoencephalography (pIR HEG) refers to the use of one or two infrared detection sensors to reflect temperature changes in the cortex. Biofeedback practitioners usually use sensors on a headband with the sensors placed over the forehead. The patient observes a feedback screen that shows a line graph with temperature on the y axis and time along the x axis. The assumption is that the temperature recorded reflects changes in the metabolic activity of the frontal lobes of the brain. (Hershel Toomim argued convincingly that it was not just surface blood flow that was measured.) If this hypothesis is true, then an increase in temperature should correlate with increased metabolic activity and beta activation in the prefrontal area. If this occurs, then executive functions may be expected to improve. This effect has been reported by Carmen (2009), who claims to have found improvements in focused attention, memory, organizational ability, mood, planning, and judgment. He has also published on positive effects in the treatment of more than 100 patients who had migraine headaches.

In this form of operant conditioning, the client is asked to find a mental state or "zone" that corresponds to calm, relaxed-yet-intense focus. At the ADD Centre/Biofeedback Institute of Toronto, when pIR is used, we usually combine the pIR feedback with BFB, most often heart rate variability (HRV) training, and NFB. The pIR feedback may be for the first 10 minutes of a NFB session, or it may be integrated into the whole 55-minute session as one part of the NFB-BFB screens.

History

The term 'hemoencephalography' was first used by Hershel Toomim in 1997 (Toomim, 2002) to describe his process of the Near Infrared Spectrophotometry system. The Passive Infrared Hemoencephalography system (pIR HEG) evolved from Carmen's application of infrared technology to peripheral thermal biofeedback (Carmen, 2004). Both systems, pIR HEG and NIRS HEG, respond to blood flow in the cerebral hemispheres. The effects of HEG training on EEG at Cz have been published (Sherrill, 2004), but there have not been many publications since that time. With either instrument the practitioner is making the assumption that an area of the brain that is more active will give off more heat and have greater blood flow and blood oxygenation than a brain area that is less active. Both systems appear to be relatively free from artifacts that are seen in NFB and which are produced by eye movement and surface EMG. This factor is particularly helpful for giving feedback concerning activity in the prefrontal region.

Hypothesized Mechanisms for the Effects of pIR HEG on Migraine *(reference: Jeffrey Carmen, website)*

The positive effects on migraine management using a pIR HEG system on the forehead may relate to the inhibitory and regulatory functions of the prefrontal cortex (Goldberg, 2001). The inhibitory hypothesis is based in part on the observation that patients with migraine that are treated using pIR HEG consistently report that they are experiencing an improvement in their sustained attention to tasks. They also report a decrease in distractibility to both internal and external stimuli. In addition, Carmen notes that patients being treated for migraine headaches using pIR HEG report that their emotional responses are more appropriate; that is, less rapid and less strong to stimuli that should not call for a rapid and strong emotional response.

Carmen warns us that the first few sessions may have an immediate but temporary aftereffect of a change in behavior that includes 'disinhibition' with difficulty attending to task, more severe migraine, and even rage that may last for the rest of the day. He notes that typically on the following day after a good night's sleep, inhibitory functions are normal or better than normal for that person. He suggests that this might represent some sort of an exercise/rest/recovery mechanism. At the ADD Centre, at the time of writing, **we have not seen these** negative changes. However, we always do other forms of NFB and BFB during these sessions because we still view pIR to be experimental at this time.

Expected Changes

It seems reasonable to expect that increased temperature should correlate with increased metabolic activity and beta activation in the prefrontal area. It seems reasonable to expect that this would improve executive functions, as has been observed by Carmen. The frontal lobes are involved with focused attention, memory, organizational ability, mood, planning, and judgment, as has been described under Section VIII, Brodmann areas and functional neuroanatomy. Carmen has reported on positive changes in some symptoms seen in persons who are experiencing a number of disorders including: migraine headaches, depression, anxiety, symptoms resulting from a Traumatic Brain Injury (TBI), ADD, ADHD, autism, Asperger's syndrome, and Tourette. However, at this time of writing, there are no randomized, controlled, multicenter studies to substantiate this. Results with migraine headaches, in a case series format, have been summarized by Carmen and are available on his website. Results with other conditions will require research. As with other forms of NFB, there is a possibility that results may be long-term in some people. Others may return for further sessions.

Instrumentation

In the past we have used NIRS HEG instruments designed by Toomim and adapted by Frank Diets to run with the Focus Technology F1000. At the present time we use an instrument designed by Thought Technology (TT) for pIR that runs using the Infiniti by TT This instrument for passive infrared heat detection uses two sensors on a headband. The sensors are normally placed over the forehead. On the feedback screen, one of the graphs shows temperature on the y axis and time along the x axis. As previously noted, the assumption is that the temperature recorded reflects changes in the metabolic activity of the frontal lobes of the brain. The Thought Technology Infiniti screen display and sound are active when the patient increases this temperature, which is assumed to correlate with metabolic activity. The client is asked to find a mental state or "zone" that corresponds to calm, relaxed yet intense focus. This is operant conditioning in which the client is rewarded for the behavior of increasing forehead temperature. At the ADD Centre/Biofeedback Institute of Toronto, we usually combine the pIR feedback with BFB that includes heart rate variability (HRV) training, and NFB. The sessions are done in 5- to 10-minute segments for 50 minutes.

Carmen says that, typically, change occurs in one session, but results may fade over the next few days. At the Biofeedback Institute we observe that the patients like the sessions and gradually improve over time as we have reported in the literature. We are not able to say the degree of improvement that can be attributed specifically to the pIR, but our impression is that to date it has been free of side effects though mild headaches have been reported by some practitioners. It is our impression that pIR may be an additional contribution to the good patient outcomes that are experienced at the ADD Centre. As with all the biofeedback/NFB approaches, the positive changes tend to last longer with each session and appear to be cumulative over time.

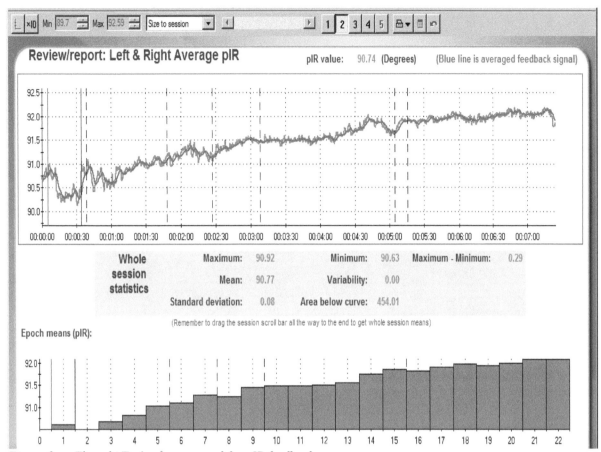

Figure from Thought Technology manual for pIR feedback.

In the above example from the Thought Technology manual, the patient showed a steady increase in temperature. This is seen both in the line graphs and in the histogram for epochs. The time length for epochs can be set in the program. Here they are set at 20 seconds.

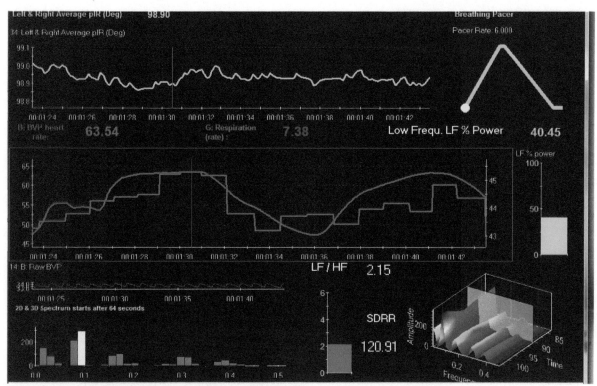

Typical screen used at the Biofeedback Institute of Toronto

In the above screen, the top line graph shows pIR - passive infrared heat detection - using two sensors on a headband. Temperature is shown on the y axis and time along the x axis. In the 20 seconds shown here one would not expect to observe an increase in temperature. Below the pIR display is HRV data. When a second monitor is available it can show a histogram of epochs over a longer time frame. This screen shows the SDRR as 120.91. (SDRR is the standard deviation of the beat to beat interval measured using a blood volume pulse – or BVP - sensor on a finger.) To get SDNN, the practitioner can, at the end of the session, use a program such as CardioPro from Thought Technology. (SDNN is standard deviation of the normalized beat to beat interval, i.e., after correction for artifacts.) The display also shows the ratio of LF/HF numerically and in a bar graph. This helps the practitioner to decide on what amount of parasympathetically controlled high frequency (HF) they wish to encourage for a particular patient. The bar graph on the right shows the percent of power that is LF. Most clinicians in North America encourage a predominance of low frequency (LF), because this reflects a healthy balance of the sympathetic and parasympathetic systems.

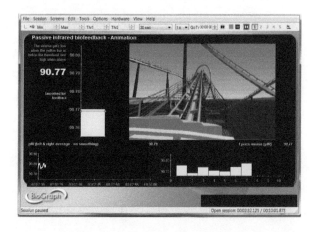

The screen shown above depicts a roller coaster that will only move when the patient is maintaining or increasing the temperature under the sensors. Progress over time is shown by the histogram of temperature levels per epoch in the lower right corner of the screen . The time frame for each bar can be set by the practitioner. We use approximately 20 seconds.

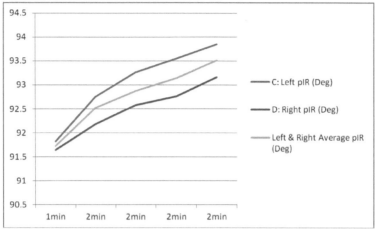

This figure shows an Excel graph from a patient at the ADD Centre. These learning curves are printed out at the end of every session by exporting statistics to excel and printing a graph. This was done at the beginning of an NFB session for ADHD and shows a one-minute baseline followed by four two-minute pIR training screens.

Client Name: SD   Jan. 11, 2012
Trainer: Andrea

	1min	2min	2min	2min	2min
C: Left pIR (Deg)	91.82	92.75	93.26	93.55	93.84
D: Right pIR (Deg)	91.64	92.18	92.58	92.76	93.16
Left and Right Average pIR (Deg)	91.73	92.52	92.87	93.14	93.51

The patient discusses what may have resulted in a change at the end of each of these segments.

Conclusion

There does appear to be good potential for pIR to, at the very least, be a significant adjunctive technique when combined with NFB and other peripheral BFB methods. Most clients can quickly learn to increase temperature measured on the forehead. Success with this kind of biofeedback encourages them regarding the slower process of learning to control the EEG.

SECTION XXII
Slow Cortical Potential (SCP) Neurofeedback

Definition and Description

Slow cortical potentials are slow, event-related, direct-current (DC) shifts of the electroencephalogram, originating from the upper cortical layer. They last from 0.3 seconds up to several seconds. SCPs are not oscillatory in nature, but occur as a consequence of external or internal events. The EEG we work with when we are doing NFB rides on top of these slow cortical potentials. Usually we use a high-pass filter at about 2 Hz when doing NFB that, for the most part, only shows waves above 2 Hz on the screen. This filter means that the EEG we view is not riding on the slow potential and going off the viewing screen. If the EEG is measured with a frequency range that goes below 1 Hz, it will appear wavy.

It has been shown that SCP shifts in the negative direction reflect the depolarization of large cortical cell assemblies, reducing their excitation threshold. Thus, a negative shift is associated with excitation – a more active cortex. Slow cortical potentials are a way of looking at brain activity that differs from the EEG as we use it in NFB. Our work is primarily concerned with frequencies above 1 Hz. We measure an alternating current (AC) that is created by the polarization and depolarization of large groups of pyramidal cells in the cortex. Slow cortical potentials are much slower than 1 Hz, and are influenced by glial cells rather than pyramidal cells. The glial cells are the main influence that creates polarity shifts between negative and positive charges. The polarity shift begins much faster and lasts much longer than the EEG changes. The shift can begin in only 1ms, as compared to about 40ms for the EEG. The shift is between positive and negative.

Thus, slow cortical potentials are due to a **massive depolarization** in layers or sheets of neurons. A negative or positive direct current (DC) shift has physiological meaning, with negativity reflecting activation. The rhythms of the EEG, on the other hand, reflect momentary excitability cycles in populations of neurons (De Ridder et al., 2007).

However, whether a positive peak is excitation or inhibition is not known by simply looking at the polarity of the surface EEG.

Clinical Examples

Two examples of conditions in which SCP differences from normal are found are seizures and ADHD. **With seizures, SCP negativity is too high,** reflecting increased excitation in the cortex. The cortex is said to be more easily 'kindled.' **With ADHD, SCP negativity is too low**; i.e., is positive. Thus, in patients who have seizures, the practitioner might encourage a shift to more positive, while in patients who have a diagnosis of ADHD, the practitioner might encourage the patient to learn to shift the polarity to be negative more of the time. Regulation of frontocentral negative slow cortical potentials is thought to affect the cholinergic-dopaminergic balance. This balance affects the ability of children to adapt to task requirements flexibly. Heinrich (2004) was the first to report feedback of slow cortical potentials (SCPs) for children with ADHD. Extensive work with SCP training for ADHD has been done at the University of Tuebingen by Ute Strehl and her group. Children with **attention problems can have reduced cortical negativity** at all electrode positions in anticipation of a task. This may suggest that a failure to engage specific cortical networks may contribute to their poor performance in academic tasks. Under activation (more positive SCP) may also relate to lack of inhibition of motor neurons, which is related to hyperactivity.

Training to be able to increase negativity to treat ADHD is the opposite training to that done with seizure disorder patients. It has been shown that SCP shifts in the negative direction reflect the depolarization of large cortical cell assemblies, reducing their excitation threshold (Strehl, Trevorrow, Veit, Hinterberger, Kotchoubey, Erb &

Birbaumer, 2006).. In patients with **epilepsy, large negative potential shifts have been observed** seconds before a seizure. In addition, there are observed to be shifts toward electrical positivity immediately after a seizure. **Suppression of negative SCP shifts has been demonstrated to significantly decrease the incidence of seizures**.

When children who have ADHD are trained using SCP, it is hypothesized that it works because they have impairments in their ability to regulate cortical excitation thresholds. These **children with ADHD are taught to increase negativity** of the SCP. As they learn to do this, there are significant improvements in the symptoms related to ADHD. All these positive changes in Strehl's study proved to be stable at follow-up six months after the end of training. In contrast, the effects of a combination of medication and parental counseling did not continue after medication washout (Fuchs, Birbaumer, Lutzenberger, Gruzelier & Kaiser, 2003).

It has also been shown in several studies that voluntary control of SCPs can be acquired by healthy populations.

Methodology

In Europe, Ute Strehl has used a very simple feedback screen on which a little circle called a ball appears at short intervals. Feedback consisted of this one-inch diameter ball appearing either at a higher or lower level on the screen. The ball (or a rocket ship) moves proportionally to the cortical electrical shift upward (negativity) or downward (positivity). Negative is up, and it reflects higher arousal/excitation. Ball movements start on the left edge of the screen and move across the screen to the right edge. A smiley face appears on the screen each time the child succeeds in having the ball move up or down as instructed. Smiley faces are exchanged at the end of the session for prizes. In addition, Strehl gives auditory feedback. This is a high-pitched (negative) or a low-pitched (positive) tone. A harmonious jingle was introduced as positive reinforcement when the result was correct. During the feedback, the child may be told to have the position of each ball a little higher or a little lower than the last ball on the screen, depending on whether the goal is to increase negativity (upward) or positivity (downward). The child must find the mental state that will do this. There is no unique cognitive strategy for the task, but the trainer may give examples of strategies that have worked for other children. The children in Strehl's research work were encouraged to verbalize any successful strategies that they used. At the ADD Centre, we have always highly recommended encouraging verbalizing of strategies whenever possible because this encourages introspection about one's mental state. We also use tokens awarded with praise for positive performance. The tokens act as secondary reinforcement and are tracked in a 'bank account,' with the child deciding after each neurofeedback session whether to spend the tokens for a prize or save them.

For a **seizure disorder**, at the University of Tuebingen, the researchers may do eight-second feedback trials. They may begin instructing the patient to move the ball 50 percent of the time each way. This will change during training to 30/70 negative-to positive-trials if the patient is being trained to produce more positive deflections of the slow wave. They have 25 percent of the trials as **"transfer" trials** with no feedback during the training phase, but they do show success scores at the end of each trial. They note that for this SCP work that 'online' correction of electrooculogram (EOG for eye movements) is essential. Clearly, the emphasis is on the client learning self-regulation, and conscious recognition of how they are self-regulating is encouraged.

Methods

Strehl has published extensively on her work. Building on the work of Niels Birbaumer, the equipment she has used a slow-wave filter with a 500-millisecond interval moving window. The slow-wave amplitude immediately before the active phase of a trial serves as a baseline, and is therefore set to 0. During the active phase, the slow-wave amplitude is calculated every 62.5 milliseconds as an average of the preceding 500 milliseconds. The position of the feedback signal (cursor, "ball") corresponds to the difference between every 500-millisecond amplitude in the active phase and the amplitude during the baseline. In Strehl's clinic, the feedback to the patient is continuously corrected online for eye movements. Thought Technology equipment and screens for doing SCP training are available (www.thoughttechnology.com) for those who are interested in adding this methodology to their practice. If you do add this type of training, you must be particularly careful about obtaining good impedance readings.

Effectiveness

Interestingly, this type of NFB is found to be **just as effective as the EEG neurofeedback (NFB)** that is more commonly used in North America. NFB normally uses frequencies above 1 Hz, and with children, often encourages a decrease in slow waves and an increase in fast waves in order to decrease the symptoms of ADHD. For seizures, NFB will usually involve increasing sensorimotor rhythm (13-15 Hz) over the sensorimotor strip. The approaches are theoretically similar because, whether one is training SCP or EEG, one is basically affecting activation in the cortex. SMR production is associated with inhibition in the cortex.

To recap, glial cells are the source of slow cortical potentials. It is a DC current, and some would argue that this represents the 'mind,' whereas the neural (EEG) activity is the brain. The DC current can turn on or off the ion channels of the neurons. In the literature it has been concluded that whether using regular NFB or SCP neurofeedback, "The long-term effects of neurofeedback can be considered as a major advantage of this treatment compared with pharmacological treatment" (Gani, Birbaumer & Strehl, 2008).

PART SIX

Assessment and Intervention With Specific Disorders and Syndromes

Additions to the First Edition of *The Neurofeedback Book*

In this section, the authors have added updated information about methods for the assessment and treatment of common disorders. While these assessments emphasize 19-channel QEEG, evoked potentials and other types of assessment that require considerable investment in terms of equipment and programs, the authors wish to remind the readers that very considerable success has been achieved over many years of practice using single-channel QEEG assessment and interventions that combine NFB with BFB and metacognitive strategies. The essential key to success still remains the learning – learning to shift EEG patterns and other physiological measures – and that learning is based on the trainer-client (therapist-patient) interaction as well as on the use of feedback that employs electronic measurements of the client's psychophysiology. Fancy machines and advances in scientific knowledge have not replaced the essentials of good clinical practice with an educational focus on learning self-regulation skills. As a clinician, you learn from every client because each person is unique. Hence our work is never boring.

SECTION XXIII
Concussion/Mild Traumatic Brain Injury (TBI)

Goal

This short section is written to provide, first, a brief description of the biochemical and biomechanical effects of a concussion, including diffuse axonal injury, and second, an overview of how various assessment and treatment modalities are combined when addressing complex patients. In our experience, patients who present following Traumatic Brain Injuries (TBI), especially those who have postconcussion syndrome, can benefit enormously from neurofeedback. Keep in mind that neurofeedback for postconcussion syndrome would only have Level 3 efficacy ("probably efficacious"), because it is supported by case studies, but randomized controlled trials are are said to be lacking (Yucha & Montgomery, 2008). However, controlled and randomized studies using NFB have more recently been summarized in a chapter by Foster and Thatcher (Foster & Thatcher, 2015).

Introduction

Patients who have suffered a mild TBI, also called concussion, are assessed and can be, for the most part, successfully treated, even a number of years postinjury, using neurofeedback. This summary is written from a clinical rather than a research perspective, and is based on both the published literature and work carried out at the ADD Centre/Biofeedback Institute of Toronto. Success is measured in terms of the patient having sufficient amelioration of cognitive and emotional symptoms to be able to resume and succeed in their social, academic and/or employment pursuits. To date, this has been achieved in all cases where the person did sufficient training. Here we summarize the most common presenting symptoms and the assessment process, including a rationale for the use of 19-channel quantitative electroencephalogram (QEEG), evoked potentials (also called event-related potentials or, simply ERPs), heart rate variability (HRV) measures, biomedical analysis and neuropsychological testing. Also included is a short description of the treatment approach. Interventions may include: single-channel neurofeedback, LORETA Z-score neurofeedback (LNFB), HRV training, tDCS, pIR HEM biofeedback, metacognitive strategies, and counseling regarding diet and supplements, sleep hygiene, and exercise. We have other modalities for treatment available at the clinic, such as audiovisual stimulation (AVS) and cranial electrical stimulation (CES), but at the time of writing there was insufficient evidence to include these among the standard interventions for TBI discussed in a basic textbook. NeuroField, a system that uses pulsed electromagnetic stimulation, has received positive comments from other practitioners, and though we are investigating the use of this approach at our Centre, we have insufficient data to report on its possible contribution to the treatment of TBI at this time. This method involves stimulation and not just feedback, and there are no controlled studies at this time of writing. There are other devices and equipment on the market that we do **not** have at our clinic for which both positive and negative outcomes and occasional negative side effects have been reported. The exhibit hall at AAPB and ISNR meetings provides an opportunity for practitioners to explore a wide range of equipment. Those meetings also afford the practitioner who wishes to carry out 'evidenced-based-practice' an opportunity to discuss various approaches with experienced professionals.

After a concussion, most people will have many symptoms resolve in a week to ten days if they rest. Rest means both physical and mental rest – no reading, television, or computer activities. In those people whose symptoms do not resolve or, in some cases, worsen, we refer to their problems as post-concussion syndrome. These people, in whom spontaneous recovery has stalled, can be helped with a combination of NFB and HRV training.

Symptoms

There is nothing "mild" about the impairments that may result from a "mild traumatic brain injury" (TBI), which is commonly called a concussion. A lack of objective pathognomonic signs of concussion means that a physician's diagnosis often rests, for the most part, on the patient's report of subjective symptoms, such as headaches, that are also common in the nonconcussed population and known to vary day to day. The most frequent symptoms include being forgetful, irritable, impatient, anxious, depressed, and having poor attention span and concentration. Less frequent persisting symptoms include: fatigue, insomnia, poor sleep that is not restorative, headaches, sensitivity to sound and light, blurred vision, double vision, restlessness, nausea, and even an inability to feel warm.

Common presenting symptoms in the emergency department include short-term symptoms of dizziness, nausea, headaches, balance deficits, and blurred vision. The longer-term problems include fatigue, sleep disturbance, slowed reaction time, plus cognitive and emotional changes that may produce difficulties in their family relationships or at work (If they are still able to work). The cognitive problems include slowed processing speed, poor sustained attention, poor concentration, impaired short-term and long-term memory, and learning problems. Deficits in attention are one of the most common symptoms after suffering a brain injury (Van Zomeren & Brouwer, 1987; Dockree, Kelly, Roche, Hogan, Reilly & Robertson, 2004). Language problems, including word-finding difficulties as well as poor comprehension, are also frequently observed among patients following TBI. There is often a tendency to blurt things out observed in individuals with frontal lobe damage (Shallice & Burgess, 1991). Emotional changes may include depression, anxiety, labile mood, irritability, and anger ("a short fuse"). After a concussion, a person is three times more likely to develop depression. Unfortunately, in some cases, the family and friends notice a major personality change to which the patient may be oblivious. It is not at all uncommon for someone to present years after the injury with these difficulties present to varying degrees. Most people who suffer a first concussion will have symptoms resolve within a week to 10 days. The chance of symptoms persisting increases with the severity of the TBI and with the number of head injuries. In a vulnerable person, such as someone who has not fully recovered from previous TBI, even a mild injury, such as a whiplash from a rear-end collision without the head hitting anything and without loss of consciousness, can result in profound memory, cognitive, and even interpersonal relationship difficulties that do not spontaneously resolve.

In the emergency department, the experienced physician will use the Sport Concussion Assessment Tool, called the SCAT 3, and the Glasgow Coma Scale. The reader can get reprints from http://journals.bmj.com/cgi/reprintform and download the scale from bjsm.bmj.com. It is published in the *British Journal of Sports Medicine* (2013, 47: 259) as the Sport Concussion Assessment Tool SCAT 3 (Guskiewicz et al., 2013). We advise the interested reader to download this scale. It includes a symptom checklist and a number of tests for the emergency physician to use with patients who present with a concussion.

Below is an example of one of the many tests included in the scale:

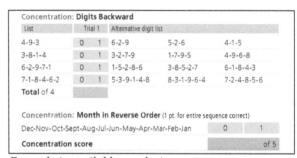

Example is available on the internet.

ADD Centre Assessment (Short Summary)

The initial interview is carried out by the most senior clinical staff, the Director or the Clinical Coordinator. This is a lengthy interview that looks into the patient's chief complaints, medical, family, social, academic, and work history. It will gather data concerning medications, diet, sleep, and exercise. Of prime importance are the patient's goals and the goals that important family members harbor for the patient. This can lead to some discussion of what may be realistic under the circumstances. Questionnaires include ADHD inventories. With adults, the Beck Depression Checklist and the Beck Anxiety Inventory are used in addition to the Zung scales that are included in an online neuropsychological battery from CNS Vital Signs. This inventory includes quite difficult neuropsychological testing. It can be done at our Centre or by the patient in their own home. It is described later in this chapter. In our office there is administration of the IVA and T.O.V.A. continuous performance tests. Previous assessments and report cards, if available, are reviewed. Dr. Lynda

Thompson always uses a low-tech, high yield measure called Draw-a-Person. Artistic attributes are not the focus, but these drawings have always, in our experience, reflected accurately changes in the patient as they progress through training. Human figure drawings and the context in which they are placed help with hypotheses about emotional functioning, as distinct from intellectual, academic, or vocational functioning.

The assessment then proceeds with single-channel EEG at Cz referenced to the ear, followed by a 19-channel quantitative electroencephalogram (QEEG). This is done twice, replicating the EEG using two different amplifiers and assessment programs. The second program also includes evoked potentials (ERPs), a continuous performance measure, heart rate variability measures, and an extensive, well-researched questionnaire. With selected patients, a biomedical analysis may be done using an outside service. Separate comprehensive neuropsychological testing may also be carried out. In the future, we may also add equipment to test balance using a force plate and virtual reality glasses as will be described later in this section.

Pathology of a TBI

A variety of neuropathological processes can be triggered by damage caused when brain matter collides with the rough, ridged inside edges of the skull or the tough falx cerebri, which is an infolding of dura matter between the hemispheres. With rapid deceleration, as in an automobile crash, or when a football player is running at full speed and is tackled, there is a coup injury where the brain first hits the skull, and a contra-coup injury as it rebounds and hits against the opposite wall of the skull. The results of this type of injury are usually detected in the hospital and show up with the more commonly used imaging techniques. These techniques may include: computerized tomography (CT scan), originally called computerized axial tomography or CAT scan, positron emission tomography (PET), single-photon emission computed tomography (SPECT), magnetic resonance imaging (MRI), and functional MRI (fMRI).

More Detail on Scans

As previously described in the section on tDCS, a PET scan measures glucose metabolism. In simple terms, PET provides a visual display of brain activity by detecting where a radioactive form of glucose goes while the brain performs a given task. SPECT is similar to PET in its use of radioactive tracer material and detection of gamma rays. In contrast with PET, however, the tracer used in SPECT emits gamma radiation that is measured directly, whereas PET tracer emits positrons that interact with electrons up to a few millimeters away, causing two gamma photons to be emitted in opposite directions. A PET scanner detects these emissions "coincident" in time, which provides more radiation event localization information and, thus, higher resolution images than SPECT (which has about 1 cm resolution). You can think of the PET scan in simple terms as a visual display of brain activity that detects where a radioactive form of glucose goes while the brain performs a given task. SPECT scans, which track perfusion (blood flow), are significantly less expensive when compared with PET scans, in part because they are able to use longer-lived and more easily-obtained radioisotopes than PET. Both these measures are 'invasive' in the sense that radioactive isotopes must be used.

In contrast, the MRI and fMRI have the advantage of not requiring any injection or ingestion of radioactive material, but the MRI tube is a noisy environment and requires the patient to lie very still in the tube while the magnetic imaging is being done. In addition, the subject cannot have any metal object in their body and may not have a pacemaker. MRI relies on the way hydrogen atoms, which make up nearly two-thirds of the human body, absorb and then give off magnetic energy at radio frequencies. The nuclei of these hydrogen atoms are like tiny spinning magnets. They respond to changes in magnetic fields, which are controlled by the machine. These responses can be recorded and three-dimensional anatomical images of the brain can be formed by the computer. The fMRI procedure is similar to MRI, but uses the change in magnetization between oxygen-rich and oxygen-poor blood as its basic measure and the subject is required to do a mental task during the recording, hence the term 'functional' MRI. Thus, MRI scans image anatomical structure and give a three-dimensional picture of that structure, whereas fMRI images metabolic activity within these structures. The machines are expensive to buy and expensive to maintain. In Canada, a scan costs the healthcare system about $2,000.

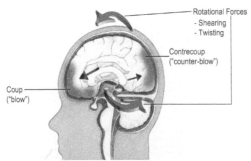

Example of sites of injury in TBI
Drawing by Maya Berenkey using original figures by Amanda Reeves.

Why Use EEG and ERP Evaluations?

The various scans, including PET, SPECT, MRI and CAT, do not generally detect diffuse axonal injury (DAI); that is, the twisting and stretching of axons. (For more information on detection of DAI, the reader is referred to Thatcher, 1989, 1998). Damage resulting from these mechanisms is most easily demonstrated using the QEEG (quantitative electroencephalography). These rotational types of injury, which the QEEG can detect, are illustrated in the figure of a pyramidal cell below.

Diffuse Axonal Injury (DAI)

In addition to gross injury to the brain described above, another type of injury occurs even more commonly in mild and moderate concussion. It is called diffuse axonal injury (DAI). It occurs both with hitting the head and also when there is rapid acceleration/deceleration and/or quick rotation of the brain such as may occur with a whiplash or a blast injury. DAI occurs from mechanically induced twisting, stretching, shearing, or tearing of nerve fibers, particularly axons of neurons. An axon is a nerve connection that goes from the cell body of one neuron to the dendrite or cell body of another neuron. It connects to the next neuron at a chemical junction called a synapse. When a neuron fires, the electrical impulse travels down the axon, which can be a long distance, to the end of the axon, where it causes the release of a chemical called a neurotransmitter. This neurotransmitter enters the synaptic space between the end of the axon and the receptor sites on the dendrite of the neuron that is receiving the message. The most common excitatory neurotransmitter is glutamate.

Medications can change this transmission and the reception of neurotransmitters in the synaptic space. Also, damage to an axon will interfere with this transmission of messages from one neuron to another. DAI is a primary pathologic feature of brain injury in concussions at all levels of severity (Thompson, 2012; Kushner, 2001). The biochemical and structural changes caused by DAI may not be detectable by MRI, but the QEEG is able to detect electrical results of these changes.

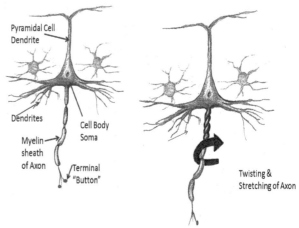

Structure of a neuron, pyramidal cell, drawn by Maya Berenkey.

The terminal button (also called "terminal bouton") is the presynaptic link to the dendrite of the next neuron. This synaptic transmission is disrupted when there is DAI. (Note that this figure shows a pyramidal cell. Only pyramidal cells produce the extracellular currents that summate to produce the EEG.) The next figure shows a schematic representation of a synaptic junction.

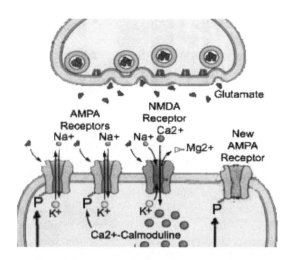

Schematic diagram of a synaptic junction to show, at the top, the presynaptic ending of the axon ('button') and the postsynaptic receptor sites for glutamate: the α-amino-3-hydroxy-5-methyl-4-isoxazolepropionic acid receptor (AMPA), which is a glutamate receptor site, and the NMDA (N-methyl-D-aspartate) receptor site, which is initially blocked by magnesium ions.

This figure is often used to illustrate long-term potentiation. It is used here to show the synapse. Thanks are extended to Wikipedia Images: Bruno Dubuc, Website of Canadian Institutes of Health Research; Institute of Neurosciences; Mental Health and Addiction (NMHA); and to Copyleft - The content of the site, The Brain from Top to Bottom, is under copyleft. ☺ This figure is also used in Nature Reviews, Neuroscience, *Nature Publishing*, 2005.

DAI: Glutamate Release

After a blow to the head that has resulted in a DAI, there is set off a flood of neurotransmitter release that includes glutamate. Glutamate excess will stimulate neurons to fire 'incessantly.' This results in an increase in neuronal permeability with an influx of sodium and calcium and an release of potassium. Both sodium (Na^+) and calcium (Ca^{2+}) excess can interfere with mitochondrial function.

More Detail Concerning Mitochondria

Each cell contains about a thousand mitochondria. The mitochondria produce a substance called adenosine triphosphate (ATP). ATP's main purpose is to transport chemical energy within cells for metabolism. When the membrane of the receptor is altered by a neurotransmitter in the synapse, ions such as sodium and calcium (Na^+ and Ca^{++}) are allowed to flow from the synaptic space on to the receptor surface of the next neuron's dendrite. The electrical potential across the cell membrane changes, and an electrical impulse goes along the dendrite to the neuron cell body. However, the **electrical potential across the receptor site membrane must immediately be restored, and this requires energy**. (This 'active transport system' requires ATP. It is described below and also in Section XXV on Asperger's syndrome.)

Energy: ATP

Metabolism in the mitochondria involves adenosine triphosphate (ATP) to produce this energy and restore normal membrane potential. ATP stores energy in the chemical bonds that hold it together, and when it breaks down into its component parts it releases that energy to the cell and it becomes adenosine diphosphate (ADP). In most cases, ADP is immediately recycled in the mitochondria, is recharged and exits the cell as ATP in a metabolic process that requires fatty acids, sugars such as glucose, and oxygen. In DAI, the twisting or stretching of the axon causes a mechanical deformation (Gennarelli, 1998) that results in disrupted neuronal membranes (Giza & Hovda, 2001).

In addition, there is an opening of K^+ channels ($K+$ efflux) while the Ca^{++} influx is taking place. Glutamate (EAA) is released, resulting in further depolarization, which triggers energy-requiring membrane pumps (Shaw, 2002). Relatively rapidly there is an **exhaustion of stored adenosine triphosphate (ATP)**. The increase in glycolysis to produce needed ATP results in an **increase of lactic acid** production (a byproduct of glycolysis), and a decrease in lactic acid metabolism (Giza & Hovda, 2001).

Lactic acidosis (a decrease in pH to <7 with an increase, therefore, in hydrogen ion concentration) will further inhibit normal mitochondrial respiration and metabolism (Hillered et al., 1984). To keep firing, the neurons employ anaerobic glycolysis, a less efficient means of generating energy, which leads to strained ion pumps (mechanisms requiring energy). This process, in turn, leads to a cascade of interdependent events (e.g., derivation of free radicals, extracellular glutamate accumulation, calcium influx, etc.).

Hypermetabolism and an energy crisis lead to a secondary reaction with increased regional cerebral blood flow (rCBF), which occurs after about one minute, to meet increased energy demands needed to restore ionic membrane balances (Meyer et al., 1970; Nilsson & Nordstrom, 1977; DeWitt et al., 1986). This is followed by a subsequent decrease of up to 50 percent in rCBF for two to four weeks, depending *upon the severity of injury (Shaw, 2002). There is an increase in the amount of endothelium-derived contractile factors (EDCFs), and a dramatic decrease in the amount of endothelium relaxing factors (EDRFs). These soluble vasoactive factors, derived from vascular endothelium, modulate the tone of underlying smooth muscle cells and thus blood pressure and blood flow. The decrease in cerebral glucose metabolism is global. With the uncoupling of the normal system, which is driven by metabolic demand, glucose utilization increases and cerebral blood flow (CBF) decreases.*

With normal blood flow and nutrition, the mitochondria can produce energy in the form of adenosine triphosphate (ATP) by means of the citric acid cycle (Krebs Cycle), which is shown below. Energy is essential to life and it is required in the cell membrane for the transport of ions against a concentration or electrical gradient. **Without this energy source, the cell cannot depolarize and repolarize normally.** Cell death is the eventual result. With a shutdown of the mitochondrial production of ATP there is a reduction of energy production and an 'energy crisis.' To keep firing, the neurons demand extra energy.

More Detail on the Krebs Cycle

Krebs, a German biochemist, in 1937 described the citric acid cycle. He published his work as: The History of the Tricarboxylic Acid Cycle (Perspect. Biol. Med., 14: 154-170 (1970).

*Briefly, after the glycolysis (**breakdown of carbohydrates**) results in the formation of pyruvic acid and hydrogen ions (H+) in the cell's cytoplasm, the pyruvic acid molecules travel into the interior of the mitochondrion. The **pyruvic acid molecules undergo oxidation** (lose electrons and combine with oxygen) in the mitochondria and the Krebs cycle begins.*

Thus, once the pyruvic acid is inside the mitochondria, the Krebs cycle is initiated. In this process, carbon dioxide is enzymatically removed from each three-carbon pyruvic acid molecule to form acetic acid. An enzyme then combines the acetic acid with an enzyme, coenzyme A, to produce acetyl coenzyme A, *also known as acetyl CoA.* **Once acetyl CoA is formed, the Krebs cycle begins.** *The cycle is split into eight steps that are shown in the diagram.*

Why Include a Krebs Cycle Diagram

It is not important for the reader to know this biochemistry, but the authors have included the diagram **to give an indication of the many vitamins and minerals** that are needed even just to complete this one process. This emphasizes the need for practitioners to have the patient consult with a professional who can analyze that patient's unique nutritional deficiencies and requirements and make dietary and supplement recommendations.

During this Krebs cycle, three major events occur. One molecule of GTP (guanosine triphosphate) is produced, which eventually donates a phosphate group to ADP to form one ATP; three molecules of NAD (nicotinamide adenine dinucleotide) are *reduced; and one molecule of FAD (flavin adenine dinucleotide) is reduced. Although one molecule of GTP leads to the production of one ATP, the production of the reduced NAD and FAD are said to be more significant in the cell's energy-generating process. This is because NADH and FADH$_2$ donate their electrons to an electron transport system that generates large amounts of energy by forming many molecules of ATP.*

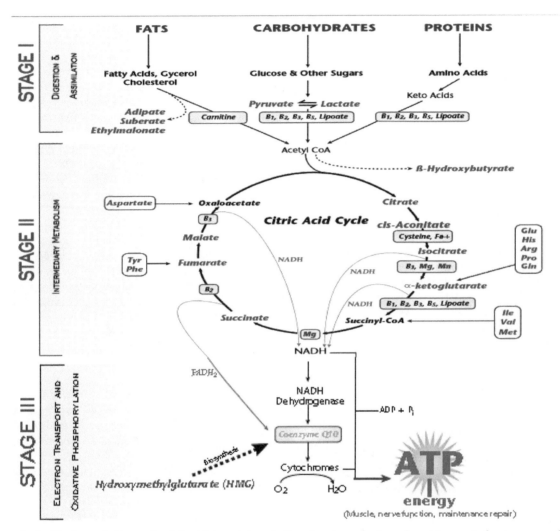

In the Krebs cycle figure from the Metametrix website, vitamin and mineral requirements for enzyme cofactors are shown in light blue boxes. Amino acids are shown in white oval boxes.

Citric Acid Cycle/Krebs Cycle

(Reference the website of: Metametrix Clinical Laboratories Inc. 3425 Corporate Way, Duluth, GA 30096, USA, at www.metametrix.com)

In the Krebs cycle figure from Metametrix, vitamin and mineral requirements for enzyme cofactors are shown in light blue boxes. Amino acids are shown in white oval boxes. Blood analysis done by specialized groups such as Metrametrix (ION profile) can demonstrate elevations of metabolites before each step in the cycle. These 'elevations' indicate **functional deficiency of specific nutrients**. They can also analyze for vitamin and mineral deficiencies. With this information, a specialist, such as a dietitian or a Ph.D. nutritionist, can prescribe an appropriate diet and necessary supplements for the specific needs of your client.

Krebs requires basic nutrients: The important point for this text is for the reader to note that in the process leading up to the production of ATP, NADH, and $FADH_2$, the mitochondria require basic nutrients: carbohydrates, proteins with all the amino acids, fats, and many vitamins and minerals. The above diagram of the Krebs cycle emphasizes this point. Counseling of patients always includes consideration of their diet and may require a comprehensive analysis of their needs by a commercial company such as Metametrix.

More Detail Concerning Nutrition

Below we list some of the essential components of a good nutritional analysis.

Vitamins
For this brief discussion, we will merely list necessary vitamins and give one area of importance for each vitamin:
- Vitamin E (also A, C, and nutrient Q10) are important antioxidants
- Vitamin A: vision
- Vitamin C: protection from illness
- Vitamin E: protection from harmful molecules that degrade cell membranes
- Coenzyme Q10: Energy production
- Vitamin B1 thiamine, B2 niacin, B5 pantothenic acid: breakdown and metabolism of food (carbohydrate amino acids). Stress increases the patient's need for these B vitamins.
- Vitamin B2 riboflavin: breakdown of carbohydrate, fat, protein.
- Vitamin B6 pyridoxine: to utilize/metabolize amino acids (AA).
- Vitamin B12 and Folic acid: to help decrease risk of disease (anemia, chronic fatigue, etc.); important to lower risk of heart disease and cancer
- Biotin
- Lipoic acid: protection of cell membranes

Minerals
- Mg magnesium, Fe iron, K potassium: These are important for nerve transmission, digestion, antibodies, metabolism of nutrients. They are involved in catalytic processes of many chemical reactions (e.g. Krebs cycle). These minerals can be measured in blood cells.
- Ca calcium, Mg magnesium: important in bone and red blood cell (RBC) metabolism.
- Cu copper: important in the metabolism of iron, energy production, and many enzyme reactions
- Mn manganese: important for skin, bone, cartilage, blood glucose control, antioxidant reactions, and many enzyme reactions
- K potassium: crucial in water balance, muscle and nerve cell functions and in nerve conduction.
- Se Selenium: a trace element important in the production of glutathione, which is an important antioxidant
- Zn zinc: More than 100 enzymes require zinc

Essential Amino Acids
An essential amino acid is an indispensable amino acid. It is an amino acid that cannot be synthesized de novo by the organism. These amino acids must be supplied in the diet. Amino acids are required for the formation of every tissue in body. They are necessary for the regulation of muscle and hormone activity. Stress will decrease all energy pathways and increase requirements for amino acids as well as other nutrients. A list of amino acids is given below.

Essential	Nonessential
Histidine	Alanine
Isoleucine	Arginine*
Leucine	Aspartic acid
Lysine	Cysteine*
Methionine	Glutamic acid
Phenylalanine	Glutamine*
Threonine	Glycine*
Tryptophan	Proline*
Valine	Serine*
	Tyrosine*
	Asparagine*
	Selenocysteine

*Essential only in certain conditions. References and list are from Reeds, P.J. (July 2000). Dispensable and indispensable amino acids for humans (*Journal of Nutrition, 130*(7), 1835S-1840S); Fürst, P., Stehle, P. (June 2004). What are the essential elements needed for the determination of amino acid requirements in humans? (*Journal of Nutrition, 134* (6 Suppl): 1558S-1565S)

In blood analysis, a few of these amino acids that often may be found to be deficient include:

- Tryptophan: This is important in the production of serotonin and is critical for mood, behavior, appetite, sleep, and may be deficient in syndromes such as fibromyalgia.
- Arginine: This is required for the production of nitric acid, which lowers BP and is important in the immune system.
- Glycine propionyl-L-carnitine: This is required in order to use fatty acids for energy.
- N-acetylcysteine: This is important in prevention of oxidative damage.

But other amino acids are also important, and only a careful analysis will highlight which ones are needed for a particular patient.

Fatty Acids
Fatty acids are essential for many important structures, including cell membranes and hormones. Good fats are found in: fish oil, flax seed, walnut and olive oils, avocado.

The good fatty acids include:
- OMEGA 3 polyunsaturated: reduce platelet stickiness (clots)
- OMEGA 6 polyunsaturated: linoleic for cell membranes (but too much favors inflammation and cardiovascular disease)
- Monounsaturated fats: Oleic is needed to build cell membranes

Practical Approach:
Given that nutrition is a complex area requiring an extensive background in biochemistry and physiology, how can the neurofeedback practitioner be of practical assistance to the patient? The simplest route is to have the blood and urine work ordered by their family practitioner and the results forwarded to a nutrition specialist. At the Biofeedback Institute, we usually do one of the following:

Blood and Urine Analysis
1. For regular clients (for example: with ADHD), we recommend OHIP (Ontario Canada health insurance plan) covered blood and urine work as recommended by a Ph.D. nutritionist/dietitian; this is ordered by the family practitioner and the results forwarded to the nutrition specialist.
2. For complex clients (depression, PTSD, TBI), we may recommend the TRIAD Profile from Metametrix, which is then interpreted for us by specialists at Evoke Neuroscience, New York. This analysis can go in progressively more advanced stages. It will look at markers for all the above nutrients, plus toxins and all major "imbalances."

Imbalances are far too biochemically detailed to include for this text. We will therefore just give one brief example to elucidate this process.

Example of an "Imbalance"
With feelings of stress, such as occur frequently in clients who present with anxiety, depression, TBI and so on, pro-inflammatory cytokines are released by macrophages. Their release will coincide with a change in tryptophan metabolism. The metabolism of tryptophan is changed with stress, depression, TBI, Alzheimer's, and perhaps autism.

Tryptophan Pathways in Vitamin B6 Deficiency and Inflammation
The following figures and discussion are from www.metametrix.com

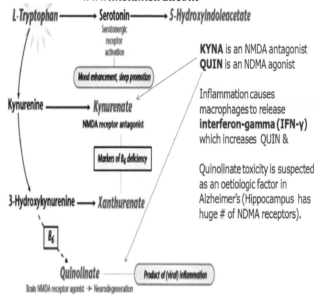

Figure adapted from Metametrix

The NMDA (N-methyl-D-aspartate) receptor is a glutamate receptor. (The reader is reminded of the earlier discussion of long-term potentiation, or LTP.) It is the predominant molecular device for controlling synaptic plasticity and memory function. It is a brain receptor activated by the amino acid glutamate that, when excessively stimulated, may cause cognitive defects such as are found in **Alzheimer's** disease. However, there may be some methods emerging in animal studies to counter the neurotoxicity induced by glutamate excess (Molz et al., 2005).

Important **markers of neurological inflammation** are kynurenate and quinolinate. The kynurenine (KYN) pathway is a major route of L-tryptophan catabolism. KYN operates as a mechanism of defense against intracellular pathogens and as a mediator of stress response signals to the brain. Tryptophan can go down the KYN pathway and be excreted as kynurenate (KYNA) or quinolinic acid (QUIN) – or it can be used to produce serotonin (5-HT), which is excreted as 5-hydroxyindoleacetate (5-HIA). Depending on the timing and juxtaposition of signals, a balance between the neurotoxic and neuroprotective metabolites of KYN serve to either activate or inhibit neuronal responses. The reader will recall from the discussion of TBIs that if glutamatergic neurons are overstimulated, they can degenerate. This degeneration has been noted in **stroke**, end stages of **HIV infection**, and **Alzheimer's** disease. It is also now postulated to be a factor in **autism**.

Inflammation can be an important factor in **depression**, and it may relate to onset of **dementia** (Leonard et al., 2006). Shifts in the kynurenine and serotonin pathways of tryptophan metabolism are important in the etiology of depression (Miura et al., 2008, 2009).

In addition, **QUIN elevation** is a metabolic event with the potential for precipitating brain developmental disruptions similar to those seen in regressive autism. QUIN is an excitotoxin that provides a critical link between the immune system and the brain. QUIN is known to be involved in the pathogenesis of several major inflammatory neurological diseases.

Stimulation of the inflammatory response causes release of interferon-gamma (IFN-γ) by macrophages. There is a tight, positive association between QUIN and IFN-γ. QUIN can bind NMDA receptors of glutamatergic neurons that respond to **pain** and other peripheral signals. Within the brain, the hippocampus is an area rich in NMDA receptors that is sensitive to the neurotoxic effects of QUIN.

Depression has been associated with immune activation and increased concentrations of **pro-inflammatory cytokines**, which have been found to affect the metabolism of **brain serotonin**. Since the gut is frequently a source of chronic inflammatory signal induction via INF-γ, there is reason to suspect that QUIN elevation could indicate both inflammatory bowel conditions and neuronal degeneration. In inflammatory diseases, the ratio of QUIN/KYNA is frequently found elevated (>2.0), so that neurotoxicity must be suspected (Heyes et al., 1992). This ratio can help to provide essential information about the pathway.

Tryptophan is catabolized to KYN by the enzyme indoleamine 2,3-dioxygenase (IDO). Pro-inflammatory cytokines (IFNc, TNFa, IL-1 and IL-2) enhance IDO under **stress, which can promote the KYN pathway, thus depriving the 5-HT pathway of tryptophan and reducing 5-HT** synthesis. This can be important in depression (Myint & Kim, 2003; Leonard & Myint, 2006). In stroke and degenerative diseases the ratio of QUIN/KYNA is very high. QUIN is also high in chronic fatigue. The widespread exposure to phthalates in plastic products has been shown to have potential for enhancing QUIN production.

KYNA is increased in inflammatory responses. One therapeutic intervention for excess QUIN is to supply magnesium, because it will compete with QUIN for the NDMA receptor sites and thus prevent calcium entry into the neuron. Magnesium and opioids have both been used as treatments to stop overstimulation by glutamate.

The key compounds, KYNA and QUIN can both be assessed by companies such as Metametrix (Organix Profile). Their profiles are one good way to inquire about the cellular function of the inflammatory pathway, as well as monitor treatments that can affect that pathway.

The NFB practitioner should be aware that diet and supplements can have a significant impact on the same patients that are being treated with NFB + BFB. It makes sense to have these patients referred to a nutrition specialist who would have the patient's doctor order the appropriate blood and urine analysis. The nutrition specialist could then prescribe the correct diet and supplements for the patient.

Summary of Metabolic Effects of TBI

In addition to the direct interference with ATP formation in the mitochondria, due to the influx of sodium and calcium, and a lactic acidosis, there is also an additional problem due to the vasoconstriction caused by the excess release of potassium. Excess potassium causes constriction of the blood vessels. This will decrease blood flow to the compromised area in the brain. The mitochondrial Krebs cycle requires fuel to keep running. In simple terms, this fuel includes glucose, fats, amino acids for coenzymes, and vitamins for the enzymatic processes. However, the constriction of the blood vessels caused by the excessive release of potassium limits the supply of these necessary nutrients and oxygen to the cell.

Thus, the high energy demand, restricted blood flow, and oxygen debt **create an energy crisis** that exhausts the neurons. This will inevitably lead to symptoms such as mental confusion and failed memory, symptoms commonly seen in people who have had a concussion. The brain will take varying amounts of time to recover. The time to restore the chemical balance will vary among people and will depend on variables including age, general health, history of previous head injuries, and many other factors, such as nutrition. Long-term potentiation

(LTP) is decreased, and long-term depression (LTD) is increased. The membrane excitability is reduced.

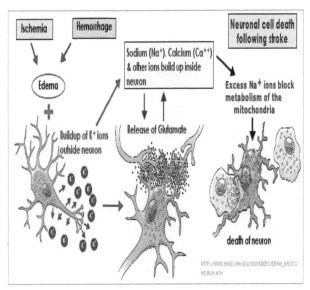

Figure to illustrate steps contributing to neuronal cell death. NIH government document in public domain. Although originally used by NIH to illustrate changes after a stroke, this diagram applies equally well to the energy crisis that ensues after TBI. The next figure summarizes some of the factors that contribute to this energy crisis.

Timelines for Metabolic Recovery

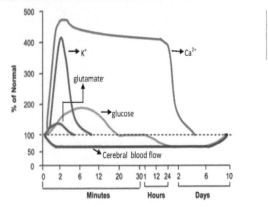

*Figure: Chemical and metabolic changes post TBI. Adapted from information contained in David A. Hovda, Ph.D. **UCLA Brain Injury Research Center** and from a drawing in: Giza and Hovda (2001), after Giza and Hovda (2000).*

This diagram is based on information from the UCLA Brain Research Center. It implies that in about 6-10 days one may observe neurochemical recovery. However, timelines predicted by different authors do show quite a range. Giza and his colleagues suggest that, due to changes and subsequent derangements in cerebral glucose metabolism, persons suffering from even minimal traumatic brain injury (mTBI) may remain vulnerable to second impact injury for periods of up to a few months (Giza and Hovda, 2001). Other authors note that hypermetabolism and an energy crisis lead to a secondary reaction with increased regional cerebral blood flow (rCBF) (occurs after about one minute) to meet increased energy demands needed to restore ionic membrane balances (Meyer et al., 1970; Nilsson & Nordstrom, 1977; DeWitt et al., 1986). Shaw notes that this initial increase in rCBF is followed by a subsequent decrease of up to 50 percent in rCBF for two to four weeks, depending upon the severity of injury (Shaw, 2002). Due to factors that include the gradual death of some neurons after diffuse axonal injury and long-term decrease in cerebral blood flow so that the energy crisis in cells cannot be resolved, it is possible for symptoms to worsen a few weeks after the concussion so that they are more serious than symptoms seen after initial neurochemical recovery during the first week.

Second Impact Syndrome

The most serious consequence of having a second concussion before recovery from the first one is complete involves acute swelling in the brain that can lead to death or serious lasting impairment. This is called second impact syndrome and accounts for one or two deaths per year in the USA in football alone, often in young players under the age of eighteen. Adolescents and females appear to be more vulnerable to the effects of a concussion, perhaps because the strength and girth of their necks is less, so that injuries like whiplash or from taking headers in soccer are greater (Tator, 2013).

Persons who receive a second mTBI during the symptomatic (injured) period are reinjured at a time when they already have disturbances in mitochondrial functioning and cerebral glucose metabolism. They may **remain vulnerable to second impact injury for periods of up to a few months** (Giza & Hovda, 2001). Athletes are particularly prone to having a second, or even third, concussion. This may, in part, be due to timing difficulties resulting from alternate pathways for information processing not being as fast or effective as the primary injured pathways in the brain. With each injury the deficits typically increase and the period of time for recovery may become significantly longer. Second impact syndrome has even resulted in deaths. It is estimated that in football players in the United States there are one to two deaths per year due to second impact syndrome.

Assessment

General Considerations

In mild to moderate TBI (mTBI), since DAI may be missed by conventional imaging techniques, and because conventional neuropsychological testing and symptom checklists, such as those traditionally used with injured athletes, have been found to normalize within a few days to a few weeks, a more advanced paradigm is needed to detect this type of injury (Thompson, J., 2006: Ph.D. thesis, Penn State University). A summary of the conclusions for mild TBI reached by James Thompson in his doctoral research, which involved development of a new metric for assessing the sequelae of head injury and measurement of recovery in varsity athletes, encompassed five points:

1. **SR (self-report)** is sensitive within **24 hours**.
2. **NP** (neuropsych.) testing typically normalizes after one week.
3. **Postural and EEG**-sensitive testing is sensitive beyond one week.
4. **Combining modalities** is more sensitive and reliable than any single method.
5. LORETA is effective to localize cortical areas affected by the head injury.

Thompson found that no one test was sufficient and he recommended the use of a testing paradigm that combines the most sensitive tests from each modality.

Comprehensive Assessment

Based on his research and that of others in this field, he and David Hagedorn, a neuropsychologist who has worked with servicemen and -women in all branches of the military, most recently the Navy and Marine Corps., developed a method for the comprehensive measurement of brain functioning. The primary tool for data collection is the Brain Injury Assessment Tool (BIAT), which is now called eVox, and this is used along with computerized neuropsychological testing and biochemical analysis (www.Evokeneuroscience.com). The eVox (formerly called the BIAT) instrument measures 19-channel EEG, QEEG, ERP (evoked potentials) and HRV, while, in addition, doing a continuous performance test (CPT) and a standardized questionnaire. When requested, a force-board/virtual reality balance (vestibular function) test can be added. Biochemical testing of blood and urine is added as deemed necessary. This is the methodology now used at the ADD Centre/Biofeedback Institute in Toronto. In addition to the comprehensive report by Evoke Neuroscience, we analyze the 19-channel data with NeuroGuide and LORETA. (We have also used in the past both SKIL and WinEEG). We additionally administer TOVA and IVA continuous performance tests and our usual battery of questionnaires. This testing is uniformly revealing abnormalities in our clients even years after the primary injury.

In summary, the assessment process will ideally take into consideration all of the networks primarily affected by a minimal TBI or concussion. These include: the Executive networks (attention, deductive and inductive cognitive processing functions, verbal comprehension, and word finding); the Affect networks including the autonomic nervous system functioning; the vestibular networks; and aspects of the biochemical functioning of the patient.

Role of Quantitative Electroencephalogram (QEEG) in Assessment

The principle findings in a QEEG, even months to years after a concussion, can include slowing, decreased power, episodic discharges, and changes in coherence. A slowed peak frequency, also referred to as the dominant frequency or peak alpha, is a common finding after a TBI. Normal adult peak frequency, measured in the occipital region with eyes closed, is about 10 Hz. With a concussion it may drop into the low alpha range, perhaps 8 Hz, or even fall into the theta range. A dominant frequency below 8 Hz in an adult is considered abnormal. Another common finding is reduced overall power. The most consistent power findings after a concussion are reduced power in the higher frequency bands (8-40 Hz), and increased slow waves (1-4 Hz) (Slobounov, Sebastianelli & Simon, 2002; Thatcher, 2009; Haneef, Levin, Frost & Mizrahi, 2013).

There are long-term changes associated with an increase in the 1-4 Hz/8-12 Hz ratio (Korn, Golan, Melamed, Pascual-Marqui & Friedman, 2005; Slobounov, Cao & Sebastianelli, 2009). In particular, there is an initial increase in low alpha, 8-10 Hz, but a decrease in fast alpha 10.5-13.5 Hz and fast beta 20-35 Hz (Haneef et al., 2013, after Tebano et al., 1988). A change (increase) in this 1-4 Hz/8-12 Hz ratio activity is negatively correlated with patient outcome six months postinjury (Leon-Carrion et al., 2009).

Early research by Thatcher demonstrated that increased frontal and temporal coherence and decreased phase, increased anterior-posterior amplitude difference, and reduced posterior power could discriminate normals from post-TBI patients (Thatcher, Walker, Gerson & Geisler, 1989). Thatcher has suggested that power changes in the EEG following concussion result from dysfunctional ionic channels of neuronal membranes (e.g., Na+, K+, Ca++) and reduced average current flux (Thatcher et al., 2001). Deceased and/or increased coherence is a common finding. Coherence reflects communication between different sites in the brain and represents the amount of phase-locked activity in a particular frequency band. TBI will disrupt coherence. The most common finding is a decrease in coherence that is colloquially called a "disconnect." Diffuse axonal injury (DAI) has also been shown in other studies to be associated with disturbances in connectivity between different areas of the brain (Meythaler, Peduzzi, Eleftheriou & Novack, 2001). A less frequent but important finding may be episodic discharges. A dramatic example reported by Sterman was that one of his patients dropped his six-month-old infant and was accused of abuse. Sterman felt that it was clear that the cause of the accident in this very conscientious parent was, in all probability, a sudden episodic discharge with momentary loss of muscle tone (Sterman, personal communication).

Despite all these QEEG findings, it should always be remembered by the practitioner that paroxysmal EEG changes, including brief, but not persistent, high-amplitude bursts of any frequency, may not be noticed as an increase when average amplitude is looked at with the quantitative EEG analysis. The practitioner must look at the whole raw EEG and would be advised to do single-point analysis (available using LORETA in the NeuroGuide program) of these isolated bursts. If their source correlates with the patient's symptoms, then these can be targeted for training when doing LNFB. An example is shown in the next figure. The blue vertical line can be moved through a short horizontal distance to different points on the high-amplitude wave in the frequency band you are interested in. In this case we told the program to look at high beta 25-30 Hz.

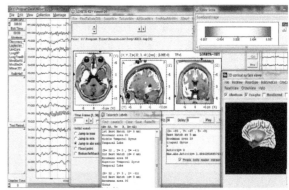

A.N. Male, age 35. Single point analysis of spindling beta (25-30 Hz band) observed at F8 shows the right uncus as the source.

Addition of Evoked Potentials (ERPs) to Assessment Measures

Information from evoked potentials, also called event-related potentials (ERPs) because they are time-locked to a specific event (sight or sound or even touch), can be used following concussion to determine the intactness of cortical pathways and speed of processing within these pathways (De Beaumont et al., 2009; Dupuis et al., 2000). ERPs reflect the neural activities associated with cognitive/behavioral demands, and thus provide access to an improved understanding of brain functioning that goes beyond either neuropsychological testing or motor testing (Gosselin et al., 2010). ERPs have been shown to be resistant to practice effects and are therefore a hearty diagnostic assessment tool (Mendez et al., 2005) that can be repeated as often as necessary.

The population who have suffered a traumatic brain injury (TBI) demonstrate reduced amplitudes for the P3 components (P3a and P3b) of their event-related potentials (see Polich, 2007, for a detailed explanation of ERPs). 'P' refers to positive-going waves and 'N' to negative-going waves. The number, such as 300 (sometimes shortened to just 3), refers to the number of milliseconds after the stimulus when you expect the response to occur. Reduced amplitude after an mTBI ('m' is used to denote mild) can indicate diminished attention, because the amplitude of the P3 has been shown to be related to the amount of attention allocated to information processing (Bernstein, 2002; Duncan et al., 2004, 2005). In addition to lower amplitude, the latency of the P3 response has been found to be greater, and this delay in conscious processing of information is called a slowed P3. This increased latency in the P3 response correlates with slowed cognitive processing (Lavoie et al., 2004). Earlier ERP responses are observed at less than 250 ms, which is before there is conscious awareness of the stimulus. They reflect early sensory processing that occurs before conscious awareness of the stimulus.

In our experience, patients with TBI generally have sensory components (P1 and N2) that are longer latency and lower amplitude compared to non-TBI patients. P3 also tends to be longer latency and lower amplitude. As previously mentioned, P3a is an attention-orienting frontal component (Fz), while P3b reflects sustained attention and monitoring and is more central-parietal, with the largest amplitude being observed at Pz. These observations of longer latencies with the P300 ERPs may help us understand why there are persistent problems with attention and concentration after concussion.

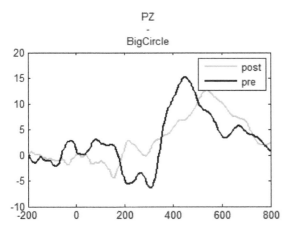

The above figure shows the ERPs for the target stimulus (a big circle). Pre (black) and post (green) P300b ERP recordings from the same subject comparing a baseline measurement and a post-injury measure. Latency of the P3 wave is delayed in the postconcussion test, and amplitude of the P3 wave is reduced following injury. ("Big Circle" refers to the target stimulus in this ERP test, which is in the program from Evoke Neuroscience.) This is just an example figure to emphasize that the P300 after a concussion tends to be 'low' (amplitude) and 'slow' (latency). Thanks to Evoke Neuroscience for this example.

Addition of Heart Rate Variability (HRV) to Assessment Measures

The observation that traumatic brain injury (TBI) impairs cardiac function has become increasingly apparent through work done with relatively young athletes and with clients who have had vehicular accidents (car, motorcycle, bicycle).

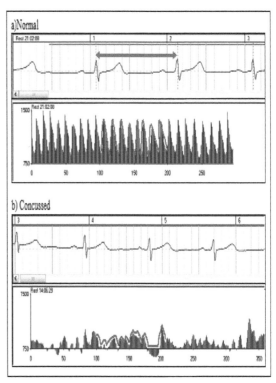

HRV profile from a healthy adult and from a concussed athlete from Thompson, J. & Hagedorn, D. (2012). Multimodal Analysis: New Approaches to the Concussion Conundrum. Journal of Clinical Sport Psychology, *6, 22-46.*

The foregoing figure contrasts heart rate variability in a healthy adult with that of an athlete who has suffered a concussion. (Thompson & Hagedorn, 2012). It demonstrates that after a concussion, even a healthy athlete can show important HRV changes. As previously explained in the section on HRV, the top EKG shows a three-second sample and below it is a tracing of variations in heart rate over a 300-second time period, both from the same healthy 36-year-old male. These demonstrate what one would expect in a healthy person: heart rate variations are even with high amplitude and steady rhythmicity. The figure below it shows data from a highly trained male athlete two weeks postconcussion. It demonstrates low amplitude and poor rhythmicity of heart rate variations. The amplitude of the variations in heart rate and the standard deviations of the heart rate variations after artifacts have been removed (SDNN) were also both lower than would be expected in a young athlete. These data provide an illustration of how brain injury affects not just the brain but also the heart, specifically that concussion has a negative effect on heart rate variability (HRV). Similar observations have been published by researchers

(Baguley et al., 2006) and their results were summarized in the section on HRV training.

Neuropsychological Testing

In a clinical, rather than research setting, testing must be confined to that which can be useful to assist in diagnosis, and more importantly, to design an appropriate treatment program, and track the patient's progress during rehabilitation. The Wechsler Intelligence Scale may be used with other testing carried out as appropriate for a particular patient. In adolescents and adults with a TBI presentation, a computerized neuropsychological test battery coordinated through Evoke Neuroscience is used at the ADD Centre. You can customize your own battery of tests through the company CNS Vital Signs. In the interest of brevity, only a brief example figure showing one-third of the first page of the report is given in the figure below. This test takes the patient a little over an hour to do, and the full report is on our computer within an hour after the testing is completed. The objective testing of aspects of response time and memory can be quite difficult, and it does appear, for the most part in our experience, to pinpoint the difficulties the patient is experiencing. This computerized methodology is simple to administer and the results can be useful, in a clinical setting, to follow progress during treatment. The cost has been <$50 per administration of the battery. The objective test procedures in the version that we use of this computerized NP test are followed by questionnaires. These include: an Alertness Rating Scale, the Zung Anxiety and Depression Self-Report Scales, an Adult ADHD Self-Report Scale, a Neuropsych Questionnaire Short Form (a series of questions about the patient's clinical state including mood, attention, sleep, fatigue, aggressiveness, and so on), and a 16-item Medical Outcomes Survey.

This computerized NP test battery is a cost-effective way to obtain objective and subjective information about the patient and track changes during training. It may be particularly useful for professionals who do biofeedback but who are not trained in the administration of psychological testing. It is also very helpful in the assessment of patients who could not afford individual testing by a psychologist.

Patient Profile	Percentile Range				> 74	25 - 74	9 - 24	2 - 8	< 2
	Standard Score Range				> 109	90 - 109	80 - 89	70 - 79	< 70
Domain Scores	Subject Score	Standard Score	Percentile	Valid Score**	Above	Average	Low Average	Low	Very Low
Neurocognitive Index (NCI)	NA	77	6	Yes				X	
Composite Memory	73	48	1	Yes					X
Verbal Memory	37	42	1	Yes					X
Visual Memory	36	68	2	Yes					X
Psychomotor Speed	137	69	2	Yes					X
Reaction Time*	720	78	7	Yes				X	
Complex Attention*	7	98	45	Yes		X			
Cognitive Flexibility	45	92	30	Yes		X			
Processing Speed	52	84	14	Yes			X		
Executive Function	46	93	32	Yes		X			
Simple Attention	40	108	70	Yes		X			
Motor Speed	84	70	2	Yes				X	

Report from CNS Vital Signs: *Domain Dashboard: Above average domain scores indicate a standard score (SS) greater than 109 or a percentile rank (PR) greater than 74, indicating a high-functioning test subject. Average is an SS 90-109 or PR 25-74, indicating normal function. Low average is an SS 80-89 or PR 9-24, indicating a slight deficit or impairment. Below average is an SS 70-79 or PR 2-8, indicating a moderate level of deficit or impairment. Very low is an SS less than 70 or a PR less than 2, indicating a deficit and impairment. Reaction times are in milliseconds. An asterisk (*) denotes that "lower is better," otherwise higher scores are better. Subject scores are raw scores calculations generated from data values of the individual subtests.*

VI** - Validity Indicator: Denotes a guideline for representing the possibility of an invalid test or domain score. "No" means a clinician should evaluate whether or not the test subject understood the test, put forth his best effort, or has a clinical condition requiring further evaluation.

Verbal Memory Test (VBM)	Score	Standard	Percentile	
Correct Hits - Immediate	7	55	1	Verbal Memory test: Subjects have to remember 15 words and recognize them in a field of 15 distractors. The test is repeated at the end of the battery. The VBM test measures how well a subject can recognize, remember, and retrieve words e.g. exploit or attend literal representations or attribute. "Correct Hits" refers to the number of target words recognized. Low scores indicate verbal memory impairment.
Correct Passes - Immediate	13	79	8	
Correct Hits - Delay	4	54	1	
Correct Passes - Delay	13	81	10	
Visual Memory Test (VSM)	**Score**	**Standard**	**Percentile**	
Correct Hits - Immediate	7	54	1	Visual Memory test: Subjects have to remember 15 geometric figures, and recognize them in a field of 15 distractors. The test is repeated at the end of the battery. The VIM test measures how well a subject can recognize, remember, and retrieve geometric figures e.g. exploit or attend symbolic or spatial representations. "Correct Hits" refers to the number of target figures recognized. Low scores indicate visual memory impairment.
Correct Passes - Immediate	11	93	32	
Correct Hits - Delay	6	67	1	
Correct Passes - Delay	12	100	50	
Finger Tapping Test (FTT)	**Score**	**Standard**	**Percentile**	
Right Taps Average	55	90	25	The FTT is a test of motor speed and fine motor control ability. There are three rounds of tapping with each hand. The FTT test measures the speed and the number of finger-taps with each hand. Low scores indicate motor slowing. Speed of manual motor activity varies with handedness. Most people are faster with their preferred hand but not always.
Left Taps Average	29	56	1	
Symbol Digit Coding (SDC)	**Score**	**Standard**	**Percentile**	
Correct Responses	53	83	13	The SDC test measures speed of processing and draw upon several cognitive processes simultaneously, such as visual scanning, visual perception, visual memory, and motor functions. Errors may be due to impulsive responding, misperception, or confusion.
Errors*	1	102	55	

Correct Passes means the subject correctly inhibited response to a nontarget (NO-GO).

This 22-year-old man ("BR") had a TBI a number of years before being seen at the ADD Centre. He was bicycling when he was hit by a truck. Following the accident he was in coma for one month, and he then spent a number of months in hospital recovering from his physical injuries. On assessment at the ADD Centre, he demonstrated all the symptoms of ADHD and there was also emotional dysregulation. He was articulate and verbally bright. He had trouble falling asleep and had multiple awakenings. His chief complaints were his memory difficulties. In addition, during the assessment, the therapist was concerned about his social awareness, particularly with respect to the effects his comments might have on others. The quick neuropsychological test (illustrated above), accurately revealed his major problems with memory and speed of mental processing. Interestingly, in the ERP test, his P3a and P3b latencies were both quite slow and the anterior posterior gradient was abnormal, as shown in the following figure.

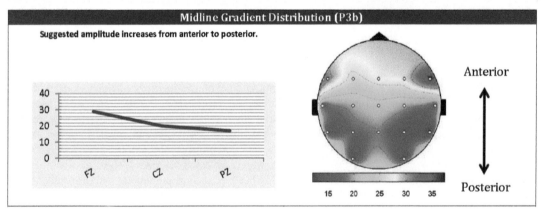

ERP Component	Waveform	Result	Reference Range
Visual processing	O1-Checkerboard	Indiscernible	< 250 ms
Auditory processing	CZ-White Noise	224 ms -16.6 uV	< 250 ms
Attention/Vigilance (P3a)	CZ-Checkerboard	712 ms 22.0 uV	< 450 ms
Information processing / Working memory (P3b)	PZ-Large Circle	712 ms 16.8 uV	< 450 ms

ERP report from Evoke Neuroscience. Note: You would normally expect P3a to be shorter latency than P3b but it is possible to have them the same or the reverse if you remember that: P3a and P3b are: 1) From different tasks; 2) Different brain mechanisms (attention vs. memory); 3) From different generators (frontal vs. temporal-parietal). Thus in some cases if frontal was abnormal but temporal-parietal was normal, then frontal would be delayed but temporal-parietal not (or not as much).

Force-Board "Vestibular Function" Assessment

At the ADD Centre we are considering adding an assessment tool that uses a force plate to quantify how well a person can balance. It was among the measures used by James Thompson when he did his doctoral research at Penn State University and developed a combined metric (four measures: brain mapping using QEEG, NP testing, balance testing, and virtual reality) for assessing varsity athletes who suffered concussions. It enables the practitioner to assess vestibular functioning. It can be a very sensitive tool that, like QEEG, will pick up functional deficits well after the time that conventional instruments (self-report questionnaires and psychometric testing) have returned to the normal range.

The subject has their balance tested in various modes (both feet, one foot, eyes open and closed) while standing on a small rectangular force plate that sends measurements to the computer, which evaluates subtle movements.

Virtual Reality

Another sophisticated measure used in Thompson's research was a virtual reality room (virtual reality glasses can also be used) where vertical lines on the wall can be moved back and forth, and the subject is asked to sway back and forth in time with that movement. Recordings from a vest worn by the subject are synchronized with the virtual room movement. A time-frequency graph shows how long it takes for the person to achieve the same rhythm as the room's movement.

Posture Assessment Tools
Neurocom Smart balance monitor

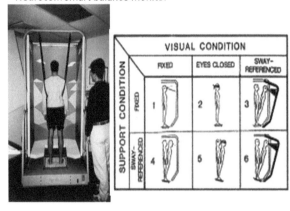

This figure shows how the balance board works. The subject has the EEG recorded while standing, balancing, in various positions. The virtual reality room (or glasses) lean the subject then lean the room then both (Slobounov & Newell, 2009; Slobounov et al., 2008; Thompson, J. et al., 2005; Slobounov et al., 2002).

Postural Data

Injured subjects are unable to integrate visual information with kinesthetic output. Images from Thompson. J.W.G. (2007). "Concussions in Sport: Investigation of Assessment Measures, Neural Mechanisms and Functional Deficits," Pennsylvania State University Libraries, http://etda.libraries.psu.edu/theses/approved/WorldWideIndex /ETD-1885/index.html. In Press. ProQuest, Ann Arbor, MI). Event-related synchronization occurs much later at 0.15 Hz infra-slow-wave EEG that modulates the higher frequency EEG (red, hot).

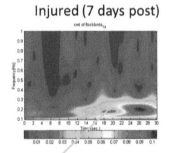

20 seconds before he is able to move in-phase with VR

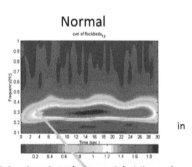

Moving in-phase with Virtual Reality room is almost immediate.

Figure from James Thompson, Ph.D. thesis research to develop a metric for assessing head injury recovery. Reprinted with permission.

After a concussion, postural adjustment was very slow. In most subjects it took more than 20 seconds to assess room virtual movement and adjust their own movements to be in phase with the room after concussion. This is in contrast to a nonconcussed athlete who achieves in-phase movement almost instantly, completely in synch after about two seconds. Further research at Penn State has confirmed these findings (Slobounov & Newell, 2009; Thompson et al., 2005; Slobounov et al., 2002). In the figure above, the y axis is frequency. This is the athlete's oscillation frequency swaying back and forth. The room was moving at 0.2 Hz. therefore, when the athlete got into a 0.2 Hz frequency zone, then they reached maximum power (the red part of the FFT that is at same height as the 0.2 Hz on the y axis). In this example, the concussed athlete did not get to max power at 0.2 Hz until 22 seconds had elapsed. That is very slow to sway in sync with the movement of the room. It indicates a delay in visual-motor response that could be particularly dangerous if an athlete returned to play in a fast moving sport like hockey. It may help explain why an athlete who has suffered one concussion is more at risk for future concussions.

Initial Intervention

If there is subdural bleeding, then the person will require acute medical care or even neurosurgery. Usually a person who has not suffered loss of consciousness is sent home with instructions to rest from both physical and mental activity for a week or until symptoms have resolved. A relative or friend will be told to check their pupils for any changes in the next few hours and days and to watch out for signs of grogginess or loss of balance that could indicate bleeding in the brain. Rest is the only universally agreed upon "intervention" after a head injury and even that is becoming a bit controversial in terms of how long a period of rest is required before one can start gentle exercise, such as going for a short walk wearing sunglasses. Something which can be done immediately (but is not always mentioned by physicians) is taking supplements, in particular, a fish oil supplement so that there is a sufficient supply of omega-3 essential fatty acids to provide the building blocks for myelin production and repair. Coconut oil is also recommended by some.

A Neurophysiological Approach to Effective Treatment

Basic rehabilitation procedures, as are carried out in hospital settings and in rehabilitation settings should be adhered to. However, these are often necessary but not sufficient. The neurofeedback/biofeedback practitioner and nutritionists have important techniques to apply that are now being demonstrated to be significant and life changing.

Normalizing the QEEG deviations from a normal database is a first step. This intervention must be designed while keeping in mind that not all deviations will correspond to symptoms. Indeed we have important examples of genius correlating with significant deviations from a normal database. For example, a musician/medical doctor in a workshop demonstrated amplitudes in the left auditory cortex far outside the database Z-score norms. This person with deviations in the auditory cortex who has perfect pitch does not have the goal of 'normalizing' that aspect of his auditory processing. The cautious practitioner will only include Brodmann areas with Z-score deviations that correlate with a particular patient's symptoms of concern and will carefully assess during each session whether the patient's changes, when meeting the feedback criteria, correlate with symptom relief. Feedback should not induce any decrement in cognitive functioning. This training, that addresses shifting functioning toward what is expected for the person's age, is often done using LORETA Z-score NFB. Nevertheless, we must note that we have had excellent results using single-channel training in past years. As with other disorders, we also do HRV training during each training session. HRV is severely affected by concussion with marked decreases evident in SDNN and power. HRV training may prove to be an important component of treatment in TBI. Part of the reason for this opinion comes from our knowledge of the inflammatory response that occurs after a TBI.

Within minutes of a traumatic impact, a robust inflammatory response is elicited in the injured brain. Inflammation is a stereotyped response that is a protective attempt by the organism to remove the injurious stimuli and to initiate the healing process. There is an activation of glial cells, microglia, and astrocytes, and the infiltration of blood leukocytes.

This is followed by the secretion of immune mediators called Cytokines. Cytokines are small, short-lived proteins produced by blood leukocytes and glial cells. They are quickly released in response to TBI. Cytokines that initiate or propagate an inflammatory response are said to be pro-inflammatory, while cytokines that inhibit the inflammatory response are called anti-inflammatory. While Interleukin-10 is regarded to be primarily an anti-inflammatory cytokine, **Intrleukin-6 (IL-6) is a cytokine related to inflammation.** It is produced by microglia and astrocytes. A 100-fold increase in IL-6 can be readily measured in serum following TBI (Woodcock, 2013). IL-6 regulates inflammation, immunity, bone metabolism, hematopoiesis (formation of blood cellular components), and neural development, and is implicated in aging, osteoporosis, autoimmune disease, Alzheimer's disease, and brain injury. Gevirtz notes that heart rate resonance training could influence the ***parasympathetic cytokine control system*** that modulates immunity through interleukins and interferons. Important for biofeedback practitioners, **this pro-inflammatory cytokine can be reduced by HRV training**. Gevirtz states that the efferent vagus nerve inhibits the release of pro-inflammatory cytokine and protects against systemic inflammation and termed this vagal function the 'cholinergic anti-inflammatory pathway' (Gevirtz, 2013). HRV training has been paired with neurofeedback for nearly every client at our Centres, and these recent findings provide one more reason to do so.

Published case studies also support the effectiveness of NFB for individuals with TBI to improve learning and memory (Reddy, Jamuna, Indira & Thennarasu, 2009). Researchers have seen improvements in functioning in small case series of people who had suffered brain injuries (Ibric, Dragomirescu & Hudspeth, 2009). At the ADD Centre, LORETA NFB is done now with people who present with post-concussion syndrome. Sometimes we alternate one regular neurofeedback session and one LORETA NFB session each week. Both are combined with HRV training. The initial assessment using LORETA in the majority of patients who have suffered a TBI typically shows both positive and negative deviations (>2 SD) at multiple sites, often in central midline areas, accompanied by both coherence and phase differences compared to database norms. LORETA NFB allows the practitioner to address up to 24 parameters, correcting the activity at various sites at the same time for amplitude and/or coherence and/or phase.

Do Patients Recover?

At the time of writing we have had more than 50 patients with postconcussion symptoms treated with a combination of NFB + HRV. These were all patients who came a considerable time after all spontaneous recovery had plateaued. We are able to state that all the postconcussion patients who completed a neurofeedback plus biofeedback program at the ADD Centre have been able to return to work and/or activities of daily living and have been successful in the amelioration of most cognition and affect symptoms. Clients have included: an author who could barely compose a paragraph when first seen two years post-injury who subsequently authored six books, one of them entitled *Concussion is Brain Injury* (Jeejeebhoy, 2012); a student who had to take a greatly reduced course load all through his undergraduate years after suffering a head injury in his first year at university, who completed a demanding graduate program in actuarial science with absolutely no accommodations while concurrently doing neurofeedback; and a Ph.D. candidate who had been stalled in his thesis work and unable to teach for two years after a whiplash injury who returned to his studies and subsequently completed his doctoral work on artificial intelligence, sending us a copy of his graduation certificate along with a note crediting his recovery to the neurofeedback intervention. These are three examples of what we mean when we say that training can be life-changing.

More Detailed Case Example

The longest time elapsed between injury and successful recovery among our clients was over fifty years. This man came to the ADD Centre with the goal of improving his concentration, explaining that he was sixty but knew he would have to work for many more years because he could not afford to retire. He was self-diagnosed with ADHD and said, to keep working, he needed to focus better. He had started work at age sixteen in construction, worked as a brick-layer and then as a truck driver. For the past six years he had been employed as a dispatcher for a taxi company, but just night shift work so there would be minimal distractions and no supervisor. In asking about ADHD symptoms in childhood, the second author (Lynda Thompson) learned that he had always felt in a fog at school and things got even worse after his mother died when he was eleven. By age fourteen he was in reform school. He did not read a book till he was 20 and it was a biography of Martin Luther King, Jr. Thereafter he became a regular user of the library. He said that he was not close to his family, did not have friends, and had never married: books allowed him an escape from his everyday existence. The most important part of his story, however, concerned a TBI he suffered in childhood.

At age four, while he was playing in front of his farmhouse near the road, a car lost control on newly laid gravel and struck a hydro pole that toppled and struck him. He did not recall the accident (doubtless he suffered retrograde amnesia) but he did recall awakening in hospital with both legs in traction and being very frightened and upset. His family had emigrated when he was two but English was not spoken at home and he had not started school, so he did not understand what anyone said to him and had no idea what had happened or where he was. His parents, when told he was out of coma, thought that his survival was a miracle. It seems they never connected his subsequent problems with learning and behavior in school with his accident, and nor had he until we discussed the long-term effects of head injuries during his intake assessment. He had done some therapy as an adult in the 1970s regarding the emotional trauma of awakening in hospital and being frightened and disoriented and unable to communicate in English but he had not realized that being a "slow learner," having a short fuse, getting into trouble with the law, suffering mood swings, and leading a solitary life with no close friends could be related to his early head injury. It was a revelation to him that he was probably not inherently a "stupid kid" or a "bad kid" (labels he recalled from childhood) but, rather, a child who had suffered brain damage. With hope that he might discover who he really was if neurofeedback could heal his brain, he embarked on training that combined neurofeedback and biofeedback.

We did recordings at Cz and determined initial training parameters to improve his concentration (his stated goal) by reducing his excessive slow-wave activity and increasing sensorimotor rhythm. We also put an inhibit on "busy-brain frequencies" in the 23-35 Hz range to help even out his mood. For biofeedback we initially did some finger temperature training for relaxation and then had him practice achieving respiratory sinus arrhythmia (RSA), getting his diaphragmatic breathing in synchrony with his heart rate changes. Today we would be doing LORETA Z-score neurofeedback plus HRV training but this was before we had those capabilities.

When we did our progress testing, which we always do at no charge after 40 sessions are completed in order to have pre- and post-data for our own tracking purposes and also to provide a bonus to the client for completing that amount of training, he was a transformed individual. The standard measures of TOVA, IVA and questionnaire data were all in the normal (non-ADHD) range. The reduction in anxiety symptoms was particularly marked. His single-channel EEG showed a reduction in busy brain activity above 20 Hz and his busy-brain/SMR ratio was fine. The 9 Hz peak that had been the most extreme deviation in his single-channel EEG profile was now just a very slight blip up, and the theta/beta ratio was fine. No longer working at night as a taxi dispatcher in order to avoid interaction with people, he was now a friendly and gregarious person who was running his own small moving business. He was not only exercising regularly, which was already a good habit upon intake, he was now passionate about doing volunteer work to bring fitness to the elderly. He had been to Europe to visit relatives for the first time in his life and had taken along a copy of his ADD Centre Initial Assessment report so that his aunts, uncles, and cousins could read it and understand how his childhood injury had led to him being so reclusive for so many decades. He even thoughtfully brought back souvenirs for the staff who had worked with him. At the age of 61 he was no longer experiencing depression or social anxiety and was, in fact, feeling that the best years of his life were ahead of him. He decided to do a further 20 sessions on a once a week basis as an optimal performance client.

SECTION XXIV
Affect Networks - Depression

Depression

Previously, the neuroanatomical connections and physiology that can underlie anxiety and responses to stress were described. This section will demonstrate that many of the same areas and systems that are part of the Affect network are also, as one might suspect, involved in various types of depression. This is particularly true when depression is accompanied by anxiety and a sense of being under stress. The major structures that are critical for our work using NFB are the central midline structures plus the insula and hippocampus. These are also structures that were keys to understanding the effects of traumatic brain injuries (TBI). This overlap perhaps helps to explain why, after a head injury, a person is three times more likely to become depressed. The following diagram of the central midline structures has been used previously in this book and is repeated here for the reader's convenience.

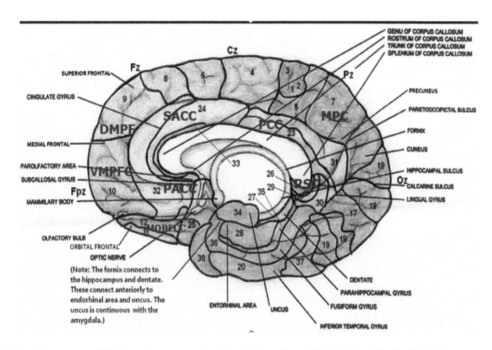

Central Midline Structures (CMS): An Anatomical and Functional Unit (after Northoff, 2006)
Original illustrations by Amanda Reeves with coloring and numbering of areas by Bojana Knezevic.

The bold letter abbreviations in the above figure indicate areas that are considered to be important central midline structures involved in neural networks. These abbreviations are as follows:

DMPF = *dorsomedial prefrontal cortex (BA 9)*
VMPFC = *ventromedial prefrontal cortex (BA 10, 11)*
MOPFC = *medial orbital prefrontal cortex (BA 11, 12)*
SACC = *supragenual anterior cingulate cortex (BA 24, 32)*

PACC = *pre- and subgenual anterior cingulate cortex (BA 24, 25, 32)*
PCC = *posterior cingulate cortex (BA 23, 31)*
RSC = *retrosplenial cortex (BA 26, 29, 30)*
MPC = *medial parietal cortex (BA 7, 31)*
We suggest that the hippocampus should also be included in this discussion of CMS, including the hippocampal region, the entorhinal-uncus area, and the insula in both hemispheres.

In the figure above, the septal area and the nucleus accumbens (now considered to be part of the basal ganglia) could have been shown inferior to the subcallosal gyrus. Also important, and not shown, is the fact that the gray matter of the amygdala is continuous with the uncus. The central nucleus of the amygdala is inhibited by the medial frontal cortex and by the fusiform gyrus and the superior temporal cortex (Porges, 2007). The septal region governs consciousness and memory. Acetylcholine is the neurotransmitter to the hippocampus from the septal region.

Anterior Cingulate

One of the key structures for us to consider in the central midline structures is the anterior cingulate cortex (ACC). As noted in the section on anxiety and the stress response in the Heart Rate Variability (HRV) section, the **ACC** has direct links to other key cortical areas that are involved in Affect networks. These areas include the medial and orbital prefrontal cortex, the insulae (left insula and right insula), and key basal ganglia areas such as the amygdala and the hippocampus. Changes in ACC activity will therefore also affect the hypothalamus and, through it, the autonomic nervous system and the hypothalamic-pituitary-adrenal (HPA) axis. We have described in previous sections how these connections play a major role in the human stress response. (See the section on HRV). All of these structures are also important in understanding the clinical presentation of the depressed patient, and are therefore important in understanding how we can intervene successfully using a combination of NFB + HRV (Thompson & Thompson, 2007; Paquette et al., 2009).

Specific Central Brain Areas Involved in Depression

In patients with major depression, PET Scans (*positron emission tomography*), which can demonstrate an increase in cerebral blood flow (CBF), show hyperactivity of ACC BA 25. These scans may also show decreased activity in the dorsolateral prefrontal cortex (DLPFC) BA 9, 46, 6, and the ACC BA 24, anterior insula and the medial prefrontal cortex (MPC) BA 10 (Mayberg et al., 2005). However, different central midline areas have different primary functions and therefore are related to different aspects, executive and vegetative, of depressive symptomatology. The reader may also wish to look up other fMRI studies. One such study was done with euthymic bipolar patients during a counting Stroop interference task and published in the *American Journal of Psychiatry* (Strakowski et al., 2005).

A simple diagrammatic overview of central midline structures involved in depression was shown previously but is shown again to assist the reader. Patients who are depressed demonstrate symptoms that relate to both Executive and Affect networks. A reasonable overview of this is illustrated in the following diagram:

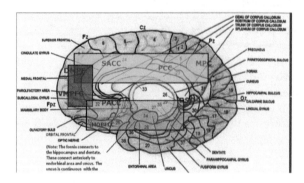

*The three colored boxes highlight aspects of the Executive and Affect networks involved in **depression** (after de Ridder, 2010 ISNR presentation):*

*Symptoms related primarily to **Executive/Cognitive** Network areas = **GREEN** (BAs 9 and 46, 24, 31, 40). Symptoms related primarily to **Affect/Vegetative** Network = **ORANGE** (BA 25, anterior insula, hippocampus, hypothalamus). **Integration and connection** of the above Cognitive and Vegetative areas = **RED** (BA 24, 9, 10, 11)*

In depression, the **vegetative/autonomic affect** network symptoms involve a ventral compartment (e.g., sleep disturbance, loss of appetite and libido). This compartment includes the hypothalamic-pituitary-adrenal axis, hippocampus, insula (BA 13), subgenual cingulate (BA 25), and the brainstem.

The **executive components** of the Affect network in depression are found in the dorsal compartment, which modulates attention and sensory-cognitive symptoms (e.g., apathy, attention, and executive deficits). This compartment includes the dorsolateral prefrontal cortex (BAs 9 and 46), the dorsolateral anterior cingulate (area 24b), posterior cingulate (BA 31), the inferior parietal lobe (BA 40), and the striatum.

The dorsal prefrontal components, including the anterior cingulate cortex (ACC), are also involved in the cognitive control of emotion, including reappraisal, evaluation, and explicit reasoning concerning emotional stimuli. Information concerning cognition and emotion from the two compartments (dorsal and ventral) is integrated by the rostral anterior cingulate (BA 24), medial frontal cortex (BAs 9 and 10), orbital frontal cortex (BA 11), and frontopolar areas (see figure). In addition to the foregoing description of involved cortical areas, the reader may recall that selection of activation of particular pathways can involve the cortex-basal ganglia-thalamus to cortex connections, as was described in the discussion of functional neuroanatomy and Brodmann areas. Different treatments may have their primary effect on different aspects of these networks. For example, **cognitive behavior therapy** (CBT) may be expected to focus more on the executive aspects of depression and therefore influence the areas that are outlined in green in the above figure.

Vegetative aspects of depression may respond more to antidepressant medications and thus affect the areas that are outlined in orange above. LORETA NFB has the potential to influence all of these structures. Even conventional surface NFB has been shown to change all the major involved areas (Paquette et al., 2009). A more invasive procedure, deep brain stimulation, has also been shown to have effects in all the critical central midline areas of both the Executive and Affect networks (Mayberg et al., 2005; Kennedy et al., 2011).

In Section VII modules and hubs were described. It was stated that a neural **network** is a number of **Nodes and Modules with functions in common** connected by links (Sporns, 2011). The links may be thought of as comprising the commissures and fasciculi and other white matter connecting pathways. The diagram in Section VII from *Gray's Anatomy* is shown below.

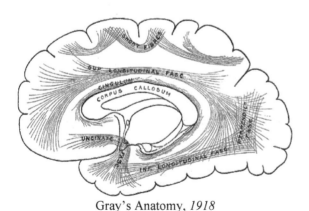

Gray's Anatomy, *1918*

Affect networks contain areas that are within what has been called the limbic system. The schematic diagram below shows Brodmann areas connected by fasciculi within each hemisphere of the brain. In this diagram the cingulum and the uncinate fasciculus are two central components of the Affect networks that we can influence when we use the combination of LORETA NFB and Heart Rate Variability training. The cingulum has connections with medial frontal, parietal, occipital, and temporal lobes, and the cingulate cortex. These connections are well illustrated in Catani and De Schotten's *Atlas of Human Brain Connections* (2012).

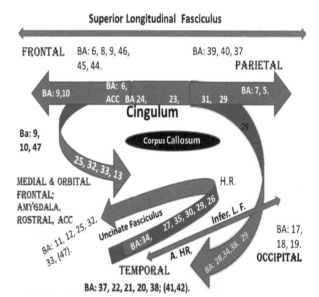

Diagram: BAs connected by white matter fasciculi
H.R. - Hippocampal region
A. – Amygdala
Infer. L. F. – Inferior Longitudinal Fasciculus
ACC – Anterior Cingulate Cortex
Major Reference: Jones, D.K., Christiansen, K.F., Chapman, R.J. & Aggleton, J.P. (2013). Distinct subdivisions of the cingulum bundle revealed by diffusion MRI fibre tracking: Implications for neuropsychological investigations.
Neuropsychologia, *51(1), 67–78.*

NFB does increase both both gray and white matter density and this was shown in controlled research done with healthy universitay students (Ghaziri et al., 2013). Anatomically, the superior longitudinal fasciculus (SLF) runs between superior frontal and parietal areas. This pathway is believed to provide a means enabling the prefrontal cortex to supervise the allocation of attentional resources. Ghaziri and his colleagues found that simultaneous NFB enhancement of 15-18 Hz activity at F4 and P4 stimulated neuronal communication between frontal and parietal regions within the right hemisphere and enhanced visual and auditory sustained attention performance. They demonstrated both increased volume in white matter pathways using fractional anisotropy and increased gray matter volume detected with MRI, in cerebral structures involved in sustained attention.

At the ADD Centre we are finding that addressing connectivity by targeting coherence and phase between relevant areas for a particular patient is turning out to be an important component in effective treatment of emotional dysregulation.

Treatments

Cognitive Behavior Therapy

Cognitive Behavior Therapy appears to have its effects on the more dorsal and executive aspects of depression by increasing activity in the ACC, BA 24. It may also decrease activity in the "connecting" areas including the MPFC, BA 10, and the orbital frontal cortex (OFC), BA 11 (Goldapple, 2004).

Medications

Mayberg has published on the regional effects of some **medications** (Mayberg et al., 2000). Essentially, they found that medications could increase the activity of the more ventral, vegetative areas such as the prefrontal cortex (PFC), BA 9, and the brain stem, and decrease activity in BA 25 and the hypothalamus. In 2003, her group noted the regional metabolic effects of fluoxetine (Prozac) in major depression and the relation to clinical response (Mayberg et al., 2003**)**.

Alpha Asymmetry NFB for Depression

An alpha asymmetry NFB protocol was developed by Rosenfeld and used by Elsa Baehr. It was modeled after the work of Richard Davidson (Davidson, 1995, 1998; Baehr et al., 1999). The anterior regions of the left and right cerebral hemispheres have been posited to be specialized for expression and experience of approach and withdrawal processes, respectively. Left frontal cortex activation was found by Davidson to be associated with positive emotions and approach behavior, whereas right frontal activation was found in association with negative thoughts and avoidance behavior. The most seriously depressed patients avoid social contact and may not even want to get out of bed. Much of the evidence supporting this hypothesis has been obtained by measuring the anterior asymmetry in electroencephalographic alpha activity; that is, comparing amplitudes of EEG activity in the 8-12 Hz range at F3 and F4. (In some individuals the asymmetry might be seen at F7 and F8). Baehr notes that, in people with more serious depression, the asymmetry may be a left frontal increase in the theta band (Baehr, personal communication). In most of this research, however, motivational direction has been confounded with affective valence such that, for instance, approach behavior is found in conjunction with positive affect. Research done by Harmon-Jones and Allen (1998) tested the hypothesis that dispositional anger, an approach-related motivational tendency with negative valence, would be associated with greater left than right anterior activity. Results supported the hypothesis, suggesting that the anterior asymmetry may vary mainly as a function of motivational direction (approach-avoidance) rather than affective valence (positive-negative).

At the ADD Centre, we do look for a frontal theta or alpha asymmetry in people who have symptoms of depression, but more often there are other findings from the 19-channel QEEG assessments with LORETA. Most of the clients presenting with depression at our Centrer are also experiencing anxiety. Perhaps this is one reason why the majority of depressed clients we have seen show excessive high frequency beta (spindling beta in frequencies >20 Hz) and low SMR, rather than an alpha or theta frontal asymmetry.

NFB + HRV Training for Depression with Anxiety

We have previously described, when discussing Brodmann areas, the importance of BA 25 in the ventral rostral anterior cingulate cortex in very serious, often intractable, depression. Brodmann area 25 is also called the subgenual area, area subgenualis (from Latin "below the knee," here referring to the knee of the corpus callosum), or subgenual cingulate. It may be influenced by work done with NFB at Fz and Cz by virtue of its importance in networks that also involve the more superior aspect of the cingulate gyrus under Cz and the medial aspect of the frontal lobe under Fz. BA 25 is very deep in the midline cortex, and LORETA NFB may have a much more direct effect than can be achieved with single-channel training. This region contains serotonin transporters. It influences the hypothalamus and brainstem, which are involved in responses to stress, including changes in HRV and appetite and sleep. This region also connects with the amygdala and insula, which affect mood, including anxiety; the hippocampus, which plays an important role in memory formation; and some parts of the frontal cortex responsible for self-esteem. All of these are functions that are seriously affected in deep depression. Of importance to the practitioner who uses a combination of NFB + BFB, the solitary nucleus in the medulla has direct connections to BA 25. Thus, HRV training will send vagal afferent feedback to the medulla, solitary nucleus, and be potentially helpful in influencing BA 25 and thus, perhaps, the symptoms of depression related to dysfunction in this area of the cortex.

Placebo Effect

Herbert Benson, M.D., a physician at the Mind/Body Medical Institute, Harvard Medical School, Division of Behavioral Medicine, Deaconess Hospital, Boston, Massachusetts, studied the placebo effect and suggested that, instead of the term Placebo Effect, which can have negative connotations in some professional circles, we use the term "Remembered Wellness" (Benson & Friedman, 1996). This effect can demonstrate beneficial clinical results in a very high percentage of patients who suffer from serious physical complaints including angina, asthma, herpes, and even ulcers. Benson noted, as have others, that positive expectations and beliefs and memories of when one felt well could have a remarkably positive effect on the symptoms that these patients experience.

In psychiatry, when dealing with anxiety and depression, physicians have always attempted to help their patients evoke positive memories to replace their negative ruminating. Cognition can alter affect and result in a change in behavior. Research on addiction was reviewed at the 2010 International Society for Neurofeedback and Research (ISNR) meeting. Dr. De Ridder, a neurosurgeon then in Belgium and now in New Zealand, showed an example in the treatment of addictions where sham repetitive transcranial magnetic stimulation (rTMS) could result in the same effect for the patient as a course of real rTMS. However, the neuroanatomical networks involved were different in the real treatment as compared to the placebo effect, as shown in the illustration below. An important point being made by this diagrammatic outline of the networks is that the placebo network emphasizes an initial hippocampal region activation, which we suggest is a good way of validating the concept of "remembered" wellness. The placebo effect involved hippocampus to the orbital frontal cortex (OFC) connections. De Ridder notes that this network is evoked with sham rTMS. With real rTMS a different neural network is activated. It goes from the ACC to the OFC.

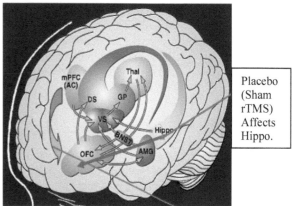

rTMS affects AC and mPFC
2010 - With thanks to Dirk De Ridder for permission to use this illustration from his presentation. He credits this illustration from his lecture to George Koob, Ph.D., Professor and Chair of the Committee on the Neurobiology of Addictive Disorders at the Scripps Research Institute.

Index: *AMG = Amygdala; Hippo = Hippocampus; OFC = Orbital Frontal Cortex; MPFC = Medial Prefrontal Cortex; AC = Anterior Cingulate; AMG = Amygdala; DS = Dorsal Striatum (includes caudate and putamen); VS = Ventral Striatum (includes olfactory cubicle and nucleus accumbens); mPFC(AC) = Medial Prefrontal Cortex; OFC = Orbital Frontal Cortex; GP = Globus Pallidus; Thal = Thalamus; Hippo = Hippocampus; BNST = Bed Nucleus of the Stria Terminalis. BNST is a basal forebrain structure involved in many motivational processes. It is closely linked to stress regulation.*

Deep Brain Stimulation

Deep brain stimulation requires a surgical procedure to access the ventral anterior cingulate gyrus in order to implant a pacemaker into BA 25. In 2011, Kennedy and Mayberg and collaborators described how they successfully treated a number of depressed people – individuals virtually catatonic with depression despite years of talk therapy, drugs, even shock therapy - with pacemaker-like electrodes for deep brain stimulation in Brodmann area 25 (Mayberg et al., 2005; Kennedy et al., 2011). BA 25 is said to be metabolically overactive in treatment-resistant depression, and this appears to correlate with poor therapeutic response to traditional treatments for depression. Their implanted pacemakers have been shown to decrease activity (PET scan follow-up) in BA 25 and OFC BA 11. These pacemakers also increase activity in the DLPC BA 9, 46 and the ACC BA 24 and PCC BA 31. In addition, they can increase parietal activity in BA 40. However, it only has these effects in responders. Interestingly, these are the same areas that were changed by NFB in Paquette's study and were only changed in those patients who respond to NFB. In Paquette's NFB study, the patients were also all persons who did not respond to medications (Paquette et al., 2009).

Neurofeedback

Vincent Paquette states correctly that NFB does not require any invasive procedure. In his study, one month after the end of NFB treatment, absolute power of high-beta (18-30 Hz) activity showed a significant reduction in the orbitofrontal cortex (BA 11/47), insula (BA 13), amygdala/parahippocampal cortex (BA 36/37), temporal pole (BA 38), lateral prefrontal cortex (BA 10 and BA 6/8), and subgenual cingulate cortex (BA 25). In addition, those who responded to the treatment with a reduction in depressive symptoms demonstrated an increased amplitude of high-beta activity in the bilateral precuneus/posterior cingulate cortex (BA 40/31) (Paquette et al., 2009). These findings are summarized in the figure below.

An interesting additional fact is that abnormally low high-beta activity has been found in these cortical areas (BA 40/31) in individuals with major depressive disorder (MDD) by Pizzagalli et al. (2002). Pizzagalli has also reported on functional, but not structural, prefrontal subgenual abnormalities and stress-related hedonic capacity blunting (Pizzagalli et al., 2008). Pizzagalli has many other papers on depression. (See Pizzagalli et al., 2003, 2004, 2005, 2007).

In addition, increased activity in the precuneus/posterior cingulate cortex has been shown to correlate with symptom remission following pharmacological treatment (Mayberg, 2003), or with interpersonal therapy (Martin et al., 2001). Since the highest level of cortical glucose metabolism during resting state occurs in these brain regions in healthy participants (Raichle et al., 2003), it is a plausible hypothesis that pharmacological treatment, interpersonal therapy and the psychoneurotherapy utilized by Paquette may all contribute to restoring appropriate functioning in the default mode of the brain.

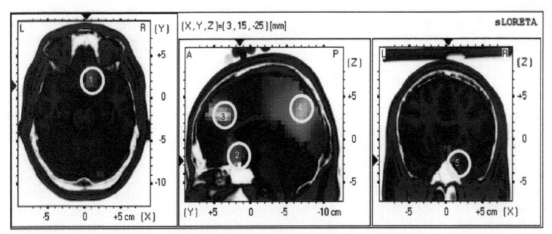

1) Right orbitofrontal cortex-BA 11/47 (r = .46), 2) Right subgenual cingulate cortex-BA25 (r = .45), 3) Right medial prefrontal cortex/dorsal anterior cingulate cortex-BA9/32 (r=.45), 4) Bilateral precuneus/Posterior cingulate cortex-BA31/40 (r=.-48), 5) Right temporal pole-BA38 (including Right parahippocampal cortex)-BA36 (r = .45)

■ The higher the reduction of high-beta activity after treatment, the higher the reduction of depressive symptoms
■ The higher the increase of high-beta activity after treatment, the higher the reduction of depressive symptoms

Figure from Paquette et al., 2009

To recap some of the functional neuroanatomical findings, we note the functions of a few of the areas that demonstrated significant EEG changes as follows:

- The **ventromedial PFC** and rostral/subgenual portions of the ACC may have an important role in the regulation of emotion, which is different from the evaluation and monitoring role of the dorsomedial PFC and dorsal ACC.

- The **amygdala and insula hyperactivation** in patients reflects excessive engagement of fear or negative emotion related circuitry.

- **Hyperactivation** of these structures in anxiety disorder patients is related to symptoms of hyperarousal and **hypervigilance.** These are symptoms shared across anxiety disorders.

- This finding is important because it may identify a core phenotype of Post Traumatic Stress Disorder (**PTSD**) in addition to other Disorders in which generalization of anxiety or profound dysregulation of emotion are prominent.

- **Hypoactivation was found in the dorsal ACC and dorsomedial PFC,** as well as in ventral portions of the ACC and the ventromedial PFC. These hypoactivations were significantly more common in **PTSD** than in the other two anxiety disorders.

- While all of these disorders appear to have in common abnormalities in functioning in the amygdala or the insula, only those in which anxiety is generalized and emotion is more widely dysregulated (PTSD and generalized anxiety disorder, GAD), have prominent dysfunction in both medial prefrontal nodes of the circuit.

- A reference for the above considerations on PTSD and other anxiety disorders is found in *Functional Neuroanatomy of Anxiety: A Neural Circuit Perspective.* It is on the internet in a chapter by Amit Etkin (2009).

Conclusion

Conventional psychiatry has tried to deal with depression through the use of psychotherapy, cognitive behavior therapy, electroconvulsive therapy (ECT), and medications. Some patients respond, but the long-term benefits for many have not been overly impressive. In addition to these therapies, diet and exercise interventions have shown promise with some depressed patients. More recently, neurofeedback, transcranial direct current stimulation (discussed elsewhere in this text), and deep brain stimulation, BA 25, have shown some very good results. Note that the various interventions, when used in a clinical setting, are not stand-alone: one would combine encouragement of healthy diet (especially increased intake of fish and other sources of omega-3 essential fatty acids) and increased exercise with neurofeedback training and counseling/cognitive behavior therapy for clients with depression. Medications can also be combined with neurofeedback. As is the case in clients with ADHD, there is a pills-skills balance: that is, the amount of medication can be reduced as self-regulation of emotion is improved with NFB training. Our preference is to try the less invasive, less expensive interventions that are supported by research (even if there is not yet much randomized, controlled research) before resorting to more invasive and expensive techniques such as ECT or deep brain stimulation. This follows both the Principle of Parsimony and the guideline to "do no harm."

SECTION XXV
Asperger's Syndrome and Autistic Spectrum Disorders
Additional material with an emphasis on newer approaches including
LORETA NFB and tDCS)

Introduction

We have chosen to discuss Asperger's syndrome despite DSM-5 deleting Asperger's disorder as a diagnosis. Our decision is due to the very different clinical pictures and the differences that are observed in the EEG profiles in those with Asperger's as compared to those with autism.

When working with autistic spectrum disorders (ASD), a multimodal approach including diet, medications, speech therapy, psychotherapy, behavior modification and education is typically used (Green et al., 2006). In the past 15 years, two new and promising approaches have been added to the interventions. These are EEG biofeedback, also called neurofeedback (NFB), and peripheral biofeedback (BFB), especially heart rate variability training (HRV). Both NFB and HRV training appear to have long-term effects when applied to other disorders such as NFB for attention-deficit/hyperactivity disorder (Lubar, 1995; Gani et al., 2008), and HRV for stress management (Lehrer, 2007). NFB + BFB can be effective for ASDs and Asperger's syndrome (Thompson et al., 2010; Pineda et al., 2008; Solnick, 2005; Martinez, 2003; Thompson & Thompson, 1995). Medications, on the other hand, have limited positive effects and may result in side effects (Sloman, 2005).

Autism and Asperger's Syndrome

Autism is a pervasive developmental disorder characterized by impaired social interaction and communication, and by restricted stereotypical behavior patterns. It may be called a spectrum condition, which means that its symptoms can be manifested in various forms and with different intensity. Its etiology is multifactorial, i.e., both genetic and environmental. Although exact causes have not yet been determined, contemporary studies link autism to biological and neurological malfunction of the brain.

Early diagnosis and early intervention are considered to be important in order to ensure an autistic child's optimal development. Traditional methods of treating autism may include pharmacology as well as behavioral techniques, such as speech therapy, sensory therapy, music therapy, and hippotherapy (the use of movements of a horse to provide carefully graded motor and sensory input). More modern methods include hyperbaric oxygen therapy (Sampanthavivat et al., 2012; Rossignol et al., 2007). Despite a large number of therapeutic methods commonly used with autistic children, no therapy at present produces satisfactory results. New names are appearing for therapeutic approaches we have used for half a century. One of these is now being called "floortime" therapy and is said to derive from the Developmental, Individual Difference, Relationship-based model (DIR), as discussed by child psychiatrist Stanley Greenspan. However, real credit should be given to Milada Havelkova, M.D., and Dorothy Chandler, R.N. of the West End Child and Family Clinic, Toronto, Canada, for their work through the 1940s to the 1970s (Thompson & Havelkova, 1983).

In recent years, neurophysiological treatment has been showing promise, in particular neurofeedback, but also older adjunctive procedures such as the Tomatis method, and new variations in listening therapy (Porges, 2009; 2014 in press).

Symptoms – A Brief Overview

People with Asperger's have deficits in the same three areas as do those with autism (social communication including social awareness, repetitive and/or intense special interests and differences in imagination), but the presentation is quite different. Patients with Asperger's want to interact but do not do so smoothly or appropriately. They are highly egocentric and tend to talk about their own special interests too much and fail to read the other person's nonverbal cues of boredom or annoyance. They may use pedantic phrases or a voice that is monotone and lacks prosody (intonation, loudness variation, pitch, rhythm). The latter symptoms may be termed a motor aprosodia (Ross, 1981), and in the EEG we may observe differences from the normal database at the F6 electrode position of the expanded 10-20 electrode placement system (Chatrian, Lettich & Nelson, 1985). F6 is over an area in the nondominant (right) frontal lobe that is the homologous site to F5, which is near Broca's area in the left hemisphere. These frontal sites are also identified as mirror neuron areas.

EEG

In autistic children we may often see EEG differences, as compared to a normative database, at both F3 and F4, which are the 10-20 sites closest to frontal mirror neuron areas. The 19-channel EEG of a 10-year-old male client diagnosed with autism is shown below. Some aspects of the pattern, especially the spindling beta, may also be seen in clients with Asperger's syndrome or with anxiety disorders. In both Asperger's and ASDs, both motor and sensory aprosodia symptoms (see the section on Brodmann areas) may be observed. Often slow-wave activity, theta, and/or low-frequency alpha is observed in the right hemisphere.

Anxiety and Anterior Cingulate

Anxiety is a key symptom in autistic spectrum disorders (ASD). Some adults with Asperger's are initially diagnosed with anxiety or panic disorder. Their anxiety may be most apparent with any transition or change in routine. Using LORETA (Pascual-Marqui, Esslen, Kochi & Lehmann, 2002), Brodmann area 24 in the anterior cingulate is often identified as the source of QEEG findings that differ from a normative database (as is shown in a figure below). In our clients, this has usually been in the 2-7 Hz range and/or 13-14 Hz, and high-frequency beta above 20 Hertz (figure). In clients with an ASD in addition to anxiety, these findings may correlate with other important symptoms, including poor modulation of emotions, inattention, and executive functioning difficulties.

Attention Span

The neural basis for momentary lapses in attention has been studied and it will be interesting to see how this develops, and whether our observations of theta and/or low alpha bursts at Cz reflect these networks (Wiessman et al., 2006). The Cz electrode site in children and FCz in adults have, in the early years at our Centre, been the primary sites for beginning NFB training in these clients. This has been both because of hypothesized influence on the anterior cingulate, which lies below these sites, and because these sites are relatively artifact-free and have been used successfully when doing NFB with children who have Attention Deficit Hyperactivity Disorder. Children who have an autistic spectrum disorder may show severe learning difficulties in addition to all the symptoms found in children with ADHD. The autistic children, in contrast to the children with Asperger's syndrome, have a severe language delay and much more severe problems with emotionally connecting to other people.

Neuroanatomical and Assessment Findings

Aprosodia

Right frontal and right parietal-temporal junction EEG differences may correlate with aprosodias as mentioned above; that is, deficient expression of emotion in voice, facial expression, and gestures (motor aprosodia) plus difficulty reading social cues, gestures, and tone of voice (sensory aprosodia; nondominant temporal-parietal junction). .

Orbital Frontal Cortex

In addition, Shamay-Tsoory and his colleagues (2005) have hypothesized that prefrontal brain damage may result in impaired social behavior, especially when the damage involves the orbitofrontal and/or ventromedial areas of the prefrontal cortex (but not dorsolateral areas in their research). These authors note that prefrontal lesions resulted in significant impairment in the understanding of irony and faux pas. In contrast to the patient who has damage to the amygdala, who cannot correctly understand the significance of another person's anger or aggressive behavior, patients with orbital frontal damage recognize the significance of other people's emotions, but may fail to modulate their behavior as the social situation changes.

This kind of impairment could lead to difficulty in correctly recognizing the intentions of others and thereby lead to inappropriate behavior (Bachevalier & Loveland, 2006). Dysfunction in the orbital-frontal cortex can lead to an inability to understand the effect of one's own behavior on other people. In their paper, Bachevalier and Loveland posit that developmental dysfunction of the orbitofrontal-amygdala circuit is a critical factor in ASD. The orbitomedial frontal cortex (BAs 11, 12, 47, 13) is responsible for evaluating the value of rewards (Kringelbach, 2005). The anterior portion is more involved in complex rewards than the posterior portion. Dysfunction in this area may be quite important for understanding the differences in everyday functioning of these persons. Parents are often encouraged by professionals to use rewards with their children, but when they try to do this with children in the ASD group, they may find that their child simply does not respond the way their other children do. At our training center, we use tokens that can be saved to purchase rewards, and this is a major motivating force for all the children except those with Asperger's or autism. These ASD children usually do not appear interested in our tokens and rewards. They get tokens but seem then to both not really be interested in purchasing a toy (ASD) or they cannot decide what reward they want (Asperger's), so their tokens just build up. In those with Asperger's it seems related to anxiety. They fear making the wrong choice, and then have trouble making decisions. Parents of children with Asperger's often report that their child will even have extraordinary difficulty deciding what to order in a restaurant. It will be interesting to see if we can alter this behavior by targeting this area with LORETA NFB.

Areas Connected by Uncinate Fasciculus

Imaging studies have shown differences in Asperger's and autistic children, as compared to neurotypical children, in the density of gray matter at the junction of the amygdala, hippocampus, and entorhinal cortex. These are areas connected by the uncinate fasciculus, as pointed out in the section on the functions of Brodmann areas. In our clients, LORETA consistently shows EEG abnormalities in these regions (Thompson, Thompson & Reid, 2010). The fusiform gyrus and the superior temporal lobe are noted in many papers to be involved in ASD, and Porges' polyvagal theory specifically notes that these areas may not be appropriately inhibiting the central nucleus of the amygdala and this could lead to an increase in sympathetic drive and a decrease in myelinated vagus activity. In addition, there may be dysfunction in the medial aspect of the prefrontal cortex that may also mean that there is less effective inhibition of the central nucleus of the amygdala (Thayer et al., 2012).

Polyvagal Contributions

Myelinated vagal activity is associated with a person feeling that they are in a safe environment. Both hearing (affected by the stapedius muscle in the middle ear) and facial expressiveness are influenced by vagal innervation, and Porges postulates that vagal dysfunction could therefore result in some of the symptoms observed in these clients, such as not listening appropriately and lack of showing emotions through facial expression (Porges, 2003, 2004, 2007, 2009, 2014). The importance of noting the above areas that are reported as deviant in the general literature on ASD is that clinical observations using EEG show these areas to be outside the database norms using LORETA and the NeuroGuide Z-score database (Thompson & Thompson, 2010).

ANXIETY & ASDs

A frequently found pattern at our centre is high amplitude 2-5Hz, low amplitude 8-10 Hz, high amplitude 11-12 Hz & 12-16 Hz, plus high amplitude higher frequency beta at various frequencies between 17–36 Hz. The specific frequency ranges & sites vary from child to child. Below: DU, Laplacian montage ec, spindling synchronous beta at F4 (25 Hz 5 SD) Fp2, & F7 (20 Hz 3.4 SD) possible orbital-frontal), C4 (25 Hz 3 SD)   and high amplitude slow wave at T6.

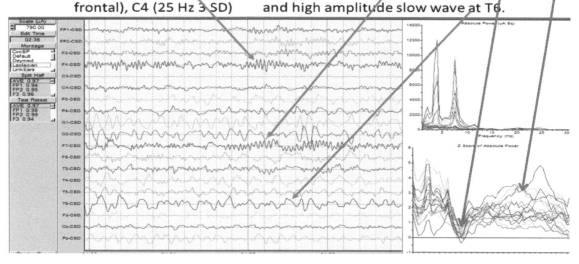

"DU," an autistic male, aged 10 years. 9 months, Laplacian montage ec, raw EEG and comparison to the NeuroGuide database. Note high standard deviations (y axis) for theta, 12-15 Hz beta, and high-frequency beta. Very low-amplitude 8-10 Hz has been observed in all our patients who have a diagnosis of autism. Usually it is –1 to –3 SD, but in this patient it was just very low relative to the high standard deviations of other frequencies. These observations of high-amplitude low theta (3-6 Hz), low-amplitude, low-frequency alpha (8-10 Hz), high-amplitude low beta (13-17 Hz), and high-amplitude, high-frequency beta have also been made by others (Murias et al., 2007; Chan et al., 2007; Thatcher, 2009).

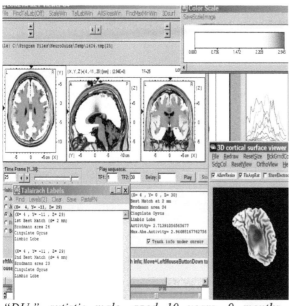

"DU," autistic male, aged 10 years, 9 months, LORETA image shows one source of 25 Hz spindling beta activity is BA 24, anterior cingulate. It is 2.95 SD from the NeuroGuide database means.

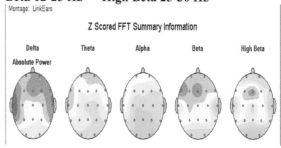

"DU," autistic male, aged 10 years, 9 months; linked ears montage ec. Red is >4 SD from the NG database. The spindling beta had a source in the anterior cingulate gyrus, as shown in the LORETA image. Note that delta 1-4 Hz and theta (4 and 5 Hz) is >4 SD between F7 and T3. With LORETA, the 2-5.5 Hz activity was observed in the orbital frontal area BA 11, the uncinate gyrus, and the parahippocampal gyrus. One may assume that the uncinate fasciculus is involved.

Coherence:
Delta 1-4 Hz   Theta 4-8 Hz   Alpha 8-12 Hz
Beta 12-25 Hz   High Beta 25-30 Hz

"DU," autistic male, aged 10 years. 9 months, linked ears montage, eyes closed (ec), Red is >3 SD from the NeuroGuide database normals. This figure shows hypercoherence (red lines) for the most part between anterior and posterior sites. In contrast, the hypocoherence (blue line C4 to F8) is between hemispheres.

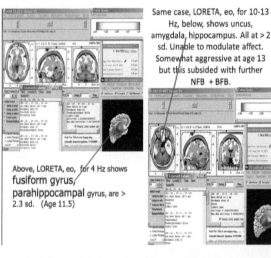

Same case, LORETA, eo, for 10-13 Hz, below, shows uncus, amygdala, hippocampus. All at > 2 sd. Unable to modulate affect. Somewhat aggressive at age 13 but this subsided with further NFB + BFB.

Above, LORETA, eo, for 4 Hz shows **fusiform gyrus, parahippocampal** gyrus, are > 2.3 sd. (Age 11.5)

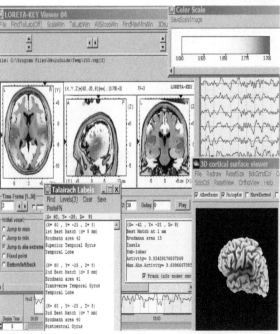

3 Hz is 3.7 SD at temporal-parietal junction BA 42, 41 (auditory cortex), BA 40 (supramarginal gyrus and postcentral gyrus), and the insula BA 13.

The autistic boy depicted in the foregoing figures has steadily progressed over the last three years with neurofeedback sessions. Progress has taken a long time, but has been steady. Unlike our former experience using behavior modification, psychotherapy, and medications, the results with neurofeedback include a significant change in the ability of this boy to react appropriately emotionally to different situations involving others. He has moved from a contained classroom special education placement to the general stream, where he is functioning well in the classroom situation. We cannot draw any conclusions concerning the relative contributions to his progress that are due to the various components of his treatment, except to emphasize that the manner in which he relates emotionally to others is very different than our past experience with hundreds of autistic children when neurofeedback was not available.

Involvement of Cuneus – A CMS: 21 year old autistic (HF) boy

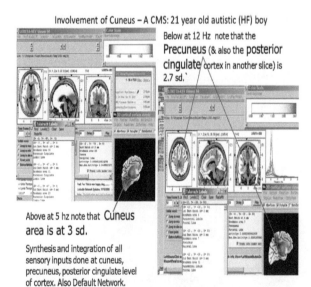

Below at 12 Hz note that the **Precuneus** (& also the **posterior cingulate** cortex in another slice) is 2.7 sd.

Above at 5 hz note that **Cuneus** area is at 3 sd.

Synthesis and integration of all sensory inputs done at cuneus, precuneus, posterior cingulate level of cortex. Also Default Network.

In most of the patients with an autistic spectrum illness, deviations from the normal database in the posterior cingulate, cuneus and precuneus are common findings. It seems likely that these findings are related to their difficulties with the synthesis, integration, and filtering of new information and their social difficulties (partly due to possible dysfunction in the Default network).

Phase

Several years ago, an interesting but as yet unexplained observation was made in New Jersey by Dr. James Neubrander, who runs a large center for children with ASD, and James Thompson, Ph.D., who was helping him introduce QEEG assessment methods and neurofeedback treatment at his center. They noted a distinct pattern at Pz in phase delay. This is shown in the following figure. We have seen this phase-delay pattern repeatedly in our patients with autism, and we have not seen it in other patients.

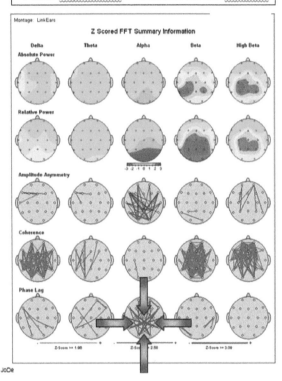

NeuroGuide figure is an average from about 10 autistic children. With thanks to Dr's J. Neubrander and J. Thompson for this figure.

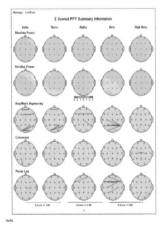

NeuroGuide figure is an average figure from >10 normal and recovered autistic children. With thanks to Dr's J. Neubrander and J. Thompson for this figure.

150 Asperger's and Nine Severe Autistic

In 2010, the authors published their findings from a case summary of 150 clients with severe Asperger's (the majority of these cases might be called by others "high-functioning autism") and nine severe autistic cases. This review extended into cases seen in the 1990s, when LORETA was not being used at the ADD Centre. Therefore, the following findings reported in that review are from those cases assessed more recently when NeuroGuide and Z-score LORETA analyzes were being carried out. More recent cases done since that publication have supported the reported findings. The following is just a list of those areas that we are finding outside the NeuroGuide database norms for most of these patients.

Assessment Findings Summary

As previously reported (Thompson et al., 2010), **eleven areas of the brain** are repeatedly found, in our work using LORETA, to differ from database Z-score norms. These areas include, in both hemispheres:

1. **Prefrontal** cortex (mirror neuron areas)
2. **Hippocampus and parahippocampal** gyrus
3. **Amygdala** (and uncus), with its connections to the orbital and medial frontal areas of the brain, the limbic system, the HPA axis, and to the brainstem (vagus).
4. **Fusiform gyrus**
5. **Superior temporal** gyrus including the auditory cortex
6. **Insula**
7. **Anterior cingulate** and insula are both part of the limbic system (the emotional brain). The central portion of the **distress network** (de Ritter describes in ISNR talk on tinnitus) consists of amygdala, **anterior cingulate,** anterior insula, BA 10 prefrontal cortex (De Ridder, 2009).
8. **Frontal and parietal-temporal mirror** neuron areas
9. Right and left **angular and marginal** gyrus areas
10. **Orbital and medial** aspects of the frontal lobes
11. **Posterior cingulate, cuneus, precuneus**

Theories for Understanding the Symptoms of ASD Support NFB Interventions

Mirror Neuron System (MNS)

The MNS is postulated to be involved in the imitation of movements, and perhaps also in copying appropriate social interactions. It is important for making critical assessment of social situations, and to understanding and predicting the behavior of others. MNS areas in the left hemisphere (there are corresponding areas on the right) include: frontal cortex near F5, the temporal pole, the temporal-parietal junction, the anterior insula, and the anterior cingulate gyrus. Each area is postulated to have mirror functions that correspond to the functions of that area of the cortex.

Mirror neurons have strong representation in the limbic system including the anterior cingulate (AC) (Iacoboni & Dapretto, 2006). The cingulate and the insular cortices both contain mirror neuron cells (Ramachandran & Oberman, 2006). A functional magnetic resonance imaging (fMRI) study demonstrated that activity of the MNS is correlated with empathic concern and interpersonal competence (Pfeifer, Iacoboni, Mazziotta & Dapretto, 2005). It has also been shown that children with ASD have reduced activity in MNS regions during tasks that require the child to mirror facial expressions of different emotions (Dapretto et al., 2006).

The reason for reminding the reader of this system is that our findings with 19-channel QEEG and LORETA source locations show these areas to be the source of EEG activity that differs from normal children (>2 SD from database means) in our clients with ASD (Thompson & Thompson, 2010). Interestingly, there is a lack of Mu suppression in children with ASD when they view videos of children moving their fingers and hands in order to catch a ball. Mu, which is measured at C3 and C4 across the motor strip, is normally found in conjunction with motor quiescence of the hands and normally would be attenuated when there is movement of the contralateral hand, or thinking of moving the hand, or watching someone else moving their hands.

Theories

Theory of Mind (ToM) (Hill & Frith, 2003) Theory of mind (which is sometimes referred to as "theory of others' minds") involves the ability to "mentalize about both the self and others" (Abu-Akel, 2003). This model implicates the posterior brain (parietal and temporal regions) in representational thinking. It involves the prefrontal regions, in addition to the medial prefrontal cortex (anterior paracingulate cortex), the temporal-parietal junction, and the temporal poles for the application and execution of theory of mind. ToM proposes that a fault in any component of these aspects of the social brain can lead to an inability to understand aspects of social communication.

Intuitive understanding of others, especially understanding what they are feeling or thinking, has always been understood to be a core deficit of those with autistic spectrum disorders (Thompson & Havelkova, 1983). The reader will note that these areas involved in ToM are also areas referred to in the above discussion of mirror neurons and in our discussion of the areas related to the **uncinate fasciculus. The amygdala** is also implicated (Adolphs, 2003; Adolphs et al., 2005) and the reader will see the overlap here with the **salience-landscape** theory (Ramachandran & Oberman, 2006). Ramanchandran also mentions findings of less connectivity between the occipital and temporal regions. Abnormal connectivity between these regions is a finding that we observe in the EEG with these subjects using coherence analysis, which is a measure of phase-locked connectivity.

Weak Central Coherence

The weak central-coherence theory seeks to explain some symptoms that are not subsumed under the above theory of mind. These include the need for '**preservation of sameness**' and also the **special interests** and talents of those with AS. The child with ASD seems to be flooded with **inconsequential details** and/or memories, without grasping the context or the Gestalt. They get stuck on details and do not see the big picture; in other words, they do not have a coherent view of things. This led to the term 'weak central coherence.' **Defensive behaviors** may include rigidity, repetitive movements and obsessive and even perseverative behaviors. It seems possible that weak central-coherence involves a lack of appropriate connectivity between areas of the brain. Connectivity in this discussion refers particularly to connections between the posterior sensory processing areas of the brain (including lingual gyrus) and the frontal areas that modulate responses to the sensory input ('top-down' modulation).

Failure of Developmental Pruning

Hill states that one cause of this weak central coherence may be a failure of normal developmental 'pruning' in early life that eliminates certain brain connections and optimizes the coordination of neural functioning. Resulting **perceptual overload** may, in turn, theoretically be partly responsible for autistic withdrawal. A reasonable, researchable hypothesis is that our findings of gross EEG coherence abnormalities with clients who have an ASD and our observations of symptomatic improvements corresponding to normalization of these coherence abnormalities might support Hill's weak central coherence theory.

Coherence and phase training has become more precise now that we are able to do this training between specific Brodmann areas deep within the central midline cortex and other areas of the cortex using LORETA Z-score NFB. In the future we may add training related to phase shift, phase lock, and phase reset (Thatcher, 2012). In addition to coherence training, we nearly always increase sensorimotor rhythm (13-15 Hz). This may be carried out at C4 or C3 when the assessment shows high-amplitude low beta 13-16 Hz at Cz (often with a source in the ACC). We hypothesize that once the thalamus is properly gating the incoming sensory information, the ASD client may be less overwhelmed by sensory stimuli. Proper gating is activated through sensorimotor rhythm (SMR) training changing thalamocortical firing patterns.

Executive Dysfunction

This third cognitive theory was advanced to explain features that do not appear to be subsumed under the former two theories. The Executive and Affect networks have been described in the section on functions of the Brodmann areas. Executive functioning (including attention, planning, inhibition, and mental flexibility) appears to be impaired in those with ASDs. The Tower of London Test (ToL) can evaluate many aspects of executive functioning, and those with an ASD score poorly on this test. The ToL involves ten trials of changing the position of colored rings by placing them on three pegs in order to copy a pattern shown by the examiner. It requires the subject to inhibit immediate responses, plan, shift mental set, use working memory, initiate a thought-out response, and then monitor and evaluate the results of that response. The required cognitive functions all depend on good prefrontal functioning. Part of this functioning involves the dorsolateral prefrontal cortex near F5, an area seen to be outside EEG database norms in many of our clients with ASDs. Improvement in performance on ToL at the ADD Centre has been reported to occur in children with Asperger's who received neurofeedback training (Knezevic, 2007). We hypothesize that in some of the children, these improvements could be related to training over the anterior cingulate cortex (ACC) at Cz and FCz. The anterior cingulate has numerous connections, including links to the left dorsolateral frontal cortex between F3 and F7 in an area that is also noted to contain mirror neurons. Improvement in executive functions using Cz NFB to decrease 4-8 Hz and increase 12-15 Hz has been reported by Kouijzer et al. (2009). They measured NFB improvements using the ToL and the Milwaukee Card Sorting Tests.

Polyvagal Theory

As previously mentioned, Stephen Porges has noted that flat facial expression, poorly modulated tone of voice, and poor listening skills are related to the neural pathways that regulate the striated muscles of the face and head. He also notes that reduced muscle tone in this circuit correlates with less expressiveness in voice and face, less eye contact (eyelids droop), and slack middle ear muscles (stapedius) that make distinguishing human voices from background noise more difficult. In addition, Porges has discussed the neurophysiological interactions between what he terms the Social Engagement System and the hypothalamic-pituitary-adrenal (HPA) axis, the neuropeptides of oxytocin and vasopressin, and the immune system (Porges, 2003, 2004, 2007; Gray et al., 2004). He has noted that training of myelinated vagal tone should be helpful in treating the symptoms of ASDs (Porges, 2003).

Systems Theory of Neural Synergy
We would add that biofeedback (BFB) training synergistically combines with our NFB training to potentially decrease at least one key symptom of ASDs, namely anxiety.

Due to the fact that HRV training gives feedback to the nucleus of the solitary tract (NTS) in the brain stem medulla, and this connects to many areas involved in Executive, Affect, Default, and Placebo networks, this combined approach using NFB with BFB may also improve other ASD symptoms. This hypothesis was central to an oral presentation titled "A Systems Theory of Neural Synergy" given by the authors at the International Society for Neurofeedback and Research (ISNR) annual conference in 2008. It has now been published in a more comprehensive format as "Heart–Brain Connections: Neuroanatomy Underlies Effective Neurofeedback plus Biofeedback Interventions" (Thompson et al., 2013b). NFB training over the ACC, which influences the entire limbic system and the brainstem nuclei, may also affect the autonomic system through feedback loops to the insula. Recall that the right insula helps regulate sympathetic drive and the left insula relates to parasympathetic calming. We hypothesize that **heart rate variability (HRV) training** or, in the early days, just teaching effortless, diaphragmatic breathing, may have been important in our success with clients along the autistic spectrum. Porges' polyvagal theory supports our rationale for adding the biofeedback to neurofeedback.

Neurofeedback Training
Training is based on an understanding of the functional significance of different areas of the brain, as elucidated in the small booklet that correlates Brodmann areas to 10-20 sites (Thompson, Thompson & Wu, 2008), and in this second edition of *The Neurofeedback Book*. When this is combined with theories concerning brain dysfunction in those with ASD, as outlined above, and 19-channel QEEG assessment, plus, in adolescents and adults, a stress assessment of psychophysiological variables, it has led to interventions that combine EEG feedback with other biofeedback modalities. Normalizing the EEG involves either using one- or two-channel NFB training, such as training at FCz (to influence the ACC) and at T6 (to decrease the sensory aprosodia symptoms) and/or LORETA Z-score NFB. Sometimes we do one session of LORETA NFB and one of regular NFB with a client each week.

Importance of Anterior Cingulate Connections
The ACC influences the entire limbic system, the amygdala-hypothalamic-pituitary-adrenal (AHPA) axis, and the autonomic nervous system. It also links to frontal areas of the cortex influencing attention and executive functioning. Training at the Cz and FCz sites is hypothesized to influence the AC, and therefore its connections to the Affective, Executive, Attention, Memory, Salience, and Default networks. The connections from the AC to functionally corresponding areas of the basal ganglia and thalamic neuron groups would then be involved in feedback loops affecting functionally related cortical areas. In addition, we should recall that the *cingulum* runs through the cingulate cortex and has connections with the frontal, parietal, and temporal lobes. This may help explain why **good results were achieved** in early years, before LORETA was available, with most clients with training at a **single site. Indeed, this type of training may have advantages** in that it allows the brain to appropriately readjust the **balance of activity** between widely dispersed areas rather than us quite 'artificially' influencing numerous sites all at the same time to be within database norms.

Frequencies Targeted with NFB
Usually, slow frequencies (2-5 Hz, 3-7 Hz or 4-8 Hz, depending on which frequency range was highest on the individual client's initial EEG profile) and high-frequency beta (20-35 Hz) are reduced, while low-=frequency beta and even low alpha may be *either* increased or decreased, depending on the QEEG assessment findings. However, the frequency bands and the decision to enhance or inhibit are based on the QEEG and on observing a correlation between the client's symptoms and the targeted area and frequency band. The decision around how to handle high-amplitude, low-frequency beta 13-16 Hz in the FCz to PCz region is a difficult one, and must be handled with careful assessment and observation. The practitioner, as previously mentioned, will wish to increase SMR 13-15 Hz. When low-frequency spindling beta is present centrally, the source is

commonly found, with LORETA, to be the ACC. When this is high-amplitude it may need to be decreased. To simultaneously raise SMR, one must, in addition to HRV training, find a site with low-amplitude 13-15 Hz along the sensorimotor strip. This may be C3 or C4. Only a careful assessment can help the practitioner decide.

NFB Combined with BFB

At the ADD Centre, we must also take into account that many of the clients have had biofeedback training to encourage effortless diaphragmatic breathing and, more recently, heart rate variability training. The vagal feedback through the medulla to the limbic system, including the thalamus and the anterior cingulate gyrus (ACC), could theoretically be an additional important factor accounting for the positive outcomes. (Note also the discussion under the Heart Rate Variability Training section in this edition of *The Neurofeedback Book*.) The combination of NFB affecting cortex-basal ganglia-thalamus-cortical loops, with peripheral BFB augmenting the NFB effects on these functional networks, fits our systems theory of neural synergy (Thompson & Thompson, 2008).

Raw EEG Findings

In the section above on neuroanatomical and assessment findings, several figures are shown. In the first figure, note that there is spindling synchronous beta at F4, Fp2 (25 Hz 5 SD) and F7 (20 Hz 3.4 SD) and high-amplitude slow-wave activity at T6. The general pattern shown by this child is a frequently found pattern in both anxiety disorders and ASDs (but not in Asperger's cases, despite the DSM5 regarding Asperger's as indistinguishable from ASDs). That pattern is excess slow-wave activity, a dip in the low frequency alpha range (8-10 Hz), and higher-than-expected beta, in particular, high 2-5 Hz, and high amplitudes at various frequencies between 12-36 Hz. The specific frequency ranges and sites vary from person to person. In general, the very high delta and theta are anterior (frontal).

In Asperger's, a different pattern is often observed with high–amplitude, low-frequency alpha at frontal and central sites. The peak is often at 8 or 9 Hz. **Asperger's is the opposite to autism** in that the **autistic patients show extremely low-amplitude, low-frequency alpha** (often –1 to –2 SD), which is generally observed in more posterior locations (Pz).

Alpha

A function that correlates with alpha oscillations is disengagement from processing sensory information that is not relevant. Some brains, such as people with ASD, do not disengage appropriately - they are overloaded by sensory information and we do not see alpha oscillations, but rather variants of beta. In addition to biofeedback (HRV) and increasing C3 and/or C4 SMR, these EEG findings suggest intervention using two-channel training at FCz and Cz to decrease low-frequency delta-theta and high-frequency beta, which may be a spindling beta (not the usual desynchronized beta) with source derivation in the anterior cingulate gyrus. In addition, it also suggests that one should increase low-frequency alpha at Pz while being careful to avoid dissociation occurring if 9 Hz is increased. Visualizing out–of-body experience is an area that has been studied with respect to brain anatomy and EEG (de Ridder et al., 2007).

Overview of Treatment

As noted above, EEG training can be combined with BFB, usually heart rate variability training (Lehrer et al., 2007; Gevirtz, 2010). We usually increase sensorimotor rhythm (SMR, 12-15 Hz or 13-15 Hz), which may have a stabilizing effect on a cortex that is unstable and easily kindled (Sterman, 2000). When enhancing SMR, we are, as previously noted, careful to identify a site on the sensorimotor strip where there is no high-amplitude spindling beta in the 13-15 Hz range. In addition to the decreased activation observed at T6, another factor that may, in future, prove to be a helpful 'marker' for ASD could be the Mu rhythm response. In ASD there is evidence of a reduction in Mu rhythm suppression during action observation (Oberman et al., 2005).

There have been other reports of the efficacy of NFB for autism (Jarusiewicz, 2002, 2006).

Coben (2005, 2007) has reported on normalizing coherence using single-channel, sequential training of the sites found to be deviant on QEEG analysis. Sequential (bipolar) training may have an influence on the phase relationship between EEG activity in the targeted frequency ranges at the two sites. We have observed normalization of coherence after amplitude training. Recently, we have added the use of Z-score feedback in selected cases, but we now usually use this with LORETA Z-score NFB. Theoretically, this may be a promising approach. However, these clients must be carefully watched to make sure we are not inadvertently training to normalize compensatory sites that, perhaps, should be outside the database norms in that client.

Treatment Summary

The approach used at the ADD Centre begins with an emphasis on diet, sleep, exercise, and parenting techniques. Gluten-free and dairy-free diets improve symptoms in some clients with autism. In Asperger's, fussy eaters abound, but elimination diets are less likely to result in amelioration of symptoms. We then prescribe a combination of NFB and BFB and metacognitive strategies. This will initially address the symptoms that interfere with the child being able to interact constructively with caregivers, including, in the following order: anxiety, impulsivity, attention span, executive functions, and finally, understanding and responding to social interactions.

Intervention focuses on the four core symptoms

(Thompson & Thompson, 2009):
1. Anxiety – (anxiety is the keystone to understanding behavior and to successful intervention)
2a. Impulsivity problems
2b. Attention span
3. Difficulties with social interactions - motor and sensory aprosodia
4. Executive functions

Each of these symptoms relates to a NETWORK - Affect, Executive (attention), Default (social) and Placebo networks. The 'counseling' in the figure below includes, but is not limited to, guiding the patient with respect to diet, sleep, exercise, and aspects of social relatioships. Diet is particularly important in some clients. A number of our staff have professional qualifications in counseling and/or psychotherapy. For these staff members, counseling may also include helping their patient consider their behavior through clarification and gentle confrontation, but never interpretation. Clients who require psychotherapy and family therapy are referred outside the clinic.

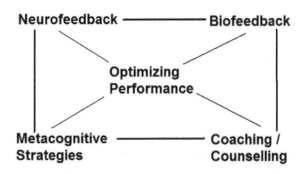

As sessions progress and the client becomes relaxed, calm, and focused, the NFB will be combined with cognitive and social strategies that place more emphasis on executive functions and social awareness. The latter may involve some training near T6 (activation) and F6 (when motor aprosodia is observed), and in the mirror neuron areas such as F5. Coherence abnormalities that remain after amplitude training has been done would then be addressed. The reader should note that some practitioners, such as Coben, might address coherence first. At the ADD Centre, we often now address amplitude, coherence, and phase differences from the normal database simultaneously using LORETA Z-score NFB. Not all children with Asperger's or an ASD can, initially, tolerate wearing the cap for LORETA NFB. For that reason we usually start with single-channel NFB. The segments of each training session (and for a series of sessions) are always graphed. This is highly motivating for both children and adults. Transcranial direct current stimulation (tDCS) may be added to the treatment regimen in selected patients for specific purposes, such as improving executive functions. The tDCS is treated as 'experimental' and fully discussed as such if it is tried. At our Centres, a few minutes of tDCS may be given to prime an area of the cortex prior to doing NFB. At the time of writing, this technique has not been used on a sufficient number of patients to be able to evaluate its possible effects.

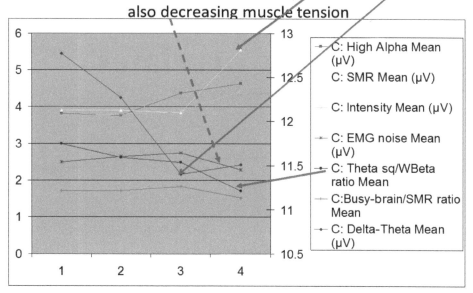

This graph demonstrates how a client can make major shifts in important variables during a session. It also raises the important point that the 'trainer' must note 'EMG noise' (here reflected in 52-58 Hz amplitude) caused by muscle tension, before concluding that the client has really raised SMR.

Understanding Apparent Regression

Some clinicians have reported on the emergence of difficult behaviors when providing treatment based purely on QEEG Z-score findings. In some cases this may be due to the clinician inadvertently countering compensatory changes in the EEG. But other possibilities do exist. We have observed that a small number of patients may appear to regress in the early stages of NFB treatment. Two possible reasons for this may be considered. First, in children with abnormal development, deviant amplitude and coherence Z-scores might, in part, reflect compensatory mechanisms. Thus care should be taken when attempting to "normalize" QEEG findings that each of these findings does correlate with negative symptoms.

Second, the autistic child has arrested development. Treatment allows these children to **begin to progress through the normal stages of development that should have been negotiated at an earlier age.** As these children move through the equivalent of rapprochement they may enter what has been termed an "aggressive-depressed" stage (Thompson & Patterson, 1986; Thompson, 1990). The child may begin to test limits. At this juncture, the caregivers must be careful not to reverse the child's forward movement in development. The caregiver, while carefully setting appropriate limits, should reinforce the child's sense of independence while still meeting their needs for dependence. These children may be going through what is commonly called The Terrible Twos but at a much later age, making their behavior more difficult to deal with because they are much bigger and stronger and even more determined and emotionally vulnerable (higher anxiety). Thus, when a child moves forward in stages of Separation-Individuation he will appear to be acting out, but really he is exploring autonomy and power and control in the world. One should not put the child down, but **rather join and then redirect**. You join in what he is doing, then introduce what you are now going to do together. **Meet their dependency needs while allowing some independence and control.**

SECTION XXVI
Attention Deficits and Their Brain Correlates
(See also Section V)

The assessment and intervention aspects of Attention Deficit Hyperactivity Disorder have been covered in an earlier section of this book. A few of the main efficacy studies will be briefly summarized here, and the reader is encouraged to read a meta-analysis (Arns et al., 2009) that argues that NFB should be given the highest level of efficacy for the treatment of ADHD (Level 5: efficacious and specific), and publications by Monastra et al. (1999, 2002, 2005) and Gevensleben et al. (2009). The latter paper reports on an excellent randomized, controlled, multicenter study. Research support for the efficacy of neurofeedback for the treatment of Attention Deficit Hyperactivity Disorder (ADHD) has increased, and in November 2012, the American Academy of Pediatrics gave biofeedback a Level 1 efficacy rating in its review of efficacious psychosocial interventions. That is the highest level in their ranking system, and the same level accorded to stimulant medications for efficacious medical interventions. Some regard NFB as having higher efficacy than medications due to its lasting effects when treatments have been discontinued (Gani et al., 2008). Medications are only effective when they are at the appropriate level in the bloodstream of the patient.

Before the research review, however, we begin this addition to our discussion of ADHD with two questions. First: Why do we achieve virtually 100 percent success with children who have simple ADHD who come from supportive, involved, positive families when we use single-channel Cz training plus BFB and metacognitive strategies for at least 40 sessions? Second: When should we do 19-channel QEEG and target other areas of the cortex?

To answer these questions, we first consider the importance of attention in cognition and the networks involved in these processes.

As stated in the first part of the functional neuroanatomy section, trying to describe attention networks in detail is not possible in a basic textbook because attention is complex and involves most areas of the brain. However, an overview of the networks contributing to our ability to attend to a specific internal or external stimulus while disengaging from other, not-so-relevant stimuli will be attempted here. A major network involved in this type of activity is called the Salience network.

Attention is the basic function that permits an appropriate selection of stimuli. It is the basis for maintaining concentration. Many other cognitive systems can operate somewhat independently of one another. For example, the child who exhibits dyslexia can still speak, reason, and visualize things without impairment. However, none of these functions could operate at a normal level without the ability to appropriately attend to stimuli. Without attention, coherent thought itself would not be possible. Attention is a process whereby one can focus and concentrate on one item while inhibiting distracting influences of nonrelevant material.

For attention to operate, an adequate level of **arousal is essential**. The core of the arousal system includes the mesencephalic reticular formation and thalamus (reticular and intralaminar nuclei), which are a part of the ascending reticular activating system (ARAS). Other components of this system include the substantia nigra (source of dopamine to basal ganglia), locus coeruleus (source of norepinephrine), ventral tegmental area (dopamine to the anterior cingulate cortex, or ACC, and the prefrontal cortex), Raphe nuclei (source of serotonin), and basal forebrain nuclei (source of acetylcholine).

Entering a focused-attention state while doing attention-demanding tasks has been found, using PET scans, to markedly increase regional cerebral blood flow (rCBF) in these areas that are related to arousal (Kinomura, Larsson, Gulyas & Roland, 1996).

The brain is able to monitor many stimuli with what is termed *divided attention*, but it cannot consciously consider more than one thing at one instant in time. The way this is often illustrated to students is by showing the Rubin Vase. The student is asked what he/she sees. They will discover that they can see either the vase or the two faces, but they are not able to see both at the same time. This figure-ground illusion was developed around 1915 by the Danish psychologist Edgar Rubin and published in his two-volume work in the Danish-language *Synsoplevede Figurer* ("Visual Figures").

The Rubin Vase

The brain uses what is called 'selective attention' to focus on the major stimulus of choice, and both the Salience network and inhibitory processes come into play to allow this singular focus. These inhibitory processes include the cortex-basal ganglia-thalamus-cortex loops with lateral inhibition within the striatum, in addition to the interplay of indirect, hyperdirect, and direct pathways as described in Part 1 of the Functional Neuroanatomy section in Section VI of this book. In that section we used the example of a football quarterback changing his decision about where to send a pass in order to introduce the concept of the indirect and hyperdirect pathways that 'excite' the globus pallidus. The hyperdirect and the indirect pathways allowed new information to be considered, and the quarterback's brain was able to inhibit the first short-pass choice in order to focus and attend to the long-pass touchdown opportunity. Because this process is important in any discussion of focused attention, we have repeated the illustration here.

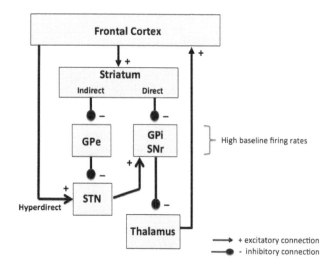

Simplified frontal-subcortical circuits illustrating selected aspects of the direct, indirect, and hyper-direct paths.

Activation of striatal neurons in the 'direct' pathway leads to inhibition of neurons in the internal segment of the globus pallidus (GPi) and substantia nigra pars reticulata (SNr). Activation of striatal neurons in the indirect path leads to inhibition of tonic activity in the external segment of the globus pallidus (GPe). This inhibition of GPe disinhibits the subthalamic nucleus (STN), which then provides an excitatory drive to the GPi/SNr. The frontal cortex can also excite STN cells directly via the 'hyperdirect' pathway.

The diagram illustrates a means for inhibition of nonrelevant cortical areas, Brodmann areas and networks, as was explained in the section on neuroanatomy. Another mechanism for stopping nonrelevant cortical areas and highlighting relevant functions is the Salience network. The key element in that network is the anterior insula (AI). The anterior insula can engage relevant systems while disengaging systems that are not immediately task-relevant (Menon & Uddin, 2010).

In the example, the quarterback has used what has been called 'top-down' Executive network, cognitive decision-making to redirect the pass. In this terminology, **top-down** involves posterior-parietal, prefrontal, limbic motivational (insula and anterior cingulate) connections to focus and modulate the so-called **'bottom-up'** competition amongst neuronal representations in the sensory association cortex regions of sensory stimuli (visual, auditory, somatosensory), to which he is allotting his attention and concentration. The disengagement of attention from his initially intended and well-practiced short pass involves the parietal cortex. It must also engage the thalamic pulvinar 'spotlight' that allows focus on one particular thing. The re-engagement to a different target will also involve the superior colliculus and the frontal eye fields (FEF). Here the reader may recall, from Part 1 in the section on functional neuroanatomy, a visual salience network described in 2006 by Bisley. In this Salience network, the lateral intraparietal area (LIP) connects to the frontal eye field (FEF) (as does the upper auditory cortex), superior colliculus, pulvinar nucleus, and inferior temporal area (BA20 - visual pattern recognition). This network feeds into the Executive aspects of the frontal lobes, which will determine the significance (salience) of new visual information and decide on which pass to execute. The vigilance network sustains this attention and maintains alertness.

However, in discussing choosing the 'salience' of new information, regardless of whether it is from internal or external sources, more emphasis should be placed on the role of the anterior portion of the insula and the anterior cingulate gyrus than seems to be done in Bisley's description of visual salience determination. As Menon and Uddin (2010) have said, "The **anterior insula is an integral hub** in mediating dynamic interactions between other large-scale brain networks involved in externally oriented attention and internally oriented or self-related cognition." These authors postulate that the **insula is sensitive to salient events,** and "Its core function is to mark such events for additional processing and initiate appropriate control signals." They reinforce the now-accepted view that the **anterior insula and the anterior cingulate cortex are the major elements that form a Salience network.** This network functions to segregate the most relevant among internal and extrapersonal stimuli in order to guide behavior. As noted above, they consider the anterior insula (AI) to be the hub of the Salience network. This is particularly important in our work, where we combine NFB with HRV and where we have observed that many of our patients who have suffered a TBI, or have an anxiety disorder, or have a diagnosis of autistic spectrum disorder, show activity in the left and/or right insula to be outside the database norms. For example, **hyperactivity** of the right insula might correlate with misinterpretation of the salience of mundane events and lead to anxiety. On the other hand, **hypoactivity** in someone with an ASD might correlate with a lack of attention to social communication. In these patients we have also observed marked changes in HRV statistics, as discussed in the section of this book on HRV.

In the diagram illustrating a Salience network connection (below), the reader can see that a bottom-up salient event is detected by the anterior insula and ACC. The insula is particularly sensitive to salient events. The insula can then communicate important events to other networks, including those for sustaining attention, executive/cognition functions, working memory functions, Affect networks, and motor activities. With the posterior insula, it will regulate the autonomic nervous system, and, as we have previously discussed, this will have an effect on HRV and other autonomic nervous system (ANS) functions through connections to the amygdala, locus coeruleus, and the hypothalamus. Thus, the **insula is a key link** between incoming bottom-up stimuli and the brain regions, such as the ACC, that are involved in monitoring this information.

Cognitively demanding tasks have the effect of decreasing activity in areas of the brain that we have previously described as being within the Default network, while increasing activity in Executive networks. The central Executive network (CEN) includes the dorsolateral prefrontal cortex (DLPFC), and the posterior parietal cortex (PPC), as previously described. The Default mode network (DMN) includes the ventromedial prefrontal cortex (VMPFC), posterior cingulate cortex (PCC), medial temporal region, and the angular gyrus. (See Sections VI, VII, VIII, on neuroanatomy, Brodmann areas, and networks.) The CEN is important for working memory, and for judgment and decision-making for goal-directed behavior.

The PCC is involved in self-referential thinking and autobiographical memory (Buchner et al., 2008) and the VMPC is important in thinking about self and others (Amodio & Frith, 2006).

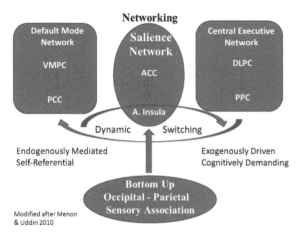

Figure to illustrate major Salience network connections
VMPC = ventral-medial prefrontal cortex; PCC = posterior cingulate cortex; ACC = anterior cingulate cortex; A Insula = anterior insula; DLPC = dorsolateral prefrontal cortex; PPC = posterior parietal cortex

This diagram emphasizes the central importance of the Salience network in the dynamic switching between activity generated by the central Executive network (CEN) and the internal self-referential activity of the Default mode network (DMN). Menon and Uddin suggest that the Salience network can be conceptualized as having a hierarchy of salience filters that can amplify inputs to different degrees. There are direct connections between the intraparietal sulcus and the insula, and this may be one important link for determining the salience of incoming sensory stimuli.

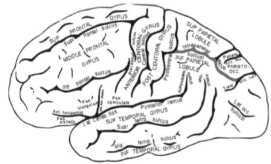

Gray's Anatomy *726 Wikimedia Commons to show intraparietal sulcus*

Menon and Uddin also note, as Juri Kropotov (2009) described (see our discussion of ERPs), that ERPs can be useful in following this process. Recall that at about 150 ms poststimulus, primary sensory areas detect a deviant stimulus as indexed by the mismatch negativity (MMN) component of the evoked potential. In the next stage, this bottom–up sensory MMN signal is transmitted to other brain regions, including the AI and the ACC. This is followed by top-down (control) signal at about 200-350 ms, and the N2 and P3a waves may be observed (P3a is best observed at Fz). These are probably, in part, generated by activity in the anterior insula and the ACC.

In the next stage, at about 300-450 ms poststimulus, the P3b is observed. It is usually maximum amplitude at Pz. We have noted earlier that this signal is likely a response to the attention shifting, and comes from frontal-temporal-parietal involvement. This can be followed by cingulate activity and response selection that may involve the motor cortex. Each of these steps from the bottom-up sensory stimulus through the Executive, Salience, and finally the Motor networks (in this example) may be thought of as being 'indexed' by an ERP.

Our emphasis in this section has been on the insula as a 'key' to salience functioning. It is important in bottom-up saliency detection. However, the reader is reminded that the insula is involved in many other areas, too. For example, the right insula is important in the subjective awareness of pain and in both positive and negative feelings, including: anger, disgust, sexual arousal, empathy, and judgments of trustworthiness.

Menon and Uddin state that the basic mechanisms of the insula are: "(1) bottom-up saliency detection, (2) switching between other large-scale networks to facilitate access to attention and working memory when a salient event occurs, (3) interaction of the anterior and posterior insula to modulate physiological reactivity to salient stimuli, and (4) access to the motor system via strongly coupling with the anterior cingulate cortex (ACC)."

Perhaps importantly, the AI and the ACC share a unique neuronal feature, namely, the von Economo neurons (VENs). The VENs have large axons. Conduction speed is higher in axons with a large diameter. This means that transmission of AI and ACC signals to other cortical regions is very rapid.

Gitelman has said, "Attention thus emerges as a combination of feed-forward signals from the visual cortices (bottom up) and feed-back signals from the parieto-frontal-limbic network (top down). There is said to be 'competition' amongst neuronal representations (visual, auditory, and somatosensory) in the sensory association cortical regions. The combined interactions are thought to bias positively neurons representing the attended target(s) to 'prevail' in their interactions" (Gitelman & Darren, 2003).

We illustrate these processes in the following manner:

Attention Networking

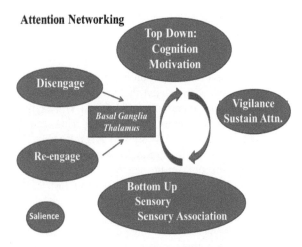

The top-down control modulates the sensory input. In a way of speaking, it alters the 'sensitivity' to sensory input. This function would appear to be deficient in people with ASD.

Attention Networking

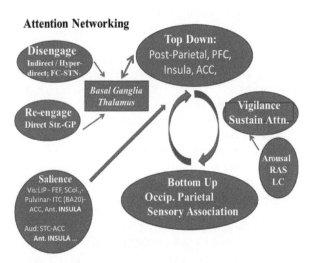

Areas: posterior-parietal (PP), prefrontal cortex (PFC), anterior cingulate cortex (ACC), frontal cortex (FC), lateral intraparietal (LIP), subthalamic nucleus (STN), globus pallidus (GP), striatum (Str), frontal eye field (FEF), superior colliculus (SCol), inferior temporal cortex (ITC), superior temporal cortex (STC), occipital (Occip.), attention (Attn), reticular activating system (RAS), locus coeruleus (LC).

In this diagram, the insula is the 'key or 'hub' for 'salience' functions, and it is therefore in bold capital letters.

The **Vigilance network** is an essential component for salience. The Vigilance network comprises the RAS, LC, the right posterior parietal cortex (RPP), and the right prefrontal cortex (RPFC). Note that the right hemisphere is particularly important in networks for sustaining attention.

Alternative Ways of Conceptualizing and Illustrating Salience Functioning and Attention Networks

These areas have been illustrated in a different way by Amir Raz, and his diagrams are available online.

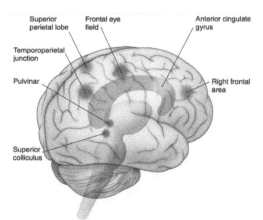

Raz, Amir (2004). Anatomy of Attentional Networks. (Visual) The Anatomical Record Part B: The New Anatomist, 281B(1), Available online: 22 Nov., 2004. Also Raz, Amir & Sapiro, Theodore (2002). Hypnosis and Neuroscience: A Cross Talk between Clinical and Cognitive Research. Archives of General Psychiatry, 59, 85-90.

Raz calls the thalamic pulvinar activation "The **Alerting Network**," the parietal activations "The **Orienting Network**," and the anterior cingulate activation "The **Conflict Network**." Raz has said that these are **distinct attentional networks** that can be identified using fMRI.

Another way of conceptualizing this is as **alerting, orienting, and executive functions**. This is really emphasizing the same processes. It is the conceptualization offered by Posner and his colleagues and it is also available on the internet. It is illustrated as follows:

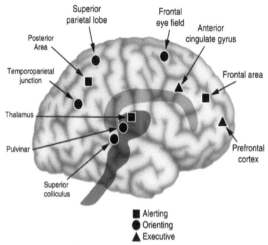

Posner, M.I., Sheese, B.E., Odludas, Y. & Tang, Y. (2006). *Analyzing and shaping human attentional networks.* Neural Networks, 19, 1422-1429. (www.sciencedirect.com).

Posner et al. highlight alerting, orienting, and executive aspects of attention. The alerting function involves the structures already mentioned under vigilance, plus the pulvinar nucleus in the thalamus. The pulvinar nuclei are lateral and posterior. It is the largest area in the thalamus. The orienting functions require that the frontal eye fields work in conjunction with the superior and inferior parietal areas. These areas are sometimes referred to as the cortical nodes of the orienting network. In this model the third important set of functions involve the anterior cingulate cortex (ACC), which is the key for executive control. It is involved in the resolution of conflict between neural systems, which may have given rise to Raz using the term 'conflict network.' The ACC functions to regulate thoughts and feelings that are competing for control of consciousness or output. In this manner, the ACC regulates the activity in other brain networks.

Notes concerning the thalamic pulvinar nuclei: The lateral and inferior pulvinar connect to visual cortical areas. The inferior (medial and lateral) aspect of the pulvinar has input from the superior colliculus for the regulation of visual attention. The dorsal-lateral aspect connects to the posterior parietal cortex. The medial aspect connects with the prefrontal and premotor cortex, cingulate, and posterior parietal cortical areas. Temporal attention involves the posterior aspects of the pulvinar nuclei. Temporal attention differs from visual/spatial attention. Temporal attention is more left-hemisphere than spatial attention, and can be defined as a process aimed at allocating brain resources on the predicted onset or timing of an incoming event.

Another way of conceptualizing Attention networks is to think of interconnected ventral and dorsal networks.

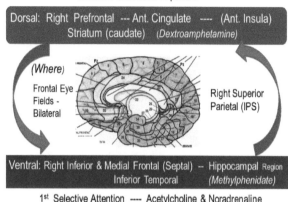

The ventral network can be thought of as answering the question, "***What*** am I seeing?". It comprises the temporoparietal junction (TPJ) and the ventral frontal cortex (VFC), (mainly right hemisphere). It typically responds when behaviorally relevant stimuli occur unexpectedly. For example, it responds when these stimuli appear outside the cued focus of spatial attention.

The dorsal network may be thought of as answering the question, "***Where*** is it?" The dorsal Attention network is organized bilaterally and comprises the intraparietal sulcus (IPS) and the frontal eye fields (FEF). These areas are active when attention is overtly or covertly oriented in space, saccade (eye fixation processes) planning, and visual working memory (Vossel, S., Geng, J.J. & Fink G.R., 2014). Dorsal and Ventral Attention Systems: Distinct Neural Circuits but Collaborative Roles. *The Neuroscientist, 20*(2), 150–159). The dorsal and ventral networks are summarized online: Montoya, E.R. (2009). Interactions between Dorsal and Ventral Attention Networks. *Faculty of Medicine (*Master thesis).

In addition to being important in attention to *what* is occurring, the Ventral Attention network can orient attention to internal or introspective processes and is part of the Default network that is activated in the so-called resting state (Swanson, J., Baler, R.D. and Volkow, N.D., 2011. Understanding the Effects of Stimulant Medications on Cognition in Individuals with Attention-Deficit Hyperactivity Disorder: A

Decade of Progress.. *Neuropsychopharmacology, 36*(1), 207–226).

The ventral and dorsal Attention networks act in sequence. The ventral Attention network activates first, followed by the dorsal Attention network, and they are anti-correlated. That is, as the ventral becomes active the dorsal is deactivated and vice versa. Therefore, an attention study of phase reset between the Brodmann areas of these two systems may show a similar phenomenon. The ventral attention network leads the dorsal network and the phase difference shortens (Thatcher, personal communication).

Now, let us return to our original questions. First: Why do we achieve virtually 100 percent success in children with simple ADHD, who are from good, involved, positive families, using single-channel Cz training plus BFB and metacognitive strategies? The answer is twofold. First, the child learns what it feels like to pay attention by resetting the thalamocortical loops as she changes her brain wave patterns during training. Rhythmic activity in the theta or alpha range produced in the thalamus is appropriately reduced. Second, there is activation of particular networks to strengthen them.

Theta/Beta Ratios and Notes About Training
In ADHD without other complicating factors such as a learning disability, affect symptoms, and/or Asperger's syndrome, single-channel QEEG will show the spectral frequencies that are extremely high- or low-amplitude. In the most common EEG pattern seen in those with ADHD, the theta/beta power ratio may be well outside the norms for the age of the client.

Here we must note that research done by Marijn Arns (SABA presentation May 2015) was reviewed by Sandra Loo (presentation at SABA conference, May, 2014) has demonstrated that the difference in theta/beta ratios between normally developing children and those with ADHD is now quite small compared to that found fifteen years ago when the norms were published by Monastra et al. (1999). The reason appears to be a marked increase in the theta amplitude in the 'normal' control group. This will require further research and publications but it is something for the NFB practitioner to keep in mind.

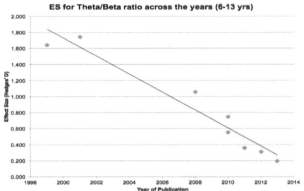

Graph from SABA 2014 presentation given by Sandra Loo (Director of Pediatric Neuropsychology and Associate Professor of Psychiatry in the Semel Institute for Neuroscience and Human Behavior at the UCLA David Geffen School of Medicine). It shows a steady decrease in effect size for theta/beta ratios from 1998 to 2014. This ratio appears to be losing its sensitivity for identifying ADHD in 6- to 13-year-old children.

One can speculate that higher theta amplitudes in the general population are due to children getting less sleep, spending more time on electronic devices, exposure to environmental factors (like exposure to Wi-Fi and other sources of electricity), change in nutrition, or some combination of these influences. With some clients, the low-frequency alpha may show an extremely high-amplitude, and/or some range within the high-frequency beta (21-35 Hz) range may show very high-amplitude peaks. These high-frequency beta amplitude peaks may correlate with that individual multitasking and/or ruminating when they had intended to focus on a single task. Our job is relatively easy. The trainer works to help the client achieve a calm, relaxed, focused state while the client practices self-regulation/control and remains concentrating on a single task. The frequencies to inhibit and/or enhance are based on the EEG assessment profile at the site chosen for feedback (often Cz, initially). A high-theta subtype will have different frequency bands chosen than a high-alpha or -beta subtype. This task, during some of the training, will be carefully chosen to correlate with the type of work, such as reading or mathematics for a young child, that that child needs to improve in school. The electrode at Cz or FCz may influence the anterior cingulate gyrus. The goal of the trainer is to help the child get into and sustain a pleasant positive affect while in a calm, relaxed, focused, state of concentration on the task at hand.

The reader will immediately observe that the role of the trainer and the relationship (therapeutic alliance) between the trainer and the client is absolutely crucial to achieving the highest level of positive outcomes in these training sessions. The trainer is assisting the client to move into the best state that needs to be reinforced for, first, the Affect network, and, second, the Executive network. The single site that has a chance of doing both is Cz, or FCz in older adolescents and adults, because the placement moves slightly forward as the frontal lobes become increasingly myelinated and thus a bit larger.

It is important that the trainer identify the network he wants to influence and chooses tasks accordingly. For example, to learn, remember and recall, new information, one child may be almost entirely visual-spatial whereas another child may be almost entirely verbal. Although boys tend to the visual right-brain, and girls somewhat more to the verbal left-brain, there are many exceptions. Indeed, much research and even popular books have been written on this subject, such as *Right Brained Children in a Left Brained World* by Jeffrey Freed. Certainly, ADHD boys often do present with significant difficulties in verbal areas while at the same time exhibiting even extraordinarily high IQ levels in nondominant right-brain tasks. The trainer would do well to clearly identify how each of their clients learns. Then, using appropriate metacognitive strategies, reinforce and expand that client's techniques for learning new material. It should go without saying that the trainer must be cognizant of her own preferred learning methods and not impose their own preferences on their client, though providing a different approach to a problem is often helpful; for example, teaching 'adding on' to find the difference as a method for subtraction to a child who often made errors when counting back. In the main, the trainer should assist his client to develop and reinforce the Executive network while using the learning method that works best for the client. An example would be teaching a child with spatial reasoning weaknesses and verbal strengths how to talk his way through math problems rather than expecting him to see the steps.

Case Example

Although 19-channel assessment and LORETA neurofeedback are widely used at our ADD Centre, we follow the Principle of Parsimony – do the least expensive, least invasive intervention that has a probability of success first. Thus we most often proceed with single-channel EEG assessment and training for students with ADHD, including those who have "ADHD +," as was the case with AB, an eleven-year-old boy with a history of ADHD plus learning and behavioral problems. This child's parents had limited funds and were in a crisis situation, so spending on training rather than on a more detailed full-cap assessment made sense.

At the time of assessment, "AB" was a source of frustration to the adults in his life, both at home, and at school. At home he often fought with his younger brother and absolutely refused to do homework. At school his limited work output and lack of acquisition of basic skills in math, reading comprehension and written work had led to low grades. In school there were also frequent problems interacting with peers. He had mastery of phonics and was at grade level for reading individual words and for spelling. His vocabulary was quite strong, and he often used adult-sounding phrases. His overall average range IQ score masked significant differences among subtests on the WISC-IV, on which Similarities (verbal reasoning skills) tested at the 95th percentile rank, whereas Block Design (spatial reasoning) was at the 9th percentile rank. Note that this is the opposite to the pattern that we usually observe in hyperactive ADHD children who do not have any ASD symptoms. The other eight subtests were average range (25th to 75th percentile range). Thus, his pattern was like that of students with a Nonverbal Learning Disorder (NLD). NLD is a close cousin to Asperger's syndrome, and he had the typical problems with organizational skills and with boundaries. He was a socially naïve boy who could be left holding the bag when other students with more "street smarts" put the blame on him for an altercation. One area of good functioning was sports, and he played on soccer and hockey teams. He sometimes had trouble getting a sense of the game, but he carefully followed the coaches' directives. Sports provided an outlet for his restlessness and high activity level.

To treat his problems, his doctor had prescribed increasingly high doses of Vyvanse. Side effects included appetite suppression and rebound hyperactivity and labile mood, often with tearfulness, as the amphetamine wore off each evening. The prescribing physician had added Zoloft to treat the mood problems. (Zoloft is an antidepressant, a selective serotonin reuptake inhibitor, or SSRI, that is not approved for pediatric use except for obsessive compulsive disorder.) Problems at home and school were escalating as he got older, despite the medical management using drugs.

The initial assessment at the ADD Centre involved:
1. Interviews with AB and his parents
2. Four parent questionnaires: ADD-Q (from *The A.D.D. Book* by Sears & Thompson), Conners 3

GI – Parent version, SNAP version of the DSM criteria for ADHD, Australian Scale for Asperger's Syndrome (from Tony Attwood's book, *Asperger's Syndrome*)
3. Two computerized tests: Test of Variables of Attention (TOVA) and Integrated Visual Auditory continuous performance test (IVA)
4. Single-channel EEG data collection at the vertex (Cz) that was artifacted and yielded statistics for single Hz bins, as well as ratios of interest. A graph showing mean amplitude for each frequency from 2-60 Hz was generated to show parents his EEG profile at the central location. NOTE: He had been off Vyvanse medication for three days so that we could get a baseline off the stimulant. He was seen on a Monday and parents had given him a drug holiday on the weekend.
5. A practice training session that compared his baseline and four segments of training, each lasting two minutes, to show his response to neurofeedback.

AB was deemed a good candidate for training, based on the findings and his ability to cooperate with the procedures. Questionnaires were all in the significant range for a diagnosis of ADHD, and the ASAS showed elevated scores, albeit not as high as seen in children with a diagnosis of Asperger's (now under DSM-5 criteria called autism spectrum disorder without intellectual impairment or language impairment, Level 1 severity). His overall ADHD score on the TOVA was -2.87, and his IVA scores indicated both inattention and impulsivity were problems. His EEG pattern showed a classic pattern of high theta/beta ratio (5.95, so above the cutoff score of 5 for children ages 6-11). His pattern also showed spindling beta above 20 Hz, which matched with symptoms of anxiety and 'busy-brain.' When there is anxiety, there is often a poor response to stimulants and more side effects.

Parents remarked that they had never seen him sitting still for so long as he did during the practice training session and were hopeful about the results of training, though they knew this was not a quick fix and would take at least five months of twice-weekly sessions (40 sessions). In addition to the neurofeedback, we gave advice regarding diet. We suggest for nearly all clients with ADHD to take an omega-3 fish oil supplement and vitamin D3 and have *serum ferritin* level checked to see if an iron supplement is needed. Parents are also referred to Chapter 10 in *The A.D.D. Book*, "Feeding the Child with A.D.D.," and Chapter 8, "Neurofeedback." In AB's case, Chapter 9, "Medications," was also recommended, as were the sections on behavior management. Every family is given a copy of this book at the initial assessment.

Parents were not happy with the high dose of stimulants and the questionable use of Zoloft, and they found the pediatrician to be uninformed about nondrug approaches. (It had been AB's grandfather who had learned of neurofeedback through a television show that featured it.) We complied with their request for us to assist in their search for a physician who was knowledgeable about ADHD and its treatments. They decided on a family physician who made referrals to our Centre and knew how to titrate medications, reducing the dose as children gained self-regulation skills through neurofeedback and biofeedback. This physician also knew how to relate to children who had ADHD, and AB's parents said she was the first physician to speak directly to AB and treat his input with respect.

AB started training in April. Training used the eight-channel Biograph Infiniti equipment that allows you to combine neurofeedback with heart rate variability training. His customized inhibit frequency range for the excess slow-wave activity was 3-9 Hz, and for his anxiety/busy brain the inhibit range was 21-35 Hz. The enhance frequencies were SMR in the 12-16 Hz range. At first we just taught "belly breathing" but, once his fidgeting was reduced after the first eight sessions, we introduced use of the respiration belt and blood volume pulse sensor (plethysmograph) on his thumb to do HRV training more precisely. We also spent part of each session teaching and then having him practice metacognitive strategies. For the most part, these strategies were related to reading comprehension, mathematics, and written work. A few of the strategies were related to reading social cues. These used social stories. Coaching in metacognitive strategies was done while EEG was monitored to ensure that he maintained a mental state that was calm and focused while doing academic tasks. In the classroom or when doing homework, when a child begins to use a strategy that was learned while doing NFB, the child will automatically enter the same focused concentrating mental state that he/she had been in while doing their training. In this manner the strategies assist in transferring the ability to maintain calm focus on tasks to outside situations and environments.

After 40 sessions, progress testing was done. AB had just started Grade 6 and was attending school. He had been successfully weaned off all medications over the summer, with careful monitoring by his new physician. The year had started well. He was attentive in class and doing his homework independently and without any argument. He and his brother could now play together, usually without arguments erupting. Positive parenting strategies were being used. These strategies had been gleaned from *The A.D.D. Book* and from discussion with trainers and sometimes with the supervising psychologist ("LT") at the end of training sessions. His soccer coach had noticed improvement in his play, especially with respect to sustaining focus throughout a game.

Objective test results on continuous performance tests showed all scores to be within the average range. Scores for Attention were in the upper half of the average range on IVA, and his ADHD score on the TOVA was now +0.70, which is an impressive gain of more than three standard deviations. Parents' ratings of symptoms on questionnaires showed significant reductions, too, and the Conners scale, on which the T-score had been >90 initially, now had a T-score of 63 - just below the cut-off score for inclusion in the ADD group, which is 65.

His drawing of a person (LT's favorite low-tech, high-yield measure) showed dramatic changes in details, including the following: blank eyes now had pupils and an appropriate gaze (reflects seeing the world around him and paying attention), sharp pointed heels on the shoes that looked very aggressive were gone and there was a solid stance with flat shoes and laces (feeling more grounded and safer, therefore less aggressive), there were ears, too (improved listening, as also seen on the IVA).

The theta/beta ratio had dropped to 5.1 - still a bit high for his age. The decision was made to continue on a once-a-week basis for 20 more sessions to consolidate gains regarding maintaining a calm and alert mental state, and to work more on metacognitive strategies to help him catch up in academic areas. Parents were delighted with the changes and the happier environment at home. They said they had been spreading the word about a nondrug approach, and took our brochures to the school principal to help build awareness of a nondrug approach. Asked how he felt about the results of training, AB said. "I feel normal." This was a brief but wonderful statement because it reflects his new self-image. If he *feels* normal, chances are that he will *act* in a normal way with respect to behavior and learning.

Second Question
Our second question was: "When should we do 19-channel QEEG and target other areas?" This question requires that we consider both adults and children who have more complex problems. These may be relatively simple difficulties, such as a single learning problem, for example, dyslexia. On the other hand, the client's problems may be quite complex, such as in persons who have many difficulties with executive functions including memory and learning; affect difficulties including anxiety and depression; and social difficulties in relating to others. Very complex difficulties are found, for example, in persons with Asperger's syndrome, or in postconcussion syndrome after a TBI when you see acquired (as contrasted to inborn) problems with attention.

In the first example of ADHD and dyslexia, initial single-channel training at Cz and C4 for the attention, hyperactivity and impulsivity can reduce these symptoms. Then, single-channel training at Wernicke's area or two-channel coherence training between Wernicke's and Broca's areas may have remarkable results, as we have seen at the ADD Centre. On the other hand, individuals with Asperger's and/or TBI may show many areas outside database norms, and they may benefit from simultaneous training at multiple sites, such as can be achieved with LORETA Z-score NFB. However, in our experience, no matter which NFB methodology is used, the best results are achieved when NFB is combined with HRV (and other BFB modalities as necessary), and with careful on-task work by the trainer to assure the client is practicing and reinforcing the neural networks that are most important for them to achieve their best outcome.

Event-Related Potentials (ERPs) and ADHD
Children diagnosed with ADHD and those diagnosed with an ASD both show difficulties with attention span. Sinzig et al. (2009) found that 53 percent of their sample of 83 children diagnosed with autism spectrum disorder fulfilled DSM-IV criteria for ADHD. In our experience with hundreds of children in the ASD category and many more with ADHD, we find that continuous performance tests such as the TOVA and IVA do show marked differences from the norms in both groups of clients. Nevertheless, there are also differences between these two groups in ERP results. We have mentioned previously the work of Kropotov in the section on ERPs and noted that he found differences from normal patterns in those with ADHD, and his group with ADHD demonstrated improvements after beta training (Kropotov, 2009). Sokhadze and his colleagues (2012a) have looked at the differences in ERP measurements between normal children, children with ADHD, and children diagnosed with autism spectrum disorders. They also used continuous

performance tests. In their study there were no differences in reaction times, but the ADHD and the ASD groups had more errors than the controls, with the ASD group showing the highest commission errors. With the ERPs, they demonstrated that the P200 and P300 in the ADHD group demonstrated longer latencies than either the ASD group or the controls. They also replicated Thatcher's findings that children diagnosed with ADHD show local hypoconnectivity and long-range hyperconnectivity as shown by coherence measures. This is the opposite pattern to that observed in those with ASDs. In addition, it has been shown that patients with autism, as compared to neurotypical individuals, have larger brains, but a smaller corpus callosum. At the opposite end of the spectrum, ADHD patients have smaller brains than the autistic population but have a larger corpus callosum (Casanova et al., 2009).

A Selection of Studies on Efficacy

A study by Arns et al. (2012) demonstrated that NFB treatment can result in improvements in patients who have insomnia associated with ADHD, which is a large subgroup of these patients. In addition, the American Academy of Pediatrics in 2012 placed NFB at the highest level among psychosocial interventions for effective treatment of ADHD. A major study they considered in coming to this decision was that of Beauregard et al. (2006), done at the Université de Montreal, Canada, that demonstrated changes in activation using fMRI after 40 sessions of NFB. More recently, that group has shown microstructural changes in both white and grey matter after neurofeedback training (Ghaziri, Tucholka, Larue, Blanchette-Sylvestre, Reyburn, Gilbert, Lévesque & Beauregard. 2013).

Effectiveness and Efficacy

Even professionals confuse the terms *effectiveness* and *efficacy*. We are therefore going to share our definition of these terms and give ADHD studies as examples for each. The reason for this is that efficacy studies require funds that ordinary clinicians do not possess. However, it is important for clinicians to recognize that they can do excellent case series *effectiveness* studies. Clinicians can do this type of study within the constraints of everyday practice, without the large costs of University-based research, which would use randomized control design and other costly procedures. These effectiveness studies can be published as case series, but more importantly, they can aid the clinician in making useful adjustments within his or her own practice.

Effectiveness: As noted above, clinical settings can do quality reviews of outcomes using their interventions. Such reviews should state precisely what is done in a manner that can be replicated by another center. They should also use standardized tests before and at the end of the intervention to demonstrate measurable change when this occurs. These standardized instruments compare the client to norms of the same age and sex. Therefore, any significant improvement can be assumed to be real and objective. A significant improvement is a statistical calculation and, if significance for pre-post testing is set at .05, that usually means that only 5 out of 100 persons would demonstrate that amount of change purely by chance. A good example of a standardized test is the Wechsler Intelligence Scale for Children (WISC).

At the ADD Centre, we have completed and published two large reviews demonstrating significant positive changes. The first was done on 111 consecutive clients with ADHD (Thompson & Thompson, 1998) and the second on 150 clients with Asperger's, plus nine clients with autism (Thompson, Thompson & Reid, 2010). Both these reviews have clearly demonstrated 'effectiveness' of the NFB + BFB interventions at the ADD Centre. These interventions have combined NFB, BFB, and the teaching of metacognitive strategies. The objective measures used demonstrated significant changes in clients' performance on pre-post measures. These tests included the Wechsler Intelligence Scales, the Wide Range Achievement Test, and Continuous Performance Tests, including the Test of Variables of Attention (TOVA) and the Intermediate Visual Auditory (IVA) in addition to subjective measures, namely questionnaires and adult self-ratings. EEG ratios were also reported. In our ADHD published review, 85 percent of the subjects, who had started the training with NFB while on medication, no longer required stimulant medication after just 40 training sessions.

A third, less rigorous, review was also published comparing the differences in EEG patterns between 165 adult and 96 child clients with ADHD

(Thompson and Thompson, 2006). A fourth publication, a single case study, demonstrated the effectiveness of combining specific NFB techniques with BFB to allow a patient with severe Parkinson's and dystonia to control her dystonic symptoms and her Parkinsonian 'freezing' symptom, reduce the dosage of medication, and restore her ability to read novels and to do fine arts and crafts (Thompson & Thompson, 2002). A fifth publication was a case study of a child with nonverbal LD and ADHD (Santhirasegaram & Thompson, 2013). The important point is that all of these publications are from a small clinical center that has no research money whatsoever. It is just a matter of systematic data collection done before and after training and then time spent reviewing the data to do pre-post comparisons. (And, yes, the data entry and statistical analysis, plus the literature review and the write-up do take time!)

A number of authors at other centers have also demonstrated the effectiveness of NFB in the treatment of ADHD in clinical settings. Perhaps the most important reports have come from the University of Tennessee and from the private practice of Joel and Judy Lubar in Knoxville, Tenn., and from the private practice of Vincent Monastra in New York state (Monastra et al., 1999, 2002, 2005).

These clinical case series and case studies provide excellent pilot data that can then be used when designing controlled research.

Efficacy: Efficacy refers to the determination of a training or treatment effect derived from a systematic evaluation obtained in a controlled clinical trial (La Vaque et al., 2002).

Research demonstrating efficacy has been done at a number of research centers. Studies for both ADHD and for seizure disorders have met the necessary criteria for a very high research rating of Level 4: Efficacious. The Arns et al. meta-analysis mentioned at the beginning of this section concluded that efficacy for ADHD reached the highest level, Level 5.

Level 4, Efficacious, means that these studies have met the research criteria for this level laid out by professional groups such as: the American Psychological Association, the Association for Applied Psychophysiology and Biofeedback, the American Academy of Child Psychiatry, and the International Society for Neurofeedback and Research. These criteria are:

a. In a comparison with a no-treatment control group, alternative treatment group, or sham (placebo) control utilizing randomized assignment, the investigational treatment is shown to be statistically significantly superior to the control condition, or the investigational treatment is equivalent to a treatment of established efficacy in a study with sufficient power to detect moderate differences; and

b. The studies have been conducted with a population treated for a specific problem, for whom inclusion criteria are delineated in a reliable, operationally defined manner; and

c. The study used valid and clearly specified outcome measures related to the problem being treated; and

d. The data are subjected to appropriate data analysis; and

e. The diagnostic and treatment variables and procedures are clearly defined in a manner that permits replication of the study by independent researchers; and

f. The superiority or equivalence of the investigational treatment has been shown in at least two independent research settings.

With respect to ADHD treated with NFB, at the time of writing there were three studies comparing NFB with a wait list control where subjects were randomly placed in the two groups. These studies all demonstrated significant improvement only in the NFB treated group. One of the studies (a controlled, multi-center study by Gani, Birbaumer & Strehl (2008) has demonstrated long-term effects, as there was a two-year follow-up. The Gani et al. study looked both at regular NFB to decrease slow wave and increase fast wave and at NFB using slow cortical potentials, and found them equally efficacious.

In addition, at the time of writing, there have been three studies that have demonstrated that NFB is equivalent or superior to a bone-fide treatment (stimulant medications) that has been shown to alter the symptoms of ADHD (Rossiter & La Vaque, 1995). There has been one randomized, controlled study with a large cohort and blind rating of improvements that demonstrated the efficacy of NFB compared to one group that received computerized cognitive training and another regular community treatment group. This was a particularly impressive study because it was done in school settings and showed the feasibility of school-based intervention using neurofeedback. Medication dosage was also monitored, and the neurofeedback group did not require any significant increase in medications, whereas the other two groups did. (Steiner et al., 2014).

Monastra has carried out a number of studies on efficacy (Monastra et al., 1999; Monastra, 2005). One of the studies is of particular interest to parents and clinicians (Monastra et al., 1999). One hundred children with ADHD were placed on a program that included parenting classes, school consults, and stimulant medication. However, 50 of these children were also given NFB. On the completion of 40 sessions of NFB training, children were taken off stimulant medication. After a 'washout' period of one week, retesting was carried out. The 50 children who had received NFB retained the gains they had made over the period of the study. All of the children who did not receive NFB returned to their premedication baseline state after medication was withdrawn. The subjects in such a large clinical study could not be randomly assigned to the groups, so this has to remain a small but valid criticism from a scientific point of view (Monastra, 2005).

Effectiveness

At the ADD Centre, at the present time, we are focusing on "effectiveness" rather than "efficacy" studies. There are good clinical reasons for this choice. The reader may have noted that researchers are often forced to use protocols in order to make the intervention exactly the same for each subject. However, it can be argued that a clinical setting should not use a 'one-size-fits-all' protocol, but that every client should be given **an individualized treatment plan that corresponds to their specific assessment findings**. At the ADD Centre, every client is given neurofeedback corresponding to their individual assessment EEG findings and **'one-size-fits-all' protocols are not used**. In addition, NFB is always combined with HRV, other BFB modalities as required, metacognitive strategies, and other interventions (often involving advice on diet, sleep, and exercise) when these may be helpful. Parenting suggestions are handled by giving every family a copy of *The A.D.D. Book* by Sears and Thompson (1998). This multimodal approach is important in responsible clinical or educational work where the client's training is based on their unique individual assessment and everything is done to include **all treatments for that client that have a reasonable possibility of improving the well-being of the client.** This is why clinical setting effectiveness reviews will continue to differ from efficacy studies that attempt to tease out the specific effects of a single intervention.

Examples of NFB Research "Protocols"
Protocols used in ADHD research studies include:

#1: Site: C3 or C4
SMR 12-15 Hz enhance
Theta 4-7 or 4-8 Hz inhibit

#2: Site: C4 for ADHD-Combined Type
SMR 12-15 Hz enhance
Beta 22-30 Hz inhibit
(Fuchs, 2003 used 1 and 2)

#3: Sites: Cz – ear, FCz – ear, Cz-Pz sequential, and C3 – ear
Theta 4-7 or 4-8 Hz inhibit
Beta 16-20 Hz enhance

Very Brief Summary of Some of the Studies Demonstrating that NFB is Efficacious for ADHD that Used the Above Protocols

Rossiter & La Vaque (1995)
These authors compared the use of stimulant medication (methylphenidate or dextroamphetamine) to NFB. There were 23 subjects assigned to the two groups, but not with random assignment. The NFB group received 20 sessions of NFB (Protocol #1 or #3). They found that as long as the drug group was on sufficient medication, both groups demonstrated similar and significant improvements as measured by the Test of Variables of Attention (TOVA), and a questionnaire, the Behavior Assessment System for Children (BASC), which showed reduced hyperactivity and externalizing behavior problems in conjunction with an increase in attention span. There was no significant difference between the two treatments while the patients assigned to the drug group remained on the stimulant medication.

Linden, Habib & Radojevic (1996)
These authors examined clients with ADHD with or without a learning disability (LD). They compared NFB treatment with a waiting list control group. Eighteen subjects were randomly assigned to each group. For NFB they used Protocol #3. They demonstrated that only the NFB group showed an average 10-point IQ gain (Kaufman Brief Intelligence Test) and only the NFB group demonstrated a significant reduction in inattention symptoms.

Carmody (2001)
This smaller study also compared NFB treatment with a waiting list control group. Eight subjects were assigned to two groups. The NFB protocol used was Protocol #1 for between 36-48 sessions. Only the NFB group improved on the continuous performance task and in improved attention span. They did not find consistent changes in electroencephalogram (EEG) amplitudes.

Monastra, Monastra & George (2002)
One hundred ADHD subjects were assigned to two groups. Both groups received what was termed Comprehensive Clinical Care (CCC) that included parenting classes, school consults, and medication. One group of 50 subjects also had NFB training sessions. The assignment to each group could not be completely random because some subjects lived a long distance from the treatment center. The NFB protocol used was Protocol #3. This study demonstrated that both groups had a reduction in ADHD symptoms and a significant improvement on a continuous performance test, the Test of Variables of Attention (TOVA). In addition, the NFB group demonstrated a significant improvement on an EEG measurement of cortical slowing.

This study went an important step further. They did a one-week washout of medication in both groups (all 100 children). They retested everyone and showed that 100 percent of the group that only received CCC (no NFB) relapsed. However, the NFB group maintained their gains on the TOVA, in the EEG, and their behavior gains in school.

Fuchs, Birbaumer, Lutzenberger, Gruzelier & Kaiser (2003)
These prominent authors are from Germany, Great Britain, and the United States. They recruited subjects diagnosed with ADHD and divided them (not random) into two groups. The first group consisted of 12 subjects who were on methylphenidate (MPH). The second group comprised 22 subjects who were given NFB for 36 sessions. Protocol #2 was used for the NFB group, who were diagnosed as ADHD-H (hyperactive type) and ADHD-C (combined type), and Protocol #3 was used for those diagnosed as ADHD-I (inattentive type).

Results of this study showed, with the TOVA and behavior rating scales, that there was no significant difference between groups as long as the medication group remained on an adequate dosage of mediation. In other words, the efficacy of NFB equaled that of medications.

Haywood & Beale (2003)
This small study involved seven clients diagnosed with ADHD. It is important in that it was a crossover design, comparing NFB treatment with a placebo. The placebo consisted of NFB with random bandwidths reinforced or inhibited.

Results of this study showed a significant improvement using NFB on a behavior rating scale (BRS) and in neuropsychological indicators in the five subjects who completed training.

deBeus, Ball, deBeus & Herrington (2003, 2006)
This is a particularly interesting study of subjects diagnosed with ADHD. Subjects were randomly assigned to groups. The first group were 26 children with a diagnosis of ADHD-C, with 14 of the subjects on stimulant medication. The second group were ADHD-I, n = 26, with 14 subjects on stimulant medication. This study was double-blind where neither the subject nor the trainer knew if the client was receiving real NFB or a 'sham' placebo NFB (play station) treatment. Those subjects who received real NFB got Protocol #1 or #3. All subjects received 40 sessions.

Results of this study demonstrated significant gains only in the real NFB group. Gains were measured using a continuous performance test, behavior ratings, and the quantitative electroencephalogram (QEEG). The QEEG showed that theta decreased, while sensorimotor rhythm (SMR) or beta increased.

Levesque, Beauregard & Mensour (2006)
This study took the entire field of measurement a step further by using not just EEG but also fMRI measurements before and after the NFB training. The clients diagnosed with ADHD (n = 20) were randomly assigned to two groups: 15 to NFB and 5 to a control group. The NFB group received 40 sessions of NFB. The second group remained as a wait-list control. The NFB group received Protocol #1 (or #3 if the diagnosis was ADHD-I).

Results demonstrated significant gains on a behavior rating scale (BRS), on a continuous performance test (CPT) and with neuropsychological measures. However, in addition, this study demonstrated that pretreatment compared to posttreatment there were fMRI findings demonstrating activation of important areas in the brains of the ADHD group that received NFB as compared to baseline, including the right anterior cingulate, right ventrolateral prefrontal cortex, left caudate nucleus, left thalamus, and left substantia nigra. There was no such activation found in the control subjects.

Functional neuroanatomical notes: Levesque, Beauregard and Mensour noted that prefrontal areas and the anterior cingulate are involved in attentional processes. They also noted how the capacity to inhibit behaviors or responses that are inappropriate can be studied using Go/No-Go tasks, in which the participant is required to refrain from responding to designated items within a series of stimuli. They have listed fMRI studies that have shown that several prefrontal regions, including the anterior cingulate cortex (ACC), dorsolateral prefrontal cortex, orbitofrontal cortex, ventrolateral prefrontal cortex, and the striatum, are all involved in this response inhibition.

Mario Beauregard, Johanne Levesque and their colleagues concluded that this study demonstrated that NFB has the capacity to functionally normalize the brain systems mediating selective attention and response inhibition in ADHD clients. Beauregard, who was in charge of the fMRI data collection, noted that seeing activation of the substantia nigra (source of dopamine) was particularly exciting because very few imaging studies have documented changes in that part of the brain.

Zhang, Zhang & Shen (2006)
This study was a randomized clinical trial. There were 22 children who were treated with NFB, 20 of whom were methylphenidate (MPH) nonresponders. These were compared with 22 MPH responders who were treated using that stimulant medication. The NFB included 4-8 Hz suppression, while 16-20 Hz was enhanced. There were 3-5 sessions per week for three months, for a total of 35-40 sessions. Evaluations were carried out at one month, three months, and also six to twelve months for follow-up.

Results reported in this study using the Conners Parent questionnaire and a timed trail-making speed test demonstrated that at three months, NFB appeared equally effective compared to MPH. At six months, the NFB group did better on the trail-making tests and on the Connor's Parent Rating Scale (CPRS).

Meta-analysis of Nine Controlled Studies of EEG Power Ratio: Snyder & Hall (2006)
This meta-analysis (data from a number of studies are grouped together and analyzed) looked at studies that included a total of 1498 subjects with ADHD comparing them to healthy controls. The grouped data had an effect size (0.8) and sensitivity (few false positives - accuracy of picking out real cases) and specificity (few false negatives – real cases not discarded) of approximately 94 percent for the theta/beta power ratio. This meant that the **diagnosis of ADHD could be accurately confirmed** using the EEG power ratio. They found that the QEEG matched or exceeded Child Behavior Checklist or Behavior Assessment System for Children in differentiating those with ADHD from healthy controls. The sensitivity of rating scales is only about 50-82 percent, and their specificity about 40-70 percent. In addition, Snyder and Hall found that reliability was substantially higher for QEEG than the inter-rater reliability of rating scales.

Steiner, Frenette, Rene, Brennan & Perrin (2014)
This landmark study, showing feasibility of neurofeedback training in school settings, was designed and carried out by a prominent pediatrician, Naomi Steiner, who is an Assistant Professor at Tufts University School of Medicine. It was published in the *Journal of Developmental and Behavioral Pediatrics, 35*(1), 18-27. This study randomly assigned 104 children with a diagnosis of ADHD from second and fourth grade to NFB (n=34), cognitive training (CT) (n=34), and control groups (n=36). Changes were assessed using Connors 3-Parent, Behavior Rating Inventory of Executive Function (BRIEF), teacher reports (Swanson et al., SKAMP), Conners 3-Teacher scale, and systematic classroom observations using Behavior Observation of Students in Schools (BOSS) done by raters blind to which group the children were in.. Training was provided in public schools in the Boston area.

The children who received NFB showed significant improvement compared to controls on the Conners 3-P attention, executive functioning and global index, all BRIEF indices, and BOSS motor/verbal off-task behavior. Children who received CT showed no significant improvement compared to controls. Children who received NFB showed significant improvements compared to those in the CT group on Conners 3-P Executive Functioning, all BRIEF summary indices, SKAMP Attention, and Conners 3-P attention subscales. Stimulant medication dosage in methylphenidate equivalencies significantly increased for children in the CT (8.54 mg) and control (7.05 mg) conditions, but not for those in the NFB condition (0.29 mg).

This is an extraordinarily carefully done study that clearly shows the improvements that can be obtained when NFB is added to standard treatment (medications) for the treatment of ADHD. It also demonstrated the feasibility of doing NFB in a school setting.

The above is only a very brief overview of a few studies. The discerning reader should check the original references for precise data. The exact references for these studies and many more can be obtained on the website www.isnr.org under Bibliography. Thanks are due to Cory Hammond for maintaining this Bibliography of articles pertaining to the use of NFB for various disorders and conditions.

Long-Term Effectiveness
Lubar (1995) published a 10- to 20-year follow-up in a textbook chapter. The telephone follow-up, done by an independent research entity (not Lubar's clinic), indicated that children between the ages of 8-12 years who demonstrated a change in their EEG patterns during their training still showed their gains when they were in college (Lubar, 1995). At the ADD Centre, Wence Leung oversaw a follow-up study, and presented at the 2011 ISNR annual meeting the initial findings from a very large, long-term follow-up that is being completed as her Ph.D. thesis at the time of writing. Her preliminary report also found that the results of NFB training are maintained years after the training has stopped.

Long-Term Efficacy

Long-term effects have been demonstrated in a controlled, multicenter study by Gani, Birbaumer, and Strehl, (2008). The Gani et al. study looked both at regular NFB to decrease slow-wave and increase fast-wave activity, and at NFB using slow cortical potentials. Beneficial gains were being maintained at the time of a two-year follow-up; indeed, half of those followed up in the NFB group no longer qualified for a diagnosis of ADHD. This study concludes by saying: "The stability of changes might be explained by normalizing of brain functions that are responsible for inhibitory control, impulsivity, and hyperactivity."

Further Comment and Critique

A carefully done review of some of these studies and others has been published by Lofthouse et al. (2012). His review gave NFB for ADHD a 'probably efficacious' rating because it embraced the skepticism that still exists among traditional researchers who are wanting double-blind, placebo-controlled studies. The authors recommend that the reader obtain this paper. However, it is our opinion that these reviewers are rather stuck on the pharmaceutical companies' blinded, placebo-control methodology. Running a clinical center, we find this approach to be inappropriate because the good clinician cannot be blind to the condition of their patient. They would know immediately from eye blink or movement artifact if the person were receiving appropriate neurofeedback or sham feedback. Automatic thresholding and the therapist being 'blind' to the treatment does not lead to appropriate patient care and will not correctly reflect NFB outcomes achieved in careful clinical practice. The double-blind design (some critics have even suggested triple-blind!) does not reflect what occurs in good clinical practice. The trainer should be working carefully and knowledgeably with the patient and educating them about their EEG patterns. Research designs used in the fields of physiotherapy and exercise would be more appropriate than research designs used in drug studies. With interventions like diet or exercise, neither the therapist nor the patient can be blind to the condition. We also suggest that authors critical of results found with neurofeedback and enamored of the established efficacy of medications look more carefully at follow-up results of medication use, such as in the large MTA study (multisite treatment approach) funded by NIMH. Longer term (more than three years), the effectiveness of medication was less impressive. Of course medication can be helpful, especially for the symptom of hyperactivity, but for those who are non-responders, have unacceptable side effects, or do not like the prospect of years on medication, neurofeedback provides an attractive alternative. Neurofeedback can also be combined with medication, and then it is usually possible to reduce the dose of medication as self-regulation skills gained with neurofeedback training increase. Decrease the pills as you increase the skills.

SECTION XXVII
Seizure Disorders

The diagnosis and treatment of seizure disorders remains essentially unchanged from the first edition of *The Neurofeedback Book*. However, there has been one important meta-analysis, done by Gabriel Tan and his colleagues, that is briefly summarized below (Tan et al., 2009). In addition, there has been a summary chapter by the 'father' of the field, Barry Sterman, and by Lynda M. Thompson (2014), and a chapter by Steriade (2005) that included discussion of the origins of SMR. Steriade also discussed another frequency, 35-45 Hz, that may accompany SMR.

The meta-analysis published by Gabriel Tan (Tan et al., 2009) notes that about one-third of patients with epilepsy do not have seizures well controlled with medical treatment. For these patients, EEG biofeedback is a viable adjunctive technique. The meta-analysis analyzed every EEG biofeedback study on epilepsy indexed in Medline, PsychInfo and Psychlit databases between 1970 and 2005 that provided information on seizure frequency changes in response to EEG biofeedback. They note that 63 studies had been published, 10 of which provided enough outcome information to be included in their meta-analysis. The subjects in all of these studies consisted of patients whose seizures were not well controlled by medical therapies, which is an important factor to keep in mind when interpreting the results. Nine of 10 studies reinforced sensorimotor rhythms (SMR), while one study trained slow cortical potentials (SCP). All of these studies reported an overall mean decreased seizure incidence following treatment, and 64 out of 87 patients (74 percent) reported fewer weekly seizures in response to EEG biofeedback. Based on this meta-analysis, EEG operant conditioning was found to have clinical efficacy in terms of producing a significant reduction in seizure frequency and severity. This finding is especially noteworthy given that the patients were individuals who had been unable to control their seizures with medical treatment.

Theoretical Discussion Concerning Mechanisms of Seizure Control

Theoretical Underpinnings of NFB to Raise SMR
The simplest form of epileptiform activity is the **interictal spike, a synchronized burst** of action potentials in a localized pool of pyramidal neurons. Spikes can be generated by phasic hyperexcitability in cells. These spikes are associated with strong axonal efferent activation. This activation leads to a subsequent period of feedback hyperpolarization due to excitation of collateral recurrent interneurons that **impose transient inhibition** on the pyramidal neurons that were the origin of the spikes. This inhibition gives rise to **the "wave" component** of spike-wave patterns. Even in the absence of spike discharges, cortical hyperexcitability and resulting inhibition produces a slowed EEG rhythm typically in the 4-8 Hz range (Gloor et. al., 1979). Accordingly, patterns of prominent 4-8 Hz "theta" slow waves can serve as a signature for focal epileptogenic sites in the cerebral cortex. Also, there has been some investigation of hemispheric effects, including left temporal slow-wave activity associated with seizures that originate in the temporal region.

Origin of Seizure Activity

There is great diversity in the pathologies leading to, and clinical manifestations of, the epileptic syndrome. It is thought, however, that the actual generation of many of the different types of seizure may occur through common cellular mechanisms and networks. Although practically every part of the brain may generate an epileptic seizure, the investigation of the cellular and network mechanisms of epilepsy over the last several decades has focused largely on three structures: the cerebral cortex, the hippocampus, and the thalamus. There are multiple factors contributing to the epileptogenicity of these structures, including the presence of massive recurrent excitatory connections, reliance upon inhibition for the regulation of excitability of this recurrent network, the ability of synaptic connections to strengthen or weaken with repetitive activation, and the presence of intrinsically burst-generating cells.

Possibly one of the most important factors in the generation of seizure activity is the **loss of regulating inhibitory modulation**. There is convergent evidence that **neurofeedback therapy can impact this deficit.**

Attenuation of efferent motor and afferent somatosensory activity can initiate production of rhythmic activity in the 12-15 Hz range that Sterman called sensorimotor rhythm (SMR). SMR oscillations can also be influenced by nonspecific neurotransmitter (cholinergic and monoaminergic) modulation of the neurons. These **other factors can affect excitability levels** both in thalamic relay nuclei and in the cortical areas receiving the relayed signals.

During waking activity, these other influences (neuromodulators), as well as cortical projections, normally keep ventral basal (VB) cells depolarized, and this will suppress rhythmic bursting patterns. On the other hand, when the subject is not motorically active, oscillations at SMR frequency may be observed.

SMR, the Dominant "Standby" Frequency

SMR constitutes the dominant "standby" frequency of the thalamocortical somatosensory and somatomotor pathways. Therefore, it is posited that operant training of SMR should result in improved control over excitation in this system. **Raised thresholds for excitation are thought to underlie the clinical benefits of SMR training** in epilepsy and other disorders characterized by cortical and/or thalamocortical hyperexcitability.

For instance, SMR training has also been shown to be an effective treatment for ADHD. For recent reviews see Monastra et. al. (2005) and Arns et. al. (2009) as discussed in the section of this book on Attention Deficit Disorder. Arns has suggested that the major effect of SMR training that is responsible for positive effects in at least some ADHD patients may be directly related to improvements in sleep (personal communication, also Arns & Kenemans, 2014). Furthermore, this training has been documented to result in a reduction of impulsive response tendencies in healthy volunteers (Egner & Gruzelier, 2001, 2004).

Thalamic Reticular Nucleus

The thalamic reticular nucleus is part of the ventral thalamus that forms a capsule around the thalamus laterally. **Thalamic reticular cells** are GABAergic, and thus **are inhibitory**. The oscillations between cells in the thalamus and the reticular nucleus are responsible for oscillatory rhythms such as theta and SMR. The thalamic reticular nucleus is the only thalamic nucleus that has **no efferent output to the cerebral cortex.** It receives input from the cerebral cortex and the dorsal thalamic nuclei. Most input comes from collaterals of fibers passing through the thalamic reticular nucleus. Primary thalamic reticular nucleus efferent fibers project to dorsal thalamic nuclei, but as noted above, never to the cerebral cortex. The **reticular nucleus modulates the information from other nuclei in the thalamus.** Its function is modulatory on signals going through the thalamus.

Generation of EEG Oscillations

When there is a **reduction of afferent somatosensory input, the cells in the ventral basal nucleus (nVB) hyperpolarize**. However, instead of remaining at a stable level of inhibition, these nuclei **gradually depolarize**. This is mediated by a slow calcium influx that causes the nVB neurons to **discharge in a 'burst'** of spikes. These bursts are **relayed to the sensorimotor** cortex and simultaneously to the thalamic reticular nucleus (nucleus reticularis thalami or nRT) neurons.

Stimulation of the reticular nucleus of the thalamus will, in turn, lead to a burst discharge of GABAergic neurons, imposing reciprocal inhibition on the original pool of VB relay cells, thus **returning them to a hyperpolarized state and initiating a new cycle** of slow depolarization. In this way, the interplay between neuronal populations in nVB, nRT, and sensorimotor cortex results in reciprocal rhythmic thalamocortical volleys and consequent cortical **EEG oscillations** that in the quiet state under normal circumstances appear as 12-15 Hz SMR rhythms (Sterman, 1996; Sterman & Bowersox, 1981).

Sterman notes that convincing evidence suggests that all normal rhythmic patterns in the EEG may arise from a similar process, but may be **expressed at different frequencies based on anatomical differences in thalamocortical projection pathways** (Hughes & Crunelli, 2007; Sterman & Thompson, 2014). Frequencies also vary as a function of background arousal (Bouyer et al., 1974,

1987; Rougeul et al., 1972), and consequent degrees of thalamic neuronal hyperpolarization (McCormack & Huguenard, 1992).

To summarize: SMR 12-15 Hz EEG activity is generated in the ventrobasal nucleus (nVB) of the thalamus, which has the function of relaying afferent somatosensory information. Inhibitory influences are enhanced with operant conditioning; that is, the client is rewarded for producing the SMR rhythm in the EEG during NFB training. This process is posited to elicit firing pattern changes in cells of the ventral posterior lateral (VPL) nucleus of the thalamus caused by a reduction of afferent somatosensory and cortical input. This results in membrane hyperpolarization. Instead of remaining at a stable level of inhibition, however, a gradual depolarization, mediated by a unique slow calcium influx, causes the VPL neurons to emit a brief burst of spikes, which is relayed to sensorimotor cortex and simultaneously to thalamic reticular nucleus (nRT) neurons. Stimulation of the latter in turn leads to a burst discharge of GABAergic neurons, imposing reciprocal inhibition on the original pool of VPL relay cells, thus returning them to a hyperpolarized state and initiating a new cycle of slow depolarization.

It is the interplay between neuronal populations in nVB, nRT, and sensorimotor cortex that results in **reciprocal rhythmic thalamocortical volleys** and recurrent cortical EEG oscillations that **appear as 12-15 Hz SMR** rhythms (Sterman & Bowersox, 1981; Sterman & Thompson, 2014). These rhythms are thus associated with a quiet (motorically still) yet alert state. Recall the cats in Sterman's original experiments who were waiting for a signal before pressing a bar to get a food reward.

Figure from original research, with thanks to Barry Sterman. Sterman observed EEG spindles when cat was waiting for a signal before making bar press for food. Next, cats were trained to produce sensorimotor rhythm (SMR) to get food.

Brief Discussion of a Possible Relationship between SMR and 40 Hz Rhythms

12-15 Hz Sensorimotor 'Expectancy Rhythm'

In Sterman's original research, his cats entered an alert, motionless, focused state and produced bursts of rhythmic activity very similar to the spindles observed in sleep. He coined the term "sensorimotor rhythm" to describe this 12.0-15.9 Hz spindling beta rhythm. In 2005, Steriade called 14 Hz spindles, observed over the sensorimotor cortex, an 'expectancy rhythm.' He, like Sterman, noted that this expectancy rhythm occurs when the alert animal is focused and immobile. Steriade gives the example of a cat watching a hole where a mouse may appear (Steriade & McCarley, 2005). Given these observations, it is not a surprise that either raising SMR or increasing negativity in slow cortical potential (SCP) feedback works to improve attention span and decrease unwanted hyperactivity and impulsivity in people with ADHD.

SMR activity is produced by the ventral posterior lateral nuclei of the thalamus. The amplitude of SMR is known to fluctuate with the individual's degree of vigilance (Steriade, 2005). Rougeul-Buser in 1983, then Steriade in 2005, termed the 14 Hz spindles observed over the sensorimotor cortex an expectancy rhythm. Rougeul-Buser in 2011 makes no reference to Sterman's work, and calls this rhythm 'Mu.' Steriade, on the other hand, gives full credit to Sterman's work on SMR. The expectancy rhythm discussed by Rougeul-Buser and by Steriade occurs when the awake animal is immobile.

35-45 Hz 'Focused Attention' Rhythm

Rougeul-Buser in 2011 and Steriade in 2005 both noted that a second rhythm may accompany the 14 Hz expectancy rhythm; namely, rhythmic activity in the gamma range. This gamma rhythm occurs along with the bursts of 14 Hz activity when a cat is immobile and watching a mouse that it can see but cannot attack due to a transparent plastic shield. They call this 35-45 Hz spindling a **'focused attention' rhythm** (Steriade, 2005; Rougeul-Buser & Buser, 1983, 2011). They note that this gamma frequency arises from the medial part of the posterior thalamus.

Other 'Functions' of the 35-45 Hz Rhythm

Interestingly, this 35-45 Hz range contains the same frequency band, 38-42 Hz, that is referred to as 'Sheer rhythm' by neurofeedback practitioners. It was named after Daniel Sheer who trained 40 Hz and found increasing these frequencies was associated with improvements in children with **learning disabilities** (Sheer, 1976). He referred to 40 Hz as a binding rhythm that helped coordinate different mental functions.

There is also a 40 Hz response when **regaining balance** if a person leans over on a balance board as far as they can without tipping (Thompson, 2005; Slobounov, 2008). Training can be conducted with high frequencies in the gamma range, such as 40 Hz, if this neurofeedback training is approached with extreme care so that EMG artifact from muscle tension is not mistaken for true 40 Hz activity of cortical origin.

Training Both SMR and the 40 Hz Rhythms

At the time of writing, a few of our clients are currently receiving this gamma training and have anecdotally reported it as being associated with feeling "extremely clear-headed." One client described feeling true empathy and feeling at one with the world. Given these positive responses, we will be considering this training with more of our clients in the future.

However, at the time of writing, we do not know of any research that has investigated combining training of both these rhythms in seizure disorders. Certainly it would make sense to train both simultaneously in learning disabilities.

Another interesting feature of gamma activity in the 40 Hz range is that it has been reported to be higher amplitude in Buddhist monks who are advanced meditators. Mental states of empathy were associated with high amplitude gamma production in the 30-40 Hz range, and amplitudes were higher in senior monks as compared with novices.

Left Temporal Lobe Seizures and Slow-Wave Activity

Spikes with after-following slow waves are not the only pattern seen in people with seizure disorders. Seizures have been reported in patients where the typical high amplitude spike with following slow-wave EEG pattern is not observed. The following short discussion concerns other high-amplitude bursts of slow-wave activity that may be associated with seizures.

Englot et al. (2010) observed significantly increased delta-range 1-2 Hz slow wave activity in bilateral frontal and parietal neocortices during complex-partial as compared with simple-partial seizures. (Recall that a simple seizure does not involve change in consciousness, whereas a complex seizure does.) They investigated 63 partial seizures in 26 patients who had surgically confirmed medial temporal lobe epilepsy. They noted altered consciousness, and stated that this can occur with bilateral or dominant temporal lobe seizure involvement. These authors postulate that with bilateral temporal lobe seizures, a network inhibition of subcortical arousal systems occurs that disrupts brainstem-diencephalic arousal systems, and this in turn leads to depressed cortical function and impaired consciousness. They also state that the signal power of frontoparietal slow-wave activity is correlated with the temporal lobe seizure activity in each hemisphere.

The clinician must be careful, however, not to over-interpret apparent abnormal EEG activity. Indeed, with any abnormal EEG waves, the NFB practitioner should refer the client to a neurologist who specializes in EEG interpretation. Severely 'epileptogenic' EEGs may be recorded from patients with infrequent or controlled clinical seizures. The EEG abnormalities do not necessarily reflect the severity of the epileptic disorder. Panayiotopoulos (2005) notes that more than 10 percent of normal people may have nonspecific EEG abnormalities, and approximately one percent may have 'epileptiform paroxysmal activity' without seizures. This author states that the prevalence of these abnormalities is higher in children, with two to four percent of children having functional spike discharges. Panayiotopoulos goes on to state that paroxysmal epileptiform activity can be high in patients with nonepileptic, neurological or medical disorders, or with neurological deficits. For example, children with congenital visual deficits frequently have occipital spikes, and patients with migraine have a high incidence of sharp paroxysmal activity and other abnormalities (Panayiotopoulos, 2005).

The following is a brief overview of types of slow-wave activity, but the practitioner should consult a neurologist who specializes in EEG interpretations if these are observed.

Abnormal slow rhythms can be focal or generalized, rhythmic (monomorphic) or arrhythmic (polymorphic), intermittent or continuous, and in terms of the dominant frequency, delta or theta. Focal polymorphic delta arrhythmic slow waves can vary in frequency, amplitude, and morphology. These waves are said to require involvement of white matter. They are usually associated with focal attenuation of normal rhythms.

Continuous Polymorphic Delta (CPD) may be correlated with structural lesions. However, seizures may occur in more than 50 percent of cases of CPD in which no structural lesion is present. Transient ischemic attacks may be observed with CPD. CPD may also be observed in postictal states (after a seizure), and this can be lateralizing. Migraine and trauma can also produce CPD. Boro (2014) notes that a structural lesion is less likely when focal slowing is intermittent, involves substantial theta, or disappears with eye opening or in sleep. Boro also notes that ischemic stroke, hemorrhage, and tumors can all cause overlapping and occasionally indistinguishable EEG findings. However, the EEG can usually indicate which hemisphere is involved. Frontal and parietal lesions often produce delta activity that has its highest amplitude, and can have phase-reversals, over the temporal regions. In addition, in the elderly, slowing (theta) of the EEG may correlate with cerebrovascular disease.

Overview of Slow Waves
(from Boro, on Medscape 2014; recommend also:
Fisch and Spehlmann's EEG Primer, 1999)
Again, the authors emphasize that these should NOT be interpreted by an NFB therapist. NFB practitioners should refer the client to a neurologist who is an EEG specialist if abnormal brain waves are observed.

Boro states that **"Frontal Intermittent Rhythmic Delta Activity (FIRDA) and Generalized Intermittent Rhythmic Delta Activity (GIRDA)** are usually associated with global cerebral dysfunction due to metabolic disarray." Boro goes on to note that, on the other hand, a small amount of FIRDA, especially when it is restricted to drowsy states, can be a normal finding in elderly subjects. FIRDA is rarely due to a subcortical lesion or elevated intracranial pressure.

Triphasic waves typically occur in the setting of more profound metabolic disarray than FIRDA. They are classically associated with hepatic encephalopathy.

Occipital Intermittent Rhythmic Delta Activity (OIRDA) is encountered most often in pediatric patients. Like FIRDA, OIRDA can be a consequence of diffuse cerebral dysfunction and is rarely a sign of increased intracranial pressure. It often occurs in children with absence seizures (petit mal epilepsy).

Temporal intermittent rhythmic theta activity is most often encountered in drowsiness and is called **rhythmic midtemporal theta** bursts of **drowsiness (RMTD)** as well as psychomotor variant. RMTD is not an ictal pattern, and is **not associated with epilepsy.**

Temporal Intermittent Rhythmic Delta Activity (TIRDA) is both less common and more specific for temporal lobe epilepsy than temporal polymorphic delta. It occurs in fewer than one percent of EEGs done for clinical indications. However, in one study, this pattern was observed in 28 percent of patients who had **temporal lobe epilepsy** (Geyer et al. 1999).

SECTION XXVIII
Movement Disorders and Fibromyalgia

Earlier in this text we discussed Parkinson's with dystonia, Tourette syndrome and fibromyalgia. One factor in common to these conditions and their treatment with NFB and BFB is the involvement of muscle spindles. Muscle spindles detect muscle length and are involved in muscle movement and tone. They have a double innervation, cholinergic and sympathetic (Passatore et al., 1985). Since both of these systems can be operantly conditioned, it was hypothesized that neurofeedback plus biofeedback, through their effects on muscle spindle activity, could result in a decrease in the undesirable muscle bracing, unplanned movement and faulty muscle tone that characterizes movement disorders.

Hypotheses concerning the pain in fibromyalgia suggest that tension in the muscle spindles may be related to trigger points. Indeed, one can have a relaxed muscle as shown by the surface EMG recordings, but have major tension in the muscle spindles within the muscles, as shown by inserted needle recordings.

The brain is constantly informed of the position of a muscle through the sensory pathways. The intrafusal fibers of the muscle spindle receive input from gamma motor efferent connections from the red nucleus in the midbrain, and send sensory information back to pathways that eventually influence red nucleus activity (Sterman, 2000). This reflects the feedback loop organization of the CNS. Sterman noted that when SMR is raised, the firing of the red nucleus decreases. Our hypothesis was that if we used operant conditioning to increase SMR, there would be less activity in the red nucleus and the result might be a decrease in the symptoms of dystonia, Tourette syndrome, and even fibromyalgia as the muscle spindles relaxed.

In addition to this gamma motor feedback loop, the muscle spindle also has innervation from the sympathetic nervous system (Grassi, Filippi, Passatore, 1986). It was therefore logical to decrease sympathetic tone using biofeedback methods such as heart rate variability training (in particular the component of effortless diaphragmatic breathing) to decrease the symptoms of these disorders.

Parkinson's Disease

The positive effects of this combined training on a patient with severe Parkinson's and dystonia has been described earlier in this book, and was published as a case study (Thompson & Thompson, 2002).

In groundbreaking research concerning nondrug approaches to treating Parkinson's Disease (PD), Ingrid Philippens' group at the Biomedical Primate Research Centre in the Netherlands trained marmoset monkeys to produce sensorimotor rhythm (12-14 Hz). They did half-hour training sessions with rewards for bursts of SMR production. The positive reinforcement used was miniature marshmallows. Philippens demonstrated that nonhuman primates can learn voluntary operant control over the sensorimotor rhythm, essentially replicating Sterman's early work in cats and rhesus monkeys. She also demonstrated that chemically induced Parkinson's symptoms did not occur in the monkeys who were first trained to produce SMR. Just as Sterman's cats were resistant to developing seizures when exposed to hydrazine, Philippens' monkeys were resistant to developing tremors when PD was chemically induced. (See Philippens, I. 2011, "Neurofeedback training on sensorimotor rhythm in marmoset monkeys." *NeuroReport*, 31 March 2010, *21*:5, p 328-332).

From a theoretical point of view, she notes that thalamocortical circuits are affected in Parkinson's disease (PD). She hypothesized that SMR training might improve the motor functions in PD patients (such as bradykinesia, or slowed movements, and resting state tremor) and/or decrease the L-dopa-induced dyskinesia that occurs with long-term medication use. In addition, she notes that SMR training will increase the metabolic activity in the striatum. Since a decline in metabolic activity can activate microglia involved in the maintenance of the neurodegenerative processes, she concludes that it may follow that an activation of metabolic activity by SMR training could be beneficial in preventing decline in neurodegenerative diseases such as PD.

A recent review of the literature on Parkinson's disease and neurofeedback discusses the neural networks involved in PD with an emphasis on the basal ganglia thalamocortical circuits (Esmail & Linden, 2014). The paper reviews six studies, beginning with our case study from 2002. One of the authors, David Linden, has shown in previous research the feasibility of doing fMRI neurofeedback in healthy subjects. Given that research interest, it is not surprising that two of the five controlled studies reviewed involved upregulation of the BOLD signal in the supplementary motor area (SMA). One controlled, nonrandomized study with 10 subjects, all of whom had PD, showed that only the neurofeedback group (n=5) showed successful upregulated SMA plus significant improvement in Unified Parkinson's Disease Rating Scale and finger tapping scores. The second study, however, had different results. In one patient with PD and three healthy controls, the fMRI neurofeedback to upregulate SMA resulted in slowed movements. One study involved training slow cortical potentials for increased negativity. It was a controlled trial with 10 subjects with PD and 11 healthy controls. Good NFB performance resulted in increased early *Bereitschaftspotential* (readiness potential), which is a slow cortical potential that represents brain activation. It precedes planned movement by about two seconds. It has been found to be lower in patients with PD who are off medication, and it increases on dopamine medication. The other two studies used regular neurofeedback. One design trained to increase 8-15 Hz and decrease 4-8 Hz using auditory feedback for 24 sessions. Decreasing 4-8 Hz caused subjective worsening of symptoms and was stopped. Neither the experimental (n=5) or the control group (n=4) showed improvements in dyskinesia or PD symptoms. The other study, however, got positive results. It was a randomized, controlled study with eight subjects in the NFB group and eight in the sham control group. The training was to increase 12-15 Hz and decrease 4-7 Hz at O1-O2, and videogame feedback was used. EEG changes were achieved in the NFB group only, and this was associated with improvements in static and dynamic balance. The conclusion drawn was that research to date is promising but not yet sufficient to support the use of NFB in everyday clinical practice.

The exciting thing is that the two studies, one using fMRI NFB and one regular NFB, were controlled studies that had positive, measurable outcomes. Numbers were small, yet results were significant, meaning the effect size was large. Since this is a non-drug, noninvasive approach that can be combined with medication and potentially reduce symptoms, it deserves further study.

Tourette Syndrome

Positive results have been demonstrated using neurofeedback (and specifically increasing SMR) in the treatment of tic disorders including Tourette syndrome (Dopfner & Rothenberger, 2007; Poncin et al., 2007; Tansey, 1986). Results for Tourette should be evaluated a year or more after completing treatment, as tics will wax and wane even without intervention. A follow-up study was done at the ADD Centre by Blair Aronovitch and Andrea Reid with 20 male patients age 8.5 to 23 years who had received treatment at the ADD Centre. This study was reported at the 2012 ISNR annual meeting. These patients had received 40 or more sessions of NFB over the sensorimotor strip to increase SMR (12-15 or 13-15 Hz) and decrease theta. The range of theta was individualized according to each patient's assessment. Diaphragmatic breathing to achieve synchrony between changes in breathing and their heart rate was also practiced. (This was respiratory sinus arrhythmia (RSA) in the 1990s, and it evolved into heart rate variability training.) Tics had disappeared at discharge for all clients. At follow-up, which was done from 1-17 years after the completion of training, 8 of the 20 subjects who could be contacted completed an interview about their current symptoms. Of those eight, two clients reported no long-term change in tics, though their other reasons for attending the training, including ADHD symptoms and learning problems, had improved and those results had lasted. The other six patients did report that their decrease in tics was sustained. No clients experienced any increase in tics, which is significant because the usual course of this illness is that it will progress.

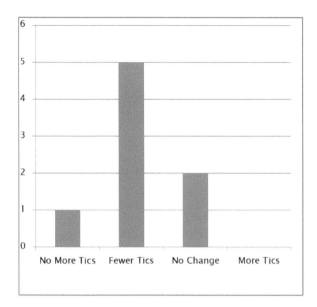

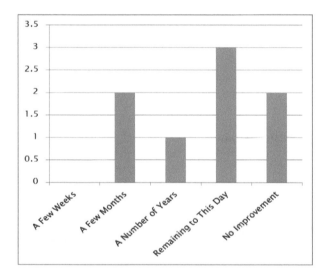

The follow-up research conclusion was that NFB training to increase SMR combined with HRV training does appear to be a promising approach for clients who present with Tourette syndrome, and that the results obtained do last in the majority of people. Perhaps a return to training would have helped the two who had an initial remission of tics but an eventual return to baseline. The finding of long-term positive results in most clients is consistent with follow-up studies concerning treatment for ADHD that show long-term efficacy without further training.

The person who remained totally without tics was a heartwarming story because he had been the most severe case initially. He had been expelled from his high school for hurling a string of obscenities at a teacher who told him to remove his hat while walking in the hall. His vocal tics were advanced enough at age seventeen to include coprolalia (involuntary swearing), but the fact that this behavior was a manifestation of a neurological disorder did not prevent the expulsion. He also had ADHD symptoms, which is a common comorbidity with Tourette syndrome, and at that point his parents sought help at the ADD Centre. Neurofeedback to increase SMR frequencies at C3 and C4, plus biofeedback to achieve breathing and heart rate synchrony, was begun. (This was before HRV training was the usual method used for training, and we were training its forerunner, RSA - respiratory sinus arrhythmia.) Both vocal and motor tics had stopped completely before 40 sessions of training were completed. Academic achievement was higher than it had ever been. On follow-up seven years later, he had graduated high school and was attending art school. He said that whenever he saw the newsletter of the Tourette Society, he remembered his difficult teen years and was so grateful to be tic-free.

The reader may also wish to read the following two articles:
- Esmail, S. & Linden, E.J. (2014). Neural Networks and Neurofeedback in Parkinson's Disease. *Neuroregulation, 1*(3-4), 240-272.
- Philippens, Ingrid (2008). Non-Human Primate Models for Parkinson's Disease. *Drug Discovery Today: Disease Models, 5*(2), 108-111.

SECTION XXIX
Post-Traumatic Stress Disorder (PTSD)
David Hagedorn, Ph.D., BCN

Diagnostic terminology for behavioral and physiological dysfunction following horrific exposure to trauma or repeated exposure to traumatic and stressful experiences has varied over the millennia. The groupings of symptoms have varied and have had many names, such as "shell-shock" during WWI, and "post-traumatic stress disorder" or PTSD in 1980 with the release of the DSM-III. Because the symptoms can be so shocking, they are seen by some as grossly pathological and not simply dysfunctional. With a deeper understanding of the pathophysiology, however, a more apt conclusion is that these post-traumatic changes are a series of normal nervous system and behavioral responses to circumstances that are themselves abnormal.

From the writings of the Greek historian Herodotus, one can infer that PTSD symptoms were associated with the Battle of Marathon in 490 B.C. In this battle an Athenian soldier lost his vision following battle, but not as a result of a primary physical injury to eye or head. Conceivably, what he witnessed in battle was so horrific that his brain disengaged the ability to see – a self-protective response. Other historical evidence describes warriors with PTSD symptoms at least as far back as 1300 B.C in Mesopotamia (now Iraq). A thousand years later, around 300 B.C., Macedonia's King Cassander, who was one suspect in the death of Alexander the Great, was described as experiencing frightful and debilitating flashbacks. By 1700, the Swiss military defined this cluster of symptoms, along with others that included tachycardia, as acute post combat *nostalgia*. German physicians referred to this pattern of behavioral and physiological findings as *Heimweh* (homesickness). In the early 1700s, French physicians included gastrointestinal symptoms, fever, emotional numbing, and depression in their post-trauma sequelae, labeling it *maladie du pays*.

The vast majority of PTSD cases are not related to combat but to other forms of traumatic exposure. Samuel Pepys related in 1666 that survivors of the Great Fire of London experienced nightmares, insomnia, fear of going to sleep, emotional outbursts of anger, and sudden fear reactions triggered by stimuli such as chimney smoke and fire. Childhood abuse, sexual assault, hindered attachment, and other traumas where a moral violation occurs during neurodevelopment, have added complicating effects on the still-maturing neurohormonal and neurophysiological systems. There is ample support from both animal and human research that structural changes to the developing brain stem, limbic system, and nervous systems may result from early trauma. It is hypothesized that early trauma results in an inefficient stress-response system and increased susceptibility to later PTSD. Of the many post combat PTSD patients I have treated, an early childhood trauma component was usually present in those with more complex symptoms, and was associated with greater resistance to treatment effects. The long-term medical abnormalities cataloged and quantified in the 17,000-subject Adverse Childhood Experiences (ACE) study punctuates this important phenomenon and hopefully brings attention to common medical conditions like headache and insomnia that may well have an early trauma genesis (Anda et al., 2010; Chapman et al., 2011).

Psychiatrist Dr. Jonathan Shay's research on the central nervous system and his subsequent work for the Department of Veterans' Affairs inspired seminal written works providing greater clarity concerning parallels between Vietnam combat veterans with central and autonomic nervous system dysfunctions and those warriors described in Homer's *Illiad* and *Odyssey*. Dissatisfied with the term PTSD, Dr. Shay, to his credit, recognized the cluster of behavioral and physiological changes as "psychological injury," a general term for what happens when bad experiences change how someone's mind and body works to protect itself in the present and future. It is an important distinction that **symptoms subsumed** under the term PTSD do not constitute a disease, but rather a **compensatory response pattern** involving the complex human nervous system. Dr. Shay further identifies what has come to be called a "moral injury" whereby victims of trauma are exposed to an added element of insult or betrayal when an immoral act is perpetrated by someone in a position of real or perceived authority. It is as if the permanence of societal mores are overturned, leaving the person with a great sense of looming danger and great insecurity so that the nervous system has but one choice to maintain survival – immediate and **sustained heightened nervous system vigilance.**

One of the more complex patients I have treated experienced loss of memory from a pivotal date and time during combat that extended back 12 years. This 12-year time period included his wedding and the birth of a daughter, now gone from memory as if those most joyful life events never happened. As a well-trained and experienced Navy Corpsman, he engaged in a close quarters firefight where he eventually resorted to the use of his side arm to shoot an aggressor. Immediately upon eliminating the threat to himself and his fellow Marines, he proceeded to deliver to the downed combatant all of his medical and combat lifesaving skills. The apparent conflict between self-protection under extreme stress and a duty and mission to save lives triggered some self-protective element in his nervous systems, rendering a period of amnesia.

Regardless of the time in history, but keeping in mind the "dose" or mechanism of trauma, the pathophysiology of electrical and chemical changes in the body has arguably remained the same. Further, the changes consistently manifest in observable dysfunctions that include both visible overt behaviors and less visible physiology markers detected by chemistry labs, new imaging technologies, and millisecond-resolution electrophysiology measures.

Autonomic Nervous System

The autonomic nervous system (ANS) is a good place to start a review of the changes to the human condition following traumatic exposure; however, at the outset it is important to recognize that the autonomic and central nervous systems and neuroendocrine system are inextricably linked to provide a synergistic process of keeping the body healthy and safe. The **parasympathetic** system originates in the medial medulla (brain stem) and is **modulated by the hypothalamus and insula**. This system offers efferent vagus nerve activation that reduces sympathetic arousal and the associated release of the hormone and neurotransmitter norepinephrine. Under a normal rest condition, vagal tone (tonic parasympathetic activation) is the predominant autonomic state that prevails, rather than the faster-to-respond sympathetic tone that is ready and waiting to mobilize the body for action. Under normal conditions the ANS maintains a homeostatic or self-regulatory state that is ever ready to permit a momentary transition to and from a "fight or flight" condition. This sympathetically driven fight-or-flight condition involves the release of hormones, muscle fiber activation, and other automatic chemical and electrical events throughout the body. Of particular concern in the pathogenesis of PTSD is the change from a flexible ANS able to rebound to maintain balance to a dysfunctional and ultimately deadly process that adversely affects nearly all bodily systems and functions.

Borrowing from the medical dose-response curve concept, trauma may be viewed as a "dose" and the nervous system reactivity and dysfunction as a "response" along a predictable curve from mild to most severe. Trauma dose is further complicated by it representing a composite of loosely measurable components: lethality, moral violation, predictability, duration, repetitiveness, intensity, trigger/stimulus generalization, and relative adverse impact. The concept of relative adverse impact may be understood by way of an example. An 18-month-old infant taken from his mother's embrace never to be seen again is arguably a relatively massive "trauma dose" compared to a 35-year-old survivor of a violent automobile accident. Ernst Gellhorn (1967) demonstrated that, under traumatic exposure conditions that exceed some "dose" threshold, the nervous system becomes stuck in a sympathetic dominant and parasympathetic diminished state. Should the "trauma dose" become even higher, not only is there a loss of vagal tone with elevated sympathetic arousal, but the dorsal (unmyelinated) vagal nerve can effect a temporary freeze or shutdown. In animals this is sometimes referred to as death feigning; for example, the mouse caught in the cat's teeth may go limp and appear lifeless. In such an instance, the otherwise depressed parasympathetic tone spikes and overwhelms the sympathetic dominant condition. In humans there ensues a cognitive collapse where the person has great difficulty speaking, making rational decisions, expressing emotions, moving small and large muscles, dissociates, and potentially may die from cardiac arrest. This recognized "tonic immobility" (Marx et al., 2008) does not always recover spontaneously and may even be involved in a negative feedback loop where the nervous system is so heightened that additional stressors are perceptively more prominent, thereby continually re-engaging the stress response pattern.

It appears that some individuals have environmental and genetic features that inoculate them from a "trauma dose" that would, in others, render a nervous system shutdown. While this feature is referred to as resilience, there is, at the other extreme, vulnerability; that is, individuals who are more susceptible to a seemingly small and insignificant trauma dose. It is unclear what all of these **protective factors and risk factors** are, but early history of trauma appears to be one risk factor for PTSD. Overtraining with realistic stimuli, as seen among military special operations groups, appears to be protective. Still other possibilities exist regarding protective factors, and these include cognitive function and nervous system differences.

Early interventions to restore a self-regulatory process may very well be one solution to avert tonic immobility or to restore balance. **Autonomic instability, measurable using heart rate variability** calculations (see the graph below), may well be reversed through tapping into the effectiveness of vagal efferent modulation of heart rate variability (HRV). Using self-regulatory HRV biofeedback and paced breathing, the person is able to learn to increase parasympathetic tone and decrease sympathetic drive, with lasting results. The respiratory sinus arrhythmia (RSA) amplitude measure is a good indicator of how the vagus nerve modulates or has efferent impact on heart rate and HRV under stress. Deep, slow breathing (diaphragmatic rather than thoracic breathing) reduces the heart rate, reduces blood pressure, and increases heart rate variability. Like yoga, which also has a breathing component in its practice, the benefits are more far-reaching than once thought (Gard et al., 2014).

The added relationship between HRV biofeedback and the Affect network in the brain extends also to the insula, a structure integral to autonomic and endocrine restorative balance. Combining central nervous system training with heart rate variability biofeedback has consistently shown positive results among those with PTSD symptoms. In addition to the restorative cardiac benefits of paced breathing, there is innervation to brain structures. The self-regulation dynamic of biofeedback may very well itself be an essential and salient healing component for those exposed to moral injuries where one's ability to maintain self-protection and personal regulation was removed. The dynamics of body awareness and purposeful self-improvement seen in yoga, mindfulness, and biofeedback appear to be an essential aspect of PTSD healing. Feelings if self-efficacy and control can be enhanced by successful acquisition of self-regulation skills learned through bio- and neurofeedback.

Figure shows 10-minute resting ECG measures in a 36-year-old male diagnosed with PTSD. Heart rate variability power spectra reflect elevated heart rate, low variability (SDNN should be >50 ms), and poor balance between sympathetic and parasympathetic branches of the nervous system. (Highest amplitude should be in the low frequency (LF) range).

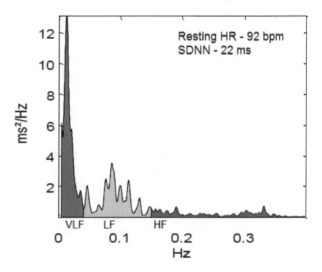

Central Nervous System

With direct connectivity to the ANS, the hippocampus, amygdala, and limbic system are integral to the brain's capacity to experience emotion, and are all involved in the interpretation of and response to traumatic events. In addition, the PTSD-related atrophy of the **anterior cingulate, hippocampus, and amygdala** results in symptoms of inattention, hindered working memory, and frontal or executive function deficits. Cognitive deficits seen with PTSD have long been considered to be a component of comorbid depression, but more recent meta-analysis results suggest that attention and working memory dysfunction are more specific to PTSD and to the brain structures known to be implicated in PTSD (Scott et al., 2015).

Due to dysfunction related to pathways to the reticular activating system, the brain orchestrates high arousal and heightened alert states. The degree of dysfunction in this arena is based on past history and the presence of current associated stimuli. The central nervous system further coordinates our momentary external and internal stimuli in order to prepare for a response, learn new information to be applied under similar conditions at a later time, and even to modify cortical connectivity to improve efficiency for survival and longevity.

The medial prefrontal cortex and anterior cingulate are involved in decision making and emotional regulation. These functions are essential to efficiently attend to the environment, quickly decipher the meaning or implications of the available stimuli, and, based on prior experience, inhibit or engage in an appropriate emotional and cognitive response. This response will also engage the neuroendocrine system. However, when the anterior regions of the brain are dysfunctional, as occurs among those with PTSD, then the stress response is poorly regulated. In these circumstances the patient will demonstrate over-arousal and incongruent emotional responses. Perhaps worse, networks such as the Salience, Affect, Executive, and Default Mode networks may be over-activated not only in response to salient or generalized trigger stimuli, but these same networks appear to remain hyperactive even in the absence of any triggering stimuli.

Mentioned earlier in the context of the parasympathetic nervous system involvement, the insula additionally serves to engage the executive system and Affect network (emotional brain including the limbic system) or, conversely, the Default mode network (DMN). The right insula is involved in self-awareness and is one of the structures assumed under the DMN. The DMN connectivity is of particular interest when considering PTSD and Reactive Attachment Disorder, since sufferers have great difficulty with self-identity, self-reflection, self-perceptions, and, by extension, self-image. It may be the case that the DMN is altered due to what may be a self-protective mechanism to block imaginal re-exposure to the sensory and visual image memories stored in the brain, and thus block associated emotions. Brain regions implicated in those with PTSD are not specific to the DMN but share other networks. The insula, anterior cingulate, and amygdala, for example, are part of the Salience network (SN), which is involved in balance between introspection and attention to external stimuli. Those with PTSD have been found to have hindered SN connectivity with either hyper- or hypo-arousal.

Based on clinical experience treating PTSD with eye movement desensitization and reprocessing (EMDR), it was repeatedly observed that, as the patient allowed introspection and physical sensations with what appeared to be time-locked emotions, symptom resolution quickly followed. In some cases it was as if a heavy cloak was lifted, and once-fitful sleep gave way to restorative sleep.

During one therapy session, a Marine began to vomit. This Marine had previously been exposed to an improvised explosive device overpressure wave. During the EMDR procedure where he visualized the blast, he began to sweat, and reported the temperature in the 72 degree office felt more like 110 degrees, the same temperature as the day of the explosion that took the life of his friend. He recalled and re-experienced the olfactory sensations of the explosive and other burnt material of that event and then became physically sick just as he had at the time of the traumatic event. Following this, he described a sense of self-awareness and calm. He later said that the previously ever-present sensation of rage and flashbacks to the memory of that blast had dissipated. This sensory and memory interaction with the associated body sensations in his chest and gut was a pivotal change in his ability to again apply introspection and self-reflection. The result was a marked recovery.

Trauma expert Bessel van der Kolk says that the traumatized person needs to engage their cingulate, insula, and medial prefrontal cortex in the context of focusing on internal experiences that involve integration of cognitive, sensory, and emotional functions. Experience appears to be demonstrating that the procedural elements of EMDR therapy embody these prescriptive nervous system capacities, perhaps more than other treatments. The application of yoga and mindfulness meditation also appears to aid in the connection and joint activation of sensory and emotional brain regions believed to be therapeutic in PTSD.

Others (Bluhm et al., 2009) have found reduced function in the DMN and SN among those with early childhood trauma. The involved DMN structures in particular include the posterior cingulate, precuneus, and medial prefrontal cortex. In addition, as it involves emotional processing and stimulus identification, the fusiform gyrus (BA 37) is part of the DMN and is also implicated in PTSD. The insula (BA 13, 43) has also been repeatedly implicated in PTSD imaging studies (Lazar et al., 2005), and, with its integral connection to autonomic nervous system balance, it is easy to see how PTSD is jointly an autonomic and central nervous system dysregulation condition. Other imaging research further supports the relationship between PTSD and the posterior cingulate, precuneus, anterior cingulate, and right amygdala (Lanius, et al., 2010). Compared to normal controls, imaging results of two nodes of the DMN (PCC and mPFC) suggest that those with PTSD have increased connectivity between the posterior cingulate and the frontal gyrus (BA 10) plus the mPFC and left parahippocampal gyrus (BA 35) (Daniels et al., 2010).

Not all published QEEG biomarkers thought to be common to PTSD appear replicable or specific. One finding that I have seen fairly often, but not always, in cases of PTSD is right parietal faster frequency alpha (Rabe et al., 2006) (see the figure below). Source localization using Brodmann areas or x, y, z coordinates to targeted voxels of interest have the added benefit of MRI imaging studies, but there is also support for scalp surface EEG amplitude findings specific to PTSD. Surface EEG data combined with analysis using LORETA mathematics provides a much simpler and less expensive imaging technique than MRI, and it has been validated against MRI. Vertex frontal (Cz) and parietal (Pz) single-channel sequential montage has the advantage of being easy to set up, and activity from those sites reflects several networks, namely the Affect, Sensory, Executive, Salience, and Default networks. Combining this simple, single EEG channel training with HRV biofeedback has certain advantages, good clinical logic, and fMRI random control trial research support.

From a Brodmann classification perspective, PTSD appears to involve, at a minimum, the following regions of interest: insula (BA 13, 43), posterior cingulate (BA 23, 31), precuneus (BA 7), medial prefrontal cortex (BA 8-12, 24, 32), anterior cingulate (BA 24, 25, 32, 33), fusiform gyrus (BA 37), and perirhinal cortex (BA 35 [hippocampus] and BA 36 [fusiform]). Some approaches to source localization neurofeedback select the list of Brodmann areas correlated with PTSD. Others opt for a correlation with subjective symptoms and match symptoms to Brodmann areas that are likely involved, and still other approaches select regions of interest based on comparison electrophysiology mapping (see the figure below titled "Regions of interest for sLORETA neurofeedback in a 36-year-old male with PTSD).

LORETA NFB can address amplitudes, coherence, and phase at and between involved Brodmann areas. This may prove to be a helpful addition to interventions when we consider that many of the areas involved in PTSD are central and ventral. Ideally, it is combined with heart rate variability (HRV) training that can influence some of the same Brodmann areas. As previously noted, the medial prefrontal cortex and anterior cingulate are involved in decision making and emotional regulation. These functions are essential to efficiently attend to the environment, quickly decipher the meaning or implications of the available stimuli, and, based on prior experience, inhibit responses or engage in an appropriate emotional and cognitive response. This response will also engage or modulate the neuroendocrine system. However, as emphasized above, when the anterior regions of the brain are dysfunctional, as occurs among those with PTSD, then the stress response is poorly regulated, the patient demonstrates over arousal, and the Affect, Executive, Salience, and Default Mode networks may be over- or underactivated even in the absence of any triggering stimuli. HRV training can have an effect on all these systems and has the added advantage of being easy to practice at home and at work.

It appears that treatments that are effective permit the person to learn and re-experience a renewed capacity to auto-regulate. For example, teaching selected patients to reduce 8-12 Hz amplitude at Pz with neurofeedback can result in significant symptom reduction. However, it may not be the directionality of the alpha power training but the ability to increase or decrease this alpha frequency in a self-regulatory manner that is bringing about the improvement.

Brain maps of a 36-year-old male with PTSD, surface amplitude (eyes closed). Deviations from healthy reference group norms are seen in elevated 12-13 Hz activity in parietal and occipital regions.

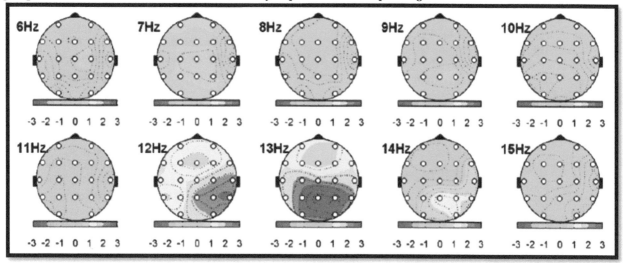

Report image used by permission of Evoke Neuroscience, Inc.

Regions of interest for sLORETA neurofeedback in a 36-year-old male with PTSD

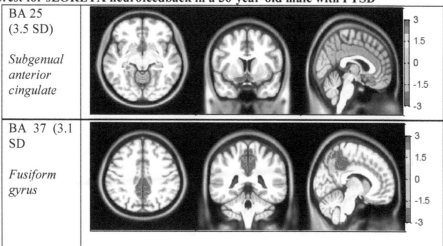

Note: Report image used by permission of Evoke Neuroscience, Inc.

Conclusion

PTSD has structural and symptom overlap with other anxiety and affective disorders; however, it does appear to differ in terms of having nervous system markers linked to how the body experiences or synthesizes the natural world and regulates emotion. The positive and negative symptoms and diagnostic criteria of PTSD are relatively simple to identify when the etiology and pathophysiology is understood. The trauma survivor with PTSD conceptualizes and experiences the present world through the lens of the past (no matter how long ago) physiological sensations, emotions, and verbal and nonverbal rules.

More complex is the delivery of effective interventions. Sufferers are often limited to medications (benzodiazepines, beta blockers, GABAergic agents, SSRIs, alpha blocking agents) that come with certain symptom benefits yet are devoid of healing properties. Secondary therapies often include clinical staff with varied skill levels and experience to provide effective exposure therapy or, ,more commonly, cognitive-behavior therapy (CBT) methods. As a result of their repeatedly documented efficacy, additional treatments like EMDR, mindfulness, and yoga therapies are being made available. Still too scarce are those trained in applied neuroscience where electrophysiological assessments and individualized heart rate variability biofeedback and brain computer interface training can be integrated into a therapeutic program designed to return the person to duty or to a productive life post-military service. True brain healing that recruits neuroplasticity can involve surface EEG amplitude neurofeedback or coherence training, or, with 19-channel data collection and source localization using LORETA, one can train specific Brodmann areas or whole networks. (See section in this book on LORETA Z-score neurofeedback.)

This brief section provided a brief clinical conceptualization of PTSD, the nervous systems involved, and some introductory direction on biofeedback modalities and added therapies that can be applied to treat those with PTSD and like conditions. Not discussed in this section, but no less important, is the research and clinical usefulness of event-related potentials that contribute further insights about PTSD (Clark, 2009). ERPs offer sensitive sensory and cognitive measures that add value to assessment(s) when used in isolation or in addition to a QEEG. Also not covered in this section is the use of neuromodulation techniques for PTSD. Transcranial direct current stimulation (tDCS) and repetitive transcranial magnetic stimulation (rTMS) are showing some clinical promise but remain experimental rather than evidence-based at this point.

Recovery is complex, and effective use of any of the above interventions should, at a minimum, include a cohesive peer support element. It is naïve to believe that peripheral biofeedback and neurofeedback are alone ample to treat or bring lasting reparative relief to those with PTSD. The peer element may be one reason group yoga interventions have been effective in the treatment of PTSD. Likewise, it is naïve to think that a treatment approach without self-regulation of the nervous system is complete. The complicated nature of moral injury and adaptive nervous system changes following traumatic experiences necessitates a multifactorial approach that includes nutrition, quality reparative sleep, psychotherapy (perhaps paired with EMDR or exposure therapy), acupuncture, exercise, peer group interaction, biofeedback, and neurofeedback.

SECTION XXX
Optimal Performance Training

In the last decade, heart rate variability training has been added to regular biofeedback, using EMG, and peripheral modalities, in the training of high-performance athletes. However, the original description of the golfer that was given earlier (and which appeared in the first edition of this book) still stands as a model. It shows how you establish the sequence of brain and body physiology associated with successful performance, and then train those parameters. The successful performance profile will differ according to different sports and even according to differing tasks within a sport. For example, during biathlon, rifle shooting differs from the skiing segment, Making a free throw in basketball is a different mental state than required when dribbling the ball and having to sense the position and readiness of every member of your own team and of the opposition. The results now with Olympic athletes are impressive. With professional athletes, Bruno Demichelis developed the Mind Room in 2009 and added neurofeedback to the biofeedback training that he was already doing for AC Milan football (soccer) players. AC Milan subsequently won the Champions League European Cup in the 2006-2007 season. The next year, four of its players were on the Italian World Cup winning team. Demichelis then moved on to work with Chelsea in England. As of early 2014, he was back with AC Milan. Regarding Olympic athletes, Tim Harkness, a South African Sports Psychologist, using NFB and BFB, trained India's first Olympic gold medal winner, who won in rifle shooting in Beijing, besting the Russian world champion and the Chinese competitor who was favored to win. Canadian freestyle skiers and short-track speed skaters received biofeedback plus neurofeedback prior to the 2010 Winter Olympics in Vancouver. The first gold medal for Canada won on Canadian soil was won by one of those skiers, Alexandre Bilodeau. He was quoted in an interview for *Maclean's Magazine*. He said that the training reduced his tension and increased his focus. We feel honored that Tim Harkness, Bruno Demichelis, and Marge Dupee (who worked under Penny Werthner to train the Canadian freestyle skiers prior to the 2010 Winter Olympics in Vancouver), all came to our Stoney Lake workshops where Vietta Sue Wilson has been a regular featured lecturer. Dr Wilson is a sports psychologist who has used NFB and BFB in her training of athletes, including Olympic athletes, for a wide range of sports, including track and field, tennis, archery, weight lifting, and, with the exception of cricket, virtually all major team sports.

At this juncture, it is important to note that a sports psychologist is a highly trained specialist whose knowledge encompasses a much larger field of expertise than is typically found in neurofeedback providers who are attempting to assist athletes.

Dr Wilson in workshops explains it thus: "A sport psychologist is typically a licensed psychologist and is credentialed with a sport psychology association. The person has knowledge of sport, specialized academic work, and placement in the sport psychology area. They additionally have training in the skills found important for performance enhancement (attentional focusing, goal setting, stress management, imagery, etc.). Additionally, there are individuals who have specialized in the application of psychological skills to sport who are not licensed psychologists. They often are called mental performance trainers, sport psychology consultants, or simply a mental training coach. Their emphasis is on assisting athletes only in developing their mental game for sport and do not do traditional counseling or therapy."

The methods used by sport psychologists who use NFB and BFB generally follow the basic training for athletes developed by Sue Wilson and used by her, as noted above, with athletes of every stripe, including Olympic athletes. Her methods and approach are well described in the book *Biofeedback and Neurofeedback Applications in Sports Psychology*, edited by Ben Strack, Michael Linden, and Vietta Wilson (2010).

Perhaps a description of training given by Tim Harkness at the 2009 AAPB Annual Conference in Albuquerque can be helpful in giving an overview of this kind of work. He described how biofeedback helped Abhinav Bindra win India's first individual Olympic gold medal in rifle shooting. He started with BFB: breathing, then skin conductivity and finger temperature control. Tim noted that Abhinav had the highest degree of body awareness that he had come across in an athlete. They did over 150 hours of training on the various modalities. For EEG training they found alpha training at T3 useful, as was rewarding 15-18 Hz at Cz and squashing 26-30 Hz. They also trained at Pz and Oz. They worked on finding a state that had him physically relaxed, but still allowed him the sharpness of reaction to trigger at the right moment. Tim notes that you train the athlete to prepare to execute the skill. This is a state prior to skill execution (3-5 seconds before) as contrasted with the slightly different state of mind and body during skill execution. Skill execution itself is short, subtle, and instinctive. You are only training the athlete to lay the foundation for optimal skill execution. Harkness found that Abhinav showed a T3 event-related desynchronization (ERD) in the alpha range when triggering. However, Abhinav wanted an event-related synchronization (ERS) in alpha at triggering, which is like being on autopilot in the alpha zone when executing the overlearned response, so they trained for that.

Tim noted that HRV was very useful and stated that the best predictor of a bad shot was heart rate and breathing being out of phase. He found that Abhinav would breathe at 2.5-3.5 breaths per minute (which is the rate Bruno Demichelis with his years of martial arts training also uses), or he would breathe at about eight breaths per minute, and then breath-hold for 40 seconds, while controlling his heart rate. The point in describing the above is to emphasize that there is no such thing as a "protocol." The practitioner addresses the needs of the individual athlete.

Dr. Wilson found that with one of her athletes, an archer, keeping high-frequency beta bursts at the central location (Cz) completely suppressed was the best predictor of a bull's-eye shot. Keeping higher frequency beta at a low amplitude may have reflected turning off any busy-brain worries or ruminations and being in the moment. The lesson here is that theory is only just theory, and with an elite athlete you must follow the athlete and measure psychophysiological parameters (EEG, breathing, heart rate, finger temperature, skin conduction, EMG) and learn what works best for that client in his or her particular sport.

There are many instruments, well advertised on the internet, for 'peak-achievement-training.' We are experimenting with some of them, but we will not be discussing these 'instruments' in this basic text because they are not yet evidence-based. Our view is that intervention without prior assessment is inappropriate because it could lead to training the wrong parameters for some individuals. In this section and, indeed, throughout this book, a simple message should be apparent: Resist the flashy gadgets and promises of quick fixes. Learn and practice using the basic fundamental principles of applied psychophysiology and biofeedback.

PART SEVEN

Efficacy, Statistics, Research Design and Multiple Choice Questions

SECTION XXXI
Efficacy Criteria Used in the Field of Neurofeedback

Efficacy Criteria

Yucha and Montgomery (2008) summarized the criteria, as listed below, for the five levels of efficacy recommended by a Joint Task Force and adopted by the Boards of Directors of the Association for Applied Psychophysiology (AAPB) and the International Society for Neuronal Regulation (ISNR), later called the International Society for Neurofeedback and Research. From weakest to strongest (Level 1 through to Level 5), these levels are: 1. Not empirically supported, 2. Possibly efficacious, 3. Probably efficacious, 4. Efficacious, 5. Efficacious and specific.

Level 1: Not empirically supported. This designation includes applications supported by anecdotal reports and/or case studies in nonpeer-reviewed venues. Yucha and Montgomery (2008) assigned eating disorders, immune function, spinal cord injury, and syncope to this category.

Level 2: Possibly efficacious. This designation requires at least one study of sufficient statistical power with well-identified outcome measures but lacking randomized assignment to a control condition internal to the study. Yucha and Montgomery (2008) assigned asthma, autism, Bell's palsy, cerebral palsy, COPD, coronary artery disease, cystic fibrosis, depression, erectile dysfunction, fibromyalgia, hand dystonia, irritable bowel syndrome, PTSD, repetitive strain injury, respiratory failure, stroke, tinnitus, and urinary incontinence in children to this category.

Level 3: Probably efficacious. This designation requires multiple observational studies, clinical studies, wait-list controlled studies, and within subject and intrasubject replication studies that demonstrate efficacy. Yucha and Montgomery (2008) assigned alcoholism and substance abuse, arthritis, diabetes mellitus, fecal disorders in children, fecal incontinence in adults, insomnia, pediatric headache, traumatic brain injury, urinary incontinence in males, and pain associated with vulvar vestibulitis (vulvodynia) to this category.

Level 4: Efficacious. This designation requires the satisfaction of six criteria:

i. In a comparison with a no-treatment control group, alternative treatment group, or sham (placebo) control group using randomized assignment, the investigational treatment is shown to be statistically significantly superior to the control condition, or the investigational treatment is equivalent to a treatment of established efficacy in a study with sufficient power to detect moderate differences.

ii. The studies have been conducted with a population treated for a specific problem, for whom inclusion criteria are delineated in a reliable, operationally defined manner.

iii. The study used valid and clearly specified outcome measures related to the problem being treated.

iv. The data are subjected to appropriate data analysis.

v. The diagnostic and treatment variables and procedures are clearly defined in a manner that permits replication of the study by independent researchers.

vi. The superiority or equivalence of the investigational treatment has been shown in at least two independent research settings.

Yucha and Montgomery (2008) assigned Attention Deficit Hyperactivity Disorder (ADHD), anxiety, chronic pain, epilepsy, constipation (adult), headache (adult), hypertension, motion sickness, Raynaud's disease, and temporomandibular joint dysfunction to this category.

Level 5: Efficacious and specific. The investigational treatment must be shown to be statistically superior to credible sham therapy, pill, or alternative bona fide treatment in at least two independent research settings. Yucha and Montgomery (2008) assigned urinary incontinence (females) to this category. More recently Arns et al. (2009) argued persuasively that ADHD should be in this category. The superiority of neurofeedback as compared to the established bona fide treatment of stimulant medication is based on a higher response rate, long-term benefits after discontinuation of treatment, and no negative side effects.

Outcome Studies On Single Components of a Treatment Program *DO NOT* Adequately Reflect the Effect of Combined Approaches

Before starting the ADD Centre, the authors had decades of experience using individual approaches. Dr. Lynda Thompson even owned and ran learning centers. It is our opinion that research that studies just one component, be it neurofeedback, HRV, metacognitive strategies, parent training, counseling, diet, sleep, exercises or other influences like computer games and electronic media of all kinds, does not capture the effect of combinations of these that occur in real life. It would be like only prescribing exercise to someone recovering from a major medical illness, with nothing done about diet or sleep or medications. Guidelines for medication use in people with ADHD always suggest it be combined with other approaches. Multimodal approaches, as outlined in *The A.D.D. Book* by Sears and Thompson, should be the norm. These include consideration of child development and family dynamics, lifestyle factors of sleep, exercise, and diet (it is of great importance to add fish oil supplements and have a healthy breakfast with sufficient protein, about 20 grams), educational interventions, neurofeedback and medications when appropriate. "Medications when necessary but not necessarily medications," is how it is phrased in that book.

The reader will note that in our description of good clinical practice, neurofeedback is never a stand-alone procedure. We recognize that researchers must, at this time, do research on individual components in order to match the paradigms used in drug research but we feel they should clearly state in their reporting that they are not doing research on neurofeedback training as it is done in standard clinical practice. In practice it is always combined with the work of a good therapist/teacher/coach/trainer who is certainly not 'blind' as to what is taking place and who adds specific metacognitive strategies and tasks, plus appropriate other biofeedback techniques such as heart rate variability training. There is also consideration of other factors such as diet, sleep, and exercise and family and social factors. The implications of positive or negative research findings concerning neurofeedback should be taken in this context.

To understand how any individual component changes the effectiveness of a total therapeutic approach to a problem, new research methodologies will be required.

At this juncture, large case series that include both subjective measures (such as questionnaires, school reports, work reports and so on) and objective measures (such as computerized testing and individualized standardized testing, like the WISC or WAIS) may better reflect the value of these combined approaches to the clients.

Addendum: Comment on the Appropriate Research Design for Studies Using Neurofeedback

There is pressure on NFB researchers to do 'placebo'-controlled studies (that is, using sham feedback) that are double-blind, or even triple-blind, in order to establish efficacy. On the other hand, the World Health Organization maintains that the best comparison is not against placebo but against an established efficacious treatment. As an example, efficacy studies concerning neurofeedback for ADHD would thus involve comparison of neurofeedback with stimulant drug treatment. Critics might argue that nonspecific effects would be greater for biofeedback than for a pill. Thus, studies using sham feedback continue to be done. In this context, recent research showing differences in brain activation with 'real' and 'false' (simulated) feedback is of interest.

Such investigations have been carried out at the Research Institute of Molecular Biology and Biophysics, Siberian Branch of the Russian Academy of Medical Sciences in Novosibirsk, Russia by Mark Shtark and colleagues (Shtark et al., 2014). Using normal young adults as the subjects, their research design involved video game feedback in which the rate of descent of a diver hunting for treasure could be controlled by decreasing one's heart rate in the 'real' feedback group. The heart rate was monitored by measurements of the interbeat RR intervals. The activation periods were 20-90 seconds long with one-minute rest periods between each trial. The subjects had to develop new and efficient self-regulation strategies in order to slow heart rate and thus win the game by collecting treasures from the bottom of the ocean.

They found that the lack of true feedback in the false feedback group resulted in an increase in overall activation as subjects searched in vain for a solution. It also resulted in an increase in the spread of RR values in the false feedback group.

The brain areas that were activated were identified and measured using fMRI, which can localize activity based on increased local blood supply to the brain tissue. Real feedback mainly activated BAs 7, 37 and 40. The simulated feedback showed more activation in BAs 7, 19, 37, 39, and, unlike real training, significantly increased activation in BAs 5 and 6. Clusters of activity in BAs 19, 22, 39, 40 and 42 occurred more clearly during the 'false' or 'simulation' training.

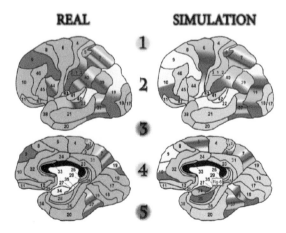

Figure from Mark Shtark et al. 2014 The five experimental stages show the progression of

activation of brain areas across the series of trials within one session. For detail the reader is referred to the original article, which is made available online by the Journal of Behavioral and Brain Science (http://www.scirp.org/journal/jbbs), and can be downloaded directly at http://dx.doi.org/10.4236/jbbs.2014.41008.

Clearly, brain activation is different under real feedback conditions than it is with false feedback. More networks become activated when no real control is possible, perhaps because the person keeps trying different things to control the speed of the game and perhaps also because of frustration. Real feedback is associated with fewer areas being activated and thus perhaps reflects a more efficient activation of appropriate brain areas. During real feedback concerning heart rate, there appears to be a sequence of involvement of brain areas that would reflect learning of the behavior. (In this experiment the reinforced behavior was a decrease in heart rate.) False feedback induces more widespread activation. The areas for both real and simulated (false) feedback overlap, but more areas were involved during the false feedback condition and the quantitative degree of activation was significantly higher during the simulated feedback trials. This research suggests that real feedback can optimize brain efficiency during the learning of self-regulation skills.

Relevance for Neurofeedback Training

Both bio- and neurofeedback typically involve visual displays for feedback. The main areas that showed activation in both real and sham conditions were, not surprisingly, areas involved in visual processing. Here is a review of the functions of the three main areas that showed activation in the fMRI images; namely, Brodmann areas 7, 37 and 40.

BA 7 is a somatosensory association cortex. It is a key area in several networks including the Default network, and it has motor, memory, attention, and both cognitive and emotional understanding functions. Other functions that may be important in carrying out the NFB task include, in the nondominant hemisphere, conscious awareness of visuospatial events, mental rotation, stereopsis, and judgments or estimates that involve visual and spatial aspects of the images, and sustained attention, particularly with respect to visual-motor attention. In the dominant hemisphere it is important in episode recall, handwriting, imaging, and shifting attention.

BA 37 is important for visual fixation and high-level visual processing of complex stimuli. It involves the parahippocampal gyrus and short-term memory networks with episode encoding. It has Executive network functions related to deductive reasoning and intentions. It is a part of the right hemisphere network for sustained attention.

BA 40 contains the supramarginal gyrus. In the dominant hemisphere it may be part of networks for the conscious recollection of previously experienced events and the ability to plan or complete motor actions that rely on semantic memory. In the nondominant hemisphere, the supramarginal gyrus is important for visual grasping and in locating places in an area (or map), locating spatial position, visualizing spatial organization, and understanding context. In both hemispheres this area is important in deductive reasoning and the ability to sustain attention. These areas are also important in other cognitive activities, such as distinguishing same and different, and conflict detection.

Activation in these areas is important for sustaining attention and for learning. This helps explain why biofeedback and neurofeedback are so effective when the goal for many of our patients is to learn self-regulation skills.

SECTION XXXII
Brief Overview of Statistics and Research Design

Introduction
In order to gain knowledge, make advancements, and improve credibility, all disciplines must have a sound scientific backing, the cornerstone of which is research. Although not a necessary requirement for all clinical practices, research is an integral component of any science, and biofeedback is no different. Indeed, biofeedback practitioners can be justifiably proud that their field is one that has been built on the solid foundations of scientific research.

Scientific Papers

Getting Started: Planning and Failure to Plan...
The old maxim "failing to plan is planning to fail" holds especially true in conducting scientific research. The solid foundation set out by a good plan allows the researcher to easily conduct his or her research, avoid pitfalls and backtracking, and ensure credibility.

The key to the planning process is familiarizing yourself with the current knowledge in your field. This is the basis of the planning stage in which you gather information to help form and mold your hypothesis. As well, this familiarization with recent work will help to inform the researcher of current methods that have been successful or unsuccessful.

The planning process can be divided into four steps:

1. Write the Problem Statement
The problem statement is the backbone of the paper. It defines the area to be investigated and guides the entire research plan. The problem statement covers five main areas:

- *Workability* – Is the research feasible given the time and resources available?
- *Critical Mass* – Is the problem being studied of sufficient proportions to justify a study?
- *Interest* – Is the field (and are you, yourself) interested enough in the area to warrant a research study?
- *Theoretical Value* – Will this research fill a gap in the current literature?
- *Practical Value* – Is there any practical value to this study?

When these questions have been satisfied, researchers can feel confident that their research and purpose will be of value to the field. The problem statement needs to be continuously consulted, assessed and refined as the researcher continues to learn about his or her specific field of study.

2. Consult Secondary and Primary Sources
This is the background research stage - the time when you truly learn how little you know about the field you are an "expert" in. It begins with the basic searches through preliminary sources. These include: reviews, encyclopedias, abstracts and bibliographies. Once the topic and sources of information have been narrowed and keywords have been identified, secondary and primary sources can be consulted. Journals, original studies, theses and dissertations fall into this category. These sources will give you more detailed information regarding the field and will be a huge help in guiding your research. With the advent of the internet, information is so readily available that it can become overwhelming. The key is to keep the search specific. Be aware of the keywords, phrases and descriptors you use. Also, use the search engine limits to avoid getting hits that are not relevant to your search. Lastly, be sure to only utilize sources that are trustworthy. Some sites that have proven their worth in the scientific field include: PsycINFO/PsycLIT (www.apa.org/pubs/databases/psycinfo/), MEDLINE (www.nlm.nih.gov/bsd/pmresources.html), ERIC (eric.ed.gov/) and PubMed (www.ncbi.nlm.nih.gov/pubmed).

3. Read and Record the Literature

By this point you no doubt have stacks of literature very relevant to your area of interest. Now comes one of the most time-consuming, yet also the most fun and interesting parts of the planning process: reading and recording the current literature. In the interest of being provident, you should keep an electronic record of all literature you found online, as well as hard copies of all literature you are going to read. This will avoid the tedious process of re-finding articles at a later date, and will help to avoid confusion as to who said what, where.

The reading and recording work is a time for an engaging challenge. It is your turn to do more than just learn. It is an opportunity to outsmart your colleagues in the field. You get to be in the driver's seat and play the role of the critic. To say "this was done well, yet I could improve on it by doing this!" There is no such thing as being too critical of a peer's work at this juncture. Whatever flaws you find and can avoid in your research will only improve your chances of success. Whatever improvements you make to something that worked previously in the field will only serve to increase the likelihood of publication of your research. It is of the utmost importance to take note of specific features of previous work. Some of the most important areas to take note of are previous hypotheses, designs, methods, subject selection and exclusion, equipment used, limitations, and suggestions for future research.

You will exit this stage of the planning process with more ideas for conducting research than you even thought possible. It may even be the case that your entire problem statement has been rewritten! The ability to learn and adapt will serve you well at this stage and throughout the research process.

4. Write the Review

This final step of writing up the literature review may at first seem redundant, or even unnecessary. It seems counterintuitive to write on a subject before performing your research; however, in the later stages you will be able to use this as the basis for the introduction of your paper or dissertation. You will have therefore already written a major portion of your paper before even testing a subject. What a feeling of accomplishment!

The review should be divided into three very basic parts: the "hamburger," if you will.

- The introduction, which defines the area and familiarizes the reader, makes clear the problem you wish to study and attracts the reader's attention.

- The body (organized around your three or four, major points), which summarizes and is critical of past research, is detailed about explaining the need for your further investigation in this field.

- The summary and conclusions, which point out gaps in current literature, thus lending support for the need for your research, give the direction of your research, defines your problem statement and your hypothesis.

Putting the Plan into Action: Contents of a Scientific Paper

Overview of Putting the Plan into Action

The research process will be described following a typical journal format divided into four sections: introduction, methods, results and discussion. The introduction will introduce the reader to the subject area and the research that will follow. The methods section will outline the scientific process and will discuss areas such as subjects and the design of a study. The third section will be the results section where the data collected is analyzed using statistics. Some basic statistics used in research will be discussed. The final section is the discussion section. The important aspects of a discussion, and how to write and organize this section, will be broken down and explained.

1. Introduction

The purpose of an introduction is simple. It raises the reader's interest in the topic, gives background information and introduces the problem. The manner in which this purpose is fulfilled is not as simple, however. There are four mutually exclusive, yet equally important, components to the introduction.

The Problem Statement

First, there is the problem statement. This is a brief description of the area to be studied. It is an identification and description of the variables. *Independent variables* (causes of changes and often the variables controlled by the experimenter), *dependent variables* (effects), *categorical variables* (e.g., sex and age), *controls* (e.g., time of day of testing) and *extraneous variables* (things that cannot be controlled) are all discussed in brief. At this stage the researcher should also mention assumptions, limitations and delimitations.

Assumptions are what the researchers must assume during the experiment, for instance, that the subjects will give their best effort and follow instructions. *Limitations* are shortcomings inherent with this research, such as the *generalizability* of results. *Delimitations* are limitations that are under the control of the researcher; ones that the researcher has imposed. These include such things as size of the sample and tests used.

The Hypothesis

Second is the hypothesis. This is the predicted outcome of the study, the expected effects. In contrast to this is the *null hypothesis*. The null hypothesis is ultimately what you will be testing. This will be further explained in the results section. For now it will suffice to say that it is the inverse of your hypothesis. For example, if your hypothesis was that 20 sessions of neurofeedback would improve a student's theta/beta ratio by five percent, then the null hypothesis would be that 20 sessions of neurofeedback would not cause any significant change in the theta/beta ratio. In formulating a hypothesis, keep in mind the SMART principle: **S**pecific, **M**easurable, **A**ttainable, **R**ealistic time frame, **T**angible. These guidelines will help the researcher to formulate a testable hypothesis.

Operational Definitions

Third, and directly following the hypothesis, comes the *operational definition* of terms. If you are measuring *intelligence,* for example, the definition you operate with will be "what an IQ test measures," and you will specify which test. This is an observable phenomenon and a concrete description of the expected effects. It enables the evaluation of results and a discussion of the findings. It is most important to operationally define the dependent variable(s). What is stated is how the expected effects will be manifested in the subjects. As with the hypothesis, these are very specific. For instance, in neurofeedback, it may be the ability to hold the magnitude (average amplitude) of beta frequencies between 15.0-18.0 Hz, above 3 µV for two minutes while performing a silent reading task with material at a Grade 8 level of difficulty. This is the standard against which you will measure.

Justification

Fourth, but certainly not of least importance, is the *justification* of the study. The purpose of this section is to draw attention to the differences between this study and others in the same area. This component will (a) address contradictory findings currently in the literature, (b) explain how this research paper will fill gaps in existing knowledge, (c) describe existing theories and how this research will refute or support them, and (d) the practical applications of the findings to the field.

After looking over the components of the introduction, it should be clear how all the work that went into the planning process and writing the literature review will aid you in expediting the writing of the introduction. Each step builds on the previous one in a logical progression when performing a research study. The thorough and successful completion of each step helps to ensure success in the steps that follow.

2. Methods

Purpose

The methods section is a very detailed and a somewhat dry section of the paper. It is a succinct description of the components of the study. It has but one purpose: to enable the *reproduction/replication* of the study. That *is* the sole purpose of the Methods section. There are four subsections within the Methods section. These are: subjects, instruments and tests, procedures, and design and analysis.

Subjects

The section on subjects is a straightforward description of the subjects chosen and how they were chosen for the study.

First comes the description of the number of subjects who participated in the study. It states the number of subjects originally recruited for the study and any dropouts or changes in numbers during the study. This is important to mention right at the onset, since sample size contributes to a very large extent to ensuring statistical results are relevant and reliable. The number of subjects used is a major contributor to the *power* of the study. It is a relatively simple calculation and can be found in almost any basic statistics book. By definition, power is the probability that you will correctly reject a false Null Hypothesis. The equation is $1 - Beta = Power$. Calculations of Power, Beta and other statistical measures will be discussed in the next section. Do not be surprised if, during your study, subjects drop out or numbers change due to other extenuating circumstances. These eventualities occur in all studies. Initial numbers and changes in numbers need to be mentioned at the conclusion of the section describing subjects and their selection.

Second, the researcher should describe who the subjects are. Were they chosen from a specific grade level, city or country? Are they male or female, old or young? Details of subjects' sex, age ranges, race, socioeconomic standing and other characteristics relevant to the study being performed, such as IQ or experience in a sport, need to be described.

The selection process is one of the most important elements of a study and is described in the Methods section. Proper selection techniques and assignment to groups of subjects is essential in setting up a study that will be considered valid in its findings. It also has a bearing on the generalizability of the findings. There are five concepts to be considered here.

1. *Random Sampling* from Population
2. *Stratified Random Sampling*
3. *Systematic Sampling*
4. *Sectioned Sampling:* areas are selected randomly, then cities, suburbs, etc...
5. *Random Assignment of Study Groups*

Different areas of study bring with them their own sampling difficulties. It is, for example, extremely difficult to define a control group in a study of ADHD. This is because many children with ADHD are extremely bright and creative and are functioning near the top of their class, albeit possibly functioning below their actual potential. Therefore, in a study of people with ADHD, not fitting the DSM criteria is a necessary but not a sufficient criterion to be included as a control subject for a study that is looking at EEG differences.

It is also important that you mention how permission and cooperation were obtained. This is simply a mention of the informed consent form used and how it was administered, Internal Review Board permissions, and, in animal studies, how the animals will be protected during the study.

Instruments and Tests (Validity and Reliability)

This section, like the Subjects section, is a detailed description. Here it is a list of the type, make, model, year, etc. of all the equipment used to carry out the study. Anything used for set-up, measurement, penalties or rewards must be described in detail. In considering the design of this section, the researcher should take into account the *validity* and *reliability* of equipment and tests. Please note that the distinction between validity and reliability is very important; they are NOT synonymous. *Validity* questions whether the researcher has measured what they intended to measure. *Reliability* addresses whether your measures were accurate, that is, could they be repeated with the same result measured at another time. Also under consideration should be the difficulty of obtaining the instruments, difficulty in taking the measurements, the range of scores expected, and the time needed to administer the tests and conduct the experiments.

Procedures

The procedures description is the meat of the methods section. Here, a step-by- step account of the process is detailed to the reader. It is like the recipe book of the experiment. There is a description of how the tests are carried out and of how the data are obtained. This section is extremely important in allowing readers to determine the credibility and validity of the study and in allowing for the replication of the study. First is the order of steps. This is important in ensuring there is a logical flow and ensuring that there are no learning effects (and if there are, they are addressed). The timeframe is also an important aspect of the procedures. Is there fatigue in the subjects (mental and/or physical)? Is there a loss of interest and therefore a decrease in test performance? Finally, the exact instructions given to the subjects must be detailed. Wording, location, time before trials, individual or in a group and many other factors can contribute to how subjects will react under test conditions. Colleagues and interested readers will scrutinize this section to the letter. If the procedures used are not clear, logical and replicable then the experiment will lose all credibility and will be deemed a failure even if results were found.

Design and Analysis

This portion of the Methods section is particularly important in experimental (quantitative) studies. It is a description of the variables (independent, dependent and extraneous) and how each is related and controlled. For example, the fact that an EEG will vary within subjects based on time of day needs to be addressed. In this section these external factors are described and the means by which they are controlled are discussed. The cause-effect relationship is also addressed here. The researcher needs to convince the reader that changes in the dependent variable(s) are explainable only on the basis of the treatments (the independent variable). There needs to be shown a method of agreement (why does A cause B?) as well as a method of disagreement (why does B not happen without A?).

The analysis portion of this section provides a description of the statistical tests that will be used to determine if an effect of statistical significance occurred. It is a description of the data processing and a description of the statistics to be used and why those particular ones were chosen. For instance, why is a parametric statistical test, such as a t-test, being performed on the data? It would need to be justified by stating that interval data was being processed, and that the data possessed normal distribution with an equal variance between samples on the variable being tested, and that the observations were independent of one another. These criteria would justify the use of a parametric test. Also, it would need to be noted that differences between two groups were being tested for, thus justifying the use of a t-test over another type of parametric test such as ANOVA. This statistical jargon will all be explained in the Results section, which follows.

The Methods section is often viewed as the 'dry section,' where list after list appears. Although the mundane nature of the writing of this section cannot be denied, the content is truly exciting. This section determines everything! Without a thorough Methods section, an experiment will never be successful. Attention to detail is the key to conducting successful research, and nowhere is this more important than in the set-up of the study. The aforementioned adage, *failing to plan is planning to fail,* has never held so true as in this section. It is here that your planning shows.

In summary, the Methods section has four main content areas: (1) information on the subjects (2) a description of the instruments, tests and apparatus used (3) an outline of the procedures and (4) a brief description of the design and analysis.

Well, with that under wraps, it is time to get to the fun stuff - the Results and Discussion sections. Did we find an effect? Why or why not? To answer those questions, get ready to dive into the wonderful world of basic research statistics.

3. Results
Overview (Significance and Meaningfulness)
Let me begin with a plea to all not to roll your eyes at the mention of the word *statistics*. As researchers, statistical calculations are our friends. They will prove the worth of a well-conducted study and its findings, while filtering out research that has been improperly conducted. This Results section is merely an introduction to some essential terms, concepts, and statistical tests that one needs to know in order to competently read research papers and prior to embarking on a research study.

Statistics tell us two very important things in research. These are: (a) the strength of an effect or relationship and (b) the significance of an effect or relationship.

It is very important to note the distinction between the two words *strength* and *significance*. A finding can be significant yet not be meaningful – the reverse does not hold true. Significance is simply a measure of reliability. It merely answers the question "did a change occur?" Meaningfulness, on the other hand, is a more difficult distinction to make. It questions whether the change that occurred is worth the input that induced the measured change. An amusing example of this is Popeye (the old cartoon character). For him, eating only one can of spinach elicited unsurpassed strength for a short period of time. For him, the change is meaningful. However, what if he had had to eat 50 cans of spinach in a row to get the same result? The experiment to deduce this would have been significant because a change did occur, but the expense, time and effort that was incurred to elicit this change (the purchasing and eating, in this case the input) is not worth the response (the increased strength for a short duration of time) and is therefore not meaningful. This distinction is unique to all research and is determined by the experimenters and their peers.

Describing the Data
For a discussion on describing the data, see the Statistics section, below.

4. Summary, Discussion and Conclusion
The discussion section actually is comprised of a summary of the findings and some discussion and conclusions that might be drawn from the findings. It is your wrap-up, and you want it to leave an impression. It is a chance to answer the questions: So what? Where is the relevance to the broader picture? Typically you interpret the information from the results, where findings and statistical significance were reported objectively but without comment. This summary is more subjective and allows for comment on how the findings relate to the original hypothesis. There is some discussion of the limitations of the study (small sample size, certain variables not controlled for, like time of day, etc.) and the

generalizability of the findings. Finally, you draw conclusions. A typical concluding sentence might run thus: Neurofeedback, when administered in such and such a manner for this many sessions, appears to be effective in improving 'x' (or decreasing 'y') in a particular population (conservatory of music students with performance anxiety, children with ADHD, athletes who have suffered concussions, or whatever group is of interest to you).

Statistics

Describing the Data

Measures of Central Tendency
Mean, Median, Mode
One of the most important basic concepts in statistics is the description of the data. Essentially there are two concepts to be discussed here: *measures of central tendency* and *measures of variability*.

Central tendency has three measures. These are the mean, median and mode. The *mean* is probably very familiar to all. It refers to the average of all the scores and is the sum of all scores divided by the number of scores. The *median* is the middle score. For instance, if your scores were 3, 6, 7, 9 and 13, then the median is 7. The term *mode* refers to the score that occurs most frequently. An example here is a set of scores of 10, 15, 16, 17, 17, 21 and 22. The mode in this example would be 17. The mean is considered to be the most accurate reflection of the entire group score.

Measures of Variability

Standard Deviation, Variance, Range
Variability is defined in statistics as the degree to which each individual score differs from the central tendency score. As with central tendency, there are three measures of variability: variance, standard deviation (here referred to as SD) and range. *Variance* and SD both measure the variability, or spread of scores around the mean. Variance is the square of the SD. SD is calculated by dividing the sum of all the squared differences between each score and the mean score by the number of scores (n) minus one, and then taking the square root of that number.

$$SD = \sqrt{\frac{\Sigma(x-mean)^2}{N-1}}$$

$$Variance = SD^2$$

The *range* in a set of data is simply the distance between the lowest and the highest score. It is the difference (subtraction) between the high score and the low score.

The concept of variance in statistics is an extremely important one. As previously mentioned, it tells about the distribution of scores around the mean. If most of the scores are very close to the mean, then you have a fairly homogeneous group with respect to the variable being measured and you will calculate a low variance within the group or between the groups. If you have a very diversified group on the variable being tested, then you will end up calculating a high inter- or intra-group variance due to the diverse nature of the scores.

Types of Data and Tests

Nominal, Ordinal, Interval
In research there are three types of data: nominal, ordinal and interval. As an aside, remember that *data* is plural. The singular is *datum.ua'* In Latin, 3rd declension nouns change their ending from '_um' to '_a' endings for the plural, as in medium (singular) to media (plural). *Nominal data* are data that are categorical, for example, males and females. The data fit into categories set by the researcher. Ordinal data refers to data that are ranked. If, for instance, the best score of ten participants was 13.5/15 and the lowest score was 5.25/15, then the highest of the ten scores may be ranked as 1 and the lowest may be ranked as 10. In this type of data scoring the order is significant, yet the quantitative distance between scores is not meaningful. The third type of data is *interval data*. This is the type of data we most often see in scientific or quantitative research. With this type of data the relative distance between data points **is** meaningful. An example could be in the long jump where a score that is .25 meters greater than another score shows a quantitative difference in ability.

Categories of Tests
The explanations of data types should elicit in you the question, When in statistics does it matter what type of data we have? The answer lies in the tests we use. There are two general categories of statistical tests: Nonparametric and Parametric. Nonparametric tests are tests that are used on data that are either nominal or ordinal in nature. Parametric tests, on the other hand, are used when data is of an interval nature and three assumptions regarding the data are met.

These assumptions are:

1. The population data are *normally distributed* (this can be inferred from your sample of the population). This means that the curve plotted by the recorded data is a normal (or *bell*) curve. It is also referred to as a Gaussian distribution, after the German mathematician, Carl Friedrich Gauss.

   A normal curve meets specific criteria:
   - The mean, median and mode are all at the same point (have the same value)
   - 68% of all the scores fall within 1 standard deviation, 95% of scores fall within 2 SDs and 99% are within 3 SDs.

2. The samples being compared have *equal variance*. Normal curves can be drawn with different heights and widths, referred to as *kurtosis*. The curves of the samples being compared must be equal in height and width.

3. Observations made during the research are *independent* (one variable cannot affect another). The influence of one variable cannot have an effect on another variable being measured. For example, if a first pill is taken and results are recorded, then a second pill is taken, and results are recorded again, *independent* means the first pill can in no way have an effect on the outcome from taking the second pill. If the first pill did affect the results of taking the second pill, the outcome of the second pill would be said to be dependent on the first and this would negate the third assumption of independent observations.

Hypotheses

Null Hypothesis

All research is based on hypotheses. A *hypothesis* is a statement whose validity the research is being conducted to test. In all research there are least two hypotheses, the (primary) hypothesis (H_1) and the null hypothesis (H_0). The hypothesis is the statement of what the researcher believes will be the treatment effect. The *null hypothesis* is the inverse of the hypothesis. It states that there will be no effect found. For example, if the research hypothesis is that taking a pill three times a day for one year will permanently turn blond hair to red, then the null hypothesis would simply be that taking the specified pill three times a day for one year will not turn blond hair permanently red. The null hypothesis is important in that it is the statement against which research results are tested.

Based on the research findings, the null hypothesis will either be accepted or rejected. If the null hypothesis is proven to be true (the treatment has no effect), then we say that we accept the null hypothesis. If a treatment effect is found, it is expressed as a rejection of the null hypothesis.

Probability and Alpha Level

Statistical tests allow one to draw conclusions that are based on results found in samples to infer things about the population that sample is assumed to represent. As with any inference, there is a chance of making an improper assumption. Statistics uses the term *alpha* to define what the researcher thinks is an acceptable probability of making the error of finding a nonexistent effect. In other words, alpha is the chance that your research will find an effect in the sample even though there is no true effect in the population.

Let's go back to our discussion of hypotheses to tie this together. By accepting or rejecting the null hypothesis, we are making a statement about the effects of the treatment. Two types of error can be made. If H_0 is actually true but is rejected, then we have committed a *type 1 error (false positive, alpha)*. If, on the other hand, H_0 is not true but is accepted, a *type 2 error (false negative, beta)* has been committed.

In the literature, it is the alpha value that is always stated as the accepted error level. Ordinarily, alpha is set at either 1% (.01) or 5% (.05); however, there are times when a researcher will set alpha at 10% (.1). These alpha levels or probability statements are called the *P-value*. For an alpha level of five percent, the P-value would be written as p<.05. That p<.05 setting for alpha means that the probability of finding an effect in the sample, given that there is no effect in the population, is smaller than five percent. In other words, there is a one in twenty chance that the effect you found to be significant is not real.

A simple chart will help you remember these terms and their meanings.

	H_0 true	H_0 false
Accept H_0	correct decision	Type 2 error (false negative, beta)
Reject H_0	Type 1 error (false positive, alpha)	correct decision

Power

The beta value discussed in the chart above is used in determining the power of the study. Power is defined as the probability that you will correctly reject H_0 (probability that you will find an existing effect). Power is calculated as 1-beta. This is logical, since, by definition, beta is the chance of making a Type 2 error (conclude that there is no effect when there actually is one). If you have a low chance of making a Type 2 error (beta is low), then the power of your study will be high. Generally, an acceptable level of power for a study is 0.8. The power of a study can be increased by manipulating one or all of three major factors. These are:

1. Increasing sample size
2. Decreasing overall variance (make the within-group variance of both groups smaller)
3. Increasing alpha from .05 to .1 (which means you are testing at a lower confidence limit and increasing your chance of making a Type 1 error from 1 in 20 to 1 in 10. The effect of this is a decrease in the chance of making a Type 2 error)

Effect Size

Now I'm sure you're wondering where this all leads. Here it is. The effect size is the number that a prudent researcher needs to calculate to ensure that they are "set for success." The effect size is the standardized value of the practical importance, or meaningfulness, of a study.

The *effect size* is calculated by measuring the difference between groups based on means and standard deviation. The calculation for effect size is:

$$ES = \frac{(M1 - M2)}{SD}$$

M1 – the mean of group 1
M2 – the mean of group 2

SD – the standard deviation of the control group. (It should be the case that the SDs of the two groups are the same since it is an assumption of parametric statistics that the within-group variances of the test groups are the same.)

A calculated effect size of 0.2 is considered small, 0.5 is considered moderate and an ES of 0.8 is considered large.

Merit of a Study

The interaction between effect size (ES), sample size (N) and power is best represented by a chart.

Schematic Representation of the Sample Size, Effect Size, and Power Relationship

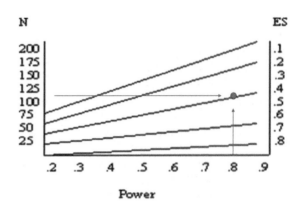

As you can see, having a small effect size will necessitate a larger sample size to maintain a high power for the study.

In conclusion, the merit of a study is determined by four factors:
1. Alpha level being used (p-value)
2. The effect size (ES)
3. Power
4. Sample size (N)

A Short List of Related Concepts and Terms

This section provides a list of additional terms that the reader may encounter when reading the literature in this field. Only a brief definition or description is given for each of these terms. For further information it is recommended that the reader refer to any standard undergraduate textbook on statistics.

Alpha level: The probability of making a Type I error, that is, a false positive, or rejection of the null hypothesis when there really is no difference or effect.

Beta level: The probability of making a Type II error, that is, a false negative. Your study, for example, concluded that neurofeedback did not have an effect on your sample when it actually does have an effect in the larger population.

Correlation: The degree to which two variables vary together. It is usually calculated as a correlation coefficient designated as *r*.

Sensitivity: The ability of a test to identify true cases. For example, the sensitivity of the theta/beta ratios for identifying ADHD is .86 (Monastra et al., 1999). That means that 86 percent of true cases are detected by that ratio. The other 14 percent of people with ADHD have a different pattern, such as excess alpha, rather than excess theta.

Specificity: The ability of a test or procedure to correctly distinguish one diagnosis from another. For example, specificity of the theta/beta ratio for distinguishing between a subject being in the ADHD population rather than the non-ADHD population is .98. That is, 98 percent of the time, the person whose score exceeds the cutoff will truly have ADHD.

Test Power: The probability that a Type II error will not be committed. This is calculated as *1-beta*. It addresses the question of whether a test has the ability (power) to detect a difference that exists.

Pre- and Post-Testing: In this text the authors have recommended that all practitioners do pre-testing (baseline) and post-testing appropriate to the client's goals and the type of BFB work that they are doing. You want to be able to document changes, not just have the person say that they feel better. In NFB work, the practitioner should keep in mind the following variables, although some of the suggestions made here may be impractical in a busy clinical practice. Any objective testing is, for the most part, better than no testing. Objective tests commonly used with clients who have ADHD include: continuous performance tests (Conners, IVA, TOVA), EEG measures (theta/beta ratios), intelligence tests, and academic achievement measures. Subjective measures would include parent questionnaires and also, with adolescents and adults, self-reports.

- Testing should be at the same time of day. In NFB work this will help control for the influence of circadian rhythms on theta.

- Most clients will be more consistent in their attention if testing is done early in the morning.

- Be sure to control for, or at least ask about, variables that could conceivably affect the results including: caffeine, meals, smoking, medications and sleep. If stress factors differ from pre- to post-testing, then you should note that when recording results.

Terms Related to Test Characteristics

Criterion-Referenced Test: Allows its users to make score interpretations in relation to a functional performance level, as distinguished from those interpretations that are made in relation to the performance of others. A grade-equivalent score is criterion referenced.

Norm-Referenced Test: This is a test that allows its users to make score interpretations based on the comparison of a test-taker's performance to the performance of other people in a specified group. Intelligence tests, for example, are age-normed; each score is compared to the norms for that person's age group.

Reliability

- **Reliability:** The degree to which test scores are consistent, dependable, or repeatable; that is, the degree to which they are free of errors of measurement.

- **Reliability Coefficient:** A coefficient of correlation between two administrations of a test. The conditions of administration may involve variation in test forms, raters or scorers, or passage of time.

  These and other changes in conditions give rise to qualifying adjectives being used to describe the particular coefficient including:
  o *Parallel form* reliability
  o *Inter-rater* reliability
  o *Test-retest* reliability

Validity

- **Validity:** The degree to which a certain inference from a test is appropriate or meaningful. Does the test measure what it purports to measure?

- **Validity Coefficient:** A coefficient of correlation that shows the strength of the relation between predictor and criterion. If you were checking the validity of a new intelligence measure, you might compare the scores with those obtained on the Wechsler scale.

- **Concurrent Validity:** This refers to having evidence of criterion-related validity in which predictor and criterion information are obtained at approximately the same time. For example, someone's self-report is that they have not been drinking alcohol in the last 12 hours and this is validated by a blood level for alcohol that measures zero.

- **Content-Related Evidence of Validity:** Evidence that shows the extent to which the content domain of a test is appropriate relative to its intended purpose. This would require, for example, that one have a representative sample of content and a clear exclusion of content outside that domain.

- **Criterion-Related Evidence of Validity:** Evidence that shows the extent to which scores on a test are related to a criterion measure. For example, Showing that SAT scores predict grades in college.

Differential Prediction: The degree to which a test that is used to predict people's relative attainments yields different predictions for the same criteria among groups with different demographic characteristics, prior experience, or treatment.

Psychometric: The measurement of psychological characteristics, such as abilities, aptitudes, achievement, personality traits, skill, and knowledge.

Error of Measurement: This term refers to the difference between an observed score and the corresponding true score. The error of measurement gives the range in which the true score is found.

Standard Error of Measurement: This is the standard deviation of errors of measurement that is associated with the test scores for a specified group of test-takers. This may be thought of as an indication of just how variant scores would be if the test were given to the same subjects an infinite number of times. For example, if a test using standard scores with a mean of 100 and a standard deviation of 15 has a standard error of measurement of + or −5, then an obtained score of 110 means that the true score would be in the range between 105 and 115.

True Score: In classical test theory, this refers to the average of the scores earned by an individual on an unlimited number of perfectly parallel forms of the same test.

Measures of Dispersion

- **Range:** The lowest to the highest score. (Calculated as Highest - Lowest + 1.)

- **Standard Deviation (SD):** This is a unit of measurement. With normal (bell curve) distributions, scores can be expressed in terms of standard deviation units from the mean. (It is useful to note that:

Over 68 % of scores fall between ±1 SD from the mean
Over 95 % .... ±2 SD from the mean
Over 99 % .... ±3 SD from the mean

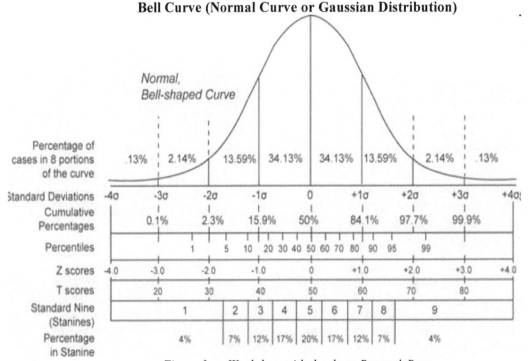

Figure from Workshop with thanks to Roger deBeus

- **Variance:** This is a measure of variability. It is the average squared deviation from the mean (the square of the standard deviation).

Note the difference in usage between the term 'variability' as it is used in statistics and how it may be used in NFB work. An example is looking at the variability of the amplitude of theta over time. Variability is the standard deviation/mean x 100. This measurement is useful to help compare measurements of a variable that changes with age, such as the amplitude of theta, because it relates the amplitude to the mean. A child has much higher amplitudes of theta and much wider deviations from the mean as compared to an adult. Note that in most instruments you must define a standard time and a standard averaging or smoothing factor before you can compare variability in one subject over several sessions. In NFB work with ADHD, the client attempts to decrease both the amount of theta and its variability. (For example, using the old F1000 equipment manufactured by Focus Technology, we used 30 seconds and a smoothing factor of 5 and had both adults and children attempt to get more than 15 out of 20 consecutive 30-second segments to have a variability of less than 35. This can be an extraordinarily sensitive measure of a client's ability to self-regulate focus and attention. At the beginning of training, most clients meet this criterion on only one or two segments out of 20. By the end of training, most clients usually score 20 out of 20. Variability in this usage is a helpful measure to track progress, but it is not the same as the term used in standard statistics developed for measuring significance of outcomes.

Normative Scores

Scores of Relative Standing

- **Percentile Rank:** The percentage of scores in a specified distribution that fall at or below the point at which a given score lies. If a child scores at the 98th percentile rank, his score exceeds or equals 98 percent of the population. He is in the top two percent.

- **Stanine:** A form of standard score that is based on a distribution of the raw scores into nine equal parts. If a stanine score is five, it is in the middle of the distribution of scores.

- **Standard Score**: A score that describes the location of a person's score within a set of scores in terms of its distance from the mean in standard units. Standard scores, such as those used to measure IQ, usually have a mean of 100 and a standard deviation of 15 and are normally distributed; that is, match a bell curve.

- **T-Score:** This is a derived standard score on a bell curve that has a mean score of 50 units and a standard deviation of 10 units.

- **Z-Score:** A type of standard score scale with Gaussian distribution (bell curve) in which the mean equals zero and the standard deviation equals one unit. Understanding the meaning of Z-scores has become extremely important in neurofeedback because database comparisons provide Z-scores.

SECTION XXXIII
Multiple Choice Questions

Note 1: These questions have been written as a teaching tool. To this end, the questions often have a longer stem than would be found in a typical exam. Many questions may also often be rather self-evident to a knowledgeable reader. These are included as a review of the basics.

Note 2: Selected references are included in the answer section that follows the questions. All questions are based on text found in the book, but selected questions, where we felt the reader might want to do further reading, are referenced further. The questions are ordered, for the most part, according to sections of the Biofeedback Certification Institute of America (BCIA) blueprint. BCIA-required hours as of 2015 are noted for most sections for the convenience of those intending to become certified. Not all of the sections have written questions. Some sections are more appropriate for demonstration and oral examination. For more details on certification, consult the BCIA website (www.bcia.org).

Orientation to EEG Biofeedback
Required Hours: 4

1. Electrical activity in the mammalian brain was first measured, using a string galvanometer, by this scientist, with the findings being published in 1875:
   a. Richard Caton
   b. Marie Curie
   c. James Maxwell
   d. Hans Berger

2. In the 1920s, a German psychiatrist recorded a pattern of uniform electrical waves from the human scalp that he labeled first-order waves. This came to be known as *alpha rhythm,* with reference to the first letter of the Greek alphabet. He also observed periods when these waves were absent and the pattern of waves was smaller and desynchronous. This pattern became known as *beta*. His name was:
   a. Richard Caton
   b. Hans Aufreiter
   c. Heinrich Hertz
   d. Hans Berger

3. In the 1950s, an American psychologist, using careful scientific methods, demonstrated that a person could correctly identify when they were producing alpha waves but subjects were not able to say precisely how they were making that discrimination. His name was:
   a. Barry Sterman
   b. Joe Kamiya
   c. John Basmajian
   d. Tom Budzynski

4. An American psychologist, working with cats in the late 1960s at the University of California Los Angeles, demonstrated that they could be trained using a method called "operant conditioning," to increase a specific spindle-like brain wave pattern that ran at a frequency between 12 and 19 cycles per second. He gave the spindle-like activity between 12 and 15 Hz the name *sensorimotor rhythm (SMR)*. Closely following on this discovery was his observation that cats who had increased their SMR activity were resistant to seizures. His name was:
   a. Barry Sterman
   b. Joe Kamiya
   c. Joel Lubar
   d. Tom Budzynski

5. It has been established that measuring the theta/beta ratio was helpful in differentiating between normal and ADHD clients. The first researcher to propose and investigate this was:
   a. Barry Sterman
   b. Keith Conners
   c. Joel Lubar
   d. Vince Monastra

Learning Theory Relevant to EEG Biofeedback

6. Operant conditioning or instrumental learning is based on _____, which can be simply stated thus: *When you reward behavior, you increase the likelihood of its recurrence.* This law was first stated by Edward Thorndike in 1911. He mainly studied cats in puzzle boxes. This law is also called trial and error learning. Which "law" fits the blank above?
   a. Thorndike's Law
   b. Watson's Law
   c. The Law of Effect
   d. The Law of Consequences

7. Classical conditioning is a term that refers to a type of learning. It was originally described by_____ as a conditioned, or learned, reflex. Who was this investigator?
   a. Thorndike
   b. Watson
   c. Skinner
   d. Pavlov

8. In what kind of learning does the conditioned stimulus *automatically elicit* a conditioned response after it has been paired a sufficient number of times with an unconditioned stimulus (that elicits an autonomic response)?
   a. Operant conditioning
   b. Classical conditioning
   c. Associative learning
   d. Conditioned response learning

9. In classical conditioning, motivation is:
   a. Very important
   b. Somewhat important
   c. Usually downplayed
   d. For the most part, irrelevant

10. What term usually refers to conditioning by successive approximations? (Rewarding a behavior or a sequence of neurophysiological occurrences, _____ the contributing components of that sequence in a way that results in an increased frequency of that sequence occurring.)
    a. Generalizing
    b. Shaping
    c. Conditioning
    d. Associating

11. What kind of learning occurs when things get unintentionally paired with reinforcers? (For example, a colored light that indicates an artifact, such as eye-blink or muscle activity, on a neurofeedback screen.)
    a. Classical learned response
    b. Incidental or associative learning
    c. Operant conditioning
    d. Unintentional learning

12. In its simplest form, what term means that what the client learns in the office doing neurofeedback will also occur at other times, places and with other people and tasks? (The ability to do this is severely impaired in some disorders such as autism.)
    a. Conditioned learning
    b. Operant learning
    c. Shaping
    d. Generalization

13. If a student who is doing neurofeedback learns to use metacognitive strategies during the feedback session when he is in a mental state that is producing positive rewards for low slow-wave activity and higher beta activity (15-18 Hz), then using those strategies while doing homework will be:
    a. Less likely to elicit focus and concentration
    b. More likely to elicit focus and concentration
    c. Irrelevant in terms of eliciting focus and concentration
    d. Produce interference with respect to memorizing new material

14. The rationale for pairing diaphragmatic breathing at a rate of about six breaths per minute with the production of SMR rhythm in patients who have dystonia and Parkinson's disease has been reported to be based on the fact there is a double innervation of the muscle spindles by:
    a. The parasympathetic nervous system and gamma motor efferents
    b. The parasympathetic nervous system and corticobulbar efferents
    c. The sympathetic nervous system and gamma motor efferents
    d. The sympathetic system and pyramidal efferents

15. Coaches may have the athlete use self-cueing of some kind, like a word or an image that has been paired with the production of the desired state, to produce:
    a. A classical response
    b. Shaping
    c. Generalization
    d. High beta rhythm

16. What may occur when the conditioned stimulus is no longer paired with the unconditioned stimulus over a number of trials? (In operant conditioning it may occur when a behavior is no longer reinforced.)
    a. Shaping
    b. Extinction
    c. A classically conditioned response
    d. An operantly conditioned response

Research (Blueprint Area IV)

17. Central tendency has three measures. These are the mean, median, and mode.
    The term mode refers to:
    a. The sum of all scores divided by the number of scores
    b. The score that occurs most frequently
    c. The most accurate reflection of the entire group score
    d. The middle score

18. Variability is defined in statistics as the degree to which each individual score differs from the central tendency score. As with central tendency, there are three measures of variability: variance, standard deviation (here referred to as SD) and range. Variance and SD both measure the variability, or spread of scores around the mean.

    Variance is:
    a. The sum of all the squared differences between each score and the mean score, divided by the number of scores (n) minus one, and then taking the square root of that number
    b. The square of the SD
    c. The sum of all the squared differences between each score and the mean score, divided by the number of scores (n) minus one
    d. The average of the standard deviations divided by the mean times 100

19. There are two general categories of statistical tests: Nonparametric and Parametric. Nonparametric tests are tests that are used on data that are either nominal or ordinal in nature. Parametric tests are used when data is of an interval nature. One or more assumptions regarding the data are made. Which of the following is/are included as an assumption?
    a. The population data is normally distributed
    b. The samples being compared have equal variance.
    c. Observations made during the research are independent (one variable cannot affect another).
    d. All of the above

20. A normal curve meets which specific criteria?
    a. The mean, median and mode are at different points (have different values)
    b. 88% of all the scores fall within 1 standard deviation
    c. 95% of scores fall within 2 SDs
    d. All of the above

21. The null hypothesis:
    a. Is a statement of what the researcher believes will be the treatment effect.
    b. States that there will be no effect found.
    c. Is a statistically invalid proposition
    d. States that no hypothesis is possible

22. Statistics uses the term *alpha* to define what the researcher thinks is:
    a. An unacceptable probability of finding an existent effect
    b. An acceptable probability of finding a non-existent effect.
    c. The probability of an event occurring
    d. The probability of an event not occurring

23. If H_0 is actually true but is rejected then we have:
    a. A Type 2 error (false negative, Beta)
    b. A Type 1 error (false positive, alpha).
    c. An inconclusive result
    d. A misuse of statistics

Basic Neurophysiology and Neuroanatomy (Blueprint Area II)
Required Hours: 4

Neuroanatomy

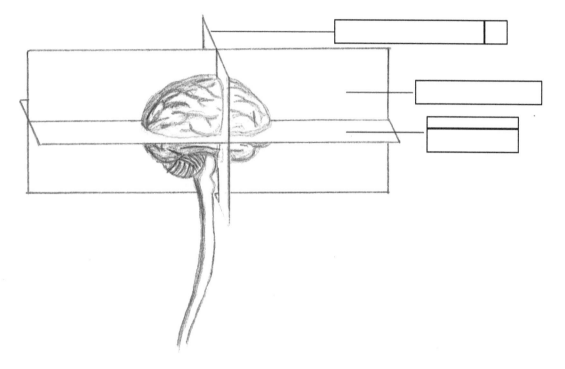

24. Place the following names on the correct lines in the above diagram:
    a. Sagittal Plane
    b. Horizontal Plane
    c. Transverse Plane

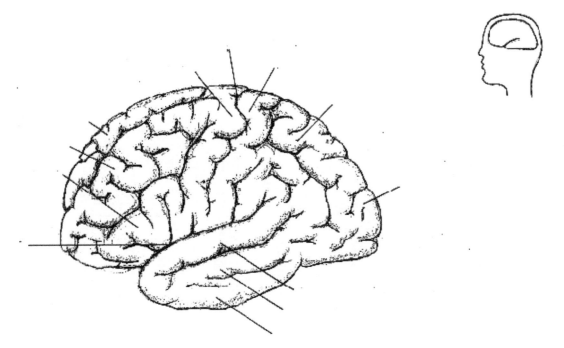

Note that there are more lines on the drawing than labels listed below. When you are labelling be sure that it is clear to which line you have attached your label.

25. Put the following labels (or the correct number) next to the appropriate line on the above drawing:
    1. Parietal lobe
    2. Postcentral gyrus
    3. Inferior temporal gyrus
    4. Superior frontal gyrus
    5. Lateral sulcus
    6. Central sulcus
    7. Inferior frontal gyrus
    8. Superior temporal gyrus
    9. Occipital lobe

    Then outline, shade and label approximately where you think the following areas would lie:
    - Wernicke's
    - Broca's
    - Exner's
    - Auditory cortex

Sketch of a Midsagittal Section of the Human Brain

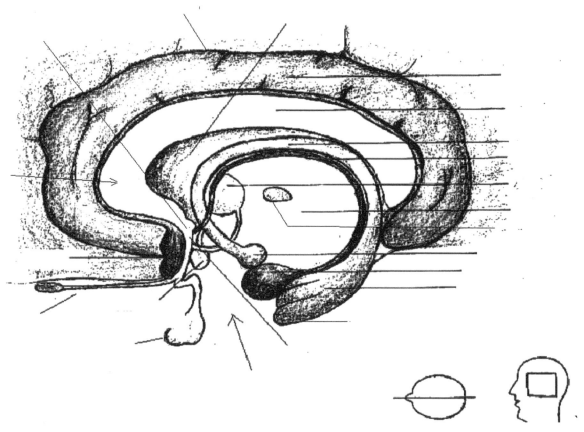

This is counted as two questions with two bonus marks available. One correct answer for five correct labels and a second correct answer for a further five correct labels. A bonus mark for a further five and a second bonus mark for 18 or more correct labels. Note that there are more lines on the drawing than labels listed below. When you are labelling be sure that it is clear to which line you have attached your label.

26. Put the following labels (or the correct number) next to the appropriate line on the above drawing:
    1. Olfactory tract
    2. Anterior thalamic nucleus
    3. Thalamus
    4. Septum
    5. Amygdala
    6. Massa intermedia
    7. Hypothalamus (on this sketch it is small, roundish and immediately posterior to the lamina terminalis and the septal area)
    8. Lamina terminalis
    9. Entorhinal cortex
    10. Pituitary
    11. Anterior commissure
    12. Fornix
    13. Ccorpus callosum
    14. Stria terminalis
    15. Cingulate gyrus
    16. Sulcus cinguli
    17. Mammillary body
    18. An arrow pointing to the approximate site of the anterior perforated area

27. In the spinal cord and peripheral nervous system, the white matter is on the outside and grey matter on the inside – the reverse of the cerebral hemispheres, where the grey matter is on the outside. This statement is:
    a. True
    b. False

28. When there is a head injury, the grey and white matter move at different speeds due to their different densities and the resultant sheer forces lead to diffuse axonal injury (DAI). This type of injury can be detected in the EEG.
    a. True
    b. False

29. The area responsible for interpretation of the meaning of language and the assemblage of letters in phonemes is called:
    a. Wernicke's
    b. Broca's
    c. Exner's
    d. Penfield's speech area

30. The area responsible for assemblage of information into intelligible speech is called:
    a. Wernicke's
    b. Broca's
    c. Exner's
    d. Penfield's speech area

31. Wernicke's area is in the:
    a. Right parietal-temporal cortex
    b. Left parietal-temporal cortex
    c. Basal ganglia
    d. Left frontal region

32. Broca's area is in the:
    a. Right parietal-temporal cortex
    b. Left parietal-temporal cortex
    c. Basal ganglia
    d. Left frontal region

33. Executive attention is believed to be, in part, organized by the:
    a. Right parietal-temporal cortex
    b. Anterior cingulate gyrus
    c. Thalamus
    d. Red nucleus

34. The *anterior perforated area* (APA) is an area that is perforated by small blood vessels that supply the corpus striatum. This area forms the inferior medial part of the innominate substance of the lenticular nucleus. It borders posteriorly with the _____. The _____ is continuous anteriorly with the innominate and with the putamen. It is also connected to the *uncus*. In part, the grey matter of the uncus is continuous with gray matter of the _____. Anterior to the anterior perforated area, the lentiform nucleus joins the caudate.

    The one correct nucleus to fill all of the blank spaces is the:
    a. Substantia nigra
    b. Locus coeruleus
    c. Claustrum
    d. Amygdala

35. Fibers connect the anterior perforated area, the septum, and the amygdala to the _____. The _____ lies below (inferior to) the thalamus and is continuous with the septal area and with the anterior perforated substance.

    The one correct term to fill all of the blank spaces is:
    a. Substantia nigra
    b. Locus coeruleus
    c. Hypothalamus
    d. Caudate

36. The _____ is critical in the regulation of blood pressure, pulse rate, body temperature and perspiration. It is involved in all the body homeostatic mechanisms. It is important in hunger, thirst, water balance, sexual behavior and lactation. It is important for our biological clock, and thus our circadian rhythms. It is, with the amygdala, an integral player in our fight or flight responses.

    The one correct term to fill the blank is:
    a. Substantia nigra
    b. Locus coeruleus
    c. Hypothalamus
    d. Caudate

37. The cortex (gray matter) of the _____ has only three layers as compared to the neocortex, which has six layers. At its anterior end, this archicortex connects with the amygdala and the anterior perforated area. Posteriorly, it is continuous with what is termed *entorhinal cortex* in the wall of the _____ sulcus, and this gives rise to the fibers of the fornix that eventually connect with the hypothalamus. The *fornix* or 'arch' also connects the _____ with the anterior thalamic nuclei and the mammillary body. In addition, the _____ has connections with the frontal, parietal and temporal lobes.

    The best term to fill all of the blank spaces is:
    a. Substantia nigra
    b. Claustrum
    c. Hippocampus
    d. Putamen

38. The _____ has connections to the areas concerned with emotions, the autonomic nervous system, the endocrine system and consciousness. It plays a key role in the laying down of memories. It is the _____ that allows the animal to compare the present situation with past memories of similar situations. This is an important function for survival.

    The best term to fill all of the blank spaces is:
    a. Substantia nigra
    b. Claustrum
    c. Hippocampus
    d. Putamen

39. The subcallosal gyrus + the parolfactory area + the cingulate gyrus + the parahippocampal gyrus + the hippocampal formation together form the _____.

    The best term to fill the blank space is:
    a. Substantia nigra
    b. Claustrum
    c. Limbic lobe
    d. Putamen

40. The hippocampal cortex connects with the entorhinal cortex anterior to it, and to the bulb of cortex there called the uncus, which is continuous with the amygdala and the anterior perforated area. This entorhinal cortex is the origin of the _____. The _____ contains **commissural fibers** that cross to the other cerebral hemisphere allowing (hippocampal commissure) for connections between the right and left hippocampal formations.

    The best term to fill the blank space is:
    a. Corpus callosum
    b. Superior longitudinal fasciculus
    c. Uncal fasciculus
    d. Fornix

41. The cortex has **extrapyramidal motor** areas (premotor). The pathways from these areas have many intervening synapses in different areas of the motor system. These areas include the nuclei within the _____, including the lenticular nucleus (putamen and globus pallidus), which sends fiber bundles to the *subthalamic* nucleus. The first bundle pierces the posterior limb of the internal capsule (lenticular fasciculus), and the second courses around the anterior border of the internal capsule (the ansa or loop lenticularis).

    The best term to fill the blank space is:
    a. Cingulate gyrus
    b. Hippocampus
    c. Fornix
    d. Corpus striatum

42. The neocortex has **extrapyramidal motor** areas (premotor). The pathways from these areas have many intervening synapses in different areas of the motor system. They course through some of the nuclei of the basal ganglia and descend to connect, in the midbrain, with the _____, the *substantia nigra,* and the *reticular formation* and from there descend to influence (extrapyramidal) the striated muscles.

    The best term to fill the blank space is:
    a. Red nucleus
    b. Hippocampus
    c. Fornix
    d. Corpus striatum

43. The voluntary motor pathways descend through the internal capsule to the *crus cerebri* on the lateral aspect of the diencephalon to become the *basis pedunculi* of the midbrain. The pathways run in this basal or ventral position through the pons segment to become the _____ on the lateral surface of the medulla, from whence they enter the spinal cord and *cross over* to the opposite side to become the *lateral corticospinal tract.*

    The best term to fill the blank space is:
    a. Inferior longitudinal fasciculus
    b. Hippocampal formation
    c. Fornix
    d. Pyramid

44. The _____ are involved in executive functions: the voluntary control of attention, the inhibition of inappropriate and/or unwanted behavior, the planning of actions and executive decision making. They have a role in the maintenance of arousal and the temporal sequencing of complex entities such as the expression of compound sentences.

    The best term to fill the blank space is:
    a. Inferior posterior portion of the temporal lobes
    b. Prefrontal areas of the cortex
    c. Anterior portion of the parietal lobes
    d. The pyramid

45. The _____ is involved in speech, articulation and writing. It has a major role in auditory and verbal representation. Object naming and word recall requires activation in this area. It is also involved in the representation of visual images evoked by auditory input.

    The best term to fill the blank space is:
    a. Left cerebral hemisphere
    b. Right cerebral hemisphere
    c. Fornix
    d. Medial portion of the right temporal lobe

46. The _____ is important in sensing the Gestalt (image of something) and in parallel processing. It attends to spatial relationships. It is to a large extent responsible for the representation of geometric forms. It is also responsible for orientation in space and holistic perceptions.

    The best term to fill the blank space is:
    a. Left cerebral hemisphere
    b. Right cerebral hemisphere
    c. The angular gyrus
    d. Medial portion of the left temporal lobe

47. _____ dominance is usually thought of as being characterized by distractibility, stimulus seeking behavior, seeking novelty and change, being emotionally involved and expressive and extroverted. There may tend to be external locus of control. There may be more of a tendency for hysterical, impulsive and even manic behaviors. Processing tends to be fast and simultaneous, in keeping with the Gestalt or holistic tendencies.

    The best term to fill the blank space is:
    a. Left cerebral hemisphere
    b. Right cerebral hemisphere
    c. Left temporal lobe
    d. Right temporal lobe

48. For the most part, the _____ controls the emotional aspects of language, such as the information conveyed by intonation.

    The best term to fill the blank space is:
    a. Left cerebral hemisphere
    b. Right cerebral hemisphere
    c. Left angular gyrus
    d. Medial portion of the right occipital lobe

49. The area responsible for auditory processing and, on the medial aspect, for short-term and working memory is:
    a. Prefrontal areas of the frontal lobes
    b. Hippocampal formation
    c. Fornix
    d. Temporal lobes

50. _____ dominance is usually thought of as being characterized by a lack of emotion, introversion, goal directed thinking and action with an internal locus of control. Processing may tend to be slow and serial in keeping with the tendency to be careful and sequential. The aptitudes tested on IQ tests are largely _____. The dominant neurotransmitter is dopamine.

    The best term to fill the blank spaces is:
    a. Left cerebral hemisphere
    b. Right cerebral hemisphere
    c. Left temporal lobe
    d. Right temporal lobe

51. Damage to the **right frontal lobe** can result in euphoria and emotional indifference. The affected person seems to ignore, and even be completely oblivious to, paralysis of their left side. The term used to describe this problem is:
    a. Anosognosia
    b. Associative visual agnosia
    c. Prosopagnosia
    d. Apperceptive agnosia

52. _____ means 'without action.' It is the inability to execute a learned movement. Constructional _____ refers to an inability to construct or even to draw an object. It is due to lesions in the right parietal cortex.

    The best term to fill the blank space is:
    a. Agnosia
    b. Apraxia
    c. Alexia
    d. Abulia

53. Understanding speech involves the inferior parietal lobe and the auditory association area of the superior temporal gyrus. If there is damage to this area, speech will still be fluent with good grammar, but the person will **speak nonsense.** This area, which is important for understanding speech, is called:
    a. Jenson's area
    b. Wernicke's area
    c. Broca's area
    d. Exner's area

54. Damage to _____ interferes with the individual's ability to instruct the motor cortex and results in defects in the articulation of speech. Damage to this area and areas to which it is linked may result in the person being able to understand what has been said to him and being conscious of what he wants to say, but being **unable to say it**. This person may substitute staccato nouns and verbs so that the flow is lost and it sounds like a telegram. Ungrammatical speech and writing errors may be observed.

    The best term to fill the blank space is:
    a. Jenson's area
    b. Wernicke's area
    c. Broca's area
    d. Exner's area

55. The _____ is a region of one of the lobes of the brain. This region sits at the *occipital-parietal-temporal* junction behind (posterior to) the lateral fissure. It involves *visual-spatial-language* skills and therefore **visual word recognition.** This region in the left hemisphere, called the _____, is required for reading. Damage to this area can result in **alexia** (inability to read) and in **agraphia** (inability to write).

    The best term to fill the blank space is:
    a. Medial aspect of the orbital cortex
    b. Inferior lateral temporal gyrus
    c. Angular gyrus
    d. Cingulate gyrus

56. When a client has dyslexia, the neurofeedback (NFB) training usually involves increasing the activation of certain areas of the cerebral hemispheres. This is done by increasing fast-wave activity (low beta often between 13 and 19 Hz) and decreasing the dominant slow-wave activity (theta or thalpha).

    Two types of dyslexia, dysphonetic (trouble sounding out) and dyseidetic (trouble with visual processing of the letters) are approached differently in NFB. It is commonly suggested that, for dyseidetic dyslexia you might first place your active electrode(s) at which of the following sites?
    a. Fp2 or F4 (sequential try F4-P4 or F8-P8 (T6)
    b. F3 or F5 (sequential try F3 – P3 or F7-P7 (T5)
    c. P3 or P5 (sequential try F3-P3 or F7-P7 (T5)
    d. O2 or T6 (sequential try P4-O2 or F4-O2

57. When a client has dyslexia, the neurofeedback (NFB) training usually involves increasing the activation of certain areas of the cerebral hemispheres. This is done by increasing fast-wave activity (low beta often between 13 and 19 Hz) and decreasing the dominant slow-wave activity (theta or thalpha).

    Two types of dyslexia, dysphonetic (trouble sounding out) and dyseidetic (trouble with visual processing of the letters) are approached differently in NFB. It is most commonly suggested that for dysphonetic dyslexia you place the active electrode(s) at which of these sites?
    a. Fp2 or F4 - sequential try F4-P4 or F8-P8 (T6)
    b. F3 or F5 - sequential try F3 – P3 or F7-P7 (T5)
    c. P3 or P5 - sequential try F3-P3 or F7-P7 (T5)
    d. O2 or T6 - sequential try P4-O2 or F4-O2

58. With injury to the visual association cortex, the person can still see but will be unable to recognize objects unless they are able to touch them.

    Injury to another portion of the association cortex can result in difficulties perceiving the shapes of objects that cannot be seen. Drawing or even following a map may become difficult. What is this part of the association cortex called?
    a. Limbic
    b. Temporal
    c. Somatosensory
    d. Supramarginal

59. In normal people, the _____ gyrus specializes in the recognition of human faces. This _____ gyrus lies on the medial aspect of the temporal lobe between the parahippocampal gyrus above, and the inferior temporal gyrus below.

    The name of this gyrus (and the best term to fill the blank spaces) is the:
    a. Superior frontal gyrus
    b. Inferior orbital gyrus
    c. Fusiform gyrus
    d. Supramarginal gyrus

60. The fusiform gyrus lies on the medial aspect of the temporal lobe between the parahippocampal gyrus above and the inferior temporal gyrus below. The fusiform gyrus has been shown to be important for:
    a. Handwriting
    b. Speech
    c. Recognition of faces
    d. Integration of auditory inputs

61. The _____ comprises a bundle of long association fibers that lie within the cingulate and parahippocampal gyri. It is at the core of the limbic lobe. It is involved, therefore, in the conscious perception of emotions. A surgical procedure to remove tissue in this area (cingulotomy) could tame a wild animal. In humans cutting certain fibers in this area (cingulectomy) has been used to treat depression, anxiety, and obsessive behavior.

    The best term to fill the blank space is:
    a. Inferior longitudinal fasciculus
    b. Superior longitudinal fasciculus
    c. Uncal fasciculus
    e. Cingulum

62. Connections between the ventral tegmental area of the midbrain through the medial forebrain bundle to the nucleus accumbens and the cortex are most often thought of as being involved in the experience of _____. (The nucleus accumbens is a region of the basal forebrain adjacent to the septum and anterior to the preoptic area.)

    The best term to fill the blank space is the:
    a. Pain
    b. Pleasure (reward)
    c. Displeasure (aversion)
    d. Rage

63. Neurofeedback studies have demonstrated that increasing alpha activity in the _____ compared to the other hemisphere, the so-called alpha asymmetry protocol, has a positive effect on depression. In this approach, the alpha is used as an inverse indicator of activation.

    The best term to fill the blank space is the:
    a. Left frontal cortex
    b. Right frontal cortex
    c. Left parietal cortex
    d. Right parietal cortex

64. The arterial supply from the vertebral basilar system is linked with the internal carotid system in a 'circle' at the base of the midbrain and diencephalon.
    a. True
    b. False

65. The _____ nervous system is involved in processes that activate and expend energy (catabolic). The cell bodies of the _____ system are located in the thoracic and lumbar regions of the spinal cord. The axons exit through the ventral roots of the spinal cord. The majority if these preganglionic fibers synapse in ganglia in the _____ chain or trunk or in ganglia of the prevertibral plexus.

    The best term to fill the blank spaces is:
    a. Parasympathetic
    b. Extrapyramidal motor
    c. Corticospinal
    d. Sympathetic

66. For the most part, the _____ nervous system is involved in processes that inhibit and those that produce or conserve energy (anabolic). The nerves of the _____ system leave the CNS in cranial nerves 3, 7, 9 and 10 and in sacral nerves 2, 3 and 4. The cell bodies or ganglia of the _____ system are located in close to the organ that they supply.

    The best term to fill the blank spaces is:
    a. Parasympathetic
    b. Extrapyramidal motor
    c. Corticospinal
    d. Sympathetic

67. The synapses within the _____ ganglia are acetylcholinergic. However, the terminal 'buttons' on the target organs are noradrenergic. An exception to this, which you might predict, is the innervation of the sweat glands, which is acetylcholinergic.

    The best term to fill the blank spaces is:
    a. Parasympathetic
    b. Extrapyramidal motor
    c. Corticospinal
    d. Sympathetic

68. The terminal buttons of both the pre- and postganglionic fibers of the _____ system are acetylcholinergic.

    The best term to fill the blank spaces is:
    a. Parasympathetic
    b. Basal ganglia
    c. Locus coeruleus
    d. Sympathetic

69. Episodic memory, or memory of a recent happening, is easier to recall if associated with emotions. These conscious memories are stored like films, and are thought to be encoded and held for about two to three years in the _____. (The _____ is part of the limbic system. It is situated in the medial temporal lobe. Profound stress has been shown to have a detrimental effect on the _____. _____ memory includes working memory.

    The best term to fill the blank spaces is:
    a. Putamen
    b. Hippocampus
    c. Amygdaloid
    d. Tegmentum

70. Alzheimer's dementia is due primarily to deterioration of the_____. Recent memory is gone and these patients often can't find their way around.

    The best term to fill the blank spaces is:
    a. Putamen
    b. Hippocampus
    c. Amygdaloid
    d. Tegmentum

71. The term _____ is applied to a condition found in alcoholics where there is an inability to form new memories (*anterograde amnesia*). The person can still recall old, long-term memories. It is associated with deterioration of the mammillary bodies and the dorsomedial nuclei of the thalamus.

    The best term to fill the blank spaces is:
    a. Senile dementia
    b. Korsakoff's
    c. Alzheimer's
    d. Amyotropic lateral sclerosis (ALS)

72. Some authors have hypothesized the following theoretical framework for the production of excess theta in ADHD. Decreased blood flow to - and metabolic activity in -, the cells in the frontal/prefrontal areas (including the **motor cortex)** may *lead to r*educed excitation by the cortex (Layer VI) of the inhibitory cells in the **putamen.** Two parallel pathways are then postulated. The first involves a direct effect of the putamen on the substantia nigra, while the second is more indirect, involving an external globus pallidus – subthalamus –(internal globus pallidus)-substantia nigra pathway. Either way, the substantia nigra would be released to increase its inhibition of which thalamic nucleus or nuclei?
    a. Ventral lateral (VL)
    b. Ventral anterior (VA)
    c. Centromedian (CM)
    d. All of the above

73. One hypothesis concerning a source of theta comes from animal (rat) studies and shows involvement of limbic areas. In these studies, theta is said to be paced by _____ pathways that project from the septal nuclei to the hippocampus. If true in humans, this finding would correspond to the observed involvement of theta in memory retrieval. Recall of words corresponds to synchronization of theta. This is thought to represent hippocampal theta. Which type of neurotransmitter is, for the most part, found in this pathway?
   a. Noradrenergic
   b. Acetylcholinergic
   c. Dopaminergic
   d. Serotonergic

Neurophysiology

74. For the most part, the process of myelinization in the cortex is generated by:
   a. Schwann cells
   b. Stellate cells
   c. Oligodendroglia cells
   d. Basket cells

75. Norepinephrine arises principally from neurons located in the locus coeruleus. It has connections through the medial-forebrain bundle to the hypothalamus. Its primary excitatory function in the central nervous system (CNS) is related to arousal and attention. It is released during stress and may be a part of the fight or flight response. It is also thought to have a role in learning and the formation of memories. Too little norepinephrine may be associated with depression and too much with:
   a. Mania
   b. Anxiety
   c. Fear
   d. All of the above

76. Serotonin (5-hydroxy-trypamine [5-HT]) is produced in the brainstem and released by the Raphe nuclei. It is primarily an inhibitory neurotransmitter. It is involved in the regulation of pain, mood, appetite, sex drive and in falling asleep. It may also be involved in memory. It is a precursor for melatonin, which in turn is important in biological rhythms. Low levels of serotonin are thought to be related to a number of psychiatric disorders including: obsessive compulsive disorder (OCD), aggression and:
   a. Mania
   b. Depression
   c. Schizophrenia
   d. ADHD

77. The amino acids group of neurotransmitters includes two inhibitory transmitters: gamma amino butyric acid (GABA) and glycine. It also includes glutamate and aspartate, which are an excitatory transmitters. The anxiolytic medications (benzodiazepines), alcohol, and barbiturates may exert their effects by potentiating the responses of GABA receptors. GABA will open both potassium and chloride channels and thus:
   a. Hypopolarize the neuron
   b. Hyperpolarize the neuron
   c. Neutralize the neuron
   d. None of the above

78. Glutamate is important in learning and memory and in an important process called *long-term potentiation (LTP)*. LTP is the process whereby which of the following is postulated to be occurring?
   a. The post synaptic cell changes (is enhanced) in response to a constant stimulus such that it can depolarize more in response to a neurotransmitter
   b. Depolarization of the postsynaptic membrane due to glutamate
   c. An influx of Ca2+ results in the activation of other 'messenger' pathways and the release, by the postsynaptic cell, of a 'paracrine'
   d. All of the above

79. Neuropeptides are short chains of amino acids. They are responsible for mediating sensory and emotional responses. Endorphins are neuropeptides. They function at the same receptors that receive heroin and morphine and are thought of as naturally occurring analgesics and euphorics. They are found in the limbic system and the midbrain. The ventral tegmental area of the midbrain and the nucleus accumbens in the frontal lobe have this type of receptors called opiate receptors.

   Another neuropeptide, *substance P,* mediates the perception of pain. Measurements of substance P in the cerebral spinal fluid (CSF) are assisting clinicians in the diagnostic work-up of persons suffering from pain associated with:
   a. Phantom limb
   b. Stroke
   c. Fibromyalgia
   d. Charcot Marie Tooth

80. Amphetamines and cocaine are catecholamine agonists. Their effect in the nucleus accumbens may be important in understanding the excitatory effects of these drugs and their ability to result in a chemical 'high.' (Alcohol, nicotine and caffeine can also increase one of the following neurotransmitters in this nucleus.) Which of the following statements is most true concerning the effect of amphetamines and cocaine?
    a. They increase the reuptake of dopamine and noradrenaline
    b. They block the reuptake of dopamine and noradrenaline
    c. They increase the effects of GABA (gamma amino butyric acid)
    d. They block the effect of serotonin at the post synaptic membrane

81. When cats are trained to increase their 12-15 Hz activity (SMR), which of the following is **NOT** true?
    a. There is an increase in the SMR in the baseline EEG
    b. There is a reliable increase in sleep spindle density and decreased awakenings during non-REM sleep
    c. Activity in the polysynaptic digastric muscle is reduced (this is the jaw opening muscle and it lacks muscle spindle afferents)
    d. Neck muscle activity is suppressed in strict association with emergence of the sensorimotor rhythm (SMR)

82. The suppression of neck muscle activity in cats, without posture change and occurring abruptly just prior to the appearance of SMR bursts (12 to 15 Hz.), suggests there is:
    a. A specific reduction in muscle tone
    b. A change in length of the innervated muscle
    c. A change in gamma motor neuron activity
    d. a and c

83. It has been demonstrated in cats that, during SMR activity;
    a. Cell discharge rates in the red nucleus are suppressed
    b. Ventrobasal thalamic nucleus firing bursts increase
    c. The thalamus produces synchronized, recurrent activity bursts that are observed in the EEG as 12-14 Hz spindles
    d. All of the above

84. It has been demonstrated in cats that during SMR activity:
    a. Ventrobasal (VB) thalamic relay cells change their behavior to recurrent, oscillatory bursting
    b. Bursting discharge of the ventrobasal thalamic cells results in hyperpolarization of the relay cells
    c. There is attenuation (decrease) of the conduction of somatosensory information to the cortex
    d. All of the above

85. During SMR activity there is:
    a. Decreased motor excitability
    b. Decreased motor reflex excitability
    c. Decreased muscle tone
    d. All of the above

86. Motionlessness in the context of attention produces altered motor output to thalamus and brainstem resulting in:
    a. Decreased red nucleus activity
    b. Decreased stretch reflex activity
    c. Decreased muscle tone
    d. All of the above

87. Operant conditioning to increase SMR (12-14 Hz activity) can:
    a. Produce sustained changes with respect to increased sleep spindle density
    b. Destabilize sleep states
    c. Inhibit ruminative thinking
    d. All of the above

88. Epileptic syndromes have in common the favoring of recurrent abnormal and excessive synchronous discharge in neuronal populations. Therefore, effective intervention, such as operant conditioning of SMR (12 to 14 Hz), might be expected to:
    a. Reduce neuronal excitability in relevant tissues
    b. Blunt the impact of transient neuronal discharge
    c. Stabilize state characteristics
    d. All of the above

89. When an action potential depolarizes the membrane of an axon's presynaptic terminal, it triggers the influx of a cation that initiates a series of steps culminating in the release of neurotransmitter into the synaptic space. This cation is:
    a. K+
    b. Na+
    c. Ca+
    d. Mg+

90. If you think that you may be observing the wicket shape of a Mu rhythm at C3 or C4, you can test to see if it can be suppressed by voluntary movement such as closing the fist on the opposite side of the body from the observed rhythm or by asking the client to open his eyes and watching to see if the rhythm disappears. Mu doesn't disappear when the client opens his eyes.

    Testing to see if the rhythm is Mu or alpha is important because:
    a. Mu rhythm indicates a seizure disorder
    b. Mu rhythm indicates a space occupying lesion
    c. If the rhythm doesn't disappear and is of high amplitude it could be alpha and correspond to difficulties with attention span
    d. Mu rhythm is usually observed at Pz and is not observed at C3 and C4

91. A wicket shape rhythm at frequency of about 7-11 Hz at C3 and/or C4, which is suppressed by voluntary movement such as closing the fist on the opposite side of the body from the observed rhythm and which is not suppressed by asking the client to open their eyes and watching to see if the rhythm disappears, is called:
    a. Lambda
    b. K-complex
    c. Alpha
    d. Mu

Neurophysiology – Nerve Transmission - Origin of the EEG

92. The speed of transmission in a nerve fiber is increased by:
    a. Increased axon diameter
    b. Salutatory conduction
    c. Myelinization
    d. All of the above

93. For the most part, the process of myelinization in the cortex is generated by:
    a. Schwann cells
    b. Stellate cells
    c. Oligodendroglia cells
    d. Basket cells

94. In the spinal cord, brain stem and midbrain (as opposed to the cerebral hemispheres) the white matter is:
    a. Medial to the gray matter
    b. Lateral to the grey matter
    c. Mixed fairly evenly with the gray matter
    d. None of the above

95. Comparing EEG and magnetic resonance imaging (MRI) techniques, which of the following is true?
    a. The EEG has better temporal resolution than MRI
    b. MRI has less spatial resolution
    c. Traditional EEG procedures and MRI have about the same spatial resolution
    d. None of the above

96. If thalamocortical connections are cut, the brain is most likely to produce:
    a. Delta
    b. Theta
    c. Alpha
    d. Beta

97. Changes in cortical loops (patterns of firing - connecting - among neurons) can change the firing rate of thalamic pacemakers and hence change their intrinsic firing pattern. It is also noted (Lubar) that changes in this intrinsic firing pattern may correspond to changes in mental state. This process may be facilitated by:
    a. Learning
    b. Emotion
    c. Neurofeedback
    d. All of the above

98. The pyramidal cells and their surrounding support cells (stellate and basket cells) are arranged in groups. Each vertical column contains hundreds of pyramidal cells. The columns are parallel to each other and at right angles to the surface of the cortex. Many adjacent groups may receive the same afferent axonal input and thus fire in unison, allowing for a sufficiently large potential to be measurable at the surface of the scalp. These 'groups' may be called:
    a. Pyramidal formations
    b. Cortical loops
    c. Macrocolumns
    d. Oligodendroglial resonant groups

99. A standard electrode on the scalp measures the activity of what area of cortex underneath it?
    a. 0.5 cm^2
    b. 2 cm^2
    c. 6 cm^2
    b. 12 cm^2

100. The cortex works in terms of three major resonant loops. They are:

Local: This loop of electrical activity is between macrocolumns. It appears to be responsible for high-frequency (>30 Hz) gamma activity.

Regional: This loop of electrical activity is between macrocolumns which are several centimeters apart. It appears that this activity is in the range of intermediate frequencies: alpha and beta.

Global: This loop of electrical activity is between widely separated areas, for example, frontal-parietal and frontal-occipital regions. Areas can be 7 cm apart. The activity produced is in the slower frequency range of delta and theta activity.

All three of these resonant loops:
a. Cannot operate independently of thalamic pacemakers
b. Can operate spontaneously or be driven by thalamic pacemakers
c. Only operate as independent entities without thalamic pacing
d. Can only operate with the influence of thalamic and other diencephalic and midbrain pace-making activity

101. The wicket shape of a Mu rhythm is usually observed at
a. F3 and/or F4
b. C3 and/or C4
c. P3 and/or P4
d. O1 and/or O2

102. If you observe the wicket shape of a Mu rhythm at C3 or C4, you can check if it is truly a Mu rhythm by observing if it can be:
a. Increased in amplitude when the client closes their fist on the opposite side of the body to the observed rhythm
b. Decreased in amplitude when the client closes their fist on the same side of the body to the observed rhythm
c. Increased in amplitude when the client closes their fist on the same side of the body to the observed rhythm
d. Decreased in amplitude when the client closes their fist on the opposite side of the body to the observed rhythm

103. In broad general terms, the approximate percent of communication activity in the human brain is:
a. 50% between cortical areas and 30% thalamocortical
b. >90% between cortical areas and 5% thalamocortical
c. 20% between cortical areas and 60% thalamocortical
d. 45% between cortical areas and 45% thalamocortical

104. On the recorded EEG, similarly appearing waves occurring at the same moment in time at nonadjacent sites probably originate from the same generator. If there is a time delay, then this probably indicates:
a. A synapse and therefore a different cellular origin
b. Cellular damage
c. A space-occupying lesion
d. An arteriovenous malformation

105. It has been noted (Lubar), "Neocortical states associated with strong corticocortical coupling are called hypercoupled and are associated with global or regional resonant modes." Hyper-coupling means that there are large resonant loops. Biochemically, it appears that the dominant neurotransmitter in this type of coupling is:
a. Dopamine
b. Acetylcholine
c. Serotonin
d. Noradrenalin

106. Corticocortical hypercoupling is said to be appropriate for states such as:
a. Focused attention
b. Reading and math
c. Hypnosis and sleep
d. None of the above

107. Corticocortical hypocoupling is associated with small regional and local resonant loops and thus higher frequencies. Biochemically it appears that the neurotransmitters principally involved in this type of coupling are:
a. Acetylcholine, norepinephrine and dopamine
b. Acetylcholine and gamma amino butyric acid (GABA)
c. Serotonin, GABA and dopamine
d. Noradrenalin, glycine and serotonin

108. Corticocortical hypocoupling is said to be appropriate for:
    a. Complex mental activity and increased attention
    b. Rapid eye movement (REM) sleep
    c. Hypnosis and meditation
    d. Stage III sleep

Waveforms and the EEG Spectrum - Correspondence of Bandwidths to Mental States

109. The transformation from the *time-related domain* of the raw EEG to the *frequency domain* for statistics is carried out by a mathematical calculation that is based on the fact that any signal can be described as a combination of sine and cosine waves of various phases, frequencies and amplitudes. The breaking apart of a complex wave into its component sine waves may be called a:
    a. Nyquist principle
    b. Fast Fourier Transform (FFT)
    c. Common average referencing
    d. Helmholtz solution

110. Observing the spectrum of an EEG recording (Fast Fourier Transform, or FFT) without the raw EEG may be deceptive because:
    a. The FFT shows too many artifacts
    b. A few very high-amplitude waves may give the same value as extensive lower-amplitude bursts of the same frequency
    c. Low amplitudes on the FFT may correspond to a combination of muscle and eye-blink artifacts
    d. All of the above

111. Morphology of a waveform refers to:
    a. The amplitude of a wave
    b. The power of a wave
    c. The shape of a wave
    d. The ratio of positive to negative power of the wave

112. A complex wave form is:
    a. Usually monophasic but not triphasic
    b. A sequence of two or more waves that is repeated and recurs with a reasonably consistent shape
    c. A sequence of six or more waves that is repeated and recurs with a reasonably consistent shape
    d. A sequence of two or more waves that is repeated and recurs with an inconsistent shape and amplitude

113. A "transient" wave is one that stands out as:
    a. Similar to the background activity of most of the EEG
    b. Pathological
    c. Different against the background EEG
    d. Evidence of a seizure disorder

114. Waves of similar morphology and frequency that are not in phase may be said to have a time delay. This delay may be expressed as a *phase angle*. If the peaks point in opposite directions it would be called a *phase reversal* and the peaks would have a phase angle of:
    a. 45°
    b. 90°
    c. 180°
    d. 360°

115. If the same kind of waves occurred simultaneously on both sides of the head (different channels) and in a constant time relationship, they would be said to be:
    a. Out of phase and asynchronous
    b. In phase and bisynchronous
    c. Asymmetrical
    d. Lateralized

116. Normal alpha rhythm frequency should exceed 8 Hz. Its amplitude may be higher on the right, but the right to left difference should not exceed:
    a. 1.5 times
    b. 2 times
    c. 2.5 times
    d. 3 times

117. If alpha never exceeds 8 Hz in an adult, this is probably abnormal and a difference of 1 Hz in frequency of the dominant alpha rhythm between the two hemispheres indicates an abnormality:
    a. In the basal ganglia
    b. In the hemisphere with the higher frequency
    c. In the hemisphere with the lower frequency
    d. In the ipsilateral locus coeruleus

118. Normal alpha rhythm is found predominantly in which region?
    a. Anterior (frontal)
    b. Central
    c. Lateral (temporal)
    d. Posterior (occipital)

119. Beta activity is found >13 Hz in adults. It is almost always a sign of normal cortical functioning. Asymmetry in beta between hemispheres should be no more than a certain percentage of the amplitude of the side with the higher amplitude. If the difference is greater than this, then the side with the lower amplitude may be abnormal. What percent difference is the maximum allowable before it may be considered abnormal?
    a. 10%
    b. 22%
    c. 35%
    d. 50%

120. Amplification is rated in terms of *sensitivity* and *gain*. Thus amplifiers have known *sensitivities* that are recorded as a number of μv/mm. The need to use a higher sensitivity indicates:
    a. Low-amplitude recording
    b. High-amplitude recording
    c. Higher frequency waves will be exaggerated
    d. The low-pass filter will function abnormally

121. 'Gain' refers to a ratio of the voltage of a signal at the output of the amplifier to the voltage of the signal at the *input* of that amplifier. Thus a gain of $10V/10\mu V$ = 1 million. You will see *gain* mentioned in the specifications for your amplifier. It will probably be recorded in *decibels*. A gain of 1 million (or 10^6) is:
    a. 20 decibels
    b. 60 decibels
    c. 120 decibels
    d. 200 decibels

122. Wave forms characteristic of an absence (petit mal) seizure are:
    a. Bursts of high-amplitude spikes followed by spike and wave complexes
    b. Focal bursts of two-per-second spike and wave 'pairs'
    c. Generalized three-per-second spike and wave 'pairs'
    d. Uniform six-per-second spike and wave 'pairs'

123. Sharp waves do not have as sharp a point as spikes. They have a duration of:
    a. 30-70 ms
    b. 70-200 ms
    c. 100-300 ms
    d. 200-400 ms

124. *Complexes* containing spikes and sharp waves that are repeated may represent *interictal* (between seizures) epileptiform activity.
    a. True
    b. False

125. The term 'paroxysmal hypnogogic hypersynchrony' refers to synchronous, slightly notched, sine waves that are higher in amplitude than surrounding waves and run at about 3-5 Hz. The burst may last for a couple of seconds.

    This type of wave:
    a. Indicates a seizure disorder
    b. Is a warning of a space-occupying lesion
    c. May be observed in normal children when they are sleepy
    d. Is a particularly prominent finding in Asperger's syndrome

126. Sleep spindles look like SMR spindles and are in the same frequency range: 12-15 Hz. Like SMR, they are maximal over the central regions. Apparently they arise from a different generator. They are frequently observed:
    a. Only in sleep (Stage II)
    b. During sleep and during physical exercise
    c. Only in rapid eye movement (REM) sleep
    d. In autistic children both when awake and asleep

127. K complexes are sharp, negative, high-amplitude waves followed by a longer duration positive wave. They are seen:
    a. In rapid eye movement (REM) sleep
    b. In Stage II sleep
    c. In ADHD children when awake
    d. As interictal activity in seizure disorders

128. The tentorium is a large infolding of dura matter (connective tissue) separating:
    a. The two cerebral hemispheres
    b. The cerebellum below from the cerebrum above
    c. The corpus callosum from the diencephalon
    d. The precentral gyrus from the postcentral gyrus

129. Deep lesions (e.g., in the internal capsule) generally demonstrate:
    a. Focal delta
    b. Hemispheric or bilateral delta.
    c. Focal beta spindling
    d. Diffuse beta spindling

130. In an alert, resting adult, delta waves may appear intermittently in bursts and be higher in amplitude than surrounding wave forms. This phenomenon may be due to:
    a. Diffuse white matter damage
    b. Diffuse gray matter damage
    c. Dopamine overactivity
    d. Noradrenergic overactivity

131. In relative terms, the communication going on in our brain between cortical areas compared to thalamocortical connections is, in very general terms, approximately:
    a. Cortical-cortical 5% - thalamocortical 90%
    b. Cortical-cortical 50% - thalamocortical 40%
    c. Cortical-cortical 25% - thalamocortical 70%
    d. Cortical-cortical 90% - thalamocortical 5%

132. Which of the following is a mathematical process that looks at surface EEG information and infers what activity is occurring in areas a little deeper in the cortex?
    a. PET (Positron Emission Tomography)
    b. MRI (Magnetic Resonance Imaging)
    c. LORETA (Low-Resolution Electromagnetic Tomographic Assessment)
    d. SPECT (Single Photon Emission Computerized Tomography)

133. An event related potential (ERP) is a measure of:
    a. Brain electrical activity that occurs as a response to a specific stimulus
    b. Spontaneous and ongoing activity of the brain
    c. A brain response unrelated to any specific stimulus
    d. A brain response that is time-independent

134. With Attention Deficit Disorder, the slowing of the EEG parallels decreased _____ recorded by Positron Emission Tomography (PET scans) and decreased blood flow on Single Photon Emission Computerized Tomography (SPECT) scans in the frontal region.

    The best phrase to fill the above blank is:
    a. PCO_2 levels
    b. Glucose metabolism
    c. Adenosine triphosphate (ATP) levels
    d. Creatinine metabolism

EEG and Electrophysiology

135. A microvolt is:
    a. One tenth of a volt
    b. One hundredth of a volt
    c. One thousandth of a volt
    d. One millionth of a volt

Wave forms - Morphology

136. In which age group is theta the dominant wave form?
    a. 3-7 years
    b. 8-16 years
    c. 17-23 years
    d. 24-30 years

137. Match the appropriate frequency band in Column 1 (letter "a" to "j") with the best description of expected corresponding mental state in Column 2 by putting the appropriate letter from Column 1 beside the answer in Column 2.

Example:

| a. Butter is placed on: | _(b)_ Hamburgers |
| b. Relish is placed on: | _(a)_ Bread |

Column 1	Column 2
a. Delta activity, 0.5-3 Hz	___ Found in sleep and also in conjunction with learning disabilities and brain injury
b. Theta waves, 4-7 Hz	___ Seen in drowsy states, which are also states in which some quite creative thoughts may occur
c. Low alpha, 8-10 Hz	
d. High alpha, 11-12 Hz	___ May correspond to jaw, scalp, neck or facial muscle activity
e. Sensorimotor rhythm frequencies, 13-15Hz over the sensorimotor cortex	___ Found in dissociative states, some kinds of meditation, and tuning out from external stimuli (daydreaming)
f. Low beta waves, 13-15 and 16-20 Hz	
g. Higher beta frequencies (20-23 Hz)	___ Often called 'Sheer' rhythm or a 'binding' rhythm. Corresponds to cognitive activity
h. 24-33 Hz	
i. 38-42 Hz	___ Can be found associated with creative reflection as well as relaxed calm states of optimal performance
j. 52-58 Hz	
Note: There is overlap in the frequency bands (theta is often defined as 4-8 Hz, for example, but experts often use 4-7 or 3-7 Hz) and there are also shifts with age, moving to the right along the spectrum. In young children, for example, alpha wave forms may be found at a frequency of 7 Hz.	___ Implies being motorically calm with reflection before action
	___ Associated with singular focus, external orientation and problem-solving
	___ May be elevated with anxiety
	___ May be associated with rumination

138. Increased activity around 40 Hz (39-41 Hz) has been found experimentally to be associated with:
    a. Dissociation
    b. Catching one's balance
    c. Intense emotion
    d. Advancing age

139. The beta frequency range, 38-42 Hz, is sometimes referred to as 'Sheer' rhythm after David Sheer who did some work in the 1970s concerning enhancing 40 Hz activity. One of the effects of raising this particular rhythm has been observed to correspond to:
    a. Decreased learning capacity
    b. Improving attention where the subject is bringing together different (binding) aspects of an object into a single percept
    c. Increased appetite
    d. Low muscle artifact activity

140. 'Beta spindling' refers to bursts of beta waves in a rising and falling spindle-like pattern. Although it can be less than 20 Hz, most spindles involve fast beta, above 20 Hz. They can be associated with epileptic auras. It may be due to a disease process and cortical irritability. It can be seen in ADHD and in obsessive compulsive disorders. When neurofeedback (NFB) is used for this pattern, the logical NFB intervention would be to:
    a. Counterbalance the increased high-beta amplitude on one side of the head by training up the same frequency of beta at the corresponding site on the opposite side
    b. Train down that range of beta in the area involved
    c. Nullify the effects of this disproportionate beta activity by training up theta rhythm on the same site
    d. Counterbalance the effects of this disproportionate beta activity by training up alpha rhythm on the same side.

141. **Mu** waves may fool you. They look like alpha waves and are usually found in the 7-11 Hz frequency range. They are usually observed at C3 and C4. When the subject opens her eyes:
    a. Both alpha and Mu are blocked
    b. Alpha is reduced, but Mu will remain in the central region
    c. Frontal alpha is increased and Mu disappears from the central region
    d. Occipital alpha remains steady and Mu will increase in the central region

142. **Spikes:** These waves have a duration of:
    a. 10-20 ms
    b. 20-70 ms
    c. 70-100 ms
    d. 90-120 ms

143. *Focal epileptiform* activity often consists of localized spikes and sharp waves. This activity may be surrounded by irregular slower waves or followed by an aftergoing slow wave. Focal spike and sharp waves may appear before and after a generalized discharge. These spikes and sharp waves are:
    a. Always generalized and seen throughout all channels
    b. Are generally just observed in a single channel
    c. Are often seen just in a few neighboring electrodes
    d. Are most often seen at Cz and Fz

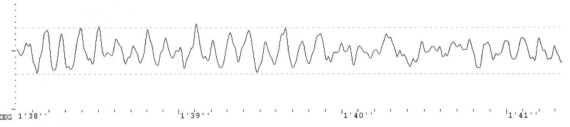

144. The dominant wave form in the first one-second section of the above EEG graph is:
    a. Delta
    b. Theta
    c. Alpha
    d. Beta

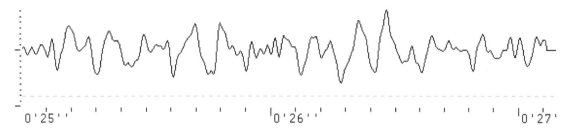

145. The dominant wave form in the first one-second EEG recording above is:
    a. Delta
    b. Theta
    c. Alpha
    d. Beta

Eyes open looking at a dot

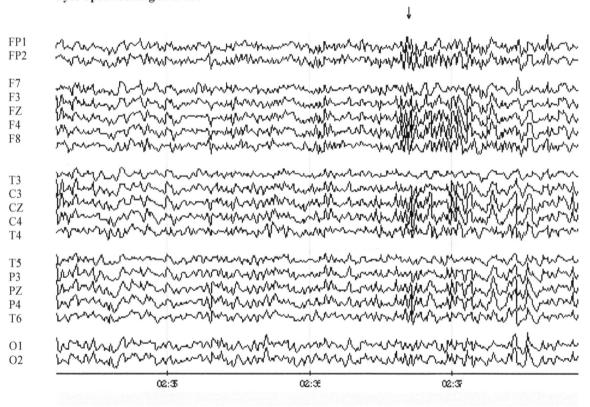

146. In the figure above, the group of waves in the prefrontal and frontal leads, under the arrow, that are higher in amplitude than the foregoing EEG, are called:
    a. Alpha
    b. Spike and wave
    c. Sharp waves
    d. Sheer rhythm

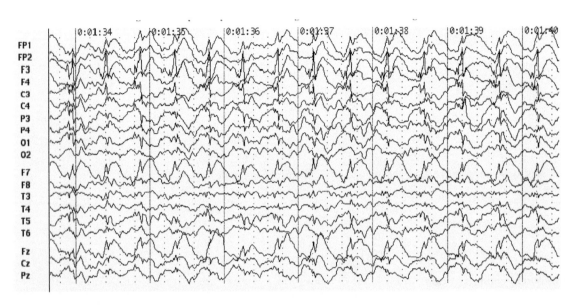

147. The group of waves seen best at F3 in the EEG graph above are called:
    a. Alpha
    b. Spike and wave
    c. Sharp waves
    d. Beta

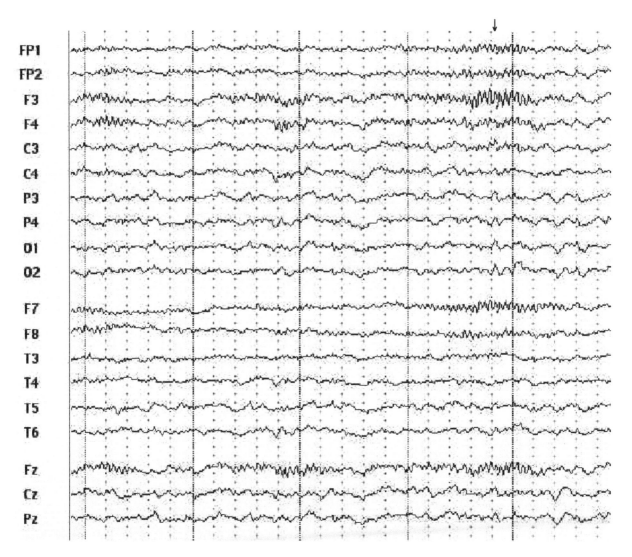

148. The wave group seen best at F3, under the arrow in the EEG graph above, is termed:
    a. Spike and wave
    b. Epileptiform spike paroxysm
    c. Alpha transient
    d. Beta spindling

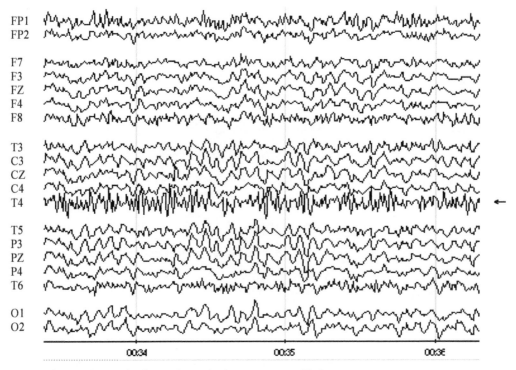

149. The waves observed at T4 in the EEG graph above are most likely:
    a. Epileptiform spikes
    b. Absence seizure waves
    c. Muscle artifact
    d. Paroxysmal burst of fast beta

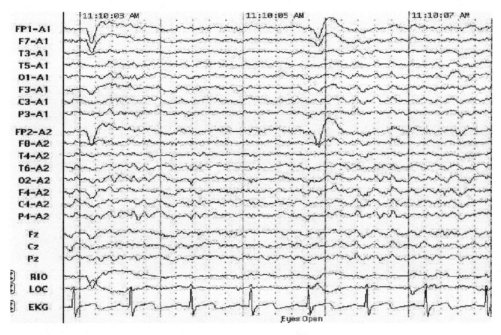

150. What has most likely caused the large excursions in the frontal leads in the EEG graph above?
    a. Muscle tension
    b. Raising the eye brows
    c. Eye blink
    d. Myoclonic seizure bursts

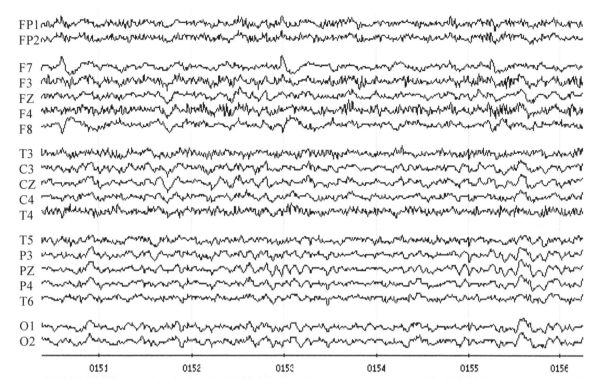

151. In F7 and F8 above, large deviations of the wave pattern are seen just prior to 0151 and again at 0153. These are most likely due to:
   a. Eye blink
   b. Isolated absence seizure (petit mal) bursts
   c. Lateral eye movement when reading
   d. Digastric muscle movement

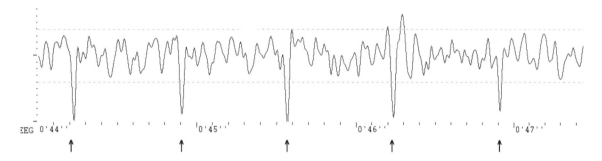

152. The large amplitude waves (arrows) shown above are most likely:
   a. Absence seizure activity
   b. Eye movement
   c. Cardiac artifact
   d. Muscle activity when speaking

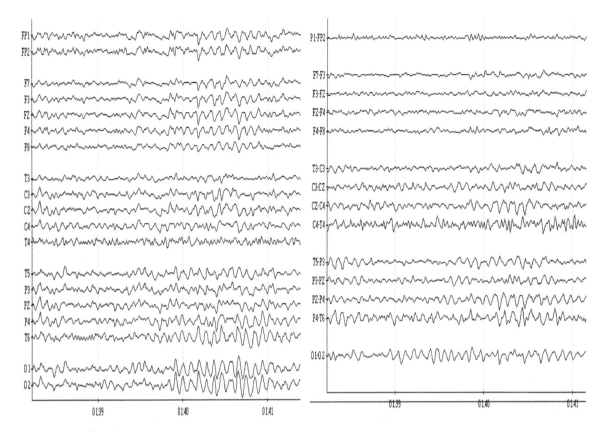

153. The referential and sequential montages above show the same patient's EEG at the same moment in time. The origin of the activity observed in the FP1 and FP2 leads in the left diagram, most likely should be interpreted as being evidence of:
   a. Prefrontal theta
   b. Prefrontal alpha
   c. Temporal alpha
   d. Paroxysmal alpha bursting

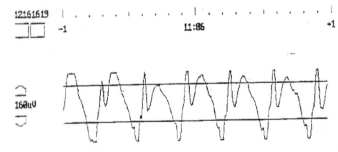

Amplitude >160 μV, two seconds of recording in a single-channel referential recording, Cz to the left ear with right-ear ground.

154. The wave form observed above is most likely the result of:
   a. Grand mal seizure
   b. Partial complex seizure
   c. Absence seizure
   d. Evenness suggesting an electrical artifact

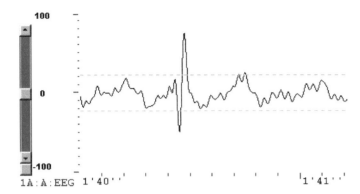

The electrode was placed 1 cm above a point, 1/3 of the way along a line drawn from external ear canal to the lateral canthus of the eye in a patient with a known partial complex seizure disorder. (This is an appropriate placement for picking up interictal activity). There were no similar waves in the next 10 seconds of the recording. The reference was to linked ears.

155. The very large amplitude wave is most likely:
   a. Spike indicating a possible seizure disorder
   b. Sharp wave transient
   c. Lateral rectus muscle spike
   d. Wave representing myoclonic seizure activity

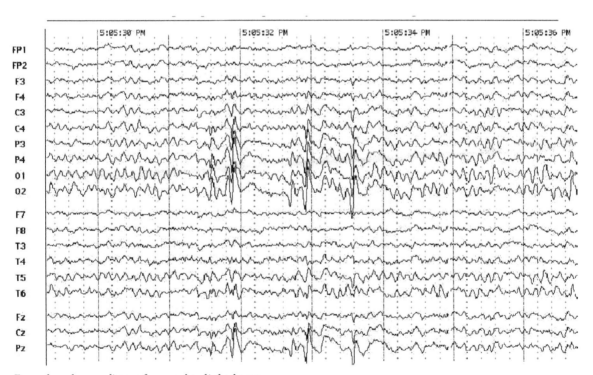

Eyes closed recording referenced to linked ears.

156. The bursts of higher amplitude waves shown above are most likely:
   a. Absence seizure complexes
   b. Normal alpha wave bursts
   c. Epileptiform paroxysmal discharge
   d. Lateral rectus muscle spikes

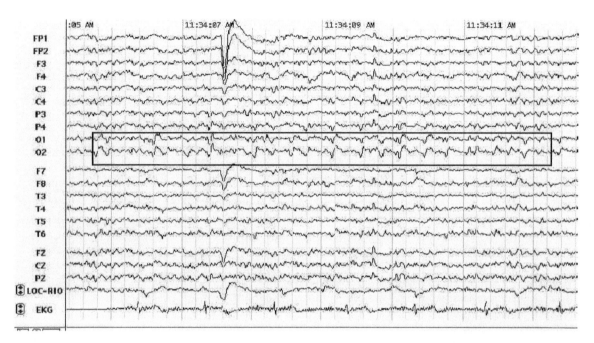

157. In the figure above the box encloses an area where repeated downward (positive) sharp deflections are seen repetitively in the occipital leads. Which of the following is the best term that could describe these?
    a. Mu waves
    b. Absence seizure complexes
    c. Eye movement artifact
    d. Lambda

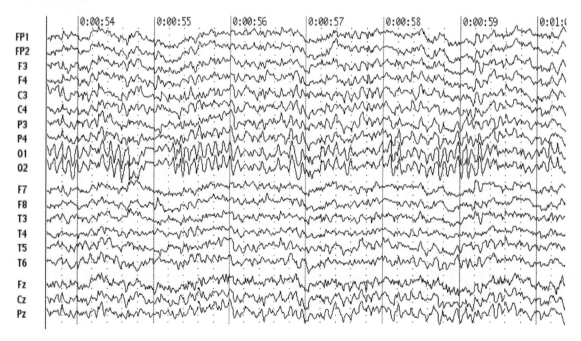

158. In this recording made with the Lexicor Neurosearch 24 using a linked-ear reference and the high-pass filter on, the large rolling excursions observed particularly in the prefrontal leads are most likely due to:
    a. Raising the eyebrows
    b. Sweat artifact
    c. Repetitive eye blinking
    d. Eye movement with reading

Laplacian Montage

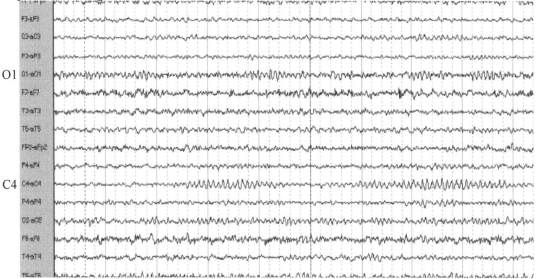

159. In the EEG (Laplacian montage) above, note the wicket shape activity at C4. This type of activity is termed:
    a. Lambda
    b. Mu
    c. Sharp wave transient
    d. Paroxysm

160a. Deep lesions (e.g., in the internal capsule) are likely to produce:
    a. Focal delta
    b. Hemispheric or bilateral delta
    c. Focal beta spindling
    d. Diffuse beta spindling

160b. Superficial lesions in the cerebral cortex are likely to produce:
    a. Focal delta
    b. Hemispheric or bilateral delta
    c. Focal beta spindling
    d. Diffuse beta spindling

161. In the awake subject, what would sudden, brief, increases in frontal 1-2 Hz activity usually represent?
    a. Muscle tension in the jaw or neck muscles
    b. Increased problem-solving cognitive activity
    c. Eye blink
    d. A meditative state

162. What frequency range would you first expect to see increasing with muscle tension in the jaw, neck, or forehead?
    a. 4-8 Hz
    b. 8-12 Hz
    c. 13-17 Hz
    d. 18-25 Hz

163. When a sequential placement is used with electrodes at Site A and Site B on the scalp, a decrease in 4-8 Hz (theta) activity, measured in microvolts, could result from:
    a. A decrease in 4-8 Hz voltage of the same magnitude at both sites, A and B
    b. A disproportionate increase in 9-12 Hz (alpha) activity
    c. Site A is more in phase with Site B
    d. Movement artifact affecting both sites equally

164. Which sites, using the 10-20 electrode placement system, would show the highest levels of sensorimotor rhythm (SMR) activity during eyes-open conditions?
    a. F3 and F4
    b. C3 and C4
    c. P3 and P4
    d. O1 and O2

165. Which sites, using the 10-20 electrode placement system, would show the highest levels of alpha rhythm activity (8-12 Hz) during eyes-closed conditions?
    a. F3 and F4
    b. C3 and C4
    c. P3 and P4
    d. O1 and O2

166. When you reinforce 12-15 Hz activity at C4, what types of brain wave activity are you encouraging?
    a. Sensorimotor rhythm (SMR), high alpha and gamma
    b. High alpha and epsilon spindles
    c. Sensorimotor rhythm (SMR), high alpha and low beta
    d. Sensorimotor rhythm (SMR), 'J' complexes and desynchronous beta

167. Fifteen minutes of sound and light stimulation in an alpha frequency may result in the following frontal EEG changes 30 minutes after the stimulation has ceased:
    a. Increased theta 4-8 Hz
    b. Increased alpha 8-12 Hz
    c. Decreased alpha 8-12 Hz
    d. Increased beta 13-21 Hz

168. Fifteen minutes of sound and light stimulation in twice the dominant frequency (usually beta 20-21 Hz), has been shown to result in the following frontal EEG changes 30 minutes after the stimulation has ceased:
    a. Increased theta 4-8 H
    b. Increased alpha 8-12 Hz
    c. Decreased alpha 8-12 Hz
    d. Increased beta 13-21 Hz

Electrophysiology

169. In the resting state, neurons have what is called a *resting potential*. The resting potential represents the potential difference between the inside and the outside of the cell. When at rest, this potential inside the cell is about how many millivolts (mV)?
    a. –20 mV
    b. +20 mV
    c. –70 mV.
    d. +70 mV

170. *The EEG is defined as the difference in voltage between two different recording locations plotted over time.* The pyramidal cell postsynaptic potentials from large groups of cortical pyramidal cells form an *extracellular dipole layer*. The EEG is generated by the synchronous activity of postsynaptic:
    a. Inhibitory potentials
    b. Excitatory potentials
    c. Inhibitory and excitatory potentials
    d. Action potentials generated by the pyramidal cells

171. Action potentials that travel down the axons or dendrites of these cortical cells have a very short time duration (1ms). Postsynaptic potentials have a long time duration (15-200 ms). These potential changes summate, and the EEG records the potential (+ve or –ve) directed towards the electrode on the surface of the scalp.

    The charge will differ depending on whether an excitatory postsynaptic potential (EPSP) or inhibitory postsynaptic potential (IPSP) has been generated in the area of the cortex beneath the electrode. To obtain a reliable scalp EEG recording how many cm^2 of cortex with predominantly synchronous activity is necessary?
    a. $0.5\ cm^2$
    b. $2\ cm^2$
    c. $6\ cm^2$
    d. $10\ cm^2$

172. *The influx of sodium into the neuron at the postsynaptic membrane site makes for what is known as* an *active sink at the level of the synaptic input from another cell's axon.*

    If the simultaneous discharges of axon terminals result in excitatory postsynaptic potentials (EPSPs) at the distal end of a sufficient number of adjacent pyramidal cell dendrites, what charge would be detected by a scalp electrode placed over that area of the cortex and referenced to a second neutral site?
    a. Negative
    b. Positive

173. The pyramidal cells and their surrounding support cells and stellate and basket cells are arranged in groups called macrocolumns. In a macrocolumn, neurons are arranged in a group of cells several mm in diameter and 4-6 layers deep.

    Each vertical column contains hundreds of pyramidal cells. The columns are parallel to each other and at right angles to the surface of the cortex. Each pyramidal cell may have thousands of synapses. The EEG electrode is detecting electrical activity from:
    a. Single pyramidal cells
    b. The extracellular charge surrounding one macrocolumn
    c. The action potentials produced by the pyramidal cells of many adjacent macrocolumns that receive the same afferent axonal input and thus fire in unison
    d. The extracellular charge surrounding the tips of the dendrites (near the scalp) of many adjacent macrocolumns

174. Anatomically, it has been found that, of the fibers entering the gray matter of a hemisphere in the cerebral cortex, the percentage that arise from the thalamus is:
    a. Less than 5%
    b. About 10%
    c. About 15%
    d. Greater than 25%

175. Coherence reflects phase-locked activity between two sites. Which of the following statements about coherence is NOT correct?
    a. Coherence is a measure of shared activity at two different locations
    b. Coherence between two sites can be evidence of a functional connection between cortical areas
    c. As coherence increases within an interhemispheric region, it will also increase between the intrahemispheric regions
    d. Decreased frontal coherences may be related to increased cortical differentiation

176. Hypocoupling favors small regional and local cortical loops that lead to faster frequencies, including beta and gamma (especially 40 Hz) activity, which are said to be important for detailed information acquisition, complex mental activity and increased attention. Such hypocoupling may be related to an increase in which neurotransmitter(s)?
    a. Acetylcholine, norepinephrine and serotonin
    b. Norepinephrine, dopamine and serotonin
    c. Acetylcholine, dopamine and norepinephrine
    d. Acetylcholine, dopamine and serotonin

177. Hypercoupling of large resonant loops that leads to slower frequencies, sleep spindles, theta and delta activity and decreased attention, is related to an increase in which neurotransmitter?
    a. Serotonin
    b. Dopamine
    c. Acetylcholine
    d. GABA (gamma amino butyric acid)

178. Very low coherence between two sites (lower than a recognized statistical database) implies that these areas are:
    a. Functionally highly connected
    b. Functionally disconnected
    c. Not differentiating related functions properly
    d. Connected by nonmyelinated fibers

179. Very high coherence (greater than a recognized statistical database) between two sites that are expected to be related in function suggests:
    a. The individual has high intelligence
    b. The two areas are functionally disconnected
    c. The two areas are not differentiating related functions properly
    d. The individual is highly focused on a task

180. Known resonances of brain electrical activity include:
    a. Local, between macro columns of cells producing EEG frequencies above 30 Hz (gamma)
    b. Regional, between macro columns that are several centimeters apart, producing EEG frequencies in the alpha and beta range
    c. Global, between widely separated areas (e.g., frontal and occipital) producing frequencies in the delta and theta range
    d. All of the above

181. An event-related potential (ERP) is a measure of brain electrical activity which occurs as a response to a specific stimulus. It is usually defined as being *time-locked* to a specific stimulusThere are exceptions to this. For example an ERP can be found at the exact time when a stimulus was expected but when there was no actual external stimulus present.
    a. True
    b. False

182. Most often event related potentials (ERP)s are measured at FZ, CZ, and PZ and amplitude and scalp distribution are among the variables measured. Amplitudes are usually:
    a. Highest frontally and lowest in the parietal region
    b. Highest in the parietal region and lowest frontally
    c. The same frontally and parietally
    d. Highest in the central region

183. Research has shown that event-related potentials (ERP)s can distinguish between different clinical conditions, so they are used in diagnosis. The most common application is the use of ERPs by:
    a. Dentists to identify dead nerve tissue
    b. Dermatologists to identify malignancy
    c. Audiologists to test hearing
    d. Ophthalmologists to detect retinal lesions

184. Most event-related potential (ERPs) are only made visible by averaging many, many samples (at least 20, and sometimes several hundred samples). Specific ERPs, in a given individual, come a set time interval after the stimulus and are always:
    a. Spike and wave morphology
    b. The same wave form
    c. A different wave form
    d. Three-per-second waves

185. It is acceptable to think of four types of event-related potentials (ERP): sensory, motor, long-latency potential and steady-potential shifts (as discussed by Vaughn). The sensory ERPs are those evoked by sight, sound, smell and touch. Auditory ERPs occur with a negative peak at about 80-90 ms, and a positive peak at about 170 ms after the stimulus. It is called the N1-P2 complex. This is most easily observed:
    a. In the frontal cortex, Fz
    b. In the auditory cortex in the temporal lobe
    c. In the parietal cortex at Pz
    d. In the occipital cortex

186. Motor event-related potentials (ERP)s precede and accompany motor movement and are proportional in amplitude to the strength and speed of muscle contraction. They are seen in the:
    a. Limbic cortex
    b. Precentral area
    c. Prefrontal area
    d. Temporal lobe

187. Long-latency potentials reflect subjective responses to expected or unexpected stimuli. They run between 250 and 750 ms after the stimulus. The most often mentioned event-related potential (ERP) is a positive response called the P300. It comes approximately 300 ms after a/an _____, although it can be later, depending on variables such as age and processing speed. A/an _____ refers to a stimulus that is different than the other stimuli in a series. It indicates that the brain has noticed something.

    The best term to fill the blanks is:
    a. Red alert
    b. Odd-ball stimulus
    c. Unexpected sentence ending
    d. Semantic deviation

188. One important negative long-latency potential is the N400. It occurs as a response to:
    a. Unexpected endings in sentences or other semantic deviations.
    b. Auditory stimulation
    c. Visual stimulation
    d. An "odd-ball" stimulus

189. An example of a *steady-potential shift* is one that occurs after a person is told that they must wait after a signal (warning) and then respond to an event. It is a kind of *anticipation* response. It is seen as a negative shift that occurs between the warning signal and the event. This type of steady-potential shift is called:
    a. An early warning stimulus (EWS)
    b. An abulic response (AR)
    c. A contingent negative variation (CNV)
    d. An odd-ball contingent

190. An event related potential (ERP) can help to distnguish response differences in ADHD children and normals. In a 'go' condition, the subject performs an action in response to a cue. A green light is an example of a cue that says you can cross the street. A 'go' stimulus produces alpha desynchronization. In a 'nogo' condition, the subject withholds acting in response to a cue that indicates he is not to act. A red light at a street corner is an example of a cue that says you must not cross the street. In this case the subject must suppress a prepared action. There is motor inhibition. Following a 'nogo' stimulus there is:
    a. An initial desynchronization followed by synchronization in the frontal and occipital areas
    b. An initial synchronization followed by desynchronization in the frontal and occipital areas
    c. An initial desynchronization followed by synchronization in the central and parietal areas
    d. An initial synchronization followed by desynchronization in the central and parietal areas.

191. 'Go' and 'no-go' ERP responses are impaired in ADHD. The ERP amplitudes are higher in normal subjects. It has been demonstrated that 20 sessions of beta training, in subjects diagnosed with ADHD, can result in:
    a. No increase in the ERP response
    b. A dramatic increase in the ERP response
    c. A dramatic decrease in the ERP response
    d. No change in ERP

192. The observation that increased cognitive or sensory workload results in a decrease in rhythmic slow-wave activity and an increase in _____ beta activity is termed:
    a. Post-reinforcement synchronization (PRS)
    b. Event-related potential (ERP) response
    c. Event-related desynchronization (ERD)
    d. Positron emission tomography (PET)

193. When the task is completed, there is _____ of the EEG. M. Barry Sterman describes these phenomena in his work with pilots. He notes that this _____ phase is self-rewarding. The best term to fill in the blanks are:
    a. Postreinforcement synchronization (PRS)
    b. Event-related potential (ERP) response
    c. Event-related desynchronization (ERD)
    d. Slow cortical potential (SCP)

194. _____ are very slow waves that indicate a shift between positive and negative. These shifts underlie the electrical activity we are usually measuring. There is great interest in this meticulous work. Birbaumer has been able to teach subjects with severe amyotrophic lateral sclerosis (ALS or Lou Gehrig's disease), who could not speak or move or otherwise communicate, to consciously make a response.
    a. Slow cortical potentials (SCP)
    b. Event-related desynchronizations (ERDs)
    c. Postreinforcement synchronizations (PRSs)
    d. Event-related potentials (ERPs)

Patterns Associated with Different Clinical Presentations

195. In a right-handed client with ADD, in which cortical area would you reasonably expect to see EEG slowing (increased theta and/or alpha activity) when carrying out a boring cognitive task?
    a. Occipital cortex
    b. Left parietal cortex
    c. Right central cortex
    d. Frontal cortex

196. Which frequency range does one usually train down (decrease) in children with Attention Deficit Disorder without hyperactivity (ADHD: Inattentive Type)?
    a. 4-8 Hz
    b. 8-12 Hz
    c. 13-17 Hz
    d. 18-25 Hz

197. Which of the following frequency ranges would you wish to reward (increase) in children with Attention Deficit Disorder without hyperactivity (ADHD: Inattentive Type)?
    a. 4-8 Hz
    b. 8-12 Hz
    c. 12-15 Hz
    d. 15-18 Hz

198. Which frequency range would you usually choose to reward (increase) in Attention Deficit Disorder with hyperactivity and impulsivity (ADHD: Combined Type)?
    a. 4-8 Hz
    b. 8-12 Hz
    c. 13-15 Hz
    d. 15-18 Hz

199. What may the sudden appearance of an extremely high-amplitude spike and wave form (at approximately 3 Hz), represent?
    a. Muscle tension in the jaw or neck muscles
    b. Increased problem-solving cognitive activity
    c. Eye blink
    d. Absence seizure (petit mal epilepsy)

Sample of Eyes-Closed Resting EEG – Linked-Ear Montage

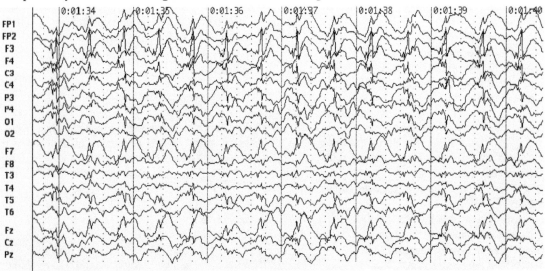

200. The EEG pattern above is most likely consistent with:
   a. Generalized Absence Seizure activity
   b. Partial seizure activity particularly in the left frontal region
   c. Generalized grand mal seizure activity
   d. Electrical artifact

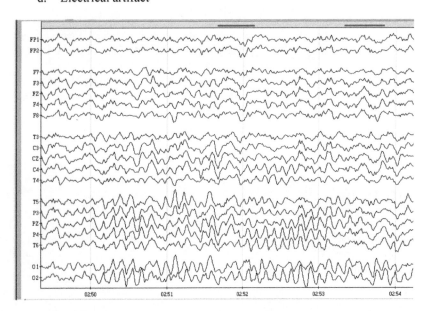

201. In the above client's EEG, the most consistent pattern observed was low-frequency alpha at P4 and T6. (T6 is oriented for spatial and emotional contextual comprehension, and nonverbal memory.) The subject had symptoms of one of the following disorders. Which is the most likely?
   a. Depression
   b. Wernicke's aphasia
   c. Asperger's syndrome
   d. Anxiety with panic attacks

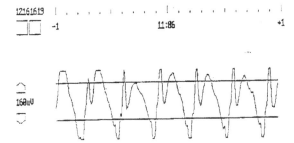

202. The EEG pattern above is most likely consistent with:
   a. Absence seizure activity
   b. Partial complex seizure activity particularly in the left frontal region
   c. Generalized grand mal seizure activity
   d. Electrical artifact

Instrumentation
BCIA Required Hours: 4

203. Impedance is measured using what type of current and bandwidth?
   a. Direct current in theta range
   b. Direct current in alpha range
   c. Alternating current in beta range
   d. Alternating current in alpha range

204. The formula: Voltage (V or E) = Current (I) x Impedance (z)

   Impedance is a more complex construct than resistance because its calculation requires measurements of not only the resistance of the conductor but other factors. These factors are:
   a. Capacitance, inductance, and the frequency of the alternating current
   b. Capacitance, and the frequency of the alternating current but not inductance
   c. Inductance and the frequency of the alternating current but not capacitance
   d. Capacitance and+ inductance but not the frequency of the alternating current

205. Current flows due to the potential difference between source (-ve site) and destination (+ve site).
   c. True
   d. False

206. Opposition to electron flow is high in substances such as rubber, where most of the atoms contain outer electron layers that are full. This increases the resistance of the electron in the outer layer to being dislodged. These substances make good insulators and very poor conductors.
   a. True
   b. False

207. Current measures the rate of transfer of electric charge from one point to another.
   e. True
   f. False

208. The electrons in the outermost layer of the atom are responsible for electricity. This electron layer may be incompletely filled. When this is the case, the electrons in that layer are less tightly held in place and collisions may dislodge them. Imagine that a loose electron acts like a billiard ball. It collides with another electron, is captured by the atom it collided with but that atom's electron is sent off on a different course to strike the next atom, and so on in a chain reaction. It is this sequence that is responsible for what we know as:
   g. An electric current
   h. Potential difference
   i. Capacitance
   j. Inductance

209. To measure the amplitude of a wave, we usually measure from the top of the positive part of the wave to the top of the negative part of the wave and we term this a:
   k. Root mean square measurement
   l. Peak-to-peak amplitude measurement
   m. Logarithmic measurement
   n. Power measurement

210. *Calibration* of a full-cap EEG instrument is done by applying a standard voltage to all input channels. This ensures that the voltage read is accurate and that all inputs are amplifying and filtering the signal in the same way.
   a. True
   b. False

211. The differential amplifier detects and amplifies differences between two inputs. It amplifies changes in the signals to each input to the same degree. The amplifier only amplifies this difference between the two inputs and not what is in common. Hence, it is correctly called a differential amplifier. In this way, the amplifier is said to 'reject' signals that are in common to both inputs. This is called common-mode rejection. For the most part, it cancels out common inputs by:
   a. Conducting any common amplitude to the ground
   b. Taking all waves that are in phase and putting them through a high input resistance
   c. Reversing the polarity of the second input so that the two currents are effectively subtracted
   d. Amplifying the first input by 100,000, but not amplifying the second input.

212. The *common-mode rejection ratio* is the ratio of the common-mode input voltage divided by the output voltage. This ratio should be >10,000, and often it is higher than 100,000 in newer instruments. In the neurofeedback practitioner's work, failure of this system to eliminate external common-mode artifact is most likely due to:
    a. Similar impedances between the two electrodes and/or a poor 'ground' connection
    b. A difference in impedance between the two electrodes and/or a poor 'ground' connection
    c. A faulty amplifier input ground
    d. A failure of the ground wire to connect with the wall socket electrical ground connection

213. A high-pass filter is not an all-or-none type of filter. It will gradually attenuate the amplitude of waves that come in at a frequency:
    o. Above its cut off point
    p. Below its cut off
    q. Over 60 Hz
    r. Over 50 Hz

214. A low-pass filter is not an all-or-none type of filter. It will attenuate the amplitude of waves that come in at a frequency:
    a. Above its cut off point
    b. Below its cut off
    c. Over 60 Hz
    d. Over 50 Hz

215. The term *band-pass filter* usually refers to a filter that allows:
    a. A defined frequency range (e.g., 4-8 Hz) to pass, while frequencies above and below the range are attenuated
    b. A filter that keeps out high-frequency artifact
    c. A filter that keeps out low-frequency artifact
    d. All of the above

216. A band-pass filter:
    a. Must be >2 Hz
    b. Must be <8 Hz
    c. Lets pass a range of frequencies
    d. All of the above

217. The maximum frequency that can be accurately reconstructed in a filter is determined by the *Nyquist* principle. This frequency is what proportion of the sampling rate?
    a. One-tenth
    b. One-quarter
    c. One-half
    d. Three quarters

218. A sampling rate that is too slow will make the analog signal incorrectly appear to be running at a slower frequency than it actually is. This effect is called
    a. The Nyquist effect
    b. Shaping
    c. Aliasing
    d. Retarding

219. The number of bits refers to the number of amplitude levels that can be resolved. An 8-bit analog-to-digital conversion (ADC) will have 2^8 or 256 amplitude levels. This would be ±128 discrete voltage levels in the voltage range allowed by that ADC. Too few bits means that:
    a. Small increases in voltage will be overemphasized
    b. Small increases in voltage will be underemphasized
    c. Large increases in voltage will be overemphasized
    d. None of the above

220. The FIR filter computes a moving average of digital samples. The number of points that are averaged is termed the:
    a. Range of the filter
    b. Order of the filter
    c. Amplitude of the filter
    d. Bits of the filter

221. When you use a FIR filter to sample a certain frequency range, say 4-8 Hz, the frequencies outside that range are attenuated but not entirely eliminated. In particular, the frequencies in each end of the range will get through to a certain extent due to the shoulders (degree of slope) on either end of the range.
    a. True
    b. False

222. Montages refer to different types of 'spatial' filtering. Sequential filtering is best to view highly localized activity and filter out coherent waveforms of similar amplitude and phase (Fisch, p 73).

Match the description given in Column 1, below, to the correct term in Column 2. Write the letter from Column 1 (a, b, etc,) beside the term to which the description best corresponds in Column 2.

Column 1	Column 2
a. Reference of an 'active' electrode to an average of all the other electrodes b. Reference of an 'active' electrode to the average of the electrodes immediately surrounding it c. Reference of an 'active' electrode to an ear or linked ears d. Adjacent pairs of electrodes are referenced to each other. This may be transverse or longitudinal **Note:** Each *montage* is just a mathematical reworking of the data, which can easily be done by the computer	_____ *Laplacian montage (Hjorth, 1980)* _____ *Sequential (bipolar) montages* _____ *Average reference montage* _____ *Common electrode reference montage*
Ref: topographical EEG interpretation, Amer. J. EEG Technol. 20: *121-132)*	

223. The *bipolar* and *Laplacian* montages are good for different procedures than the *common reference montage.* The common reference montage is often considered better for:
   a. Detecting localized activity
   b. Locating the origin of seizures activity
   c. Analyzing asymmetry and detecting artifacts
   d. Finding a small space-occupying lesion

Psychopharmacology Considerations
BCIA Blueprint Section V: Required Hours: 2

224. QEEG measures of theta (4-8 Hz) and beta (13-21Hz) taken at 19 sites in children with Attention Deficit Hyperactivity Disorder when on and off stimulant medication (Ritalin) demonstrate that on medication there is:
   a. Increased alpha activity in the occipital area
   b. Very large decrease in theta activity in the frontal regions
   c. Only small changes in theta/beta power ratio at any site
   d. A very large decrease in theta/beta ratio in the frontal area

225. Sedatives and tranquilizers produce an increase in which EEG frequency range:
   a. Delta
   b. Theta
   c. Alpha
   d. Beta

226. Selective serotonin reuptake inhibitors (**SSRIs**) are most likely to:
   a. Increase beta activity and may decrease alpha.
   b. Decrease beta activity and increase alpha
   c. Increase theta and delta activity
   d. None of the above

227. **Chronic** excessive alcohol consumption can:
   a. Increase beta (usually above 20 Hz) and decrease "thalpha" and alpha
   b. Decrease high-frequency beta
   c. Decrease high-frequency beta but increase theta and alpha
   d. None of the above

228. The use of lithium for the management of bipolar disorder can result in:
   a. Raising the frequency band for alpha
   b. Generalized asynchronous slowing, increased theta, and some slowing of alpha
   c. Increase beta (usually above 20 Hz) and decrease thalpha and alpha
   d. Decreased SMR 13-15 Hz with an increase in 12 Hz alpha

229. Tricyclic antidepressants may produce:
    a. Decreased SMR, 13-15 Hz, with an increase in 12 Hz alpha
    b. Generalized asynchronous slow waves and spike-and-wave discharges. Although they decrease alpha and also perhaps low beta, they may increase high beta.
    c. A rise in the frequency band for alpha
    d. Decreased high-frequency beta but increased theta and alpha

230. Marijuana will typically:
    a. Increase beta
    b. Increase 38-42 Hz activity
    c. Increase alpha
    d. Decrease alpha

Effects of Drugs on Learning Tasks

231. State-dependent learning refers to which of the following?
    a. Information (or skill) acquired in one physiological state is best recalled when again in that state
    b. The finding that when in a drugged state people cannot learn
    c. The observation that a repeated operant will produce a consistent response
    d. The repeated finding since the 1970s that children learn better on Ritalin

Treatment Planning
BCIA Blueprint; VII. Developing Treatment Protocols – 6 hours
VIII. Treatment Implementation – 6 hours

232. It has been hypothesized that in ADD there is reduced dopamine in the fronto-mesolimbic system in the left hemisphere. The type of cognitive processing that is affected and deficient is that which requires slow, serial effort. It is this kind of processing that may be improved with stimulant medication. Training using neurofeedback in the left frontal area for clients with ADHD may be done to:
    a. Increase SMR
    b. Increase high beta (26-34 Hz)
    c. Decrease high alpha (11-13 Hz)
    d. Decrease the dominant slow wave

233. Dopaminergic overactivity is thought to be associated with blunting of affect, excessive intellectual ideation, and introversion. Pathologically, it may underlie disorders including: paranoid states, anxiety, and hallucinations; psychosis, including the +ve symptoms of schizophrenia such as paranoia; Tourette syndrome; overly excited states including euphoria and mania; and also in:
    a. Attention Deficit Disorder
    b. Obsessive compulsive disorder (OCD)
    c. Parkinson's disorder
    d. None of the above

234. Too little dopamine has been reported to be associated with: Parkinson's disease with its tremor and inability to start movement; with the negative symptoms of schizophrenia including lethargy, misery, catatonia and social withdrawal; addictions; and with which of the following?
    a. Obsessive compulsive disorder
    b. Paranoid states
    c. Adult ADHD
    d. Anxiety

235. In obsessive compulsive disorder (OCD), the pathophysiology is hypothesized to involve an overactive *loop* of neural activity between the **orbital prefrontal cortex,** which is involved in *feeling that something is wrong,* to the **caudate nucleus,** which gives the *urge to act on personal memories or on* instincts *such as cleaning or grooming,* to the **cingulate,** which is important in registering conscious emotion and *which can keep focus or attention fixed on the feeling of unease.* The caudate nucleus is involved in automatic thinking. For example, when you check that you closed the fridge, turned off the stove, locked the door. Overactivity of this circuit means constantly checking and rechecking. Brain scans show that when a person with OCD is asked to imagine something related to their compulsion (such as dirt if it is someone who compulsively cleans) their caudate and prefrontal cortex light up. In NFB we tend to:
    a. Decrease high-amplitude beta activity frontally
    b. Increase 24-29 Hz activity frontally
    c. Decrease delta frontally
    d. Increase Sheer rhythm 38-42 Hz activity frontally

236. In Tourette's syndrome, the **putamen** is overactive. The putamen is related to the *urge to do fragments of* preprogrammed *motor skills*. The putamen is linked to the **premotor cortex,** which governs the production of the actual movements. Instead of the relatively complete procedures seen with compulsions, the motor and vocal tics appear in Tourette's syndrome as fragments of known actions. In neurofeedback and biofeedback work we will attempt to reduce and/or modulate the firing of the fusiform fibers in the muscle spindles by:
    a. Increasing sympathetic drive and decreasing SMR
    b. Decreasing sympathetic drive and increasing SMR
    c. Decreasing RSA and increasing high beta (22-32 Hz)
    d. Increasing EDR and decreasing theta

237. Which of the following areas is involved in the general maintenance of attention and arousal that is said to be "wide" to extrapersonal space? This area regulates information processing, which requires peripheral vision, spatial location and rapid shifts in attention. These aspects of attention appear to involve the noradrenergic system.
    a. The left cerebral hemisphere
    b. The right cerebral hemisphere
    c. The diencephalon
    d. The basal ganglia

238. In ADHD there might be excessive locus coeruleus norepinephrine production and excess noradrenergic stimulation to which area? This has been suggested because clonidine affects α_2 (inhibitory) receptors (locus coeruleus) and helps to decrease the symptoms of ADHD.
    a. Left cerebral hemisphere
    b. Right cerebral hemisphere
    c. The diencephalon
    d. The hypothalamus

239. People with ADD seem to have automatic processing that is fast and simultaneous. This style of attention is biased towards novelty and change. It has been posited that the positive effect that Dexedrine has on some people with ADHD is due to its effects on alpha2 inhibitory receptors in the locus coeruleus. It has also been noted that overactivation of the noradrenergic system in which of the following areas is associated with extroversion, histrionic behavior, impulsivity and manic behaviors?
    a. Left cerebral hemisphere
    b. Right cerebral hemisphere
    c. The diencephalon
    d. The hypothalamus

240. Stimulants, in animal studies, have been found to do which of the following?
    a. Decrease right hemisphere processing speed
    b. Facilitate release of dopamine (but not norepinephrine) from the striatum (which includes the caudate)
    c. Increase left hemisphere processing speed
    d. All of the above

241. A model of cortico-thalamic feedback that may be relevant to understanding and treating ADHD clients with neurofeedback suggests that decreased blood flow to, and metabolic activity in, the cells in the frontal/prefrontal areas (including the motor **cortex**) may *lead to r*educed excitation by the motor cortex (Layer VI) of the inhibitory cells in the **putamen,** which, in turn, may affect the activity of the substantia nigra (either directly or indirectly via the external globus pallidus and the subthalamus). Either way, the substantia nigra would increase its inhibition of areas of the thalamus, resulting in the thalamic relay cells producing bursts of activity, and then, by projection to the anterior association cortex (Layer IV), a wave form and rhythm of activity that we pick up in the EEG. What is this rhythm?
    a. Delta
    b. Theta
    c. High alpha
    d. Beta

242. A model of cortico-thalamic feedback that may be relevant to understanding and treating ADHD clients with neurofeedback suggests that decreased blood flow to, and metabolic activity in, the cells in the frontal/prefrontal areas (including the motor **cortex**) may lead to reduced excitation by the motor cortex (Layer VI) of the inhibitory cells in the **putamen,** which, in turn, may reduce its inhibition of the inhibitory cells in the **substantia nigra,** which would result in the substantia nigra being released to increase its inhibition of the **thalamus.** The net result of this model is thought to be increased inhibition of the ventral-lateral (VL), ventral-anterior (VA) and the centromedian (CM) nuclei of the thalamus. These cells would depolarize, then repolarize in a slow rhythmic manner. This would begin an oscillatory process that would be conveyed to the cortex. The end result of this cycle in the EEG would be:
   a. Delta
   b. Theta
   c. Beta
   d. Sheer rhythm

243. A commonly "inhibited" frequency range in ADHD students may sometimes represent constructive brain activity and not just tuning out. This wave frequency may be paced by cholinergic pathways that project from the septal nuclei to the hippocampus. This would correspond to the observed involvement of this wave form in memory retrieval. Recall of words can correspond to synchronization of this frequency wave. This is thought to represent hippocampal production of this wave. Which of the following frequencies is this discussion referring to?
   a. Higher frequency delta
   b. Theta
   c. High alpha (12-14 Hz)
   d. Low beta (15-18 Hz)

244. This area of the cortex and prefrontal cortex is said to be involved in inhibiting inappropriate action. Also, in tests of intellectual and/or attentional functions, this area appears to have decreased functional activity. The most likely area referred to in the above statements is:
   a. Right temporal cortex
   b. Medial posterior cortex
   c. Left orbitofrontal cortex
   d. Exner's area in the left frontal cortex

245. In which area of the brain have SPECT scans of subjects with ADHD demonstrated a decrease in blood flow during a continuous performance task? (Found in about 65% of the ADD subjects, compared to 5% of the controls in one study.)
   a. Broca's cortex
   b. Prefrontal cortex
   c. Supramarginal gyrus
   d. Hippocampal cortex

246. The amygdala strongly connects to an area of the cortex where emotions are experienced and meaning is bestowed on perceptions. Which area is this?
   a. Prefrontal cortex
   b. Ventromedial frontal lobe cortex
   c. Ventrolateral frontal lobe cortex
   d. Right parietal cortex

247. This area of the cortex focuses attention and facilitates tuning in to one's own thoughts. Increased blood flow on the right side of this area may correspond to attention being focused on internal events. This area distinguishes between internal and external events and is underactive in schizophrenia where the subject is unable to distinguish their own thoughts from outside voices. These statements most likely apply to:
   a. The left parietal lobe
   b. The substantia nigra
   c. The anterior cingulate
   d. The red nucleus

248. The area that is thought to be principally involved when you **hold** a thought in mind, **select** thoughts and perceptions to attend to, **inhibit** other thoughts and perceptions, **bind** the perceptions into a unified whole, endow them with **meaning, conceptualize, plan** and **choose** is:
   a. Left parietal lobe
   b. Substantia nigra
   c. Anterior cingulate
   d. Dorsolateral prefrontal cortex

249. People who have ADD may have a genetic alteration in the dopamine D4 receptors. These differences may lead to reduced dopaminergic activity (LaHoste et al., 1996). Deficiency in dopamine-related functions in the frontal-striatal system is the basis of some theories as to the emergence of ADHD symptoms (Charcot et al., 1996; Malone et al., 1994; Sterman, 2000). Stimulant medication such as Ritalin (methylphenidate) is thought to:
   a. Increase dopamine production
   b. Block the re-uptake of dopamine
   c. Increase receptors for dopamine
   d. Decrease the need for dopamine

Understand Procedures/Initial Assessment

250. Place in the following two figures the abbreviations (e.g., O1, P4) for the 10-20 sites indicated by ovals and circles using the old or the new site designations.

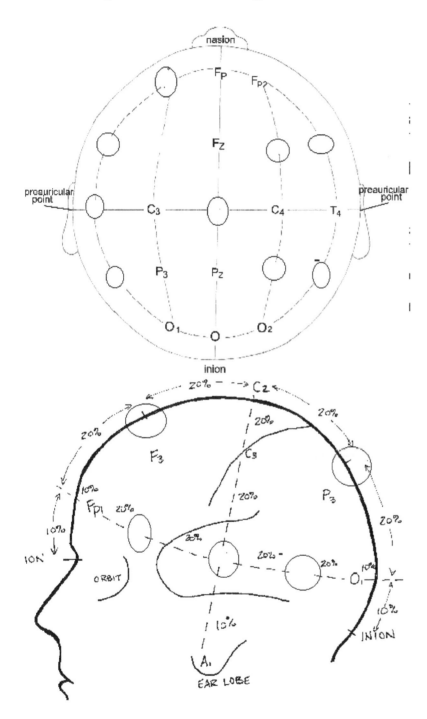

Artifacting the EEG

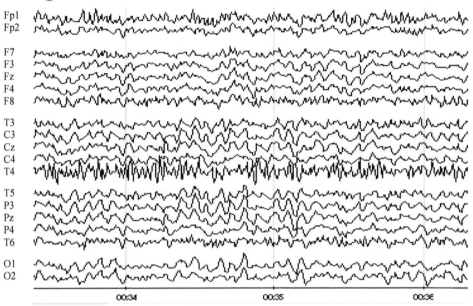

251. The above tracing at T4 is an example of:
    a. Eye flutter
    b. Spiking epileptiform activity
    c. Muscle artifact
    d. Severe beta spindling

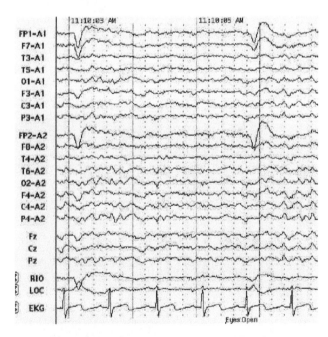

252. In the above figure, the two large downward excursions followed by a wave in the tracings at Fp1-A1 and Fp2-A2 represent:
    a. Eye flutter
    b. Spike and wave epileptiform activity
    c. Muscle artifact
    d. Eye blink artifact

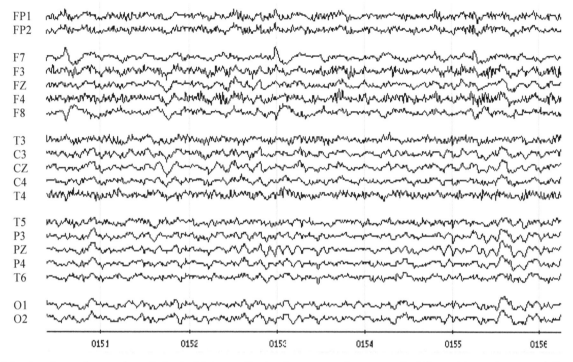

253. In the above figure, the relatively high-amplitude sharp excursion upwards followed by a downward flowing wave at F7 just prior to 0151 and again at 0153 represents:
   a. Eye-blink artifact
   b. Cardiac artifact
   c. Single epileptiform spike and wave
   d. Lateral eye movement

254. In the above figure, the five large downward excursions most likely represent:
   a. Eye-blink artifact
   b. Cardiac artifact
   c. Single epileptiform spike and wave
   d. Lateral eye movement

255. In the figure to the right, from the EEG (eyes open, referenced to linked ears) from an 11-year-old boy, the waveform seen at Fp1 and Fp2 most likely represent:
   a. Repeated eye blink artifact
   b. Delta activity associated with a learning disability
   c. Eye flutter
   d. Frontalis muscle artifact

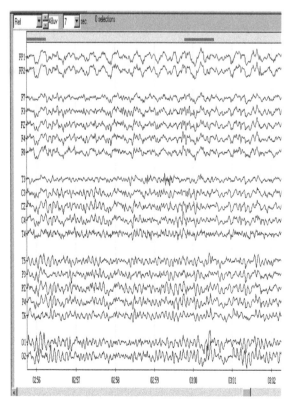

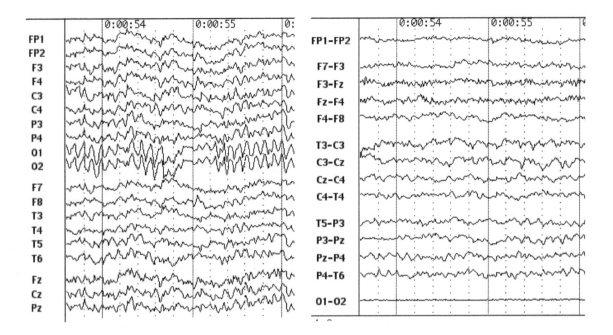

256. In the above recording (above two figures), displayed with two different montages, the activity seen in the left-hand montage at O1 and O2 most likely represents:
   a. High-amplitude alpha at sleep onset
   b. "Salt bridge" artifact
   c. Normal occipital alpha that is the same in both channels
   d. Occipitalis muscle artifact

QEEG (Single-Channel QEEG and Multi-Channel Brain Mapping)

Interpretation: databases, montages, comparison to databases using parameters such as frequency, amplitude, coherence, phase, and asymmetry.

257. In the figure to the right, the 19-channel recording is displayed in which montage?
   a. Longitudinal sequential
   b. Transverse sequential
   c. Linked-ear referential
   d. Laplacian or Hjorth

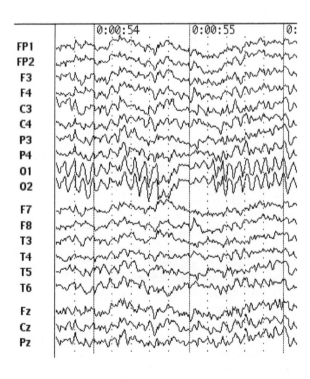

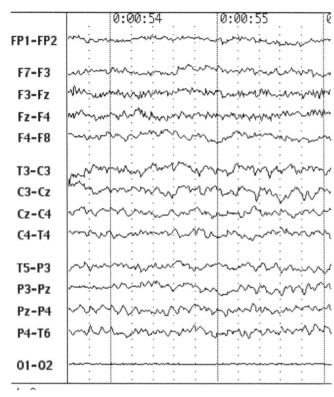

258. In the above figure, the 19-channel recording is displayed in which montage?
    a. Longitudinal sequential
    b. Transverse sequential
    c. Linked-ear referential
    d. Laplacian or Hjorth

259. Electrodes record electrical activity produced by the neurons (nerve cells) in the brain. Which is the best statement of what an EEG instrument measures?
    a. The voltage under each electrode
    b. The current under each electrode
    c. The potential difference between pairs of electrodes
    d. The potential difference between the electrode and ground

260. The term 'LORETA' refers to:
    a. Electromagnetic resolution of neuronal orientation
    b. A mathematical process that looks at surface EEG information and infers what activity is occurring in areas a little deeper in the cortex
    c. Glucose metabolic activity detected by positron emission from a radioactive form of oxygen
    d. Blood flow perfusion detected by a scan procedure

Treatment Protocols

261. Biological markers for Attention Deficit Hyperactivity Disorder include:
    a. Increased glucose metabolism in frontal and certain subcortical regions as recorded in PET scans
    b. Increased slow-wave activity (4-8 Hz) in frontal and central cortical regions
    c. Increased blood flow as recorded in SPECT scans
    d. All of the above

262. Positive features found in most people with Attention Deficit Disorder do NOT generally include:
    a. Spontaneity
    b. Ability to hyperfocus
    c. Constant reflection
    d. Creativity

263. In the United States in the mid 1990s, the proportion of the school-age population taking stimulant drugs, such as Ritalin (methylphenidate), for Attention Deficit Hyperactivity Disorder approximated:
    a. Less than 1%
    b. 3%
    c. 7%
    d. 10%

264. Research supports the finding of long-term benefits after cessation of treatment, measured by behavioral questionnaires (sampling behavior, attitude, homework completion, grade averages and relationships), in children with Attention Deficit Hyperactivity Disorder for:
    a. Stimulant drug therapy
    b. Behavior modification
    c. EEG biofeedback
    d. Nutritional supplements

265. Students who have Attention Deficit Disorder when expected to sustain attention during a cognitive task (such as listening to the teacher or doing homework) often experience:
    a. A decrease in EDR (electrodermal response)
    b. An increase in peripheral skin temperature
    c. Deep diaphragmatic breathing
    d. An increase in 28-35 Hz EEG activity at Cz

266. Successful EEG biofeedback training for students who have Attention Deficit Hyperactivity Disorder is **NOT** associated with significant increases in:
   a. Reading comprehension
   b. Intelligence test scores
   c. Attention span (measured by continuous performance tests)
   d. Theta activity (3-7 Hz)

267. After 40 sessions of EEG biofeedback, students who have Attention Deficit Disorder were found to demonstrate the following pattern on Wechsler Intelligence Scale testing:
   a. An increase in overall IQ but no significant improvement on subtests for arithmetic, coding, digit span, information.
   b. Subtest scores for arithmetic, coding, digit span, and information increase significantly but remain lower than other subtest scores.
   c. Subtest scores for arithmetic, coding, digit span, information increase significantly and are no longer lower than other subtest scores.
   d. Subtest scores for arithmetic, coding, digit span, information get worse, though other scales do make a significant improvement.

268. EEG biofeedback for Attention Deficit Disorder usually requires:
   a. More sessions for students with comorbidity of specific learning disabilities
   b. Fewer sessions for students who have associated hyperactivity
   c. The addition of intense individual psychotherapy
   d. Fewer sessions for students who have Asperger's syndrome.

269. Successful EEG biofeedback training for depression has utilized:
   a. At Cz, decreasing theta 4-8 Hz activity and increasing SMR (12-15 Hz) activity
   b. At F3, decreasing dominant alpha relative to F4 alpha in the same frequency band
   c. At T6, increasing low-frequency beta, 13-16 Hz activity and decreasing dominant slow-wave activity
   d. At P5, decreasing dominant theta activity and increasing beta 15-18 Hz activity

270. Successful EEG biofeedback training for Asperger's syndrome that may help to improve interpretation of social cues and abstract thinking may utilize:
   a. At Cz, decreasing theta 4-8 Hz activity and beta 20-24 Hz activity
   b. At F3, decreasing dominant alpha relative to F4 alpha in the same frequency band
   c. At T6, increasing low-frequency beta (13-16 Hz activity), and decreasing dominant slow-wave activity
   d. At P5, decreasing dominant theta activity and increasing beta 15-18 Hz activity

271. Successful EEG biofeedback training in conjunction with appropriate teaching strategies for dyslexia has utilized:
   a. At Cz, decreasing theta 4-8 Hz activity and increasing SMR (12-15 Hz) activity
   b. At F3, decreasing dominant alpha relative to F4 alpha in the same frequency band
   c. At T6, increasing low-frequency beta, 13-16 Hz activity and decreasing dominant slow-wave activity
   d. At P5, decreasing dominant theta activity and increasing beta 15-18 Hz activity

272. Successful EEG biofeedback training for athletic performance, as reported in professional meetings, has utilized:
   a. At Cz, decreasing theta 4-8 Hz activity and increasing SMR (12-15 Hz) activity
   b. Decreasing high EMG activity, diaphragmatic breathing and increasing peripheral skin temperature, while increasing 11-13 Hz and 13-15 Hz activity at PCz and decreasing 21-32 Hz activity in central and frontal regions
   c. At T6, increasing low-frequency beta, 13-16 Hz activity, and decreasing dominant slow-wave activity
   d. At P5, decreasing dominant theta activity and increasing beta 15-18 Hz activity

273. Successful EEG biofeedback training for dystonia combined with Parkinson's disease has been reported utilizing:
    a. At C4, decreasing 24-32 Hz beta spindling activity
    b. At F3, decreasing dominant alpha relative to F4 alpha in the same frequency band
    c. Promoting diaphragmatic breathing at 6 BrPM and RSA while concomitantly increasing 13-15 Hz in central regions (C3, Cz, C4)
    d. At P5, decreasing dominant theta activity and increasing beta 15-18 Hz activity

274. Successful EEG biofeedback training for seizure disorders has utilized:
    a. At Cz, decreasing theta 4-8 Hz activity and increasing SMR (12-15 Hz) activity
    b. Decreasing the dominant slow-wave activity near the seizure site while concomitantly increasing SMR (13-15 Hz) activity in the central regions (C3, C4)
    c. At T6, increasing low-frequency beta, 13-16 Hz activity and decreasing dominant slow-wave activity while doing diaphragmatic breathing at 6 BrPM
    d. Increasing beta 16-22 Hz activity near the seizure site while concomitantly increasing SMR (13-15 Hz) activity in the central regions (C3, Cz, C4)

275. Successful EEG biofeedback training to improve musical performance (musicality and emotional interpretation) has been reported using:
    a. At Cz, decreasing theta 4-8 Hz activity and increasing SMR (12-15 Hz) activity
    b. Increasing high frequency theta activity at Pz to the point where theta amplitude just exceeds alpha activity at the same site
    c. At T6, increasing low-frequency beta, 13-16 Hz activity and decreasing dominant slow-wave activity while doing diaphragmatic breathing at 6 BrPM
    d. Increasing beta 16-22 Hz activity at Fz while concomitantly increasing SMR (13-15 Hz) activity in the central regions (C3, Cz, C4)

276. Successful EEG biofeedback training to decrease anxiety and tension has been reported using:
    a. At Cz, decreasing theta 4-8 Hz activity and increasing SMR (12-15 Hz) activity
    b. At T6, increasing low-frequency beta, 13-16 Hz activity and decreasing dominant slow-wave activity
    c. Increasing high alpha (11-13 Hz) and SMR (13-15 Hz) while decreasing high beta (20-23Hz and 24-32 Hz), plus hand warming and diaphragmatic breathing at 6 BrPM
    d. Increasing beta 16-22 Hz activity at Fz while concomitantly increasing SMR (13-15 Hz) activity in the central regions (C3, Cz, C4)

Other Treatment Techniques
Psychotherapy

277. Mechanisms of defense are mental processes that are defensively motivated in an attempt to obtain relief from emotional tension or anxiety, and to resolve conflict. They are:
    a. Conscious but automatic
    b. Unconscious and automatic
    c. Conscious efforts to relieve emotional pain
    d. Require thought and motivation to enact but become automatic

278. A detached and intellectual awareness and understanding of feeling, emotions and behavior of another person is termed:
    a. Sympathy
    b. Empathy
    c. Transference
    d. Psychic awareness

279. The _____ is a combination of defense mechanisms including projection, repression, identification and rationalization. It is that unconscious process (defense mechanism) whereby one has an extreme negative reaction to another person who has characteristics that are really disowned aspects of one's own self (physical, behavioral, and/or personality characteristics). These are aspects of self that are unacceptable to one's own self-image and that one does not consciously recognize. These aspects of self are seen in (projected onto) the other person and adamantly reacted to without any recognition of the inappropriateness of that reaction.

This process is important for trainers and therapists to recognize when they have a very strong reaction to someone they are working with. It is called:
   a. Displacement
   b. Transference
   c. Countertransference
   d. The King David's Reaction

280. _____ is a defense mechanism outside of conscious awareness in which emotional feelings are 'transferred' from their appropriate (now 'internal object') to an external one. (phobia is an example). Substitution _____ may result in minor similarities between a person in one's present life and a significant other in earlier life, to cause reactions towards the person in the here-and-now "as if" they were the person from the past. The defense mechanism that fits in the above blank spaces is called:
   a. Rationalization
   b. Suppression
   c. Repression
   d. Displacement

281. The unconscious transfer to another person of feelings that were originally felt towards another (important) person earlier in life is called:
   a. Countertransference
   b. Projection
   c. Reaction formation
   d. Transference

282. The transfer (partially or wholly unconscious) of emotional responses of a therapist to a patient, that may in part be due to the patient's transference and/or based on the therapist's early relationship with an important other, is called:
   a. Transference
   b. Projection
   c. Countertransference
   d. King David's Reaction

283. A special kind of collaborative relationship between therapist and client to promote meaningful progress in psychotherapy is called:
   a. Countertransference
   b. Therapeutic alliance
   c. Reaction formation
   d. Transference

284. A mental mechanism operating outside of and beyond conscious awareness by which the emotional significance and affect is separated and detached from an idea, situation or object is called:
   a. Countertransference
   b. Dissociation
   c. Reaction formation
   d. Transference

285. _____ is unconscious and defensively intended to help maintain the lack of awareness of one's rejected and disowned ideas, wishes, impulses, and motives. The best term to fill in the blank is:
   a. Countertransference
   b. Repression
   c. Reaction formation
   d. Transference

286. _____ directs outwardly and attributes to others one's own disclaimed and objectionable character traits, thoughts, feelings, attitudes, motives and desires. It distorts one's understanding of the outside world and enables people to remain blind to important dynamic factors in their own personality.
   a. Countertransference
   b. Suppression
   c. Reaction formation
   d. Projection

287. A mechanism beyond conscious awareness in which the ego justifies or attempts to make otherwise unacceptable impulses, needs, feelings, behavior and motives into ones which are consciously tolerable and acceptable is called:
   a. Countertransference
   b. Therapeutic alliance
   c. Reaction formation
   d. Rationalization

288. Unconscious mental mechanism by which needs, complexes, attitudes and motives are developed in large areas of the personality that are the reverse of consciously disowned ones. Personally and socially acceptable goals are developed that are the antithesis of the hidden, previously disowned ones. This is called:
   a. Countertransference
   b. Therapeutic alliance
   c. Reaction formation
   d. Transference

289. _____ is the conscious effort to subjugate (remove from conscious recognition) unacceptable thoughts or desires. One directs attention away from undesirable thoughts, objects, or feelings and is aware of so doing.
   a. Countertransference
   b. Repression
   c. Reaction formation
   d. Suppression

Cognitive Interventions

290. A metacognitive strategy is:
   a. An unconscious strategy learned by repetition of an exercise
   b. An executive thinking function that consciously monitors learning and planning
   c. A mathematical exercise usually learned in calculus
   d. A philosophical term applied to advanced theological learning

291. Research has demonstrated that metacognitive strategies are:
   a. Used by good students and not by poor students
   b. Not a differentiating factor between good and poor students
   c. Often accompanied by increased 32-35 Hz activity
   d. Improved in high-stress situations

Regular Biofeedback

292. The human normal reaction to stress involves a release of norepinephrine from the locus coeruleus, which causes the hypothalamus to increase its output of:
   a. ACTH
   b. Acetylcholine
   c. Cortisol
   d. Serotonin

293. An increase in peripheral skin temperature is correlated with:
   a. Peripheral vasoconstriction
   b. A decrease in sympathetic nervous system arousal
   c. Increased heart rate
   d. Anxiety

294. Increase in electrodermal response (EDR) is correlated with:
   a. Parasympathetic nervous system arousal
   b. An increase in skin resistance measured in ohms
   c. An increase in skin conduction measured in 'mho'
   d. Decreased sweat gland secretion

295. Which of the following physiological changes might you expect to observe in a client who feels very tense and anxious?
   a. EMG decrease in the trapezius muscle groups
   b. Diaphragmatic breathing pattern
   c. Synchrony between changes in heart rate and the inspiration and expiration phases of breathing.
   d. Decreased peripheral skin temperature

Relaxation Techniques

296. There are six standard exercises in a classic form of rel training. They are limb heaviness, limb warmth, cardiac, respiration (diaphragmatic breathing), solar plexus warmth and forehead cooling. The term used for this is:
   a. Progressive relaxation
   b. Guided imagery
   c. Autogenic training
   d. Six-step meditation

****

Answers for the Multiple Choice Questions

All questions are covered in the text with the exception of the questions on mechanisms of defense. Additional references are mentioned for a few of the questions.

1. a
2. d
3. b
4. a
5. c
6. c
7. d
8. b
9. d
10. b (Sterman, 2000)
11. b
12. d
13. b
14. c (Thompson & Thompson JNT, Nov. 2002)
15. c
16. b
17. b
18. b
19. d
20. c
21. b
22. b
23. b
24. See diagram in text, p 69
25. See diagram in text, p 69
26. See diagram in text, p 84
27. a
28. a
29. a
30. b
31. b
32. d
33. b (Amen, 1997)
34. d (Smith)
35. c
36. c
37. c
38. c
39. c
40. d
41. d
42. a
43. d
44. b
45. a
46. b
47. b (Malone et al.)
48. b (Carlson from Tucker, 1977)
49. d
50. a
51. a (Carlson, p 674, after Babinski, 1914)
52. b
53. b
54. c
55. c
56. c (Reference: Joel Lubar workshop handout of applications, supported by peer-reviewed journal papers)
57. b (Reference: Joel Lubar workshop handout of applications, supported by peer-reviewed journal papers)
58. c
59. c
60. c
61. d
62. b (Olds & Milner, 1954)
63. b (Rosenfeld, 1997)
64. a (Smith)
65. d
66. a
67. d
68. a
69. b
70. b
71. b
72. d (Sterman, 2000; DeLong, 1990)
73. b (Klimesch et al., 1999; Thatcher, personal communication)
74. c
75. d
76. b
77. b
78. d (Silverthorn & Unglaub, 1998)
79. c
80. b
81. c (Sterman, 2000)
82. d (Sterman, 2000)
83. d (Sterman, 2000)
84. d (Sterman, 2000)
85. d (Sterman, 2000)
86. d (Sterman, 2000)
87. a (Sterman, 2000)
88. d (Sterman, 2000)
89. c (Campbell, 1996, p 971)
90. c
91. d
92. d (Campbell, 1996)
93. c
94. b
95. a
96. a (Sterman)
97. d (Joel Lubar, 1997, after Nunez)

98. c
99. c (Dyro, 1989)
100. b (Lubar)
101. b
102. d
103. b
104. a (Fisch, 1999, p 16)
105. c (Lubar)
106. c
107. a
108. a (Lubar)
109. b (Fisch, p 125)
110. b
111. c
112. b (Fisch, p 145)
113. c
114. c
115. b (Fisch, p 152)
116. a (Gibbs, F.A. & Knott, J.R., 1949)
117. c (Fisch, p 185, 187)
118. d
119. c (Fisch, p 181)
120. a (Fisch, p 45, p 149)
121. c
122. c
123. b
124. a
125. c
126. a
127. b
128. b
129. b (Fisch, p 349)
130. b
131. d
132. c
133. a
134. b
135. d (Schwartz, 1987, p 76)
136. a (Niedermeyer & Da Silva)
137. See table in text, page 10
138. b
139. b
140. b
141. b
142. b
143. c (Fisch, p 271)
144. c
145. b
146. c
147. b
148. d
149. c
150. c
151. c
152. c
153. c
154. c
155. c
156. d
157. b
158. b
159. b
160a. b (Fisch, p 353)
160b. a (Fisch, p 352)
161. c
162. d
163. c
164. b (SMR by definition is only over the central area)
165. d
166. c
167. d (Lubar)
168. d (Lubar)
169. c
170. c (Fisch, 1999)
171. c (Dyro, 1989)
172. a
173. d
174. a (Thatcher, 1986, 64: 123-143)
175. c (Thatcher, 1986, 64: 123-143)
176. c (Lubar, 1997)
177. a (Lubar, 1997)
178. b
179. c
180. d (Lubar, 1997)
181. a (Sutton, Teuting, Zubin & John, 1967)
182. b
183. c
184. b
185. b (Vaughn and Arezzo, 1988).
186. b
187. b
188. a (Kutas & Hillyard 1980).
189. c (Walter, Cooper, Aldridge, McCallum & Winter 1964)
190. a
191. b (Grin-Yatsenko, Kropotov, 2001)
192. c
193. a
194. a
195. d. (Mann, 1992)
196. a (Lubar, 1995; Thompson & Thompson, 1998)
197. d (Lubar, 1995; Thompson & Thompson, 1998)
198. c (Lubar, 1995; Thompson & Thompson)
199. d
200. b
201. c
202. a
203. d
204. a

205. a
206. a
207. a
208. a
209. b
210. a
211. c
212. b
213. b
214. a
215. a
216. c
217. c
218. c
219. a
220. b
221. a
222. b
    d
    a
    c
223. c (Lubar)
224. c
225. d (Fisch, p 414)
226. a
227. a
228. b
229. b
230. c
231. a
232. d
233. b
234. c
235. a
236. b
237. b (Malone, 1994)
238. b
239. b (Tucker)
240. d (Malone)
241. b (Sterman, 2000; DeLong, 1990)
242. b (Sterman, 2000; DeLong, 1990)
243. b (Klimesch et al., 1999)
244. c (Sterman, Child Study, 2000)
245. b (Amen et al., 1997)
246. b
247. c
248. d
249. b
250. See diagram in text, page 268
251. c
252. d
253. d
254. b
255. c
256. b
257. c
258. b
259. c (Lubar, Thompson)
260. b
261. b (Zametkin, A.J., 1990; Mann, 1992, Amen)
262. c (Thompson & Thompson, 1998)
263. b (Thompson & Thompson, 1998, after Safer, 1996)
264. c (Lubar in Schwartz)
265. a (Thompson & Thompson, 1998)
266. d (Thompson & Thompson, 1998; Lubar, 1995)
267. b (Thompson & Thompson, 1998)
268. a
269. b
270. c (Thompson, 2003; Jarascewicz, 2002)
271. d
272. b
273. c (Thompson, 2002)
274. b (Sterman, 2000)
275. b (Grazelier)
276. c
277. b (Introduction to Psychiatry. George Washington University Medical School, Washington D.C.)
278. b
279. d (Laughlin, 1967)
280. d
281. d
282. c
283. b
284. b
285. b
286. d
287. d
288. c
289. d
290. b
291. a (Thompson & Thompson, 1998, after Cheng, 1993)
292. a (Smith-Pellettier)
293. b (Schwartz, 1987, p 98)
294. c (Schwartz, 1987, p 98)
295. d (Schwartz, 1987, p 98)
296. c

New Questions for the 2nd Edition of *The Neurofeedback Book*

These questions apply, for the most part, to Sections II, IV, V and VI.

Multiple choice and index words for the second edition of *The Neurofeedback Book* and the *Companion to the Neurofeedback Book* (Europe, Asia, Africa)

Give the *most correct* answer; at times more than one answer may be partially correct.

Neuroanatomy, Neurophysiology, Brodmann Areas and Networks

1. Brodmann areas (BAs) delineate cortical areas that have specific functions that differ from other Brodmann areas.
    a. True
    b. False

2. Inhibition of cortical areas and networks is extremely important. One inhibitory mechanism that is available involves links from the cortex to subcortical structures that are broadly labeled the "basal ganglia." These connections can focus action while simultaneously inhibiting areas that are not necessary for a particular task. This mechanism is often described as a sequence:
    a. Cortex, putamen, thalamus, globus pallidus, cortex
    b. Cortex, locus coeruleus, solitary nucleus, cortex
    c. Cortex, putamen, globus pallidus, thalamus, cortex
    d. Cortex, caudate, nucleus ambiguous, thalamus, cortex

3. The lentiform nucleus comprises:
    a. The globus pallidus laterally, the innominate and anterior perforated area superiorly, and the subthalamus medially
    b. The putamen laterally, globus pallidus medially and the innominate substance, which contains the anterior perforated area, inferiorly
    c. The nucleus accumbens laterally, the putamen medially, the thalamus inferiorly, and the solitary nucleus posteriorly
    d. The thalamus medially, the nucleus accumbens laterally, the innominate, which contains the anterior perforated area, inferiorly

4. The relationship between the putamen, globus pallidus, and thalamus is:
    a. The putamen activates the globe is pallidus
    b. The putamen activates the globus pallidus, which can inhibit the thalamus
    c. The putamen inhibits the globus pallidus; the globus pallidus inhibits the thalamus
    d. The putamen inhibits the globus pallidus; the globus pallidus activates the thalamus

5. The dominant hemisphere is:
    a. The right hemisphere in left-handed persons
    b. The left hemisphere only in right-handed persons
    c. The left hemisphere in right-handed and many left-handed persons
    d. The right hemisphere in all left-handed persons and some right-handed persons

6. The anterior cingulate cortex is a major component of the following networks:
    a. The Affect network but not the Executive or Salience networks
    b. The Executive but not the Salience or Affect networks
    c. The Salience and Executive but not the Affect network
    d. The Affect, Executive, and Salience networks

7. The lobes of the brain that are usually considered crucial for all executive functions including the voluntary control of attention, the inhibition of inappropriate and/or unwanted behavior, the planning of actions and executive decision-making are:
    a. The temporal lobes
    b. The frontal lobes
    c. The parietal lobes
    d. The occipital lobes

8. Which of the following areas is most involved in handwriting?
   a. Broca's area
   b. Wernicke's area
   c. Gerstman's area
   d. Exner's area

9. Tactile apperceptive agnosia refers to an inability to:
   a. Visually identify items in their immediate environment
   b. Recognize objects based on their shape
   c. Recognize faces
   d. Execute a learned action with the hands

10. Constructional apraxia refers to an inability to:
    a. Name an object even though the person can recognize it
    b. Move a limb voluntarily
    c. Construct or even draw an object
    d. Have in mind the concept of an object

11. Frontal lobe damage to the ability to perceive a "pattern" (such as knowing that you are looking at a car and not a tree or a plane) is called:
    a. Constructional apraxia
    b. Anomia
    c. Aphasia
    d. Associative agnosia

12. After a stroke, a person may have nonfluent speech but good comprehension. This may be called:
    a. Broca's aphasia
    b. Anomia
    c. Wernicke's aphasia
    d. Apraxia

13. The following are true concerning the term "mirroring" as it is used in psychology and medicine:
    a. Infants learn by mirroring their mothers
    b. Mirroring was discovered in monkeys near the International 10-20 System Site F5
    c. Mirroring functions are a characteristic of an area near the temporal-parietal junction
    d. All of the above

14. Sensory aprosodia is term that is used to describe:
    a. Nonfluent speech but good comprehension
    b. Fluent speech, poor repetition, poor comprehension of both written and spoken propositional language
    c. Inability to plan movement related to an object because the patient has lost the perception of the object's purpose
    d. Unable to detect the meaning of nuance, innuendo, and tone of voice that reflects emotions

15. Deductive arguments are:
    a. Arguments that can be evaluated for validity where the premises provide absolute grounds for accepting the conclusion
    b. Arguments that must be evaluated for plausibility or reasonableness
    c. Never valid but entirely reliable
    d. Primarily a function of the right prefrontal areas

16. Nondominant (usually right) hemisphere dominance is thought of as being characterized by:
    a. Stimulus seeking behavior, seeking novelty and change, extroverted
    b. Excellent verbal comprehension
    c. Inability to easily comprehend novelty
    d. Relying on stable patterns and routines

17. The dominant neurotransmitter in:
    a. The dominant left frontal lobe is noradrenaline
    b. The dominant left frontal lobe is dopamine
    c. The nondominant right frontal lobe is dopamine
    d. Both frontal lobes is serotonin

18. The major functions usually attributed to the dominant (usually left) frontal lobe include:
    a. Sensing the gestalt
    b. Attention to spatial relationships
    c. Letter and word perception and recognition
    d. Synthesis and integration of sensory inputs

19. One major connecting link between neurons in the entorhinal area with the orbital surface of the frontal lobe is called the:
    a. Cingulum
    b. Uncinate fasciculus
    c. Superior longitudinal fasciculus
    d. Arcuate fasciculus

20. Overactivity of the right prefrontal area is often thought to be related to:
    a. Difficulties with expressive language
    b. Inability to interpret nuance and innuendo
    c. Irritability, impulsivity, tactless, and even manic behavior
    d. Approach behavior and positive thoughts

21. The nucleus basalis of Meynert (NB) is a diffuse grouping of neurons within the basal forebrain that projects to widespread areas of the neocortex and represents the main source of:
    a. Noradrenaline
    b. Serotonin
    c. Neocortical acetylcholine (Ach)
    d. Dopamine

22. The nucleus basalis of Meynert and the cholinergic enervations to the hippocampus and temporal lobes are the first to be affected by:
    a. Picks disease
    b. Alzheimer's
    c. Huntington's
    d. Asperger's

23. The ventromedial prefrontal cortex is an important area for:
    a. Regulating auditory memory
    b. Shifting attention from interpreting gestures to comprehending emotional tone of speech
    c. Executive decision-making including planning and monitoring
    d. Tonic inhibition of the amygdala (also reactivating past emotional associations and events, therefore essentially mediating pathogenesis of PTSD)

24. The right insula can:
    a. Decrease heart rate
    b. Influence the parasympathetic side of the autonomic nervous system
    c. Influence the sympathetic side of the autonomic nervous system
    d. Decrease peristalsis

25. The anterior cingulate cortex Brodmann area 24:
    a. Is of primary importance in the default Affect network
    b. Is a key structure involved in speech production
    c. Is important for clear handwriting
    d. Has motor control of the hands

26. In depression, the **vegetative/autonomic affect** network symptoms involve:
    a. A ventral compartment (e.g., sleep disturbance, loss of appetite and libido), which includes the hypothalamic-pituitary-adrenal axis, hippocampus, insula (BA 13), subgenual (SG) cingulate (BA 25), and the brainstem
    b. A dorsal compartment, which modulates attentional and sensory-cognitive symptoms (e.g., apathy, attentional and executive deficits), which includes the dorsolateral prefrontal cortex (BA 9 and 46), the dorsolateral anterior cingulate (area 24b), posterior cingulate (BA 31), the inferior parietal lobe (BA 40), and the striatum
    c. The left, dominant, dorsolateral prefrontal cortex
    d. The right parietal temporal junction, and the right supramarginal gyrus (BA 40) and angular gyrus

27. The left frontal operculum (F5) (Broca's) area is the key area for:
    a. Verbal comprehension
    b. Speech articulation
    c. Understanding innuendo
    d. Spatial perception

28. Encoding of visuospatial patterns activates:
    a. The left inferior prefrontal cortex and the left medial temporal lobe
    b. The right inferior prefrontal cortex and the right medial temporal lobe
    c. The right superior temporal lobe (BA 41)
    d. The left superior temporal lobe (BA 41)

29. The integration of auditory, visual, perceptual and memory inputs that can assist reading can include: phonemic (letter to sound), lexical (sight words), and semantic (meaning) inputs to:
    a. Areas near the right temporal parietal junction
    b. Areas near the left prefrontal cortex
    c. Areas near the right prefrontal cortex
    d. Areas near the left temporal parietal junction

30. The auditory association area can include:
    a. BA 20 and BA 38 of the temporal lobe
    b. BA 41, BA 42 and BA 22 of the superior temporal gyrus
    c. BA 1, BA 2 and BA 3 of the sensory cortex
    d. BA 44 and BA 45 of the frontal cortex

31. A client who has speech that is fluent with good grammar but whose speech doesn't seem to make sense, who speaks without understanding what she is saying and substitutes inappropriate or wrong words in sentences in a jargon that is just nonsense, will most likely suffer from dysfunction in:
    a. The right dorsolateral prefrontal cortex
    b. The inferior parietal lobe and the auditory association area of the superior temporal gyrus in the nondominant hemisphere
    c. The inferior parietal lobe and the auditory association area of the superior temporal gyrus in the dominant hemisphere
    d. In the precuneus and posterior cingulate (BA 31) of the dominant hemisphere

32. A region of the brain that is involved in *visual-spatial-language* skills, visual word recognition, and can be involved in learning disabilities including alexia (inability to read), agraphia (inability to write), difficulties with: *calculations, left-right discrimination, and technical competence* is the:
    a. Angular gyrus in the dominant hemisphere
    b. Ventral rostral cingulate gyrus
    c. Parahippocampal gyrus in the dominant hemisphere
    d. Hypothalamus

33. A region of the brain that is involved in spatial-emotional-contextual perception, symbol recognition, ability to copy emotional tones, and the comprehension of innuendo, nuance, and the emotional aspects of both gesture and verbal communications is the:
    a. Angular gyrus and the supramarginal gyrus in the dominant hemisphere
    b. Angular gyrus the supramarginal gyrus in the nondominant hemisphere
    c. Parahippocampal gyrus and the fusiform gyrus in the dominant hemisphere
    d. Parahippocampal gyrus and the fusiform gyrus in the nondominant hemisphere

34. An area of the cortex that plays an important role in autobiographical memories and, in particular, spatial memories is the:
    a. Dominant hemisphere, entorhinal cortex (EC) BA 34 and 28 - hippocampus system
    b. Nondominant hemisphere, entorhinal cortex (EC) BA 34 and 28 – hippocampus system
    c. Dominant hemisphere, left frontal operculum (F5) (close to Broca's area BA 44, 45)
    d. Nondominant prefrontal area BA 9, BA 10 and BA 46

35. The most important areas for encoding and consolidation but not storage of long-term memories lie in:
    a. The dominant hemisphere parahippocampal gyrus
    b. The septal area of the frontal lobe
    c. The hippocampus
    d. The structures in the entire Papez circuit

36. Extensive damage to which area of the cortex will result in a neglect of the contralateral (opposite side) of space and the body and a lack of awareness of the loss, which is termed *anosognosia*?
    a. The left/dominant temporal cortex
    b. The right/nondominant temporal cortex
    c. The left/dominant parietal cortex
    d. The right/nondominant parietal cortex

37. One or more of the following symptoms: dysgraphia, dyscalculia, finger agnosia, left-right disorientation, are most likely related to dysfunction in the:
    a. The left/dominant temporal cortex
    b. The right/nondominant temporal cortex
    c. The left/dominant parietal cortex
    d. The right/nondominant parietal cortex

38. Persons who are unable to perceive the visual field as a whole (simultanagnosia), have difficulty in fixating the eyes (ocular apraxia), and are unable to move the hand to a specific object by using vision (optic ataxia) have a syndrome named after a neurologist, Bálint. These difficulties are most often related to lesions in:
    a. Both temporal lobes
    b. The dominant prefrontal cortex
    c. The right parietal lobe
    d. Parietal-occipital area bilaterally

39. A person's cortex may have a cortical stream for information flow that provides the "what" of information to the individual's:
    a. Occipital-temporal-frontal pathway
    b. Occipital-parietal-frontal pathway
    c. Dominant temporal-nondominant frontal pathway
    d. Nondominant parietal-dominant temporal-frontal pathway

40. The cuneus and precuneus are found in:
    a. The superior temporal cortex
    b. The medial superior portion of the parietal lobe
    c. The inferior lateral aspect of the parietal lobe
    d. The anterior portion of the occipital lobe

41. The cuneus and precuneus with the posterior cingulate gyrus are areas that are responsible for the:
    a. Comprehension of verbally presented material including phonemic (letter to sound), lexical (sight words), and semantic (meaning)
    b. Organization, integration and synthesis of auditory, visual and kinesthetic inputs, and for orientation in space
    c. Spatial-emotional-contextual perception, symbol recognition, ability to copy emotional tones, and the comprehension of innuendo, nuance, and the emotional aspects of both gesture and verbal communications
    d. Primary import in the inhibition of cortical networks

42. Nondominant parietal lobe functions include:
    a. Spatial position, opposite-side body awareness, visualization of spatial organization, and context
    b. Cognitive reasoning, attention, spelling, verbal short-term memory, and imagination
    c. Comprehension of verbally presented material including phonemic (letter to sound), lexical (sight words), and semantic (meaning)
    d. Organization, integration and synthesis of auditory, visual and kinesthetic inputs

43. Lateral inferior parietal (LIP) neuronal activity is necessary for:
    a. The organization, integration and synthesis of auditory, visual, and kinesthetic inputs
    b. Disengaging attention from a current stimulus in order for focus to change to a different stimulus
    c. Interpretation of metaphors
    d. Pattern recognition

44. Visual information comes to the occipital lobes from the:
    a. Medial geniculate
    b. Lateral geniculate
    c. Posterior lateral nucleus of the thalamus
    d. Pulvinar

45. Brodmann area 24 is located in:
    a. The temporal cortex
    b. The parietal cortex
    c. The anterior cingulate gyrus
    d. The posterior cingulate gyrus

46. Long-term potentiation is a process wherein:
    a. A postsynaptic cell is enhanced so that it can depolarize more in response to a neurotransmitter
    b. A presynaptic membrane is encouraged to inject more neurotransmitter into the synaptic space
    c. The presynaptic axon develops more receptors
    d. The postsynaptic dendrite increases its capacity to transmit Mg^{++} ions

47. In the membrane of the postsynaptic receptor site:
    a. Ca^{++} blocks the NMDA (N-methyl-D-aspartate) receptor sites
    b. Na^{++} blocks the NMDA (N-methyl-D-aspartate) receptor sites
    c. Mg^{++} blocks the NMDA (N-methyl-D-aspartate) receptor sites
    d. Mg^{++} blocks the APMA (alpha-amino-3-hydroxy-5-methyl-4-isoxazolepropionic acid) receptor sites

48. A unique property of the NMDA (N-methyl-D-aspartate) receptor is its voltage-dependent activation. This means that when glutamate causes an influx of sodium into the AMPA (alpha-amino-3-hydroxy-5-methyl-4- isoxazole-propionic acid) postsynaptic terminal, an electrical voltage depolarization change takes place that allows the:
    a. Mg^{2+} block of the NMDA receptor site to be released
    b. Ca^{2+} block of the NMDA receptor site to be released
    c. Ca^{2+} to pass through the receptor site, which then allows the Mg^{2+} block of other AMPA receptor sites to be released
    d. Na^+ to pass through the receptor site, which then allows the Ca^{2+} block of other AMPA receptor sites to be released

END OF BRODMANN AREA MONOGRAPH QUESTIONS

SUMMARY OF ANSWERS
to the New Multiple Choice Questions
Added to the 2nd Edition of the NFB Book
Part 1 – Brodmann Area Monograph Questions

For some of the questions that are more than just neuroanatomy, an author's name may be given in order to facilitate finding a reference in the reference list.

1. b
2. c
3. b
4. c
5. c
6. d
7. b
8. d
9. b
10. c
11. d
12. a
13. d
14. d (Ross, 1981)
15. a
16. a
17. b
18. c
19. b
20. c
21. c (Divac, 1975; Mesulam & Van Hoesen, 1976)
22. b
23. d (Thayer et al., 2012)
24. c (Oppenheimer, 2001; Oppenheimer et al., 1992)
25. a
26. a (Paquette et al., 2009)
27. b
28. b
29. d
30. b
31. c
32. a
33. b
34. b
35. d
36. d
37. c
38. d (Ropper & Brown, 2005)
39. a (Ungerleider & Mishkin, 1982)
40. b (Smith, 1962)
41. b
42. a (Thatcher, 2012)
43. b (Bisley et al., 2006)
44. b (Smith, 1962)
45. c (Thompson et al., 2007)
46. a (Cooke & Bliss, 2006)
47. c (Cooke & Bliss, 2006)
48. a (Bear, 2004)

Procedures for Assessment and Treatment Planning

49. Independent component analysis in QEEG assessment procedures is primarily used to assist the practitioner to:
    a. Identify the frequencies of waves at a given international 10-20 sites
    b. Remove high-amplitude artifacts from specific international 10-20 sites
    c. Clarify which of two 10-20 sites should be inhibited
    d. Pinpoint the one 10-20 site that is responsible for a particular symptom

50. An evoked potential (ERP) is:
    a. An electrical method for removing high-amplitude artifacts from specific international 10-20 sites
    b. A method for identifying the source correlation for specific brain wave frequencies
    c. The desired result when using transcranial direct current stimulation
    d. A brain response that shows a stable time relationship to actual or anticipated stimuli

51. When recording evoked potentials, if a new stimulus is detected, attentional processes govern a change or updating of the stimulus representation that is concomitant with:
    a. N100
    b. P200
    c. P300
    d. P450

52. P300 may have two components, P3a and P3b. For these components, which of the following statements is true:
    a. Frontal attention corresponds to P3a and may be a dopaminergic activity
    b. Temporal-parietal memory-related processes correspond to P3b and may be related to locus coeruleus norepinephrine (LC-NE) activity
    c. P300 and its underlying subprocesses could reflect rapid neural inhibition of ongoing activity to facilitate transmission of stimulus/task information from frontal (P3a) to temporal-parietal (P3b) locations
    d. All of the above

53. Episodic memory operations primarily alter:
    a. Slow alpha (1-9.5 Hz) activity
    b. Fast alpha (10-12 Hz) activity
    c. Sensorimotor rhythm (SMR) (13-15 Hz)
    d. High-frequency beta (23-35 Hz) activity

54. NoGo-N2 (200 ms) and NoGo-P3 (300-600ms) are different time points of the response inhibition process of the:
    a. Prefrontal cortex
    b. Orbital-frontal cortex
    c. Occipital cortex
    d. Anterior cingulate cortex

55. Which of the following statements are true concerning slow cortical potentials?
    a. They are slow event-related direct-current (DC) shifts of the electroencephalogram
    b. They last from 0.3 seconds up to several minutes
    c. They are alternating current AC oscillations
    d. All of the above

56. In patients with ADHD, the clinician who does slow cortical potential (SCP) training will be asking the patient to:
    a. Raise cortical negativity because cortical negativity is too low
    b. Increase cortical positivity because negativity is too high
    c. Do another form of training because these patients have labile cortical potentials
    d. Neither raise or lower negativity because neither approach will show any clinical changes in ADHD patients

57. In seizure disorders, the clinician who does slow cortical potential (SCP) training will be asking the patient to:
    a. Decrease cortical negativity because cortical negativity is too high
    b. Increase negativity because cortical negativity is too low
    c. Do another form of training because these patients have labile cortical potentials
    d. Neither raise or lower negativity because neither approach will show any clinical changes in patients who have a seizure disorder

58. LORETA (low resolution brain electromagnetic tomography):
    a. Is an inverse solution
    b. Is used to find locations (sources) of electrical activity in the cortex
    c. Is not used to find sources of electrical activity in the thalamus
    d. All of the above

59. In a typical psychophysiological stress test, the stressors such as a Stroop color test or mathematics, will cause:
    a. Skin temperature to decrease
    b. Skin conduction to decrease
    c. Electromyogram activity to decrease
    d. Heart rate to decrease

Heart Rate Variability

60. In heart rate variability training, the most commonly used interbeat interval measurement in a biofeedback practitioners' practice is usually:
    a. Q to Q intervals of the QRS complexes
    b. R to R intervals of the QRS complexes
    c. Intervals between the start of an increase in heart rate and the beginning of a decrease in heart rate
    d. Intervals between low frequency (LF) and high frequency (HF) heart rates

61. When any resonant system (such as the heart) is rhythmically stimulated at its resonant frequency, this external stimulation will:
    a. Magnify the persistent oscillations, thus greatly increasing total variability
    b. Decrease the persistent oscillations, thus greatly decreasing total variability
    c. Cause labile oscillations to occur and thus greatly increase total variability
    d. Have absolutely no effect on the rate of oscillations

62. Vagal afferents from the heart and afferents from the carotid and aortic baroreceptors go first to the:
    a. Nucleus ambiguous
    b. Nucleus coeruleus
    c. Nucleus solitarius
    d. Parabrachial nucleus

63. Parasympathetic vagal afferents to the heart will:
    a. Increase heart rate
    b. Increase blood pressure
    c. Occur during inspiration
    d. Decrease heart rate

64. Hyperventilation, now called 'overbreathing,' can result in:
    a. Seizures
    b. Angina
    c. Anxiety
    d. All of the above

65. The Bohr effect is a:
    a. Decrease in the ability of hemoglobin to release oxygen
    b. Direct result of an increase in PCO2
    c. Result of excess beta EEG activity
    d. Result of vagal parasympathetic drive

Interventions that Can Be Combined with NFB and BFB

66. Transcranial direct current stimulation (tDCS) requires the application of a DC electrical current between anode and cathode of:
    a. 0.2 -2 milliamperes
    b. Less than 0.2 milliamperes
    c. 10-20 milliamperes
    d. 50-100 milliamperes

67. Transcranial direct current stimulation (tDCS) uses:
    a. A negative anode pad of about 28-38 cm^2
    b. A positive anode pad of about 28-38 cm^2
    c. A cathode that results in cortical stimulation
    d. An anode that results in cortical inhibition

68. In transcranial direct current stimulation (tDCS), the:
    a. Anode electrode induces a lack of positive ions (negativity, depolarization) at the basal part of neuronal membrane
    b. Anode electrode induces an excess of positive ions (positivity, hyperpolarization) at the basal part of neuronal membrane
    c. Cathode induces a negativity at the basal part of neuronal membrane
    d. Effects of tDCS usually last for more than a week after the first session

69. Passive infrared hemoencephalography (pIR HEG) refers to the use of one or two infrared detection sensors to:
    a. Detect EEG rhythm changes in the cortex
    b. Reflect temperature changes in the cortex
    c. Measure blood flow in the cortex
    d. Decrease cortical activity in the cortex

70. An important assumption that is made when using passive infrared hemoencephalography (pIR HEG) with patients is that an increase in temperature correlates with:
    a. Decreased cerebral blood flow in the underlying cortex
    b. Increased metabolic activity in the underlying cortex
    c. Decreased beta frequency activity in the underlying cortex
    d. Increased alpha frequency activity in the underlying cortex

71. Passive infrared hemoencephalography (pIR HEG) with sensors placed over the forehead has been reported to have a positive effect in decreasing the symptoms of:
    a. Attention Deficit Hyperactivity Disorder (ADHD)
    b. Visual acuity
    c. Migraine
    d. Dyslexia

Specific Disorders Interventions and Efficacy

72. The highest levels of efficacy (Level 5) for neurofeedback `reported in the literature by 2012 was for:
    a. Depression
    b. Dyslexia
    c. Attention Deficit Hyperactivity Disorder (ADHD)
    d. None of the above

73. Disorders, as of 2012, for which neurofeedback had been shown to reach and efficacy at Level 4 are:
    a. Depression and migraine
    b. ADHD and seizure disorder (grand-mal)
    c. Anxiety and ADHD
    d. Asperger's syndrome and seizure disorder (grand-mal)

74. In addition to the frontal alpha-asymmetry QEEG subtype of depressive disorder, a second QEEG subtype has been reported in the neurofeedback literature. It reflects more abnormality in:
    a. Parietal theta asymmetry
    b. Temporal alpha asymmetry
    c. Increased high-frequency beta in frontal, temporal, and cingulate cortices
    d. Low-amplitude delta in the right parietal-temporal junction

75. Concussion can result in diffuse axonal injury (DAI). This can most easily be detected using:
    a. PET scans
    b. fMRI
    c. QEEG
    d. SPECT scans

76. After a concussion, an athlete's return to play should be based on:
    a. Neuropsych testing + QEEG + questionnaires
    b. Objective postural (balance) testing + neuropsych testing
    c. Detailed questionnaires + QEEG + neuropsych testing
    d. Objective postural (balance) testing + neuropsych testing + QEEG + questionnaires

77. Concussion can result in diffuse axonal injury (DAI), which can result in:
    a. An intracellular Ca^{++} increase that impairs mitochondrial metabolism
    b. An increase in adenosine triphosphate (ATP)
    c. A decrease in of lactic acid production (byproduct of glycolysis)
    d. An increased regional cerebral blood flow (rCBF)

78. Second Impact syndrome is a term that refers to the fact that postconcussion (post-traumatic brain injury), an individual is more vulnerable and easily reinjured due to:
    a. Psychological factors that make them less likely to avoid reinjury
    b. Softened scalp skeletal structures
    c. Continued disturbances in mitochondrial functioning
    d. Decreased reaction time due to muscle motor injury

79. The principle QEEG post concussion findings include:
    a. Slowed peak alpha
    b. Increased high frequency power
    c. Decreased power in slow (delta and theta) frequency bands
    d. Increased coherence between frontal lobes

80. A high-functioning person who has difficulties that lie within the autistic spectrum disorders may use pedantic phrases or a voice that is monotone and lacks prosody (intonation, loudness variation, pitch, rhythm). This difficulty may be called motor aprosodia and be associated with disturbed functioning close to:
a. F5
b. P7
c. F6
d. P8

81. A high-functioning person who has difficulties that lie within the autistic spectrum disorders may appear unable to interpret nuance, innuendo, gestures, and the emotional tone of other people's speech. This may be termed sensory aprosodia and be associated with disturbed functioning close to:
a. F5
b. P7
c. F7
d. P8

82. Impaired social behavior, which includes an inability to understand the effects of one's own behavior on other people, in the autistic spectrum disorders can correlate with dysfunction in the:
a. Orbital surface of the left frontal lobe
b. Substantia nigra
c. Nucleus solitarius
d. Locus coeruleus

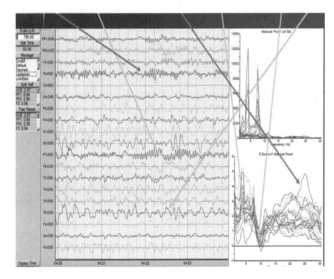

83. The two red arrows in the chart above point to a high-amplitude 25-27 Hz wave that is observed at F4.

The morphology of this wave is termed:
a. Desynchronous alpha
b. Desynchronous high beta
c. Spindling beta
d. Sheer rhythm

84. Intervention(s) for intractable depression in some studies have found, in responders to treatment, an effect on multiple Brodmann areas. These effects include:

Decreased activity (shown by PET scan) in BA 25, and the orbital frontal cortex BA11; increased activity (PET scan) DLPC BA 9/46; ACC BA 24; PCC BA 31; and, in responders to treatment, increased activity in BA 40 (parietal). These results occurred with which of the following treatment(s)?
a. SSRI medications
b. Cognitive behavior therapy
c. Deep brain stimulation
d. All of the above

85. In cases of serious depression that did not respond to treatment using antidepressant medications, those cases who responded positively to neurofeedback that included reducing frontal high-amplitude high frequency beta (referred to here as high "activity"), showed which of the following Brodmann area (BA) results?
a. Decreased activity in BA 25 and BA 31
b. No change in activity of BA 25
c. Decreased activity BA 25 and increased activity BA 31
d. Increased activity BA 25 and decreased activity BA 31

86. Treatments for major depression can include antidepressant medications and/or cognitive behavior therapy (CBT). Which of the following statements is most often correct concerning their effects on Brodmann areas (BA)?
a. Medications increase activity in ventral areas such as BA 25
b. Medications decrease activity BA 25 while CBT increases activity BA24
c. Medications increase activity BA 25 while CBT decreases activity BA24
d. None of the above

87. An important effect of a traumatic brain injury/concussion is calcium influx into the neurons. One important outcome of this process is an "energy crisis" in which there is:
    a. An increase mitochondrial activity to produce too much adenosine triphosphate (ATP)
    b. A decrease in the Krebs cycle production of ATP
    c. A large increase in GTP (guanosine triphosphate), which stimulates the citric acid cycle
    d. An increase in the Krebs cycle production of nicotinamide adenine dinucleotide (NAD)

88. Vitamins and minerals are essential for the mitochondrial, Krebs cycle, production of adenosine triphosphate (ATP).
    a. True
    b. False

89. After a traumatic brain injury/concussion, hypermetabolism and an energy crisis leads to a secondary reaction, with increased regional cerebral blood flow (rCBF) that occurs after about one minute to meet increased energy demands needed to restore ionic membrane balances (Meyer et al., 1970; Nilsson and Nordstrom, 1977; DeWitt et al., 1986). This is followed by:
    a. No further change in regional cerebral blood flow (rCBF)
    b. Immediate increase of up to 50 percent in rCBF usually lasting for 24 hours
    c. a subsequent decrease of up to 50 percent in rCBF
    d. None of the above

SUMMARY OF ANSWERS
to the New Multiple Choice Questions
Added to the 2nd Edition of the NFB Book
Part 2

49. b (Kropotov, 2009)
50. d (Kropotov 2009)
51. c (Polich, 2007)
52. d (Polich, 2007)
53. b (Hanslmayr et al., 2007; Klimesch, 2003)
54. d (Sokhadz, Stewart, Hollifield &Tasman, 2008)
55. a
56. a (Fuchs, Birbaumer, Lutzenberger, Gruzelier & Kaiser, 2003)
57. a (Birbaumer, Elbert, Canavan & Rockstroh, 1990)
58. d
59. a
60. b
61. a (Lehrer, 2007)
62. c (Porges, 2007)
63. d (Oppenheimer, Gelb, Girvin & Hachinski, 1992
64. d (Thompson, M., Thompson, L., Budzinski)
65. a (Lehrer, 2007)
66. a (Angelakis & Liouta, 2011)
67. a (Angelakis & Liouta, 2011)
68. a (Angelakis & Liouta, 2011)
69. b (Carmen, 2005)
70. a (Carmen, 2005)
71. c (Carmen, 2005)
72. c (Arns, De Ritter, Strehl, Breteler & Coenen, 2009)
73. b (Tan, Thornby, Hammond, Strehl, Canady, Arnemann, Kaiser & DeBakey, 2009; Arns, De Ritter, Strehl, Breteler & Coenen, 2009)
74. c (Paquette, Beauregard, Beaulieu-Prévoste, 2009)
75. c (Thompson, J.W.G., Hagedorn, 2012)
76. d (Thompson, J.W.G., Hagedorn, 2012; Thompson, Sebastianelli, Slobounov, 2005)
77. a (Meyer et al., 1970; Nilsson & Nordstrom, 1977; DeWitt et al., 1986; Shaw, 2002)
78. c (Giza & Hovda, 2001)
79. a (Slobounov, Sebastianelli & Simon, 2002; Thatcher, 2009)
80. a (Ross, 1982; Thompson, 2003)
81. d (Ross, 1982; Thompson, 2003)
82. a (Shamay-Tsoory, Tomer, Berger, Goldsher, & Aharon-Peretz, 2005; Thompson & Thompson, 2010
83. c
84. c (Mayberg, Lozano, Voon, McNeely, Seminowicz, Hamani, Schwalb & Kennedy, 2005)
85. c (Paquette, Vincent, Beauregard, Mario, Beaulieu-Prévoste & Dominic, 2009)
86. b (Goldapple, Segal, Garson, Lau, Bieling, Kennedy & Mayberg, 2004; Mayberg, Brannan, Tekell, Silva, Mahurin, McGinnis & Jerabek, 2000)
87. b (Thompson, & Hagedorn, 2012)
88. a (Kauslendratripathi, 2013; Metrametrix)
89. c (Shaw, 2002; Yuan, Xiao-Quan, Prough, Smith, Thomas & Dewitt, 1988; Meyer et al., 1970; Nilsson and Nordstrom, 1977; DeWitt et al., 1986; Pascual-Marqui, Michel & Lehmann, 1994)

References

Note: A helpful listing of articles is the Comprehensive Neurofeedback Bibliography organized and kept current by D. Corydon Hammond. It is arranged according to disorders (epilepsy, ADHD, anxiety disorders, depression, etc.), and is accessible under the heading Neurofeedback Archive on the website of the International Society for Neuronal Regulation, www.isnr.org.

Abu-Akel, A. (2003). A Neurobiological Mapping of Theory of Mind. *Brain Research Reviews, 43*, 39-40.

Achard, S., Salvador, R., Whitcher, B., Suckling, J. & Bullmore, E. (2006). A resilient, low-frequency, small-world human brain functional network with highly connected association cortical hubs. *J Neurosci., 26*(1), 63-72.

Adolphs, R. (2003). Is the human amygdala specialized for processing social information? *Annals of the New York Academy of Sciences, 985*, 326 - 340.

Adolphs, R., Gosselin, F., Buchanan, T.W., Tranel, D., Schyns, P. & Damasio, A.R. (2005). A mechanism for impaired fear recognition after amygdala damage. *Nature, 433*(7021), 68-72.

Aguirre-Perez, D.M., Otero-Ojeda. G.A., Pliego-Rivero, F.B. & Ferreira-Martinez, A. (2007). Relationship of working memory and EEG to academic performance: a study among high school students. *Internat J. Neurosci 117*(6), 869-882.

Alexander, G.E., DeLong, M.R. & Strick, P.L. (1986). Parallel organization of functionally segregated circuits linking basal ganglia and cortex. *Annual Review of Neuroscience, 9*, 357-81.

Alhambra, M.A., Fowler, T.P. & Alhambra, A.A. (1995). EEG biofeedback: A new treatment option for ADD/ADHD. *Journal of Neurotherapy, 1*(2), 39-43.

Amen, D.G. (1998). *Change Your Brain: Change Your Life*. New York: Times Books division of Random House, Inc.

Amen, D.G. (2001). *Healing ADD*. New York: G. P. Putnam's Sons.

Amen, D.G., Carmichael, B.D. & Thisted, R.A. (1997). High resolution SPECT imaging in ADHD. *Annals of Clinical Psychiatry, 9*(2), 81-86.

American Psychiatric Association (2000). *Diagnostic and statistical manual of mental disorders - 4th edition – Text Revision*. Washington, DC.

Amodio, D.M. & Frith, C.D. (2006). Meeting of minds: the medial frontal cortex and social cognition. *Nature Reviews Neuroscience, 7*, 268-277.

Anda, R., Tietjen, G., Schulman, E., Felitti, V. & Croft, J. (2010). Adverse childhood experiences and frequent headaches in adults. *Headache. 50*(9), 1473-81.

Anderson, P. & Anderson, S.A. (1968). *Physiological basis of the alpha rhythm*. New York: Appleton.

Andreassi, J.L. (1995). *Psychophysiology, Human Behavior & Physiological Response - Third Edition*. New Jersey: Lawrence Earlbaum Associates.

Andreassi J.L. (1995). *Op. cit.* p. 327, after Blanchard, E.B., Theobald, D.E., Williamson, D.A., Silver, B.V. & Brown, D. A. (1978). A controlled evaluation of temperature biofeedback in the treatment of migraine headaches. *Archives of General Psychiatry, 41*, 121-127.

Andreassi, J.L. (1995). *Op. cit.*, after Ekman, P., Levenson, R.W., & Friesen, W.V. (1983). Autonomic nervous system activity distinguishes among emotions. *Science, 22*, 1208-1210.

Andreassi, J.L. (1995). *Op. cit.*, after Sakai, L., Baker, L. & Dawson, M. (1992). Electrodermal lability: Individual differences affecting perceptual speed and vigilance performance in 9 to 16 year-old children. *Psychophysiology, 29*, 207- 217.

Andreassi, J.L. (1995). *Op.cit.*, p 326, after Cohen, J. & Sedlacek, K. (1983). Attention and autonomic self regulation. *Psychosomatic Medicine, 45*, 243-257.

Andreassi, J.L. (1995). *Op. cit.*, p 318, after Denkowski, K.M., Denkowski, G.C., Omizo, M.M. (1984). Predictors of success in the EMG biofeedback training of hyperactive male children. *Biofeedback and Self-Regulation, 9*, 253-264.

Andreassi, J.L. (1995). *Op. cit.*, p 156, after Fridlund, A.J., Cottam, G.L., Fowler, S.C. (1984). In search of a general tension factor: Tensional patterning during auditory stimulation. *Psychophysiology, 19*, 136-145.

Andreassi, J.L. (1995). *Op. cit.,* p 176, after Yuille, J.C. & Hare, R.D. (1980). A psychophysiological investigation of short term memory. *Psychophysiology, 17,* 423-430.

Andreassi, J.L. (2007). *Psychophysiology: Human behavior and physiological response.* Hillsdale, NJ: Erlbaum.

Angelakis, E. & Evangelia, L. (2011). Transcranial Electrical Stimulation: Methodology and Applications. *Journal of Neurotherapy, 15,* 337-357

Angelakis, E., Lubar, J.F., Stathopoulou, S. & Kounios, J. (2004). Peak alpha frequency: an electroencephalographic measure of cognitive preparedness. *Clinical Neurophysiology, 115*(4), 887-897.

Antal, A., Brepohl, N., Poreisz, C., Boros, K., Csifcsak, G. & Paulus, W. Transcranial direct current stimulation over somatosensory cortex decreases experimentally induced acute pain perception. *Clin J Pain, 24*(1), 56-63.

Applehans, B.M. & Luecken, L.J. (2006). Attentional processes, anxiety, and the regulation of cortisol reactivity. *Anxiety, Stress & Coping: An International Journal, 19*(1), 81-92.

Arena, J., Bruno, G. & Brucks, A. (1997). The use of EMG biofeedback for the treatment of chronic tension headache. *The Biofeedback Foundation of Europe,* www.bfe.org.

Arena, J., Bruno, G., Hannah, S. L. & Meader, K.J. (1995). Comparison of frontal electromyographic biofeedback training, trapezius electromyographic biofeedback training, and progressive muscular relaxation therapy in the treatment of tension headache. *Headache, 35*(7), 411- 419.

Arns, M., De Ridder, S., Strehl, U., Breteler, M. & Coenen, A. (2009). Efficacy of neurofeedback treatment in ADHD: The effects on inattention, impulsivity and hyperactivity: A meta-analysis. *Clinical EEG and Neuroscience; 40*(3), 180-189

Arns, M. & Kenemans, J.L. (2012). Neurofeedback in ADHD and insomnia: Vigilance Stabilization through sleep spindles and circadian networks. *Neuroscience and Biobehavioral Reviews.* doi: 10.1016/j.neubiorev. 2012.10.006

Arnsten, A.F. (2001). Dopaminergic and noradrenergic influences on cognitive functions mediated by prefrontal cortex. In Solanto, M.V., Arnsten, A.F.T. & Castellanos, F.X. (eds.) *Stimulant Drugs and ADHD Basic and Clinical Neuroscience.* New York: Oxford University Press, p. 186.

Aron, A.R., Robbins, T.W. & Poldrack, R.A. (2004). Inhibition and the right inferior frontal cortex. *Trends in cognitive sciences, 8*(4), 170-177.

Arul-Anandam A.P. & Loo C. (2009). Transcranial direct current stimulation: a new tool for the treatment of depression? *J Affect Disord., 117*(3):137-145.

Asperger, H. (1944). Die "Autistischen Psychopathen" im Kindesalter. *Archiv fuer Psychiatrie und Nervenkrankheiten, 117,* 76-136. English translation (1991) Autistic psychopathy in childhood. In Frith, U. (ed. & trans.) *Autism and Asperger's syndrome* (37-92). Cambridge, United Kingdom: Cambridge University Press.

Astin, J.A., Beckner, W., Soeken, K., Hochberg, M.C. & Berman, B. (2002). Psychological interventions for rheumatoid arthritis: A meta-analysis of randomized controlled trials. *Arthritis and Rheumatism. 47*(3), 291-302.

Astur, R.S. & Constable, R.T. (2004). Hippocampal dampening during a relational memory task. *Behav Neurosci, 118,* 667- 675.

Astur, R.S., St. Germain, S.A., Baker, E.K., Calhoun, V., Pearlson G.D. & Constable, R.T. (2005). FMRI hippocampal activity during a virtual radial arm maze. *Appl Psychophysiol Biofeedback, 30,* 307-317.

Attwood, T. (1997). *Asperger's Syndrome: A Guide for Parents and Professionals.* London: Jessica Kingsley Publications.

Augustine, J.R. (2008). *Human Neuroanatomy: An Introduction.* Academic Press. San Diego, California.

Ayers, M.E. (1995). Long-term follow-up of EEG neurofeedback with absence seizures. *Biofeedback and Self-Regulation, 20*(3), 309-310.

Babiloni, C., Babiloni, F., Carducci, F., Cincotti, F., Vecchio, F., Cola, B. Rossi, S., Miniussi, C. & Rossini, P.M. (2004). Functional Frontalparietal Connectivity During Short-Term Memory as Revealed by High-Resolution EEG Coherence Analysis. *Behav Neurosci, 118*(4), 687-697

Bachevalier, J. & Loveland, K.A. (2006). The orbitofrontal-amygdala circuit and self-regulation of social-emotional behavior in autism. *Neuroscience and Behavioral Reviews, 30,* 97-117.

Baehr, E., Rosenfeld, J.P., Baehr, R. & Earnest, C. (1998). Comparison of two EEG asymmetry indices in depressed patients vs. normal controls. *International Journal of Psychophysiology, 31,* 89-92.

Baehr, E., Rosenfeld, J.P., Baehr, R. & Earnest, C. (1999). Clinical use of an alpha asymmetry neurofeedback protocol in the treatment of mood disorders. In Evans, J.R. & Abarbanel, A. (eds.) *Introduction to Quantitative EEG and Neurofeedback.* San Diego: Academic Press.

Baehr, E., Rosenfeld, J.P. & Baehr, R. (2001). Clinical use of an alpha asymmetry neurofeedback protocol in the treatment of mood disorders: Follow-up study one to five years post therapy. *Journal of Neurotherapy*, *4*(4), 11-18.

Baguley, I.J., Heriseanu, R.E., Felmingham, K.L. & Cameron, I.D. (2006). Dysautonomia and heart rate variability following severe traumatic brain injury. *Brain Injury*, *20*(4), 437-444.

Baker, L.D., Frank, L.L., Foster-Schubert, K., Green, P.S., Wilkinson, C.W., McTiernan, A., Plymate, S.R., Fishel, M.A., Watson, G.S., Cholerton, B.A., Duncan, G.E., Mehta, P.D. & Craft, S. (2010). Effects of aerobic exercise on mild cognitive impairment: a controlled trial, *Arch Neurol.* 67(1), 71-79.

Banks, S.L., Jacobs, D.W., Gevirtz, R. & Hubbard, D.R., (1998). Effects of autogenic relaxation training on electromyographic activity in active myofascial trigger points. *Journal of Musculoskeletal Pain*, *6* (4), 23-32.

Barnell Loft, Ltd., *Multiple Skills Series – Second Edition.* (1990) New York: SRA Division of MacMillan/McGraw-Hill School Publishing Co.

Barry, R.J., Clarke, A.R., Johnstone, S.J., McCarthy, R. & Selikowitz, M. (2009). Electroencephalogram theta-beta ratio and arousal in attention-deficit-hyperactivity disorder: Evidence of independent processes. *Biological Psychiatry, 66*, 398-401.

Barry, R.J., Johnstone, S.J. & Clarke, A.R. (2003). A review of electrophysiology in attention-deficit/hyperactivity disorder: II. Event-related potentials. *Clinical Neurophysiology, 114*, 184-198.

Basmajian, J.V. (1989). *Biofeedback: Principles and Practice for Clinicians - Third Edition.* Baltimore: Williams and Wilkins.

Basmajian, J.V. (1989). Anatomical and physiological basis for biofeedback of autonomic regulation. In Basmajian, J.V., *Op cit.*, 33-48.

Basmajian, J.V. (1989). *Op. cit.*, p 77, after Morgan, W.P. (1896). A case of congenital word blindness. *British Medical Journal*, *2*, 1612.

Basmajian, J.V. (1989). *Op.cit.,* p 207, after Mitchell, K.R. & Mitchell, D.M. (1971). Migraine: An explanatory treatment application of programmed behavior therapy techniques. *Psychosomatic Research*, *15,* 137-157.

Basmajian, J.V. (1989). *Op cit.*, p 179, after Stoyva, J.M. (1986). Wolfgang Luthe: In memoriam. *Biofeedback and Self-Regulation*, *11*, 91-93.

Bauer, G. & Bauer, R. (1999). EEG drug effects and central nervous system poisoning. In Niedermeyer, E. & Da Silva, F.L. (eds.) *Electroencephalography: Basic Principles, Clinical Applications and Related Fields.* Baltimore: Williams & Wilkins, 671-691.

Bear, D.M. & Fedio, P. (1977). Quantitative analysis of interictal behavior in temporal lobe epilepsy. Arch Neurol., 34(8), 454-67.

Beasley, C.L., Pennington, K., Behan, A., Wait, R., Dunn, M.J. & Cotter, D. (2006). Proteomic analysis of the anterior cingulate cortex in the major psychiatric disorders: Evidence for disease-associated changes. Proteomics, 6(11), 3414-3425.

Beauregard, M. & Levesque, J. (2006). Functional Magnetic Resonance Imaging Investigation of the Effects of Neurofeedback Training on the Neural Bases of Selective Attention and Response Inhibition in Children with Attention-Deficit/Hyperactivity Disorder. *Applied Psychophysiology and Biofeedback, 31*(1), 3-20.

Beauregard, M. & Paquette, V. (2006). Neural correlates of a mystical experience in Carmelite nuns. Neuroscience Letters, 405, 186-190.

Benedetti, F., Mayberg, H.S., Wager, T.D., Stohler, C.S. & Zubieta, J.K. (2005). Neurobiological Mechanisms of the Placebo Effect. *Journal of Neuroscience, (25)*45, 10390-10402.

Beniamini, Y., Rubenstein, J.J., Zaichkowsky, L.D. & Crim, M.C. (1997). Effects of high intensity strength training on quality-of-life parameters in cardiac rehabilitation patients. *Am J. Cardiol, 80*, 841-846.

Bergeron, S., Binik, Y.M., Khalifé, S., Pagidas, K., Glazer, H.L., Meana, M. & Amsel, R. (2001). A randomized comparison of group cognitive-behavioral therapy, surface electromyographic biofeedback and vestibulectomy in the treatment of dyspareunia resulting from vulvar vestibulitis. *Pain, 91*(3), 297-306.

Berman, M.G., Peltier, S., Nee, E.D., Kross, E., Deldin, P.J. & Jonides, J. (2011). Depression, rumination and the default network. *SCAN, 6*, 548-555

Bernstein, D.M. (2002). Information processing difficulty long after self-reported concussion. *Journal of the International Neuropsychological Society, 8*(5), 673-682.

Bhandari, T., Thompson, L. & Reid-Chung, A. (2013). Treating post-concussion syndrome using neurofeedback: A case study. *Biofeedback, 41*(4).

Bigger, J.T., Jr., Kleiger, R.E., Fleiss, J.L. Rolnitzky, L.M., Steinman, R.C. & Miller, J.P. (1988). Components of heart rate variability measured during healing of acute myocardial infarction. Am J Cardiol, 61, 208-215.

Bisley, J.W. & Goldberg, M.E. (2006). Neural correlates of attention and distractibility in the lateral intraparietal area. *J Neurophysiol., 95*(3), 1696-1717.

Bluhm, R.L., Williamson, P.C., Osuch, E.A., Frewen, P.A., Stevens, T.K., Boksman, K., Neufeld, R.W.J., Théberge, J. & Lanius, R.A. (2009). Alterations in default mode network connectivity in posttraumatic stress disorder related to early-life trauma. *Journal of Psychiatry and Neuroscience*, 34(3), 187-94.

Blum, K. & Cummings, D. (1996). Reward deficiency syndrome. *American Scientist, 84.*

Blumberg, H.P. (2000). Increased Anterior Cingulate and Caudate Activity in Bipolar Mania. *Biol Psychiatry, 48,* 1045-1052.

Blumer D. (1975). Temporal lobe epilepsy and its psychiatric significance. In: Benson, D.F., Blumer, D. (eds.), *Psychiatric Aspects of Neurological Disease.* Grune & Stratton; New York, 171-198.

Blumer, D. & Benson, D.F., Psychiatric manifestations of epilepsy. In Blumer, D., Benson, D.F. (eds.), *Psychiatric aspects of neurological disease, volume II,* New York, Grune & Stratton, 1982.

Blumer, D. & Walker, A. (1967). Sexual behaviour in temporal lobe epilepsy. *Arch. Neurol., 16,* 31-43

Boddaert, N. & Chabane, N. (2002). Temporal lobe dysfunction in childhood autism. *Journal of Radiology, 83,* 1829-1833.

Bogacz, R., Wagenmakers, E. J., Forstmann, B. U. & Nieuwenhuis, S. (2010). The neural basis of the speed-accuracy tradeoff. *Trends in Neurosciences, 33*(1), 10-6.

Boggio, P.S., Ferrucci, R., Rigonatti, S.P., Covre, P., Nitsche, M., Pascual-Leone, A. & Fregni, F. (2006). Effects of transcranial direct current stimulation on working memory in patients with Parkinson's disease. *J Neurol Sci., 249*(1), 31-38.

Boggio, P.S., Rigonatti, S.P., Ribeiro, R.B., Myczkowski, M.L., Nitsche, M.A., Pascual-Leone, A. & Fregni, F.A. (2008). Randomized, double-blind clinical trial on the efficacy of cortical direct current stimulation for the treatment of major depression. *Int J Neuropsychopharmacol. 11*(2), 249-254.

Boggio, P.S., Sultani, N., Fecteau, S., Merabet, L., Mecca, T., Pascual-Leone, A., Basaglia, A. & Fregni, F. (2008). Prefrontal cortex modulation using transcranial DC stimulation reduces alcohol craving: a double-blind, sham-controlled study. *Drug Alcohol Depend., 1;92*(1-3), 55-60.

Bonanni, L., Franciotti, R., Onofrj, V., Anzellotti, F., Mancino, E., Monaco, D., Gambi, F., Manzoli, L., Thomas, A. & Onofrj, M. (2010). Revisiting P300 cognitive studies for dementia diagnosis: Early dementia with Lewy bodies (DLB) and Alzheimer disease (AD). *Neurophysiologie Clinique/Clinical Neurophysiology, 40,* 255-265.

Bonelli, R.M. & Cummings, J.L. (2007). Frontal-subcortical circuitry and behavior. *Dialogues in Clinical Neuroscience, 9*(2), 141.

Boro, A.D. (2014). Benbadis, S.R. (chief ed.), Focal EEG Waveform Abnormalities. Medscape online resource.

Bottini, G., Corcoran, R., Sterzi, R., Paulesu, E., Schenone, P., Scarpa, P., Frackowiak, R.S.J. & Frith, D. (1994). The role of the right hemisphere in the interpretation of figurative aspects of language A positron emission tomography activation study. *Oxford Journals, Medicine, Brain, 117*(6), 1241-1253.

Boulay, C.B., Sarnacki, W.A., Wolpaw, J.R. & McFarland, D.J. (2011). Trained modulation of sensorimotor rhythms can affect reaction time. *Clinical Neurophysiology: Official Journal of the International Federation of Clinical Neurophysiology, 122*(9), 1820-6.

Bouyer, J.J., Detet, L., Konya, A. & Rougeul, A. (1974). Convergence de trios systems rythmiques thalamo-corticaux sur l'aire somesthesique du chat et du babouin normaux. *Rev. Electroenceph. Clin. Neurophysiol. 4,* 397-406.

Boyd, W.D. & Campbell, S.E. (1998). EEG biofeedback in the schools: The use of EEG biofeedback to treat ADHD in a school setting. *Journal of Neurotherapy, 2*(4), 65-71.

Brodmann, K. (1909), *Brodmann's Localisation in the Cerebral Cortex* (Garey, L.J., trans., ed.) Springer. 2006.

Brotman, P. (1989). Low-intensity transcranial stimulation improves the efficacy of thermal biofeedback and quieting reflex training in the treatment of classical migraine headache. *American Journal of Electromedicine, 6*(5), 120-123.

Brown, J.W. & Braver, T.S. (2005). Learned predictions of error likelihood in the anterior cingulate cortex. *Science* (New York, N.Y.), *307*(5712), 1118-21.

Brown, J.W., Bullock, D. & Grossberg, S. (2004). How laminar frontal cortex and basal ganglia circuits interact to control planned and reactive saccades. *Neural Networks : The Official Journal of the International Neural Network Society, 17*(4), 471-510.

Brucker, B.S. & Bulaeva, N.V. (1996). Biofeedback effect on electromyography responses in patients with spinal cord injury. *Archives of Physical Medical Rehabilitation. 77*(2), 133-137.

Brunoni, A.R., Valiengo, L., Baccaro, A., Zana, T.A., de Oliveira, J.F., Goulart, A., Boggio, P.S., Lotufo, P.A., Bensen, I.M. & Fregni, F. (2013). The Sertraline vs Electrical Current Therapy for Treating Depression Clinical Study Results from a Factorial, Randomized, Controlled Trial. *Jama Psychiatry Published Online,* February 6, 2013.

Buchanan, T.W., Tranel, D. & Adolphs, R., (2009). In *The Human Amygdala,* Whalen, P.J. & Phelps, E.A. (eds.) 289-318, Guilford, New York.

Buckner, R.L., Andrews-Hanna, J.R. & Schacter, D.L. (2008). The Brain's Default Network: Anatomy, Function, and Relevance to Disease. *Ann. N.Y. Acad. Sci., 1124,* 1-38.

Budzynski, T.H. (1979). Biofeedback and the twilight states of consciousness. In Goleman, D. & Davidson, R. J. (eds.) *Consciousness in Brain States of Awareness and Mysticism.* New York: Harper & Row.

Bullmore, E. & Sporns, O. (2012). The economy of brain network organization. *Nature Reviews Neuroscience 13,* 336-349.

Bullock, D., Tan, C.O. & John, Y.J. (2009). Computational perspectives on forebrain microcircuits implicated in reinforcement learning, action selection, and cognitive control. *Neural Networks : The Official Journal of the International Neural Network Society, 22*(5-6), 757-65.

Buysse, D.J., Germain, A., Moul, D.E., Franzen, P.L., Brar, L.K., Fletcher, M.E., Begley, A., Houck, P.R., Mazumdar, S., Reynolds, C.F., III & Monk, T.H. (2011). Efficacy of Brief Behavioral Treatment for Chronic Insomnia in Older Adults. *Arch Intern Med. 171*(10), 887-895.

Buzsaki, G. (2006). *Rhythms of the Brain.* Oxford University Press.

Byers, A.P. (1998). *The Byers Neurotherapy Reference Library – Revised Second Edition.* Wheat Ridge, Colorado; Association for Applied Psychophysiology and Biofeedback.

Campbell, N.A., Reece, J.B., Michell, L.G. (1996). *Biology - Fifth Edition.* New York: Addison Wesley Longman, Inc., 960-974.

Campbell, T.A., Winkler, I. and Kujala, T. (2007). "N1 and the mismatch negativity are spatiotemporally distinct ERP components: Disruption of immediate memory by auditory distraction can be related to N1". *Psychophysiology* 44: 530–540.

Cannon, R., Congedo, M., Lubar, J. & Hutchens, T. (2009). Differentiating a network of executive attention: LORETA neurofeedback in anterior cingulate and dorsolateral prefrontal cortices: Mapping a Network of Executive Attention. *International Journal of Neuroscience, 119* (3), 401-441.

Cannon, R. & Lubar, J.F. (2007). EEG Spectral Power and Coherence: Differentiating Effects of Spatial-specific Neuro-Operant Learning (SSNOL) Utilizing LORETA Neurofeedback Training in the Anterior Cingulate and Bilateral Dorsolateral Preftontal Cortices. *Journal of Neurotherapy 11* (3), 25-44.

Cannon, R.L., (2012). *Low Resolution Electromagnetic tomography (LORETA): Basic Concepts and Clinical Applications.* DMED Press, Corpus Christi, Texas , p 160.

Cantor, D.S. (1999). An Overview of Quantitative EEG and Its Applications to Neurofeedback. In J. R. Evans & A. Abarbanel, *Introduction to Quantitative EEG and Neurofeedback.* New York: Academic Press.

Carlson, J.G., Chemtob, C.M., Rusnak, K., Hedlund, N.L. & Muraoka, M.Y. (1998). Eye movement desensitization and reprocessing (EMDR) treatment for combat-related posttraumatic stress disorder. *Journal of Traumatic Stress, 11*(1), 3-24.

Carlson, N.R. (1986). *Physiology of Behavior - Third Edition.* Toronto: Allyn and Bacon Inc.

Carmen, J. (2004). Passive infrared hemoencephalography: four years and 100 migraines. *Journal of Neurotherapy, 8* (3), 23-51.

Carmody, D.P., Radvanski, D.C., Wadhwani, S., Sabo, M.J. & Vergara, L. (2001). EEG biofeedback training and ADHD in an elementary school setting. *Journal of Neurotherapy, 4*(3), 5-27.

Carter, R. (1998). *Mapping the Mind.* London: Weidenfeld & Nicolson.

Casanova, M.F., El-Baz, A., Mott, M., Mannheim, G., Hassan, H., Fahimi, R. & Farag, A. (2009). Reduced gyral window and corpus callosum size in autism: Possible macroscopic correlated of a minicolumnopathy. *Journal of Autism & Developmental Disorders, 39,* 751-764.

Castellanos, F.X. (2001). Neuroimaging Studies in ADHD children on Stimulant Drugs. In Solanto, M.V., Arnsten, A.F.T. & Castellanos, F.X. (eds.) *Stimulant Drugs and ADHD Basic and Clinical Neuroscience.* New York: Oxford University Press, 243-258.

Catani, M. & de Schotten, M.T. (2012). *Atlas of Human Brain Connections.* Oxford University Press, NY.

Catani, M. & Jones, D.K. (2005). Perisylvian language networks of the human brain. *Annals of neurology, 57*(1), 8-16.

Cattaneo Z, Pisoni A. & Papagno C. (2011). Transcranial direct current stimulation over Broca's region improves phonemic and semantic fluency in healthy individuals. *Neuroscience, 2*(183), 64-70.

Cavanna, A.E &, Trimble, M.R. (2006). The precuneus: a review of its functional anatomy and behavioural correlates. *Brain 129* (Pt 3), 564-583.

Chabot, R.J., di Michele, F., Prichep, L. & John, E.R. (2001). The clinical role of computerized EEG in the evaluation and treatment of learning and attention disorders in children and adolescents. *Journal of Neuropsychiatry & Clinical Neurosciences. 13*(2), 171-86.

Chabot, R.J., Orgill, A.A., Crawford, G., Harris, M.J. & Serfontein, G. (1999). Behavioural and electrophysiological predictors of treatment response to stimulants in children with attention disorders. *Journal of Child Neurology, 14*(6), 343-351.

Chabot, R.J. & Serfontein, G. (1996). Quantitative electroencephalographic profiles of children with attention deficit disorder. *Biological Psychiatry, 40*, 951-963.

Chapman, D.P., Wheaton, A.G., Anda, R,F., Croft, J.B., Edwards, V.J., Liu, Y., Sturgis, S.L. & Perry, G.S. (2011). Adverse childhood experiences and sleep disturbances in adults. *Sleep Medicine, 12,* 773-79.

Chatrian, G.E., Lettich, E., Nelson, P.L. (1985). Ten percent electrode system for topographic studies of spontaneous and evoked EEG activity. *American Journal of EEG Technology, 25,* 83-92.

Chauhan, N.B., Siegel, G.J. & Lee, J.M. (2001). Depletion of glial cell line-derived neurotrophic factor in substantia nigra neurons of Parkinson's disease brain. *Journal of Chemical Neuroanatomy, 21*(4), 277-288.

Cheng, P. (1993). Metacognition and Giftedness: The state of the relationship. *Gifted Child Quarterly, 37* (3).

Chiesa, V., Gardella, E., Tassi, L., Canger, R. & Lo Russo, G. (2007). Age-related gender differences in reporting ictal fear: Analysis of case histories and review of the literature. *Epilepsia, 48*(12), 2361-2364.

Choo, I.H., Lee, D.Y., Lee, J.H., Kim, K.W., Jhoo, J.H., Ju, Y-S., Yoon, J.C., Kim, S.G., Ha, J. & Woo, J.I. (2009). The prevalence of cognitive impairment with no dementia in older people: The Seoul study. *International Journal of Geriatric Psychiatry, 24*(3), 306-312.

Choo, I.H., Lee, D.Y., Oh J.S., Lee J.S., Lee D.S., Song I.C., Youn J.C., Kim S.G., Kim K.W., Jhoo J.H. & Woo J.I. (2010). Posterior cingulate cortex atrophy and regional cingulum disruption in mild cognitive impairment and Alzheimer's disease. *Neurobiol. Aging 31,* 772-779.

Christopoulos, G.I., Tobler, P.N., Bossaerts, P., Dolan, R.J. & Schultz, W. (2009). Neural Correlates of Value, Risk, and Risk Aversion Contributing to Decision Making under Risk. *J Neurosci 26* (24), 6469-6472.

Civier, O., Bullock, D., Max, L. & Guenther, F.H. (2013). Computational modeling of stuttering caused by impairments in a basal ganglia thalamo-cortical circuit involved in syllable selection and initiation. *Brain and Language, 126*(3), 263-78.

Clark, C.R., Galletly, C.A., Ash, D.J., Moores, K.A., Penrose, R.A. & McFarlane, A.C. (2009). Evidence-based medicine evaluation of electrophysiological studies of the anxiety disorders. *Clinical EEG and Neuroscience, 40*(2), 84-112.

Clarke, A.R., Barry, R.J., McCarthy, R. & Selikowitz, M. (2001). Electroencephalogram differences in two subtypes of attention-deficit/hyperactivity disorder. *Psychophysiology, 38,* 212-221.

Coben, R. (2005). Assessment guided neurofeedback for autistic spectrum disorder. Paper presented at the 13th Annual Meeting of the Society for Neuronal Regulation , Denver, Colorado.

Coben, R. (2006). Autism Spectrum Disorders. Panel presentation at the 36th Annual Meeting of the Association for Applied Psychophysiology and Biofeedback, Portland, OR.

Coben, R. (2007). Connectivity-guided neurofeedback for autistic spectrum disorder. *Biofeedback, 35*(4), 131-135.

Coben, R. & Pudolsky, L. (2007). Infrared imaging and neurofeedback: Initial reliability and validity. *Journal of Neurotherapy, 11*(3), 3-13.

Cohen, B. (1989). Basic biofeedback electronics for the clinician. In Basmajian, J.V., *Biofeedback Principles and Practice for Clinicians - Third edition.* Baltimore: Williams and Wilkins.

Collins, D.L., Neelin, P., Peters, T.M. & Evans, A.C. (1994). Automatic 3-D intersubject registration of MR volumetric data in standardized Talairach space. *J. Comput. Assist. Tomogr., 18* (2), 192-205.

Congedo, M. (2006). Subspace projection filters for real-time brain electromagnetic imaging. *IEEE Transactions on Bio-Medical Engineering, 53,* 1624-1634.

Congedo, M., Lubar, J.F. & Joffe, D. (2004). Low-resolution electromagnetic tomography neurofeedback. *IEEE Trans Neural Syst Rehabil Eng., 12*(4), 387-397.

Congedo, M., Ozen, C. & Sherlin, L. (2002). Notes on EEG resampling by natural cubic spline interpolation. *Journal of Neurotherapy, 6*(4), 73-80.

Contreras, D., Destexhe, A., Sejnowski, T.J. & Steriade, M. (1997). Spatiotemporal patterns of spindle oscillations in cortex and thalamus. *Neuroscience, 17*(3), 1179-1196.

Cooke, S.F. & Bliss, T.V. (2006). Plasticity in the human central nervous system. *Brain, 129* (Pt 7), 1659-1673.

Coull, J.T., Nobre, A.C. (1998). Where and When to Pay Attention: The Neural Systems for Directing Attention to Spatial Locations and to Time Intervals as Revealed by Both PET and fMRI. *The Journal of Neuroscience, 18*(18), 7426-7435.

Courchesne, E., Karnes, C.M., Davis, H.R., Ziccardi, R., Carper, R.A., Tigue, A.D., Chisum, H.J., Moses, P., Pierce, K., Lord, D., Lincoln, A.J., Pizzo, S., Schreiban, L., Haas, R.H., Akshoomoff, N.A. & Courchesne, R.Y. (2001). Unusual brain growth patterns in early life in patients with autistic disorder: An MRI study. *Neurology. 57*(2), 245-54.

Cowen, M.J., Pike, K.C. & Budzynski, H.K. (2001). Psychosocial nursing therapy following sudden cardiac arrest: Impact on two year survival. *Nursing Research, 50*(2), 68-76.

D'Argembeau, A., Collette, F., Van der Linden, M., Laureys, S., Del Fiore, G., Degueldre, C., Luxen, A. & Salmon, E. (2005). Self-referential reflective activity and its relationship with rest: a PET study. *NeuroImage, 25,* 616-624.

Damasio, A. (2003a). Feelings of Emotion and the Self. *Annals of the New York Academy of Sciences 1001,* 253-261.

Damasio, A. (2003b). Mental self: the person within. *Nature, 423* (6937), 227-227.

Daniels, J.K., McFarlane, A.C., Bluhm, R.L., Moores, K.A., Clark, C.R., Shaw, M.E., Williamson, P.C., Densmore, M. & Lanius, R.A. (2010). Switching between executive and default mode networks in posttraumatic stress disorder: alterations in functional connectivity, *Journal of Psychiatry and Neuroscience, 35*(4) 258-266.

Dapretto, M., Davies, M.S., Pfeifer, J.H., Scott, A.A., Sigman, M., Bookheimer, S.Y. & Iacoboni, M. (2006). Understanding emotions in others: mirror neuron dysfunction in children with autism spectrum disorders. *Nature Neuroscience, 9*(1), 28-30.

Davidson, R.J. (1995). Cerebral asymmetry, emotion and affective style. In Davidson, R.J. & Hugdahl, K. (eds.) *Brain Asymmetry.* Cambridge, MA: MIT Press, 369-388.

Davidson, R.J. (1998). Anterior electro-physiological asymmetries, emotion, and depression: Conceptual and methodological conundrums. *Psychophysiology, 35,* 607-614.

Davidson, R.J. (2000). Cognitive neuroscience needs affective neuroscience (and vice versa). *Brain Cogn. 42,* 89-92.

Davidson, R.J. 2002. Anxiety and affective style: role of prefrontal cortex and amygdala. *Biolog. Psychiat., 51* (1), 68-80.

Davidson, R.J., Abercrombie, H., Nitschke, J.B. & Putnam, K. (1999). *Current Opinion in Neurobiology, 9,* 228-234.

Davidson. R.J., Pizzagalli, D., Nitshke, J.B. & Putnam, K. (2002). Depression: perspectives from affective neuroscience. *Annual Rev. Psychol., 5*(3), 545-574.

De Beaumont, L., Theoret, H., Mongeon, D., Messier, J., Leclerc, S., Tremblay, S. & Lassonde, M. (2009). Brain function decline in healthy retired athletes who sustained their last sports concussion in early adulthood. *Brain, 132*(Pt 3), 695-708.

De Ridder, D. (2009). An evolutionary approach to brain rhythms and its clinical implications for brain modulation. *Journal of Neurotherapy, (13)*1, 69-70.

De Ridder, D., (2010). Alcohol Addiction: A Clinical Pathophysiological Approach. *Proceedings of the ISNR 18th Annual Conference,* Denver, Colorado.

De Ridder, D., (2011). Limbic dysrhythmia and delta pathologies. *ISNR Presentation* 2011.

De Ridder, D., Van Laere, K., Dupont, P., Menovsky, T. & Van de Heyning, P. (2007). Visualizing Out-of-Body Experience in the Brain. *N Engl J Med, 357,* 1829-1833.

De Ridder, D., Vanneste, S., Adriaenssens, I., Lee, A.P.K., Plazier, M., Menovsky, T., van der Loo, E., Van de Heyning, P.& Møller, A. (2010). Microvascular Decompression for Tinnitus: Significant Improvement for Tinnitus Intensity Without Improvement for Distress. A 4-Year Limit. *Neurosurgery, 66*(4), 656-660.

deBeus, R. (2005). Efficacy of Attention Training for Children with ADHD: A Randomized Double-Blind Placebo-Controlled Study. Presented at: *The Annual Meeting of the International Society for Neurofeedback and Research.* Atlanta, GA, USA, 7-10 September 2006.

Dedovic, K., Duchesne, A., Andrews, J., Engert, V. & Pruessner J.C. (2009). The Brain and the stress axis: the neural correlates of cortisol regulation in response to stress. *Neuroimage, 47*(3), 864-871.

Dedovic, K., Renwick, R., Mahani, N.K., Engert, V., Lupien, S.J. & Pruessner, J.C. (2005). The Montreal imaging stress task: using functional imaging to assess the effects of perceiving and processing psychosocial stress in the human brain. *J Psychiatry Neuroscience, 30*(5), 319-325.

Deepak, K.K. & Behari, M. (1999). Specific muscle EMG biofeedback for hand dystonia. *Applied Psychophysiology and Biofeedback, 24*(4), 267-280.

Dehaene, S., Le Clec'h, G., Poline, J. LeBihan, D. & Cohen, L. (2002). The visual word form area. A prelexical representation of visual words in the fusiform gyrus. *NeuroReport, 13*, 321-325.

DeLong, M.R. (1990). Primate models of movement disorders of basal ganglia origin. *TINS, 13*(7), 281-285.

Devinsky, O., Morrell, M. & Vogt, B. (1995). Contributions of Anterior Cingulate Cortex to behavior, *Brain, 118*, 279-306.

DeWitt, D.S., Jenkins, L.W., Wei, E.P., Lutz, H., Becker, D.P. & Kontos, H.A. (1986). Effects of fluid-percussion brain injury on regional cerebral blood flow and pial arteriolar diameter. *Journal of Neurosurgery, 64*(5), 787-794.

Ding, J.B., Guzman, J.N., Peterson, J.D., Goldberg, J.A. & Surmeier, D.J. (2010). Thalamic gating of corticostriatal signaling by cholinergic interneurons. *Neuron, 67*(2), 294-307.

Dobbs, D. (2006). Turning off depression. *Scientific American Mind, 17*(4), 26-31.

Dockree, P.M., Kelly, S.P., Roche, R.A., Hogan, M.J., Reilly, R.B. & Robertson, I.H. (2004). Behavioural and physiological impairments of sustained attention after traumatic brain injury. *Cognitive Brain Research, 20*(3), 403-414.

Dohrmann K., Weisz N., Schlee W., Hartmann T. & Elbert T. (2007). Neurofeedback for treating tinnitus. *Prog Brain Res., 166*, 473-85.

Doidge, N. (2007). *The Brain that changes itself*. Viking Press.

Doidge, N. (2015). *The Brain's Way of Healing*. Viking Press.

Donegan, N.H., Sanislow, C.A., Blumberg, H.P., Fulbright, R.K., Lacadie, C., Skudlarski, P., Gore, J.C., Olson, I.R., McGlashan, T.H. & Wexler, B.E. (2003). Amygdala hyperreactivity in borderline personality disorder: implications for emotional dysregulation. *Biological Psychiatry, 54* (11), 1284-1293.

Dopfner, M. & Rothenberger, A. (2007). Behavior Therapy in Tic-Disorders with Co-Existing ADHD. *European Child & Adolescent Psychiatry, 16*(Suppl.1), 89-99.

Dougherty, M.C., Dwyer, J.W., Pendergast, J.F., Boyington, A.R., Tomlinson, B.U., Coward R.T. Duncan, R.P., Vogel, B. & Rooks, L.G. (2002). A randomized trial of behavioral management for continence with older rural women. *Research in Nursing and Health, 25*(1), 3-13.

Dranias, M.R., Grossberg, S. & Bullock, D. (2008). Dopaminergic and non-dopaminergic value systems in conditioning and outcome-specific revaluation. *Brain Research, 1238*, 239-87.

Duffy, F.H., Hughes, J.R., Miranda, F., Bernad, P. & Cook, P. (1994). Status of quantitative EEG (QEEG) in clinical practice. *Clinical Electroencephalography, 25,* vi-xxii.

Duffy, F.H., Iyer, V.G. & Surwillo, W.W. (1989). *Clinical Electroencephalography and Topographic Brain Mapping: Technology and Practice*. New York: Springer-Verlag.

Duncan, C.C., Kosmidis, M.H. & Mirsky, A.F. (2005). Closed head injury-related information processing deficits: an event-related potential analysis. *International Journal of Psychophysiology, 58*(2-3), 133-157.

Dupuis, F., Johnston, K.M., Lavoie, M., Lepore, F. & Lassonde, M. (2000). Concussions in athletes produce brain dysfunction as revealed by event-related potentials. *Neuroreport, 11*(18), 4087-4092.

Dyro, F.M. (1989). *The EEG Handbook*. Boston: Little, Brown and Co., p 18.

Egner, T. & Gruzelier, J.H. (2001). Learned self-regulation of EEG frequency components affects attention and event-related brain potentials in humans. *Neuro. Report, 12*(18), 4155-4160.

Egner, T. & Gruzelier, J.H. (2003). Ecological validity of neurofeedback modulation of slow wave EEG enhances musical performance. *NeuroReport, 14*(9), 1221-1224

Egner, T. & Gruzelier, J.H. (2004). EEG Biofeedback of low beta band components: Frequency-specific effects on variables of attention and event-related brain potentials. *Clin. Neurophysiol., 115*, 131-139.

Elliott, R., Sahakian, B.J., Matthews, K., Bannerjea, A., Rimmer, J. & Robbins, T.W., (1997). Effects of Methylphenidate on Spatial Working Memory and Planning in Healthy Young Adults. *Psychopharmacology, 131*, 196-206.

Englot, D.J., Yang, L., Hamid, H., Danielson, N., Bai, X., Marfeo, A.,Yu, L. Gordon, A., Purcaro, M.J., Motelow, J.E., Agarwal, R., Ellens, D.J., Golomb, J.D., Shamy, M.C.F., Zhang, H., Carlson, C., Doyle, W., Devinsky, O., Vives, K., Spencer, D.D., Spencer, S.S., Schevon, C., Zaveri, H.P. & Blumenfeld, H. (2010). Impaired consciousness in temporal lobe seizures: role of cortical slow activity. *Brain, 133*, 3764-3777.

Erlandsson, S.I., Rubinstein, B. & Carlsson, S. G. (1991). Tinnitus: Evaluation of biofeedback and stomatognathic treatment. *British Journal of Audiology, 25*(3), 151-161.

Esmail, S. & Linden, D.E.J. (2014), Neural Networks and Neurofeedback in Parkinson's Disease, *NeuroRegulation, 1*(3-4), 240-272.

Evans, A.C., Collins, D.L., Mills, S.R., Brown, E.D., Kelly, R.L. & Peters, T.M. (1993). 3D statistical neuroanatomical models from 305 MRI volumes. Proc IEEE-Nuclear Science Symposium and Medical Imaging Conference, 1813-1817.

Ewing, D.J., Neilson, J.M. & Travis, P. (1984). New method for assessing cardiac parasympathetic activity using 24 hour electrocardiograms. *British Heart Journal 52*, 396-402.

Farwell, L.A. & Donchin, E. (2000). The truth will out Interrogative Polygraphy ("Lie Detection") With Event-Related Brain Potentials. *Psychophysiology*, 28, 531-547.

Fehring, R.J. (1983). Effects of biofeedback-aided relaxation on the physiological stress symptoms of college students. *Nursing Research, 32*(6), 362-366.

Ferrucci, R., Mameli, F., Guidi, I., Mrakic-Sposta, S., Vergari, M., Marceglia, S., Cogiamanian, F., Barbieri, S., Scarpini, E. & Priori, A. (2008). Transcranial direct current stimulation improves recognition memory in Alzheimer disease. *Neurology, 71*(7), 493-498.

Fertonani, A., Rosini, S., Cotelli, M., Rossini, P.M. & Miniussi, C. (2010). Naming facilitation induced by transcranial direct current stimulation. *Behav Brain Res., 208*(2), 311-318.

Filipek, P.A., Semrud-Clikeman, M., Steingard, R.J., Rendshaw, P.F., Kennedy, D.N. & Biederman, M.D. (1997). Volumetric MRI analysis comparing attention-deficit hyperactivity disorder and normal controls. *Neurobiology, 47*, 618-628.

Fink, A. & Neubauer, A.C. (2006). EEG alpha oscillations during the performance of verbal creativity tasks: the differential effects of sex and verbal intelligence. *Int J Psychophysiol, 62*(1), 46-53.

Fisch, B.J. (1999). *Fisch and Spehlmann's EEG Primer: Basic Principles of Digital and Analog EEG - Third revised and enlarged edition.* New York: Elsevier.

Flöel, A., Rösser, N., Michka, O., Knecht, S. & Breitenstein, C. (2008). Noninvasive brain stimulation improves language learning. *J Cogn Neurosci., 20*(8), 1415-1422.

Flor, H., Hagg, G. & Turk, D.C. (1986). Long-term efficacy of EMG biofeedback for chronic rheumatic back pain. *Pain, 27*(2), 195-202.

Fox, M.D., Snyder, A.Z., Vincent, J.L., Corbetta, M., Van Essen, D.C. & Raichle, M.E. (2005).The human brain is intrinsically organized into dynamic, anticorrelated functional networks. *Proceedings National Academy of Science, USA, 102*(27), 9673-9678.

Frank, M.J. (2006). Hold your horses: A dynamic computational role for the subthalamic nucleus in decision making. *Neural Networks : The Official Journal of the International Neural Network Society, 19*(8), 1120-36.

Frank, Y. & Pavlakis, S.G. (2001). Brain imaging in neurobehavioral disorders (Review). *Pediatric Neurology, 25*(4), 278-87.

Freed, J. & Parsons, L. (1997). *Right-Brained Children in a Left-Brained World.* New York: Simon & Schuster.

Freedman, J. (1993).*Failing Grades.* Society for Advancing Educational Research: Full Court Press Inc.

Fregni, F., Boggio, P.S., Nitsche, M., Bermpohl, F., Antal, A., Feredoes, E., Marcolin, M.A., Rigonatti, S.P., Silva, M.T., Paulus, W. & Pascual-Leone, A. (2005). Anodal transcranial direct current stimulation of prefrontal cortex enhances working memory. *Exp Brain Res, 166*(1), 23-30.

Fregni, F., Boggio, P.S., Nitsche, M.A., Marcolin, M.A., Rigonatti, S.P., Pascual-Leone, A. (2006). Treatment of major depression with transcranial direct current stimulation. *Bipolar Disord, 8*(2), 203-204.

Fregni, F., Gimenes, R., Valle, A.C., Ferreira, M.J., Rocha, R.R., Natalle, L., Bravo, R., Rigonatti, S.P., Freedman, S.D., Nitsche, M.A., Pascual-Leone, A. & Boggio, P.S. (2006). A randomized, sham-controlled, proof of principle study of transcranial direct current stimulation for the treatment of pain in fibromyalgia. *Arthritis Rheum, 54*(12), 3988-3998.

Fried, I., Wilson, C., MacDonald, K. & Behnke, E. (1998). Electric current stimulates laughter. *Nature, 391*(6668), 650

Fried, R. (1987). *The Hyperventilation Syndrome: Research and Clinical Treatment.* Baltimore: Johns Hopkins University Press.

Friedman, M. & Rosenman, R.H. (1974). *Type A Behavior and Your Heart.* New York: Knopf.

Frith, U., Frith, C.D. (2003). Development and neurophysiology of mentalizing. *Phil. Trans. R. Soc. Lond. B. 358*(1431), 459-473.

Fritz, G. & Fehmi, L.G. (1982). *The Open Focus Handbook.* Princeton, NJ: Biofeedback Computer.

Frodl, T., Shaub, A., Banac, S., Charypar, M., Jäger, M., Kümmler, P., Bottlender, R., Zetzsche, T., Born, C., Leinsinger, G., Reiser, M., Möller H-J & Meisenzahl, E.M.(2006). Reduced hippocampal volume correlates with executive dysfunctioning in major depression. *J Psychiatry Neurosci, 31*(5), 316-323.

Fuchs, T., Birbaumer, N., Lutzenberger, W., Gruzelier, J. & Kaiser, J., (2003). Neurofeedback treatment for attention-deficit/hyperactivity disorder in children: a comparison with methylphenidate. *Journal of Applied Psychophysiology and Biofeedback, 28*(1), 1-12.

Fuller, G.D. (1984). *Biofeedback: Methods and Procedures in Clinical Practice.* San Francisco: Biofeedback Press.

Fürst, P. & Stehle, P. (2004). What are the essential elements needed for the determination of amino acid requirements in humans? *Journal of Nutrition, 134* (6 Suppl): 1558S–1565S.

Gani, C., Birbaumer, N. & Strehl, U. (2008). Long term effects after feedback of slow cortical potentials and of theta-beta-amplitudes in children with attention-deficit/hyperactivity disorder (ADHD). *International Journal of Bioelectromagnetism, 10*(4), 209-232.

Gardea, M.A., Gatchel, R.J. & Mishra, K.D. (2001). Long-term efficacy of biobehavioral treatment of temporomandibular disorders. *Journal of Behavioral Medicine, 24*(4), 341-359.

Garin, P., Gilain, C., Van Damme, J.P., de Fays, K., Jamart, J., Ossemann, M. & Vandermeeren, Y. (2011). Short- and long-lasting tinnitus relief induced by transcranial direct current stimulation. *J Neurol., 258*(11), 1940-1948.

Geda, Y.E., Roberts, R.O., Knopman, D.S., Christianson, T.J.H., Pankratz, S., Ivnik, R.J., Boeve, B.F., Tangalos, E.G., Petersen, R.C. & Rocca, W.A. (2010). Physical Exercise, Aging, and Mild Cognitive Impairment. *Journal American Medical Association Neurology (formerly Archives Neurology, 67*(1), 80-86.

Gellhorn, E. (1967). The tuning of the nervous system: physiological foundations and implications for behavior. *Perspect. Biol. Med. 10*, 559–591.

Gennarelli, T.A., Thibault, L.E., Adams, J.H., Graham, D.I., Thompson, C.J. & Marcincin, R.P. (1982). Diffuse axonal injury and traumatic coma in the primate. *Annals of Neurology, 12*(6), 564-574.

Gennarelli, T.A., Thibault, L.E. & Graham, D.I., (1998). Diffuse Axonal Injury: An Important Form of Traumatic Brain Damage. *Neuroscientist, 4*(3), 202-221.

Gevensleben, H., Holl, B., Albrecht, B., Vogel, C., Schlamp, D., Kratz, O., Studer, P., Rothenberger, A,, Moll, G.H. & Heinrich, H. (2009). Is neurofeedback an efficacious treatment for ADHD? A randomised controlled clinical trial. *Journal of Child Psychology and Psychiatry, 50*, 780-789.

Gevins, A., Smith, M.E., McEvoy, L. & Yu, D. (1997). High-resolution EEG mapping of cortical activation related to working memory: effects of task difficulty, type of processing, and practice. *Cereb Cortex, 7*, 374-385.

Gevirtz, R. (2010). Autonomic Nervous System Markers for Psychophysiological, Anxiety, and Physical Disorders, Chapter 9, in *Integrative Neuroscience and Personalized Medicine,* Gordon, E. & Koslow, S.H. (eds.), Oxford Press, 164-181.

Gevirtz, R. (2013). The Nerve of that Disease: The Vagus Nerve and Cardiac Rehabilitation. Biofeedback, 4, Issue 1, 32-38.

Ghaziri, J., Tucholka, A., Larue, V., Blanchette-Sylvestre, M., Reyburn, G., Gilbert, G., Lévesque, J. & Beauregard, M. (2013). Neurofeedback training induces changes in white and gray matter. *Clin EEG Neurosci., 44*(4), 265-272.

Giardino, N.D, Lehrer, P.M. & Edelberg, R. (2002). Comparison of finger plethysmograph to ECG in the measurement of heart rate variability. *Psychophysiology, 39*(2), 246-253.

Gibbs, F.A. & Knott, J.R. (1949). Growth of the electrical activity of the cortex. *Electroencephalography and Clinical Neurophysiology,* 223-229.

Gibson, T. & O'Hair, D. (1987). Cranial application of low level transcranial electrotherapy vs. relaxation instruction in anxious patients. *American Journal of Electromedicine, 4*(1), 18-21.

Gilula, M. & Kirsch, D. (2005). Cranial Electrotherapy Stimulation Review: A Safer Alternative to Psychopharmaceuticals in the Treatment of Depression. *Journal of Neurotherapy, 9*(2):7-26.

Gitelman, D.R. (2003). Attention and its disorders. *British Medical Bulletin, 65*, 21-34.

Giza, C.C. & Hovda, D.A. (2000). *Ionic and metabolic consequences of concussion;* in: Cantu RC, Cantu RI. Neurologic Athletic and Spine Injuries. St Louis, MO: WB Saunders Co; 80-100.

Giza, C.C. & Hovda, D.A. (2001), The Neurometabolic Cascade of Concussion. *Journal of Athletic Training, 36*(3), 228-235.

Glascher, J. & Adolphs, R. (2003). Processing of the arousal of subliminal and supraliminal emotional stimuli by the human amygdala. *Journal of Neuroscience, 23*, 10274-10282.

Gloor, P., Olivier, A., Quesney, L.F., Andermann, F. & Horowitz, S. (1982). The role of the limbic system in experiential phenomena of temporal lobe epilepsy. *Ann Neurol, 12,* 129-144.

Gloor, P., Pellegrini, A. & Kostopoulos, G.K. (1979). Effects of changes in cortical excitability upon the epileptic bursts in generalized penicillin epilepsy of the cat. *Electroencephalogr Clin Neurophysiol., 46*(3), 274-289.

Goela, V. & Dolana, R.J. (2004). Differential involvement of left prefrontal cortex in inductive and deductive reasoning. *Cognition, 93*(3), 109-121.

Golby, A.J., Poldrack, R.A., Brewer, J.R., Desmond, J.E., Aron, A.P. & Gabrieli, J.D.E. (2001). Material-specific lateralization in the medial temporal lobe and prefrontal cortex during memory encoding. *Brain, 124*(9), 1841-1854.

Goldapple, K., Segal, Z., Garson, C., Lau, M., Bieling, P., Kennedy, S. & Mayberg, H. (2004). Modulation of cortical-limbic pathways in major depression: Treatment-specific effects of cognitive behavior therapy. *Arch. Gen. Psychiatry, 61*, 34-41.

Goldberg, E. (2001). *The executive brain: Frontal lobes and the civilized mind.* New York: Oxford University Press (paperback 2002).

Goldberg, I., Harel, M. & Malach, R. (2006). When the brain loses its self: prefrontal inactivation during sensorimotor processing, *Neuron, 50*(2), 329-339.

Goleman, D. (1995) *Emotional Intelligence.* New York: Bantam Books.

Gorman, J.M., Kent, J.M., Sullivan, G.M. & Coplan J.D. (2000). Neuroanatomical hypothesis of panic disorder revised: *Amer J Psychiatry, 157*(4), 493-505

Gosepath, K., Nafe, B., Ziegler, E. & Mann, W.J. (2001). Neurofeedback in therapy of tinnitus. *Hals-Nasen-Ohrenärzie, 49*(1), 29-35.

Gosselin, N., Saluja, R.S., Chen, J.K., Bottari, C., Johnston, K. & Ptito, A. (2010). Brain functions after sports-related concussion: insights from event-related potentials and functional MRI. *The Physician and Sportsmedicine, 38*(3), 27-37.

Gotman, J., Grova, C., Bagshaw, A., Kobayashi, E., Aghakhani, Y. & Dubeau, F. (2005). Generalized epileptic discharges show thalamocortical activation and suspension of the default state of the brain. *Proc Natl Acad Sci USA, 102*(42), 15236-15240.

Grafman, J. (2002). *The structured event complex and the human prefrontal cortex.* In Stuss, D.T. & Knight, R.T. (eds.), The frontal lobes (292-310). New York: Oxford University Press.

Grassi, C., Filippi, G.M. & Passatore, M. (1986). Postsynaptic alpha 1- and alpha 2-adrenoreceptors mediating the action of the sympathetic system on muscle spindles in the rabbit. *Pharmacological Research Communications, 18*(2), 161-170.

Gray, H. (1918). *Anatomy of the Human Body.* Lea & Febiger, Philadelphia.

Gray, L. (2013). The University of Texas Medical School at Houston, *FreeScience - Books – Neuroscience Online*, Chapter 10, Vestiblar System: Structure and Function; *NeuroScience 2nd edition.* Chapter 11: Vestibular System: Pathways and Reflexes.

Green, F. & Green, A. (1977). *Beyond Biofeedback.* New York: Knoll Publishing.

Green, V.A., Pituch, K.A., Itchon, J., Choi, A., O'Reilly, M. & Sigafoos, J. (2006). Internet survey of treatments+ used by parents of children with autism. *Research in Developmental Disabilities, 27*, 70-84.

Greenberg, B.D., Malone, D.A., Friehs, V., Gerhard M., Rezai, A.R., Kubu, C.S., Malloy, P.F., Salloway, S.P., Okun, M.S., Goodman, W.K. & Rasmussen, S.A. (2006). Three-Year Outcomes in Deep Brain Stimulation for Highly Resistant Obsessive-Compulsive Disorder. *Neuropsychopharmacology, 31*, 2384-2393.

Grin'-Yatsenko, V.A., Kropotov, J.D., Ponomarev, V.A., Chutko, L.S. & Yakovenko, E.A. (2001). Effect of biofeedback training of sensorimotor and beta-sub-1 EEG rhythms on attention parameters. *Human Physiology, 27*(3), 259-266.

Grippo, A.J. & Johnson, A.K. (2009). Stress, depression and cardiovascular dysregulation: A review of neurobiological mechanisms and the integration of research from preclinical disease models. *Stress, 12*(1), 1-21.

Gruber, G. (1986). *Dr. Gary Gruber's Essential Guide to Test Taking For Kids.* New York: William Morrow and Company.

Gruzelier, J. (2002). Neurofeedback training to enhance musical performance. *Proceedings of the AAPB Annual Meeting*, Las Vegas, NV, March 2002.

Gruzelier, J. & Egner, T. (2003). Theta/Alpha neurofeedback training to enhance musical performance. *Proceedings of the combined annual meetings of the European chapter of iSNR and the Biofeedback Foundation of Europe*, Udine, Italy, February 2003.

Guberman, A. (1994). Hyperkinetic movement disorders. In *An Introduction to Clinical Neurology.* New York: Little, Brown & Co.

Gunkelman, J., Personal communications. jay@brainsinternational.com

Gurtubay, I.G., Alegre1, M., Labarga, A., Malanda, A. & Artieda, J. (2004). Gamma band responses to target and non-target auditory stimuli in humans. *Neuroscience Letters, 367,* 6-9.

Gusnard, D.A., Akbudak, E., Shulman, G.L., Raichle, M.E. (2001). The medial prefrontal cortex and self-referential mental activity: relation to a default mode of brain function. *Proc Natl Acad. Sci USA, 98*(7), 4259-4264

Hadhazy, V.A., Ezzo, J., Creamer, P. & Bergman, B.M. (2000). Mind-body therapies for the treatment of fibromyalgia. A systematic review. *Journal of Rheumatology, 13*(3), 487-492.

Hagmann, P., Cammoun, L., Gigandet, X., Meuli1, R., Honey,C.J., Wedeen, V.J. & Sporns O. (2008). Mapping the Structural Core of Human Cerebral Cortex. *PLoS Biology, 6(*7), 1479-1493.

Halgren, E., Squires, N.K., Wilson, C.L., Rohrbaugh, J.W., Babb, T.L. & Crandall, P.H. (1980). Endogenous potentials generated in the human hippocampal formation and amygdala by infrequent events. *Science, 210,* 803-805.

Hamer, M. & Chida, Y. (2009). Physical activity and risk of neurodegenerative disease: a systematic review of prospective evidence. *Psychol Med., 39,* 3-11.

Hammond, D.C. & Gunkelman, J. (2001). *The Art of Artifacting.* Merino, CO: Society for Neuronal Regulation.

Hampson, M., Driesen, N.R., Skudlarski, P., Gore, J.C., Constable, R.T. (2006). Brain Connectivity Related to Working Memory Performance *The Journal of Neuroscience, 26(*51), 13338-13343.

Haneef, Z, Levin, H.S., Frost, J.D. & Mizrahi, E.M. (2013). Electroencephalography and quantitative electroencephalography in mild traumatic brain injury. *J Neurotrauma, 30*(8), 653-656.

Harmon, K.G. (1999). Assessment and management of concussion in sports. *American Family Physician 60*(3), 887-892.

Harmon-Jones, E. & Allen, J.J.B. (1998) Anger and frontal brain activity: EEG asymmetry consistent with approach motivation despite negative affective valence. *Journal of Personality and Social Psychology, 74*(5), 1310-1316.

Hassett, A.L, Radvanski, D.C., Vaschillo, E.G., Vaschillo, B., Sigal, L.H., Karavidas, M.K., Buyske, S., Lehrer PM. (2007). A Pilot Study of the Efficacy of Heart Rate Variability (HRV) Biofeedback in Patients with Fibromyalgia. *Appl Psychophysiol Biofeedback, 32,* 1-10.

Hatcher, R.T. (2007). Applied Neuroscience Inc. *NeuroGuide Delux, 2.3.7 (www.appliedneuroscience.com)*

Hauri, P.J., Percy, L., Hellekson, C., Hartmann, E. & Russ, D. (1982). The treatment of psychophysiologic insomnia with biofeedback: A replication study. *Biofeedback and Self-Regulation, 7*(2), 223-235.

Hawkins, R.C., II, Doel, S.R., Lindseth, P., Jeffers, V. & Skaggs, S. (1980). Anxiety reduction in hospitalized schizophrenics through thermal biofeedback and relaxation training. *Perceptual & Motor Skills. 51*(2), 475-482.

Hebb, D.O. (1949). *The Organization of Behavior.* New York: Wiley-Interscience.

Heffernen, M. (1997). The effect of variable microcurrents on EEG spectrum and pain control. *Canadian Journal of Clinical Medicine, 4*(10), 2-8.

Heinrich, H., Gevensleben, H., Freisleder, F.J., Moll, G.H. & Rothenberger, A. (2004). Training of Slow Cortical Potentials in Attention-Deficit/Hyperactivity Disorder: Evidence for Positive Behavioral and Neurophysiological Effects. *Biol Psychiatry, 55,* 772-775.

Heller, W., Etienne, M.A. & Miller, G.A. (1997). Patterns of regional brain activity differentiate different types of anxiety. *Journal of Abnormal Psychology, 104,* 327-333.

Henderson, R.J., Hart, M.G., Lai, S.K. & Hunyor, S.N. (1998). The effect of home training with direct blood pressure biofeedback of hypertensives: A placebo controlled study. *Journal of Hypertension, 16*(6), 771-778.

Herbert, M. (2012). Findings in Autism (ASD) Consistent with Electromagnetic Fields (EMF) and Radiofrequency Radiation (RFR). BioInitiative Working Group: TRANSCEND Research Program, Massachusetts General Hospital, Harvard Medical School, Sage Associates, Santa Barbara, CA, USA.

Hermann, C. & Blanchard, E. B. (2002). Biofeedback in the treatment of headache and other childhood pain. *Applied Psychophysiology and Biofeedback, 27*(2), 143-162.

Heyes, M.P., Saito, K., Crowley, J.S., Davis, L.E., Demitrack, M.A. & Tourtellotte, W.W. (1992). Quinolinic acid and kynurenine pathway metabolism in inflammatory and non-inflammatory neurological disease. *Brain, 115*(5), 1249-1273.

Heymen, D., Jones, K.R., Ringel, Y., Scarlett, Y. & Whitehead, W.E. (2001). Biofeedback treatment of fecal incontinence: A critical review. *Diseases of the Colon and Rectum, 44*(5), 728-736.

Heywood, C. & Beale, I. (2003) EEG Biofeedback vs. placebo treatment for Attention-Deficit/Hyperactivity Disorder: A Pilot study. *Journal of Attention Disorders, 7*(1), 43-55.

Hickok, G. (2001). Functional Anatomy of Speech Perception and Speech Production: Psycholinguistic Implication. *Journal of Psycholinguistic Research, 30(*3), 225-235.

Hill, E.L. & Frith, U. (2003). Understanding Autism: Insights from mind and brain. Theme Issue *'Autism: mind and brain'*. *Phil. Trans. of The Royal Society London, Bulletin, 358*, 281-289.

Hillered, L., Ernster, L. & Siesjö, B.K. (1984). Influence of in vitro lactic acidosis and hypercapnia on respiratory activity of isolated rat brain mitochondria. *J. Cereb Blood Flow Metab., 4*(3), 430-437.

Hjorth, B. (1980). Source derivation simplifies topographical EEG interpretation. *American Journal of EEG Technology, 20,* 121-132.

Holmes, T.H., Rahe, R.H. (1967). The Social Readjustment Rating Scale. *J Psychosom Res 11*(2): 213-8.333

Holtzheimer, P.E. & Mayberg, H.S. (2011). Stuck in a rut: Rethinking depression and its treatment. *Trends in Neurosciences, 34*(1), 1-9.

Horwitz, B., Rumsey, J.M., Grady, C.L. & Rapoport, S.I. (1988). The cerebral metabolic landscape in autism: Intercorrelations of regional glucose utilization. *Archives of Neurology, 45*(7), 49-55.

Hughes, J.R. (2002). The Mozart Effect. *Proceedings of the 10th Annual Conference of the Society for Neuronal Regulation.* Scottsdale, AZ, September 2002.

Hughes, J.R. & John, E.R. (1999). Conventional and quantitative electroencephalography in psychiatry. *Journal of Neuropsychiatry and Clinical Neuroscience, 11* (2).

Hughes, S. & Crunelli, V. (2007). Just a phase they're going through: The complex interaction of intrinsic high-threshold bursting and gap junctions in the generation of thalamic alpha and theta rhythms. *Int. J. Psychophysiol., 64,* 3-17.

Humphreys, P.A. & Gevirtz, R. (2000). Treatment of recurrent abdominal pain: Components analysis of four treatment protocols. *Journal of Pediatric Gastroenterological Nutrition, 31*(1), 47-51.

Hynd, G.W., Hern, K.L., Novey, E.S. & Eliopulos, D. (1993). Attention deficit hyperactivity disorder and asymmetry of the caudate nucleus. *Journal of Child Neurology, 8,* 339-347.

Iacoboni, M. (2006). Understanding emotions in others: mirror neuron dysfunction in children with autism spectrum disorders. *Nature Neuroscience, 9*(1), 28-30.

Iacoboni, M. & Dapretto, M. (2006). The mirror neuron system and the consequences of its dysfunction. *Nature Reviews and Neuroscience, 7*(12), 942-951.

Ibric, V.L., Dragomirescu, L.G. & Hudspeth, W.J. (2009). Real-time changes in connectivities during neurofeedback. *Journal of Neurotherapy, 13*(3), 156-165.

Introduction to Psychiatry. Washington, D.C.: George Washington University Medical School.

Jääskeläinen, I.P., Ahveninen, J., Belliveau, J.W., Raij, T. & Sams, M. (2007). Short-term plasticity in auditory cognition. *Trends Neurosci. 30,* 653-661.

Jääskeläinen, I.P., Ahveninen, J., Bonmassar, G., Dale, A.M., Ilmoniemi, R.J., Levänen, S., Lin, F.-H. May, P., Melcher, J., Stufflebeam, S., Tiitinen, H. & Belliveau, J.W. (2004). Human posterior auditory cortex gates novel sounds to consciousness. *Proc. Natl. Acad. Sci. U.S.A. 101*(17), 6809-6814.

Jacobs, J., Kahana, M.J., Ekstrom, A.D., Mollison, M.V. & Fried, I. (2010). A sense of direction in human entorhinal cortex. *Proc Natl Acad Sci U S A, 107*(14), 6487-6492).

Jacobson, E. (1970). *Modern treatment of tense patients.* Springfield, IL: Charles C. Thomas.

Jantzen, T., Graap, K., Stephanson, S., Marshall, W. & Fitzsimmons, G. (1995). Differences in baseline EEG measures for ADD and normally achieving pre-adolescent males. *Biofeedback and Self-Regulation, 20*(1), 65 - 82.

Jarusiewicz, E. (2002). Efficacy of neurofeedback for children in the autistic spectrum: A pilot study. *Journal of Neurotherapy, 6*(4), 39-49.

Jasper, H. (1958). Report of the committee on methods of clinical examination in electroencephalography. *EEG and Clinical Neurophysiology, 10, 374.*

John, E.R. (1989). The role of quantitative EEG topographic mapping or 'neurometrics' in the diagnosis of psychiatric and neurological disorders: The pros. *Electroencephalography and Clinical Neurophysiology, 73,* 2-4.

John, E.R., Prichep, L.S. & Easton, P. (1987). Normative data banks and neurometrics: Basic concepts, methods and results of norm constructions. In Gevins, A.S. & Remond, A. (eds.), *Handbook of Electroencephalography and Clinical Neurophysiology. Vol. 1.*

John, E.R., Prichep, L.S., Fridman, J. & Easton, P. (1988). Neurometrics: Computer-assisted differential diagnosis of brain dysfunctions. *Science, 239,* 162-169.

Johnson, V.E., Stewart, J.E., Begbie, F.D., Trojanowski, J.Q., Smith, D.H. & Stewart, W. (2013). Inflammation and white matter degeneration persist for years after a single traumatic brain injury. *Brain, 136*(Pt 1), 28-42.

Jones, D.K., Christiansen, K.F., Chapman, R.J. and Aggleton, J.P. (2013). Distinct subdivisions of the cingulum bundle revealed by diffusion MRI fibre tracking: Implications for neuropsychological investigations. *Neuropsychologia, 51*(1), 67–78.

Jones, G.E., Leonberg, T.F., Rouse, C.H. & Scott, D.M. (1986). Preliminary data exploring the presence of an evoked potential associated with cardiac visceral activity, *Psychophysiology,23*, p. 445.

Joseph, R. (2000). *Neuropsychiatry, Neuropsychology,* BA 34, Uncus, Temporal Lobe, *Clinical Neuroscience* 3rd Edition. Academic Press, New York; (available online: BrainMind.com).

Kaiser, D.A. & Othmer, S. (2000). Effects of Neurofeedback on Variables of Attention in a Large Multi-Center Trial. *Journal of Neurotherapy, 4*(1), 5-15.

Kamiya, J. (1979). Autoregulation of the EEG alpha rhythm: A program for the study of consciousness. In Peper, E., Ancoli, S. & Quinn, M. (eds.) *Mind Body Integration: Essential Readings in Biofeedback.* New York: Plenum Press, 289-298.

Karavidas, M.K., Lehrer, P.M., Vaschillo, E., Vaschillo, B., Marin, H., Buyske, S., Malinovsky, I., Radvanski, D. & Hassett, A. (2007). Preliminary results of an open-label study of heart rate variability biofeedback for the treatment of major depression. *Applied Psychophysiology and Biofeedback, 32*(1), 19-30.

Kaufmann, T., Vögele, C., Sütterlin, S., Lukito, S. & Kübler, A. (2012). Effects of resting heart rate variability on performance in the P300 brain-computer interface. *Int J Psychophysiol, 83*(3), 336-341.

Kennedy, S.H., Giacobbe P., Rizvi, S.J., Placenza, F.M., Yasunori, N.Y., Mayberg, H.S. & Lozano, A.M. (2011). Deep Brain Stimulation for Treatment-Resistant Depression: Follow-up after 3 to 6 years. *Am J Psychiatry; 168*(5), 502-510.

Kim, M.J., Gee, D.G., Davis, F.C. Loucks, R.A. & Whalen, P.J. (2011). Anxiety dissociates dorsal and ventral medial prefrontal cortex functional connectivity with the amygdala at rest. *Cerebral Cortex, 7*, 1667-1673.

Kim, M.J., Loucks, R.A., Palmer, A.L., Brown, A.C., Solomon, K.M., Marchante, A.N. & Whalen, P.J. (2011). The structural and functional connectivity of the amygdala: from normal emotion to pathological anxiety. *Behavioral Brain Research, 223*, 403-410.

Kinomura S., Larsson J., Gulyas B. & Roland P.E. (1996). Activation by attention of the human reticular formation and thalamic intralaminar nuclei. *Science, 271,* 512-515.

Kirchhoff, B.A., Wagner, A.D., Maril, A. & Stern, C.E. (2000). Prefrontal-Temporal Circuitry for Episodic Encoding and Subsequent Memory. *The Journal of Neuroscience, 20*(16), 6173-6180.

Kirsch, D. (2002). *The science behind cranial electrotherapy stimulation – second edition.* Medical Scope Publishing. Edmonton, Alberta, Canada.

Klabunde, R.E. (2013). *Cardiovascular Physiology Concepts.* Available online: www.cvphysiology.com.

Klimesch, W. (1996). Memory Processes, Brain Oscillations, and EEG Synchronization. *New Advances in EEG & Cognition; Special Issue of International Journal of Psychophysiology, 24*(1-2), 61-100.

Klimesch, W. (1999). EEG alpha and theta oscillations reflect cognitive and memory performance: a review and analysis. *Brain Research. Brain Research Reviews, 29*(2), 169-195.

Klimesch, W., Pfurtscheller, G. & Schimke, H. (1992). Pre- and post-stimulus processes in category judgment tasks as measured by event-related desynchronization (ERD). *Journal of Psychophysiology, 6*, 185-203.

Klimesch, W. Schack, B. & Sauseng, P. (2005). The functional significance of theta and upper alpha oscillations . *Exp. Psychol., 52*(2), 99-108.

Kneeland, T.W. & Warren C.A.B. (1994). Pushbutton Psychiatry: A history of electroshock in America. Westport, CT: Praeger.

Knezevic, B. (2007). Pilot Project to Ascertain Utility of Tower of London Test (ToL) to Assess Outcomes of Neurofeedback in Clients with Asperger's Syndrome. Student Award paper presented at the International Society for Neurofeedback and Research Annual Meeting, San Diego, CA.

Knezevic, B., Thompson, L. & Thompson, M. (2010). Pilot project to ascertain the utility of Tower of London Test to assess outcomes of neurofeedback in clients with Asperger's Syndrome. *Journal of Neurotherapy, 14*(3), 3-19.

Knoch, D., Gianotti, L.R.R., Pascual-Leone, A., Treyer. V., Regard, M., Hohmann, M. & Brugger, P. (2006). Disruption of Right Prefrontal Cortex by Low-Frequency Repetitive Transcranial Magnetic Stimulation Induces Risk-Taking Behavior. *Journal of Neuroscience, 26*(24), 6469-6472.

Koelega, H.S. (1993). Stimulant drugs and vigilance performance: A review. *Psychopharmacology, 111*, 1-16.

Koenigs, M. & Grafman, J. (2009). The functional neuroanatomy of depression: Distinct roles for ventromedial and dorsolateral prefrontal cortex. *Behavioural Brain Research, 201*, 239-243.

Konishi, S., Chikazoe, J., Jimura, K., Asari, T. & Miyashita, Y. (2005). Neural mechanism in anterior prefrontal cortex for inhibition of prolonged set interference. *Proc Natl Acad Sci U S A, 102*(35), 12584-12588.

Korn, A., Golan, H., Melamed, I., Pascual-Marqui, R. & Friedman, A. (2005). Focal cortical dysfunction and blood-brain barrier disruption in patients with Postconcussion syndrome. *Journal of Clinical Neurophysiology, 22*(1), 1-9.

Kouijzer, M., de Moor, J.M.H., Gerrits, B.J.L., Congedo, M. & van Shie, H.T. (2009). Neurofeedback improves executive functioning in children with autistic spectrum disorders. *Research in Autistic Spectrum Disorders 3*, 145-162.

Kringelbach, M. & Rolls, E.T. (2005). The Human Orbitofrontal Cortex: Linking Reward To Hedonic Experience. *Nature Reviews Neuroscience, 6*(9), 691-702.

Kropotov. J.D. (1997). Striatum *Russian Journal of Physiology, 83*, 45-51.

Kropotov, J.D. (2009). *Quantitative EEG, Event-Related Potentials and Neurotherapy.* San Diego, CA: Academic Press, Elsevier, 292-309.

Kropotov, J.D., Alho, K., Näätänen, R., Ponomarev VA, Kropotova, O.V., Anichkov, A.D. & Nechaev, V.B. (2000). Human auditory-cortex, mechanisms of preattentive sound discrimination. *Neurosci Lett, 280*(2), 87-90.

Kropotov, J.D. & Etlinger (1999). *International Journal of Psychology, 31*, 197-217.

Kropotov, J.D., Grin-Yatsenko,V.A., Ponomarev,V.A., Chutko, L.S., Yakovenko, E.A. & Nikishena, I.S. (2005). ERPs correlates of EEG relative beta training in ADHD children. *International Journal of Psychophysiology, 55*, 23-34.

Kuo, M.-F., Grosch, J., Fregni, F., Paulus, W & Nitsche, Michael A. (2007). Focusing Effect of Acetylcholine on Neuroplasticity in the Human Motor Cortex. *The Journal of Neuroscienc, 27*(52), 14442-14447.

Kurova, N.S. & Cheremushkin, E.A. (2007). Spectral EEG characteristics during increases in the complexity of the context of cognitive activity. *Neurosci Behav Physiol, 37*(4), 379-385.

Kushner, D.S. (2001) Concussion in sports: minimizing the risk for complications. *American Family Physician, 64*, 1007-1014.

Kwak, Y.S., Um, S.Y., Son ,T.G. & Kim, D.J.(2008). Effect of regular exercise on senile dementia patients. *Int J Sports Med., 29*(6), 471-474.

La Rovere, M.T., Pinna, G.D. & Raczak, G. (2008). Baroreflex Sensitivity: Measurement and Clinical Implications. *Annals of Noninvasive Electrocardiology*, *13*(2), 191-207.

Lachaux, J.P., Rodriguez, E., Martinerie, J., Adam, C., Hasboun, D. & Varela, F. (2000). Gamma-band activity in human intracortical recordings triggered by cognitive tasks, *Europ. J. Neuroscience, 12*, 1-15.

LaHoste, G.L., Swanson, J.M., Wigal, S.B., Glabe, C., Wigal, T., King, N. & Kennedy, J.L. (1996). Dopamine D 4 receptor gene polymorphism is associated with attention-deficit hyperactivity disorder. *Molecular Psychiatry, 1*, 121-124.

Laird, A.R., Fox, P.M., Eickoff, S.B., Turner, J.A., Ray, K.L., McKay, D.R., Glahn, D.C., Beckmann, C.F., Smith, S.M. & Fox, P.T. (2011). Behavioral Interpretations of Intrinsic Connectivity Networks. *Journal of Cognitive Neuroscience*, 23(12), 4022-4037.

Landers, D.M., Petruzzello, S.J., Salazar, W., Crews, D.J., Kubitz, K.A., Gannon, T.L. & Han, M. (1991). The influence of electrocortical biofeedback on performance in pre-elite archers. *Medicine and Science in Sports and Exercise, 23*(1), 123-128.

Lanius, R.A., Bluhm, R.L., Coupland, N.J., Hegadoren, K.M., Rowe, B., Theberge, J., Williamson, P.C. & Brimson, M. (2010). Default mode network connectivity as a predictor of post-traumatic stress disorder symptom severity in acutely traumatized subjects. *Acta Psychiatrica Scandinavica, 121*(1), 33-40.

Laughlin, H.P. (1967). *The Neuroses.* Washington: Butterworth Press, p 76.

LaVaque, T.J., Hammond, D.C., Trudeau, D., Monastra, V., Perry, J., Lehrer, P., Matheson, D. & Sherman, R. (2002). Template for developing guidelines for the evaluation of the clinical efficacy of psychophysiological interventions. *Applied Psychophysiology and Biofeedback, 27*(4), 273-281. Reprinted in *Journal of Neurotherapy, 6*(4), 11-23.

Lavoie, M.E., Dupuis, F., Johnston, K.M., Leclerc, S. & Lassonde, M. (2004). Visual p300 effects beyond symptoms in concussed college athletes. *Journal of Clinical and Experimental Neuropsychology, 26*(1), 55-73.

Lawrence, A.D., Sahakian, B.J. & Robbins. T.W. (1998). Cognitive functions and corticostriatal circuits: insights from Huntington's disease. *Trends in Cognitive Sciences, 2*(10), 379-388.

Lazar, S.W., Kerr, C.E., Wasserman, R.H., Gray, J.R., Greve, D.N., Treadway, M.T., McGarvey, M., Quinn, B.T., Dusek. J.A., Benson. H., Rauch. S.L., Moore. C.I. & Fischl. B. (2005). Meditation experience is associated with increased cortical thickness. *Neuroreport, 16*, 1893-1897.

Lehrer, P. & Eddie, D. (2013). Dynamic processes in regulation and some implications for biofeedback and behavioral interventions. *Applied Psychophysiology and Biofeedback. 38*(2), 143-155.

Lehrer, P., Smetankin, A. & Potapova, T. (2000). Respiratory sinus arrhythmia biofeedback therapy for asthma: A report of 20 unmedicated pediatric cases using the Smetankin method. *Applied Psychophysiology and Biofeedback, 25*(3), 193-200.

Lehrer, P. & Vaschillo, E. (2008). The future of heart rate variability (HRV) biofeedback. *Biofeedback, 36*, 11-14.

Lehrer, P., Vaschillo, E., Lu, S.-E., Eckberg, D., Vaschillo, B., Scardella, A. & Habib, R. (2006). Heart rate variability biofeedback: effects of age on heart rate variability, baroreflex gain, and asthma. *Chest, 129*, 278-284.

Lehrer, P.M. (2007). Biofeedback Training to Increase Heart Rate Variability. In Lehrer, P.M., Woolfolk, R.L. & Sime, W.E. (eds.) *Principles and Practice of Stress Management: Third Edition* (227-148). New York: Guilford Press.

Lehrer, P.M., Carr, R., Sargunaraj, D. & Woolfolk, R.L. (1994). Stress management techniques: Are they all equivalent, or do they have specific effects? *Biofeedback and Self-Regulation, 19*(4), 353-401.

Lehrer, P.M., Karavidas, M.K., Lu, E.E., Feldman, J., Kranitz, L., Abraham, S., Sanderson, W. & Reynolds, R. (2008). Psychological treatment of comorbid asthma and panic disorder: a pilot study. *Journal of the Anxiety Disorders, 22,* 674-683

Leins, U., Goth, G., Hinterberger, T., Klinger, C., Rumpf, N. & Strehl, U. (2007). Neurofeedback for Children with ADHD: A Comparison of SCP and Theta/Beta Protocols. *Applied Psychophysiology & Biofeedback, 32*(2):73-88

Leon-Carrion, J., Martin-Rodriguez, J.F., Damas-Lopez, J., Barroso y Martin, J.M. & Dominguez-Morales, M.R. (2009). Delta-alpha ratio correlates with level of recovery after neurorehabilitation in patients with acquired brain injury. *Clinical Neurophysiology, 120*(6), 1039-1045.

Leonard, B.E. & Myint, A. (2006). Inflammation and depression: is there a causal connection with dementia? *Neurotoxicity research, 10*(2), 149-160.

Levesque J., Beauregard M. & Mensour B. (2006). Effect of neurofeedback training on the neural substrates of selective attention in children with attention-deficit/hyperactivity disorder: a functional magnetic resonance imaging study. *Neuroscience Letters, 394*(3), 216-221.

Levine, B., Black, S.E., Cabeza, R., Sinden, M., Mcintosh, A.R., Toth, J.P., Tulving, E. & Stuss, D.T. (1998). Episodic memory and the self in a case of isolated retrograde amnesia. *Brain, 121*(Pt 10), 1951-1973.

Li, C-S.R., Huang, C., Constable, R.T. & Sinha, R. (2006). Imaging Response Inhibition in a Stop-Signal Task: Neural Correlates Independent of Signal Monitoring and Post-Response Processing.*The Journal of Neuroscience, 26*(1), 186-192.

Lin, G., Xiang, Q., Fu, X., Wang, S., Chen, S., Shao, L., Zhao, Y. & Wang, T. (2012). Heart rate variability biofeedback decreases blood pressure in prehypertensive subjects by improving autonomic function and baroreflex. *Journal of Alternative and Complementary Medicine, 18*(2), 143-152.

Linden, M. (2006). *Autism Spectrum Disorders.* Panel presentation at the 36th Annual Meeting of the Association for Applied Psychophysiology and Biofeedback, Portland, OR.

Linden, M., Habib, T. & Radojevic, V. (1996). A controlled study of the effects of EEG biofeedback on cognition and behavior of children with attention deficit disorder and learning disabilities. *Biofeedback and Self-Regulation, 21*(1), 106-111.

Lisanby, S.H., Luber, B., Perera, T. & Sackeim, H.A. (2000). Transcortical magnetic stimulation: applications in basic neuroscience and neurpsychopharmacology. *Int J Neuropsychopharmacol 3,* 259-273.

Lofthouse, N., Arnold, L.E., Hersch, S. & Hurt, E. (2012). Current status of neurofeedback for attention-deficit/hyperactivity disorder. *Current Psychiatry Report*, 14(5), 536-542.

Lofthouse, N., Arnold, L.E., Hersch, S., Hurt, E. & Debeus, R. (2012). A review of neurofeedback treatment for pediatric ADHD. *Journal of Attention Disorders*, 16, 351-372.

Lofthouse, N., McBurnett, K., Arnold, L.E. & Hurt, E. (2011). Biofeedback and neurofeedback treatment for ADHD. *Psychiatric Annals*, 41 (1), 42-48.

Loo, C.K., Alonzo, A., Martin, D., Mitchell, P.B., Galvez, V. & Sachdev, P. (2012). Transcranial direct current stimulation for depression: 3-week, randomised, sham-controlled trial. *British Journal of Psychiatry, 200*, 52-59.

Lopez, C., Blanke, O. & Mast, F.W. (2012). The human vestibular cortex revealed by coordinate-based activation likelihood estimation meta-analysis. *Elsevier, Neuroscience, 212*(14), 159-179.

Love, A.J. & Thompson, M.G.G. (1988). Language disorders and attention deficit disorders in a child psychiatric outpatient population. *American Journal of Orthopsychiatry, 58*(1), 52-64.

Lubar, J.F. (1995). *Neurofeedback for the management of attention-deficit/hyperactivity disorders.* In Mark S. Schwartz and Associates, Biofeedback: A Practitioner's Guide - Second Edition. New York: Guilford Press, 493-522.

Lubar, J.F. (1997). Neocortical dynamics: Implications for understanding the role of neurofeedback and related techniques for the enhancement of attention. *Applied Psychophysiology and Biofeedback, 22*(2), 111-126.

Lubar, J.F. & Lubar, J. (1999). *Neurofeedback assessment and treatment for ADD/ hyperactivity disorder.* In Evans, J.R. & Abarbanel, A., Introduction to Quantitative EEG and Neurofeedback. San Diego: Academic Press.

Lubar, J.F. & Shouse, M.N. (1976). EEG and behavioral changes in a hyperkinetic child concurrent with training of the sensorimotor rhythm (SMR): A Preliminary report. *Biofeedback and Self-Regulation, 3*, 293-306.

Lubar, J.F., Swartwood, M.O., Swartwood, J.N. & O'Donnell, P.H. (1995). Evaluation of the effectiveness of EEG neurofeedback training for ADHD in a clinical setting as measured by changes in TOVA scores, behavioral ratings, and WISC-R performance. *Biofeedback and Self-Regulation, 21*(1), 83-99.

Lubar, J., White, J.N., Swartwood, M.O. & Swartwood, J.N. (1999). Methylphenidate effects on global and complex measures of EEG. *Pediatric Neurology, 21*, 633-637.

Lundstrom, B.N., Ingvar, M. & Petersson, K.M. (2005). The role of precuneus and left inferior frontal cortex during source memory episodic retrieval. *NeuroImage, 27*, 824-834.

Lutz, A., Greischar, L.L., Periman, D. & Davidson, R.J. (2009). BOLD signal in insula is differentially related to cardiac function during compassion meditation in experts vs. novices. *Neuroimage, 47*, 1038-1046.

Lutz, A., Greischar, L.L., Rawlings, N.B., Ricard, M. & Davidson, R.J. (2004). Long-term meditators self-induce high amplitude gamma synchrony during mental practice. *Proceedings of the National Academy of Sciences USA 101*, 16369-16373.

Mabbott, D.J., Rovet, J., Noseworthy, M.D., Smith, M.L. & Rockel, C. (2009). The relations between white matter and declarative memory in older children and adolescents. *Brain Research, 1294*, 80-90.

Macintosh, K.E & Dissanayake, C. (2004). Annotation: The similarities and differences between autistic disorder and Asperger's disorder: A review of the empirical evidence. *Journal of Child Psychology & Psychiatry, 45*(3), 421-434.

MacKinnon, S., Gevirtz, R., McCraty, R. & Brown, M. (2013). Utilizing Heartbeat Evoked Potentials to Identify Cardiac Regulation of Vagal Afferents During Emotion and Resonant Breathing. *Applied Psychophysiology and Biofeedback, 38*(4), 241-255.

MacLeod, A.K., Buckner, R.L., Miezin, F.M., Petersen, S.E. & Raichle, M.E. (1998). Right anterior prefrontal cortex activation during semantic monitoring and working memory. *Neuroimage, 7*, 41-48.

Malenka, R. & Bear, M. (2004). LTP and LTD: an embarrassment of riches. *Neuron, 44*(1), 5-21.

Malone, M.A., Kershner, J.R. & Swanson J.M. (1994). Hemispheric processing and methylphenidate effects in attention-deficit/ hyperactivity disorder. *Journal of Child Neurology, 9*(2), 181-189.

Mann, C.A., Lubar, J.F., Zimmerman, A.W., Miller, C.A. & Muenchen, R.A. (1992). Quantitative analysis of EEG in boys with attention-deficit/hyperactivity disorder: Controlled study with clinical implications. *Pediatric Neurology, 8*(1), 30-36.

Marchie, A. & Cusimano, M.D. (2003). Bodychecking and concussions in ice hockey: Should our youth pay the price? *Canadian Medical Association, 169* (2), 124-128.

Marosi, E., Harmony, T., Sánchez, L., Becker, J., Bernal, J., Reyes, A., Díaz de León, A.E., Rodríguez, M. & Fernández, T. (1992). Maturation of the coherence of EEG activity in normal and learning-disabled children. *Electroencephalography and Clinical Neurophysiology, 83*, 350-357.

Martinez, Y. (2003). Unpublished thesis done as part of the requirements for an Honours B.A. degree. Copy on file at the ADD Centre, Toronto.

Marx, B.P., Forsyth, J.P., Gallup, G.G. & Fusé, T. (2008). Tonic immobility as an evolved predator defense: implications for sexual assault survivors. *Clini. Psychol. Sci. Prac. 15*, 74-90.

Matsuura, M., Yamamoto, K., Fukuzawa, H., Okubo, Y., Uesugi, H., Moriiwa, M., Kojuma, T. & Shimazomo, Y. (1985). Age development and sex differences of various EEG elements in healthy children and adults – Quantification by a computerized waveform recognition method. *Electroencephalography and Clinical Neurophysiology, 60*, 394-406.

Matthews, S.C., Paulus, M.P., Simmons, A.N., Nelesen, R.A. & Dimsdale, J.E. (2004). Functional subdivisions within anterior cingulate cortex and their relationship to autonomic nervous system function. *Neuroimage, 22*, 1151-1156.

Mayberg, H.S. (2003). Modulating dysfunctional limbic-cortical circuits in depression: towards development of brain-based algorithms for diagnosis and optimised treatment. *British Medical Bulletin, 65*, 193-207.

Mayberg, H.S., Brannan, S.K., Tekell, J.L., Silva, J.A., Mahurin, R.K., McGinnis, S. & Jerabek, P.A. (2000). Regional metabolic effects of floxetine in major depression: Serial changes and the relationship to clinical response. *Biol.Psychiatry, 48*(8), 830-843.

Mayberg, H.S., Lozano, A.M., Voon, V., McNeely, H.E., Seminowicz, D., Hamani, C., Schwalb, J.M. & Kennedy, S. H. (2005). Deep brain stimulation for treatment-resistant depression. *Neuron, 45*(5), 651-660.

Mayer, E, Martory, M.D., Pegna, A.J., Landis, T., Delavelle, J. & Annoni, J.M. (1999). A pure case of Gerstmann syndrome with a subangular lesion. *Brain, 122*(6), 1107-1120.

McClintic, J. (1978). Physiology of the Human Body, p.103. New York: John Wiley & Sons Inc.

McCormick, D.A. & Huguenard, J.R. (1992). A model of the electrophysiological properties of thalamocortical relay neurons. *J. Neurophysiol., 68*, 1384-1400.

McCrory, P., Meeuwisse, W.H., Auby, M. Cantu, B., Dvo ák, J., Echemendia, R.J., Engebretsen, L., Johnston, K., Kutcher, J.S., Raftery, M., Sills, A., Benson, B.W., Davis, G.A., Ellenbogen, R.G., Guskiewicz, K., Herring, S.A., Iverson, G.L., Jordan, B.D., Kissick, J., McCrea, M., McIntosh, A.S., Maddocks, D., Makdissi, M., Purcell, L., Putukian, M., Schneider, K., Tator, C.H. & Turner, M. (2013). Consensus statement on concussion in sport: The 4th International Conference on Concussion in Sport held in Zurich, November 2012. *British Journal of Sports Medicine, (47)*, 250-258.

McKeith, I.G., Galasko, D., Kosaka, K., Perry, E.K., Dickson, D.W., Hansen, L.A. et al. (1996). Consensus guidelines for the clinical and pathologic diagnosis of dementia with Lewy bodies (DLB): report of the consortium on DLB international workshop. *Neurology, 47*, 1113-1124.

Mcrone, J. (2002). The first word. *Lancet Neurol. 1*(1), p 72.

Mehta, M.A., Owen, A.M., Sahakian, B.J., Mavaddat, N., Pickard, J.D. & Robbins, T.W. (2000). Methylphenidate enhances working memory by modulating discrete frontal and parietal lobe regions in the human brain. *Journal of Neuroscience, 20*, 1-6

Mehta, M.A., Sahakian, B. J. & Robbins, T.W. (2001). Comparative psychopharmacology of methylphenidate and related drugs in human volunteers, patients with ADHD and experimental animals. In Solanto, M.V., Arnsten, A.F. & Castellanos, F.X. (eds.) *Stimulant Drugs and ADHD Basic and Clinical Neuroscience.* New York: Oxford University Press, 303-331.

Mendez, C.V., Hurley, R.A., Lassonde, M., Zhang, L. & Taber, K.H. (2005). Mild traumatic brain injury: Neuroimaging of sports-related concussion. *The Journal of Neuropsychiatry and Clinical Neurosciences, 17*(3), 297-303.

Menon, V. & Uddin, L.Q. (2010). Saliency, switching, attention and control: a network model of insula function. *Brain Struct Funct, 214*, 655-667

Mesulam, M. & Guela, C. (1988). Nucleus basalis (Ch4) and cortical cholinergic innervation in the human brain: Observations based on the distribution of acetylcholinesterase and choline acetyltransferase. *Journal of Comparative Neurology, 275*(2), 216-240.

Metcalfe, J. & Shimamura, A.P. (eds.) (1996). *Metacognition.* Cambridge, MA: MIT Press.

Metrametrix Clinical Laboratories Inc. 3425 Corporate Way, Duluth, GA 20096, USA.

Meyer, J.S., Kondo, A., Szewczykowski, J., Nomura, F. & Teraura, T. (1970). The effects of a new drug (Hexobendine) on cerebral hemodynamics and oxygen consumption. *Journal of the Neurological Sciences, 11*(2), 137-145.

Meythaler, J.M., Peduzzi, J.D., Eleftheriou, E. & Novack, T.A. (2001). Current concepts: Diffuse axonal injury—Associated traumatic brain injury. *Archives of Physical Medicine and Rehabilitation, 82*, 1461-1471.

Middaugh, S.J., Haythornwaite, J.A., Thompson, B., Hill, R., Brown, K.M., Freedman, R.R. Attanasio, V., Jacob, R.G., Scheier, M. & Smith, E.A.(2001). The Raynaud's treatment study: Biofeedback protocols and acquisition of temperature biofeedback skills. *Applied Psychophysiology and Biofeedback, 26*(4), 251-278.

Mikosch, P., Hadrawa, T., Laubreiter, K., Brandl, J., Pilz, J., Stettner, H. & Grimm, G. (2010). Effectiveness of respiratory-sinus-arrhythmia biofeedback on state-anxiety in patients undergoing coronary angiography. *Journal of Advanced Nursing, 66*(5), 1101-1110.

Minamimoto, T., Hori, Y. & Kimura, M. (2009). Roles of the thalamic CM-PF complex-basal ganglia circuit in externally driven rebias of action. *Brain Research Bulletin, 78*(2-3), 75-9.

Minshew, N.J., Luna, B. & Sweeney, J.A. (1999). Oculomotor evidence for neocortical systems but not cerebellar dysfunction in autism. *Neurology, 52*, 917-922.

Mitchell, K.R. & Mitchell, D.M. (1971). Migraine: An explanatory treatment application of programmed behavior therapy techniques. *Psychosomatic Research, 15*, 137-157.

Miura, H., Ozaki, N., Sawada, M., Isobe, K., Ohta, T. & Nagatsu, T. (2008). A link between stress and depression: shifts in the balance between the kynurenine and serotonin pathways of tryptophan metabolism and the etiology and pathophysiology of depression. *Stress, 11*(3), 198-209.

Miura, H., Ozaki, N., Shirokawa, T. & Isobe, K. (2008). Changes in brain tryptophan metabolism elicited by aging, social environment, and psychological stress in mice. *Stress, 11*(2), 160-169.

Miura, H., Shirokawa, T., Isobe, K. & Ozaki, N. (2009). Shifting the balance of brain tryptophan metabolism elicited by isolation housing and systemic administration of lipopolysaccharide in mice. *Stress, 12*(3), 206-214.

Molz, S., Decker, H., Oliveira, I.J., Souza, D.O. & Tasca, C.I. (2005). Neurotoxicity induced by glutamate in glucose-deprived rat hippocampal slices is prevented by GMP. *Neurochem. Res., 30*(1), 83-89.

Monastra, V.J., Lubar, J.F., Linden, M., VanDeusen, P., Green, G., Wing, W., Phillips, A. & Fenger T.N. (1999). Assessing attention deficit hyperactivity disorder via quantitative electroencephalography: An initial validation study, *Neuropsychology, 13* (3), 424-433.

Monastra, V.J., Lynn, S., Linden, M., Lubar, J.F., Gruzelier, J. & La Vaque, T.J. (2005). Electroencephalographic biofeedback in the treatment of attention-deficit-hyperactivity disorder. *Applied Psychophysiology and Biofeedback, 30*, 95-114.

Monastra, V.J., Monastra, D. & George, S. (2002). The effects of stimulant therapy, EEG biofeedback and parenting on primary symptoms of ADHD. *Applied Psychophysiology and Biofeedback, 27*(4), 272-250.

Moore, L.E. & Wiesner, S.L. (1996). Hypnotically-induced vasodilation in the treatment of repetitive strain injuries. *American Journal of Clinical Hypnosis, 39*(2), 97-104.

Moran, J.M., Wigg, G.S., Adams, R.B., Janata, P. & Kelly, W.M. (2004). Neural correlates of humor detection and appreciation. *Neuroimage, 21*(3), 1055-1060.

Moreland, J.D., Thomson, M.A. & Fuoco, A.R. (1998). Electromyographic biofeedback to improve lower extremity function after a stroke: A meta-analysis. *Archives of Physical Medical Rehabilitation, 79*(2), 134-140.

Morin, C.M., Hauri, P.J., Espie, C.A., Spielman, A.J., Buysse, D.J. & Bootzin, R.R. (1998). Nonpharmacological treatment of chronic insomnia. An American Academy of Sleep Medicine review. *Neuroscience and Behavior Psychology, 28*(3), 330-335.

Morrow, L., Urtunski, P.B., Kim, Y. & Boller, F. (1981). Arousal responses to emotional stimuli and laterality of lesion. *Neuropsychologia, 19*, 65-72.

Moss, D. & Gunkelman, J. (2003). Task Force on Methodology and Empirically Supported Treatments: Introduction. *APB, 7*(4). Reprinted, *Journal of Neurotherapy, 6*(4), 7-10.

Moss, D. & Shaffer, F. (2007). Psychophysiology & General Health Heart Rate Variability (HRV) Version, *HRV Clinical Guide,* Thought Technology.

Mourot, L., Bouhaddi, M., Perrey, S., Cappelle, S., Henriet, M., Wolf, J.-P., Rouillon, J.-D. & Regnard, J. (2004). Decrease in heart rate variability with overtraining: assessment by the Poincaré plot analysis. *Clin Physiol Funct Imaging, 24*, 10-18.

MTA Cooperative Group. (2004), National Institute of Mental Health Multimodal Treatment Study of ADHD Follow-up: Changes in Effectiveness and Growth After the End of Treatment, *Pediatrics*, 113:762-769.

Muel, S., Knott, J. R. & Benton, A. L. (1965). EEG abnormality and psychological test performance in reading disability. *Cortex, 1,* 434.

Mueller, H.H., Donaldson, C.C., Nelson, D.V. & Layman, M. (2001). Treatment of fibromyalgia incorporating EEG-driven stimulation: A clinical outcomes study. *Journal of Clinical Psychology, 57*(7), 933-952.

Munoz, D.P., Hampton, K.A., Moore, K.D. & Armstrong, I.T. (1998). Control of saccadic eye movements and visual fixation in children and adults with ADHD. *Proceedings of the Society for Neurosciences*, Annual meeting, Los Angeles, CA.

Murias, M., Swanson, J.M. & Srinivasan, R. (2007). Functional connectivity of frontal cortex in healthy and ADHD children reflected in EEG coherence. *Cerebral Cortex, 8*, 1788-1799.

Murias, M., Webb, S.J., Greenson, J. & Dawson, G. (2007). Resting state cortical connectivity reflected in EEG coherence in individuals with autism. *Biological Psychiatry, 62*(3), 270-273.

Myint, A. & Kim, Y. (2003). Cytokine-serotonin interaction through IDO: a neurodegeneration hypothesis of depression. *Medical Hypothesis, 61*(5-6), 519-525.

Näätänen, R., Kreegipuu K. (2011).The mismatch negativity (MMN) as an index of different forms of memory in audition. In: Bäckman, L., Nyberg. L. (eds.), *Memory, aging, and brain. A Festschrift in honour of Lars-Göarn Nilsson*. Psychology Press, 287-299.

Näätänen, R., Paavilainen, P., Rinne, T., Alho, K. (2007). The mismatch negativity (MMN) in basic research of central auditory processing: a review. *Clin. Neurophysiol. 118*, 2544-2590.

Näätänen, R., Pakarinen, S., Rinne, T. & Takegata, R. (2004). The mismatch negativity (MMN): towards the optimal paradigm. *Clin Neurophysiol, 115,* 140-144.

Nagai, M., Hoshide, S. & Kario, K. (2010). The insular cortex and cardiovascular system: a new insight into the brain-heart axis. *Journal of the American Society of Hypertension 4*(4), 174-182.

Nature's Children (1985) Publisher: J. R. DeVarennes, USA: Grolier Limited.

Nee, D.E., Kastner, S. & Brown, J.W. (2011). Functional heterogeneity of conflict, error, task-switching, and unexpectedness effects within medial prefrontal cortex. *NeuroImage, 54*(1), 528-40.

Netter, F.H. (1997). *Atlas of Human Anatomy.* 2nd Edition. Havas Medi Media. Friesens Corporation. Canada.

Neuner I., Podoll K., Janouschek H., Michel T.M., Sheldrick A.J. & Schneider F. (2009). From psychosurgery to neuromodulation: deep brain stimulation for intractable Tourette syndrome. *World J Biol Psychiatry,* 10(4 Pt 2), 366-376.

Nilsson, B. & Nordström, C.H. (1977). Experimental head injury in the rat: Part 3: Cerebral blood flow and oxygen consumption after concussive impact acceleration. *Journal of Neurosurgery, 47*(2), 262-273.

Nitsche, M., Cohen, L., Wassermann, E., Priori, A., Lang, N., Antal, A., Paulus, W., Hummel, F., Boggio, P., Fregni,F. & Pascual-Leone, A. (2008). Transcranial direct current stimulation: State of the art. *Brain Stimulation, 1,* 206-23.

Nitsche, M.A., Boggio, P.S., Fregni, F. Pascual-Leone, A. (2009).Treatment of depression with transcranial direct current stimulation (tDCS): A Review. *Experimental Neurology 219,* 14-19.

Nitsche, M.A., Fricke, K., Henschke, U., Schlitterlau, A., Liebetanz, D., Lang, N., Henning, S., Tergau, F. & Paulus, W. (2003). Pharmacological modulation of cortical excitability shifts induced by transcranial direct current stimulation in humans. *J Physiol., 553*(Pt 1), 293-301.

Norris, L.S. & Currier, M. (1999). Performance enhancement training through neurofeedback. In Evans, J.R. & Abarbanel, A. (eds.), *Introduction to Quantitative EEG and Neurofeedback.* San Diego: Academic Press.

Northoff, G., Heinzel, A., de Greck, M., Bermpohl, F., Dobrowolny, H. & Panksepp, J. (2006). Self-referential processing in our brain—A meta-analysis of imaging studies on the self. *NeuroImage 31,* 440-457.

Nusslock, R., Almeida, J.R.C., Forbes, E.E., Versace, A., Frank, E., LaBarbara, E.J., Klein, C.R. & Phillips, M.L. (2012). Waiting to win: elevated striatal and orbitofrontal cortical activity during reward anticipation in euthymic bipolar disorder adults. *Bipolar Disorder, 14,* 249-260.

Nutt, D.J. (2008), Relationship of neurotransmitters to the symptoms of major depressive disorder, *Journal of Clinical Psychiatry, 69* Suppl E1, ISSN:0160-6689, 4-7.

Oberman, L.M., Hubbard, E.M., McCleery, J.P., Altschuler, E.L., Ramachandran, V.S. & Pineda, J.A. (2005). EEG evidence for mirror neuron dysfunction in autistic spectrum disorders. *Brain Research & Cognitive Brain Research, 24,* 190-198.

Ochsner, K.N. & Gross, J.J. (2005). The cognitive control of emotion. *Trends in Cognitive Sciences,* 9(5), 242-249.

Odom, J.V., Bach, M., Barber, C., Brigell, M., Marmor, M.F., Tormene, A.P., Holder, G.E. & Vaegan, (2004).Visual Evoked Potentials Standard, *Documenta Ophthalmologica, 108,* 115-123.

Ohn, S.H., Park, C.I., Yoo, W.K., Ko, M.H., Choi, K.P., Kim, G.M., Lee, Y.T. & Kim, Y.H. (2008). Time-dependent effect of transcranial direct current stimulation on the enhancement of working memory. *Neuroreport, 19*(1), 43-47.

Olds, M.E. & Milner, P. (1954). Positive reinforcement produced by electrical stimulation of septal area and other regions of the rat brain. *Journal of Comparative and Physiological Psychology, 47,* 419-427.

Olejniczak, P. (2006). Neurophysiologic Basis of EEG, *J Clin Neurophysiol, 23,* 186-189.

Ongur, D. (2000). The Organization of Networks within the Orbital and Medial Prefrontal Cortex of Rats, Monkeys and Humans. *Cerebral Cortex, 10*(3), 206-219.

Oppenheimer, S. (2001) Forebrain lateralization and the cardiovascular correlates of epilepsy. *Oxford Journals, Medicine, Brain, 124*(12), 2345-2346.

Oppenheimer, S.M., Gelb, A., Girvin, J.P. & Hachinski, V.C. (1992). Cardiovascular effects of human insular cortex stimulation, *Neurology, 42,* 1727-1732.

Othmer, S.F. & Othmer, S. (1991). EEG biofeedback training for ADD, specific learning disabilities and associated conduct problems. Encino, CA: EEG Spectrum, Inc.

Pacak, K., Palkovits, M., Kvetnansky, R., Matern, P., Hart, C., Kopin, I.J. & Goldstein, D.S. (1995). Catecholaminergic inhibition by hypercortisolemia in the paraventricular nucleus of conscious rats. *Endocrinology, 136*(11) 4814-4819.

Palincsar, A. S., Brown, D. A. (1987) Enhancing instructional time through attention to metacognition. *Journal of Learning Disabilities, 20*(2).

Palsson, O.S., Pope, J.D., Ball, M.J., Turner, S.N. & DeBeus R. (2001). Neurofeedback videogame ADHD technology: Results of the first concept study, Abstract, Proceedings of the 2001 Association for Applied Psychophysiology and Biofeedback Meeting, Raleigh-Durham, NC.

Panayiotopoulos CP. (2005). *The Epilepsies: Seizures, Syndromes, and Management:* Chapter 2, Optimal Use of the EEG in the Diagnosis and Management of Epilepsies. Oxfordshire (UK): Bladon Medical Publishing (available on the internet).

Panu, N. & Wong, S. (eds.) (2002). *MCCQE Review Notes & Lecture Series.* Toronto: University of Toronto Press.

Papez, J.W. (1937). A proposed mechanism of emotion, *Archives of Neurology & Psychiatry, (38),* 725-743.

Paquette, V., Beauregard, M. & Beaulieu-Prévoste, D. (2009). The effect of psychoneurotherapy on brain electromagnetic tomography in individuals with major depressive disorder. Psychiatry Research: *Neuroimaging, 174,* 231-239.

Pascual-Marqui, R. (2000). *Proceedings of the annual meeting of the Society for Neuronal Regulation,* Minneapolis, MN, October 2000.

Pascual-Marqui, R.D. (2002). The sLORETA method: Standardized low-resolution brain electromagnetic tomography (sLORETA): technical details. *Methods and Findings in Experimental and Clinical Pharmacology, 24*(Supplement D), 5-12.

Pascual-Marqui, R.D., Esslen, M., Kochi, K. & Lehmann, D. (2002). Functional Imaging with Low Resolution Electromagnetic Tomography (LORETA): A review. *Methods & Findings in Experimental & Clinical Pharmacology, 24C,* 91-95.

Pasqual-Marqui, R.D., Michel, C.M. & Lehmann, D. (1994). Low resolution electromagnetic tomography: a new method for localizing electrical activity in the brain. *International Journal of Psychophysiology, 18*(1), 49-65.

Passatore, M., Grassi, C. & Filippi G.M. (1985, Dec.). Sympathetically-induced development of tension in jaw muscles: the possible contraction of intrafusal muscle fibres. *Pflugers Archiv - European Journal of Physiology.* 405(4),297-304.

Pasternak O., Koerte I.K., Bouix S., Fredman E., Sasaki T., Mayinger, M., Helmer, K.G., Johnson, A.M., Holmes, J.D., Forwell, L.A., Skopelja, E.N., Shenton, M.E. & Echlin, P.S. (2014). Hockey Concussion Education Project, Part 2. Microstructural white matter alterations in acutely concussed ice hockey players: a longitudinal free-water MRI study. *J. Neurosurgery, Feb. Internet ahead of print.*

Patrick, G.J. (1996). Improved neuronal regulation in ADHD: An application of fifteen sessions of photic-driven EEG neurotherapy. *Journal of Neurotherapy, 1* (4), 27-36.

Paulesu, E., Frith, U., Snowling. M., Gallagher, A., Morton, J., Frackowiak, S.J. & Frith, C.D. (1996). Is develpmental dyslexia a disconnection syndrome? Ebidence from PET scanning. *Brian, 118,* 143-157.

Pavlakis, F.Y. (2001). Brain imaging in neurobehavioral disorders. (Review) *Paediatric Neurology, 25*(4), 278-287.

Pavlov, V.R. & Tracey, K.J. (2005). The cholinergic anti-inflammatory pathway. *Brain, Behavior, and Immunity* 19(6), 493-499.

Penfield, W. & Jasper, H. (1954). *Epilepsy and the functional anatomy of the human brain.* Little Brown, Boston.

Penfield, W. & Perot, P. (1963). The brain's record of auditory and visual experience. A final summary and discussion, *Brain, 86,* 595-696. (also *Brain, 128*(3), 449-450).

Peniston, E.G. & Kulkosky, P.J. (1989). Alpha-theta brainwave training and beta-endorphin levels in alcoholics. *Alcoholism: Clinical and Experimental Research, 13*(2), 271-279.

Peniston, E.G. & Kulkosky. P.J. (1990). Alcoholic personality and alpha-theta brainwave training. *Medical Psychotherapy, 3,* 37-55.

Peniston, E.G., Marrinan, D.A., Deming, W.A., & Kulkosky, P.J. (1993). EEG alpha-theta brainwave synchronization in Vietnam theater veterans with combat-related post-traumatic stress disorder and alcohol abuse. *Advances in Medical Psychotherapy,* 6, 37-50.

Pentacost, H. (1991). A kind of murder. In Robinson, K. (ed.) *Scholastic Scope Literature,* New York: Scholastics Ltd., 730 Broadway, New York, N.Y., 10003.

Peper, E. Shaffer, F. & I-Mei, L. (2010). Garbage in; Garbage out—identify blood volume pulse (BVP) Artifacts before analyzing and interpreting BVP, blood volume pulse amplitude, and heart rate/respiratory sinus arrhythmia data. *Biofeedback, Spring, 38*(1), 19-23.

Pepper, E. & Tibbetts, V. (1997). Electro-myography: Effortless diaphragmatic breathing. *The Biofeedback Foundation of Europe,* www.bfe.org.

Perez, J.J., Guijarro, I.E. & Barcia, I.J.A. (2005). The cardiac electric field artifact from the heart action evoked potential. *Med. Biol. Eng. Comput., 43,* 572-581.

Perminder, S. (2012). Transcranial direct current stimulation for depression: 3-week, randomised, sham-controlled trial, *British Journal of Psychiatry, 200,* 52-59.

Persinger, M.A. & Saroka, K.S. (2012). Protracted parahippocampal activity associated with Sean Harribance, *Int J Yoga, 5,* 140-150.

Petersen, S.E., Fox, P.T. Snyder, A.Z. & Raichle, M.E. (1990). Activation of extrastriate and frontal cortical areas by visual words and word-like stimuli. *Science, 249*(4972), 1041-1044.

Peterson, G. (2000). Operant Conditioning. *Proceedings of the Society for Neuronal Regulation annual meeting,* October 2000, Minneapolis.

Peterson, L.L. & Vorhies, C. (1983). Reynaud's syndrome: Treatment with sublingual administration of nitroglycerin, swinging arm manoeuvre, and biofeedback training. *Archives of Dermatology, 119*(5), 396-399.

Petrofsky, J.S. (2001). The use of electromyogram biofeedback to reduce Trendelenburg gait. *European Journal of Applied Psychophysiology, 85*(5), 491- 495.

Pfeifer, J.H., Iacoboni, M., Mazziotta, J.C. and Dapretto, M. (2008). Mirroring others' emotions relates to empathy and interpersonal competence in children. *NeuroImage, 39,* 2076–2085.

Pierce, K., Muller, R.-A., Ambrose, G., Allen, G. & Courchesne, E. (2001). Face processing occurs outside the fusiform 'face area' in autism: evidence from functional MRI. *Brain. 124,* 2059-2073.

Pineda, J.A., Brang, D., Hecht, E., Edwards, L., Carey, S., Bacon, M., Futagaki, C., Suk, D., Tom, J., Birnbaum, C. & Rork, A. (2008). Positive behavioral and electrophysiological changes following neurofeedback training in children with autism. *Research in Autistic Spectrum Disorders, 2,* 557-581.

Pizzagalli, D.H., Bogdan, R., Ratner, K.G. & Jahn, A.L. (2007). Increased perceived stress is associated with blunted hedonic capacity: potential implications for depression research. *Behavior Research & Therapy, 45*(11), 2742-2753.

Pizzagalli, D.A., Iosifescu, D., Hallett, L.A., Ratner, K.G. & Fava, M. (2008). Reduced hedonic capacity in major depressive disorder: Evidence from a probabilistic reward task. *Journal of Psychiatric Research, 43*(1), 76-87.

Pizzagalli, D.A., Nitschke, J.B., Oakes, T.R., Hendrick, A.M., Horras, K.A., Larson, C.L., Abercrombie, H.C., Schaefer, S.M., Koger, J.V., Benca, R.M., Pascual-Marqui, R.D., Davidson, R.J. (2002). Brain electrical tomography in depression: the importance of symptom severity, anxiety, and melancholic features. *Biological Psychiatry, 52*(2), 73-85.

Pizzagalli, D.H., Oakes, T.R. & Davidson, R.J. (2003). Coupling of theta activity and glucose metabolism in the human rostral anterior cingulate cortex: an EEG/PET study of normal and depressed subjects. *Psychophysiology, 40*(6), 939-949.

Pizzagalli, D.A., Oakes, T.R., Fox, A.S., Chung, M.K., Larson, C.L., Abercrombie, H.C., Schaefer, S.M., Benca, R.M. & Davidson, R.J. (2004). Functional but not structural subgenual prefrontal cortex abnormalities in melancholia. *Mol Psychiatry, 4*(325), 393-405.

Pizzagalli, D.A., Sherwood, R.J., Henriques, J.B. & Davidson, R.J. (2005). Frontal Brain Asymmetry and Reward Responsiveness: A Source-Localization Study. *Psychological Science, 16(*10), 805-813.

Pohunková, D., Sulc, J. & Vána, S. 1989. Influence of thyroid hormone supply on EEG frequency spectrum. Endocrinol Exp., 23(4), 251-258.)

Polich, J. (2004). Clinical application of the P300 event-related brain potential. *Phys Med Rehabil Clin N Am, 15*(1), 133-161.

Polich, J. (2007). Updating P300: An integrative theory of P3a and P3b. *Clinical Neurophysiology, 118*(10), 2128-2148.

Poncin, Y., Sukhodolsky, D., McGuire, J. & Scahill, L. (2007). Drug and Non-Drug Treatments of Children with ADHD and Tic Disorders. *European Child & Adolescent Psychiatry. 16*(Suppl.1), 78-88.

Pope, A.T. & Palsson, O.S. (2001). Helping Video Games "Rewire Our Minds," NASA Langley Research Center

Poreisz, C., Boros, K., Antal, A. & Paulus, W. (2007). Safety aspects of transcranial direct current stimulation concerning healthy subjects and patients. *Brain Res Bull., 72*(4-6), 208-214.

Porges, S.W. (2003). Social engagement and attachment: A phylogenetic perspective. *Annals of the New York Academy of Sciences, 1008,* 31-47.

Porges, S.W. (2004). The Vagus: A mediator of behavioral and physiologic features associated with autism. In Bauman, M.L. & Kemper, T.L. (eds.) *The Neurobiology of Autism.* (65-78). Baltimore: Johns Hopkins University Press.

Porges , S.W. (2007). The Polyvagal Perspective . *Biological Psychiatry***,** *74,* 116-143.

Porges, S.W. (2009). Music Therapy & Trauma: Insights from the Polyvagal Theory. Stewart, K. (ed.), Symposium on Music Therapy & Trauma: Bridging Theory and Clinical Practice. New York: Satchnote Press.

Porges, S.W., Bazhenova, O.V., Bal, E., Carlson, N., Sorokin, Y., Heilman, K.J., Cook, E.H. & Lewis, G.F. (In Press for 2014). Reducing auditory hypersensitivities in autistic spectrum disorders: Preliminary findings evaluating the Listening Project Protocol. Frontiers in Pediatrics.

Porges, S.W., Macellaio, M., Stanfill, S.D., McCue, K., Lewis, G.F. Harden E.R., Handelman, M., Denver, J., Bazhenova, O.V. & Heilman, K.J. (2013). Respiratory sinus arrhythmia and auditory processing in autism: Modifiable deficits of an integrated social engagement system? *International Journal of Psychophysiology, 88*(3), 261-270.

Posner, M.I. & Petersen, S.E. (1990). The attention system of the human brain. *Annual Review of Neuroscience, 13,* 25-42.

Posner, M.I., Sheese, B.E., Odludas, Y. & Tang, Y. (2006). Analyzing and shaping human attentional networks. *Neural Networks, 19,* 1422-1429. (www.sciencedirect.com) 2006 Special Issue.

Prather, A.A., Marsland, A.L., Hall, M., Neumann, S.A., Muldoon, M.F. & Manuck, S.B. (2009). Normative variation in self-reported sleep quality and sleep debt is associated with stimulated pro-inflammatory cytokine production. *Biological Psychology, 82,* 12-17.

Prather, A.A., Rabinovitz, M., Pollock, B.G. & Lotrich, F.E. (2009). Cytokine-induced depression following IFN-☐ treatment: the role of IL-6 and sleep quality. *Brain, Behavior, and Immunity, 23,* 1109-1116.

Prichep, E.S., Mas, F., Hollander, E., Liebowitz, M., John, E.R., Almas, M., DeCaria, C.M. & Levine, R.H. (1993). Quantitative electroencephalographic subtyping of obsessive-compulsive disorder. *Psychiatry Research: Neuroimaging,* 50, 25-32.

Prinsloo, G.E., Rauch, Laurie, H.G., Lambert, M.I., Muench, F., Noakes, T.D. & Derman, W.E. (2011). The effect of short duration heart rate variability (HRV) biofeedback on cognitive performance during laboratory induced cognitive stress. *Applied Cognitive Psychology, 25*(5), 792-801.

Rabe, S., Beauducel, A., Zollner, T., Maercker, A. & Karl, A. (2006). Regional brain electrical activity in posttraumatic stress disorder after motor vehicle accident. *Journal of Abnormal Psychology, 115*(4), 687-698.

Raichle, M.E. (2010). Two views of brain function. *Trends in Cognitive Sciences, 14*(4), 180-190.

Raichle, M.E., MacLeod, A.M., Snyder, A.Z., Powers, W.J., Debra A. Gusnard, D.A. Shulman, G.L. (2003). A default mode of brain function, National Academy of Sciences, *Proceedings of the National Academy of Sciences of the United States of America, 98* (2), 676-682.

Railo, H., Koivisto, M. & Revonsuo, A. (2011).Tracking the processes behind conscious perception: A review of event-related potential correlates of visual consciousness. *Consciousness and Cognition, 20,* 972-983.

Ramachandran, V.S. & Oberman, L.M. (2006). Broken Mirrors. *Scientific American, 295*(5), 62-69.

Ramos, F. (1998). Frequency band interaction in ADD/ADHD neurotherapy. *Journal of Neurotherapy, 3*(4), 27-36.

Raz, A. (2004). Anatomy of Attentional Networks. (Visual) *The Anatomical Record Part B: The New Anatomist, 281B*(1), Available online: 22 Nov., 2004.

Raz, A. & Sapiro, T. (2002). Hypnosis and Neuroscience: A Cross Talk between Clinical and Cognitive Research. *Archives of General Psychiatry, 59,* 85-90.

Read, D. (1981). Solving deductive-reasoning problems after unilateral temporal lobectomy. *Brain and Language, 12*:116-127.

Reddy, R.P., Jamuna, N., Indira, D.B. & Thennarasu, K. (2009). Neurofeedback training to enhance learning and memory in patient with traumatic brain injury: A single case study. *The Indian Journal of Neurotrauma, 6*(1), 87-90.

Reeds, P.J. (2000). Dispensable and indispensable amino acids for humans, *Journal of Nutrition, 130*(7), 1835S-1840S.

Reid, A. (2005). *Autistic Spectrum Disorders: Assessment and Intervention Results after Neurofeedback in 146 Cases.* Student Award Presentation at the Annual Meeting of the International Society for Neuronal Regulation, Denver, Colorado. (See publication 2010)

Reid, A., Nihon, S., Thompson, L. & Thompson, M. (2013). The effects of heart rate variability training on sensorimotor rhythm: A pilot study *Journal of Neurotherapy:* Investigations in Neuromodulation, Neurofeedback, and applied Neuroscience, *17*:1, 43-48.

Reiner, R. (2008). Integrating a portable biofeedback device into clinical practice for patients with anxiety disorders: results of a pilot study. *Applied Psychophysiology and Biofeedback, 33,* 55-61.

Reynaud's Treatment Study Investigators (2000). Comparison of sustained-release nifedipine and temperature biofeedback for treatment of primary Reynaud's phenomenon. Results from a randomized clinical trial with 1-year follow-up. *Archives of Internal Medicine, 160*(8), 1101-1108.

Rice, B., Kalder, A.J., Schindler, J.V. & Dixon, R.M. (2001). Effect of biofeedback-assisted relaxation therapy on foot ulcer healing. *Journal of the American Podiatric Medical Association, 91*(3), 131-141.

Rice, K.M., Blanchard, E.B. & Purcell, M. (1993). Biofeedback treatments of generalized anxiety disorders: Preliminary results. *Biofeedback and Self-Regulation, 18*(2), 93-105.

Richard, J.M. & Berridge, K.C. (2011). Nucleus Accumbens Dopamine/Glutamate Interaction Switches Modes to Generate Desire versus Dread: D_1 Alone for Appetitive Eating But D_1 and D_2 Together for Fear. *The Journal of Neuroscience, 31*(36), 12866-12879;

Rivera, E. & Omizo, M. M. (1980). The effects of relaxation and biofeedback on attention to task and impulsivity among male hyperactive children. *The Exceptional Child,* 27, 41-51.

Robbins, J. (2000). *A Symphony in the Brain.* New York: Atlantic Monthly Press.

Robbins, T.W. & Everitt, B.J. (2009). Limbic-striatal memory systems and drug addiction. *Neurobiol. Learn Mem, 78*(3), 625-636

Robinson, R.G., Kubos, K., Starr, L.B., Rao, K. & Price, T.R. (1984). Mood disorders in stroke patients: Importance of location of lesion. *Brain, 107*, 81-93.

Romero, J.R., Anschel, D., Sparing, R., Gangitano, M. & Pascual-Leone, A. (2002). Subthreshold low frequency repetitive transcranial magnetic stimulation selectively decreases facilitation in the motor cortex. *Clinical Neurophysiology, 113*(1), 101-107.

Ropper, A. & Brown, R. (2005). *Adam's and Victor's Principles of Neurology, 8th Edition* McGraw-Hill Companies Inc., United States of America, 417-430.

Rosenfeld, J.P. (1997). EEG Biofeedback of frontal alpha asymmetry in affective disorders. *Biofeedback, 25*(1), 8-25.

Rosenfeld, J.P. (2000a). An EEG biofeedback protocol for affective disorders. *Clinical electroencephalography, 31*(1), 7-12.

Rosenfeld, J.P. (2000b). Theoretical implications of EEG reference choice and related methodological issues. *Journal of Neurotherapy, 4* (2), 77-87.

Rosenfeld, J P., Baehr, E., Baehr, R., Gotlib, I. & Ranganath, C. (1996). Preliminary evidence that daily changes in frontal alpha asymmetry correlate with changes in affect in therapy sessions. *International Journal of Psychophysiology, 23*, 241-258.

Ross, E.D. (1981). The aprosodias: Functional-anatomic organization of the affective components of language in the right hemisphere. *Archives of Neurology, 38*, 561-569.

Rossignol, D.A., Rossignol, L.W., James, S.J., Melnyk, S. & Mumper, E. (2007). The effects of hyperbaric oxygen therapy on oxidative stress, inflammation, and symptoms in children with autism: an open-label pilot study. Biomed. Central (BMC) *Pediatrics, 7(*36), 1-13.

Rossiter, T.R. (1998). Patient-directed neurofeedback for AD/HD. *Journal of Neurotherapy, 2*(4), 54-63.

Rossiter, T.R. (2004). The effectiveness of neurofeedback and stimulant drugs in Treating AD/HD: Part II. Replication. *Applied Psychophysiology & Biofeedback, 29*(4), 233-243.

Rossiter, T.R. & LaVaque, T.J. (1995). A comparison of EEG biofeedback and psychostimulants in treating attention deficit hyperactivity disorders. *Journal of Neurotherapy, 1*(1), 48-59.

Rougeul, A., Letalle, A. & Corvisier, J. (1972). Activite rythmique du cortex somesthesique primair in relation avec l'immobilite chez le chat libre eveille. *Electroencephalogr. Clin. Neurophysiol., 33*, 23-39.

Rougeul-Buser A., Bouyer J.J., Montaron M.F. & Buser P. (1983). Patterns of activities in the ventro-basal thalamus and somatic cortex S1 during behavioral immobility in the awake cat: focal waking rhythms. *Exp. Brain Res.,* Suppl 7: 67-87.

Rougeul-Buser, A & Buser, P. (2011). Attention in cat revisited. A critical review of a set of brain explorations in fully alert animals *Archives Italiennes de Biologie,* 149 (Suppl.), 204-213,

Rourke, B.P. & Tsatsanis, K.D. (2000). Nonverbal learning disabilities and asperger's syndrome. In Klin, A., Volkmar, F.R. & Sparrow, S.S. (eds.) *Asperger Syndrome.* New York: Guilford Press.

Rozelle, G.R. & Budzynski, T.H. (1995). Neurotherapy for stroke rehabilitation: A single case study. *Biofeedback and Self-Regulation, 20*(3), 211-228.

Sacks, O. (1985). *The Man Who Mistook His Wife for a Hat.* New York: Summit Books.

Sampanthavivat, M., Singkhwa, W., Chaiyakul, T., Karoonyawanich, S. & Ajpru, H. (2012). Hyperbaric oxygen in the treatment of childhood autism: a randomised controlled trial. *Diving Hyperb Med, 42*(3), 128-133.

Santhirasegaram, L., Thompson, L., Reid, A. & Thompson, M. (2013). Training for Success in a Child with ADHD. *Biofeedback* 41(2), 75-81.

Sapolsky, R. (2010). This is your brain on metaphors, Exclusive on line commentary from the times, *Opinionator, The Stone*.

Sara, S.J. & Bouret, S. (2012). Orienting and reorienting: The locus coeruleus mediates cognition through arousal. *Neuron*, 76(1), 130-41. doi:10.1016/j.neuron.2012.09.011

Sarkar, P., Rathee, S.P. & Neera, N. (1999). Comparative efficacy of pharmacotherapy and biofeedback among cases of generalized anxiety disorder. *Journal of Projective Psychology & Mental Health.* 6(1), 69-77.

Sarter, M., Givens, B. & Bruno, J.P. (2001). The cognitive neuroscience of sustained attention: where top-down meets bottom-up. *Brain Research Reviews, 35,* 146-160

Sauseng, P., Hoppe, J., Klimesch, W., Gerloff, C. & Hummel, F.C. (2007). Dissociation of sustained attention from central executive functions: local activity and interregional connectivity in the theta range. *European Journal of Neuroscience, 25,* 587-593. (Theta increase with activation of memory based functions which require visuomotor integration.)

Sauseng, P., Klimesch, W., Gruber, W., Doppelmayr, M., Stadler, W., Schabus, M. (2002). The interplay between theta and alpha oscillations in the human electroencephalogram reflects the transfer of information between memory systems. *Neuroscience Letters, 324*(2), 121-124.

Saxby, E. & Peniston, E.G. (1995). Alpha-theta brainwave neurofeedback training: An effective treatment for male and female alcoholics with depressive symptoms. *Journal of Clinical Psychology, 51*(5), 685-693.

Schandry, R., Sparrer, B. & Weitkunat, R. (1986). 'From the heart to the brain: a study of heartbeat contingent scalp potentials'. *Int. J. Neurosci., 30*, 261-275.

Schenk, L. & Bear, D. (1981). Multiple personality and related dissociative phenomena in patients with temporal lobe epilepsy. *American Journal of Psychiatry, 138*, 1311-1316.

Schleenbaker, R.E. & Mainous, A.G. (1993). Electromyographic biofeedback for neuromuscular re-education in the hemiplegic stroke patient: A meta-analysis. *Archives of Physical Medical Rehabilitation, 74*(12). 1301-1304.

Schneider, H.D., Hopp, J.P. (2011). The use of the Bilingual Aphasia Test for assessment and transcranial direct current stimulation to modulate language acquisition in minimally verbal children with autism. *Clin Linguist Phon.*, (6-7), 640-654.

Schultz, R.T., Romanski, L.M. & Tsatsanis, K.D. (2000). Neurofunctional models of autistic disorder and Asperger syndrome, clues from neuroimaging. In Klin, A., Volkmar, F.R. & Sparrow, S.S. (eds.) *Asperger Syndrome.* New York: Guilford Press.

Schwartz, M.S. and Associates (1995). *Biofeedback: A Practitioner's Guide - Second Edition.* New York: Guilford Press.

Schwartz, M. (1987). *Biofeedback: A Practitioner's Guide.* New York: Guilford Press.

Scott, J.C., Matt, G.E., Wrocklage, K.M., Crnich, C., Jordan, J., Southwick, S.M., Krystal, J.H. & Schweinsburg, B.C. (2015). A quantitative meta-analysis of neurocognitive functioning in posttraumatic stress disorder. *Psychological Bulletin, 141*(1), 105-40.

Sears, W. & Thompson, L. (1998). *The A.D.D. Book: New Understandings, New Approaches to Parenting Your Child.* New York: Little, Brown and Co.

Seeley, W., Menon, V., Schatzberg, A., Keller, J., Glover, G., Kenna, H., Reiss, A. & Greicius, M. (2007). Dissociable Intrinsic Connectivity Networks for Salience Processing and Executive Control. *The Journal of Neuroscience, (27)*9, 2349-2356.

Seger, C.A., Stone, M. & Keenan, J.P. (2004). Cortical Activations during judgments about the self and another person. *Neuropsychologia, 42,* 1168-1177.

Sella, G.E. (1997). Electromyography: Towards an integrated approach of sEMG utilization: Quantitative protocols of assessment and biofeedback. *The Biofeedback Foundation of Europe*, www.bfe.org/protocol/prol3eng.htm.

Selye, H. (1976). *The Stress of Life - Revised Edition.* New York: McGraw-Hill.

Shaffer, Fred (2002). The Neuron. Workshop presentation at the AAPB annual meeting. Los Vegas, NV.

Shah, S.G., Klumpp, H., Angstadt, M., Nathan, P.J. & Phan, K.L. (2009). Amygdala and insula response to emotional images in patients with generalized anxiety disorder. *J Psychiatry Neurosci, 34*(4), 296-302.

Shain, R.J. (1977). *Neurology of Childhood Learning Disorders - Second Edition.* Baltimore: Williams & Wilkins.

Shallice, T.I.M. & Burgess, P.W. (1991). Deficits in strategy application following frontal lobe damage in man. *Brain, 114*(2), 727-741.

Shamay-Tsoory, S.G., Tomer, R., Berger, B.D., Goldsher, D. & Aharon-Peretz, J. (2005). Impaired "Affective Theory of Mind" is associated with right ventromedial prefrontal damage. *Cognitive & Behavioral Neurology, 18*(1), 55-67.

Shaw, N. (2002). The neurophysiology of concussion. *Progress in Neurobiology, 67*(4), 281-344.

Shealy, N., Cady, R., Culver-Veehoff, D., Cox, R. & Liss, S. (1988). Cerebrospinal fluid and plasma neurochemicals: Response to cranial electrical stimulation. *Journal of Orthopedic Medical Surgery, 18*, 94-97.

Shealy, N., Cady, R., Veehoff, D., Houston, R., Burnetti, M., Cox, R. & Closson, W. (1992). The neurochemistry of depression. *American Journal of Pain Management, 2*(1), 13-16.

Sheer, D.E. (1977). Biofeedback training of 40 Hz EEG and behavior. In Burch, N. & Altshuler, H.I. (eds.) *Behavior & brain electrical activity.* New York: Plenum.

Sheline, Y.I., Barch, D.M., Price, J.L., Rundle, M.M., Vaishnavi, S.N., Snyder, A.Z. Mintunn, M.A., Wang, S., Coalson, R.S., & Raichle, M.E. (2009). The default-mode network and self-referential processes in depression. *Proc Natl Acad Sci USA, 106*(6), 1942-1947.

Sherin, L. (2003). Recovery after stroke, a single case study. *Proceedings of the AAPB annual meeting,* Jacksonville, Florida.

Sherman, R.A., Davis, G.D. & Wong, M.F., (1997). Behavioral treatment of exercise-induced urinary incontinence among female soldiers. *Military Medicine, 162*(10), 690-704.

Sherrill, R. (2004). Effects of hemoencephalographic (HEG) training at three prefrontal locations upon EEG ratios at Cz. *Journal of Neurotherapy, 8*(3), 63-76.

Shouse, M.N. & Lubar, J.F. (1979). Sensorimotor rhythm (SMR) operant conditioning and methylphenidate in the treatment of hyperkinesis. *Biofeedback & Self-Regulation, 4*, 299-311.

Shtark, M.B., Mazhirina1, K., Rezakova, M., Savelov, A., Pokrovskiy, M. & Jafarova, O. (2014). Neuroimaging Phenomenology of the Central Self-Regulation Mechanisms. *Journal of Behavioral and Brain Science, 4*, 58-68.

Sichel, A.G., Fehmi, L.G. & Goldstein, D.M. (1995). Positive outcome with neurofeedback treatment in a case of mild autism. *Journal of Neurotherapy, 1*(1), 60-64.

Sieb, R.A. (1990) *Medical Hypotheses, 33*, 145-153.

Siever, D. (1999). *David User's Guide*. Edmonton, Alberta: Computronic Devices Limited.

Siever, D. (2013). Transcranial DC Stimulation. *Neuroconnections,* Spring, 33-40.

Silverthorn, D.U. (1998). *Human Physiology, An Integrated Approach*. New Jersey: Prentice Hall.

Simpson, D. (2004). Asperger's syndrome and autism: Distinct syndromes with important similarities. In Rhode, M. & Klauber, T. (eds.) *The many faces of Asperger's syndrome. The Tavistock Clinic Series. xviii, 302,* 25-38. London, England: Karnac Books.

Sinzig, J., Walter, D. & Doepfner M. (2009). Attention deficit-hyperactivity disorder in children and adolescents with autism spectrum disorder: Symptom or syndrome? *Journal of Attention Disorders, 13*, 117-126.

Slobounov, S., Cao, C. & Sebastianelli, W. (2009). Differential effect of first versus second concussive episodes on wavelet information quality of EEG. *Clinical Neurophysiology, 120*(5), 862-867.

Slobounov, S., Hallett, M., Cao, C. & Newell, K. (2008). Modulation of cortical activity as a result of voluntary postural sway direction: An EEG study. *Neuroscience Letters, 442,*(3), 309-313.

Slobounov, S.M., Halletta, S., Stanhopec, H. & Shibasakia, W. (2005). Role of cerebral cortex in human postural control: an EEG study. *Clinical Neurophysiology, 116*, 315-323.

Slobounov, S. & Newell, K. (2009). Balance and Posture: Human. *Encyclopedia of Neuroscience*, Oxford: Academic Press. volume 2, 31-35.

Slobounov, S., Sebastianelli, W. & Simon, R. (2002). Neurophysiological and behavioral concomitants of mild brain injury in collegiate athletes. *Clinical Neurophysiology, 113*(2), 185-193.

Sloman, L. (2005). Medication use in children with high functioning pervasive developmental disorder and Asperger syndrome. In Stoddart, K. (ed.) *Children, Youth and Adults with Asperger Syndrome: Integrating multiple perspectives.* London: Jessica Kingsley Publishers.

Smith, C.G. (1962). *Basic Neuroanatomy.* Toronto, Canada: University of Toronto Press.

Smith, R. (1999). Cranial electrotherapy in the treatment of stress related cognitive dysfunction with an eighteen month follow-up. *Journal of Cognitive Rehabilitation, 17* (6), 14-19.

Smith, R. (2006). *Cranial Electrotherapy Stimulation; Its First Fifty Years, Plus Three - A Monograph.* Self published.

Smith, Y., Surmeier, D.J., Redgrave, P. & Kimura, M. (2011). Thalamic contributions to basal ganglia-related behavioral switching and reinforcement. *The Journal of Neuroscience : The Official Journal of the Society for Neuroscience, 31*(45), 16102-6.

Smith-Pellettier, C. (2002). The Hypothalamic-Pituitary Adrenal Axis. Paper presented at the Canadian Medical Association annual meeting.

Snowdon, D. (2001). *Aging with Grace: What the Nun Study Teaches Us About Leading Longer, Healthier, and More Meaningful Lives.* New York: Bantam Books.

Snyder, S.M. & Hall, J.R. (2006). A Meta-analysis of Quantitative EEG Power Associated With Attention-Deficit Hyperactivity Disorder. *Journal of Clinical Neurophysiology, 23*(5), 441-456.

Sokhadze, E.M., Baruth, J.M., Sears, L., Sokhadze, G.E., El-Baz, A.S. & Casanova, M.F. (2012). Prefrontal neuromodulation using rTMS improves error monitoring and correction function in autism. *Appl Psychophysiol Biofeedback, 37*(2), 91-102.

Sokhadze, E.M., Baruth, J.M., Sears, L., Sokhadze, G.E., El-Baz, A.S., Williams, E.L., Klapheke, R. & Casanova, M.F. (2012). Event related potential study of attention regulation during illusory figure categorization task in ADHD, autism spectrum disorder, and typical children. *Journal of Neurotherapy, 16*, 12-31.

Solanto, M.V., Arnsten, A.F.T. & Castellanos, X.F. (eds.) (2001). *Stimulant Drugs and ADHD Basic and Clinical Neuroscience.* New York: Oxford University Press.

Solnick, B. (2005). Effects of electroencephalogram biofeedback with Asperger's syndrome. *International Journal of Rehabilitation Research, 28*(2), 159-163.

Soltani, M. & Knight, R.T. (2000). Neural origins of the P300. *Crit Rev Neurobiol, 14*(3-4), 199-224.

Spargo, E. (ed.) (1989). *Timed Readings in Literature.* Providence, Rhode Island: Jamestown Publishers.

Sridharan, D., Levitin, J. & Menon, V. (2008). A Causal Role for the right Fronto-Insular Cortex in Switching between Executive-Control and Default-Mode Networks. *Proceedings of the National Academy of Sciences, 105*(34), 12569-12574.

Steiner, N.J., Frenette, E.C., Rene, K.M., Brennan, R.T. & Perrin, E.C. (2014). Neurofeedback and cognitive attention training for children with attention-deficit hyperactivity disorder in schools. *J Dev Behav Pediatr., 35*(1), 18-27.

Steriade, M., Gloor, P., Llinas, R.R., Lopes da Sylva, F.H. & Mesulam, M.M. (1990). Basic mechanisms of cerebral rhythmic activities. *Electroencephalography and Clinical Neurophysiology, 76*, 481-508.

Steriade, M. & McCarley, R. W. (2005). *Brain Control of Wakefulness and Sleep.* Springer, New York.

Sterman, M.B. (1996). Physiological origins and functional correlates of EEG rhythmic activities: Implications for self-regulation. *Biofeedback and Self-Regulation, 21*, 3-33.

Sterman, M.B. (1999). *Atlas of Topometric Clinical Displays: Functional Interpretations and Neurofeedback Strategies.* New Jersey: Sterman-Kaiser Imaging Laboratory.

Sterman, M.B. (2000a). Basic concepts and clinical findings in the treatment of seizure disorders with EEG operant conditioning. *Clinical Electroencephalography, 31*(1), 45-55.

Sterman, M.B. (2000b) EEG markers for attention deficit disorder: pharmacological and neurofeedback applications. *Child Study Journal, 30*(1), 1-22.

Sterman, M.B. & Bowersox, S.S. (1981). Sensorimotor EEG rhythmic activity: A functional gate mechanism. *Sleep, 4*(4):408-422.

Sterman M.B. & Thompson, L.M., (2014). *Neurofeedback for Seizure Disorders: Origins, Mechanisms, and Best Practice*; Chapter 12 in Cantor and Evans/Clinical Neurotherapy: Application of Techniques for Treatment. Elsevier Ltd., The Boulevard, Langford Lane, Kidlington, Oxford, OX5 1GB, United Kingdom

Sterman, M.B. & Wyrwicka, W. (1967). EEG correlates of sleep: Evidence for separate forebrain substrates. *Brain Res. 6*, 143-163.

Stoyva, J.M. (1986). Wolfgang Luthe: In Memoriam. *Biofeedback and Self-Regulation, 11*, 91-93.

Strack, I.B., Linden, M. & Wilson, V. (eds.) (2010). *Biofeedback and Neurofeedback in Sport.* Wheat Ridge Co.; Association of applied Psychophysiology and Biofeedback.

Strakowski, S.M., Adler, C.M., Holland, S.K., Mills, N.P., DelBello, M.P., Eliassen, J.C. (2005) Abnormal fMRI brain activation in euthymic bipolar disorder patients during a counting Stroop interference task. *American Journal of Psychiatry.* 162 (9):1697-1705.

Strehl, U., Leins, U., Goth, G., Klinger, C., Hinterberger, T. & Birbaumer, N. (2006). Self-regulation of Slow Cortical Potentials: A New Treatment for Children with Attention-Deficit/Hyperactivity Disorder. *Pediatrics, 118*, 1530-1540.

Strehl, U., Trevorrow, T., Veit, R., Hinterberger, T., Kotchoubey, B., Erb, M. & Birbaumer, N. (2006). Deactivation of brain areas during self-regulation of slow cortical potentials in seizure patients. *Appl Psychophysiol Biofeedback, 31*(1), 85-94.

Sullivan, M.W. (1970) *Comprehension Readers.* Box 577, Palo Alto, California, 94302:

Sung, M.S., Hong, J.Y., Choi, Y.H., Baik, S.H. & Yoon, H. (2000). FES-biofeedback versus intensive pelivic floor muscle exercise for the prevention and treatment of genuine stress incontinence. *Journal of Korean Medical Science, 15*(3), 303-308.

Supekar, K., Uddin, L.Q., Prater, K., Amin, H., Greicius, M.D. & Menon, V. (2010). Development of functional and structural connectivity within the default mode network in young children, *J. Neuroimage 52*, 290-301.

Suthana, N., Haneef, Z., Stern, J., Mukamel, R., Behnke, E., Knowlton, B.; Fried, I. (2012). Memory Enhancement and Deep-Brain Stimulation of the Entorhinal Area. *New England Journal of Medicine, 366*, 502-51

Sutton, S., Tueting, P., Zubin, J. & John, E. R. (1967). Information delivery and the sensory evoked potential. *Science, 155*, 1436-1439

Swanson, J.M., McBurnett, K., Wigal, T., Pfiffner, L.J., Williams, L. Christian, D.L., Tamm, L., Willcutt, E., Crowley, K., Clevenger, W., Khouam, N., Woo, C., Crinella, F.M. & Fisher, T.M. (1993). The effect of stimulant medication on children with attention deficit disorder: A "Review of Reviews." *Exceptional Children, 60*(2), 154-162.

Swanson, J., Baler, R.D. & Volkow, N.D., 2011. Understanding the Effects of Stimulant Medications on Cognition in Individuals with Attention-Deficit Hyperactivity Disorder: A Decade of Progress. *Neuropsychopharmacology, 36*(1), 207–226).

Sykora, M., Diedler, J., Rupp, A., Turcani, P., Steiner, T., (2009). Impaired baroreceptor reflex sensitivity in acute stroke is associated with insular involvement, but not with carotid atherosclerosis. *Stroke, 40*, 737-742.

Talairach, J. & Tournoux, P. (1988). *Co-Planar Stereotaxic Atlas of the Human Brain: 3-dimensional proportional system: An approach to cerebral imaging.* New York, Thieme Medical Publishers.

Tamm, L., Willcutt, E., Crowley, K., Clevenger, W., Khouam, N., Woo, C., Crinella, F.M. & Fisher, T.M. (1993). The effect of stimulant medication on children with attention deficit disorder: A "Review of Reviews." *Exceptional Children, 60*(2), 154 - 162.

Tan, C.O. & Bullock, D. (2008). A dopamine-acetylcholine cascade: Simulating learned and lesion-induced behavior of striatal cholinergic interneurons. *Journal of Neurophysiology, 100*(4), 2409-21.

Tan. G., Thornby, J., Hammond, D.C., Strehl, U., Canady B., Arnemann, K., Kaiser, D.A. (2009). Meta-analysis of EEG biofeedback in treating epilepsy. *Clin EEG Neurosci. 40*(3),173-179.

Tansey, M. (1986). A Simple and a Complex Tic (Gilles de la Tourette's Syndrome): Their Response to EEG Sensorimotor Rhythm Biofeedback Training. *International Journal of Psychophysiology, 4*, 91-97.

Tansey, M.A. (1993). Ten year stability of EEG biofeedback results for a hyperactive boy who failed fourth grade perpetually impaired class. *Biofeedback & Self-Regulation, 18*, 33-44.

Task Force of The European Society of Cardiology and The North American Society of Pacing and Electrophysiology (1996). Heart rate variability Standards of measurement, physiological interpretation, and clinical use. *European Heart Journal, 17*, 354-381.

Tator, C.H. (2013). Concussions and their consequences: Current diagnosis, management and prevention. *Canadian Medical Association Journal, (185)*11, 975-979.

Taylor, D.N. (1995). Effects of behavioral stress-management program on anxiety, mood, self-esteem, and T-cell count in HIV positive men. *Psychological Reports. 76*(2), 451-457.

Tebano, M.T., Cameroni, M., Gallozzi, G., Loizzo, A., Palazzino, G., Pezzini, G. & Ricci, G.F. (1988). EEG spectral analysis after minor head injury in man. *Electroencephalogr.Clin. Neurophysiol. 70*, 185-189.

Tekin, S. & Cummings, J.L. (2002). Frontal-subcortical neuronal circuits and clinical neuropsychiatry: An update. *Journal of Psychosomatic Research, 53*(2), 647-54.

Tesche, C.D. & Karhu, J. (2000). Theta oscillations index human hippocampal activation during a working memory task. *Proc Natl Acad Sci USA, 97*, 919-924.

Thatcher, R.W. (1999). EEG data base-guided neurotherapy. In Evans, J.R. & Abarbanel, A., *Introduction to Quantitative EEG and Neurofeedback.* New York: Academic Press.

Thatcher, R.W. (2009). *EEG evaluation of traumatic brain injury and EEG biofeedback treatment.* Elsevier, Inc.

Thatcher, R.W. (2012). *Handbook of quantitative electroencephalography and EEG biofeedback: Scientific Foundations and Practical Applications.* www.appliedneuroscience.com; self-published; p 35, 305, in prepublication Adobe format supplied by author.

Thatcher, R.W., Biver, C., Gomez, J.F., North, D., Curtin, R., Walker, R. A. & Salazar, A. (2001). Estimation of the EEG power spectrum using MRI T2 relaxation time in traumatic brain injury. *Clinical Neurophysiology, 112*(9), 1729-1745.

Thatcher, R.W., Biver, C., McAlaster, R., Camacho, M. & Salazar, A. (1998). Biophysical linkage between MRI and EEG amplitude in closed head injury. *NeuroImage, 7*(4), 352-367.

Thatcher, R.W., Biver, C., McAlaster, R. & Salazar, A. (1998). Biophysical linkage between MRI and EEG coherence in closed head injury. *Neuroimage, 8*, 307-326.

Thatcher, R.W., Krause, P.J. & Hrybyk, M. (1986). Cortico-cortical associations and EEG coherence: A two-compartmental model. *Electroencephalography and Clinical Neurophysiology, 64*, 123-143.

Thatcher, R.W., Walker, R.A., Gerson, I. & Geisler, F.H. (1989). EEG discriminant analysis of mild head trauma. *Electroencephalography and Clinical Neurophysiology, 73*(2), 94-106.

Thayer, J.F., Åhs, F., Fredrikson, M., Sollers, J.J., III & Wager, T.D. (2012). A meta-analysis of heart rate variability and neuroimaging studies: Implications for heart rate variability as a marker of stress and health. *Neuroscience & Biobehavioral Reviews, 36*(2), 747-756.

Thayer, J., Hansen, A., Saus-Rose, E. & Johnson, B.H. (2009). Heart rate variability, prefrontal neural function, and cognitive performance: The neurovisceral integration perspective on self-regulation, adaptation, and health. *Annals of Behavioral Medicine Publication of the Society of Behavioral Medicine, 37*(2), 141-153.

Thayer, J.F. & Lane, R.D. (2009). Claude Bernard and the heart-brain connection: further elaboration of a model of neurovisceral integration. *Neurosci. Biobehav. Rev., 33*, 81-88.

Thompson, J. & Hagedorn, D. (2012). Multimodal analysis: New approaches to the concussion conundrum. *Journal of Clinical Sport Psychology, 6*, 22-46.

Thompson, J., Sebastianelli, W. & Slobounov, S. (2005). EEG and postural correlates of mild traumatic brain injury in athletes. *Neuroscience Letters 377*(3), 158-163.

Thompson, L. & Thompson M. (1995). Exceptional Results with Exceptional Children. *Proceedings of the Society for the Study of Neuronal Regulation.* Annual Meeting: Scottsdale, Arizona.

Thompson, L. & Thompson M. (1998). Neurofeedback combined with training in metacognitive strategies: Effectiveness in students with ADD. *Applied Psychophysiology and Biofeedback, 23*(4), 243-263.

Thompson, L. & Thompson, M. (2006). *Autism Spectrum Disorders.* Panel presentation at the 36th Annual Meeting of the Association for Applied Psychophysiology and Biofeedback, Portland, OR.

Thompson, L. & Thompson, M. (2011). Beginning a Neurofeedback Combined with Biofeedback Practice, *NeuroConnections*, Fall, 32-38.

Thompson, L., Thompson, M. & Reid, A. (2010) Neurofeedback Outcomes in Clients with Asperger's Syndrome. *Journal of Applied Psychophysiology and Biofeedback, 35*(1) 63-81.

Thompson, L.M. (1979). *The Effect of Methylphenidate on Self-concept and Locus of Control in Hyperactive Children.* A Thesis submitted in conformity with the requirements for the Degree of Doctor of Philosophy in the University of Toronto.

Thompson, M., Thompson, J. & Wu, W. (2008). Brodmann Areas (BA) correlated with 10-20 sites and their primary functions. San Rafael, CA: International Society for Neurofeedback and Research.

Thompson, M. & Thompson, L. (2006) Improving Attention in Adults and Children: Differing Electro-encephalograhy Profiles and Implications for Training. *Biofeedback*, Fall 2006

Thompson, M. & Thompson, L. (2002) Biofeedback for Movement Disorders (Dystonia with Parkinson's Disease): Theory and Preliminary Results. *Journal of Neurotherapy, 6*(4), 51-70.

Thompson, M. & Thompson, L. (2003a). Neurofeedback for Asperger's Syndrome: Theoretical Rationale and Clinical Results. *The Newsletter of the Biofeedback Society of California, 19*(1).

Thompson, M. & Thompson, L. (2003b). *The Neurofeedback Book: An Introduction to Basic Concepts in Applied Psychophysiology*, Wheat Ridge, CO: Association for Applied Psychophysiology. Polish translation 2012.

Thompson, M. & Thompson, L. (2007a). Autistic Spectrum Disorders including Asperger's syndrome: EEG and QEEG findings, results, and neurophysiological rationale for success using neurofeedback training. Presented at the 11th Annual Meeting of the Biofeedback Foundation of Europe, Berlin, Germany. Abstract reprinted in *Applied Psychophysiology and Biofeedback, 32*(3-4), 213-214.

Thompson, M. & Thompson, L. (2007b). Neurofeedback for Stress Management. In Lehrer, P., Woolfolk, R. & Sime, W. (eds.) *Principles and Practice of Stress Management, 3rd Edition, 249-287.* New York: Guilford Publications.

Thompson, M. & Thompson, L. (2009a). Asperger's Syndrome Intervention: Combining Neurofeedback, Biofeedback and Metacognition. In Budzynski, T., Budzynski, H., Evans, J. & Abarbanel, A. (eds.), *Introduction to Quantitative EEG and Neurofeedback: Advanced Theory and Applications* (2nd ed., 365-415). New York: Academic Press/Elsevier.

Thompson, M. & Thompson, L. (2009b) Systems Theory of Neural Synergy: Neuroanatomical Underpinnings of Effective Intervention Using Neurofeedback plus Biofeedback. *Journal of Neurotherapy, 13*(1), January-March 72-74.

Thompson, M. & Thompson, L. (2010). Functional Neuroanatomy and the Rationale for Using EEG Biofeedback for Clients with Asperger's Syndrome. *Journal of Applied Psychophysiology and Biofeedback, (35)*1, 39-61.

Thompson, M., Thompson, L. & Reid, A. (2014). LORETA z-score Neurofeedback Combined with Heart rate Variability Training. In Thatcher, R. & Lubar, J. (eds.), EEG Z Score Neurofeedback: Clinical Applications, Academic Press.

Thompson, M., Thompson, L., Thompson, J. & Hagedorn, D. (2011). Networks: A Compelling Rationale for Combining Neurofeedback, Biofeedback and Strategies. *NeuroConnections*, Summer, 8-17.

Thompson, M., Thompson, L., Thompson, J. & Reid, A. (2009). Biofeedback Interventions for Autistic Spectrum Disorders: An Overview. *Neuroconnections*, Fall, 9-14,

Thompson, M., Thompson, L., Thompson, J., Reid, A. & Hagedorn, D. (2013b). Heart-Brain Connections Underlie Effective Neurofeedback plus Biofeedback Interventions: Complex Case with Depression. *NeuroConnections,* Winter, 11-26.

Thompson, M., Thompson, L., Reid-Chung, A. & Thompson, J. (2013). Managing TBI: Appropriate Assessment and a Rationale for Using Neurofeedback and Biofeedback to Enhance Recovery in Post-concussion Syndrome. *Biofeedback, 41*(4).

Thompson, M., Thompson, L. & Reid-Chung, A. (2015). Treating Post-Concussion Syndrome with Loreta Z-score Based Neurofeedback and Heart Rate Variability Biofeedback: Neuroanatomical/ Neurophysiological Rationale, Methods, and Case Examples. *Biofeedback, 43*(1).

Thompson, M., Thompson, L., Thompson, J. and Reid-Chung, A. (2015). Biofeedback for Autism Spectrum Disorders in *Biofeedback, Fourth Edition,* Schwartz, M. and Andrasik, F. (eds.). Guilford Press, New York, N.Y.

Thompson, M.G.G. (ed.) (1979). (Editorial Board - Dr. S. Woods, Los Angeles; Dr. D. Langsley, Cincinnati; Dr. M. Hollander, Tennessee; Dr. F. Lowy, Toronto; Dr. H. Prosen, Manitoba; Dr. K. Rawnsley, Great Britain; Dr. R. Ball, Australia) *A Resident's Guide to Psychiatric Education.* New York: Plenum Publishing.

Thompson, M.G.G. (1990). Developmental Assessment of the Preschool Child. In Stockman, J.A. (ed.) *Difficult Diagnoses in Pediatrics,.* W. D. Saunders Co., Harcourt Brace Jovanovich Inc. Philadelphia, Chapt. 2, 15-27.

Thompson, M.G.G. & Havelkova, M. (1983). Childhood Psychosis. In Steinhauer, P. & Rae-Grant, Q. (eds.) *Psychological Problems of the Child in the Family*. New York: Basic Books, Inc.

Thompson, M.G.G. & Patterson, P.G.R. (1986). The Thompson-Patterson scale of psychosocial development: I - Theoretical Basis. *Canadian Journal of Psychiatry, 31*(5).

Toomim, H. (2002). Hemoencephalography (HEG): The study of regional cerebral blood flow. C*alifornia Biofeedback*, Summer. 17-21.

Tower of London Test (1982). Colorado Assessment Tests, www.catstests.com/prod03.htm.

Tranel, D., Rudrauf, D., Vianna, E.P.M. & Damasio, H. (2008). Does the clock drawing test have focal neuroanatomical correlates? *Neuropsychology, 22*(5), 553-562.

Traub, R.D., Miles, R. & Wong, R.K.S. (1989). Model of the origin of rhythmic population oscillations in the hippocampal slice. *Science, 243*, 1319-1325.

Tucker, D.M., Watson, R.T. & Heilman, K.M. (1977). Affective discrimination and evocation in patients with right parietal disease. *Neurology, 27*, 947-950.

Tucker, D.M. & Williamson, P.A., (1984) Asymmetric neural control systems in human self regulation. *Psychological Review, 91*, 185-215.

Uecker, A., Barnes, C.A., McNaughton, B.L. & Reiman, E.M. (1997). Hippocampal glycogen metabolism, EEG, and behavior. *Behav Neurosci* 111:283-291.

Ullman, M. (1975). In A. M. Freedman, I. Kaplan, & B.J. Sadock, *Comprehensive Textbook of Psychiatry – Second Edition, Vol. 2*, 2552-2561. Baltimore: Williams and Wilkins.

Ungerleider, L.G. & Mishkin, M. (1982). Reported in Kim, M.S., Robertson, L.C. (2001). Implicit representations of space after bilateral parietal lobe damage. *J Cogn Neeurosci., 13* (8), 1080-1087.

van der Kolk, B.A. (2006). Clinical implications of neuroscience research in PTSD. *Annals of the New York Academy of Sciences,* 1071, 277-93.

Van Kampen, M., De Weerdt, W., Van Poppel, H., De Ridder, D., Feys, H. & Baert, L. (2000). Effect of pelvic-floor re-education on duration and degree of incontinence after radical prostatectomy: A randomized controlled trial. *Lancet, 355*(9198), 98-102.

Van Zomeren, A.H. & Brouwer, W.H. (1987). Head injury and concepts of attention. *Neurobehavioral recovery from head injury*, New York, Oxford University Press, 398-415.

Vanathy, S., Sharma, P.S.V.N. & Kumar, K.B. (1998). The efficacy of alpha and theta neurofeedback training in treatment of generalized anxiety disorder. *Indian Journal of Clinical Psychology, 25*(2), 136-143.

Vernon, D., Egner, T., Cooper, N., Compton, T., Neilands, C., Sheri, A. & Gruzelier, John. (2003). The effect of training distinct neurofeedback protocols on aspects of cognitive performance. *International Journal of Psychophysiology 47*, 75-85

Vertes, R., Hoover, W., Szigeti-Buck, K. & Leranth, C. (2007). Nucleus reuniens of the midline thalamus: Link between the medial prefrontal cortex and the hippocampus. *Brain Research Bulletin 71*(6), 601-609.

Vines, B.W., Nair, D. & Schlaug, G. (2008) Modulating activity in the motor cortex affects performance for the two hands differently depending upon which hemisphere is stimulated. *Eur J Neurosci., 28*(8), 1667-1673.

Vlaeyen, J.W., Haazen, I.W., Schuerman, J.A., Kole-Snijders, A.M. & van Eek, H. (1995). Behavioural rehabilitation of chronic low back pain: Comparison of an operant treatment, an operant-cognitive treatment and an operant-respondent treatment. *Clinical Psychology, 34*(1), 95-118.

Vossel, S., Geng, J.J. & Fink G.R. (2014). Dorsal and Ventral Attention Systems: Distinct Neural Circuits but Collaborative Roles. *The Neuroscientist, 20*(2), 150–159.

Wadhwani, S., Radvanski, D.C. & Carmody, D.P. (1998). Neurofeedback training in a case of attention deficit hyperactivity disorder. *Journal of Neurotherapy, 3*(1), 42-49.

Wagner, T., Fregni, F., Fecteau, S., Grodzinsky, A., Zahn, M. & Pascual-Leone, A. (2007). Transcranial direct current stimulation: a computer-based human model study. *Neuroimage; 35*(3), 1113-1124.

Wang, J., Rao, H., Wetmore, G.S., Furlan, P.M., Korcyowski, M., Dinges, D.F. & Detre, J.A. (2005). Perfusion functional MRI reveals cerebral blood flow pattern under psychological stress. *Proc Natl Acad Sci USA, 102*(49) 17804-17809.

Weidmann, G., Pauli, P., Dengler, W., Lutzenburger, W., Birbaumer, N. & Buckkremer, G. (1999). Frontal brain asymmetry as a biological substrate of emotions in patients with panic disorders. *Archives of General Psychiatry, 56,* 78-84.

Weins, W.J. (1983). Metacognition and the adolescent passive learner. *Journal of Learning Disabilities, 16*(3).

Weissman, D.H., Roberts, K.C., Visscher, K.M. & Woldorff M.G. (2006), The neural bases of momentary lapses in attention, *Nat Neurosci,* 9(7), 971-8.

Westmoreland, B.F., Espinoa, R.E. & Klass, D. W. (1973). Significant prosopo-glosso-pharyngeal movements affecting the electroencephalogram. *American Journal of EEG Technology, 13,* 59-70.

Westmoreland, B.F. & Klaus, B. W. (1998). Defective alpha reactivity with mental concentration. *Journal of Clinical Neurophysioogy, 15,* 424-428.

Whalen, P.J. (2003). Remembering people: Neuroimaging takes on the real world. *Learning & Memory, 10*(4), 240-241.

White, N.E. (1999). Theories of the effectiveness of alpha-theta training for multiple disorders. In Evans, J.R. & Abaranel, A. (eds.), *Introduction to Quantitative EEG and Neurofeedback.* San Diego: Academic Press.

Wilson, V.E., Thompson, J., Thompson, M. & Pepper. E. (2010). Using the EEG for enhancing performance: Arousal, Attention, Self Talk, and Imagery, In Strack, I.B., Linden, M. & Wilson, V. (eds.) *Biofeedback and Neurofeedback in Sport.* Wheat Ridge Co.; Association of Applied Psychophysiology and Biofeedback.

Wing, L. (2001). *The Autistic Spectrum: A parent's guide to understanding and helping your child.* Berkeley, CA: Ulysses Press.

Woodcock, T. & Morganti-Kossmann, M.C. (2013). The Role of Markers of Inflammation in Traumatic Brain Injury. *Front Neurol,* 4(18).

Wyrwicka, W. & Sterman, M.B., (1968). Instrumental conditioning of sensorimotor cortex EEG spindles in the waking cat. *Physiology and Behavior, 3,* 703-707.

Yamadera, H., Kato, M, Tsukahara, Y., Brandeis, D. & Okuma, T. (1997). Zipiclone versus diazepam effects on EEG power maps in healthy volunteers. *Acta Neurobiol Exp., 57,* 151-155.

Yang, T.T., Simmons, A.N., Matthews, S.C., Tapert, S.F., Bishoff-Grethe, A., Frank, G.K. Arce, E., & Paulus, M.P. (2007). Increased amygdala activation is related to heart rate during emotion processing in adolescent subjects. *Neurosci Lett. 428*(2-3), 109-114.

Yocum, D.E., Hodes, R., Sundstrom, W.R. & Cleeland, C.S. (1985). Use of biofeedback training in treatment of Reynaud's disease and phenomenon. *Journal of Rheumatology, 12*(1), 90-93.

Yucha, C. & Montgomery, D. (2008). *Evidence-Based Practice in Biofeedback and Neurofeedback.* Wheat Ridge, CO: Association for Applied Psychophysiology and Biofeedback.

Yucha, C.B., Clark, L., Smith, M., Uris, P., Lafleur, B. & Duval, S. (2001). The effect of biofeedback in hypertension. *Applied Nursing Research, 14*(1), 29-35.

Zaehle, T., Rach, S., Herrmann, C. (2010). Transcranial Alternating Current Stimulation Enhances Individual Alpha Activity in Human EEG. *PlosOne, 5* (11), 1-7.

Zametkin, A.J., Nordahl, T.E., Gross, M., King, A.C., Semple, W.E., Rumsey, J.H., Hamburger, S. & Cohen, R.M. (1990). Cerebral glucose metabolism in adults with hyperactivity of childhood onset. *New England Journal of Medicine, 323*(20), 1361-1366.

Zheng, X., Alsop, D.C. & Schlaug, G. (2011). Effects of transcranial direct current stimulation (tDCS) on human regional cerebral blood flow. *Neuroimage, 58*(1), 26-33.

Zhou, Y., Dougherty, J.H. Hubner, K.F., Bai, B., Cannon, R.L. & Hutson, R.K. (2008). Abnormal connectivity in the posterior cingulate and hippocamus in early Alzheimer's disease and mild cognitive impairment. *Alzheimer Dement., 4*(4), 265-270.

Zoeller, R.F., Jr. (2007). Physical activity: Depression, anxiety, physical activity, and cardiovascular disease: What is the connection? *American Journal Lifestyle Med, 1*, 175-180.

Zöllei, E., Paprika, D. & Rudas L. (2003). Measures of cardiovascular autonomic regulation derived from spontaneous methods and the Valsalva maneuver, 103(1-2), 100-105.

Index

Figures and tables are denoted respectively by *f* and *t* after the page number.

A

A620 assessment program vs. Monastra norms, 270, 284, 324, 395, 397
AAPB (Association for Applied Psychophysiology and Biofeedback), assessment training, 16, 324, 357
AAPB/SNR joint guidelines, 14
abnormal amplitude, 44–45
abnormal beta, 44
abnormal EEG patterns, 45–50
    absence seizures, 47, 47*f*, 350–351
    generalized seizures, 46
    non-spike and wave patterns, 48
    partial seizures, 46, 48*f*, 49*f*
    seizures, 46–50
    SKIL comparison, 50*f*
    spike-and-wave patterns, 46–48, 47*f*
absence seizures (petit mal epilepsy), 47, 47*f*, 350–351
absolute band values, 33
absolute power vs. relative power, 433
abulia, 99, 121
academic tasks and NFB, 415–417
acalculia, 92, 122
ACC. *See* anterior cingulate cortex
accommodation (right brain), 88
ACE (Adverse Childhood Experiences) study, 684
acetylcholine (Ach), 24, 182, 201
acetylcholinergic synapses, 102
acidosis, 347
acquired lefties, 95
ACTH (adrenocortico-tropic hormone), 102
Action authorship, 234
action inhibition, 176
action potentials, 24, 26–28
actions, anticipating and understanding, 187

active (+ve) lead placement, 4
active electrode effects on neurons, 594
AC vs. DC circuits and impedance, 317
ADC (analog to digital) converters, 54, 55
ADD. *See* Attention Deficit Disorder
The A.D.D. Book (Sears and Thompson), 672, 694
ADD Centre
    ADD Centre Planner and Distraction Dissipater (ADD-PADD), 490
    ADD Centre Store, 390*f*
    assessment, concussion/mild traumatic brain injury (TBI), 621–622
    assessment training, 357
    effectiveness studies, 670–671
    HRV and SMR study, 584–585, 585*f*
    individualized treatment plans, 672
addictions, 14, 25
addition and subtraction study techniques, 507
ADD-PADD (ADD Centre Planner and Distraction Dissipater), 490
adenosine triphosphate (ATP), 23, 587, 593–594, 607–608, 624–625, 626*f*
ADHD. *See* Attention Deficit Hyperactivity Disorder
adrenocortico-tropic hormone (ACTH), 102
Adrian, 7
Adverse Childhood Experiences (ACE) study, 684
affect circuit. *See* anterior cingulate cortex (ACC) circuit
affect disorders, HRV training and, 580
Affect Network
    amygdala and, 199–200
    BA 38, 203–204
    CMS areas and, 146, 146*f*
    depression, 641–647
    executive components, 217
    insula and, 211–213
    temporal lobe medial aspect, 201
    ventral portion and ACC, 215

afferent and efferent ANS connections between brain stem and cardiovascular system, 576f
after-effects with training, 458
age, dominant frequencies and, 34
aggression, serotonin and, 25
aggressive-depressed state in ASD, 659
Aging with Grace (Snowdon), 17
agnosia, 91, 92, 122
agraphesthesia, 186
agraphia, 92, 95, 121, 231, 233
"A Kind of Murder" (Pentecost), 502–503
akinesia, 99
akinetic mutism, 141–142
alcoholism
    19-channel (full-cap) QEEG assessment, 355
    alcohol, data collection on effects, 345
    dopamine activity and, 25
    as experimental application of NFB, 14
    high beta waves, 38
    Korsakoff's syndrome and, 107
    Peniston on, 15, 475
Aldrich, Robert, 607
Alerting network, 664
alertness, 110–111, 266, 495
Alexander, G.E., 140
Alexander techniques, 481
alexia, 93, 121, 231, 233
aliasing, 55
Allen, J.J.B., 644
all-or-none training, 561
alpha and beta reciprocal effects, 344
alpha asymmetry protocol, 99, 644
alpha desynchronization produced by 'go' stimulus, 6
alpha frequencies/rhythms
    bilateral decrease in, 44
    defined and described, 37
    as first-order brain waves, 7
    IQ correlations, 34
    rhythms and asymmetries, 34
    slowing of frequency, 44
    subject awareness of, 7
    unilateral reduction of, 45
    visual system and, 34

alpha synchronization, 6–7, 473
alpha-theta therapy, 475–482
    alpha-theta therapy addition, 480–481
    BFB and NFB, 477
    diaphragmatic breathing and relaxation, 477
    EEG assessment, 476
    enhanced alpha and theta, 481–482
    hypnagogic state, 15
    Ideal Me visualization, 479
    imagery and desensitization, 477–478
    imaging, 481–482
    inhibits, use of, 481
    negative ruminations, 478
    NFB training objectives, 477
    physiologically and psychologically relaxed state, 480
    relaxation-desensitization technique, 478
    scene visualization, 479
    stress assessment, 476
    stress desensitization, 479
    training overview, 477–478
ALS (Lou Gehrig's disease), SCP work, 7
alternating current (sine wave), 52
aluminum, 347
Alzheimer, Alois, 131
Alzheimer's dementia
    19-channel (full-cap) QEEG assessment, 354
    acetylcholine deficiency, 24
    BA 38, 203
    effect on NBM, 182
    ERC and, 204–205
    hippocampal region and, 191–192
    NFB treatment, 107
    slow-wave activity increase, 354
    tDCS and, 597
ambidexterity, 95
ambient electrical activity, 38–39
Amen, Dan, 112, 344–345, 496
American Academy of Child Psychiatry, efficacy criteria, 671
American Academy of Pediatrics, ratings of efficacious psychosocial interventions, 660, 670
American Biotech, assessment training, 357
American Clinical Neurophysiology Society, 272

American Journal of Psychiatry,
  on euthymic bipolar patients, 642
American Psychological Association, efficacy criteria, 671
Amini, Fari, 71
amino acids as neurotransmitters, 25
amitriptyline, 345
amotivational syndrome, 214, 344
amphetamines, 25, 113, 346
amplification, EEG, 52–53
amplitude, 4, 36, 51
amygdala
    Affect Network and, 146, 199–200
    as "gear shift" to thalamus, 199
    illustrations, 200*f*
    location and function, 73–74, 78, 82
    negative emotions generated in, 97
    PTSD and, 686
analogies, understanding, 173
analog to digital (ADC) converters, 54, 55
analytical processing, 164
Anderson, P.F., 29
Anderson, S.A., 29
angel dust, 346
anger control, 168, 177
angular gyrus
    Brodmann areas, 236*f*
    as connecting auditory and visual link, 198
    disconnection from left visual cortex, 94
    location and function, 93–94, 237, 238*f*
    supramarginal gyrus and, 238*f*
anodal activation of neurons, 593, 596
anode and cathode in tDCS, 590, 591*f*
anomia, 91, 121
anosognosia, 91, 92, 122, 225–226, 237
ANS. *See* autonomic nervous system
anterior and hippocampal commissures, 80*f*
anterior cerebral arteries, 100
anterior choroidal arteries, 100
anterior cingulate connections, 216–217
anterior cingulate cortex (ACC)
    Affect network and, 141–142, 218–220
    in ASD and Asperger's syndrome, 656
    disease-specific protein changes in, 215
    effects on other structures, 642
    fixing attention, 112
    location and function, 146, 214–216
    networks influenced by, 656
    PTSD and, 686–687
anterior commissure, 80, 85
anterior communicating artery, 100
anterior insula and speech, 176
anterior intraparietal (AIP) area, spatial perceptions, 225
anterior nuclei, pain perception in, 97
anterior nuclei of metathalamus, 75
anterior nucleus of thalamus, 86
anterior perforated area (APA), 78, 81–82, 100
anterior portion of brain, 68
anterior thalamic nuclei, 73
anterograde amnesia, 107, 192, 201, 205
An Anthropologist on Mars (Sachs), 108
antibiotics, 346
antidepressants, 9, 345
antihistamines, 347
Anton's agnosia, 239, 242
anxiety disorders
    19-channel (full-cap) QEEG assessment, 355
    ACC and, 218
    in ASD and Asperger's syndrome, 649, 651*f*–652*f*
    Brodmann area 10, 173
    CES for, 602
    high beta waves, 38
    LORETA images, 547*f*
    metacognitive strategies and, 484
    NFB and BFB combinations, 133
    norepinephrine and, 25
    right ventral prefrontal cortex and, 177
    training caution, 404
anxiolytic medications (benzodiazepines), 25, 37, 344
aphasia, 46, 92, 122, 198, 231
apperceptive agnosia, 91
applied psychophysiology,
  as client-controlled physiology, 2
approach behaviors, 173, 174, 419
appropriate behaviors, 140–141
apraxia, 92, 121, 233
aprosodia, 649

arachnoid membrane, 70
archicortex, 85
arcuate fasciculus, 466
Arns, Marijn, 666, 670–671, 677, 694
Aronovitch, Blair, 682
artifact effects
    assessment training, 358
    automatic artifacting, 350
    cardiac activity, 333–334, 333f, 334f, 338f, 339f, 340f
    client's typical artifacts, 320–321
    common artifacts, 326–341
    common artifactual currents, 62
    comparisons, NFB and BFB intervention fundamentals, 465
    correcting, 570
    defined, 326, 336
    in EEGs, 33, 59–60
    in EKGs, 570
    electrical artifacts, 62, 326, 392–393
    electrode bridges, 348
    electrode movement, 329, 330f
    electrode pop, 335f, 336, 348
    EMG interference, 39
    evoked potentials, 348
    eye blink, 330, 330f
    eyelid flutter, 334, 335f
    eye movement, 331–332, 331f, 332f, 342f
    frequency bands, effects on, 332, 333f
    hearing aids, 340f
    hyperventilation, 348
    importance, 326
    metal in mouth, 341f
    minimizing, 321–322
    muscle activity, 304f, 326–329, 327f, 337f, 392–393, 393f
    myogenic spike biphasic morphology, 42
    prevention, 348–350
    removal for assessments, 330
    rules for detecting, 326
    Sheer rhythm and, 38
    single motor unit muscle fasciculation, 338f
    special assessment and artifacting screen, 322–326
    spectral arrays and, 35
    tongue and swallowing, 334
    troubleshooting, 390–392
The Art of Artifacting (Hammond and Gunkelman), 336, 349, 358
ASD. *See* Autism Spectrum Disorder
Asperger's syndrome
    ACC and, 218. *see also* Autism Spectrum Disorder (ASD)
    adolescents, working with, 119–120
    attention span in, 649
    attention to external stimuli, 250
    brain side dominance, 88–89
    children, working with, 118–119
    child's drawings, 121f
    concrete thinking, 174
    deleted as diagnosis in DSM-5, 648
    as developmental deficiency, 120
    motor aprosodia, 176
    NFB interventions, 117–120
    promising clinical reports for NFB, 14
    reading comprehension, 501–502
    right hemisphere underactivation, 117
    soft encouragement, 403–404
    symptoms, 648–649
"Assessing Attention Deficit Hyperactivity Disorder via Quantitative Electro-encephalography" on theta/beta ratios, 284
assessment, concussion/mild traumatic brain injury (TBI), 630–633, 632f
assessment and intervention, 248–358
    absence seizures, 350–351
    alert mental state, 266
    ANS and EEG assessment, 265
    assessment data and bandwidths, 400–401
    assessment goals, 265–267, 270, 356
    assessment summaries, 373–375
    attention to external stimuli, 250
    calmness under stress, 266–267
    candidates for, 249, 269
    closed head injuries (CHIs), 351
    common factors in assessment, 250–251
    decision pyramid, 248–249, 249f
    differential amplifiers, 290

display screens, 320
EEG assessment, 265, 270–273
EEG normalization, 251, 267
electrode placements, 271–273, 272f, 285–286
environment, awareness of, 266
epilepsy, 250, 351, 352f
false impressions and misinterpretations, 349, 350f
full-cap findings, 314–315
global average Laplacian montage, 290
hemispheres, differences between, 286
hypoglycemia, 351
impedances in two channel approach, 289
inhibit and reward frequencies, 275f
internal reflection, 266
interpretation of findings, 314–315
linked-ear referential montage, 290
local average Laplacian or Hjorth montage, 290
LORETA, 355–357, 357f
montages, 290
narrow focus, 267
parsimony principle, 269–270
partial seizures, 351
pathological activity, 350–354
performance optimization, 265–267
problem-solving concentration, 267
push-pull amplifiers, 290
QEEG assessment, 267–273
reasons for, 248
relaxed mental and physical states, 265–266
seizures, 250
sequential (bipolar) montages, 290
sequential (bipolar) QEEG assessment, 274–275
single-channel assessment bandwidths findings, 275–284
single-channel assessments to 62 Hz, 314–315
single channel EEG assessments, 273–275
single channel QEEG assessment, 273–275
stability, 250–251
subtypes of disorders, 353–354
theta/beta ratios in ADHD, 285
three-electrode referential placement, 273–274
training in assessment procedures, 357–358, 358f
two-channel findings, 314
two-channel QEEG assessments, 285–289
units of measurement, 284–285
unusual analyses, 352–353, 353f
See also 19-channel (full-cap) QEEG assessment; artifact effects; case studies/case examples; data collection
assimilation (left brain), 88–89
association cortex, 72, 96
association fibers, 80–81
Association for Applied Psychophysiology and Biofeedback, 16, 18, 250, 671, 693
Association for Applied Psychophysiology and Biofeedback (AAPB), efficacy research methodology standards, 16
associative cortex, speech and language functions, 91
associative learning, 13, 446
associative visual agnosia, 91
astereognosis (tactile agnosia), 190, 227, 234
asthma, HRV training effect on, 579
astrocytes, 70
asymmetries, QEEGs and, 34
athletes, optimal performance training for, 15, 19, 250, 254–255, 324, 364, 402
Atlas of Human Brain Connections (Catani and De Schotten), 643
atonic seizures, 46
ATP. See adenosine triphosphate
attention
    attention span problems, frontal lobe injury and, 164
    BA 10 and, 173–174
    as basic function, 660
    cortex areas involved, 111–112
    to external stimuli, 250
    importance in cognition, 660–669
    in parietal lobes, 237
    in right frontal cortex, 166
Attention Deficit Disorder (ADD), 108–113
    alertness and movement correlation, 110–111
    anterior cingulate cortex (ACC), 111
    arousal (attention theory), 111
    attention findings, 111–113
    attention to external stimuli, 250
    basal ganglia and, 99

dorsolateral prefrontal cortex, 111–112
excessive theta waves in, 37
focus (attention theory), 111
genetic differences in dopamine receptors, 112
high-beta ADD, 273–274
hippocampus-septal nuclei-prefrontal cortex circuit, 112
with hyperactivity, 405–407
left frontal cortex, 111
math skills and, 513
measuring technique correlation, 5–6
movement for ADD children, 110–111
MRI studies, 112
neurotransmitters, 112
orbitofrontal cortex, 111
orientation (attention theory), 111
prefrontal cortex, 111
reticular formation, 112
right hemisphere deficiencies, 112
right hemisphere deficiencies in, 112
secondary reinforcers for, 13
shaping in, 13
single channel NFB and BFB, 165–166
SMR seizure studies, 8
Sterman's hypothesis on theta cortex-putamen-substantia nigra-thalamus, 110–111
theta sources, 110
three-element theory of attention, 111
as validated application for NFB, 14
ventromedial cortex, 111
without hyperactivity, 405
See also metacognitive strategies

Attention Deficit Hyperactivity Disorder (ADHD)
19-channel (full-cap) QEEG assessment, 355
ACC and, 216
alertness and movement correlation, 110–111
attention to external stimuli, 250
CZ site for, 219
dopamine activity and, 24
EEG changes with stimulation, 30
event-related potentials and, 669–670
frontal lobe dysfunction, 88
GP and, 74
high beta waves, 38
hypnotizability, 383
impaired go/no-go paradigms in, 6
incorrect diagnoses, 15
and LD, two-channel assessment and training, 418
mathematic study strategies, 513
NFB treatment, 113–115
owl or monkey pattern, 308f
SCP and, 616–617
single channel NFB and BFB, 165–166
SKIL comparison plot, 305f, 306f
SMR seizure studies, 8
stimulant medications, 113–115
testing results, 669–670
theta/beta ratios in, 285

attention deficits and brain correlates, 660–675
attention importance in cognition, 660–669
comments and critiques, 675
efficacy studies, 670–675
long-term efficacy, 675
See also Attention Deficit Disorder; Attention Deficit Hyperactivity Disorder (ADHD)

Attention Network, 133, 149, 663–665, 664f
audiology, ERP in, 5–6
auditory cortex, 91, 96, 191, 192, 194
auditory functions, 95–96, 194, 195, 522
auditory odd-ball paradigm, 195
auditory-verbal memory, 178

Autism Spectrum Disorder (ASD), 648–659
ACC and, 216, 649, 651f–652f, 656
accommodation (right brain) and, 88–89
aprosodia, 649
assessment findings summary, 654
attention span in, 649
attention to external stimuli, 250
autism studies, 262–263
autism vs. Asperger's, 648
developmental pruning failure, 655
EEG differences in, 121, 649
executive dysfunction theory, 655
frequencies targeted with NFB, 656–657
fusiform gyrus activation, 96
mirror neuron system and, 175–176, 187, 654

neuroanatomical and assessment findings, 649–653
NFB for, 656–659
orbital frontal cortex, 649–650
phase in, 653, 653f
polyvagal theory, 650, 655–656
raw EEG findings, 657
regression, 659
single channel NFB and BFB, 165–166
soft encouragement, 403–404
symptoms, 648–649
systems theory of neural synergy, 656
tDCS and, 597
theory of mind (ToM), 654–655
traditional treatments, 648
treatment overview, 657–659, 659f
uncinate fasciculus connections, 650
weak central coherence theory, 655
See also Asperger's syndrome
autobiographical/declarative/episodic memories, ERC-hippocampus system, 204
autobiographical memory, 168
autobiographical 'Sense of Past Self,' NDH and, 166
Autogen A620 assessment program, 270, 284, 324, 395, 397
autogenic training, 364, 480
automatic artifacting, 350
automatic motor training, 12
automatic thresholding, 436–437
autonoetic self-awareness, 178
autonomic instability, 686, 686f
autonomic nervous system (ANS)
    ANS/EMG feedback, 268
    control, insula and, 212
    flags for, 268
    insulae and, 212
    measurable elements, 359
    nicotinic receptors in, 27
    overview, 101–102, 102f
    parasympathetic nervous system (PSNS), 102–105
    PTSD and, 685–686, 686f
    sympathetic nervous system (SNS), 102
autonomic nervous system (ANS) and skeletal muscle tone (EMG) assessment, 359–382
    ANS and EEG assessment, 265, 377–380
    assessment goals, 360, 369
    assessment summaries, 373–375
    associated therapeutic techniques, 364
    autogenic training, 364
    BFB effectiveness, 369–371, 380–382, 380f
    compartmentalizing, 365–366
    data examination, 375–377, 378–379
    electrodermal response (EDR), 360–361
    electromyography (EMG), 362–364
    generalization, 365
    heart rate (HR), 361
    imagery, 364
    light-sound stimulation, 366
    measurable elements, 359
    office environment, 379
    peripheral skin temperature, 360
    psychophysiological profiles, 371–377, 375f, 376–377t
    quick stress test, 372–373
    relaxation and stress charts, 369f
    respiration, 361
    respiratory sinus arrhythmia, 361–362
    SAMONAS Sound Therapy, 366–369
    sensor application, 377–378
    shaping (reinforcement technique), 365
    stress assessment, 371–372, 378
    subliminal alpha, 366
    summary graph, 374f
    systematic desensitization, 365
    theory and objectives, 379
    Tools for Optimal Performance States, 370–371
    training sessions, 379–380
average reference montage, 57
averaging and continuous feedback, 472–473
averaging factor, 407–408
avoidance behavior, 173, 174, 418
axon connections, 72
axon hillock, 26
axons with Schwann cells, 28f

B

backup equipment, 61

Baehr, Elsa, 19, 177, 257, 419, 644
Baguley, I.J., 580
balance rhythm, regaining, 679
Bálint, Rezso, 228
Bálint's syndrome, 92, 228
Bancaud's phenomenon, 34, 45
band-pass filters, 54
bandwidth-mental state correlation, 36–45
    abnormal amplitude, 44–45
    abnormal beta, 44
    alpha waves, 37
    ambient electrical activity, 38–39
    beta waves, 37–38
    bilaterally synchronous slow waves, 44
    continuous irregular data, 44
    delta waves, 36
    electromyogram (EMG), 38–39
    generalized asynchronous slow waves, 44
    K complexes, 43–44
    localized slow waves, 44
    Mu waves, 40, 41*f*
    overview, 9, 10*t*
    paroxysmal discharge, 43, 43*f*
    paroxysmal hypnagogic hypersynchrony, 43
    sensorimotor rhythm, 37–38
    sharp waves, 43
    Sheer rhythm, 38
    sleep spindles, 43–44
    slow waves, 44
    spikes and waves, 42, 42*f*
    square waves, 44
    theta waves, 36–37
    V waves, 43–44
bandwidths, defined, 35
bandwidths to enhance, 401–404
barbiturates, 37, 344
Barnell Loft Multiple Skills Series, 501
baroreceptor stimulation, 173, 575
baroreflex gain (BRS), 212, 581
Barron, 6
BAs. *See* Brodmann areas
basal ganglia, 74, 74*f*, 99, 135
basal ganglia thalamocortical circuits, 682

basal nucleus of the amygdala, 83
basal zone of cerebral hemispheres, 89
basilar arteries, 100
basis pedunculi, 79, 86
basket cells, 20, 23, 134
Bauer, 344
Bauman, Margaret, 262
Beauregard, Mario, 670, 674
Beck Anxiety Inventory, 621
Beck Depression Checklist, 621
Behavioral Interpretations of Intrinsic Connectivity Networks, 126
behavior changes through EEG use, 18–19
behavior inhibition, 176, 184
behaviorism, coined by Watson, 12
Bell's phenomenon, 330
Benson, Herbert, 645
benzodiazepines (anxiolytic medications), 25, 37, 344
Bereitschaftspotential (readiness potential), 528, 682
Berger, Hans, 7, 17, 35
Bertelli, Marco, 527
beta activity
    as absence of alpha waves, 7
    beta waves, defined and described, 37
    beta waves, increase in, 44
    desynchronization, PRS patterns in, 6–7
    minima pattern, 295
    QEEGs and, 34
BFB. *See* biofeedback
BFE (Biofeedback Foundation of Europe), 324, 357
BFE Sterman Suite, 408–409
bicoherence, 442–443
Bikson, Marom, 596
bilaterally synchronous slow waves, 44
billiard game technique, 502
Bilodeau, Alexandre, 250, 691
binding rhythm, 38
Bindra, Abhinav, 691–692
Binet, Simon, 486
biofeedback (BFB)
    assessment, 369–371
    autonomic nervous system, measures of, 3–4
    defined, 2–4

effectiveness, 381–382
epidermal response and NFB for ADD, 108
HRV plus peripheral BFB, 576–577, 577f
HRV training and, 162
and two-channel NFB, 462–463, 463f
used first in NFB and BFB intervention fundamentals, 447
variables, monitoring, 380–381, 380f

Biofeedback and Neurofeedback Applications in Sports Psychology (Strack, Linden, and Vietta Wilson, eds.), 691

Biofeedback Federation of Europe (BFE), 408, 429, 526
biogenic amines (catecholamines), 24
BioGraph program, 7, 33, 270, 584
BioInitiative Working Group, 120
biphasic waves, 33
bipolar disorder, 57, 217, 314, 355, 462–463
Birbaumer, Nils, 7, 617, 671, 675
Bisley, J.W., 662
bispectrum, 443
bisynchronous waves, 34
bit numbers, 55
blindsight, 242
blood-brain barrier, 70
blood-oxygen-level dependent (BOLD), 593, 682
blood supply, brain, 70, 99–101, 101f
bodily self-awareness, insula and, 212
body orientation, 187
Bohr effect, 588
BOLD (blood-oxygen-level dependent), 593, 682
Bonanni, L., 541
borderline personality disorder, 215
Boro, A.D., 679
bottom-up decision making, 661–662, 663f
boxing technique, 503
bradykinesia (slowed movement), 140
brain activation patterns, 153
brain activity in tDCS, 593–596
brain blood supply, 70, 99–101, 101f
brain brightening in elderly, 14
brain changes with NFB, 14
brain-derived neurotrophic factor (BDNF), 182–183
Brain Injury Assessment Tool (BIAT). *See* eVox program

brain lobes. *See* lobes of brain
brain maps, 18–19
BrainMaster equipment, 324, 395
brain positions, 73f
brains, information processing speed, 521–523
brain stem (medulla oblongata), 77
brain stem-amygdala-hypothalamus pituitary-adrenal axis (AHPA), stress response and, 103–105
brain structures, basic, 68–71, 69f, 71f, 99–101, 101f
The Brain's Way of Healing (Doidge), 134, 583, 607
The Brain that Changes Itself (Doidge), 134
brain wave patterns alteration with EEG, 18
breathing/breaths per minute (BrPM), 187, 447
Breslaw, Jon, 284–285
British Journal of Sports Medicine, Glasgow Coma Scale, 621
broad total awareness, 266
Broca, Paul, 176
Broca's Area, 92, 93, 94f, 165, 175, 232f
Brodmann, Korbinian, 125, 130, 131–132
Brodmann areas (BAs)
BA 1, 186
BA 1, 2, 3, 186f
BA 2, 186
BA 3, 187
BA 4, 187
BA 5, 189f
BA 6, 185f, 188–190, 188f
BA 7, 226f, 229f
BA 8, 169f, 174f, 215
BA 9, 215, 217
BA 9 both hemispheres, 170–171, 171f, 174f
BA 10, NDH, 173–174
BA 11, 181, 217
BA 12, 182
BA 13, 211–213
BA 17, 239f, 240, 240f
BA 18, 239f, 241, 241f
BA 19, 239f, 241, 241f
BA 20, 197–198, 197f
BA 21, 196f
BA 22, 194f, 195, 196f
BA 23, 217, 220f, 221, 221f

BA 24, 214f, 215, 218, 218f
BA 24,' 217
BA 25, 214f, 219, 219f
BA 26, 223–224, 223f
BA 27, 208f
BA 28, 204–206, 204f
BA 29, 223–224, 223f
BA 30, 223f
BA 31, 217, 220, 220f
BA 32, 214f, 215, 218, 218f
BA 32,' 217
BA 33, 214f, 215, 220
BA 34, 206–207
BA 35, 208f
BA 36, 207f, 208f
BA 37, 207f, 236f
BA 37, DH and NDH, 207
BA 38, 203–204, 203f
BA 39, 233f, 236f
BA 40, 217, 234f, 235–236, 235f
BA 41, 194f
BA 42, 194f
BA 43, 196–197, 196f
BA 43 (near), 211–213
BA 44, dominant hemisphere, 175–176
BA 45, 177f
BA 46, 174f, 177f, 178, 185f, 217
BA 47, 178f
BA 52, 197, 197f
BAs 9 and 10, 169, 169f
BAs 10 and 11, 172f
BAs 44, 45, DH, 176
BAs 44, 45, NDH, 176
BAs 46, 8, 9, DH, 174
BAs 47 and 12, NDH right inferior frontal region, 180
child vs. adult brain, 128
in efficacy studies, 695–696
F4: BAs 46, 8, 9, NDH, 174
F5: BAs 44 and 45, 175f
F5-F7: BAs 44 and 45, 175f
F7: BAs 38, 44, 45, 46, 47, DH and NDH, 177–178
functional correlates, 125–126
left Superior Frontal gyrus, BA 9, 171, 173

for LNFB, 552–555, 552f, 554f, 555f
lobes of brain and, 156–166, 161t
locations on cortex, 127–128
location summary, 158, 246t
networks and, 152–153, 243t
overlapping functions, 126, 156, 160, 161t
phase-reset, 152–153
primary functions and locations, 159f
regions of interest, 244t
Brodmann areas and NFB training, 133–137
    BA functions overlap, 133
    basal ganglia, 135
    coronal section with putamen, globus pallidus, and thalamus, 135f
    cortex, inhibition in, 134
    cortex-basal ganglia-thalamus networks, 134
    LORETA images, 135f
    LORETA Z-score NFB, 134
    midsagittal section with thalamus position, 135f
    networks, importance of, 134
    nuclei of thalamus, 138f
    single-channel training, 133–134
    thalamus and general overview, 136f
    transverse section through right cerebral hemisphere, 136f
Brown, Val, 417
BRS (baroreflex gain), 581
Budzynski, Tom
    on dominant frequency, 34
    frontal lobe computer analogy, 88
    on RAD children, 251, 404, 465
    theta state for rescripting, 475
    twilight learning state, 477
    on white matter in brain, 70
busy brain, 424
Byers, Ralph, 250
The Byers Neurotherapy Reference Library, 16

C

caffeine, 25, 346
calibration, EEG, 52
calmness under stress, 266–267

Cammoun, Leila, 153
Canon, Rex, on ACC, 215
capacitance, defined, 64
capacitors, 57
carbamazepine, effect on tDCS, 590
cardiac activity as artifact, 333–334, 333*f*, 334*f*, 338*f*, 339*f*, 340*f*, 394–395, 394*f*–395*f*
cardiac functioning, HRV training, effect on, 579
cardioballistic (EKG) artifact, 321
CardioPro program (Thought Technology), 566, 566*f*, 570, 573*t*
cardiopulmonary lecture gridding task, 496–497
Carlson, N.R., 76
Carmen, Jeffrey, 611–612
Carmody study, 673
Carter, Rita, 112
cartoon technique, 500
case studies/case examples
    AB (ADHD, LD, and behavioral problems), 667–669
    Allison (Hunter Mind traits), 255
    anxiety and Asperger's, 556–562, 556*f*–562*f*
    Asperger's syndrome, 282–284, 283*f*
    autistic girl, high beta in P4, 230
    Barry (non-ADD), 277*f*
    Ben (severe Asperger's symptoms, low IQ), 258–259, 260*f*–261*f*
    Benny (memory techniques), 501
    BR (TBI with ADHD), 634–635, 634*f*–635*f*, 636*f*
    Brad (mild Asperger's symptoms), 257*f*, 258, 282, 282*f*
    Brent (ADD with hyperactivity), 405–407
    BS (severe Asperger's plus ADHD), 530–534, 530*f*–534*f*
    C. (ADHA with severe LD), 484
    case series by Thompson, Thompson, and Reid-Chung, 118
    Christian (LD, social awkwardness, dysphoria), 301–302, 301*f*, 302*f*
    concussed hockey player, 538–540, 539*f*–540*f*
    Daniel (Asperger's syndrome), 380–381
    Dave (race car driver), 451, 451*f*
    David (bipolar disorder), 462–463, 463*f*
    Dawn (closed-head injuries), 299–300, 299*f*, 300*f*, 301*f*
    dominant hemisphere temporal – parietal area injury, 221, 222*f*
    DU (autistic male), 651–652, 651*f*–652*f*
    dystonia, SMR treatment, 116
    EEG assessment summary, 264–265
    encephalitis patient, 192–193, 263, 264*f*
    Harry (performance optimization), 451–456, 452*f*, 453*f*, 454*f*, 456*f*, 458
    head injury, 306, 306*f*, 307*f*
    high school student with Tourette's syndrome, 683
    HRV training for post concussion symptoms, 587–588
    Jacob (recall and retention), 500
    Jane (ADD without hyperactivity), 405
    Jane (memory problems), 490
    Jane (performance optimization), 252
    Jane (reading disability), 302–303, 303*f*
    Jason (ADHD), 252–253
    Jason (math skills), 509
    Jason (NFB results), 115–116, 118–119, 412
    Jason (short-term memory and CAP problem), 383–384, 384*t*, 385, 386, 386*f*, 392, 396–397, 397*f*, 398–400, 399*f*, 400*f*
    Joan (eye-blink artifact), 391
    Joan (professional athlete), 254, 279, 280*f*, 448–450, 449*f*
    John (ADHD, combined type), 277, 278*f*
    John (dysphoria and anxiety), 287–289, 288*f*
    John (math skills), 512–513
    John (performance optimization), 252
    John (professional athlete), 254–255
    John (stress), 373–375, 378–379
    Johnny (multiple drug treatments), 114–115
    Judy (debate organization), 504–505
    long term TBI with ADHD, 639–640
    Lorraine (depression), 257, 281, 281*f*
    male with TBI, 2 years post-injury, 535–538, 535*f*–538*f*
    Marine with PTSD, 687
    Mary (Parkinson's with dystonia), 253–254, 279*f*, 459–461, 460*f*, 461*f*

    Mike (ADHD and anxiety), 277f, 384–385, 386–387, 387f
    Mike (metacognitive skills), 256
    Navy Corpsman with amnesia, 685
    Peter (complex symptoms), 421–423
    Peter (high functioning Asperger's syndrome), 263
    Philip (ADD without hyperactivity), 276f
    Rod (ADHD; schizophrenia), 355
    Roger (reading disability), 90, 92
    Sam (learning disability), 253
    Sam (stroke), 439
    A. (scanning and packaging techniques), 505–506
    Sean (ADD, LD, and depression), 293–295, 294f, 295f
    success story, 469
    Susan (gridding exercise), 497–498, 498t
    Troy (autism), 90, 257–258
    Vince (concentration problems), 491
Catani, M., 643
catecholamine agonists, 25
catecholamines. *See* biogenic amines
categorization, 177, 179
cathodal stimulation, 598
cathodal suppression of neurons, 593
Caton, Richard, 7, 35
cat operant conditioning study, 7–8, 8f, 29, 468, 678, 678f
caudal division of cingulate cortex, 220–222
caudate nucleus, 78, 99, 107
CB radio booster artifact, 58
CBT (cognitive behavior therapy) in depression treatment, 643
cell column distance in cortex, 30
cell walls as capacitors, 64
central autonimic network (CAN), nucleus of tractus solitarius (NTS) and, 575
central canal of spinal cord, 70
central gray matter, 98
central midline structures (CMS), 133, 147, 147f, 191, 641f, 642f
central midline theta, 110
central midline ventral structures, LORETA NFB for, 209, 209f
central nervous system (CNS), 70, 634, 686–689, 689f
central nucleus of amygdala, 83

central regions, 185–190
    C3 and C4, 185–188
    cerebellum, 187
    cortical homunculus, 186f
    operculum areas, 188f
    premotor cortex, 188–190
    somatosensory association cortex, 189
central sites, training effect on networks, 150
central sulcus, 72, 225f
centrencephalic theory, 46
cerebellum, 77, 187
cerebral arteries, 100
cerebral cortex damage, 90–92
cerebral cortex role in epileptic syndrome, 676–677
cerebral hemispheres, 71–72, 80f, 88–89
cerebrospinal fluid (CSF), 70–71, 78
cerebrovascular disease, 19-channel (full-cap) QEEG assessment, 354
cerebrum, physiology of, 98
CES. *See* cranial electrical stimulation
chaining (reinforcement technique), 11
Chandler, Dorothy, 648
changing memories, 107
channel (derivation), 34
channel choices. *See* 19-channel (full-cap) QEEG assessment; single-channel EEG; two-channel EEG assessment and training
chaos and nonlinear mathematics, 443–444
Chartier, Dan, 324, 458
checkerboard (novel infrequent distractor), 524f
Checktrode, 391
CHIs (closed head injuries), 14, 351
chloride ions in IPSP, 24
chlorinated hydrocarbons, 347
chlorpromazine, 344
cholinergic innervations, 541
choroid plexus, 70
chronic hyperstimulation, 206
cingulate, OCD and, 107
cingulate cortex, 217–218
cingulate gyrus, 214–224
    anterior cingulate affect networks, 218–220
    anterior cingulate connections, 216–217

anterior cingulate cortex (ACC), 214–216
caudal division of cingulate cortex, 220–222
cingulate cortex, 217–218
communication fibers for limbic system, 80
mammillothalamic tract and, 85–86
posterior cingulate gyrus, 223–224
cingulectomy, 97
cingulotomy, 97
cingulum, 80, 81f, 97, 202
citric acid cycle. *See* Krebs cycle
citrulline effect on tDCS, 590
Claparède, Édouard, 82
classical conditioning, 12–13, 471, 473–474
claustrum, 78
clients
client doing different tasks, 428f–429f
client experiences (EEGs), 65
goal review, 388, 389
production of specific frequencies, 411–412
questions for, 438
typical artifacts, 320–321
clinical EEGs, neurology vs. medical uses, 8
Clinical Electroencephalography (January 2000) as recommended reading, 8
clinical practice vs. double-blind studies, 675
clinical studies on CES, 603f, 603t, 604f, 605f
clinical tips, 30, 286–287
closed head injuries (CHIs), 14, 351
CMS. *See* central midline structures
CNS (central nervous system). *See* central nervous system
Coben, R., 657
cocaine, 25, 98, 345–346
cognition, VMPC and, 183
cognitive behavior therapy (CBT) in depression treatment, 643–644
cognitive control of emotion, dorsal prefrontal regions and, 215–216
cognitive function improvement, 596, 605
cognitive functions, Executive network and, 217
coherence, defined, 291–292
coherence measures, 292
coherence training, 19

color-coded wiring, 57
"Columbo" questioning technique, 502
commissural fibers, 85
commissures, 72, 79–80, 202f
common electrode reference montage, 57
common-mode rejection
in amplifiers, 53
cardiac artifacts, 394
common-mode rejection ratios, 63–64
electrical artifacts, 39, 58–59
electrode movement artifacts, 321, 329
communication linkages, 30–31
communication patterns in brain, 19
comodulation, 292, 295–297, 296f, 297f, 298, 310f
"Comparative Psychopharmacology of Methylphenidate and Related Drugs in Human Volunteers, Patients with ADHD and Experimental Animals" (Mehta et al.), 113
compartmentalizing, 365–366
complex cases, two-channel assessment and training, 421–423
complexity, understanding, 173
complex mental activity, hypocoupling and, 31
complex object features, BA 20, 197–198
complex partial seizures, 46
complex waves, 33
"Comprehensive Neurofeedback Bibliography," 16
Comprehensive Textbook of Psychiatry (Freedman, Kaplan, and Sadock), 476
computer connection troubleshooting, 59
concentration, increasing, 404
concentration gradient for sodium and potassium, 23
concept organization, 168
Concerta (methylphenidate), 345
concrete thinking, 174
Concussion is Brain Injury (Jeejeebhoy), 639
concussion/mild traumatic brain injury (TBI), 620–640
ADD Centre assessment, 621–622
assessment, 630–633, 632f
diffuse axonal injury (DAI), 623–629, 623f
EEG and ERP evaluations, 623
EEG assessments, 622
ERPs in, 632–633, 633f, 636f

force-board vestibular function assessment, 636
glutamate release in DAI, 624–625
HRV in, 579–580, 579f, 580f, 633–634, 633f
intervention, 637–638
Krebs cycle, 626f
neurophysiological treatment, 638–640
neuropsychological testing, 634–635, 634t, 635t
QEEG role in, 630–633
recovery rates, 639
scan types used, 622–623
second impact syndrome, 630–631
symptoms, 621–622, 621f
TBI metabolic effects, 629–630, 630f
TBI pathology, 622–623
tryptophan pathways in vitamin B6 deficiency, 628f
virtual reality, 636–638
conditioned stimuli/conditioned responses, 12, 98, 470
conducted phase, 291
conduction aphasia, 195
Conflict network, 664
Congeto, Marco, 550
connecting principles, 503
conscious awareness, loss of, 194
consciousness, states of, 35
constructional apraxia, 92, 167, 234
context-dependent behavior pattern regulation, ACC and, 218
contextual cues, 171
contingent negative variation (CNV), 6
continuous irregular data, 44
continuous performance tests, 522–523
continuous polymorphic delta (CPD), 680
continuous reinforcement schedule, 11
continuous sound, 408
contralateral side of head, 40
convergent thinking, frontal lobe injury and, 164
Cook, Edwin, 262
coronal sections, 68
corpus callosum, 72, 79
corpus striatum, 78–79
cortex
    BAs and functional areas, 126f–127f, 127–128
      cortex-basal ganglia-thalamus networks, 134

cortex pathways, 187
inhibition in, 134
pathways to spinal cord, 87
cortical-basal ganglia circuits, 139f
cortical connectivity, 153–155, 154f, 155f
cortical electrical communication, 30–31
cortical functioning disorders, 90–96
    agnosia, 91
    agraphia, 95
    angular gyrus, 93–94
    anomia, 91
    apraxia, 92
    association cortex injuries, 96
    Broca's area, 93, 94f
    cerebral cortex damage, 90–92
    dyslexia, 94–95
    emotion recognition and expression, 90–91
    hemispheric interactions, 92–93
    inter-lobe connections, 93
    left cerebral cortex damage, 91
    left-handedness, 95
    limb apraxia, 92
    motor apraxia, 92
    neurologically based disorders, 94–95
    neurophysiology study hints, 95–96
    occipital lobe damage, 96
    parietal lobe damage, 91–92
    reading and writing functions, 92–93
    reading disorders, 94–95
    right cerebral cortex damage, 90–91
    speech and language functions, 92–93
    temporal lobe damage, 96
    Wernicke's area, 93, 94f
cortical homunculus, 186f
cortical information flow, 226
cortical surface connection maps, 153, 154f
cortical-to-cortical connections, 30
corticobulbar tract, 86
corticocortical coupling, 31
corticopontine tract, 86
corticothalamic communication, 29f
counting tricks, 508
Courchesne, Eric, 262

Covey, 445
cps (Cycles-per-second), 4
cranial electrical stimulation (CES), 599–610
    background, 599
    clinical studies, 603–605, 603f, 603t, 604f, 605f
    current developments, 606, 606f
    electromagnetic stimulation approaches, 607–610
    electronics of, 601–602, 601f
    frequencies, 602–603, 603f, 603t, 606
    household power and, 602
    laser therapy, 607–609
    LENS and NeuroField, 609–610
    physiological basis, 599–601, 600f
    physiological studies, 602–603, 603f
    tDCS vs. tACS, 606
    wave forms, 606, 606f
Crick, Francis, 111
crus cerebri, 79, 86
cryptogenic epilepsy, 46
CSF (cerebrospinal fluid), 70–71, 78
current, 51
current control in tDCS, 593
current detection, conditions for, 22
cycles-per-second (cps), 4
cyproterone acetate, 347
cytoarchitectonics of human brain (Brodmann), 130f, 131
cytokines
    cytochrome and, 608
    cytokine IL6 and sleep disturbance, 217
    inflammatory cytokines and HRV, 638
    laser light and, 608
    parasympathetic cytokine control system, 638
    pro-inflammatory cytokines, 628–629

D

dACC (dorsal anterior cingulate cortex), as switch with insula, 148–149
DAI (diffuse axonal injury), 27–28, 623–629, 623f
Damásio, António, 147, 184
da Silva, Alex, 344, 596
data collection, 315–350
    alcohol and, 345
    aluminum and, 347
    antibiotics and, 346
    antidepressants and, 345
    antihistamines and, 347
    artifact effects, 320–322
    barbiturates and, 344
    benzodiazepines and, 344
    caffeine and nicotine and, 346
    chlorinated hydrocarbons and, 347
    data consistency and accuracy, 464
    data examination, 378–379
    data recording (statistics), 375–377
    differential amplifiers, 315–316
    electrode bridges, 320f
    electrode site preparation, 316
    encephalopathies and, 347–348
    hallucinogens and, 346
    haloperidol and, 345
    heroin and morphine and, 346
    hormones and, 347
    impedances, 316–319
    independent component analysis (ICA), 342–344, 343f
    lead and, 347
    lithium and, 345
    marijuana use and, 344–345
    math tasks, 322
    measurements, voltage and amplification, 315
    medication effects on EEG, 344–347
    mercury and, 347
    methanol and, 347
    neuroleptics and, 344
    organophosphates and, 347
    phenothiazines and, 345
    rauwolfia derivatives and, 345
    site preparation, 315
    solvents and, 347
    special assessment and artifacting screen, 322–326
    stimulant medications and, 345–346
    teaching and assessment screen, 322–326, 323f
    thyroid hormone levels and, 347
    time of day effects, 322
    toxic materials and, 347
    tranquilizers and, 344

tuberculostatics and, 347
withdrawal and, 347
See also artifact effects
Davidson, Richard
    alpha asymmetry NFB protocol, 644
    avoidance behavior, 212
    depression studies, 19
    emotions and left frontal activation, 173
    on insular cortex, 212
    NFB alpha-asymmetry protocol, 177
daydreaming, metacognitive strategies and, 484–485
day planners, 490–491
DBS. *See* deep brain stimulation
DC vs. AC circuits and impedance, 317
deBeus, Ball, deBeus & Herrington ADHD study, 673
Decade of the Brain (1990s), 17
decision making process, 173
decision pyramid, 248–249, 249f
decreasing mental capacities, 354
deductive reasoning, 164–165, 174, 175, 178, 187, 234
deep brain stimulation (DBS), 141, 219, 643, 645–646
Default-mode network (DMN)
    BA 23 and, 221
    BA 30 and, 223
    cognitive tasks and, 662–663, 663f
    connections to, 179–180
    dACC (dorsal anterior cingulate cortex), as switch with insula, 148–149
    defined, 134
    depression and, 217
    insula and, 212
    PTSD and, 687
defined mental states through EEG use, 18
déjà vu, 192, 202
Delight Pro (Mind Alive) CES device, 600
delta waves, 36
Dement, William, 45
Dementia, 269, 347, 354, 435, 523, 528, 535, 541 609, 629
Demichelis, Bruno, 691
dendrites, 22, 23
depolarization, 23–24, 26

depression, 641–647
    19-channel (full-cap) QEEG assessment, 355
    abnormal neurochemical levels, 603–604
    alpha asymmetry NFB, 644
    anterior cingulate cortex, 642
    BA 25 and LNFB, 553–556
    central brain areas involved, 642–644, 643f
    central midline structures, 641f, 642f
    CES for, 602–604
    cognitive behavior therapy, 644
    deep brain stimulation, 645–646
    EEGs to determine, 354
    in epilepsy patients, 210
    Executive and Affect network areas, 216f
    HRV training, effect on, 579, 581
    LORETA NFB and, 545f–546f, 549–550
    medications, 644
    medication vs. CES treatment, 604t
    NFB and HRV training, 644–645
    NFB as noninvasive, 646–647, 646f
    placebo effect, 645, 645f
    studies, 25
    superior longitudinal fasciculus, 643f
    tDCS and, 597
    treatments, 644–647
    two-channel assessment and training, 419–420, 420f
    vegetative aspects, 643
    vegetative/autonomic Affect network, 217
    vmPFC and dlPFC relationship, 603–604
de Ridder, Dirk, 598, 598f, 645, 645f
derivation (channel), 34
De Schotten, M.T., 643
"design a school" gridding task, 496
desired mental state, 409, 446
desynchronization in thalamic cells, 29
desynchronized EEG activity in biofeedback, 29
desynchronous beta waves, 403
detail discrimination, 195
Developmental, Individual Difference, Relationship-based model (DIR), 648
developmental pruning failure in ASD and Asperger's syndrome, 655
DeyMed equipment, 289

DH (dominant hemisphere), 163–166
diadochokinesis, 188
dialysis encephalopathies, 347
diaphragmatic breathing, 404, 447, 473, 477
Dickens, Charles, 46
diencephalon, 71–72, 75, 84*f*, 86*f*
Dietz, Frank, 324, 611, 612
differential amplifiers, 53, 62–63, 63*f*, 290, 315–316
diffuse axonal injury (DAI), 27–28, 623–629, 623*f*
Diffuse Lewy Body disease (DLB), 205
digital (FFT) filtering, 56*f*
digital or FFT filtering, 56*f*
dipoles, 20–21, 548–549, 549*f*
directionality in pyramidal cells, 22
directional terms, 135
direct pathways, 142–146, 142*f*, 661, 661*f*
disconnection syndromes, conduction aphasia and, 195
Discovery (Brainmaster) equipment, 289
discrete feedback, 408–409
discrete sounds, 408
discriminative stimulus, 98
disease-specific protein changes within ACC, 215
disorder improvement, tDCS and, 597
disorders appropriate for NFB, 269
display options, NFB and BFB intervention fundamentals, 447–448
display screens, 320
distal position similarity in pyramidal cells, 22
diurnal variations in alpha waves, 37
divergent thinking, frontal lobe injury and, 164
divided attention, 661
DLB (Diffuse Lewy Body disease), 205
DLPFC (dorsolateral prefrontal cortex), 145, 178, 596
DMN (Default-mode network), 148
DMPFC (dorsomedial prefrontal cortex), 147*f*, 215
dog conditioning studies, 12
Dogris, Nicholas, 609
Doidge, Norman, 17, 134, 242, 583, 607
dominance, right or left brain, 88–89
dominance types, 95–96
dominant hemisphere (DH), 163–166
dopamine, 24–25, 98, 164, 344
dopaminergic activity in left hemisphere, 109

dorsal (head), 68
dorsal and lateral aspects of BA 7, 227
dorsal anterior cingulate cortex (dACC), as switch with insula, 148–149
dorsal hypothalamus, defensive or flight responses, 98
dorsal lateral aspect, 226–228
dorsal network, 666
dorsal periaqueductal gray matter, 199
dorsal stream, Where Pathway as, 240
dorsolateral prefrontal cortex (DLPFC), 141, 145, 178, 596
dorsomedial nuclei, pain perception and, 97
dorsomedial prefrontal cortex (DMPFC), 147*f*, 215
double-blind studies, 600–601, 675
Draw-a-Person assessment tool, 621–622
driving distractions and internal focusing, 401
Drug Discovery Today, "Non-Human Primate Models for Parkinson's Disease" (Philippens), 683
Dubin, Mark, 127
Duffy, Frank, 19, 32, 270, 304, 354
Dupee, Marge, 691
dura mater (connective tissue), 70
dyscalulia, 122
dyslexia, 94–95, 122, 407
dysphoria, 97, 251, 465
dystonia, 14, 74, 116

E

early child development, 117–121
    control of environment, 118
    inborn deficits, 118
    temperament and stages of development, 117
    See also Asperger's syndrome; Attention Deficit Disorder; Attention Deficit Hyperactivity Disorder; Autism Spectrum Disorder
echolalia, 93, 233
Edison, Thomas, 36, 62
Edison Syndrome, 255, 256*f*
EDR. *See* electrodermal response
EEG biofeedback (neurofeedback), defined, 2–3
"EEG Findings in Selected Neurological and Psychiatric Conditions" (Hughes), 354

EEG frequencies and waveforms. *See specific frequencies and waveforms*

EEGs (electroencephalgrams)
    abnormal EEG patterns, 45–50
    advantages, 18
    for alpha-theta therapy, 476
    in ASD and Asperger's syndrome, 649, 657
    biofeedback operant conditioning, 12
    capacitance, 64
    changes with stimulation, 30
    client experiences, 65
    DAI detection, 28
    defined and described, 4, 17, 20, 269
    and ERP evaluations, concussion/mild traumatic brain injury (TBI), 623
    filtering, 52–53, 55–57, 269
    frequencies, 2–3
    grounding, 59–60, 62
    histogram or graph presentation, 270–271
    history, 7–8
    ictal EEG activity, 46
    impedances, 63–66
    inductance, 64–65
    interical EEG activity, 46
    interpretation, 114
    learned normalization of EEG patterns, 3
    low-power EEGs, 432–435
    medications as artifacts, 344–347, 349
    method, 271–273
    named by Berger, 7
    normalization, 251, 267
    origins, 17–31
    outline, 270–271
    pre- and post-EEG comparisons, 311–312
    QEEGs and, 32–34, 269
    quality EEGs, achieving, 63–66
    rationale for, 265
    raw EEG findings in ASD and Asperger's syndrome, 657
    raw EEGs, 385
    reasons for use, 18–19
    self-regulation, learning, 11
    signals, 52–54
    synchronous EEG activity, 29, 31
    temporal resolution, 5

topographic maps, 33, 294
    See also abnormal EEG patterns; instruments and electronics; 19-channel (full-cap) QEEG assessment; single-channel EEG; two-channel EEG assessment and training
EEG training, HRV training and, 585*f*
effectiveness studies, 670–671
effect-size statistical analysis, 603
efferent connections, 79
efficacy, 16, 671, 671–672
efficacy criteria, 693–696
Einstein, Albert, 28
EKG (electrocardiogram), 2, 321, 570
electrical artifacts, 58, 326, 392–393
electrical charge, 51
electrical housekeeping, 273
electrical outlets, 57
electrocardiogram (EKG), 2, 321, 570
electrodermal response (EDR), 90–91, 265–266, 360–361, 447
electrodes
    electrode bridges, 320*f*, 348
    electrode paste, 62
    electrode pop as artifact, 335*f*, 336, 348
    electrode positions, 129*f*
    electrode sponge, dampness of, 590, 591, 596, 601
    homologous (interhemispheric) electrodes, 348
    linked, 287
    linked-ground electrodes, 406
    local frequency output as artifact, 58–59
    materials and care, 59
    movement artifacts, 329, 330*f*, 393–394, 394*f*
    O1 active electrode site, 481
    placement for feedback, 404–407
    placement for two channels, 285–286
    placement in ADD with hyperactivity, 405–407
    placement in ADD without hyperactivity, 405
    placements for assessment, 273
    polarization, 317–318
    site changes for new objectives, 399–400
    site locations, 271–272, 272*f*
    site preparation, 316
    in tDCS, 591–592, 592*f*
    unlinked, 286
electroencephalograms (EEGs). *See* EEGs
electroencephalographs, 4, 17
Electroencephalography (Neidermeyer and da Silva), 344
electromagnetic fields (EMFs), 120, 607–610
Electromedical Products International, Inc., CES studies, 604, 604*f*, 604*t*
electromyograms (EMGs)
    artifact minimization, 38–39, 392–393, 393*f*
    induced artifact, 13
    interference, 39
    muscle relaxation benefits, 362–364
electrooculogram (EOG for eye movements), SCP and, 617
Elefix, 17
eLORETA, 544
EMDR (eye movement desensitization and reprocessing), 687
EMF (electromagnetic field), 120, 607–610
EMGs. *See* electromyograms
emotion recognition and expression
    amygdala and, 199
    auditory cortex, 196
    BA 7 and, 227
    BA 34 and, 206–207
    BA 38 and, 203–204
    BA 47 and, 178
    cingulum, 202
    CMS and, 147
    control of in depression, 642–643
    emotional conditioning, defined and described, 12
    emotional cues, 171
    emotional disorders, HRV training in, 579
    emotional expressiveness, lack of, 176
    emotional lability, 168
    emotional stability, HRV training in, 579
    emotional value of different reinforcers, 182
    insula and, 211–212
    limbic system and, 98–99
    medial frontal cortex and, 215
    right cerebral hemisphere damage and, 90–91
    temporal lobe medial aspect, 202
    VMPC and, 183–184

empathic concern, 175, 181–182, 679
emptying of the mind, 589
enabling objectives, 265
encephalitis patient QEEG findings, 193f
encephalopathies, 347–348
encoder troubleshooting, 59
encoding new images, 166–167
endocrine system, 212
endorphins, 26
energy levels, atomic, 51
Englot, D.J., 679
enhanced alpha and theta, 481–482
enhanced thalamic inhibition, 142
entorhinal cortex (ERC), 85, 146, 204
environment, awareness of, 266
EOG (electrooculogram) for eye movements, SCP and, 617
epilepsy
    absence seizures (petit mal epilepsy), 47, 47f, 350–351
    assessment and intervention, 250
    depression in, 210
    epileptiform activity, 250
    epileptiform paroxysmal discharge, 43f
    high beta waves, 38
    SCP work, 7, 617
    SMR studies, 8
    temporal lobe epilepsy, 192
    temporal lobe functions and, 209–210
    tonic-clonic seizures (grand mal epilepsy), 46, 351, 352f
    wave forms, 43
    See also seizure disorders
episodic contextual associations, 179
episodic memory, 106, 179
EPSP (excitatory postsynaptic potential), 20, 21, 23–24
equipment troubleshooting, 58–61
ERC (entorhinal cortex), 145, 204
ERC-hippocampus system, 204–205
ERPs. See event-related/evoked potentials
error processing, 173
Esmail, S., 683
Essential Guide to Test Taking for Kids (Gruber), 501

Etkin, Amit, 647
euphoria, dopamine activity and, 25
The European Heart Journal, heart measurements defined, 573
eustress physiological state, 484
Event-related desynchronization (ERD), defined, 6–7
event-related/evoked potentials (ERPs), 521–541
    in ADHD, 6
    in ADHD and ASD, 669–670
    as artifact, 332, 348
    auditory ERPs, 522
    checkerboard (novel infrequent distractor), 524f
    concussion/mild traumatic brain injury (TBI), 632–633, 633f, 636f
    defined and described, 5–6, 521
    ERP graph, 522f
    exogenous ERPs, 524–525
    explanations to patients, 540–541, 541f
    heart evoked potentials (HEP), 541–542
    injury effects and, 6
    Kropotov's study, 541
    long-latency ERPs, 528–541
    long-latency potential ERPs, 6, 528
    measurement, 521–522
    for mental disorders, 354
    mismatch negativity (MMN), 525–528, 526f, 527f
    motor ERPs, 6, 528
    N1-P2 complex ERP, 6, 525
    N1-P2 complex ERPs, 525
    N2 ERP, 525
    P3a and P3b, 529
    P300 components, 528–529
    sensory ERPs, 6, 524–528
    steady-potential shift ERPs, 6
    in very slow frequencies, 321
    visual ERPs, 522–524, 522f
Evoke Neuroscience equipment, 5
eVox program (Evoke Neuroscience)
    for ERPs, 521, 522f
    eyes open and closed capabilities, 321
    HRV training with NFB and BFB, 584
    for LORETA NFB, 549, 551
    multiple assessment capabilities, 289

sLORETA and, 544
Excel program for trend reports, 464
excitatory neurons, stellate cells as, 23, 28
excitatory postsynaptic potential (EPSP), 20, 21, 23–24
executive circuit. See dorsolateral prefrontal cortex circuit
Executive network
    ACC and, 217
    BA 24 and, 219
    BA 32 in, 218
    BA 38 and, 203–204
    CEN and cognitive tasks, 662–663, 663f
    DLPFC and, 217
    executive attention, 169
    executive components of the Affect network in depression, 642
    executive dysfunction theory in ASD and Asperger's syndrome, 655
    executive functions, 163, 173, 178
    insula and, 212
    medial frontal cortex and, 215
executives, optimal performance training for, 15
exercise value, 182, 183, 581
Exner's area (handwriting), 188
exogenous ERPs, 524–525
Expanded 10-20 System, 272
expectancy rhythm, 586, 678
experimental vs. validated applications, 14
external capsule (white matter capsule), 79
external open awareness, 445–446
external stimuli, 250
extinction of reinforced behavior, 11, 14
extracellular dipole layers, 20
extrapyramidal motor areas (premotor), 87
extrastriate cortex, BAs 18 and 19, 241
eye blink artifact, 330, 330f, 392, 394f, 403
eyelid flutter artifact, 334, 335f
eye movement artifact, 331–332, 331f, 332f, 342f
eye movement desensitization and reprocessing (EMDR), 687

F

F9, F10, and orbital surface, 178–179
F1000 (Focused Technology)
    cables for, 52
    digitally tunable low-pass filter, 54
    electrode placement, 405–407
    feedback rate, 407
    HRV training with NFB and BFB, 584
    impedance in, 318, 324, 395
face and name association (BA 11), 182
facial recognition (BA 37), 207
false frontal slow-wave activity, 349, 350f
false impressions from artifacts, 333–334, 390
false memories, 107
falx cerebri, 70
Family Ruler (long division), 510
Farmer mind, 255, 256f
Farwell, L.A., 528–529
Fast Fourier Transform (FFT), 32–33, 35, 55, 56f, 57, 291
fast waves, 39
FDA (Food and Drug Administration) on CES, 599
fear, temporal lobe activation and, 206
fear conditioning (BA 23), 221
fear response study, 82–83
feedback criteria, option to change, 553
feedback inhibits, 468
feedback screens, 410–411, 410f, 412f, 413f
feedback speed, 407–408
feedback types and cueing, 477–478
FEFs. See frontal eye-fields
Fehmi, Les, 446, 467, 477
FFT. See Fast Fourier Transform
fibromyalgia, 459, 597, 681–683
fight or flight (stress) response, 25, 199
filing cabinet vs. garbage bag mind, 495–496
filtering, EEG, 52, 53, 55–57, 269
finite impulse response (FIR), 55, 55f
first letters as acronyms technique, 501
Fisch, B.J., 34, 358
Fisch and Spehlmann's EEG Primer, Basic Principles of Digital and Analog EEG (Fisch), 358
fish oil supplements for concussion, 637–638
fissures, 69, 72

flexibility problems, frontal lobe injury and, 164
flexible shifting of mental state, 446
flicker-fusion-focus, 445
focal epileptiform activity, 46
focus functions
    and concentration, NFB and BFB intervention fundamentals, 465
    focal phase reversal, 291
    focal waves, 34
    focused attention rhythm (35-45 Hz), 678–679
    focused attention SMR rhythm, 586, 678–679
    sustaining, slow-wave activity and, 406
Food and Drug Administration (FDA) on CES, 599
foramen of Magendie, 70
foramen of Monro (interventricular foramen), 78, 85
foramina of Luschka, 70
force-board vestibular function assessment, 636
forebrain development, 77–86
    anterior and hippocampal commissures, 80f
    anterior commissure, 80, 85
    anterior perforated area (APA), 81–82
    association fibers, 80–81
    cerebral hemispheres and diencephalon transverse section, 80f
    cingulum, 80, 81f
    commissures, 79–80
    communication links, 79–80
    corpus callosum, 79
    corpus striatum, 78–79
    development, 77–78, 78f
    diencephalon and forebrain, 84f, 86f
    fear response study, 82–83
    forebrain described, 71–72, 72f
    fornix, 85
    head of caudate, 83f
    hippocampal commissure, 80, 85
    hippocampus, 85
    hypothalamus, 82
    inferior longitudital fasciculus, 81, 81f
    internal capsule, 79
    lamina terminalis, 81
    linkage development, 85
    long association fibers, 80
    mammillothalamic tract, 85–86
    sagittal section, 84f
    septal area, 81–82
    septum, 82
    superior longitudital fasciculus, 81, 81f
    ventricles, 78
fornix, 73, 85, 201
Fourier, Jean Baptiste, 33
Fourier analysis, 33
Fp1,Fp2, 172f
Fp2, right prefrontal area, 173
fractions study techniques, 510–514, 512f
Franz, Margarite, 131
Freed, Jeffrey, 88, 418, 667
Freeze-Framer device (HeartMath), 446
Fregni, Felipe, 596
frequencies
    bandwidths, 35–36
    in CES, 606
    dominant frequency and age, 34
    effects on artifacts, 332, 333f
    effects on brain waves, 602–603, 603f, 603t
    frequency bands, 395–396
    ranges, 35
    targeted with NFB in ASD and Asperger's syndrome, 656–657
frequency domain. *See* EEGs
Freud, Sigmund, 15
frontal-central disconnect pattern, 311f
frontal cortex, lateral and medial views, 174f
frontal eye-fields (FEFs), 169, 188, 195, 237
frontal intermittent rhythmical delta (FIRDA)activity, 44, 680
frontal lobes, 163–184
    anterior insula and speech, 176
    deductive reasoning, 164–165
    Default mode network, 179–180
    DH (left) frontal lobe functions, 163–164
    encoding new images, 166–167
    Executive functioning, 163
    F9, F10, and orbital surface, 178–179
    Fp1,Fp2, 172f
    Fp2, right prefrontal area, 173

frontal cortex, lateral and medial views, 174f
frontal lobe injury, 164
frontal lobe lesions, 167
hippocampal theta and memory, 179–180
inductive reasoning, 165–166
inferior/basal area of dominant hemisphere, 168
inferior frontal gyrus in DH, 179
inferior view, 168f
injuries, 164
lateral and medial aspects, 169–180
memory, 166
nondominant (right) frontal functions, 167
nondominant (right) hemisphere, 170, 170f
orbital frontal cortex and temporal lobe links, 168
orbital sulci, 179f
overview, 88–89
P. Broca and, 176
prefrontal cortex, lateral fiew, 163f
reasoning, 164–166
uncinate fasciculus connections, 166f, 168
uncus of Parahippocampal Gyrus, 168f
See also Brodmann areas

The Frontal Lobes and Voluntary Action (Passingham), 140
frontal-subcortical circuits, 139–146, 142f
    affect network, 146f
    anterior cingulate cortex (ACC) circuit, 141–142
    cortical-basal ganglia circuits, 139f
    direct pathways, 142–146
    dorsolateral prefrontal cortex circuit, 141
    hyperdirect pathways, 142–146
    indirect pathways, 142–146
    motor circuit, 140
    motor network, 145f
    oculomotor circuit, 140–141
    open-loop integration, 145–146
    orbitofrontal cortex circuit, 141
    pathways and SMR production, 144–145
    simplified frontal-subcortical circuits, 142f
    spatial network, 145f
    visual network, 146f
frontotemporal dementia (Pick's disease), 184, 205, 354
Fuchs, Birbaumer, Lutzenberger, Gruzelier & Kaiser ADHD study, 673
full cap assessments. *See* 19-channel (full-cap) QEEG assessment
Functional Magnetic Resonance Imaging (fMRI), brain activity measurements, 593
functional modules, 155
functional neuroanatomy, 124–137
    Brodmann areas illustrated, 124f–125f
    electrode sites illustrated, 124f–125f
    International Society for Neurofeedback and Research, 125
    lateral view of brain, 124f
    lobes of brain illustrated, 124f–125f
    midsagittal section of brain, 125f
Functional Neuroanatomy of Anxiety: A Neural Circuit Perspective, chapter on depression, 647
fusiform cells, 72
fusiform gyrus, 96
Fz placement, 215

G

GABA (gamma amino butyric acid), 24, 25, 441–442, 590
Gage, Phineas, 178, 181–182
Galen (Greek physician), 599
galvanic cells, 62
galvanic skin response (GSR), 266
games for feedback training, 413–414
gamma waves as high beta waves, 38
Gani, C., 14, 114, 671, 675
garbage bag vs. filing cabinet mind, 495–496
gates, sodium and potassium, 26
gating mechanisms, 142–146
Gauss, Johann Carl Friedrich, 355–356
Gellhorn, Ernest, 685
generalization, 13–14, 365, 446, 475
generalized anxiety disorder (GAD), 14, 647
generalized asynchronous slow waves, 44
generalized conditioned reinforcers, 13
generalized seizures, 46
generalized waves, 33–34
generalizing techniques to home, school, and work, NFB and BFB intervention fundamentals, 446
A General Theory of Love (Lewis, Amini, and Lannon), 71

Gerstmann syndrome, 92, 227, 233–234
Gestalt, 88, 174, 177, 195
Gevirtz, Richard, 586, 638
Ghaziri, J., 643–644
Gigandet, Xavier, 153
Gilula, M., 604
Gitelman, D.R., 663
Giza, C.C., 629–630
Glasgow Coma Scale, 621, 621f
glial cells, 20, 28, 70
global average Laplacian montage, 290
global resonant loops, 30
globus pallidus (GP), 74, 78, 142
glutamate, 25, 590, 624–625
glutamate activation, 595
glycine, 24, 25
goal-directed behavior, ACC and, 217–218
goal setting with client, 369, 387–388
Gogh, Vincent Van as epileptic, 46
Goleman, D., 486
go/no-go paradigms in ERPs, 6
go/no-go pathways, 142–146
go/no go tasks, 180, 522
Gorman, J.M., 218
GP (globus pallidus), 74, 78, 142
grade school math tips, 506–508
grammar, 177
grand mal epilepsy (tonic-clonic seizures), 46, 351, 352f
granular cells, 72
graphing training sessions, 437–439
gray matter, 27, 70, 72
Great Depression gridding task, 497
Green, 32, 355–356
Greenberg, B.D., 219
Greenspan, Stanley, 648
gridding and organizing problems, 496–498, 497t, 498t
grid technique, 496–499
Grippo, A.J., 580, 581
grounding (EEGs), 59–60, 62
groups times people study technique, 507, 510f
Gruber, Gary, 501
Gruzelier, John, 7, 15, 250, 401, 475

Gunkelman, Jay
    arousal level with stimulants, 346
    The Art of Artifacting, 336, 349, 358
    on Brodmann areas, 127
    medication and toxin artifacts, 344
Gustatory Cortex (BA 43), 196–197
gyrus, 69, 72

H

habit inhibition, NDH and, 166
habitual bracing, 496
Hagedorn, David
    Brain Injury Assessment Tool (BIAT), 631
    communication patterns in brain maps, 19
    EEG interpretation, 127
    eVox workshops, 550
    HRV feedback screen, 577–578
    normative databases, 32
    side effects of tDCS, 596
Hagmann, Patric, 153
Hall, M., 674
hallucinogens, 24, 346
haloperidol, 345
hamburger or 3-3-3 method, 504–505, 504f
Hammond, Cory
    The Art of Artifacting, 336, 349, 358
    on hypnosis, 475
    NFB literature suggestions, 16, 250
handwarming, 480
Hardt, Thomas, 7
Harmon-Jones, E., 644
Harness, Tim, 692
Hartmann, Thom, 15, 255, 451
Harvard Medical School, cortical surface connection maps, 153
Havelkova, Milada, 648
Haywood & Beale ADHD study, 673
headache, 381–382
headbands, 329, 393
head of caudate, 83f
hearing aids as artifact, 340f
heart as oscillating system, 571f, 572

"Heart–Brain Connections" (Thompson and Thompson), 656
heart evoked potentials (HEP), 541–542, 586–587, 587f
heart rate (HR), 361, 447
heart rate variability (HRV) training, 565–589
    ADD Centre HRV and SMR study, 584–585, 585f
    affect disorders, 580
    afferent and efferent ANS connections between brain stem and cardiovascular system, 576f
    artifact corrections, 570
    biofeedback from sympathetic nervous system, 2
    clinical conclusions, 589
    in combined therapy, 549
    in concussion/mild traumatic brain injury (TBI), 633–634, 633f
    definitions, 570
    depression, 581
    EEG training and HRV, 585f
    exercise, 581
    goal, 570
    heart evoked potentials (HEP), 586–587, 587f
    HR displays; smooth line or steps, 577–578, 578f
    inflammatory cytokines and, 638
    injury effects and, 579–580, 580f
    insulae (BA 13) and, 580–581
    literature on, 579
    measurements used, 570–571, 571f
    method, 566–567, 567f
    neuroatomical links between HRV and NFB, 581–583, 582f, 583f
    and NFB, reasons for, 579–581, 579f, 580f
    NFB and, 162
    NFB and BFB intervention fundamentals, 446
    with NFB and LNFB, 584
    oscillations in heart-brain connections, 571–573, 572f, 573t
    overbreathing (hyperventilation), 588–589
    P300 and HRV, 585–586
    peripheral psychophysiological (biofeedback) measurements, 565–566
    plus peripheral BFB, 576–577, 577f
    pNN50 defined, 573
    in relaxation periods, 447
    root mean square of the standard deviation of the RR interval (RMSSD), 573
    SMR origin of term, 585–586
    standard deviation of NN interval (SDNN), 573
    standard deviation of the average NN intervals (SDANN), 573
    stress assessment profile, 567f
    top-down and bottom-up training, 151–152, 151f–152f
    training method, 567–570, 568f, 569f
    vagal control of HRV, 574–575, 575f
    vagal efferent output, 212
    Wiggers diagram, 574f
hedonic (pleasure) responses, 183
Heffernen, M., 606
Heinrich, H., 616
hemispheric interactions, 92–93
hemoencephalography, 611–615
HEP (heart evoked potentials), 541–542, 586–587, 587f
hepatic encephalopathies, 345
heroin and morphine, 346
Hertz (Hz), defined, 4
Hertz, Heinrich R., 4, 17, 35
Heschl's gyri (transverse temporal gyri), 194
high alpha waves, 37, 401
high-amplitude beta, 113
high beta waves, 38
high-functioning autism, clinical reports for NFB, 14
high-pass filters, 53–54, 321
hindbrain, 77
hippocampus
    Affect Network and, 146
    defined and described, 85
    hippocampal commissure, 80, 85
    hippocampal stimulation, 207
    hippocampal theta and memory, 179–180
    hippocampal theta waves, 36
    memories and, 97
    PTSD and, 686
    role in epileptic syndrome, 676–677
Hjorth, B., 57
Hodgkin cycle, 26, 27f, 593
homeostasis, 103, 211–212

homologous (interhemispheric) electrodes, 348
Honey, Christopher J., 153
hormones, 347
Hospital for Sick Children (HSC), Toronto, 346
household power, 602, 602t
HR (heart rate), 361, 447
HR displays; smooth line or steps, 577–578, 578f
HRV. *See* heart rate variability training
hubs, defined, 155, 155f
Hudspeth, William, 19, 32, 270, 304
Hughes, 354
Hunter Mind, 15, 255, 451
Huntington's chorea, 99
hydrazine seizures, 8
hydrocephalus, 71
hyperactivity and impulsivity, 405–407
hyperalgesia, 185
hypercomodulation, 294
hypercoupling, 31
hyperdirect pathways, 142–146, 142f, 661, 661f
hyperfocus, 473
hyperpolarized membranes, 23–24, 593
hypertension, 579
hyperventilation (overbreathing), 348, 568, 588–589
hypervigilance, 230, 235
hypnagogic state, 15
hypnopompic state, 15
hypnosis, 31, 458–459, 475
hypoactivation, 559
hypocapnia, 588
hypocoupling, 31
hypoglycemia, 351
hypothalamic-pituitary-adrenal (HPA) axis, 656
"The Hypothalamic-Pituitary Adrenal Axis" (Smith-Pelletier), 103–105
hypothalamus, 76–77, 82, 97
hypothesis generation, 182
Hz (Hertz), defined, 4

I

IBIs (interbeat intervals), 570–571, 571f
ICA (independent component analysis), 342–344, 343f
ictal EEG activity, 46
Ideal Me visualization, 479
ideomotor apraxia, 190, 227, 234
idiom understanding, 171, 182
idiopathic epilepsy, 46
IIR (infinite impulse response), 55
IIR Butterworth filter, 56f
imagery, 364
imagery and desensitization, 381–382, 477–478
imagined events, 166
imaging for alpha-theta therapy, 481–482
imipramine, 345
impedances
    AC vs. DC circuits and, 317
    in CES, 599–600
    client's perspective, 318–319
    defined, 63–64
    definition and measurement, 316–319, 600
    electrical artifacts and, 58–59, 390–391
    high readings, 318
    impedance meter, 317
    measuring for quality readings, 65–66
    site preparation, 315
    in two channel approach, 289
impulsivity, 168, 173, 404
in-born deficits/delays in early child development, 118
incidental learning, 13
increased attention, hypocoupling and, 31
independent component analysis (ICA), 342–344, 343f
indirect pathways, 142–146, 142f, 661, 661f
individualized treatment plans, 672
indoleamines, 24
inductance, 64–65
inductive reasoning, 165–166
inductive resistance, 65
inferences, making, 173
inferior/basal area of dominant hemisphere, 168
inferior frontal gyrus in DH, 179
inferior longitudital fasciculus, 81, 81f
inferior temporal gyri (IT BA 20), 145
infinite impulse response (IIR), 55
Infiniti/Biograph (Thought Technology), 395, 398

Infiniti instruments (Thought Technology), 52, 53–54, 270, 407, 584
inflammation, 608–609, 638
    See also cytokines
information acquisition, hypocoupling and, 31
information context, 174
information retrieval, 179
inhibit all feedback if above threshold setting, 397
inhibit and reward frequencies, 275*f*
inhibit frequency bands, 396–397
inhibition in cortex, 134
inhibitory neurons, 23, 24, 26, 28
Inhibitory Postsynaptic Potential (IPSP), 20, 24
inhibit-over-threshold, 472
inhibits, use of for alpha-theta therapy, 481
inion (posterior) of head, 271
innominate substance, 78
insomnia, CES for, 604
instability (mood), 250–251
Institute for Personality and Ability Testing Depression Scale Questionnaire (IPAT Depression Index), 605
instrumental learning. *See* operant conditioning
instruments and electronics (EEG), 51–67
    amplification, 52–53
    artifact effects, 59–60
    band-pass filters, 54
    capacitors, 57
    computer connection troubleshooting, 59
    digital or FFT filtering, 56*f*
    EEG signals, 52–54
    electrical artifacts, 58
    electrical outlets, 57
    electrodes, 58–59
    encoder troubleshooting, 59
    equipment troubleshooting, 58–61
    filtering, 53–57
    FIR Blackburn filter, 55*f*
    IIR Butterworth filter, 56*f*
    impedances, 58–59
    interference factors, 58
    montage reformatting, 57
    NFB system, 56*f*
    offsets, 58–59
    optical isolation, 58
    sampling rates, 54–55, 55*f*
    See also electroencephalograms)
instruments for biofeedback, 446
insulae
    (BA 13) and HRV training, 580–581
    Brodmann areas and, 211–213, 211*f*
    in corpus striatum, 78
    dyslexia and, 95
    in mediating dynamic interactions, 149
    PTSD and, 686–687
    Salience network and, 662–664, 664*f*
    as switch with dACC, 148–149
integrating center of neuron, 26
integration of inputs, NFB in dominant temporal lobe, 192
interactive video games, 473
interbeat intervals (IBIs), 570–571, 571*f*
interference factors, 54, 58
interhemispheric (homologous) electrodes, 348
interictal EEG activity, 46
interictal epileptiform activity, 43, 272
interictal spikes, 676
interleukin-6 (IL-6), 638
inter-lobe connections, 93
intermittent reinforcement to prevent extinction, 14
internal capsule of neocortex, 79
internal capsule of spinal cord, 87
internal carotid arteries, 70, 99–100
internal reflection, 266
internal speech, 176
International 10-20 System of electrode placement, 132, 156, 161*t*, 271, 272*f*
International Journal of Psychophysiology, on LORETA, 543
International Society for Neurofeedback and Research (ISNR), 16, 357, 645, 656, 671, 693
International Society for Neuronal Regulation, 16
interpretation of findings, 314–315
intervention fundamentals, 383–444
    absolute power vs. relative power, 433
    academic tasks and NFB, 415–417
    ADD Centre Store, 390*f*
    ADHD and LD, 418

artifact effects, 390–392
assessment data and bandwidths, 400–401
automatic thresholding, 436–437
averaging factor, 407–408
bandwidths to enhance, 401–404
bicoherence, 442–443
cardiac artifacts, 394–395, 394f–395f
chaos and nonlinear mathematics, 443–444
client doing different tasks, 428f–429f
client goal review, 388, 389
client's production of specific frequencies, 411–412
complex cases, 421–423
continuous sound, 408
depression, 419–420, 420f
discrete feedback, 408–409
discrete sounds, 408
electrical artifacts, 392–393
electrode movement artifacts, 393–394, 394f
electrode placement for feedback, 404–407
electrode placement in ADD with hyperactivity, 405–407
electrode placement in ADD without hyperactivity, 405
electrode site changes for new objectives, 399–400
EMG artifact minimization, 392–393, 393f
eye blink artifact minimization, 392, 394f
feedback screens, 410–411, 410f, 412f, 413f
feedback speed, 407–408
frequency bands, 395–396
games for feedback training, 413–414
graphing training sessions, 437–439
inhibit frequency bands, 396–397
Laplacian montage brain maps, 433f–434f
low-power EEGs, 432–435
Mind-Mirror screen, 422f
music feedback, 409
negative z-scores and relative power, 434
neuromodulation, 444
objective measurements with training screens, 435–436
one site's effect on others, 405
operant and classical conditioning with NFB, 414–415
percent over threshold, 397–398, 397f
phase reset, 440–441, 441f, 442f
phase shift and phase lock, 439–440, 440f, 442f, 443f
referential vs. sequential amplitude comparison, 426f
resting waves, 414
reward frequency bands, 397f, 398–399, 399f, 400f
reward frequency ranges, 404
rewarding 11-13 Hz, 401–402
rewarding beta 16-20 Hz or 13-15 Hz, 403–404
rewarding SMR (12-15 Hz or 13-15 Hz), 402–403
seizure disorders, 420
setting thresholds, 395–396
single-channel assessment procedures, 424–432, 425f, 430f, 431f
slow- and fast-wave production control, 412
sound feedback, 408–409
sustained feedback, 409
training adults, 384–387, 387f, 388t
training children, 383–384, 384f, 386, 386f
training screen, 436f
training session graphs, 437f
training sessions, beginning, 385
training session with formatting axis, 427f
training session without formatting axis, 427f
transferral of NFB to home and school, 414–415
two-channel assessment and training, 417–423, 417f
two-channel feedback for ruminations and impulsivity, 423f
wide vs. narrow bandwidths, 400
intervention fundamentals using NFB and BFB, 445–466
   artifact effects comparisons, 465
   BFB and two-channel NFB, 462–463, 463f
   biofeedback first, 447
   bipolar disorder, 462–463
   business executive case study, 451–457, 452f, 453f, 454f, 456f, 462–463
   data consistency and accuracy, 464
   desired mental state linked with learning, 446
   display options, 447–448
   dysphoria and depression, 465
   external open awareness, 445–446
   fibromyalgia, 459
   flexible shifting of mental state, 446
   focus and concentration, 465

generalizing techniques to home, school, and work, 446
HRV with respiration, 446
instruments for BFB, 446
LD and dyslexia, 465–466
left-brain feedback, 465–466
mental state, ability to hold, 464
mind-body changes for optimal performance, 450
movement disorders, 459–462
muscle spindle disorders, 459–462
NFB and BFB parameters, 448, 460f
one-channel feedback, 465–466
Parkinson's disease with dystonia, 459–460, 461f
performance optimization, 451–452
physiological and brain state changes, 449–450
points vs. percent-over-threshold tracking, 463–464
professional golfer case study, 448–450, 449f
race car driver case study, 451, 451f
right-brain feedback, 465
screen design, 449
self-regulation, 445–446
sequential (bipolar) training, 466
within-session tracking, 463–464
side effects, 458–459
similar conditions comparisons, 464–465
success sequence, 457t
time of day effects, 465
tracking client progress, 463–464
training screens, 453–456
trend reports, 464–465
untoward client experiences, 457–458
intervention in concussion/mild traumatic brain injury (TBI), 637–638
intervention summary, 471–478
    averaging and continuous feedback, 472–473
    classical conditioning, 473–474
    feedback types and cueing, 477–478
    generalization, 475
    learning paradigms, 474
    motivating children, 476–477
    operant conditioning, 471–473
    reinforcing feedback, 472
    screens and screen sequences, 475–476
    sessions, steps in, 475–476
    setting up for success, 471
    tracking trends, 476–477
    training session components, 471–475
interventricular foramen (foramen of Monro), 78, 85
ionotropic receptor sites, 27
iproniazid, 345
IPSP (inhibitory postsynaptic potential), 20, 21, 24
IQ correlations, 167
IQs and peak alpha frequencies, 34
irregular waves, 33
Isaacs, Julian, 476, 481
ISNR. *See* International Society for Neurofeedback and Research
isocarboxazid, 345
isonicotinic acid hydrazide (INH), 347

J

jamais vu, temporal lobe medial aspect, 202
Jarusiewicz, Betty, 263
Jasper, Herbert, 186
Jeejeebhoy, 639
Joan of Arc, 46, 202
John, E. Roy
    communication patterns in brain maps, 19
    EEG databases, 270, 304
    imaging techniques for mental disorders, 354
    medication response prediction, 9, 19
    neurometric approach, 18
    normative databases, 32
    psychiatric and neurological disorder research, 354
    seizure disorder subtypes, 353
joining and redirecting in ASD, 659
Jones, 587
Journal fuer Psychologie und Neurologie, 131
Journal of Brain Research, 131
Journal of Development and Behavioral Pediatrics, Steiner ADHD study, 674
Journal of Neurotherapy, efficacy research methodology standards, 16
judgment, BA 46 and, 174
Jungian techniques, 478

K

Kahn, Fred, 607–608
Kaiser, David, 32, 127
Kamiya, Joe, 7
Kaufmann, T., 586
K complexes, 43–44
Kenemans, J.L., 677
Kennedy, S.H., 645
Kennerly, 602
Kim, Slava, 607
kinesthetic dominance, 95–96
kinetic apraxia, BA 6 and, 187–188
Korsakoff's syndrome, 107, 205
Krebs cycle, 625–626, 626f
Kropotov, Juri
    communication patterns in brain maps, 19
    ERPs and, 6, 127, 663
    on frontal lobes and executive function, 140
    Independent Component Analysis, 342, 344
    on LORETA, 544
    memory traces and auditory cortex, 196
    mismatch negativity (MMN), 526, 526f
    normative databases, 32
    time relationships and evoked potentials, 197–198, 198f
    visual stimulus study, 242
    WinEEG program, 342
Kübler, A., 586
kynurenine (KYN) pathway, 628–629

L

labile mood, 177
lactic acidosis, 624–625
Laird, A.R., 126
lamda waves, 39, 40f
lamina terminalis, 81
Landau-Kleffner syndrome, 46
Lander, 250
language comprehension, 127, 178, 179
language processing, 173
Lannon, Richard, 71
Laplacian montage, 41f, 57, 433f–434f, 548f
Largus, Scribonius, 599
La Rovere, M.T., 581
Larsen, Stephen, 609
"Laser Fundamentals" (Aldrich), 607
laser therapy, 607–609
latency in brain speed, 523
lateral (brain), 68
lateral and medial aspects of cortex, 169–180
    anterior insula and speech, 176
    BA 4, dominant hemisphere, 175–176
    BA 9 both hemispheres, 170–171
    BA 46, 178
    BAs 44,45, DH, 176
    BAs 44,45, NDH, 176–177
    BAs 46,8,9, DH, 174
    Default-mode network connections, 179–180
    F4, NDH, 174
    F7 location and, 177
    F8, 177–178
    F9, F10, and orbital surface, 178–179
    FP1-F3, FP2-F4, 169–170
    inferior frontal gyrus in DH, 179
    left, dominant hemisphere, 171–174
    NDH right inferior frontal region, 180
    P. Broca, 176
    right, nondominant hemisphere, 170
    See also Executive network
lateral aspect, cerebral hemisphere, 69f
lateral aspect of brain, 71f
lateral corticospinal tract, 87
lateral fissure, 72
lateral geniculate nucleus of metathalamus, 75
lateralized waves, 34
lateral pulvinar nucleus, orienting and focusing, 112
lateral ventricles, 78
laughter, in superior frontal gyrus, 171–172
Law of Effect (Thorndike), 3, 11
LBD (Lewy body dementia), 541
lead, 347
leakage (sodium and potassium), 23
learned normalization of EEG patterns, 3
learning, limbic system and, 98–99
learning disabilities (LDs)
    19-channel (full-cap) QEEG assessment, 355

attention to external stimuli, 250
and dyslexia, NFB and BFB intervention fundamentals, 465–466
as experimental application of NFB, 14
metacognitive strategies and, 485–486
learning strategies, 488
LeDoux, Joseph, 82
Leduc, 599
left (dominant) temporal lobe, 192
left-brain feedback, NFB and BFB intervention fundamentals, 465–466
left cerebral cortex damage, 91
left cerebral hemisphere, 88–89, 108, 128f, 129f, 208f
left-handedness, reading and writing functions, 95
left hemisphere, frontal portion and positive emotions, 98
left hemispheric inactivation of insula, 581
left hemispheric underactivation, in ADD, 109
left lateral and medial orbitofrontal cortex, 183
left motor precentral gyrus, arm movement difficulty, 92
left temporal lobe seizures and slow-wave activity, 679–680
left visual cortex, disconnection from angular gyrus, 94
Lehrer, P., 568
lentiform nucleus, 78, 135
lesions, 167, 169, 178
lethargy and low alertness, rewarding beta, 403
Leung, Wence, 675
Levesque, Beauregard, & Mensour ADHD study, 673–674
Levesque, Johanne, 674
Lewis, Thomas, 71
Lewy body dementia (LBD), 541
lexical categorization, BA 20 and, 197–198
lexical functions, 177, 192, 596
LEXICOR equipment, 53, 54, 251, 289
lie detection, ERPs in, 6, 529
light-sound stimulation, 366
limb apraxia, 92
limbic cortex, 72
limbic system
    ACC and, 656–657
    described, 73, 96–99
    learning and, 98–99
    orbitofrontal cortex and, 141
    physiology, 96–97
Linden, David, 682
Linden, E.J., 683
Linden, Habib, & Radojevic ADHD study, 672
Linden, Michael, 32, 263, 691
lingual gyrus, visual sensory area, 96
linkage development, 85
linked-ear reference, 287f, 290, 406
linked-ground electrodes, 406
Liss Pain Suppressor, 602–603, 603f
listening training, 583
lithium, 345
"living environment for another world" gridding task, 497–498
Lloyd, Dan, 127, 133
lobes of brain
    basal zone of cerebral hemispheres, 89
    cerebral hemispheres, 88–89
    dominance, right or left brain, 88–89
    frontal lobes, 88–89
    functional significance, 88–89
    functions, described, 161t
    International 10-20 sites and, 156
    left cerebral hemisphere, 88–89
    medial zone of cerebral hemispheres, 89
    occipital lobes, 89
    parietal lobes, 89
    right cerebral hemisphere, 88–89
    temporal lobes, 89
local average Laplacian or Hjorth montage, 290
Localisation in the Cerebral Cortex (Brodmann), 131
localized slow waves, 44
local resonant loops, 30
lockjaw, glycine and, 25
locus coeruleus (LC), 77, 141
locus coeruleus norepinephrine production in ADD, 109
Lofthouse, N., 675
logical information attention, 171
long association fibers, 80
long distance connectivity, 153
long division study techniques, 510
longitudinal cerebral fissure, 69

longitudinal stria, 82
long-latency ERPs, 528–541
long-latency potential ERPs, 6, 528
long-term memory, 106, 205–206
long-term potentiation (LTP), 25, 183, 594–596
Loo, C.K., 597
LORETA (low resolution electro-magnetic tomography assessment)
    assessment and intervention, 355–357, 357f
    for BA 25, 219
    criteria for using, 562
    defined, 5
    EEG information and, 28
    eVox program (Evoke Neuroscience) and, 289
    HRV training and, 162
    images, 135f
    and low-power EEGs, 432
    for mental disorders, 354
    TBIs and, 638–639
LORETA mathematics in assessment, 543–564
    conclusions, 562–564
    definition and overview, 543–544
    dipoles, 548–549, 549f
    explanations to patients, 551
    Laplacian montage LORETA analysis, 548f
    LORETA Z-score (LNFB), 549–556, 552f, 554f, 555f
    magnetoencephalography (MEG), 548
    site choices (BAs), 551–555, 553f–555f
    sLORETA and LORETA, 544
    training examples, 556–562, 556f–562f
    typical images, 545f–547f
LORETA Z-score (LNFB)
    Brodmann areas and, 156
    case example, 554f, 555f
    HRV training with NFB and BFB, 584
    for multiple midline and ventral structures, 549–556, 552f
    for simultaneous parameter training, 134
losing the reward, 409
Lou Gehrig's disease (ALS), 7
Loveland, K.A., 650
low alpha waves, 37
low beta waves, 38
Low Energy Neurofeedback System (LENS), NeuroField and, 609–610
low-pass filters, 54
low-power EEGs, 432–435
low resolution electromagnetic tomography assessment. See LORETA
Low Voltage Fast activity, 435
LSD (lysergic acid diethylamide), 24, 346
LTP (long-term potentiation), 25, 183, 594–596
Lubar, Joel
    ADD research, 14, 19
    ADHD NFB study, 32
    Autogen A620 assessment program, 324
    cell column distance in cortex, 30
    on dysphonetic dyslexia, 95
    effectiveness studies, 671
    electrode placement for training, 405
    follow-up study, 675
    LORETA use for NFB, 356, 550
    NeuroGuide workshops, 550
    OCD treatment, 107
    self-regulation for ADHD, 269–270
    single-channel assessment study, 314
    SMR seizure studies, 8
    on strong corticocortical coupling, 31
    task learning options, 4
    ten-year follow-up, 114
    on thalamic pacemakers, 29
    "thalpha" range, 398
    theta/beta ratio norms study, 18, 113
Lubar, Judith
    effectiveness studies, 671
    SMR seizure studies, 8
Lukito, S., 586
lysergic acid diethylamide (LSD), 24, 346

M

MacKinnon, S., 587, 587f
Maclean's Magazine, Bilodeau on optimal performance training, 691
macrocolumn (pyramidal cells), 20
macroelectrodes, 20

magnetic resonance imaging (MRI) for mental disorders, 354
magnetoencephalography (MEG), 548
magnitude, defined, 36
major depressive disorder (MDD), 597, 646
Malone, Molly, 109
mammillary body, 73
mammillothalamic tract, 85–86
mania, 25
manic behavior in NDH, 167
The Man Who Mistook His Wife for a Hat (Sachs), 91, 241
MAO (monoamine oxidase) inhibitors, 345
mapping technique, 500
Mapping the Mind (Carter), 112
marijuana use, 37, 344–345
massa intermedia, 70
mathematic study strategies, 506–516
    addition and subraction study techniques, 507
    ADHD children and math, 513
    difficult multiplication study techniques, 508–509
    fractions study techniques, 510–514, 512f
    grade school math tips, 506–508
    groups times people study technique, 507, 510f
    long division study techniques, 510
    multiplication and division study techniques, 507
    multiplication concept, 509
    reciprocal rule study technique, 508
    word problem study techniques, 514–516, 516f
math tasks, 322
Matthews, S.C., 7
Mayberg, Helen S., 219, 644, 645
MDD (major depressive disorder), 597, 646
medial (brain), 68
medial aspect, inferior temporal lobe, 204–210
medial aspect of occipital and posterior temporal cortex bilaterally, 91
medial forebrain bundle (MFB), electrical stimulation as reward, 98
medial frontal cortex, 215
medial geniculate nucleus of metathalamus, 75
medial geniculate to auditory cortex information path, 197f
medial hypothalamus, irritability and aggression, 98
medial intraparietal (MIP), 225

medial lemniscus, 87
medial orbital prefrontal cortex (MOPFC), 147f
medial parietal cortex (MPC), 147f
medial parietal lobe/posterior cingulate, 223
medial prefrontal cortex, 89
medial zone of cerebral hemispheres, 89
medication response prediction with QEEGs, 9
medications as artifacts, 344–347, 349
medications for depression, 644
meditation, 37, 679
Meditech Laser Therapy, 608
medulla oblongata (brain stem), 77
MEEG (magneto-encephalography) for mental disorders, 354
MEG (magnetoencephalography), 548
melody generation, 176
memory disorders, 105–107
memory functions
    amygdala and, 199
    BA 6 and, 188
    BA 21, 196
    BA 38, 203
    changing memories, 107
    and emotion, 107
    ERC-hippocampus system, 204–205
    memory areas, 166
    memory retrieval, 106–107, 169, 174, 267
    memory trick for adolescents, 119
    Papez circuit, 201
    parahippocampal gyrus, 206, 208
    and recall techniques, 500
    temporal lobe medial aspect, 201
memory problems, frontal lobe injury and, 164
memory strategies, 172, 489–491
memory traces, auditory cortex, 196
meninges, 70
Menon, V., 149, 662–663
mental disorders, brain imaging techniques for, 354
"The Mental Edge" (Robbins), 15
mental processing, HRV training, effect on, 579
mental rotation, 175
mental state, ability to hold, NFB and BFB intervention fundamentals, 464

mental state-band width correlation, 9, 10t
mercury, 347
mescaline, dopamine receptors and, 24
mesencephalon (midbrain), 77
Meshalkin, 607
meta-analysis of controlled studies of EEG power ratio, 674
metabolism, metabotropic receptors and, 27
metabotropic receptors, 27
metacognitive strategies, 483–519
    3-3-3 method, 504–505, 504f
    addition and subraction study techniques, 507
    ADHD, mathematic study strategies, 513
    alertness and motivation, 495
    anxiety, 484
    boxing technique, 503
    cardiopulmonary lecture gridding task, 496–497
    cartoon technique, 500
    daydreaming, 484–485
    day planners, 490–491
    "design a school" gridding task, 496
    difficult multiplication study techniques, 508–509
    fractions study techniques, 510–514, 512f
    garbage bag vs. filing cabinet mind, 495–496
    grade school math tips, 506–508
    Great Depression gridding task, 497
    grid technique, 496–499
    groups times people study technique, 507, 510f
    habitual bracing, 496
    hamburger (3-3-3) method, 504–505, 504f
    learning disabilities, 485–486
    learning strategies, 488–499
    "living environment for another world" gridding task, 497–498
    long division study techniques, 510
    mathematic study strategies, 506–516
    memory and recall techniques, 500
    memory strategies, 489–491
    metacognition, defined, 483
    multiplication and division study techniques, 507
    multiplication concept, 509
    Nelson vs. Napoleon exercise, 492–494, 494f, 496t
    organization of ideas, 502
    organized thinking, 495
    organizing and gridding tasks, 496–498, 497t, 498t
    packaging technique for studying, 505–506
    project timeline/flow sheet, 499
    reading, 491–496
    reading comprehension, 501–503
    reading/listening basic principles, 494–496
    reciprocal rule study technique, 508
    red thread, 499–500
    reporter's questions technique, 503
    Roman Room technique, 500
    self-stimulated-recall, 502
    short story and novel techniques, 502–503
    single picture technique, 500
    at start of NFB and BFB sessions, 133
    study hall, 516–517
    synthesis of data, 502
    three-key-facts technique, 489–490
    Titanic exercise, 500
    verbal recall techniques, 500–501
    word problem study techniques, 514–516, 516f
    writing technique, 504–505
metal in mouth as artifact, 341f
metaphorical thinking, 174
metaphors, understanding, 170, 173
metencephalon, 77
methanol, 347
methylphenidate, 113, 345–346
Meuli, Reto, 153
MFB (medial forebrain bundle), electrical stimulation as reward, 98
microelectrodes, 20
microglia, 70
microvolts (mV), 4, 284
midbrain (mesencephalon), 77
middle cerebral arteries, 100
middle frontal gyrus, memory strategies and, 172
midline nuclei of metathalamus, 75
midsagittal section of the telencephalon and the diencephalon, 97f
midsagittal section with thalamus position, 135f
migraine treatment, passive infrared hemoencephalography (pIR HEG), 612
mild cognitive impairment (MCI), 183

mild head injury, 19-channel (full-cap) QEEG assessment, 354
Milwaukee Card Sorting Test, 655
Mind-Mirror screen, 422f
Mind Room, 691
Minshew, Nancy, 262
MIP (medial intraparietal) area, 225
mirror neuron system (MNS), 175, 176, 187, 229, 654
mismatch negativity (MMN), 525–528, 526f, 527f
mitochondria, ATP production in, 624–625
Mitsar equipment, 251, 289, 342, 521
Monastra, Monastra & George ADHD study, 673
Monastra, Vince
    ADHD NFB study, 14, 32
    effectiveness studies, 671
    efficacy studies, 672
    single-channel assessment study, 314
    SMR for ADHD, 677
    theta/beta ratio norms study, 18
Monastra norms vs. A620 assessment program, 284
monoamine oxidase (MAO) inhibitors, 345
monophasic waves, 33
montage reformatting, 57
montages in full-cap assessment, 290
Montgomery, D., 693
Montoya, E. R., 665
Montreal Neurological Institute (MNI) atlas, 544
mood, depressed, as experimental application of NFB, 14
mood regulation, 177, 203, 223
MOPFC (medial orbital prefrontal cortex), 147f
moral injury and PTSD, 685
morphine and heroin, 346
morphological reading, 94
morphology (waveform), 33
motivating children, 476–477
motivation, limbic system and, 98–99
motivational direction in dispositional anger, 177
motor apraxia, 92
motor aprosodia, 176, 649
motor association area, 187–188
motor circuit, 140
motor control, ACC and, 218
motor ERPs, 6, 528

motor network, 145f
motor pathways of cortex to spinal cord, 86–87
motor-speech functions, 176
mouth and taste functions, Gustatory Cortex, 196–197
movement disorders, NFB and BFB intervention fundamentals, 459–462
movement disorders and fibromyalgia, 681–683, 683f
movement functions, BA 6 and, 187–188
MPC (medial parietal cortex), 147f
MRI (magnetic resonance imagery), 5, 354
Mu activity, 40, 41f, 185–186, 657
multichannel encoders, troubleshooting, 61
multi-infarct dementia, 354
multimodal treatment approaches, 694
Multimodal Treatment of Attention Deficit Hyperactivity Disorder study, 346
multiplication and division study techniques, 507
multiplication concept, 509
multiplication study techniques, 508–509
multitasking, 111
Municipal Mental Asylum, Frankfurt, 131
muscarinic receptors, 27
muscle activity as artifact, 326–329, 327f, 337f
muscle spindle disorders, 459–462, 681
mushin (no-mindedness), 445
music feedback, 409
music performance, optimal performance training, 15
myelencephalon, 77
myelinated fibers (white matter), 17
myelinated vagus system, vagal control of HRV, 574–575
myelinization, 27, 70
myelin sheath, 70
myoclonic activity, 347
myoclonic seizures, 46, 352
myogenic spike biphasic morphology, 42

N

N1-P2 complex ERP, 6, 525
N2 ERP, 525
N400 negative long-latency potential, defined, 6
narcolepsy, 45
narrow focus, 266–267

nasion (anterior) of head, 271
National Institutes of Health (NIH), biofeedback and neurofeedback, 19
natural neurofeedback, 3
Nature's Children series, 503
nausea and vertigo, BA 22 and, 237
NBM (nucleus basalis of Meynert), 182–183
NDH (nondominant hemisphere), 163
near infrared spectrophotometry hemoencephalography (NIRS HEG), 610–615
    definition and history, 611
    expected changes, 612
    instrumentation, 612–615
near infrared spectrophotometry system, 611
negative bracing, 515
negative emotions and thoughts, 98, 148, 173
negative ruminations for alpha-theta therapy, 478
negative z-scores and relative power, 434
Nelson vs. Napoleon exercise, 492–494, 494f, 496t
neocortex, 72, 72f
nerve cell resting potential, 22–23
nervous system vigilance, 685
networks, defined, 152, 153
networks, importance of, 133, 134
Neubrander, James, 653
neural foundation of sense of self, 147
neuritic outgrowth, 182–183
neuroanatomical/biochemical changes, 595–596
neuroanatomical structures, 71–77
    amygdala, 73–74
    basal ganglia, 74, 74f
    brain positions, 73f
    cerebral hemispheres, 71–72
    diencephalon, 71–72, 75
    forebrain, 71–72
    hindbrain (metencephalon, myelencephalon), 77
    hypothalamus, 76–77
    limbic system, 73
    midbrain (mesencephalon), 77
    neocortex, 72f
    nuclei of thalamus, 76f
    right cerebral hemisphere, transverse section, 75f
    striatum, 74
    telencephalon, 71–72
    thalamus, 75, 76f
neuroatomical links between HRV and NFB, 581–583, 582f, 583f
Neurobiologishes Laboratorium, Berlin, 131
Neurocybernetics system (EEG Spectrum), 324
neurofeedback (NFB)
    in ASD and Asperger's syndrome, 656–659
    and BFB, in ASD and Asperger's syndrome, 657
    and BFB for alpha-theta therapy, 477
    and BFB parameters, 448, 460f
    defined, 4
    and HRV training, for depression, 644–645
    hypocoupling and, 31
    long-term results, 113–114
    as noninvasive, 646–647, 646f
    over dominant temporal lobe, 198
    results at single site, 133
    subject awareness of mental state, 7
    system, 56f
    training objectives, for alpha-theta therapy, 477
    training regimens, QEEGs for, 9
    See also metacognitive strategies
"Neurofeedback training of sensorimotor rhythm in marmoset monkeys" (Philippens), 681–682
NeuroGuide database, 270, 304, 321
NeuroGuide Program (Applied Neuroscience), 33, 290, 549, 550–554, 609
neurohormones and brain's message to body, 2
neuro-inhibition in P3a and P3b, 529
neuroleptics, 344
neurological inflammation in depression, 628–629
neurologically based disorders, 94–95
neurology vs. medical uses for EEG, 8
neurometric approach, accuracy of, 18
neuromodulation, 444
neuromodulators and brain's message to body, 2
neuronal cell death, 629–630, 630f
NeuroNavigator equipment, 54, 251, 289, 299
neuron communication, 22–26
    acetylcholine, 24
    amino acids, 25
    biogenic amines, 24

    dopamine, 24–25
    excitatory postsynaptic potential (EPSP), 23–24
    GABA, 25
    glutamate, 25–26
    glycine, 25
    inhibitory postsynaptic potential, 24
    nerve cell resting potential, 22–23
    neuropeptides, 26
    neurotransmitters, 24–26
    norepinephrine, 25
    postsynaptic potentials, 23–24
    serotonin, 25
    summation of EPSPs and IPSPs, 24
    synapses, 23
neurons, integrating centers of, 26
neuropeptide Y (NPY)/ polypeptide YY (PPYY), 26
neurophysiological profile (NP) test battery, 634
neurophysiological treatment, 638–640
neurophysiology and NFB interventions, 105–121
    Alzheimer's dementia, 107
    anterograde amnesia, 107
    Asperger's syndrome, 117–120
    changing memories, 107
    episodic memory, 106
    Korsakoff's syndrome, 107
    long-term memory, 106
    memory and emotion, 107
    memory disorders, 105–107
    memory retrieval, 106–107
    movement disorders, 116
    obsessive compulsive disorder, 107–108
    procedural memory, 106
    results, 120–121
    reward deficiency syndrome, 108
    semantic memory, 106
    senile dementia, 107
    Tourette syndrome, 107–108
    See also Attention Deficit Disorder; Attention Deficit Hyperactivity Disorder (ADHD)
neurophysiology study hints, 95–96
neuroplasticity, 17
neuropsychological testing, concussion/mild traumatic brain injury (TBI), 634–635, 634*t*–635*t*

NeuroPulse equipment, 251, 289
Neuroregulation, "Neural Networks and Neurofeedback in Parkinson's Disease" (Esmail & Linden), 683
Neurotherapy Reference Handbook (Byers), 250
neurotoxins, 347
neurotransmitters, 2, 24–26
neurotrophic factors, 182
Newton, Isaac, as epileptic, 46
NFB. *See* neurofeedback
nicotine, 25, 346
nicotinic receptors, 27
Niedermeyer, 344
nigral toxins, 183
19-channel (full-cap) QEEG assessment, 289–314
    ADHD, 355, 669
    advantages and disadvantages, 312–313
    alcoholism, 355
    Alzheimer's dementia, 354
    anxiety disorders, 355
    assessment training, 358
    bipolar disorder, 355
    cerebrovascular disease, 354
    common electrode reference montage, 57
    defined, 5
    dementia, 354
    depression, 355
    full-cap findings, 314–315
    lamda waves in, 39
    learning disabilities, 355
    mild head injury, 354
    montages and, 290–291
    as more complete, 314–315
    obsessive compulsive disorder (OCD), 355
    psychiatric and neurological disorders, 354–355
    schizophrenia, 354–355
    See also quantitative electroencephalogram
NIRS HEG (near infrared spectrophotometry hemoencephalography). *See* near infrared spectrophotometry hemoencephalography
Nissl, Franz, 131
Nitsche, M., 596, 597
NLD (Nonverbal Learning Disorder), 667
nociceptive area, 218

nodes, defined, 155
nodes of Ranvier, 27, 28
noise (EEGs), 64
nondominant (right) frontal functions, 167
nondominant (right) hemisphere, 163, 170, 170f
nondominant supramarginal gyrus, 234
non-N-methyl-D-aspartate receptors, 25–26
non-spike and wave patterns, 48
nonverbal communication, 476
nonverbal learning abilities, 418
Nonverbal Learning Disorder (NLD), 667
noradrenalin in NDH, 167
noradrenergic buttons on target organs, 102
noradrenergic stimulation to the right cerebral hemisphere in ADD, 109
norepinephrine, 24, 25
normalization of EEG, 267
normal patterns, targeting with operant conditioning, 19
normative databases, QEEGs and, 32
notch filters, 53
novelty, comprehension of, 167
NTS (nucleus tractus solitarius), 147, 575
nuclei of thalamus, 76f, 138f
nucleus accumbens, pleasure response and, 98
nucleus basalis of Meynert (NBM), 182–183, 541
nucleus reticularis, 29
nucleus tractus solitarius (NTS), 147, 575
numerical relationships, parietal lobes, 225–226
Nun Study, 17
Nuprep, 316
nutritional balance, 627–628
Nyquist principle, 54

O

O1 active electrode site, 481
Oasis Pro (Mind Alive), 600
objective measurements with training screens, 435–436
obsessive compulsive disorder (OCD)
    19-channel (full-cap) QEEG assessment, 355
    ACC and, 218
    dopamine activity and, 25
    NFB treatment, 107–108
    promising clinical reports for NFB, 14
    serotonin and, 25
    tDCS and, 597
occipital intermittent rhythmic delta activity (OIRDA), 680
occipital lobes, 89, 96
occipital regions, 239–242, 239f, 240f
OCD. See obsessive compulsive disorder
Ochs, Len, 609
ocular apraxia, 92, 228
oculomotor circuit, 140–141
oddball paradigm, 522
oddball stimulus, 6
OFC (orbitofrontal cortex), 146
office environment, 379
offsets, 58–59, 390–391
Ohm, Georg, 51
OIRDA (occipital intermittent rhythmic delta activity), 680
oligodendroglia cells, 27, 70
Omniprep, 316
one-channel assessment bandwidths findings, 275–284
    See also single-channel EEG
one-channel feedback, NFB and BFB intervention fundamentals, 465–466
one site's effect on others, 405
one-trial learning, 383
open awareness, 266, 401–402, 477
open focus, 446, 467, 477
open-loop integration, 145–146
operant and classical conditioning with NFB, 414–415
operant classes, 11
operant conditioning, 3, 11, 13, 98, 471–473
operculum areas, 188f
optical isolation, 58
optic ataxia, 92, 122, 228
optic chiasm, 76
optimal performance training
    assessment and intervention, 265–267
    in athletic competition, 691–692
    as experimental application of NFB, 15
    high alpha waves and, 37
    mental flexibility and, 404
    mind-body changes for golf, 450
    narrow focus and, 267–268

NFB and BFB intervention fundamentals, 451–452
    with operant conditioning, 19
orbital frontal/inferior frontal cortex, 181–184
    in ASD and Asperger's syndrome, 649–650
    nucleus basalis of Meynert (NBM), 182–183
    orbital frontal association cortex, 181
    orbitofrontal cortex, 183–184
    Phineas Gage, 181–182
    and temporal lobe links, 168
    ventromedial prefrontal cortex (VMPFC), 183–184
orbital prefrontal cortex circuit, 107
orbital sulci, 179f
orbitofrontal cortex (OFC)
    acetycholine and, 182
    in Affect network, 146
    social functions, 183–184
    socially appropriate behavior, 111–113, 141, 168
organic delusional states, 354
organized thinking, metacognitive strategies and, 495, 502
organophosphates, 347
orientation problems, frontal lobe injury and, 164
Orienting network, 664
orthography-phonology links BAs 37 and 39 and, 237
oscillations in heart-brain connections, 571–573, 572f, 573t
overactive behavior, 173
overbreathing (hyperventilation), 348, 568, 588–589
overlapping training components, 467
overriding questions, red thread and, 499
oversampling, 55
overview illustration, 194f
owl or monkey pattern, 308f
oxytocin neuropeptide, 656

P

P3a and P3b, 529
P8, 237
P300, 5–6, 528–529, 585–586
Paced Auditory Serial Addition Task (PASAT), 373
packaging technique for studying, 505–506
pain anticipation, 186
pain functions, BA 5 and 7 and, 188
pain perception, substance P and, 26

pairing, mental states and academic tasks, 12–13
Palsson, Olaf, 473
Panayiotopoulos, C.P., 679
panic, 173
panic disorders, ACC and, 218
Papez, J.W., 85
Papez circuit, 201, 203, 205–206
    See also limbic system
Paquette, Vincent, 646, 646f
parabrachial nucleus (PBN), 147
paracrines, 25
parahippocampal cortex, 208
parahippocampal gyrus, 80, 206
parahippocampal place area (PPA), 206
paralimbic cortex, 204
paranoia, 24–25
paraphrasias, 195
parasympathetic cytokine control system, 638
parasympathetic nervous system (PSNS), 2, 102–105, 212, 685
paraventricular nucleus of hypothalamus, 575
parental support for NFB and BFB, 383
parietal cortex, orienting and focusing, 112
parietal high alpha training, 89
parietal lobes, 225–238
    anatomical boundaries, 225
    angular gyrus, 236f
    angular gyrus and supramarginal gyrus, 238f
    BA 7, 229f
    BA 37, 236f
    BA 39, 233f, 236f
    BA 40, 234f, 235–236, 235f
    Bálint's syndrome, 228
    Broca's and Wernicke's areas, 232f
    central sulcus, 225f
    cortical information flow, 226
    dorsal and lateral aspects of BA 7, 227
    dorsal lateral aspect, 226–228
    functions, 225–226
    Gerstmann syndrome, 227
    lateral and inferior parietal lobes, 235–237
    lateral aspect, 228, 228f
    medial dorsal aspect, 229–230

midsagittal section, 226f, 229f
P8, 237
parietal lobe damage, 91–92
per Gray's Anatomy, 226f
somatosensory association cortex, 226–228
supramarginal gyrus, 236f
temporal-parietal junction, lateral aspect, 230–235, 230f
vestibular connections, inferior parietal-temporal BA 22, 237
visual acuity and, 89
Wernicke's area, 231
parietal spindling beta, 236
Parkinson's disease
   basal ganglia and, 74
   dopamine and, 24–25, 99
   with dystonia, NFB and BFB intervention fundamentals, 459–460, 461f
   high STN activity, 143
   nigral toxins and, 183
   promising clinical reports for NFB, 14
   research, 681–682
   tDCS and, 597
parolfactory area, 80, 85, 89, 97
paroxysmal discharge, 43, 43f
paroxysmal hypnagogic hypersynchrony, 43
parsimony principle, 269–270, 667
pars opercularis ("part that covers"), 175
partial lefties, 95
partial reinforcement, 11
partial seizures, 46, 48f, 49f, 351
PASAT (Paced Auditory Serial Addition Task), 373
Pasqual-Marquis, Roberto
   current detection, conditions for, 22
   on dipoles in pyramidal cells, 20
   LORETA development, 5, 356, 543–544
Passingham, Richard, 140
passive infrared hemoencephalography (pIR HEG), 611–615
   defined and described, 611
   expected changes, 612
   history, 611
   instrumentation, 612–615, 613f, 614f, 615f
   migraine treatment, 612

pathological activity, assessment and intervention, 350–354
pathways and SMR production, 144–145
pattern encoding, temporal lobes and, 191
pattern recognition, 164, 188, 241, 268
Pavlov, V.R., 12
Paxil, 345
PBN (parabrachial nucleus), 147
PCP (Phencyclidine or Angel Dust), 346
Peak Achievement Trainer (PAT), 324
peak alpha frequency, 37
Peak Performance Mind (Wise), 417
Pearson psychological tests, 171
pEMF (pulsed electromagnetic field) stimulation, 609
Penfield, Wilder, 186
Peniston, William, 15, 475
Peniston protocol, 475, 480–481
Pentecost, Hugh, 502–503
percent over threshold, 397–398, 397f, 464
percent-over-threshold tracking vs. points, 463–464
perception defects, 91
Perez, J.J., 586–587
performance IQ, 167
performance optimization. See optimal performance training
periaqueductal gray matter, 77
peripheral neuropathies, 347
peripheral psychophysiological (biofeedback) measurements, 565–566
peripheral skin temperature, 360
perisylvian language zone, 466
Perminder, S., 609
perseveration, 122, 164
personality change, frontal lobe injury and, 164
Pervasive Developmental Disorder (PDD), 117, 250
PET (positron emission tomography) scan, defined, 5
Peterson, Gail, 408
petit mal epilepsy (absence seizures), 47, 47f, 350–351
PET studies, 112
phase
   in ASD and Asperger's syndrome, 653, 653f
   defined, 291
   phase angles, 34

phase-locked activity, coherence as, 152
phase of waveforms, 4
phase-reset, 152–153, 440–441, 441f, 442f
phase reversals, 34, 353
phase shift and phase lock, 439–440, 440f, 442f, 443f
phasic attention abilities, 109
phenothiazines, 345
Philippens, Ingrid, 681–682, 683
phonemes, 177
phonemic input, NFB in dominant temporal lobe, 192
phonetic knowledge, 465–466
phonological reading, 94
phonology, 171
phosphorylation, 596
phthalates in plastic products, 629
physician assessment, 269
physiological and brain state changes, NFB and BFB intervention fundamentals, 449–450
pia mater, 70
Picks disease (frontotemporal dementia), 184, 205, 354
picowatts (pW), 4, 284
Pierce, Karen, 262
Pinna, G.D., 581
pIR HEG. *See* passive infrared hemoencephalography
piriform lobe, 215
Pizzagalli, D.A., 646
placebo effect in depression, 645, 645f
placebo network, 570
planes of the brain, 69f
planning and controlling, 174
PNN50, 573
points vs. percent-over-threshold tracking, 463–464
polarization of ventral basal (nVB) nucleus cells, 677–678
Polich, John
    ERP case examples, 530–531, 535
    visual continuous performance test, 522–523
poly spike waves, 42
polyvagal contributions in ASD and Asperger's syndrome, 650
polyvagal theory (Porges), 574, 655–656
pons, 77
Pope, Alan, 473

Poreisz, C., 596
Porges, Stephen
    HRV training, 583
    on inability to inhibit defensive reaction responses, 216
    polyvagal theory, 574, 650, 655–656
    on unmyelinated vagus, 199
positive emotions, left hemisphere and, 98–99
positron emission tomography (PET), 354, 593
Posner, M.I., 664–665, 665f
post-concussion syndrome., 620
posterior (head), 68
posterior cerebral arteries, 100
posterior cingulate cortex (PCC), 147f, 220–222
posterior cingulate gyrus, 223–224
posterior communicating arteries, 100
posterior parietal cortex (PPC BA 7), 145
posterior part of SACC, 142
post-reinforcement synchronization (PRS), 6–7
postsynaptic membranes, 23
postsynaptic potentials (PSPs), 20–24, 21f, 26
post-traumatic stress disorder (PTSD), 684–690
    autonomic nervous system (ANS), 685–686, 686f
    Brodmann areas, 688
    central nervous system (CNS), 686–689, 689f
    as compensatory response pattern, 685
    conclusions, 690
    core phenotype, 647
    in history, 684
    LORETA NFB for, 688
    VMPC and, 183–184
posture, 473
potassium gates, 26
potassium resting potential, 23
potential differences, 4, 51
power spectrums, 32–33
PPA (parahippocampal place area), 206
preamplifiers, 52, 53
pre-and subgenual anterior cingulate cortex (PACC), 141, 147f
preauricular notch, 271
precuneus, 229–230
prefrontal cortex, 112–113, 163, 163f
prefrontal lobes, 88

premotor cortex, 108, 188–190
preoptic area, 76
presbycusis, 597
present input/past experience comparison, BA 22 and, 195
prestriate cortex, visual functions and, 241
presynaptic autoreceptors, 409
Prichip, Leslie, 9, 18, 353, 354
primary process thinking, 15
problem-solving beta, 38
problem-solving concentration, 267, 404
problem solving difficulty, frontal lobe injury and, 164
procedural memory, 106, 237
ProComp+/BioGraph equipment
    bar-graph screens, 395
    feedback rate, 407
    filter type and order, 55
    high-pass filters, 53–54
    preamp configuration, 52
    screen example, 323–324
    single-channel assessments, 270
ProComp+ or Infiniti (Thought Technology), impedance in, 318
projection fibers, 85
project timeline/flow sheet, 499
The Promise of Sleep (Dement), 45
propagated phase, 291, 291*f*
propagation of nerve impulses, 24
proprioception, 185, 235
prosody, 91, 176
prosopagnosia, 91, 96, 192, 207
protein analysis, 215
protein receptor sites, 23
protein synthesis effect on tDCS, 590
proximal position similarity in pyramidal cells, 22
Prozac, 345
PSNS (parasympathetic nervous system), 102–105, 212
psychiatric and neurological disorders, 19-channel (full-cap) QEEG assessment, 354–355
psychiatric disorders, ACC and, 215
psychiatric syndromes, brain maps for, 19
psychological injury, PTSD as, 684–685
psychophysiological stress profiles, 249, 371–377, 375*f*, 376–377*t*

psychosis, dopamine activity and, 24–25
psychotherapy, NFB as adjunct procedure, 15
pulse artifact, 333
pulsed electromagnetic field (pEMF) stimulation, 609
punishment vs. reward sensitivity, 182
push-pull amplifiers, 290
putamen, 78, 107–108, 135–136
pyramid (forebrain), 79
pyramidal cells (macrocolumn), 17, 20, 22, 72
pyramidal motor pathways, 86
pyramid on medulla, 86–87

Q

quantitative electroencephalogram (QEEG)
    amplitude and power, 32
    assessment, 267–273
    assessment training, 357
    asymmetries, 34
    beta activity, 34
    defined, 32
    described, 4
    dominant frequency and age, 34
    vs. EEGs, 269
    EEG spectrum, 32–33
    for mental disorders, 354
    normative databases and, 32
    phases, 34
    rhythms, 34
    role in concussion/mild traumatic brain injury (TBI), 630–633
    sequential (bipolar) QEEG assessment, 274–275
    synchrony, 34
    uses, 8–9
    wave forms, 33
    See also 19-channel (full-cap) QEEG assessment
quick stress test, 372–373
quinolate (QUIN) elevation, 628–629

R

racial EEG studies, 18
Raczak, 581

RAD (reactive attachment disorder), 216, 251, 404
radio frequency radiation (RFR), 120
Rae-Grant, Naomi, 269
rage, LORETA images and, 548f
Raphe nuclei, 25
RAS (reticular activating system), 112
rauwolfia derivatives, data collection, effects on, 345
raw EEG findings in ASD and Asperger's syndrome, 657
Raz, Amir, 664
rCBF (regional cortical blood flow), 594
RCZ (rostral cingulate zone), 141
reactive attachment disorder (RAD), 216, 251, 404
reading
    metacognitive strategies and, 491–496
    reading and writing functions, 92–93
    reading comprehension, metacognitive strategies and, 501–503
    reading disorders, 94–95, 406–407
    reading/listening basic principles, 494–496
Reali, Gualtiero, 526–527
reasoning, defined, 164–166
recall of episodic information, PCC and, 221
receptor sites, 27f, 595
reciprocal rhythmic thalamocortical volleys, 678
reciprocal rule study technique, 508
red nucleus, 77, 86
red thread, 499–500
referential montage, 314
referential placement, 4, 273
referential vs. sequential amplitude comparison, 426f
refractory period, 26
regional cortical blood flow (rCBF), 594
regional resonant loops, 30
region(s) of interest (ROI), 551
regression, 560, 659
regular waves, 33
regulating inhibitory modulation loss, 677
Reid-Chung, Andrea, 118, 550, 585, 585f, 682
reinforced behavior, limbic system and, 98
reinforcement schedules, 11
reinforcing feedback, 472
relative power vs. absolute power, 433
relaxation and stress charts, 369f

relaxation-desensitization technique for alpha-theta therapy, 478
relaxation methods, 373
relaxed mental and physical states, assessment and intervention, 265–266
relaxed state for alpha-theta therapy, 480
relay cells, thalamic cells as, 28–29
relevance, 178
religiosity in temporal lobe, 46
Remembered Wellness (placebo effect), 645
renaming of sites, 272
repetitive transcranial magnetic stimulation (rTMS), 592, 592f
reporter's questions technique, 503
rescripting, 475
research design for NFB studies, 695–696, 695f
resistance (impedance), 51
    See also impedances
resonant cortical loops, 30
respiration, 361
respiratory sinus arrhythmia (RSA), 268, 361–362, 446, 570
resting potential in nerve cells, 22–23
resting state, 26
resting waves, 414
reticular activating system (RAS), 112
reticular formation, 77
reticular nuclei of metathalamus, 75
retrosplenial cortex (RSC), 147f
reward deficiency syndrome, NFB treatment for, 108
reward frequency bands, 397f, 398–399, 399f, 400f
reward frequency ranges, 404
rewarding 11-13 Hz, 401–402
rewarding beta 16-20 Hz or 13-15 Hz, 403–404
rewarding SMR (12-15 Hz or 13-15 Hz), 402–403
reward network, 181
reward system, medial forebrain bundle involvement, 98
reward vs. punishment sensitivity, 182
RFR (radio frequency radiation), 120
Rhett's syndrome, 47
rhythmic midtemporal theta, 680
rhythmic waves, 33
right (nondominant) temporal lobe, 192–193

Right-Brained Children in a Left-Brained World (Freed and Parsons), 88, 418, 667
right-brain feedback, NFB and BFB intervention fundamentals, 465
right cerebral cortex damage, 90–91
right cerebral hemisphere, 75*f*, 88–89
right frontal lobe damage, 90–91
right frontal lobes as heteromodal, 167
right hemisphere, frontal portion and negative emotions, 98
right hemisphere overactivation in ADD, 109
right hemispheric inactivation of insula, 581
right inferior prefrontal cortex, 184
right insula, 147
right parietal faster frequency alpha, PTSD and, 688
risk aversion, 180
risk versus benefit analysis, 173
Ritalin, 113, 345–346
Robbins, Jim, 7, 15, 150
ROI (region(s) of interest), 551
Roman Room technique, 500
root mean square of standard deviation of RR interval (RMSSD), 573
Rosenfeld, Peter, 6, 177, 419, 529, 644
Roshi system, 324
Rossiter & La Vaque ADHD study, 672
rostral cingulate zone (RCZ), 141
rostral portion of brain, 68
Rougeul-Buser, 586, 678–679
Rouxeau, 599
Royal College of Music (RCM), optimal performance training, 15, 250, 481
RSA (respiratory sinus arrhythmia), 361–362, 446, 570
RSC (retrosplenial cortex), 147*f*
rTMS (repetitive transcranial magnetic stimulation), 592, 592*f*
Rubin, Edgar, 661
Rubin vase, 661, 661*f*

S

Saab, Marc, 611
SACC (supragenual anterior cingulate cortex), 141, 147*f*, 215
saccades, 240
Sachs, Oliver, 91, 241
sagittal section, 68, 84*f*
salience-landscape theory, 654
Salience network, 148–149, 150*f*, 212, 660–663, 663*f*, 687
salutatory conduction, 27, 28
same and different discrimination, 182
SAMONAS Sound Therapy, 366–369, 479, 583
sampling, 52
sampling rates, EEG, 54–55, 55*f*
Sams, Marvin, 5
sandwich technique, 475
Sapolsky, Robert, 170
sarcasm in parahippocampal gyrus, 208
scene visualization, 479
Schandry, R., 587
schizophrenia, 7, 24–25, 354–355
Schwann cells, 27, 28*f*
scientific papers, writing, 697–702
    collecting material, 697–698
    content organization, 698–701
    results and summary, 701–702
screen design, 449
screens and screen sequences, 475–476
Sears, W., 672, 694
secondary reinforcement, 11, 13
second impact syndrome, 630–631
second order conditioning, 12
seizure disorders, 46–50, 676–680
    assessment and intervention, 250
    complex-partial vs. simple-partial seizures, 679
    continuous polymorphic delta (CPD), 680
    convergent analysis, 309*f*
    diaphragmatic breathing caution, 404
    EEG oscillations, 677–678
    EEG spindles, 678*f*
    expectancy rhythm (12-15 Hz), 678
    focused attention rhythm (35-45 Hz), 678–679
    frontal intermittent rhythmical delta (FIRDA) activity, 680

left temporal lobe seizures and slow-wave activity, 679–680
occipital intermittent rhythmic delta activity (OIRDA), 680
origin of seizure activity, 676–677
rhythmic midtemporal theta, 680
SCP and, 616, 617
seizure control mechanisms, 676
seizure resistance with SMR, 8
simple partial seizures, 46, 48f, 49f, 202
slow waves overview, 680
SMR and 40 Hz rhythms, 678
SMR as standby frequency, 677
temporal intermittent rhythmic delta activity (TIRDA), 680
temporal intermittent rhythmic theta activity, 680
temporal lobe epilepsy, 680
temporal lobe medial aspect, 202
thalamic reticular nucleus, 677
training SMR and 40 Hz rhythms, 679
triphasic waves, 680
two-channel assessment and training, 420
as validated application for NFB, 14
selective attention, 176–177, 661
selective serotonin reuptake inhibitors (SSRIs), 25
self-control, 178
self-in-relation-to-others, 182
self-referenced tasks, 216
self-referential processing, CMS and, 216
self-reflection, 148
self-regulation, 11, 445–446, 471–472
self-stimulated-recall, 502
self-will experiment, 112
semantic associations, BA 29 and, 223
semantic categorization, 173
semantic dementia, 106
semantic input, NFB in dominant temporal lobe, 192
semantic memory, 106, 179
semantics understanding, 171
senile dementia, NFB treatment for, 107
    See also Alzheimer's dementia
sensitivity and gain, defined, 36
sensor application, 377–378

sensorimotor rhythm (SMR)
    and 40 Hz rhythms, 678
    for ADHD, 677
    BFE Sterman Suite, 408–409
    defined and described, 37–38
    for dystonia, 116
    epilepsy studies, 8
    origin of term, 585–586
    rewarding (12-15 Hz or 13-15 Hz), 402–403
    as standby frequency, 677
    Sterman cat study, 7–8, 18–19
    for Tourette's, 108
sensor-motor strip, 37–38
sensory ERPs, 6, 524–528
sensory information synthesis, BA 23 and, 221
sensory input and integration, BA 31 and, 220–221
sensory networks, temporal lobes and, 191–192
sensory pathways, spinal cord to cortex, 87, 87f
septal area, 81–82
septal hippocampal area, directionality, 97
septum, 82
septum pellucidum, 82
sequencing, 174
sequential (bipolar) montages, 290
sequential (bipolar) QEEG assessment, 274–275
sequential (bipolar) training, NFB and BFB intervention fundamentals, 466
sequential processing, 164
sequential recordings, 57
sequential vs.referential amplitude comparison, 426f
Sergievskii, 607
serial sevens math task, 322, 372
series retention, hippocampal region and, 192
serotonin (5-hydroxy-trypamine, or 5-HT), 24, 25, 167
serotonin selective reuptake inhibitors (SSRIs), 115, 345
sessions, steps in, 475–476
set-up-for-success principle, 120
sexual arousal, 170
Shaffer, Fred, 574–575, 575f
Shamay-Tsoory, S.G., 649–650
shaping (reinforcement technique), 3, 11, 13, 365
sharp waves, 43
Shay, Jonathan, 684–685

Shealy, N., 602–603, 603f
Sheer, David, 38, 329, 586, 679
Sheer rhythm, 38, 270, 586, 679
shell-shock. *See* post-traumatic stress disorder (PTSD)
shifting mental set, 173
short story and novel techniques, 502–503
short term memory, 178, 208, 218
Shouse, Margaret, 8
Shtark, Mark, 695, 695f
side effects with NFB and BFB, 458–459
Siever, Dave, 599, 606, 606f
similar conditions comparisons, 464–465
simple partial seizures (SPS), 46, 48f, 49f, 202
simultagnosia, 92, 122
simultaneous discharge in pyramidal cells, 22
single-channel EEG
    single-channel assessment procedures, 424–432, 425f
    single-channel assessment screen, 430f, 431f
    single-channel assessments to 62 Hz, 315
    single channel EEG assessments, 273–275
    single-channel findings, 314
    single-channel neurofeedback, BA 24 and, 218–219
    single channel QEEG assessment, 273–275
    single-channel QEEG for ADHD, 666
    single-channel training, 133–134
single-ended amplification, 53
single motor unit muscle fasciculation as artifact, 338f
single picture technique, 500
sink (pyramidal cells), 20–21
Sinzig, J., 669
site preparation, 59, 315
skeletal muscle tone (EMG) assessment, measurable elements, 359
SKIL (Sterman Kaiser Imaging Laboratory) database
    3 SD comparison, 50f
    ADHD client comparison to, 305f
    analysis, 293
    database norms, 270
    high-pass filters for, 321
    theta comparison, 49, 49f
    time of day effects, 307f
Skinner, B.F. "the Great Behaviorist," 11, 13, 415
Skinner Boxes, 11

skin reactions with tDCS, 591
skin temperature, controlling, 446–447
sleep, 31, 45
sleep apnea, 45, 403
sleep disturbance, depression and, 217
sleep spindles, 43–44
sleep-wake regulation, 168
SLF (superior longitudinal fasciculus), 93
sLORETA and LORETA, 544
slow- and fast-wave production, control of, 412
slow cortical potentials (SCPs), 7, 616–618
slow waves, 39, 44, 406, 680
SMIRB (Stop My Irritating Ruminations Book), 385, 478, 491
Smith, Ray, 603, 603t, 605, 605f
Smith-Pelletier, Carolyn, 103–104
smoothing factor, 407
Snowdon, David, 17
SNS (sympathetic nervous system), 102, 212, 574–575
Snyder, S.M., 674
social anxiety, 168
social avoidance, 173
social circuit. *See* orbitofrontal cortex
social context, parahippocampal gyrus, 208
Social Engagement System, 656
social interactions, dorsal CMS lesions and, 216
socially appropriate behavior, orbitofrontal cortex and, 141
Society for Neuronal Regulation (SNR), 16, 250, 358
sociopathic personality disorders, ACC and, 218
Socratic questions, 502
sodium gates, 26
sodium-potassium pump, 23
soft encouragement, 403–404
soft-eyes state, 445
solitary nucleus, BA 25 and, 219–220
solitary tract (NST), HRV training and, 212
solvents, 347
somatosensory association cortex, 96, 189, 226–228
sound feedback, 408–409
sound localization, 195
sound perception, right temporal lobe and, 192
sound rewards, 397
source (cell current), 20–21

Soutar, Richard, 127
space, localization in and perception of, 91
Spargo, Edward, 501
spatial information, 174
spatial memory, 173
spatial network, 145f
spatial-object memory retrieval, 174
spatial perceptions, parietal lobes and, 225–226
spatial summation, 24
special assessment and artifacting screen, 322–326
spectral arrays, 35
spectral density, 292–293
spectral magnitude, 292–293
SPECT studies, 112, 354, 496, 593
speech and language functions
    auditory cortex, 192
    BA 6 and, 188
    BA 20 and, 197
    BA 21 and, 196
    BA 22 and, 195
    BA 30 and, 223
    BA 38 and, 203–204
    BAs 21 and 22 and, 231
    BAs 37 and 39, 237
    Broca's Area, 93, 233–234
    NFB for, 92–93
    Wernicke's area, 93, 232
speech initiation, 169
spike-and-wave patterns, 42f, 43, 46–48, 47f
spikes, 42, 42f
spinal cord at medulla obongata, 87, 87f
spindle-like wave forms, 38
spiritual functions, temporal lobe activation, 206
Sporns, Olaf, 153
Sport Concussion Assessment Tool (SCAT 3), 621
SPS (simple partial seizures), 46, 48f, 49f, 202
square waves, 44
SSRIs (serotonin reuptake inhibitors), 115, 345
stability (mood), 250–251
standard deviation of NN interval (SDNN), 573
standard deviation of the average NN intervals (SDANN), 573
standardized testing, 170–171

Stanford-Binet IQ test, 486
state changes, 350
State-Trait Anxiety Inventory (STAI), 605
static electricity, 52, 58
statistics, collecting and evaluating, 702–707
    concepts and terms, 704–707, 706f
    data description, 702
    data types, 702–703
    effect size, 704
    hypothesis generation, 703
    power, 704
    probability and alpha level, 703
status epilepticus, 344
steady-potential shift ERPs, 6
Steiner, Frenette, Rene, Brennan & Perrin ADHD study, 674
Steiner, Naomi, 674
stellate cells, 20, 23, 28
stems of posterior cerebral arteries, 100
Stens Corporation assessment training, 357
Steriade, M., 31, 586, 678
Sterman, Maurice Barry
    alpha activity after task completion, 267
    on anatomical differences in thalamocortical projection pathways, 677–678
    cat operant conditioning study, 7–8, 8f, 29, 468, 678, 678f
    communication patterns in brain maps, 19
    on comodulation, 296, 298
    EEG interpretation, 127
    epileptiform activity, 250
    full-cap demonstration, 358f
    HRV training study, 585, 585f
    math tasks, 321–322
    normative databases, 32
    owl or monkey pattern, 308f
    patient with myoclonic seizures, 352
    seizure disorders, 14, 251, 419
    sensorimotor rhythm (SMR), 38
    SKIL artifact display, 332
    SKIL database, 292–293
    SMR cat studies, 585–586
    SMR for movement disorders, 116
    SMR training, 408–409

SMR training graphs, 431
theta/beta ratio norms study, 113
Theta Cortex-Putamen-(Globus Pallidus)-Substantia Nigra-Thalamus hypothesis, 110–111
Thorndike's Law of Effect, 415
Top Gun pilots and, 6–7, 52
Sterman Kaiser Imaging Laboratory. *See* SKIL (Sterman Kaiser Imaging Laboratory) database
Stimulant Drugs and ADHD, 113
stimulant medications, 109–110, 345–346
stimuli evaluation, BA 31 and, 220
STN (subthalamic nucleus), 142
Stop My Irritating Ruminations Book (SMIRB), 385, 478, 491
Strack, Ben, 691
Strehl, Ute, 616–617, 671, 675
strength-duration curve, 600, 600f
stress
    assessment, 371–372, 378, 476, 567f
    desensitization for alpha-theta therapy, 479
    stress control, NFB and HRV training for, 216
    stress response, neurophysiology of, 103, 103f
striatal medium spiny neurons (MSNs), 145
striate cortex, BA 17 and, 241
stria terminalis, 73
striatum, 74, 217
strokes and TIAs, 186, 580–581
Stroop task, 143–144, 218, 567
study hall, 516–517
subarachnoid space, 70
subcallosal gyrus, 82
subclinical electrographic seizure patterns, 43
subcortical cells, 28
subcortical rhythmic influences, 30
subdural bleeding, 637–638
subgenual area, 219
subiculum of hippocampal region, 205
subliminal alpha, 366
substance P, 26
substantia nigra, 74, 77
subthalamic nucleus (STN), 142
subthalamus, 86

subtypes of disorders, assessment and intervention, 353–354
success, setting up for, 471
successive approximations, 395
sulcus, 69, 72
Sullivan Readers, 501
summation of EPSPs and IPSPs, 24
superior (head), 68
superior colliculus, orienting and focusing, 112
superior frontal gyrus for laughter, 171–172
superior longitudinal fasciculus (SLF), 81, 81f, 93, 643f
superior temporal gyri (ST BA 22), 145
superior temporal lobe of auditory cortex, 195
supragenual anterior cingulate cortex (SACC), 141, 147f, 215
supramarginal gyrus, 236f
SURA (sharp, unique, reliable, ambient) feedback criteria, 408
sustained attention, 174, 227, 409
sustained feedback, 409
Sütterlin, S., 586
Sutton, S., 6
Sweeney, John, 262
sympathetic and parasympathetic systems, balance between, 467
sympathetic nervous system (SNS), 102, 212, 574–575
sympathetic stimulation measurements, 102
A Symphony in the Brain (Robbins), 7, 150
symptomatic epilepsy, 46
synapses, 23
synaptic terminals, 23, 26
synaptic transmission, 182–183, 409
synchronous EEG activity, 29, 31
synchrony (coherence) between oscillators, 445–446
synchrony, defined, 292
Synsoplevede Figurer (Rubin), 661
syntactic working memory, 176
synthesis of data, metacognitive strategies and, 502
systematic desensitization, 365
systems theory of neural synergy, 565, 656
"A Systems Theory of Neural Synergy" (Thompson and Thompson), 656

T

tACS vs. transcranial direct current stimulation (tDCS), 601, 606
tactile localization, 190
Talairach coordinate system, 543–544
talking inhibition loss, right temporal lobe and, 192
Tan, Gabriel, 14, 250, 676
tardive dyskinesia, 344
"Task Force Report on Methodology and Empirically Supported Treatments: Introduction" (Moss & Gunkelman), 16
Tattenbaum, Rae, 15
TBI. *See* traumatic brain injury
tDCS. *See* transcranial direct current stimulation
teaching and assessment screen, 322–326, 323*f*
technical support for equipment problems, 60–61
tectum of mesencephalon, 77
tegmentum, 77, 86
telencephalon. *See* forebrain development
telepathy, parahippocampal gyrus and, 208
"Template for Developing Guidelines for the Evaluation of the Clinical Efficacy of Psychophysiological Interventions" (La Vaque & Hammond), 16
temporal coherence in music, 179
temporal context recognition, 190
temporal intermittent rhythmic delta activity (TIRDA), 680
temporal lobes
    auditory association cortex, 91
    damage, 96
    described, 191–193
    encephalitis patient graphs, 193*f*
    functions, 89
    inferior temporal lobe, 193
    left (dominant) temporal lobe, 192
    medial aspect, 201–202
    right (nondominant) temporal lobe, 192–193
    seizures, 46
    temporal lobe epilepsy, 192, 209–210, 680
    visual functions, 206–207
temporal lobes lateral aspect, 194–198
    auditory cortex, 194
    connections, 198
    NFB over dominant temporal lobe, 198
    overview illustration, 194*f*
    transverse temporal gyri, 194
temporal-parietal junction, lateral aspect, 230–235, 230*f*
temporal pole, 203–204
temporal resolution with EEG, 5
temporal summation, 23–24
tentorium, 44, 70
Test of Variables of Attention (TOVA), 19
thalamo-cortical-basal ganglia loops, 150
thalamocortical oscillations, 144–145
thalamus
    cell synchronicity in, 22
    described, 75, 76*f*
    as hub of nervous system, 73–74
    overview, 136*f*
    role in epileptic syndrome, 676–677
    sensorimotor rhythm (SMR), 38
    thalamic inhibitory interneurons, 29
    thalamic pacemakers, 28–29
    thalamic reticular nucleus, 677
    theta waves, 38
thalamus-basal ganglia-cortical inter-changes, EEG patterns in, 3
Thatcher, Robert
    ADHD testing results, 670
    on brain modules and hubs, 153, 155
    on Brodmann areas, 127
    communication patterns in brain maps, 19
    EEG amplitude ratios, 434
    EEG databases, 270, 304
    on Fourier analysis spectrum, 435
    on Laird research, 126
    location importance, 545, 547
    LORETA Z-score (LNFB), 550
    NeuroGuide system power, 33
    NeuroGuide workshops, 550
    normative databases, 32
    on phase shift and phase lock, 153, 442
    psychiatric and neurological disorder research, 354
    QEEG findings for TBI, 632
    on relative power, 433
    Z-Score Neurofeedback, 554
Theory of Mind (ToM), 173, 216, 234, 654–655

theta/beta ratios, 14, 18, 284–285, 666–667, 666f
theta waves, 15, 36–37, 110
Thimersol-autism controversy, 120
thioridazine, 344
third ventricle (lamina terminalis), 78, 85
Thompson, James
    case studies, NFB for Asperger's, 118
    communication patterns in brain maps, 19
    The A.D.D. Book, 672, 694
    eVox workshops, 550
    force-board balance test, 636, 637f
    "Heart-Brain Connections," 656
    ISNR Brodmann area booklet, 125
    normative databases, 32
    phase delay pattern in ASD children, 653, 653f
    TBI testing research, 631
    virtual reality room, 636–637, 637f
Thompson, Lynda
    at ADD Centre Store, 390f
    Draw-a-Person assessment tool, 621–622
    "Heart-Brain Connections," 656
    as learning center owner, 694
Thompson, Michael, 125
Thompson/Hagedorn Evoke Neuroscience, EEG databases, 270
Thompson Setting-up-for-Clinical-Success Suite, single channel assessment screen, 429
Thorndike, Edward, 3, 11
Thorndike's Law of Effect, 11, 415
Thought Technology assessment training, 357
3-3-3 method, 504–505, 504f
three-electrode referential placement, 273–274
three-key-facts technique, 489–490
threshold for excitation, 26
threshold potentials, 24
thresholds, setting, 395–396
thyroid hormone levels, 347
thyroxine, 347
Timed Readings in Literature (Spargo, ed.), 501
time of day effects, 307f, 322, 465
time-related domains, 32–33
time sequence after visual stimulus, 198f
tinnitus, 195, 597–598, 598f

TIRDA (temporal intermittent rhythmic delta activity), 680
Titanic exercise, 500
TLE (temporal lobe epilepsy), 209–210
TMS (transcranial magnetic stimulation), 592
ToL (Tower of London Test), 170, 178, 655
ToM (Theory of Mind). *See* Theory of Mind
Tomatis Sound Therapy, 583
tomography, defined, 543
tonal sequence recognition, 192
tongue and swallowing as artifact, 334
tonic-clonic seizures (grand mal epilepsy), 46, 351, 352f
tonic immobility, 685–686
tonic inhibition, 183
tonic processing, 108
Tools for Optimal Performance States (TOPS), 370–371
Toomim, Herschel, 611–612
top-down decision making, 661
Top Gun pilots, 6–7, 52, 268, 327
topographic maps (EEGs), 33, 294
topometric analysis, 295
Tourette syndrome
    ADHD symptoms with, 459
    basal ganglia and, 74
    dopamine activity and, 25
    NFB and BFB treatment, 462
    NFB treatment, 107–108
    promising clinical reports for NFB, 14
    SMR training for, 279, 682–683, 683f
    tDCS and, 597
TOVA (Test of Variables of Attention), 19
TOVA (Test of Variables of Attention), NFB results, 120
Tower of London Test (ToL), 170, 178, 655
toxic materials, data collection, effects on, 347
tracking client progress, 463–464
tracking trends, 476–477
training adults, 384–387, 387f, 388t
training children, 383–384, 384f, 386, 386f
training in assessment procedures, 357–358, 358f
training screen, 436f
training screens, NFB and BFB intervention fundamentals, 453–456
training session components, 471–475
training session graphs, 437f

training sessions, 379–380, 385
training session with formatting axis, 427f
training session without formatting axis, 427f
training SMR and 40 Hz rhythms, 679
tranquilizers, 344
transcranial direct current modulation, 591
transcranial direct current stimulation (tDCS), 590–598
    active electrode effects on neurons, 594
    anodal stimulation, 596
    anode and cathode, 590, 591f
    brain activity, effects on, 593–596
    brain activity measurements, 593
    calcium influx, 595–596
    cathodal stimulation, 598
    clinical effects, 596–598
    cognitive function improvement, 596
    current control, 593
    depression, 597
    disorder improvement, 597
    dosage, 590
    electrodes and effects, 591–592, 592f
    glutamate activation, 595
    long-term potentiation (LTP), 594–596
    in NDH, 171
    neuroanatomical/biochemical changes, 595–596
    neuronal membrane changes, 593–594
    overview, 132
    prior to NFB, 165–166
    receptor sites, 595, 595f
    side effects, 596
    vs. tACS, 601, 606
    time length of effects, 594
    tinnitus, 597–598, 598f
transcranial magnetic stimulation (TMS), 592
transcutaneous electric nerve stimulation (TENS), 599
transference cures, 13, 469
transferral of NFB to home and school, 414–415
Transfer Trials, 428
transient (change in kind), 332
transient waves, 33, 290
transition frequencies, 37
transverse section of brain, 68
transverse section through right cerebral hemisphere, 136f

transverse temporal gyri, 194
trauma dose, 685–686
traumatic brain injury (TBI)
    cardiac health, effect on, 579–580, 579f
    as experimental application of NFB, 14
    metabolic effects, 629–630, 630f
    pathology, 622–623
traumatic memories, amygdala and, 199
treatment vs. training, 3
trend reports, 464–465
Trial and Error Learning (Law of Effect), 3, 11
tricyclic antidepressants, 345
triphasic waves, 33, 680
true lefties, 95
truth telling, BA 20 and, 198
tryptophan, 24
tryptophan pathways in vitamin B6 deficiency, 628f, 629
tuberculostatics, data collection, effects on, 347
tumors, wave forms and, 44
tuning out, rewarding, 403
twilight learning state, 477
2D and 3D spectrums, 431–432
two-channel EEG assessment and training, 285–289, 314, 417–423, 417f, 423f
tyrosine, 24

U

Uddin, L.Q., 149, 662–663
uncinate fasciculus, 81, 166f, 168, 650
unconditioned stimuli/unconditioned responses, 12, 98, 470
unconscious memories, amygdala and, 199
uncus, 73, 82, 96, 206–207
uncus of Parahippocampal Gyrus, 168f
underlying principles, 499
undershoots, 26
understanding others' intentions, 170, 176
Unified Parkinson's Disease Rating Scale, 682
units of measurement, assessment and intervention, 284–285
University of Indiana, cortical surface connection maps, 153
University of Lausanne, Switzerland, cortical surface connection maps, 153

unmyelinated vagus, vagal control of HRV, 574–575
untoward client experiences, NFB and BFB intervention fundamentals, 457–458
unusual analyses, assessment and intervention, 352–353, 353f
upper auditory cortex, stimuli integration, 112

V

vagal afferent connections, 212
vagal control of HRV, 574–575, 575f
valences in pyramidal cells, 22
validated vs. experimental applications, 14
value-based behavior, 168
van der Kolk, Bessel, 687
Van Poppel, H., 17, 155
variable reinforcement schedule, 11
vasopressin, 347, 656
Vaughn, 5–6
vegetative/autonomic Affect network, 217, 642
ventral (head), 68
ventral amygdalofugal pathway, 73
ventral network, 665–666
ventral stream, What Pathway as, 240
ventral tegmental area, 77
ventricles, 70, 78
ventrolateral nucleus of metathalamus, 75
ventrolateral prefrontal cortex (VLPFC), 145
ventromedial prefrontal cortex (VMPFC), 147f, 183–184
ventromedial prefrontal region, 89
ventroposterior nucleus of metathalamus, 75
verbal comprehension, BA 7 and, 227
verbal encoding, temporal lobes and, 191
verbal fluency, 177, 596
verbal IQ, 167
verbal recall techniques, 500–501
Vergleichendes Lokalisationslehre der Grossgehirn-rinde (Brodmann), 131
Vernon, D., 586
vertebral arteries, 70, 99
vestibular connections, inferior parietal-temporal BA 22, 237
vestibular input to cortex, 187

Vigilance network, 664
virtual reality, concussion/mild traumatic brain injury (TBI), 636–638
vision functions, BA 6 and, 188
visual agnosia, 96, 241
visual association cortex, injury effects and, 96
visual cortex, 72
visual dominance, 95–96
visual ERP, 522–524, 522f
visual functions, 207, 241
visual hallucinations, temporal lobe medial aspect, 202
visualization, hypercoupling and, 31
visual memory decay time, 96
visual network, 146f
visual processing, BA 20 and, 197
visual-spatial comprehension, BA 7 and, 227
visual-spatial-language skills, electrode placement for training, 406–407
visual-spatial processing, right NDH somatosensory cortex, 190
visual word recognition, 93, 194
vitamin B6 deficiency, 628f, 629
VLPFC (ventrolateral prefrontal cortex), 145
VMPFC (ventromedial prefrontal cortex), 147f, 183–184
Voffset, 62
Vögele, C., 586
Vogt, Cecelia, 131–132
Vogt, Oskar, 131–132
voltage and amplification measurements, 315
voltage detection, 20
voltage divider models, 66–67
voltage measurements, 315
voltage ranges, 55
voltage-sensitive change, 26
volume conductors, 31
voluntary control of motor pathways, 86–87
voluntary eye movements, 140–141
von Economo neurons (VENs), 663
voxels, 356, 543–544
V waves, 43–44

W

Wagner, T., 591
Walker, Jonathan, 127
Watson, John, 12, 416
wave component of spike-wave patterns, 676
wave forms, 33–34, 606, 606f
   See also bandwidth-mental state correlation
wave phases, 34
WCS (Wisconsin Card Sort), 170
weak central coherence theory in ASD and Asperger's syndrome, 655
websites for assessment training, 358
Wechsler Intelligence Scale (WISC, WAIS), 19, 170, 173, 486, 634
Wedeen, Van J., 153
Wernicke, Carl, 231f
Wernicke's aphasia, 91, 196
Wernicke's area
   language comprehension, 127, 191
   learning dysfunction and, 193, 231
   reading training and, 93, 94f, 407
   as temporal-parietal junction, 230f
What Pathway, ventral stream as, 240
"When in doubt, throw it out," 332, 337f, 350
Where Pathway, dorsal stream as, 240
White, Nancy, 476
white matter, 27, 70
WHO (World Health Organization), on efficacy studies, 695
Wide Range Achievement Test (WRAT), 120
wide vs. narrow bandwidths, 400
Wiggers diagram, 574, 574f
willpower, 181
Wilmarth, Eric, 475
Wilson, Vietta (Sue), 432, 445, 691–692
WinEEG database, 321
WinEEG program, 342, 344, 521
WISC, WAIS (Wechsler Intelligence Scale), 170, 173, 486
Wisconsin Card Sort (WCS), 170
Wise, Anna, 417
withdrawal, data collection, effects on, 347
within-cluster connectivity, 153
within-session tracking, 463–464
word deafness, 93, 195, 232

word problem study techniques, 514–516, 516f
working memory
   BA 5 and 7, 188
   BA 9 and, 170
   BA 10 and 11 and, 173
   BA 47 and, 179
   defined, 106
   F7 location and, 177
   tDCS on DLPFC, 596
work-rest function in instruments, 411, 472
World Health Organization (WHO), on efficacy studies, 695
WRAT (Wide Range Achievement Test), NFB results, 121
writing technique, 504–505
Wu Wenqing, 125
W-W-H-W paradigm, 488, 492–493

Y

Yucha, C.B., 693

Z

Zametkin, A.J., 112
Zen Buddhism, 7, 467
Zhang, Zhang, & Shen ADHD study, 674
Zoloft, 345
Z-Score Neurofeedback (Thatcher), 554
z-tunes-style feedback, 561
Zubin, J., 6
Zung scales (CNS Vital Signs), 621